Satellites, Packets, and Distributed Telecommunications

A COMPENDIUM OF SOURCE MATERIALS

Roy Daniel Rosner, Editor

LIFETIME LEARNING PUBLICATIONS
Belmont, California
A division of Wadsworth, Inc.

London, Singapore, Sydney, Toronto, Mexico City

To
Rachel, Matthew, Stuart,
Deborah, Ira,
and especially
Judith

Text & Jacket Designer: Rick Chafian

Copy Editor: Laurel Cook

Composition: Bi-Comp, Inc.

Printed in the United States of America

1 2 3 4 5 6 7 8 9 10—86 85

Library of Congress Cataloging in Publication Data
Main entry under title:

Satellites, packets, and distributed telecommunications.

  Bibliography: p.
  Includes index.
  1. Packet switching (Data transmission) 2. Computer
networks.   3. Artificial satellites in telecommunication.
I. Rosner, Roy D., 1943–
TK5105.S28 1983   621.38′0422   83-18761
ISBN 0-534-97924-6

Contents

iii

Preface

WHAT THIS BOOK DOES

Divided into ten major sections, this book brings together some of the most fundamental and important journal and conference articles in the fields of packet switching, satellite networks, and distributed data communications networks. In addition, a number of the articles relate to the evolution of voice communications within these fields, which led to the formation of large, distributed hybrid networks serving both voice and data.

Specifically, the book covers the following topics:

Introduction to networking

Development of packet-switching technology

The introduction of broadcast satellite technology

The integration of these concepts into large, multifunctional, multivendor networks

Security and privacy considerations

The commercial services and systems currently available

This volume contains some forty-eight articles drawn from technical journals and conference proceedings over the last fifteen years, and is a direct outgrowth of my previous two books, *Packet Switching: Tomorrow's Communications Today* and *Distributed Telecommunications Networks via Satellites and Packet Switching*, as well as of my own involvement in large-scale telecommunications networks projects for the U.S. Government. If you have not read those earlier books, you should note that many of the present articles presume a level of technical knowledge of the field imparted by books such as those. While the introduction to each section provides some basic overview of the underlying technology, it may be useful before trying to understand fully the reference articles within this volume to review the more elementary material in a comprehensive text.

As a minimum, this volume will save many hours of searching through old professional journals and conference proceedings trying to locate copies of these outstanding references. More important is the fact that these articles will, in the manner in which they have been selected and grouped, help to provide a concise, in-depth insight into the allied aspects of packet networks, satellite broadcasting, and distributed systems.

INTENDED AUDIENCE

This book is intended both as a learning tool and a reference for systems analysts, computer scientists, engineers, teleprocessing systems planners, and managers of organizations that supply communications services to the public at large or to their own organizations. It is meant to assist such professionals and anyone else who needs an in-depth knowledge of the design considerations, alternatives, and techniques for meeting present and future telecommunications requirements. It is also of use in understanding fundamental principles needed to evaluate the performance and cost-effectiveness of services and facilities supplied on a lease, purchase, or tariffed basis from the rapidly expanding providers of telecommunications services. In that context it can be of great use to systems analysts, communications staff officers, and managers of network-user complexes in selecting and providing the most economical and efficient mix of services to their organizations.

STRUCTURE AND ORGANIZATION OF THE BOOK

Section I provides a broad overview of networking, ranging from architectural and structural issues to cost and legal issues. The unique requirements placed upon networks by non-voice, interactive data-users are contrasted, for example, with the demands of basic telephone users.

Section II consists of two papers that introduce the concept of packet communications. Section III extends the ideas contained in those papers to terrestrial networks, showing just how such networks are structured, designed, and implemented. Section IV examines the concept of protocols for packet networks and how the protocols that define user-interaction with the networks operate in conjunction with the protocols that define how the networks themselves function.

Section V provides a general introduction to satellite communications, with particular emphasis placed on the ability of satellites to provide multi-point "connectivity" between a large number of users. Section VI specifically examines the employment of satellites as part of the fundamental mechanism of a packet-based network.

The discussion in Section VII focuses on local area networks, wherein large amounts of information are exchanged among users within a fairly limited geographic area via a broadcast radio channel or cable medium.

Section VIII provides an amalgamation of all of these concepts into networks that can combine voice and data users with satellite, terrestrial, and local access facilities so as to provide large, efficient user-service networks. This section also covers a range of issues from technological and structural to economic and practical implementation concerns.

Section IX explains the techniques available for protecting the integrity and privacy of all information. The same technology that encourages the sharing of resources among many users also renders information highly susceptible to interception and disruption. It is thus imperative that provision be made for the protection of network-based information.

Section X describes the vendor marketplace, outlining some of the key vendors and their services available to network users, designers and implementers. While it is impossible to keep such information wholly up to date in this dynamic marketplace, the industry structure and service types actually change more slowly than the prices and services themselves.

As noted, this volume provides a careful sampling of key articles and papers relating to this information technology. Readers are encouraged to follow up on these secondary sources in order to fully grasp the complex issues, applications, and technology of advanced communications networking.

ACKNOWLEDGMENTS

I am delighted to acknowledge the cooperation of the dozens of authors who have granted me permission to reproduce their work in this volume. In some cases it has been quite a challenge to locate some of these individuals after a number of years since their works first appeared, but I am pleased to have had the opportunity to renew some old acquaintances in the process. I would also like to thank the professional societies and groups—in particular the Institute of Electrical and Electronics Engineers (IEEE), the American Federation of Information Processing Societies (AFIPS), the International Council for Computer Communications (ICCC)—for their assistance and their permission to reprint materials from articles and conference proceedings prepared under their auspices. Finally, I would like to thank those people working for commercial publications who have kindly granted permission for some of their articles to be assembled in this volume. I believe that the amount of expertise as represented by this diverse group of individuals and organizations will provide an extremely useful body of information leading to the further development of large, sophisticated, distributed telecommunications networks.

Roy Daniel Rosner

Introduction

Even as this is being written, the telecommunications industry and marketplace are undergoing very sweeping and fundamental changes. Motivated by the anti-trust charges of the United States Justice Department, the divestiture agreement, negotiated by the American Telephone and Telegraph Company (AT&T), is effectively completing a process begun more than fifteen years ago. That process, introducing, stimulating, and nurturing competition in the field of telecommunications within the United States, has provided both opportunity and anxiety for the communications consumer. The opportunity lies in the consumer's ability to get a wider range of efficient and innovative services from a variety of providers. The anxiety lies in deriving services from a mixture of competitive, often incompatible vendors without loss of reliability, responsiveness, and cost efficiency.

The AT&T divestiture agreement separates the local service-providing telephone companies (operating under local public-utility commission regulation and control) from the interstate and interarea long-distance carriers (regulated by the Federal Communications Commission for basic services, and largely unregulated for a wide range of specialized services). No single company has a monopoly to provide end-to-end user communications service. Conversely, no company is inhibited from providing end-to-end service, particularly when that service is anything beyond basic, dial-up universal telephone service. There will continue to be regulatory and legal argument about the ability of nationwide providers to "bypass" the services of the local telephone company for voice-only services; however, it is clear that a variety of legal bypass possibilities exist for non-voice and specialized services.

Of interest as well is the fact that, outside of the United States and in countries where telecommunications have traditionally been provided by government owned and operated monopolies, experiments have been proceeding to introduce competitive services or to denationalize the operations of the telecommunications providers. Telecommunications and its tight intercoupling with the computer industry is the technological foundation for much of today's information-based society. Fostering competition is one way to stimulate innovation and development in a high-tech industry that is essential to each country's survival in the world marketplace.

My two previous books, *Packet Switching: Tomorrow's Communications Today* and *Distributed Telecommunications Networks via Satellites and Packet Switching,* anticipate these events in the telecommunications industry and attempt to equip the reader with an understanding of the array of new services, facilities, and technologies that are available to meet specialized needs. Packet switching, in various shapes, forms, and structures is the foundation of most nonvoice communications today, whether that communication takes place in sophisticated value-added networks or just between two devices sending data across a telephone line. In *Distributed Telecommunications Networks* the ideas of packet technology are combined with the high visibility of synchronous communications satellites to provide extremely flexible, large-scale networks, potentially servicing hundreds or thousands of diverse, geographically dispersed users. Both of these books provide a fundamental introduction to these exciting and challenging areas of telecommunications and find their natural extension in the publication of this present volume.

Introduction to Networking

This section:

■ provides a comprehensive overview of telecommunications networking

■ characterizes many of the unique properties of the different classes of network users

■ defines, describes, and compares various switching techniques that can be used in networks

■ introduces some legal and regulatory issues that must be considered, in addition to technology, in developing networks.

Many technological prognosticators and philosophers theorize that the world is moving from the industrial age into the information age. In this new age, communications systems might be considered the blood stream of the societal organism and computer hardware and software its brain and nervous system. Already, the combination of computer facilities and telecommunications resources, resulting in the rapid, accurate, and low-cost movement of information, has changed the way people work, shop, play, travel, and exchange ideas.

The evolution of total information systems, however, has followed some bumpy and irregular roads, resulting in very interesting dichotomies in the implementation of their technology. One of the most significant of these dichotomies is the large discrepancy between the cost of processing information and the cost of transmitting that same information. In many applications, telecommunications services are one of the major cost elements of a modern business establishment, although rapid changes in the technology and marketing of communications services are beginning to change this situation.

As the communications marketplace has evolved from being monopolized by regulated common carriers to being open to the competition of a wide range of high-technology companies, it has become increasingly important for telecommunications users to be technically knowledgable about the range of services and facilities available to meet their organizational needs. The distributed telecommunications networks required to meet the information transfer needs of even very modestly sized organizations must be synthesized from an array of components, devices, facilities, and services that are drawn from the total marketplace. To do less can result in user costs often many times as great as the "optimum" solution.

NETWORKING ALTERNATIVES

While historically used for voice communications among people, telecommunications networks have become increasingly important for transporting non-voice information between people and their machines as well as between and among machines—with little or no human intervention. Although it is true at the present time that about 90% of the revenues expended on telecommunications is for voice-based services, a significantly larger percentage of the world's commerce is implemented, managed, or controlled via non-voice communications. The reliance of our economy on the flow of nonverbal, digital information can be seen in the worldwide use of computer-based banking and electronic funds transfer networks, the nation's air traffic control system, the military command and control systems and networks, and the computer networks that support the marketing and manage the activities of most businesses.

While the inventory of worldwide voice networks has been marginally satisfactory in meeting many of the non-voice demands of users, serious deficiencies in cost, performance, and overall suitability quickly become evident to these users. As the number of suppliers and the sophistication of the available technology have both increased, many alternative approaches have become practical. The approach that users most commonly select is to combine the use of public voice-networking facilities with private data-networking facilities, or else to continue to use the public voice networks in an attempt to meet all needs.

Two factors make either of these approaches less than satisfactory from both an efficiency and effectiveness viewpoint. First, the operational procedures at each end of the communications link must be much more highly structured and controlled for communications to proceed between non-voice users (e.g., computers and terminals) than for voice users. Secondly, the usage statistics, in terms of average usage rates, message lengths, and information content of voice users and non-voice users, differ markedly. These two distinctions are at the core of the information presented in the articles that comprise this section.

APPROACHES TO DATA NETWORKS

The first paper on network architectures and protocols by Paul E. Green, Jr. of IBM provides an overview of the issues associated with non-voice networks. His primary emphasis is on generalized computer-based networks, where intelligent, programmable machines are the primary communicating element of the network. First, a set of functions that the network must perform for the users' equipment are defined and described. These functions, having evolved into essentially universal characteristics, have been standardized into a multilayered protocol hierarchy. Because the seven-layer protocol hierarchy of the International Standards

Organization (ISO) is fundamental to the successful implementation of data communications networks, it will be reintroduced in many of the papers appearing throughout this volume. This first paper shows in some detail how two of the major suppliers of data-processing equipment, IBM and Digital Equipment Corporation (DEC) have mapped these functions into their own networking architectures (SNA and DECNET).

CHARACTERISTICS OF DATA COMMUNICATIONS TRAFFIC

The next two papers in this section, both authored by researchers at Bell Telephone Laboratories, provide a lasting insight into the statistical differences between voice and non-voice communications users. These papers, which first appeared more than a decade ago, have become classic for their definitions and characterizations of the parameters influencing the use of a data communications circuit.

The paper by A.L. Dudick, E. Fuchs, and P.E. Jackson deals with a particular class of computer-based users—one whose primary input is via a simplified keyboard (principally the telephone tonepad) and whose computer response is via a voice-answerback mechanism. The system operation is modeled in such a way that the relative importance of the various phases of user activity are explicitly shown. In such systems, not only does the user transmit to the computer rather slowly and inefficiently, but considerable periods of idleness occur while the user "thinks" about what to do, and while the computer processes information before responding. In addition to reporting the analytical techniques developed by these authors, this paper presents the results of applying these techniques to four sample systems. These analyses provided very early indications ("early" in the development of advanced data communications networks) of the vast differences in the statistics of data users compared to telephone voice users. Table 2 of this paper shows the typically short call-holding time (ranging from 20 to 40 seconds) compared to the typical 300 seconds for a voice call. Figure 2 of the paper graphically displays the very large components of delay and idle time as part of the typical data "transaction" occurring even during the typically short holding times associated with data calls.

The paper by P.E. Jackson and C.D. Stubbs develops similar analytical models for the usage characteristics of communications circuits serving fairly sophisticated multiaccess, time-sharing computer applications. Three sample systems are studied, two doing scientific processing, and the third doing business processing. User input is derived from full keyboard-display units.

It is important to note that for these systems typical operation has the users' terminal devices connected to the computer over an extended period of time. During this time period a long succession of transactions flow between the user and the computer. The overall length of the session becomes the "holding time", as used in this article, and is typically 30 to 60 minutes for the systems studied by these researchers. What they have determined, however, and what has become the basis for new development of data systems, is that 60 to 85% of the holding time is "idle" time—i.e., time during which no information is being sent, in either direction, on the communications channel. From the data collected, they were able to derive a theoretical model for a "hypothetical short holding time system," which they show as Figure 6 in their paper. Such a system would have to have the ability to use idle periods of one user's circuit to meet the needs of others—an ability which we shall see in later sections is achieved by packetized operation. In such a system, with terminals operating at 100 characters per second (nominally 1200 bits per second), average holding times would drop to approximately 15 seconds, as also shown in Figure 6 of this paper.

Other investigators have made similar measurements, further reinforcing the understanding of the differences between voice and non-voice traffic statistics. In Figure I-1, we have plotted the distribution of call holding times for about 27,000 potential and actual data users within the U.S. Department of Defense. The holding time was based on the actual transmission time of the various messages and transactions, where the transmission rate was the bit rate of operation for the data terminal involved. In other words, the holding time, measured in seconds, depends not only on the length of the transaction (in bits), but also on the speed of the terminal (in bits/second). For example, a 600-bit inquiry will take only one-quarter of a second using a 2400-bit/second terminal whereas the same 600-bit inquiry will take 8 seconds using a 75-bit/second teleprinter terminal. Figure I-1 thus reflects not only the distribution in user transaction lengths, but also the distribution in user terminal characteristics. In fact, as user terminal devices become more capable, with more common operation at higher speeds, the curve in Figure I-1 will shift upward and to the left, further reducing the median and average holding times for typical data users.

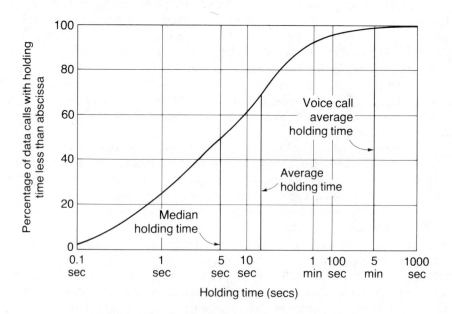

Figure I-1   Distribution of data call holding times.

These two articles, as well as more recent information suggest several conclusions. First, voice and data communication are vastly different from each other. Second, trying to put data communications through a voice network that has been optimized over the years for the unique characteristics of person-to-person voice communications may not be very efficient or very economical. Finally, the very short duration of most data calls and data communications transactions implies that a network would have to act quickly in response to data-oriented calls.

INTRODUCING TELEPHONE SWITCHING TECHNOLOGY

The next two papers in this section, both by Fellows of the Institute of Electrical and Electronics Engineers (IEEE), provide a fundamental explanation of the functions, applications, and implementation of the switched telephone networks and their associated plant and facilities. This basic knowledge is needed before the reader can appreciate the information given in subsequent papers, which describe the alternative approaches that are better suited to data users and draw comparisons among the various techniques.

The paper by Amos E. Joel, Jr. provides a functional definition and operational description of the switching function, and places that introductory information into the context of nationwide and worldwide networks. He then provides a historical perspective tracing the switching technology from the earliest switching patents (1879 and 1891) to the fully electronic goliaths of today. In addition to introducing the basic communications functions provided by the switches, the author develops a basic understanding of the technical approaches used to meet those functions and the special service features made available to the individual users.

The paper by J. Gordon Pearce extends the ideas introduced by Joel's paper, and shows a possible evolutionary path for large, telephony networks, through the incorporation of new technology. The advent of low-cost electronic circuit components has resulted in the application of digital techniques to the switching systems, which, together, have resulted in the combination of switching and transmission elements within single equipment structures. Pearce shows that the longer-term objective, made possible by low-cost memory and electronic gate circuits, is to combine switching, transmission, and terminal devices and incorporate them into multifunctional, integrated voice and data networks. This practice is already occurring, and Section 8 of this volume will deal extensively with the issue of integrated networks.

COMPARING SWITCHING TECHNOLOGIES

Having now developed an understanding of switching techniques and the fundamental differences in telecommunications user characteristics, it is possible to appreciate the paper by Roy D. Rosner and Ben Springer. This paper provides a comparison of circuit (telephone-type) switching and packet switching, where the packet switching has been optimized for data users. In this study, a large common user network for data communications was investigated. Networks were configured using packet switching and compared to circuit-switching-based networks for functionally equivalent performance. Under the assumptions of the study, packet switching showed decided advantages over circuit-switched implementations, as measured by transmission efficiencies and overall network costs. The sensitivity of the results to the assumptions was also included to make sure that the conclusions were applicable over a broad range of interest. (Although this paper is quite comprehensive, describing fundamental packet-switching operations at a level of detail needed to understand the analytical comparisons being made, some aspects of the packet network operation may not be totally apparent—from this paper or, for that matter, from the preceding papers. These features will be covered in much more detail in the two papers reprinted in Section II of this volume.)

The paper by G.J. Coviello and R.D. Rosner extends these ideas further, by investigating the sensitivities of a large packet-based network to key design parameters of such a network. Of particular significance are the architectural design of the network, the data rates used on the trunk lines of the network, the balance of transmission and switching costs, and the life-cycle model used. In this paper, we show the interplay of these factors to arrive at a network optimization based upon user requirements and system data transmission rate. In addition, this paper provides a unique methodology for modeling and projecting future costs of the digital transmission facilities that are fundamental to the operation of packet-based data networks or, ultimately, to integrated voice and data systems.

NONTECHNICAL CONSIDERATIONS IN NETWORKS

The last paper in this section by Stuart L. Mathison deals with a number of principally nontechnical aspects of data networks and packet communications. It begins with the recognition that, historically, telecommunications has been a heavily regulated, monopoly-based, public utility whereas data processing has not. Since packet switching begins to blend aspects of each, difficult legal and regulatory issues are raised. The earliest commercial applications of packet communications within the United States were based upon the concept of "value-added service". Packet services were not seen by the regulatory commissions as being fundamentally competitive with basic communications because packetizing operations, through various technical enhancements including error protection, format and speed conversions, session control, and the like, "added value" to the services provided.

While the United States has taken a more pro-competitive approach to telecommunications, most of the other countries around the world continue to provide services through a monopoly-franchised, (often publicly owned) entity. Thus, while some of the issues raised in this paper are now being solved by the U.S. marketplace on a national level, they are still very important from the perspective of worldwide network operation. Some countries (Great Britain, Canada, and Japan) have begun to permit telecommunications competition, but the major issues of regulation and competition discussed in this paper are still largely unresolved.

An Introduction to Network Architectures and Protocols

PAUL E. GREEN, JR., FELLOW, IEEE

(Invited Paper)

Abstract—This tutorial paper is intended for the reader who is unfamiliar with computer networks, to prepare him for reading the more detailed technical literature on the subject. The approach here is to start with an ordered list of the functions that any network must provide in tieing two end users together, and then to indicate how this leads naturally to layered peer protocols out of which the architecture of a computer network is constructed. After a discussion of a few block diagrams of private (commercially provided) and public (common carrier) networks, the layer and header structures of SNA and DNA architectures and the X.25 interface are briefly described.

I. INTRODUCTION

EVER since computer users began accessing central processor resources from remote terminals over 25 years ago, computer networks have become more versatile, more powerful, and inevitably, more complex. Today's computer networks [1]-[6] range all the way from a single small processor that supports one or two terminals to complicated interconnections in which tens of processing units of various sizes are interconnected to one another and to thousands of terminals, often with various forms of special multiplexors and controllers in between.

As this evolution has proceeded, so have attempts to replace ad hoc methods of network design with systematic ways of organizing, understanding, and teaching about computer network details. Today there is a way of looking at networks in terms of *layered architectures* that all the experts use, but which is replete with its own jargon, and unclear and seemingly conflicting definitions, which often make it difficult to follow what is going on.

This paper aims at providing an introduction to how computer networks work from two perspectives: by briefly tracing the historical evolution of network implementations, and by summarizing some of today's layered architectures. These architectures are the rulebooks upon which the implementations are based, consisting of a collection of *protocols,* the rules by which physically separated entities interact.

In the following section, we analyze a list of the basic functions that the network provides in putting the parties that the network serves into communication with one another. This sets the stage for later discussion of layered architectures. Following these basic functions is a discussion of the evolution of private network implementations. Next, the closely related *standards* of the common carriers are introduced. Finally, matters previously presented are re-presented in terms of the underlying layered architectures and their component protocols.

II. A FRAMEWORK FOR DISCUSSING NETWORKS: THE TOTAL ACCESS PATH BETWEEN END USERS

A. Characterizing the Network

The basic function to be performed by any computer network is the provision of *access paths* by which an *end user* at one geographical location can access some other end user at another geographical location. Depending on the particular circumstances, the pair of end users might be a terminal user and a remote application program he or she is invoking, two application programs interacting with one another, one application program querying or updating a remote file, and so forth. By *access path* we mean the sequence of functions that makes it possible for one end user not only to be physically connected to the other, but to actually *communicate* with the other in spite of errors of various types and large differences in the choices of speed, format, patterns of intermittency, etc., that are natural to each end user individually.

There are many ways of characterizing networks, as, for example, the following: 1) according to the particular application (banking, time sharing, etc.), 2) according to geography (in-plant, out-plant), and 3) according to ownership (public, private), and so forth. Another way of characterizing different network types is to examine the topological character of the web of transmission lines that connect together the nodes at which the different end users are located and/or which perform some connection and message forwarding function. Here, a *node* is a physical box such as a computer, controller, multiplexor, or terminal and an *end node* is one where an end user resides. Thus, we have the various network types shown in Fig. 1.

None of these four approaches really reveals what the network is actually doing. A much better scheme is to examine the total repertoire of functions that the network must provide in making up an effective access path between two end users (Table I). By doing this in an ordered way, one is in a good position to characterize the important features of both common carrier networks (of the leased, dial, fast circuit-switched, and packet-switched types that we shall define later in this paper) and the network designs of computer manufacturers. (Two typical examples of the latter are the Systems Network Architecture (SNA) of IBM and DECNET of the Digital Equipment Corporation.)

B. Access Path Requirements

First, someone must make sure that a set of physical transmission resources (lines) exist that run from the origin node to the destination node, possibly by way of intermediate nodes.

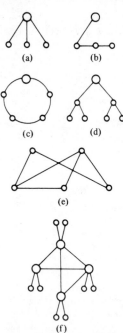

Fig. 1. Six network topologies. (a) Star. (b) Multidrop. (c) Loop. (d) Tree. (e) Mesh. (f) Mesh of trees.

TABLE I
ACCESS PATH REQUIREMENTS

To give a user access to processor-based resources, someone must:

Make sure a transmission path exists.	Using	Common carrier provided links.
See that it talks in bits.	Using	Modems.
Move individual messages.	Using	DLC's.
Provide economies for intermittent use.	Using	Dial-up; multidropping, multiplexing, packet and fast circuit switching.
Send messages to correct node and correct subaddress within node. Bypass failed line or station.	Using	Addressing, routing.
Accommodate buffer size; avoid need to resend long messages.	Using	Packetizing-depacketizing.
Resolve mismatches between actual and accommodatable flow rates.	Using	Buffering, flow control.
Accommodate end-user intermittency patterns.	Using	Datagram, transaction, or session dialogue management.
Accommodate end-user format, code, language requirements.	Using	Protocol conversions.

In out-plant situations (beyond one contiguous set of customer premises), this is done by *common carrier* provided links, either terrestrial or satellite.

Then one must see that the two ends of each line talk in bits using waveforms whose energy lies in a frequency range accommodated by the lines. *Modems* (modulator–demodulator units) provide this function [7]. One modem at the sending end of the line converts bits to analog waveforms, and a second modem at the receiving end converts analog waveforms back to bits.

A capability must also be provided for making sure that the

bit stream received is an error-free replica of the bit stream transmitted. This is one of the functions of *data link control* elements [8], [9], one at each end of the line, that see to it that successive groups of bits (frames) all arrive successfully at the receiving node. This is done by checking at the receiver after each frame to see whether there has been a violation of an error check of information bits against redundant bits that were added to each frame at the transmitter. If a frame is found to be in error, a retransmission is requested.

The art of Data Link Controls (DLC's) has advanced considerably from the simple but inefficient and inflexible asynchronous (start–stop) DLC's in which precious line capacity was wasted in adding to each character fixed bit patterns for synchronization. *Synchronous* character-oriented DLC's (such as BISYNC) alleviate many of the problems with start–top, but have proved to retain several disadvantages, notably that the same alphabet set (for example, ASCII or EBCDC) and the same positions in a frame are used for line control characters, text characters, and device control characters. Thus, a character of text could be spuriously converted by noise into a character that signals the end of a frame, for example. Another disadvantage of having line control characters drawn from the same alphabet as device control and text characters is that every time a new choice of alphabet is made for the peculiar needs of some particular end user, a new and different variant of the line control results. These difficulties as well as bit efficiency problems and other problems were alleviated in the new "bit-oriented" DLC's, such as the High Level Data Link Control (HDLC), Advanced Data Communication Control Protocol (ADCCP), as well as IBM's Synchronous Data Link Control (SDLC) Burroughs Data Link Control (BDLC), and Univac Data Link Control (UDLC), which are subsets of HDLC and ADCCP. In these protocols, line control information always occurs at its own same place in a frame. Thus, the time origin of the entire frame must be knocked out of line in order for link control and data to become confused, a much less likely circumstance than to have a character in error. The line control commands are specified as bit patterns that have nothing to do with any alphabet set. High Level Data Link Control is the standard being developed by the International Standards Organization; Advanced Data Communication Control Protocol is the standard of the American National Standards Institute.

The next problem to be faced is to exploit the intermittent ("bursty") nature of most end user traffic by sharing the capacity of one line across many such users. If each end user were to send bit streams at a constant rate, networks made entirely of simple point-to-point lines of the right capacity would be appropriate solutions. But since this is, in practice, almost never the case, a wide variety of *multiple access* schemes have been invented to realize economies of sharing the transmission medium across many stations [10]. For example, with multiple nodes per leased line comes the need to add to the DLC certain link address fields and control characters that are used by the DLC elements at each node to avoid conflicting attempts to use the line. Multistation DLC's thus perform *time division multiplexing* or interleaving of traffic from various stations on the same line. For rarely used connections, dial-up links offer a solution. Even more attractive economically are the new *fast circuit-switched* and *packet-switched* services with minimum billing times down to a fraction of a second and fractional-second time to connect. In *circuit switching,* the common carrier commits a path through his system until the users finish and break the connection. In *packet switching* [11], which aims at dynamically sharing intermittently used transmission resources, the user sends properly addressed frames or packets to the common carrier who delivers them individually.

The action taken in response to the addressing information is, of course, the *routing* operation. We have just encountered this addressing/routing requirement on a single link connecting several stations. When the nodes at which the end users are located are separated by not just one line but by one or more intervening nodes and links, addressing and routing become quite elaborate, particularly if there is a multiplicity of possible routes between the two end user nodes [12]. In such a topologically complex network, upon failure of a node or link, alternate path rerouting provides a powerful tool for recovery.

Before leaving the subject of addressing and routing, it should be noted that a line connected to a node often carries traffic to or from more than one location within that node. To resolve the ambiguity, an intranode addressing and routing function is required in such cases.

The next function that must be provided is the buffering of incoming messages from the line until they can be serviced, and the buffering of outgoing messages until they can be carried away by the transmission line. Limitations on available buffer size and the desire for fast response time, together with the aforementioned need to do error checking on a frame-by-frame basis (while seeing to it that the inevitable retransmissions do not take too long), lead to the need to segment (packetize) outgoing bit streams into elements of reasonable size and similarly to reassemble (depacketize) incoming bit streams.

Next, the rate of flow of outgoing packets has to be regulated so as neither to overflow the buffers at the receiving station nor to leave the receiving end user waiting for more traffic. This can be accomplished by feeding back along part or all of the access path from the receiving node to transmitting node special *pacing* or *flow control* signals. There are many options here. One may need to control rate of flow on an individual internode link to protect a buffer dedicated to that link at the same time that a completely different mechanism is controlling end-to-end flow to protect a buffer dedicated to an end user. The flow control signals sent from receiver to sender may simply turn off and on the emission of packets, they may tell the latter how many more packets can for the moment be safely sent, or there may be other strategies [13].

The next function needed is a way for the end user to use all the functions just listed to conduct a dialogue with the end user at the other end of the access path. The access path must be managed so that the dialogue between end users has the *pattern of intermittency* that the end users require. For example, the pair of users might be such that a single packet should flow in just one direction. This simplest case has been termed the *datagram* type of dialogue (actually a monologue) [14]. Or there might be a tightly structured *transaction* form of dialogue in which, for example, a single packet in one direction elicits a fixed number of reply packets in the other

direction. A third possibility is a *session* between end users in which the flow of packets back and forth is part of a related series of transactions. In analogy with a telephone conversation, it would be as though an access path were set up for each word, each sentence and its response, or for an entire telephone call, respectively. In managing the dialogue, there is the need not only to set up and take down the dialogue, but while it is in progress, to associate related packets with one another, and to decide when an end user should listen and when it should talk.

The last function required is to make sure, for each end user at a node, that the access path accommodates his pecularities with respect to such things as format, character code, device control, and database access conventions.

Once all the elements just listed are provided, the access path can be considered complete. This is shown in Fig. 2 where the actions just discussed are listed in order. Character streams typed in by the terminal user undergo a protocol conversion, then have sequence numbers and other information added for managing the dialogue, are then provided with addresses, are arranged in packets, and so forth. Two interesting things are immediately obvious: the elements occur in pairs, and the two members of each pair talk essentially only to each other. For example, one modem talks to the other, ignoring both details of the transmission link and the meaning of bits it is handling. As another example, a DLC element ignores what its modem is doing about modulation and demodulation and also what the information field within a frame contains. A DLC interacts only with the DLC at the other end to convey the frame successfully from one sending node to the next receiving node on the same line, and so forth.

C. Peer Interaction and Interfaces

This pairwise interaction, or *peer interaction,* of the functions we have enumerated is summarized in Fig. 3, which is derived directly from Fig. 2. Another way of thinking of Fig. 3 is that it is in a sense the inverse of one end user's view of a network. Thus, instead of showing one end user at the center of his network, we show the transmission facilities at the center and the two end users at the periphery. The access path across the network is depicted at the bottom for illustrative cases of zero and two intermediate nodes. Note that when the access path goes through intermediate nodes, in each intermediate node it goes no higher in the layered structure than the routing operation.

Layers up through (roughly) addressing-routing-packetizing provide a portion of the access path that hides from the end user much of the network detail. Functions outside this layer in Fig. 3 are often referred to as *higher level* functions.

Several caveats are in order about this seemingly tidy picture. For example, some generic functions can occur in more than one layer. Consider, for example, *multiplexing,* the interleaving of several traffic streams as they flow through the same path. We have already met this function in data link control. It also occurs (invisibly to the nodes) buried within the common carrier's transmission system. Moreover, several end users can be multiplexed on one transmission path, and as one proceeds from a set of end users at a sending node inward in the con-

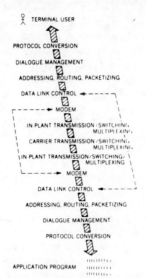

Fig. 2. Access path elements with dashed lines showing two examples of peer interaction.

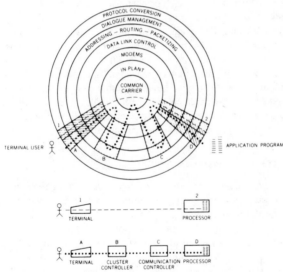

Fig. 3. Peer pairs of access path elements. (The modem may be absent in local in-plant connections.)

centric circles of Fig. 3, there is a choice of options as to the layer at which this merging might take place.

Another complication is that there is some interlayer communication of control information within the same node. This weakens the prior statements to the effect that the two peer-related members of a given layer at the two ends of the access path ignore the contents of the bit stream handed down by the next higher layer and are also not involved in the service provided to them by the next lower layer. For example, in an intermediate node, the routing function must supply to the DLC function an address it can use in forwarding a message to the proper choice of several stations on the same link.

The term *interface* has been widely used to describe the interactions *between adjacent layers* of Fig. 3, and we shall adopt

that convention here. For example, between the modem and the DLC level is the *electrical* (or *physical*) *interface* [15].

D. Network Control

Not shown in Figs. 2 and 3 is *network control* [16], the set of functions that do the activation and deactivation of the various portions of the access path shown, provide some of the control parameters required in their operation, and manage recovery. Network control can to various degrees be centralized (in one node) or decentralized (no single node dominant). The many network control functions that are required in forming the access path can be classified into several phases. One such rough classification is the following.

1) Establishing the electrical transmission path between nodes. This may involve dial-up, which requires that appropriate telephone numbers be supplied to a participating node.

2) Assigning data link addresses of stations, designating who is primary or secondary, and activating the DLC-level function.

3) Establishing and updating routing tables that tell each node where to forward a message. If the message must proceed on to another node, the table must say which outgoing link to use.

4) Establishing and updating directories of all end users in the network, and providing name-to-address conversion.

5) Establishing and later disestablishing the datagram, transaction, or session connection out to the end users. Parameters must be supplied at each end to set up the specific dialogue convention required by the end user at that end. Queues of requests and responses within a session must be managed.

6) Providing an interface to the human network manager. This includes problem determination functions, such as error reporting, testing, sending traces, and making measurements.

In this section, we have introduced the notion of layers of function as they occur in peer-related pairs to form an access path through the network. We have also mentioned the control of these functions. Before discussing how these ideas are manifested in specific network protocols of the computer manufacturers and the public common carriers, let us return to a topological view of things and examine in a little more detail what computer networks look like from that standpoint.

III. NETWORKS OF COMMERCIALLY PROVIDED ACCESS PATHS

A. Early Systems

In order to discuss the rationale of access path implementations that have been of most interest, it is instructive to sketch the historical evolution of private networks since the 1960's. Let us look first at what has happened with large computers, then minicomputers, and then common carrier computer network services.

The earliest systems were single-processor batch systems that later evolved to support a few local terminals. True teleprocessing (remote access of a terminal end user to an application program in a processor) came with systems such as that shown in Fig. 4, of which a typical example was the IBM System/360 running the Basic Telecommunications Access Method (BTAM). Essentially all the processing was concentrated in the central host processor, as befitted the technology available at that time. Of the various access path functions we have enum-

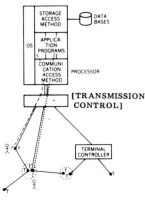

Fig. 4. Typical teleprocessing system of the 1960's such as System/360; dotted and dashed lines are access paths.

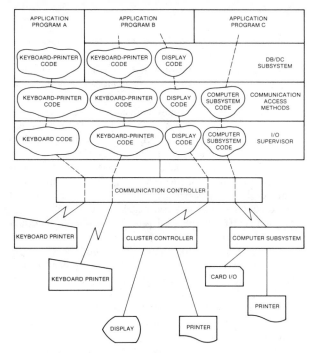

Fig. 5. Distribution of terminal-specific code in an early teleprocessing system.

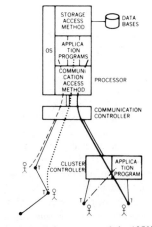

Fig. 6. Typical teleprocessing systems of the 1970's such as System/370 with SNA.

erated in Table I, only elementary DLC-level functions were performed outboard of the host, specifically in a *transmission control unit,* which was often hard-wired and not programmable. The other functions were never cleanly layered, as in Fig. 3, but were so spread out among the different software systems (as shown in Fig. 5) that a change in the configuration of a line or its attached terminal required reprogramming in all these software systems. Terminal cluster controllers performed the device control functions, but essentially none of the communication access path functions. What proved to be a particularly inconvenient restriction was the lack of *line sharing* or *terminal sharing.* By this is meant that, since a given line and all the terminals on it were part of the access path to only one and the same application program, if a user wanted to access two different applications (e.g., savings accounts and credit checking), he required two terminals and two lines.

The next step in commercially provided networks came around 1974, with systems such as that of Fig. 6, of which a typical example was the System/370 with software and hardware releases referred to as Systems Network Architecture (SNA) generations 1 and 2 [17], [5]. The transmission control unit gave way to a programmable *communication controller* that handled all data link control and a great deal more. In the communication controller code, the host communication access method code, and the cluster controller code, a significant attempt was made to delineate function into layers, as in Fig. 3. Thanks to the availability of microcomputers and the lowered cost of main and secondary storage, it began to be possible to execute limited application code, including that involving significant databases, in the cluster controllers, and (for some non-IBM realizations) in the communication controller. Most significantly, this design allowed terminals to share a line to separate applications located in the same host and to do the same thing with applications in the cluster controller. Moreover, it allowed access paths between host application programs and cluster controller application programs.

B. Computer Networking

It was soon clear that functions of commercially provided networks did not go far enough. The ARPANET [18], developed under U.S. Defense Department sponsorship, had shown how a number of *resource sharing* functions could be provided, and it was soon found that such functions were needed by customers of the computer manufacturers. Specifically, many commercial network users had multiple processors individually serving *tree networks* such as that of Fig. 6. These networks could not intercommunicate. A given terminal user frequently wanted an access path to an application *in a different host* from the one that normally served him, and it was either uneconomical or infeasible to run a second copy of that application in his own host just to provide this service. Moreover, it became desirable for one application to talk to a remote other application. These capabilities were needed for sharing processor resources among locations and for improving system availability through remote backup. These requirements led to the *computer networking* solution shown in Fig. 7(c), realized, for example, by IBM in Systems Network Architecture with Advanced Communication Function (SNA/ACF), which is also known as SNA-3 [19]. The most recent version (SNA 4.2) is described in [20]. In this arrangement, any terminal can gain an access path to any of the applications in any of the hosts. Application-to-application access paths are also supported. Fig. 7(c) shows several of the tree structures of Fig. 6 (schematized in Fig. 7(b), just as Fig. 7(a) abbreviates Fig. 4) connected together into a mesh of trees [as in Fig. 1(f)] by physical paths between communication controllers. Thus, an SNA tree network can be characterized as a *hierarchical* network with network control centralized in the processor (actually in a module called the System Services Control Point (SSCP) located in the communications access method). SNA/ACF is a *hybrid* peer-hierarchical structure, that is, hierarchical within each tree or *domain* (with its own SSCP), but with peer interconnection between trees at the level of the host-attached communication controllers. Not shown in the diagram is the *multitail* capability of communication controllers, in which one such controller can support more than one processor. (Also, one processor can support several controllers.)

In the world of minicomputers, networks have evolved somewhat differently. Originally, minicomputers were used individually for stand-alone, real-time, or batch processing or for supporting a few simple terminals. When the need developed for connecting these together, it was found desirable to do this in a strictly *peer* style of interconnection rather than the peer-plus-hierarchical pattern just discussed. Peer connection had been used in the ARPANET, and the flexibility of this mode of operation undoubtedly had a strong influence on minicomputer networking. In the peer mode of interconnection, no one computer does network control for the other; there is no master/slave distinction and there need be no identifiable central control point. Network control steps are managed in each node more or less symmetrically. In principle, this allows a wide range of topologies to be implemented, but requires special procedures for managing routing tables, flow control, directory functions, and recovery operations, especially when the network consists of a large number of nodes.

One of the better known of the peer computer network designs is the set of DECNET offerings of the Digital Equipment Corporation based on Digital Network Architecture (DNA) [21]. The DECNET design has been implemented not only for the minicomputers of the DEC product line (e.g., PDP-11), but also for the high end (e.g., DECSYSTEM-20). The ultimate objective is to connect the machines together in a mesh [as in Fig. 1(e)] or in a hierarchy [as in Fig. 7(d)], or other arrangements. In fact, a natural user evolution for minicomputer users has been for independent users to start with stand-alone minicomputers of roughly

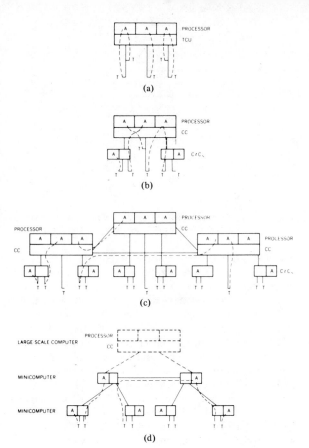

Fig. 7. Schematization of access paths. (a) abbreviates Fig. 4. (b) abbreviates Fig. 6. (c) Top-down network of trees. (d) Bottom-up approach of DECNET.

equal power, later to connect them together, and still later to connect this set to a single large host. This *bottom-up* evolutionary pattern may be contrasted with the *top-down* pattern of network growth experienced by many users of large machines, as just described.

The ARPANET, which had a great influence on all succeeding computer networks, whether commercially or carrier provided, embodied [22] a mesh-connected backbone network of many small Interface Message Processors (IMP's), connected together by a packet switching (Section IV-B) IMP-to-IMP protocol. Most computers were connected into the network by means of the Host-to-IMP protocol, very roughly equivalent to the later X.25 Interface, which we shall describe in the next section. (The ARPANET Host-to-Host protocol is roughly equivalent to a *virtual circuit*.) Specially augmented IMP's, called Terminal Interface Processors (TIP's) provided the additional terminal handling software to allow terminal connection into the backbone network. Special higher level software was provided in the hosts, for example, to support interactive terminals (TELNET) or to effect bulk file transfers (FTP).

IV. NETWORKS OF ACCESS PATHS PROVIDED BY CARRIERS

In commercially provided networks, such as the IBM and DEC offerings just described, the physical transmission-level function between nodes in the network is, of course, provided by the common carriers. The carriers have been investigating whether there is any technical reason why other functions of Fig. 3 at a higher level than the transmission level might not also be provided by them—for example, protocol conversion. Several references, e.g., [23], detail the status of common carrier offerings and data network interfaces.

A. Fast Circuit Switching

The common carriers are, in fact, taking steps not only to improve service at the transmission level, but to provide higher level services. At the transmission level, an urgent need of the data processing community has been to have dial-up service with much faster connect times and much shorter minimum billing increments than ordinary voice grade dial-up service provides. There has also been the need to improve the space-division modem-to-machine interface, such as V.24 (known in the United States as RS-232C), by providing a combined space- and time-division interface of wider generality. These needs have been met partially by the X.21 Recommendation [24] of the international standards body, the International Consultative Committee for Telephony and Telegraphy (CCITT). The 21 (or fewer) wires of V.24, each performing one and only one function, are replaced in X.21 by up to eight wires, of which one is used in each direction to send bit patterns for specific control functions. By this means, the repertoire of control functions is flexible and expandable. But the real significance of X.21 is not as an electrical interface, but as a peer protocol below the DLC level, to be used for dialing and disconnecting at data processing bit stream speeds, thus serving as the basis of *fast circuit-switching* common carrier networks.

B. Packet Switching

Packet switching [11] seems to have been inspired by the idea of sharing communication channel capacity across a number of users by implementing the same time-slicing philosophy that had earlier proved so successful in sharing the execution power of a single processor across many user processes. Every user node that connects to a packet-switched common carrier makes a contract with the carrier (i.e., follows standard protocols) to hand him bit streams already segmented (packetized) as we have described in the beginning of this paper, with each packet supplemented with a header saying, among other things, to which other user node he wishes the packet delivered. Widespread interest in packet switching on the part of the carriers has led them to standardize this contract in the form of the CCITT Recommendation X.25 [25], [26], [14], which is discussed further in the next section.

The contract includes an agreement on the electrical interface, the data link control, how the remote user is to be addressed, packet size, how the flow of packets toward and out of the carrier's network is to be regulated, and how certain recovery actions are to be effected. The contract also includes some network control functions such as protocols for establishing and disestablishing the access path. Thus, two user nodes (say A and B) each agree to exchange packets with the carrier network using the X.25 standard, and the carrier agrees to deliver to B properly addressed packets from A and vice versa. The combined actions of: 1) the X.25 interface of A to the network, 2) the X.25 interface of B to the network, and 3) the network provide a full duplex path, termed a *virtual circuit*, between the higher level function at the two nodes.

Actually, if one adopts the definition of an *interface* we have been using (Section II-C), then X.25 is not, strictly speaking, an interface (although often called one), but a set of layered *peer* interaction protocols, by which a machine talks to a packet network (as we shall see in Section V-D).

There is currently some debate over whether a degenerate form of virtual circuit, called the *datagram* mode of operation and referred to earlier in this paper, should be supported under X.25. There, the duration of the contract is essentially only one packet long [14].

Fast circuit switching and packet switching both offer the user the economies of paying for the transmission service only to the extent that it is used. Fast circuit switching has the particular advantage over packet switching that once the transmission path has been set up, it is totally transparent. That is, except for uncontrollable random errors, the bit stream out is the same as the bit stream in for a period of time whose duration is up to the user. Packet switching, although highly nontransparent (since the user is required to adhere to what the contract says about packet length, rate of flow, header structure, etc.) does allow the carrier to offer the user more of the access path function discussed earlier in this paper than does fast circuit switching, and it allows him many freedoms in buffering, delayed delivery, etc.

V. NETWORK ARCHITECTURES AND PROTOCOLS

A. Architecture Versus Implementation

The precise definition of the functions that a computer network and its components should perform is its *architecture*. Exactly by what software code or hardware these functions are actually performed is the *implementation*, which is supposed to adhere to the architecture. Both the data processing and carrier communities have expressed their network ideas in layered peer architectures that in one way or another resemble Fig. 3. Communication architecture is different from processor architecture or storage subsystem architecture in that it usually involves a *pairwise* interaction of *two* parties (although there are a few exceptions, such as routing [12] or distributed directory protocols [27], in which more than two parties are involved). For example, as we have said earlier in this paper, a DLC element in one node interacts with a DLC element in another; the flow control functions in two nodes interact specifically with each other, and so forth. The set of agreements for each of these pairwise interactions may be termed a protocol, and thus we find network architecture specified in terms of *protocols for communication between pairs of peer-level layers*. A network protocol consists of the following three elements: 1) *syntax*—the structure of commands and responses in either *field-formatted* (header bits) or *character-string* form; 2) *semantics*—the set of requests to be issued, actions to be performed, and responses returned by either party; and 3) *timing*—specification of ordering of events.

We shall now briefly discuss SNA, DNA, and X.25 from this point of view, saying something about semantics and syntax, but nothing about timing. All three of these structures make strict definitions of protocols between the two members of a pair of functions at the same level (although in different nodes), but usually leave details of interaction of adjacent layers in the same node (interfaces) to be decided by the implementer. They are all slightly different in the way they assign functions to the different layers, in spite of the fact that these assignments may at first glance appear to be equivalent. The SNA and DECNET architectures are different in kind from X.25. The former two manage the access path from end to end. On the other hand, as originally conceived, X.25 is not an end-to-end protocol, but a node-to-packet network protocol; it manages the access path from a user node to the immediately adjacent node internal to the packet network. End user to end user functions are built up by a concatenation of the two X.25 paths between each user and the network, plus the internal network paths, as noted in Section IV-B. Recent work has strengthened the end-to-end message integrity provisions of common carrier networks that use X.25 into and out of the network [26].

B. SNA

Fig. 8 shows the layers of two SNA nodes. No intermediate nodes are shown, but in practice, one or more of these could exist along the access path. Furthermore, the layers at one end could be in more than one physical box. For example, at the host end, all functions could be in the host (as in the System/370, Models 115 and 125), or functions roughly corresponding to Data Link Control and Path Control could be in the software of a separate communication controller and the rest in the host. Or it might be possible to move almost all the access path functions out to a *front-end communications processor*, leaving the host processor freer to concentrate its resources on application processing. At the terminal end, all the functions shown might be in the same box in the case of an "intelligent terminal," or almost all except the upper layer might be in the cluster controller that supports a number of "dumb" terminals.

The functions of the SNA protocol layers are as follows [5], [20], [28].

1) *Data Link Control* (DLC) transfers packets intact across the noisy transmission facility. For every line attached, there is one instance of DLC or *DLC Element* (DLCE).

2) *Path Control* [29] (PC) routes incoming packets to the appropriate outgoing DLCE or to the correct point within its own node. It allows alternate path routing between nodes and the use of several sets of parallel DLC facilities ("transmission groups") between a node pair for better reliability and throughput. It also does packetizing of outgoing and depacketizing of incoming messages. There is one instance of PC per node. The pair of PC's at the two end nodes provide to the higher protocol layers a set of eight *virtual routes* upon which sessions may be built, with flow control within each virtual route.

3) *Transmission Control* (TC) manages pacing (flow control for an individual session), helps manage session establishment/disestablishment, and performs a number of other functions on behalf of one of the end users. There is one instance of TC, namely, a Transmission Control Element (TCE), per session per end user. Each TCE can be thought of a one end user Session's "front office" to the communication network.

4) *Data Flow Control* [30] (DFC) has the function of accommodating the idiosyncrasies of message direction and intermittency demanded by the end user. Such idiosyncrasies include, for example, whether a user wants to communicate duplex or half-duplex or whether the separate messages (RU's) are parts of larger units of work as seen by the end user. For example, different RU's might represent different lines of text that make up a single display screen of text. A screen full of lines of text would be handled by SNA as a *chain*. A structured set of related screens, plus the messages flowing in the other direction, would be handled by SNA as a *bracket*, i.e., a set of chains. There is one instance of DFC per end user session.

5) *Presentation Services* [30] (PS) define the end user's port into the network in terms of code, format, and other attributes. The pair of PS realizations in the pair of nodes have the job of accommodating, for example, the totally different interfaces seen by a terminal end user (and his supporting device control hardware or code) and the application that is being accessed. The PS layer (and other layers as well) are designed for flexibility as to the fraction of the complexity that lives at each member of the peer pair. It is thus possible to have a small or even null PS function in a simple terminal while doing most of it in the processor. As has been mentioned, it is possible optionally to have not just one, but a number of concurrently operating "sub-end users" for each end user (as we have employed the term *end user*), so that a form of multiplexing (using so-called *FM headers*) takes place at the PS level that is roughly analogous to that at the DLC level.

There are a number of other SNA functions that have to do with network control [16], but which are too detailed for a discussion here. These network control functions involve a separate family of access paths that emanate from the System Services Control Point, which might be in some other node not shown, and terminate in modules (also not shown) that control the various functions shown in Fig. 8.

The function of the various headers is shown in Fig. 8.

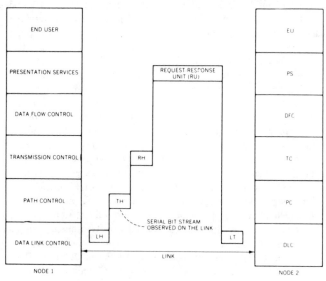

Fig. 8. SNA architectural layers. Compare with Fig. 3. The Request-
Response Unit (RU) is usually converted user information.

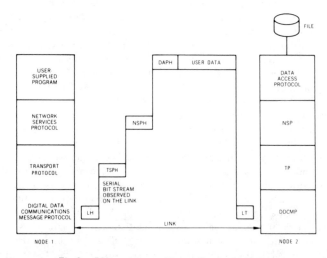

Fig. 9. DNA architectural layers. Compare with Fig. 3.

The zig-zag strip shows the bit stream that would be observed on the line. On an outbound message, TC adds to the RU a Request/Response Header (RH) on behalf of itself and DFC, PC adds a Transmission Header (TH), and DLC adds a Link Header (LH) and Link Trailer (LT). Inbound, each layer strips off its appropriate header (and trailer) and forwards what is left. If there is multiplexing within PS, there is still another header, namely, the Function Management (FM) header, not shown. All of this illustrates the following important property of peer protocols: it is by means of the *header* that belongs to a given layer of the protocol that the interaction of the peer pair constituting that layer takes place.

C. DNA

The architecture on which the DECNET implementations are based is DNA (Digital Network Architecture) [20]. In the DNA set of protocols, illustrated in Fig. 9, there are four basic layers, of which the bottom three are vendor-provided, and the top one is a user implementation, or (in the case of file access) vendor-supplied. The bottom three layers of DNA correspond roughly to the bottom three layers of SNA, as shown in Fig. 8, but with some interesting differences. The Data Link Level is exactly the DCL level of Figs. 3 and 8, the preferred realization being the DEC line control, Digital Data Communications Message Protocol (DDCMP). DDCMP is character-oriented (like BISYNC), but has many of the characteristics of bit-oriented DLC's. As in HDLC, for example, control and data characters are distinguished positionally.

The Transport Layer uses the Transport Packet Header (TPH) for its peer communication. Each packet, with its associated TPH, is handled as a datagram, i.e., each packet is handled by the *Transport Protocol* (TP) as a stand-alone unit, and TP guarantees only a "best effort" to deliver the packet. It will, however, guarantee that if the packet has not arrived by a certain time, then it will never arrive. Successive packets may follow different routes as the TP routing algorithm in each node responds to changing connectivity conditions in the network, so that packets may arrive out of order. Packet loss can occur due to temporary line outages, due to action of TP flow control (which allows for the relief of congestion by discarding packets), or because the packet had exceeded the age limit without being delivered and had to be subjected to involuntary retirement. The seeming disadvantages of uncertainty of delivery and ordering buys considerable simplicity in Transport Protocol, compared, for example, to SNA Path Control, where tight control is maintained over connectivity, sequentiality, and guaranteed arrival.

In DNA, sequentiality and guaranteed arrival are restored (if required by the user) in the Network Service Layer, using sequence numbers, acknowledgments, and *timeouts*, much as with Data Link Controls. That is, if a packet having a given sequence number arrives successfully, this is acknowledged to the sending *Network Services Protocol* (NSP), and transmission proceeds to the next packet; but if the sender has to wait longer than a certain delay before hearing from the receiver, it retransmits the missing packet. Other NSP functions include end-to-end control and packetizing/depacketizing.

The basic access path provided by the two NSP's at two end nodes (using a TP protocol pair per hop, in turn supported by a DDCMP protocol pair per hop) is called a *logical link*. The Application Layer of DNA provides a means for a number of concurrent user processes to communicate with partners across the network, each using a separate logical link. Usually this layer is user-implemented, but a number of file access and distributed file management options can be built using the venter provided Data Access Protocol (DAP), as illustrated in Fig. 9.

The Network Control functions mentioned in Section II-D are almost completely decentralized in DNA. For example, logical links are activated and deactivated by commands to the NSP from the Application level process. In SNA, a session between end users at separate nodes is set up and taken down by a third party, the System Services Control Point, which might be in one of the two nodes or might be in a third node.

D. X.25

The X.25 protocol is illustrated in Fig. 10. The X.21 protocol, defined earlier in this paper as an interface, is used as a peer protocol for providing the electrical connection between the user node and the nearest Data Circuit Termi-

nating Equipment (DCE) node owned by the carrier. The X.25 specification allows for use of X.21 bis (in which the interface appears to each user as a V.24 interface) as an interim solution. In Fig. 10, stations 1 and 2 are the Data Terminal Equipments (DTE's), i.e., the user's end nodes. Packets P1 and P3 are intended for station (DTE) 2 and packet P2 is intended for some other station. The Frame Level protocol, which manages error-free transfers of strings of packets to and from the packet network, is equivalent to the DLC layer of SNA and the Physical Link layer of DNA. The Frame Level protocol uses one of two variants of HDLC. The preferred one at the moment appears to be "Link Access Protocol B," specified as the full-duplex Asynchronous Balanced mode of HDLC. Here each of the two DLC stations is neither solely a primary station nor a secondary station, but a "combined" station that is able to take responsibility unilaterally for transmission and recovery.

The Packet Level protocol produces the Virtual Circuits (VC's) referred to earlier. There may be one or many (as in Fig. 10) VC's multiplexed onto one access line. These may be *fixed* (assigned upon initial subscription to the service and always in place) or *switched* (invoked as needed). (A switched VC is also known as a *virtual call*.) These virtual circuits have end-to-end aspects during setup or takedown of the VC and end-to-network aspects otherwise. For example, flow control operates only to regulate traffic between the user node and the network. After a VC is initially set up, the addressing is between each end node and the network, not between end users. These are clearly end-to-network functions. But in initially establishing the VC, the end-user node must know how to address the other end-user node. This is clearly an end-to-end function.

As Fig. 10 shows, there are two X.25 interfaces between each of the two customer-owned end nodes and the network. The packet carrier appears in this diagram in roughly the position where a single intermediate node would appear in Figs. 8 and 9. If an SNA or DECNET system operates across an X.25 packet carrier facility, there are some divided responsibilities. For example, the SNA and DECNET implementations have specific rules about packet size, addressing/routing, flow control, internal multiplexing of flows, and recovery from error and lost- or duplicated-message conditions. When X.25 services are used, these responsibilities may overlap with those that the carrier is willing to undertake. There is a growing literature (e.g., [31]) discussing how these overlaps may be resolved.

F. Open System Architecture[32]

Before ending this brief review of network protocols, architectures, and implementations, it should be mentioned that there is considerable interest and activity in the standards bodies that have defined HDLC, X.21, X.25, etc., in standardizing even higher level functions than those represented by the Packet Level of X.25. The object is to allow any-to-any (i.e., *open*) interconnection capability for communication products. This is being attempted by adding four more layers

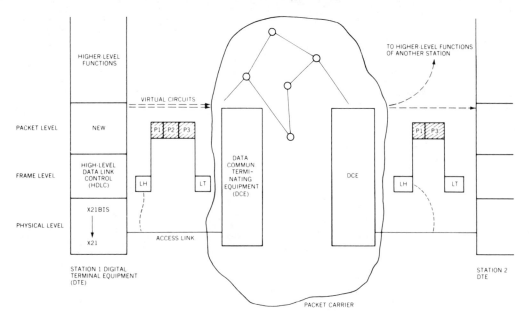

Fig. 10. Layers and virtual circuits in X.25.

above the X.25 Packet Level, making seven in all. (The bottom four layers provide end-to-end access path function roughly equivalent to the bottom three levels of DNA shown in Fig. 9 and the bottom three levels of SNA shown in Fig. 8.) This ambitious undertaking rapidly becomes as much of a data processing end-user issue as a communications issue, and because of the bewildering variety of special end-user needs to be accommodated, it is the author's opinion that this difficult task will succeed only very slowly.

VI. CONCLUDING REMARKS

Even though networks have been growing more complicated, they should be getting easier to dissect and understand as systematic formalization and layering become more pervasive in the implementations. One reason for persistence of complexity is that, until now, the architects have carried a heavier burden than is commonly realized of maintaining compatibility with individual software and hardware product offerings that antedated the evolution of systematic, clearly layered sets of network protocols. These earlier offerings are gradually disappearing or in later releases are adhering more and more to the strict terms of the architecture. The modularization means that new ideas ought to be more easily incorporated without producing system-wide disruptions. Continuing research will provide such new ideas.

ACKNOWLEDGMENT

The author thanks R. J. Cypser, A. Endres. R. F. Steen, and the reviewers for their helpful comments.

REFERENCES

[1] D. W. Davies and D. L. A. Barber, *Communications Networks for Computers.* New York: Wiley, 1973.
[2] P. E. Green Jr., and R. W. Lucky, Eds., *Computer Communications.* New York: IEEE Press, 1975
[3] L. Kleinrock, *Queuing Systems, Vol. II* New York: Wiley, 1976.
[4] M. Schwartz, *Computer Communication Network Design and Analysis.* Englewood Cliffs, NJ: Prentice-Hall, 1977.
[5] R. J. Cypser, *Communications Architecture for Distributed Systems.* Reading, MA: Addison-Wesley, 1978.
[6] D. W. Davies, D. L. A. Barber, W. L. Price, and C. M. Solomonides, *Computer Networks and their Protocols,* New York: Wiley, 1979.
[7] J. R. Davey, "Modems," *Proc. IEEE,* vol. 60, pp. 128–1292, Nov. 1972. Reprinted in [2].
[8] J. W. Conard, "Character-oriented data link control protocols," this issue, pp. 445–454.
[9] D. E. Carlson, "Bit-oriented data link control procedures," this issue, pp. 455–467.
[10] F. Tobagi, "Multi-access protocols in packet communication systems," this issue, pp. 468–488.
[11] R. E. Kahn, Ed., *Special Issue on Packet Communication Networks, Proc. IEEE,* vol. 66, Nov. 1978.
[12] M. Schwartz and T. E. Stern, "Routing techniques in communication networks," this issue, pp. 539–552.
[13] L. Kleinrock and M. Gerla, "Flow control algorithms: A comparative survey," this issue, pp. 553–574.
[14] H. C. Folts, "X.25 transaction oriented features-datagram and fast select," this issue, pp. 496–500.
[15] H. V. Bertine, "Physical level protocols, this issue, pp. 433–444.
[16] J. P. Gray, "Network services in systems network architecture," *IEEE Trans. Commun.,* vol. COM-25, pp. 104–116, Jan. 1977.
[17] J. H. McFadyen, "Systems network architecture: An overview," *IBM Syst. J.,* vol. 15, no. 1, pp. 4–23, 1976. See also three companion papers in the same issue.
[18] L. G. Roberts and B. D. Wessler, "Computer network development to achieve resource sharing," in *1970 AFIPS Conf. Proc (SJCC),* vol. 36, pp. 543–549.
[19] *Introduction to Advanced Communication Function,"* Order No. GC30-3033, IBM Data Processing Div., White Plains, NY, 10504.
[20] J. P. Gray and T. B. McNeill, "SNA multiple-system networking," *IBM Syst. J.,* vol. 18, no. 2, pp. 263–297, 1979.

[21] S. Wecker, "DNA: The digital network architecture," this issue, pp. 511–526.

[22] See, for example, the papers on ARPANET reprinted in [2].

[23] J. Halsey, L. Hardy, and L. Powning, "Public data networks: Their evolution, interface, and status," *IBM Syst. J.*, vol. 18, no. 2, pp. 223–243, 1979.

[24] H. C. Folts, "Procedures for circuit-switched service in synchronous public data networks," this issue, pp. 489–496.

[25] A. Rybczynski, B. Wessler, R. Despres, and J. Wedlake, "A new communication protocol for accessing data networks—The international packet mode interface," in *AFIPS Conf. Proc. (NCC)*, vol. 45, June 1971, pp. 477–482.

[26] A. Rybczynski, "X.25 interface and end-to-end virtual circuit service characteristics," this issue, pp. 500–510.

[27] J. Bremer, and O. Drobnik, "Specification and validation of a protocol for decentralized directory management,"IBM Research Ctr., Yorktown Hts., NY, Tech. Rep. RC-7800, Sept. 25, 1979.

[28] *SNA Format and Protocol Reference Manual*, Order No. SC30-3112, IBM Data Processing Div., White Plains, NY 10504.

[29] J. D. Atkins, "Path control: The transport network of SNA," this issue, pp. 527–538.

[30] V. L. Hoberecht, "SNA function management," this issue, pp. 594–603.

[31] F. P. Corr, and D. H. Neal, "SNA and emerging international standards," *IBM Syst. J.*, vol. 18, no. 2, pp. 244–262, 1979.

[32] H. Zimmermann "The ISO model for open systems interconnection," this issue, pp. 425–432.

COMMUNICATIONS SYSTEMS

A.L. DUDICK, E. FUCHS and P.E. JACKSON

Bell Telephone Laboratories, Holmdel, New Jersey, USA

This paper reports the results of traffic studies of data stream parameters of a class of inquiry—response computer communications systems. Systems in this class employ TOUCH-TONE® telephone signals for input of digital data; responses are generated by voice answerback units under computer control. It is shown that, while traffic models of these systems preserve some of the tailor-made properties of the specific installation studied, we are able to describe the important traffic variables in terms of robust, analytically tractable random variables. In a few cases, where the observed physical behavior of the traffic parameters of interest cannot be described adequately by single random variables, mixtures of analytically tractable random variables are found to yield statistically accurate representations. While such mixtures do not as readily lend themselves to straightforward analytical treatment in system studies, they are easily accommodated in simulation studies.

INTRODUCTION

This is the third of a series of papers that report the results of studies of traffic characteristics of multiaccess computer communication systems. The first two papers [1,2] described traffic models of systems with long call duration used for on-line programming and problem solving. This paper reports similar results for a limited class of systems with short call durations used for inquiry—response applications. This class of systems employs TOUCH-TONE telephones or TOUCH-TONE pad adjuncts to dial telephones for input of digital data inquiries. Responses are generated by voice answerback units under computer control.

Data for this study were collected from four systems; three of the systems operate in the telephone network while the fourth employs a private switcher. To acquire the data, specially designed logic equipment was placed at the interfaces between two computer communication ports and the data sets (or modems) on the communications lines of each system. The purpose of these logic units was to enable us to capture the communication characteristics and timing relationships of the traffic traversing the lines without violating the privacy of the transaction or the anonymity of the users. The logic units transferred the observed information via private lines to a small, dedicated computer at Bell Laboratories where the captured information and timing relationships were recorded on magnetic tape for further processing.

Each system was monitored continuously (during all hours the system was in service) for two to three weeks. During this period, data describing 5,000 to 10,000 calls were recorded for each system.

The vehicle used for analysis of the data and presentation of results is a data stream model, which is described in the next section. Following that, the relationships among the model variables are presented. Finally the observed data and statistical analyses are presented.

Sufficient data were collected and analyzed to ensure that the distributions presented as descriptive of the model variables are robust in the following senses: First, the distribution forms are stationary over time; scale and location parameters may change but the shapes do not. Second, the forms of the distributions remain the same for different system operating loads, different system applications and for systems employing different computer hardware. Thus only the parameters of the distributions need be changed in system studies and analyses to reflect different operating environments.

DATA STREAM MODEL

The model used to depict the user—computer data stream is illustrated in fig. 1. The first division of the data stream is into user and computer "segments" or "times". Each such segment is the longest period of time during which either the user or computer transmits information to the other party with no intervening periods of reception of information. The user time is composed of an initial inactive period called the "think time" followed by the user input, a succession of distinct characters (TOUCH-TONE signals). The time for each character is represented in the model as a "character width" or "character time," and the associated interval between successive characters as an "inter-character time". The computer time consists of an initial period of inactivity called the "idle time" followed by an "audio response time". In the time from the last audio response of a call to disconnect, one of two distinct events occurs. In a normal call, this time is referred to as "dropoff time". In some calls, the computer program is not equipped to complete the transaction and a credit manager takes over in place of the computer. In this case, the time is called the "referral time". Additional variables necessary to characterize the data stream completely are the length of each call (connect—disconnect) called the "holding time", the "inter-call time",* the "number of characters per user segment", the "number of user segments per call", and the "number of computer segments per call".

RELATIONSHIPS AMONG DATA STREAM MODEL PARAMETERS

Let the following notation be introduced:

τ = holding time of a call (seconds)
W = character width (seconds)
G = intercharacter time or gap (seconds)
N^w = number of characters per user segment
I = idle time (seconds)
T = think time (seconds)
A = audio response time (seconds)
R = referral time (seconds)
D = dropoff time (seconds)
N^c = number of computer segments per call
N^u = number of user segments per call
N^r = number of referrals per call ($N^r = 0$ or 1).

The two indices of sumation to be used are: i to designate the ith segment (user or computer), j to designate the jth character in a user segment.

The sum of these components is the holding time for a call:

* The inter-call time observable at a computer port is a function of the number of ports provided in the system for a given amount of originated traffic. Thus it is a variable that can be changed at the will of the system provider and measurements of it cannot be generalized.

$$\tau = \sum_{i=1}^{N^u} \left[\sum_{j=1}^{N_i^w} W_{ij} + \sum_{j=1}^{N_i^w - 1} G_{ij} + T_i \right]$$
$$+ \sum_{i=1}^{N^c} (I_i + A_i) + N^r R + (1 - N^r)D. \quad (1)$$

Knowing the distributions for the 11 parameters in the expression for holding time, it is theoretically possible to solve directly for the distribution of τ. Except for very restricted cases, however, the mechanics of finding the solution are prohibitive.

One particular case for which a solution is readily obtainable is where the discrete variables used as the limits of summation are geometrically distributed and where the continuous variables being summed are exponential. For this case, Feller [3] has shown that the resulting random variable is exponentially distributed. For the long holding time systems examined earlier [2], it was shown that the discrete random variables were geometrically distributed and that the continuous random variables were gamma distributed but could be approximated by the exponential distribution for many cases of interest. Hence, the holding time was approximately exponentially distributed for those cases.

In later sections of this paper, distributions fitted to samples of each of the important random variables including holding time are displayed. In addition to examining the distributional form of the holding time components, the mean values of these components are of interest in performing systems analyses.

Taking expected values of both sides of (1) where we assume that the random variables are stationary and mutually independent,** we obtain

$$H = N_u[N_w W + (N_w - 1)G + T]$$
$$+ N_c[I + A] + N_r(R - D) + D$$

where H is the average value of holding time and the symbol for each of the other random variables without a subscript of i or j and with u, c, w, and r used as subscripts rather than as superscripts indicates its expected value.

** While this assumption is an obvious approximation to reality, it is a meaningful approximation for the analyses of system behavior discussed in this paper.

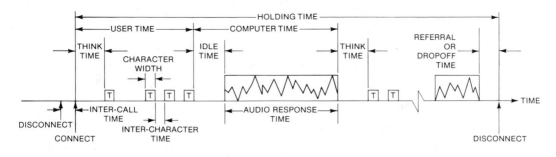

Fig. 1. TOUCH-TONE remote access data stream model.

For further analysis, this expression for H may be separated into five parts each having its own functional significance:

(a) User send time, U_s (the total amount of time during which user characters are being transmitted): $U_s = N_u N_w W$

(b) Computer send time, C_s (the total amount of time during which the computer is transmitting): $C_s = N_c A$

(c) User delay time, U_d (the sum of all inactive periods during user segments): $U_d = N_u(T + (N_w - 1)G)$

(d) Computer delay time, C_d (the sum of all inactive periods during computer segments: $C_d = N_c I$

(e) Call termination time: $C_t = N_r R + (1 - N_r)D$

Finally, we can derive expressions for the total send time, S (the total time data is being transmitted during a call) and for total delay time, Δ (the total inactive time during a call) by:

$$S = U_s + C_s$$
$$\Delta = U_d + C_d + C_t$$

OBSERVED DATA AND ANALYSIS

In this section, we present the quantitative results of analyses of data on a large number of calls to each of four short-holding time, multiaccess computer systems. These systems are labelled A, B, C, and D. Systems A and B have the same user applications with different equipment configurations and data input modes. Systems A, C, and D have the same equipment configuration but different applications. Systems A and B serve occasional users relatively unsophisticated in their use of the systems, whereas the users of systems C and D are more proficient from frequent use of the systems. These systems and user characteristics are summarized in table 1. None of these systems were compute-bound or in any way heavily loaded. In fact, the inquiry-response traffic used only a small portion of the real-time of each of the systems.

Table 1
Monitored systems' characteristics

	System A	System B	System C	System D
Computer Type	IBM 360/40	Honeywell H2206	IBM 360/40	IBM 360/50
Audio Response Unit Type	IBM 7770	Cognitronics 285–8D	IBM 7770	IBM 7770
Primary Application	Credit Bureau	Credit Bureau	On-Line Banking	Production Control
Data Input Mode	Manual TOUCH–TONE and Card Dial	Manual TOUCH–TONE	Manual TOUCH–TONE	Manual TOUCH–TONE
User Proficiency	Low	Low	Moderate	High

Table 2
Mean values of holding time components

		Systems		
	A	B	C	D
H – Holding Time (seconds)	40.	43.	28.	20.
T – Think Time (seconds)	2.1	2.1	.28	1.3
W – Character Width (seconds)	.17	.16	.24	.14
G – Intercharacter Time (seconds)	.42	.41	.37	.31
I – Idle Time (seconds)	1.7	1.8	.22	.89
A – Audio Response Time (seconds)	4.4	8.8	17.	2.0
R – Referral Time (seconds)	77.	33.	N/A	N/A
D – Dropoff Time (seconds)	13.	9.6	0.	.97
N^w – No. Characters/User Segment	28.	20.	18.	5.2
N^u – No. User Segments/Call	1.1	1.5	1.0	3.1
N^c – No. Computer Segments/Call	1.0	1.3	.92	2.9
N^r – No. Referrals/Call	0.13	0.28	N/A	N/A

The data gathered from the four systems were laundered to remove obvious measurement errors and ambiguities (such as two contiguous "connects" without an intervening disconnect) and then partitioned into sets describing each of the 11 variables of the model and the holding time. For each variable for each system, means and variances of the variable were estimated and several well-known distribution functions were tested for goodness of fit. Two techniques were used to test goodness of fit. First, the set of unimodal distributions discussed by Fuchs and Jackson [2] were tested using the new goodness of fit test developed by Jackson (see Appendix in [2]), In each case, the parameters of the hypothesized distribution (that distribution being tested) were adjusted so that the mean and variance for a two-parameter distribution were the same as the sample mean and sample variance. For a single-parameter distribution, the mean of the distribution was equated to the sample mean.

Although the unimodal distributions were sufficient to describe the variables of the long-holding-time systems investigated earlier, several variables could not be described by a single, unimodal distribution for the short-holding-time systems. Our initial analyses of these data indicated that in some cases a mixture of the same common distributions was appropriate. Consequently Pack developed a new procedure for fitting mixtures of distributions to empirical data [4]. Using Pack's procedure we first estimated the mixture of distributions which most closely fit the data under the criterion of minimum mean squared error and then tested goodness of fit using the Kolmogorov-Smirnov test.

Table 2 summarizes the average values of each of the variables of the traffic models for each of the four systems. Table 3 summarizes the results of fitting the data for each of the variables to standard unimodal and sums of unimodal distributions. It must be emphasized that the objective of the distribution fitting exercise is not to find the best fit or the fit which may be hypothesized with the highest confidence to be a model of the source process. Rather, a model is desired that describes the statistical data at least at the five percent level of significance, that has properties that reasonably well fit the physical process modelled, and that best lends itself to traffic analyses and systems studies.

Examining table 2, we find that the holding times range from 20 sec to 43 sec with a grand mean of 33 sec and a standard deviation about

Table 3
Results of goodness of fit tests
acceptable* distributions**

Random Variable	Systems			
	A	B	C	D
H — Holding Time	Γ	Γ	$\Sigma\Gamma$	Γ
T — Think Time	$\Sigma\Gamma$	Γ	Γ	Γ
W — Character Width	Γ	Γ	Γ	Γ
G — Intercharacter Time	$\Sigma\Gamma$	$\Sigma\Gamma$	Γ	Γ
I — Idle Time	$\Sigma\Gamma$	$\Sigma\Gamma$	Γ	Γ
N^w — No. Characters/User Segment	NR	NR	NR	G
N^u — No. User Segments/Call	G	G	G	G
N^c — No. Computer Segments/Call	E	E	B	E

* Acceptable at the five percent level of significance.

** Γ — gamma distribution,
$\Sigma\Gamma$ — sum of two gamma distributions,
E — exponential distribution,
B — Binomial distribution,
G — geometric distribution,
NR — nonrandom process such as a fixed response time.

that mean of 9.5 sec.* Two components of the holding time that only serve to obscure the interpretation of the data flow timing are the Dropoff Time, the time from the end of the last transaction to disconnect, and the referral time. For example, System C allows only one transaction per call; the computer therefore disconnects immediately at the termination of its response, resulting in a dropoff time which is identically zero for all calls. For the other systems, more than one transaction per call is allowed. Therefore, the computer initiates a timeout interval (which differs from system to system) after each response, waiting for either an indication of terminal disconnect (carrier-off) or additional input. At the end of the time-out interval, the computer disconnects. The dropoff time is therefore a race between near-end and far-end disconnect and is peculiar to the selected time-out interval for each system and the communication configura-

tion in which it is imbedded. We therefore find it useful to eliminate the dropoff time from our model by subtracting its average value from the average holding time to arrive at a modified holding time, and by removing it from our set of random variables. Although its magnitude and causal factors are of interest to the telephone companies, it is not a relevant component of our systems or traffic studies.

Similarly, the referral time represents a non-data use of the communications channel and is not relevant to the flow of information from the data terminal to the computer. In addition, the fraction of systems employing referrals in their operations is declining as the number of fully automated short holding time data systems increases.

Our examination of the data and our understanding of the processes underlying the audio response time variable have led to the conclusion that it should not be characterized as a random variable. Neglecting for a moment calls aborted during an audio response, we find that the several systems have one or two- at most three-standard length response types, the difference between responses of constant length being solely in the choice of words at various opera-

* It should be noted that the estimated mean holding time does not necessarily equate to the sum of the estimated means of the holding time components. The difference occurs because the test used for removing outlier data points from the data set for each variable is specific to the characteristics of that data set, and each data set was measured independently.

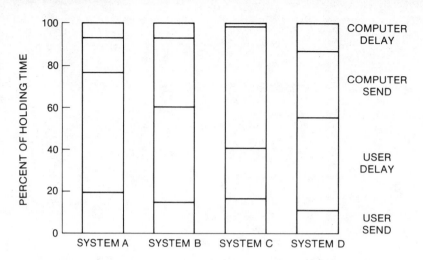

Fig. 2. Holding time of components as percent of adjusted holding time.

tive positions in the phrases of the response. Thus, the observed data on response length is distributed as one or two or three impulses, modified only by aborted calls and measurement noise.

Indeed the audio response time given for System C is not even meaningful. It results from an overlap of transmission of the last TOUCH-TONE digit of an inquiry and the computer-generated audio response. Our measurement equipment was designed to be faithful to our half-duplex model. Hence, it suppressed recording the beginning of the audio response until the end of the TOUCH-TONE digit transmission.

Fig. 2 is a graphic characterization of the data shown in table 2. The macroscopic variables are displayed as a percentage of the modified holding time. We focus our attention first on the average values of the model variables and make the following observations.

The human-factor determined averages (think time, character width and intercharacter time) for the two systems with the same applications are virtually identical. Further, the interarrival times of characters (a derived variable in the model, being the sum of character width and intercharacter time) exhibits a marked and important robustness between all systems. The grand mean character interarrival time for characters within user segments is 0.553 sec with a standard deviation of 0.042 sec. This variable is an important one; it is employed in statistical multiplexing analyses that are directed at har-

nessing efficiently the statistical properties of multiaccess communications.

The length of each user segment varies from a low of 5.2 characters for the moderately interactive System D to a high of 28 characters for system A. Systems A, B, and C were noninteractive as each normal transaction had only one user segment and one computer segment. System C allowed only one transaction per call while Systems A and B allowed multiple transactions. Notice, however, that the average number of transactions per call was close to one for these systems. In contrast, the long holding time systems had a mean of 41 user segments per call (called "Burst Segments" by Jackson and Stubbs [1]).

In contrast to the extensive generalizations that were made for long holding time systems based upon or derived from the average values of the data stream model variables, there is little more of general significance that one can draw from the data on the mean values of the parameters of short holding time systems. The short duration of each transaction preserves the tailor-made properties of the individual systems in most of the observed communication characteristics. For example, the average total delay per call (sum or user delay and computer delay) as a percent of adjusted holding time varied from a low of 25% of System C to a high of 64% for System A and was greater than 50% for three of the systems. In addition, as each of the computers was lightly loaded, the idle times were all short and conclusions about system loading cannot be made similar to those for the long holding time systems.

To some lesser extent, this property of individuality of system characteristics is also carried over to the more important characterization of the distributional form of the model variables.

The continuous random variables in the model can all be described by either the gamma distribution or by a sum of gamma distributions. So, while the model retains much of the robustness of the model for long holding time systems, here we found it necessary to fall back on a mixture of distributions (the sum of two gamma distributions). Thus the simplicity of the long holding time model is somewhat diminished and the comments by Fuchs and Jackson [2] on approximating certain gamma distributions by the exponential distribution must be re-evaluated. For many types of analyses, though, the sum of gamma distributions is tractable for analysis and it certainly is useful for studies of large-scale networks using simulation techniques.

The discrete variables exhibited considerably more variability in the form of the acceptable distributions. The number of characters per user segment was nonrandom in three of the systems as the data given to the computer consisted of a fixed-length character string. Only in System D could N^w be described as a random variable, where it could be described by a geometric distribution.

Both the discrete exponential distribution and the geometric distribution have exponential tails, but the discrete exponential distribution has mass at the origin (a nonzero probability of the random variable having the value zero) while the geometic distribution has no mass at the origin. The geometric distribution could be used to characterize the number of user segments per call, while the number of computer segments per call is described by the discrete exponential distribution for every system but one. These results are consistent with the long-holding time system results.

SUMMARY

The results presented in this paper provide to those engaged in research and design of inquiry-response computer communication systems esti-

mates of significant system random variables. It has been shown that for most cases, assumptions ordinarily made in system studies about the behavior of input processes are supportable within tolerable error bounds.

REFERENCES

[1] P.E. Jackson and C.D. Stubbs, A study of multiaccess computer communications, AFIPS Conference Proceedings, Volume 34.
[2] E. Fuchs and P.E. Jackson, Estimates of distributions of random variables for certain computer communications traffic models, Commun. ACM, (December, 1970).
[3] W. Feller, An Introduction to Probability Theory and its Applications, Vol. II, (Wiley, New York, 1966) 53–54.
[4] C.D. Pack, A constrained curve-fitting procedure based on Prony's method, unpublished paper.

APPENDIX

The "Constrained Prony Procedure" is a curvefitting procedure based on the classical Prony method.* Given, the sampled data and a hypothesized distribution family (e.g., exponential) the problem is to determine the value of the parameters associated with the distribution (at least) two alternatives are available for determining the parameters:

(1) estimation by such means as maximum likelihood, moment matching (as used by Jackson's "Fourier Series Tests"), etc.,

(2) attempting to find the member of the given family of distributions (say exponential) which is the best fit in some sense, to the empirical histogram formed from the sampled data.

Either method determines estimates of the parameters associated with the given distributional form. It is the method (2) which is used by the Constrained Prony (Pack's) Procedure. Constraints must be imposed in order to ensure that the parameters are real** and are within allowed ranges (e.g., the exponent of the exponential must be negative). The procedure allows for fitting mixtures of distribution from a large selection of classical distributions of the form

$$f_x \approx \sum_{i=1}^{K} C_i \mu_i^x, \quad x = 0, 1, \ldots, N-1$$

where ‡

$$C_i = P_i D(\lambda_i)$$

$$\mu_i = B(\lambda_i)$$

λ_i = parameter of distribution i of a given family

P_i = mixing proportion of distribution i.

Included in the distributions of this form are the binominal, negative binominal, Poisson, geometric, Erlangian, gamma, normal, and exponential.

The problem is then to

$$\min_{\{C_i\},\{\mu_i\}} \left\| f_x - \sum_{i=1}^{K} C_i \mu_i^x \right\| \quad \text{(in some sense)}$$

such that,

$F_i(x) = 1 - \exp(-\lambda_i x)$, $D(\lambda_i) = \lambda_i$, $B(\lambda_i) = \exp(-\lambda i)$.

$$P_i = C_i/D(\lambda_i) \geqslant 0$$

$$\underline{\lambda} \leqslant \lambda_i = D^{-1}(C_i/P_i) \leqslant \bar{\lambda}$$

$$\sum_{i=1}^{K} P_i = 1$$

$$P_i, \lambda_i \in R'.$$

This nonlinear programming problem is solved by applying the Prony transformation and some theorems from numerical analysis to obtain a sequential (and suboptimal) solution to the problem in either a sum of least squares or sum of absolute values sense.

Having obtained estimates of the parameters, classical (Kolmogorov–Smironov) goodness of fit tests are applied in the standard manner.

---

* See F. Hildebrand, Introduction to Numerical Analysis (McGraw-Hill, 1956).

** Because the Prony procedure involves transformation extraneous, nonreal roots are often created.

‡ For the exponential distribution

A study of multiaccess computer communications

by P.E. JACKSON and CHARLES D. STUBBS

Bell Telephone Laboratories, Incorporated
Holmdel, New Jersey

INTRODUCTION

The communications characteristics of multiaccess* computing are generating new needs for communications. The results of a study of multiaccess computer communications are the topic of this paper. The analyses made are based on a model of the user-computer interactive process that is described and on data that were collected from operating computer systems. Insight into the performance of multiaccess computer systems can be gleaned from these analyses. In this paper emphasis is placed on *communications considerations.* For this reason, the conclusions presented deal with the characteristics of communications systems and services appropriate for multiaccess computer systems.

The problem

Digital computers requiring communications with remote terminals exhibit a set of communications needs which, in some respects, are different from those of both voice traffic and other record communications. It is important for the providers of data communications to have an understanding of the broad characteristics of this communication process so that new, more appropriate offerings can be designed to satisfy these needs.

Previous studies** by the manufacturers and providers of multiaccess computer systems have begun to characterize both the computer systems and their users. The principal interest of these studies, however, has been computer and/or user performance rather than data communications.

---

* The word "multiaccess" is chosen to avoid confusion over the use of the word "time-shared" which is often used synonymously but which has a specialized meaning in some contexts.

** For example, see References 1, 2, 3 and 4.

There are several reasons why the quantitative characterization of the communications process is timely but intricate. First, multiaccess computing is still in its infancy. Therefore, computer system design is going through a trial and error process with a high rate of change of system characteristics. Lacking a unified, well-tested body of technical knowledge applicable to the problems of multiaccess computing, systems designers have been led to heuristic solutions to system organization. Certain specific problems such as scheduling algorithms for single and multiple central processors have been studied in detail. No intensive, overall, general system studies, however, have been reported with the constraints of total cost minimization including the effects of system characteristics on communications costs and human factors such as reduction in efficiency due to long turn-around times.

Second, the rate of change of the size of the user community, the number of systems in operation, and the introduction of new equipment and operating systems is high. In fact, most systems are changing so rapidly that a detailed characterization of any one will probably be outdated before it is completed. The insight to be gained from such studies, however, far outweighs the drawback of obsolescence. Indeed, this situation calls for continued study and review.

Third, the applications of time-sharing are diverse. Where one of the parties in the transaction is a person, uses range from inquiry-response systems with short call durations of a minute or less, to scientific problem-solving and certain types of business information systems with call durations of 10 to 30 minutes, to computer aided learning with long call durations of one to two hours or more. Where the transaction involves an automatic terminal such as a telemetry device, call durations may be measured in milliseconds. Also, the volume of information exchanged in a computer-to-computer or computer-to-data-logger interaction varies widely from a small number of bits in polling, meter reading and some banking and credit services, to a large number of bits in CRT displays, information retrieval and file manipulation. The speed of transmission is wide-ranging from the low bit rates of supervisory and control terminals to megabits per second for CRT displays.

Fourth, the data required for such studies are microscopic in nature. Unlike voice traffic, which can be characterized by measures of holding times, arrival rates and other parameters independent of a call's content, the characterization of calls to a computer requires some information about a call's content, e.g., timing information interrelating the transmission times of data characters is essential for the design of an efficient time division data multiplexer. An additional factor is that some of the desired statistics on these data have very skewed distributions. Thus, large data samples are required. The implications of these considerations upon our study are that:

a. new data gathering procedures and equipment are needed,
b. data analysis procedures must be capable of handling very large quantities of data,[5]
c. legal, ethical, and business requirements related to communications and computing privacy must be satisfied.

The problem, then, is to provide communications services to a rapidly growing market of multiaccess computer systems and their terminals. These exhibit diverse and changing communications requirements. The study described below is directed at this problem.

The modus operandi for this study is an in-depth analysis of selected multiaccess computer communications systems. The subset of system types chosen for detailed study is composed of computer service providers whose systems are representative of multiaccess computer installations. Besides representativeness, additional prerequisites for the choice of a system to study were that the use of multiaccess computing be advanced, and that the provider of the particular system be knowledgeable in the communications area. By "advanced in multiaccess usage," we mean that the system be fully operational on a daily basis with the initial break-in period accomplished. A final prerequisite for inclusion in the study is the willingness of the computer service provider to participate in the study.*

---

* Part of the study reported herein involved the collection of data from three operating multiaccess computer systems. In every case these data were obtained on the premises of the computer service provider and with this full permission and cooperation. To ensure the privacy of the three systems under discussion, however, they are not identified by name.

To ensure that a cross section of on-line systems was included in the study, the characteristics of such systems were classified as shown in Table I.

Table I—*Classification characteristics for multiaccess computer systems for communications study*

---

1. Computer Type
2. I/O Device Type
3. Loading (Number of Simultaneous Users)
4. User's Applications
5. User Community (In-House or Utility)
6. Error Control (e.g., Echoplex)
7. Holding Time

In the table, by "computer type" we mean the manufacturer and model number of the central processor and the system configuration, i.e., whether or not a separate communications computer is used. Not all models of all manufacturers can be covered, but at least two large manufacturers were included for each application. I/O device types include teletypewriter-like terminals and TOUCH-TONE® telephones. Loading is the average percentage of ports that are active. User's applications include scientific and business programming, inquiry-response systems, extended file retrieval and maintenance, message switching and mixtures of these. Both in-house and utility systems were included. Error control includes systems which retransmit each character back to the terminal (Echoplex) and those which do not. The systems selected for examination include short holding time systems with average call durations on the order of one or two minutes or less and long holding time systems with average holding times of 20 to 30 minutes.

From the systems selected for detailed analyses, measurements of three different categories were obtained. The first category included telephone facilities measurements such as occupancies and overflow counts on computer access lines (port hunting groups) and pen recordings of call durations from several terminal lines. The second category of measurements was made by the computer service providers within their computer systems by identifying the arrival and departure times of calls, the amount of central processor time used, the serving port and an identifier of call type. Distributions of call holding time, call interarrival time, CPU usage, and port loading can be obtained from these data. The third category of measurement was the collection of data at computer ports describing the characteristics of such microscopic statistics as intercharacter time. The first two categories of data are being used to formulate traffic and engineering practices. These will be used by telephone company personnel to provide appropriate computer communications by properly configuring existing telephone company equipments.

The third category of data is being employed in the analyses reported here. These data are required to investigate new systems and service characteristics such as the desirability of various transmission speeds or multiplexing methods, as they include detailed information on the timing relationships within a call.

An analytic model of the communications process between a multiaccess computer and a user at a remote console is the vehicle being used to conduct these analyses. The model describes the communications process in terms of random parameters which give the times between characters transmitted through the communications network. All of the parameters are measurable at the communications interface to the computer, i.e., none requires the gathering of data on internal computer processes such as the length of various queues.

The model is used to focus on the user-computer communications process and to exhibit how the characteristics of the computer and of the user affect communications requirements. It is also used to study the converse, i.e., how the constraints of the communications medium affect the user and the computer. The model does not directly represent the detailed characteristics of the computer system or its organization or the internal operation of the user's console. Rather, it reflects the effect of these on the characteristics of the communications signals entering and leaving the computer. From the characteristics of the communications process, however, it is possible to employ the model to predict the effects of changing system characteristics such as improving computer scheduling algorithms or increasing the computer's transmission rate. The following two sections further discuss this model.

The data stream model

The next two sections develop the data stream model, the analytical model used to describe the stochastic interactive communications process between user and computer. In this section, the basic parameters of the model are defined. In the next section, the relationships among the parameters are described and an expression for the holding time of the process is developed where holding time is the duration of a user-computer session.

Figure 1 illustrates the data stream model. A "call" (or a connect-disconnect time period) is represented as the summation of a sequence of time periods during which the user sends characters without receiving, interleaved with time periods during which he receives characters

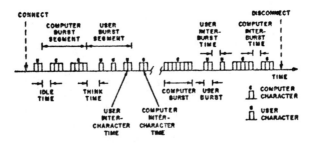

Figure 1—The data stream model

without sending. (This implies half-duplex operation. Simple modifications to the model would allow the accommodation of full-duplex operation.) The periods during which the user is sending characters to the computer are defined as user burst segments. The periods during which he is receiving characters sent from the computer are computer burst segments. A user burst segment, by definition, begins at the end of the last character of the previous computer burst segment. Similarly, a computer burst segment begins at the end of the last character sent by the user. The first burst segment of a call begins when the call is established and the last burst segment ends when the call is terminated as measured at the computer interface.

Within a given burst segment, there are periods of line activity and of line inactivity. The first inactive period of a user burst segment is defined as think time. That is, think time is the time that elapses from the end of the previous computer character until the beginning of the first user character in that burst segment. In most cases, think time is employed by the user to finish reading the previous computer output and to "think" about what to do next. The corresponding inactive period in a computer burst segment is called idle time. In some systems idle time represents time during which the user waits for the return of "line feed" after sending "carriage return"; in other systems, idle time represents time during which the user's program is being processed or is in queue. The remaining inactive periods within a burst segment are called intercharacter times and interburst times. A prerequisite for their definition is the definition of a "burst."

Two consecutive characters are defined as belonging to the same burst if the period of inactivity between the characters is less than one-half character width. Thus, each "burst" is the longest string of consecutive characters where the period of inactivity between any two consecutive characters is less than one-half character width. All of the characters in a burst must, of course, be transmitted from the same party (user or computer). For example, every character of an unbroken string of characters sent at line speed is in the same burst.

For characters within the same user burst, an inactive time between two consecutive characters is called a user intercharacter time. The corresponding parameter for computer bursts is computer intercharacter time. For bursts within the same user (computer) burst segment, the inactive time between two consecutive bursts is called a user (computer) interburst time. Five final parameters of the data stream model are number of user bursts per burst segment, number of computer bursts per burst segment, number of characters per user burst, number of characters per computer burst, and temporal character width (time from start to end of one character).

For a given user-computer environment, a knowledge of the distributions of the parameters defined above allows the calculation of some interesting measures. Examples are distributions for (a) holding time, (b) percent of holding time during which the communication channel carries data, and (c) amount of delay introduced by the computer. The next section shows how some of these distributions can be calculated from the parameters.

Relationships among data stream model parameters

Let the following notation be introduced:

τ = holding time of call (seconds)
S = number of burst segments in call
T = think time (seconds)
I = idle time (seconds)
B = interburst time (seconds)
N = number of bursts per burst segment
M = number of characters per burst
W = character width (seconds)
C = intercharacter time (seconds)

The lower case letters "c" and "u" will be used as superscripts to B, N, M, W, and C to represent "computer" and "user" respectively. For example, N^c will represent the number of computer bursts per computer burst segment. The three indices of summation to be used are:

i—to designate the ith burst segment,

j—to designate the jth burst of a given burst segment,

and

k—to designate the kth character of a given burst.

In summing expressions over these indices, the primary index will be shown as a subscript and the secondary index (or indices), if any, will be enclosed in parentheses.

Using this notation, it is possible to construct an equation relating the holding time of a call to its component parts in the following manner:

a. In burst segment 2i, in the jth burst, the amount of time required by the kth user character is $W_k^u(2i, j)$. Summing over all k such characters in the burst, the time required is

$$\sum_{k=1}^{M_j^u(2i)} W_k^u(2i, j).$$

Summing over all bursts in the burst segment gives

$$\sum_{j=1}^{N_{2i}^u} \sum_{k=1}^{M_j^u(2i)} W_k^u(2i, j).$$

If one defines all burst segments where i is even as user burst segments and assumes that the number of burst segments per call (S) is always even, the total contribution of user character times to total holding time is

$$\sum_{i=1}^{S/2} \sum_{j=1}^{N_{2i}^u} \sum_{k=1}^{M_j^u(2i)} W_k^u(2i, j).$$

Since S is usually large the error introduced by assuming S even, even when it is not, is small. Assuming the odd numbered burst segments are computer burst segments, the corresponding contribution of computer character times is

$$\sum_{i=1}^{S/2} \sum_{j=1}^{N_{2i-1}^c} \sum_{k=1}^{M_j^c(2i-1)} W_k^c(2i - 1, j).$$

b. A corresponding argument shows the total contributions of user intercharacter times are

$$\sum_{i=1}^{S/2} \sum_{j=1}^{N_{2i}^u} \sum_{k=1}^{[M_j^u(2i)-1]} C_k^u(2i, j)$$

and of computer intercharacter times are

$$\sum_{i=1}^{S/2} \sum_{j=1}^{N_{2i-1}^c} \sum_{k=1}^{[M_j^c(2i-1)-1]} C_k^c(2i - 1, j).$$

c. The total amount of user interburst time is

$$\sum_{i=1}^{S/2} \sum_{j=1}^{[N_{2i}^u-1]} B_j^u(2i),$$

and the total computer interburst time is

$$\sum_{i=1}^{S/2} \sum_{j=1}^{[N_{2i-1}^c-1]} B_j^c(2i-1).$$

d. Total think time is

$$\sum_{i=1}^{S/2} T_{2i},$$

and total idle time is

$$\sum_{i=1}^{S/2} I_{2i-1}.$$

The sum of these components is the holding time for a call. That is, the time of each burst segment summed over all burst segments is the holding time. The time of a burst segment equals the sum of the durations of all bursts, interburst times, and the think (or idle) time in that burst segment. The duration of a burst is equal to the sum of the character times and the intercharacter times contained therein. That is, the holding time of call l, τ_l, is

$$
\begin{aligned}
\tau l = \sum_{i=1}^{S/2} \Bigg\{ & \Bigg[T_{2i} + \sum_{j=1}^{[N_{2i}^u-1]} B_j^u(2i) \\
& + \sum_{j=1}^{N_{2i}^u} \sum_{k=1}^{M_j^u(2i)} W_k^u(2i,j) \\
& + \sum_{j=1}^{N_{2i}^u} \sum_{k=1}^{[M_j^u(2i)-1]} C_k^u(2i,j) \Bigg] \\
& + \Bigg[I_{2i-1} + \sum_{j=1}^{[N_{2i-1}^c-1]} B_j^c(2i-1) \\
& + \sum_{j=1}^{N_{2i-1}^c} \sum_{k=1}^{M_j^c(2i-1)} W_k^c(2i-1,j) \\
& + \sum_{j=1}^{N_{2i-1}^c} \sum_{k=1}^{[M_j^c(2i-1)-1]} C_k^c(2i-1,j) \Bigg] \Bigg\}.
\end{aligned}
\tag{1}
$$

Knowing the distributions for the 12 parameters in Equation (1), it is theoretically possible to solve directly for the distribution of holding time. The mechanics of

finding the solution are prohibitive, however, except for very restricted cases. One method of solving (1) is to find the moments of holding time rather than the complete distribution. This approach will be used here and, in fact, it will be sufficient for our purposes to solve merely for the mean value of holding time. In order to arrive at the solution, we assume that the random variables are stationary and mutually independent.*

Taking the expected value of both sides of (1), we obtain

$$
\begin{aligned}
\tau = (S/2)[& T + B^u(N^u - 1) + N^u M^u W^u \\
& + N^u C^u(M^u - 1) + I + B^c(N^c - 1) \\
& + N^c M^c W^c + N^c C^c(M^c - 1)]
\end{aligned}
\tag{2}
$$

where the symbol for each variable without a subscript implies its mean value. For further analysis, Equation (2) may be separated into four parts each having its own functional significance:

a. user send time (the total amount of time during which user characters are being transmitted)

$$= (S/2)(N^u M^u W^u),$$

b. computer send time (the total amount of time during which computer characters are being transmitted)

$$= (S/2)(N^c M^c W^c),$$

c. user delay (the sum of all inactive periods during user burst segments)

$$= (S/2)[T + B^u(N^u - 1) + N^u C^u(M^u - 1)],$$

d. computer delay (the sum of all inactive periods during computer burst segments)

$$= (S/2)[I + B^c(N^c - 1) + N^c C^c(M^c - 1)].$$

The sensitivity analysis performed in the next section is an investigation of the properties of these four parts and includes a discussion of how their interrelation affects holding time. If any of these parts can be reduced without increasing others, holding time can be reduced leading to possible cost savings.

---

* Analyses of these assumptions have exposed their limitations. However, these assumptions have been shown to be reasonable for the analyses and conclusions of this paper.

Before discussing such analyses, it may be well to indicate what values have been observed for each of the 12 parameters. This will allow us to concentrate our attention on those parameters and those measures of parameters that promise to be the areas of greatest possible holding time reduction.

Collected data and sensitivity analyses

During the current study, data have been gathered on a large number of calls to each of several multiaccess computer systems. For each system, the data have been partitioned into sets representing each of the 11 random parameters (the twelfth parameter, character width, is a constant). Probability density functions have been fitted to the data collected on each parameter from each system.[5]

Data from three of the systems are discussed in this paper. These systems are labeled A, B, and C. Systems A and B have the same computer equipment and basically the same mix of user applications (programming—scientific). System C has computer equipment different from that of the other two systems and its mix of user applications is primarily business oriented. All three systems serve low-speed teletypewriter-like terminals. System B is rather heavily "loaded" compared to Systems A and C. Table II summarizes these characteristics for Systems A, B, and C.

Table II—Characteristics of systems studied

	System A	System B	System C
Computer Type	Brand X	Brand X	Brand Y
Transmission Speed (Characters/sec)	10	10	15
Primary Application	Scientific	Scientific	Business
Load	Moderate	Heavy	Moderate

Table III summarizes the measured values of the model parameters. To ensure the privacy of the three systems under discussion, these values are not shown on a per system basis. Rather, for each parameter,* an average value $\hat{\mu}$ is given where $\hat{\mu}$ is the average of the

---

* As the character widths W^u and W^c are treated in the model as random variables, they are included in Table III for completeness. In the three systems discussed, however, they were constant as can be derived from Table II.

three system averages for the given parameter. The numbers in the column headed $\sigma\hat{\mu}$ are the standard deviations of the three numbers averaged in $\hat{\mu}$ for each parameter. The analyses subsequently reported, however, are based on the actual per system average values of these parameters.

Table III—Average parameter values

		$\hat{\mu}$	$\sigma\hat{\mu}$
S	—No. of Burst Segments	82.	37.
T	—Think Time (sec.)	4.3	3.4
I	—Idle Time (sec.)	.65	.48
B^u	—User Interburst Time (sec.)	1.6	.90
B^c	—Computer Interburst Time (sec.)	16.	25.
N^u	—No. of Bursts/User Burst Seg.	11.	3.1
N^c	—No. of Bursts/ Computer Burst Seg.	3.3	2.8
M^u	—No. of Characters/ User Burst	1.1	.12
M^c	—No. of Characters/ Computer Burst	47.	27.
$W^u(=W^c)$	—Character Width (sec.)	.089	
C^u	—User Inter- character Time (sec.)	.00021	.00023
C^c	—Computer Inter- character Time (sec.)	.00030	.000090

One characteristic of the data summarized in Table III deserves further comment. It is that measures which should be most sensitive to computer characteristics seem to just that. For example, the users of both Systems A and B have predominantly programming-scientific applications and the average numbers of characters per user burst segment ($N^u M^u$) are 9.2 and 10.7 for Systems A and B, respectively, versus 13.8 for the primarily business oriented users of System C. Such relationships prevail in spite of widely different average computer delays. The average amount of time spent per computer burst segment in interburst delay, (N^c-1)B^c, is 1.4 seconds in System A and 35.8 seconds in System B.

	System A	System B	System C
Average Holding Time, τ			
Minutes	17.	34.	21.
Average User Send Time, $(S/2)(N^u M^u W^u)$			
Minutes	0.50	0.45	0.96
% of τ	3%	1%	5%
Average Computer Send Time, $(S/2)(N^c M^c W^c)$			
Minutes	5.7	4.5	7.5
% of τ	33%	13%	35%
Average User Delay, $(S/2)(T + B^u[N^u - 1] + N^u C^u[M^u - 1])$			
Minutes	10.	12.	11.
% of τ	58%	35%	53%
Average Computer Delay, $(S/2)(I + B^c[N^c - 1] + N^c C^c[M^c - 1])$			
Minutes	0.95	17.	1.5
% of τ	6%	51%	7%

Table IV summarizes the macroscopic characteristics of these data as they contribute to holding time. An inspection of the table leads to the following observations:

Observation 1: The average holding time for the heavily loaded system (System B) is considerably larger than for the lightly loaded Systems A and C (94 percent and 60 percent larger).

Observation 2: For the lightly loaded systems, A and C, Computer Delay is less than 10 percent of the total holding time but for System B Computer Delay accounts for over half of the total holding time.

Observation 3: User Delay is a significant component of holding time in all three systems, and in each case, is between 10 and 12 minutes.

Observation 4: User Send Time is less than 5 percent of total holding time in each system and is not a significant contributor to holding time.

Observation 5: Computer Send Time is smallest in both absolute value and in percent holding time for the heavily loaded System B.

These five observations lead to three broad areas of interest that are discussed in the next three sections. The first area is the relationships between holding time and computer delay (Observations 1 and 2). The second is the relationships between holding time and user characteristics (Observations 3 and 4). The third is the relationship between holding time and computer send time (Observation 5).

Relationships between holding time and computer delays

The Computer Delay times shown in Table IV indicate a large variability among computer systems. This section investigates this variability.

There are a number of convenient measures that can be used to describe "computer load" and "computer delays." The "load" on a computer is a function of the number of simultaneous users who are "active" (in queue waiting for the computer to run their program or output to them), and the characteristics of user programs. For the purposes of the present discussion, data availability requires the use of "simultaneous users" as our measure of computer load. The manner in which a computer system reacts to a fluctuating load is a function of many additional variables, including characteristics of the scheduler.

Average computer delay may be calculated as the average total amount of computer delay per call, i.e., the sum of all the idle times, computer interburst times, and computer intercharacter times in a call.* This method was used in Table IV to demonstrate the effects of total computer delay on the average holding times of the three computer systems. It is also beneficial to examine the individual components of total computer delay. For example, it appears reasonable to divorce our measure of computer delay from the number of burst segments in

---

* Symbolically, average total computer delay per call is $(S/2)$ $(I + B^c[N^c - 1] + N^c C^c[M^c - 1])$.

Table V—Components of average computer delay per computer burst segment
(all times in seconds)

	System A	System B	System C
Average Computer Delay per Computer Burst Segment (Δ)	1.7	37.	1.4
Average Idle Time per Computer Burst Segment	.33	1.2	.41
% of Δ	19%	3%	28%
Average Interburst Time per Computer Burst Segment	1.4	36.	.99
% of Δ	79%	97%	69%
Average Intercharacter Time per Computer Burst Segment	.03	.02	.04
% of Δ	2%	—	3%

a call by considering computer delay per burst segment. This new measure is reasonable because the number of burst segments per call appears to be highly sensitive to user application type. Thus, calculating computer delay per burst segment reduces the dependence of our results on user application. We now consider contributions from average idle time (I), from average intercharacter times ($N^c C^c [M^c - 1]$), and from average interburst times ($B^c [N^c - 1]$). These three components of average computer delay per computer burst segment are shown in Table V.

As can be observed, the delays introduced by interburst times are the majority of all computer delay components in each system. The explanation for the relative sizes of these three components is as follows:

a. The characteristics of user programs are such that two or more quanta of execution time are required for a run to completion but output is generated by each quantum; and

b. The combination of system load, output buffer size, and characteristics of the scheduler preclude the immediate availability of additional output when the transmission of a computer burst is completed.

Because average *InterBurst Time* is the largest single contributor to average computer delay, it will be denoted by a special symbol, I-B-T. Figure 2 shows the relationships between holding time and I-B-T for each of the three computer systems. The three points in Figure 2 associated with computer systems are the observed values of holding time and I-B-T for those systems. The three lines are generated by changing I-B-T for each of the systems while holding all other parameters fixed. The slope of each line is equal to one-half the number of burst segments per call, i.e., S/2, because when every other factor in Equation (2) is held constant, it becomes

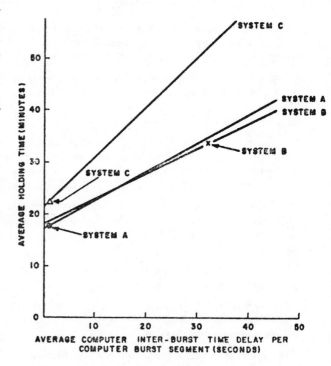

Figure 2—Average holding time versus average computer interburst time delay per computer burst segment

$$\text{Holding Time} = \text{Constant} + (S/2)\,(\text{I-B-T}). \quad (3)$$

Note that the lines for Systems A and B are close together. As these two systems have similar configurations of hardware and software, support is given to the conjecture that increasing the load on System A would lead to values of I-B-T and holding time comparable to those of System B. Conversely, deloading System B should result in these parameters having values comparable to those of System A.

Next, as it appears that holding time is a function of I-B-T and I-B-T is in turn a function of the loading on the computer and hence on the number of simultaneous users, an expression relating I-B-T and number of active users can be established.

To establish the relationship between I-B-T and number of simultaneous users, both quantities were measured on the systems as a function of time of day. The solid curve in Figure 3 indicates the average number of simultaneous users of System A for 15-minute periods of the day. The average I-B-T's were calculated for hourly periods on data from System A and a least squares fit of a variety of curve types was investigated. For these data the best fit is

$$\text{I-B-T} = (0.18)\exp(0.13\ u) \tag{4}$$

where

$$u = \text{number of simultaneous users,}$$

or

$$u = (7.7)(1.7 + \ln \text{I-B-T}). \tag{5}$$

The dashed curve in Figure 3 is a plot of u versus time of day by using Equation (4) and the actual measurements of I-B-T versus time of day. This fit seems to re-flect the major characteristics of the data as shown by the solid line in the sense that at least the morning and afternoon busy periods are reflected along with the intervening noontime lull.

Using (4), a plot can also be made relating I-B-T and the number of simultaneous users. Figure 4 shows this relationship as well as three curves showing the effects of I-B-T on holding times in Systems A, B, and C. These latter curves are plotted from Equation (3) after substitution of (4). Figure 4 indicates that above some threshold (represented by the knees of the curves) the computer's grade of service deteriorates rapidly as additional users are accepted.*

Relationships between holding time and user characteristics

Tables VI and VII, below, are used to illustrate several relationships between holding time and user characteristics in this section and between holding time and computer send times in the next section.

---

* Here an analogy can be drawn between the manner in which a multiaccess computer reacts to increasing loads and the fashion in which a telephone switching system reacts to overloads. As the link occupancy increases in a telephone office, the probability that a path through the switching networks cannot be found for an incoming call increases in a manner similar to the computer delay curve in Figure 4 (Reference 6). Such failures to complete connections can cause the common control equipment to generate additional attempts that consume additional real time. In computer systems, the analogous work is the "swapping" of user programs into and out of central processor core. For a further discussion of this computer problem see Scherr,[1] Raynaud,[2] Greenberger,[7] and Coffman.[6]

Table VI—Send time information
(all times in minutes)

	System A	System B	System C
Average Holding Time (τ)	17.	34.	21.
Average Total Send Time $(R) = (S/2)(N^u M^u W^u + N^c M^c W^c)$	6.2	5.0	8.4
% of τ	36%	15%	40%
Average User Send Time $(S/2)(N^u M^u W^u)$	.50	.45	.96
% of τ	3%	1%	5%
% of R	8%	9%	11%
Average Computer Send Time $(S/2)(N^c M^c W^c)$	5.7	4.5	7.5
% of τ	33%	13%	35%
% of R	92%	91%	89%

Table VI shows, for each system, the average holding time (τ), the average total send time** (R), and the aver-age user and computer send times. These quantities are measured both in minutes and, for the latter three categories, as a percentage of holding time.

Table VII is constructed identically for delay quantities rather than for send time quantities.

** Average total send time is the sum of average user send time and average computer send time.

Table VII—Delay information
(all times in minutes)

	System A	System B	System C
Average Holding Time (τ)	17.	34.	21.
Average Total Delay (D)	11.	29.	13.
% of τ	64%	85%	60%
Average Total User Delay	10.	12.	11.
% of τ	58%	35%	53%
% of D	92%	40%	88%
Average Total Computer Delay	.95	17.	1.5
% of τ	6%	51%	7%
% of D	8%	60%	12%

Table VI indicates that (a) in all three systems average user send time accounts for less than five percent of average holding time and less than 12 percent of average total send time, and (b) the users of System C inputted about three times† as many characters as the users of the other two systems. A conclusion that can be drawn from (a) is that user send time is an insignificant contributor to holding time. Even if average user send time increased by a factor of three, the increase in holding

---

† Recall that terminals in System C operated at 15 characters per second versus 10 characters per second in the other two systems.

time would be only one to two minutes, assuming that total user delay remains fixed. The reasons for (b) are probably a combination of, first, the business oriented applications of many of the system's users and, second, the rather low computer delays experienced in System C. It is possible that this increase in user input volume in System C is encouraged by the small computer delays experienced in that system. The same factor could also be partially responsible for the greater degree of on-line interaction in System C which when compared to Systems A and B had about double the average number of burst segments per call.

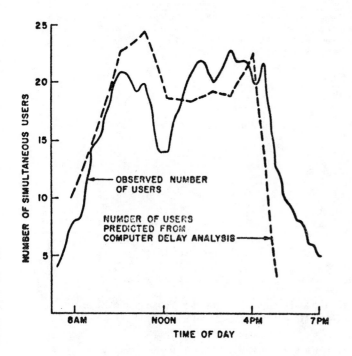

Figure 3—Number of simultaneous users in System A versus time of day

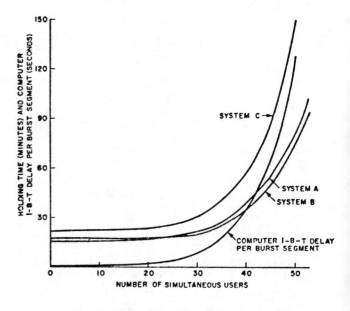

Figure 4—Average holding times and computer interburst time (I-B-T) delay per burst segment versus number of simultaneous users

Table VII indicates that average total delay D (the sum of average user and computer delays) accounts for more than half of average holding time. The lowest percentage is for System C where average total delay is 60 percent of average holding time; the highest is System B with 85 percent. Of these delays, users contributed from 40 percent (System B) to 92 percent (System A). Another observation is that the absolute value of average user delay is between 10 and 12 minutes in all three systems. This consistency is rather-remarkable when one considers the diversity of other parameters that affect user delay.

One might conjecture further about how user send time and delay characteristics would be affected by different transmission speeds, different program applications, different levels of user sophistication, and many other variables. For example, the user who inputs his prepared program from punched tape eliminates the user delay introduced by the user who performs this function by the hunt and peck method. An analogous "user" characteristic is computer send time which is determined almost entirely by the amount of computer output requested by the user. The next section discusses how this measure affects holding time.

Relationships between holding time and computer send time

Table VI shows that the system with the highest load, i.e., System B, has the smallest user send times and computer send times in both absolute value and in percent of holding time. This fact may be partially caused by the user's tendency to limit the amount of on-line I/O when he experiences long computer delays. The second, and more useful, observation that can be made from Table VI is that in systems which are not heavily loaded, average total send time may be on the order of 35 percent to 40 percent of average total holding time. Of this time, approximately 90 percent is computer send time.

We may infer from these data that, barring changes in user patterns and other influencing factors, holding time may be materially reduced by providing a high-speed channel from the computer to the user and a high-speed printer, or other display device at the user's location. This system redesign would enable a decrease in W^c, computer character width, with a corresponding decrease in

$$\text{Holding Time} = \text{Constant} + (S/2)\ (N^cM^cW^c). \quad (6)$$

Figure 5 is a plot of average holding times versus computer-to-user channel speed assuming all other factors remain constant. Of the three systems, two of them, viz., Systems A and B, have computer channel speeds of 10 characters per second. System C transmits at 15 characters per second. Note that if System C transmitted at 10 characters per second, its average holding time would be expected to increase to about 25 minutes from 21 minutes.

If computer channel speed were infinite, the average holding times for Systems A, B, and C would be 11.7, 29.3, and 13.7 minutes, respectively. To approach these minima within ten percent would require computer channel speeds of about 60 ch/sec for System A, 20 ch/sec for System B and 100 ch/sec for System C. If a channel speed of 360 char/sec were available, the computer send times would be less than one half minute and less than 2.5 percent of average holding time in each system.

At this point one might conjecture that there are at least two components of holding time that are likely to increase if computer transmission rates are increased. The first is think time because, in at least some instances, the user utilizes computer send time to read the output he receives. Hence, if the computer outputs the same number of characters in a much shorter time interval, the user may increase his think time in order to do the same amount of reading and thinking. This interplay of responses and transmission rates has effects on holding time in that it suggests the existence of some upper bound on computer transmission rate beyond which decreases in computer send time are matched by equal increases in user think time. The result is that average holding time cannot be further reduced by increasing computer transmission rates. In order to attach quantitative significance to this conjecture let us assume that the average user employs currently 10 percent of computer send time for reading and thinking.* This assumption

---

* Ten percent is certainly a high figure for the user who is listing his program to provide a paper copy for filing purposes but is low for the user who is checking every comma and parenthesis in order to find a program bug.

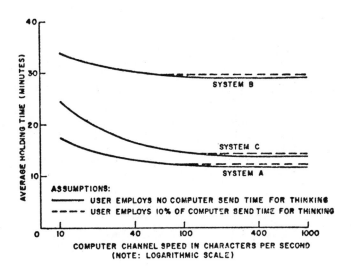

Figure 5—Average holding time versus computer transmission rate for three long holding time systems

implies that a tenfold increase in computer transmission rate will result in minimum average holding time. For Systems A and B this rate is 100 characters per second (1000 words per minute) and yields holding time of about 12.5 minutes and 29.5 minutes, respectively. These average holding times are 28 percent and 13 percent less than the current holding times and seven percent and one percent greater than the theoretical minima shown in Figure 5 for Systems A and B. For System C, the corresponding transmission rate is 150 characters per second. This would reduce the average holding time in System C by 31 percent to about 14.5 minutes which is six percent greater than the theoretical minimum of 13.7 minutes. The minimum holding times implied by this assumption are shown as dashed lines in Figure 5.

The second variable that may naturally increase if computer transmission rates increase is the quantity of output requested by the user. For example, if the computer transmitted at 360 characters per second, computer send time would be less than 2.5 percent of average holding time in all three systems, and users may find it quite convenient to request two or three times as much output as they do presently. This increase in output will not severely affect average holding times, however, as even a tripling of computer send time would increase average holding time by less than a minute assuming a 360 character per second computer rate.

It should be noted that these figures are heavily dependent on such factors as user application. For example, if the application is not scientific programming but rather inquiry-response one might anticipate drastically shorter holding times. Consider the telephone directory assistance operator who makes a five second query of a computer that responds in one second with the beginning of a 1000 character transmission to be displayed with a video terminal. Assume that "holding time" is defined to be

$$\tau = \text{operator keying time}$$
$$+ \text{ computer response time}$$
$$+ \text{ computer transmission time}$$

or

$$\tau = K + I + 1000 \, W_c.$$

For the numbers we have assumed,

$$\tau = 6 \text{ seconds} + 1000 \, W_c,$$

---

his program to provide a paper copy for filing purposes but is low for the user who is checking every comma and parenthesis in order to find a program bug.

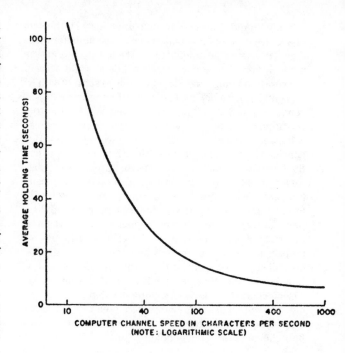

Figure 6—Average holding time versus computer transmission rate for hypothetical short holding time system

and at 10 characters per second, this results in a 106 second holding time of which 94 percent is computer send time. At 360 characters/second, $\tau = 8.8$ seconds but computer send time is still 32 percent of τ. At 4000 characters/second, $\tau = 6.25$ seconds of which four percent is computer send time. Figure 6 shows a graph of holding time versus computer channel speed for the short holding time example we have assumed.

SUMMARY AND CONCLUSIONS

The results of a study of the communications considerations of serving multiaccess computer systems have been presented. A model of user-computer interaction, as observed at the communications interface, has been developed. Summary data from three "long holding time" computer systems have been given for the parameters of this model.

Examination of these data has revealed that

a. Computer introduced delays can be a large component of holding time and, above some threshold, are acutely sensitive to the number of simultaneous users. The largest component of computer delay occurs during those periods when the computer is outputting to remote users. The conclusions to be drawn from these findings stem from the consideration of holding time as being composed of periods of computer outputting activity, and periods of no computer outputting. Of the inactive periods, some time is due to user-dependent delays and some to computer-dependent delays, some, such as execution time, may not be reducible and others, such as delays due to overhead, can be reduced. It should be noted that changes in the computer system such as changes to the scheduling algorithm,[7,8,9] or changes to the communications control unit[10] can strongly influence computer delays. Thus, it is within the *computing* system that some reductions in holding time may be made resulting in communications economies. As not all computer delays can be eliminated in a heavily loaded system, the technical and economic feasibility of employing data multiplexers at the computer to decrease the number of access lines should be explored.

b. The average number of characters sent by the computer to the user is an order of magnitude greater than the number of characters sent by the user to the computer. If other parameters did not change drastically, the availability of higher transmission rates for computer outputting would effect significant reductions in average holding time. A computer transmission rate of 360 characters per second would reduce total computer send time below one half minute per call. To achieve this rate requires either high speed or asymmetric data sets and correspondingly higher speed output terminals.

c. Delays introduced by the user are a significant contributor to average holding time. The average user delays in the three systems reported are remarkably close in absolute values. As user delays are appreciable, the multiplexing of inputs from user terminals which are geographically clustered appears attractive and is being studied.

The data analyzed in this report are from systems having primarily scientific and business problem-solving applications. The users of such systems demonstrate a wide range of sophistication in their use of these systems. As users educate themselves in the efficient use of multiaccess computers and their terminals, the data traffic characteristics of these systems will change.

Studies of multiaccess computer communications systems are continuing. Data are being collected from systems with markedly different terminal types, average holding times, and user applications. Analyses of these data will allow the characterization of such systems in a manner analogous to that reported above for "long holding time" systems.

Implications of results

The implications of the results of this study extend into several aspects of computer communications. The study has produced quantitative indications of the degree to which computer operations can influence such communications parameters as holding time. The developers of computer systems, in turn, have noted that the provision of computer hardware and software to accommodate data communications is a major problem area.[11,12] In order to jointly optimize the computation-communication solution to the problem, it is apparent that closer coordination between the computer and communications systems designers would be extremely fruitful in terms of economic and technological improvements to overall systems design.

One of the impediments to finding rapid and robust solutions to the problems of multiaccess computer communications has been the unavailability of data descriptive of the user-computer interaction process. The acquisition of the data reported here is a contribution toward the removal of that obstacle.

These data are currently being used in systems engineering studies at Bell Telephone Laboratories to further define the systems requirements for new systems and services to satisfy the needs of the multiaccess computer community.

The analyses made of these data support for the first time in more than qualitative terms some proposals proffered in the past as solutions to the data communications problems associated with multiaccess computer systems. The delays which are introduced by both user and computer suggest the possibilities for effective employment of multiplexing techniques. For example, it has been shown for the systems studied that the average total send time (the sum of user send time and computer send time) is as little as five to nine minutes. This corresponds to 15 to 40 percent of average holding time. For the other 60 to 85 percent of average holding time the communications channel is idle. One method of obtaining higher utilization of these facilities is by time division multiplexing.* The following assumes a multiplexing technique in which the user channel is independent of the computer channel. Only one to five percent of average holding time (or three to eight percent of the average user burst segment) is user send time. Thus, for 95 to 99 percent of an average call, the user-to-computer channel is idle and could be made available to additional users. Average computer send times for the three systems were from 13 to 35 percent of average holding time

---

* Multiplexing is not always economical, of course, despite the large idle times. Other important considerations involve the geographical placement of terminals and computers and several statistical traffic characteristics other than average occupancy.

40

(21 to 86 percent of the average computer burst segment) indicating that higher usage of the computer-to-user channel may be realized by the use of appropriate multiplexing techniques.

The asymmetric nature of the data flow in multiaccess computer systems suggests that different transmission treatments may be appropriate for computer-to-user versus user-to-computer transmissions. The large volumes of computer-to-user data are an order of magnitude greater than volumes in the opposite direction. The provision of computer transmission rates of 100 to 200 characters per second could reduce average holding times up to 30 percent. As the user is capable of generating characters for transmission at a much slower rate, the application of asymmetric channels or data sets receives quantitative support and is now being studied. Provision for higher computer transmission rates would require, of course, user terminals with accordingly higher input rates.

The final conjecture receiving quantitative support from the above analyses is that users themselves contribute substantially to the communications costs of their real-time computer access calls by introducing delays. Some of these delays are likely to decrease as users gain proficiency. Others are due to the inveterate characteristics of human users. As they pertain to the use of communications and to the use of computers[1,3] these characteristics are being intensively studied to enable the design of versatile and responsive computer/communications systems.

ACKNOWLEDGMENTS

Many people have contributed their efforts to various parts of this study. Data acquisition was accomplished with the considerable help of the American Telephone and Telegraph Company and the Bell System Operating Companies. Contributions to the model and the analyses and many helpful criticisms were made by Messrs. E. Fuchs, R. J. Price and R. J. Roddy, all of Bell Telephone Laboratories.

Our special thanks are extended to the companies whose computer systems are being studied. Without their full permission and very helpful cooperation these analyses would not be feasible.

REFERENCES

1 A L SCHERR
An analysis of time-shared computer systems
Massachusetts Institute of Technology Project MAC Thesis
MAC–TR–18 June 1965
2 T G RAYNAUD
Operational analysis of a computation center
Operations Research Center MIT July 1967
3 H SACKMAN
Experimental investigation of user performance in time shared computer systems retrospect, prospect, and the public interest
System Development Corporation May 1967
4 G E BRYAN
JOSS: 20.000 hours at the console, a statistical summary
Rand Corporation August 1967
5 P E JACKSON
A fourier series test of goodness of fit
To be published
6 ANON.
Switching systems
AT & TCo 1961
7 M GREENBERGER
The priority problem
MIT November 1965
8 E G COFFMAN JR
Analysis of two time sharing algorithms designed for limiting swapping
J A C M July 1968
9 E G COFFMAN JR   L KLEINROCK
Computer scheduling methods and their countermeasures
Proc S J C C 1968
10 L J COHEN
Theory of the operating system
The Institute for Automation Research Inc 1968
11 W F BAUER   R H HILL
Economics of time shared computing systems—Part II
Datamation Vol 13 No 12 41–49 December 1967
12 W E SIMONSON
Data communications: The boiling pot
Datamation Vol 13 No 4 22–25 April 1967
13 W W CHU
A study of the technique of asynchronous time division multiplexing for time-sharing computer communications
Proc of the 2nd Hawaii International Conference on Systems Sciences January 1969

What Is Telecommunications Circuit Switching?

AMOS E. JOEL, JR., FELLOW, IEEE

Abstract—This article is an attempt to provide one with no knowledge of telecommunications circuit switching with some insight into its complexity and technological progress. This subject has for many practicing telecommunication engineers been treated as enigmatic. The problem is that switching is a combination of many details. The power of modern technology has added to the complexity of the features and services that can be offered economically by these systems. Perhaps in no other area of telecommunication does the evaluation of the choice of technology and system architecture so obfuscate the true purpose of the system. The functional requirements of a switching system are relatively simple and easily understood, but the service and feature requirements are growing and being implemented at a rate greater than ever before. With its growth has emerged a new era of switching. It is the desire of the author to place the past and current technology, as well as the fundamentals into prospective so that the reader may be prepared for the new era in telecommunication circuit switching.

I. Introduction

AS WITH MOST technical subjects, especially those with long evolving histories, telecommunication switching has grown from a simple need to a complex and most important element of public and private telecommunications. The field has grown so complex that to avoid slighting one at the expense of the other it has been necessary to divide the field into two principal but not independent segments. "Circuit switching" applies to switching where a continuous two-way path is established in space, time, or frequency of the bandwidth suitable to the calling and called station transducers for the duration of the call. The other area of switching is its application to some forms of data transmission where a continuous path is not established but where memory is used to store the message to be transmitted awaiting the availability of transmission facilities. This form of switching is known generally as "store and forward switching" and includes a current popular form known as "packet switching." (The definitions and differences in these forms of switching are covered in more detail in Section XI of this paper.) Combined circuit and store and forward switching systems have been developed for some applications.

There are different ways to approach the study of switching. For this article two different approaches are included, historical and philosophical. These were chosen since they both tend to classify the field of switching. However, classification of the field is only to illuminate many specific topics for further study.

II. Centralization of Switching

It is well known that if a system is to serve n stations $n(n-1)/2$ two-way transmission facilities or lines can serve them (see Fig. 1). This type of telecommunication system requires what is known as "noncentral circuit switching" using $(n-1)$ switches per station as shown for the lower right station. The switches used to complete the establishment of

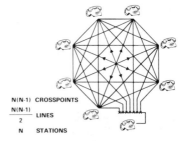

Fig. 1. Noncentral switching.

N(N-1) CROSSPOINTS
$\frac{N(N-1)}{2}$ LINES
N STATIONS

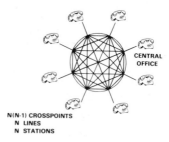

Fig. 2. Centralized switching. Remote controlled network.

N(N-1) CROSSPOINTS
N LINES
N STATIONS

CENTRAL OFFICE

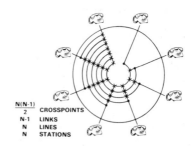

Fig. 3. Centralized switching. Nonblocking network.

$\frac{N(N-1)}{2}$ CROSSPOINTS
N-1 LINKS
N LINES
N STATIONS

paths at both ends of the transmission facilities are known in the art as "crosspoints." Therefore, this noncentral system requires a total of $n(n-1)$ crosspoints. Noncentral systems have only limited application, generally where n is small.

From the beginnings of telephony the concept of centralized switching was obviously more efficient (Fig. 2). Here only one two-way line is required per station but the number of crosspoints has not changed. The only difference is that the crosspoints must be remotely operated, manually or automatically. But central switching can be more efficient in the number of crosspoints shown in Fig. 3, only half as many crosspoints ($n(n-1)/2$) are required. This is because being centralized only one crosspoint is needed for each connection. This is known as a single stage nonblocking switching network.

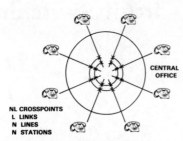

Fig. 4. Centralized switching. Blocking network.

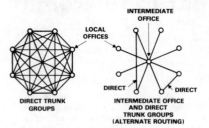

Fig. 5. The intermediate office.

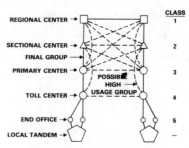

Fig. 6. Hierarchial switching plan.

There is an additional advantage to central switching. The nonblocking network provides for all stations to be simultaneously involved in some connection. By providing fewer than $n/2$ simultaneous connections even fewer crosspoints are required. As shown in Fig. 4 providing for less than $n/2$ connections introduces the concept of lines within the central switching system or office known as "links". The concept of blocking, that is when all links are in use connecting $2L$ stations, the remaining $(n - 2L)$ stations cannot be connected until one of the links becomes available. The number of links L is determined by the desired "grade of service" to be given to the offered load.

These concepts are the start of two important topics in switching—*switching center network topology* and *teletraffic theory and engineering.*

III. NETWORKING

Public telecommunications means serving many stations over long distances. Deploying a plurality of central switching systems known as "offices" or "entities" and located in "wire centers" provides a trade of switching for transmission. With a single central office n lines replaced $n(n - 1)/2$ lines. Not only are there less lines (transmission facilities) but they are on the average shorter.

If each office were to be connected by transmission facilities with every other office then all stations could communicate with one another. These facilities are known as "trunks." Obviously more than one or a "group" of trunks is needed if the blocking is to be held to a reasonable value so that satisfactory service is given to the load expected between offices. The trunks may be one- or two-way, that is, used for calls originating in one office or both. Obviously two one-way groups are required, one in each direction.

Interconnected central offices are known collectively as a "network." The offices are sometimes spoken of as "nodes" in this network.

This network of transmission facilities and central switching offices is different than the in-office switching or link network. Where necessary to distinguish between trunk or facilities networks the in-office network of links will in this paper be called "switching center network" (SCN).

Rather than directly interconnecting a plurality of central offices by trunk groups it is possible to have an "intermediate office" that interconnects trunks (see Fig. 5). Depending upon more practical tariff and administrative considerations these generic offices may be given more specific names like "tandem," "transit," "toll," "gateway." In this context central offices connected to stations are known as "local" or "end" offices. Intermediate offices may also be connected by trunk groups so that more than one may be involved in a call.

A popular concept in networking is providing more than one "route" for a call. A "direct" trunk group may connect two offices. Trunks in this group are selected first if idle. If they are all busy and if both offices have a trunk group to the same intermediate office then an "alternate route" may be used if a combination of idle trunks between the local and intermediate offices is found.

With a plurality of intermediate offices in a network including direct trunk groups, the way calls are routed to provide the desired degree of blocking is also the subject of teletraffic theory and engineering. Added factors are the connectivity discipline and routing must include provision to avoid calls passing through the same intermediate office more than once and, in some networks, insuring that calls are not dependent upon a single intermediate office for survivability in case of total office failure. Connectivity disciplines and routing result in networks known as "hierarchical" and "symmetrical" or "polygrid."

Fig. 6 shows the hierarchical public telephone network of North America [1]. The switching plan assigns classes to the offices with alternate routing permitted up the originating "ladder" and down the terminating "ladder." There are two types of trunk groups, the "final" groups between the intermediate offices or next class within the ladder and between ladders at class 1 (regional) and so called "high usage" groups, from which alternate routing may occur, between offices of all classes within and between each ladder whenever call volumes justify. The engineering of this network is very sophisticated and the network is amorphous, ever changing and growing.

In addition to the office types shown, there are also intermediate offices between end offices that connect to the same or similar class 4 office within the same exchange. These are usually known as "local tandem" offices. In this network a particular switching system might serve several functional classes simultaneously including the end office. Therefore, some system features are known as "local/toll," "local/tandem," etc.

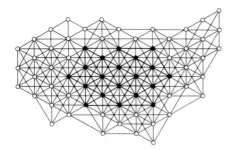

Fig. 7. Polygrid network.

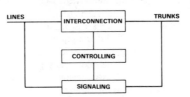

Fig. 8. Basic functions.

A symmetrical or polygrid network is shown in Fig. 7 [2]. Here each office or node is of the same rank and can be used as an end or intermediate office. In the network shown in the figure, the one used by the AUTOVON leased private military service, there are generally a minimum of two two-way paths connecting any two offices in the network.

IV. BASIC FUNCTIONS

The basic functions are those that telephony requires to provide a useful universal (public) service. With the exception of remote control of the switching center network, these functions apply to both manual and automatic switching. Since, except for special services, manual switching has become obsolete this paper deals only with automatic switching, with the exception of Section IX-A. The nine basic functions are shown on Fig. 8. There are three categories of basic functions: signaling, interconnecting with the link or switching center network, and controlling. Fig. 9 shows their basic interrelationship.

While each of these functions is simple in concept, the description of their interaction during the progress of a call may be complex. Furthermore, depending upon the technology employed, the organization or architecture of these functions, and the services to be rendered or the features to be provided by the switching system as a whole can become very complex. Each new generation of switching system introduced into a network generally must work with and initially continue to provide most of the services and features of the previous generation [3].

The following is a brief explanation of each of the basic functions taken in the order in which they generally occur during the progress of a call in a modern switching system.

Attending—This is the reception by a central office of a request for service from a station or another office, i.e., a call origination.

Signal Reception—After the central office responds to the request it receives information, usually numerical, to address the desired called station.

Interpreting—Determination of the action required based on the received signal information.

Path Selecting—Determining an idle link or series of links or channels through the switching center network.

Route Selecting—Determining the trunk group to which a path is to be established including interoffice calling.

Busy Testing—Determining that a link or trunk is in use or reserved for use on another call. When links or trunks are busy successive testing of trunks or links is known as "hunting."

Path Establishment—Control of the elements of the switching center network to establish a channel for use on the call thereby making the desired interconnection. This function in

Fig. 9. Interrelation among categories of basic functions.

a circuit switching requires some form of memory to retain or remember the connection for the duration of the call. In older systems the memory may be retaining a physical position as in a mechanical switch.

Signal Transmission—On interoffice calls transmission of the address of the call for which a connection is to be established in the distant office.

Alerting—Informing the called station or office that a call is being sent to it. On calls to stations this is called "ringing." On interoffice calls it is the transmission of the attending signal.

Supervising—To detect when the connection is no longer needed and to effect its release. Supervision is also required for other purposes such as for call service features.

V. CALL SERVICE FEATURES (FIG. 10)

Universality is a basic service objective of a public switched-telecommunication network, i.e., that each station be able to reach every other station to the extent dictated by the desired economics of the service. The universality may but does not always extend to the (numerical) address identifications of the station. If a truly "closed" numbering plan is used, universality is achieved in this respect. In many networks the numbering plan is "open" meaning that different length addresses are transmitted to reach a given station depending upon the location of the calling station. In the North American public network the numbering plan is "quasi-closed," that is, within a numbering plan area the same number of digits is transmitted and the same number prefix plus an area code is used on calls originating in other areas.

A practical service requires progress tones such as "dial tone" to indicate that the central office is ready to receive address signals, "busy tone" to indicate that the called station is busy, a tone to indicate that the called station is being rung, and "reorder tone" to indicate that all links or trunks by which the call might be routed are busy, that is the call is blocked.

```
NUMBERING PLAN
    OPEN
    CLOSED
    QUASI CLOSED (NORTH AMERICAN)
CALL PROGRESS TONES AND ANNOUNCEMENTS
    DIAL
    BUSY
    RINGING
    RECORDER
    RECEIVER OFF-HOOK
    RECORDED ANNOUNCEMENT
SUPERVISE
    CALL ABANDONMENT
    NO ADDRESS TRANSMITTED
    PARTIAL ADDRESS RECEIVED
    UNASSIGNED ADDRESS RECEIVED
    UNEQUIPPED ADDRESS RECEIVED
```

Fig. 10. Call service features.

```
PROVISIONING
GROWTH
RELIABILITY
MANAGEMENT
    MANUAL
    AUTOMATIC
OPERATOR ASSISTANCE
MAINTAINABILITY
CHARGING
    CASH - (COIN)
    CREDIT - (BILLED)
```

Fig. 11. Operational service features.

It is also necessary that a practical service include provision for supervising the call from the moment of orgination until either or both parties indicate that the call has ended and that the connection, if established, is no longer required. This function also implies that before a call attempt is answered it is continuously supervised for abandonment thereby eradicating any action taken that might affect future calls.

As a corollary, it is possible that, at any point in the processing of a call, the expected information is not received, or the information received does not meet the established address format, or the address is not assigned or equipped. These conditions in the jargon have become known as "permanent signal," "partial dial" (address), etc. To provide a viable service requires that these conditions also be detected, generally by timing for a predetermined time and then taking a specific action which is usually equivalent to establishing a connection, such as to a "reorder" tone or "receiver-off-hook" tone or a recorded announcement.

VI. OPERATIONAL SERVICE FEATURE (FIG. 11)

In the broader aspects of the service, certain features have become recognized and expected features of a switched public telecommunication service.

The degree of blocking in networks is a judgment factor that determines the overall grade or quality of service. The tools provided by the application of modern teletraffic theory are important for engineering the quantity of equipment and facilities to be provided. But the complexity of modern networks and systems makes it almost impossible to predict with certainty the quality of service any particular user will encounter.

Nevertheless it is incumbent upon an administration to set and maintain a grade of service, such as one call blocked in 500 (written $p = 0.002$), as an objective to be used in deciding the quantity of equipment to be provided. It takes time to provide (install and test) working combinations of equipment and facilities. The time factor means that future load and needs must be estimated and also sufficient equipment and facilities made available for expected growth before the next installation. The grade of service is therefore an average value based on the center or end of an engineering interval.

Not only must an adequate quantity of equipment and facilities be provided at the appropriate time but features must be built into the system to measure the load offered to the system, the number of calls of each type that are processed, and the number of times all trunks are busy or a channel cannot be established in part of the network. Facilities to obtain traffic measurement data and periodically pass it along to other systems for processing can be designed into the system [4].

The reliability of the switching system and its operation must be taken into account in provisioning. As an extreme example

with electronic switching it is usually possible that a single control unit can be designed to provide all of the offered service needs of the largest office. However, to ensure continuity of service at least one additional control unit is installed [5]. If the reliability of the components is inadequate or the time to repair a fault is excessive, a simple calculation might predict that more than two control units would be needed.

Allowance needs to be made for out-of-service times that might be longer than the normal repair time when a unit is being modified. The modification may be due to changing or correcting the design of the hardware or software [6].

Once a network of offices is in service, it must be kept in service such as by the detection, location, and repair of troubles in the equipment or facilities which are bound to occur regardless of the reliability of the individual components. Some troubles are man-made. Features are designed into systems to enable the system to be maintained, which includes equipment to test routinely individual portions of the service providing equipment or to determine the overall quality of service as seen by the customers [7].

Other features are required to make the best use of the equipment and facilities that are available. These are known as "network management" features and they may be implemented automatically or manually [8]. One such feature is alternate routing in a polygonal network which steers calls around a wire center that cannot process calls. When an excess number of calls reaches reorder on certain routing, or when certain end offices cannot be reached, routing automatic changing features may be provided.

In the North American and some other public networks, "operators" are accessible by users to aid in completing their calls or to answer other questions concerning the service [9]. Features are required in the switching system design to recognize requests for or the need to connect to an operator, to distribute calls to operator positions, and to provide operators with special signaling and switching features.

Another category of operational service features are those required in the switching system to provide the administration with information for charging for the services rendered. These are divided into cash or coin services and credit or billed measured services.

In summary there are many considerations in the design and deployment of switching systems that are required besides those for call processing to provide a high-quality service acceptable to the public. Any compromises on these factors and elimination of these features in a modern system would be to deny the user the high-quality service that it is possible to offer.

Note that in network management, maintenance, traffic measurement, and charging systems are now in an environment of inputs and outputs (I/O's) in addition to the lines and trunks that are the traditional inputs and outputs of a switching system. These operational service features in many respects resemble modern information processing systems I/O's. They are a different category of functional units of a switching system. They may be magnetic recording units or data links to other information processing systems. Data links are also being used for call signaling between offices (see Section XII).

VII. TELECOMMUNICATION-SERVICES

The features described in the previous sections are for the most part required for administration and maintenance of a switching system. But to be useful it must provide one or more services. The basic call processing service and the functions for their implementation have been described for an ordinary call or what has been known as a "POTS—plain old telecommunication service" call.

However, a sophisticated telecommunications minded public demands many types of services. In particular business users and residence users require distinct services. For business there is another class of switching system known as a private branch exchange (PBX). In the United States, Japan, and France a service known as CENTREX is offered generally from the central office [10].

In addition to coin service, there is also shared or party line service in the United States and a few other countries [11].

These are the most familiar services but there are many others such as wide area telecommunication service (in the U.S. [12]) or abbreviated dialing [13]. A "service" as distinguished from a feature is defined as one of which the customer is aware when using the switching facilities. Some administrative features may be needed to implement a service. It is not uncommon to include a "class of service" with each facility to indicate the services to which a customer is entitled from a given line. A class may also be associated with a trunk to indicate the type of calls it can accept or delivers. The number of service combinations that may apply to a given line in a modern system may be in the hundreds. Counting specific features and services it is not unusual for a switching system to be designed to support many hundreds (see Fig. 12).

This concludes the introduction to the field of telecommunication circuit switching. While the requirements for individual features and services appear to be easy to understand by themselves they become most complex when they must function cooperatively in the same switching entity. This combinational aspect of the many requirements of a switching system is one of the chief factors that makes it difficult to understand and to formalize.

VIII. APPROACHES TO UNDERSTANDING

While it was stated at the beginning of this paper that of the many possible approaches to the understanding of switching only the philosophical and historical approaches would be covered, the author believes that in view of the tremendous amount of literature on the subject, the various approaches which might apply to any complex subject, should be mentioned.

This paper deals with the switching system and its functions. This might be called the "inward" look. Many study switching systems in the context of the environment of the facilities network or the "outward" look. When there are so many combinations of variables the requirements for switching systems becomes an important factor in the design. Many have tried to

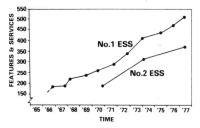

Fig. 12. Growth in features and services.

Fig. 13. Approaches to the study of switching.

formalize this process, but to date only the natural language has been used, and this is subject to misinterpretation.

The requirements are interpreted in the context of a specific switching system architecture and at this point might be considered a development specification. Many different systems could be designed to meet a specific set of requirements, particularly with different techniques. It has been largely a designers choice with innovation, economics, and salesmanship ultimately determining which of the many possible designs are manufactured and placed in service.

The needs of the student of switching has tended to determine the most prevalent method of approach to the subject and this has been the "descriptive" approach. While this approach is most useful and effective for many with an immediate need to know, there are many other approaches that have been used to a lesser degree. These are listed on Fig. 13 and are briefly discussed below:

Descriptive—This approach covers "how" a specific system(s) works, the organization, technology, call and other processing in terms of either hardware or software or both.

Historical—This approach is related to descriptive approach but it covers many systems in chronological order.

Philosophical—The philosophical approach is one that examines these events and designs and tries to assess them from the standpoint of their relative contributions and how they meet current and possible future needs.

Principles—There are many principles in switching. Many are lost due to the morasses of detail necessary for their implementation. However, a careful study of the art of switching as it has developed reveals certain principles some of which have become accepted. This approach to the study in essence is the classification and categorization of switching system information from which may usually be extracted common elements that might be called "principles."

The combination of an historical, philosophical, and principles approach results from the study of the evolution of switching and is, in general, the one employed in this paper.

SWITCHING PLAN
TRANSMISSION PLAN
CHARGING FACILITIES
OPERATOR ACCESS
ROUTING PLAN
NUMBERING PLAN
DIALING PROCEDURE
SIGNALING COMPATIBILITY

Fig. 14. Switching system facilities network environments.

Environmental—This approach is largely one of requirements for features and services. It is "what" the system has to do rather than the "how." However, it generally goes beyond the system itself looking outwardly to the facilities network into which the system is to be inserted. Fig. 14 lists the important environmental factors many of which have already been discussed in the previous sections.

Statistical—This approach is largely one of recording existing and/or projected quantities of systems, offices, or the degree of utilization of the equipment and its features.

Theory—This is the formalization of the principles [14]. To date very little theory exists about the structure of switching systems. For the most part theory has been applied only to link and trunk networks and to a lesser degree the system control capacity [15].

Operational—To the practicing engineer this is probably the most important approach to switching. It is the description of systems viewed from the point of view of an administrator charged with the responsibility of providing and maintaining service. Two related factors become the focal point of this approach, they are serviceability and costs. How good is or will be the service rendered and how much will the system cost is not just initial cost but also consideration of the annual cost of operation, the expected annual revenues, and the cost of growth. While this is information that is much sought, it is difficult to obtain such information for comparison purposes due to the differences in system capabilities.

IX. The Evolution of Switching

There have been three principal eras of switching to date based mostly upon the technology: manual, electromechanical, and electronic. Initial dates of invention, trial or commercial service are usually easy to establish, but an end date is indefinite. A peak year in terms of number of lines in service perhaps is the best measure, since it establishes the start of the decline of an era. Using this definition manual switching peaked about 1938. Today about 2.5 percent of the world's telephone lines are still switched manually. Although we consider ourselves in the electronic era that started in 1965, only about 6 percent of the lines are switched by electronic systems. Electromechanical switching is not expected to peak world-wide until about 1985.

Although electronic offices are being placed in service by most administrations for growth, it will be some time before the number of lines served by electromechanical offices will peak since in many administrations it is still more economical to extend existing installations rather than start new units. Also replacement of electromechanical systems will take some time since in some administrations much of this equipment is functioning satisfactorily and not fully depreciated.

Each of these eras, including the electronic era to date, have made many unique contributions, too numerous for complete coverage in this paper. In this section only the evolution of the basic functions will be covered.

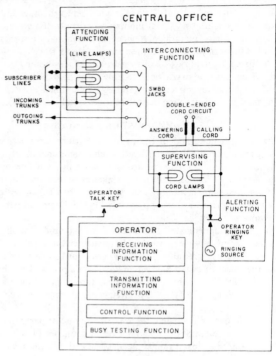

Fig. 15. Basic functions of a manual central office.

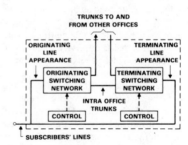

Fig. 16. Separation of traffic trunks to and from other offices.

A. The Manual Era

By 1878 the first switchboards designed for telephone switching were placed in service. The basic functions were as discernable then as now. They are shown with respect to the plug and jack technology in Fig. 15. Much effort was spent during this era to improve the reliability of the basic apparatus and to modify and extend the operator switchboards to improve the efficiency of their operation. Of principal importance for large metropolitan areas was separating the originating and terminating flow of traffic with the use of separate switchboards [16]. This idea of separate switching center networks carried over into early electromechanical switching systems as shown in Fig. 16.

Another concept, one that is basic to switching center networks, introduced into switchboards in the interest of efficiency was the multiple appearance of lines and trunks [17]. Lines appeared before several operators so that any one could serve a call and the trunks were multipled to appear before all positions so that any operator could reach all trunks that might be used to switch a particular call.

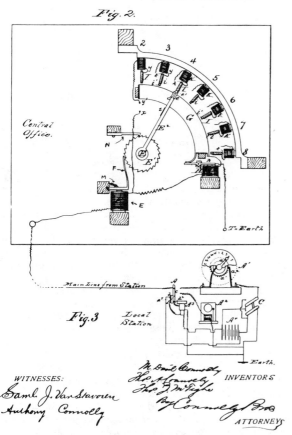

M. D., & T. A. CONNOLLY & T. J. McTIGHE.
Automatic Telephone-Exchange.
No. 222,458.   Patented Dec. 9, 1879.

Fig. 17. M. D. & T. A. Connolly & T. J. McTighe Patent No. 222 458.

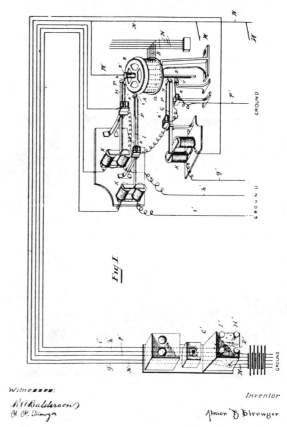

A. B. STROWGER.
AUTOMATIC TELEPHONE EXCHANGE.
No. 447,918.   Patented Mar. 10, 1891.

Fig. 18. A. B. Strowger Patent No. 447 918.

Also developed early during this era were the attending and alerting functions along with the simultaneous development of the telephone line and station [18]. Initially batteries were provided locally at each telephone station. An important innovation and improvement was the centralization of the battery, known generally as "common" battery, that is, the same battery may be used to power all station transmitters without crosstalk. Accompanying this development was the use of two-wire, thereby, eliminating longitudinal unbalance that induced electrical noise into connections switched on a one-wire basis.

B. Electromechanical Switching Era

1) Direct Control—The electromechanical era started out and was largely maintained by a proliferation of inventions of switching network devices. A device technology evolved unique to switching. The names of systems were based on the names of the network device. The other two categories of the basic system functions, signaling and control, were initially also inherent in the switching device design, but innovations borne of experience and changing requirements brought change to these functions without dependence on the device characteristics.

The principal objective of the first device and system was remote control of a selective device. Fig. 17 shows the first automatic switching system in 1879, almost 100 years ago, and Fig. 18 shows the first Strowger two-motion step-by-step switch of 1891. (The first commercial step-by-step switch of 1892 was different from this but retained the two-motion idea [19]). Each of these switches was controlled differently: the first by a dial-like pulse generator over a single wire such as then used with crude teleprinter developments of the day and the other by push bottons, each using a separate wire. In both cases they were controlled "directly" from the station by "calling devices."

2) Gross Motion Switches—The motion required by these mechanisms was supplied by electromagnetically operated ratchets that consumed considerable power supplied by the station local battery. The operating power was related to the state of mechanical precision that could be used to define contact size and spacing and the contact force needed to provide a reliable, relatively noise free transmission path. Because of the space traversed between terminals, switches of this type are classified as "gross motion."

These problems concerned the early designers for over a decade. A reasonable two-motion switch size emerged that was

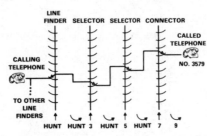

Fig. 19. Progress of a call in a 10 000 line SXS office.

Fig. 20. Step-by-step switches.

Fig. 21. Switching network functions.

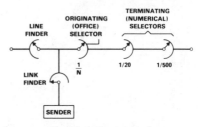

Fig. 22. Indirect control.

conveniently 10 × 10 or 100 terminals in capacity, so that it fitted the decimal numbers then being assigned to telephone addresses. After another five years a switching center network design evolved using these switches in stages interconnected by links, so that more than 100 stations could be accessed.

Fig. 19 shows how a series of three stages of switches, labeled "selectors" and "connectors," are used to reach one of 10 000 terminals. This was known as "progressive" stage-by-stage or "step-by-step" switching.

The other stage labeled "line finder" also an innovation of this era is a single stage of "concentration" so that each line need not have its own series of 1 + 10 + 100 = 111 switches to reach 10 000 terminations. Fig. 20 shows a group of modern step-by-step line finders.

Not only were the selection stages arranged serially to reach any line but the links between stages were shared so that other switches in the succeeding stages did not have to stand idle when a connection was made through others. Each selector makes a one out of ten selection with access after each selection to ten links to the next stage. While this increased the number of stages to achieve selection or "expansion" as it is known more generally, the "distribution" function was added. The switching center network therefore takes on the more general internal functions as shown in Fig. 21. This was the first automatic system and also to date the most successful. During the first two decades methods were devised for accomplishing

and integrating most of the basic functions that were not initially carried over from manual switching.

3) Indirect Control—Interconnecting is the one function that had been the focal point of technological change until the electromechanical era. During the early development of the step-by-step system telephone growth accelerated and it became obvious that to serve large cities would require many offices and many stages of decimal (one out of ten) selection. In 1906 E. C. Molina proposed that nondecimal selection stages would be possible if the customer dialed into a storage device or register that would convert the dialed decimal digits into nondecimal digits and then "send" pulses to the selectors [20]. Nondecimal selectors with capacities as much as five times (500 terminals) the step-by-step selector were devised. Since the "sender" as it was called, was required only during the addressing phase, a concentration switching network using the same or other type of selectors was inserted between the line concentration and the first selector stage. When the selections were completed (or the caller abandoned) the sender could then be released for reuse on other calls. Thus an element of the control was divorced from the connection path and became the genesis of a most important concept (see Fig. 22).

The larger selector capacity brought with it new problems that were solved with the available technology. To cover more terminals gave rise to the need for greater speed in selection, particularly since an interval was required by the "sender" to recode what was dialed into nondecimal form. New selectors were driven by a friction coupling to a constant motor-driven shaft. A new generation of "power-driven" gross motion switches were born. Some of these moved the wipers, or wipers known as "brushes" that were tripped when selections started, in a straight line such as the "panel" selector [21] (see Fig. 23) or rotary directions [22], [23]. Power was also applied hydraulically. Brush tripping was used for terminal group selection in place of different directions of motion although two-motion power driven selectors were also developed. The size of the switch made it possible to engineer links or trunks into groups of different sizes depending upon the traffic needs.

Since the number and speed of pulses to set a selector increased, more accurate pulsing (address signaling) methods were developed. One popular arrangement was "revertive" pulsing where the selector in moving generated pulses that were received by the "sender" and compared with the number

Fig. 23. Panel switch frame.

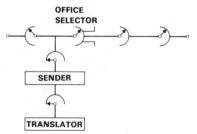

Fig. 24. Common office code translator.

Fig. 25. Coordinate switch.

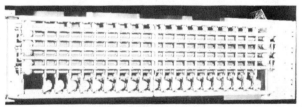

Fig. 26. Crossbar switch.

desired [24]. When they matched the selector was stopped by opening the trunk. This also permitted the more delicate pulsing elements to be concentrated in the sender.

While there was a period when systems of this type were used in conjunction with manual switching (called "semi-automatic" "semimechanical" and "automanual"), fully automatic systems of this type were placed into production just prior to 1920. Later the indirect control principle was extended. It was possible to further translate the digits dialed to designate the central office so that the terminals of the early stages of switching could be used efficiently and independently of the particular digits or names selected to identify the desired offices. As shown in Fig. 24, the central office code digits, referred by such plurality as "translators" or "decoders", accessed through another concentration network to act on each call to determine the location of trunks to the desired office on the early selector stages of the network [25]. By concentrating the nonfixed translation it was possible to make changes readily on a few translators so that it would affect all calls originating and established in a particular office. The holding time of a translator could be only a few hundred milliseconds per call. This is the first application of bulk memory in information processing.

4) Fine Motion Switches—The success of indirect control led to other control technique improvements. The logic required for sequencing and storing information in these systems

was carried out with electrical relays. As a result this technology saw many advances and improved reliability. Some designers even used them for crosspoints [26]. Switches with coordinate contact arrays (Fig. 25) were proposed for many years as a possible replacement for the panel switch [27]. They had the advantage of "fine" rather than gross motion and, as compared with a matrix of relays, did not require a magnet or coil for each crosspoint.

The crossbar switch (Fig. 26) was the first coordinate switch to use this principle in combination with reliable relay-type contacts. The contacting elements were noble metals such as silver, gold, platinum, or their alloys. As with most modern relays the contact springs were bifurcated and have two independent sets of contacts [28]. Prior to this development all switches used phosphor–bronze "base" metal contacts that were inexpensive and could be formed by stamping. Fig. 27 shows the contact selection mechanism of a crossbar switch.

5) Common Control—Early crossbar switches had 10 X 10 arrays of crosspoints and were used in direct control systems, in effect replacing the step-by-step selectors [29].

During the coordinate switch development the concept of using these simple crosspoints without individual switch con-

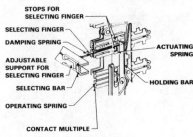

Fig. 27. Crossbar switch crosspoint.

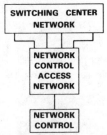

Fig. 28. Common control.

Fig. 29. Crossbar switch frames.

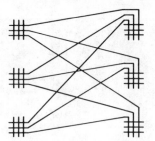

Fig. 30. Coordinate link principle.

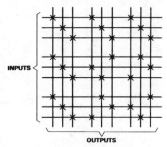

Fig. 31. Graded multiple.

trols was proposed. The idea was to provide a common circuit that would control the operation of selected crosspoints and then be available to select and operate upon other crosspoints for other calls. This gave rise to another network, a network control access network (NCAN) (see Fig. 28). (The photograph, Fig. 29, shows multicontact relays at the top of the frame. These constitute the NCAN.) The activated crosspoints would be electrically held for the duration of the call by a supervisory circuit. (Note that the electrical holding of the activated crosspoint link like the positional holding of a gross motion switch is a "memory" of the connection.)

When the smaller crossbar (10×10 or 10×20) switch was proposed for this type of operation, it was possible to carry the selection process further, keeping in mind the general advantage of making larger selections. From this grew the "coordinate link" principle as shown in Fig. 30 [30]. Here each input can reach many outputs in the second stage which are also accessible, with blocking, to the other inputs. This principle may be extended in many ways as described in another paper in this issue [31]. Other methods for extending both gross and fine motion switch selectivity are beyond the scope of this paper. Mention should be made however of the "graded multiple" invented in 1907 by E. A. Gray [32]. By restricting accessibility of each input to an orderly pattern of a fixed number of outputs it is possible to obtain an output group that is larger than the accessibility of any one switch (see Fig. 31). This was particularly important as gross motion switches were used in large heavy traffic offices and for fine motion switching with comparatively expensive crosspoints.

Once adequate accessibility within link networks using small access devices was possible and combined with indirect control and common network control it was possible to free the entire network control process from the progressive establishment of connections through the switching center network. With common control, an idle desired output of the network is found first. Then a set of idle links through the network is found. Finally the crossbar switches are operated and a check made of the talking and holding paths. This concept of common control is known as "look ahead." As a result an idle trunk to a suitable destination, including possible alternate routes, can be selected before selecting the switching center network path. On terminating calls "look ahead" gave two

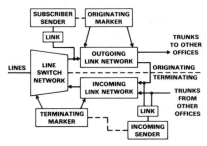

Fig. 32. No. 1 Crossbar system diagram.

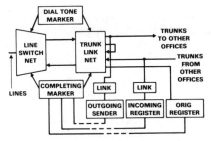

Fig. 33. No. 5 Crossbar system diagram.

ASSOCIATED

    DIRECT CURRENT   LOOP (ON/OFF) (HIGH/LOW)
                       COMPOSITE, SIMPLEX

      SUPERVISORY

      PULSING

         DIAL (DP), REVERTIVE (RP)

    ALTERNATING CURRENT

      SUPERVISORY   SINGLE FREQUENCY
                       INBAND
                       OUT-OF-BAND

      PULSING   MULTIFREQUENCY (MF)

             SINGLE FREQUENCY (DP, RP)

    DIGITAL

      PER CHANNEL

      COMMON CHANNEL

  DISASSOCIATED

    COMMON CHANNEL

Fig. 34. Signaling.

advantages. First, it was possible to determine if the called line was busy or to select an idle line in a (PBX) group without establishing a connection from the incoming trunk. Second, it is possible to place lines on the switches according to traffic requirements and independent of their directory numbers. A translator (bulk memory) is used to indicate the correspondence between directory and "equipment" numbers. A switching center network consists only of crossbar switches and associated NCAN multicontact relays. The network controls have been called "markers." Generally one marker will not handle all offered call attempts, and a plurality is provided to operate simultaneously on different sections of the network.

In early systems the switching center network was separated into originating and terminating networks with a group of markers for each. In the No. 1 Crossbar system of the Bell System and other similar systems (see Fig. 32), separate originating line concentration and terminating line selection stages were made unnecessary by combining them into one stage known as a "line switch" [33]. This stage had a separate network control for originating calls.

Later crossbar systems, like the No. 5 Crossbar system (see Fig. 33 combined the outgoing and incoming networks [34]. The switching center networks were also used as the link for connecting to the originating register (equivalent of the signal receiving portion of a sender). The markers that originally served to establish all network connections were later divided into two groups; one for connecting calling lines "dial tone") to originating registers and one for completing all other network actions.

C. Signaling (Fig. 34)

As electromechanical systems progressed from directly controlled gross motion switches to common control with fine motion switches much progress was made in the technology and techniques for the basic functions. Signaling was one

function that improved greatly not only in reliability but also in range. Initially the improvements were in dc pulsing such as dial pulsing repeated from the originating line to interoffice trunks. Special techniques were developed to extend this signaling to offices from 50 to 100 mi away using techniques known as "simplex" and "composite" signaling [35].

Originally trunk facilities to distant cities were insufficient to meet peak traffic demands. All calls were recorded by operators who processed them as facilities became available [36]. As sufficient facilities were built to meet the demand dial operation on a no-delay basis became feasible. To implement the dialing and supervision of calls over long distances tone signaling was used. Many types were developed. Most important was multifrequency (MF) pulsing where a combination of frequencies represent a digit, so that in effect address signaling could be speeded up by almost an order of magnitude [37].

The signaling function takes place from stations to the switching system in which case it is known as "station signaling." Between switching offices it is known as "interoffice signaling."

In each case dc or ac signals are used. In the case of station signaling care must be taken that ac signals cannot be easily simulated by the voice. Therefore, different dual-tone multifrequency (DTMF) signaling is used rather than the MF tone used for interoffice signaling.

Tone supervisory signals were sent out of the speech band to avoid interference by voice but later they were placed in the speech band but on a "tone on" basis for idle circuit as an on-hook indication.

More recently it has become possible, with the introduction of digital transmission facilities, to allot separate bits for signaling. If the bits are on a per trunk basis then this is another form of out-of-band signaling.

Out-of-band signaling is either ac or digital on a per channel basis [38]. A further form of associated out-of-band signaling occurs when separate but parallel signaling paths are used. The best example is in Europe where one channel in a digital transmission system is used exclusively for the signaling associated with 30 other channels [39].

The principles of switching, like many other areas have a habit of returning after a generation or two of new technology. In the early days of manual switching separate associated channels were used for both station signaling and intra- and interoffice signaling [40]. Now after several generations of over the channel out-of-band and in-band signaling a new generation of disassociated signaling known as "common channel signaling" is beginning to appear [41].

Common channel signaling uses modern data transmission channels to interchange supervisory and address signaling generally between processor controlled switching systems (see Section IX-D4). In the past due in part to the slowness of the transmission and individual channel complexity, the specific signals have been held to the minimum required to support the basic functions. With common channel signaling not only can these signals be sent and received more rapidly, but also additional information relative to the call may be interchanged. In the North American message network the time for establishing a connection across the country is 10 s on the average. With the full implementation of common channel signaling this average will be reduced to about 3 s.

Specific arrangements for the international use of common channel signaling have been standardized by the International Consultative Committee Telephone and Telegraph (CCITT) [42].

D. Electronic Era

The successful application of electronics to reduce the costs and extend transmission distance encouraged switching engineers to look at this technology. Prior to World War II vacuum tube electronic beam selectors were devised and some used in the war effort [43]. Gas discharge tubes were improved in reliability sufficiently to be used in subscriber station equipment [44]. Switching engineers experimented with the application of these devices primarily to replace relays for logic but after the war also as possible crosspoints.

The Bell System's first electronic switching system used small gas diodes with a negative impedance characteristic as crosspoints [45]. In other systems they were used to assist in the operation of crossbar [46] and reed relay crosspoints [47]. Reed relays are hermetically sealed contacts in a glass envelope and inserted into electromagnetic coils. They were also a product of war-time technology and were used to avoid troubles that could occur in some open contact crosspoint devices such as crossbar switches [48]. Some reed crosspoints magnetically hold or "latch." These bistable crosspoints require a reselection to effect a release [49].

1) Time-Division Networks—Another idea from war-time transmission developments and theory is time sharing or time multiplexing. These developments showed that sampled speech signals could be multiplexed over broad-band channels. Internal to switching systems, broad-band links are easier to apply than over distances. Furthermore, the sample, if digitized may be transmitted over parallel channels at lower pulse rates.

All previous switching center networks used the separate physical paths in space and have been called "space division." The new concept, the switching of speech samples with electronic crosspoints that are simultaneously shared by other calls, is called "time division." (Frequency division switching, analogous to analog carrier transmission has also been investigated for switching but to date no practical scheme has been devised.) In switching the applications of electronics lagged because of the cost of power as well as life of hot cathode vacuum tubes that would be required in large quantities for the logic and network of a large switching system. It was quite natural that the first applications of electronics was to time share the elements in the network portion of switching systems. Many experimental [50] and production [51] systems used time-division

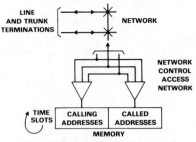

Fig. 35. Time-multiplex switch control.

networks. Initially the switched samples were of variable amplitude. More recently digitized or coded samples have been switched [52].

So it was no surprise that the invention at Bell Laboratories of the transistor, a device that required no appreciable standby power, was found to be an ideal new technology for application to switching. At the same time study and exploratory development was being applied to the bulk (large) electronic memory such as storage tubes and later magnetic devices. These events ushered in the electronic switching era.

2) Electronic Switching Center Networks—Electronic switching center networks use electronic crosspoints. In time division the crosspoints operate much faster, but many fewer are required. As mentioned earlier with space-division switching memory is required. In electronic space-division networks the memory is usually built into the crosspoint electrically so that it will hold once it is operated. Semiconductors p-n-p-n diode crosspoints have been developed with these characteristics [53]. In some systems the memory function is centralized in bulk storage where it is known as the "network map."

In time-division switching center networks a cyclic memory is separate from the crosspoint (see Fig. 35). During the cyclic readout of each address of the memory, known as a "timeslot," the calling and called terminals are connected together long enough for the transference of the representation of the speech sample. Received signals eventually pass through a low-pass filter to restore the original continuous signal envelope. Early networks had 64 time slots, whereas current technology permits as many as 512 time slots.

As time-division networks are extended to serve more terminals, either of two types of stages may be used to extend the network capacity. One is the storage of the samples either analog or digital, independent of a time slot. With the sample in memory it is possible to place it onto the same or another common medium or "highway" in a different time slot from which was placed in the memory. This is known as "time-slot interchange" (TSI). The other technique is to provide more than the single stage of time-shared crosspoints, that is more than one common medium. This then is a space division network that is switched at a rate necessary to provide sufficient time slots. It has become known as a "time-multiplexed switching (TMS) network." The TSI and TMS stages, briefly known as *T* or *S* stages are subsets and assembled in various combinations to form time-division switching center networks (see Fig. 36) [54].

3) Time-Division Controls—The control of switching systems became quite sophisticated with the introduction of common control for networks and the centralization of translation.

The availability of a technology (solid state) that was at least three orders of magnitude faster and potentially more reliable

LEVELS OF SWITCHING	CONFIGURATIONS
1	T/S
2	ST/TS
3★	STS/TST
4★	STTS/TSST
5★	SSTSS/TSSST
6★	STSSTS/TSSSST

★ OTHER CONFIGURATIONS POSSIBLE AT THESE LEVELS

Fig. 36. Time-division digital switching configurations. The time-space pyramid.

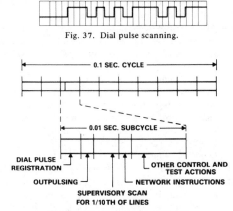

Fig. 37. Dial pulse scanning.

Fig. 38. The time-sharing principle.

WIRED LOGIC

STORED PROGRAM

SINGLE PROCESSOR

SEPARATE PROGRAM AND DATA STORES

SINGLE STORE

MULTIPROCESSOR

TRAFFIC DIVISION

FUNCTIONAL DIVISION

DISTRIBUTED CONTROL

ACTION TRANSLATORS

Fig. 39. Electronic switching controls.

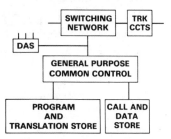

Fig. 40. Stored program system.

than that being used to implement these common arrangements appeared to have considerable potential to greatly reduce the quantity of equipment required for this basic function. Generally register-senders were required for approximately one per 100 terminals and markers or other controls, one per 1000. (These quantities vary widely depending upon the number and class of call attempts offered to the office.) With the application of electronics it became possible for a single control or "processor" as they are called, to perform all of the control functions of an office, formerly requiring a plurality of controls, some of them competing for the access to the same switching center network. With a single control the same principles used to extend networks into the time domain were applied.

Time-shared electronic logic with a common electronic memory made it possible to reduce the complexity of control accessibility. In particular input signals may be examined one at a time in a process known as scanning (Fig. 37) while the control functions as a whole including "scanning" are scheduled serially or time shared (Fig. 38).

4) Stored Program Control (Fig. 39)—As mentioned in Section XII, each generation of switching systems incorporates in its controls those items needed for new services and features that were added peripherally after the initial design. The burden of this task was becoming formidable after the second generation of crossbar systems were modified for the new service and features requirements of the late 1950's and early 1960's. It was realized that for initially designing and meeting changing requirements needs, electronic logic for the complex control would be difficult to administer. In 1955 a new flexible con-

cept of stored program control for the application to switching systems using general purpose electronic logic and bulk memory was considered [55]. This concept adapted the real-time needs of a switching systems to the programming ideas then being developed for computers [56]. The real-time difference was and is still an important difference. For the first time the memory was used for the control logic as well as the translations. The bulk memory is also used to record temporarily other information about each call.

By removing the system logic from the "hardware" of the control a new degree of design independence became possible. The common control, called "central control" because of its singularity, became the heart of general purpose processors for switching systems (see Fig. 40) [57]. Not only are the controls general purpose and time shared but the input and output functions as related to call processing are also time shared and general purpose. The output information is conveyed by "distributors" and the input information accessed by "scanners," known collectively as the distributing and scanning (DAS) functions.

Much of the emphasis in the design of system control switched from hardware to "software," particularly by the continuing design after the system was installed. Software had always been present in system design but not recognized as separate from the hardware design [58].

In a broad sense "software" not only includes the programs written for switching systems and computers that aid in the design and engineering but also in the detailing of system service and feature requirements, action sequences, the theoretical call capacity determinations of stored program systems, and the other items listed in Fig. 41 [59].

With more recent developments in large-scale integrated circuits, READ-ONLY memories (ROM) with programmed logic, shown as wired program logic (WPL), are being used for the logic needs that remain in many systems. This class of systems

1. FEATURE AND
   SERVICE REQUIREMENTS
2. FLOW CHARTING—
   CALL SEQUENCES
3. PROGRAM LANGUAGES AND
   ORDER STRUCTURE
4. PROGRAMMING AND CODING
5. SIMULATION
6. CALL PROCESSING CAPACITY

Fig. 41. The software dimension.

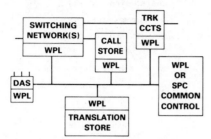

Fig. 42. Distributed logic system.

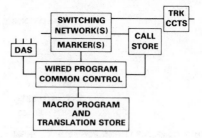

Fig. 43. Action translation system.

is being called "distributed logic" systems (see Fig. 42) but they still require a central control for directing or sequencing the information flow. As described below, these peripheral functions of centrally controlled systems can also be described as almost autonomous units making possible another approach to distributed control.

As these concepts were successfully applied and systems of the stored program type placed in service, the maximum size for which the system was designed became larger. The maximum size of the No. 5 Crossbar system was 35 000 lines. The latest metropolitan electronic switching system, the No. 1A ESS, has a capacity of more than 128 000 lines [60]. While some have developed these systems with advanced technology so that a single processor (a pair for service continuity) can service this large number of traffic sources, others have solved this need for larger systems with a plurality of processors, known as "multiprocessing" [61].

The processors may be used as a plurality of common controls, as in electromechanical systems, each capable of dealing with a portion of the total traffic offered [62] (called "traffic division") or with each control serving specific functions of a system (called "functional division") such as call originations, translation, charging, etc., the basic functions and operation service features [63]. This division of functions is related to the program sections rather than just the peripherals. Some system proposals have been made to relate distributed processing and functional multiprocessing [64]. In the latter case a characteristic is a changeable stored program and a common call memory record.

As the general flexibility of the system control has improved, so too has the memory technology and accompanying reliability. Originally programs and translations were stored in separate ROM's called "program stores." Later it was possible to store programs and translations in electrically writable stores with magnetic tape backup, the same stores used for call and other data [65].

Another form of system control that is not fully stored program control is a variety closer to wired logic known as "action translator" systems (see Fig. 43) [66]. Systems of this type are a compromise and may involve distributed control, traffic, or functional division. Action translators usually employ special purpose controls with class of service and other tabularized data stored in bulk memory. In addition, macro call sequences are stored in a translator so that call state actions or changes may be read from a table.

The deployment choice between full stored program control (FSPC), distributed control, action translator control, and wired logic is dependent upon the objective of the installing and operating administration [67]. If large numbers of new services and features are expected to be added then the FSPC-type system on a life cycle basis will probably prove to be least costly. Where simple POTS calls are the only ones expected, the action translator or even simpler wired logic systems might suffice. Distributed control system designs at this time are in a state of flux so that while the intentions of some designers are to retain stored program control flexibility in the central control, others are distributing the functions and with it the flexibility. This means distributed changes when they are required which may prove more costly than centralized changes. If the distributed functions are implemented with programmable ROM's (PROM's) then they may at some expense be reused if changed. If they are implemented with nonprogrammable ROM's they are wired program logic (WPL) and must be disposed of if changes are required. Of course electrically changeable ROM's are most desirable for these applications.

X. THE FUNCTIONAL APPROACH

The evolutionary (see Fig. 14) approach provided greater insight into the way the basic functions are implemented in practical switching systems. Principles are best understood by abstracting the specific relationships found in different systems. An approach classifying these details has been devised [68] and its use demonstrated [70]. As a result, the more general diagram of Fig. 9 may now be replaced with one (Fig. 44) that more accurately illustrates general system architecture. Here the functional elements are given more appropriate names to better delineate their purposes.

The more specific aspects of the basic signaling function in the processing of calls are known collectively as "call signaling processing (CSP)" and include service request detection, address (pulsing), and supervisory signaling. Each line and most trunk circuits contain a CSP element.

The controlling function usually consists of a minimum of two distinguishable parts, the "network control (NC)" and the "call information processing (CIP)." Call information processing consists of buffering (storage), gathering, translating, and interpreting information related to calls. Separate CIP elements

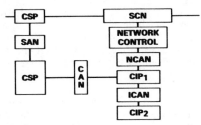

Fig. 44. Addition of access networks to connect basic functions.

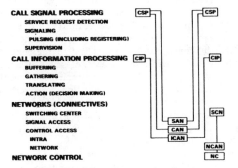

Fig. 45. Switching system functions.

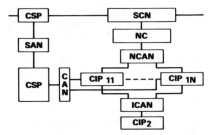

Fig. 46. Switching system with traffic division.

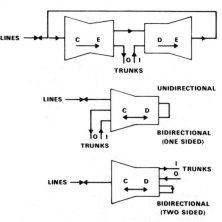

Fig. 47. Functional applications of switching networks.

NUMBER OF LINES,
TRUNK GROUPS, TRUNKS

CALLING RATE (ATTEMPTS)

HOLDING TIME (USAGE)

COMMUNITY OF INTEREST -
(INTEROFFICE VS. INTRAOFFICE)

Fig. 48. Traffic characteristics.

This functional symbolism may be most useful in delineating architectural difference between switching systems [70]. It is unfortunate that standardization of nomenclature has been most difficult even within equipment of one manufacturer let alone among manufacturers. Using these symbols with a nomenclature translation table aids in the study of system architecture.

A. Switching Center Network Characteristics (Fig. 47)

All networks described so far in this paper might be classified as two-sided bidirectional, the latter meaning traffic may originate from either side. Single-sided bidirectional networks are also useful where the intraoffice (line-to-line) traffic is high and where trunk-to-trunk calls are to be accommodated economically. Fig. 47 shows these networks as well as one using only unidirectional networks. The C, D, and E represent the concentrations, distributions, and expansion functions. To form a switching center with a unidirectional network at least two portions are required although only one need be unidirectional (see top of figure). Combinations of these networks may also be used.

There is also a duality in network design between space and time division. Generally there is an equivalence of every time division network in a space division representation. It has not been proven that the reverse is always true [71].

For every network design many factors must be considered. Fig. 48 lists some of more important traffic factors. The methods of engineering offered load to carried load (service) differ between switching center and access networks. Load is defined as usage—the total occupied time. The lines, service circuits, and the trunks associated with the terminations and control are engineered on a demand basis while within the system the access connectives are usually engineered on a "delay" basis. The processing and control functions are engineered on the basis of

may occur in the system for separately performing these aspects of the basic function or for different system architectures.

In addition to the various call signal and information processing, the main switching center network, and its control, it is obvious that additional functional elements are needed to interrelate them. Fig. 45 shows the possible interrelations. These are known collectively as "connectives" or "access networks." They range from register or sender links in indirect control systems to scanners in electronic switching systems.

Figs. 44 and 46 include the addition of four access networks as well as more specific naming of the basic functions. In general the access networks are named for what they connect to in the initial flow for call utilization. In a specific system a function may occur more than once. For example, in the No. 1 Crossbar system [69] there are both originating and terminating markers. A different subscript is used to designate each different function. Also the identical function might appear as in a traffic divided (or multiprocessor) system, as in Fig. 46, where the same or similar designation is used for each appearance of multiply specified items. Generally connectives are only required where there is a multiplicity of functional units although in the trivial case it may be a single gate or relay.

calls carried. The application of probability theory to teletraffic has matured with the developments in switching. The theoretical results coupled with modern computer aids and have enabled excellent engineering methods to be developed [72].

XI. SWITCHING DEFINITIONS

This paper has been confined to cover switching that was first introduced for the telephone and now is known more generally as "circuit switching." It has been difficult to obtain agreement on switching nomenclature definitions. There is an IEEE Standard on the basic terms as used in the United States [73] (it has been updated). Here circuit switching is defined as providing through connections for the exchange of messages. Circuit switching can also be defined as applied to a system with a switching center network used for two-way transmission with no chance of transmission delay due to passage through the network. The messages may be voice, data (including Telex, Teletypewriter), or video. Digitally encoded voice or data may even pass through space division networks.

Time-division switching center networks were discussed in Section IX-D1. When time division switching is used with digital data messages it is sometimes called "virtual" switching to distinguish it from space division circuit switching.

It is obvious that the switching center network may be implemented to carry any type of information delivered to it in electrical form. It is possible for the control portion of the system to operate simultaneously with more than one switching center network, for example, one for voice and one for video [74]. Also the transmission may represent different types or speeds of messages and that as such several types may be carried through the same switching center network simultaneously. In this case the network must be designed to carry the bandwidth required by the highest bandwidth message or to take advantage of the bandwidth of the network several channels, and as for example low speed data, may use the same network path simultaneously [75]. This is a form of combined voice and data switching.

The other generic form of switching applies in general to one-way transmission that may be delayed. This is known as "store and forward switching." Generally this implies that no switching center network is employed, but from the principles given in the preceding section it is obvious that a signal access network is required.

The storage aspect of this type of switching varies widely with different types of applications. The delay is the reason for storage and may relate to one of the following: the lack of liberally engineered or common user group of idle transmission facilities at the moment the beginning of message reaches the switching office; the need for multiple transmissions of the same message also affects the engineering of facilities and the need for storage; the called station or terminal being busy; the speed of capability of the message transmission facilities or medium which may be less than the received rate; or the formatting or coding of the transmitted message being different from the way it was received [76]. The length of delay depends upon the quantity of memory provided in a switching center and the transmission capacity (number and speed) available for use in routing the message to its destination. Generally in these networks the desire is to employ switching memory at the expense of more adequate transmission facilities.

The message itself may be divided into smaller, usually uniform size segments. This form of store and forward switching

is known as "packet switching." In this case each packet must be preceded by a repetition of the called address. It has the advantage that the delays are shared by all users rather than serving complete messages in the order of receipt. For this reason it is used for interactive data transmission where the messages tend to be short.

XII. GROWTH OF SERVICES AND FEATURES

After the introduction of each new generation of switching technology there is considerable effort expended adapting it to new features and offering new services, usually extending over many years and peaking just before the deployment of another technology. As a result systems of each generation are or become more complex than those of previous generations. Furthermore, as needs develop some features and services are implemented by functional adjuncts added to the periphery of the system. In the past the initial design of each new generation of system has incorporated these improvements so that the peripherals added to the previous generation are absorbed into the basic switching functions. It is an important philosophical point that new features and services successfully introduced in one system are generally included as basic features and services for succeeding system designs.

The extent of feature and service development varies widely around the world as well as with the general area of application of the switching system; PBX, local central office, intermediate office, store and forward, etc. Fig. 12 shows the development that has taken place in the 11 and 7 years, respectively, since No. 1 and No. 2 ESS's were introduced to meet the sophisticated needs of the Bell System [77]. History teaches us that it is only a matter of time before similar features and needs appear in other administrations. The high-speed interchange of information between offices made possible by the application of common channel signaling will bring to the entire public network of telecommunication offices the same advantages stored program control brought to the switching offices themselves. Many new services will become available to customers relative to their own needs when large portions of their calls are dealt with in the same way with interaction available over the common signaling channels [78]. For example, knowing the source of a call, special ringing signals or reverse charging would be possible if desired by the called subscriber. The new services and features introduced by these methods will further enhance the great expansion already made feasible by stored program control.

XIII. CONCLUSION

Circuit switching is such a pervasive aspect of modern telecommunications that the public has grown highly dependent upon it. This paper is only an introduction to its history and principles. Many important and even critical topics have been omitted. For example, missing are discussions on methods for charging for calls, equipment methods to test for and locate trouble, and the introduction of redundancy to maintain service during periods of trouble and growth. Others are operator and business services, traffic measurement and control, and network management.

Much has been written on the subject of switching and its technology. It is the writer's opinion that much more is needed before the subject is understood by telecommunication engineers specifically and public generally.

[1] "Notes on distance dialing," AT&T, 1975.
[2] J. W. Gorgas, "Polygrid network for Autovon," *Bell Labs. Rec.*, vol. 46, p. 222, 1968.
[3] F. J. Singer, "Compatibility in communications," *Elec. Eng.*, vol. 81, p. 94, 1962.
[4] J. A. Grandle and R. E. Machol, "Engineering and administrative data acquisition system for processing and analyzing traffic data," in *IEEE Nat. Telecommun. Conf. Rec.*, vol. 1, pp. 5-25, 1975.
[5] W. B. Rohn and T. F. Arnold, "Design for low expected downtime in real-time control systems," in *Rec. IEEE Int. Conf. Communications*, pp. 16-25, 1972.
[6] R. J. Hass, "On line retrofitting of electronic switching systems," *IEEE Trans. Commun. Technol.*, vol. COM-17, p. 99, 1969.
[7] R. C. Clark and C. A. Martin, "Quality control through service observing," *Auto. Electric Labs. J.*, vol. 8, p. 154, 1963.
[8] W. B. Macurdy, "Network management in the United States—Problems and progress," in *6th Int. Teletraffic Congr. Rec.*, p. 621/1, 1973.
[9] R. J. Jaeger and A. E. Joel, Jr., "TSPS no. 1 systems organization and objectives," *Bell Syst. Tech. J.*, vol. 49, p. 2417, 1970.
[10] P. D. Shea, "Centrex service—A new design for customer group telephone service in the modern business community," *Trans. AIEE*, vol. 80, pt. I, p. 474, 1961.
[11] K. B. Miller, *Telephone Theory and Practice*. New York: McGraw-Hill, 1933, vol. 2, ch. V.
[12] R. J. Murphy, "Expansion of step-by-step switching facilities for data (MTWX) and for bulk-type (WATS) traffic," AIEE paper CP-62-1286, 1962.
[13] M. Takenaka, K. Yamagishi, and M. Chiba, "Centralized memory systems for abbreviated dialing," *Japan Telecommun. Rev.*, vol. 12, p. 239, 1970.
[14] R. Syski, *Introduction to Congestion Theory in Telephone Systems*, Edinburgh, Scotland: Oliver and Boyd, 1960.
[15] J. E. Brand and J. A. Warner, "Processor call carrying capacity estimation for stored program systems," this issue, pp. 1342-1349.
[16] *Ibid.* [11, ch. X].
[17] *Ibid.* [11, ch. XI].
[18] Rhodes, *Beginnings in Telephony.* New York: Harper and Brothers, 1929.
[19] A. B. Smith, "History of the automatic telephone," *Sound Waves*, vol. XIV, no. 5, Oct. 1907 (to telephony, vol. XVIII, no. 20, Nov. 1909).
[20] U.S. Patent 1 083 456.
[21] E. B. Craft, L. F. Morehouse, and H. F. Charlesworth, "Machine switching telephone system for large metropolitan areas," *AIEE Trans.*, p. 320, Apr. 1923.
[22] G. Deakin, "The 7A machine switching system—Rotary system," *Elec. Commun.*, vol. 3, p. 153, 1925.
[23] "LM Ericsson automatic telephone systems," *Post Office Elec. Eng. J.*, vol. 18, p. 128, 1925.
[24] C. Breen and C. A. Dahlbom, "Signaling systems for control of telephone systems," *Bell Syst. Tech. J.*, vol. 39, p. 1381, 1962.
[25] Ibid. [11, vol. 3, ch. IV].
[26] H. W. Dipple, "Relay system of automatic switching," *Inst. Elec. Eng. (London)*, paper 84, 1921.
[27] U. S. Patents 1 131 734 and 1 525 787.
[28] H. W. Wagar, "The U type relay," *Bell Labs. Rec.*, vol. XV, p. 300, 1938.
[29] B. Bjurel, "Rural automatization in Swedish telephone system," *Trans. AIEE*, vol. 73, pt. I, p. 552, 1954.
[30] C. Jacobaeus, "Employment of crossbar selectors in shunt circuit systems," *Ericsson Rev.*, p. 30, 1945.
[31] M. J. Marcus, "The theory of connecting networks and their complexity: A Review," this issue, pp. 1263-1271.
[32] U. S. Patent 1 002 388.
[33] F. J. Scudder and J. N. Reynolds, "Crossbar dial telephone switching system," *Bell Syst. Tech. J.*, vol. 18, p. 76, 1939.
[34] R. C. Davis, "No. 5 post-war crossbar," *Bell Labs. Rec.*, vol. 27, p. 85, 1949.
[35] H. A. Sheppard, "A signaling system for intertoll dialing," *Bell Labs. Rec.*, vol. 18, p. 337, 1940.
[36] Bell Telephone Labs., Inc., *A History of Engineering and Science in the Bell System—The Early Years 1875-1925*, p. 615, 1976.
[37] *Ibid.* 24.
[38] *Ibid.* 24.
[39] W. New and A. Kuendig, "Switching, synchronizing, and signaling in PCM-exchanges," in *Int. Colloquium for Electronic Switching Rec.*, p. 513, 1972.
[40] Ibid. [36, p. 730.]
[41] C. A. Dahlbom, "Common channel signaling—A new flexible interoffice signaling technique," in *Int. Switching Symp. Rec.*, p. 421, 1972.
[42] J. J. Bernard, "CCITT international signaling no. 6," *Telecommunication J.*, vol. 41, no. 2, p. 69, 1974.
[43] U. S. Patent 2 122 102.
[44] L. F. Stacey, "Vacuum tube improves selective ringing," *Bell Telephone Labs. Rec.*, vol. 15, p. 111, 1936.
[45] M. A. Townsend, "Cold cathode gas tubes for telephone switching systems," *Bell Syst. Tech. J.*, vol. 36, p. 775, 1957.
[46] Z. Kujasa and Katsunuma, "Experimental telephone exchange system using Parametron," *J. Inst. Elec. Commun. Eng. Jap.*, vol. 42, p. 225, 1959.
[47] W. A. Mathaner and H. E. Vaughn, "Experimental electronically controlled automatic switching system," *Bell Syst. Tech. J.*, vol. 31, p. 443, 1952.
[48] P. W. Renaut and C. E. Pollard, "The seated contact family," *Bell Labs. Rec.*, vol. 48, p. 203, 1970.
[49] A. Feiner, "Ferreed," *Bell Syst. Tech. J.*, vol. 43, pt. 1, p. 1, 1964.
[50] L. R. F. Harris, V. E. Mann, and P. W. Ward, "Highgate Wood experimental electronic telephone exchange system," *Inst. Elec. Eng. Proc.*, vol. B107, suppl. 20, p. 200, 1960.
[51] W. A. Depp and M. Townsend, "Electronic PBX telephone switching system," *IEEE Trans. Commun. Electron.*, vol. CE-83, p. 329, 1964.
[52] "No. 4 ESS—Long distance switching for the future," *Bell Telephone Labs. Rec.*, vol. 51, p. 226, 1973.
[53] T. Feldman and J. W. Rieke, "Application of breakdown devices to large multistage switching networks," *Bell Syst. Tech. J.* vol. 37, p. 1421, 1958.
[54] M. J. Kelly, "Introduction to PCM switching," *Auto. Elec. Tech. J.*, vol. 12, p. 234, 1971.
[55] U. S. Patent 2 955 165.
[56] *Ibid.* [41]
[57] J. A. Harr, F. F. Taylor, and W. Ulrich, "Organization of the no. 1 ESS central processor," *Bell Syst. Tech. J.*, vol. 43, p. 1845, 1964.
[58] J. A. Harr, E. S. Hoover, and R. B. Smith, "Organization of the no. 1 ESS stored program," *Bell Syst. Tech. J.*, vol. 43, p. 1923, 1964.
[59] R. W. Ketchledge, "Development of development methods," in *Int. Switching Symp. Rec.*, p. 412, 1974.
[60] J. S. Nowak, "No. 1A—A new high capacity switching system," in *Int. Switching Symp. Rec.*, p. 131-1, 1976.
[61] K. Katzeff and U. Jerndal, "L. M. Ericsson transit exchange system," in *Rec. IEE Conf. Switching Techniques for Telecommunication Networks*, publ. no. 52, p. 230, 1969.
[62] S. Kobus, A. Kruthof, and L. Vrellevoye, "Central control philosophy for the Metaconta telephone switching system," in *Int. Switching Symp. Rec.*, p. 509, 1972.
[63] A. H. Doblmaier and S. M. Neville, "No. 1 ESS signal processor," *Bell Labs. Rec.*, p. 120, 1969.
[64] S. Pitroda, W. A. Fechalos, and C. J. Stehman, "The microprocessor controlled 580 digital switching system," in *Int. Switching Symp. Rec.*, p. 213/4, 1976.
[65] R. E. Staehler and R. J. Walters, "No. 1A processor—An ultra-dependable common control," in *Int. Switching Symp. Rec.*, p. 423/1, 1976.
[66] R. C. Long and G. E. Gorringe, "Electronic telephone exchanges: TXE2—A small electronic exchange system, *Post Office Elec. Eng. J.*, vol. 62, p. 12, 1969.
[67] A. E. Joel, Jr., "Realization of the advantages of stored program control," in *Int. Switching Symp. Rec.*, p. 143, 1974.
[68] ——, "Classification and unification of switching system functions," in *Int. Switching Symp. Rec.*, p. 446, 1972.
[69] *Ibid.* [32].
[70] A. E. Joel, Jr., Ed., *Electronic Switching: Central Office Switching Systems of the World.* New York: IEEE Press, 1976.
[71] M. J. Marcus, "Space–time equivalents in connecting networks," in *Int. Switching Symp. Rec.*, p. 35/25, 1970.
[72] R. I. Wilkinson, "Comparisons of load and loss data with current teletraffic theory," *Bell Syst. Tech. J.*, vol. 50, p. 2807, 1971.
[73] "Trial-use standard definitions of terms for communication switching," *IEEE Trans. Commun. Technol.*, Oct. 1970.
[74] P. N. Burgess and J. E. Stickel, "The Picturephone system: Central office switching," *Bell Syst. Tech. J.*, vol. 50, p. 533, 1971.
[75] A. F. Connery, "System philosophy and parameters—Broadband switching," *IEEE Trans. Commun. Electron.*, p. 687, 1963.
[76] "No. 1 ESS arranged for data features," *Bell Syst. Tech. J.*, vol. 49, p. 2733, 1970.
[77] W. O. Fleckenstein, "Development of telecommunication switching in the United States of America," in *Int. Switching Symp. Rec.*, Keynote Address, p. 121, 1976.
[78] C. R. Jacobsen and R. L. Simms, "Toll switching in the Bell System," in *Int. Switching Symp. Rec.*, p. 132/4, 1976.

The New Possibilities of Telephone Switching

J. GORDON PEARCE, FELLOW, IEEE

Abstract—Switching system architectures have evolved to be responsive to the needs of the user. Their design has been constrained by the existing environment which treats terminals, transmission channels, and switching entities as separate "black boxes." Telephone switching circuits have used relatively expensive discrete components, and hence, system designers used common control techniques to minimize the system cost. This was done by providing the circuits in common whenever this was possible. Such common circuits were associated with a switched path for the period of time for which the functions of the circuit were required. This was followed by "one at a time" operation and by the use of stored program controls. The advent of low-cost electronic circuit components has resulted in the application of digital techniques to switching systems. For the first time, the combination of switching and transmission is possible. This is the near-term objective. The long-term objective, made possible by forecasts of low-cost memory and electronic gate circuits, is the combination of switching, transmission, and the terminals. This will result in much more complex terminals.

INTRODUCTION

TO PEOPLE who are not involved in the industry, the telephone is simply a device (which has the same significance for them as a switch which they operate to use electric current). In either case, they are particularly interested with what happens as the result of the action, and not how the results are accomplished. However, they expect that when the action is made (be it operating a switch or signaling the requirements for a telephone call by dialing or keying) the results they desire are achieved rapidly and without error, 24 hours a day, and seven days a week.

A telephone system to accomplish this requires

1) transducers to convert speech signals to electrical signals;
2) signal generator to identify the required recipient (station equipment);
3) means to transmit such signals (outside plant);
4) means of automatically identifying the required recipient (interface with outside plant);
5) machines capable of providing a path between the transmission means to the required destination (switching centers).

The switching center associates equipment which can identify and select a path to the caller's telephone terminal (either directly as an output from the switching center or via other switching centers). Fig. 1 shows the general arrangements.

The telephone terminal includes speech and signaling transducers and may also include data modems for communication between information processing equipment or video facilities. The telephone user expects to be able to obtain access to compatible equipment located anywhere else in the world. The bandwidth usable for this communication depends on the transmission media and the characteristics of the terminals.

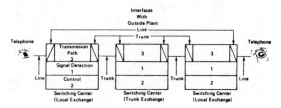

Fig. 1. General arrangements of switching centers.

THE BEGINNINGS OF TELEPHONY

The art of telephone switching goes back to 1876 when two inventors made applications for telephone patents on the same day: Alexander Graham Bell [1] and Elisha Gray [2]. Prior to this, in 1774, a primitive method of transferring electrical signals had been devised by a Swiss inventor using keys and buzzers, and in the succeeding years, telegraph systems were developed. Bell's patent (which was allowed) meant that a voice transducer was available which could be applied to telegraph lines.

The next required step was to make the selective connections (switching centers).

BASIS FOR A SWITCHING SYSTEM

A switching system can be divided into three parts:

1) equipment to detect the signals sent out by a subscriber and convert these into a code so that the call can be established to the required destination;
2) a control element which, in response to the information from the subscriber, sets up the appropriate path through the system;
3) a transmission path which connects the two subscribers for the duration of the call.

Systems performing these functions have taken many different forms and have used various media for control purposes. However, the introduction of any new switching technology into the existing telecommunications network has always been constrained by the interfacing requirement for the existing environment. All switching centers must conform to their environment. In essence, they are parts of a jigsaw puzzle—the national network of which they form part—Fig. 2. A switching center must meet the parameters set by its position in the network such as connecting lines and trunks. In addition, it must conform to the local requirements for transmission, signaling, numbering, etc. This need for compatibility with existing equipment sets a formidable restriction on innovation—which has largely been related to the switching center architecture.

At the present time, information, whether it be in the form of voice or data, is transferred to central switching points over

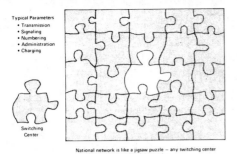

Typical Parameters
- Transmission
- Signaling
- Numbering
- Administration
- Charging

Switching Center

National network is like a jigsaw puzzle — any switching center has to match its own piece of the puzzle

Fig. 2. National network.

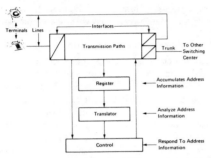

Fig. 3. Principle of switching center.

REGISTER	TRANSLATOR
1	NORTH AMERICA
716	UPSTATE N.Y.
461	STROMBERG-CARLSON
6453	EMPLOYEE

Fig. 4. Principle of number translation.

dedicated connections. The switch centers (Fig. 3) analyze and act on the address signals to provide the necessary channels to interconnect the addressee (the called party) and the caller (the calling party). This requires an organization of switching entities which are controlled by address and supervisory signals from the calling party. The signals from the calling terminal are stored in registers and are translated into the necessary address information to locate the desired party by a translator. The control then establishes the channel to connect the parties' terminals. The dialed address code, which can consist of 11 digits, provides the address information as shown in Fig. 4. The digits dialed will always consist of the exchange code (in this case 461) and the main station code (in this case 6237). The other digits 716 (area code) and 1 (world zone number) will be dialed if the locations of the calling and called parties are in different areas.

Telecommunication systems have taken many different forms and have used various media for control purposes depending upon the available materials and techniques. The first manual exchanges combined a space-division appearance of subscriber lines on a jack field with the time-shared common control of an operator.

AUTOMATIC TELEPHONE EXCHANGES

The interconnection of the subscriber's lines is done at a switching center (a telephone exchange). There are many such exchanges which are interconnected by transmission media (trunks). It is necessary to know which exchange a subscriber is connected to, and hence, what number has to be signaled to reach a subscriber. This requires a telephone directory, and hence, a calling device to signal the required number. The commonly used calling device is the telephone dial which is based on decimal selection.

The telephone dial transmits a series of one to ten impulses as selected by the caller. The required number is indicated to the switching equipment by successive dialed digits. The stream of impulse trains is signaled to an apparatus which is capable of selecting a path to the required subscriber.

The telephone dial was designed to set the automatic selecting equipment directly (Fig. 5). The earliest automatic systems employed the direct control principle, using counting equipment based on electromechanically controlled ratchet and pawl mechanical devices. It was convenient and economical to provide such counting devices on a per switching selector basis.

This has had a major effect on all system development even though the needs for nationwide direct dialing rapidly made the direct control principle inconvenient. This was due to the requirement that the same number be dialed by the telephone customer anywhere in the network. This requires varying the routing, and hence, varying the digits to set the selectors depending upon the points of origination and termination. This requirement is best met by equipment at the switching center which interprets the dialed number and generates the signals relevant to the destination. The need for the analysis of the dialed digits before setting up the connection led to the provision of separate circuits for counting and analyzing, hence the principle of indirect control emerged. These control functions only occupied a small proportion of the total holding time of the call and thus were provided in common. Once the need for direct control of the switches had vanished, less complex switching devices, requiring no compatibility with the dial, were developed. Such a system can employ any kind of calling customer's signaling method such as a tone calling key set. The need for compatibility between the dial and the selectors disappeared. The dial was now used only because it was cost effective.

An outline of an indirectly controlled switching system is shown in Fig. 6. In this figure, the switching matrix is subdivided into the line selection, transmission bridge, and trunk selection functions.

The telecommunications switching systems that have been developed and described in the literature have followed an evolutionary pattern which has largely been related to the availability of materials and the necessity to progress by evolution, rather than by revolution, in system design. New telephone switching systems have evolved to improve the service and provide a greater degree of flexibility in the options available to the user.

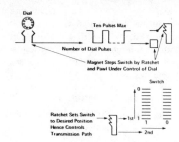

Fig. 5. Direct control.

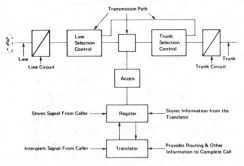

Fig. 6. Indirect control.

The collection and processing of the address information requires the association of equipment for a period long enough to effect the desired function, and which is then released from the connection and made available for another call. This principle is called common control.

Over the years, many techniques had been derived for controlling systems. In the beginning, systems used circuits composed of combinations of electromechanical relays.

The inherent ability of electronic circuitry to perform such control functions, virtually without wear, is important. This made controls available which could operate rapidly enough to handle the operation of the entire systems. Thus common control has been the basis of all system development to date, culminating in "one-call-at-a-time" system operation.

The electronic industry has been busily developing computers. Bell Telephone Laboratories applied data processing techniques to switching system program control (Morris, IL, was the first installation).

PROGRAM CONTROL

The objective of an automatic electronic switching system is to provide telephone service facilities to users at an acceptable cost, and to provide such service on an around-the-clock basis. Provided that this basic concept can be achieved, then the means employed are immaterial.

Any automatic telephone system operates according to a program. It is the way in which the program is accomplished that determines the classification of the system: wired program or stored program.

STORED PROGRAM CONTROL

The peak of achievement in the common control of telephone switching systems is the application of stored program

control. This provides one-at-a-time operation with the aid of digital processors.

The use of stored program, resident in a system, offers possibilities of flexibility of system operation and the ease of conversion of data. However, the practical operation of such systems is really a function of the ability to store and transfer information from memory areas (since this represents an economical way of processing the information). If we, however, analyze the degree to which flexibility of operation is utilized in systems, then it becomes apparent that the technique is simply another means to the same end, which is to provide reliable service at minimum cost.

The design of the system is, in effect, words of programs which should make an easily modified system that contains less hardware. It is, of course, the hope that programming will become easier as programmers become more familiar with their machines and the tasks associated with telephone switching.

SOME PROS AND CONS OF CONTROL METHODS

An individual circuit providing all the functions needed in its role in establishing and maintaining a path through a switching circuit can be inherently less reliable than a common circuit. This is because its failure only reduces the capacity of the system. The provision of many circuits providing the same functions in a system results in the graceful degradation of the system rather than a dependence upon limited redundancy.

The alternative concept means that the circuits forming part of the switched transmission path are simpler circuits, but the associated common control circuits have become more and more complex. It has been necessary to provide extensive maintenance features to diagnose faults in these circuits. This includes the use of monitoring circuits to ensure that the equipment is performing, or capable of performing, its assigned tasks. This may mean that large portions of the switching system programs have to do with the ability of the system to provide continuing service.

THE COMMON CONTROL PRINCIPLE

A number of decades ago, the common control principle was applied extensively in the machine shops of our forebears. These were based upon an awe inspiring system of belts, shafts, and pulleys driven by single electric or other rotating prime movers. Each of the machines used took its power from the shaft system and this was considered an ideal solution, in that only one common control power source was provided. There were obvious disadvantages to this, such as the necessity to maintain an elaborate belt drive system and the resultant inefficiency of the transference of power, and eventually the system was replaced by one in which, instead of having common motors driving many machines, individual motors were provided for each machine. This was purely a matter of cost effectiveness but if one had attempted to say to the people at the time (when the common belt drive was the only compromise solution) that this was not the way to go, their reaction would have been one of disbelief.

It is interesting to reflect that in the mechanical aspects of system construction, the common power drive of selectors which has been a feature of some systems for several decades, has given way to a system in which individual motors power individual selectors. Whether this was not a cost effective point at the time the system was designed or not is not relevant. The point is that this kind of common control, which looks so good (and surely had its protagonists) gave way to a simpler, maybe less elegant, method of power driving, but inherently a much more reliable one.

The minimization of the number of controls provided has been regarded as the ultimate in system architectural construction. Indeed, the protagonists of early electronic switching systems made a claim for the merits of one-at-a-time operation with electronic control instead of numerous separated markers used in some Crossbar systems.

It is easy to regard one-at-a-time operation as being the end and not the means to the end, and it is easy to fall into the trap of supposing that all systems must follow this concept in order to be effective. In practice, of course, this is not the case. The only reason that the equipment has been provided in common has been because this is the most economical way to do it, with the techniques and materials available at the time.

The minimization of equipment by common control has stemmed from the necessity to minimize the number, and hence, the cost of discrete components used in the system. The minimization of equipment by common control may be an invalid concept in the future. This is due to the possibility of assembling very elaborate circuits, capable of performing many functions, on small chips of semiconductor material. The opportunities of reconfiguring systems with the decentralized provision of circuit function are quite large, and care has to be taken not to be constrained by the requirements of older technologies to minimize componentry, and hence, to repeat and duplicate common equipment using techniques of this type.

ALTERNATIVE APPROACHES: EVOLUTIONARY

Common control systems are designed to set up only a very limited number of calls at one time. These systems are no more economical than those using other control techniques. It is time that a careful examination be made of the alternatives. One such alternative may be found in a deviation from the centralized control approach. A system might be proposed which incorporates subsidiary controls arranged to control segments of the network autonomously and which provide functions under the control of a central processor. Another method may be wired logic implementation of the call setup process with the control provided as an integral part of each switching matrix inlet. This technique would provide much greater protection against overloads caused by too many simultaneous call attempts. The problem of locating faults in such a system might be overcome by built-in or centralized monitoring techniques.

The low cost of implementing control capability with integrated circuits that now are both very inexpensive and reliable offers attractive possibilities.

The use of such integrated circuits for distributed controls could result in a system which is more reliable than one in which minute instructions must be transmitted from a central processor control to peripheral units—and also one which is less expensive.

The provision of many of the functions on an individual circuit basis would obviate the need for regular and standby controls with the accompanying diagnostic routines, and the need to ensure that both contain the same information.

Each of the individual circuits could then be so designed that it has a built-in testing means to check the correct operation of the circuit at regular intervals. This would determine if it is operative or not, and would automatically take it out of service and indicate when the circuit is unusable. This would obviate the need for standby circuits since only a small portion of the total equipment is unusable.

A system based on high-density electronic circuits—where many functions are available on small chips—is an extension of the techniques used in the hand electronic calculator and the electronic watch industries. With such techniques, the function can be performed at an acceptable cost, individually, as compared to the cost of common circuitry. This avoids groups of equipment becoming unusable due to a faulty condition in shared equipment. A system organization using such techniques may provide the following: the first stage, an analysis of the local code to check whether the call can be dealt with locally or not. It would also provide the relevant translation function. This would be justified purely on a cost effective basis.

Exchange code translation can vary, and would present problems in effecting changes at many points if it was provided on an individual basis. However, systems with translation spread over many devices, such as the early director systems, are in operation. Hence the extent to which the translation also can be delegated to individual circuits tends to be mainly one of the cost of providing the information on an individual basis. For instance, the question of modifying the individual circuits is simply a matter of processing each one of them in turn when the changes have to be made.

Register sender operation can be provided in the same way; however, there may be specific problems having to do with analog circuitry, such as tone signal detection which may require shared equipment.

APPLICATION OF MULTICONTROLS

We have to be careful not to be carried away with the idea that we can revert to such a system as "electronic step-by-step." This, obviously, is impractical because of the need to collect, store, and process data into areas outside the switching system. Any system will require an interface at some point with common processing equipment, and such processing equipment should be available as required on a demand basis. However, its purpose should be simply to provide the collection and storage of data which has to be processed by central computers and which is used to maintain or administer the system. We still find ourselves in the situation where, as in many systems in the past, we need to develop a system which is adaptable to our planned requirements. It should not necessitate major changes in construction as new requirements develop, or as changes are required to correct errors found during the testing and field trial periods.

The stored program technique would appear to offer a very attractive intermediate design step for such a system (in essence, modeling the behavior of the system in software). The next step would be to analyze those parts of the system which do not require change, such as portions of the executive program and some of the intermediate line-connecting, data-processing, and call-handling requirements. This could then be embodied in firmware in the form of integrated circuits, as could be the register sender and other functions. Circuits such as register senders would be provided on a call basis (even if they are required only during part of a call). There is no reason why they should not be there all the time, if this does not impose a cost which the system cannot bear. The resultant system would then be a compromise making the best use of whatever techniques were relevant.

We are, therefore, in the area of a paradox. The basic question is as follows. Is stored program, which seems to have so much to offer in the construction of telephone switching systems, doomed to be replaced by other techniques as its cost of operation and its needs of maintenance become more complex? Or, conversely, will the programming techniques become simpler and the maintenance routines much less complex?

There's no doubt that the ability to manipulate data in memory offers great advantages, but this can be applied, in turn, without having to control the many logic elements of a system. Would the system be any better if there was more firmware, more wired program, and less stored program control? How far is the stored program control simply carrying the one-at-a-time operation to its climax? Is one-at-a-time operation in a system what we really want? If it is, is it only accomplished through large shared central controls?

There is no reason at all why cooperating controls on a decentralized basis with adequate redundancy could not operate to some extent on a concurrent basis, bidding for circuits in the selection path and setting them up in advance of the system operation.

Again, the question boils down to the basis of engineering, which is essentially one of compromise between the advantages and disadvantages of competing techniques or solutions. The conclusion can, therefore, be drawn that:

inevitably, many circuits can be provided on an individual basis using high-density integrated circuits;

data storing (and especially that part of data storage which has to do with subscriber number information and charging information) could well be provided by stored programs or other common processing means.

It is difficult to see (although again we may be falling into common control reasoning) that the functions of call charging, etc., can be incorporated into low-cost chips on a per circuit basis.

The question is simply one of economics, basically a low enough component cost. It is necessary to provide a means, if a means is required, of updating the decentralized information from common sources and maintaining the information accurately in a number of locations. Again, this is solely a function of cost; the cost of providing the feature, the cost of distribution, and the cost of diagnostics.

Another advantage of circuitry based on high-density electronics is that supervisory tone control, traffic evaluation, and the determination of charges for calls can also be decentralized. Common equipment could be restricted to circuits which are not directly involved in call handling, such as test circuits and common channel signaling circuits. Under these circumstances, a common channel is simply a means of access for transferring data, and does not involve in any way the evaluation of such data; many decisions could be made in the individual circuits of the subsystems.

SEPARATION OF SWITCHING AND TRANSMISSION

For the greater part of the telephone era, the transmission channels and switching systems have been separated by the MDF into independent entities. Hence switching system architects designed equipment which interconnected cable pairs, which carried analog signals. All switching techniques that have been introduced, whether they use cross points spa-

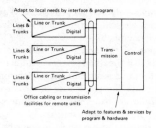

Fig. 7. Basic digital switching system structure.

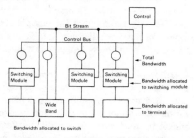

Fig. 8. System adapts to bandwidth.

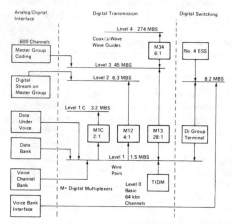

Fig. 9. The digital hierarchy.

tially divided or time-divided multiplexing of the voice path, have acknowledged the need to switch analog inputs to analog outputs. This has required digital carrier systems to be separated into individual channels, and demodulated into an analog form at switching centers.

In the past several years, a great deal of attention has been given to digital switching. This technique reduces the need for maintaining amplitude and other basic transmission signal parameters, without distortion, in its areas of application. It encodes the voice signals in a binary way, and hence has the following advantages:

it is regenerative;
it is noise resistant;
it uses simple cross points;
it is compatible with digital computers;
it exploits the continuing cost erosion of logic and memory.

The principle of operation of a digital switching system is shown in Fig. 7. The system interfaces with analog inputs in the same way as other switching systems. However, since it uses digital multiplexed signals in a bit-stream mode direct-digital interfacing of submultiples of the bit-stream bit rate can be made (Fig. 8). This opens up new possibilities for inter-facing digital signaling systems at various levels. Fig. 9, which shows the Bell System digital hierarchy, illustrates one possi-bility of such a higher level interface with the Bell No. 4 ESS digital switching center.

A number of manufacturers have announced their plans for the introduction of switching systems employing digital tech-niques. No doubt other manufacturers will, at the time appro-priate for them, announce their own products in this field.

DIGITAL SWITCHING SYSTEM

Digital switching systems are intended to provide a cost effec-tive means of switching telecommunications traffic. Many authors have described the advantages of a completely digital network as the ultimate technique; however, the systems that are being introduced must work in the analog environment (MDF to MDF) on a cost effective basis. The system's ulti-mate ability to switch digital transmission systems without demodulation offers a combination of transmission and switching. The problems anticipated with network synchroni-zation of cooperating systems, operating in the digital mode, is receiving a great deal of attention. The next decade will see a gradual increase in the number of digital switching systems in service, and also a start in the combination of the transmission and switching functions into a common system.

APPLICATION OF NEW TECHNOLOGY

Digital Telephone Transducers

A possible application for new technology is the extension of the digital format towards the station. Digital telephone transducers have been studied for some time, and no doubt, technology will ultimately provide cost effective solutions which will translate the voice signals directly into an encoded bit stream. However, no viable solutions have been proposed to date. The proposed system does not anticipate such a direct encoding technique, but is rather aimed at providing a terminal which will convert any input signals into a digital form, through the use of the relevant transducers and coders.

Information Processing Terminals

The technology to design such terminals exists at present; its application to a product and viable entry into the market de-pends upon a low-cost per logic gate or memory bit which in turn depends upon component technology. It is interesting to note, that costs of 1/10 of a cent, per gate or memory bit, have been forecast for the next decade (Fig. 10).

In the meantime, it would appear appropriate to examine the possibilities that arise if the customer terminal can be made more elaborate so that it contains functional program-mable logic, resulting in a simplification of the switching sys-tem and a reduction in the amount of outside plant.

One application of this is in the more elaborate terminals be-ing proposed, e.g., Bell Laboratories visual communications terminal, Fig. 11. It uses a microprocessor and has a flat-screen ac plasma-display panel, television camera, telephone data set, keyboard, and a light pen for the interchange of sketches, text, and continuous-tone pictures. It uses the existing switched telephone network, slow scan techniques, and storage to pro-duce a picture every 8 s. A display of 8 in X 8 in has about 250 000 discrete light points.

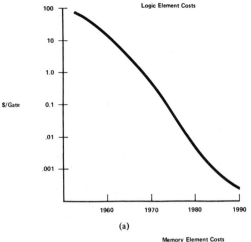

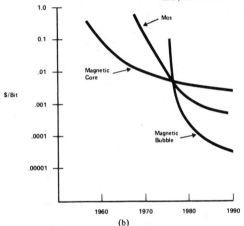

Fig. 10. Projected costs for logic and memory.

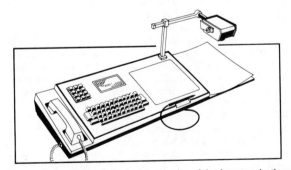

Fig. 11. Picturephone—Bell laboratory version of visual communications terminal from *Electronics*/April 1, 1976).

It would appear appropriate to examine the possibilities that arise if the customer terminal can be made more elaborate, so that it contains functional programmable logic, resulting in a simplification of the switching system and a reduction in the amout of outside plant.

The system assumes low-cost logic and memory and inexpen-sive programming and firmware inputs. It also assumes non-

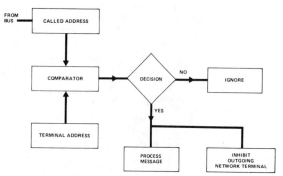

REGISTER	COMPARATOR	TRANSLATOR
1	NO	NORTH AMERICA
716	NO	UPSTATE NY
461	MAY BE	STROMBERG-CARLSON
6453	MAY BE	EMPLOYEE

Fig. 12. Routing by number translation.

Fig. 13. Principle of comparison.

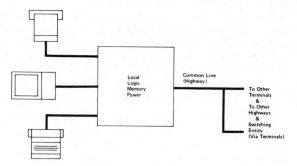

Fig. 14. Decentralization.

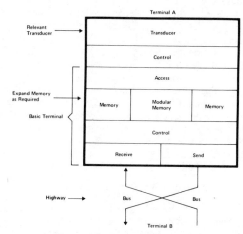

Fig. 15. Association of terminals.

dedicated transmission media—in the sense that a path is provided for the duration of the conversation. No matter whether such a path is metallic or time divided, it will not be provided on a per terminal basis. What is considered is the allocation of the transmission path only when a message is being sent. This opens up the possibility of developing coding technology to minimize the amount of signal information and to take advantage of the fact that speech contains many intervals of silence. The signaling system would convert all information (voice as well as data) into a series of data packages.

Another important basis for the system is that the terminal extracts only the information which is intended for it. This differs from the present switching system which translates a code address into a physical location. The terminal would send the code address, and this is screened at all terminals and matched at the desired terminal. Fig. 12 shows the way in which a dialed number would be processed. The dialed number is stored in the originating terminal. It is encoded in binary code and sent to all terminals connected to the bus. The terminal on the bus which has this address compares the number on the bus with its own number. A match occurs at the desired terminal. Such matching would take place on the basis of storing information, either in a terminal for which it is intended, or within the terminal for subsequent retransmittal into another highway or the outside network. Fig. 13 shows a simplified flow chart of the matching function.

A reallocation of the system functions results in a decentralized system (Fig 14). In this system, the facilities provided at the terminal include the means to perform the processing of data at the terminal. The terminal (Fig. 15) connects to a common highway. It consists of a basic module and the rele-

vant transducer for the type of information being transferred. All information is converted into data packages. The highway is allocated to a connection only when a data package is being sent. Such data packages contain routing and control, as well as message information. The routing information includes the called address. This is screened by all terminals and matched at the desired terminal. This results in the storing of information, either in the addressed terminal or within a terminal, for subsequent retransmittal into another highway or the outside network.

Such a system could operate on a synchronous basis in which each terminal is assigned a recurring interval of time on the highway or it could operate asynchronously. An asynchronous method is preferred since more terminals can be served. However, the impact of a nondedicated random-access time-assignment speech channel requires experimentation and simulation to determine the degree to which delays in transmission can be tolerated by a speaker or listener.

BASIS FOR INFORMATION TRANSFER

The system would only utilize the transmission facilities when information is actually being transferred. In this respect, it has a similarity with the TASI principle (talk spurts) and an affinity with package switching systems (short data messages). The concept is one which is aimed at only sending meaningful information; that is, actual signals relating to speech (and not

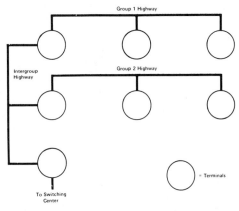

to the intervals between speech). It will also send only the signals corresponding to black, grey, etc., and not white as in the case of facsimile transmission. Thus the various terminals will automatically store for transmission—as data packages—only customer input which contain meaningful information. The noninformation intervals will be recreated when the package arrives at its destination. This enables the reconstruction on a real time basis of the actual message.

A typical package will contain about 1024 bits and the voice processor will utilize talk spurts of about 1 s. It is believed that low-speed visual display will serve most needs. If memory cost of $0.001 per bit can be obtained in the next decade, a cost of less than $100 could provide enough storage for a 9 × 11 picture.

System Outline

General Concept

The customer terminals are connected to common wideband channel highways (Fig. 16). The highway could utilize any transmission media including optical cables. A number of terminals are served by a pair of highways (send bus and receive bus). The highways are interconnected via a special terminal. Access to the national network from such a highway group is provided via a special terminal.

The terminal circuits convert input information to a digital bit stream, organize it into labeled packages, and vice versa.

Highway System

The highway buses replace the switching stages of conventional telecommunications exchanges. In addition to the customer terminals, there are interface terminals which provide access to other highway bus systems and the national network, and a monitor terminal extracts administrative information and provides diagnostics.

The interconnecting bus is passive and each terminal that has information to transmit bids for the bus, and if the bus is free, applies the necessary busy signals and the message.

Highway Capacity

The ability of the system to handle traffic is a function of the number of packages of information which can be sent over the highway bus, within a time interval when the signal transducers are able to convert the signals into meaningful information.

As an example of the capacity of the highway, let us assume that information is transmitted at 3.088 Mbits/s and the capacity of the bus is sufficient to provide the equivalent of 48 T1 carrier channels.

If the system is used by audio customers, the capacity can be effectively doubled by the application of the TASI principle. (The equivalent of 96 channels will carry about 80 erlangs of traffic.) It can be shown that it is possible to accommodate about 1024 terminals on a single bus using this technique.

The number of terminals served by a highway depends upon the type of information being sent, for example, in the case of low-speed video terminals then the amount of data necessary to transmit a picture will utilize much of the highway capacity. In a similar way, interconnection of data sets will reduce the number of terminals per highway.

It should be pointed out, however, that the actual bit rate in use for the bus can be very much higher than the example.

Fig. 16. Bus/terminal association.

If fiber-optical cables are being used, then the possibility for connecting many terminals, irrespective of the information mode, increases.

Allocation of Communication Channels

There is no dedicated allocation of a channel to a communication between two terminals. The communication consisting of a series of messages interspaced by other messages. In order that a given called party shall only receive the messages relevant to the conversation, the called terminal receives both the called and calling numbers with the initializing message and stores these in its own memory banks. When a further message is addressed to that number, then the message is only accepted if it originates from the original caller. If some other terminal is attempting to communicate at the same time, then the terminal indicates a busy condition by sending the appropriate message to the relevant caller. The removal of the terminal busy condition is only effected when the call is terminated either by the caller or the called party.

Use of Dedicated Channels

The system can also be based on a time-divided carrier system with individual drops per terminal. However, the number of terminals will always be significantly larger than the number of channels if random rather than cyclic access to channels is used.

Terminals

General

The terminal contains the necessary transducers to encode and decode visual and audible signals and also, if necessary, produce hard copy outputs. The terminal processing logic assembles segments of the information into data packages and transmits them, with the relevant information relating to their destination and point of origin, to the highway. It also converts digital information intended for it into the relevant output format.

Customer Terminal

Fig. 17 shows the general arrangement of a terminal. The terminal consists of a basic interface and optional interfaces for various transducers. The basic terminal circuitry includes

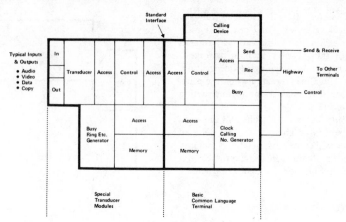

Fig. 17. Block diagram terminal.

the send and receive gating circuits—a control and memory and the access circuitry for the optional transducers. The relevant transducers (visual, audible, and hard copy) and their controls are provided as needed for the terminal.

The terminal automatically generates its own address, indicates the relevant information input modes, and encodes the information.

The terminal generates signals to initiate and terminate ringing and ringing tones, and to initiate the generation of whatever other audible or visual signals are necessary.

All such tones are generated in the terminal to which they are addressed. The highway is only used to indicate the start and finish of the tone signal.

The accumulation of information for generating telephone bills could be done in the terminal monitor.

Audio Terminal

The send portion of the terminal uses a telephone transmitter and digital encoding circuitry. This circuitry is activated whenever a talk spurt occurs. The equivalent bit stream is stored and assembled into packages of approximately 1024 bits and the appropriate header address and control information is inserted. Access to the highway including the check of the busy condition of the bus is carried out by control sequences. These provide the necessary instructions for the generation of the messages and their acknowledgment. The calling device has its own digitizer which stores the dialed information.

The receive portion of the terminal is driven by the incoming pulses derived from the bit stream. The signals are regenerated. The identity of the calling party is checked (to prevent interrupted messages) and the relevant information signals are generated.

Interbus Terminal

Terminals which interconnect highway buses have a checking system for identifying packages with addresses originating and terminating on the bus. Any outgoing call using the bus which is not in this category and all incoming calls to the bus will be repeated to the relevant highway or the national network.

Administration Terminal

The information relating to charging for calls and for monitoring traffic is extracted by an administrative terminal. This

monitors each package and it can record in its memory the point of origin, the point of destination, and the time of arrival. This information can be collated over a period of time and used in the preparation of material for message accounting.

The advantage of monitoring in this way is that it reduces the number of diagnostic routines that are necessary to check the operation of a particular terminal. This can be done by sending a specially coded address to it and receiving a specially encoded message from it.

MESSAGE FORMAT

The message is made up of packages of say 1024 bits. Each message contains information for steering the call to its destination such as the called directory number. It also includes the calling directory number and an identification of the sequence of the package so that the information can be regenerated in the correct order. Additional information relating to the type of information being transmitted is sent to initiate the appropriate transducer action at the desired terminal. The initial message which locates the desired terminal will contain the calling and called terminal numbers and other information pertaining to the nature of the call. A signal acknowledging receipt of the message will be sent with an identifying code. The information message, which is being stored in memory, is now sent as a series of packages with information relating to the identifying code and the sequential position of the message.

ROUTING OF CALLS

Intrahighway Routing

When a terminal wishes to send information to another terminal, it seeks access to the highway and, when it obtains it, sends an initializing message. After receiving the acknowledgment message, it proceeds to send its information. The required terminal is identified by the address and the information is stored by it. It is then processed and applied to a transducer of the appropriate signal type. Messages are exchanged by the two terminals, and at the end of the information transfer, disconnect messages are sent.

Interhighway Routing

If the connection is to a terminal on a different highway, then interhighway switching is used. This is accomplished by using the relevant interhighway terminal. This terminal stores the information for retransmittal. This arrangement has the advantage that there is no need for any kind of translation of the called address into a location, nor is there any need to establish any kind of switching path since the highway buses are passive, and all the control functions are in the terminals which interconnect the various buses. At some point in time, however, the call may be routed outside the highway system. When this is necessary, then the information is restored to the appropriate form and sent over conventional transmission media. Register sender functions will be provided at this point to repeat the called number into the network.

It appears that the present interest on alternatives to transportation such as teleconferencing, etc., will accelerate the development of an information transfer system. The philosophy of information transfer, when applied, will transform our present transmission and switching systems into an integrated telecommunications information exchange system. Fig. 18 shows the three stages of evolution (today to 1985 can be regarded as near future). It seems certain that there will be more elaborate terminals providing a full range of information access in visual as well as audio form. There may well be data processing and computing capabilities provided as part of the service. A system providing these features has been described in general terms only. The system is based on practical materials as they exist today. The present rapid escalation of technology could well produce entirely different solutions, as only the future can tell.

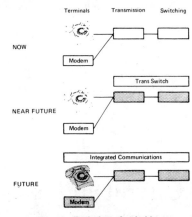

Fig. 18. Evolution of switching.

Circuit and packet switching
A cost and performance tradeoff study

Roy D. Rosner
Defense Communications Engineering Center,
1860 Wiehle Ave., Reston, VI 22090, U.S.A.

Ben Springer
UNIVAC, Dallax, TX 75240, U.S.A.

In this paper a large common user network for data communications is investigated in an operational environment where about half the traffic was in the form of short messages and the other half was in the form of long ones, where short messages carry the implicit requirement to transit the network with very short time delays. A number of hypothetical networks were designed for the purpose of comparing circuit and packet switching technologies with respect to efficiency, cost, and throughput. The sensitivity of the results to certain assumptions and certain parameters of network performance were also examined in order to make the results applicable over a broad range of interest. Under the assumptions of this study, packet switched implementations showed decided advantages over circuit switched implementation, as measured by transmission utilization efficiencies and overall network costs.

Keywords: Circuit switching, Packet switching, Switching comparisons, Switching networks, Network costs and comparisons, Computer network implementations, Data networking concepts, Data network cost comparisons, Data network cost/structure/performance comparisons, Computer and data network analysis, Large scale networks, Cost comparisons for data networks, Performance comparisons for data networks.

1. Introduction

Rapid increases in the quantity of digital data communications flowing through the telecommunications systems of the world have simulated considerable interest in developing more efficient methods of handling this form of communications traffic. Recent efforts have focused attention on the switching and networking aspects of data communications and have led to the widespread acceptance of packet switching as an effective method of handling data communications traffic in a switched network environment. However, agreement is not universal, nor is a decision easily arrived at, as to the conditions under which packet switching is more efficient or less costly as a network technique compared to other possible techniques, particularly circuit switching. In fact, it has been shown elsewhere [7,10] that the efficiency of the network switching technique is closely related to the length and transaction frequency statistics of the data messages. Simply stated, it is more efficient to packet switch short messages and circuit switch long messages, or groups of short messages sent frequently enough to justify holding the connection. The crossover between short and long message-length is heavily dependent on the technology used to implement the network, but appears to be in the range of 2000 to 50,000 bits per message.

Suppose however, a large common-user network for data communications was to be built in an operational environment where about half the traffic was in the form of short messages and the other half was in the form of long ones, where short messages carry the implicit requirement to transit the network with very short time delays (a few seconds, at most). What then would be the best switching technique to use in the single network, circuit switching or packet switching? In the study described in this paper a number of hypothetical networks were designed for the purpose of examining this problem. The sensitivity of the results to certain assumptions and certain parameters of network performance were also examined in order to make the results applicable over a broad range of interest.

2. Data switching techniques

Circuit switching of data provides for a direct circuit connection through the network in exactly the same way as for voice subscribers. This has the obvious advantage of compatibility with voice switching concepts and can utilize the same switching hardware. The call establishment procedure is common to voice. However, in order to achieve efficient transmission utilization, circuit switches would have to be engineered to provide much shorter connection delay times, and the high arrival rates associated with much interactive data communications would result in signaling and processing levels substantially higher than those presently encountered in circuit switched networks. Circuit switching requires complete code, speed, and format compatibility of the end terminals since no in-network processing is performed, and the network is subject to blocking either within the switched network or on the local loop to a busy terminal.

Classical store and forward switching relays the entire message through the network while storing the message in full at each relay point. This concept permits code, speed, or format conversions to take place during the in-switch processing, and appears essentially nonblocking to the subscribers. Powerful processors with large storage capacities are required at each node of the data network. There is a very large variance in delivery delay, especially for messages which find themselves in queue behind a very long one, leading to poor responsiveness for interactive traffic.

Packet switching, a concept pioneered by the Advanced Research Projects Agency (ARPA) and various contractors, established not only a conceptual mode of operation for interactive data but also a testbed network on which to prove the new concepts and ideas. Packet switching as a concept refers to the mode of handling data-oriented communications, but it does not imply the particular distributed network concept embodied in the experimental ARPA network.

In the packet switch concept small processors provide formatting and buffering of subelements (i.e., packets) of the full message. Address information, appended to each packet, permits flow through the network to the processor nearest the addressee. The packet switching concept provides the capability of exchanging short messages despite the presence of long messages in the system and provides the same service features in terms of conversions and journaling available in store and forward networks. Packet switching entails the use of a fairly large number of small processors, and therefore requires complex control and routing procedures.

Packet switching and circuit switching can be viewed as the extremes of a spectrum of techniques available for data communications switching. In actuality, a continuous range of techniques which blend elements of each are available for meeting specific problems. In this paper, however, attention is focused on pure packet switching and pure circuit switching.

3. Assumptions

In order to perform a specific comparison between circuit and packet switching, a fairly extensive set of assumptions on network parameters must be made. The comparison of switching technologies has been made on the basis of the performance measures of delay, blocking, throughput, channel utilization, and reliability, as governed by the fundamental cost elements of the network, (network capability as determined by the number of channels and channel miles in the backbone network, and the number of nodal processors). It is clear that many of the assumptions which had to be made to make the comparative analysis possible could have a significant impact on the results. The fact that the comparison is focused on the backbone portion of the network (as opposed to access lines and terminals), and that the network was fixed in size at 70 nodes, certainly impact on the overall credibility of the comparative results. The impact of the assumptions in general, and these two facts in particular, is not trivial, and is discussed in detail in a later section, but let it suffice for the moment to make the following points. The analysis performed as part of this study was aimed at the comparison of different networks with fundamentally different backbone structures, and not aimed at developing a single optimized data network. Thus, the comparison of the technologies is only minimally affected by many of the assumptions, and further, the backbone portion of the network is the focus of major attention.

Current data communications is predicated on the lease of circuits from common carriers, and the continuity of this policy is assumed. The availability of such facilities was, is, and will continue to be dependent on the financial, technological, and political pressures for or against development, expansion, or enhancement of existing or new facilities and capabilities. Where needed, previously developed projections of data circuit costs [3,4] are used to estimate future costs of leased facilities.

A predominant cost factor in today's data networks is the cost of the long haul, high volume circuits. Although a significant reduction in the per-mile cost of long haul circuits is a definite possibility, due to advances in technology, such a decrease will probably be slow due to the size of the investment required for new plant facilities. Also advancing technology will probably reduce some of the other network cost factors, particularly the processor and software more rapidly than the tariffed costs of long haul data circuits. Thus the ratio of the circuit costs to the other costs will decrease even more slowly than the reduction in circuit costs. Hence, the predominance of network cost by circuits costs is presumed.

The cost of network access from the user terminals to the network nodes is assumed to be equivalent for both techniques and is therefore ignored.

The network size is chosen to be 70 nodes in a nonhierarchical network with no imposed minimum or regular (lattice) connectivity. A previous study [5] has shown that a minimum of approximately 50 nodes in a nationwide distributed network is necessary to provide a cost-effective network with an acceptable level of network redundancy and reliability, for a total projected traffic level and user community as defined for this study. Other work [11]

indicates that for the 70-node configuration under study the costs of hierarchical and distributed networks are within 5% of each other if the number of links is between 150 and 300.

The network is sized to serve a user community composed of about 2500 host computers and 25,000 user terminals with a traffic load 1,600,000 busy hour call originations and 3.27×10^{10} bits of busy hour data flow [10]. About 2500 different geographic locations are assumed and the traffic flow distribution is largely proportioned according to the user population associated with each physical location. By aggregating the traffic by location to the nearest of the 70 network nodes a 70×70, to/from traffic-flow matrix with nonuniform terms is generated, such that the sum of all the entries is equal to the total projected traffic flow.

For the purposes of the study, the circuit switch is assumed "perfect." It has the properties of zero signaling delay, zero connect time, zero cross-office delay, and unlimited termination capacity. Also the access lines provided to the users are assumed to be errorless. This precludes the necessity of incrementing the user's traffic load by some debatable scale factor to allow for user error control. These assumptions yield a circuit-switched network lower bound cost determined only by the traffic load and the blocking grade-of-service offered by the finite circuit capacity.

The model packet switch is based largely on ARPANET operating experience. Specific packet switching assumptions are a circuit bit-error-rate of 1 in 10^5, and an average packet length of 1144 bits, with 128 bits of routing and error control and 16 bits of user information (nondata overhead) in each 1144-bit packet. The packet-switching nodal processors are capable of handling 1000 packets per second.

Since this study was basically concerned with the comparison of circuit- and packet-switching techniques for data networking, the assumptions are uniformly selected to favor or bias the results toward circuit-switching; since preliminary analyses indicated that packet switching was the more desirable technique. The impact of the assumptions are further analyzed in Section 9 of this paper.

4. Method of analysis

The quantitative comparison of the use of circuit switching versus packet switching to handle the flow of traffic in a large common-user data network has been reduced by the stated assumptions to primarily a comparison of the network channel kms required to support each switching technique under equal traffic load and comparable user service demands. The network topologies and performance evaluations for each technique were computed using an IBM 370/155. Over 120 separate runs, with an accumulated cost of about 10 hours CPU time, were made to provide sufficient range in topologies and performances for the comparative and sensitivity analyses.

The problem of network synthesis, i.e., the design of the network capacity and connectivity to satisfy a given level of user demand, is a nonlinear capacity flow problem which, if solvable, is not tractable by the mathematical methods and capable of being implemented on digital computers. However, a proper set of assumptions and heuristic approximations allow the problem solution to be approximated by iteratively solving a nested sequence of related-integer linear capacity flow equations, where the capacity of the links is approximated by simple additive super-position (linear independence assumed) of the various simplex flows and capacities of each origin destination node pair [1]. The required assumptions for the synthesis model are:

Point-to-point traffic originates according to Poisson process.

The system is in statistical equilibrium

Link blocking probabilities are statistically independent of each other.

Overflow and carried traffic distributions are adequately characterized by their intensity of flow.

Blocked traffic is lost and is not retried in the circuit-switched model but is queued in the packet-switched model.

The circuit switches have infinite termination capacity.

Call processing (signaling and interconnection) is instantaneous in the circuit switched network.

All channels are full duplex.

There is only one class of traffic.

All routing is fixed to one preferred path for each origin-destination pair.

The holding time of calls (circuit switch) and the packet lengths (packet switch) can each be described by a negative exponential distribution.

The channels of the circuit-switched network are errorless while those of the packet-switched network are not, but the error effect can be adequately approximated by an appropriate incremental increase in the original offered traffic.

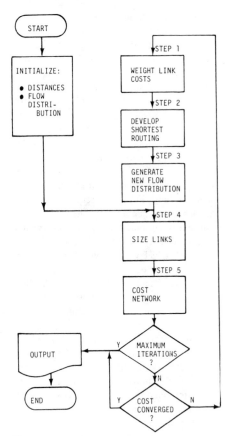

Fig. 1. Flow chart: Minimum cost network design algorithm.

Even though some of these assumptions are not realistic for a practical network, experience with design studies of existing DoD networks, such as AUTO-VON and AUTODIN, has shown that network designs produced by the network algorithm based on them are reasonably sound.

The algorithm, shown in Fig. 1, is initialized by the input of parameters, node locations, and point-to-point offered-traffic loads. A matrix of the direct distances between all node pairs is generated from the input node locations. An initial flow distribution matrix is created which is actually a copy of the point-to-point offered load. These initial arrays are provided as input to Step 4 of the main iterative loop. They force Step 4 to generate and size the trunks of a fully connected network which has adequate direct connection for the traffic between any node pairs which provides the starting point for the connectivity reduction—cost reduction phases of the algorithm. This procedure generates the initial link matrix (channel size and implied connectivity) also required as input for Step 1 of the main iterative loop. Initialization proceeds to Step 4.

Step 1 of the design algorithm uses the link matrix, the distance matrix. and the flow distribution matrix to generate a weighted cost matrix which represents the marginal cost to increase the link capacity to accommodate unit of flow increase. This Step is the heart of the algorithm because it allows the nonlinear relationship between flow and capacity (with performance fixed) to be approximated by a sequence of linear relationships. For the circuit switch, the weighing of each link is derived from the ERLANG B (blocked traffic lost) formula for the channel size and flow distribution of that link as found by the previous iteration. The ERLANG B value at which the marginal cost is to be determined is selected by an input parameter. For the packet-switched model the link weighing has the effect of a cost penalty on the less-efficient small-capacity links in the routing algorithm of Step 2.

Step 2 of the design algorithm uses the weighted cost matrix, the link matrix, and the offered traffic matrix to generate a new routing matrix. This step reformulates the minimum cost capacity-flow problem as a multiterminal shortest (least costly) path problem. The cheapest path for each point-to-point flow is found independently of other flows using Floyd's algorithm [6] and the connectivity established by the previous iteration. This reformulation provides the short solution times which make the problem computationally feasible for networks of practical size.

Step 3 uses the new routing matrix to create a new flow distribution matrix from the offered traffic matrix. The traffic for each point-to-point pair in the network is distributed over the links encountered in the routed path with the link loading being cumulative. The process assumes zero blocking and no alternate paths. Processes not limited by these assumptions have been found to be too costly in computation time for only small increases in accuracy. Where precision is necessary, an alternative method using blocking and alternate routing consideration can be used to fine tune the network produced by this simplification [8].

Step 4 uses the new flow distribution matrix to generate a new link matrix (trunk size) with the restriction that the blocking or delay (depending on circuit- or packet-switched case) the performance of the link must equal or exceed that specified by the input parameter mentioned in Step 1. In this step, certain links will appear which, because of the weighting factor in Step 1. were not included in any routing path by Step 2. Since they have zero load, they will be sized to zero. They are eliminated from further consideration (nonproductive mode of oscillations due to reintroduction of traffic on an unused link) by setting the corresponding distance matrix element to a computationally infinite value. The links with non-zero load are sized by the Erlang relations between capacity and offered load. Erlang B is used for the circuit-switched model providing for overflow traffic to be lost from the network, and Erlang C is used for the packet-switched model, where overflow traffic is queued.

Step 5 uses the new link matrix and the distance matrix to cost the new network. Since for the purposes of this study the channel km. costs of the network are assumed to be the dominant cost factor, the cost per channel km. was set to one unit and all other tariff costs were set to zero. The cost (channel km.) of the new network is checked against the cost of the previous iteration. If the savings are less than a specified minimum, the design phase is ended, otherwise another iteration is begun at Step 1. At the end of the design phase, all the netoork matrices are saved for u use in the performance evaluation phase.

For the circuit-switched model. the performance analysis consists of a procedure similar to that of the design phase but uses a more complex method of calculating the effects of blocking combined with use of alternate paths in the steady-state flow. The output of Floyd's [6] shortest path algorithm is used to select the three shortest alternate routes in ascending order. (If the origin is connected to less than three nodes, there will be less than three alternate routes.) For each origin-destination pair the offered load is distributed among the possible routes by an iterative procedure which is based on the link blocking probabilities of each successive route as determined by the flow distribution on the previous iteration (link capacities being constant). The superposition of all newly determined point-to-point flow distributions yields a new flow distribution for the entire network including values for link blocking losses, overall blocking loss, and average number of links traversed by completed calls. The new flow distribution and blocking losses are used to calculate a new matrix of link blocking probabilities, which is used to calculate

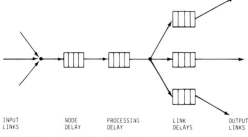

INPUT   NODE   PROCESSING   LINK   OUTPUT
LINKS   DELAY   DELAY   DELAYS   LINKS

Fig. 2. Packet-switched node model.

a new matrix of blocking probabilities, which . . ., etc., until the mean blocking probability for all point-to-point flows converges. Outputs selected from the program include the grade-of-service (total traffic), defined as the ratio of total blocked (and lost) traffic to total offered traffic, the overall trunk occupancy, defined as the traffic weighted average of the occupancy for each trunk group, and the average number of links traversed per call.

For the packet-switched model, the performance analysis is based on determination of point-to-point and link delay statistics. Given the connectivity and the number of channels in each link of the synthesized network, the model, shown in Fig. 2, computes the mean expected delay of each link and node by using the queueing theory results of Burke [1] and Cobham [2]. Mean point-to-point delays are computed by summing the node processing delays and link delays in the routing path for each origin-destination pair, which is possible because of the assumptions of Poisson statistics holding throughout the network. The overall mean delay for the total network is found by taking a traffic-weighted average of the point-to-point delays. As a final step, the overall and individual link utilizations are found by using Little's formula [9] together with the delay results for a m/m/n queueing system. In contrast to the circuit-switched model, this model allows no alternate routes in evaluating performance.

5. Sizing the circuit-switched network

The use of a circuit switch in performing a data switching function causes an inherent problem with respect to traffic management and sizing. Since circuit switching fully allocates network resources (end-to-end transmission path) to a particular pair of users

for the duration of a call, the level of traffic is a function of the call arrival rate (e.g., calls/sec. or calls/hr.) and the average length of the call. In addition, the connection results in a full duplex path between subscribers, such that information can flow in both directions at the same time.

These concepts are very straightforward when the circuit switch user is a voice user. Certain fundamental changes in point of view occur when the user is a data subscriber. In general, a data subscriber is characterized as having a number of data bits to be sent from location A to location B. It is normally unidirectional transfer, where even an interactive data session can be characterized by a sequence of unidirectional data transactions. The duration of a particular unidirectional data transaction is dependent upon not only the length of the transaction (in bits) but also on the speed of the circuit interconnecting the users. The speed of the circuit, in turn, is often determined by the operating characteristics of the terminal equipment integral to the circuit. Since the analysis of circuit-switched networks is inherently time-sensitive (e.g., the call holding time measured in seconds or minutes) and is well documented in terms of Erlangian theory (where arrival rates multiples by holding time results in traffic intensity measured in erlangs), the task is then one of creating a circuit switch-network concept and operating procedures which make efficient use of the network assets, and yet can be analyzed realistically.

In order to do this, the circuit-switched network was looked at for three distinctly different cases. In all three the data communications traffic load in bits (in the busy hour) was held constant at 3.27×10^{10} bits and 1,600,000 transactions which is a projection of DoD data-traffic in the late 1960's [10]. This level of traffic is also used for the packet-switched network sizing and analysis so that direct comparisons are possible. The three cases examined are characterized as:

Case A. "Perfect" Circuit Switches – Buffered Terminals;
Case B. "Perfect" Circuit Switches – Unbuffered Terminals;
Case C. Rapid Dial with Limited Circuit Holding.

In Case A, the circuit switches are assumed to be "perfect" with a zero connection time delay and an infinite call handling capability. Thus, each data transaction can be handled as an independent call origination with no user impact. All user terminals are buffered so that they enter the network at an 8000 bits/sec. line rate.[1] These assumptions lead to the minimum theoretical traffic load on the circuit switched network.

Under this operating procedure, the equivalent circuit-switched load in Erlangs is calculated by

$$E = \frac{V}{C} = \frac{V}{3600R} \qquad (1)$$

where E = Traffic intensity in erlangs, V = Busy hour traffic volume (bits), C = Channel capacity in bits per hour, R = Channel rate (bits/sec).

For Case A, $V = 3.27 \times 10^{10}$, and R is 8000.

$$E_A = \frac{3.27 \times 10^{10}}{3600(8000)} = 1135 \text{ erlangs.}$$

Since each data transaction would result in a new call origination, this traffic intensity rejects a total of 1,600,000 call originations, as reflected by the assumed requirements.

In Case B, the switches are still assumed to be "perfect", as in Case A. However, the terminals are unbuffered and enter the network at their own charactertistic rate, which in general is less than the 8000 bits/sec. line rate. Thus a 300 bits/sec terminal would take 8 times as long to complete a transaction of a given length as would a 2400 bits/sec terminal. The traffic requirements used to generate the traffic flow matrix characterize the data flow in terms of low-speed terminals (nominally 450 bits/sec), high-speed terminals (3600 bits/sec), and computer connections at the backbone line rate (8000 bits/sec). The traffic flow in each category is respectively 6.97×10^9 bits/hr, 5.5×10^9 bits/hr, and 20.25×10^9 bits/hr (total of 3.27×10^{10} bits/hr).

The traffic intensity for this case is calculated in a similar fashion to Case A, except it is composed of three components, one for each class of terminal.

$$E = \sum_{i=1}^{3} \frac{V_i}{C_i} = \frac{1}{3600} \sum_{i=1}^{3} \frac{V_i}{R_i} \qquad (2)$$

where

$V_1 = 6.97 \times 10^9$ bits/hr., $R_1 = 450$ bits/sec., $\}$ low speed terminals ,

---

[1] This study uses line rates drawn from the 8000 N ($N \geqslant 1$) hierarchy of digital data rates. The rate 8000 bits/sec is used for the circuit-switched case since such channel rates could, if needed, be readily accommodated by normal voice grade channels.

$V_2 = 5.5 \times 10^9$ bits/hr., $R_2 = 3600$ bits/sec., $\}$ high speed terminals ,

$V_3 = 20.25 \times 10^9$ bits/hr., $R_3 = 8000$ bits/sec., $\}$ computer-to-computer .

Substituting the values of V_i and R_i in equation (2) results in

$$E_B = 5428 \text{ erlangs.}$$

As in Case A this traffic intensity is the result of 1,600,000 individual call originations.

In Case C, an attempt is made to define a circuit-switched operating procedure which significantly reduces the number of call originations seen by the circuit switches. Data users "dial up" and are permitted to hold the circuit-switched connections for periods of time which are significantly longer than the actual data transmission time, but are short relative to present operational systems. Low-speed terminals are assumed to conduct 4 data transactions per call origination with an average holding time of 50 seconds per origination. High-speed terminals are presumed to conduct 2 data transactions per origination, where in general the query and the response would be combined in a single origination, with an average holding time of 30 seconds per origination. Computer-to-computer traffic is assumed to lead to the equivalent of a full period connection to on the average one other computer during the busy hours, which for calculation purposes is the equivalent of one call per hour with a one hour holding time.

Using the fundamental definition of traffic intensity (call rate times holding time) the traffic intensity for this case is calculated. At the same time the revised number of call originations for each user type is calculated.

Since the lines are being held for longer periods of time than the actual data transmission time, the Erlang load (E_c) must be calculated from the basic definition of arrival rate multiplied by holding time. Thus

$$E_c = \frac{1}{3600} \sum_{i=1}^{3} T_i C_i H_i \qquad (3)$$

where T_i = Number of terminals of ith type, C_i = Calls per hour per terminal = (transactions/hr)/(transactions/call), H_i = Holding time for average call (seconds).

$T_1 = 20042$ terminals, $C_1 = 10$ calls/hr., $H_1 = 50$ seconds, $\}$ low speed terminals ;

T_2 = 7836 terminals,
C_2 = 6 calls/hr., $\left.\right\}$ high speed terminals ;
H_2 = 30 seconds,

T_3 = 2560 terminals,
C_3 = 1 call/hr., $\left.\right\}$ computer-to-computer .
H_3 = 3600 seconds,

Substituting the values in equation (3) results in

E_c = 2784 + 392 + 2560 = 5736 erlangs .

The number of call originations is given by

$$C = \sum_{i=1}^{3} T_i C_i = 249,996 .$$

The circuit-switched network parameters are summarized in Table 1.

Another significant parameter in the design of a circuit-switched network is the end-to-end grade-of-service experienced by the users. When a voice call is attempted which cannot be completed due to the lack of sufficient resources, the attempt is generally cleared from the network, and the user usually retries the call at a later time. Data networks generally work on a delay grade-of-service, in which calls are queued and delayed when network resources are overutilized. In order to compare the different switching technologies, it was judged that a grade-of-service of **P**.01 (1% of all call attempts blocked) would lead to the same level of user-satisfaction as an average delay through the network of $\frac{1}{2}$ second per transaction. While this judgement is somewhat subjective, it is shown below that the network size and cost is a fairly slowly varying function of either blocking or delay for a fixed traffic level and fixed line rate.

The application of the network design programs to the required traffic levels resulted in separate detailed network designs for each traffic level. The program output consisted of the needed node-to-node connectivity, the number of channels on each link of the network, the total number of channel km, and the

total number of switch terminations. Fig. 3 shows the network connectivity design for a traffic level of 1135 Erlangs and a grade-of-service of **P**.01. Fig. 4 shows a circuit switched network design sized to 5428 Erlangs, and a grade-to-service of approximately **P**.01. Note that except for the number of links, both networks have the same basic structure, with most of the capacity well distributed among the different nodes. Note also the use of links which are generally short (1600 km or less) relative to the coast-to-coast distances, which minimizes the use of long, expensive, highly vulnerable trunk routes. Under the set of given input conditions, that is traffic matrix, node locations, and required grade-of-service, these networks are optimum in the sense of utilizing the minimum number of channel miles to achieve the required grade-of-service.

Fig. 5 shows the number of links required for the optimum circuit-switched network design as a function of the traffic intensity. A link is a transmission path between two nodes, and may consist of one or more communications channels. Curves are shown for grades-of-service of **P**.01 (judged nominally equivalent to a $\frac{1}{2}$ second end-to-end delay) as well as **P**.10 which represents ten times as much network blocking. It is noteworthy that the better grade of service, **P**.01 results in fewer links than the poorer grade of service, **P**.10. This is due to the fact that a better grade of service is achieved, most economically, by the use of fewer, high capacity, higher density links. As shown in Fig. 6 and 7 below, the number of channels and channel km required increases as the grade of service improves.

Fig. 6 plots the number of channels in the network, again as a function of traffic intensity, at blocking levels of **P**.01 and **P**.10. The number of channels is a significant parameter of the network design since the cost of switches is at least partially dependent on the number of channel terminations at the switch. In addition, most common-carrier tariffs include a channel-connection charge, as well as a channel-km charge. The number of channel km as a function of traffic intensity is shown in Fig. 7, again for grades-of-service of **P**.01 and **P**.10. It should be noted that both the number of channels and channel km are nearly linear functions of the traffic intensity. This is intuitively satisfying since for a large network such as that under consideration, very high levels of channel utilization (efficiency) are achieved such that an incremental increase of traffic results in a nearly linear increase in required assets.

Table 1.
Circuit-switched network parameters.

	Traffic intensity	Originations
Case A (perfect, buffered)	1135 erlangs	1,600,000
Case B (perfect, unbuffered)	5428 erlangs	1,600,000
Case C (dial and hold)	5736 erlangs	249,996

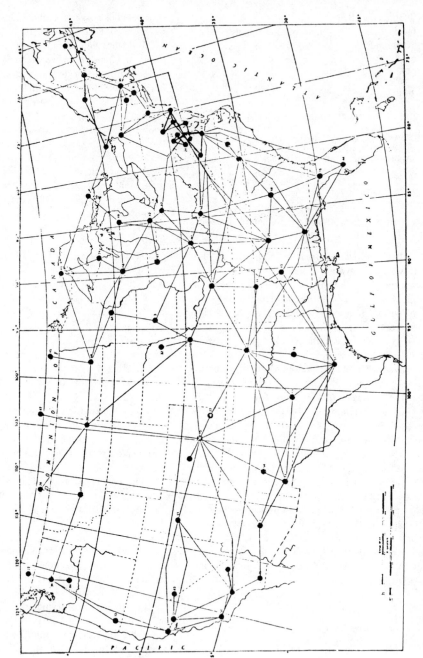

Fig. 3. Optimized circuit-switched network connectivity for traffic intensity of 1135 erlangs.

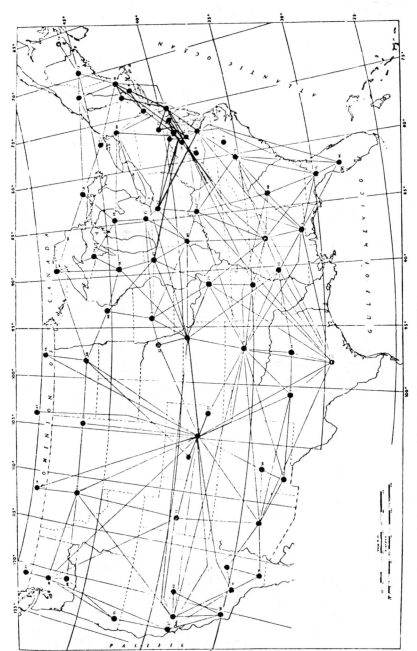

Fig. 4. Optimized circuit-switched network connectivity for traffic intensity of 5428 erlangs.

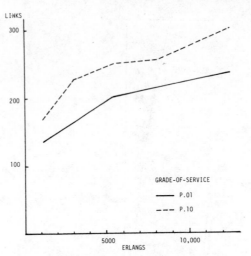

Fig. 5. Number of links versus traffic intensity (70-node optimized circuit-switched network).

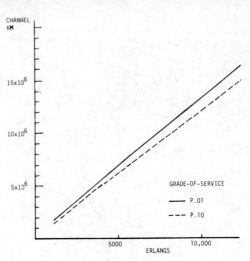

Fig. 7. Number of channel km versus traffic intensity (70-node optimized circuit-switched network).

Because of the somewhat subjective equivalence of circuit-switch blocking factors to transnetwork delays of a packet-switched network, the impact of grade-of-service on network size is shown in Fig. 8. It is clear that for any traffic intensity, the number of channel miles is a rather slowly varying function of the grade-of-service. Fig. 8 shows grade-of-service up to **P**.10 (one call blocked out of each 10 attempts) which is

considered the minimum acceptable level a subscriber would tolerate.

6. Sizing the packet-switched network

Totally packet-switched data network designs were developed for a large number of different configurations

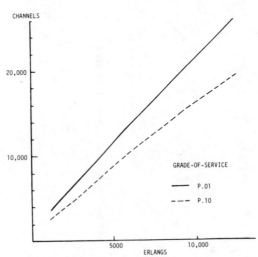

Fig. 6. Number of channels versus traffic intensity (70-node optimized circuit-switched network).

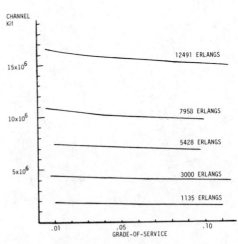

Fig. 8. Sensitivity of channel km to grade-of-service (70-node optimized circuit-switched network).

tions and parameters. In all cases the total magnitude of the data traffic was held fixed at the busy hour value of 3.27×10^{10} bits used in designing and sizing the circuit-switched configurations. In the packet-switched cases, however, the concept of a "perfect" switch was not used, and approximately 15% overhead traffic flowed in the network in addition to the basic data flow requirement.

Since packet switching is basically a form of statistical multiplexing of the switch-to-switch channels, and because it is a system where blocked traffic is queued and delayed, the network design is not made on the basis of erlangs as in the circuit-switch case. Network sizing is based on the actual bit flow requirement during the busy hour, subject to the delay constraints of the traffic. In the circuit-switching case, the basic rate of the individual channels impacts the conversion of bits to erlangs, and likewise in the packet-switching case, it heavily impacts the sizing of the network. For that reason most packet-switched sizing analyses were made at channel bit rates ranging from 8000 to 56,000 bits/sec. Comparison of the packet-switch and circuit-switch designs is made at the 8000 bits/sec rate. However, it should not be interpreted that 8000 bits/sec results in the optimum packet-switched network. In fact 8000 bits/sec channels provide just barely marginal performance for a packet-switched network, and rates considerably higher would be more desirable, all other things being equal. In addition, using identical channel bit rates for every line in the network is an analytical fiction, used to reduce the dimensionality of the network analysis/synthesis problem. In actual implementations, it would be expected that channel bit rates available from the tariffed offerings of common carriers, which most closely match the link requirements would be used. Assuming that carrier offerings include some economy of scale as channel rate increases, the actual network costs for a packet switched implementation would actually be lower than for the case analyzed, where all channel rates throughout the network are assumed identical.

The application of the packet-switched network design program to the required data traffic flow resulted in different network configurations, depending on the line data rate, delay constraints, and permissible line utilization. A typical network configuration, employing 149 links, is shown in Fig. 9. The basic node and link structure of the network is quite similar to the circuit-switched data network designs.

The results of various packet-switched network sizing analyses are displayed in Figs. 10–13. Fig. 10 shows the number of links needed to optimize the packet-switched network design as a function of channel bit rate. In this, and the figure that follow, curves are presented showing 100% and 80% maximum line capacity utilization. Since, in queueing systems, delays become excessive when a facility approaches capacity, it is generally desirable to limit the nominal design level to that no element of the system operates at traffic levels close to 100% utilization under normal system loading. Design curves are therefore shown at this 80% level along with the 100% utilization for comparison.

It is noteworthy, as seen in Fig. 10, that the number of links in the optimized network configuration at any particular bit rate increases as the channel bit rate increases. This somewhat counterintuitive result is due to the fact that at higher bit rates the incremental capacity of adding an additional channel to an existing link is such that the capacity is not as efficiently used as placing that capacity in a new location, e.g. establishing a new link in some other part of the network, which can make more effective total use of the incremental capacity. At lower bit rates, the increments of capacity can be added in small enough units so that the capacity on the links can closely approximate the desired design maxima (e.g. 80% or 100% maximum for the cases of this study).

Fig. 11 shows the number of channels, and Fig. 12 the number of channel km required in the optimum packet-switched network design. Fig. 13 shows the peak and average delay actually achieved in the optimum network design. The peak delay is the delay encountered along the longest, most circuitous user-to-user path in the data network, while the average delay is the achieved delay averaged over all user-to-user pairs. This definition of peak delay leads to the rather peculiar behavior of the delay curves shown in Fig. 13. Since the network is configured and sized based on an average delay constraint, the peak delay, can have, for certain implementations, a fairly high value, without affecting the overall sizing of the network. In other words, a high peak delay between the most distant user pair is not considered "bad" by the program, provided that the delay, averaged over all user pairs, is within the defined design range. The somewhat anomalous behavior of the peak delay curve of Fig. 13, at 100% possible line utilization and 48,000 bits/sec line rate is due to this phenomenon, coupled with the fact that as any element of a queueing system approaches 100% utilization delays rise

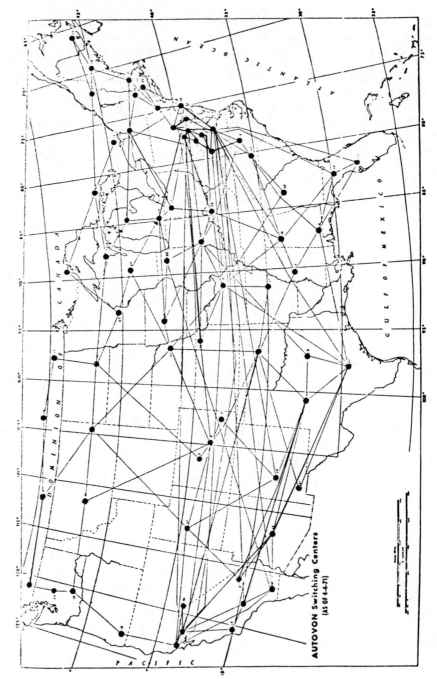

AUTOVON Switching Centers
(AS OF 6-4-71)

Fig. 9. Optimized packet-switched network connectivity employing 149 links.

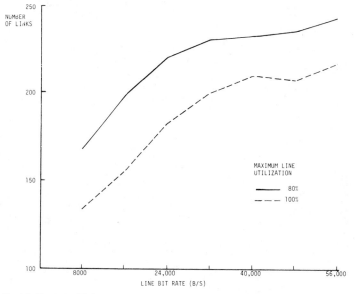

Fig. 10. Number of links versus line rate (70-node optimized packet-switched network).

rapidly with utilization. The high value of peak delay in the region of 48,000 bits/sec indicates that at this value of channel rate, at least one link in the path between the most remote user pair was filled to nearly 100% occupancy, this leading to a sharp increase in the peak delay at this value. At higher and lower channel bit rates, the channel rate modularity led to achieved occupancies which were sufficiently lower

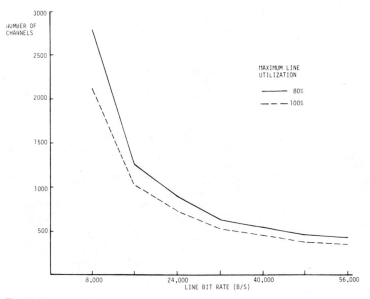

Fig. 11. Number of channels versus line rate (70-node optimized packet-switched network).

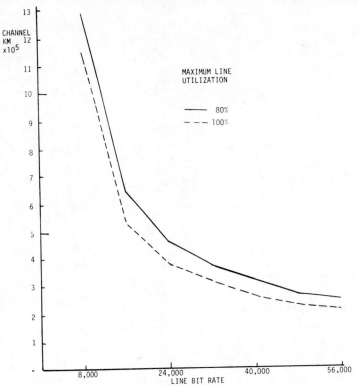

Fig. 12. Number of channel miles versus line rate (70-node optimized packet-switched network).

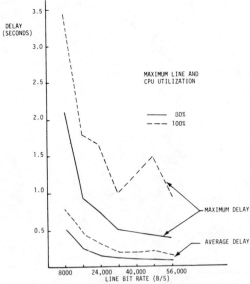

Fig. 13. Network delay (peak and average) versus channel rate (70-node optimized packet-switched network).

than 100% to still achieve lower peak delay than the 48,000 bits/sec case.

7. Comparison of sizing results

Using the results of the above analyses, a direct comparison of circuit- and packet-switching networks, which are handling the same total traffic, can be made. In addition, the techniques can be compared on the basis of overhead efficiency, and on the basis of relative potential cost differences of the network designs, based on present tariffs and estimated future tariffs.

The ability of each network structure to efficiently use the available network resources is reflected by the number of lines and quantity of transmission facilities (channel miles) required by the network design to meet the specified user requirement. The analytical data developed allows for direct comparison of the network designs as shown in Table 2. The table used the analytically derived figures, under

Table 2.
Network design comparisons.

Network	Circuit switched networks						packet switched network
	Case A		Case B		Case C		
	perfect/buffered		perfect/unbuffered		Dial & hold		
GOS	$P.10$	$P.01$	$P.10$	$P.01$	$P.10$	$P.01$	(0.5 sec. delay)
Links	170	135	240	180	250	200	165
Channels	2600	3600	9400	12500	9900	13000	2800
Channel km	1.53×10^6	1.92×10^6	6.72×10^6	7.52×10^6	7.04×10^6	7.88×10^6	1.29×10^6

the condition of 8000 bits/sec transmission lines, a 70-node network design, and both $P.10$ and $P.01$ blocking probability for the circuit-switched network. The achieved delay for the packet-switched network is 0.5 second average transnetwork delay.

Note that only in the case of the perfect, buffered circuit-switched network does the overall network efficiency approach that of the packet-switched network. Even at this level, the circuit-switched network requires about 20% more channel km than the packet-switched network. If the figures for a perfect, circuit-switched network operating at $P.01$ blocking were used (3600 channels and 2,000,000 channel km), a differential of nearly 50% would exist between the "perfect" circuit-switched network and the packet network.

Before proceeding further, it is perhaps important to reiterate the fundamental differences between the theory of operation of the two network concepts which lead to these results. The circuit-switched network, a blocking network, must, during busy periods, prevent users from completing calls (blocking), forcing them to place their calls again at a later time. During the less busy periods, the capacity required to keep the blocking percentage at acceptable levels is idle, leading to a certain amount of network inefficiency. On the other hand, the packet network, a queueing network, can hold on to "blocked" or excess traffic during busy periods and use the traffic so queued to "smooth" out the statistical traffic flow and achieve a higher average level of resource utilization. There are other factors in the network operation, such as the more efficient use of full duplex lines in the packet-switched network, which lead to its higher efficiency, but the fundamental operating principles of the networks provide the basic explanation for the packet networks superior performance. There are

techniques, such as the introduction of statistical multiplexers or concentrators to a circuit-switched network, which can "smooth" out the traffic flow and improve the network efficiency, but these devices begin to introduce elements of packet switching into the network, such that one can no longer consider the network design "pure".

In addition to the topological aspects of the network comparison, an important consideration in comparing packet- and circuit-switched network designs is their relative efficiencies in establishing the communications path through the network, especially when considering circuit switches which are less than "perfect". The network sizing establishes the network design from the point of view of its traffic carrying capacity or capability, but the overhead efficiency is a measure of how flexibly the network capacity can be utilized, via the switching facilities, to establish the constantly changing user-to-user paths.

In the circuit-switched network, the overhead consists of the time it takes to establish the end-to-end path from one subscriber to the other, and in general consists of the switch-to-switch signaling time and the switch processing and operating time, known as "cross-office" time, In the packet network, the overhead consists of the header and error control bits that must be transmitted with each packet to control its flow from subscriber to subscriber. In Fig. 14 the packet- and circuit-switched network overheads are graphically compared by plotting the connection overhead time (in milliseconds) versus the data transaction length (in bits). A very conservative packet-network design is used, employing 8000 bits/sec lines, 1000 text bits and 144 overhead bits per packet. It is assumed that the typical user-to-user path employs three nodes (originating, terminating, and one tandem). Common channel interoffice signaling (CCIS) is

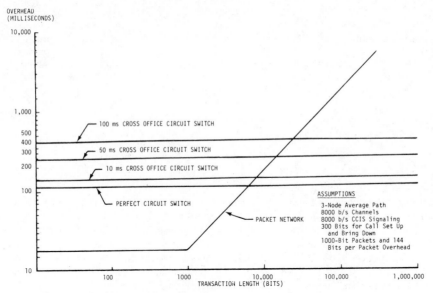

Fig. 14. Overhead comparison of circuit and packet switching.

used at the 8000 bits/sec line rate, and a total of 300 bits are assumed to set up and later disconnect a circuit switched call.

Since the amount of overhead needed to establish a circuit-switched call is independent of the transaction length, the circuit switch overhead appears as horizontal lines in Fig. 14. Overhead curves are shown for perfect circuit switches, as well as circuit switches with 10, 50, and 100 millisecond cross office times. The overhead of the packet-switched network is constant until the first packet is filled, and then is a step function of the transaction length (approximated in Fig. 14 by a continuous linear function) as the transaction is composed of more and more packets.

The intersection of the packet network curves with the circuit-switched network curves defines a crossover between the preferred handling techniques for data transactions based on overhead considerations. Even in the case of "perfect" circuit switches, packet switching would be the preferred transaction handling technique for transactions shorter than 6500 bits. The crossover point is up to 25,000 bits for a more realistic 100 msec cross-office-time circuit switch.

8. Potential economic impact

The absolute measure of the difference between the packet-switched and circuit-switched network designs is the potential cost difference that would result from the greater transmission efficiency of one network design versus the other. While detailed cost analysis of the network designs developed in the course of this study is a highly complex undertaking and was not performed, it is possible to get a significant estimate of the magnitude of the cost differences which result from the different choices of network technology. In so doing, attention is again focused only on the backbone network of switching nodes and channels, and channel miles of backbone trunking. Even though substantial resources are consumed in the access portions of the network, consideration of those resources is neglected on the basis that the access facilities of the packet network and the circuit-switched network are equivalent. (This, in fact, is a bias in favor of circuit switching as discussed in Section 9.) Further, it is presumed that the cost of a node is independent of the technology employed (e.g. either circuit or packet switching), and dependent only upon the number of terminations.

In a recent study, Coviello and Rosner [4] developed general cost factors useful in cost estimating for large data networks. Cost factors from that study for the 1975–76 time frame applicable to the backbone portion of the data network are nominally $ 0.75 per channel km per month (8000 bits/sec lines), and $ 150 per month per termination, a fixed charge. The nominal cost for switching is $ 1500 per termination (purchase cost). This equates to $0.16 \times 1500 = \$\,240$ per year per termination, assuming a 10-year useful switch life, a 10% per year allowance for operations and maintenance, and a 10% present worth factor on annualized investment. The total annual costs are then $\$\,2040$ per termination and $\$\,9.00$ per channel km. The cost difference between any two technologies can then be expressed as

$$D = 2040[T_2 - T_1] + 9[M_2 - M_1]$$

where D = cost differential (yearly, dollars), T_i = number of terminations, M_i = number of channel km.

Since each channel has two ends, the number of terminations is twice the number of channels, thus

$$D = 4080[C_2 - C_1] + 9[M_2 - M_1]$$

where C_i = number of channels.

As an example, the circuit switch Case A with **P**.10 blocking is compared with the packet switched case (from Table 2) as

$$D = 4080[2600 - 2800] + 9[1,530,000 - 1,290,000]$$

$$= 4080[-200] + 9[240,000]$$

$$= -816,000 + 2,160,000$$

$$= \$1,344,000.$$

Thus, the annual cost of the circuit-switched network is approximately $\$\,1.3$ million more than the equivalent packet-switched network, even under the most ideal assumptions.

The value of D can be obtained for a number of different cases, and plotted against the circuit-switched traffic load in erlangs. This was done and is shown in Fig. 15. The figure shows that the cost difference grows from an annual difference of $\$\,1.2$ million for the perfect, buffered circuit-switched network (1135 erlangs) to nearly $\$\,80$ million for the dial and hold mode of circuit-switched network operation (5736 erlangs) in the **P**.10 blocking case.

If present network cost factors dictate a clear cost advantage for packet switching over circuit switching for a data network, is it possible that future cost reductions could obviate this advantage. It is widely held that the cost of digital services will be substantially reduced in the next decade, and in their paper Coviello and Rosner project in fact, on a technological (rather than competitive) basis, a 50! decrease in the cost of digital transmission services in the decade from 1975 to 1985. However, rather than speculate on how much future digital costs may drop, Fig. 16 relates the cost advantage of packet switching over

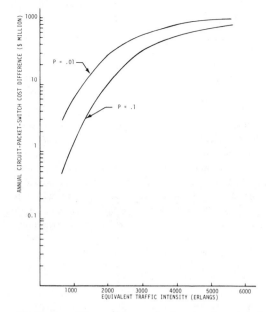

Fig. 15. Cost difference between circuit- and packet-switched networks.

circuit switching for the case of a perfect, buffered circuit-switched network. If a cost difference of $\$\,100,000$ per year is considered negligible, future costs would have to drop by a factor of 12 relative to today's costs to achieve this level of negligibility. In

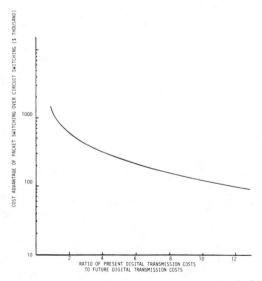

Fig. 16. Cost advantages – packet network versus "perfect" circuit switched network.

the cases of the nonperfect of nonbuffered networks, even more dramatic cost reductions would be required to achieve cost comparability.

9. Impact of the assumptions

In any technical study, and this is no exception, a number of assumptions are made to make the problems tractable. The assumptions are chosen so as to have little or no impact on the results, at least not upon the fundamental nature of the results. Since this study was basically concerned with the comparison of circuit- and packet-switching techniques for data networking, the assumptions are uniformly selected to favor or bias the results toward circuit switching, since preliminary analyses indicated that packet switching was the more desirable technique. Several of the assumptions are now examined to indicate how they actually did impact the results.

The impact of choosing a 70-node network (as opposed to a smaller network) is potentially significant in the comparison of switching technology. The choice of the number of nodes in the network design primarily impacts the balance of network cost and complexity between the backbone and access portions of the network design. If very few nodes are used, there are many more channel km of access connectivity required, while if the number of nodes is oversized, the efficiency of the node-to-node links begins to diminish, since the amount of traffic per node pair begins to diminish. It is generally agreed that the 70-node network used in the analysis is as large as any practical data network would have to be, but the question is whether a choice of a substantially smaller number of nodes would change the basic comparative results of the two network designs. A good measure to use for that consideration is the average utilization of the links in each technology, since reducing the number of nodes would tend to increase the average utilization of the links. Fig. 17 plots the average utilization for the packet network design, and shows about 55% average utilization at 8000 bits/sec. In the Case "A" circuit-switched network, the average link has 15 channels, which at P.01 level of blocking yields an average utilization of 54%, nominally the same as the packet network. Reducing the number of nodes is therefore likely to affect the link efficiencies in the two networks similarly, and thus have no impact on the basic comparison of the two technologies.

The assumption that the access sizing and costs for both the packet network and circuit network are equivalent was used in order to permit the focus of attention on the backbone networks. Actually the currently available techniques permit efficiencies in the access areas of packet networks which are presently unattainable in circuit-switched networks. For example, if a large computer is simultaneously supporting a number of different remote terminals via the network, it will require one access line for each terminal to give the computer the physical connection via the circuit switch to each remote terminal. In a packet network, large computers can logically multiplex, in a time shared fashion, a number of virtual or logical connections to a large number of individual terminals via a single access line. This actually leads to a conclusion that the cost of the packet network

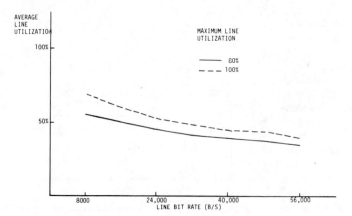

Fig. 17. Average line utilization versus channel bit rate for a packet switched network.

access capability would be smaller than that for the circuit-switched network.

The sensitivity of the network comparisons to traffic levels and total data flow requirements is minimal, since in terms of the relatively large network being considered, most of the aspects of the network design are fairly linear. As seen from most of the previous figures, the channels and channel km for the networks are relatively linear functions of traffic intensity. Thus changes of data traffic load by factors of 2, 5, or even 10 would change the absolute magnitude of the network designs, but would not alter the comparative advantages of the two network technologies.

The sensitivity of the design comparison to the network performance measures can be viewed in terms of the information provided in Fig. 8 and Fig. 18. Figure 8 shows that the network size (channel km) is reduced about 20% (1,900,000 to 1,500,000 channel km) goving from **P**.01 to **P**.10 level of blocking. From Fig. 18 the size of the packet network is reduced by about 25% (1,300,000 to 950,000 channel km) going from 0.5 sec to about 1.5 sec average delay through the network. The potential savings for both the packet and circuit networks tend to bottom out as constraints are relaxed, so that a point of diminishing returns is clearly presented in both cases. Since saturation occurs for both techniques for savings of 20–25% over the nominal design points, it is apparent that the comparison of the two networks is substantially independent of the actual performance level chosen. It is significant, however, that the "perfect" circuit-switched network performance saturation (for every high blocking levels) is about 1,400,000 channel km, while the packet network blocking level, (for arbitrarily long delays) is about 950,000, a nominal 50% advantage for packet switch-

ing over circuit switching with even the most relaxed performance constraints.

10. Conclusions

The conclusion of this study is clearly that a common user-data-network can be more efficiently and cost effectively implemented using packet switching rather than circuit switching. This result appears to be valid over a broad range of assumptions and traffic loads. It is emphasized, however, as shown in other studies [7,10] as well as Fig. 14 of this paper, that because of the overhead requirements of packet switching, long messages are best handled in circuit-switched networks. An area now ripe for exploration is the determination of how best to mix various switching technologies to achieve a hybrid structure which is highly efficient under all types of requirements.

References

[1] Burke, P.J., "The Output of a Queueing System," Operations Research, 4, 1956, pp. 699–704.

[2] Cobham, A., "Priority Assignment in Waiting Line Problems," J. Operations Research Soc. Am., 2, (1954), pp. 70–76.

[3] Coviello, G.J. and Rosner, R.D., "System Considerations in Evaluation of Speech Digitization Techniques," Proceedings of the Communications Systems and Technology Conference, Dallas, Texas, April 1974, page 81.

[4] Coviello, G.J. and Rosner, R.D., "Cost Considerations for a Large Data Network," Proceedings of ICCC '74, Stockholm, Sweden, August 1974, page 289.

[5] Defense Communication Engineering Center, DCASEF TR 33–73, "CONUS/CSN Cost and Worth Study," July 1973.

[6] Floyd, R.W., "Algorithm 97: Shortest Path," Communication of ACM, 5(6), 345, 1962.

[7] Itoh, K., Kato, T., Hashida, O., and Yoshida, Y., "An analysis of Traffic Handling Capacity of Packet Switched and Circuit Switched Networks," Proceedings of Datacomm '73, Third Data Communications Symposium, St. Petersburg, Florida, November 1973, page 29.

[8] Katz, S.S., "Trunk Engineering of Non-Hierarchical Networks," Sixth International Teletraffic Congress, 1970.

[9] Little, J.D.C., "A Proof of the Queueing Formula: L = λW," Operations Research, 9, No. 3, (1961), pp. 383–387.

[10] Rosner, R.D., "A Digital Data Network Concept for the Defense Communications System," Proceedings of the National Telecommunications Conference, NTC '73, Atlanta, Georgia, November 1973, page 22c-1.

[11] Rosner, R.D., "Large Scale Network Design Considerations," Proceedings of ICCC '74, Stockholm, Sweden, August 1974, page 189.

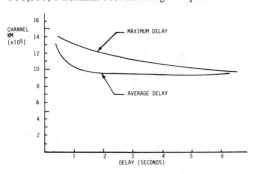

Fig. 18. Sensitivity of channel miles to permitted peak and average delay (70-node packet-switched network).

COST CONSIDERATIONS FOR A LARGE DATA NETWORK

©Coviello, G J, Defence Communications Agency, Reston, Virginia, USA
©Rosner, R D, Defence Communications Agency, Reston, Virginia, USA

ABSTRACT

The 10-year cost sensitivity, as function of channel bit rate, for a wide geographically dispersed mix of variable size data customers is developed based on current and projected costs for leased transmission service and purchased packet switching processors. To achieve this objective, a model for the projection of leased transmission costs into the future is proposed. The results are compared to a previously derived result for digital voice networks and an interesting implication for possible cost advantages of integrating data and voice networks is introduced.

1. INTRODUCTION

Since the advent of the U.S. ARPA (Advanced Research Projects Agency) network, the use of packet-switching to handle data and computer interactive traffic has been receiving widespread attention. This trend is of increasing importance as the rapidly decreasing costs of computer and switching hardware is thereby elevating the residual communication cost to a more critical role in the cost structure analysis. Hence, the efficient and cost-effective utilization of available transmission facilities becomes a prime objective in order to achieve minimum over-all costs. It has been estimated that cross-over between the domination of computer versus communications costs occurred around 1969 [1]. The purpose of this paper is to propose a model for the expected transmission costs over the nominal 1976-1986 time frame, particularly with respect to the variation of network cost as a function of the channel bit rate. The network structure which is treated differs markedly from the ARPA-type network which serves a relatively small number of highly concentrated traffic sources. Rather, a large modified hierarchical network, designed to serve a geographically wide distributed mix of high and low density users, will be examined.

The customer requirements postulated for the network to be discussed, and the resulting structure which is proposed as an effective solution to these requirements have been described by Rosner [2] in a companion paper to this conference. The requirements are intended to be somewhat typical of what might be expected in a large diversified private usage network typified by the Defense Communications System of the U.S. Department of Defense. The type of traffic to be handled is a mixture of computer-to-computer and man-to-computer data-exchange, interactive and query response modes. Although voice communications are also required between these same locations, these particular requirements are not discussed until the last section of this paper.

The data users of the proposed network are considered to be concentrated in approximately 2400 locations whose geographic distribution and total transmission requirements are assumed to be described by table 1 below.

TABLE 1
Description of Data Installations

Descriptor	Total Number	Transmission Requirements
Very Large	131	90 Kb/s
Large	404	60 Kb/s
Medium	479	30 Kb/s
Small	667	12 Kb/s
Very Small	777	4 Kb/s

To serve the above requirements, the backbone structure shown in figure 1 was derived, consisting of 70 Regional Switches of which 7 are elevated to the higher hierarchical position of Tandem Switches by virtue of the fact that these 7 are fully interconnected while the other Regional Switches may be connected to as few as 2 other Regional Switches and at least 1 Tandem Switch. The inter-connection of all 70 switches is prohibitive since the number of switch-to-switch lines, K_s as a

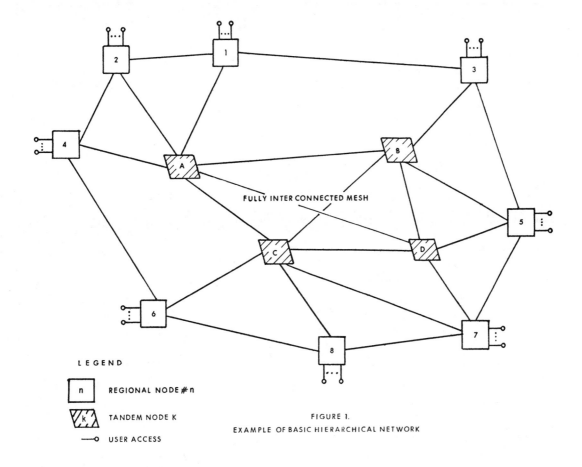

LEGEND

n	REGIONAL NODE #n
k	TANDEM NODE K
⊸	USER ACCESS

FIGURE 1.
EXAMPLE OF BASIC HIERARCHICAL NETWORK

function of number of switches, N_S is given by:

$$K_S = \frac{N_S (N_S-1)}{2} \qquad (1)$$

For a 70-switch network, K_S = 2415. The proposed hierarchical network, on the other hand, has been sub-optimized at only 149 links of which 21 are used to provide the fully-interconnected mesh existing between the 7 Tandem Switches. The proposed structure still provides a reasonable degree of security should any switch or link fail, and also insures that no more than 4 switches are ever needed in the principal route for any packet transfer. With this structure, the response through the network can be bounded as illustrated in figure 2.

In order to conduct a cost analysis of the above network, it is necessary to first perform a sizing of the network in order to determine the number and total length of transmission channels required. As indicated, a total of 149 links are used to interconnect the Regional and Tandem Switches. To determine the actual number and length of channels required as a function of channel bit rate, a computer analysis was utilized to satisfy the load requirements while maintaining the response times of figure 2. The number of backbone channels, $N_B(R)$ and the total length, $M_B(R)$ are plotted in figure 3. Likewise the number and lengths of access channels, $N_A(R)$ and $M_A(R)$ required to satisfy the traffic load indicated in Table 1 were derived and are plotted figure 4.

2. SYSTEM COST ELEMENTS

Transmission Cost Model

The transmission requirements for the network are

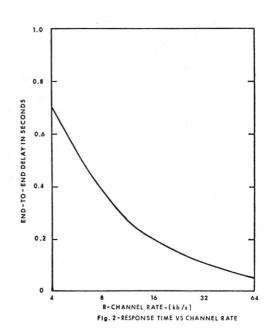

Fig. 2-RESPONSE TIME VS CHANNEL RATE

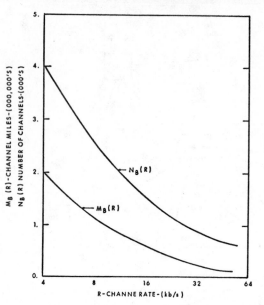

FIG.3 BACKBONE CHANNELS, $N_B(R)$, AND TOTAL MILEAGE, $M_B(R)$

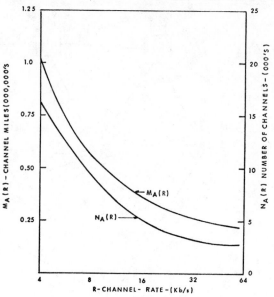

FIG.4- ACCESS CHANNELS, $N_A(R)$, AND TOTAL MILEAGE, $M_A(R)$

presumed to be met entirely by commercial leased services and hence are governed by the approved tariff schedules. Monthly leased costs per link, L , as presented in these schedules are related to:

$$L_m(R) = C_t(R) + C_m(R)M \qquad (2)$$

where $C_t(R)$ represents a termination charge attributable to multiplex and line terminating equipments and $C_m(R)$ represents a per mile charge based on the length of the link, m. $C_t(R)$ and $C_m(R)$ are both functions of the leased channel bit rate, R. In addition, they are functions of geographic locations with the lesser charges being levied within the more highly populated areas where the resulting higher concentration of trunks leads to lower costs. Estimates for the above charges for the 1976 time period were derived from present existing tariffs and from proposed, trial, and sample rates publicized by various U. S. Common Carriers for future near-term digital transmission services. A best estimate was derived by a weighted average of the available data for values of R varying from 4 to 64 Kb/s. Rates below 4 Kb/s were not considered since the response time of the network would be greater than the 1 second which was considered a threshold requirement of the network. Separate cost figures are also derived for both high-density and low-density areas. The resulting cost estimates for the 1976 time base are plotted in figure 5. As would be expected, $C_m(R)$ is more sensitive to R than $C_t(R)$ since the costs of digital circuitry to terminate links are relatively insensitive to the channel bit rate, R, while the percentage of transmission bandwidth (and hence percentage of link cost) is nearly in direct proportion to the channel bit rate, R.

For costing out the transmission network in later sections, the high-density cost estimates of figure 5 will be used for the 149 backbone links inter-connecting the Regional and Tandem Switches since they are generally assumed to be located at or nearby highly populated areas. On the other hand, the access links which interconnect remote data installations to Regional Switches will generally be located in less populous areas and will, therefore, be costed at the low-density rates of figure 5.

Projection of Costs into the Future

There is very little basis available from which to project tariff costs for digital transmission services into the future due to the unpredictable behavior of competitive and legislative pressures, inflationary factors and technological development. Hence, an approach is proposed here based on a hueristic technological argument. This approach establishes a model for comparing the cost of a 56 Kb/s PCM channel to an analog voice band channel based on relative bandwidth occupancy, and then using this as a cost basis for digital rates as a function of an assumed projection for analog costs.

The rationale for comparing the PCM to analog voice channel is developed in the following manner. With present technology, the use of wideband modems to transmit a PCM channel displaces a bandwidth equivalent to about four analog voice band channels and hence carries a cost penalty of about 4:1. This situation generally prevails today since about 90% of the common carrier facilities have capabilities for such wideband modem interfaces but essentially no purely digital long haul facilities. The gradual replacement of analog with digital transmission and multiplex equipments will proceed only as the effective cost becomes advantageous. It is, therefore, postulated that the costs for an analog versus a PCM channel would be equivalent at that time. It is further postulated based on present estimates that the implementation of digital facilities will follow an "S" shaped growth curve over the 1975 to 1999 time fram which is asymptotic to the 75% level of digital capability. The increasing proportion of digital facilities will, therefore, result in a decreasing average cost for PCM and proportionally lower digital services as compared to analog. To establish a cost basis for analog voice service it is assumed that although the cost elements for the transmission of analog bandwidth may change (e.g., downwards due to technology advances, upwards due to inflationary factors), these elements tend to counter balance each other and keep the overall costs relatively constant over the next 10-20 years.

Putting these factors together results in the projection of digital costs according to the curve of figure 6. As illustrated, the 1975 cost is projected to decrease 50% by 1981-82 and ultimately a level of less than 25% of the 1975 cost by the end of this century. It is again emphasized that this projection is based on a technological basis only and does not reflect the impact of competition, legislation or regulatory agency rulings.

Switch Cost Model

Costing of the Regional and Tandem Switches of the backbone is related only indirectly to the channel bit rate (R) due to the cost being related primarily to the number of terminations on that switch. Letting N_i represent this number for the ith switch, the total system cost for switching is given by:

$$P_S = \sum_{i=1}^{70} [C_S(N_i)]_i \qquad (3)$$

where $C_S(N_i)$ represents a cost relationship to be derived. It is clear that since each access channel terminates on only one switch and each backbone channel terminates on two switches, the total terminations in the system, N, is given by:

$$N = \sum_{i=1}^{70} N_A(R) + 2N_B(R) \qquad (4)$$

Hence, the total switching cost will be a function of the channel rate R. To complete the cost model, a relationship for $C_S(N_i)$ will now be developed.

Estimates of core capacity, number of line terminations and, throughput required of each size category were used as the basis for obtaining cost data. A number of manufacturers and models were surveyed and cost estimates were obtained for processor, memory and line termination hardware needed to implement each size. The number of manufacturers and the number of models built by each manufacturer made it impossible to compile a complete survey of all machines which might be suitable. However, the machines selected were representative of the hardware which could be used for the various size switches and the resulting averages are shown in table 2.

TABLE 2
Comparative Switch Costs

No. of Term.	Processor (&)	Core (&)	Termination Hardware ($)	Total (&)
60	16K	23K	129K	167K
110	38K	39K	146K	223K
225	199K	165K	596K	960K
450	470K	431K	1230K	2131K

These costs generally fit an exponential relationship as a function of the number of terminations given by:

$$C_S(N_i) = \$1150 \, (N_i)^{1.21} \qquad (5)$$

for $N_i >$ approximately 50

The exponent being greater than one implies that it is less expensive to buy two smaller switches rather than one large switch for the same number of terminations. This is the basic rationale for the present widespread success of mini-computers and multiprocessor configurations for communications applications. However, the relatively recent development by a number of vendors of new exclusively communications oriented processor systems indicates that in the 1976-1982 period this cost structure

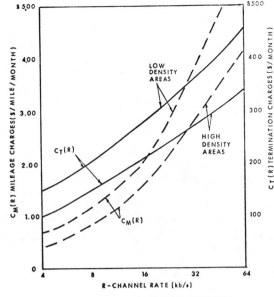

FIG.5 TRANSMISSION COST STRUCTURE DATA

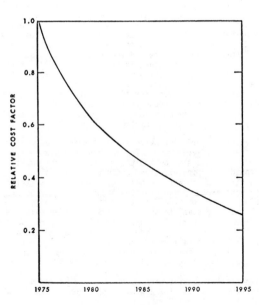

FIG.6 TRANSMISSION COST PROJECTION

will change. It is anticipated that mini- and medium-sized multiprocessor communications processor systems, which are expandable to a large number of terminations (approximately 500), will become commonly available at substantially reduced costs. This trend is represented by the exponent in the above cost relationship decreasing to some value less than one and the proportionally factor increasing slightly. Based on these considerations it is estimated that a more refined and yet still conservative cost relationship for communications oriented switches in the 1976-1982 time frame will be given by:

$$C_S(N_i) = \$1500 \ N_i \qquad (6)$$

By substitution of this relationship into equation 3 yields:

$$P_S = \sum_{i=1}^{70} 1500 \ N_i = 1500[N_A(R)+2N_B(R)] \qquad (7)$$

Hence, P_S becomes independent of the variations in size from one switch to another. Rather it is a direct function of the total terminations in the network. This relationship is plotted in figure 7.

3. EVALUATION OF SYSTEM COST

The development of system costs as a function of the channel rate, R, is somewhat complex due to the fact that both lease (on a yearly basis) and acquisition (initial one-time) costs are involved. A useful method of combining both aspects fairly is to accumulate all costs over a fixed time period, say 10 years after discounting all annual costs to present value. For example, the cost of leasing in the 10th year is reduced to the amount of investment it would take in the first year to build up to that needed amount in the 10th year. For this paper, the period 1976-86 was chosen for this base period and a 10% discount factor was used.

To determine total lease costs, therefore, the annual lease cost for 1976, the initial year, must first be established. This is readily obtained by the application of equation 2 using backbone and access data provided in figures 3, 4, and 5. Actual lease costs for each succeeding year are then determinable by application of the lease cost projection curve contained in figure 6. The 10% annual discount factor is subsequently applied to derate future costs to present dollars. It is interesting to note that the lease appropriation for the 10th year is approximately 10% of the first year due to approximately a 25% reduction in actual tariffs and a 40% derating factor. The total 10-year lease costs are plotted in figure 8.

The cost figures for switching must also be augmented to reflect a 10-year cost. In addition to the actual procurement costs, an additional one-time cost must be expended for installation to cover such items as construction, power equipment, and labor for installation and checkout. A conservative figure that is generally used is 100% of procurement costs. Annual costs for operation and maintenance must also be added which generally amounts to an additional 10% of procurement costs. This annual charge is also derated according to the 10% discount factor. Hence, the total 10-year cost for switching is derived as the sum of these three components and is also plotted in figure 8.

The final 10-year system cost, obtained by adding the corresponding total lease and procurement costs, is also plotted in figure 8.

4. DISCUSSION OF RESULTS

The final results presented in figure 8 would a first appear to indicate that a channel rate somewhat below 32 Kb/s would result in a minimum cost network, at least for the model and cost factors derived for this study.

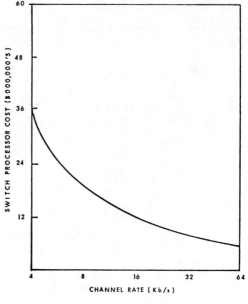

FIG.7
SWITCHING COST

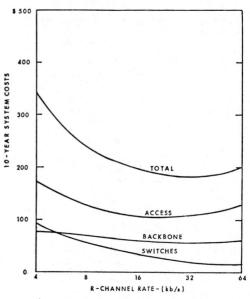

FIG. 8-10-YEAR SYSTEM COST COMPARISON

Before we prematurely accept such a simplistic conclusion, however, we should probe deeper into the components of the cost model and apply some judgement as to the sensitivity of this result with respect to our assumptions.

First of all, reviewing the transmission aspects, it is noted that the trunk cost is relatively flat. A major cause of this is due to the relatively high interconnectivity of the backbone network and the fact that the number of links for each channel rate has been optimized so that the total network capacity is always efficiently utilized. This tends to maintain a flat cost curve and is an important advantage of a highly interconnected network.

In contrast, the access cost curve exhibits a definite minimum essentially caused by the following factors. At high channel rates, many installations are too small to utilize the higher capacities efficiently which leads to higher unused capacity, which is reflected in higher attendant cost. At low channel rates, the number of channels required to handle the larger installations rises dramatically and causes excessive charges for channel terminations which because of the relatively shorter access channels become predominant over the mileage charges. Upon further examination, it becomes quite clear that the true optimum solution here is to match the channel rate to capacity for each particular data installation since there is no real requirement for over-all uniformity in access channel rate. The cost evaluation of this approach was conducted with the dramatic results that the total 10-year access costs reduced to $66.6 million as compared with the 104.6 million obtained as the previous minimum at 16 Kb/s. It is seen therefore, that with a properly designed network, the total transmission costs can be held relatively insensitive to channel rate.

With respect to switching costs, it is noted that this element does tilt the economic advantages to the higher channel rates. It should be noted, however, that the switching costs represent the smallest component of the cost model and if we were to project these costs further into the future it is generally recognized that the switch costs would decrease at a much faster rate than the transmission costs, and hence become an even less significant factor.

Results of this study have shown that in an efficiently designed data network, the sensitivity of cost with channel bit rate is very minimal over a significantly broad range of interest, and further indicates that this result is not expected to change with time. Hence, one can proceed with reasonable confidence towards a system design that is expected to remain cost-effective for a long time span.

There is one final point that we would like to make. In many cases, a data network such as we have described will interconnect facilities that also require voice communications and hence will have need for a voice network as well. An example of this is the U.S. Defense Communications System. For a number of reasons, it is being planned to ultimately convert voice signals to digital and replace the present analog system with a digital one. The opportunity is present therefore, to achieve significant cost savings by combining the two needs into a single integrated system. A previous paper by Coviello and Rosner [3] presented an example of a systems cost trade-off similar to that treated in this paper except it was oriented to a voice network alone. These results by contrast, indicated a definite minimization of system cost in the 8 Kb/s region. Hence, if one were to design an integrated system using common transmission channels, it is seen that the use of an 8 Kb/s channel would minimize the cost of the voice portion of the network without incurring any cost penalty for the data portion. This of course assumes that the network response time which is obtained at 8 Kb/s, as indicated in figure 2, is sufficient to meet system requirements.

5. AREAS FOR FUTURE INVESTIGATIONS

It has often been noted in past papers [4,5,6], that typical voice channels are used somewhat less than 50% of the time, since in general only one party is talking at any one time. It is interesting therefore to consider an approach whereby packets of data could be inserted into the voice network during such "quiet" periods. Since a 1000-bit packet at 8 Kb/s would only require time slot durations of 125 milliseconds through such a phantom network the above approach might possibly be used to "eliminate" a significant portion of the separate data network entirely. Although the feasibility of such an approach has not yet been investigated it does serve to point out one factor. In the rush to design new exotic data networks one should not ignore the total communication picture, and instead be receptive to methods of reducing costs on a totally integrated basis. It is yet too early to decide whether data networks should evolve as a separate entity from voice networks or whether the optimal solution is an ultimate marriage of the two.

6. REFERENCES

[1] L. G. Roberts, "Data by The Packet", IEEE Spectrum, pp. 46-51, Vol. 11, No. 2., Feb. 1974.

[2] R. D. Rosner, "Large Scale Network Design Considerations", 1974 ICCC, these proceedings.

[3] G. J. Coviello, R. D. Rosner, "System Considerations in Evaluation of Speech Digitization Techniques", Record of the Communications Systems and Technology Conference, Dallas, Texas, April 1974.

[4] J. A. Sciulli and S. J. Campanella, "A Speech Predictive Encoding Communication System for Multichannel Telephone". IEEE Trans on Communications, Vol. COM-21, pp. 827-835, July 1973.

[5] K. Bullington and M. M. Fraser, "Engineering Aspects of TASI". Bell System Tech. J., Vol. 38, pp. 353-364, March 1959.

[6] K. Amano and C. Ota, "Digital TASI System in PCM Transmission", 1969 International Conference on Communications, June 9-11, 1969.

Commercial, Legal, and International Aspects of Packet Communications

STUART L. MATHISON, MEMBER, IEEE

Invited Paper

Abstract—Packet switching technology emerged rapidly in the 1970's as another viable mode of communications switching, along with circuit and message switching. Since packet switching offers economical and versatile data communication capabilities in a multiuser environment, it is particularly well suited for furnishing public data communication network services.

Public packet networks are now established or being developed in most industrialized countries, and the introduction of these networks has raised policy issues relating to the structure and regulation of the national networks, and the interconnection of national networks into an international packet switching system.

This paper reviews these issues and concludes that public packet switching network services will continue to be regulated in all cases; that competitive packet networks will coexist in the U.S. and in Canada, but that only one national packet network will exist in each of most other countries; that packet networks will aggravate the problem of distinguishing nonregulated data processing services from regulated data communication services; that international interconnection of public packet networks based upon CCITT standards will occur rapidly over the next several years; and that a unified international packet switching system will eventually emerge similar to today's international telephone and telex systems.

I. INTRODUCTION

DURING the mid-1970's public packet switching networks emerged in a number of countries, offering computer users and communicators a highly cost-effective and versatile means of transferring digital data between terminals and computers. The introduction of these new services has raised several national and international policy issues, the resolution of which will affect computer users, computer equipment manufacturers, communication common carriers, and the general communicating public.

This paper is intended to introduce the reader to these policy issues, to review the principal arguments expressed on each issue, and to indicate the likely policy resolution in the future. The paper is organized into two major sections, the first covering the structure and regulation of packet switching services at the national level, and the second section covering the structure, pricing, and standards-making activities relating to packet networks at the international level.

A typical packet switching network consists of many distributed store-and-forward switching centers, multiply interconnected. Such a network is similar to many private data communication networks in that it may be implemented by leasing the communication channels from the traditional communication common carriers—i.e., the telephone companies—for the links between the switching centers. However, packet switching networks, by their nature, unlike traditional tree-structured data communication networks, are well suited to provide data communication services in a multiuser environment. The characteristics of packet networks which make them suitable in such an environment are as follows.

1) The distributed architecture of a packet network makes it possible to attach user computers to the network at virtually any network node. In contrast, most private data communication network architectures are "tree-structured"; that is, they are designed to service one or more computers located at a central site.

2) Packet networks are "fully switched" networks, permitting any network station (terminal or computer) to establish communications with any other network station on a demand basis. In contrast, typical private data communication networks provide fixed linkages between data terminal entry points and the centralized server computer.

3) Packet network architecture lends itself to rapid, yet flexible and orderly, expansion in response to growth in data traffic and demand for service, while private networks are generally limited by their initial design.

4) Packet network architecture permits a variety of interface arrangements to be established between the network and the end user equipment (data terminals and computers). With the private networks' rigid design, a change in the type of user equipment could entail an attendant redesign of the entire network.

Multiuser communications services offered to the public on a for-hire basis are generally offered by Postal Telephone and Telegraph (PTT) administrations or (in North America) by government-regulated private common carrier organizations, in order to take maximum advantage of economies of scale, to facilitate communication among all users, to ensure equitable treatment of all users, and to serve national interests. Public packet switching network services, now offered in several countries on a common carrier basis, are subject to the same types of government regulation.

As of early 1978, there were several public packet networks in operation and many more under development in various countries. In the United States, Telenet Communications Corporation operates a public packet network, and Tymnet Inc. operates a somewhat similar public data network utilizing a character-oriented packet technique. ITT Domestic Transmission Systems Inc. (ITTDTS) and Graphnet Systems Inc. have also been authorized by the Federal Communications Commission (FCC) to establish and operate public packet networks, and both firms at this writing are in the process of implementing their networks. Both the ITTDTS and Graphnet networks initially intend to offer message and facsimile communication service, and later, computer and data communications service. In addition, ATT has petitioned the FCC for authority to introduce its packet network, the Advanced Communications Service (ACS). At this writing the FCC has not acted upon the ATT petition.

In Canada, the Trans-Canada Telephone System (TCTS) operates a public packet network, similar to the U.S. Telenet, called Datapac; Canadian National/Canadian Pacific Telecommunications (CN/CPT) is also developing a public data network, called Infoswitch, which will utilize a hybrid circuit/packet switching technique.

In Western Europe, several public packet networks are currently being developed. The French PTT is nearing comple-

tion of its Transpac network, scheduled for operation in late 1978. The British Post Office, which operates all public telecommunications facilities in the UK, has been operating the Experimental Packet Switching Service (EPSS) for several years and is now beginning the development of a permanent public packet network. Several other European countries, including the Scandinavian countries, West Germany, the Netherlands, Belgium, Switzerland, and Italy, are also beginning plans for the development of public packet networks.

In Spain, the Spanish National Telephone Company (CTNE) has been operating a hybrid packet and message switching network, primarily oriented to computer-communications applications in banking, for several years.

In addition to these domestic packet networks, nine Common Market countries (Belgium, Denmark, France, Germany, Ireland, Italy, Luxembourg, the Netherlands, and the United Kingdom) have joined together to develop and operate a multinational public packet network called EURONET. This network will initially operate on a semi-public basis, providing terminal access to selected government and research computer centers and data bases, but will later become a general-purpose public data network.

In Japan, Nippon Telephone and Telegraph (NTT) is developing a public packet network, while the Australian telecommunications administration is also planning to develop a similar network.

In virtually all of these public packet networks, the respective telecommunications carrier has implemented (or is planning to implement) the network in conformity with packet network interface standards established by the Consultative Committee for International Telephone and Telegraph (CCITT), the international standards-making body for telecommunications. Consequently, as these national networks become interconnected with one another, a unified world-wide packet switched data communications system will gradually emerge, providing similar data communications services on a world-wide basis. The services offered and the standards utilized in these national networks are discussed in "Public Packet Switched Data Networks, International Plans and Standards" by P.T.F. Kelly, in this issue of the PROCEEDINGS.

In addition to these public packet networks, several other multiuser packet networks in operation or under development should be noted. First is the recognized "grandfather" of all packet networks—the ARPA network (ARPANET), which was first put into operation in September 1969 under sponsorship of the Advanced Research Projects Agency (ARPA) of the U.S. Department of Defense. It is currently managed by the Defense Communications Agency (DCA).

Second, the international SITA network, a private network operated on a nonprofit basis on behalf of the international airlines, links the many airline reservation and message switching computers through a common packet switching communications subnetwork.

The development of these public and quasi-public packet networks has raised a series of policy questions which are discussed in this paper. Due to variations in industry structure and philosophy, the answers to these questions are often different in the United States than in other countries. In many cases, definite answers are not possible at the present time. This paper discusses the issues involved with these questions.

1) Should several public packet networks be permitted or encouraged to coexist within one country, or should such services be limited to a single monopoly supplier?

2) Should public packet network services be offered only by

authorized common carriers, or should any organization be permitted to furnish public packet network services?

3) If government regulation is required, what form should such regulation take?

4) To the extent that entities offering public packet network services utilize the transmission facilities of underlying telephone carriers, what steps are required to ensure equitable competition in the event that the underlying carrier undertakes to furnish similar packet network services itself?

5) Where private organizations, such as computer service companies, operate packet switching networks, to what extent and under what conditions should such firms be permitted to offer the use of their networks to the general public without first becoming authorized carriers?

6) On what basis should carriers offering public packet network services furnish customer-site equipment (i.e., should customer-site equipment be "bundled" with network services; should such equipment furnished by carriers, in competition with similar equipment furnished by noncarriers, be tariffed)?

7) How should domestic public packet networks be interconnected with domestic packet networks in other countries, from a technical, organizational, and commercial point of view?

8) How should public packet network services be priced domestically and internationally? Should such services be priced high to protect existing service offerings, or should they be priced low in order to stimulate new data communication applications and to convert users from private line facilities to "shared" public packet network facilities?

9) What types of interface standards should be utilized and to what extent should these interface protocols be supported by carriers and equipment suppliers?

10) Should the greatly increased flow of data across international boundaries, stimulated in part by the introduction of international packet communication services, be restricted to protect privacy and national interests?

These issues are discussed in this paper in the context of packet communications, the subject of this Special Issue of the PROCEEDINGS. However, most of these issues apply as well to public computer-based message switching services that utilize either a single centralized message switching host computer, or a distributed network of store-and-forward message switching computers. In fact, the principal distinction between a public packet network and a distributed public message switching network is the transit delay between source and destination. (Typically the delay is hundreds of milliseconds for a packet network and many minutes, or even hours, for a message switching network.)

II. Domestic Industry Structure and Regulation

The industry structure and the regulatory framework for the provision of public packet switching services varies somewhat from country to country. In the United States several private firms have been authorized by the FCC to furnish such services. These firms thus operate within a regulated but competitive framework.

In Canada, both the telephone company (TCTS) and the telegraph company (CN/CPT) have been authorized by the Canadian Radio and Television Commission to furnish such services; this has resulted in a regulated, but competitive, duopoly structure.

In most other countries, all telecommunications services are furnished by government agencies or government-owned cor-

porations, and public packet services are furnished on a monopoly basis.

In all countries, public packet network services are furnished as common carrier services, either by a government entity or under government regulatory control.

The industry structure that has evolved within each country reflects the local circumstances. The following section discusses the evolution of the U.S. approach to regulating packet network services, the nature of certain changes that have been proposed, and the pros and cons of each; then similar issues in other countries are discussed.

A. The United States

1) Communications Act of 1934: U.S. policy regarding packet network services is based upon the Communications Act of 1934, which provides for FCC regulation of communication services that are common carrier in nature, under Title II of the Act. While the Act does not specify when a communication service is "common carrier in nature," it endorses the historical common law definition of common carriage. Accordingly, all communication services offered to the general public for hire are typically regarded as common carrier services subject to FCC regulation.

Title II of the Communications Act requires that carriers furnish services at reasonable charges upon reasonable request, that carriers obtain FCC authorization prior to constructing or leasing new lines or curtailing service, and that carriers file nondiscriminatory tariffs showing all charges, practices, classifications, and regulations for interstate services offered to the public. Technological developments and institutional changes in the communications industry have impelled Congress to review the Communications Act of 1934. This review may result in new policies regarding the amount of competition in the communications industry, the manner in which public packet carriers are regulated, and the amount of discretion available to the FCC.

2) First Computer Inquiry: In 1966 the FCC initiated the first of two rule-making inquiries concerning policy problems presented by the growing interdependence of computers and communications. The FCC's proceedings in this area became known as the "Computer Inquiries I and II" and addressed several questions affecting the regulatory status of packet communication services.

First, the FCC examined the questions of whether data processing services involving communications should be regulated, and if not, to what extent and in what manner regulated carriers should be permitted to furnish data processing services. In 1971, the FCC concluded that data processing services should not be regulated, and could only be offered by carriers through a separate data processing subsidiary. To enforce this ruling it was necessary for the FCC to distinguish data processing from communication services; accordingly the Commission adopted an *ad hoc* procedure known as "the primary purpose test."

Under this approach, the FCC would examine a "hybrid" data processing/data communications service offering to determine which component was "primary" in nature and which was only "incidental," and would forego or impose regulation accordingly. Consequently, a firm could offer a computer-based service involving some communications capability, and so long as the communications element of the overall service was "integral and incidental" to the data processing element, the firm could avoid subjecting itself to FCC regulation.

Few firms actively sought an FCC determination of whether

their particular service offerings were data processing or communications; the primary purpose test worked reasonably well, however, in that most organizations made their own judgments as to what they believed was the primary nature (data processing versus communications) of their business and acted accordingly.

One particular class of service—store-and-forward message switching—was specifically considered by the FCC and was found to be a common carrier service when offered to the public, whether or not the provider of the service furnished both the terminal access lines and the message switching computer, or just the message switching computer (in which case the users furnished their own access lines). The FCC reaffirmed this ruling in the context of message switching services offered by several international record (telegraph) carriers, one offering the service under tariff, the other not. Here, the FCC concluded that both carriers must furnish the message switching service under tariff.

This conclusion indicated that the emerging public packet network services would probably be treated by the FCC as common carrier services subject to Title II regulation.

A second issue raised by the FCC in the first computer inquiry involved the sharing and resale of leased communication lines. Sharing was defined as the "joint use" of a line by two or more organizations on a nonprofit basis; resale was distinguished from sharing in that a reseller made a profit from the activity, either by acting simply as a broker of a wide-band channel, or by offering a "value-added" communications service.

While the FCC did not issue any new rules pertaining to sharing and resale in the first computer inquiry, the strong support among users for greater latitude to "share" leased communications channels led ATT to revise its private line tariff to permit sharing of voice-grade and smaller bandwidth channels.

3) Specialized Carrier Inquiry: While the Communications Act of 1934 provided a mechanism for FCC authorization of new common carriers to offer new services, there were actually few significant new carriers offering new communication services between the time the Act was passed and the late 1960's. This was due largely to the determined opposition of the telephone and telegraph companies, which opposed most such attempts at their outset.

This pattern changed in 1969, when the FCC, over strong telephone company opposition, authorized Microwave Communications Inc. (MCI) to build and operate a small 200-mile microwave system between Chicago and St. Louis that would offer private line communication services in competition with the telephone and telegraph companies. The MCI decision was later reaffirmed and broadened in an FCC "specialized common carrier" inquiry in 1971. In this inquiry the FCC concluded that the introduction of new specialized communications services and competitive carriers into the marketplace was beneficial to the public and would be permitted in the future. As a result, manifold new specialized carriers were authorized, each of which proposed to build and operate microwave facilities in competition with similar facilities operated by the telephone and telegraph companies. All but one of these "specialized carriers" offered primarily private line voice-grade point-to-point analog communication channels. Data Transmission Company (DATRAN), now defunct, offered private line and fast-connect circuit-switched digital transmission channels.

The FCC's Specialized Carrier decision, strongly opposed by the established carriers, marked a turning point in U.S. communications policy. Rather than maintain the then current monopoly structure for all public communication services, the FCC had decided to permit competition within the communication industry, in carefully selected areas where such competition was feasible and publically beneficial.

4) Value-Added Carriers: In the early 1970's, shortly after the technical feasibility of packet switching had been demonstrated by the ARPA network (see "The Evolution of Packet Switching" by L. G. Roberts, in this issue of the PROCEEDINGS), two companies applied to the FCC for authorization to establish and operate public packet networks. The first application, submitted by Packet Communications Inc. (PCI), sought a determination of whether it was even necessary to obtain FCC authorization to offer such services. Since, according to PCI, all communication lines in the proposed packet network were to be leased from ATT and such lines had already been authorized by the FCC, it was not clear whether any further FCC authorization was necessary. After considering the PCI application, the FCC concluded that the offering of public packet network services was a common carrier undertaking, subject to FCC regulation, and that PCI's particular proposal would be approved. Although approved by the FCC, PCI was unable to raise sufficient venture capital to proceed with its network plans, and subsequently went out of business.

The second packet network application was filed with the FCC by Telenet Communications Corporation in 1973 and was granted by the Commission in 1974. After sucessfully raising the necessary capital, Telenet proceeded to implement its public packet network.

Telenet, and each of the "value-added carriers" following it, utilize transmission lines leased from the telephone companies. In part for this reason, the telephone companies did not oppose the proposals of PCI, Telenet, or other subsequent value-added packet carriers. In fact, the telephone companies made specific revisions in their tariffs to permit value-added carriers to "resell" the telephone company communications lines. Such resale had been, until that time, strictly prohibited by the telephone company tariffs.

For some time after Telenet began operations, an anomaly existed in the U.S. data communications industry. Both Telenet and Tymshare Inc. were offering "value-added" network services, utilizing somewhat similar packet-switching techniques, but Telenet was furnishing service as an FCC-approved common carrier, while Tymshare Inc. was operating without FCC approval, under ATT's "joint use" tariff provisions.

This issue was raised before the FCC and as a result, Tymshare subsequently formed a separate subsidiary, Tymnet Inc., which submitted an application to become a common carrier. The action was apparently taken in order to be able to furnish message switching services which could not be provided under the "joint use" tariff provisions. The application was approved by the FCC in 1976.

5) ATT's Transaction Network Service and the ACS: In 1976 ATT introduced, on an intrastate basis in the state of Washington, a new service called Transaction Network Service (TNS). TNS is the first packet communications type of service offered by ATT. It consists of a packet switch, with high-speed lines linking customer host computers to the switch, and two types of terminal access facilities for terminals—dedicated polled access lines and dial-in facilities.

TNS is specifically designed for users who need to interchange large volumes of short transactions between terminals and data base computers. Typical applications include credit verification, check authorization, and other banking and financial transaction systems.

ATT furnishes only the communications facilities and, at the user's option, terminals. ATT does not furnish the data base computers.

TNS is classed as a packet communications type of service since it operates by switching short packets (maximum packet length is 128 characters), utilizes the same general type of host computer interface employed in most packet switching networks, and performs communications processing functions within the switching computer (protocol conversion, error detection and correction, polling, etc.).

At this writing, TNS is offered in two states, Washington and Minnesota. ATT has not yet sought or received FCC approval to offer TNS service on an interstate basis.

The future of TNS is unclear at this time. While presently offered by ATT as a tariffed communications service, the Commission on Electronic Fund Transfers, established by Act of Congress in 1974, recommended in its final report that communications services for electronic funds transfer applications, such as TNS, be furnished on a nonregulated nontariffed basis. Since ATT is prohibited by a 1956 antitrust consent degree from furnishing nonregulated services, the enactment of the EFT Commission recommendation would prevent ATT from continuing to furnish TNS.

TNS has also encountered only limited acceptance in the marketplace, partly because it furnishes only communications services, while other EFT systems furnish both communications and data processing services.

Of considerably greater significance than TNS, both from a policy and an industry point of view, is another value-added packet switching service which ATT plans to introduce in the 1979-1980 time frame. Although not yet approved ATT has applied to the FCC for authority to offer a new service called ACS. This anticipated service has also been referred to as the Bell Data Network.

Although all the details of the ACS offering have not yet been announced, the policy issues that will be raised by this offering can be anticipated. First, ATT will become both a supplier of transmission circuits to value-added carriers (such as Telenet, Tymnet, ITT, and Graphnet), while at the same time furnishing value-added packet network services itself. This problem is discussed in Section V below.

The second issue highlighted by ACS concerns the distinction between unregulated data processing and regulated data communications services. ATT can only furnish the latter, due to its antitrust consent decree, but the distinction between data processing and data communications has never been satisfactorily established. This issue is discussed further below.

6) *Resale/Sharing Inquiry:* In 1974 the FCC instituted another related inquiry addressing the questions: Under what circumstances should organizations be permitted to share and/ or resell communication lines leased from telephone and telegraph carriers? And, if resellers are to be regulated, what form should such regulation take? Most of the more than 60 parties in this proceeding were content with variations on the *status quo*, whereby nonprofit sharing of lines was permitted without regulation, and for-profit resale was limited to authorized carriers. Several organizations, however—including the Office of Telecommunications Policy (OTP) within the executive branch (this office was subsequently abolished), and the U.S. Department of Justice—argued that any organization, carrier or not, should be permitted to engage in unlimited sharing and resale

of leased communications lines. Adoption of this proposal would enable any company to establish and operate a public packet network without any FCC regulatory control.

The FCC concluded that sharing of a leased communication channel by several organizations was permissible, but only if such organizations shared the line on a nonprofit cost-sharing basis. The FCC also concluded that resale of leased channels, when performed on a for-profit basis, was a common carrier service and subject to full Title II regulation. Since public packet carriers in the U.S. are engaged in resale on a for-profit basis, this FCC ruling reaffirmed that such activities were common carrier in nature. IBM and several other parties appealed this ruling to the U.S. Court of Appeals. In January 1978 the Court affirmed the FCC in all respects, concluding that the Communications Act requires the regulation of any firm which "undertakes indifferently to provide communications service to the public for hire, regardless of the actual ownership or operation of the facilities involved," and that the FCC cannot abdicate this responsibility [1].

Given that resellers are to be regulated, the FCC also asked how they should be regulated. Telenet Communications Corporation argued that the full range of Title II regulations, which were originally intended for the regulation of telephone and telegraph companies, were inappropriate for value-added carriers which leased their transmission facilities and operated within a competitive environment. Specifically, for example, Telenet argued that it should be unnecessary for the FCC to authorize every circuit used by a value-added carrier over ten miles in length, as presently required under FCC regulations.

The FCC however decided to continue to apply the full range of common carrier regulations to value-added carriers. In fact, the FCC concluded that the term "value-added" has no regulatory significance—that is, a value-added carrier and an underlying carrier are equivalent from a regulatory point of view.

7) *The Second Computer Inquiry:* In 1976 the FCC initiated the second of its two computer inquiries, for the purpose of establishing a more precise distinction between data processing and communications. The Commission proposed to define data processing as "the electronically automated processing of information wherein: a) the information content, or meaning, of the input information is in any way transformed, or b) where the output information constitutes a programmed response to input information. Part b) of the proposed rule would bar carriers, including packet communications carriers, from offering users the ability to input "commands" which request status information or instruct the network to establish parameters concerning routing of calls, mode of communication (half or full-duplex), insertion of padding characters, etc., since such commands produce a programmed response to input information.

Comments regarding the proposed definitions were submitted by several dozen organizations in the computer and communications industry, and by several government agencies. These comments indicated an almost universal consensus that the definitions proposed by the Commission were unworkable, would inhibit development of new services and technologies, and would be unenforceable. Most commenting parties recommended a continuation of the "primary purpose" test, which was originally proposed to the FCC by IBM some 12 years ago.

Significantly, in the Second Computer Inquiry IBM proposed an entirely new definition of what services constitute common carrier communications. IBM proposed that only "pure transmission" services be subject to regulation. Under this proposal, according to IBM, "a regulated service would be permitted to make signal conversions for purposes of transmission. It would, however, be required to deliver information in its original data format, code, or protocol to the addressee. If a service does not deliver information in its original data format, code, or protocol, it would be performing more than the transportation of information from place to place," (See reference [2]).

The IBM proposal, if adopted, would bar packet communications carriers from offering customers "translation" capability —the ability to interface terminals and host computers utilizing different protocols to the networks. For example, Telenet and other public packet carriers would be barred from using the CCITT Interactive Terminal Interface for linking teleprinter and CRT terminals and the CCITT X.25 protocol for linking host computers to the networks, and permitting such terminals and computers to intercommunicate—unless these carriers set up separate unregulated data processing subsidiaries. And then it would be necessary to distinguish those communication services which utilized the packet network "transparently" from those that involved conversions from one speed, code, or protocol to another, since presumably the former would be unregulated while the latter would be regulated. IBM's proposal would also bar ATT from providing the same type of conversion capability in its TNS and ACS. Significantly, the ability of public packet networks to tailor interfaces to customer site equipment and to enable dissimilar customer systems to intercommunicate is regarded by many as one of the most important features of public packet network services.

While the January 1978 U.S. Court of Appeals ruling affirming the FCC's sharing and resale policy, referred to above, appears to have disposed of IBM's argument that only pure transmission is the proper province of the common carriers, the question of how far a carrier can go in furnishing "communications processing" services is certain to be raised again in the context of ATT's anticipated.

8) The Proposed Communications Act of 1978: In June 1978, a bill entitled the "Communications Act of 1978" was introduced in the U.S. House of Representatives by the Chairman and ranking minority member of the Communications Subcommittee of the Interstate and Foreign Commerce Committee. The proposed bill, if enacted, would repeal and replace the Communications Act of 1934, discussed above.

Provisions in the proposed bill which would affect packet communication services include the following.

First, ATT would be freed from the terms of its 1956 antitrust consent decree which currently restrict the company to common carrier communication services, and prohibit it from offering unregulated computer services. Under the proposed bill, ATT would be free to furnish data processing services, and thus could furnish enhanced packet services containing data processing features which would otherwise be impermissible.

Second, the bill would permit carriers offering competitive services to install new facilities without prior regulatory approval.

Finally, and potentially most significantly, the proposed bill would give the regulatory agency the discretion to totally deregulate any class of carrier, and the agency might exercise such discretion in the case of "resale" services which are competitive in nature. Under this provision, firms such as Telenet

and Tymnet might be deregulated while ATT and its ACS service might continue to be regulated.

The proposed bill will undoubtedly undergo much revision, but at least some portions of it are expected, by this author, ultimately to be introduced into law. Since the overall thrust of the bill is towards deregulation, even if the bill is not passed, it is likely to influence the FCC to relax regulatory controls wherever possible and to rely more heavily upon competitive market forces.

B. Canada

As noted earlier, packet communication services in Canada are furnished by the established telephone and telegraph carriers, the TCTS and CN/CPT. At this writing no "value-added" carriers have applied for nor been authorized to furnish packet communications service in Canada.

In the early 1970's Canada undertook an extensive communications policy analysis under the auspices of a government group called the Telecommission. This group, and a subsequent task force focusing specifically on computer communications, reached conclusions similar to those of the First FCC Computer Inquiry in the U.S. Additionally, the Canadian Computer Communications Task Force concluded that it was desirable for the common carriers to establish a common user data communications network to facilitate the exchange of financial transaction data among banks, retail establishments, and other financial service organizations. Interestingly, this conclusion is diametrically opposed to the recommendation of the U.S. Electronic Funds Transfer Commission, as noted above.

Regarding the regulatory status of packet communication services in Canada, it appears to be a well-established fact and accepted policy that such services, when furnished as a public offering, constitute common carrier services subject to regulation by the Canadian Radio and Television Commission, the Canadian equivalent of the FCC.

C. European Policies

In Europe, telecommunications services are generally furnished by a government agency or government-owned corporation—the postal telephone and telegraph administration, or PTT. As in the U.S. and Canada, the PTT's regard public packet communication services as common carrier services subject to PTT control. However, unlike the U.S. and Canada, the PTT's have determined that such services should be furnished on a monopoly basis. Thus, in the United Kingdom, the British Post Office operates the only public packet network service (on an experimental basis at this time), and in France the PTT will begin operating the only such public network in France in late 1978. The PTT's of nine European countries are also cooperating in the implementation of the supranational Euronet packet network, mentioned earlier, which will link these countries together in early 1979.

D. Industry-Oriented Networks

In the U.S. and in other countries there are a number of industry-oriented networks, some of which utilize variations of packet communications and/or message switching technology. In the U.S., Aeronautical Radio Inc. (ARINC) operates an extensive message switching network linking many airline reservation systems and message switching computers. Also, two U.S. networks—the Fed Wire and BankWire—are used to transfer funds among several hundred banks whose computers are linked to these networks. In Europe, SITA operates a similar network linking the computers of many international airline organizations. Another European network, called the SWIFT network, links together the computers of many banks and financial institutions for fund transfers and message communications.

In most of these cases the organization operating the network furnishes service on a nonprofit basis to its members only. The various regulatory authorities and telecommunications administrations have recognized that the services which these networks offer are not available from the common carriers, are highly specialized in nature, and are not offered in order to derive a profit. Therefore, such networks have been granted special exceptions from the general requirement that common-user communication services be furnished on a common carrier basis under tariff. In Europe (see Section III-B, "International Tariffs") a surcharge may be imposed upon such multiuser industry-oriented network operators.

E. Value-Added Carriers and Transmission Carriers

In the U.S. both value-added carriers (Graphnet, ITT Domestic Transmission Services, Telenet, and Tymnet), and the principal underlying carriers (ATT and Western Union) furnish packet and message switching services. The value-added carriers, however, depend upon the underlying carriers for transmission facilities necessary to their operation. This raises the potential problem of unfair competition between the value-added and the underlying carriers.

The underlying carrier operating its own network service has the ability to utilize whatever long haul and local circuit facilities it chooses, and has considerable discretion to allocate costs and establish rates in a fashion which strengthens its competitive position. Moreover, the underlying carrier may utilize circuit facilities not available to the value-added carriers. (In fact, certain of the low-cost access facilities which ATT utilizes for its Transaction Network Service are not part of the standard tariff of facilities available to the general public.) Therefore, unless the underlying monopoly carrier is required by the regulatory authorities to furnish competitive services (defined as services involving communications processing) through separate subsidiaries that obtain circuits on the same basis as the value-added carriers, unfair competition is likely to result. Recognizing the growing importance of this problem, and the historic difficulty of dealing with cross-subsidies between monopoly and competitive service within ATT, the FCC has indicated its intention to conduct a public inquiry concerning the manner in which monopoly carriers will be permitted to furnish competitive services, such as private line and communications processing services, in the future.

F. Value-Added Carriers and Vendors of Data Processing Services and Equipment

The principal policy issue affecting both packet carriers and vendors of data processing services and equipment is one of definition. What types of services are regarded as communications and may be furnished by common carriers? For example, a common carrier offering a packet network service could furnish, as part of a terminal interface protocol, the ability to "edit" terminal input data. Should this function be defined as data processing and excluded totally from the carrier's service offering, which would thereby reduce the utility of the service to the end user? Or should the regulatory authority look to the primary purpose of the service, recognize it as communications, and permit incidental additional features which enhance the service for the end user? (In fact, the Communications Act of 1934 defines communications as including ". . . all instrumentalities, facilities, apparatus, and services (among other things, the receipt, forwarding, and delivery of communications) incidental to transmission") [3]. On the other hand, if an organization is primarily furnishing a text editing service, with transmission as an incidental part, its offering might be considered data processing.

While the primary purpose test approach may involve administrative delays in those cases where firms are unable to determine the nature of their offerings or are unwilling to proceed without official determination by a regulatory authority, it permits carriers to enhance their services to the public, permits the data processing firms to incorporate communications capabilities in their services, and provides the regulatory authority with the latitude to consider the public interest in evaluating hybrid data processing/data communications service offerings.

The commonly used argument in support of deregulation of all communication processing services, and regulation of only "pure transmission" services, is the claim that communication processing services are not subject to economies of scale, are not natural monopolies, and therefore should not be regulated. However, communication processing network services are, in fact, subject to economies of scale (although not to the same degree as in the physical telephone plant). Moreover, the desirable goal of enabling any user to communicate with any other user requires either a single universal service, or the interconnection of many services. Either option requires regulatory control.

Another argument in support of deregulation is the claim that new communication services would be introduced by firms unwilling to submit to regulation. Certainly this would occur to some extent; however, another possible result might be the domination of the data communications field by one or more computer manufacturers.

For example, in considering the application of Satellite Business Systems Inc. (SBS), a joint venture of IBM, Comsat, and Aetna Insurance Co., the FCC took note of the concerns expressed by other data processing equipment manufacturers and other domestic satellite carriers. These organizations claimed that IBM might utilize its computer marketing resources, its established customer base, and various product packaging and bundling practices to extend its domination of the computer field into the data communications field.

Apparently agreeing with these arguments, the FCC imposed strict constraints upon SBS. The FCC required that: 1) SBS establish a wholly separate marketing organization from that of IBM; 2) IBM salesmen be barred from proposing integrated IBM/SBS systems; and 3) SBS offer only "transparent" communication services utilizing industry standard interface protocols for linking user equipment to the SBS system.

If the definition of communications proposed by IBM (namely that only "pure transmission" services be regulated) were adopted, IBM and other computer manufacturers would be free to undermine the intent of the FCC's SBS ruling by the simple expedient of leasing communication channels (terrestrial or satellite), adding communications processing services, and offering "nontransparent" services on a totally unregulated basis, perhaps bundled with other terminal or computer equipment. Not only would such a result reduce competition among communication carriers, but it would also strengthen and extend IBM's control over the data processing equipment marketplace.

III. International Issues

Packet communication services are potentially more significant internationally than nationally because such services offer very substantial economies where circuit costs are high (e.g., overseas channels), provide built-in mechanisms for overcoming differences in national data communication standards (e.g., CCITT and Bell System modem standards), and thereby facilitate the flow of information, computer data, and person-to-person communications among nations. Greater international information flows, similar to increased international commerce in goods and materials, contribute to the transfer of technology and know-how, increase the well-being of all nations, and increase the amount of cooperation and interdependence among nations. It is therefore advisable to ensure that institutional, political, and technical obstacles do not impede the development of a worldwide public packet communication system, similar in scope and benefit to the worldwide public telephone and telex communication systems.

The several areas of policy and practice which will influence the evolution of international packet communication services over the next several years are as follows:

1) the industry structure–i.e., the manner in which the national and international packet networks are interconnected;

2) the pricing structure and level, and the associated regulations concerning the use of international telecommunication facilities;

3) the development of and degree of adherence to international standards;

4) the establishment of national "data privacy laws" which may inhibit the free flow of information and data across international borders.

The nature of the policy issues in each of these areas, and the possible courses of development, are discussed in the following paragraphs.

A. International Industry Structure

1) Interconnection Arrangements: As discussed above, packet communications is evolving toward an industry structure consisting of several competing public carrier networks in the U.S., two competing networks in Canada, one network in each of several other countries in Western Europe and elsewhere, and a supranational network serving the common market countries. In the case of contiguous national networks, such as Datapac in Canada and Telenet in the U.S., the networks have been directly interconnected to one another based upon bilateral agreements between the respective national carriers.

However, interconnection between national networks in the U.S. and national networks overseas is evolving somewhat differently, due to the presence of several domestic carriers in the U.S., and several U.S. international record communication carriers that have historically provided most record/message/data communications service between the U.S. and the administrations overseas. These administrations typically have separate international departments, which work closely with the U.S. international record carriers.

In 1976 Telenet and the British Post Office (BPO) planned to directly interconnect Telenet's national network with an international gateway packet switching exchange operated by the BPO in London. Telenet submitted an application to the FCC for the necessary authority, proposing to utilize low-cost satellite channels leased directly from Comsat on a carrier-to-carrier basis. Although the application was vigorously opposed by all of the U.S. international record carriers (IRC's), the FCC approved it. The FCC also approved applications to furnish overseas packet service subsequently filed by the several U.S. international telegraph carriers–Western Union International, ITT World Communications Inc., and RCA Global Communications Inc. Each of the IRC's proposed to provide service between their gateway exchanges (primarily in the New York City area) and one or more overseas gateway exchanges operated by the respective PTT. Each of the IRC's also proposed to serve as the link between the national networks in the U.S. (primarily Telenet and Tymnet) and the overseas networks. The reason why the IRC's limited their proposals to their gateway cities is discussed below.

The PTT's became concerned that FCC policies potentially would lead to dozens of national public packet networks in the U.S. over the next several years and that establishing a precedent of direct interconnection with national U.S. networks would eventually force the PTT's to deal with dozens of U.S. network operators and to furnish overseas "half-circuits" to each.[4]

As an alternative, the PTT's could refuse to directly interconnect with multiple national U.S. packet carriers, and could require such carriers to interconnect their national networks to gateway exchanges operated by the IRC's. Each PTT would then interconnect only with the IRC's with which it traditionally dealt, rather than with both the IRC's and the several national packet carriers as well.

With the encouragement of the IRC's, the PTTs have generally chosen the latter course. Consequently, both Telenet and Tymnet in the U.S. interconnect with each of IRC's, who in turn interconnect with many PTT gateway exchanges in Europe and elsewhere.

This type of industry structure potentially reduces the number of U.S. organizations with which each PTT has to deal, but it increases the amount of equipment required, the number of switching exchanges through which packets must pass, the end-to-end delay, and the cost to the end user. Also, the intervening facilities utilized by the IRC's may constrain the services which otherwise would be offered internationally. Only when there are, in fact, a large number of U.S. public networks seeking interconnection with PTT networks, will the "concentration function" of the IRC's result in economies over the direct interconnection approach.

2) U.S. Gateway Restrictions: Another aspect influencing interconnection arrangements is Section 222 of the Communications Act of 1934, as amended, which provides that Western Union shall be limited to furnishing telegraph service in the contiguous domestic states, and that the international record carriers shall be limited to operating between several U.S. gateway cities (New York City, Washington, DC, San Francisco, New Orleans, and Miami) and overseas points. This provision resulted from the separation of Western Union's domestic and international operations into two completely separate companies—Western Union and Western Union

International, respectively. It was intended to ensure that Western Union could not direct overseas telegraph traffic through its own international department at the expense of the other international carriers, which did not have domestic telegraph operations. The gateway rule has, effectively, resulted in a cartelization of the U.S. telegraph industry into domestic and international sectors.

The gateway restrictions on Western Union and the IRC's are increasingly under pressure for change, by both Western Union, which wishes to interconnect directly with overseas PTT's, and by the IRC's, who wish to extend their facilities throughout the U.S. In fact, ITT has established a subsidiary to build and operate a domestic public packet network and has sought authority to interconnect this network with its international carrier subsidiary. The FCC has refused to grant such authority, because the interconnection would constitute a possible extension of ITT's gateways and would enable ITT to extend its privileged position as an international carrier into the domestic field.

Although the FCC is currently examining the gateway restrictions, elimination of this rule would require revision of the Communications Act. (The proposed Communications Act of 1978, if enacted, would eliminate the gateway restriction.)

3) Sharing and Resale of International Circuits: One final issue affecting international links between national networks is the matter of whether (and to what extent) international links can be shared and/or resold by noncarriers. The FCC has established a new policy permitting sharing of domestic channels by users on a nonprofit basis, and resale of such circuits on a profit-making basis by authorized carriers only. In reaching this conclusion, the FCC initially indicated it would apply the same rules to international circuits.

However, the foreign PTT's, which would have to concur in any revised regulations pertaining to international circuits, strongly oppose any form of generalized sharing of circuits, either within their respective countries or between them. Such sharing would undermine the PTT's international pricing structures, result in the creation of "pseudo-carriers," and fragment the integrated international communication service approach. Thus the FCC has limited the application of its resale/sharing ruling to domestic circuits. The FCC has, however, indicated its intention to conduct further inquiries into resale and sharing of international circuits in cooperation with the PTT's.

The prohibition against both sharing and resale of international circuits prevents the establishment of common user international packet networks, and bars private international networks from carrying "third party" traffic. (Third party traffic is communication in which the private network operator has no direct business interest in the content of the messages/data transported.) Thus international common user packet network services may only be offered by common carriers. Exceptions to this regulation require approval of the appropriate national authorities and have been granted in special cases, such as in the case of the SITA network.

B. International Tariffs

1) Basis for Charges: Many factors combine to determine the structure and level of tariffs for packet communication services—the underlying cost structure for the facilities, competitive alternatives (in countries with competing networks), impact upon other communication services offered by the carrier, national policy, and the desire for international uniformity and simplicity in tariffs.

In large-scale packet networks, cost elements are basically 1) access lines, 2) access ports (line termination equipment at switching centers), and 3) backbone network switching equipment and communication lines. Access line and port charges are basically domestic charges and vary by country. Backbone network usage depends on the peak volume of data (packets) transmitted; the cost for such packet transmission, within a given national network, is largely independent of the distance traveled by the data packets. Therefore, most national packet networks have established a flat charge per kilopacket of data transmitted. (A kilopacket equals one thousand packets, each of which contains a maximum of 64, 128, or 256 characters of user data.)

For international packet communications, where overseas links and gateway nodes are typically required, different packet charges apply for communication between each pair of countries. Such charges may be expressed in terms of kilopackets, kilocharacters, or kilosegments (discussed below).

While many aspects of international packet tariffs are established, certain items remain unresolved, including reverse charging, proper charging where packet size is converted in passing from one net to another, and volume discounts. A related issue, involving much controversy, is the question of whether volume-sensitive pricing should be applied to leased international lines.

2) Reverse Charging: Both Telenet and Datapac, in their respective national networks, serve large numbers of terminal users, who dial into local network access nodes and establish network connections to remote host computers. The network does not require the user to identify himself, but "reverses the charges"; that is, it bills the host computer for network usage from dial-in terminals. (Generally, such hosts require the user to "log-in" in order to establish the terminal user's identity, and may, if they wish, rebill the communication charge to the user.)

Reverse charging practices have not yet become accepted for international packet service, although it is expected that as institutional arrangements evolve, reverse charging will become available, as "collect calls" have become available between most countries in international telephony.

3) Packet Size Conversion: At this writing, international charges are based upon either kilopackets or kilocharacters. Where charges are based upon kilocharacters they are unaffected by the "packet size" used in each of the national networks. However, if charges are based upon kilopackets, it is necessary to establish a policy for international charges where the interconnected networks utilize different packet sizes (assuming that a single end-to-end charge for kilopackets applies). The concept of a standard international unit of measure for packet communications traffic, called the "segment," is evolving, as a result of the efforts of a CCITT study group concerned with international tariffs, to handle this situation. A segment contains up to 64 characters of user data.

4) Volume Discounts: Volume discounts for telecommunication services are commonplace within the U.S. for both conventional telephone and private line facilities as well as for public packet networks. They are generally not offered by PTT's in their respective countries nor over international circuits, however, because the PTT's believe such discounts discriminate against the small user in favor of the large user. Therefore, international packet services are highly attractive to the small to medium-scale user, but may not be competitive in instances where a firm has sufficient traffic in one location to install a dedicated concentrator. In order for international packet network services to handle the majority of international data traffic, it will be necessary to incorporate volume discounts, or very low basic packet charges in international tariffs.

5) Volume-Sensitive Line Pricing: Organizations operating wholly private packet switching networks in the U.S. or in other countries generally obtain leased line facilities at fixed monthly rates. However, organizations operating multiuser packet or message switching networks may be charged a surcharge, above normal leased line rates, which is based upon traffic volume and is intended to recover the revenues which the PTT would have otherwise obtained, had the network customers utilized public PTT services exclusively. Such volume-based surcharges are being applied in the case of the SWIFT and SITA networks mentioned earlier. Volume sensitive charges have also been proposed for all use of international leased lines by several telecommunications administrations.

These administrations allege that many firms leasing international circuits are using the circuits to carry third party message traffic. Data processing service companies, in particular, have been accused of such activities. Given the difficulty of distinguishing data processing from communications, as discussed above, this allegation is extremely difficult to validate.

Telecommunications authorities have, in the past, permitted special-purpose multiuser networks where it was clear that the public telecommunications administration could not immediately furnish the service required (e.g., the SITA and the SWIFT networks) although in these cases they also imposed a surcharge.

In the future, with the advent of public packet networks in many countries, the telecommunications administrations may seek to establish tariffs which discourage use of leased lines and encourage use of "shared" public packet network facilities. Volume-sensitive charging for use of leased line facilities is one mechanism for accomplishing this. Needless to say, large user organizations are opposed to such pricing. At this writing, several administrations have taken a small first step toward volume-sensitive pricing by introducing speed-sensitive pricing on international voice-grade leased lines—e.g., a base rate for 2400 bit/s, and a surcharge for 4800 and 9600 bit/s over the same channel facility.

Volume-sensitive pricing for leased lines has, at this writing, been suggested only in the context of international circuits. Some user organizations have suggested that the underlying purpose of such pricing is to serve as a new type of tariff barrier preventing primarily U.S. data processing firms from competing in foreign markets and encouraging the establishment of local data processing centers in each country. This political and economic undercurrent, barriers to international computer-communications, is discussed again below under a somewhat different banner—"protecting individual and national information and data."

6) Uniform International Charges: One objective in establishing rates and tariffs for public communication services is simplicity and uniformity. Often, however, different jurisdictions institute different charging structures and levels.

A significant first step in the direction of establishing uniform international rates for packet communication services is an agreement between the nine European PTT's developing Euronet to establish a common European-wide tariff. Proposed tariff rates are approximately $2.00 per kilosegment, and approximately $2.50 per hour for the use of dial-in network access ports. Charges for service between Europe and the U.S. will be several times higher, but still approximately an order of magnitude lower than rates for traditional international circuit switching services (dial telephone and telex).

Since packet network costs are relatively independent of distance, it is not inconceivable that a single traffic charge would apply in the future, for all traffic from a given point to any destination in the world.

C. International Standards

In the past several years, there has been remarkable progress in the establishment of standards for public packet networks, primarily in the area of customer-to-network and network-to-network interfaces. The CCITT has adopted, for example, the X.25 protocol as the international standard for connecting computers and other programmable devices to public packet networks. This important standard may have an even greater impact upon the data communications field than the vendor-proprietary protocols and standards embedded in IBM's System Network Architecture and DEC's DECNET architecture.

The CCITT has also adopted several related standards—X.3, X.28, and X.29—for interfacing interactive keyboard terminals to public packet networks.

Work is also underway on the establishment of standards for network-to-network interconnection (X.75), and for additional computer-to-network interface protocols. In addition, various parties are also developing a number of new higher level protocols, for computer-to-computer communication in general, and for specific applications such as file transfer and message communications. Considerable attention will continue to be devoted to the development and standardization of communication protocols in the years ahead.

It is not possible in this paper to cover adequately the standards relating to packet communications. These standards are mentioned here only because 1) they reflect the remarkable degree of international cooperation that has occurred—and is still occurring—in this field, and because 2) they indicate that a unified, worldwide public packet switching system is evolving very rapidly. Further discussion of packet network standards and protocols may be found in "Trends in Public Packet Switched Data Networks," by P. T. F. Kelly; in "Tutorial on Protocols," by L. Pouzin and H. Zimmermann; and in "Issues in Packet Network Interconnection," by V. G. Cerf and P. T. Kirstein, all in this issue of this PROCEEDINGS.

D. International Data Flow

While rapid progress is being made in packet communications, the attendant benefits, in facilitating international information flow, may be thwarted to some extent by a wave of "data protection laws" already passed or under consideration in several countries. These are being advocated in the name of "individual privacy," "national sovereignty," or "protection of infant industries." Typical of such laws is Sweden's requirement that a Data Inspection Board must approve any export of computer files or personal data.

While such laws may have valid purposes, the implications of such regulations do not, in this author's view, appear to have been fully considered. Some industry observers have suggested that the underlying purpose of the data protection laws is to force the U.S. and other multinational firms to set up local subsidiaries for handling data within national borders [5]. However, even the U.S. Congress is considering passage of data protection laws. One such proposal, House of Representatives bill number 1984, (the number is purely coincidental) provides that any information system that includes personal data cannot transfer information outside the U.S. without the individual's consent or a treaty guaranteeing the foreign government or organization receiving the data will protect it as the U.S. would.

These new regulations, which restrict the free flow of information across international boundaries, are similar to commodities tariffs, which restrict the free flow of goods in international commerce. Such commodities tariffs increase the overall costs of goods, reduce the overall standard of living, and impinge upon individual freedom. Limitations on the flow of information in today's "information society" may have equally serious effects—reducing productivity, limiting technology transfer, and fragmenting organizations into uneconomic units.

Typical of our times, there appears to be a growing, superstitious overreaction to computerized information systems containing data on individuals, firms, and national economies. Thus we will probably see the passage of many regulations that will hinder the free flow of information among nations and will, as a result, retard the creation and development of multinational institutional fabrics which bind society together.

Perhaps later, when the impact of these regulations is recognized, international efforts to "eliminate barriers to information flow" will be established, similar to the international efforts now aimed at eliminating the import duties which restrict the free flow of goods in international commerce.

IV. CONCLUSIONS AND PROGNOSIS

Forecasts are invariably wrong in some respect, and forecasts involving political issues are based upon particularly weak foundations. Nevertheless, some general comments on the manner in which the policy issues discussed here will be handled in the future is worthwhile. The reader is reminded that these comments reflect only the opinions of the author, and are not based upon empirical studies, quantitative analyses, or other scientific forecasting techniques.

In the U.S., it is anticipated that public packet network services will remain regulated for the foreseeable future. Present statutory policy, as interpreted by the FCC, mandates regulation of such carriers. Revision of the Communications Act of 1934, as amended, is likely to occur during the next several years, but the arguments presented in favor of total deregulation of value-added carriers in the U.S. are not persuasive, in the author's view, particularly with the introduction of ATT's packet network services.

It is also expected that the FCC will continue to foster an environment of competition among packet network carriers, authorizing new carriers and attempting to ensure fair competition among new and established carriers. Due to the large capital investment required to develop and operate public packet networks, the number of packet carriers is likely to remain relatively small for the foreseeable future.

It is expected that the FCC will fail to establish a clear demarcation line between data processing and data communication services, and will revert to *ad hoc* application of the primary purpose test which, despite its shortcomings, appears to have worked satisfactorily over the past several years.

It is expected that ATT's ACS, when introduced will stretch the definition of data communications to the limit. Extensive FCC deliberations are likely to be involved, with both computer equipment manufacturers and computer service companies arguing that ATT's ACS performs functions ordinarily considered to be data processing.

After considerable policy debate, it is expected that ATT will be authorized to proceed with ACS under a set of constraints intended both to ensure fair competition among carriers and to restrict ATT to primarily data communications functions. ATT may be required, for example, to maintain separate books and accounts for ACS, or to set up a separate subsidiary offering ACS, in order to ensure compensatory pricing.

Eventually, in the opinion of this writer, all customer premises data terminal equipment will be furnished on a competitive, nontariffed basis. This approach will require ATT to either divest its Teletype Corporation subsidiary or to be freed from the constraints of the consent decree, which limits it to the provision of tariffed services.

In Canada, it is expected that packet network services will continue to be regulated, and that the primary suppliers will be TCTS and CN/CPT. While the entry of value-added carriers is possible in Canada, the limited size of the market and the capital requirements are likely to deter new entrants.

In regard to the distinction between data processing and data communications, the regulatory authorities in Canada are likely to follow the basic position adopted by the FCC, but to allow the Canadian carriers greater discretion in furnishing services that contain a high component of data processing functions.

Outside of North America, public packet network services are likely to be furnished on a monopoly basis by the PTT's. Moreover, sharing of private line circuits will generally not be permitted except in rare cases, and in those instances, surcharges above normal leased line rates will apply.

Although carriers outside of North America may have the option of furnishing data processing services, as a practical matter they will not represent major suppliers of data processing services in the long run since they bring few competitive advantages into the data processing marketplace. Industry pressures are likely to force PTT's to liberalize any restrictions on the use of PTT facilities for data communications by computer service companies. For example, those countries (such as Mexico), which prohibit the use of dial telephone circuits for data transmission, are likely to rescind such restrictions.

International packet network services will almost certainly remain regulated for the foreseeable future. Moreover, the sharing of international circuits will not be permitted to the same extent that sharing of leased line facilities is permitted in the U.S.

For the immediate future, interconnection arrangements

between U.S. packet networks and overseas networks will continue to utilize the facilities of international record carriers. Similarly, in the near term, the U.S. gateway regulations restricting the international record carriers to acceptance and delivery of traffic in gateway cities only are likely to remain in force. However, in the long run, it is the opinion of this writer that two major changes in overseas communications will occur: 1) the domestic gateway restrictions on the IRCs and 2) the PTT requirement that overseas interconnections between U.S. packet networks and non-U.S. packet networks transit IRC facilities will both disappear.

These changes are significant and, if they occur, will require legislative revisions in the U.S. and substantial shifts in policy outside the U.S. The impetus for direct overseas interconnection and abolition of the U.S. gateway rule may be a future attempt by ATT to interconnect its ACS service to overseas points. Since ATT generally interconnects directly with foreign carriers, and since no gateway restrictions apply to such arrangements, ATT is likely to seek similar arrangements for international packet network interconnection.

Tariffs for public packet network services will consist of the same structural elements in each country. National tariffs, and in some cases regional international tariffs, will generally be distance-independent. (The Euronet tariff charges are likely to be distance-independent, for example.) Most international tariffs will vary according to the particular pair of countries or regions.

Charges for international packet network services will generally be one to two orders of magnitude lower than alternative circuit switched telephone and telex services, particularly between overseas points. In the long run, these international packet network charges are likely to decline even further toward the domestic packet network rates.

Volume-sensitive rates for international leased communication circuits, proposed in 1977, are not likely to be adopted for many years, due to strong U.S. industry and government opposition.

The recent and timely establishment of international standards for public packet network interfaces by the CCITT is greatly accelerating the deployment of such networks, and will stimulate rapid growth of packet network services in the future. In this writer's view, the establishment of these standards represents a step forward fully as significant as the development of packet switching technology itself, and these standards will influence the architecture of computer software systems to a substantial degree in the future.

Finally, international packet network services are likely to aggravate the problem of trans-national data flow by significantly reducing the costs of international data communications and thereby stimulating increased traffic. Those countries concerned that transborder data flows may compromise individual privacy and national economic interests have therefore adopted or are considering legislation to restrict transnational data communication. In this writer's view, such legislation is premature, and generally aimed at problems that have not yet occurred and are not likely to occur to any considerable degree. In many cases, transnational data legislation is politically attractive, however, and statutes are likely to be passed in many countries. Eventually, these statutes will be recognized for what they are—namely, impediments to the free flow of information among nations and therefore impediments to economic progress.

Packet communications technology is most cost-effective today for the transmission of computerized data. However, the transmission of speech through packet communications systems is also technically feasible today and is likely to become increasingly attractive in the future. (See L. G. Roberts, "The Evolution of Packet Switching," in this issue.)

Packet switched voice communications systems will initially be most cost-effective in overseas communications where very high transmission costs can be minimized through dynamic allocation of transmission bandwidth. ATT has, since 1960, employed on certain overseas channels a technique called Time Assignment Speech Interpolation (TASI) which, like packet switching, dynamically allocates transmission capacity. TASI dynamically allocates capacity by using the listening time and other natural pauses in one speaker's conversation to transmit portions of other conversations. Packet technology, in conjunction with continually improving techniques for the digital encoding of speech, is likely to provide greater cost effectiveness than TASI in the future, particularly since packetized speech can be efficiently interspersed with computerized data in the same packet network.

The policy implications of packetized speech are severalfold. First, the prospect of packetized voice communication services diminishes the possibility that regulatory and congressional bodies in the U.S. will deregulate all packet communication carriers. Second, questions concerning the monopoly status of switched telephone services may eventually have to be faced. And third, international issues concerning interconnection policies, standardization, and tariff rates for packetized speech may become important in the foreseeable future.

Acknowledgments

The author wishes to acknowledge the assistance of P. M. Walker, B. D. Wessler, B. Sincavage, and C. Kinsey in the preparation of this paper. All opinions expressed, however, are the sole responsibility of the author.

The author would also like to express his appreciation to Dr. L. G. Roberts, whose creativity and leadership have profoundly changed the course of communications development. His contributions, both technical and organizational, to the evolution of packet communications systems, cannot be overestimated.

References

[1] U.S. Court of Appeals (2nd Circuit), p. 6546, Jan. 26, 1978.
[2] Response of IBM FCC Docket No. 20828 (The Second Computer Inquiry), p. 42, June 6, 1977.
[3] Communications Act of 1934, 47 U.S.C., paragraph 153 (a)–(b).
[4] Overseas channels between carriers consist of two imaginary "half circuits," each carrier furnishing its respective half circuit from its network to the midpoint of the channel.
[5] Brendan McShane, quoted in "The Coming Information War," by John Eger, *The Washington Post*, Jan. 15, 1978, pp. B1–B2.

Bibliography

[6] R. P. Blanc and I. W. Cotton, Eds., *Computer Networking*. New York: IEEE Press, 1976.
[7] W. W. Chu, Ed., *Advances in Computer Communication*, 2nd ed. Dedham, MA: Artech House, 1976.
[8] D. W. Davies and D. L. A. Barber, *Communication Networks for Computers*, New York: Wiley.
[9] S. L. Mathison and P. M. Walker, *Computers and Telecommunications: Issues in Public Policy*. Englewood Cliffs, NJ: Prentice-Hall, 1970.

[10] S. L. Mathison and P. M. Walker, "Regulatory and Economic Issues in Computer Communications," *Proc. IEEE*, vol. 60, pp. 1254–1272, Nov. 1972.

[11] P. M. Walker, "Regulatory developments in data communications—The past five years," *Proc. 1972 Spring Joint Computer Conf.*, AFIPS Press, 1972.

[12] *IEEE Trans. Communications*—Special Issue on Computer Communications, Jan. 1977.

REGULATORY RULINGS

[13] *Specialized Common Carriers*, 29 FCC2d 870(1971).

[14] *First Computer Inquiry*, Notice of Inquiry, 7 FCC2d 11(1966); Tentative Decision, 28 FCC2d 291(1970); Final Decision, 28 FCC2d 267(1971).

[15] *Second Computer Inquiry*, Notice of Inquiry, 61 FCC2d 103(1976). Supplemental Notice of Inquiry, 64 FCC2d 771(1977).

[16] *Resale and Shared Use of Common Carrier Services*, Notice of Inquiry, 47 FCC2d 644(1974); Report and Order, 60 FCC2d 261(1976); Memorandum Opinion and Order on Reconsideration, 62 FCC2d 588(1977).

[17] *Re Packet Communications Inc.*, Memorandum, Opinion, Order and Certificate, 43 FCC2d 922(1973).

[18] *Re Telenet Communications Corp.*, Memorandum, Opinion, Order and Certificate, 46 FCC2d 680(1974).

[19] *Re Tymnet, Inc.*, Memorandum, Opinion, and Certificate, 65 FCC2d 247(1976).

The Concept of Packet Communications

This section:

- traces the history and evolution of packet network operation

- provides simplified descriptions of packet network functions and operations

- shows how design choices made in early packet network implementations were improved through experience, invention, and technological enhancement.

This short section includes just two papers; however, if it were possible to capture the essence of packet communications in one hour's worth of reading, these two papers would do it. The authors, Dr. Lawrence G. Roberts and Dr. Leonard Kleinrock, are, without dispute, the recognized worldwide experts on packet switching. But, more importantly, they jointly and individually are largely responsible for the conception, development, and commercialization of packet technology.

THE "INVENTION" OF PACKET SWITCHING

Dr. Roberts' paper traces his own work, the work of J.C.R. Licklider, and the work of Donald Davies from the early 1960s to late in the decade when the first packet network implementations occurred in the United States and in the United Kingdom. Experimental systems were built in both countries, with the U.S. version, known as the ARPANET, becoming the principal prototype for most packet networks. The Advanced Research Projects Agency (ARPA) of the U.S. Department of Defense initiated and funded the development of the ARPANET to meet the desperate need for linking the dozens of very large computer installations that were supporting fundamental research throughout the country. With no existing commercial capability available, ARPA, primarily through initial contracts with Bolt, Beranek and Newman, Inc. and later contracts with a large number of companies and universities, developed the needed capabilities. The ARPANET served two very important functions: As desired, it successfully provided the inter-computer communications resource that was needed, and, equally important, it provided a nationwide, live test bed upon which entirely new concepts of communications could be developed, tested, and improved.

COMMERCIALIZATION OF PACKET SWITCHING

The transition of packet communications from an experimental, government resource to a commercially viable set of services was largely attributable again to the efforts of Dr. Roberts. Leaving federal government service, he joined with Bolt, Beranek and Newman in late 1972 to form Telenet Communications Corporation (now GTE-Telenet Communications Corporation). It was at this time that the concept of value-added services and the processing role of the packet switches were recognized, making such a service feasible and legal in the then highly structured and regulated public communications environment of the United States. Commercial development of packet technology by Telenet, other U.S. entries such as Tymnet, and counterpart organizations in Canada, United Kingdom, and France led to rapid development of international standards for packet network interface, and culminated in the development of the CCITT X.25 standard in 1976 and its refinement in 1980. The X.25 standard and its related standards for terminal interface, inter-network operation, international numbering, and so forth have permitted worldwide packet networks to evolve in a commercially practical and profitable environment in an incredibly short period of time.

THE FUTURE OF PACKET SYSTEMS

Dr. Roberts' paper concludes with a section devoted to his view of some of the important future developments in packet technology. The paper deals with applications of satellite technology to packet networks, packet radio (foreshadowing local area networking technology) and integration of voice services into packet networks. Each of these areas will be treated in detail in later sections of this volume (Sections V, VII and VIII, respectively).

TECHNICAL EVOLUTION OF PACKET NETWORKS

While Dr. Roberts was largely responsible for the initial development and commercialization of packet switching, the author of the second paper in this section, Dr. Leonard Kleinrock, was largely responsible for the studies, measurements, and analyses that promoted our technical understanding of network operation and the practical improvements in the operation of such networks. As a professor of Computer Science at the University of California at Los Angeles (UCLA), he had contractual responsibility, under research contracts with ARPA, for the Network Measurement Center for the ARPANET. Working together with a number of graduate students, many of whom have since become renowned experts in their own right, Dr. Kleinrock made major contributions to the protocols, operations, and management of large-scale data networks, including not only packet switches but also satellite channels, radio networks, and local area networking facilities.

Dr. Kleinrock's paper traces the early history of the ARPANET and provides a somewhat more technical description than we had in the earlier paper. He then shows how design problems, inherent in the early implementations, were discovered, analyzed, and corrected within the ARPANET structure as well as in other network implementations. Much of the introductory information presented in this paper was first discovered through the ARPANET Network Measurement Center and became the fundamental basis for many of the later papers, particularly in the development of network protocols described in Sections III and IV.

The Evolution of Packet Switching

LAWRENCE G. ROBERTS, MEMBER, IEEE

Invited Paper

Abstract—Over the past decade data communications has been revolutionized by a radically new technology called packet switching. In 1968 virtually all interactive data communication networks were circuit switched, the same as the telephone network. Circuit switching networks preallocate transmission bandwidth for an entire call or session. However, since interactive data traffic occurs in short bursts 90 percent or more of this bandwidth is wasted. Thus, as digital electronics became inexpensive enough, it became dramatically more cost-effective to completely redesign communications networks, introducing the concept of packet switching where the transmission bandwidth is dynamically allocated, permitting many users to share the same transmission line previously required for one user. Packet switching has been so successful, not only in improving the economics of data communications but in enhancing reliability and functional flexibility as well, that in 1978 virtually all new data networks being built throughout the world are based on packet switching. An open question at this time is how long will it take for voice communications to be revolutionized as well by packet switching technology. In order to better understand both the past and future evolution of this fast moving technology, this paper examines in detail the history and trends of packet switching.

THERE HAVE ALWAYS been two fundamental and competing approaches to communications: pre-allocation and dynamic-allocation of transmission bandwidth. The telephone, telex, and TWX networks are circuit-switched systems, where a fixed bandwidth is preallocated for the duration of a call. Most radio usage also involves preallocation of the spectrum, either permanently or for single call. On the other hand, message, telegraph, and mail systems have historically operated by dynamically allocating bandwidths or space after a message is received, one link at a time, never attempting to schedule bandwidth over the whole source-to-destination path. Before the advent of computers, dynamic-allocation systems were necessarily limited to nonreal time communications, since many manual sorting and routing decisions were required along the path of each message. However, the rapid advances in computer technology over the last two decades have not only removed this limitation but have even made feasible dynamic-allocation communications systems that are superior to preallocation systems in connect time, reliability, economy and flexibility. This new communications technology, called "packet switching," divides the input flow of information into small segments, or packets, of data which move through the network in a manner similar to the handling of mail but at immensely higher speeds. Although the first packet-switching network was developed and tested less than ten years ago, packet systems already offer substantial economic and performance advantages over conventional systems. This has resulted in rapid worldwide acceptance of packet switching for low-speed interactive data communications networks, both public and private.

A question remains, however. Will dynamic-allocation techniques like packet switching generally replace circuit switching and other preallocation techniques for high-speed data and voice communication? The history of packet switching so far indicates that further applications are inevitable. The following examination of the primary technological and economic trade-offs involved in the growth of the packet switching communications industry should help to trace the development of the technology toward these further applications.

EARLY HISTORY

Packet switching technology was not really an invention, but a reapplication of the basic dynamic-allocation techniques used for over a century by the mail, telegraph, and torn paper tape switching systems. A packet switched network only allocates bandwidth when a block of data is ready to be sent, and only enough for that one block to travel over one network link at a time. Depending on the nature of the data traffic being transferred, the packet-switching approach is 3–100 times more efficient than preallocation techniques in reducing the wastage of available transmission bandwidth resources. To do this, packet systems require both processing power and buffer storage resources at each switch in the network for each packet sent. The resulting economic tradeoff is simple: if lines are cheap, use circuit switching; if computing is cheap, use packet switching. Although today this seems obvious, before packet switching had been demonstrated technically and proven economical, the tradeoff was never recognized, let along analyzed.

In the early 1960's, preallocation was so clearly the proven and accepted technique that no communications engineer ever seriously considered reverting to what was considered an obsolete technique, dynamic-allocation. Such techniques had been proven both uneconomic and unresponsive 20–80 years previously, so why reconsider them? The very fact that no great technological breakthrough was required to implement packet switching was another factor weighing against its acceptance by the engineering community.

What was required was the total reevaluation of the performance and economics of dynamic-allocation systems, and their application to an entirely different task. Thus, it remained for outsiders to the communications industry, computer professionals, to develop packet switching in response to a problem for which they needed a better answer: communicating data to and from computers.

THE PIONEERS

Rand

The first published description of what we now call packet switching was an 11-volume analysis, *On Distributed Communications*, prepared by Paul Baran of the Rand Corporation in August 1964 [1]. This study was conducted for the Air Force, and it proposed a fully distributed packet switching system to provide for all military communications, data, and voice. The study also included a totally digital microwave system and integrated encryption capability. The Air Force's primary goal was to produce a totally survivable system that contained no critical central components. Not only was this goal achieved by Rand's proposed packet switching system, but even the economics projected were superior, for both voice and data transmissions. Unfortunately, the Air Force took no follow-up action, and the report sat largely ignored for many years until packet switching was rediscovered and applied by others.

ARPA I

Also in the 1962–1964 period, the Advanced Research Projects Agency (ARPA), under the direction of J. C. R. Licklider (currently at M.I.T.), sponsored and substantially furthered the development of time-sharing computer systems. One of Licklider's strong interests was to link these time-shared computers together through a widespread computer network. Although no actual work was done on the communication system at that time, the discussions and interest Licklider spawned had an important motivating impact on the initiators of the two first actual network projects: Donald Davies and me.

As previously indicated, the development of packet switching was primarily the result of identifying the need for a radically new communications system. Licklider's strong interest in and perception of the importance of the problem encouraged many people in the computer field to consider it seriously for the first time. It was in good part due to this influence that I decided, in November 1964, that computer networks were an important problem for which a new communications system was required [2]. Evidently Donald Davies of the National Physical Laboratory (NPL) in the United Kingdom had been seized by the same conviction, partially as a result of a seminar he sponsored in autumn 1965, which I attended with many M.I.T. Project MAC people. Thus, the interest in creating a new communications system grew out of the development of time-sharing and Licklider's special interest in the 1964–1965 period.

National Physical Laboratory

Almost immediately after the 1965 meeting, Donald Davies conceived of the details of a store-and-forward packet switching system, and in a June 1966 description of his proposal coined the term "packet" to describe the 128-byte blocks being moved around inside the network. Davies circulated his proposed network design throughout the U.K. in late 1965 and 1966. It was only after this distribution that he discovered Paul Baran's 1964 report.

The first open publication of the NPL proposal was in October 1967 at the A.C.M. Symposium in Gatlinburg, TN [3]. In nearly all respects, Davies' original proposal, developed in late 1965, was similar to the actual networks being built today. His cost analysis showed strong economic advantages for the packet approach, and by all rights, the proposal should have led quickly to a U.K. project. However, the communications world was hard to convince, and for several years, nothing happened in the U.K. on the development of a multinode packet switching network.

Donald Davies was able, however, to initiate a local network with a single packet switch at the NPL. By 1973 this local network was providing an important distribution service within the laboratory [4], [5]. This project, plus the strong conviction and continued effort by those at NPL (Davies, Barber, Scantlebury, Wilkinson, and Bartlett), did gradually have an effect on the U.K. and much of Europe.

ARPA II

In January 1967, I joined ARPA and assumed the management of the computer research programs under its sponsorship. ARPA was sponsoring computer research at leading universities and research labs in the U.S. These projects and their computers provided an ideal environment for a pilot network project; consequently, during 1967 the ARPANET was planned to link these computers together.

The plan was published in June 1967. The design consisted of a packet switching network, using minicomputers at each computer site as the packet switches and interfacing device, interconnected by leased lines. By coincidence, the first published document on the ARPANET was also presented at the A.C.M. Symposium in Gatlinburg, TN, in October 1967 [6] along with the NPL plan. The major differences between the designs were the porposed net line speeds, with NPL suggesting 1.5 Mbit/s lines. The resulting discussions were one factor leading to the ARPANET using 50-kbit/s lines, rather than the lower speed lines previously planned [7].

During 1968, a request for proposal was let for the ARPANET packet switching equipment and the operation of the network. The RFP was awarded to Bolt Beranek and Newman, Inc. (BBN) in Cambridge, MA, in January 1969. Significant aspects of the network's internal operation, such as routing, flow control, software design, and network control were developed by a BBN team consisting of Frank Heart, Robert Kahn, Severo Ornstein, William Crowther, and David Walden [8], [9], [10]. By December 1969. four nodes of the net had been installed and were operating effectively. The network was expanded rapidly thereafter to support 23-host computers by April 1971, 62 hosts by June 1974, and 111 hosts by March 1977.

The ARPANET utilized minicomputers at every node to be served by the network, interconnected in a fully distributed fashion by 50-kbit/s leased lines. Each minicomputer took blocks of data from the computers and terminals connected to it, subdivided them into 128 byte packets, and added a header specifying destination and source addresses; then, based on a dynamically updated routing table, the minicomputer sent the packet over whichever free line was currently the fastest route toward the destination. Upon receiving a packet, the next minicomputer would acknowledge it and repeat the routing process independently. Thus, one important characteristic of the ARPANET was its completely distributed, dynamic routing algorithm on a packet-by-packet basis, based on a continuous evaluation within the network of the least-delay paths, considering both line availability and queue lengths.

The technical and operational success of the ARPANET quickly demonstrated to a generally skeptical world that dynamic-allocation techniques—and packet switching in particular—could be organized to provide an efficient and highly responsive interactive data communications facility. Fears that packets would loop forever and that very large buffer pools would be required were quickly allayed. Since the ARPANET was a public project connecting many major universities and research institutions, the implementation and performance details were widely published [11], [12], [13],

[14], [15]. The work of Leonard Kleinrock and associates at UCLA on the theory and measurement of the ARPANET has been of particular importance in providing a firm theoretical and practical understanding of the performance of packet networks. (See "Principles and Lessons in Packet Communications" by L. Kleinrock, in this issue pp. 1320–1329.)

Packet switching was first demonstrated publicly at the first International Conference on Computer Communications (ICCC) in Washington, DC, in October 1972. Robert Kahn of BBN organized the demonstration. He installed a complete ARPANET node at the conference hotel, with about 40 active terminals permitting access to dozens of computers all over the U.S. This public demonstration was for many, if not most, of the ICCC attendees proof that packet switching really worked. It was difficult for many experienced professionals at that time to accept the fact that a collection of computers, wide-band circuits, and minicomputer switching nodes—pieces of equipment totaling well over a hundred—could all function together reliably, but the ARPANET demonstration lasted for three days and clearly displayed its reliable operation in public. The network provided ultra-reliable service to thousands of attendees during the entire length of the conference.

The widespread publicity the ARPANET demonstration earned contributed greatly to the task of introducing modern dynamic-allocation technology to a preallocation trained world. However, during the same period in the early 1970's many other dynamic-allocation techniques were being developed and tested in private networks throughout the world. Hopefully, the extensive publications on the ARPANET have not *oversold* the particular variety of packet switching used in this first major network experiment.

SITA

The Societe Internationale de Telecommunications Aeronautiques (SITA) provides telecommunications for the international air carriers. In 1969 SITA began updating its design by replacing the major nodes of its message switching network with High Level Network nodes interconnected with voice-grade lines—organized to act like a packet switching network. Incoming messages are subdivided into 240-byte packets and are stored and forwarded along predetermined routes to the destination. Prestored distributed tables provide for alternate routes in the event of line failures [16].

TYMNET

Also in 1969, a time sharing service bureau, Tymshare Corporation, started installing a network based on minicomputers to connect asynchronous timesharing terminals to its central computers. The network switches, which are interconnected by voice-grade lines, store and forward from node to node data characters for up to 20 calls packaged in 66-byte blocks. The data is repackaged at each node into new blocks for the next hop. Routing is not distributed, but is accomplished by a central supervisor on a call-by-call basis [17].

CYCLADES/CIGALE

In France the interest in packet switching networks grew quickly during the early 1970's. In 1973 the first hosts were connected to the CYCLADES network, which links several major computing centers throughout France. The name CYCLADES refers to both the communications subnet and the host computers. The communications subnetwork, called CIGALE, only moves disconnected packets and delivers them in whatever order they arrive without any knowledge or concept of messages, connections or flow control. Called a "datagram" packet facility, this concept has been widely promoted by Louis Pouzin, the designer and organizer of CYCLADES. Since a major part of the organization and control of the network is imbedded in the CYCLADES computers, the subnetwork, CIGALE, is not sufficient by itself. In fact, Pouzin himself speaks of the network as "including" portions of the host computers. The packet assembly and disassembly, sequence numbering, flow control, and virtual connection processing are all done by the host. The CYCLADES structure provides a good testbed for trying out various protocols, as was its intent; but it requires a more cooperative and coordinated set of hosts than is likely to exist in a public environment [18].

RCP

Another packet network experiment was started in France at about the same time by the French PTT Administration. This network, called RCP (Reseau a Commutation par Paquets), first became operational in 1974. By this time the French PTT had already decided to build the public packet network, TRANSPAC, and RCP was utilized primarily as testbed for TRANSPAC. The design of RCP, directed by Remi Despres, differed sharply from that of the other contemporary French network, CYCLADES. Despres' design was organized around the concept of virtual connections rather than datagrams. RCP's character as a prototype public network may have been a strong factor in this difference, since a virtual circuit service is more directly marketable, not requiring substantial modifications to customers' host computers. In any case, the RCP design pioneered the incorporation of individually flow-controlled virtual circuits into the basic packet switching network organization [19].

EIN

Organized in 1971 and originally known as the COST II Project and later as the European Informatics Network (EIN) is a multination-funded European research network. The project director is Derek Barber of NPL, one of the original investigators of packet switching in the U.K. Given freedom from the red tape of multinational funding, this project would have been one of the earliest pace-setters in packet networks in the world. As it happened, however, EIN was not operational until 1976 [20], [21].

Public Data Networks

The early packet networks were all private networks built to demonstrate the technology and to serve a restricted population of users. Besides those early networks already mentioned, which were the most public projects, many private corporations and service bureaus built their own private networks. Generally these private networks did not make provision for host computers at more than one location, and thus their organization usually developed into a star network.

All these networks were the result of a basic economic transition, which occurred in 1969 [22] when the cost of dynamic-allocation switching fell below that of transmission lines. This change made it economically advantageous to build a network of some kind rather than to continue to use direct lines or the circuit switched telephone network for interactive data communications. Universal regulatory conditions in all countries restricted "common carriage" to the government or government-approved carriers, and thereby led to the development of many private networks instead of a competitive market of public networks.

However, the extensive private network activity in the early 1970's encouraged some of these public carriers to make plans for building their own packet networks, although all public networks and plans for future networks were based on preallocation techniques until about 1973. Many plans to provide public data service arose; some were even under way, like the German EDS system; but all were based on circuit switching until that time. The shift in economics in the late 1960's that made packet switching more cost-effective, instigated more rapid change in communications technology than had ever before occurred.

The established carriers and PTT's took their time reacting to this new technology. The United Kingdom was the first country to announce a public packet network through the British Post Office's planned Experimental Packet Switched Service (EPSS) [23]. Donald Davies' 1966 briefings with the BPO on packet switching clearly played a strong role in the U.K.'s early commitment to this new technology.

In the United States the dominant carrier, American Telephone and Telegraph (AT&T), evidenced even less interest in packet switching than many of the PTT's in other countries. AT&T and its research organization, Bell Laboratories, have never to my knowledge published any research on packet switching. ARPA approached AT&T in the early 1970's to see if AT&T would be interested in taking over the ARPANET and offering a public packet switched service, but AT&T declined. However, the Federal Communications Commission (FCC), which regulates all communications carriers in the U.S., was in the process of opening up portions of the communications market to competition. Bolt Beranek and Newman, the primary contractor for the ARPANET, felt strongly that a public packet switched data communications was needed. The FCC's new policies encouraged competition, so BBN formed Telenet Communications Corporation in late 1972. In October 1973 Telenet filed its request with the FCC for approval to become a carrier and to construct a public packet switched network; six months later the FCC approved Telenet's request. (See "Legal, Commercial, and International Aspects of Packet Communications," by S. L. Mathison in this issue, pp. 1527–1539.)

In France in November 1973, the French PTT announced its plans to build TRANSPAC, a major domestic packet network patterned after RCP [24]. The next year, in October 1974, the Trans-Canada Telephone System announced DATAPAC, a public packet network in Canada [25]. Also during this period, the Nippon Telegraph and Telephone Corporation announced its plans to build a public packet switched data network in Japan [26].

Thus, only four years after the building of the first experimental networks, the concept of data communications networks began to move into the public arena. Still, the networks were only planned and had yet to be built; most PTT's and carriers adopted a wait and watch attitude toward these first public networks.

INTERNATIONAL STANDARDIZATION AND ACCEPTANCE

CCITT X.25

With five independent public packet networks under construction in the 1974–1975 period, there was strong incentive

for the nations to agree on a standard user interface to the networks so that host computers would not have unique interfacing jobs in each country. Unlike most standards activities, where there is almost no incentive to compromise and agree, carriers in separate countries can only benefit from the adoption of a standard since it facilitates network interconnection and permits easier user attachment. To this end the parties concerned undertook a major effort, to agree on the host–network interface during 1975. The result was an agreed protocol, CCITT Recommendation X.25, adopted in March 1976.

The X.25 protocol provides for the interleaving of data blocks for up to 4095 virtual circuits (VC's) on a single full-duplex leased line interface to the network, including all procedures for call setup and disconnection. A significant feature of this interface, from the carriers' point of view, is the inclusion of independent flow control on each VC; the flow control enables the network (and the user) to protect itself from congestion and overflow under all circumstances without having to slow down or stop more than one call at a time. In networks like the ARPANET and CYCLADES which do not have this capability, the network must depend on the host (or other networks in interconnect cases) to insure that no user submits more data to the network than the network can handle or deliver. The only defense the network has without individual VC flow control is to shut off the entire host (or internet) interface. This, of course, can be disastrous to the other users communicating with the offending host or network.

Another critical aspect of X.25, not present in the proposals for a datagram interface, is that X.25 defines interface standards for both the host-to-network block transfer and the control of individual VC's. In datagram networks the VC interface is situated in the host computer; there can be, therefore, no network-enforced standard for labeling, sequencing, and flow controlling VC's. These networks are in the author's opinion, not salable as a public service since they must offer individual terminal interfaces, as well as host interfaces, to provide complete host–terminal communications; to sell these interfaces requires knowing how to interface to one VC as well as to a host.

The March 1976 agreement on X.25 and on virtual circuits as the agreed technique for public packet networks marked the beginning of the second phase of packet switching: large interconnected public service networks. In the two years since X.25 was adopted, many additional standards have been agreed on as well, all patterned around X.25. For example, X.28 has been adopted as the standard asynchronous terminal interface; X.29, a protocol used with X.25 to specify the packetizing rules for the terminal handler, will be the host control protocol. More recently X.75, the standard protocol for connecting international networks has been defined.

Public Data Network Services

Capitalizing on BBN's ARPANET experience, TELENET introduced the first public packet network service in August 1975. Initially TELENET consisted of seven multiply interconnected nodes. By April 1978 the network had grown to 187 network nodes which used 79 packet switches to provide 156 U.S. cities with local dial service to 180 host computers across the country, with interconnections to 14 other countries. Originally TELENET supported a virtual connection host interface similar to X.25. However, shortly after the specification was adopted, X.25 was introduced into TELENET as the preferred host interface protocol.

In early 1977 both EPSS in the U.K. and DATAPAC in Canada were declared operational. Also, in the U.S., TYMNET was approved as a carrier and began supplying public data services. EPSS, having been designed long before X.25 was specified, is not X.25 compatible, but the U.K. intends to provide X.25 based packet service within the next year.

DATAPAC was X.25 based from the start of commercial service since the development was held until X.25 was approved. Using X.25 lines, DATAPAC and TELENET were interconnected in early 1978. This connection demonstrated the ease of international network linking, once a common standard had been established.

In France, TRANSPAC is due to become operational later this year (1978); in Japan, the NTT packet network, DX-2, should become operational in 1978 or 1979. A semipublic network, EURONET, sponsored by nine European Common Market Countries, is due to become operational in late 1978 or 1979. Many other European countries, like Germany and Belgium, are making plans for public packet networks to start in 1979. These networks are all X.25 based and therefore should be similar and compatible.

Datagrams versus VC's

As part of the continuing evolution of packet switching, controversial issues are sure to arise. Even with universal adoption of X.25 and the virtual circuit approach by public networks throughout the world, there is currently a vocal group of users requesting a datagram standard. The two major benefits claimed for datagrams are reliability and efficiency for transaction-type applications.

Reliability: It is claimed that datagrams provide more reliable access to a host when two or more access lines are used, since any packet could take either route if a line were to fail. This reflects a true deficiency in X.25 as currently defined—the absence of a reconnect facility on the call request packet. If, when a call is initiated, a code number for the call is placed in the call request packet, the X.25 network (or host) can save the code number. If the line over which the call is placed fails, the network simply places a new call request, marked as a reconnect, over another line and supplies the original code number to insure reconnection to the correct VC. Since packets on each VC are sequence numbered, this reconnection can be accomplished with no data loss and usually just as quickly as rerouting of the packets in a datagram interface. If the network uses VC's internally, the same reconnect capability is used to insure against connection failures.

Cost: It is often assumed that datagrams would be cheaper for networks to provide than packets on VC's. However, the cost of memory and switching have fallen by a factor of 30 compared to transmission costs over the last nine years resulting in the overhead of datagram headers becoming a major cost factor. A datagram packet or end-to-end acknowledgment requires about 25 bytes of packet header in addition to the actual data (0–128 bytes) whereas only 8 bytes of overhead are required for similar packets on a virtual circuit. In the unique case of a single packet call, the overheads are the same. For all longer calls, datagram overhead adds 13–94 percent to the cost of all transmission costs, both long haul and local. Originally, this increase in transmission cost was more than offset by increased switch costs but with modern microprocessor switches very little of the increase is offset. Thus, with this radical shift in economics, datagram packets are now more expensive than VC packets.

CONTINUING TECHNOLOGICAL CHANGE

A decade ago computers had barely become inexpensive enough to make packet switching economically feasible; computers were still slow and small, forcing implementers to invent all sorts of techniques to save buffers and minimize CPU time. Computer technology has progressed to the point where microcomputer systems have now been especially designed for packet switching, and there is no shortage of memory or CPU power. This development has been partially responsible for the shift from datagrams to virtual connections and has also eliminated buffer allocation techniques (which cost transmission bandwidth to save memory). The modularity and computational power of today's microprocessors has made it economical and practical to provide protocol conversion from X.25 to any existing terminal protocol, polled or not.

As a result of these improvements, packet networks are rapidly becoming universal translators, connecting everything to everything else and supplying the speed, code, and protocol conversions wherever necessary. As this trend continues, it is almost certain that the techniques in use today will have to be continually changed to respond to the changing economics and usage patterns.

For example, one major change that will be required in the next few years is an increase in the backbone trunk speed from 56 kbit/s to 1.544 Mbit/s (the speed of "T1" digital trunks). Both Paul Baran and Donald Davies in their original papers anticipated the use of T1 trunks, but present traffic demand has not yet justified their use. As the traffic does justify T1 trunks, many aspects of network design will change by a corresponding order of magnitude. Packet networks have always incorporated a delay in the 100–200 ms range. This delay has so strongly affected both the system design and the choice of applications that it is hard to remember which decisions depended on this delay factor.

With T1 carrier trunks, the transit delay in the net will drop to around 10 ms plus propagation time requiring a complete reexamination of network topology and processor design issues. The number of outstanding packets on a 2500 mile trunk will increase from around 3 to 75, requiring extended numbering, and perhaps, new acknowledgment techniques. The user will be most strongly affected by a 10–30 ms net delay; his whole strategy of job organization may change.

Of course, there will be a significant price decrease accompanying this change. This, combined with the short delay, will make many new applications attractive; remote job entry (RJE) and bulk data transfer applications through public packet networks will probably be economically and technically feasible, even before T1 trunks are introduced; but if not before, certainly afterwards, when the packet price reflects the new trunk speed. Dynamic-allocation permits savings over pre-allocation by a factor of four in line costs for RJE, and by a factor of two for bulk data transfer. As the switch cost continues to fall far more rapidly than the line cost, dynamic-allocation techniques will be used for RJE, batch transfer and even voice applications.

FUTURE

Packet Satellite

One change which will clearly occur in packet networks in the next decade is the incorporation of broadcast satellite facilities. ARPA has sponsored extensive research into packet satellite techniques and, over the past few years, has tested these techniques between the U.S. and England. (See "General Purpose Packet Satellite Networks" by I. M. Jacobs, R. Binder, and E. Hoversten in this issue, pp. 1448–1467.) Fundamentally, a satellite provides a broadcast media which, if properly used, can provide considerable gains in the full statistical utilization of the satellite's capacity. Using ARPA's techniques, a single wideband channel (1.5 Mbit/s–60 Mbit/s) on a satellite provides an extremely economical way to interconnect high bandwidth nodes within a packet network.

With the current cost of ground stations ($150K–$300K), it appears to be marginally economic to install separate private ground stations at major nodes of a domestic packet network rather than to lease portions of commercial ground stations and trunk the data to the packet network nodes. However, either way, the cost of ground station facilities are such that the use of satellites only becomes economic compared to land lines when the aggregate data flow exceeds about 100 packets/s (100 kbit/s) to and from a node or city. Furthermore, satellite transmission has an inherent one-way delay of 270 ms; therefore, the packet traffic must logically be divided between two priority groups—interactive and batch. Only batch traffic can presently be considered for satellites, since the 270 ms delay is unacceptable for interactive applications, at least if any other options are available, even at a somewhat higher price. With current economics, the long-haul land line facilities only add about $0.50/hr to the price of interactive data calls, which is far too little a cost to encourage the acceptance of slower service. Therefore, interactive service will almost always require ground line facilities in addition to satellite facilities at all network nodes.

This introduces another factor that limits the potential satellite traffic: land lines can easily carry 10–25 percent batch traffic at a lower priority, using a dual queue, without any significant increase in cost. Further, if ground lines are required and satellite facilities are optional, the full cost for the satellite capability, must be compared with the incremental cost of simply expanding the land line facilities. All these factors considered, it is probable that satellites will be used by public data networks within the next five years for transmissions between major nodal points, but that ground facilities will be used exclusively for transmissions between smaller nodes.

Packet Radio

Since local distribution is by far the most expensive portion of a communications network, ground radio techniques are of considerable interest to the extent that they can replace wire for local distribution. Packet radio is another area where ARPA has been sponsoring research in applying dynamic-allocation techniques. The basic concept in packet radio is to share one wide bandwidth channel among many stations, each of which only transmits in short bursts when it has real data to send. (See "Advances in Packet Radio Technology" by R. E. Kahn *et al.*, in this issue, pp. 1468–1496.) This technique appears to be extremely promising for both fixed and mobile local distribution, once the cost of the transceivers has been reduced by, perhaps, a factor of ten. Considering the historical trend of the cost of electronics, this should take about five years; from that point onward packet radio should become increasingly competitive with wire, cable, and even light fibers for low to moderate volume local distribution requirements.

One important consequence would be the use of a simple packet radio system inside buildings to permit wireless communication for all sorts of devices. Clearly, as electronic devices multiply throughout the home and office, low-power packet radio would permit all these devices to communicate among themselves and with similar devices throughout the world via a master station tied into a public data network.

Voice

The economic advantage of dynamic-allocation over pre-allocation will soon become so fundamental and clear in all areas of communications, including voice, that it is not hard to project the same radical transition of technology will occur in voice communications as has occurred in data communications.

Digitized voice, no matter what the digitization rate, can be compressed by a factor of three or more by packet switching since in normal conversation each speaker is only speaking one third of the time. Since interactive data traffic typically can be compressed by a factor of 15, voice clearly benefits far less from packet switching than interactive data. This is the reason why packet switching was first applied to data communications. However, modern electronics is quickly eliminating any cost difference between packet switches and circuit switches, and thus packet switching can clearly provide a factor of three cost reduction in the transmission costs associated with switched voice service.

Probably there will be many proposals, and even systems built, using some form of dynamic-allocation other than packet switching during the period of transition. The most likely variant design would be a packetized voice system that does not utilize sum checks or flow control. Of course, this would be just a packet switch with those options disabled. If the similarity to present packet switching were not recognized, the packetized voice system might be built without providing these essential capabilities and would be useless for data traffic. However, the obvious solution would be an integrated packet switching network that provides both voice and data services.

On further consideration, it becomes apparent that the flow control feature of packet switching networks can provide a substantial cost reduction for voice systems. Flow control feedback, applied to the voice digitizers decreases their output rate when the network line becomes momentarily overloaded; as a result, peak channel capacity required by users can be significantly reduced.

In short, packet switching seems ideally suited to both voice and data transmissions. The transition to packet switching for the public data network has taken a decade, and still is not complete; many PTT's and carriers have not accepted its viability. Given the huge fixed investment in voice equipment in place today, the transition to voice switching may be considerably slower and more difficult. There is no way, however, to stop it from happening.

REFERENCES

[1] P. Baran *et al.*, "On distributed communications, vols. I–XI," RAND Corporation Research Documents, Aug. 1964.
[2] T. Marill and L. G. Roberts, "Toward a cooperative network of time-shared computers," *Proc. FJCC* pp. 425–431, 1966.
[3] D. W. Davies, K. A. Bartlett, R. A. Scantlebury, and P. T. Wilkinson, "A digital communications network for computers giving rapid response at remote terminals," ACM Symp. Operating Systems Problems, Oct. 1967.
[4] R. A. Scantlebury, P. T. Wilkinson, and K. A. Bartlett, "The de-

sign of a message switching centre for a digital communication network," in *Proc. IFIP Congress 1968, vol. 2—Hardware Applications,"* pp. 723-733.

[5] R. A. Scantlebury, "A model for the local area of a data communication network—Objectives and hardware organization," ACM Symp. Data Communication, Pine Mountain, Oct. 1969.

[6] L. G. Roberts, "Multiple computer networks and intercomputer communication," ACM Symp. Operating System Principles, Oct. 1967.

[7] L. G. Roberts and B. D. Wessler, "Computer network development to achieve resource sharing," *Proc. SJCC 1970,* pp. 543-549.

[8] F. E. Heart, R. E. Kahn, S. M. Ornstein, W. R. Crowther, and D. C. Walden, "The interface message processor for the ARPA computer network," in *AFIPS Conf. Proc. 36,* pp. 551-567, June 1970.

[9] R. E. Kahn and W. R. Crowther, "Flow control in a resource-sharing computer network," in Proc. Second ACM/IEEE Symp. Problems, Palo Alto, CA, pp. 108-116, Oct. 1971.

[10] S. M. Ornstein, F. E. Heart, W. R. Crowther, S. B. Russell, H. K. Rising, and A. Michel, "The terminal IMP for the ARPA computer network," in *AFIPS Conf. Proc. 40,* pp. 243-254, June 1972.

[11] H. Frank, R. E. Kahn, and L. Kleinrock, "Computer communications network design—Experience with theory and practice," in *AFIPS Conf. Proc.,* vol. 40, pp. 255-270, June 1972.

[12] L. Kleinrock, "Performance models and measurement of the ARPA computer network," in *Proc. Int. Symp. Design and Application of Interactive Computer Systems,* Brunel University, Uxbridge, England, May 1972.

[13] L. Kleinrock and W. Naylor, "On measured behavior of the ARPA network," in *AFIPS Conf. Proc.,* vol. 43, NCC, Chicago, IL, pp. 767-780, May 1974.

[14] L. Kleinrock and H. Opderbeck, "Throughput in the ARPANET—

[15] Protocols and measurement," in *Proc. 4th Data Communications Symp.,* Quebec City, Canada, pp. 6-1–6-11, Oct. 1975.

[15] H. Frank and W. Chou, "Topological optimization of computer networks," *Proc. IEEE,* vol. 60, no. 11, pp. 1385-1397, Nov. 1972.

[16] G. J. Chretien, W. M. Konig, J. H. Rech, "The SITA network," in *Computer Communication Networks,* R. L. Grimsdale, F. F. Kuo, Eds. Noordhoff: NATO Advanced Study Institute Series, pp. 373-396, 1975.

[17] L. R. Tymes, "TYMNET—A terminal oriented communication network," *Proc. AFIPS, SJCC,* pp. 211-216, May 1971.

[18] L. Pouzin, "Presentation and major design aspects of the CYCLADES network," in *Third Data Communications Symp.,* Tampa, FL, pp. 80-85, Nov. 1973.

[19] R. F. Despres, "A packet network with graceful saturated operation," in *Proc. ICCC,* Washington, DC, pp. 345-351, Oct, 1972.

[20] D. L. A. Barber, "The European computer network project,"*Proc. ICCC,* Washington, DC, pp. 192-200, Oct. 1972.

[21] ——, "A European informatics network: Achievement and prospects," *Proc. ICCC,* Toronto, Canada, pp. 44-50, Aug. 1976.

[22] L. G. Roberts, "Data by the packet," *IEEE Spectrum,* Feb. 1974.

[23] R. C. Belton, J. R. Thomas, "The UKPO packet switching experiment," *ISS Munich,* 1974.

[24] A. Danet, R. Depres, A. LesRest, G. Pichon, S. Ritzenthaler, "The French public packet switching service: The TRANSPAC network," *Proc. ICCC,* Toronto, Canada, pp. 251-260, Aug. 1976.

[25] W. W. Clipsham, F. E. Glave, M. L. Narraway, "DATAPAC network overview," *Proc. ICCC,* Toronto, Canada, pp. 131-136, Aug. 1976.

[26] R. Nakamura, F. Ishino, M. Sasaoka, M. Nakamura, "Some design aspects of a public packet switched network," *Proc ICCC,* Toronto, Canada, pp. 317-322, Aug. 1976.

Principles and Lessons in Packet Communications

LEONARD KLEINROCK, FELLOW, IEEE

Invited Paper

Abstract—After nearly a decade of experience, we reflect on the principles and lessons which have emerged in the field of packet communications. We begin by identifying the need for efficient resource sharing and review the original and recurring difficulties we had in achieving this goal in packet networks. We then discuss various lessons learned in the areas of: deadlocks; degradations; distributed control; broadcast channels; and hierarchical design. The principles which we discuss have to do with: the efficiency of large system; the switching computer; network constraints; distributed control; flow control; stale protocols; and designers not yet experienced in packet communications. Throughout the paper, we identify various open issues which remain to be solved in packet communications.

I. Introduction

WHAT IS IT WE now know about packet communications that we did not know a decade ago? What made the problem difficult, and why were the solutions not immediately apparent to us in the late 1960's? Whereas the answers to these questions may entice the system designer (indeed, I, for one, delight in such investigations), why should the network user care a whit? To most users (and, alas, to many designers), communications is simply a nuisance and they would just as soon ignore those problems and get on with the "real" issues of information processing!

In this paper (and in conjunction with the other papers in this Special Issue of the PROCEEDINGS), we hope to answer some of these questions. We will identify the need for resource sharing, explain why the problem of efficient resource sharing is hard, and why it must be understood, review some of the lessons we learned (mostly from the three ARPA packet networks), and then, finally, state some principles which have evolved from the study and extensive use of packet communications.

II. Resource Sharing

A privately owned automobile is usually a waste of money! Perhaps 90 percent of the time it is idly parked and not in use. However, its "convenience" is so seductive that few can resist the temptation to own one. When the price of such a poorly utilized device is astronomically high, we do refuse the temptation (how many of us own private jet aircraft?). On the other hand, when the cost is extremely low, we are obliged to own such resources (we all own idle pencils).

An information processing system consists of many poorly utilized resources. (A resource is simply a device which can perform work for us at a finite rate.) For example, in an information processing system, there is the CPU, the main memory, the disk, the data channels, the terminals, the printer, etc.

One of the major system advances of the early 1960's was the development of multiaccess time-sharing systems in which computer system resources were made available to a large population of users, each of whom had relatively small demands (i.e., the ratio of their peak demands to their average demands was very high) but who collectively presented a total demand profile which was relatively smooth and of medium to high utilization. This was an example of the advantages to be gained through the smoothing effect of a large population (i.e., the "law of large numbers") [1]. The need for resource sharing is present in many many systems (e.g., the shared use of public jet aircraft among a large population of users).

In computer communication systems we have a great need for sharing expensive resources among a collection of high peak-to-average (i.e., "bursty") users [1]. In Fig.1 we display the structure of a computer network in which we can identify three kinds of resources:

1) the terminals directly available to the user and the communications resources required to connect those terminals to their HOST computers or directly into the network (via TIPS in the ARPANET, for example—this is an expensive portion of the system and it is generally difficult to employ extensive resource sharing here due to the relative sparseness of the data sources;

2) the HOST machines themselves which provide the information processing services—here multiaccess time sharing provides the mechanism for efficient resource sharing;

3) the communications subnetwork, consisting of communication trunks and software switches, whose function it is to provide the data communication service for the exchange of data and control among the other devices.

The HOST machines in 2) above contain hardware and software resources (in the form of application programs and data files) whose sharing comes under the topic of time sharing; we dwell no further on these resources. Rather, we shall focus attention on those portions of the computer communications system where packet communications has had an important impact. Perhaps the most visible component is that of the communications subnetwork listed in item 3) above. Here packet communications first demonstrated its enormous efficiencies in the form of the ARPANET in the early 1970's (the decade of computer networks). The communication resources to be shared in this case are *storage capacity* at the nodal switches (these switches are called IMP's in the ARPANET), *processing capacity* in the nodal switches, and *communications capacity* of the trunks connecting these switches. Packet switching in this environment has proven to be a major technological breakthrough in providing cost effective data communications among information processing systems. Deep in the backbone of such packet-switched networks there is a need for long-haul high-capacity inexpensive communications, and it is here where we

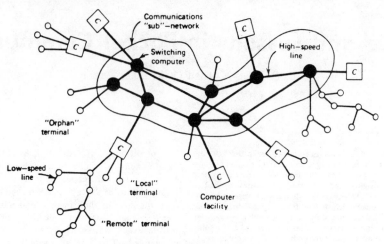

Fig. 1. The structure of a computer-communication network.

see the second application of packet communications for resource sharing in the form of satellite packet switching; elsewhere in this PROCEEDINGS [2] you will find a description of the SATNET, an ARPA-sponsored research network connected to the ARPANET. The third application may be found in the local access problem stated in item 1) above which also lends itself to the use of packet switching to provide efficient communications resource sharing; this takes the form of the use of a multiaccess broadcast channel in a local environment, commonly known as ground radio packet switching. Here too, ARPA has sponsored an experimental system, and its description may be found in this PROCEEDINGS [3]. The common element running through all these systems is the application of the smoothing effect of a large population to provide efficient resource sharing, an exquisite example of which is provided by packet communications.

Having described the environment and the resources of interest, let us now discuss the performance measures which permit us to evaluate the effort of resource sharing in a quantitative way. Indeed, there are basically four measures that both the system designer and network user apply in evaluating the service provided in a communications environment. These are *throughput, response time, reliability*, and *cost*. Before packet networks came into existence, the obvious solution for providing communications between two devices was to lease or dial a line between the two. In such a case the user was able to associate precise quantitative values to the four measures listed above. On the other hand when one attaches to a packet network, the user cannot get deterministic answers to the same quantities as he has in the past. He must accept probabilistic statements regarding throughput, delay, and reliability (and alas, sometimes even cost). Moreover the quantities so prescribed can seldom be measured in a straightforward fashion. This is the state of affairs to which we have evolved today! It is to the credit of those who developed packet communications in the last decade that the system design was carefully studied and well-analyzed *prior* to and during the systems implemention; this certainly has not, in general, been the history in the information processing industry.

III. WHY THE PROBLEM IS HARD

Back in 1967, when the concept of the ARPANET was first taking form, we found ourselves entering the uncharted terrain

of packet switching. Let us trace our initial confusion regarding that project briefly. Certainly, there existed at that time some communication networks, but they were mostly highly specialized networks with restricted goals. In the early 1960's Paul Baran had described some of the properties of data networks in a series of Rand Corporation papers [4]. He focused on the routing procedures and on the survivability of distributed communication systems in a hostile environment, but did not concentrate on the need for resource sharing in its form as we now understand it; indeed, the concept of a software switch was not present in his work. In 1968 Donald Davies at the National Physical Laboratories in England was beginning to write about packet-switched networks [5]; at around the same time, Larry Roberts at ARPA pursued the use of packet switching in an experimental nationwide network [6]. For a more complete history of the evolution of packet communications, see [7].

In the initial conception of a packet network, we identified some problems and looked to the technical literature for solutions to these problems. For example, how should one design the topology of a network, and how should one select the bandwidth for the various channels in such a network, and in what fashion should one route the data through the network, and what rules of procedure should two communicating processes adopt, and how much storage did one need at the multiplexing nodes of the network? These and many other questions confronted us. Indeed the general problem was how to achieve efficient resource sharing among a set of incompatible devices in a geographically distributed environment where access to these devices arose from asynchronous processes in a highly bursty fashion. Moreover, not only was the demand process bursty, it was also highly unpredictable in the sense that the instants when the demands arose and the duration of the demands were unknown ahead of time. Fortunately we were unaware of the enormity of the problems facing us and so we plunged ahead enthusiastically and with naive optimism. The remainder of this section describes why the problem was difficult, and in following sections we describe the lessons we learned and the principles we established in the development of packet communications. Our efforts have been well rewarded and the technology of packet communications has come of age and has proven itself to be a cost-effective technology.

We quickly found that many of our old techniques could not be directly applied to packet network design and that new techniques had to be developed; these new techniques turned out to be of great generality and have led to principles and to understanding which are sure to benefit us for many years to come. One of our first tasks was to develop tools which would allow us to analyze the performance of a given network. This involved evaluating the delay-throughput profile for networks. Basically, this is a queueing problem in a network environment. It had earlier been recognized [8] that the probabilistic complexities one encounters in computer networks are extremely difficult. One of the simplest analytical questions involving the solution of two nodes in tandem was first posed at that time in 1964 and has only been satisfactorily answered (in the queueing theoretic sense) within the past year [9]; this, note, is for the simplest problem. Indeed we have come to realize that an exact solution for the delay-throughput profile is probably hopeless in a complex network environment. Fortunately suitable approximations [1], [8] have been developed which permit one to predict the performance of given networks with a high degree of accuracy. More than that, these approximate tools allow us to expose and understand the phenomenological behavior of networks.

The astute reader will observe that the resource sharing problem stated above sounds very much like the problem faced in the design of time sharing systems. Surely, with time sharing, we are faced with the problem of sharing resources among asynchronous processes which behave in a bursty fashion. The major difference between the two problems, however, is that our problem exists in a *geographically distributed* environment which requires expensive communication resources in the communications and coordination functions. The implications here are strong. For example, when communication is cheap, then wide-band communications can be obtained with extremely small delays; such is the case, for example, within the resources of a local operating system connected together by a data bus. In a regional or nationwide network subject to the relatively expensive cost of telecommunications, we find that typical bandwidths are many orders of magnitude less than that in a local time-sharing environment, and the delays are many orders of magnitude greater. Furthermore, the control of these processes in the time sharing environment can be very tightly coupled if desired or left loosely coupled if there is sufficient reason; in the network environment we typically find our processes are very loosely coupled due to the difficulty of tightening the control between them (indeed, the inherent delay due to the finite speed of light is a fundamental limitation on the tight coupling of remote processes). The overhead in the time sharing system is variable and may be very high with poor system design (for example, thrashing) but may be made small with clever design. In the network environment, for a variety of reasons, we find that the overhead due to packet headers, control information and resource allocation tends to be relatively high. These comparisons are summarized in Table I.

Not only do we have extremes in communications cost between these two systems, we also have a significant difference in the reliability of that communications. Indeed, in the local time shared system, the process-to-process communication is usually assumed to be reliable and therefore the acknowledgment procedure (if any exists) is simple and tends to be invoked only under exceptional circumstances. On the other hand, in the distributed computer network environment, communications reliability is not assumed, and, therefore, an elaborate

TABLE I
ASYNCHRONOUS PROCESS-TO-PROCESS COMMUNICATION AND CONTROL
COST COMPARISON BETWEEN LOCAL PROCESSES IN A TIME-SHARED
SYSTEM AND DISTRIBUTED PROCESSES IN A NETWORK

	Multiaccess Time-Shared Systems	Geographically Distributed Computer Networks
Typical bandwidth	megabytes/sec	kilobits/sec
Typical communications delay	fractions of a micro-second	tens to hundreds of milliseconds
Process-process coupling	tight to loose	typically loose
Overhead due to system control	variable (typically low)	variable (typically high)

acknowledgment procedure (often including timeouts and other defaults) is usually invoked. We are here reminded of the "two-army" problem. This is the problem where two blue armies are each poised on opposite hills preparing to attack a single red army in the valley. The red army can conquer either of the blue armies separately but will fall to defeat if both blue armies attack simultaneously. The blue armies communicate with each other over an unreliable communications system. The problem is to coordinate the two blue armies so that they will attack simultaneously. Let us assume that Blue Army 1 (B1) sends a message to Blue Army 2 (B2) indicating that they should jointly attack at noon the next day. Clearly B1 must await a positive acknowledgment from B2 that the command was properly received; were B1 to attack without such an acknowledgment, then the possibility exists that B2 did not receive the message correctly, in which case B1 is subject to the certain annihilation of his army. If B2 properly receives the command and acknowledges it, then he must await an acknowledgment of the acknowledgment from B1, for if B1 did not receive his acknowledgment then we know B1 will not attack and B2's attack will then be doomed to failure. The argument continues in a circular fashion where acks of acks of acks . . . are continually transmitted with no action ever being taken; the two blue armies can *never* get perfectly synchronized with certainty using this unreliable communications system.

We see therefore that the new culprit in resource sharing in a distributed environment is the fact that the resources are distributed and cannot easily be assigned to the demands. Indeed, in previous resource allocation problems, which often come under the name of scheduling algorithms, we made a big assumption, namely, that the scheduling information could be passed around for free. That is, the competing demands could organize themselves into a cooperating queue at no price. Unfortunately, in the distributed environment, the cost of organizing demands into a cooperating queue may be very large, and in one fashion or another nature will make you pay a price [10].

The problem of resource sharing in a distributed environment manifests itself in the routing control and flow control problems. The problem of flow control is to regulate the rate at which data crosses the boundary of the communications subnetwork (both into and out of the network). The problem of routing control is to efficiently transport the data (which has been admitted by the flow control procedure) across the network to its destination. In 1967 we were aware of the sophistication needed in the routing procedure, but were relatively ignorant of the need for effective flow control (see below). These two problems are difficult because we are dealing with a control procedure in a distributed environment subject to random delays in passing that control information around in order to control the random demands. The purpose of both procedures is to efficiently use the network resources (IMP storage, IMP processing capacity, and communications capacity). In achieving this goal one must attempt to control congestion, route data around busy or defective portions of the network, and in general must adaptively assign capacity to the data traffic flow in an efficient, dynamic way. These are hard control problems and represent a class which has not been adequately studied up to and including the present time. We have come to learn that distributed control is a sophisticated problem. Below we return to the issues involved in distributed resource allocation and sharing. For the moment let us introduce some of the other sources of complexity in packet communications.

In any distributed communications system design one is faced with a *topological design problem*. The problem basically is, given a set of constraints to meet, find that topological design structure which meets these constraints at least cost. The field of network flow theory addresses itself to such problems and the salient feature of this theory is that most of its problems cannot be solved! To exhaustively search over all possible topological designs for a given problem is certainly not a solution since the number of possible alternatives to consider can easily exceed the number of atoms in the universe even for relatively small problems. (For example, if at some stage you must consider all permutations of 20 objects, then a computer would take more than 75 000 years to process all 20! cases even if it could examine one million cases per second.) Rather, a solution consists of elegant search procedures which are computationally efficient and which find the optimal topology for the given problem. Very few problems in network flow theory yield to such efficient algorithms. Rather, one gets around the combinatorial complexity naturally inherent in these problems by accepting suboptimal solutions. (Beware! A suboptimal solution to a problem is simply the result of a procedure which examines a subset of possible solutions and picks the best of those examined–this suboptimal solution may or may not be close to the optimal.) The trick here is to find efficient heuristic search procedures which come close to the optimal rapidly. Over the past decade, efficient procedures have been developed in many cases and new procedures are constantly being investigated for the topological design problem.

Another source of difficulty in the resource sharing problem is in defining the appropriate *performance measure*. For example, what is the capacity of a network? It is well-known in network flow theory that one can easily calculate the capacity (i.e., the throughput) between any two pairs of points. What is not straightforward is to evaluate the total data-carrying capacity of a network where throughput is measured in terms of messages successfully received at their destination. The difficulty comes about because the capacity of the network strongly depends upon the traffic matrix one assumes for the data flowing through that network. For example, if the traffic matrix were such that traffic passes only between immediate neighbors in the topological structure and in an amount equal to the capacity of the line connecting those two, then the network capacity would approach a value equal to the sum of all channel capacities in the network. This is clearly an upper bound to the capacity for any other traffic matrix. Since in general we do not know the traffic matrix for a network to be designed for future use, how is one to evaluate that capacity?

Yet another source of difficulty in the network problem is that of *interfacing* terminals and HOSTS to networks as well as one network to another network. For example, there is the general issue of *virtual circuit service* as opposed to *datagram service*. A virtual circuit service presents to the user a communications environment in which all functions appear as if she were directly connected between the two points communicating (i.e., an orderly and controlled flow is maintained whereby data is guaranteed to be correct upon delivery to the destination, comes out of the network in the same order in which it came in, and a flow control may be applied to that transmission to maintain efficient use of network resources and of terminal-HOST resources). A datagram service ensures none of these things and simply sends blocks of data (packets) across the network, not guaranteeing correct delivery (or delivery at all), not guaranteeing orderly flow in terms of sequencing (packets may arrive at the destination out of order), and not enforcing any flow control procedure at the process-to-process level. Which of those two services is desirable has become an issue of international proportions discussed elsewhere in these proceedings [7], [11]. Futhermore, the interconnection of two networks presents an enormous and rich variety of difficult problems. For example, should one apply flow control at the boundary of each network in a tandem chain of networks or should one apply flow control from the source HOST to the destination HOST across many networks simultaneously? If we consider the interconnection of networks with different packet sizes, we have the general problem of fragmentation whereby long packets get fragmented into smaller packets when crossing network boundaries. A variety of other very important issues arise in internetting (see [27] for a more detailed discussion of internetting).

Many of the problems we have just presented come about due to the distributed nature of our message sources and system resources. The problems created by distributed sources are very clearly seen in the environment of geographically dispersed message sources which communicate with each other over a common broadcast channel; in this case, access to the capacity of the channel must be carefully coordinated.

Thus in answering the question, "Why is the problem hard?" we have found the following sources of difficulty:
1) the analytic problems from queueing theory and the probabilistic complexity resulting therefrom;
2) the topological design problems from network flow theory and the combinatorial complexity resulting therefrom;
3) the price for coordinating and sharing resources in a geographically distributed environment (the new culprit) leading to problems of resource allocation, routing control, flow control, and general process-to-process communication problems.

IV. LESSONS LEARNED

After a decade of experience with packet communications it is fair that we ask what lessons have we learned and what have we come to know about the needs of the user and the questions he would like to have answered. So far as the user is concerned we shall see as we step through the lessons learned below that he cannot insulate himself completely from the underlying technology of packet communications. Indeed the service he sees is quite different from that which he has with leased lines as mentioned above. Moreover, certain decisions will either be thrust upon him or accepted by him due to the nature of the service offered; if he is unaware of the consequence of setting parameters in that decision-making process then he may seriously degrade the performance of the network due to his ignorance. Let us now list some of the lessons we learned and return to the principles in the following section.

A. Deadlocks

In [1], [12], and [13] we described in detail some of the deadlocks and degradations of which we have become aware. In this section we simply enumerate and sketch the details of the deadlocks. Simply stated, a *deadlock* (also commonly referred to as a *lockup*) is the unpleasant situation in which two (or more) competing demands have each been assigned a subset of their necessary resources; neither can proceed until one of them collects some additional resources which currently are assigned to the other and neither demand is willing to release any resource currently assigned to him. Deadlocks are one of the most serious system malfunctions possible, and one must take great care to avoid them or find ways to recover from them. It is ironic that flow control procedures by their very nature present *constraints* on the flow of data (e.g., the requirement for proper sequencing), and if the situation ever arises whereby the constraint cannot be met, then, by definition, the flow will stop, and we will have a deadlock! This is the philosophical reason why flow control procedures have a natural tendency to introduce deadlocks. In this section we briefly discuss four ARPANET deadlocks which have come to be known as: *reassembly* lockup; *store-and-forward deadlock*; *Christmas lockup*; and *piggyback lockup*.

Reassembly lockup, the best known of the ARPANET deadlock conditions (and one which was known to exist in the very early days of the ARPANET implementation), was due to a logical flaw in the original flow-control procedure. In the ARPANET, a string of bits to be passed through the network is broken into "messages" which are at most approximately 8000 bits in length. These messages are themselves broken into packets which are at most approximately 1000 bits in length. A message requiring more than one packet (up to a maximum of eight) is termed a multipacket message and each of these packets traverses the network independently; upon receipt at the destination node, these packets are "reassembled" into their original order and the message itself is recomposed, at which time it is ready for delivery out of the network. (A more complete description of the ARPANET protocols may be found in [1], [13].) Reassembly lockup occurred when partially reassembled messages could not be completely reassembled since the network through which the remaining packets had to traverse was congested, thus preventing these packets from reaching the destination; that is, each of the destination's neighbors had given all of their relay (store-and-forward) buffers to additional packets (from messages other than those being reassembled) heading for that same destination and for which there were no unassigned reassembly buffers available. Thus the destination was surrounded by a barrier of blocked IMP's which themselves could provide no store-and-forward buffers for the needed outstanding packets, and which at the same time were prevented from releasing any of their store-and-forward buffers since the destination itself refused to accept these packets due to a lack of reassembly buffers at the destination. The deadlock was simply that the remaining packets could not reach the destination and complete the reassembly until some store-and-forward buffers became free, and the store-and-forward buffers could not be released until the remaining packets reached the destination.

Store-and-forward deadlock is another example of a lockup that can occur in a packet-switched network if no proper precautions are taken [1], [13]. The case of "direct" store-and-forward lockup is simply a "stand-off." Let us assume that all store-and-forward buffers in some IMP *A* are filled with packets headed toward some destination IMP *C* through a neighboring IMP *B* and that all store-and-forward buffers in IMP *B* are filled with packets headed toward some destination IMP *D* through IMP *A*. Since there is no store-and-forward buffer space available in either IMP *A* or *B*, no packet can be successfully transmitted between these two IMP's and a deadlock situation results. One way to prevent the deadlock is to prohibit these packets in IMP *A* from occupying all those store-and-forward buffers which are needed by the packets coming in from IMP *B* (and vice versa) by the introduction of "buffer classes" as in [14]. This is accomplished by partitioning the buffers in a switch into classes, say, B_0, B_1, $\cdots$, B_k, where k is the longest path length in the network. A packet arriving at a switch from outside the net has access only to class B_0 buffers. When a packet arrives at a switch after having made k hops so far, it has access to class B_0, $\cdots$, B_k buffers, etc. Thus, the closer a packet gets to its final destination, the more access it has, and therefore the speedier its passage through the network. It can be proven [14] that this "buffer class" allocation will prevent direct store-and-forward lockup. "Indirect" store-and-forward lockup can occur when all store-and-forward buffers in a loop of IMP's become filled with packets all of which travel in the same direction (clockwise or counter-clockwise) and none of which are within one hop of their destination. Both store-and-forward lockup conditions are far less likely if, as in the ARPANET, more than one path exists between all pairs of communicating IMP's.

In December 1973, the dormant *Christmas lockup* condition was brought to life. This lockup was exposed by collecting measurement messages at UCLA from all IMP's simultaneously. The Christmas lockup occurred when these measurement messages arrived at the UCLA IMP for which reassembly storage had been allocated but for which no reassembly blocks had been given. (A reassembly block is a piece of storage used in the actual process of reassembling packets back to messages.) These messages had no way to locate their allocated buffers since the pointer to an allocated buffer is part of the reassembly block; as a consequence, allocated buffers could never be used and could never be freed. The difficulty was caused by the system first allocating buffers before it was assured that a reassembly block was available. To avoid this kind of lockup, reassembly blocks are now allocated along with the reassembly buffers for each multipacket message in the ARPANET.

Piggyback lockup is a deadlock condition which was identified by examining the flow control code and has, as far as we know, never occurred. This lockup condition comes about due to an unfortunate combination of intuitively reasonable goals implemented in the flow-control procedure. One of these goals, which we have already mentioned, is to deliver messages to a destination in the same order that the source received them. The other innocent condition has to do with the reservation of reassembly storage space at the destination. In order to make this reservation procedure efficient, it is reasonable that only the first multipacket message of a long file transfer be required to make the reservation. The ARPANET flow control procedure will then maintain that reservation for a given file transfer as long as successive multipacket messages from that file are promptly received in succession at the source IMP. We have now laid the groundwork for piggyback lockup. Assume that there is a maximum of eight reassembly buffers in each IMP; the choice of eight is for simplicity, but the argument works for any value. Let IMP A continually transmit eight-packet messages (from some long file) to some destination IMP B such that all eight reassembly buffers in IMP B are used up by this transmission of multipacket messages. If now, in the stream of eight-packet messages, IMP A sends a single-packet message (not part of the file transfer) to destination IMP B it will generally not be accepted since there is no reassembly buffer space available. The single packet message will therefore be treated as a request for buffer allocation (these requests are the mechanism by which reservations are made). This request will not be serviced before the RFNM (an end-to-end acknowledgment from the destination to source) for the previous multipacket message has been sent. When the RFNM is generated, however, all the free reassembly buffers will immediately be allocated to the next multipacket message in the file transfer for efficient transmission as mentioned above; this allocation is said to be "piggybacked" on the RFNM. In this case, the eight-packet message from IMP A that arrives later at IMP B (and which is stored in the eight buffers) cannot be delivered to its destination HOST because it is out of order. The single-packet message that should be delivered next, however, will never reach the destination IMP since there is no reassembly space available, and, therefore, its requested reservation can never be serviced. Deadlock! A minor modification removes the piggyback lockup.

These various deadlock conditions are usually quite easy to prevent once they are detected and understood. The trick, however, is to expurgate all deadlocks from the control mechanism ahead of time, either by careful programming (a difficult task) or by some automatic checking procedure (which may be as difficult as proving the correctness of programs). Those deadlocks found in the ARPANET have, to the best of our knowledge, been eliminated.

B. Degradations

A degradation is just that, namely, a reduction in the network's level of performance. (Deadlocks are, of course, an extreme form of degradation which is why we discussed them in the separate section above.) For our purposes, we shall measure performance in terms of delay and throughput. In this section we discuss four sources of ARPANET degradation, namely: *looping* in the routing procedure; *gaps* in transmission; *single-packet turbulence*; and *phasing*.

Looping comes about due to independent routing decisions made by separate nodes which cause traffic to return to a previously visited node (or, in a more general definition, causes traffic to make unnecessarily long excursions on the way to its destination). Of course any reasonable adaptive routing procedure will detect these loops (through the build-up of queues and delays perhaps) and will then break the loop and guide the traffic directly on to its destination. However, the occurrence of loops does cause occasional large delays in the traffic flow and in some applications this is quite unacceptable. It is ironic that a remedy which was introduced in the ARPANET to reduce the occurrence of loops, in fact made them worse in

the sense that whereas they occurred less frequently, when they did occur, they persisted for a longer time. Some loop-free algorithms have recently been published [15], [16].

The next degradation we wish to discuss is the occurrence of *gaps* in the message flow. These gaps come about due to a limitation on the number of messages in transit which the network will allow. Assume that between any source and destination, the network will permit n messages in flight at a time. If n messages are in flight, then the next one may not proceed until an end-to-end acknowledgment is returned back at the source for any one of the n outstanding messages. We now observe that if the round-trip delay (i.e., the time required to send a message across the network in the forward direction and to return its acknowledgment in the reverse direction) is greater than the time it takes to feed the n messages into the network, then the source will be blocked awaiting ack's to release further messages. This clearly will introduce gaps in the message flow resulting in a reduced throughput which we might classify as a mild form of degradation.

We now come to the issue of *single packet turbulence* as observed in the ARPANET. We note that "regular" single-packet messages in the early ARPANET were not accepted by the destination if they arrived out of order. Rather, they were then treated as a request for the allocation of one reassembly buffer. Therefore if, in a stream of single-packet messages, packet p arrived out-of-order (say it arrived after packet $p + 3$), then packets $p + 1$, $p + 2$, and $p + 3$ would all be discarded at the destination, and only after packet p arrived would a single packet buffer be allocated to message $p + 1$. This allocation piggybacked on the end-to-end ack for packet p, and when it arrived at the source IMP, it then caused a retransmission of the discarded packet $p + 1$ (which had been stored in the source). Of course any packets arriving at the destination after packet $p + 3$, but before $p + 1$ arrived in order, would themselves be discarded. When packet $p + 1$ finally arrived for the second time at the destination IMP it was then in order and this caused an allocation of a single-packet buffer to packet $p + 2$, etc. The net result was that only one packet would be deliverable to the destination per round-trip time along this path; had no packets been received out-of-order, then we would have been pumping at a rate close to n packets per round trip time (if the maximum number in transit n could fit into the pipe). Observe that once a single packet arrived out-of-order in this stream, then the degradation from n to 1 packets per round-trip time would persist forever until either some supervisory action was taken or until the traffic stream ceased and began again from a fresh start in the future. We refer to this effect as "single-packet turbulence," and it was observed in the ARPANET as described in [17]. The need to handle a continuous stream of traffic (e.g., packetized speech) was recognized some time ago and resulted in the definition of "irregular" packets known as type 3 packets (as contrasted to "regular" type 0 packets); these packets are allowed to be delivered out of order, receive no end-to-end acknowledgment, and are generally handled in a much more relaxed fashion.

The last degradation we discuss is known as "*phasing*." In a typical packet network, more than one resource is often required before a message is allowed to flow across that network. For example, some required resources may be: a message number; storage space at the source; storage space at the destination, etc. Tokens move around the network passing out these resources in some distributed fashion. Phasing is the phenomenon whereby enough free tokens are available in the network to permit message flow, but, the proper mix of tokens is not available simultaneously at the proper location in the net. The delay in gathering these tokens represents a degradation [1], [18].

Fortunately, the degradations here described have been remedied in the ARPANET and in later networks.

C. Lessons of Distributed Control

We have had "lessons" in two areas of distributed control. The first has to do with flow control, and it is simply the observation that flow control procedures are rather difficult to invent and extremely difficult to analyze. The deadlocks and degradations referred to the in previous subsections were principally due to flow control failures (and occasionally routing control failures). To data there is no satisfactory theory or procedure for designing efficient flow control procedures, much less evaluating their performance, proving they contain no deadlocks, and proving that they are correct. Attempts in this direction are currently under way.

An important lesson we have learned with flow control is that a packet communications system offers an opportunity for passing data between two devices of (very) different speeds. We can effectively connect a slow-speed teletype to an enormously high-speed memory channel over a packet network and apply flow control procedures which protect the two devices from each other as well as protecting the net from both. Specifically, we must not drown the teletype with a flood of high-speed input, nor must we "nickel-and-dime" a high performance HOST to death with incessant interrupts, nor must we use the network as a storage medium for megabytes of data. Flow control mechanisms provide the means to accomplish this; the trick is to do it well.

The second area of distributed control in packet communications has to do with the routing control. The ARPANET, and many of the networks which have since based their design on the packet-switching technology which emerged from the ARPANET experiment, employ an adaptive routing procedure with distributed control. In such a procedure, routes for the data traffic are not preassigned but rather are dynamically assigned when they are needed according to the current network status. Control packets (called routing update packets) which describe the state of the network to some degree are passed back and forth between neighboring IMP's in some fashion and current queue lengths and congestion measures are added to these updates by each IMP. The ARPANET employs a periodic update routing procedure whose rate depends upon channel utilization and line speed. The updates passing between IMP's have no priority in competing for the processing capacity of the CPU at the IMP's but do have priority in the queue discipline feeding the output modems between IMP's. An important lesson learned is that giving low priority to the processing of routing updates appears to be advantageous since the processing load on the CPU is rather large and prevents the further dispatching of arriving packets to output queues [19]. Another routing lesson we have learned is that frequent updates cause background congestion in a network which may be intolerably high even in the absence of other data traffic; the update procedure and update rate must be carefully chosen. A number of alternatives to periodic updates have been suggested [1] including such things as aperiodic updating (send updates only when status information has crossed certain thresholds and then send it immediately); and purely local information for routing decisions based on queue lengths within a given node and knowledge of the current topology. Furthermore, unless care is taken, there is a tendency for looping to occur in these distributed control algorithms; looping can be prevented with more sophisticated algorithms [15].

One of the lessons which is now beginning to emerge is that the most important advantage of distributed control adaptive routing is its ability to *automatically* sense configuration changes in the network; these configuration changes may be planned or accidental as for example the result of a line or IMP failure. This is important for two reasons: first because configuration changes do happen often enough so that the requirement for a centralized control evaluating new routing tables based on the current configuration would be an enormously complex task from an administrative point of view; second because it is specifically at times of configuration changes when drastic network action must be taken and only then is the adaptive routing procedure really called upon to do serious work (it is not yet clear to what extent the routing algorithm should adapt to statistical fluctuations in traffic).

Without diminishing the result of these lessons, it is fair to say that the most significant lesson learned regarding routing is that it works at all. Perhaps one of the greatest successes of the ARPANET experiment was to show that a distributed control adaptive routing algorithm would indeed converge on routes which were sufficiently good. The difficulty in proving this lies in the fact that we are dealing with a dynamic situation in a distributed control environment with delays in the feedback paths for control information flow. The empirical proof that things do work has had an important impact on network design; indeed, these distributed algorithms are currently operating successfully in a number of packet networks.

D. Lessons from Broadcast Channels

As mentioned earlier, packet communications has found important applications in the areas of satellite packet broadcasting and in ground radio packet switching. In both environments we have a situation in which a common broadcast channel is available to be shared by a multiplicity of users. Since these users demand access to the channel at unpredictable times, we must introduce some access scheme to coordinate their use of the channel in a way which prevents degradations and mutual interference. In many of the schemes described [10] we have found that "burst" communications provides efficiencies over that of "trickle" transmission. By this we mean that when a data source requires access to the channel, it should be given access to the full capacity of that broadcast channel and not be required to transmit at a slow speed using only a fraction of the available bandwidth (see Section V-A on principles regarding "bigger is better").

In examining the recent literature, we find that a number of access schemes have been invented, analyzed, and published; for a summary of many of these access schemes, see [10]. We observe that these access schemes fall into one of three categories, each with its own cost. The first of these involves random access contention schemes whereby little or no control is exerted on the users in accessing the channel, and this results in the occasional collision of more than one packet; a collision destroys the use of the channel for that transmission. Such schemes as pure ALOHA, slotted ALOHA, and (to a much lesser extent) Carrier Sense Multiple Access fall into this category. At the opposite extreme, we have the static reservation access methods which preassign capacity to users thereby creating "dedicated" as opposed to multiaccess channels. Here the problem is that a bursty user will often not use his preassigned capacity in which case it is wasted. Such schemes as Time Division Multiple Access and Frequency Division Multiple Access fall in this category. Between these two extremes are the dynamic reservation systems which only assign capacity to a user when he has data to send. The loss here is due to the overhead of implementing the demand access. Such schemes as Polling (where one waits to be asked if he has data to send), active reservation schemes (where one asks for capacity when he needs it), and Mini-Slotted Alternating Priority (where a token is passed among numbered users in a prearranged sequence, giving each permission to transmit as he receives a token) all fall in this category. Each of these schemes pays its tribute to nature as shown in Table II.

Unfortunately, at this point in time we are unable to evaluate the minimum price (i.e., a degradation to throughput and/or delay) one must pay for a given distributed multiaccess broadcast environment.

We have found that contention schemes are fundamentally unstable in that they have a tendency to drift into a congested state where the throughput decreases significantly at the same time the delay increases. Fortunately, however, we have been able to design and implement amazingly effective control schemes which stabilize these contention schemes [20]. Another lesson we have learned is that certain tempting ways of mixing two access schemes (e.g., taking a fraction of the traffic and a fraction of the capacity assigned to one access scheme, and using that capacity to handle that traffic using a second access scheme) does not give an improvement in the overall throughput-delay performance [10]. Furthermore we have found that certain capture effects exist in some of the contention schemes (e.g., a group of terminals may temporarily hog the system capacity and thereby "lock out" other groups for extended periods of time) and one must be wary of such phenomena [20].

We have also found that in a ground radio broadcast environment, a few buffers in each packet radio unit appear to be sufficient to handle the storage requirements [21]; this comes about largely due to the fact that our transceivers are half-duplex (i.e., they can either transmit or receive, but not both, at a given time). We can show (see Section IV-E) that dedicated broadcast channels have an inherent advantage over dedicated wire networks in a large (many-user) bursty store-and-forward environment [22]. Moreover, we have investigated the optimal transmission range for ALOHA networks and have found that those broadcast networks can be made quite effective when the traffic is not bursty; indeed this optimal range is chosen so that the channel utilization in the resulting local ALOHA system is $1/2e$ and then those networks need only $\sqrt{e}$ more capacity than the corresponding M/M/1 network [22].

Lastly we point out that perhaps one of the first applications of broadcast radio access schemes will be to implement these access schemes on wire networks (for example, coaxial cables or fiber-optics channels) in a local environment; an example of such an implementation is the Ethernet [23].

E. Hierarchical Design

As N (the number of nodes in a network) grows, the cost of creating the topological design of such a network behaves like N^E where E is typically in the range from 3 to 6. Thus we see that topological design quickly becomes unmanageable. Secondly, we note that as N grows, the size of the routing table in each IMP in the network grows linearly with N and this too

TABLE II
THE COST OF DISTRIBUTED RESOURCES

Access Method	Collisions	Control Overhead	Idle Capacity
Random access contention	Yes	No	No
Dynamic reservation	No	Yes	No
Fixed allocation	No	No	Yes

places an unacceptable burden on the storage requirements within an IMP. In addition, the transmission and processing costs for updating such large tables is prohibitive. Third, even were the design possible, the cost of the lines connecting this huge number of nodes together grows very quickly unless extreme care is taken in that design. In all three cases just mentioned, one finds that the use of hierarchical structures saves the day. In the design case, one may decompose the network into clusters of nodes, superclusters of clusters, etc., designing each level cluster separately. This significantly reduces the number of nodes involved in each subdesign, thereby reducing the overall design cost significantly. For example, a 100-node net would have a cost on the order of $100^4 = 10^8$ (for $E = 4$), whereas a 2-level hierarchical design with 5 clusters would cost on the order of $5(20)^4 + 5(4) \leqslant 10^5$, yielding an improvement of three orders of magnitude! The same approach may be used in routing, where names of distant clusters, rather than names of distant nodes, are used in each routing table, thereby reducing the table length down from N to a number as small as $e \ln N$ giving a significant reduction [24]. For example, a 1000-node net would give a 50-fold reduction in the routing table length when hierarchical routing is used.

In [22] we discuss the overall effect and gain to be had in the use of hierarchically designed wire networks and broadcast networks. For example, we can show that in a bursty dedicated broadcast environment, the use of hierarchical network structures (even with fixed allocation schemes), yields a system cost which is proportional to $[\log M]^2$, where M is the number of users. Comparing this to the case of wire networks where the cost is proportional to the $\sqrt{M}$, we see the significant advantages that broadcast channels have over wire networks in a bursty environment when hierarchical structures are allowed. We can see this intuitively since we assume that the cost of a broadcast channels is proportional only to capacity, but is independent of distance; if we properly select the transmission range, then the broadcast capacity can be reused spatially (i.e., it can be used independently and simultaneously in more than one area). Further, it can be shown that a 2-level hierarchy using random access in the lower level and dedicated channels in the upper level can be quite efficient in a broadcast environment; this is true since the lower level has gathered together enough traffic so that it is no longer bursty when delivered to the upper level (recall that dedicated channels do well with nonbursty traffic)

V. PRINCIPLES ESTABLISHED

This section is really a continuation of the last since there is a somewhat fuzzy boundary between lessons and principles. Indeed, one might accept the pragmatic definition that a principle is a lesson you had to learn twice.

A. Bigger is Better

The law of large numbers states that a large collection of demands presents a total demand which is far more predictable than are the individual demands. We are thus led to the consideration of large shared resources (large in the sence that we increase both the number of users—or the load presented by each user—and the capacity of the resources). Furthermore it is easy to show that the performance improves significantly as we make our systems larger. In particular we can show that a small system whose capacity is C operations per second and whose throughput is J jobs per second (with each job requiring an average of K operations per job) performs A times as slowly (i.e., the response time is A times longer) as a system whose capacity is AC and whose throughput is AJ. The lesson here is very clear, namely that bigger systems perform far better than smaller ones [25]. This is a statement about performance and not one about cost. Indeed if one is talking about communication channel capacity, then one usually also gains through an economy of scale due to the tariff pricing structure as presented by the common carriers. All the more reason, therefore, to concentrate more and more traffic on ever larger channels to gain both cost and efficiency in performance; of course one must be careful not to abuse any "resale" restrictions. Moreover, our lesson about burst communications tells us that in sharing this large channel dynamically, one should provide the full capacity to a single user on demand, rather than to preallocate fractions of the capacity on a permanent basis (omitting consideration of such channel-sharing schemes as spread-spectrum).

The "bigger is better" principle may not apply to the case of stream traffic (defined as real-time traffic which requires a low delay and moderately large throughput requirement—an example being packetized speech). Indeed, an unresolved issue recently raised by Dr. Robert E. Kahn (Editor of this Special Issue) is how effective it would be to handle stream traffic by dividing each trunk into a multiplicity of medium-capacity channels which may then be linked together to form a stream traffic path. We are currently looking at this issue.

B. The Switch

Our second principle has to do with the use of a software switch at the nodes of a network. The principle here is that it pays to place intelligence at the switching nodes of a network since the cost of that intelligence is decreasing far more rapidly than the cost of the communications resource to which it is attached. The idea is to invest some cost in an intelligent switch so as to save yet greater cost in the expensive communications resource. The ability to introduce new programs, new functions, new topologies, new nodes, etc., are all enhanced by the programmable features of a clever communications processor/multiplexer at the software node.

C. Constraints

The principle here is simply, "constraints are necessary and often are evil." Indeed some of the constraints we have seen are sequencing, storage management, capacity allocation, speed matching, and other flow and routing control functions. These "natural" constraints render us vulnerable to dangerous deadlocks and degradations. As mentioned above, if the constraint cannot be met due to some possibly unfortunate accident, than the system will stop all flow. If one is slow in meeting the constraints, then that represents a delay-throughput degradation. As a result of this principle, we see that it behooves us to provide sufficient resources in the network which then allow us to be more relaxed about assigning them. That is, the more precious is a given resource, the tighter we are in

allocating it to a demand, the more likely we are to run into a deadlock or degradation. With an ample resource, we can be more cavalier in assignment and even renege on the assignment if necessary, assuming that a backup facility (in the form of an ample resource elsewhere in the network) is provided.

D. Distributed Control

The principle here is that one must pay a price to nature for organizing a collection of distributed resources into a cooperating group. We have not yet established what that minimum price is, but we have classified the price in the form of collisions, control overhead, and idle capacity.

E. Flow Control

The "principle" here is that flow control is a critical function in packet communications and we are still naive in the invention and analysis of flow control procedures. Hopefully, cleaner code and cleaner concepts will simplify our ability to design and evaluate flow control procedures in the future. There is a "miniprinciple," which seems to be emerging from our preliminary studies [26] which states that if one wants to maximize the power in a network at fixed cost, where power is defined as throughput divided by response time, then under simple statistical assumptions on the flow, one should operate at a point where the throughput delivered is half the maximum possible and the response time is then twice the minimum (no-load) response time.

F. Stale Protocols

In our experiments in the ARPANET, the SATNET, and the packet radio network, we have occasionally attempted to adopt a protocol from one network directly over into a new network. We have found that this is a dangerous procedure and must be carefully analyzed and measured before one adopts such a procedure. Indeed, the use of old protocols in a new environment is dangerous. For example, we found that the use of the ARPANET-like RFNM end-to-end protocol was extremely wasteful of channel capacity and resulted in a capture effect between pairs of users when used in the SATNET. In a 2-user Time Division Multiple Access scheme (in which odd-numbered slots are permanently assigned to user A and even-numbered slots to user B), user A could prevent B from sending any data if he simply started transmitting first in each of his slots since this would require B to devote all of his slots to returning RFNM's to A. Time in the SATNET is divided into fixed length slots (of 30-ms duration). A slot is used for a single packet transmission even if the packet itself is tiny, as is the case of a RFNM. This inefficiency does not exist in the ARPANET since no extra bits are stuffed into ARPANET packets to artificially increase their size. Indeed, gateways have been introduced between the ARPANET and SATNET which renders these nets independent of each other's protocols and formats [2].

G. Inexperienced Designers

It is important that users recognize the difference in function, performance, and operation of a packet network as opposed to a leased line. Certain decisions regarding the parameter settings in any process-to-process communication are often left up to the user of a packet network; for example the buffer allocation he provides in his HOST to accept data from another process communicating through the network with his HOST is a decision often left to the system user. If his buffer allocation is too small, he may degrade the apparent performance of the network to an unacceptably low degree; this comes about, not because the network is slow, but rather because his allocation was too small. The principle here is that if one leaves design decisions in the hands of the users (or even network designers) then those individuals must be informed as to the effect of their decisions regarding these parameter settings; they cannot be expected to understand the consequence of their actions without being so informed.

VI. Conclusions

The purpose of this paper has been to boil down a decade of experience with packet communications and from this to extract some lessons and principles we have established. We have succeeded only in part in this endeavor; the field is still moving rapidly and we are learning new things each day. Indeed, in addition to lessons and principles, we have identified a number of open issues which require further study. Aside from the meager principles we stated in the preceding section, we feel it is necessary to make some concluding statements. First we feel that one must view packet communications as a system rather than as a trivial leased line substitute. The use of packet communications offers opportunities to the informed user on the one hand and sets traps for the naive user on the other. It is necessary that the overriding principles which we have established and others which we have yet to establish be well understood by the practitioners in the field. We must continue to learn from our experience, and alas, that experience is often gained through mistakes observed rather than through clever prediction. In all of our design procedures we must constantly be aware of the opportunity to share large resources among large populations of competing demands. We must further be prepared to incorporate new technologies and new applications as they arise; we cannot depend upon "principles" as these principles become invalid in the face of changing technologies and applications.

Lastly, we must point out that the true sharing of processing facilities in the network (i.e., the HOSTS) has not yet been realized in modern day networks. One would dearly love to submit a task to a network, ask that it be accomplished in the most efficient fashion, and expect the network to find the most suitable resources on which to perform that task. Currently, one must specify on which HOST his program should be stored, where his job should be executed, where to store his results, at which location his results should be printed, and specify when all this must happen. The next phase of networking must address this general question of automatic resource sharing among HOSTS in a distributed processing environment. Perhaps in the next special issue on packet communications we will be in a position to identify lessons and principles for true resource sharing of this type.

References

[1] L. Kleinrock, *Queueing Systems, Volume II: Computer Applications.* New York: Wiley Interscience, 1976.

[2] I. M. Jacobs, R. Binder, and E. V. Hoversten, "General purpose packet satellite networks," this issue, pp. 1448–1467.

[3] R. E. Kahn, S. A. Gronemeyer, J. Burchfiel, and R. C. Kunzelman, "Advances in packet radio technology," this issue, pp. 1468–1496.

[4] P. Baran, "On distributed communications," RAND Series Reports, Rand Corporation, Santa Monica, CA, Aug. 1964.

[5] D. Davies, "The principles of a data communication network for computers and remote peripherals," in *Proc. IFIP Congress '68,* Edinburgh, Scotland, pp. 709–714; Aug. 1968.

[6] L. G. Roberts, "Multiple computer network development to achieve resource sharing," in *Proc. ACM Symp. Operating Syst.,* Gatlinburg, TN, 1967.

[7] L. G. Roberts, "The evolution of packet switching," this issue, pp. 1307–1313.

[8] L. Kleinrock, *Communication Nets; Stochastic Message Flow and Delay.* New York: McGraw-Hill, 1964, out of print. Reprinted by Dover Publications, 1972. (Published in Russian, 1970, Published in Japanese, 1975.)

[9] O. J. Boxma, "On a tandem queueing model with identical service times at both counters, I.," University Utrecht, Dept. of Mathematics, Preprint No. 78, Mar. 1978.

[10] L. Kleinrock, "Performance of distributed multi-access computer-communication systems," in *Proceedings of IFIP Congress '77* Toronto, Canada, pp. 547–552; Aug. 1977.

[11] L. Pouzin and H. Zimmerman, "A tutorial on protocols," this issue, pp. 1346–1370.

[12] L. Kleinrock, "ARPANET lessons," in *Proc. Int. Conf. Communications,* Philadelphia, PA, pp. 20-1–20.6; June 1976.

[13] R. Kahn and W. Crowther, "Flow control in a resource sharing computer network," in *Proc. 2nd IEEE Symp. Problems in Optimization of Data Communication Systems,* Palo Alto, CA, pp. 108–116, Oct. 1971, (also reprinted in *IEEE Trans. Communications,* pp. 539–546; June 1972).

[14] E. Raubold and J. Haenle, "A method of deadlock-free resource allocation and flow control in packet networks," in *Proc. Third Int. Conf. Computer Communication,* Toronto, Canada, pp. 483–487; Aug. 1976.

[15] W. Naylor, "A loop-free adaptive routing algorithm for packet switched networks," in *Proc. Fourth Data Communications Sym.,* Quebec City, Canada, pp. 7.9–7.14; Oct. 1975.

[16] A. Segall, P. M. Merlin, and R. G. Gallager, "A recoverable protocol for loop-free distributed routing," in *Pro. Int. Conf. Communications,* Toronto, Canada, vol. 1, pp. 3.5.1–3.5.5; June 1978.

[17] H. Opderbeck and L. Kleinrock, "The influence of control procedures on the performance of packet-switched networks," in *Nat. Telecommunications Conf. Record,* San Diego, CA., pp. 810–817; Dec. 1974.

[18] W. Price, "Simulation of packet-switching networks controlled on isarithmic principles," in *Proc. Third Data Communication Symp.,* St. Petersburg, FL, pp. 44–49; Nov. 1973.

[19] W. Naylor and L. Kleinrock, "On the effect of periodic routing updates in packet-switched networks," in *Nat. Telecommunications Conf. Record,* Dallas, TX, pp. 16.2-1–16.2-7; Nov. 1976.

[20] L. Kleinrock and M. Gerla, "On the measured performance of packet satellite access schemes," in *Proc. Fourth Int. Conf. Computer Communication,* Kyoto, Japan, Sept. 1978.

[21] F. Tobagi, "Performance analysis of packet radio communication systems," in *Nat. Telecommunications Conf. Record,* pp. 12.6-2–12.6-7; Dec. 1977.

[22] G. Akavia, "Hierarchical organization of distributed packet-switching communication systems," Ph.D. Dissertation, Computer Science Department, Univ. of California, Los Angeles, Mar. 1978.

[23] R. Metcalfe and D. Boggs, "Ethernet: Distributed packet switching for local computer networks," *Communications of the ACM,* vol. 19, no. 7; pp. 395–404, July 1976.

[24] L. Kleinrock and F. Kamoun, "Data communications through large packet-switching networks," in *Proc. Int. Teletraffic Congress,* Sydney, Australia, pp. 521-1–521-10; Nov. 1976.

[25] L. Kleinrock, "Resource allocation in computer systems and computer communication networks," in *Proc. of IFIP Congress '74,* Stockholm, Sweden, pp. 11–18; Aug. 1974.

[26] ——, "On flow control," in *Proc. Int. Conf. Communications,* Toronto, Canada pp. 27.2–1 to 27.2–5, June 1978.

[27] V. G. T. Cerf and P. Kirstein, "Issues in packet network interconnection," this issue, pp. 1386–1408.

Terrestrial Packet Networks

This section:

- shows how packet networks have been modeled and analyzed

- introduces the many ways in which data can be routed through complex networks—an essential aspect of the success of a packet network

- provides examples of operational packet networks.

The ability of packet communications networks to provide rapid, efficient, and economical interconnection among a large number of users inherently depends upon the sharing of the total resources available within the network. The basic premise of sharing is that not all users of a network will want to communicate with each other at the same time. It is impossible, however, to know in advance when every user will want to communicate with some other user. The behavior of users, and therefore of the network, can be defined only in statistical terms.

The papers in this section deal with the statistical behavior of packet networks, starting with the generalized terrestrial structures introduced in the previous sections and including the traffic characteristics which, as we saw, set data users apart from voice users.

CONCEPT OF TRAFFIC INTENSITY

The principle of "sharing" in switched communications networks is dependent upon the random nature of the call- or message-arrival process. If there are a large number of users and sufficient total resources available, the randomness will tend to average out and, most of the time, users will be satisfied with the service they receive from the network. This is where the challenge of traffic analysis lies: to supply enough resources that users are generally satisfied but not so many that capacity is wasted. The essential factors can be estimated from statistical knowledge of the amount of network demand and the number of available network resources. In packet networks, the dynamics of user demand can also be com-

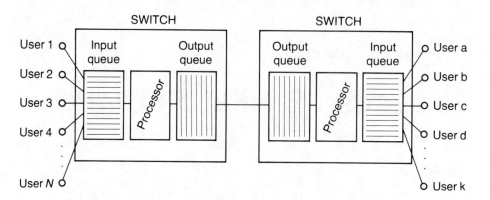

Figure III-1

bined with self-optimizing routing patterns so that all capacity in the network can be used with high efficiency, even if it is not ideally located at any particular time.

For a data communications network, the traffic intensity presented to the network is simply the product of the arrival rate of user messages multiplied by the average length of those messages. If our entire network consisted of a pair of switches connected by a single high-speed trunk, we could model it as a simple single-server queueing system, as shown in Figure III-1. Employing some simple and logical assumptions simplifies the model to that shown in Figure III-2. Application of basic queueing analysis to this model allows us to estimate the delay through the system by the equation:

$$T = \tau/(1 - \rho)$$

where τ is the average service time for a message and ρ is the average occupancy of the interswitch trunk circuit.

APPLICATION TO REAL NETWORKS

Clearly, the application of such a simple model to real networks—with multiple user-to-user paths, different classes of users, and a mixture of different service elements—is not a trivial process. The first paper in this section, authored by Fouad A. Tobagi, Mario Gerla, Richard W. Peebles, and Eric G. Manning attempts to describe this transition by presenting an overview of the modeling techniques that have been successfully applied to actual networks and later verified and validated through actual measurements on those networks.

The paper begins with an excellent introduction to the basic results that have came from queueing theory used in network models, and deals not only with average performance measures but also the variance associated with mean values. Application of such results to a network of queues, as found in networks with multiple routes through the network, is also shown. The concepts are extended to radio- and satellite-type channels and use examples from actual networks to illus-

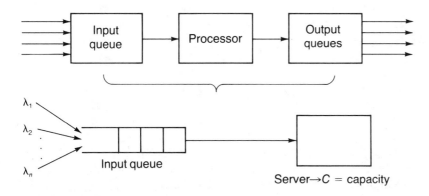

Figure III-2

trate the results. This paper is a summary and overview paper on data network modeling and, as such, contains a comprehensive list of 94 references, each of which is valuable for deriving and validating the applicability of the techniques.

THE ROLE OF ROUTING

Inherent in the analysis of packet networks is the implication that they are able to utilize capacity that may be temporarily underutilized because of the statistical variation in traffic intensity. Many routing techniques have been suggested and, in fact, tried in various packet networks. Some techniques are quite random; others are highly structured and deterministic. The paper by Mischa Schwartz and Thomas E. Stern describes and compares a number of these techniques. Specific examples are drawn from actual networks, including ARPANET, TRANSPAC, Tymnet, IBM SNA (Systems Network Architecture) and Digital Equipment Corporation Network (DECNET).

All of the techniques described in this paper derive from the concept of a shortest-path-routing algorithm. In this context, the length of the path can be defined in terms of distance, delay, or cost. In most packet applications, the path length is measured in terms of delay; that is, the network is always attempting to select a path that will minimize the transit delay of the information between a pair of users.

ROUTING ALGORITHM DETAILS

Substantial detail about the actual routing algorithm used in the ARPANET is presented in the paper by John M. McQuillan, Ira Richer, and Eric C. Rosen. Not only does this paper describe the "new" ARPANET routing algorithm introduced into the network in May 1979, but it compares the design and operation of the new algorithm with the earlier ARPANET routing algorithms. Although the new algorithm requires more processor power and memory, it provides more efficient utilization of the network connectivity—a very favorable tradeoff in light of the continuing rapid decrease in the cost of nodal processing equipment compared to the cost of network transmission facilities.

FURTHER EXAMPLES OF TERRESTRIAL PACKET NETWORKS

The final two papers of this section describe two practical, operational packet networks. M.A. Bergamo and A.S. Campos show how REXPAC, a nationwide packet network in Brazil, was designed and implemented. REXPAC is deployed to provide both a research avenue for the national universities as well as an operational utility for Brazilian industry. While the REXPAC network will employ all of the CCITT standard interfaces, a high reliability, multiprocessor node has been developed by Brazil's own research community.

D.W.F. Medcraft described the data networking plans within the United Kingdom; they include not only a nationwide packet switching service but also longer-range integrated voice/data services and, possibly, circuit-switched (transparent) data services.

Modeling and Measurement Techniques in Packet Communication Networks

FOUAD A. TOBAGI, MEMBER, IEEE, MARIO GERLA, MEMBER, IEEE,
RICHARD W. PEEBLES, AND ERIC G. MANNING, MEMBER, IEEE

Invited Paper

Abstract—Considerable advances in the modeling and measurements of packet-switched networks have been made since this concept emerged in the late sixties. In this paper, we first review the modeling techniques that are most frequently used to study these packet transport networks; for each technique we provide a brief introduction, a discussion of its capabilities and limitations, and one or more representative applications. Next we review the basic measurement tools, their capabilities, their limitations, and their applicability to and implementation in different networks, namely land based wire networks, satellite networks, and ground packet radio networks; we also show the importance of well-designed experiments in satisfying the many measurement goals. Finally we discuss briefly some open problems for future research.

I. Introduction

COMMUNICATIONS engineers have long recognized the need to multiplex expensive transmission facilities and switching equipment. The earliest techniques for doing this were synchronous time-division multiplexing and frequency-division multiplexing. These methods assign a fixed subset of either the channel bandwidth or the time frame to each of several subscribers, and are very successful for voice traffic. But the advent of computers has led to an explosive growth in data traffic and the old multiplexing techniques are not nearly so successful. We measure success by the degree of utilization of the transmission channels and switches, which is reasonable since the lower the utilization the more of these expensive resources we need to support a given level of subscriber demand. Data traffic is less effectively supported by a fixed subchannel allocation because it is more "bursty" in nature than voice traffic. This simply means that if one were to watch data traffic over a long period of time then there would be no activity at all for a while, then a flurry of transmission, no activity for another long while, and so on. That is, there is inherently large peak to average transmission rate. Burstiness is a statistical property, both the time between messages and the message length are usually random variables. Fixed subchannel allocation schemes must assign enough capacity to each subscriber to meet his peak trans-

mission rates with the consequence that the resulting channel utilization is low.

Packet-transmission networks have been developed over the past ten years in an attempt to solve this problem [16], [18], [69], [70], [73], [76], [79]. The basic idea is to allocate some or all of the system capacity (along some path between subscribers) to one customer at a time; but only for a very short period of time. Customers are required to divide their messages into small units (packets) to be transmitted one at a time. Each packet is accompanied by the identity of its intended recipient. In packet-switched networks each packet is passed from one packet switch to another until it arrives at one connected to that recipient, whereupon it is delivered. Packets arriving at a switch may need to be held temporarily until the transmission line that they need is free. The resulting queues require that packets be stored in the switches and it is not unusual that all packet buffers are occupied in a given switch. Thus both the switch capacity (processing and storage) and transmission capacity between switches is statistically multiplexed by subscribers. The designers of such packet networks are faced with the problem of choosing line capacities and topologies that will result in relatively high utilization without excessive congestion.

Another type of packet-transmission network is the "multi-access/broadcast" network typified by the ALOHA network [2], ring network [70], and the ETHERNET [73]. Here, a single transmission medium is shared by all subscribers. Again, each subscriber is given the whole resource but only for the time required to transmit a single packet. In these networks each subscriber is interfaced through a smart port that listens to all transmissions and absorbs any packets addressed to itself. These multiaccess/broadcast networks also statistically multiplex the communications channel. Queueing does not occur within the system but at the ports to it. Packet-switching networks on the contrary exhibit queues both within the system and at the ports.

All of the packet transmission schemes are designed to share a resource that is inadequate to meet the simultaneous peak demand of all subscribers. It is a reasonable assumption that such simultaneous demands are exceedingly rare events and that average traffic can be adequately supported. In the design of these systems we require that we achieve a desired balance between high average utilization and acceptable levels of congestion under peak loads. The difficulty of such a task leads us to the need for applying modeling techniques to assist us in this design.

131

We should observe that packet switching networks offer another potential advantage in addition to improved resource utilization. They offer greater reliability. In fact, that was the original motivation for suggesting the idea [4]. One of the original design goals for the ARPA network was to insure network connectivity between any pair of nodes with a downtime of less than 30 s/yr [79].

A. Modeling

A broad spectrum of mathematical tools has been applied to the design of packet transport networks. Various aspects of the theory of stochastic processes have been used to understand their behavior as average system utilization increases. Typically we are interested in developing estimates of transit delay, packet loss probability, line and buffer utilizations, network throughput and so forth. Queueing models and models of networks of queues have been used to predict the behavior of packet-switching networks with a high degree of success [40], [47], [49], [65], [94]. A recent survey paper by Kobayashi and Konheim [60] examines the application of queueing theory to communications systems in general and serves as a useful companion paper to this one. Comprehensive introductions to queueing theory can be found in many texts including [7], [13], [20], [53], [57], [78]. The theory of networks of queues was heavily influenced by the fundamental papers of Jackson [41], and Gordon and Newell [34]. Recently significant extensions of that work have been made and are reported in [5]. Markovian models and Markov decision processes have been applied to the study of multiaccess networks and flow control procedures [48], [54], [56], [61], [64], [88].

A separate discipline of mathematics has been applied to the design of packet switching networks, the theory of network optimization. Early work by Frank *et al.* [24], was used in the design of the ARPA network. The complexity of the problem has led to many enhancements of their model. Gerla [28] has applied multicommodity flow optimization to the problem of determining optimal routing algorithms. The general problem of determining the optimal structure of a packet-switching network (topology, link capacity, and routing algorithms) is NP complete (i.e., no polynomial time-bounded algorithm can be found to solve it [3]) and for networks of size above ten nodes, say, we are obliged to turn to heuristic solution techniques. The general idea of these schemes is to "guess" a good topology, apply optimization tools to determine the link capacities for that topology which minimizes the cost subject to delay constraints (or which minimizes the delay subject to cost constraints). The algorithm then makes small perturbations to the first topology and recomputes the optimal link assignments, comparing the resulting cost to previously computed values. These methods usually will converge to a "local minimum" but there is no guarantee that the best solution will be found. The heuristics are often proprietary in nature since a good one is highly marketable; some published work has been presented by Gerla [30] and Lavia [66].

There are numerous occasions where analytical techniques may also have to be abandoned. This happens because the theory is not currently able to deal with some of the "real" properties of packet networks such as state dependent transition probabilities and correlations between interarrival times and service time requirements. We move then to simulation. The application of simulation techniques to packet transport networks has frequently led to useful insights, [39], [49],

[62], [77]. Most of the major systems that have been implemented have also been studied through simulation prior to construction. Great care must be taken, however, in the use of this tool since it is often the case that the results of an experiment are based on correlated samples and the underlying assumptions of statistics are violated. A great deal of work has been done lately to resolve this problem. For example, Crane and Iglehart [14], [15] have introduced the notion of regeneration points for defining intervals in which independent samples can be taken. Unfortunately many experiments have taken no account of these problems and have reported simulation results with experimental data points supported just by single samples, without mentioning the applicable confidence intervals. This was mainly motivated by simple cost considerations; there are so many sample points to consider that it is impossible to run the simulation a large number of times for each point! However, computing costs are dramatically decreasing so that the trend is for more accurate and more reliable simulation models to support future network studies.

B. Measurements

We will not know the true operating characteristics of packet transport networks for some years. All of the modeling work is based upon assumptions about the traffic characteristics and the subnet behavior. It is inherent both in analytic and simulation models that fairly gross assumptions are made. If this is not done, the models become intractable. We expect to find an iterative procedure where initial models guide first implementation; measurements of real characteristics then are fed back into refined models and so forth. This has already occurred, and is important for improving the value of simulations.

In the remaining sections of this paper, we will examine both modeling and measurement in more detail. Section II focuses on modeling. We first consider the application of queueing theory and Markov models, then turn to the theory of optimization and to simulation techniques. In Section III we address the measurement problem: we define the measurement functions and the performance measures; we describe the tools necessary to perform the measurement tasks; and we review some of the significant results obtained in some recent experiments. Finally, in Section IV, we describe some of the important open problems.

II. MODELING TECHNIQUES

This section reviews the modeling techniques that are most frequently used to study packet transport networks. For each technique we provide a brief introduction, a discussion of its capabilities and limitations, and one or more representative applications.

We start with *queueing theory*. In Section II-A, we consider the (simple) single queue server model; we state two widely applicable results, namely, Little's result and the Pollaczek-Khinchin formula for the mean number of customers in an $M/G/1$ queueing system, and discuss their limitations. Two applications are treated: i) the analysis of an allocation strategy of buffers in a packet switch known as the complete partitioning strategy, and ii) Kleinrock's model for message delay in a packet-switched network. Next, we turn to a discussion of "network of queues" models and their application to packet network modeling in Section II-B. We state the condition under which a network of queues has a tractable solution and treat two applications: i) a derivation of the distribution of end-to-end delay for source–destination pairs in a packet-switched network, and ii) the analysis of another allocation strategy of buffers in a packet switch known as the complete sharing strategy.

A second set of powerful analytical tools is provided by the *theory of stochastic processes*. This includes renewal theory, Markov chain theory, semi-Markov and regenerative processes, and Markov decision theory. Numerous applications exist to illustrate the usefulness of these techniques, but due to the great interest we see today in multiaccess/broadcast networks, in this presentation we shall limit ourselves to examples drawn from radio communications systems. We first start, in Section II-C, by the consideration of renewal theory. We give a brief account on the (relatively recent) use of radio for data transmission and discuss the related issues, in particular, the so-called random access schemes. We then show how the assumption of an infinite population of users in conjunction with renewal theory arguments have allowed the determination of the (radio-multiaccess) channel capacity and other performance measures under various access schemes. Following that, we briefly discuss the limitations of the infinite-population/renewal-theory model for these access schemes and emphasize the need for a more accurate performance evaluation. The latter is obtained in Section II-D via Markov and semi-Markov chain models which are used to analyze slotted ALOHA and Carrier Sense Multiple Access respectively [54], [81]. These models greatly improved our understanding of the behavior of random access schemes under "static" conditions. It is important, however, to design systems which can dynamically adapt to time-varying inputs and to changes in the system state. We discuss this issue in Section II-E, in which we give a brief introduction to Markov decision theory and its most relevant results, and then proceed with a discussion of various practical control schemes and their analysis.

Some concepts of network optimization relevant to the design of packet networks are then introduced. Linear, nonlinear, and integer programming techniques are briefly reviewed, and a nonlinear programming technique, namely, the method of Lagrangian multipliers, is illustrated in an optimal capacity assignment problem in Section II-F. In the following section, the routing problem, i.e., the problem of optimally routing packets in the network is formulated and solved using a multicommodity flow approach.

Unfortunately, mathematical programming has its limitations, and heuristic approaches are often required to obtain practical solutions to network optimization problems, as discussed in Section II-H. Similarly, many of the performance models are analytically intractable, and require simulation for their solution, as discussed in Section II-I.

A. Queueing Theory

The single server queue is perhaps the simplest of all the mathematical modeling tools and has been widely applied. The model assumes that "customers" arrive at a service facility, that we know the service time distribution, the interarrival time distribution, and the order in which customers are served. The models tell us the distributions for the number of customers in the system, their waiting time, the server busy period distributions, and so forth. Single server queues are categorized according to the interarrival time and service time distributions and we speak of $M/M/1$ systems, $G/G/1$ systems, and others, where the first letter denotes the interarrival time distribution, the second denotes the service time distribution, and the third element denotes the number of servers in the

system. The letters used in this paper have the following meaning, by convention: M−exponential, D−fixed, and G−general. There have been many contributors to the theory of single server queues and readers looking for an introduction should turn to one of the many basic texts [13], [53], [57].

There are some basic formulas that we will state without derivation which are broadly applicable. The first is Little's result [68]. The result is stated as a simple formula:

$$\bar{n} = \lambda T. \tag{1}$$

Here, $\bar{n}$ is the average number of customers in the system (both queue and server). λ is the customer arrival rate, and T is the average time that a customer spends in the system (including service). The formal proof makes no assumptions about the arrival process distribution, the service time distribution, the number of servers, nor the service discipline which can be first-come-first-serve (FCFS), last-come-first-serve (LCFS), round-robin (RR), etc.

A second widely applicable result is the Pollaczek–Khinchin formula for the mean number of customers in an $M/G/1$ queueing system. This is expressed as

$$\bar{n} = \rho + \rho^2 \frac{(1 + C_b^2)}{2(1 - \rho)} \tag{2}$$

where ρ is the traffic intensity (also referred to as the server utilization if $\rho < 1$) and is defined as λ/μ with μ denoting the service rate, and C_b^2 is the squared coefficient of variance for the service time and is defined as σ_b^2/μ^2, with σ_b^2 denoting the variance of the service time. For the $M/D/1$ queueing system this formula holds with $C_b^2 = 0$.

We observe that this formula only gives us the mean value for the number of customers in the system. This is certainly useful, but it is dangerous to design systems based on mean-value estimates only. The variance of the number in the system is also of great interest: in fact one would ideally like to know the exact distribution for the number of customers in the system. This would then allow one to compute other statistics of interest. Specifically, it is often desirable to design to such criteria as: 90 percent of messages will be transmitted within 2 s and the average time will be 0.9 s. The general approach taken to the analysis of $M/G/1$ systems is to develop the Laplace transform for the distribution of the number in the system (or the waiting time). But then it is often very difficult to invert the result and the detailed distribution cannot be obtained. Nevertheless it is easy to derive the moments of the distribution of the number of customers by evaluating the derivatives of the transform at $s = 0$ [53] where s denotes the argument of the transform.

For an $M/M/1$ queue, $C_b^2 = 1$, and the expression for $\bar{n}$ reduces to

$$\bar{n} = \frac{\rho}{1 - \rho}. \tag{3}$$

In this case, the detailed state probabilities can easily be obtained and they are expressed as

$$P(j) = (1 - \rho)\rho^j. \tag{3a}$$

The $M/M/1$ model has been applied to a wide variety of problems for three reasons. First, it is so very simple that it is convenient to work with. Second, it is a very good approximation to many real systems. Third, even when the approximation is not good, it provides upper bounds to systems with $C_b^2 < 1$. Practitioners must beware, however, if the service time distribution of the real system is known to have wide variance.

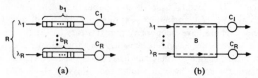

Fig. 1. Buffer management at the switch. (a) Complete partitioning (CP). (b) Complete sharing (CS).

Let us now consider two problems in the design of packet-transport networks that have been attacked with elementary queueing theory: first, the problem of buffer allocation in a packet switch and secondly, the problem of estimating packet transit delay in a packet-switched network.

a) Buffer Allocation: Each packet switch has several outgoing lines; incoming packets will queue for these lines awaiting retransmission. If there is no available storage space an incoming packet will be dropped. How can one allocate the storage of the switch in such a way that packets will not be refused entry when space is available and that no single outgoing line is able to capture all of the switch buffers?

Kamoun [46], [47] has analyzed several strategies, the simplest of which is an $M/M/1$ model with finite customer waiting room. Each of the output lines has a fixed protion of the buffer pool assigned to it, b_i for line i. The rate of external arrivals destined for line i is λ_i. The service rate is μC_i, where C_i is the capacity of line i and $1/\mu$ is the average length of the message, the latter assumed to be exponentially distributed. If the buffer assigned to line i is full, an arriving packet destined for line i is dropped and does not return. Service at a line is FCFS and there are R output lines, as shown in Fig. 1(a). In this model all of the R subsystems are independent and we can write down the equations for each separately. The finite waiting room variant of the $M/M/1$ queue also has a simple solution; (see [13] or [53].) Let n_i denote the number in subsystem i. Its distribution is given by

$$\Pr\{n_i = k\} = \begin{cases} \dfrac{1 - \rho_i}{(1 - \rho_i^{b_i+1})}\rho_i^k, & 0 \leqslant k \leqslant b_i \\ 0, & \text{otherwise} \end{cases} \quad (4)$$

where $\rho_i = \lambda_i/\mu C_i$.

The probability that a packet destined for line i is dropped is the probability that there are exactly b_i packets in subpool i at the time it arrives. Let PB_i denote that probability. We have

$$PB_i = \frac{1 - \rho_i}{1 - \rho_i^{b_i+1}}\rho_i^{b_i}. \quad (5)$$

The independence of the subsystems also means that the global state has a simple product form. Let $\tilde{n} \triangleq (n_1, n_2, \cdots, n_R)$ and let $P(\tilde{n})$ denote the probability of state $\tilde{n}$. We have:

$$P(\tilde{n}) = \left(\prod_{i=1}^{R} \frac{1 - \rho_i}{1 - \rho_i^{b_i+1}}\right)\left(\prod_{i=1}^{R} \rho_i^{n_i}\right). \quad (6)$$

The average time spent at the switch by type i packets, denoted by $\bar{t}_i$, is obtained from Little's result:

$$\lambda_i' = (1 - PB_i)\lambda_i$$
$$\bar{t}_i = \bar{n}_i/\lambda_i' \quad (7)$$

where

$$\bar{n}_i = \sum_{j=1}^{b_i} j \Pr\{n_i = j\}.$$

This simple example shows how even very modest queueing theory allows for the determination of important quantities and thus can give us useful insight into the design of packet networks. We would like to compare this partitioned storage strategy to other strategies. In particular, we like to compare it to the case where the entire buffer pool is shared by all lines. However, this other alternative is more difficult to analyze and we defer the discussion until Section II-B after we have discussed networks of queues.

b) Transit Delays: Another useful application of $M/M/1$ queues was made by Kleinrock [57] in predicting the delay of messages flowing through a network. The delay is expressed in terms of the line capacity and the traffic between ports on the network. With the aid of this expression it is then possible to adjust the line capacity to meet certain delay constraints. It was precisely this technique which was applied in the design of the ARPANET (see Sections II-F and G). The model is developed as follows.

The network is assumed to contain M links between switches of infinite storage capacity and we seek to determine T, the expected message delay in the net. This is just the average delay over all messages flowing through the net. Let N be the average number of packets in the system as a whole and let n_i be the average number of packets in the link i subsystem. Clearly

$$N = \sum_{i=1}^{M} n_i. \quad (8)$$

Then, if γ is the aggregate packet arrival rate from all sources, λ_i is the arrival rate at link i, and T_i is the expected time spent at link i, Little's result yields:

$$T = \sum_{i=1}^{M} \frac{\lambda_i T_i}{\gamma} \quad (9)$$

It remains to compute the values of λ_i and T_i. The former are obtained by considering the traffic between all source–destination pairs and the routing rules (which can be fixed or random but not adaptive). The T_i's are then obtained by treating each link as an independent $M/M/1$ queueing system. (We defer the discussion concerning the validity of this assumption to the next paragraph.) Given that link i has capacity C_i bits per seconds and the average packet size is $1/\mu$ bits, T_i is expressed as

$$T_i = \frac{1}{\mu C_i - \lambda_i} \quad (10)$$

so the expected packet delay is

$$T = \sum_{i=1}^{M} \frac{\lambda_i}{\gamma}\left[\frac{1}{\mu C_i - \lambda_i}\right]. \quad (11)$$

This simplified model can be enhanced by introducing terms that express propagation delay, processing delay, and multiple customer types. The more elaborate the model the more accurate its predictions. The reader is referred to [57] for a detailed discussion.

Why should we believe that each of the links behaves as a separate $M/M/1$ queue? There are many reasons to suspect that it may not. For example, although traffic entering the first switch of a path may be Poisson it may not be so when it leaves the switch. Furthermore, the interarrival time and service time for packets are supposed to be independent random variables but at the second server on a chain this is not the case since packets preserve their length! It is, in fact, not true that the two variables are independent in packet-switched networks but IF THERE IS SUFFICIENT MIXING AT A NODE (i.e., packets joining the queue arrive from several different input lines) then the switch behaves AS THOUGH they were independent. This is Kleinrock's celebrated "independence assumption." It is reasonable if the network topology and routing algorithms are such that the traffic from many preceding switches is mixed at any successive switch. This is true under remarkably loose constraints: a "fan in" of two or three lines seems to be sufficient for the approximation to be accurate [48a].

We turn now to a discussion of networks of queues and their application to packet network modeling.

B. Networks of Queues

Most computer-communications systems are most naturally represented as networks of queues. We have seen in the last section that a simple model of delay in a packet-switched network can be developed if we are free to treat each switch link as an independent $M/M/1$ queue. This depends upon whether or not it is reasonable to believe that the process of passing through a switch does not alter the basic Poisson nature of the traffic. Pioneering work was done in this area by Burke [9] and by Jackson [41].

Burke showed that the output of an $M/M/1$ queue is Poisson. (Limited details regarding Burke's output theorem can be found in Inose and Saito [38].) Jackson extended this work to include feedback networks of N servers as well. If $\tilde{n} \triangleq (n_1, n_2, \cdots, n_N)$ is the global state variable denoting n_i customers at server i then the equilibrium probability distribution has a simple product form:

$$P(n_1, n_2, \cdots, n_N) = P(n_1)P(n_2) \cdots P(n_N) \quad (12)$$

where $P(n_i)$ is the marginal probability of finding n_i customers at server i, and is given by the simple $M/M/1$ formula. To apply Jackson's result we must know the actual traffic arriving at server i. This is easily computed if we know the external arrival rate a_i and the customer branching probabilities, b_{ij}. This yields the set of equations:

$$\lambda_j = a_j + \sum_{i=1}^{N} \lambda_i b_{ij}, \quad j = 1, \cdots, N. \quad (13)$$

The network will reach an equilibrium state provided that none of the servers is overloaded. The interesting point in this result is that the network of queues behave as though the traffic remained Poisson in that the equilibrium state probabilities factor into the product of the marginal probabilities despite the fact that, in truth, it is not Poisson.

Jackson considered also more elaborate models, but the most general results have recently been derived by Baskett et al. [5]. They assume that there are N nodes, L classes of customers (such that each class may have different routing through the network and possibly different service time at a node), and four allowed node types which satisfy the Poisson

output property and thus guarantee a product form solution. These are: type 1—FCFS, $M/M/1$; type 2—RR, $M/G/1$; type 3—processor sharing $M/G/\infty$; type 4—LCFS, $M/G/1$. For the general service time distributions, we require that they have a rational Laplace transform; in types 2, 3 and 4 the service time distribution can vary with the customer class. Each customer class travels through the network according to a probabilistic routing (which can be fixed) specified by $b_{ij;mn}$ where the implication is that customers can also change class $(m \rightarrow n)$ during a transition from one node to another $(i \rightarrow j)$. The network can be open for some classes of customers and closed for others. (A closed network has a fixed number of customers and none leave or enter the network; an open network allows for external arrivals and departures and the number of customers in the network may vary). Again the customer arrival rate at each node is computed by using the external arrival rates and the routing information; in particular for the case where there are no class changes, we have

$$\lambda_j(c) = a_j(c) + \sum_{i=1}^{N} \lambda_i(c)b_{ij} \quad (14)$$

where c denotes the customer class. The solutions for the global state distributions are given in [5]. They are of the form

$$P(n_1, n_2, \cdots, n_N) = C \prod_{j=1}^{N} f_j(n_j) \quad (15)$$

where the f_j depends on the node type. For the type 1 nodes, for example, we have

$$f_j(n_j) = \left(\sum_c \frac{\lambda_j(c)}{\mu_{jn_j}} \right)^{n_j} \quad (16)$$

where μ_{jn_j} is the service rate of node j when there are currently n_j customers and where the summation over customer classes is assumed to be over those that are routed through the node. The constant C is a normalizing constant (i.e., is determined by the condition that $\sum_{\tilde{n}} P(\tilde{n}) = 1$). The case of a completely open system is of special interest; it can be shown that

$$P(\tilde{n}) = \prod_{j=1}^{N} P_j(n_j) \quad (17)$$

where

$$P_j(n_j) = \begin{cases} (1 - \rho_j)\rho_j^{n_j}, & \text{for nodes of types 1, 2 and 4} \\ \dfrac{\rho_j^{n_j}}{n_j!} e^{-\rho_j}, & \text{for nodes of type 3.} \end{cases}$$

That is, these rather complex systems (node types 1, 2, and 4) behave just like a set of connected but independent $M/M/1$ queues! And type 3 nodes behave like isolated $M/G/\infty$ servers. This gives us a few very powerful tools for analytic modeling.

Before we consider examples of the application of these models it is worthwhile to consider some of the important problems that cannot be handled: dynamic storage allocation at a switch which allows for variable sized blocks (this is a situation where the allowed number of packets at a node depends on their total required storage volume); the flow of multiple customer classes through a FCFS node in which the classes have different service time distributions; state dependent customer routing (thus representing adaptive routing algorithms in packet-switched networks); priorities. As a final comment we must warn potential users to consider carefully whether or not their network will support the independence assumption.

An important application of this theory has recently been developed by Wong [94]. He uses the results of Basket *et al.* as a starting point and develops the distribution of the end-to-end delay for source-destination pairs in a packet switched network. Thus Kleinrock's earlier result for the mean delay in a net (described above) has been extended in an important way. The fact that the detailed distributions are known means that we can compute variance and percentile information. The model allows for both fixed and random routing. The independence assumption is still required. The switches use a FCFS discipline on their communications links; infinite storage is assumed, and all packets are assumed to have the same length distribution, namely exponential with mean $1/\mu$. Channel i has the capacity C_i, so the service rate is μC_i. If c denotes a customer class then $a(c)$ represents its path through the network and $\gamma(c)$, its traffic rate. (For random routing this becomes a set of paths, each with a known probability of use.) We will present the formulas for the fixed routing case only. We let λ_{ic} be the mean arrival rate of class c customers to channel i. For fixed routing:

$$\lambda_{ic} = \begin{cases} \gamma(c), & \text{if } i \text{ is in } a(c) \\ 0, & \text{otherwise.} \end{cases} \quad (18)$$

We let ρ_{ic} be the utilization channel i by class c customers, so,

$$\rho_{ic} = \frac{\lambda_{ic}}{\mu C_i} \quad (19)$$

and the total utilization of channel i, $i = 1, \cdots, M$, must satisfy

$$\rho_i = \sum_{c=1}^{R} \rho_{ic} < 1. \quad (20)$$

Let the probability density function of the end-to-end delay for class c customers be denoted by $t_c(x)$, and its Laplace transform by $T_c^*(s)$. Wong shows for this model of a packet-switched network with fixed routing that

$$T_c^*(s) = \prod_{i \in a(c)} \frac{\mu C_i(1 - \rho_i)}{s + \mu C_i(1 - \rho_i)}. \quad (21)$$

Notice that each term in the product is the Laplace transform of the time spent in each queue calculated as if the queue was independent of the rest of the network (while in reality it is not). This implies that the end-to-end delay can be interpreted as the sum of independent delays along the path! The Laplace transform expression can be inverted using partial fractions (see [53]) to obtain $t_c(x)$. The mean and variance can be obtained by taking derivatives and are given by

$$\bar{T}_c = \sum_{i \in a(c)} \frac{1}{\mu C_i(1 - \rho_i)}$$

$$\sigma_c^2 = \sum_{i \in a(c)} \frac{1}{[\mu C_i(1 - \rho_i)]^2}. \quad (22)$$

This result allows us to greatly inhance our understanding of packet networks; we note, however, that the case of finite storage capacity in the switches is still not modeled. In fact, it is still an open problem.

An important tool in solving for the system state probabilities is the set of "local balance" equations. We have not as yet written down any of the system state equations because we wished to expose answers, not derivations. But when the time comes to derive similar models to those described here, these state equations will immediately arise. State equations for stable systems come in two types: global balance equations and local balance equations. The former states that, in equilibrium, the total rate of flow into any given state must equal the total rate of flow out of that state. "Flow," here, means probabilistic flow (state transitions over time). This is not surprising: if it were not true then some states would have increasing (or decreasing) probability of occurrence as time passed. The local balance equations are useful in studying networks of queues and assume that in equilibrium the flow into a state due to arrivals at server i can be equated to the flow out of that state due to departures from server i. It is known that if a solution to the local balance equations can be found then it will also satisfy the global balance equations. The local balance equations are generally much easier to solve and are thus the preferred route. What is not known are the necessary conditions for the local balance equations to have solutions. All of the conditions listed in the paper by Baskett *et al.* are sufficient conditions.

We will now look at an example of global and local balance equations: the second part of our switch buffer allocation example [46]. Recall that the problem is to allocate buffer space in a packet switch to a set of R communication links, and that we have considered a partitioned allocation where each link is assigned a fixed subpool of size b_i. This clearly has the disadvantage of blocking packets on a busy link when buffer space is available in the switch but is dedicated to other channels. Another alternative is *complete sharing*; any buffer can be used for any outgoing link; (see Fig. 1(b)). The total available storage will accommodate B packets (here each packet takes a full buffer even if it is not of full length). A state of the switch is described by the R-tuple $\tilde{n} \triangleq (n_1, n_2, \cdots, n_R)$. Let J be the set of states where $n_j = 0$. In Fig. 2, we show a portion of the state transition diagram. Each node in the figure is a possible state. The edges indicate possible transitions and are weighted by the rate of flow conditioned upon starting in the state at the tail of the arrow. The global balance equations are obtained by drawing a circle around state $(n_1, n_2, \cdots, n_R)$ and equating flows across that boundary; that is, the stochastic flows in and out of that state. Here we get two sets of equations depending on whether or not the total number of customers is less than or equal to B. For states with $\sum_i n_i < B$ we have

$$\sum_{\substack{i=1 \\ i \notin J}}^{R} \lambda_i P(n_1, \cdots, n_i - 1, \cdots, n_R)$$

$$+ \sum_{i=1}^{R} \mu_i P(n_1, \cdots, n_i + 1, \cdots, n_R)$$

$$= \left(\sum_{i=1}^{R} \lambda_i + \sum_{\substack{i=1 \\ i \notin J}}^{R} \mu_i \right) P(n_1, \cdots, n_i, \cdots, n_R). \quad (23)$$

For all states with $\sum_i n_i = B$ the equations are

$$\sum_{\substack{i=1 \\ i \notin J}}^{R} \lambda_i P(n_1, \cdots, n_i - 1, \cdots, n_R)$$

$$= \sum_{\substack{i=1 \\ i \notin J}}^{R} \mu_i P(n_1, \cdots, n_i, \cdots, n_R). \quad (24)$$

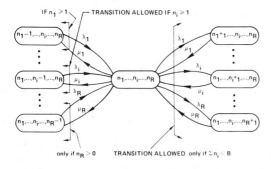

Fig. 2. State transitions for the R-link buffer pool.

The local balance equations for this system equate the flow rates due to arrivals and departures from a given channel. Specifically, we get

$$\lambda_i P(n_1, \cdots, n_i - 1, \cdots, n_R) = \mu_i P(n_1, \cdots, n_i, \cdots, n_R). \tag{25}$$

The traditional method of solving these equations is to guess the answer and try it out. With a little practice the guessing is not too hard. Here the solution is

$$P(\tilde{n}) = \begin{cases} P_0 \rho_1^{n_1} \rho_2^{n_2} \cdots \rho_R^{n_R}, & \text{for all } \tilde{n} \text{ such that } \sum_i n_i \leqslant B \\ 0, & \text{otherwise.} \end{cases} \tag{26}$$

To prove this we need only to substitute this expression back into the local balance equations. The evaluation of P_0 again follows from the normalization requirement (that the $P(n)$ be valid probabilities) so

$$P_0^{-1} = \sum_{\tilde{n}} \rho_1^{n_1} \cdots \rho_R^{n_R} \tag{27}$$

where the summation is taken over the set of all feasible states. Efficient algorithms for the evaluation of such constants can be found in [9a], [74], [93].

The complete sharing scheme has a lower probability of dropping a packet if traffic is reasonably well-balanced, but under highly assymetrical loads it tends to be unfair, i.e., favors heavily utilized channels far too much. The delay on links with low utilization becomes exorbitant. This suggests that some of the buffers should be permanently allocated to each link; but how many? Kamoun goes on to explore several other sharing strategies and concludes that no one scheme is always optimal. It is desirable to select a scheme whose delay and packet loss behavior best suit the operational constraints. When this is not possible, a scheme that dedicates some buffers to each channel and leaves some in a general pool is preferred.

The problem of buffer allocation in a packet switch has also been studied by Irland [40]. He considers a scheme that bounds each channel queue subject to the constraints that all space can be used and no queue can have more than the total space. He then develops a queueing model for the state distributions and uses this to drive a Linear Programming model that seeks the optimum assignment of queue bounds.

We have shown so far in this section that queueing theory is a powerful tool for the study of packet transport network behavior and a very broad class of related problems. We have also identified several limitations in the queueing theory approach, which force us in many cases to turn to simulation techniques. We shall defer the discussion on the use of simulation to Section II-I below. In the following sections we review the (more basic) theory of stochastic processes and discuss its applications.

C. Renewal Theory

Computer communication systems, as pointed out earlier in the paper, are characterized by unpredictable sequences of random demands on the available resources. The theory of stochastic processes (which generally includes renewal theory, Markov chain theory, semi-Markov processes and regenerative processes) thus also provides a large and effective set of analytical tools particularly suited for the modeling of these probabilistic systems. This body of theory is well known and has been established throughout many years; its applications are numerous in very many different areas. In fact, carefully examining the queueing systems for which some solution is obtainable, we realize that virtually in each of them there exists an underlying Markov or semi-Markov process. Queueing theory is a very powerful tool and was shown in the above sections to be extremely effective in the design of computer networks. However, there are situations in which queueing theory does not provide the appropriate model. The latter has to be drawn from the more basic theory of stochastic processes, thus allowing for the determination of the system's steady-state performance. Another problem which also is of great practical importance is the optimization of these probabilistic communication systems. By viewing the models from a probabilistic point of view, dynamic programming and applied probability theory have been combined to give rise to a simple and precise treatment of sequential decision theory, a result of which is the well known Markov decision theory [82]. This is found particularly useful in the design and analysis of efficient procedures for the (optimum) control of communication subsystems.

We intend here to briefly review these tools and illustrate their usefulness by calling on examples from the (relatively recent) packet radio communication systems in both satellite and ground environments. Although as pointed out in the introduction, the latter are not the only examples of applications one can give for these tools, we restrict ourselves to these here for the sake of a unified presentation.

The advantages of using radio communication for data transmission have been extensively discussed in the literature [1], [33], [43], [44], [45], [54], [63], [84]. In essence, satellite transponders in a geostationary orbit above the earth provide long-haul communication capabilities, while broadcast ground radio communications provide us with easy access to central computer installations and computer networks. The topic of interest to this discussion, common to most of these radio systems, is the sharing of a *single* radio channel by users. The difficulty in controlling a multiaccessed channel of this sort, which has to carry its own control information, gave rise to the so-called *random access* techniques. In the event of transmission overlap, these techniques suffer from destructive interference (unless a spread spectrum modulation scheme is used); acknowledgement procedures are devised to recover from errors and overlapping transmissions.

A simple scheme, known as "pure-ALOHA," permits users to transmit any time they desire [1]. Another method, referred to as "slotted-ALOHA," requires each user to start his packets only at the beginning of a slot (whose duration is equal to the transmission time of a packet) [50], [81]. These two ALOHA schemes are suitable for both satellite and ground environments. In ground radio environments, where the channel can further be characterized as a wide-band channel with a small propagation delay between any source–destination pair as compared to the packet transmission time,[1] a third scheme has proven to be efficient: it is the carrier sense multiple access (CSMA) mode. In this scheme one attempts to avoid collisions by listening to the carrier due to another user's transmission; a terminal never transmits when it senses the channel is busy [54], [84]. In the (simple) nonpersistent CSMA protocol, a terminal with a packet ready for transmission transmits the packet if the channel is sensed idle, or reschedules the (re)transmission of the packet to some later time if the channel is sensed busy. A slotted version of this scheme is also considered in which the time axis is slotted and the (mini-) slot size is τ seconds (the propagation among pairs of devices is assumed to be the same [54]). All terminals are synchronized and start transmission only at the beginning of a slot, according to the protocol described above. In addition to these and other CSMA protocols, a number of clever schemes have also appeared in the literature offering improved performance under various specific conditions such as heavy traffic, large users, etc. For more details the reader is referred to [8], [42], [43], [80].

The remainder of this subsection will be devoted to renewal theory and its application to the analysis of packet switching in radio channels. Markov chain models and Markov decision models will be treated in the following two subsections.

The Infinite Population Model and Renewal Theory: The focus here is to show how the assumption of an infinite population in conjunction with renewal theory arguments have allowed the determination of the channel capacity under the various schemes. In this presentation, we shall make sure not to overlook the importance of simulation techniques and simulation results whenever they have proven useful, be it for the validation of a model, or the determination of some performance measure hard to obtain analytically, or just the gain of insight into the behavior of the system under specific conditions.

The model assumes that the traffic source consists of an infinite number of users who collectively form an independent Poisson source with an aggregate mean packet generation rate of S packets per packet transmission time T. (We assume here that each packet is of constant length requiring T seconds for transmission.) This is an approximation of a large but finite population in which each user generates packets infrequently and each packet can be successfully transmitted in a time interval much less than the average time between successive packets generated by a given user. Each user in the infinite population is assumed to have at most one packet requiring transmission at any time (including any previously blocked packet). Under equilibrium conditions, S is also the channel throughput. Because of packet interference, the achievable throughput will always be less than 1. The traffic offered to the channel from our collection of users consists of not only new packets but also of previously collided packets: this increases the mean offered traffic rate which we denote by G (packets per transmission time T) where $G \geqslant S$. To avoid repeated conflicts, each user delays the transmission of

a previously collided packet by some random time whose mean is $\overline{X}$ (chosen, for example, uniformly between 0 and $X_{\max} = 2\overline{X}$). Two additional assumptions are introduced here.

Assumption 1: The average retransmission delay X is large compared to T.

Assumption 2: The interarrival times of the point process defined by the start times of all the packets plus retransmissions (and reschedulings) are independent and exponentially distributed.[2]

If we use $T = 1$ (for normalization), then we express τ as $a = \tau/T$ and $\overline{X}$ as $\delta = \overline{X}/T$. The throughput analysis consists of solving for S in terms of G and other system parameters (namely a). The channel capacity is then found by maximizing S with respect to G.

Renewal theory, and the theory of regenerative processes in general, relate to systems in which there exists an underlying process which probabilistically restarts itself. Perhaps the result in renewal theory which proves most useful here is the one corresponding to alternating renewal processes.[3] An alternating renewal process is one which describes a system which can be in one of two states, say on or off. Starting in the on state, the system alternates between these two states. The periods of time it spends in each are random variables which follow a common distribution for each of the two states. Let EX be the average time the system remains in the on state, and EY be the average time it remains in the off state. Let $P(t)$ be the probability that the system is on at time t; we have the following simple result: $P(t) = EX/(EX + EY)$. This result is easily extendable to any number of states that the system may be cycling through. The key element in such an analysis is to identify points in time at which the system regenerates itself: the interval of time separating two consecutive regenerative points is called a cycle; the ratio of the average time that the system spends in a given state to the average cycle time is precisely the fraction of time that the system spends in that state.

Consider for example the slotted ALOHA scheme. By the infinite population assumption and the Poisson assumption on the channel traffic, each slot boundary is a regenerative point. It is clear that e^{-G} is the probability that the slot is empty and this is also the fraction of time that the channel is idle. The probability that a slot is carrying a successful packet is clearly Ge^{-G}, the probability that a single packet is transmitted in that slot; by the same argument this is also the fraction of time that the channel is carrying successful information and thus constitutes the average throughput S.

A slightly different approach using the same type of argument can also lead to the result derived above. Considering the (slotted) time axis, it is clear that we observe a number of consecutive nonempty slots (which we refer to as a busy period (B)), followed by a number of consecutive empty slots (which we refer to as an idle period (I)). A busy period and the following idle period constitute a cycle. The idle period is geometrically distributed with mean $\overline{I} = 1/(1 - e^{-G})$. The busy period is also geometrically distributed with mean $\overline{B} =$

---

[1] Ratio on the order of 0.01 [54].

[2] It is clear that Assumption 2 is violated and that it has been introduced for analytic simplicity. However some simulation results are discussed below which show that performance results based on this assumption are excellent approximations. Moreover, in the context of slotted-ALOHA it was analytically shown that, in the limit as $X \to \infty$, Assumption 2 is satisfied [63].

[3] It will be clear from the discussion below that this constitutes a special case of the more general result obtained with regenerative processes.

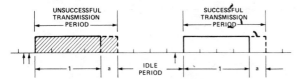

Fig. 3. Slotted nonpersistent CSMA: Transmission and idle periods.

$1/e^{-G}$. Thus the fraction of time that the channel is idle is equal to

$$\frac{\overline{I}}{\overline{I} + \overline{B}} = \frac{(1 - e^{-G})^{-1}}{(1 - e^{-G})^{-1} + e^G} = e^{-G} \quad (28)$$

Let $\overline{U}$ denote the average time during a cycle that the channel is carrying successful packets. Given that a slot is nonempty, the probability that it is successful is simply $Ge^{-G}/(1 - e^{-G})$. $\overline{U}$ is, therefore, given by

$$\overline{U} = \overline{B}Ge^{-G}/(1 - e^{-G}) = G/(1 - e^{-G}). \quad (29)$$

Taking the ratio of $\overline{U}$ to $\overline{B} + \overline{I}$, we find again that the channel throughput is precisely $S = Ge^{-G}$.

Let us consider now, as another example, the slotted non-persistent CSMA protocol, analyzed by Kleinrock and Tobagi [54], [84]. Considering the time axis, we define a transmission period (TP) to be the period of time required for transmission and reception of a packet and its (possible) over-lapping packets. Thus we observe on the time axis transmission periods separated by idle periods, as depicted in Fig. 3. The length of a TP is $1 + a$. A TP is successful if only one packet is transmitted; the probability of this occurring is

$$P_s = \frac{aGe^{-aG}}{1 - e^{-aG}}. \quad (30)$$

Due to the memoryless property of the Poisson process, the average idle period (normalized to T) is simply

$$\overline{I} = \frac{ae^{-aG}}{1 - e^{-aG}}. \quad (31)$$

Using the same renewal theory argument as above, we find that the average channel utilization is given by

$$S = \frac{P_s}{\overline{I} + 1 + a} \quad (32)$$

Substituting for P_s and $\overline{I}$ the expressions found above, we get

$$S = \frac{aGe^{-aG}}{1 + a - e^{-aG}}. \quad (33)$$

This relatively simple argument has been applied in numerous occasions to analyze the throughput and channel capacity of many other (more complex) protocols as well as the effect on system capacity of the overhead created by various acknowl-edgment schemes. For this the reader is referred to the work by Tobagi and Kleinrock [54], [84], [85], [90]. We illus-trate such results here by plotting in Fig. 4 S versus G for various random access schemes. An important question re-mains: what about packet delay analysis? Kleinrock and Lam [50] formulated an analytic model for a slotted-ALOHA channel using a uniform retransmission randomization scheme, and assuming that the channel is in equilibrium. Such a task proved more difficult for CSMA, and simulation techniques

appeared then to be the only recourse. A brief discussion of some simulation results and of the validity of the equilibrium assumption follows.

Discussion and Delay Analysis [54], [84]: The above anal-ysis is based on renewal theory and probabilistic arguments requiring independence of the random variables provided by Assumption 2. Steady state conditions are also assumed to exist. However, from the (S, G) relationships derived above (see Fig. 4) and the throughput-delay performance derived in [50] for slotted-ALOHA, one can see that steady state may not exist because of the inherent instability of these random-access techniques. This instability is simply explained by the fact that when statistical fluctuations in G increase the level of mutual interference among transmissions, then the positive feedback causes the throughput to decrease to 0. Extensive simulation runs performed on a slotted-ALOHA channel with an infinite population [63] have indeed shown that the as-sumption of channel equilibrium is not strictly speaking valid; in fact, after some finite time period of quasi-stationarity conditions, the channel will drift into saturation with prob-ability one.[4]

In the simulation models considered, [54], [63] Assump-tions 1 and 2 concerning the retransmission delay and the independence of arrivals for the offered channel traffic are relaxed. That is, only the newly generated packets are de-rived independently from a Poisson distribution. In general, simulation results obtained with moderate length runs indicate the following. For each value of the input rate λ, there is a minimum value δ_0 for the average retransmission delay vari-able such that below that value it is impossible to achieve a throughput equal to the input rate. The higher λ is, the larger δ_0 must be to prevent a constantly increasing backlog, i.e., to prevent the channel from saturating. Simulation also shows that for finite values of δ, $\delta > \delta_c$, but not large compared to 1, the system already "reaches" the asymptotic results ($\delta \to \infty$). That is, for some finite values of δ, Assumption 2 is excellent and delays are acceptable. Moreover, the comparison of the (S, G) relationship as obtained from simulation and the results obtained from the analytic model exhibits an excellent match. Thus we consider the results derived above under the assump-tion of channel equilibrium useful since they are meaningful for these finite (and possibly long) periods of time. Also they provide an accurate assessment of the channel capacity. In [54] simulation experiments were also conducted to find the CSMA "optimal" delay; that is, the value of $\delta(S)$ which allows one to achieve the indicated throughput with the mini-mum delay. "Delay" here refers to the average over all sam-ples collected in the period of time which represents the length of the simulation runs.

D. Markov Chain and Semi-Markov Chain Models

It is clear from the above discussion that the (assumed) equi-librium throughput-delay results are not sufficient to charac-terize the performance of the infinite population model. A more accurate measure of channel performance must reflect the trading relations among stability, throughput and delay. The intent here is to show how this can be done by formulat-ing a Markovian model for a population of M users, where M

---

[4] It is interesting to point out here that it was more difficult to ob-serve this behavior of saturation with the CSMA simulator because CSMA, as shown in [88] and as will be discussed later, is relatively speaking, less unstable than slotted-ALOHA; it will require an ex-tremely long run before one can observe the unstable behavior.

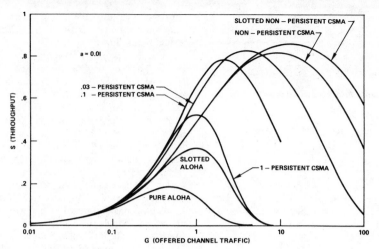

Fig. 4. Throughput versus channel traffic for various random-access schemes.

can be infinite as well. In summary, in this section we first give the definition of a Markov chain and state the Limit theorem which is most relevant to the present discussion. Then we proceed with the Markovian model for a slotted-ALOHA channel and derive its steady-state throughput-delay performance. Next, a discussion concerning non persistent CSMA under similar conditions follows which shows that, due to the dependence of the system evolution on the state of the channel (busy or idle), a simple Markov Process is not sufficient to model the system; instead, results from the theory of semi-Markov chains and regenerative processes are required. We give a brief account on this theory, and the resulting model.

1) Markov Chains: A Markov chain is a stochastic process $\{X_n, n = 0, 1, 2, \cdots\}$ with a finite or countable state-space such that for all states $i_0, i_1, \cdots, i_{n-1}, i, j$ and all $n \geqslant 0$

$$\Pr\{X_{n+1} = j \mid X_0 = i_0, X_1 = i_1, \cdots, X_{n-1} = i_{n-1}, X_n = i\}$$

$$= \Pr\{X_{n+1} = j \mid X_n = i\}. \qquad (34)$$

If $\Pr\{X_{n+1} = j \mid X_n = i\}$ is independent of n, then the Markov chain is said to possess stationary transition probabilities. In this case we let

$$p_{ij} = \Pr\{X_{n+1} = j \mid X_n = i\}. \qquad (35)$$

Perhaps the major results in the theory of Markov chains consist of the Limit Theorems (as $n \to \infty$), and in particular the following [82]

"*Theorem:* An irreducible aperiodic Markov chain belongs to one of the following two classes:

(a) either the states are all transient or all null recurrent; in this case $p_{ij}^{(n)} \to 0$ as $n \to \infty$ for all i, j and there exists no stationary distribution.

(b) or else, all states are positive recurrent, that is,

$$\pi_j = \lim_{n \to \infty} p_{ij}^{(n)} > 0. \qquad (36)$$

In this case, $\{\pi_j, j = 0, 1, 2, \cdots\}$ is a stationary distribution and there exists no other stationary distribution where $p_{ij}^{(n)} \triangleq \Pr\{X_{n+m} = j / X_m = i\}$ (the probability of reaching state j from state i in n steps)."

Consider again the slotted-ALOHA scheme and let the channel user population consist of M independent users. Each

such user can be in one of two states: blocked or thinking [56], [63], [72]. In the thinking state, a user generates and transmits a new packet in a time slot with probability σ. A packet which had a collision and is waiting for retransmission is said to be backlogged. A backlogged packet retransmits in the current slot with probability p; thus a backlogged packet incurs a retransmission delay which is geometrically distributed. Let N^t denote the total number of backlogged packets at time t. Given the memoryless property of both the generation process and the retransmission process, N^t is a Markov chain with stationary transition probabilities. The state space consists of the set of integers $\{0, 1, \cdots, M\}$. The one-step state transition probabilities of N^t are easily derived (see [56]). For finite M the Markov chain is finite, irreducible and aperiodic and all states are positive recurrent; there exists a stationary distribution $\{\pi_j\}_{j=0}^M$. The channel input rate at time t is $S^t = (M - N^t)\sigma$. The average stationary throughput is then simply given by $S = (M - \overline{N})\sigma$ where $\overline{N} = \Sigma_j j \pi_j$. The average packet delay is equal to the average backlog time plus the transmission time of the packet (which equals one slot); the average backlog time, by Little's result, is simply $\overline{N}/S$. The (true) steady-state throughput-delay performance of a slotted-ALOHA system with finite population is thus obtained.

Consider now the slotted nonpersistent CSMA protocol in which the time axis is (mini-) slotted and the slot size is τ seconds, the propagation delay. Packets, assumed to be of fixed length, require a transmission time of T slots. Just as with slotted-ALOHA above, we consider here a user population consisting of M users (terminals), all in line of sight and within range of each other. Again each such user can be in one of two states: backlogged or thinking. In the thinking state, a user generates and transmits (if the channel is sensed idle) a new packet in a (mini-) slot with probability σ. A user whose packet either had a channel collision or was blocked because of a busy channel is said to be backlogged. A backlogged user remains in that state until he successfully transmits the packet at which time he switches to the thinking state. The rescheduling delay of a backlogged packet is also assumed to be geometrically distributed, i.e., each backlogged user is scheduled to resense the channel in the current slot with a probability ν; as specified in the description of the protocol, a retransmission would result only if the channel is sensed idle.

The memoryless property of the geometrically distributed retransmission delay will permit here again a simple state description for the mathematical model. The terminal stays in the backlogged state during the transmission period.

Let again N^t be a random variable representing the number of backlogged users at the beginning of (mini-) slot t. The channel input rate at time t, defined as the average number of new packets generated by the thinking users at time t, is denoted by S^t. Assume M, σ and ν to be time invariant. In slotted-ALOHA, the action that a terminal takes pertaining to the transmission of a packet (either newly generated or rescheduled) is completely independent of the state of the channel (busy or idle). In CSMA, on the contrary, the action taken by a terminal does depend on the state of the channel. One could increase the state description to include an indicator $\mathbf{\hat{y}}$ for the channel state (busy or idle), however, with *fixed length* packets, $(N^t, \mathbf{\hat{y}})$ is still not a Markov chain. Tobagi and Kleinrock [88] show that a simpler analysis can be obtained by using results from semi-Markov and regenerative processes. But before we proceed with the model, we give here a brief account on the notion of semi-Markov and regenerative processes and present results most relevant to this discussion.

2) Semi-Markov Chains: A stochastic process which makes transitions from state to state in accordance with a Markov chain but in which the amount of time spent in each state before a transition occurs is random, is called a *semi-Markov* process. Basic results similar to those obtained for Markov chains exist for semi-Markov chains. In particular, the process $X(t_e)$ taken at times t_e defined as immediately before or after a transition is a Markov chain and is referred to as the *imbedded* Markov chain.

If a stochastic process $\{X(t), t \geqslant 0\}$ with state space $\{0, 1, 2, \cdots\}$ has the property that there exists time points at which the process probabilistically restarts itself, then it is called a *regenerative* process. A cycle is said to be completed every time a renewal occurs. Let $E[.]$ denote the expected value of the random variable following the letter E. A main result relating to regenerative processes is the following

$$p_j \triangleq \lim_{t \to \infty} \Pr\{X(t) = j\}$$

$$= \frac{E[\text{amount of time in state } j \text{ during one cycle}]}{E[\text{time of one cycle}]} \quad (37)$$

for all $j \geqslant 0$; a very simple and pleasing result.

Furthermore, we can impose a reward structure on the process in the following manner. Suppose that when the process is in state j we earn a reward at a rate $f(j), j \geqslant 0$. Because of the regenerative nature of the process, it follows that with probability one

$$\lim_{t \to \infty} \int_0^t \frac{f[X(s)] \, ds}{t} = \lim_{t \to \infty} E \int_0^t \frac{f[X(s)] \, ds}{t}$$

$$= \frac{E \int_{\text{cycle}} f[X(s)] \, ds}{E[\text{cycle length}]} \quad (38)$$

Moreover,

$$\lim_{t \to \infty} E \int_0^t \frac{f[X(s)] \, ds}{t} = \sum_{j=0}^{\infty} p_j f(j). \quad (39)$$

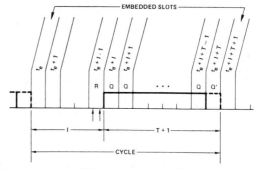

Fig. 5. The imbedded Markov-chain in slotted nonpersistent CSMA.

We now proceed with the semi-Markov model for the nonpersistent CSMA scheme. Referring to Fig. 5, which depicts transmissions and idle periods, consider the imbedded slots to be the first slot of each idle period. The intervals of time between two consecutive imbedded slots are defined as subcycles. These subcycles, of course, are of random length. Consider one such subcycle and let t_e denote the first slot. N^{t_e} denotes the state of the system at t_e. N^{t_e} is an imbedded Markov chain. Let R denote the one-step transition matrices over slot $t_e + I - 1$; Q, over slots $t_e + I$ through $t_e + I + T - 1$; and Q', over slot $t_e + I + T$. The determination of these matrices is easily done based on simple probabilistic arguments (for details see [88].) Given R, Q, and Q', the transition matrix P corresponding to the imbedded Markov chain N^{t_e} is easily calculated. For finite M, the stationary distribution $\Pi = \{\pi_0, \pi_1, \cdots, \pi_M\}$ where $\pi_i = \lim_{t_e \to \infty} \Pr\{N^{t_e} = i\}$ is obtained by solving recursively $\Pi = \Pi P$. This distribution exists for finite values of M.

It is clear that our process N^{t_e} is actually a regenerative process. Thus we can apply the above stated results. Consider an imbedded point t_e and its corresponding subcycle. Since the subcycle is entirely determined by N^{t_e} we consider N^{t_e} to be the representative state for the subcycle which is of random length $I + T + 1$. The success of a transmission in a subcycle N^{t_e} is also entirely determined by the value of N^{t_e}. Given that $N^{t_e} = n$, it is expressed as [88]:

$$P_s(n)$$

$$= \frac{(1 - \nu)^n (M - n)\sigma(1 - \sigma)^{M-n-1} + n\nu(1 - \nu)^{n-1}(1 - \sigma)^{M-n}}{1 - (1 - \nu)^n (1 - \sigma)^{M-n}}.$$
$$(40)$$

Let t_e' and t_e'' denote two successive regenerative points; they are such that $N^{t_e'} = N^{t_e''}$ and such that for any $t_e' < t_e < t_e''$, $N^{t_e} \neq N^{t_e'}$. The interval of time separating t_e' and t_e'' constitutes a cycle. π_j represents the fraction of subcycles such that $N^{t_e} = j$. We now derive the channel throughput. In each subcycle, we define a reward function $f(N^{t_e}) = P_s(N^{t_e})T$, representing the expected time during the subcycle that the channel is used without conflicts; then the throughput S can be written as (s denoting a generic imbedded point)

$$S = \lim_{t_e \to \infty} \frac{\sum_{s_e=0}^{t_e} f(N^{s_e})}{t_e} = \frac{E\left[\sum_{t_e=t_e'}^{t_e''} f(N^{t_e})\right]}{E[t_e'' - t_e']}. \quad (41)$$

The numerator is given by $\sum_{j=0}^{M} \pi_j T P_s(j)$; while the denominator is given by $\sum_{j=0}^{M} \pi_j [\bar{I}_j + T + 1]$; where $\bar{I}_j$ is the average idle period, given that $N^{t_e} = j$.

If we now define a cost function $f(N^t) = N^t$, representing the backlog at time t, then the average number of backlogged packets is given by (s denoting a generic minislot)

$$\bar{N} = \lim_{t \to \infty} \frac{\sum_{s=0}^{t} f(N^s)}{t} = \frac{E\left[\sum_{s=t'_e}^{t''_e} f(N^s)\right]}{E[t''_e - t'_e]}. \quad (42)$$

Given $N^{t_e} = j$, the knowledge of transition matrices R, Q, and G' allows the determination of [88]

$$n_j = E\left[\sum_{t=t_e}^{t_e + T + I + 1} f(N^t) \bigg| N^{t_e} = j\right].$$

The numerator is then expressed as $\sum_{j=0}^{M} \pi_j n_j$. By Little's result, the average packet delay is $\bar{N}/S$. The (true) steady-state throughput-delay performance of a CSMA channel with finite population is obtained (analytically!).

So far, we have shown how a model based on Markov chain theory and regenerative process has permitted us to derive analytic expressions for the average throughput-delay performance of some important random access schemes under a finite population constraint. This has constituted a considerable progress in our attempt to understand the behavior of these systems.

In particular, a very important observation of the results obtained from the above analysis has been the following. Even in a finite population environment (thus guaranteeing the existence of equilibrium), if the retransmission delay is not sufficiently large (i.e., the retransmission or rescheduling probability is not sufficiently small), then the stationary performance attained is significantly degraded (low throughput, very high delay), such that, for all practical purposes, the channel is said to have failed; it is then called an *unstable* channel. With an infinite population, the Markov chain is not ergodic and stationary conditions just do not exist [11]; the channel is always unstable thus confirming the results obtained from simulation as discussed in Section II-C above. For unstable channels Kleinrock and Lam [56] defined a stability measure which consists of the average time the system takes, starting from an empty state, to reach a state determined to be critical. In fact, this critical state partitions the state space into two regions, a safe region and an unsafe region. The stability measure is the average first exit time (FET) into the unsafe region (again a concept borrowed from Markov chain theory!). As long as the system operates in the safe region, the channel performance is acceptable; but then, of course, it is only valid over a finite period of time with an average equal to FET. For more details, concerning the determination of FET and the numerical results, the reader is referred to [56], [88].

E. Markov Decision Models

In the Markovian models discussed above it was assumed that the system parameters were all fixed, time invariant and state-independent. These models are referred to as *static*. Clearly, it is often advantageous to design systems that dynamically adapt to time-varying input and to system state changes, thus providing improved performance. Dynamic adaptability is achieved via dynamic control consisting of time and state dependent parameters. The basic problem is to find the control functions which provide the best system performance. If the system is Markovian in nature, then the theory known as Markov decision theory provides a basis for analysis. We first give in this section a brief introduction to Markov decision theory and the most relevant results. We then proceed with our example of packet-switching in radio channels and discuss various practical control schemes and their analysis.

Consider the process $X(t)$ and its state space δ labeled by the nonnegative integers $\{0, 1, 2, \cdots, M\}$. Let $\mathcal{C}$ be a finite set of possible actions such that corresponding to each action $\alpha \in \mathcal{C}$, a set of state transition probabilities $\{p_{ij}(\alpha)\}$ is specified and a cost $C_i(\alpha)$ is incurred. A policy f is a rule for choosing actions. Let $\mathcal{P}$ be the class of all policies. An important subclass is the class of stationary policies $\mathcal{P}_s$. A stationary policy is defined to be one which chooses an action at time t depending on the state of the process at time t. It easily follows that if a stationary policy f is employed, then the sequence of states $\{X_t, t = 0, 1, 2, \cdots\}$ forms a Markov chain with transition probabilities $p_{ij}[f(i)]$. It is thus called a Markov decision process, and it possesses stationary transition probabilities. Let us define the expected cost per unit time for $X(t)$ which was initially in state i as

$$\phi_i(f) \triangleq \lim_{\tau \to \infty} \frac{1}{\tau + 1} E_f\left[\sum_{t=0}^{\tau} C_{X(t)}[f(X(t))]/X(0) = i\right]. \quad (43)$$

For a stationary policy f such that $X(t)$ is irreducible we have the following result: $\phi_i(f)$ is simply expressed as

$$\phi_i(f) = \sum_{j=0}^{M} \pi_j(f) C_j(f) \triangleq g(f), \quad \forall i = 0, 1, \cdots, M \quad (44)$$

where $\{\pi_i(f)\}$ is the (unique) stationary distribution of $X(t)$ such that

$$\pi_j(f) = \sum_{i=0}^{M} \pi_i(f) p_{ij}(f)$$

$$\pi_j(f) \geqslant 0 \quad (45)$$

$$\Sigma \pi_i(f) = 1$$

$g(f)$ is also called the cost rate.

Another important result in the theory of Markov decision processes states that if every stationary policy gives rise to an irreducible Markov chain, then there exists a stationary policy f^* which is optimal over the class of all policies such that

$$g(f^*) = \min_{f \in \mathcal{P}} \Phi_i(f), \quad \forall i = 0, 1, \cdots, M. \quad (46)$$

This pleasing result means that, in most cases of interest, one can limit the attention to the class of stationary policies. A very efficient computational algorithm known as the Howard's policy iteration method [36], [37] exists which allows the evaluation of the cost rate $g(f)$ and which leads to an optimal stationary policy.

We continue here with our now familiar example of random access techniques. It follows from the discussion in the previous section that, if M is finite, a stable channel can be achieved by using a sufficiently small retransmission or rescheduling probability. But a smaller retransmission probability implies a larger backlog size and hence a larger packet delay! Moreover, it is noted that, for stable channels and for a given total throughput, the packet delay increases with increasing population size [56], [88]. Basically, the reason for this behavior is that the appropriate constant retransmission probability has to be small enough to overcome the degrading effect resulting from those statistical fluctuations which otherwise would drive the system into an "unsafe region." Dynamic control provides an effective solution to this problem; it enables an originally unstable channel to achieve a much improved throughput-delay performance, and conversely it improves the (otherwise high) delay performance of a stable channel with large M. With dynamic control we also allow the channel to accommodate varying input load without any disastrous effect.

Markov Decision theory has successfully been applied by Lam and Kleinrock [64] to the design and analysis of control procedures suitable to slotted-ALOHA in particular and random-access techniques in general. More precisely the objective is to decide on some practical control scheme and to derive the optimal stationary policy. Two main types of control are proposed: an *input* control procedure (ICP) corresponds to an action space consisting of either accepting or rejecting all new packets arriving in the current slot; a *retransmission* control procedure (RCP) corresponds to an action space $\{f(i)\}$ where $f(i)$ denotes the retransmission probability in a slot in which the backlog is i. A more general description of the action space, of which ICP and RCP are special cases, is as follows [64]. Let $\mathfrak{A}_1 = \{\beta_1, \beta_2, \cdots, \beta_m\}$ where $0 \leqslant \beta_1 < \beta_2 < \cdots < \beta_m \leqslant 1$ and $\mathfrak{A}_2 = \{\gamma_1, \gamma_2, \cdots, \gamma_k\}$ where $0 < \gamma_1 < \cdots < \gamma_k < 1$. Let $\mathfrak{A} = \mathfrak{A}_1 \times \mathfrak{A}_2$. A stationary policy f maps the state space $\{0, 1, \cdots, M\}$ into $\mathfrak{A}$ such that $f(i) = (\beta, \gamma)$ means that whenever the system state at slot t is $N^t = i$, then each new packet is accepted with probability β and rejected with probability $1 - \beta$ and each backlogged packet is retransmitted in that slot with probability γ.

Given a policy f, it is easy to write the one step transition probabilities $\{p_{ij}[f(i)]\}$ for the Markov decision process N^t. Under the condition that N^t is irreducible, the stationary distribution $\{\pi_i[f]\}$ exists. It remains to define the cost rates $C_i(f)$ and to determine the performance measures. For the sake of simplicity, let us restrict ourselves here to just RCP. The more general treatment can be found in [64]. Let $f = \{f(i)\}$ denote the policy. Supposing $N^t = i$, define the immediate reward $C_i(f)$ to be the expected channel throughput in the tth time slot, $\overline{S}_{out}(i, f)$. Then by (44), the reward rate becomes

$$g_s(f) = \sum_{i=0}^{M} \pi_i(f) \overline{S}_{out}(i, f). \qquad (47)$$

This is also the channel throughput rate. Consider now the stationary average packet delay. Supposing $N^t = i$, define the expected immediate cost to be $C_i(f) = i$ thus accounting for the waiting cost of i packets incurred in the tth time slot. By (44), the cost rate is

$$g_d(f) = \sum_{i=0}^{M} i \pi_i(f). \qquad (48)$$

Applying Little's result, the average "wasted" time of a packet is simply $D_w = g_d(f)/g_s(f)$. The main objective here is to find a policy which optimizes channel performance, that is, which minimizes the delay for a given stationary throughput. Fortu-

nately, this task is simple since it can easily be shown that for any stationary policy $f = \delta \rightarrow \mathfrak{A}$, we have

$$g_d(f) = -\frac{g_s(f)}{\sigma} + M \qquad (49)$$

meaning that if there exists a stationary policy which minimizes the cost rate $g_d(f)$, this policy will also maximize the reward rate $g_s(f)$ [64]. Having decided upon a channel control procedure, the optimal policy is determined via Howard's policy-iteration algorithm. This will also allow the determination of the associated optimum performance [64].

Independently, Fayolle et al. [19] give yet another treatment of the instability of slotted ALOHA channels with infinite populations and propose similar control procedures to recover stability. In particular, it was shown that, with retransmission control procedures, only policies which assure a rate of retransmission f from each blocked terminal which is inversely proportional to the number of backlogged terminals, will lead to a stable channel. An expression for the optimal policy $\hat{f}$ was also given. In their paper, Lam and Kleinrock [64] suggest that a good control policy must be of the control limit type. Their intuition was confirmed by the numerical solutions obtained from Howard's policy iteration algorithm; however, there is no rigorous proof of optimality. In [88] Tobagi and Kleinrock similarly addressed themselves to the dynamic control of the nonpersistent CSMA protocol. In essence, it was shown that one can improve the channel performance by selecting the retransmission probability which maximizes the "instantaneous" throughput, that is, the average throughput over a subcycle. The resulting overall channel performance was further shown to be then insensitive to the population size.

It is apparent to the careful reader that the preceding (exact) analysis is based on a major system assumption, namely that each user knows the exact current state of the system. Clearly, this assumption does not hold in practical situations! The channel users have no means of communication among themselves other than the multiaccess broadcast channel itself. But each channel user may individually estimate the channel state by observing the channel outcome over some period of time, and apply a control action based upon the estimate. In the context of slotted-ALOHA, Lam and Kleinrock [64] give some heuristic control-estimation algorithms which prove to be very satisfactory. With appropriate modifications and extensions, these algorithms can be applied to CSMA channels as well. The difficulty in incorporating the estimation algorithms into the mathematical model incites us again to the use of simulation techniques. The results obtained by the mathematical model assuming full knowledge of the system state represent the ultimate performance; a bound on the performance obtained via any heuristic estimation algorithms. The goodness of the latter is evaluated by comparing simulation results to these bounds. (For numerical results, the reader is referred to [64].)

F. Mathematical Programming

So far we discussed modeling tools which have been mainly used to evaluate throughput and delay performance of communications systems without paying much attention to optimization issues (except, perhaps, in the development of Markov decision models). Here, we address the optimization problem more directly, and review the mathematical programming tools commonly used in network design.

The typical problem consists of optimizing a performance measure (e.g., cost, delay, throughput), over a set of variables, meeting given performance constraints. The ability to solve the design problem depends critically on our success in expressing both objective function and constraints in analytically manageable form as a function of the design variables. Thus network models are an essential prerequisite to design.

Unfortunately, most performance expressions are rather complex, requiring approximations in the model or relaxation of the constraints in order to formulate the design problem in convenient terms and solve it with powerful mathematical programming techniques. A typical example of this approach is the linearization of discrete line costs in the minimum cost design of land based packet networks [57]. In some cases, the nature of the variables and of the constraints is so complex that a mathematically manageable formulation of the problem is not possible. An example is the topological network design to satisfy two-connectivity constraints [83]. In these cases only heuristic approaches can be of help (see Section II-H).

Design variables may be continuous (e.g., link data rates, packet length) or discrete (e.g., topology, number of switch sites).

Examples of methods commonly used in continuous variable computer network designs are: 1) linear programming; 2) Lagrangian optimization; 3) multicommodity flow optimization (discussed in the following section); 4) gradient projection method. Examples of methods used in discrete optimization are: 1) dynamic programming; 2) Lagrangian decomposition; 3) branch and bound [32].

In this section, as an example, we discuss the Lagrangian method as it applies to the Capacity Assignment (CA) problem in land based packet networks [57].

The CA problem can be formulated as follows.

Given: topology, average link data flows, $f = (f_1, f_2, \cdots, f_M)$

(where M is the number of links and f_i is the bit rate on link i and is given by $f_i = \lambda_i/\mu$),

minimize link costs: $D = \sum_{i=1}^{M} d_i(C_i)$

over the selection of link capacities $C = (C_1, C_2, \cdots, C_M)$ subject to: $\quad C \geqslant f$

$$T = \frac{1}{\gamma} \sum_{i=1}^{M} \frac{f_i}{C_i - f_i} \leqslant T_{\text{MAX}}.$$

Line capacity options are in general discrete. To simplify the problem we may approximate the discrete capacities with continuous values; furthermore, we may approximate the cost function with a linear function, i.e.:

$$d_i(C_i) = d_i C_i + d_{i_0} \tag{50}$$

To solve the linearized problem we use the method of Lagrange multipliers. To this end we construct the Lagrangian L, defined as the sum of the objective function plus the constraint function multiplied by the multiplier β:

$$L = D + \beta(T - T_{\text{MAX}})$$

$$= \sum_{i=1}^{M} d_i C_i + d_{i_0} + \beta\left(\frac{1}{\gamma} \sum_{i=1}^{M} \frac{f_i}{C_i - f_i} - T_{\text{MAX}}\right). \tag{51}$$

By setting the partial derivatives $\partial L/\partial C_i$ to zero, and choosing β so that the delay constraint is satisfied, we obtain the optimal expressions for C_i:

$$C_i = f_i + \frac{\sum_{k=1}^{M} \sqrt{d_k f_k}}{\gamma T_{\text{MAX}}} \sqrt{\frac{f_i}{d_i}}. \tag{52}$$

G. Multicommodity Flow Optimization

In earlier sections we showed how to evaluate delay performance in land based networks, assuming static routing. In most networks, however, route selection is adaptive to load pattern and to network conditions. Since the delay performance is critically dependent on the routing policy, we wish to develop models that predict delay performance also in a dynamic routing environment.

Unfortunately, the general dynamic routing problem is very complex. Network of queues theory may be used. However, the fact that transition probabilities depend on network load precludes the derivation of "product form" solutions (see Section II-B). The system may still be represented as a very general Markovian system and solved using numerical techniques. This type of solution, however, is computationally very cumbersome and offers little insight into the dependence of network performance on dynamic routing parameters, let alone their optimization.

To overcome this problem, we approximate the dynamic routing solution with the optimal static solution. More precisely, we first find the optimal static routing strategy using mathematical programming techniques. Then we verify that the dynamic strategy performs almost as well as the static strategy at steady state. This verification may be carried out using Markovian models in simple cases, and simulation in more complex situations.

The advantage of this two-stage approach is that the verification stage although computationally cumbersome needs to be carried out only for a few representative benchmarks, while the static optimization stage (which must be solved an endless number of times in a typical network design problem with several topological alternatives) is performed very efficiently using multicommodity flow techniques.

Multicommodity flow techniques are mathematical programming techniques used to optimize the distribution of "commodities" (in our case, packet flows) throughout a network, between several origin–destination pairs. The problem constraints are generally the line capacities. The objective may be the total throughput, or the delay, or another appropriate function of the flows in the network.

The multicommodity flow matrix F for a data network is an $N \times L$ matrix (N = number of nodes; L = number of links) whose entry $F(i, k)$ represents the average data flow (bits per second) on link k with final destination i. Each row of F represents the "single commodity" flow to a distinct destination.

The matrix F uniquely identifies the steady-state routing scheme. In order to find the optimal routing solution (at steady state) we just need to optimize $g(F)$, where $g(F)$ is the desired performance measure. This optimization can be carried out very efficiently using a decomposition approach. We decompose each single commodity flow into the convex sum of the flows on all possible paths to a given destination. Clearly, the number of possible paths can grow very large, but one can show that only at most N paths are included in the optimal solution. Using an iterative procedure (called flow-deviation method [26], we introduce at each step a new path that can improve performance. The selected path is the shortest path evaluated using as weight for link i the partial derivative $W_i = \partial g(F)/\partial f_i$, where f_i is the total flow (sum of all commodities) in link i. An appropriate amount of traffic is then "deviated" from the previous paths to the new shortest path. The iterative procedure terminates when the relative improvement obtained by the deviation falls below a specified tolerance, at which point we have reached a local minimum.

Multicommodity flow techniques can be used to solve a variety of problems in data network designs. Here we consider, as a specific example, the problem of finding the minimum delay routing in a land-based packet-switched network. The problem is formulated as follows.

Given:
Topology
Channel capacities
Traffic requirement matrix R.
Minimize:

$$T = \frac{1}{\gamma} \sum_{i=1}^{M} \frac{f_i}{C_i - f_i}.$$

Over the design variable:

$$f = (f_1, f_2, \cdots, f_M).$$

Subject to
a) f is a multicommodity flow induced by requirement matrix R
b) $f_i \leqslant C_i$.

It is easy to verify that the objective function $T(f)$ is convex, and, therefore, the locally optimal solution found by the Flow Deviation Algorithm is also globally optimal. The link weights used at each iteration are:

$$W_i = \frac{\partial T}{\partial f_i} = \frac{C_i}{(C_i - f_i)^2}, \qquad \forall i = 1, \cdots, M. \qquad (53)$$

In general, the key requirements for the successful solution of the routing problem are the capability of expressing the delay as a function of link flows, and the convexity of such a function. For land-based networks, we have just shown that such requirements are satisfied. One may show that the requirements are also met for mixed terrestrial and satellite networks [29]. Distributed packet radio networks, on the other hand, the delay formulation is much more complex because of the interference existing between neighboring nodes.

H. Heuristic Techniques

Heuristic and approximate solutions are often the only feasible approach to some of the more complex analysis and design problems related to large packet networks. A classification of the various heuristic methods is certainly beyond the scope of this paper. Interested readers are referred to [67] for an overview. One particularly important class of heuristics, however, will be discussed here: namely, the class of topological design algorithms.

The topological optimization of a packet network is a formidable mathematical programming problem made difficult by the combination of integer type variables (switch number and location, topology, line speeds) and multicommodity flow variables describing the routing of packets in the network.

Therefore, the only practical way of approaching medium and large network designs is via heuristics. Several procedures have been proposed, which are based on different concepts. The common philosophy of these procedures consists of identifying a condition which is necessary (although generally not sufficient) for optimality, and of achieving this condition by means of repeated topological transformations until a locally optimal solution is found. Starting from randomly chosen initial configurations, a large number of local minima is explored to enhance the probability of success of the heuristic method.

One of the most popular topological design heuristics is the Branch X-Change (BXC) method [24]. The local condition for optimality is the condition that the cost be not reduced by any BXC (i.e., the elimination of one or more old links and the insertion of one or more new links). Thus, the BXC algorithm explores exhaustively all the feasible BXC's, accepting only the X-changes that lead to cost reduction, until no more improvements are possible.

Another, more recent algorithm, the Cut Saturation (C-S) algorithm , can be viewed as a refinement of the BXC Algorithm in that only a selected subset of the possible X-changes is explored [25], [30]. More precisely, only the X-changes involving the insertion of links in the Saturated Cut (i.e., the minimal set of most utilized links that, when removed, disconnects the network) and the deletion of lightly utilized links are considered.

A third algorithm, the concave branch elimination (CBE) is based on flow optimization concepts [32]. The CBE algorithm can be applied whenever the discrete line costs can be approximated with continuous concave curves of cost versus capacity. Under these assumptions the total network cost for a given topology can be expressed as a concave function D of the link flows, namely:

$$D(f) = \sum_{i=1}^{M} \left[d_{i_0} + d_i f_i + \frac{\left(\sum_{k=1}^{M} \sqrt{d_k f_k} \right)^2}{\gamma T_{\text{MAX}}} \right]. \qquad (54)$$

We can then apply the flow-deviation algorithm to obtain a minimum. This minimum is only a local minimum since the function $D(f)$ is concave (instead of convex, as in the routing problem). In the process of finding a local minimum, it can be shown that the algorithm eliminates uneconomical links (i.e., reduces their flow to zero), and, therefore, strongly reduces the initial topology. Exploiting this locally optimal property, several local solutions are investigated starting from different initial configurations.

Several other heuristic methods have been proposed and applied with more or less success to various types of networks. The common element of all the methods is the existence of a local optimality principle and the need to randomly explore several solutions in order to improve the accuracy of the design.

I. Simulation

As we have noted at several points in our earlier discussion, analytic modeling techniques are inadequate to deal with many of the details of a system to be modeled. Simulation is then used.

Basically, simulation has two main purposes: a) the performance evaluation of network protocols that are analytically intractable; and b) the validation of analytical models based on simplifying assumptions. These purposes, however, are adequately served only if the simulation model itself is valid; and thus techniques that can be used to guarantee the validity of the model and its results are required. We discuss each of these issues in more details and supply the reader with some typical examples.

Numerous examples exist where simulation is used as a tool to evaluate alternative protocols. Unfortunately, there is no particular work that serves as a basis for a discussion here as was the case with analytic modeling. We refer the reader to applications which have appeared in the literature on the ARPANET, the Atlantic Packet Satellite Network, the NPL network and the Cigale Network [27], [39], [33], [49], [77].

A recent example of simulation applied to validate analytical models is offered by Lam's study on network congestion control [65]. In this model, each nodal switch is represented by a network of queues. The nodes in turn are interconnected by a higher level network, which is, in fact, the network topology. Due to buffer constraints and, consequently, nodal blocking, the global network of queues does not have a convenient "product form" solution, as discussed in Section II-C. However, by postulating static (i.e., state independent) blocking probabilities for the nodes, the global model was reduced to a collection of independent submodels (one for each packet switch), which were solved separately, using the product form approach. Clearly, the assumption of static blocking probabilities is of critical importance for this study. Thus it was thoroughly tested using a very accurate simulation program. Experiments showed good agreement between simulation and analytical results.

Analytic queueing models and simulation queueing models go hand in glove. We have pointed out that, when we are trying some new technique of analysis it is useful to cross check that analysis with the results of simulation runs. Also very important is the use of analytic models to validate large simulation programs. It is often the case that a simulation program will be thousands of lines of code (in Simscript, say) and the modeler is faced with the question of whether or not this enormous program actually models the system in question. Now it is often the case that the logic of the program can be tested by comparing its results with an analytic model when the program parameters are set to correspond. That is, much of the complexity that cannot be handled in analytic models may appear in the simulation as parameter settings. When the parameters are set to certain values the simulation may be modeling a Markovian network of queues for which we can analytically predict the behavior. If the simulation produces correct results for this case then one is inclined to believe that the results for other parameter settings are also correct. An example of this approach can be found in [62].

When we perform a simulation we usually gather statistics on certain variables that we wish to measure (queue lengths, transit times, ···). But these statistics are not very accurate since they are derived from highly correlated samples. Their proper interpretation requires the use of time series analysis rather than classical statistics. It is all too common, however, to find that simulation results are quoted for a single run of each desired experimental point. Simulation languages often gather statistics using models based on classical statistical theory. However, a correct approach requires the experimenter to obtain a large number of independent samples (40 to 100 samples, say) if classical analysis is to be used to estimate the mean of some measured quantity and to place a confidence interval around that estimate. This would involve thousands of computer runs each of which may be several hours long. Few people can afford this; and the typical strategy is to take the one observation as correct. There are very many problems where this will work, but it is fortuitous and modelers must beware of extrapolating such successes.

An alternative scheme is to break the simulation run down into several "subruns" in which statistics are separately gathered. These runs hopefully are independent. The problem is to be sure of this. Recent work by Crane and Iglehart [14], [15] and by Fishman [21], [22], [23] helps us here. They carefully explain the problem and suggest the use of "regeneration points" to obtain independent samples. The basic idea is that systems return periodically to certain configurations and that the behavior of the system after reaching such a regeneration state is independent of its behavior prior to entering that state. Thus by taking separate samples during the cycles between entry into the regeneration state one obtains the required independent samples that permit the use of the simple classical statistical tools. The most recent work by Fishman [23] offers practical guidance here.

Regeneration points are not a panacea. The problem is that the cycle time between reentry may be too long for practical application. Consider a system with 10 queues each with finite waiting room for 10 customers. There are 10^{10} possible states! Nevertheless, the above referenced material is important reading since it offers useful insight into the problem of establishing confidence in our simulation experiments.

These problems are one reason that analytic models are to be preferred to simulation if they can be used. They give results much more cheaply, even if several weeks are required to develop the model. Furthermore, they aid our intuition. It is much easier to comprehend the implications of even the most complex formulas than it is to comprehend the meaning of 8000 lines of Simscript code and a basketful of output tables! Both tools are needed.

Finally, a novel approach known as *hybrid simulation*, is worth mentioning. Basically, the idea is to combine both discrete-event simulation and mathematical modeling in an attempt to achieve good agreement with the results of an equivalent complete simulation model, but at a significantly lower cost. This approach is possible if the frequency of state transitions of some portions of the system is much higher than those of other portions. Then the high-frequency events are accounted for in a computationally efficient analytic submodel while the relatively infrequent (and often more complex) events are accurately simulated. Parameters are exchanged back and forth between the various submodels. Hybrid models have been successfully applied for the analysis of computer systems [10a], but have not yet been widely used in computer communications systems.

III. MEASUREMENTS

Manfred Eigen wrote: "A theory has only the alternative of being right or wrong. A model has a third possibility: it might be right but irrelevant" [17]. Indeed, most if not all of the modeling work is based on simplifying assumptions without which the analysis becomes intractable; and with these assumptions we run the risk of providing results which do not exactly conform to the real situation. "Irrelevant" is perhaps too strong a word: in the absence of a real system (that is, in the early design phase) analytic and simulation models are the only tools available to guide us in first implementations. But once the system is built, measurements allow us to gain valuable insight regarding the network usage and behavior. They provide a means to evaluate the performance of the operational protocols employed and the identification of their key parameters; they allow for the detection of system inefficiencies and the identification of design flaws; when used to improve network design, they are a valuable feedback process by which existing analytical models are validated and/or improved, and in which design deficiencies are detected and subsequently corrected. Thus contrary to early designed computer systems which did not allow sufficient freedom in experimentation, and in line with Hamming's observation that "it is difficult to have a science without measurement," elaborate measurement facilities constitute an integral part of all experimental and many operational computer communications systems of today. Basically, the measurement task consists of identifying the measurement functions with respect to the system and devising the measurement facilities required to support those functions under the constraints that the system imposes. In this section, we shall first review the basic measurement tools, their capabilities, their limitations, and their applicability to and implementation in different network environments, namely, land-based wire networks, satellite networks, and ground packet radio networks; next, we shall show the importance of well designed experiments in satisfying the many measurement goals.

A. Measurement Facilities

Although the objective of measurements is basically the same for all types of networks (wire, ground radio, or satellite), several factors exist which do not allow for a simple transfer of measurement facilities from one to the other. The techniques may very well be the same, but the implementations of these tools will have to be compatible with the particular system's design and comply with its limitations.

Most of us are now familiar with land-based wire packet-switched networks and their structure; the switches are minicomputers which provide the store-and-forward function and handle routing and error control; typical examples are the ARPANET [79] (in which the nodal switch is called the IMP), the Cigale Network [76a], TELENET [75a], DATAPAC [10b]. Satellite and ground radio networks, however, are far less common than wire networks and a brief introduction here is in order. In a satellite system (an example of which is the SATNET [33], [43], a node is basically a minicomputer switch similar to the ARPANET IMP interfacing with the satellite channel by means of satellite radio equipments. All nodes share a common satellite channel via some access scheme. In a ground radio environment, the issues are somewhat more complex. Besides the original but simple one-hop ALOHA system at the University of Hawaii, the only and prominent example of a fully distributed radio network is the ARPA Packet Radio Network (PRNET), a prototype of which has already been deployed in the San Francisco Bay Area [44], [45]. There are three basic functional components in the PRNET: (i) the packet radio terminals which are the sources and sinks of traffic, (ii) the packet radio repeaters whose function is to extend the effective radio range by acting as store-and-forward relays; and (iii) the packet radio stations which provide global control for the network of repeaters and act as interfaces between the broadcast system and other

computers or networks. The repeater is called a packet radio unit (PRU) and consists of a radio transceiver and a microprocessor. It receives and transmits packets according to dynamic routing and control information provided by the stations. For simplicity and uniformity of design the PRU is also used as the front-end of terminals and stations interfacing them with the radio net.

Although the PRNET utilizes the technique of packet-switching, the packet radio measurement facilities are unique with respect to the system constraints [87]. These constraints are in large part due to the desire to keep the components small and portable, to the limited speed of the microprocessor (which in turn is due to the assumed limited power supply available in some military applications) and to the limited available storage at each PRU. The overhead placed on the PRU is also of utmost importance in evaluating the feasibility of a measurement tool and of the collection of data in support of a measurement function. In particular, due to the broadcast nature of transmissions, the transmission of collected measurement data not only introduces overhead over its own path, but causes interference at all neighboring repeaters within hearing distance and creates additional overhead on those PRU's activities. The development of the measurement tools in the PR Net is an excellent illustration of the iterative design process involved whereby a balance is sought between supporting the measurement functions and satisfying the system constraints, thus making sure that the network communication protocols allow the implementation and proper functioning of those tools. Specific examples will surface in the sequel substantiating this statement.

We now describe the various types of statistics used, and the artificial traffic (and noise) generators needed in measurement experiments and the various techniques available for data collection.

1) Cumulative Statistics (CUMSTATS): Cumulative statistics consist of data regarding a variety of events, accumulated over a given period of time. These are provided in the form of sums, frequencies, and histograms. In the ARPANET, for example, cumstats are collected in each IMP and include a summary of the sizes of messages entering and exiting the network [12], [51], the number of IMP words, the number of control messages, etc. "Global" traffic data are also collected; they are referred to as round-trip CUMSTATS; they include the number of round-trips per possible destination and their delays.

In the PRNET, distinction is made between cumstat data collected at the PRU's (which provide information about the local environment and behavior such as traffic load, channel access, routing performance and repeaters' activity) and those collected at the end devices (which reflect more global network behavior such as user delays and network throughput). A detailed list of the data items of interest in PRNET experiments can be found in [87].

The implementer may add the flexibility of tailoring to some degree, the content of the CUMSTAT to the requirements of a specific experiment. In SATNET, for example, most of the items in the CUMSTAT message are optional and may be requested by the experimenter at the beginning of the experiments. The advantage of this solution is to make available a very large set of measurements, without the line and processor overhead usually required to construct and transmit long CUMSTAT messages.

2) Trace Statistics: The trace mechanism allows one to literally follow a packet as it flows through a sequence of nodes and thus trace the route it takes and the delays it encounters at each hop. In the ARPANET, selected IMP's whose trace parameter has been set gather data (at the IMP's themselves) on each (trace marked) packet and send this data to the experimenter at the collection point as a new packet. Trace data consists of time-stamps related to the time of arrival, the time of transmission, and the time that the acknowledgment is received [12]. While the above implementation has been possible in the ARPANET, in a PRNET, the collection of trace data at the repeaters is prohibited by the limited size of storage in the PRU. To overcome this problem, a new type of packet called the pick-up packet has been introduced [87]. Pick-up packets are generated with an empty text field by traffic generators at end devices. As these packets flow normally in the network according to the transport protocols, selected repeaters gather the trace statistics and store them within the text field of the pick-up packets themselves.

3) Snapshot Statistics: A Snapshot provides an instantaneous look at a device (IMP, PRU) showing its state with regard to various queue lengths and buffer allocation. Although this information can be obtained by a time sequence of state changes, the snapshot technique is preferable in that it reduces considerably the overhead and artifact caused by collecting and sending the statistics too frequently. In the ARPANET, snapshots also include the IMP routing table and its delay table. The correlation of these with other collected statistics will help explain abnormal or unexpected behavior. In the PRNET, the stations play a central role in providing some degree of global control for the operation of the entire network. They contain tables describing the network connectivity, the repeaters states and their parameter values. Changes in appropriate tables are time stamped and collected as the stations snapshot functions.

4) Artificial Traffic Generators: Artificial traffic generation is clearly a requirement of any experimental system. In the absence of real user traffic, the experimenter is thus given the ability to create streams of packets between given points in the net, with specified durations, inter-packet gaps, packet lenghts, and other appropriate characteristics. In the ARPANET, the IMP's message generator can send fixed length messages to one destination; in the PRNET a higher level of flexibility is implemented: each traffic source (at terminals and stations) can provide one or more streams of both "information" packets and/or pick-up packets, each with a specified packet length, frequency, destination and duration. In SATNET, each station may generate up to 10 independent streams of artificial traffic, each stream having its own characteristics (generation rate, packet length, priority, etc.). Furthermore, the generation rate of a stream may be changed to preset values at preset times during an experiment. For example, the generation rate of a station may be set to 0 in the interval $(0, T_0)$; to 1 in $(T_0, T_0 + \Delta T)$; and again to 0 in $(T_0 + \Delta T, \infty)$. The result, in this case, is a pulse of amplitude 1 and duration ΔT, at time T.

5) Emulation: In most initial experiments, the system under investigation consists of a limited number of elements, thus placing severe constraints on the experimenter in his attempt to understand the system behavior in future and more realistic environments. This makes it desirable to emulate in a single element the traffic that would be generated by several separate sources. An interesting example of multiple node emulation is

offered by SATNET. The physical configuration of SATNET consists of 4 stations, a number too small to carry out any meaningful type of stability experiments. The experimental capabilities in this direction were considerably expanded by implementing in each real station 10 "fake" stations equipped with all the essential protocols to permit their independent operation.

6) Network Measurement Center—Control, Collection, and Analysis: The need for controlling the measurement facilities, and collecting and analyzing measurement data, gives rise to the notion of a network measurement center (NMC). For the three above mentioned networks (ARPANET, PRNET, and SATNET), for example, the University of California at Los Angeles (UCLA) has been successfully playing the role of NMC. For the French Cigale network, the measurement task is being undertaken by IRIA [35].

In the ARPANET, messages are sent to a background program in the IMP to trigger the necessary parameter change; in fact, once an experiment is specified, these messages are automatically formatted and sent by a set of programs constructed at the NMC. Conversely, measurement data is gathered at the various sites, formatted and routed to UCLA-NMC where it is stored and analyzed. Similar techniques are employed for the PRNET and SATNET while using the ARPANET to communicate with the measurement facilities and to transport the data back. In the PRNET, in particular, it is through the station that the initiation and termination of measurement experiments is controlled. When provided with the appropriate commands, the station enables and disables the CUMSTAT and Pick-up packet functions at the PRU's and assigns to the various elements the intervals for CUMSTAT collections, and to the artificial traffic generators their corresponding parameters. At the present time, it is also to the station that all measurement data is destined; upon arrival at the station, the data is time-stamped and stored in a single measurement file for shipment to UCLA–NMC where off-line reduction and analysis is performed.

It is possible that if enough care is not taken, the collection of measurement data at a network measurement center will create overhead traffic on the network, and serious considerations have to be made as to the techniques used. For an illustration, let us limit the discussion here to the PRNET context, and consider specifically the collection of cumulative statistics at the station using the PRNET itself. Two ways can be thought of to achieve the transport of the data. One method, called the Automatic method, consists of having the PRU form, at the end of each CUMSTAT interval, a measurement CUMSTAT packet which is time-stamped and transmitted to the station. The second method consists of having the station issue at regular intervals executable packets (control packets with code to be executed at the destination PRU) called Examine packets, to PRU's; these collect the time-stamped CUMSTAT data and return to the station. Both methods are certainly valid; but since for analysis purposes, it is strongly desirable for the CUMSTAT data received at the station to correspond to equal length time intervals at the originating device, the automatic method in conjunction with a reliable transmission protocol between the PRU's and the station is preferred. With the Examine method, variable length intervals will occur since Examine packets, even though sent at regular intervals from the station, are subject to network random delays en route to the PRU and subject to the possibility of loss in either direction. Analysis of these collection methods and the overhead they impose appeared in [86]. Also, to alleviate the congestion around the station due to the convergence of (i) traffic normally flowing through the station, (ii) control traffic carrying information needed at the station, and (iii) measurement traffic, it is not unlikely that the future design will use a separate station remote from control stations for the sole purpose of measurements and other less vital system functions. In fact, in recent experimentations with the CIGALE network, to prevent any interference with the operational CIGALE traffic, the IRIA measurement group went to the extreme of installing a measurement laboratory isolated from the network [35]. Such a setup was perfectly adequate for their need, namely measuring the performance of line protocols in an environment of somewhat limited complexity.

7) Network Control Center—Status Report, Monitoring, and Control: In addition to the measurement effort which is primarily for experimentation and performance evaluation, certain measurement data can be of vital importance to the proper operation of a network. With the notion of network control emerges also the concept of a network control center (NCC). In the ARPANET, each IMP sends periodically a status report to the NCC containing various data such as the up-down status of each HOST and channel, a count of the number of packets entering each IMP, and other statistics regarding each channel and the traffic it carries. Data is processed at the NCC creating summary statistics and advising the operator of failures. Network monitoring and control is even more significant in a PRNET where, in order to satisfy the constraints imposed by the repeaters, one or more station have the responsibility for distributed control over an entire region of the network (reliability is achieved through redundancy). To carry out this responsibility, the station requires various indications of network activity and performance. Some of this information is acquired from incoming traffic, but much of it is specifically obtained by having monitoring procedures collect, from the various devices, a subset of the measurement items. These will assist the control and routing algorithms implemented at the stations in taking the proper actions.

B. Measurement Experiments

After having described the types of measurement tools, available in packet networks, we proceed to show how these tools are selected and coordinated to carry out a specific experiment. We first define the object of our experiment, i.e., the specific aspect of the network that we want to investigate (e.g., channel access protocol in a ground radio network), along with the goals of the experiment (e.g., verification of correct protocol implementation). Then we select the performance measures which best characterize the aspect of the network under study (e.g., in a random access experiment we may choose to monitor the number of retransmissions until success). Finally, we select the appropriate subset of measurement tools (e.g., CUMSTATS; Pick-up packets etc.) which permit us to monitor the desired performance measures and, if necessary, to follow step by step the various network operations. Thus four ingredients come together to form an experiment: 1) the measured object of the measurement; 2) the goals; 3) the performance measures; and 4) the tools used in the experiment.

The complete list of experiments proposed for a given system constitutes the so-called experiment plan. A preliminary experiment plan is generally prepared before the implementation of the network. Based on this plan, the network designer and the experimenter agree on a set of tools that are adequate to carry out the desired functions, and yet do not overtax the system. Naturally, after network implementation, the experiment plan is frequently revised using as feedback the results of previous experiments. A new set of experimental requirements may arise which were not anticipated during the preliminary plans and for which there is no adequate measurement software support. While this occurrence may be minimized by careful preplanning (possibly assisted by analytical and simulation models), a more general solution is the flexible design of the measurement software to permit modifications and expansions of existing tools to meet new demands.

An example of production interaction between measurement planning, interpretation of the results, and new measurement software design is offered by SATNET, the Atlantic experimental satellite network. In SATNET the software was developed in stages, each stage having a different set of operational capabilities and measurement tools. This allowed the upgrading of the measurement software at each step, based on previous experience. In particular, the traffic generator, a very critical element in the testing of channel access performance, underwent a remarkable evolution throughout the duration of the experiments. In the original implementation the artificial messages were subject to the ARPANET, RFNM based end-to-end protocols [33]. Early measurement results showed that this protocol structure was too restrictive in the presence of large satellite delays and would actually trigger undesired "capture" effects. A new generation, therefore, was developed to provide a separate source of packets at constant rate. Later on, during the experimentation of S-ALOHA channel access protocols, the need was identified for a time varying packet generation rate. This feature was also introduced making the evaluation of channel stability properties much more effective. More recently, the generator in each station was upgraded to generate a certain number of parallel packet streams, each stream having different rate, priority and message length characteristics. This feature was used to simulate a diversified user environment, necessary to exercise the very sophisticated, priority oriented, demand assignment protocols implemented in SATNET [42].

1) Objects of the Measurements: Let us return to the basic ingredients that define an experiment. First, we focus on the object, i.e., the aspect of the network that we want to test. Here the following classification may be found useful.

a) Experiments on communications subnet protocols: These experiments may be directed to the study of specific network protocols (e.g., routing protocol; acknowledgment protocol; source-destination node protocol [52]) or may involve the interaction of all the protocols (global performance evaluation). Generally, only artificial traffic is used in this phase to better isolate the behavior of the protocols from user related effects.

b) Experiments on user behavior: We are interested in determining user traffic characteristics (traffic pattern; message length distribution, etc.) which are independent of the subnet itself. An excellent example of this type of experiment is offered by the investigation of ARPANET user behavior by Kleinrock and Naylor [51].

c) Experiments on user performance in presence of subnet protocols: Here we want to study the performance of a Host-to-Host (more generally, user to user) connection across a packet network. Typical examples of such experiments include the evaluation of the quality of digitized speech transmitted through a packet network, and the evaluation of Host-to-Host protocols [6].

d) Experiments on nodal processor performance: We are interested in determining throughput, delay and reliability characteristics of a nodal processor which supports given protocols. This set of experiments differs from the subnet protocol experiments in a) in that we want to assess the limitations of hardware and software implementation of the protocols, rather than the intrinsic limitations of the protocols themselves. Clearly, in an operational network, the subnet performance will depend both on protocol design and hardware and software implementation. An example of experiments aimed at the nodal processor evaluation is the series of throughput tests performed on the ARPANET IMP [71].

Undoubtedly, the majority of packet network experiments reported so far in the literature fall under category a). This is explained by the fact that the packet technology is still young (especially in the satellite and ground radio areas) and there are many performance issues to be clarified at the internal network protocol level before considering higher level protocols. Generally, internal subnet performance is better studied in a controlled traffic environment (i.e., artificial traffic generators) rather than in a real traffic environment. There is a growing interest, however, in the experimental evaluation of end-to-end performance as seen by the user and of the interaction of subnet protocols and higher level protocols (i.e., type c) experiments). The recent experimental networks are promoting both subnet and end-to-end experiments. In SATNET, for example, one of the important experimental issues is the performance of the TCP protocol (a Host-to-Host level protocol) [10] in presence of different channel access schemes. In this case, TCP experiments are carried out using specialized measurement facilities in the Hosts.

User behavior (type b)) and nodal processor performance (type d)) are generally given more attention in an operational environment than in an experimental environment. In fact, the main goal of an experimental network is to demonstrate the validity of a new communications concept. This evaluation is usually carried out independent of the actual hardware implementation of the nodal processor, and assumes very general user traffic characteristics. (Clearly, after having verified the validity of the communications scheme, some analysis of the actual user behavior, and a rigorous testing of the communications hardware and software is necessary before committing to a full scale operational net).

In contrast to experimental networks, the main goal of an operational network is to provide reliable service to its users. Therefore, the network manager must constantly monitor the trends in the user load profile and must be aware of the hardware and software limitations of the existing equipment in order to plan appropriate system expansions/modifications.

2) Measurement Goals: As we mentioned earlier in this section, measurements in an experimental network are motivated by one or more of the following goals:

a) Software verification: Although the implementor generally performs a systematic checkout of each software component before release, it is advisable that the experimenter test the software after field installation, to make sure that it operates according to the specifications.

b) Performance evaluation and verification: System performance is evaluated to determine whether the system meets the original design goals, and eventually to identify the applications for which the system can be effectively used. Measured performance parameters are compared with modeling results to verify the validity of the models.

c) Feedback for system design iterations: In a complex system, some of the parameters which affect performance may be properly tuned only using experiments on the real system, since analytical and simulation models are too limiting and do not possess the required accuracy.

d) Study of user behavior and characteristics: In their simplest form, user behavior experiments are intended to assess important user attributes (e.g., traffic pattern; degree of burstiness etc.) that may impact network design. In some cases, user behavior may also be monitored to determine the effect of network protocols on user characteristics. In the SATNET packet speech experiment, for example, one of the issues currently being investigated is the effect of long satellite delays (which vary according to the access scheme) on speech statistics and, more generally, on user behavior in dialogues as well as conferencing situations [43].

3) Performance Measures: The performance measures of a system (or part of a system, or of a procedure) are those parameters, or sets of parameters which best characterize and quantify the performance of the system in a real operating environment, and which permit comparison of the system with other systems performing similar functions.

The traditional performance measures used in most data communications models and experiments are the average delay and the average throughput (under a given offered traffic pattern). These measures are always available, in one form or another, from the CUMSTATS messages and probably offer the most straightforward means of evaluation and comparison of different systems, without requiring the detailed knowledge of the internal mechanisms. For these reasons we may refer to throughput and delay as the basic measures in the set of possible performance criteria. The average delay measure may of course be refined by introducing histograms (in addition to mean values). This feature, however, often proves to be very costly in terms of nodal processor requirements, and is rarely implemented. An exception to the general rule is offered by SATNET, in which the capability to measure delay distributions was deemed essential for the thorough evaluation and understanding of priority and delay class disciplines [42]. In SATNET, however, to limit storage and processor overhead, the experimenter is required to specify for each experiment the appropriate range and quantization subintervals in which histograms must be collected.

Delay and throughput, although conceptually simple to define, are not always so straightforward to measure. For example, end-to-end delay cannot in general be measured in a distributed network due to the lack of clock synchronization among the nodes. The common substitute for end-to-end delay is the round-trip delay, i.e., the interval between transmission of the data packet and reception of the acknowledgment from the destination. This approach is currently used in ARPANET. Clearly, the round-trip measurement divided by two provides only an approximation to the one-way delay measurement, since data packets have different length and (possibly) priority than acknowledgment packets, and some additional delay may be incurred at the destination before returning the acknowledgment.

A more accurate measurement of one-way delay may be obtained with the Pick-up packet. Each packet is stamped with entry and exit time at each intermediate node. The total delay is then the sum of the times spent at each node (which are reported in the pick up packet) and the transmission times on each hop (which can be accurately computed). Some imprecision, however, may still exist due to time gaps between the actual arrival/departure of a packet and the time stamps.

An exact measure of one-way delay is possible only with time-synchronized nodes. This is generally the case of the nodes in a packet satellite network, which must be synchronized in order to properly schedule packet transmissions to the multiple access channel (an exception, of course, is the case of the pure-ALOHA satellite system, in which no synchronization is required). In particular, the nodes in SATNET have synchronous clocks and therefore permit exact one-way delay measurements [33].

Some complications may also arise during throughput measurements, especially if throughput is measured as the sum of all packets successfully received at the destination. In this case, the measured throughput will include also duplicate packets (generated by the network because of missed acknowledgments, and therefore will be higher than the actual throughput seen by the user. Duplicate detection is, of course, performed by user level protocols and, in some cases, by network protocols (e.g., ARPANET). However, artificial traffic generators and sinks in the network are generally not equipped with duplicate detection protocols. Thus the experimenter has to carefully filter out duplicates from the measured throughput using some additional information (e.g., measured channel errors etc.).

The basic delay and throughput measures have some limitations, especially in those experiments aimed at gaining insight into a specific procedure. In such cases, the basic measures must be complemented with specialized measures. A typical example of specialized measure is the count of the number of times a loop is detected on a path from origin to destination: this measure is an essential complement to average throughput and delay in the evaluation of adaptive routing policies [75]. Similarly, in the PRNET the number of additional transmissions beyond success incurred by a packet is used to measure the efficiency of the Echo Acknowledgment protocol [87].

Besides basic and specialized measures, there is another category of measures which reflect some global system properties not easily characterized solely by total throughput and delay. These measures generally require the collection of a set of different measurements during a properly designed experiment and therefore may be referred to as composite measures. A typical example of composite measure is "fairness." Fairness is the property of allocating network capacity among an arbitrary population of large and small users (i.e., users with large and small traffic requirements) in a fair manner, without favoring one class of users over another. One of the possible definitions of fairness, proposed in [59], verifies the condition that small users get a share of capacity equal to their demand, while large users are all given the same allocation (which is larger than the allocation of any small user). The boundary between large and small users is determined by the condition that the sum of the individual allocations be equal to the maximum network capacity. Clearly, the total throughput measure alone is not a sufficient representation of fairness, since a protocol may be efficient and yet unfair. For a better appraisal of fairness one has to investigate the ratios between offered rate and effective throughput for each user in a carefully designed experiment. Other examples of composite per-

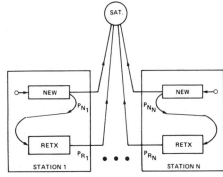

Fig. 6. S-ALOHA access protocol.

formance measures are: congestion protection, stability, robustness of network algorithms to line errors, and reliability of a network configuration with respect to component failures.

4) Designing an Experiment: The design of the experiment is probably the most critical and delicate phase of the experimentation process. Once the measurement object and the goals are defined, we must identify a meaningful set of performance measures consistent with our goals and with the tools at our disposal. Experience shows that a bad choice of performance measures and measurement facilities (e.g., inadequate traffic pattern; system parameters chosen outside of the range of interest) may produce results irrelevant to the original goals and cause a frustrating waste of efforts and resources. The design of experiments should be carefully conducted and, if possible, guided by analytical and simulation models.

To illustrate the various phases of the experiment design process we report here the highlights of the S-ALOHA control experiment carried out on SATNET in mid-1977 [59]. First, a brief description of the S-ALOHA access protocol is in order.

In the S-ALOHA channel access protocol each station maintains two output queues as shown in Fig. 6. The new queue (for new packets); and the retransmit queue (for packets that need to be transmitted because of previous conflict). All stations follow the same rules for transmission: at the beginning of a slot the station will transmit a packet from the retransmit queue with probability PR (retransmit gate). Only if the retransmit queue is empty, will the station then transmit a packet from the new queue with probability PN (new gate). If two or more stations transmit in the same slot, their packets will collide and will mutually destroy each other. The senders detect the conflict by monitoring the channel after a round-trip time and promptly return a copy of the collided packet to the retransmit queue.

It can be easily shown that if several stations have data to send at the same time and the ALOHA gate values are not properly adjusted, the system may become congested, i.e., total throughput in the system may reduce to zero [56]. The congestion situation will persist for a prolonged period of time even if the external traffic sources are removed. To overcome this congestion problem, a distributed stability control algorithm was implemented in SATNET [31]. The algorithm dynamically controls the ALOHA gates based on channel load observations, and implements a set of gate values which are optimal for the current traffic problem. One of the objectives of the S-ALOHA experiment in SATNET was to evaluate the performance of the controls, namely their performance at steady state and their ability to prevent congestion. Regarding

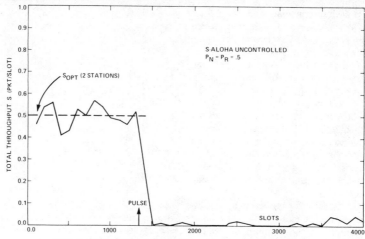

Fig. 7. S-ALOHA (uncontrolled) measurements. 10 stations.

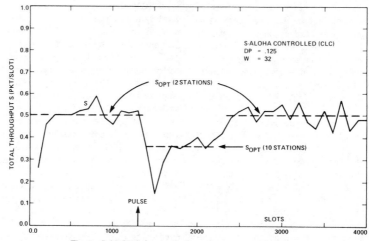

Fig. 8. S-ALOHA (controlled) measurements. 10 stations.

the network configuration, it was soon recognized that a large user population was essential in a stability experiment. Therefore, a variety of test configurations with number of stations varying from 5 to 30 were defined. As explained in Section III-A, only a few of these stations were real stations, while the remaining stations were emulated in the software of the real stations (fake stations). The best traffic environment to probe stability was found to be the superposition of a stable background pattern (involving only a small number of stations) and a traffic pulse of short duration, during which all stations are active. This traffic pattern would induce a sharp degradation of throughput immediately after the pulse, followed by a slow recovery to preexisting performance values.

In order to characterize stability, the following measures were chosen: average throughput before the pulse, time to recover after the pulse, and average throughput during the recovery period. The rationale behind this choice was the need to monitor performance both at steady state and during the recovery period, since both situations are of great concern to the user. Clearly, a simple measure of throughput averaged over the entire duration of the experiment would be grossly inadequate.

Another important set of decisions in the planning of the experiment involved the selection of the stability control parameters. These were chosen on the basis of existing analytical and simulation results [31] as well as previous experiments carried out in steady traffic conditions.

The main results of the experiments are summarized in Figs. 7 and 8 [59]. Fig. 7 shows the measured throughput performance S (packets/slot) as a function of time t (in slots) for a 10 station, uncontrolled S-ALOHA system, with fixed gates $PN = PR = 0.5$. The traffic pattern is the sum of a steady load consisting of two stations active all the time with rate $R = 1$ pkt/slot, and a pulse pattern of 20 slot duration, in which all stations are active. From Fig. 7, we notice that the average throughput before the pulse is $S = 0.5$, i.e., the theoretical optimum in a 2 station system. The introduction of the pulse, however, causes the throughput to collapse to zero, confirming the well known tendency of uncontrolled systems to become unstable.

Fig. 8 shows the measured throughput performance for a controlled S-ALOHA system using the same traffic environment as in the previous experiment. Recovery from the effect of the pulse is completed in the controlled system in 1000 time slots (analytical calculations also show that recovery in the controlled system would have required about 1000 time slots, while recovery in the uncontrolled system would have required on the order of 100 000 slots!). Performance before the pulse and after recovery (steady state) is nearly optimal.

The above experiments, and similar ones, led to important quantitative conclusions regarding the need for controls in S-ALOHA systems and the performance of specific control procedures implemented in SATNET. The success was not in little part due to the careful design of the experiments, namely, the appropriate definition of the performance measures, and the perfect matching of tools and measures to achieve the proposed goals.

We mentioned earlier that a successful network experiment is usually the result of carefully conducted modeling, simulation, and measurement activities. Analytic models assist in the specification of the experiment and the selection of parameter ranges; simulation is an extension of analysis in that it validates analytic models and permits the evaluation of analytically intractable protocols; measurements complete the experimental cycle.

The Atlantic Pachet Satellite Experiment was no exception. Analytical models for various channel access protocols were derived [33]. When analysis failed, simulation was used. In particular, the closed loop control mechanism proposed for the stabilization of the S-ALOHA protocol was tested extensively via simulation before implementation and before measurement planning. Finally, measurements were performed on the system. By this time, the experimenter already had a good idea of the results that he should expect, since the same experimental configuration had been previously evaluated via analysis or simulation. This proved to be very rewarding, and permitted tracing many of the discrepancies which arose between models and measurements back to their origin—typically, a wrong setting of the parameters, or a bug in the measurement software. More rarely, the discrepancies were found to stem from a different interpretation of the protocols by the implementer and the experimenter, respectively. This generally stimulated a productive discussion and reevaluation of the protocol among the working groups involved, leading to the adoption of the best alternative.

IV. Open Areas and Conclusion

Many advances in modeling and measurements of packet-switched networks have been made since this concept emerged in the late sixties. We have described in this paper many of the basic techniques and illustrated their use by calling on specific examples. It is clear, however, that we are far from having answered all our needs. In the following we briefly discuss some open areas, just to name a few, where more work is in order.

Random access has been thoroughly studied in single-hop environments. The performance analysis of multiaccess schemes in (the more interesting) packet radio multihop environments has proven to be a rather difficult task; no simple model exists yet for this more general problem, nor is there any obvious way to translate the results obtained in single-hop systems to multihop environments. The only analysis work relating to this that has already appeared can be found in [89], [91], [92].

In addition to the many design problems (topological, capacity assignment, routing, etc.–), network behavior is believed to be greatly affected by the flow and congestion control algorithms in use; the modeling and analysis of these techniques are still in their infancy. Perhaps the most elaborate work so far achieved consists of the analysis of a dynamic decision

process relating the number W of messages outstanding in the network to the destination buffer occupancy in an environment where the changes in W do not affect the network response time [48]. Unfortunately, such limitations in the model render it only a first approximation. In sum, there is a definite shortage of analytic work in this area. The measurement of end-to-end protocols has been also extremely limited due to the difficulty of synchronizing end devices and their measurement facilities, and the difficulty in interpreting the results which depend not only on the particular protocol implementation, but also on the characteristics of the communication subnet [6].

Although in this paper we have uniquely focused on packet switched networks suitable mainly for computer-to-computer communications, we observe today an important trend towards the design of integrated packet-circuit switching communications to satisfy a broader class of users with a large variety of traffic characteristics (interactive data, long files or facsimile, real time such as digitized voice, etc.–). The design and analysis of such systems are only at their start.

Finally, we get to the problem of network interconnection. It is all too evident that the behavior of a communication network varies with its type (land based, ground radio, satellite) as well as with the specific implementation and control techniques used. The interconnection of networks exhibits the need for a simple and accurate characterization and classification of these networks and for the development of analytic tools which help predict the performance of various interconnection topologies. Moreover, measurement facilities which allow for coordination, control and collection of simultaneous measurements in several interconnected networks in view of future internetworking experiments will be of utmost importance. These experiments include among others the evaluation of internetwork protocols and the end-to-end user performance in a multinetwork environment.

Thus we conclude here that in this exciting area of modeling and measurements of data communication networks, we are still faced with many problems of the most challenging kind.

References

[1] N. Abramson, "The ALOHA system—Another alternative for computer communications," in *1970 Fall Joint Comput. Conf. AFIPS Conf. Proc.*, vol. 37, Montvale, NJ: AFIPS Press, 1970, pp. 281–285.

[2] N. Abramson, "The Aloha system," *Computer Commun. Networks*, N. Abramson and F. Kuo, Eds. Englewood Cliffs, NJ: Prentice Hall, 1973.

[3] A. V. Aho, J. E. Hopcroft, and J. D. Ullman, *The Design and Analysis of Computer Algorithms*. Reading, MA: Addison Wesley, 1974.

[4] P. Baran, "On distributed communications," Rand Corporation, Santa Monica, CA, Rand Series Reports, Aug. 1964.

[5] F. Baskett, K. M. Chandy, R. R. Muntz, and F. G. Palacios, "Open, closed, and mixed networks of queues with different classes of customers," *J. Assoc. Comp. Mach.*, vol. 22, no. 2, pp. 248–260, April 1975.

[6] C. J. Bennett and A. J. Hinchley, "Measurements of the transmission control protocol," in *Proc. Workshop on Computer Network Protocols* (Liege, Belgium), Feb. 1978.

[7] U. N. Bhat, "Sixty years of queueing theory," *Management Sci.*, vol. 15, pp. B280–B294, 1969.

[8] R. Binder, "A dynamic packet-switching system for satellite broadcast channels," in *Proc. Int. Communications Conf.*, pp. 41-1–41-5, June 1975.

[9] P. J. Burke, "The output of a queueing system," *Operations Res.*, vol. 4, pp. 699–704, 1956.
    a) J. Buzen, "Queueing network models of multiprogramming," Ph.D. dissertation, Division of Engineering and Applied Science, Harvard Univ., Cambridge, MA, 1971.

[10] V. Cerf and R. E. Kahn, "A protocol for packet network intercommunication," *IEEE Trans. Commun.*, vol. COM-22, pp. 637–648, May 1974.

a) W. W. Chiu and W. Chow, "A hybrid hierarchical model of a multiple virtual storage (MVS) operating system," IBM T. J. Watson Research Center, Yorktown Heights, NY, Research Rep. R. C. 6947, Jan. 1978.

b) W. W. Clipshaw and F. Glave, "Datapac network review," in Int. Comput. Commun. Conf. Proc., pp. 131–136, Aug. 1976.

[11] J. W. Cohen, The Single Server Queue. New York: Wiley, 1969.

[12] G. D. Cole, "Performance measurements on the ARPA computer network," IEEE Trans. Commun., vol. COM-20, pp. 630–636, June 1972.

[13] D. R. Cox and W. L. Smith, Queues. Chapman and Hall Monographs on Applied Probability and Statistics, 1961.

[14] M. A. Crane and D. I. Iglehart, "Simulating stable stochastic systems, I: General multiserver queues," J. Assoc. Comp. Mach., vol. 21, no. 1, pp. 103–113, Jan. 1974.

[15] ——, "Simulating stable stochastic systems, II: Markov chains," J. Assoc. Comp. Mach., vol. 21, no. 1, pp. 114–123, Jan. 1974.

[16] D. W. Davies, K. A. Bartlett, R. A. Scantlebury, and P. T. Wilkinson, "A digital communication network for computers giving rapid response at remote terminals," presented at ACM Sym. Operating System Principles. Gatlinburg, TEN, Oct. 1–4, 1967.

[17] A Selection of Scientific Quotations. Collected by A. L. Mackay (M. Ebison, Ed.).

[18] W. D. Farmer and E. E. Newhall, "An experimental distributed switching system to handle bursty computer traffic," in Proc. ACM Conf. (Pine Mountain, GA), Oct. 1969.

[19] G. Fayolle, E. Gelembe, and J. Labetoulle, "Stability and optimal control of the packet-switching broadcast channels," J. Assoc. Comp. Mach., vol. 24, no. 3, pp. 375–386, July 1977.

[20] W. Feller, An Introduction to Probability Theory and Its Applications, Volume 2. New York: Wiley, 1966.

[21] G. S. Fishman, "Statistical analysis for queueing simulations," Management Sci., vol. 20, no. 3, pp. 363–369, 1973.

[22] ——, "Estimation in multiserver queueing simulations," Operations Res., vol. 22, no. 1, pp. 72–78, 1974.

[23] ——, "Achieving specific accuracy in simulation output analysis," Commun. Assoc. Comp. Mach. vol. 20, no. 5, pp. 310–314, May 1977.

[24] H. Frank, I. T. Frisch, and W. Chou, "Topological considerations in the design of the ARPA network," in Proc. AFIPS Conf. (SJCC), vol. 36, pp. 581–587, 1970.

[25] H. Frank, R. E. Kahn, and L. Kleinrock, "Computer communication network design: Experience with theory and practice," Networks, vol. 2, no. 2, 1972.

[26] L. Fratta et al., "The flow deviation method—An approach to store and forward communications network design," Networks, vol. 3, 1973.

[27] G. L. Fultz, "Adaptive routing techniques for message switching computer communications networks," UCLA School of Engineering, Rep. No. 7252, July 1972.

[28] M. Gerla, "The design of store-and-forward (S/F) networks for computer communications," Computer Science Dep., School of Engineering and Applied Science, Univ. California, Los Angeles, Tech. Rep. UCLA-ENG-7319, Jan., 1973. (Also published as Ph.D. dissertation.)

[29] M. Gerla, W. Chou, and H. Frank, "Cost-throughput trends in computer networks using satellite communications," presented at ICC 74, Minneapolis, MN, June 1974.

[30] M. Gerla, H. Frank, W. Chou, and J. Eckly, "A Cut-Saturation Algorithm for Topological Design of Packet Switched Communications Networks," in Proc. Nat. Telecommunications Conf. (San Diego, CA), Dec. 1974.

[31] M. Gerla and L. Kleinrock, "Closed loop stability controls for S-ALOHA satellite communications," Data Communications Symp., Snowbird, UT, Sept. 1977.

[32] ——, "On the topological design of distributed computer networks," IEEE Trans. Commun., COM-25, Jan. 1977.

[33] M. Gerla et al., "Packet satellite multiple access: Models and measurements," in NTC Conf. Proc. (Los Angeles, CA), Dec. 1977.

[34] W. J. Gordon and G. F. Newell, "Closed queueing systems with exponential servers," Operations Res., vol. 15, pp. 254–265, 1967.

[35] J. L. Grange and P. Mussard, "Performance measurements of line control protocols in the Cigale network," in Proc. Workshop on Computer Network Protocols (Liege, Belgium), Feb. 1978.

[36] R. Howard, Dynamic Programming and Markov Processes. Cambridge, MA: M.I.T. Press, 1960.

[37] ——, Dynamic Probabilistic Systems, Vol. 1: Markov Models, Vol. 2: Semi-Markov and Decision Processes. New York: Wiley, 1971.

[38] H. Inose and T. Saito, "Theoretical aspects in the analysis and synthesis of packet communication networks," this issue, pp. 1409–1422.

[39] M. Irland, "Simulation of Cigale 74," in Proc. 4th Data Communications Symp. (Quebec City, P.Q., Canada), Oct. 1975.

[40] ——, "Queueing analysis of a buffer allocation scheme for a packet switch," in Proc. IEEE National Telecommunications Conf. (New Orleans, LA), Dec. 1975.

[41] J. R. Jackson, "Job shop-like queueing systems," Management Sci., vol. 10, pp. 131–142, 1963.

[42] I. Jacobs et al., "C-PODA—A demand assignment protocol for SATNET," presented at 5th Data Communication Symp., Snowbird, UT, Sept. 1977.

[43] I. M. Jacobs, R. Binder, and E. V. Hoversten, "General purpose packet satellite networks," this issue, pp. 1448–1467.

[44] R. E. Kahn, "The organization of computer resources into a packet radio network," in Nat. Comput. Conf., AFIPS Conf. Proc., vol. 44, Montvale, NJ: AFIPS Press, 1975, pp. 177–186; also in IEEE Trans. Commun., COM-25, Jan. 1977.

[45] R. E. Kahn, S. A. Gronemeyer, J. Burchfiel, and R. C. Kunzelman, "Advances in packet radio technology," this issue, pp. 1468–1496.

[46] F. Kamoun, "Design considerations for large computer communication networks," Ph.D. dissertation, Computer Science Dep., Univ. California, Los Angeles, Apr. 1976.

[47] F. Kamoun and L. Kleinrock, "An analysis of shared storage in a computer network environment," in Proc. 8th Hawaii Int. Conf. System Science (Honolulu, HI), pp. 89–92, Jan. 1976.

[48] P. Kermani, "Switching and flow control techniques in computer communication networks," Ph.D. dissertation, Dep. Computer Science, Univ. California, Los Angeles, 1977.

a) L. Kleinrock, Communication Nets: Stochastic Message Flow and Delay. (New York: McGraw-Hill, 1964 (out of print). Reprinted by Dover Publications, 1972).

[49] L. Kleinrock, "Analytic and simulation methods in computer network design," in AFIPS Conf. Proc. (SJCC), vol. 36, pp. 569–579, 1970.

[50] L. Kleinrock and S. Lam, "Packet-switching in a slotted satellite channel," in Nat. Computer Conf. AFIPS Conf. Proc., vol. 42. Montvale, NJ: AFIPS Press, 1973, pp. 703–710.

[51] L. Kleinrock and W. E. Naylor, "On measured behavior of the ARPA network," in AFIPS Conf. Proc. 1974, National Computer Conf., vol. 43, pp. 767–780, 1974.

[52] L. Kleinrock and H. Opderbeck, "Throughput in the ARPANET—Protocols and measurements," in Proc. 4th Data Communications Symp. (Quebec City, P.Q., Canada), Oct. 1975.

[53] L. Kleinrock, Queueing Systems, Volume I: Theory. New York: Wiley Interscience, 1975.

[54] L. Kleinrock and F. A. Tobagi, "Packet switching in radio channels: Part I—Carrier sense multiple-access modes and their throughput-delay characteristics," IEEE Trans. Commun., vol. COM-23, pp. 1400–1416, Dec. 1975.

[55] F. A. Tobagi and L. Kleinrock, "Packet switching in radio channels: Part II—The hidden terminal problem in carrier sense multiple-access and the busy-tone solution," IEEE Trans. Commun., vol. COM-23, pp. 1417–1433, Dec. 1975.

[56] L. Kleinrock and S. S. Lam, "Packet switching in a multiaccess broadcast channel: Performance evaluation," IEEE Trans. Commun., vol. COM-23, pp. 410–423, Apr. 1975.

[57] L. Kleinrock, Queueing Systems, Volume II: Computer Applications. New York: Wiley Interscience, 1976.

[58] L. Kleinrock, W. E. Naylor, and H. Opderbeck, "A study of line overhead in the ARPANET," Commun. Assoc. Comp. Mach., vol. 19, no. 1, pp. 3–12, 1976.

[59] L. Kleinrock and M. Gerla, "On the measured performance of packet satellite access schemes," presented at Int. Conf. on Computer Communications, Kyoto, Japan, Sept. 1978.

[60] H. Kobayashi and A. G. Konheim, "Queueing models for computer communications system analysis," IEEE Trans. Commun., vol. COM-25, pp. 2–29, Jan. 1977.

[61] A. G. Konheim and B. Meister, "Service in a loop system," J. Assoc. Comp. Mach., vol. 19, no. 1, pp. 92–108, Jan. 1972.

[62] J. Labetoulle, E. Manning, and R. W. Peebles, "A homogeneous computer network: Analysis and simulation," Computer Networks, vol. 1, no. 4, pp. 225–240, 1977.

[63] S. S. Lam, "Packet switching in a multi-access broadcast channel with application to satellite communication in a computer network," Ph.D. dissertation, Dep. Computer Science, Univ. California, Los Angeles, Mar. 1974; also in Univ. California, Los Angeles, Tech. Rep. UCLA-ENG-7429, Apr. 1974.

[64] S. S. Lam and L. Kleinrock, "Packet switching in a multiaccess broadcast channel: Dynamic control procedures," IEEE Trans. Commun., vol. COM-23, pp. 891–904, Sept. 1975.

[65] S. Lam and M. Reiser, "Congestion control of store-and-forward networks by input buffer limits," in NTC 77 Conf. Proc. (Los Angeles, CA), Dec. 1977.

[66] A. Lavia and E. Manning, "Topological optimization of packet switched data communications networks," CCNG Tech. Rep. No. E-35, Oct. 1975, CCNG, Univ. Waterloo, Waterloo, Ont., Canada.

[67] S. Lin, "Heuristic programming as an aid to network design," Networks, Jan. 1975.

[68] J. Little, "A proof of the queueing formula $L = W$," Operation Res., vol. 9, pp. 383–387, Mar.–Apr. 1961.

[69] M. T. Liu and C. C. Reames, "Communication protocol and network operating system design for the distributed loop computer

network (DLCN)," *Proc. 4th Annu. Symp. Computer Architecture*, pp. 193–200, Mar. 1977.

[70] M. T. Liu, "Distributed loop computer networks," (to appear in *Advances in Computer Networks*, M. Rubinoff and M. C. Yovits, Eds. New York: Academic Press, 1978.)

[71] J. M. McQuillan, W. R. Crowther, P. P. Cossell, D. C. Walden, and F. E. Heart, "Improvements in the design and performance of the ARPA network," in *AFIPS Conf. Proc. 1972 Fall Joint Computer Conf.*, vol. 41, pp. 741–754, 1972.

[72] R. M. Metcalfe, "Steady-state analysis of a slotted and controlled ALOHA system with blocking," in *Proc. 6th Hawaii Int. Conf. System Sciences*, Univ. Hawaii, Honolulu, Jan. 1973.

[73] R. M. Metcalfe and D. R. Boogs, "ETHERNET: distributed packet switching for local computer networks," *Commun. Assoc. Comp. Mach.*, vol. 19, no. 7, pp. 395–403, 1976.

[74] F. R. Moore, "Computational model of a closed queueing network with exponential servers," *IBM J. Res. Dev.*, vol. 16, pp. 567–572, 1972.

[75] W. E. Naylor, "A loop-free adaptive routing algorithm for packet switched networks," in *Proc. 4th Data Communications Symp.* (Quebec City, P. W., Canada), pp. 7–9 to 7–14, Oct. 1975.
a) H. Opderbeck and R. B. Hovey, "Telnet-Network features and interface protocols," in *Proc. NTG-Conf. Data Networks* (Baden-Baden, West Germany), Feb. 1976.

[76] L. Pouzin, "Presentation and major design aspects of the cyclades computer network," presented at Datacom 73, ACM/IEEE, 3rd Data Communications Symposium. St. Petersburg, FL, pp. 80–87, Nov. 1973.
a) L. Pouzin, "CIGALE, the packet switching machine of the CYCLADES computer network," presented at IFIP Congress (Stockholm), pp. 155–159, Aug. 1974.

[77] W. L. Price, "Data network simulation: Experiments at the national physical laboratory 1968–1976," *Computer Networks*, vol. 1, no. 4, pp. 199–210, 1977.

[78] J. Riordan, *Stochastic Service Systems*. New York: Wiley, 1962.

[79] L. G. Roberts and B. D. Wessler, "Computer network developments to achieve resource sharing," in *Proc. AFIPS Conf. (SJCC)*, vol. 36, pp. 543–549, 1970.

[80] L. G. Roberts, "Dynamic allocation of satellite capacity through packet reservation," in *1973 Nat. Comput. Conf., AFIPS Conf. Proc.*, Vol. 42. New York: AFIPS Press, 1973, pp. 711–716.

[81] L. G. Roberts, "ALOHA packet system with and without slots and capture," *Computer Communication Rev.*, vol. 5, no. 2, pp. 28–42, Apr. 1975.

[82] S. M. Ross, *Applied Probability Models with Optimization Applications*. San Francisco, CA: Holden Day, 1970.

[83] Steiphitz *et al.*, "The design of minimum cost survivable networks," *IEEE Trans. Circuit Theory*, Nov. 1969.

[84] F. Tobagi, "Random access techniques for data transmission over packet switched radio networks," Ph.D. dissertation, Comput. Sci. Dep., School Eng. and Appl. Sci., Univ. California, Los Angeles, rep. UCLA-ENG 7499, Dec. 1974.

[85] F. Tobagi and L. Kleinrock, "Packet switching in radio channels: Part II—The hidden terminal problem in carrier sense multiple access and the busy tone solution," *IEEE Trans. Commun.*, vol. COM-23, pp. 1417–1433, Dec. 1975.

[86] F. Tobagi and S. E. Lieberson, "Measurements in packet radio systems: Methods for collection of cumstat data," *Packet Radio Temporary Note* #175, Comput. Sci. Dept., Univ. California, Los Angeles, Mar. 1976.

[87] F. Tobagi *et al.*, "On measurement facilities in packet radio systems," in *Nat. Computer Conf. Proc.* (New York), June 1976.

[88] F. Tobagi and L. Kleinrock, "Packet switching in radio channels: Part IV—Stability considerations and dynamic control in carrier sense multiple access," *IEEE Trans. Commun.*, vol. COM-25, pp. 1103–1120, Oct. 1977.

[89] F. Tobagi, "Performance analysis of packet radio communication systems," in *Proc. IEEE Nat. Telecom. Conf.* (Los Angeles, CA), pp. 12:6–1 to 12:6–7.

[90] F. Tobagi and L. Kleinrock, "The effect of acknowledgment traffic on the capacity of packet-switched radio channels," *IEEE Trans. Commun.*, vol. COM-26, pp. 815–826, June 1978.

[91] F. Tobagi, "On the performance analysis of multihop packet radio systems: Parts I–IV," Packet Radio Temporary Notes #246–249, Comput. Sci. Dept., Univ. California, Los Angeles, 1978.

[92] F. Tobagi, "Analysis of slotted ALOHA in a centralized two-hop packet radio network," in *Proc. 17th IEEE Computer Society Int. Conf.* (CompCon, Fall 78), Washington, DC, Sept. 1978.

[93] A. C. Williams and R. A. Bhandiwad, "A generating function approach to queueing network analysis of multiprogrammed computers," Mobil Oil Corp., Princeton, NJ, Apr. 1974.

[94] J. W. Wong, "Distribution of end-to-end delay in message-switched networks," submitted for publication, Dep. Computer Science, Univ. Waterloo, Waterloo, Ont., Canada.

Routing Techniques Used in Computer Communication Networks

MISCHA SCHWARTZ FELLOW, IEEE, AND THOMAS E. STERN, FELLOW, IEEE

(Invited Paper)

Abstract—An overview is provided in this paper of the routing procedures used in a number of operating networks, as well as in two commercial network architectures. The networks include TYMNET, ARPANET, and TRANSPAC. The network architectures discussed are the IBM SNA and the DEC DNA. The routing algorithms all tend to fall in the shortest path class. In the introductory sections, routing procedures in general are discussed, with specialization to shortest path algorithms. Two shortest path algorithms, one appropriate for centralized computation, the other for distributed computation, are described. These algorithms, in somewhat modified form, provide the basis for the algorithms actually used in the networks discussed.

INTRODUCTION

IN this paper, we provide an overview of routing techniques used in a variety of computer communication networks in current operation. These include the public data networks TYMNET and TRANSPAC (the former is a specialized common carrier network based in the United States, but with connections to Europe as well; the latter is the French government PTT data network), ARPANET, the U.S. Department of Defense Computer Network, and the commercial network architectures SNA (Systems Network Architecture) and DNA (Digital Network Architecture), developed by IBM and Digital Equipment Corporation, respectively. The networks are all examples of store-and-forward networks with data packets[1] moving from a source to a destination, buffered at intermediate nodes along a path. The path is defined simply as the collection of sequential communication links ultimately connecting source to destination.

The routing algorithms used in these networks all turn out to be variants, in one form or another, of shortest path algorithms that route packets from source to destination over a path of least cost. The specific cost criterion used differs among the networks. As will become apparent in the discussion following, some networks use a fixed cost for each link

in the network, the cost being roughly inversely proportional to the link transmission capacity in bits per second. For a network with equal capacity links, minimization of the path cost generates a minimum hop path. Links with measured congestion and/or high error rates may be assigned higher costs, steering traffic away from them. Costs may also vary with the type of traffic transmitted—whether interactive, asynchronous terminal type, synchronous traffic, or file transfers between computers. Other networks attempt to estimate average packet time delay on each link and use this to assign a link cost. The resultant source-destination path chosen tends to provide the path of minimum average time delay.

Since a least cost routing algorithm is used in all cases, we provide in the next section a unifying discussion of least cost routing to further demonstrate the similarities in the network algorithms.

Although the basic routing procedures are similar, differing primarily in the choice of a link cost function used to establish the minimum cost path, the routing techniques used tend to differ in implementation and the place at which the algorithms are run. The routing algorithm may be run in centralized fashion by a central supervisory program or Network Control Center, or may be carried out in a decentralized or distributed way with individual nodes in the network running the routing algorithm separately. In the former (centralized) case, global information about the network required to run the algorithm (current topology, line capacity, estimated link delays if required, condition of links and nodes, etc.) need only be kept by the central supervisor. Path setup is then accomplished through routing messages sent to each node along the path selected. In the latter (distributed) case, the required information must be exchanged among nodes in the network. This implies some means of disseminating changes in topology (nodes and links coming up or going down), congestion, and estimated time delay information if used in the algorithms. In the next section, we discuss two least cost routing algorithms, which are the basis for routing procedures in many networks.

The routing procedures adopted also differ in how dynamic they are—how rapidly and in what manner they adapt, if at all, to changes in network topology and/or traffic information. In some cases, routes are fixed during the time of a user session. (This is the length of a call from sign-on or connect time to sign-off or disconnect time.) A node or link failure during a session will then abort the call or may, in some cases, cause a

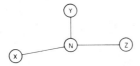

(a) CURRENT NODE AND NEIGHBORS

PACKET IDENTIFICATION	NEXT NODE ASSIGNMENT
(1,4)	X
(1,5)	Y
(2,4)	X
(2,5)	Z
(2,6)	Y
⋮	⋮

(b) ROUTING TABLE

Fig. 1.   Routing at a node in a network.

new route to be selected, transparent to the user. In other cases, paths may be changed during a session (although unknown to the user and relatively slowly to avoid stability problems).

Once the path has been determined, routing tables, set at each node, are used to steer individual packets to the appropriate outgoing link. An example appears in Fig. 1. A typical node N in a network is shown, with three neighboring nodes X, Y, Z to which it is connected. In part (b) of the figure, a partial routing table is shown, associating individual user packets with an appropriate outgoing line leading to one of the neighboring nodes. The packet identification requires two numbers. The two numbers could be source and destination address, or they could be a mapping of these two fields into a corresponding pair given by the incoming link number and a number associated with that link. (This is variously called the logical record number, the logical link number, next node indicator etc.) The source-destination addresses could also be combined into one unique network-wide virtual circuit number, although this becomes difficult to monitor and assign with large networks. If all packets routed to a particular destination follow the same path, only a destination address is required to determine the proper outgoing link. (This would be the case, for example, if the paths chosen are independent of message class, type, etc.)

It is obvious that routing procedures play an important role in the design of data networks. Together with the techniques of flow and congestion control, they are implemented as part of the transport level or end-to-end protocols of networks. In layered protocols, this is the level just above the data link level that ensures correct transmission and reception of packets between any neighboring nodes in a network.

Because of their importance to the proper operation of data networks, routing techniques have received a great deal of attention in recent years. They have been variously classified as deterministic, stochastic, fixed, adaptive, centrally controlled, or locally controlled [1].

The fixed versus adaptive classification is particularly vague, since all networks provide some type of adaptivity to accommodate topological changes (links and/or nodes coming up or

going down, new topologies being established). In the past, the distinction had been made primarily on the basis of individual packet handling. In the original ARPA routing algorithm, routing tables could be updated at intervals as short as 2/3 s [2]. Routing changes were made by individual nodes in a decentralized manner. As a result, individual packets in a message could follow diverse routing paths. The ARPA adaptive routing algorithm was adopted by a number of other networks as well [3], [4]. The French Cigale network used a related decentralized algorithm [5].

The hope was that by adapting on a packet-by-packet basis, the network could be made more responsive to changes in traffic characteristics and to topology, enabling packets to arrive at their destinations more rapidly, as well as avoiding failed links and/or nodes and regions of congestion. This was the case to some extent, yet the ARPA experience indicated some fundamental problems arising—there were problems with message reassembly at the destination, packet looping, adaptation problems ("too rapid a response to good news and too sluggish a response to bad news") [2], etc. As a result, the ARPA algorithm has been changed, making it less dynamic.

Although these routing techniques will be called adaptive in the sense of responding to network changes, the time constants are considerably longer. In the case of the new ARPA algorithm, changes may take place about every 10 s. Details appear in Section III.

If the algorithms used in most of these networks are adaptive and of the shortest path type, how then are they to be distinguished? We have already indicated that they may differ in the cost criterion used, and as to whether the computations are done centrally or on a distributed basis. The rate of adaptation is another distinguishing characteristic. This has also been noted already—the ARPA network, as an example, will change routes, if necessary, every 10 s. TYMNET makes changes from session to session only. DNA changes paths only when necessary.

Other differences arise due to the actual implementation: the size of routing tables, the routing overhead required, the time required to set up a path or change one if necessary; all of these will be found to differ in the networks to be described. Other differences will be noted during the discussion.

Interestingly, shortest path single routes turn out not to be optimum if the long-term average *network* time delay is to be minimized. In this case, multiple or bifurcated paths arise [6], [1]. Packets at a node are assigned to one of several outgoing links on a probabilistic basis. Bifurcated routing has not as yet been used in routing algorithms implemented in operating networks, although there are plans to incorporate this procedure in future routing mechanisms for the Canadian DATAPAC network [7].

In Section II, we provide a more detailed treatment of routing procedures in networks, focusing, as already noted, on shortest path (least cost) algorithms. In Section III, we then describe the routing implementation currently found in TYMNET, ARPANET, TRANSPAC, and the two commercial network architectures, IBM's SNA and Digital Equipment's DNA. In these last two cases, the routing procedures adopted are part of the overall protocol design and do not refer to a specific network implementation.

II. STRUCTURE OF ROUTING PROCEDURES IN PACKET-SWITCHED NETWORKS

Efficient utilization and sharing of the communications and nodal processing resources of a packet-switched communication network require various types of control, perhaps the most important of these being packet routing, that is, selecting paths along which packets are to be forwarded through the network. The objective of any routing procedure is to obtain good network performance while maintaining high throughput. "Good performance" usually means low average delay through the network, although many other performance criteria could be considered equally valid. Since poor routing algorithms often lead to congestion problems, and conversely, local congestion often requires at least temporary modification of routing rules, the routing problem cannot be completely divorced from that of congestion control [8]. Nevertheless, in this paper, we restrict ourselves to routing under the assumption that the better the routing algorithm, the less congestion is likely to occur.

While routing procedures can be set up within a network more or less independently of the protocols seen by the users (i.e., the devices external to the network), the choice of an appropriate routing procedure is influenced to some extent by the transport protocols operating at the network/user interface. It is convenient to classify these as either *virtual circuit-oriented* or *message-oriented*. In the former case, a device, or a process within a device (e.g., an application program within a computer), prepares to communicate with another device by exchanging a number of control messages with the network. The purpose of these messages is to determine whether the destination device is connected and ready to receive messages, to agree on certain aspects of the transmission protocol, and to set up a virtual circuit (VC) from source to destination.[2] Fig. 2 illustrates three VC's connecting terminals $T1$, $T2$, $T3$ to a host H. The VC appears to the external devices as if it were a dedicated line; under normal operation, individual data packets arrive at the destination essentially without loss or error and in the proper sequence. It is important to note, however, that *within* the network, packets from many different virtual circuits are generally sharing the same communication lines; errors, losses, and changes of packet order may occur. However, it is the function of the internal network protocols to correct for all of these effects. In the message-oriented case, communication is on a message-by-message basis. Each message or packet (often called a *datagram* in this case) must therefore contain its own destination address, but no preliminary control messages are required to set up a communication path.

A. Functions of Routing Procedures

In an idealized situation where all parameters of the network are assumed to be known and not changing, it is possible to determine a routing strategy which optimizes network performance, e.g., minimizes average network delay. The routing problem posed in this form is equivalent to the multicommodity flow problem well known in the operations research literature, and has been treated extensively in the communications network context [6], [8]–[11]. Changing situations in real networks, such as a line failure or a change in the traffic distribution, necessitate some degree of adaptivity. Any adaptive routing procedure must perform a number of functions.

1) Measurement of the network parameters pertinent to the routing strategy.

2) Forwarding of the measured information to the points(s) (Network Control Center (NCC) or nodes) at which routing computation takes place.

3) Computation of routing tables.

4) Conversion of routing table information to packet routing decisions. (This may include dissemination of a centrally computed routing table to each switching node as well as the conversion of this information to a form suitable for "dispatching" packets from node to node.)

Typical information that is measured and used in routing computation consists of states of communication lines, estimated traffic, link delays, available resources (line capacity, nodal buffers), etc. The pertinent information is forwarded to the NCC in a centralized system and to the various nodes in a distributed system. In the distributed case, two alternatives are possible: 1) forward only a limited amount of network information to each node (i.e., only that which is required for computing its local routing decisions), or 2) forward "global" network information to all nodes. (See Section III-A1) for a comparison of these two strategies in a specific network.) Based on the measured information, "costs" can be assigned to each possible source-destination path through the network. Routing assignments may be based on the principle of assigning a single path to all traffic between a given pair of source/destination nodes, or else traffic for a given source/destination pair might be distributed over several paths, resulting in the *multiple path*, or *bifurcated routing* procedure mentioned in Section I. In the latter case, single paths might still be maintained for each virtual circuit (if a VC-oriented protocol is used). This case is illustrated in Fig. 2, wherein VC_1 and VC_2 involve the same source/destination nodes, but take different paths. (Bifurcated routing on a packet basis is illustrated in Fig. 3.) While maintenance of single paths for each VC is not an optimal procedure, it has a number of practical advantages, an important one being the fact that packets always arrive at their destination in the proper order. It is therefore not surprising that most of the networks currently in operation use VC-oriented protocols with single-path routing per VC. These paths generally remain fixed for the duration of operation of the VC, unless a failure occurs.

Once one thinks in terms of single-path routing, it is natural to choose the "shortest" or, more generally, the *least cost* path whenever alternate paths exist. The path costs can, of course, be assigned using whatever cost functions seem appropriate (see above), the only essential property being that the *path* cost is computed as the sum of the costs of the *links* comprising that path. In such a case, the routing problem is equivalent to that of finding the shortest path through a graph, wherein link "length" is understood to have the more general

---

[2] VC's set up in this manner are termed "switched" VC's, in contrast to "permanent" VC's, which require no call setup procedure.

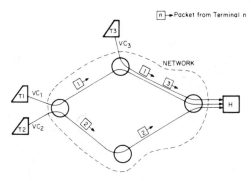

Fig. 2. Virtual circuits.

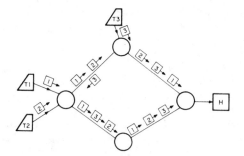

Fig. 3. Bifurcated routing.

meaning of link "cost." The set of shortest paths from all source nodes to a common destination node in a network forms a tree with the destination as root node. Thus, it is clear that if single-path routing on a source/destination basis is to be used, the path of a packet is uniquely determined by its destination alone. (Optimal bifurcated routing on a packet basis also only requires destination information.) On the other hand, single-path-per-VC routing requires either explicit or implicit VC identification for each packet; source/destination information alone is insufficient. This is because each time a new VC comes into operation, the costs determining the shortest path may be different since they generally change with time as network operating conditions change. Thus, VC's between the same source/destination pairs, established at different times, may take different paths as illustrated in Fig. 2.

B. Shortest Path Algorithms

The shortest path problem described above has received much attention in the literature. A variant of this problem, that of finding the k shortest paths between source and destination, is also applicable to the routing problem. (One is often interested in two or three alternate routes ranked in order of cost.) This too has been extensively treated. Since most operating networks use some version of shortest path routing, we discuss in this section the two algorithms most commonly used in communication network shortest path calculations. Algorithm A, due to Dijkstra [12], [13], is adapted to centralized computation, while B, a form of Ford and Fulkerson's algorithm [14], is particularly useful in dis-

tributed routing procedures. Since they are simple and intuitive, we present them informally, aided by an example.

Consider the network of Fig. 4(a) in which the numbers associated with the links are the link costs. (It is assumed for simplicity that each link is bidirectional with the same cost in each direction. However, both algorithms are applicable to the case of links with different costs in each direction.) We first use algorithm A to find shortest paths from a single source node to all other nodes. The algorithm is a step-by-step procedure where, by the kth step, the shortest paths to the k nodes closest to the source have been calculated; these nodes are contained in a set N. At the $(k + 1)$th step, a new node is added to N, whose distance to the source is the shortest of the remaining nodes outside of N. More precisely, let $l(i, j)$ be the length of the link from node i to node j, with $l(i, j)$ taken to be $+\infty$ when no link exists. Let $D(n)$ be the distance from the source to node n along the shortest path *restricted to nodes within N*. Let the nodes be indicated by positive integers with 1 representing the source.

1) Initialization. Set $N = \{1\}$, and for each node v not in N, set $D(v) = l(1, v)$.

2) At each subsequent step, find a node w not in N for which $D(w)$ is a minimum, and add w to N. Then update the distances $D(v)$ for the remaining nodes not in N by computing

$$D(v) \leftarrow \text{Min} \, [D(v), D(w) + l(w, v)] \, .$$

Application of the algorithm to the network of Fig. 4(a) is shown in Table I, and the resultant tree of shortest paths appears in Fig. 4(b), together with a *routing table* for node 1, indicating which outbound link the traffic arriving at that node should take. (It should be clear that the same algorithm can be used to find shortest paths *from* all nodes to a common *destination*.)

Now consider algorithm B. This is an iterative procedure, which we will use in the same network to find shortest paths from all nodes to node 1, considered now as the common *destination*. To keep track of the shortest paths, we label each node v with a pair $(n, D(v))$ where $D(v)$ represents the current iterate for the shortest distance from the node to the destination and n is the number of the next node along the currently computed shortest path.

1) Initialization. Set $D(1) = 0$ and label all other nodes $(\cdot, +\infty)$.

2) Update $D(v)$ for each nondestination node v by examining the current value $D(w)$ for each adjacent node w and performing the operation

$$D(v) \leftarrow \text{Min}_{w} \, [D(w) + l(v, w)] \, .$$

Update of node v's label is completed by replacing the first argument n by the number of the adjacent node which minimizes the above expression. Step 2) is repeated at each node until no further changes occur, at which time the algorithm terminates.

Table II illustrates the procedure for the network of Fig. 4(a). Two complete cycles of updates are required, after which

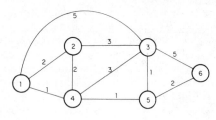

(a) NETWORK

Routing Table
for Node 1

Dest.	Next Node
2	2
3	4
4	4
5	4
6	4

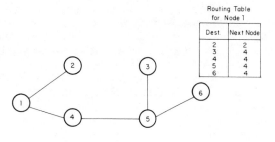

(b) TREE

Fig. 4. Example of shortest path routing.

TABLE I
ALGORITHM *A*

Step	N	D(2)	D(3)	D(4)	D(5)	D(6)
Initial	{1}	2	5	1	∞	∞
1	{1,4}	2	4	1	2	∞
2	{1,2,4}	2	4	1	2	∞
3	{1,2,4,5}	2	3	1	2	4
4	{1,2,3,4,5}	2	3	1	2	4
5	{1,2,3,4,5,6}	2	3	1	2	4

TABLE II
ALGORITHM *B*

Cycle	Node →	2	3	4	5	6
Initial		(·,∞)	(·,∞)	(·,∞)	(·,∞)	(·,∞)
1		(1,2)	(2,5)	(1,1)	(4,2)	(5,4)
2		(1,2)	(5,3)	(1,1)	(4,2)	(5,4)

no further changes occur and the iteration is complete. The tree of shortest paths generated is, of course, the same as that of Fig. 4(b). In this case, the nodes were updated in numerical order; however, any arbitrary order, cyclic or acyclic, will work. For each nondestination node, the first argument of its final label indicates the next node on the shortest path to the destination, and thus supplies the necessary routing information (for this destination *only*).

A word of comparison is now in order. Construction of routing tables based on algorithm *A* requires a shortest path tree calculation for each node in the manner described above. The tree is constructed with the particular node chosen as source (root) node, and the routing information that is generated is used to construct the table for that node as illustrated

in Fig. 4(b). The tree can then be discarded. It should be noted that tree construction for each node requires *global* information about the network. Construction of a routing table using algorithm *B* requires repeated application of the algorithm for each *destination* node, resulting in a *set* of labels for each node, each label giving the routing information (next node) and distance to a particular destination. Note that in this case, the algorithm can be conveniently implemented in a *distributed* fashion, in which case each node requires only information from its neighbors.

Evaluation of the comparative merits of the two algorithms depends on a number of factors: amount of overhead required in passing measured information to the point (s) at which computation is performed, amount of data to be stored, complexity of the computation, speed with which the algorithm can respond to changes in link costs, etc. These comparisons can only be made meaningful in the context of a specific network. See Section III-A1), ARPANET, for an example.

Finally, it should be noted that the algorithms described here have been assumed to be operating under *static* conditions of topology and link costs. (Their convergence has been proved in the literature for this case only.) In some applications, the link costs are defined to depend in some fashion on link traffic, which in turn depends, through the routing algorithm, on link cost; the result is a feedback effect. By studying the dynamics of such situations, it has been shown [15] that poor choices of link cost functions can, in fact, produce instabilities in the resultant traffic patterns. Stability can, however, be ensured by making the link costs sufficiently insensitive to link flow.

C. Packet Routing Implementation

As indicated in Section II-A, computation of the routing tables does not complete the routing procedure. These tables must be converted to a form appropriate for dispatching packets from node to node. In this section, we describe a method which underlies a number of schemes used for implementing routing on a single-path-per-VC basis in some existing or proposed systems [16], [17]. The essence of the procedure is that each VC has a *path number* (PN) associated with each link it traverses; if two VC's share a link, they obtain different path numbers on that link. Each packet carries the appropriate PN, which is updated as the packet traverses the network. The updating procedure is determined by, and replaces the routing table, at each node. The PN contains all information necessary for routing; thus, the packet need not carry a VC number. To illustrate, consider a set of four active virtual circuits traversing the network of Fig. 5. The second column of Table III indicates the node sequence for the paths chosen for these VC's. (Note that VC_1 and VC_2 have the same source/destination node pair, but different paths.) Let PN (*n*) be a path number associated with a path on a link outbound from node *n*; each link will have as many path numbers as there are distinct active VC's sharing that link. The remaining columns of Table III show how a sequence of PN's is assigned at each node, serving to identify uniquely the path to be followed by a packet on each VC. When a packet is received on an inbound line at a node, its PN must be updated, and the packet must be placed

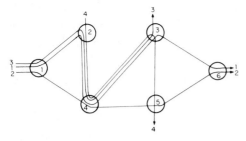

Fig. 5. Example of routing implementation.

TABLE III
PATH NUMBERS

VC#	Path	PN(1)	PN(2)	PN(3)	PN(4)	PN(5)
1	1-2-4-3-6	1	1	1	1	-
2	1-4-5-6	1	-	-	1	1
3	1-2-4-3	2	2	-	3	-
4	2-4-3-5	-	3	1	2	-

TABLE IV
ROUTING AT NODE 4

Arriving from node	Old PN	New PN	Next node
1	1	1	5
2	1	1	3
2	2	3	3
2	3	2	3

on the proper outbound line or released to its destination. At each node, a simple table lookup procedure can perform this function. In Table IV we show the necessary table for node 4. Note that it is derived directly from the routing information in Table III.

The PN used in this section is roughly equivalent to the *logical record number* used by TYMNET [16] [see Section III-A2)] and the *next node indicator* (NNI) proposed for explicit routing [17]. It might be thought that it would be simpler to tag each packet with a unique VC number rather than using the procedure outlined here. However, there will generally be far more active VC's in the network (perhaps thousands) than there are distinct VC's sharing a link (up to 256 in the case of TYMNET, for example). Thus, the PN approach is generally more efficient in memory requirement and table lookup time than any method using VC numbers.

III. EXAMPLES OF ROUTING PROCEDURES USED IN PRACTICE

A. Computer and Data Networks

1) ARPANET, A Computer Network:[3] ARPANET [18] was created in 1969 as an experiment in computer resource sharing. Beginning with four nodes in 1969, it now runs as an operational system with over 100 computers connected to 56 nodes throughout the continental United States, Hawaii, and Europe. It is a distributed network with at least two paths between any pair of nodes. Most of its lines are 50 kbit/s synchronous links. It is a store-and-forward packet-switched network in which the transport protocol is message-oriented. Messages longer than the maximum packet length are segmented into up to 8 packets at the source node and are reassembled at the destination node. This requires special

provision for buffer allocation at the nodes to prevent various types of lockups (a significant problem in the early stages of network development).

The network was originally operated with a distributed adaptive routing algorithm of the minimum cost, i.e., shortest path, type wherein link cost was evaluated in terms of measured link delay. Since the measured delays were determined by queue lengths encountered along a packet's transmission path, these quantities varied rapidly with time. Routing was on an individual packet basis where each packet was forwarded along the path that was perceived by the forwarding node to be the shortest in time to the packet's destination at the time of transmission. Since adaptivity was quite rapid, and different nodes could have different views of network conditions, perceptions of shortest paths could change during the period the packet traversed the network. The shortest path algorithm used was essentially our algorithm *B* (Section II-B), with information necessary for node updates passed among neighbors at 2/3 s intervals. Details of the algorithm can be found in [19]. A number of difficulties appeared in the algorithm, and it underwent several modifications [2] from the time it was implanted until May 1979 when it was replaced by a basically different procedure [20].

The new routing algorithm is distributed in the sense that each node independently computes its own routing tables using what is called a *shortest path first* algorithm (essentially our algorithm *A* with some modifications). That is, each node computes a shortest path tree with itself as the root node. Since algorithm *A* requires availability of global network information at the node doing the routing computation, this procedure can also be viewed as a "partially" centralized method.

Link costs are evaluated in terms of time delays on the links. Each node calculates an estimate of the delay on each of its outbound links by averaging the total packet delay (processing, queueing, transmission, retransmission, propagation time) over 10 s intervals. (One of the problems with the first algorithm was that delay estimates were obtained too frequently to be accurate.) Since all nodes must be informed of any changes in link time delays, a "flooding" technique is used for forwarding the measured delays throughout the network. Each node transmits to all its neighbors delay information for all of its outgoing links. It also acts as a repeater, broadcasting to all of its neighbors the link delay information it has received from other nodes. (Transmitting delay information back to the adjacent node from which it was received provides an automatic positive acknowledgment mechanism.) Duplicate delay information packets are dropped, so that while the information propagates to all nodes in the network, it does not circulate

---

[3] The authors are indebted to Dr. John McQuillan of Bolt Beranek and Newman for providing information on ARPANET.

indefinitely. To reduce the amount of communication overhead involved in this information exchange, the 10 s average link delay measurements are not always transmitted. Only when the *change* in link delay since the last transmission exceeds a certain threshold does a new transmission take place. The threshold is reduced as time increases since the previous transmission. (However, a change in the status of a line is reported immediately.) The total communication overhead involved in delay update exchanges is less than 1 percent.

Since a complete execution of algorithm *A* at each update requires considerable computation, the algorithm has been modified so that "incremental" computation can be performed. When a single link delay changes (or if a link or node is added or deleted from the network), each node does a *partial* computation to restructure its shortest path tree. (This, of course, implies that each node must store the most recently updated tree as a basis for future updates, imposing an additional memory requirement.) Also, to take care of the case where link or node failures cause a complete partition of the network, an indication of "age" is inserted in each delay update packet. In this way, "out of date" delay information can be recognized and discarded when lines are reconnected and routing tables are recomputed. Operational results indicate that complete processing of a routing update at a node requires several milliseconds on the average.

A series of tests were performed with the algorithm under actual operating conditions, revealing a number of its features: it responds fairly rapidly (100 ms) to topological changes (one of the problems with the earlier algorithm was that it responded too slowly to line failures); it usually does minimum hop routing, but under heavy load conditions it spreads traffic over lines with excess capacity; it can respond to congestion by choosing paths to avoid congested nodes; and it seems to be stable and free of sustained looping.

Based on the information available at this time, the new algorithm seems to show some advantages over the old in terms of speed of response to changing topology, stability, and suppression of looping. These advantages are apparently attained without undue overhead. It must be kept in mind that in going from algorithm *B* to *A*, many other aspects of the routing scheme were also changed, most importantly, the procedure for estimating and forwarding link delay information. Many of the problems encountered using algorithm *B* were due to the extremely rapid updating that was used based on information whose accuracy did not warrant such rapid adaptivity.

2) TYMNET Routing Algorithm: TYMNET is a computer-communication network developed in 1970 by Tymshare, Inc. of Cupertino, CA. It has been in commercial operation since 1971 [1], [21], [22]. Originally developed for time-shared purposes, it has more recently taken on a network function as well, and is classified by the Federal Communications Commission as a valued-added specialized carrier. As of 1978, the network had 300 nodes in operation and was growing at the rate of 2 nodes/week [16]. Almost all nodes are connected to at least two other nodes in the network, giving rise to a distributed topology with alternate path capability. The network is designed primarily to handle interactive terminal users,

although it does handle higher speed synchronous traffic as well. The lines connecting the nodes range in speed from 2400 to 9600 bits/s. The network covers the United States and Europe, with connections also made to the Canadian Datapac Network. Trans-Atlantic lines are cable with satellite backup. Satellites are avoided, where possible, for interactive users because of the substantial delay involved.

Individual user data packets or logical records, each preceded by a 16-bit header incorporating an 8-bit logical record number to be discussed below and an 8-bit packet character count, are concatenated to form a physical record of at most 66 8-bit characters, including 16 bits of header and 32 bits of checksum for error detection [1]. These data packets can range in length from a few characters to a maximum of 58 characters. (Physical records are transmitted as soon as available, without waiting for a specified size logical record to be assembled.)

TYMNET routing is set up centrally on a virtual circuit, fixed path, basis by a supervisory program running on one of four possible supervisory computers in the network.

A least cost algorithm [1], [16] is used to determine the appropriate path from source to destination node over which to route a given user's packets. The path is newly selected each time a user comes on the network, and is maintained unchanged during the period of the user connection or session. (In the event of an outage, the session is interrupted and a new routing path has to be computed. In TYMNET I, the version of TYMNET that has operated to the present, this could take up to 2.5 min as the supervisor learned of the incident and established the new topology. In the newer TYMNET II, which is gradually replacing the earlier version, rerouting in the event of an outage is carried out by the supervisor in a manner transparent to the user.) The algorithm used by the supervisor is a modification of Floyd's algorithm, a variation of our algorithm *B*.

Integer-valued costs are assigned to each link, and costs are then summed to find the path of least cost. The cost assignments depend on line speed and line utilization. Thus, the number 16 is assigned to a 2400 bit/s link, 12 to a 4800 bit/s link, and 10 to a 9600 bit/s link. A penalty of 16 is added to a satellite link for low-speed interactive users. This shifts such users to cable links, as noted above.

A penalty of 16 is added to a link if a node at one end complains of "overloading." The penalty is 32 if the nodes at both ends complain. Overload is experienced if the data for a specific virtual circuit have to wait more than 0.5 s before being serviced. This condition is then reported by the node to the supervisor. An overload condition may occur because of too many circuits requesting service over the same link, or it may be due to a noisy link with a high error rate, in which case the successive retransmissions which are necessary slow the effective service rate down as well. The penalty used in this case serves to steer additional circuits away from the link until the condition clears up.

Details of the specific algorithm used appear in [16]. In the absence of overloading, the algorithm tends to select the shortest path (least number of links) with highest transmission speed. As more users come on the network, the lower speed links begin to be used as well. In lightly loaded situations, users tend to have relatively shorter time delays through the network. The minimum hop paths, favored in the lightly loaded case, also tend to be more reliable than ones with more links. Users coming on in a busy period may experience higher time delays due both to congestion and to the use of lower speed lines. The use of the overload penalties tends to spread traffic around the network, deviating from the shortest path case, but attempting to reduce the time delay. In practice, the average response time for interactive users in 0.75 s [16].

It takes 12 ms for the supervisor to find the least cost path using this algorithm [23]. Once the path has been selected, the supervisor notifies each of the nodes along the path, assigning an 8-bit *logical record number* to each link on that path. (This allows up to 256 users or channels to share any one link. In practice, the maximum number ranges from 48 for a 2400 bit/s line to 192 for a 9600 bit/s line. In addition, one number or channel is reserved for a node to communicate with the supervisor and one channel is reserved for communications with the neighboring node.) The supervisor also associates a logical record number on an incoming link to a node with a number on the appropriate outgoing link setting entries in routing tables, called permuter tables in TYMNET terminology. This process, described in more detail later, is basically the same as the method of *path numbers* described in Section II-C. In the TYMNET II version of the network, the nodal computers themselves establish the routing table sizes and entries, as well as the buffers associated with them, relieving the supervisor of this burden.

Routing information is sent to a particular node in a 48-bit supervisory record with the usual 16 bits of logical record overhead as part of a normal physical record. The data transmission overhead due to the dissemination of this routing information is calculated, on a worst case basis, to be 1.6 percent [23]. This assumes that the circuit to be set up is 4 links long, with 5 nodes to be notified (the average path in TYMNET is 3.1 links) during a busy period in which an average of 1 user/s requests entry to the network. The supervisory overhead is taken as distributed equally over a minimum of 8 outgoing 2400 bit/s links from the supervisor. This calculation does not assign any physical record overhead to the supervisory logical record. The assumption is made that there are always data waiting to be transmitted and that the supervisory record is piggybacked onto a normal data record, as noted earlier.

Each node acknowledges receipt of the routing information, again doing this as part of a physical record. (Nodes, in addition, report any link outages to the supervisor as part of a 48-bit record transmitted every 16 s.)

The procedure at a node for forwarding an incoming data packet (logical record) to the appropriate outgoing link, or to either a host computer or terminal if at the destination node, proceeds as follows. As noted earlier, there is a routing or permuter table associated with each link at a node. Each logical record number in either direction on the link is associated with an entry in the table. That entry, in turn, corresponds to the address of a pair of buffers at the node, one for each direction of data flow (inbound and outbound). For L links at a node, L permuter tables are needed, each receiving up to 256 buffer addresses. An error-free physical record arriving at a node is disassembled into its component data packets (logical records). Each data packet is steered by the permuter table entry to its appropriate buffer. Data in buffers destined for terminals and/or computers associated with this node are then transferred to the appropriate device. This node thus represents the destination node for these logical records. Logical records waiting in transit buffers are handled differently. A physical record for a given outgoing link is created, under program control, by scanning sequentially the entries in the permuter table for that link. As each buffer address is read, a determination is made as to whether its *pair* has had data entered. If so, the data are then formed into a logical record with their corresponding new logical record number. This logical record is incorporated in the physical record and is transmitted out over the link.

A specific example appears in Fig. 6 [23]. Fig. 6(a) shows a typical two-link virtual circuit connecting nodes numbered 5, 7, and 10. In this example, terminal data enter the network via a terminal port at node 5, destined for a Host computer connected to node 10. The link connecting nodes 5 and 7 is labeled 1, as seen at the node 5 side, and 2 as seen at the node 7 side. Similarly, the link connecting nodes 7 and 10 is labeled 3 at the node 7 side and 1 at the node 10 side.

Fig 6(b) portrays the logical record number assignments and permuter table entries in detail, node by node. (Eight possible logical record numbers only have been assumed for simplicity.) The logical record numbers 4 and 6 have been assigned to this virtual circuit over the two links shown, respectively. At node 5, the entry node, the number 3 in entry 4 in the permuter table for link 1 indicates that data with logical record number 4 are to be found in buffer 2, the pair of buffer 3.

At node 7, data coming from link 2 are stored in buffers designated by the contents of the permuter table for link 2 at that node. Continuing with this example, data arriving at that link with logical record 4 are to be further transmitted over outgoing link 3 to node 10. Their outgoing logical record number is to be changed to 6. To accomplish this, note that the contents of entry 4 of permuter table 2 and entry 6 of permuter table 3 are paired together. Data arriving over incoming link 2 are stored in buffer 8. They are read out over link 3 when the entries for the permuter table for that link are scanned, entry 6 pointing to buffer 9, the pair of buffer 8. At node 10, the destination node for this virtual circuit, data arriving with logical record 6 are stored in buffer 100 of that node and are then transferred to the appropriate Host.

3) Routing in TRANSPAC:[4] TRANSPAC, the French public packet-switching service [24], began operation in December 1978 with ten nodes (soon to be expanded to twelve) in a distributed network configuration. As is the case with most public packet-switching services, the transport procedures for TRANSPAC follow the X.25 international standard protocol. Thus, this is a virtual-circuit-oriented system, and the routing procedures discussed below reflect this orientation. For pur-

---

4 The authors are indebted to J. M. Simon of TRANSPAC for providing information used in preparing this section.

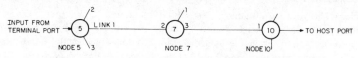

(a) TYPICAL VIRTUAL CIRCUIT: TWO LINKS, TERMINAL TO HOST PORT

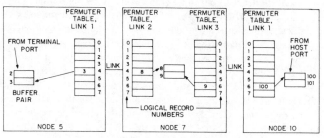

(b) PERMUTER TABLES AND LOGICAL RECORD NUMBERS

Fig. 6.   Routing example, TYMNET.

poses of reliability, there are at least two 72 kbit/s lines, following different physical paths, connecting each node to the remainder of the network. Each node consists of a control unit (CU) (a CII Mitra 125 minicomputer) to which are attached a number of switching units (SU). Each incident link is controlled by an SU, which executes all data link procedures. The SU's also execute the access protocols for customers connected to the node. Routing is handled by the CU, using information from the Network Management Center (see below).

Network control is partially decentralized through six local control points which handle a certain amount of statistics gathering and perform test and reinitialization procedures in case of node or line failures. However, general network supervision, including the bulk of the routing computation, is exercised through a single Network Management Center (NMC).

Routes in TRANSPAC are assigned on a single-path-per VC basis. The algorithm of interest to us here is that which governs the assignment of a route to a switched virtual circuit, i.e., a VC which is established temporarily in response to a "call request." The call request takes the form of a *Call Packet*, emitted by equipment connected to the originating network node, and requesting connection to a specified destination. The path that eventually will be retained by the switched VC is identical to that taken by the Call Packet as it is forwarded through the network, Routing of the Call Packet is effected through routing tables stored at each node; as indicated in Section II, the tables associate a unique outbound link with each destination node. The network as currently configured has two classes of nodes. One class is connected in a distributed fashion, with alternate route capability. The second class consists of nodes homing in via a single link to a node of the first class. Node 5 in Fig. 7 is an example of a node of the first type; node 6 is a node of the second type. Messages destined to nodes of the second type are routed to the "target" node to which they are connected. In Fig. 7, messages destined for node 6 have node 5 as a target node.

The routing tables for the network are constructed in an essentially centralized fashion, using a minimum cost, i.e.,

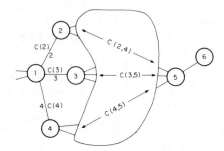

Fig. 7.   Routing example, TRANSPAC.

shortest path criterion. Link costs are defined in terms of link resource utilization. Thus, the cost assigned to a link varies dynamically with network load. We shall first describe the method of evaluation of link cost and then the routing algorithm [25], [26]. Consider a full duplex link k connected to nodes m and n. Let $C_m(k)$, $C_n(k)$ be the cost assigned to link k as perceived by nodes m and n, respectively, and let $C(k) = \text{Max}\,[C_m(k), C_n(k)]$ be the "combined" estimate of link cost. The quantities $C_i(k)$ are the basic data on which routing computation is based; they are determined locally by each node's CU which gathers estimated and measured data from its associated SU's. Link cost is defined as a function of the level of utilization of two types of resources: line capacity and link buffers. The utilization of these quantities is evaluated both by estimation (based on the parameters of the active VC's using the link) and by measurement. The cost $C_i(k)$ is set to infinity if either the link is carrying its maximum permissible number of VC's or it has exceeded a preset threshold of buffer occupancy. Otherwise, $C_i(k)$ is defined as a piecewise constant increasing function of average link flow, quantized to a small number of levels and including a "hysteresis" effect. A typical function is shown in Fig. 8, with the arrows indicating the way link cost changes as a function of changing utilization. The nodes send updated values of their $C_i(k)$'s to the NMC whenever a change occurs; these events are infrequent due to the combined effect of coarse quantization and hysteresis. At the

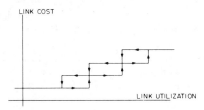

Fig. 8.   A typical link cost relation, TRANSPAC.

NMC, the costs perceived by the nodes at both ends of each link are compared to form $C(k)$ as defined above.

The major part of the routing computation takes place at the NMC, but some local information is used at each node. The procedure is illustrated by an example in Fig. 7 in which a Call Packet arriving at node 1 (which may be either the originating node or an intermediate one) is to be forwarded through one of the adjacent nodes 2, 3, 4 to the target node 5, and finally to the destination node 6. Let $C(k, n)$ (computed by the NMC) be the total cost associated with the minimum cost path between nodes k and n. Node 1 determines the "shortest" route to node 5 by choosing the value of k which minimizes $C(k, 5) +$ Max $[C(k), C_1(k)]$, $k = 2, 3, 4$. In this way, node 1 chooses the intermediate node that would have been chosen by the NMC, unless the value of $C_1(k)$ has changed recently. Ties are resolved by giving priority to the shortest hop path. Because of the way in which link costs are defined, the routing procedure becomes a minimum hop method upon which is superimposed a bias derived from the level of link resource utilization.

Although the TRANSPAC routing algorithm has many of the features of a typical centralized routing procedure, its operation departs from being purely centralized by allowing the final routing decision to be made locally, based on a combination of centrally and locally determined information. This is similar to the concept of "delta routing" suggested by Rudin [27].

By examining the current topology of the network [24], one can deduce the order of magnitude of the computational load at the NMC. The $C(k, n)$'s must be determined for all k and n belonging to the subset of all possible target nodes. Only six out of the twelve currently planned network nodes are in this category. Furthermore, rather than doing a complete shortest path computation to determine these quantities, the designers chose to limit the shortest path computation to a minimization over a prescribed subset of four or five paths joining each pair of nodes. Thus, the computation of all pertinent $C(k, n)$'s involves at most 75 path length evaluations.

At this writing, the network has recently entered its operational phase, with 300–400 subscribers as compared to an expected full load population of 1500. It is reported that the routing algorithms are operating satisfactorily, without undue overhead.

B. Commercial Network Architectures

The examples discussed thus far have all been operational networks. Specific physical implementations exist, although the networks have been steadily growing and changing their topologies. In the two examples discussed in this section, we focus on another type of distributed network architecture for which routing procedures become important. These are the network protocols introduced by most large computer system manufacturers during the past decade or so to enable private users to configure their own computer networks. Such networks are being increasingly developed to handle such diverse tasks as distributed processing, distributed database handling, and computer resource sharing. A computer manufacturer's protocol is designed to enable a user to interconnect a variety of computer systems and terminals in any desired configuration. All of these network protocols tend to follow a layered architecture, starting at the lowest level, that of setting up physical connections, continuing to the next, data link level, which controls the flow of data packets between neighboring nodes, then proceeding to the transmission or transport level, involved with end-to-end (source to destination) control of packet flow, routing, and congestion control, and finally concluding, at the highest levels, with several levels of "handshaking" between users or programs at the two ends. Other papers in this Special Issue discuss these network protocols in detail.

In this section, we describe the routing procedures defined for distributed versions of the IBM Systems Network Architecture (SNA) and the Digital Equipment Corporation's Digital Network Architecture (DNA). These are both relatively recent developments since earlier versions of both SNA and DNA were tailored primarily to star- or tree-type network configurations with no real need for routing. It will be noted that, unlike the network examples discussed previously, where networking is essentially transparent to the user, it is left to the user of either SNA on DNA to configure his own network. There is a certain flexibility in the routing procedures as well, with the user free to define his own link costs and paths to be taken. This is, of course, not the case in the earlier networks described.

1) IBM's Systems Network Architecture (SNA):[5] The early versions of SNA, appearing in 1974, were designed for single-computer system tree-type networks [28], [29]. In these networks, it is apparent that routing was not really a significant problem. Later versions of SNA allow two or more such single-system networks to be interconnected, leading to the concept of cross-domain networking [28]. Here, too, routing requirements were quite simple. IBM's latest SNA architecture, termed SNA 4.2 [30], envisions multiple computer systems interconnected to form a distributed network. Routing thus plays an important role in the architecture.

The routing procedure chosen for SNA incorporates predetermined fixed paths from source to destination. A multiplicity of possible routes is provided to increase the probability that a route will be available when needed to achieve load leveling, to provide alternate route capability in the event of node/link failures or congestion, and to provide different types of services for different classes of users [30], [31]. For example, batch traffic would normally be routed differently from interactive traffic. (Not only are the response time re-

[5] The authors are indebted to Dr. James P. Gray of IBM for help with this section.

quirements different calling usually for different capacity links, as noted earlier in discussing the TYMNET routing procedure, but one would not normally want to have batch traffic interfering with, and hence slowing down, interactive traffic. In SNA 4.2, this can be done by assigning a lower transmission priority to batch traffic.) Some traffic may require high security handling and will therefore be routed differently.

Multiple routing is provided at two levels: when first initiating a session, the user specifies a name corresponding to a particular class of service. Examples of classes of service include low response time, high capacity lines, more secure paths, etc. Associated with each class of service name is a list of possible virtual routes for use by sessions specifying that name. This list provides load balancing and backup capability. A particular session uses only one of these virtual routes at a time. This corresponds to the first level of multiple routing. Each virtual route provides a full-duplex connection between source and destination nodes, and can support multiple users or sessions. Each virtual route in turn maps into a so-called explicit route, the actual physical path from source to destination. It is this path that has been precalculated to provide the desired performance. Multiple explicit routes will exist, on a unidirectional basis, between any source-destination nodal pair. The multiple explicit routes provide the second level of multiple route control noted earlier. In the current SNA 4.2 release, up to eight explicit routes can be made available between any source-destination nodal pairs. Several virtual routes may use the same explicit route.

Although explicit routes are established by the source node on a unidirectional basis, explicit routes are used in pairs that are physically reversible. This simplifies user notification of route failure.

Up to 24 virtual routes are currently available between any pair of nodes. These are grouped into three levels of transmission priority, with eight possible virtual route numbers associated with each level. The entire set of virtual routes, each identified by a virtual route number and transmission priority, is stored in a virtual route identifier list. Class of service names are then associated with subsets of this list, in some preassigned order. A user setting up a session specifies his class of service name. He is then assigned to the first virtual route in the virtual route list that is available or can be activated. Multiple sessions may be assigned to the same virtual route. The virtual route is defined by four fields—the source and destination addresses, a virtual route number, and the transmission priority.

The explicit route corresponding to a specific virtual route is, in turn, defined by the source and destination addresses and an explicit route number. Each explicit route number represents one of the eight distinct routes possible between any source-destination nodal pair. A given explicit route is made up of a sequence of logical links connecting adjacent nodes along the path.

The term transmission group is used for logical link in the SNA terminology. Transmission groups may consist of multiple physical links. Thus, a set of parallel physical links between any two nodes can be divided into one or more transmission groups. This adds flexibility to the transmission function:

physical links may be combined in parallel to provide higher capacity, links may be dynamically added or deleted without disruption, and scheduling of links is employed to optimize the composite bandwidth or capacity available. But the use of multiple-link transmission groups means that data packets or blocks may arrive out of sequence. Out-of-order blocks must thus be reordered at the receiving end of each transmission group along the composite path.

Routing of data packets is carried out by examining the destination address and explicit route number as a packet arrives at an intermediate node along the path. An explicit routing table at each node associates an appropriate outgoing transmission group with the destination address and explicit route number. An example of such a table at a particular node appears in Fig. 9. The letters represent the transmission groups to which packets with the corresponding address, route number pair are directed. By changing the explicit route number for a given destination, a new path will be followed. This introduces alternate route capability. If a link or node along the path becomes inoperative, any sessions using that path can be reestablished on an explicit route that bypasses the failed element. Explicit routes can also be assigned on the basis of type of traffic, types of physical media along the path (satellite or terrestrial, for example), or other criteria, as already noted. Routes could also be listed on the basis of cost, the smallest cost route being assigned first, then next smallest cost route, etc.

Note that the explicit routing concept is similar to that adopted by TYMNET in its virtual circuit approach. Here the path selected may be changed by the source node, however, by choosing a new explicit route number. In essence, a variety of alternate routes is laid out in advance. This introduces the alternate route capability noted above. In the current TYMNET approach, the central supervisor must set up a new path if one is desired.

The concept of explicit routing, as first enunciated and as noted in Section II earlier, is somewhat broader than the one described here [17]. There, rather than using a fixed explicit route number, a variable "next node index" (NNI) field was proposed for the packet header. The combination of the destination address and the NNI field then directs the packet to the appropriate outgoing transmission group. The NNI is changed at the same time as well. This allows more explicit routes to be defined than through the use of a fixed explicit route number. The idea is similar to the (variable) logical record number concept used by TYMNET. In addition, some form of intermediate or local node routing capability could be introduced through the use of the NNI. For by changing the NNI locally, a new path from that point on will be followed. This makes it possible to introduce alternate route capability along the initial path chosen in the event of localized congestion or some other delaying phenomenon. The British NPL, in a series of network simulation experiments, has shown the benefits of alternate route capability [32]. Rudin has proposed as well a routing strategy that combines centralized routing with a measure of local adaptability [27]. The general idea of explicit routing thus enables centralized, distributed, and local routing strategies (or some combination of them) to be introduced into the

Fig. 9.   Explicit routing table, SNA.

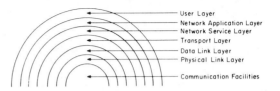

Fig. 10.   DNA layered architecture.

network. In the current IBM implementation, however, only precalculated routes are used.

Three steps are required to activate a route. The individual links of the transmission groups forming the explicit route must be brought up. The explicit route is then activated. Finally, the virtual route to be used that maps into this explicit route must be activated. Special command packets are used for this purpose [31]. For example, an explicit route is activated by transmitting a specific activate command from node to node along the path. This packet verifies the routing tables of the nodes along the path. It ensures loop-free routes by checking the routing tables of nodes along the route. It verifies that there are no packets along the path with the same source-destination address pairs. It also measures the length of the explicit route in hops. Activation of the explicit route is considered completed when verified by a reply command from the destination. If the activation of the first-choice virtual route and its associated explicit route fails, the second-choice virtual route is tried, repeating with the third choice, and so on, if necessary.

The user is involved in setting the routing tables, and hence in route definition in the SNA architecture. Thus, the user can define the routes he desires, given his physical topology, by providing table entries at system definition time. The user can also ensure that a session is established on a desired route. For unique or specialized requirements, the user can write a user-exit routine that is invoked during session establishment. This exit can assign a session to a specific route.

2) Routing in Digital's DNA:[6] Digital Equipment Corporation's network architecture is called Digital Network Architecture, or DNA for short. It provides the interfaces and protocols that enable users to create their own networks using Digital Equipment Corporation Systems. The family of network products supporting DNA is generally called DECnet. DNA and DECnet were first introduced in 1973.

As does the IBM SNA, DNA employs a layered architecture. Six levels have been defined, as shown in Fig. 10 [33]. The Transport Layer shown in the figure was sufficiently simple in the Phase II DECnet implementations introduced in 1978 so as to be encompassed within the Network Services Layer. Phase II provides for point-to-point connections with no routing capability required. The next phase, Phase III DECnet, will have provision for store-and-forward distributed topologies requiring routing, flow and congestion control, and a network management capability. For this phase of DECnet, the Transport Layer has been defined and provides the necessary routing

[6] The authors are indebted to Anthony Lauck of the Digital Equipment Corporation for providing information used in preparing this portion of the paper.

and congestion control features. In this section of the paper, we focus on routing in Phase III of DECnet.

The routing procedure adopted by Digital Equipment Corporation is based on a variant of the distributed shortest path algorithm (our algorithm *B*), with each node carrying out its own calculations. It is similar to the protocol analyzed by Tajibnapis [34] which has been implemented on the Michigan MERIT Computer Network. The routing algorithm adapts to changes in network topology (it does not use traffic flow information), and so needs to be invoked only when a link or node in the network comes up or fails. Unlike the IBM SNA approach, routing is done on a packet-by-packet or datagram basis, as contrasted to a virtual circuit service.

To carry out the least cost routing procedure, each link in the network is assigned a fixed cost. The specific cost is set by the user, but is approximately inversely proportional to link capacity. (Note that these assignments are then roughly similar to those used by TYMNET.) Paths with high capacity links are favored. These costs are used by each node to derive a routing database (or routing table) which lists the cost to each destination using each of the node's outgoing lines. An example appears in Fig. 11(a). (Each node in the network is assigned a unique address. Naming and addressing are carried out at a level higher than the Transport level.)

Packets going to a particular destination are routed to the output link with the smallest cost. In the example of Fig. 11(a), packets going to destination *C* would take output line 2 with a cost of 2 units. Those going to destination *B* would take output line 1. The listing of minimum cost outgoing lines, one for each destination, that is used in routing the packets is kept in a second database, called the forwarding database. An example appears in Fig. 11(b) for the routing table of Fig. 11(a). (A third, Boolean, database indicates whether each destination is reachable or not. This is discussed briefly later.)

As noted earlier, these tables are changed only on receipt of routing messages, triggered by a line (or node) coming up or going down. Specifically, a node, on learning that one of its links or a neighboring node has either been brought up or has failed, will update its tables. If the minimum cost to any destination has changed, the cost information is broadcast using a routing message to all the neighbors. These nodes, in turn, add to the cost forwarded for each destination the link cost for the link over which the message has arrived. The sum is then entered in the routing database. Minimization is then carried out row by row (i.e., for each destination), and the forwarding database is changed. If the resultant cost changes, this is, in turn, broadcast, using routing messages, to all neighbors. In this way, changes percolate throughout the network.

Routing messages contain 16 bits per destination, with a

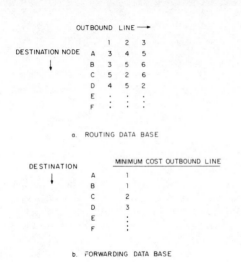

a. ROUTING DATA BASE

DESTINATION	MINIMUM COST OUTBOUND LINE
A	1
B	1
C	2
D	3
E	⋮
F	⋮

b. FORWARDING DATA BASE

Fig. 11. Typical nodal routing tables, DNA routing protocol. (Similar tables are kept at each node.)

maximum of 128 nodes allowed at present. The routing message thus consists of a maximum of 256 bytes. Of the 16 bits, 11 are used to transmit total cost information and 5 bits represent a hop count that is transmitted as well. This hop count is incremented by 1 at each node, and is used for reachability analysis. To avoid indefinite ping-ponging, one node adding 1, its neighbor adding 1, back and forth, a destination is declared unreachable if the hop count reaches a specified maximum. This maximum could be one more than the maximum path length, it could be the diameter of the network, etc.

Packets going to an unreachable destination are discarded. However, there is an option of notifying the source that a particular destination is unreachable through the use of a "return to sender" packet. This would be used on setting up a connection or initializing the operation of the network.

If a link fails, packets queued on that link are discarded as well. To maintain end-to-end (source to destination) integrity, an acknowledgment and time-out procedure is carried out by the higher level Network Service layer of DNA. (The lower level digital data communications message protocol-DDCMP-provides link, or node-to-node, error control as well. This is, of course, similar to link error control carried out by HDLC, SDLC, and other data link control protocols [1].)

How does a node know if a link is down? This is based on the number of retransmissions of packets needed. If the number 7 is reached the link is declared down. In addition, provision is made for transmitting a low-priority "Hello" message to a neighboring node that has not been heard from for a while. If there is no acknowledgment, the node is declared down.

The actual software implementation of the routing procedure involves three processes: a *decision process* which receives routing messages; an *update process* which updates the routing tables; and a packet *forwarding process* which uses the forwarding database to route the packets. Normally, the third process only is used. The first two are run only when changes in the network topology dictate changes in the routing table. Provision is made to check the routing algorithm periodically,

if desired, with the use of a timer. Such a check might be made once a minute, for example. Although the forwarding database (minimum cost paths) is normally used for routing, the entire routing database is retained as well at each node. This is required to run the distributed routing algorithm when needed. The routing database can also be used to provide alternate path capability as well, if desired, or if necessary.

Some additional factors provided by the DNA phase III Transport Layer in addition to routing include a packet lifetime control and a congestion control mechanism. The packet lifetime control is used to bound the time a packet spends in the network. A nodal visit count is kept in each data packet. If the number is too large, the packet is purged. The congestion control involved is the one analyzed by Irland [35]. The queues at each outbound link at a node are limited in size. Packets are discarded if the number queued will exceed this maximum value. Priority is, however, given to transit messages (those already in the network, as contrasted to packets originating at the node in question).

IV. CONCLUSIONS

After a brief discussion of routing in general, we have presented the basic features of routing procedures currently used in five representative packet-switched communication networks and network architectures. While the networks were chosen to represent a broad spectrum of operational characteristics, it is interesting to note that there are many similarities in their routing algorithms. At the same time, there is a great deal of diversity in the manner in which these algorithms are implemented. Most of the networks use some variation or approximation of a shortest path routing strategy. However, each network defines the "length" or "cost" of a communication link differently. Some use centralized computation, some decentralized, and some use a hybrid of the two. Adaptivity ranges from the bare minimum necessary to react to line failures to more sophisticated procedures sensing and responding to queueing delays, error rates, and line loading. Undoubtedly, a larger set of representative networks would have yielded a still richer set of alternative schemes for information gathering, routing computation, and packet forwarding. One can conclude from this survey that while the routing function is central to the smooth and efficient operation of packet-switched networks, no one scheme can be identified as "best." Many viable alternatives exist at all levels of the routing function.

REFERENCES

[1] M. Schwartz, *Computer Communication Network Design and Analysis.* Englewood Cliffs, NJ: Prentice-Hall, 1977.

[2] J. M. McQuillan, G. Falk, and I. Richer, "A review of the development and performance of the ARPANET routing algorithm," *IEEE Trans. Commun.,* vol. COM-26, pp. 1802–1811, Dec. 1978.

[3] T. Cegrell, "A routing procedure for the TIDAS message-switching network," *IEEE Trans. Commun.,* vol. COM-23, pp. 575–585, June 1975.

[4] F. Poncet and C. S. Repton, "The EIN communications sub-network: Principles and practice," in *Proc. 3rd ICCC,* Toronto, Ont., Canada, Aug. 1976, pp. 523–531.

[5] J. L. Grangé and M. I. Irland, "Thirty-nine steps to a computer network," in *Proc. 4th ICCC,* Kyoto, Japan, Sept. 1978, pp. 763–769.

[6] L. Fratta, M. Gerla, and L. Kleinrock, *The Flow Deviation Method: An Approach to Store-and-Forward Communication Network Design*, *Networks*, vol. 3. New York: Wiley, 1973, pp. 97–133.

[7] W. Older and D. A. Twyver, personal communication.

[8] J. M. McQuillan, "Interactions between routing and congestion control in computer networks," in *Proc. Int. Symp. Flow Contr. in Comput. Networks*, Versailles, France, Feb. 1979, J. L. Grangé and M. Gien, Eds., Amsterdam: North-Holland, pp. 63–75.

[9] M. Schwartz and C. Cheung, "The gradient projection algorithm for multiple routing in message-switched networks," *IEEE Trans. Commun.*, vol. COM-24, pp. 449–456, Apr. 1976.

[10] R. Gallager, "An optimal routing algorithm using distributed computation," *IEEE Trans. Commun.*, vol. COM-25, pp. 73–85, Jan. 1977.

[11] T. E. Stern, "A class of decentralized routing algorithms using relaxation," *IEEE Trans. Commun.*, vol. COM-25, pp. 1092–1102, Oct. 1977.

[12] E. W. Dijkstra, "A note on two problems in connection with graphs," *Numer. Math.*, vol. 1, pp. 269–271, 1959.

[13] A. V. Aho, J. E. Hopcroft, and J. D. Ullman, *The Design and Analysis of Computer Algorithms*. Reading, MA: Addison-Wesley, 1974.

[14] L. R. Ford, Jr. and D. R. Fulkerson, *Flows in Networks*. Princeton, NJ: Princeton Univ. Press, 1962.

[15] D. P. Bertsekas, "Dynamic behavior of a shortest path routing algorithm of the ARPANET type," presented at the Int. Symp. Inform. Theory, Grignano, Italy, June 1979.

[16] A. Rajaraman, "Routing in TYMNET," presented at the European Computing Conf., London, England, May 1978.

[17] R. R. Jueneman and G. S. Kerr, "Explicit routing in communications networks," in *Proc. 3rd ICCC*, Toronto, Ont., Canada, Aug. 1976, pp. 340–342.

[18] D. C. Walden, "Experiences in building, operating, and using the ARPA network," presented at the 2nd U.S.A–Japan Comput. Conf., Tokyo, Japan, Aug. 1975.

[19] J. M. McQuillan, "Adaptive routing algorithms for distributed computer networks," BBN Rep. 2831, May 1974.

[20] J. M. McQuillan *et al.*, "The new routing algorithm for the ARPANET," *IEEE Trans. Commun.*, vol. COM-28, pp. 711–719, May 1980.

[21] L. Tymes, "TYMNET—A terminal oriented communication network," in *1971 Spring Joint Comput. Conf.*, AFIPS Conf. Proc., vol. 38, 1971, pp. 211–216.

[22] J. Rinde, "Routing and control in a centrally-directed network," in *1977 Nat. Comput. Conf.*, AFIPS Conf. Proc., vol. 46, 1977, pp. 603–608.

[23] ——, "TYMNET I: An alternative to packet technology," in *Proc. 3rd ICCC*, Toronto, Ont., Canada, Aug. 1976, pp. 268–273.

[24] A. Danet, R. Despres, A. LaRest, G. Pichon, and S. Ritzenthaler, "The French public packet switching service: The Transpac network," in *Proc. 3rd ICCC*, Toronto, Ont, Canada, Aug. 1976, pp. 251–260.

[25] J. M. Simon and A. Danet, "Contrôle des ressources et principes du routage dans le réseau Transpac," in *Proc., Int. Symp. Flow Control in Comput. Networks*, Versailles, France, Feb. 1979, J. L. Grangé and M. Gien, Eds. Amsterdam: North-Holland, pp. 33–44.

[26] J. M. Simon, personal communication.

[27] H. Rudin, "On routing and "Delta-routing": A taxonomy and performance comparison of techniques for packet-switched networks," *IEEE Trans. Commun.*, vol. COM-24, pp. 43–59, Jan. 1976.

[28] R. J. Cypser, *Communications Architecture for Distributed Systems*. Reading, MA: Addison-Wesley, 1978.

[29] P. E. Green, "An introduction to network architectures and protocols," *IBM Syst. J.*, vol. 18, no. 2, pp. 202–222, 1979.

[30] J. P. Gray and T. B. McNeill, "SNA multiple-system networking," *IBM Syst. J.*, vol. 18, no. 2, pp. 263–297, 1979.

[31] V. Ahuja, "Routing and flow control in systems network architecture," *IBM Syst. J.*, vol. 18, no. 2, pp. 298–314, 1979.

[32] W. L. Price, "Data network simulation experiments at the National Physical Laboratory, 1968–1976," *Comput. Networks*, vol. 1, no. 4, pp. 199–210, 1977.

[33] Digital Network Architecture, General Description, AA-H202A-TK, Digital Equipment Corp., Maynard, MA, Nov. 1978.

[34] W. D. Tajibnapis, "A correctness proof of a topology information maintenance protocol for distributed computer networks," *Commun. Ass. Comput. Mach.*, vol. 20, pp. 477–485, July 1977.

[35] M. Irland, "Buffer management in a packet switch," *IEEE Trans. Commun.*, vol. COM-26, pp. 328–337, Mar. 1978.

The New Routing Algorithm for the ARPANET

JOHN M. McQUILLAN, MEMBER, IEEE, IRA RICHER, MEMBER, IEEE, AND ERIC C. ROSEN

Abstract—The new ARPANET routing algorithm is an improvement over the old procedure in that it uses fewer network resources, operates on more realistic estimates of network conditions, reacts faster to important network changes, and does not suffer from long-term loops or oscillations. In the new procedure, each node in the network maintains a database describing the complete network topology and the delays on all lines, and uses the database describing the network to generate a tree representing the minimum delay paths from a given root node to every other network node. Because the traffic in the network can be quite variable, each node periodically measures the delays along its outgoing lines and forwards this information to all other nodes. The delay information propagates quickly through the network so that all nodes can update their databases and continue to route traffic in a consistent and efficient manner.

An extensive series of tests were conducted on the ARPANET, showing that line overhead and CPU overhead are both less than two percent, most nodes learn of an update within 100 ms, and the algorithm detects congestion and routes packets around congested areas.

I. INTRODUCTION

THE last decade has seen the design, implementation, and operation of several routing algorithms for distributed networks of computers. The first such algorithm, the original routing algorithm for the ARPANET, has served remarkably well considering how long ago (in the history of packet switching) it was conceived. This paper describes the new routing algorithm we installed recently in the ARPANET. Readers not familiar with our earlier activities may consult [1] for a survey of the ARPANET design decisions, including the previous routing algorithm; readers interested in a survey of routing algorithms for other computer networks and current research in the area may consult [2].

A distributed, adaptive routing scheme typically has a number of separate components, including: 1) a *measurement process* for determining pertinent network characteristics, 2) a *protocol* for disseminating information about these characteristics, and 3) a *calculation* to determine how traffic should be routed. A routing "algorithm" or "procedure" is not specified until all these components are defined. In the present paper, we discuss these components of the new ARPANET algorithm. We begin with a brief outline of the shortcomings of the original algorithm; then, following an overview of the new procedure, we provide some greater detail on the individual components. The new algorithm has undergone extensive testing in the ARPANET under operational conditions, and the final section of the paper gives a summary of the

test results. This paper is a summary of our conclusions only; for more complete descriptions of our research findings, see our internal reports on this project [3]–[5].

II. PROBLEMS WITH THE ORIGINAL ALGORITHM

The original ARPANET routing algorithm and the new version both attempt to route packets along paths of least delay. The total path is not determined in advance; rather, each node decides which line to use in forwarding the packet to the next node. In the original approach, each node maintained a table of estimated delay to each other node, and sent its table to all adjacent nodes every 128 ms. When node I received the table from adjacent node J, it would first measure the delay from itself to J. (We will shortly discuss the procedure used for measuring the delay.) Then it would compute its delay via J to all other nodes by adding to each entry in J's table its own delay to J. Once a table was received from all adjacent nodes, node I could easily determine which adjacent node would result in the shortest delay to each destination node in the network.

In recent years, we began to observe a number of problems with the original ARPANET routing algorithm [7] and came to the conclusion that a complete redesign was the only way to solve some of them. In particular, we decided that a new algorithm was necessary to solve the following problems.

1) Although the exchange of routing tables consumed only a small fraction of line bandwidth, the packets containing the tables were long, and the periodic transmission and processing of such long, high-priority packets can adversely affect the flow of network traffic. Moreover, as the ARPANET grows to 100 or more nodes, the routing packets would become correspondingly larger (or more frequent), exacerbating the problem.

2) The route calculation is performed in a distributed manner, with each node basing its calculation on local information together with calculations made at every other node. With such a scheme, it is difficult to ensure that routes used by different nodes are consistent.

3) The rate of exchange of routing tables and the distributed nature of the calculations causes a dilemma: the network is too slow in adapting to congestion and to important topology changes, yet it can respond too quickly (and, perhaps, inaccurately) to minor changes.

The delay measurement procedure of the old ARPANET routing algorithm is quite simple. Periodically, an IMP counts the number of packets queued for transmission on its lines and adds a constant to these counts; the resulting number is the "length" of the line for purposes of routing. This delay measurement procedure has three serious defects.

1) If two lines have different speeds, or different propagation delays, then the fact that the same number of packets is

queued for each line does not imply that packets can expect equal delays over the two lines. Even if two lines have the same speed and propagation delay, a difference in the size of the packets which are queued for each line may cause different delays on the two lines.

2) In the ARPANET, where the queues are constrained to have a (short) maximum length, queue length is a poor indicator of delay. The constraints on queue length are imposed by the software in order to fairly resolve contention for a limited amount of resources. There are a number of such resources which must be obtained before a packet can even be queued for an output line. If a packet must wait a significant amount of time to get these resources, it may experience a long delay, even though the queue for its output line is quite short.

3) An instantaneous measurement of queue length does not accurately predict average delay because there is a significant real-time fluctuation in queue lengths at any traffic level. Our measurements show that under a high constant offered load, the average delay is high, but many individual packets show low delays, and the queue length often falls to zero! This variation may be due to variation in the utilization of the CPU, or to other bottlenecks, the presence of which is not accurately reflected by measuring queue lengths.

These three defects are all reflections of a single point, namely, that the length of an output queue is only one of many factors that affect a packet's delay. A measurement procedure that takes into account only one such factor cannot give accurate results.

The new routing algorithm is an improvement over the old one in that it uses fewer network resources, operates on more realistic estimates of network conditions, reacts faster to important network changes, and does not suffer from long-term loops or oscillations.

III. OVERVIEW OF THE NEW ROUTING PROCEDURE

The routing procedure we have developed contains several basic components. Each node in the network maintains a database describing the network topology and the line delays. Using this database, each node independently calculates the best paths to all other nodes, routing outgoing packets accordingly. Because the traffic in the network can be quite variable, each node periodically measures the delays along its outgoing lines and forwards this information (as a "routing update") to all other nodes. A routing update generated by a particular node contains information only about the delays on the lines emanating from that node. Hence, an update packet is quite small (176 bits on the average), and its size is independent of the number of nodes in the network. An update generated by a particular node travels *unchanged* to all nodes in the network (not just to the immediate neighbors of the originating node, as in many other routing algorithms). Since the updates need not be processed before being forwarded because they are small, and since they are handled with the highest priority, they propagate very quickly through the network, so that all nodes can update their databases rapidly and continue to route traffic in a consistent and efficient manner.

Many algorithms have been devised for finding the shortest path through a network. Several of these are based on the concept of computing the entire tree of shortest paths from a

given node, the root of the tree. A recent article [9] discusses some of these algorithms and references several survey articles. The algorithm we have implemented is based on one attributed to Dijkstra [10]; because of its search rule, we call it the shortest-path-first (SPF) algorithm.

The basic SPF algorithm uses a database describing the network to generate a tree representing the minimum delay paths from a given root node to every other network node. Fig. 1 shows a simplified flowchart of the algorithm. The database specifies which nodes are directly connected to which other nodes, and what the average delay per packet is on each network line. (Both types of data are updated dynamically, based on real-time measurements.) The tree initially consists of just the root node. The tree is then augmented to contain the node that is closest (in delay) to the root and that is adjacent to a node already on the tree. The process continues by repetition of this last step. LIST denotes a data structure containing nodes that have not yet been placed on the tree but are neighbors of nodes that are on the tree. The tree is built up shortest-paths-first—hence, the name of the algorithm. Eventually, the furthest node from the root is added to the tree, and the algorithm terminates. We have made important additions to this basic algorithm so that changes in network topology or characteristics require only an incremental calculation rather than a complete recalculation of all shortest paths.

Fig. 2 shows a six-node network and the corresponding shortest path tree for node 1. The figure also shows the routing directory which is produced by the algorithm and which would be used by node 1 to dispatch traffic. For example, traffic for node 4 is routed via node 2. Only the routing directory is used in forwarding packets; the tree is used only in creating the directory.

The two other important components of the routing procedure are the mechanism for measuring delay and the scheme for propagating information. The routing algorithm must have some way of measuring the delay of a packet at each hop. This aspect of the routing algorithm is quite crucial; an algorithm with poor delay measurement facilities will perform poorly, no matter how sophisticated its other features are.

Each node measures the actual delay of each packet flowing over each of its outgoing lines, and calculates the average delay every 10 s. If this delay is significantly different from the previous delay, it is reported to all other nodes. The choice of 10 s as the measurement period represents a significant departure from the old routing algorithm. Since it takes 10 s to produce a measurement, the delay estimate for a given line cannot change more often than once every 10 s. The old routing algorithm, on the other hand, would allow the delay estimate to change as often as once every 128 ms. We now believe, however, that there is no point in changing the estimate so often, since it is not possible to obtain an accurate estimate of delay in the ARPANET in less than several seconds. (See Section IV-B.)

The updating procedure for propagating delay information is of critical importance because it must ensure that each update is actually received at all nodes so that identical databases of routing information are maintained at all nodes. Each update is transmitted to all nodes by the simple but reliable

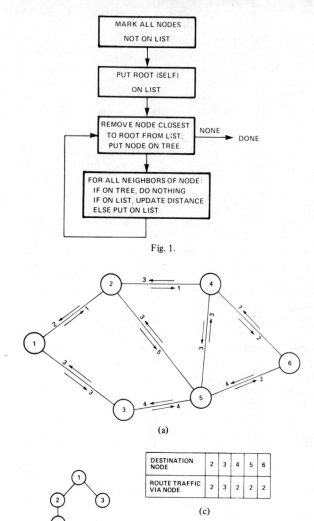

Fig. 1.

Fig. 2. (a) Example network (line lengths indicated by the numbers beside the arrowheads). (b) Shortest path tree. (c) Routing directory.

DESTINATION NODE	2	3	4	5	6
ROUTE TRAFFIC VIA NODE	2	3	2	2	2

(c)

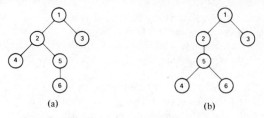

Fig. 3. (a) Shortest path tree for network of Fig. 2(a) after the length of the line 2 → 4 increase to 6. (b) Modified tree after the length of line → 5 decrease to 2.

transient loops may form for a few packets when a change is being processed, but that is quite acceptable, since it has no significant impact on the average delay in the network.

IV. DETAILED DESCRIPTION OF THE NEW ROUTING PROCEDURES

A. Routing Calculation—The SPF Algorithm

We now describe the additions to the basic algorithm of Fig. 1 which we have developed to handle various possible changes in network status without having to recalculate the whole tree. For each change described below, we assume that the shortest path tree rooted at node I prior to the change is known.

First, consider the case where the delay of the line AB from node A to node B increases. Clearly, if the line is not in the tree (i.e., not in the shortest path from that node to any other node), nothing need be done because if the line were not part of any shortest path prior to the change, then it will certainly not be used when its delay increases. If the line is in the tree, then the delay to B increases, as does the delay to each node whose route from I passes through B. Thus, the nodes in the subtree whose root is B are candidates for changed positions in the tree. Conversely, nodes not in this subtree will not be repositioned.

The first two steps for handling an increase of X in the delay from A to B are as follows.

1) Identify nodes in B's subtree and increase their delays from I by X.

2) For each subtree node S, examine S's neighbors which are not in the subtree to see if there is a shorter path from I to S via those neighbors. If such a path is found, put node S on LIST.

At the conclusion of these steps, LIST either will be empty or will contain some subtree nodes for which better paths have been found. In order to find the best paths to the nodes on LIST, a slightly modified version of SPF can be invoked. This will also find better paths, if any exist, for other subtree nodes. Fig. 3(a) shows the modification to the tree of Fig. 2 that results when the delay of the line from node 2 to node 4 increases to 6.

Now consider the case where the delay on AB decreases by X. If this line is in the tree, then paths to the nodes of the subtree which have B as its root will be unchanged because the subtree nodes were already at minimum delay, and hence the decreased delay will only shorten their distances from I. Moreover, any node whose delay from I is less than or

method of transmitting it on all lines. When a node receives an update, it first checks to see if it has processed that update before. If so, the update is discarded. If not, it is immediately forwarded to all adjacent nodes. In this way, the update flows quickly (within 100 ms) to all other nodes. The fact that an update flows once in each direction over each network line is the basis for a reliable transmission procedure for the updates. Because the updates are short and are generated infrequently, this procedure uses little line or node bandwidth (less than two percent). We have augmented this basic procedure with a mechanism to ensure that databases at nodes are correctly updated when a new node or line is installed, or when a whole set of previously disconnected nodes joins the network. This is discussed in more detail in Section IV-C.

Since all nodes perform the same calculation on an identical database, there are no permanent routing loops. Of course, equal to B's new distance from I will not be repositioned, since the node's path must reach B first in order to take advantage of the improved line. However, nodes which are not in the subtree and which are farther from I than B may have a shorter distance via one of the subtree nodes.

The algorithm must thus first perform the following steps.

1) Identify the nodes in the subtree and decrease their distances from I by X.

2) Try to find a shorter distance for each node K that is not in the subtree but is adjacent to a subtree node by identifying a path to K via an adjacent node which is in the subtree. If such a path is found, put node K on LIST.

At the conclusion of these steps, LIST will contain some (possibly zero) subtree adjacent nodes that have been repositioned. Nodes adjacent to these that are not in the subtree are also candidates for improved paths, and starting with the LIST generated in step 2) above, the basic SPF algorithm (with minor modifications) can be used to restructure the rest of the tree. Fig. 3(b) shows how the tree of Fig. 3(a) changes when the length of the link from node 2 to node 5 decreases to 2, while the length of the link from node 2 to 4 remains at 6.

If the delay on line AB improved, but AB was *not* originally in the shortest path tree, the algorithm first determines whether B can take advantage of this improvement. Since the delay from I to A cannot be improved, the delay to B using the line AB will be equal to the original distance to A plus the new delay of AB. If the new delay is greater than or equal to the former delay from I to node B, then the improved line does not help and no changes are made to the tree or to the routing table. If, on the other hand, the updated delay is less than the original delay, then the best route to B now includes AB. The first change to the shortest path tree is, therefore, to relocate B (and its subtree), attaching it to node A via line AB. Now the situation is identical to that of the previous paragraph in which the line from A to B was in the tree in the first place and its delay decreased.

Finally, a change in the status of a node—namely, the addition of a new node, the removal of a node, a node failure, or its recovery from a failure—is implicitly recognized by the change in the status of its lines. For example, if a node fails, its neighbors determine that the lines to that node have failed, and when other nodes receive this information, they calculate that the failed node is unreachable. (Of course, nodes can become unreachable even if their lines do not fail.) Thus, the algorithm need explicitly consider only line changes.

The basic SPF calculation and all of the above incremental cases are consolidated into the semiformal version of the algorithm given in the Appendix.

B. Delay Measurement

Measuring the delay of an individual packet is a simple matter. When the packet arrives at the IMP, it is time-stamped with its arrival time. When the first bit of the packet is transmitted to the next IMP, the packet is stamped with its "sent time." If the packet is retransmitted, the original sent time is overwritten with the new sent time. When the acknowledgment for the packet is received, the arrival time is subtracted from the sent time. To this difference are added the propaga-

tion delay of the line (a constant for each line) and the packet's transmission delay (found by looking it up in a table indexed by packet length and line speed). The result is the packet's total delay at that hop—the time it took the packet to get from one IMP to the next.

Every 10 s the average delay of all packets which have traversed a line in the previous 10 s is computed. Our measurements show that when we take an average over a period of less than 10 s, the average shows too much variation from measurement period to measurement period, even when the offered load is constant. There is a tradeoff here: a longer measurement period means less adaptive routing if conditions actually change; a shorter period means less optimal routing because of inaccurate measurements.

Another important aspect of the measurement technique is that the measurement periods are *not* synchronized across the network. Rather, the measurement periods in the different IMP's are randomly phased. This is an important property because synchronized measurement periods could, in theory, lead to instabilities [4], [11].

The new routing algorithm does not necessarily generate and transmit an update at the end of each measurement period; it does so only if the average delay just measured is "significantly" different from the average delay reported in the last update that was sent (which may or may not be the same as the delay measured in the previous measurement period). The delay is considered to have changed "by a significant amount" whenever the absolute value of the change exceeds a certain threshold. The threshold is not a constant but is a decreasing function of time because whenever there is a large change in delay, it is desirable to report the new delay as soon as possible, so that routing can adapt quickly; but when the delay changes by only a small amount, it is not important to report it quickly, since it is not likely to result in important routing changes. However, whenever a change in delay is long lasting, it is important that it be reported eventually, even if it is small; otherwise, additive effects can introduce large inaccuracies into routing. What is needed, then, is a scheme which reacts to large changes quickly and small changes slowly. A threshold value which is initially high but which decreases to zero over a period of time has this effect. In the scheme we have implemented, the threshold is initially set to 64 ms. After each measurement period, the newly measured average delay is compared with the previously reported delay. If the difference does not exceed the threshold, the threshold is decreased by 12.8 ms. Whenever a change in average delay equals or exceeds the threshold, an update is generated, and the threshold is reset to 64 ms. Since the threshold will eventually decay to zero, an update will always be sent after a minute, even if there is no change in delay. (This feature is needed to ensure reliability of the updating protocol under certain conditions. See Section IV-C.) It should be pointed out that when a line goes down or comes up, an update reporting that fact is generated immediately.

C. Updating Policy

We next discuss the policy for propagating the delay information needed in SPF calculations, which require identical data bases at all the nodes [12]. The updating technique must meet two basic criteria, high efficiency (i.e., low utilization of line and CPU bandwidth) and high reliability. Efficiency is important both under normal conditions and when a change is detected that requires immediate updating. Reliability means that updates must be processed in sequence, handled without loss during equipment failures, and treated correctly after failure recovery.

Rather than having separate updates for each line, each update contains information about all the lines at a particular IMP. That is, each update from a given node specifies all the neighbors of that node, as well as the delay on the direct line to each of the neighbors. This results in more efficiency (i.e., less overhead), and the simplicity of only one single serial number per node. The latter makes sequencing and other bookkeeping easier.

We considered different approaches for distributing the updates [8] and decided on "flooding," in which each node sends each new update it receives on all its lines except the line on which the update was received. An important advantage of flooding is that the node sends the same message on all its lines, as opposed to creating separate messages on the different lines. These messages are short (no addressing information is required), so that the total overhead due to routing updates is much less than one percent. A final consideration which favors flooding is that it is independent of the routing algorithm. This makes it a safe, reliable scheme.

We considered several different ways of augmenting the basic flooding scheme to ensure reliable transmission [4]. An important feature of all the schemes is that updates which need to be retransmitted can be reconstructed from the topology tables in each IMP. The protocol we have adopted uses an explicit acknowledgment which is a natural extension of the basic flooding scheme. Using flooding, there is no need to transmit an update back over the line on which it was received since the neighbor on that line already has the update. In our protocol, however, the updates are transmitted over *all* lines, including the input line. The "echo" over the input line serves as an acknowledgment to the sender; if the echo is not received in a given amount of time (measured by a retransmission timer for each line), the update is retransmitted. In order to cover the case of a missed echo, the retransmitted update is specially marked (with a "Retry" bit) to force an echo even if the update has been seen before. Note that acknowledging an update at each hop ensures that the update will be received by all nodes which have a path to the source.

One difficult problem in maintaining duplicate databases at all nodes is that some nodes may become disconnected from each other due to a network partition. For some period of time, certain nodes are unable to receive routing updates from certain other nodes. When the partition ends, the nodes in one segment of the network may remember the serial numbers of the last updates they received from nodes in the other segment. However, if the partition lasted a long enough time, the serial numbers used by the disconnected nodes may have *wrapped around* one or more times. If there has been wrap-around, it is meaningless to compare the serial numbers of new updates with the serial numbers of old updates. Some method must be developed to force all nodes to discard the prepartition updates in favor of the postpartition ones. The obvious approach of ignoring updates from unreachable nodes is not workable, since the SPF databases may temporarily be inconsistent, and different nodes may ignore different updates.

This problem is resolved by having the update packets carry around some indication of their age. There is a k-bit field in each packet, and each node has a clock which ticks once every t seconds. When an update is first generated, the "age field" is $2k - 1$. When an update is received, its age field is decremented once each tick of the clock. An update is considered "too old" when its age field has been decremented to zero. This scheme ensures that the age of an update as seen by a given node is determined by the time it has been held in the given node, plus the time it was held in any nodes from which it was retransmitted. The use of a time-out scheme like the one just described places several constraints on the parameters used by the routing scheme.

1) It should be impossible for the serial numbers of updates generated by any one node to wrap around (i.e., to get halfway through the sequence number space) before the time-out period expires.

2) The time-out period should be somewhat longer than the maximum period between updates from a single node. This means that good, recent updates from reachable nodes will not time out.

3) It should be impossible for a node to stop and be restarted within the time-out interval. This ensures that all of the node's old updates will time out before any new updates are sent.

There is one other important facet to the updating protocol. When a network line which has been down is determined to be in good operating condition, it is placed in a special "waiting" state for a period of one minute. The line is not "officially" considered to be up until the waiting period is over. While a line is in the waiting state, therefore, no data can be routed over it. However, routing updates *are* transmitted over lines in the waiting state. As we indicated in Section IV-B, each node is required to generate at least one update per minute, even if there is no change in delay. This means that while a line is in the waiting state, an update from every node in the network will traverse it; the line cannot come up until enough time has elapsed so that recent updates from all nodes have been transmitted over it. This feature is needed for three reasons.

1) In order to properly perform the routing computation, a node must have a copy of the network database which is identical to the copies in all the other nodes. Recall that the database specifies the topology of the network (i.e., which nodes are direct neighbors of which other nodes), as well as the delay on each network line. When a new node is ready to join the network, it has none of this information. It must somehow obtain the information before it can be permitted to join. Note, however, that the procedure described above ensures that a node cannot come up (because its lines cannot come up) until it has received an update from each other node. Since an update from a given node specifies the neighbors of that node, as well as the delay on the line to each neighbor, it follows that a node cannot come up until it has received enough information to construct a complete and up-to-date copy of the network database.

2) When the network is partitioned, the partition must not

be permitted to end until updates from each segment have flowed into the other. Otherwise, nodes in one segment may have copies of the database which are inconsistent with those in the other segment. Again, the procedure of having each node generate at least one update per minute, while holding a line in the waiting state before allowing it up, is sufficient to avoid this problem. Since a partition can only end when a line comes up, and a line cannot come up until updates from all nodes have traversed it, a partition cannot end until all nodes have complete, consistent, and up-to-date copies of the database.

3) There are certain peculiar cases in which flooding is not totally reliable, even when augmented by a retransmission strategy. For example, suppose a node has two lines, one of which comes up at the precise moment the other goes down. An update which is being flooded around the network might arrive at each line at a time when it is down. This means that the update may never reach that node, even though there is no instant when both of the node's lines are down. However, by ensuring that a line cannot come up until enough time has passed for updates from all nodes to have traversed it, this situation is prevented.

V. PERFORMANCE

We next describe some analytical and empirical results on the performance of the new routing algorithm. One important measure of the efficiency of the SPF algorithm is the average time required to process changes in the delays along network lines, since such changes comprise the bulk of the processing requirements. When a given node receives an update message indicating that the delay along some line has increased, the running time of the SPF algorithm is roughly proportional to the number of nodes in that line's subtree; that is, it is roughly proportional to the number of nodes to which the delay has become worse. When a given node receives an update message indicating that the delay along some line has decreased, the amount of time it takes to run the incremental SPF algorithm is roughly proportional to the number of nodes in that line's subtree *after* the algorithm is run; that is, it is roughly proportional to the number of nodes to which the delay got better. Thus, in either case, the SPF running time is directly related to the subtree size.

Since the average subtree size provides a measure of SPF performance, it is useful to understand how this quantity varies with the size of the network. Let N denote the number of network nodes, and let hi represent the number of hops on the path from the source node, $i = 1$, to node i; in other words, if the length of each line is 1, then hi is the length of the path to node i. Clearly, node i appears in i's subtree and in the subtrees of all the nodes along the path to i. Thus, hi is equal to the number of subtrees in which node i is present, so that

total number of all subtree nodes

$$= \sum_{i=2}^{N} h_i$$

and since there are $N - 1$ subtrees (the complete tree from the

source node is not considered to be a "subtree"), the average subtree size is given by

average subtree size

$$= \frac{1}{N-1} \sum_{i=2}^{N} h_i.$$

But this expression is identical to the average hop length of all paths, and thus we have the remarkable result that in any tree, the average subtree size is equal to the average hop length from the root to all nodes. This result is significant because the average hop length generally increases quite slowly as the number of nodes increases. (For a network with uniform connectivity $c > 2$, the average hop length increases roughly as $\log N/\log (c - 1)$ [3].)

To establish some estimate of the running time of the algorithm, we programmed a stand-alone version for the ARPANET nodes. We randomly assigned each line in the ARPANET a length between 1 and 20. We ran the SPF algorithm to initialize the data structure in each node. Then we picked 50 lines at random and successively gave each a new random length. Every time we changed the length of a line, we changed it by at least 15 percent. Also, some lines were brought down by being assigned a length which represented infinity. Each time we did this, we ran the SPF algorithm with each node as the source node. We obtained the following results.

1) The average time per node to run the incremental SPF algorithm was about 2.2 ms.

2) The average time per subtree node to run the incremental SPF algorithm was about 1.1 ms.

Since we calculated that the average subtree size multiplied by the probability that a line is in the tree is about 2, these two results are in agreement. Note that these are average times; actual times varied from under 1 to 40 ms.

The figures given above are for the shortest path calculation only. Processing an update invokes a routine to maintain the topology database (including the ability to dynamically add or delete lines and nodes), and a routine to determine which nodes can be reached from the root node. These modules increase the running time by about a factor of two; and the total storage requirement, including these modules, the topology database, and the measurement and updating packages, is about 2000 16-bit words.

We designed and programmed the new routing procedure over a period of about six months. We than began an extensive series of tests on the ARPANET, at off-peak hours but under actual network conditions [5]. Our tests involved a great deal more than simply turning the new routing algorithm on to see whether it would run. The tests were specifically designed to stress the algorithm, by inducing those situations which would be most difficult for it to handle well. To stress its ability to react properly to topological changes, we induced line and node failures in as many different ways as we could think of, including multiple simultaneous failures. We also generated large amounts of test traffic in order to see how the algorithm performs under heavy load. (In this respect, it should be noted that the periods during which we were testing were "off-peak" only with respect to the amount of ordinary user traffic in the network. The amount of test traffic we generated far exceeds the amount of traffic generated by users, even during peak hours.) We experimented with many different traffic patterns, in order to test the algorithm under a wide variety of heavily loaded conditions. In particular, we tried to induce those situations which would be most likely to result in loops or in wild oscillations. We also designed and implemented a sophisticated set of measurement and instrumentation tools, so that we could evaluate the routing algorithm's performance. Some of these tools enabled us to monitor the utilization of resources used by the algorithm. Others enabled us to monitor changes in delay (as measured by the routing algorithm), as well as changes in the routing trees themselves at particular network nodes. One of our most important tools was the "tagged packet." A tagged packet is a packet which, as it travels through the network, receives an imprint from each node through which it travels. When such a packet reaches its destination, it contains a list of all the nodes it has traversed, as well as the delay it experienced at each node. These packets provided us with a very straightforward indication of the routing algorithm's performance. Of course, since the network was also in use by ordinary users during our tests, we cannot claim to have performed "controlled" experiments, in the strict scientific sense. However, all our experiments were repeated many times before being used to draw conclusions. Some of our main results are as follows.

1) Utilization of resources (line and processor bandwidth) by the new routing algorithm is as expected, and compares quite favorably with the old algorithm. Line overhead is less than one percent; CPU overhead is less than two percent. We have measured these quantities repeatedly since the new routing algorithm became operational in May 1979, and we have found this result to hold even during peak hours on the network.

2) The new algorithm responds quickly and correctly to topological changes; most nodes adapt to the change within 100 ms.

3) The new algorithm is capable of detecting congestion, and will route packets around a congested area.

4) The new algorithm tends to route traffic on minimum hop paths, unless there are special circumstances which make other paths more attractive.

5) The new algorithm does *not* show evidence of serious instability or oscillations due to feedback effects.

6) Routing loops occur only as transients, affecting only packets which are already in transit at the time when there is a routing change. The few packets that we have observed looping have not traversed any node more than twice. However, the loop can be many hops long. Although packets which loop may experience a longer delay than packets which do not, there is no significant impact on the average delay in the network.

7) Under heavy load, the new algorithm will seek out paths where there is excess bandwidth, in order to try to deliver as much traffic as possible to the destination.

Of course, the new routing algorithm does not generate optimal routing—no single-path algorithm with statistical input data could do that. It has performed well, and is successful in eliminating many of the problems associated with the old routing scheme. After several months of careful testing during which both old and new routing algorithms were resident in the network and used for experiments [5], we began to operate the ARPANET with the new routing scheme in May 1979, and removed the old routing program. Since that time, we have continued to monitor the performance of the algorithm. The results obtained during our test periods have continued to hold, even during peak hours, and no new or unforeseen problems have yet arisen.

Is the new routing algorithm really better than the old? We are convinced that it is for reasons that we will summarize shortly. We would like first to point out, though, that there is no general answer to the question, "What makes routing algorithm A better than routing algorithm B?" If one could claim that algorithm A performs better in every possible situation than algorithm B does, according to some well-defined metric of performance, then one would have a good reason for preferring A. However, such a claim could never be supported for it is untestable. One might try to claim that algorithm A performs better than algorithm B in "most" situations, but that would not necessarily show that A is a better algorithm than B. A's performance in the minority of situations might be so much worse than B's that B is to be preferred. Furthermore, it is difficult to define a performance metric which is equally applicable to every possible situation. For example, in some situations it may be desirable to minimize delay; in others to maximize throughput. Yet these two intuitively desirable performance metrics are in conflict. In attempting to decide which of two routing algorithms is the better one, one cannot appeal to any procedure simple enough to be followed by rote. Rather, one must first look at particular situations which are known to give rise to performance difficulties. Then one must decide what sort of performance one would like to see in those situations (a decision often akin to a value judgment). Only then can one compare the two algorithms to see which gives the more desirable performance.

Our purpose in designing and implementing a new routing algorithm for the ARPANET was to eliminate certain problems in the performance of the old algorithm, while at the same time maintaining the strengths of the old algorithm. We believe that one of the strengths of the old algorithm was that it was distributed, in the sense that the routing computation was performed by every node. In the ARPANET environment, this makes good sense from the point of view of reliability and efficiency. The new routing algorithm retains this feature by replicating the SPF computation at every node. There is a sense, however, in which the old routing computation was a distributed computation but the SPF computation is not. In the old algorithm, the inputs to the computation at one node were the outputs of the computation at the neighboring nodes. In this sense, then, the old routing computation was a global computation, with each node performing just a piece of it. Since the nodes performed the computation in an unsynchronized manner, the output of the global computation at any instant depended more on the history of events around the network than on the traffic in the network at that instant. The SPF computation, on the other hand, is a local computation. It does depend on measurements which have been made all around the network, but the updating protocol provides these measurements to all nodes unchanged and unprocessed; the SPF computation at one node never learns of the results of the SPF computation at any other node. In this way, we have kept the advantages of distributed routing while dispensing with the disadvantages of having a distributed computation.

Another good quality of the old algorithm was its attempt to adapt to changing delay conditions in the network. We realize that there may be certain applications, where network traffic can be accurately predicted and the network can be sized to handle exactly that traffic load and pattern, in which it may not be important for the routing to be adaptive. In the ARPANET, however, nodes and trunks are frequently added or removed. These changes are primarily made for administrative or economic reasons, rather than for the purpose of optimizing traffic flows. The traffic in the ARPANET is quite unpredictable, being largely determined by the behavior of a community of researchers. Furthermore, although there are sites on the ARPANET separated by as many as 11 hops, about one-third of the messages in the network travel no more than one hop; about half travel no more than three hops. This leads to situations in which the load in the network is very nonuniform, and these are the situations in which adaptive routing is likely to be of great utility. For reasons such as these, adaptive routing seems no less important to us now than it did to the original designers of the ARPANET many years ago. There were, however, several problems in the way the old algorithm responded to changes in network delay. Most of the problems stemmed from deficiencies in the delay measurement procedure of the old algorithm (see Section II). Because of these deficiencies, the old algorithm was often incapable of detecting congestion, and would sometimes send traffic directly into a congested area, thereby causing the congestion to spread. Our tests [5] show that we have eliminated this problem, and have done so without introducing any countervailing problems, such as instability or wild oscillation of the routing patterns. In its ability to adjust to changes in delay, the new algorithm appears to dominate the old completely.

An important deficiency of the old algorithm was its slow response to topological changes. The old algorithm would take many seconds to respond properly to a node or line failure. During this period, many nodes could be directing traffic towards a failed node. Having to buffer such traffic for seconds at a time was a significant cause of congestion in the network. With the new algorithm, however, the time for all nodes to respond to topological changes is on the order of 100 ms. Since the new algorithm was installed, we have not observed any congestion arising due to slow response to line or node failures.

The updates of the old routing algorithms were over 1200 bits long. There were often as many as seven such updates per second on each line. The new routing updates average 176 bits. Even during peak periods, it is rare to see more than two updates per second per line. One of the problems with the old algorithm was the increase in the delay of ordinary data packets due to the presence of the long, frequently sent routing updates. Clearly, the new routing updates interfere much less with ordinary network traffic than did the old.

The old routing algorithm took a fixed amount of time (15-20 ms) to process an update. The new algorithm, as we have pointed out, takes a variable amount of time, with the amount of time proportional to the size of the routing change necessitated by the update. This results in a much more efficient use of the CPU.

One of the major problems of the old algorithm was that it was prone to form loops which might persist for several seconds at a time. A given packet might be trapped in such a loop for a significant amount of time. Often a large number of packets would get "sucked in" to such loops, causing congestion which began at the locus of the loop and then spread throughout the network. While the new algorithm cannot be said to be loop-free, the loops that it forms occur only as transients while the network is adapting to a routing change. The loops which do form do not persist; a packet will sometimes loop once, but we have never seen packets traveling around and around in a loop, as would sometimes happen with the old algorithm. The small amount of looping which has been observed has *never* led to congestion, or even to a significant increase in average network delay. We conclude, therefore, that looping is not a problem with the new algorithm, as it was with the old.

Someone might object that any algorithm that permits loops is seriously deficient; this point is worth commenting on briefly. It is certainly true that, other things being equal, it is better not to have looping than to have it. But other things are never equal—we know of no pair of routing algorithms that perform exactly alike, except that one permits looping and the other does not. An algorithm which does not permit looping does not necessarily result in lower delay, less variable delay, higher throughput or less congestion than an algorithm which does. We simply do not believe that a small amount of transient looping should be regarded as a problem.

Are there any ways in which the old algorithm is better than the new one? The new algorithm does take about three times the memory as the old one, but conservation of memory is not generally considered to be an important desideratum for routing algorithms. From the point of view of performance, the new algorithm seems to dominate the old one in every respect. This is not to say that our approach is appropriate for every possible application, or even that it is the only possible approach for *our* application. We do believe, though, that we have met our goal of designing a new routing algorithm which kept the known strengths of the old one while eliminating many of its known weaknesses.

APPENDIX

This Appendix gives a semiformal description of the algorithm to calculate and update the shortest path tree. SOURCE denotes the node in which the algorithm is running. The algorithm's basic data structure, LIST, is a variable-length list whose elements are ordered triples. An ordered triple T is of the form ⟨SON, FATHER, DISTANCE⟩ where SON and FATHER are nodes and DISTANCE is a member. (We use the notation SON (T) in the obvious way to mean the first element of the triple T.) Each triple represents a particular path from SOURCE to SON. The penultimate hop on this path is FATHER, and the total length of this path is DISTANCE. The algorithm we describe here has been modified so that changes to the tree can be computed incrementally, without having to recalculate those parts of the tree that do not change. Hence, it does not correspond exactly to the flow chart in Fig. 1.

SPF Algorithm

0) If no tree exists, place ⟨SOURCE, SOURCE, 0⟩ on LIST, and go to Step 4).

1) If the change was to line AB, then perform one of the following steps.

    a) If AB is in the tree, set DELTA equal to the change in distance along AB.

    b) If AB is not in the tree, set DELTA equal to the distance to node A plus the distance along AB minus the distance to B; if DELTA is greater than or equal to 0, done.

2) Identify B and all of B's descendants (both first generation and succeeding generations) as members of the subtree; increase the distances of all subtree members by DELTA.

3) For each subtree node S, perform one of the following steps.

    a) If DELTA is positive, try to find a shorter path to S via each of S's neighbors that is not in the subtree; if such an improved path is found, put the triple representing S on LIST.

    b) If DELTA is negative, try to find a shorter path to each of S's nonsubtree neighbors by attempting to route each neighbor via S; if such an improved path is found, put the triple for the neighbor node on LIST.

4) Search LIST for the triple T with the smallest DISTANCE. Remove T from LIST; place SON (T) on the shortest path tree so that its father on the tree is FATHER (T). (Exception: if SON (T) = SOURCE, place it in the tree as its root.)

5) For each neighbor N of SON (T), do *one* of the following steps.

    a) If N is already in the shortest path tree, then

        i) if its distance from SOURCE along the tree is less than or equal to DISTANCE (T) + LINE–LENGTH (SON $(T), N$), do nothing;

        ii) if its distance from SOURCE along the tree is greater than DISTANCE (T) + LINE–LENGTH (SON $(T), N$), remove N from the tree and place ⟨N, SON (T), DISTANCE (T) + LINE–LENGTH (SON $(T), N$)⟩ on LIST.

    b) If there is no triple T' on LIST such that SON (T') = N, then place the triple ⟨N, SON (T), DISTANCE (T), + LINE–LENGTH (SON $(T), N$)⟩ on LIST.

    c) If there is already a triple T' on LIST such that SON (T') = N, and if DISTANCE $(T') \leq$ DISTANCE (T) + LINE–LENGTH (SON (N), do nothing.

    d) If there is already a triple T' on LIST such that SON (T') = N, and if DISTANCE $(T') >$ DISTANCE (T) + LINE–LENGTH (SON $(T') N$), then

        i) remove T' from LIST;

        ii) place the triple ⟨N, SON (T), DISTANCE (T) + LINE–LENGTH (SON $(T), N)$⟩ on LIST.

6) If LIST is nonempty, go to step 4). Otherwise, the algorithm is finished.

REFERENCES

[1] J. M. McQuillan and D. C. Walden. "The ARPANET design decisions." *Comput. Networks*, vol. 1, Aug. 1977.

[2] J. M. McQuillan. "Routing algorithms for computer networks—A survey." presented at the 1977 Nat. Telecommun. Conf., Dec. 1977.

[3] J. M. McQuillan. I. Richer, and E. C. Rosen. "ARPANET routing algorithm improvements—First semiannual technical report." BBN Rep. 3803, Apr. 1978.

[4] J. M. McQuillan, I. Richer, E. C. Rosen, and D. P. Bertsekas, "ARPANET routing algorithm improvements—Second semiannual technical report." BBN Rep. 3940, Oct. 1978.

[5] E. C. Rosen. J. Herman, I. Richer, and J. M. McQuillan. "ARPANET routing algorithm improvements—Third semiannual technical report." BBN Rep. 4088, Apr. 1979.

[6] J. M. McQuillan, G. Falk, and I. Richer. "A review of the development and performance of the ARPANET routing algorithm." *IEEE Trans. Commun.*, Dec. 1978.

[7] J. M. McQuillan, I. Richer, and E. Rosen. "ARPANET routing study—Final report." BBN Rep. 3641, Sept. 1977.

[8] J. M. McQuillan, "Enhanced message addressing modes for computer networks," *Proc. IEEE (Special Issue on Packet Communication Networks)*, Nov. 1978.

[9] D. B. Johnson. "Efficient algorithms for shortest paths in sparse networks." *J. Ass. Comput. Mach.*, vol. 24, pp. 1–13, Jan. 1977.

[10] E. Dijkstra. "A note on two problems in connection with graphs." *Numer. Math.*, vol. 1, pp. 269–271, 1959.

[11] D. P. Bertsekas. "Dynamic models of shortest path routing algorithms for communications networks with a ring topology." in preparation.

[12] E. C. Rosen. "The updating protocol of the ARPANET's new routing algorithm: A case study in maintaining identical copies of a changing distributed data base." in *Proc. 4th Berkeley Conf. Distributed Data Management and Comput. Networks*, Aug. 28–30, 1979, pp. 260–274.

REXPAC — A Brazilian Packet Switching Data Network

M.A. Bergamo and A.S. Campos
Research and Development Center CPqD/ Telebras, Brazil

ABSTRACT

This paper describes REXPAC, the nationwide experimental packet switching data network being developed by CPqD, the research and development center of TELEBRAS, the holding company responsible for public telecommunications in Brazil. REXPAC uses internationally-standardized CCITT interfaces, following CPqD interpretations for X.3, X.25, X.28, X.29 and X.75. Its six backbone nodes with a centralized control system are connected by 9.6 and 64 kbits/s full-duplex trunks. The packet switching exchange architecture is multiprocess/multiprocessor based and two different structures with different capacity characteristics have been chosen. The higher capacity version foresees future application in the national public data network.

I. INTRODUCTION

TELEBRAS is the holding company responsible for public telecommunications in Brazil and EMBRATEL is its subsidiary in charge of providing and operating the data communication services all over the country. The Data Com munication Group of CPqD, The Research and Development Center of TELEBRAS, is in charge of designing and implementing a nationwide multi-purpose experimental packet switched network based on internationally - standardized interfaces. This network, named REXPAC, represents the scientific and technical nu cleus of the future Brazil-wide public data network that is already being specified by a group gathering TELEBRAS an EMBRATEL engineers. Although relying mostly on TELEBRAS, REXPAC might be regarded as a national joint -venture involving universities, industries and several other Brazilian corporations.

REXPAC actually constitutes a development laboratory for experiments and studies of packet switching technology envisaging its application to public data networks. Its direct fulfillment will be the network it- -self, but, more important, it is the oppor tunity offered to researchers all over the country for adequate scientific and techni-

cal progress in the data communications field.

This paper describes the architecture, inter faces, protocols and switching nodes of REXPAC.

II. NETWORK ARCHITECTURE

REXPAC will be the backbone for the local da ta networks currently being planned or devel oped by universities and R & D institutes and it supplies a natural interconnection structure for the TELEBRAS computing centers scat tered throughtout Brazil. REXPAC will also serve as an access to distributed database systems belonging to TELEBRAS and its various associates.

REXPAC is composed of six nodes located in Sao Paulo, Rio de Janeiro, Brasilia, Recife, Porto Alegre and Campinas. The necessary software and hardware facilities for a network control center are centered in the Campinas node.

The REXPAC node configuration is shown in Fig.1. The distribution of concentrators, line multiplexors and remote Packet Assembly/ Disassembly equipments (PAD) has been left to be decided lately, according to demand requirements.

REXPAC PACKET SWITCHING

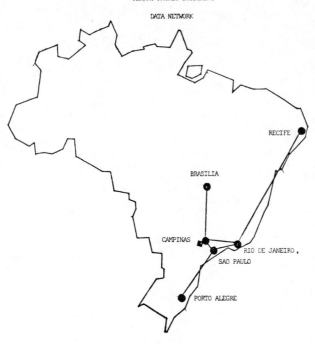

DATA NETWORK

Fig. 1 - REXPAC Architecture

The communication trunks connecting Rio de Janeiro, Sao Paulo and Campinas are full- duplex 64 kbit/s links while, the Brasilia, Recife and Porto Alegre ones are full - duplex 9.6 kbit/s links. The implementation of a multilink communication structure between pairs of nodes is also planned.

A brief description of REXPAC elements represented in Fig. 2 and of its interface protocols is presented in the next item.

III. PROTOCOLS, FACILITIES AND SERVICES

The basic service provided to REXPAC users, and available for the communication among network elements, is the Virtual Call. All the interfaces follow CCITT Recommendations (1, 2, 3).

Users can access the network by means of Packet-mode Data Terminal Equipments (DTE-P) connected directly to a node via full-duplex dedicated lines or, indirectly, via PADs, using character oriented synchronous or asynchronous Terminals (DTE-C).

The interface between a Data Circuit Terminating Equipment (DCE) and a DTE-P obeys the CCITT X.25 Recommendation and has been implemented according to a CPqD/TELEBRAS interpretation (4, 5). The physical interfaces (X.25,

level 1) follow V.24 and X.21 bis. Link Access Procedure - Balanced Mode (LAPB/ X.25 level 2) is the only procedure available for DTE-P connections to REXPAC. The CCITT recommended facilities and services (6) to be offered by REXPAC are restricted to the essential ones.

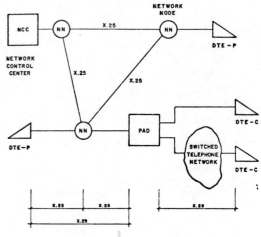

Fig. 2 - REXPAC elements

A character oriented terminal (DTE-C) is connected to REXPAC via PAD equipment. All the functions associated to code conversions, terminal control, character-to-packet adaptation and communication protocol management are performed by the PAD. The PAD is connected to a node as a standard DTE-P.

Physical connections of asynchronous DTE- Cs use full-duplex dedicated lines or the switched telephone network. The PAD/DTE-C interface follows the CCITT X.28 Recommendation . The X.20 bis option is the only physical interface available at REXPAC for PAD / DTE- C connection.

The PAD functions are defined by Recommendation X.3. At first, only the functions defined as basic will be supported by REXPAC .

Data exchange between a PAD and a remote DTE-P follows the CCITT X.29 Recommendation .

REXPAC implementation of CCITT X.3, X.28 and X.29 Recommendations obeys the CPqD/ TELEBRAS interpretations (7,8,9) .

Synchronous DTE -Cs not adapted for packet communication will access REXPAC through PADs.

These accesses are initially restricted to dedicated full-duplex 1200-2400 bit/s links and BSC terminals.

Interconnections between nodes follow the CCITT X.25 Recommendation. The physical couplings related to the 64 kbit/s links are V.36 compatible. The multilink structure and X.75 compatibility will be gradually introduced in REXPAC.

Permanent Virtual Circuits and Closed User Groups are standard facilities supplied by REXPAC. The use of these facilities is restricted to DTE-Ps. Permanent Virtual Circuit implementation is based on a network node path setup (Virtual Call) between predefined DTE-P pairs. A Virtual Call is automatically originated when the calling DTE-P is turned on.

The control and statistical information exchanged between network elements make use of a transport protocol (for message protection and consistency) and dedicated Permanent Virtual Circuits forming Closed User Groups.

IV. PACKET SWITCHINC NODE

The layer structure of the X.25 protocol makes packet switching well suited to a multiprocessor implementation (10).

Link level processing involves real-time attention to line events (transmission, reception, etc.) and a strict time discipline for the logical tasks involved. Packet level processing, otherwise, is less restrictive in terms of real-time requirements, being more concerned with the subscriber and network table handling.

The REXPAC packet switching node is based on the multiprocess approach and its most relevant characteristics are: a multiprocessor architecture; a distributed bus structure; and a operating system responsible for message communication between processors and process synchronization.

1. Multiprocess/Multiprocessor Approach

The packet switching functions are grouped according to necessary or convenient parallel configurations and an adequate software entity called process is assigned to each group.

The communication rules between processes are defined in terms of
. specific messages exchanged between processes,
. consistency checks executed with respect to type, format and activation predicates for

any transmitted or received message, and actions related to transmitted or received messages.

The process allocation to processors may occur during system initialization for some types of processes or during process generation for other ones.

The synchronization of and the message exchange between processes are based on SEND and WAIT primitives. A copy of the operating system is stored in every processor and it works in such a way that the physical localization of a given process is transparent to the application programs.

2. Packet Switching Exchange Implementation

The Packet Switching Exchange being developed at CPqD is composed of specialized processor modules communicating through a distributed bus structure. Two different bus structures are being considered, referring to different speed characteristics.

An exploratory packet switching exchange utilizing a duplicated ring structure as the basic communication path between processors is currently being implemented at CPqD to test the limitations of and permit experiments with a multiprocessor/multiprocess based technology for packet switching. The resulting medium capacity packet switching exchange will be used at the initial network nodes of REXPAC.

The multiprocessor based switching node with the ring structure (withouth showing the ring redundancies) is presented in Fig. 3.

The processes LHM, VCM, CCM, etc., defined bellow, are allocated to processors in such a way as to explore the natural existing parallelisms.

The ring has a total capacity of 1.6 M bit/s The processor access to the ring is achieved through FIFO buffers filled or emptied via DMA.

The decisions about putting a message in the ring, or taking it out, are made independently in each processor by specially designed local controllers located in the ring interfaces.

A higher speed distributed bus structure is being analysed for the high capacity packet switching exchange. This bus is composed of FIFO buffers and a per-message switching structure (Fig. 4).

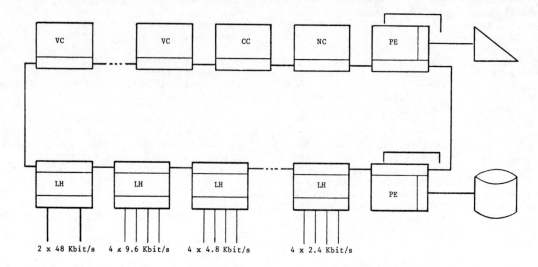

Fig. 3 - Exploratory Packet Switching Node

Each FIFO buffer is dedicated to a processor module as far as writing is concerned. After a message has been written, the buffer control, including clock signals, is completely liberated. A distributed polling mechanism transfers the message to the addressed processor.

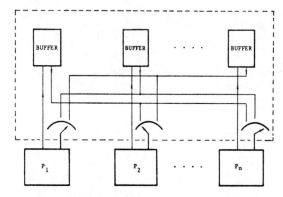

Fig. 4 - High Capacity Multiprocessor Architecture

This type of bus structure is rather simple and permits high level of parallelism (N simultaneous transmissions, with N equal to the number of FIFO buffers in the bus) and complete independence of the communications between cooperating processors.

3. Process Definitions

In the exploratory packet switching exchange the following modules were defined:
(i) Link Handling Module (LHM): it includes procedures associated with X.25 levels 1 and 2;

(ii) Call Controller Module (CCM): it includes call setup, clearing and restarting procedures for logical channel associations;

(iii) Virtual Circuit Module (VCM): it includes procedures for data switching from a logical channel to another one, after the call setup phase is completed;

(iv) Node Controller Module (NCM): responsible for the control of the switching node as a whole; and

(v) Peripheral Equipment Modules (PEM): responsible for the peripheral interfaces in general.

Each of these modules is associated at least to one process. The use of more than one process for the implementation of a given function depends on the desired reliability characteristics.

4. Call Setup and Data Transfer Phases

The call setup procedure between two X. 25 terminals connected to the same Network Node at different Link Handling Modules is represented schematically in Fig. 5.

A frame (X.25, level 2) protected CALL REQUEST packet (1) is physically received and cheked for link errors by LHM #1. This packet, after type identification, is enveloped in a internal message (2) and directed to the selected Call Controller Module (CCM). The message parameters include input link number, LHM number, CCM number, packet type identifier, logical channel number and the called Data Terminal Equipament (DTE) address. The correspondences between logical channel number and utilized CCM are kept in the LHM #1.

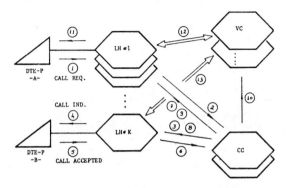

Fig. 5 - Call Setup and Data Transfer

The tables with the network link allocations, LHMs and network addresses are kept in the CCMs. The CCMs also manage the logical channel utilization.

The addressed CCM determines the output link and the LHM # k corresponding to the addressed DTE. A message ③ with the INCOMING CALL packet is transferred to LHM # k. The message parameters include output link number, LHM number, packet type identifier and the selected logical channel number.

The INCOMING CALL packet is framed ④ and transferred to the called DTE.

The CALL CONNECTED packet ⑤ is routed through the LHM # k, CCM and LHM # 1 (⑥ , ⑦) to the calling DTE ⑪ .

The CALL CONNECTED packet to the calling DTE is preceded by the preparation of the CCM selected Virtual Circuit Module ⑩ for handling the data transfer phase of the established virtual circuit. The VCM preparation is implemented by transferring the informations about the logical channels, links and LHMs involved. The LHMs are also informed about the VCM to be utilized to handle that virtual circuit (⑧ , ⑨).

These logical channels/LHMs/VCMs associations are used to switch the data packets through the network node during the data transfer phase. (⑫ , ⑬).

V. CONCLUSION

REXPAC, an experimental data communication packet switched network, is a nationwide development laboratory for the progress of Tele-Informatics in Brazil and a scientific and technical nucleus for the future national public data network. A thorough description of aspects concerning its architecture, interfaces, protocols and switching nodes has been given.

VI. REFERENCES

1. CCITT Recommendation X.25 "Interface between DTE and DCE for terminals operating in the packet mode on public data networks" , 1976.

2. CCITT Recommendation X.25, "Draft Revised Recommendation of X.25 Introduction and Packet Level", COM VII - No. 439-E, November 1979.

3. CCITT Recommendation X.25, "Proposed Revision to Level 1 and 2 of Recommendation X.25" COM VII - No. 374-E, October 1979.

4. "TELEBRAS X.25 DTE/DCE Interface for Packet Mode Terminals on Public Data Networks-CCITT X.25/Level 2 Interpretation" , CPqD/TELEBRAS Internal Report, May 1980.

5. "TELEBRAS X.25 DTE/DCE Interface Packet Mode Terminals on Public Data Networks-CCITT X.25/Level 3 Interpretation", CPqD/ TELEBRAS Internal Report, May 1980.

6. CCITT Recommendation X.2, "Proposed Revision to CCITT Recommendation X.2 and Question 2/VII, COM VII- No.414-E, November 1979.

7. "TELEBRAS X.3 Packet Assembly/Disassembly (PAD) on Public Data Networks-CCITT X.3 Interpretation", CPqD/TELEBRAS Internal Report , May 1980.

8. "TELEBRAS X.28 DTE/DCE Interface for Start-Stop Terminal for PAD Access on Public Data Networks-CCITT X.28 Interpretation" , CPqD/TELEBRAS Internal Report, June 1980.

9. "TELEBRAS X.29 Procedures for the Exchange of Control Information and User Data Between a PAD Facility and a Packet Mode DTE or Another PAD-CCITT X.29 Interpretation" , CPqD/TELEBRAS Internal Report, June 1980.

10. M.A. Bergamo and R.Gadelha P. , " The Architecture of the Telebras Packet Switching Node", presented at INTELCOM 80, Rio de Janeiro, Brazil.

11. M.A.Bergamo,"Design of Data Communication Packet Switched Data Networks", XII Congresso Nacional de Processamento de Dados-SUCESU,pg. 623-640, São Paulo, October 1979.

12. E.C. Grizendi, "An Interface for Public Packet Switching Data Networks", XII Congresso Nacional de Processamento de Dados - SUCESU, pg.649-658, Sao Paulo, October 1979.

Data Network Plans for the UK

D.W.F. Medcraft
British Post Office, UK

ABSTRACT This paper describes the data communication services offered by the British Post Office (BPO) in the United Kingdom. It also outlines the plans for modernisation of the basic telecommunications infrastructure in the United Kingdom, which are already in the first stages of implementation. The sheer size of the network and the magnitude of investment involved means that the modernisation process will take many years to achieve, ie before the advantages are available in every location in the ubiquitous telephone network. In the meantime the BPO, recognising the requirement to offer a range of services adequate to meet the needs of the 1980s, is also implementing specialised public data services; the National Public Packet Switching Service (PSS) is an example. The planning of PSS, which opened this year, is described in detail.

II.   DATA TRANSMISSION ON THE TELEPHONE NETWORK

The demand for data communication began to materialise in the 1960s since when the readily available telephone network has provided a convenient bearer for data transmission at data rates which have increased as modem technology developed. The ubiquitous telephone network provides for service in two ways; over the Public Switched Telephone Network (PSTN) and via leased lines (private circuits).

Two wire connections over the PSTN, originally designed for speech transmission, limit data rates to 1200 bit/s full duplex and 9600 bit/s half duplex (although 2400 bit/s duplex and even higher duplex operation may eventually be possible as a result of recent developments eg echo cancelling techniques). The main limitation of the present PSTN is the long call set-up time (typically 15 seconds) which is a result of the step by step, principally electro-mechanical, switching system. This disadvantage, together with bandwidth limitation, poor performance (relating to those forms of distortion and interference to which data, but not speech, is sensitive) makes the PSTN unsuitable for many modern data communication requirements.

4 wire circuits provided by leased lines enable rates of 9600 bit/s full duplex or higher to be achieved on voice circuits, whilst use of a full Frequency Division Multiplex (FDM) "group" band equivalent to 12 voice circuits (48kHz) enables 48kbit/s up to 128kbit/s to be achieved.

At present 60-70% of data customers in the UK make use of leased lines rather than the PSTN. The reasons for this are possibly:-

1. Leased lines are cheaper for many applications because of the volume of traffic involved.

2. The present day data communication systems are mainly "within house", ie belonging to single business organisations with very little need or possibility for inter system working.

3. The switching within the telephone network is inadequate for modern systems requirements.

4. The need for commercial security.

The BPO provides a range of ℝ Datel services as shown in Table 1. At 31st December 1979, there were approximately 63,000 modems in service and a forecast growth of 10-15% per annum over the next decade.

TABLE 1
BRITISH TELECOMMS DATEL SERVICES

Datel Service	Switched (bit/s)	Leased (bit/s)
0200	up to 300	up to 300
0600	up to 1200 (hd)	up to 1200
1200 Duplex	1200	1200
2400	600/1200 (hd)	2400
2400 Dial-up	600/1200/ 2400 (hd)	-
02412	1200/2400 (hd)	2400
4832	3200/4800 (hd)	4800
4800	4800/2400 (hd)	4800
9600	9600/7200/ 4800 (hd)	9600/7200/ 4800
48k	-	48k

0 Also International Services

hd = Half duplex operation only

III AN INTEGRATED DIGITAL NETWORK

Studies of the UK network requirements up to the year 2000 have clearly indicated that there would be considerable economic advantages, for the basic telephone system, if the present network, with its analogue transmission and space switching systems, were to be replaced with an Integrated Digital Network (IDN) consisting of digital switching and digital transmission equipment. The provision of an IDN would also give better service benefits including faster call set-up time, better transmission with reduced noise, greater flexibility for the introduction of new services and facilities, and improved network management facilities.

The creation of an IDN from the existing network is a mommoth task, the prime emphasis of which must be centred on meeting the needs of the telephone service ie the majority service deriving the

ℝ "Datel" is a registered BPO trademark and covers all services operating via a BPO provided modem.

greatest revenue. Nevertheless the introduction of digital switching and transmission equipment for the telephone service could have profound consequences for the manner in which data, amongst other non-voice services, will be provided in the future.

So as to implement an IDN, a family of digital switching systems known as System X is currently being developed which, by virtue of its modular nature, has the potential to respond to future needs and technology. Initially it is planned to provide in effect an overlay network giving a 64kbit/s switched service between the 30 major conurbations in the UK, by 1986/87. This will lay the foundation of an IDN which, with gradual expansion, and later extension of working over local loops to customers' premises, will enable the creation of an Integrated Services Digital Network (ISDN) on which both voice and non-voice services may be carried.

However, the development timescales for digital transmission systems are shorter than those for digital exchanges and the transmission systems can therefore be implemented in advance of the exchanges. Long haul digital transmission systems, 120 Mbit/s (1680 x 64kbit/s channels), and 140 Mbit/s (1920 x 64kbit/s channels) to the new CCITT standard, are already being installed on coaxial cable and microwave radio systems, and obsolete 24 channel FDM systems on balanced pair cables are being replaced by 120 channel 8 MHz systems. At the end of the 1980s almost all growth in long haul transmission will be on digital systems. An equivalent emphasis is being placed on the introduction of digital plant into the junction network since current economic studies show digital plant to be economic at distances down to 3 km. 2048kbit/s (30 channel) Pulse Code Modulation systems, designed to the CCITT/CEPT standards, are superseding the earlier 24 channel systems in the junction networks. These 2 Mbit/s systems will provide the basic transmission elements linking System X exchanges. It is anticipated that, by 1986, nearly all of the 6000 local exchange sites in the UK will be accessible by digital plant.

As the implementation of these plans proceeds the opportunity will be taken to provide "all digital" leased line services, for data communication, in certain locations. From the standard 2 Mbit/s system, line rates of 3.2, 6.4, 12.8 and 64kbit/s will be derived giving customer data rates of 2.4, 4.8, 9.6 and 48kbit/s, respectively, as shown in Figure 1.

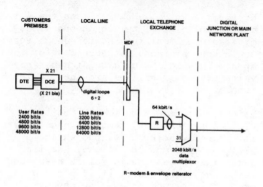

Fig. 1. DDS : INTEGRATED SERVICES – USE OF 64KBIT/S TO PROVIDE DATA SERVICES

IV NEED FOR INTERIM PUBLIC SWITCHED DATA NETWORK

There is however ample evidence that improved switching facilities will be required to meet the needs of data users well in advance of the availability of the "new" all digital integrated services network. Developments in microprocessor technology are enabling the range and quality of modems to be extended, but whilst the Datel services are expected to continue to cover the needs of the bulk of users in the early 1980s there will be an increasing requirement for intercommunication between data systems belonging to different business organisations, and many new applications emerging that will require an "open system" public data network superior to the capabilities of the existing PSTN. Such a network should have the characteristics listed in Table II.

TABLE II

REQUIREMENTS FOR A PUBLIC SWITCHED DATA NETWORK

Fast call set-up times

High signalling rates readily available

Low error rates

Full duplex operation

A range of facilities

Table III shows the present CCITT X1 Recommendation for the services that should be offered on public switched data networks. At present the BPO is concentrating on providing a Packet Switching Service offering user classes 8 to 11 for Packet terminals with access available from asynchronous terminals at data rates of 110 to 300 bit/s, 1200/75 bit/s, and 1200/1200 bit/s. Asynchronous terminals can be directly connected to the packet network or access it via dial-up over the PSTN.

User Class	Terminal mode	Data signalling rate	Address selection and service signals
1	Start-stop	300bit/s (11 units/char)	300bit/s-IA No 5
2	Start-stop	50-200bit/s (7.5 to 11 units/char)	200bit/s-IA No 5
3	Synchronous	600bit/s	600bit/s-IA No 5
4	Synchronous	2,400bit/s	2,400bit/s-IA No 5
5	Synchronous	4,800bit/s	4,800bit/s-IA No 5
6	Synchronous	9,600bit/s	9,600bit/s-IA No 5
7	Synchronous	48,000bit/s	48,000bit/s-IA No 5
8	Packet	2,400bit/s	2,400bit/s) See Rec
9	Packet	4,800bit/s	4,800bit/s) X25
10	Packet	9,600bit/s	9,600bit/s) for format
11	Packet	48,000bit/s	48,000bit/s)

TABLE III

CCITT USER CLASSES OF SERVICE IN PUBLIC DATA NETWORKS

Although there will be a capability for low speed data (User classes 1 and 2) on the modernised SPC (Stored Program Controlled) Telex network in the UK it is likely that such services can adequately and more economically be provided on the existing ubiquitous PSTN.

Studies conducted by the BPO into the feasibility of providing a dedicated synchronous Circuit Switching system covering CCITT User classes 3 to 7, indicated that it could not be economically justified in the UK in the environment of a predominantly analogue telecomms infrastructure. Therefore synchronous circuit switched services are not planned to be provided until there is adequate penetration of digital plant in the telephone network during the mid to late 1980s. With the penetration of IDN, together with the 64Kbit/s "time slot" becoming the basic element for a voice channel, a high speed circuit switched system becomes a practical and interesting possibility.

The flexibility of a packet switching system, with its ability to support a range of terminals operating at different rates and the dynamic multiplexing facilities, make it an attractive proposition in the existing analogue environment. There are also many characteristics of a packet switching system that facilitate "open system" working such that it is likely to remain a part of the range of services offered in the longer term ISDN. The establishment of a standard protocol for terminal to network also provides a convenient foundation for the building on of additional "within network" services such as mail-box, store and forward message services etc.

V   EXPERIMENTAL PACKET SWITCHING SERVICE (EPSS)

The BPO decided in the early 1970s that practical involvement was the best method of obtaining evidence to support the provision of a packet switched data service and EPSS was introduced in April 1977 with Packet Switching Exchanges (PSEs) in London, Manchester and Glasgow. Customer's access to the exchanges and inter-nodal links were based on the use of existing analogue plant and modems. EPSS has enjoyed a good measure of success as a test bed, enabling the computer industry, users and the BPO to evaluate the technical and economic aspects of packet switching. However, since EPSS was designed before work on International Standards came to fruition it has not been found economically

possible or desirable to incorporate it into an ongoing national packet switched service conforming to the latest International Recommendations. For this reason a new network was planned to supersede EPSS during 1980.

VI   NATIONAL PACKET SWITCHED SERVICE (PSS)

As indicated in Fig. 2, at opening date PSS consists of a 9 node network capable of accommodating initially 160 packet ports and 800 asynchronous ports. The PSEs are based on multiple TP4000s, a modular multi-microprocessor system, developed by Telenet and supplied in the UK by Plessey Controls Limited.

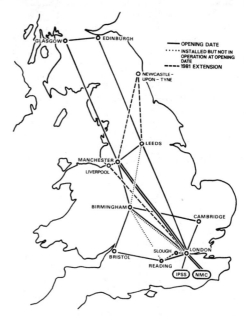

Fig. 2.   UK PSS NETWORK

PSS will support directly connected Packet mode customer terminals with CCITT X25 interfaces, operating full duplex at speeds of 2400, 4800, 9600 bit/s and 48kbit/s (CCITT User classes 8-11) and for which special distance independent tariffs will apply. Character mode access at up to 300 bit/s and at 1200/75 or 1200 bit/s full duplex is available either by direct connection or dial-up via the PSTN.

The network is controlled by main and standby Network Management Centres (NMCs) in London, each comprising a Prime 350 minicomputer equipped with disc store, magnetic tape, VDU and printer. The PSEs are designed for normally unattended operation with all test, maintenance and program loading being undertaken remotely from the NMC. Additional VDU and printer terminals will however be provided at Birmingham, Manchester and Leeds and will have access to all NMC facilities appropriate to these locations. PSE sites are interconnected via 48kbit/s trunk circuits, CCITT X75 inter-network protocol being incorporated to facilitate flexibility in future network development.

Each PSE has automatic fault diagnosis and will communicate automatically the existence of many types of hardware faults to the NMC. On-line diagnostics from a NMC or other Maintenance Control Points (MCPs) can be achieved to a module level or in some cases to a function within a module. When a TP4000 is "cold started" it initialises a route to a neighbouring switch and requests initialisation from the NMC. The NMC then sends software and tables back which are loaded by a program stored and executed in Read only Memory (ROM). The NMC also deals with call accounting information which is passed from individual PSEs on a call by call basis. The call accounting information is written to disc stores at the NMC for subsequent transfer to magnetic tape for off-line processing and production of bills.

All network nodes use primary/secondary choice routing tables to choose an optimum inter-nodal link for Call Request packets not destined for a local customer. Since for any destination address there may be a multiple choice primary and/or multiple secondary link the optimum choice at any time will depend on the instantaneous loading and relative bit rates of the associated lines.

In setting standards for delay across the system account has to be taken of future expansion whereby the introduction of more PSEs into the network could increase the number of links required to complete a connection. As PSS expands a hierarchial topology will be introduced to ensure that no more than 5 links in tandem will be necessary on a connection between any two PSEs. Over such a connection involving 5 links it is estimated that 95% of all full length data packets will be conveyed in significantly less than 500 ms. The average delay for the average connection over PSS will be much shorter than this - around 150-200 ms. Delay is a function of packet length, CPU loading and performance, and the bandwidth of the transmission links. The performance and traffic loading of the network will be carefully monitored to determine the optimum configuration as

experience of the type of usage is obtained.

The initial BPO market forecast suggested the need for a fairly small network mainly accessed by asynchronous terminals seeking facilities and use of databases on a few X25 connected host computers. However the market response to the announcement of the service and publication of tariffs indicates that an earlier than anticipated enhancement will be necessary. There are also signs that there is a greater demand than anticipated for packet to packet terminal applications, particularly at 9600 bit/s. Consequently plans are already being implemented for enhancement of the system so that during 1981 PSS will be extended to become a twelve node network with additional PSEs at Liverpool, Newcastle and Slough, and will cater for about 600 packet ports and 1400 character ports. As soon as possible PSS will be connected to the international packet switching gateway exchange that already exists in the UK, thus providing access to existing Packet Switching Services to the USA and Canada (Fig. 3). It is also planned to provide links from the international gateway to packet switching service in other countries (Transpac in France, and to the Federal Republic of Germany, the Netherlands, Belgium etc), whilst connection to EURONET will provide access to at least ten European countries together with access to European databases via DIANE, the Direct Information Access Network Europe development by the EEC, that utilizes EURONET as its communication vehicle.

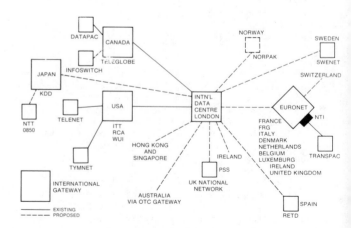

Fig. 3. UK PLANNED INTERNATIONAL PACKET SWITCHED DATA SERVICES

Planning philosophy for PSS is based on the choice of a small but modular, readily enhanceable, system. Whilst recognising the potential demand for this type of service, the BPO faced the usual dilemma when contemplating introduction of a new dedicated network. Forecasts for demand were of necessity uncertain and speculative; the capital outlay that could be authorised for a speculative venture was limited; and because of the latter, the size and geographical coverage of the initial system would also be limited. This in turn would inhibit the initial customer response to the new service. The offering of distance independent access line rentals for packet and directly connect asynchronous terminals means that PSE location is of little concern to these types of customers. In the case of dial-up aynchronous terminals the location of the nine initial nodes was chosen to give an estimated local fee dial-up access to 60% of the geographicl distribution of existing Datel customers. The BPO was constrained to install a system that would achieve an estimated pay back in a period of seven years. It was therefore crucial that the choice of system was optimum for an initial embryonic service, but capable of expansion within short lead times.

The overheads of opening new PSEs are not only in the form of fixed costs for TP equipment, but in such factors as provision of accommodation, power feeds, staffing and training. As regards line costs the most important element is probably local and junction line plant plus the HF terminal multiplexing equipment. Once it is necessary to route on the main trunk HF network distance in the UK is relatively unimportant. This suggests that PSEs should be located to serve local concentrations of customers rather than large geographical area. For example it will be economic to provide a number of PSEs in and around London within the first 5 years of service but serve customers in the remote districts of the UK from existing PSEs in the main cities. Much

depends upon the future demand for character terminal access, a cheaper simple remote PAD facility could extend the local fee area PSTN coverage, and there is also an attraction for private PADs and concentrators located on customers premises for serving a number of local terminal installation within one organisation. The use of digital plant for PSE access (2Mbit/s telephony systems to telephone exchange sites with simple baseband modem access of short local links) could significantly reduce the access costs and hence the future overall cost of the service.

VII.   SUMMARY

BPO plans for Data communication services in the UK can be summarised as:-

a.   Continuing to enhance the range and quality of Datel services offered over the Public Switched Telephone Network and leased lines.

b.   Concentration of effort into modernisation of the whole telephone network into an integrated digital switching and transmission system based on System X concepts.

c.   Because (b) will of necessity take a number of years to achieve; provide additional dedicated public data services such as PSS, the National Packet Switching Service, to complement the Datel Services during the 1980s.

d.   Adopt an implementation strategy for (b) so that various elements of the modernised system, eg digital transmission, can be utilized to offer data customers improved services over as wide an area as soon as possible.

Therefore during the 1980s a wide range of network options should be available in the UK to facilitate the development of new customer data application and also support new services such as Videotex, Teletex, electronic mail, electronic funds transfer.

Packet Switching Protocols

This section:

- provides a comprehensive overview of data communications protocols and their application to packetized communications

- describes the operational internal protocols in the ARPANET and the impact of the protocols on network performance

- introduces another type of internal network protocol which improves both throughput and resource utilization

- shows the functional applications of the Network Control Center in a packet network, and its influence on protocol operation and network performance monitoring.

To communicate through an electronic medium, both the people and the machines involved must operate according to a set of well-defined rules and procedures known as *protocols*. Such protocols are particularly critical in a data-oriented network, where much of the network usage is between "machines" at either end of the network. A large measure of agreement must exist between the network supplier and network users to insure that the operations will be mutually satisfactory over a wide range of conditions.

TYPES OF NETWORK PROTOCOLS

Data communications protocols can be generally classified as user-to-user protocols, user-to-network (or alternatively network interface) protocols, and switch-to-switch (or internal network) protocols.

User-to-User Protocols

The user-to-user protocols involve a large range of functions from authentication of the data exchange to the remote management and manipulation of user files. From a network operations viewpoint, these protocols are considered to be ADP functions within the host software or host operating system. These protocols are so complex that standards are evolving to define at least four different levels of user-to-user protocols. We will discuss this multilevel structure in more detail later in this section.

User-to-Switch Protocols

The user-to-switch (or network interface) protocols define the relationship between the host computers or user terminals at the packet switches with which they interface. These user-to-switch protocols can be adapted to deal with many different terminal devices and network facilities.

In general, these protocols are concerned with controlling the physical communications circuit between the user and the network, assuring data transparency, detecting and recovering from errors and failures, and controlling information flow.

User-to-switch protocols for packet networks fall into two broad categories—the packetized virtual circuit and the packetized datagram mode. The virtual circuit mode assures the sequencing of user data. It attempts to streamline the flow of long messages that routinely exceed the capacity of a single packet. The datagram mode tries to use the high-speed, real-time capabilities of the packet network to speed data across the network one packet at a time. It is not concerned with initiation of the data connections or the sequence of the data. This mode is particularly suited to the exchange of information that will fit within a single packet.

User-to-switch protocols have evolved to a standard definition involving three distinct levels—the physical level, the data-link level, and the network-control level.

Switch-to-Switch Protocols

The switch-to-switch protocols deal with the internal operation of the packet network and the way that the network controls the flow of information from user input port to user output port. The various ways that user information can be controlled and protected within a packet network define the switch-to-switch protocols. These protocols are, to some extent, related to the user-to-switch protocols, but there are many design decisions and tradeoffs to be made in defining the internal protocols of a packet network. As a result of the layered approach to structuring the networking protocols, a seven-level standardized definition of protocol functions has evolved.

PROTOCOL HIERARCHY OF THE INTERNATIONAL STANDARDS ORGANIZATION (ISO)

Between 1975 and 1980, international standards for data network activity evolved quickly. To facilitate the intercommunication of data-processing devices, the In-

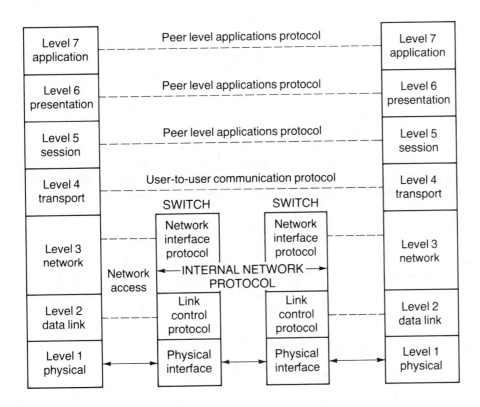

Figure IV-1 International Standards Organization (ISO) protocol hierarchy.

ternational Standards Organization (ISO) adopted the protocol hierarchy shown in Figure IV-1.

The lowest level of the ISO protocol hierarchy is the physical level, which applies previously defined standards to define the physical interface. By "physical interface" to the network we refer to the pin connections, electrical voltage levels, and signal formats between the user equipment and the network. Level 2, known as the data-link level, controls the data link between the user and the network—meaning data format, error control and recovery procedures, data transparency, and implementation of certain command sequences. For non-switched networks, or the interface of simple terminals with computers through point-to-point services, generally only levels 1 and 2 are required. Networks designed by a single manufacturer around a single product line, usually do so with a combination of level 1 and level 2 protocols.

Level 3, the network level, defines most of the protocol-driven functions of the packet network interface, or the internal network. It is at this level that the flow-control procedures are employed and where switched services are initiated through a data-call establishment procedure.

Level 4, known as the transport level, assures the end-to-end flow of complete messages. If the network requires that messages be broken down into segments or packets at the interface, the transport level assures that the message segmentation takes place and that the message is properly delivered.

Level 5, the session-control level, controls the interaction of the user software that is exchanging data at each end of the network. Session control includes such things as network log-on, user authentication, and the allocation of ADP resources within user equipment.

Level 6, the presentation level, controls display formats, data code conversion, and information going to and from peripheral storage devices.

Level 7, the user process or user application level, deals directly with the software application programs that interact through the network.

Although at levels 5, 6, and 7 the protocol is defined from a functional viewpoint, implementation of standard software that can operate at these levels has been slow. The software at all of these levels (often referred to as peer-level software) tends to be both equipment- and application-dependent. However, the layered approach to protocol development achieves a degree of isolation and modularity between the various layers, such that changes in one level can be made without changes in any other level.

THE EVOLUTION OF NETWORKING PROTOCOLS

In recent times hardly a month has gone by when the ISO protocol hierarchy was not discussed or referenced in at least one article in each of the popular trade journals on telecommunications and data communications networks. The promise of this standard protocol hierarchy is that a user will eventually be able to sign onto a network with a smart terminal or personal computer and interact with data in any data base, anywhere in the world. The layers of protocol will each, in turn, resolve any differences in form or structure among the various users.

This Utopia is, and will continue to be, difficult to reach. On the other hand, even though a universal data communications service and network would certainly be of enormous utility, by analogy, there is no reason to expect that a phone call originated by a Japanese-speaking person has to be automatically translated into French (or English, or German, or another language) by the network. For voice users, the communications network simply forces the users to agree upon a common protocol (language), which may be the language of one user or the other, or a third language common to the two users. In fact, data communications systems often take a similar approach, as, for example, the use of a "network virtual terminal" for standardizing terminal device functions.

The articles in this section provide a detailed look at protocols, their definition, and their evolution from the early ARPANET days through the definition of the various levels of the ISO hierarchy. The tutorial by Louis Pouzin and Hubert Zimmermann introduces the concept of protocols first from an intuitive viewpoint, then from a more technically rigorous approach. From the functional descriptions follow the attributes of the layered approach and the impacts on both the computer and network systems designs. The interactions between protocol structure and network performance are developed and these ideas will be treated more quantitatively in several of the subsequent papers. The tutorial concludes with a discussion of the various protocol standardization activities throughout the world.

The second paper of this section, by Leonard Kleinrock and Holger Opderbeck, provides a detailed and quantitative look at the protocol interactions and their performance impact on the ARPANET. The authors also describe the system measurement techniques, developed at the Network Measurement Center (NMC) at the University of California, Los Angeles (UCLA), as well as the internal protocols which provide the limitations on the network operation.

This is a particularly significant paper in that it treats in detail the relationship between the internal network operating protocols and their effect on overall network performance. Analytical and experimental results are used to show how subtle changes in the internal network protocols can have major impact on the integrity and throughput of a packet network. This paper also provides excellent insight into the operation of the basic ARPANET protocols, which are based upon the concept of multiple packets per user segment. This protocol means that the input information from the network users is permitted to have an overall length greatly exceeding the sizes of internal network packets. As a result, the network switching elements must break down user "segments" into a number of composite packets and reassemble them at the network output. In contrast, the next paper in the section describes the class of internal network protocols known as *single packet per segment:* i.e., each user input segment is limited in length to the amount of information that can fit into a single network packet. This protocol, which places additional burden on the user input and output devices, has become typical of the network operations under the international standards for packet interface protocols, the so-called CCITT X.25, in particular. (No single article has been included in this section on protocols dealing specifically with the CCITT X.25 packet switching standard. A good introduction to this standard can be found in Chapter 8 of *Packet Switching: Tomorrow's Communications Today* (Roy D. Rosner, Lifetime Learning Publications, 1982). The standard itself, Technical Bulletin 80-5, is available from the U.S. National Communications System, NSCTS, Washington, D.C. 20305.)

Roy Rosner, Raymond Bittel, and Donald Brown, provide in the third paper of this section, a detailed description of the operation of the single-packet-per-segment type of protocol. This protocol, which, in contrast with the ARPANET mode of operation, is more closely aligned with the operational modes of commercial packet networks, requires that network users provide their data in segments that essentially match up to the internal network packet size. This approach results in a design which, ideally, assures efficient operation for short messages and transactions that fit into a single packet. The protocol is further extended to accommodate longer messages by providing a two-tiered acknowledgement system which, although it limits the number of uncontrolled packets in the network, achieves good throughput characteristics for longer messages. This accommodation is accomplished by splitting the buffering between the origination and destination switches, and essentially eliminating the message reassembly process within

---

ASSURE NETWORK INTEGRITY

Connectivity
Reliability/availability
Restoration
Extension

MAXIMIZE TRAFFIC THROUGHPUT

Routing
Flow control

MONITOR ABUSES AND MALFUNCTIONS

Privacy
Misrouting
Disruption and spoofing
Data accuracy

PROVIDE BASIS FOR BILLING OR COST RECOVERY

Traffic quantity
Delay and precedence data

PROVIDE DATA FOR PLANNING AND ENGINEERING

Traffic growth
User sensitivities
Traffic patterns

**Figure IV-2   Roles of control in a
packet-switched network**

the network. Buffer holding times are thereby reduced, leading to much more efficient utilization of network buffers and lower probability of delaying or blocking messages because of the unavailability of buffers.

In addition to describing the actual protocol design, the authors describe various analytical techniques for estimating the performance of packet networks, and then apply these techniques to the estimation of buffer requirements, network delay, and source and destination throughput. This protocol design eventually became the core of the protocol specification for the AUTODIN II network developed by the Western Union Telegraph Company for the U.S. Department of Defense.

The final paper of this section by A.A. McKenzie, B.P. Cosell, J.M. McQuillan and M.J. Thrope describes the design, implementation, and functions of the network control center used in the ARPANET. Issues discussed in this paper are important because the development of a control and management structure for a communications network is rather imprecise as a result of the many unpredictable and statistical conditions that can affect network operation. While the actual implementation of network monitoring and control varies with the user environment, the role of control with respect to network operation is essentially uniform. Figure IV-2 summarizes the various functions that must be performed in a packet network. Some of these control activities can be achieved using passive techniques—that is, where the network control facilities merely monitor network operations and collect data such as statistical usage data and billing information. Other control functions can best be achieved through complex, active, real-time processes operating on a distributed basis in the various network switches. Figure IV-3 presents a decision-tree approach to selecting alternative modes of control implementation among the major choices of centralized or decentralized, passive or active.

As described in the fourth paper in this section, the functions of the ARPANET Network Control Center (NCC) are accomplished through a combination of active distribution control mechanisms operating in each of the network switches and passive centralized data collections at a special processor within the network. This approach has been used as a model in many practical military and commercial packet networks. A key feature of the Network Control Center operation of a packet network is that its operation be fail-safe, or at least *fail-soft*. In other words, the operation of the network should be benefited by the proper operation of the NCC but the failure of the NCC should not prevent the network from achieving continued operation—at worst, at some reduced capability. These points are well illustrated in this reference paper.

**Figure IV-3   Decision choices in packet network
control implementation**

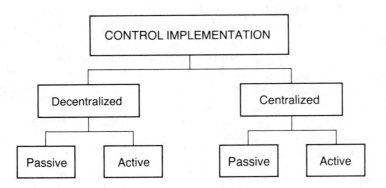

A Tutorial on Protocols

LOUIS POUZIN AND HUBERT ZIMMERMANN

Invited Paper

Abstract—Protocols are common tools for controlling information transfer between computer systems. The concept of a protocol, which grew out of experimental computer networking, is now fundamental to system design. In this paper, basic protocol functions are explained and discussed. Then, the concept of a distributed system architecture is presented. It provides the framework for layers and protocols to operate across heterogeneous systems. The purpose and functions of each protocol layer such as, transmission, transport, virtual terminal, are described. Interactions between design and performance are discussed, and typical mechanisms are reviewed. CCITT and ISO relevant standards are summarized. Finally, the similarity between protocols and programming languages is emphasized as it points to the major impact brought about by protocols in system design.

I. INTRODUCTION

EVERYONE has had the opportunity to overhear such cryptic conversations exchanged over the radio by taxi drivers, policemen, and aircraft pilots. Although upon hearing these conversations at first do not mean much to the layman, these abbreviated languages carry well-defined meanings and obey well-defined rules. Speakers give their name, ask correspondents if they are listening, confirm reception, etc. This form of conversation differs drastically from a face-to-face chat. The communication channel is shared by many speakers. To save bandwidth and reduce interferences, messages are short and coded. External noise and other interferences are common occurences, hence, repetition and confirmation are normal practice. These rules are known as *protocols*.

The term *protocol* entered the computer jargon at the turn of the 70's, when the U.S. Defense Advanced Research Project Agency set out to build a network of geographically distribute heterogeneous computers [71]. Up to that time, communication between computer programs or processes was limited to processes which were located within the same machine. Interprocess communication was accomplished through the use of shared memory and special signals exchanged through the mediation of the operating system. This technique represented the analog of a face-to-face chat between processes. Interprocess communication between geographically distant systems would have left processes with the same kind of constraints that taxi drivers encounter. They would have to interact through a potentially hostile environment with limited bandwidth, delay and unreliable transmission. In addition, the processes in the different computer systems did not even speak the same native tongue, having been created by different manufacturers.

Computer veterans remember the sinuous evolution that led from binary programming to assembly code, to Fortran, Cobol, Algol, and other high-level languages. Originally viewed as a collection of tricks and hobbies, programming languages have developed into a major branch of computer science. The evolution of protocols has followed a strikingly similar path. Indeed, *Protocols are common tools designed for controlling information transfer between computer systems.* They are made up of sequences of messages with specific formats and meanings. These messages are equivalent to the instructions of a programming language, although protocol languages are still in an early stage of ad hoc development.

Actually, progress has been very rapid. Techniques for protocol formalization and verification are already under development [25]. The primitive era is passing and a body of protocol science is emerging. It is time for computer and communications specialists to acquaint themselves with the basic concepts of protocols and to make them a part of their repertoire of design and implementation tools.

Summary

Section II of this article is a short *introduction* to protocols. It presents a simplified and condensed view of the rest of the article, with the emphasis put on examples drawn from analogies. This should help the reader to acquire an intuitive vision of the whole subject, and to understand more easily the following sections.

Section III presents the *basic elements* found in protocols. Indeed, the elementary components used for building protocols are very limited in number although they appear under different names and in different forms in the literature. The first step towards understanding protocols is to uncover the simple mechanisms used in their construction.

Section IV introduces the concept of a *distributed system architecture*. It shows how the functional system components can cooperate with one another through the use of protocols to organize an otherwise heterogeneous system into a consistent homogeneous structure. It explains the role of *protocol layering* in the rationalization of distributed system architectures. This is a crucial concept since, without layering, it would be impossible to allow local optimization of different parts of the heterogenous *system* while preserving opportunities to establish common conventions for interprocess communication, independent of the applications to be supported.

Section V contains a discussion of the *functional elements* of typical layers, for example, data transmission, transport and virtual terminal protocols. These layers form the foundations of any distributed system.

Section VI shifts the focus from design to *engineering*. It points out the advantages and disadvantages associated with the various basic protocol elements and shows how performance is affected by design choices and environmental factors.

Section VII addresses *standardization*. As in the case of programming languages, protocols will only be effective when they are widely adopted by large communities. Due to commercial interests, this is an area replete with nontechnical complications. Major data communication standards are summarized, along with work in progress within standardization bodies.

II. Protocol Overview

A. Basic Protocol Functions

In the first subsection, we introduced the intuitive notion of a protocol and provided a simple working definition. To the newcomer, it is necessary to understand intuitively the ingredients which comprise a protocol in order to acquire a general mental model of its structure and its *raison d' être*. We proceed by giving an analogy with the interaction between organizations in a business context. Admittedly, this analogy will only give a macroscopic view of the nature of protocols. Terms and concepts are used and intended to be understood generally without the need for introducing very precise definitions at this point. We have made obvious approximations and omissions for the sake of achieving clarity in presentation.

Let us assume that two corporations, SIXOUNS and SNAPSE, are considering some joint business. They both produce whisky (of course). One of the two presidents will presumably invite his colleague for lunch (no martinis, please), and agree upon a business strategy. They will then turn the matter over to their marketing vice-presidents who will arrange a meeting. Later on, plant managers may get together, etc. For our model, we assume regular discussions take place between opposite numbers, i.e., between people at the *same level* (or layer) in the corporate hierarchy. However, these discussions are highly informal and also time-consuming during the initial stages. Since the strategy is now shaping up, SIXOUNS and SNAPSE feel that they should streamline their interactions. To that effect, they design procedures for maintaining a mutual exchange of information between the various participants, without the need for actually meeting. These procedures might include such elements as rules, message or document formats, frequency of interaction, etc., which are established by a particular layer of the corporate hierarchy. They constitute the *protocol* to be used by that layer.

SIXOUNS and SNAPSE have created a common distillery. Each corporation uses its own network of retail stores to market its own brand. Orders are collected separately by SIXOUNS and SNAPSE, and transmitted to the distillery. On the other hand, deliveries are carried directly from the distillery to the retailers. Occasional mixups may happen, such as delivering a SNAPSE whisky, instead of a SIXOUNS, or miscounting the number of bottles, or dropping a carton of whisky on the floor with shattering consequences. In order to prevent disputes from arising, each corporation acknowledges each order received from its retailers, and retailers have to sign delivery vouchers. Thus, discrepancies may be traced back to their origin and corrected. We may say that order and delivery protocols include *error control*.

The volume of orders may at times vary substantially. This fluctuation causes inefficient operation for the distillery because its production capability is limited. Sometimes, orders must be delayed because of overload, while at other times there is unused distillery capacity. Thus, it was decided that SIXOUNS and SNAPSE should place production orders by *anticipation*, in order to regulate the amount of whisky produced. Of course, some care must be exercised, since the distillery has also a limited space for inventory. This tricky balance is a matter of *flow control*.

Even though considerable care is taken in handling properly the orders, deliveries and inventory, some errors may slip through, due possibly to human factors. An occasional discrepancy is acceptable so long as it has no cumulative effect. To prevent that, the books are balanced once a month, at which time all figures must be reconciled. This accounting is not completely straightforward, as there are orders or acknowledgments in the mail, and filled trucks on the road. In fact, it is not possible for all accounts to be settled simultaneously and, as a result, there is a need for a specific set of accounting rules so that all outstanding orders, deliveries, and production be clearly accounted for each month. This protocol may be called *synchronization*.

Another requirement always appears whenever an exchange of information is to occur, viz. providing the address of the intended recipient. In the whisky business, an order may be sent by mail to the following address:

SIXOUNS Corp.
Order Dept.
4/5 Bourbon St.
DOWN HATCH 70127 YU

Or the retailer may dial 435–2217, ask for extension 652, and say: Hello Sally, . . . This is Louis. I would like. . . etc. Or he may send a TWX message, or something else. In these examples, the composite address is actually a concatenation of several address elements. Some elements are meaningful to the Post Office or the public telephone system. Other elements are only meaningful within the SIXOUNS corporation. The retailer must learn them beforehand.

A different example illustrates another addressing possibility. A friend of mine happens to be in town and has previously advised me that he will be staying at the DALTON hotel. (Of course, he had to know me and my address). How can I send him a message? I may call the DALTON, and leave a message for Mr. Jim DOLITTLE with the operator. I could also ask the hotel operator for my friend's room number, and the next time I may just leave a message for room 625. Now, there could be more than one Jim DOLITTLE in the DALTON, or he may have moved to a different room, or checked out. If I really want to be sure that my message gets to the right person, I have to check further.

Things would be simpler if every person in the whole world had one portable telephone (or TWX) number which only he can use. When it answers, then we know the right person is at the other end.

The previous examples illustrate three typical methods used for conveying information to its destination. The first example corresponds to a hierarchical addressing strategy, the second example denotes a method based on dynamic allocation of temporary names, and the third example illustrates mapping. In the protocol jargon we often use the term *naming* to designate methods of composing addresses and selecting a proper correspondent. Each of the above three naming approaches, namely hierarchy, allocation and mapping will be discussed in Section III along with the other key elements of protocols, namely, error control, flow control and synchronization.

B. Distributed Systems

So far, our examples have been taken from a common business setting without any reference to computer systems. In fact, there are definite similarities between the functional organization of a computer system and an organization. A piece of software, often called an executive, allocates system resources (memory, files, channels, processor time, etc.) to subsystems (or programs) in charge of specific services. In turn, these sub-systems allocate their own resources, and control the execution of their tasks, and so on. There is some difference, however. When drawing a corporation chart, the president is normally placed at the top. When drawing a computer system functional diagram, the executive is normally at the bottom, but this is immaterial for our purposes.

When two or more computer systems are interconnected for a common purpose, they are said to make up a *distributed system*. As we have seen already by analogy, cooperation between these entities will normally be formalized into a number of protocols established at each layer of the system structure.

In the example of the whisky business, we assume that SIXOUNS and SNAPSE are similar organizations, with the same functional levels (or layers). Thus, it is rather straightforward to structure their interaction protocols by functional layers (marketing, production, etc.). Establishing a business relationship between a whisky corporation and a corn farm would require a different model, because these two organizations exhibit totally different structures. Perhaps the most sensible solution would be a contract between the farmer and the purchasing department of the whisky corporation. As we can infer, each case of business connection requires a specific solution, along with a specific set of protocols.

Even though most computer systems are structured along similar lines, there are a multiplicity of differences between systems from various manufacturers. It is also common that a single manufacturer produces several families of computer systems, with distinctive structural differences. In this case, computer systems are termed *heterogeneous*. Nevertheless, it is usually possible to work out some sensible interconnection scheme between any two computer systems [17]. One must decide how layers of each system are to be paired, and what protocols are to be used.

The development of an interconnection scheme between computers is often a significant investment, as it requires skilled computer professionals who know the intricacies of both systems. On the other hand, if most computer systems would contain very similar, or even identical layers, their interconnection could become a simple routine task. This is the case with computer systems of the same family. In this case they are termed *homogeneous*.

Many technical problems encountered in building distributed systems would disappear with homogeneous computer systems. But homogeneity cannot be a general solution. We can expect that there will always be a variety of suppliers. Even with a single supplier computer systems evolve, due to changes in technology and in user needs. On the other hand, as we have mentioned earlier, heterogeneity may introduce technical incompatibilities and require ad hoc adaptations. Is this dilemma irreconcilable? Experimental developments of distributed systems have led to a middle of the road approach, which aims to take full account of heterogeneity in the real world, while making interconnection more or less straightforward. This approach is presented below. We call it a *reference architecture*.

C. Reference Architecture

Let us assume we want to connect a cassette recorder and a radio set of arbitrary makes. We may study the electrical diagrams of both devices, and try to determine appropriate points where we might plant an electrical coupling, so that sounds received by the radio set become recorded on a cassette tape. It may be a little difficult to locate these appropriate coupling points, if the diagrams are not well-documented. Electrical signals may also require some adjustment to be performed by an intermediate black box, which we would have to build. Then, we may connect external wires with a soldering iron. If things do not work satisfactorily, we will need metering instruments and possibly some professional advice.

Interconnecting a cassette recorder and a radio set becomes much easier if both manufacturers have anticipated the need, and installed some plugging socket carrying well-defined electrical signals. In this case, we do not have to study the inner workings of a whole device. We only have to understand how to use the pins of the socket. These pins are the only peepholes into the device. We call them *visibility points*.

Computer systems are more complex than cassette recorders. Thus, interconnecting them requires more than the definition of a simple socket. As we mentioned earlier, computer systems are layered structures. Interconnecting layers requires the definition of protocols, which are the set of rules followed within each layer by the interactions between interconnected systems. Furthermore, protocol definitions assume some interactions between layers within each computer system. For example, in the SIXOUNS–SNAPSE agreement described earlier, protocols established for the marketing layer and the production layer assume some interactions between marketing and production within SIXOUNS and SNAPSE. We call an *interface* the set of rules followed by the interactions between two layers within a single computer system.

Once interfaces and protocols are defined between interconnected systems, they may work in cooperation to achieve a common purpose. They become a distributed system. If there exists a common set of definitions for all interfaces and protocols used in distributed systems, it is no longer necessary to study the inner workings of each system. Interfaces and protocols are analogous to sockets in the connection of a cassette recorder and a radio set. They make interconnection much easier.

If the construction of distributed systems required that all interconnected systems follow rigidly a precise model defining systems *in their entirety*, then distributed systems would only comprise computer systems of the same family. Indeed, not many manufacturers, if any, would accept being constrained to such an extent in the design of their own products. But if the constraints are *limited to interfaces and protocols* necessary for building a model distributed system, it may become commercially attractive for many manufacturers to accept these constraints as the counterpart of a wider market. For easy reference to the set of interfaces and protocols of a distributed system model, we introduce the term *reference architecture*. Then, we may say that a system conforms to the reference architecture when it contains the *visibility points* of the distributed system model, i.e., the interfaces and protocols necessary for interconnection.

D. Basic Protocols Layers

Most business organizations are structured along similar lines. Below the president, one can find a number of departments: finance, marketing, production, purchasing, etc. There is no proof that this structure is the best, but in practice it seems to be satisfactory. The same rationale applies to layers of a distributed system. Even though the choice of appropriate protocol layers is not yet supported by a rigorous demonstration, a certain consensus has emerged from experience. The *basic protocol layers* consist of *transmission, end to end transport*, and application oriented functions. Among the latter, the *virtual terminal* and *file transfer* protocols are the most commonly used.

The *transmission* layer includes functions pertaining to the transfer of data between geographically distinct locations. In the example of the whisky business, transmission is analogous to the transportation of whisky bottles. Whether transportation is carried out by truck, train, or helicopter, is immaterial, as long as whisky is delivered as ordered. Similarly, whether data are carried through wires, microwaves, or satellites is immaterial, as long as they are delivered without alteration at the proper destinations. New transmission techniques such as packet switching require some specific packaging of the data. This packaging is analogous to the practice of putting goods into containers for shipping. The transmission protocol consists in using properly any available transmission system.

The *end to end transport* layer (or transport for short) is in charge of checking the integrity of the transmission layer. Indeed, no transmission system is totally reliable, and some are far less than perfect. When data must travel through more than one transmission system (e.g., for international traffic), it is not unusual that each public carrier will blame the other in case of trouble. Thus, the transport protocol includes error control for assuring that data are delivered correctly, or else for having them corrected or retransmitted. In addition, the transport layer acts as a transportation bureau for the benefit of application layers.

It receives transport requests from other layers and makes the best use of available transmission systems. A similar function is carried out in the distillery. Packages for the retailers are prepared for shipment, but they are not delivered separately. Indeed, several packages may be destined to the same retailer, and delivery trucks follow established routes. Thus packages must be assigned to some trucks, depending on their route and their available capacity. The set of mechanisms carrying out the transport protocol in a distributed system is called a *transport station*.

The *virtual terminal* layer performs various adaptations between the characteristics of physical terminals and application programs. Indeed, real terminals available on the market may use different codes, keyboards, formats, etc. and new varieties of terminals are popping up constantly. Thus it is practically impossible for every application program to work properly with any terminal. For example, let us assume that an application program outputs series of lines having a length of 120 characters. Each character is binary coded in the specific code used by the computer, say EBCDIC. How can we use a display terminal working with ASCII coded characters and having a width of 80 character positions? We may define some adaptation rules:

1) translation of EBCDIC characters into ASCII characters;
2) reformatting 120-character lines into 80-character lines.

There is no universal solution for shrinking lines. A first option is to select a subset of 80 positions out of 120. Some fields will be truncated, but this may be acceptable if they contain redundant information. A second option is to fold lines, i.e. each output line will be displayed as an 80-character line followed by a 40-character line. This presentation may be acceptable for program listings. A third option is to use 3 lines on the display for 2 output lines. This is acceptable for plain text.

Assuming that we can select the most suitable adaptation, the application program works as if it were in communication with a 120-character wide EBCDIC terminal, which is not really there. This is why we call it a virtual terminal.

Actually, a virtual terminal does not have to be identical with an existing real terminal. Rather, it is a model of an ideal terminal with enough adjustable parameters so that it can look like a large number of real terminals. The virtual terminal protocol is the set of rules for setting the parameters and exchanging data with a virtual terminal.

The *file-transfer* layer is in charge of moving or copying files across computer systems. Ideally, in a distributed system, the location of a file should be immaterial. In practice, it is not so, for a number of reasons such as the following. Access is faster when file and application program are located in the same computer system. Storage costs may vary with systems. Security may not be good enough on certain systems. Thus, file transfer is frequently used in distributed systems. The file transfer protocol includes rules for opening and closing sender and receiver files, for data transfer, and for error recovery. If data translation is required, it may be performed by a specific user-provided routine, or it may be invoked as a protocol option.

In the above, we have introduced an intuitive presentation of the basic protocol layers of a distributed system. At this point, we would like to warn the reader that there is so far no universal agreement about the names and the number of layers. There is some analogy with geology. Where some people would identify only a sand layer, others would find a sand layer, a gravel layer, and a thin clay layer. No one is wrong, it is just a matter of defining sand.

E. Concepts in Distributed Systems

Some basic protocol concepts, such as error control, synchronization, etc., have been introduced informally earlier in this section. We shall have occasion to draw upon a few more conceptual terms in the body of this article. Thus, a short presentation, without rigorous definition, is included below to convey a sufficient understanding of these additional concepts, before proceeding to the next sections.

Process: A program running on a computer. It could be microcode running on a microprocessor or conventional software running on a mainframe processor. The term is most frequently used to distinguish among several running programs in time-sharing systems.

Resource: When a program is executed on a computer, it needs some area of memory, some processor time, and usually access to some input–output device, such as a teleprinter, a disk file, or a digital plotter. These equipments, or parts thereof, are called *resources*. By extension, any machinery made available to users is also called a resource. For example a data base, a compiler, a text editor, a graphics system.

Activity: One or several processes, along with appropriate resources, which work towards a common goal; e.g., a seat reservation application handling all the terminals of an airline company may be called an *activity*.

Entity: When information is exchanged within a distributed system, it is often immaterial whether we consider sources and destinations as processes, activities, files or terminals. Due to the heterogeneous nature of computer systems, a source of information might be called a process on one system, and a terminal on another system. In order to avoid misnomers, we use the generic term of *entity* to designate anything capable of sending or receiving information.

Correspondent: When two entities exchange information, each one is said to be the other's *correspondent*.

Port: Interconnected pieces of equipment are usually linked by a cable plugging onto a socket on each equipment. Each socket comprises a number of pins, which are linked to the other sockets pins through a wire in the cable. A pin has no function of its own. It is only a convenient device for speeding up interconnection, and for identifying signal channels. A *port* is analogous to a pin. It is a convenient device for establishing and identifying information channels between entities.

Association: An information channel between two ports or two entities. Also used in a general sense to mean any kind of relationship.

Liaison: A reliable association. Reliability is obtained by using a transport protocol providing for error and flow control.

To Access: To establish an association and exchange information with an entity.

Context: Some kinds of information transfers are composed of totally unrelated messages; e.g., when a point-of-sale terminal sends a message containing item number, quantity, credit card number, and salesman identification, there is no correlation with the previous or the following message. Each message is self-contained and may be interpreted and processed independently. In other cases, messages are interdependent; e.g., when accessing a time sharing system from a terminal, a user might follow a question and answer procedure, giving successively his name, password, account number, etc. Clearly, messages sent from the terminal are no longer independent. They are interpreted in taking account of the previous ones. This means that the user logging process remembers something about the ongoing conversation. The information set aside during the logging in of a user is called a *context*. Usually, the context contents evolve with the arrival of new messages, or other events related to the user activity, such as occurrence of an error, or exhaustion of allocated processor time. In a dialogue, the turn (i.e., whose turn is it to send the next message) is part of the dialogue context.

Context Identifier: In the previous example of a time-sharing system, the user logging process may be unique and in charge of logging in all incoming users. Since a logging operation may take a significant time due to slow user reactions, it would be irritating for other incoming users to have to wait until the logging process is free. Therefore a typical logging process is constructed for handling several users in parallel, thence, several contexts in parallel. Obviously, messages arriving from the incoming users terminals must be sorted out and interpreted in *the proper context*. Since a user cannot be completely identified and validated until the logging procedure is completed, sorting out messages on the basis of the originating terminal appears to be the most practical method. Therefore, the terminal (or its physical line) number is used as a search key to find the context of an incoming message. The key relating a message to its context is called a *context identifier*.

Context Initialization: Necessarily, a context has to be brought to existence. Such an operation is called *context initialization*. The initialization may be *static* (i.e., it is built in the environment, e.g., at system generation time), or it may be *dynamic*. In this latter case, some specific messages are accepted *out of context*, and the result of their interpretation is to build an initial context. Normally, other specific messages are intended for the reinitialization of an existing context, and for terminating (i.e., destroying) an existing context. Typically, statically initialized contexts are protected from destruction, but they may be reinitialized.

Context Synchronization: When two entities communicate, they are related by an association, which they have to remember. Thus there must be an association context. On a single computer system an association context could be an area of memory shared by the two entities using it. In a distributed system memory sharing is rather cumbersome, because there is no memory directly addressable from geographically distant entities. In practice, either entity maintains its own view of the association context. We may say that an association *context is distributed*. Ideally, the two context views should always be identical, but this is impossible in practice due to message transit delays, not to mention occasional failures; e.g., let us assume that two entities called PING and PONG are engaged in a dialogue, and that it is PING's turn to send a message. In either context view some variable, say TURN, should read PING. As soon as PING has sent its next message, it updates its own context view so that TURN now reads PONG. However, the TURN of PONG's context will read PING until PONG has successfully received PING's message and updated its own context view. In the meantime the two views are *inconsistent*. Should PING's message get lost, the inconsistency becomes permanent (eventually, some external action is necessary to break the deadlock). Bringing both context's views into a consistent identical state is called *context synchronization*.

C-Name, and Local Name: Putting together a collection of heterogeneous computer systems is similar to an assembly of football teams, aircraft crews, church choirs, and police squads, coming from various countries. Probably only the French would understand when AMBROISE is called, and know who he is. There are several DAVID's and PAUL's. There is also someone from Netherland, but only he can pronounce his name. Clearly, there is a need for some way to call people that should be unambiguous, and understood by everyone. A possible solution could be to give each person a badge with a unique number. In a distributed system, such global identifiers are called *C-names* (common names) as opposed to *local names* used locally within each computer system.

Cycle: Let us assume a life insurance company produces 10 000 new policies every year (only one policy for one customer). Policies are numbered sequentially for 1 to 999 999. When all policy numbers are exhausted the numbering starts again with 1. Thus it takes one hundred years to go through one *cycle* of the policy numbering scheme. The number 125, 786 is sufficient to identify uniquely a policy as long as it never happens that some customer keeps a life insurance for more than one hundred years. Communications protocols frequently use sequential numbering schemes to identify messages and detect losses and duplicates. The message numbering cycle must be long enough so as to outlive any numbered message. Otherwise it could happen that two different messages carry simultaneously the same identifier.

F. Evolution Trends

Techniques used in distributed systems are now submitted to seemingly conflicting forces. Stability and standardization are crucial for achieving worldwide interworking between all kinds of computer systems, terminals, data bases, etc. On the other hand computer and communication technologies are evolving at a swift pace. The advent of composite services which tightly integrate communications and data processing is tipping off the historical balance between the communications domain, mainly regulated and monopolistic, and the data processing domain, vigorously competitive and innovative.

Traditionally, standardization has progressed at a snail's pace, especially in the data processing field, where competing manufacturers never showed much enthusiasm for intersystem compatibility. A new spirit is now distinctly spreading. The demand for interconnection and world-wide access to computerized services has reached the point where standardization has become unavoidable. Standardization bodies are now engaged in a race with the requirements of distributed systems.

Two major standardization bodies are referred to in this article.

1) Comité Consultatif International Télégraphique et Téléphonique (CCITT) is practically a common carriers club. In most countries common carriers are government controlled monopolies often termed PTT's (Post, Telegraph and Telephone).

2) International Standard Organization (ISO) is the forum of national standardization bodies, and a few international standardization associations.

Over the past fifteen years CCITT has produced interface standards intended for data transmission over analog telephone circuits. These standards are known as the *V*-series. Since 1972, new standards intended for digital networks are being worked out. They are known as the *X*-series. They include standards for packet switched services, even though some public packet networks use mostly analog circuits.

ISO has produced some standards for the control of data transmission over circuits. These standards are known as high-level data link control (HDLC) [43], [44]. Objectives for the next few years are the definition of a standard architecture and basic protocols applicable to heterogeneous distributed systems.

Presumably, some new developments will appear that may obsolete the technical premises on which the existing standards are founded. Indeed, protocol performance and even the choice of protocols depends on environmental factors, which may vary substantially; e.g., the sizeable increase in bandwidth brought about by satellites or optical fibers may lead to a redistribution of data storage functions. The layered approach systematically adopted in distributed systems is essential for minimizing the impact of technical evolutions, since it attempts to concentrate and isolate within a specific layer all functions related to a particular system objective. Should this layer need to be redefined, the impact on other layers should be negligible. The difficulty resides in a careful definition of functionally independent system layers.

The experience acquired by a number of experimental and industrial distributed systems has uncovered some basic layers that are good candidates for a first cut at distributed systems standards (see Section II-D). They are as follows: 1) data transmission, 2) end-to-end transport, 3) virtual terminal, 4) and file transfer.

G. For the Reader

Section II was intended to be a primer to the rest of this article. We would like to bring again the attention of the reader to the approximate and intuitive nature of the material presented in this section. The following sections give more technical coverage, and at times redefine concepts from a different point of view, with another set of justifications.

The reader may wish to compare the technical descriptions in Section III with the simple intuitive notions in Section II.

III. ELEMENTS OF PROTOCOLS

The design of an automobile must take into account a minimum set of basic requirements: wheels, engine, body, steering, brakes, etc. The very same approach applies to protocols. They would be incomplete, or inappropriate, unless they properly handled a number of basic functions. This section explains these basic functions, but does not describe any specific protocol.

A. Naming

Protocols involve communications and thus transfer of information. When a piece of information must be transferred from one domain into another, some unambiguous indication of the destination must be specified. In most circumstances, the sender does not perform the physical transfer of information by itself. Typically, the sender invokes some common underlying mechanism along with a set of parameters, including names or addresses, e.g., MOVE ⟨data⟩ FROM ⟨A⟩ TO ⟨B⟩.

In homogeneous systems, names such as A or B designate well-defined entities, e.g., core locations, files, processes, physical or logical devices. Computer systems make use of various naming schemes, depending on the type of resource to which names are attached. Names may be of fixed or variable length, and follow specific conventions: alphabetic, numeric, special characters, and so on. Changing naming conventions in existing systems is an insuperable task because the conventions are usually so ingrained in the design that any change causes a major upheaval.

Thus networks of heterogeneous systems bring about new problems, as there is typically no commonly agreed naming scheme. A usual approach in such a situation is to superimpose a new network-wide scheme, while keeping existing ones inside each system. For example, telephone numbers are superimposed on persons or companies names. We shall see now how these common names can be created.

For convenience, we shall refer to names which exist in each computer system as *local names*. Network-wide names will be called *C-names* (for common names). There are a number of ways to construct C-names and relate them to local names. Three most frequent methods are discussed below: hierarchical concatenation, allocation, and mapping.

Hierarchical Concatenation: We imagine that each computer system has a set of local names $\{L_j\}$. To each set is assigned a unique C-name, $\{C_i\}$. The C-names, obviously must be assigned by global agreement to preserve their uniqueness. It is not ruled out that more than one C-name is assigned to a set $\{L_j\}$ of local names, but the same C-name would not be assigned to two different local name sets because of the potential ambiguities this would introduce. The network name space is obtained by concatenating local names with a C-name, i.e., $\{\langle C_i \times L_j \rangle\}$.

For example

Local names in system X:

> JOHN.FILE3.TEXT
> TOM.LETTERS.MARY

Local names in system Y:

> SYS COBOL BIN LINK.3
> FACTORY PARTS

Local names in system Z:

> 53409
> 42121

The corresponding C-names might be

> X TOM.LETTERS.MARY
> X JOHN.FILE3.TEXT
> Y SYS COBOL BIN LINK.3
> Y FACTORY PARTS
> Z 53409
> Z 42121

This scheme is similar to the international telephone numbering plan in which phone numbers are obtained by concatenation of a country number with an internally assigned national number.

Allocation: A second method is to permanently allocate C-names to only a few processes, for example, the "logger" process in time-sharing systems which validates remote user access to the system.

Another set of C-names is allocated to each local system for dynamic association with local processes. At least one

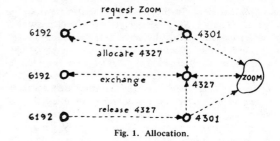

Fig. 1. Allocation.

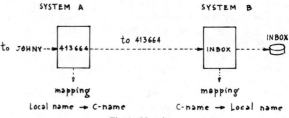

Fig. 2. Mapping.

process in each local computer system is assigned a permanent C-name and the responsibility for dynamically assigning the others allocated to the system.

For example, suppose that C-names are simply integers in the range of 1 to 9999. Let us further assume that system A has been assigned C-names 6100–6199 for its use and system B has been assigned 4300–4399. Let us further suppose that there is a network access process at system B whose permanent C-name is 4301. Finally, let a process in system A, with C-name 6192 attempt to access a process in system B whose local name in system B is ZOOM (see Fig. 1).

Process 6192 would send a request to access local process ZOOM to the network access process in system B whose C-name is 4301. The network access process locates or creates process ZOOM within system B and allocates an unused C-name for it, say, 4327, and sends a reply to process 6192 which associates C-name 4327 with local process ZOOM.

Thereafter, process 6192 exchanges information with process 4327 until such exchanges are no longer necessary, at which time process 6192 reports to the network access process at system B that the C-name 4327 can be freed again. Alternatively, the ZOOM process could make this last report.

Plainly, the originating process (6192) needed a considerable amount of external knowledge to conduct this exchange. It needed to know that there was a network access process at system B and that its C-name was 4301. It also needed to know the local name, ZOOM, as the process it wanted to communicate with.

This strategy is closely related to the method used in the ARPANET *initial connection protocol* [31], [61] associated with the network control protocol (together they constitute the ARPANET transport protocol layer).

Mapping: An alternative to concatenation is to create a set of C-names $\{C_i\}$ and statically assign them among all the processes, files, and other local entities which must be accessed on a network-wide basis (see Fig. 2). The C-names assigned are *mapped* by each local system into the associated local name. The relationship between local and global naming mechanisms is discussed further in this section. This strategy is used in the Cyclades network [65].

Other strategies besides concatenation, allocation, and mapping are possible. Indeed, network designers have tended to invent ad hoc strategies that suit their implementation constraints rather than developing network naming techniques for every general use.

Discussion: Among the three methods described above, the hierarchical concatenation method is apparently the simplest one because a C-name may be parsed easily into a system name and local name, but it is only practical when local names are rather homogeneous. Otherwise, C-names take so many different formats that protocols become unwieldy and inefficient. In particular, the introduction of a new set of local names may require modifications in a number of existing network access protocols when the characteristics of the new set have not been anticipated.

The allocation method is favored by a number of operating system-minded people. It has the advantage of fitting within conventional single computer structures. Most operating systems in use over the past ten years were designed as geo-centric objects centralizing all critical functions. Accessing entities is usually a multistep procedure beginning with a well-known log-in procedure. This vision of a rigidly partitioned universe is still taken for granted by a number of computer professionals.

An advantage of the allocation method is that users can access other computer systems almost as if there were no network. It is also contended that using a small number of C-names, for active entities only, saves overhead in network access machinery.

The deficiencies of this method are apparent from its advantages. Entities must be rigidly associated with specific computer systems and this puts a straitjacket on users who might prefer instead homogeneous access to all resources, with computer system boundaries fading out. In other words, the proper vision is a network resources rather than a network of computers. In addition, the allocation method is more complex in implementation as it requires the management of changing associations between C-names and local entities. Indeed, it implies a dynamic modification of the structure of the naming mechanism. Due to transit delays and fuzzy states associated with any distributed system, there appear transient conditions which require specific safeguards to prevent errors; e.g., a C-name may be accidentally released on one end and reassigned without the other correspondent being aware of this event. The elimination of inconsistent states takes additional mechanisms and delays the allocation of C-names. (See example in Section II-A.)

The *mapping* method provides for a homogeneous name space for accessing any network entity. It is similar to a national telephone numbering plan. Mapping C-names onto local names is a matter of local implementation. This allows both permanent and temporary associations between entities. This method is therefore more general than the allocation method, which provides only temporary associations. Any desirable access control procedure can be triggered as part of the mapping machinery, depending on accessed entities or their correspondents.

Such a facility makes it possible to offer homogeneous network-wide access protocols for specific services, which may be available on different computers, e.g., compilers, text editors, help, distributed data bases. This flexibility results from the ability to design and implement specific access protocols independently from the others, because they are attached to well-defined entities. In principle, the allocation method could offer an equivalent facility, but in practice the logger machinery is too sensitive software to be frequently adapted to new classes of users.

A criticism of the mapping method is that it takes overhead in scanning large tables of C-names when they are permanently associated with local names. However, this objection is not well substantiated. Indeed, a search is always necessary, whether the key is a C-name or a local name, and there is no reason why searching by C-names should be less efficient. Space occupied by the C-name table is not critical, as it can be on secondary storage, like any file directory.

A few examples of the flexibility inherent in the mapping method are as follows.

1) Users may choose local names according to their own symbolism, e.g., in their native tongue.

2) Several local names may map into a single C-name, for the convenience of using at will abbreviated or full local names.

3) Since local names do not have to be known remotely, there cannot occur any ambiguity when identical local names are used by several computer systems.

4) A resource may be moved to a different computer system without its users knowing it. Actually, this capability is truly effective only when the transmission system can use C-names as message destinations. Examples of such networks are CIGALE [66], DCS [29], and ETHERNET [58].

Group Names: When the total space of network names becomes very large, C-names may become very long and may generate overhead or inconvenience. A countermeasure is to partition the name-space into a hierarchy of subsets designated by group names [66]. Communications taking place between correspondents belonging to the same group require only short names. Communications crossing group boundaries require full length names. Thus there are two or more C-name formats, depending on the number of partition levels crossed by communications. This is similar to telephone dialing for local and long distance calls. However, the analogy is only partial, because a group does not carry any geographical connotation, as opposed to an area in the telephone system. Entities of a single group may be geographically scattered throughout. On the other hand, it is obviously possible to define certain groups as having geographical boundaries, but not necessarily restricted to a single piece of land.

For this technique to be effective, there must exist natural clusters of correspondents establishing a significant proportion of mutual exchanges. This is almost always the case when data communication is directly related to human users who are naturally clustered geographically (urban areas) or within a business structure (corporation, institution, affiliation).

Some refined techniques may be used for partitioning the network name space so that short names be also usable to communicate with neighbor groups [69]. When groups are not geographical, the notion of neighborhood has to be defined within some space, e.g., the space of integers representing C-names.

Port Names: When two entities communicate, they exchange information in the form of messages, following the rules of a certain protocol. It might appear sufficient that messages contain some destination field designating unambiguously the receiving entity. Thence, the name of the receiving entity could be used in the destination field of a message. However, this technique is not generally applied for the following reasons.

1) Since computer systems are mainly heterogeneous, protocols are designed so as to make a minimum of assumptions about the characteristics of the communicating entities and of their environment. In particular, they tend to use names which are independent of any entity.

2) Information exchanged between two entities may pertain to several independent channels (or associations).

Consequently, the notion of *port name* is commonly used as a substitute for entity names. Port names present the following characteristics:

3) They constitute a homogeneous network-wide name space. Therefore, they are *C*-names by definition.

4) They are defined independently of any entity, but they may be assigned to any one.

In the naming techniques presented earlier (allocation and mapping) *C*-names are in effect port names. A port name may be allocated temporarily to an entity in the allocation method, or permanently in the mapping method.

Communications protocols are then defined as operating *from port to port*. Each port to port association constitutes an independent information channel. An entity may use several port names in order to maintain simultaneously several independent channels with another entity.

Port names may be created and distributed randomly to entities of a distributed system, but a possibly time-consuming table search would then be necessary to locate them. Customarily, some simple algorithm is used to relate port name and computer system, e.g., contiguous series of integers are allocated to each system, or a port name is formed by concatenation of a computer system *C*-name and a local integer. Many variants are acceptable, as long as resulting port names are homogeneous.

Association Names: When two entities exchange information, they make up an association. In some systems, any two entities may exchange information at any time without prior arrangement, e.g., the public mail, or a message switching system. Actually, if traffic is observed from the point of view of an end user, information exchanged is usually associated with an existing relationship which may be temporary or permanent, e.g., a business connection, a complaint, an acquaintance. In order to interpret correctly the arriving information, it is necessary to relate it to some other information kept into the receiver's memory. This information already in store is called a *context*. It may contain such items as the number of the last message correctly received, the number of errors encountered, the address of a buffer for the next message, etc. The context contents are specific to the communication protocol in effect between the corresponding entities.

Relating an arriving message to the appropriate context requires that some item of information contained in the message designate this context unambiguously. Such an information is called a *context identifier*. We will examine in the following some typical methods for choosing context identifiers. There are two cases.

1) Entities cannot establish and operate more than one association at a time (Fig. 3). In this case, there is not really a need for a context identifier, because the entity can easily locate its only existing context, if any. However, in the

Fig. 3. Single association.

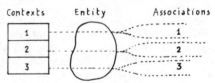

Fig. 4. Multiple association.

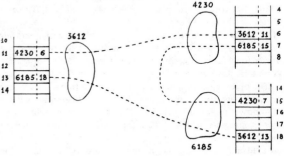

Fig. 5. Context identifiers.

absence of context identifier, any arriving message would be deemed related to the only existing association. Spurious messages might cause disruption. In order to tell valid messages from others, the protocol used with the association should include error control (see Section III-B). Schemes such as carrying a password within each message, or encrypting messages may be used as a form of error control.

2) Entities may establish and operate multiple concurrent associations (Fig. 4). In this case, messages must carry contexts identifier. Identification can be exchanged between entities when they set up an association. This is similar to the business practice of using file numbers to reference letters. It does not matter whether an entity uses its own or its correspondent's identifier when it sends a message, as long as this convention is defined by the protocol used with the association (Fig. 5). An entity engaging simultaneously in several associations is in effect *multiplexed*, according to the terminology to be introduced in Section IV-B. Logically this entity becomes visible as a set of independent instances. Thus context identifiers are extensions to the name of an entity for multiplexing purposes. For example, Fig. 5, entities 3612 and 4230 have set up an association. The association context identifier in entity 3612 is 11. Thus messages sent by entity 4230 should carry 11 as context identifier. To that effect, the association context in entity 4230 contains 3612-11, i.e., the context identifier of the same association as seen by entity 3612. At this point it is clear that entity 4230 holds all the information it needs for labeling messages properly. The same holds true for the reverse direction. Contexts are built when an association is set up by exchanging context identifiers assigned by either entity.

Context identifiers can be assigned in the same way as *C*-names. For simplicity, such identifiers may be integers which

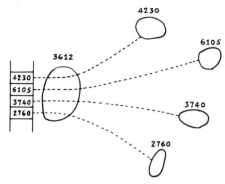

Fig. 6. C-names as context identifiers.

are assigned in monotonically increasing order. If the maximum value of the context identifier is large enough compared to the lifetime of associations and the rate at which associations are created, it should never happen that a new identifier is assigned while the same value is still assigned to an existing association context. However, this cannot always be guaranteed.

On the other hand, context identifiers might be assigned, released, and reused dynamically, but since these identifiers are shared by two processes, there may be a risk of transient conditions in which the interpretation of identifiers might become ambiguous (e.g., the crash of one computer system may cause it to lose all knowledge of previously assigned identifiers). The same problem can occur with the dynamic allocation of C-names.

In a distributed system, any dynamic allocation scheme comes with additional complexity and potential for trouble. The complexity results from the need to synchronize allocation or release. The potential for trouble derives from the vulnerability to crashes or erratic working in which the current allocation status may be corrupted. Due to the technical realities, static allocation schemes offer simpler and more robust solutions.

Assuming that entities are allocated C-names in a reliable way (e.g., by mapping), one can capitalize on this by using C-names as context identifiers (Fig. 6). Since there is no additional information, there can only be one association between two entities, however, a single entity may still set up multiple associations with other distinct entities. This restriction is not as effective as it might appear. Indeed, the real limitation is only one association for a pair of C-names. Since port names act as substitutes for entities in communication protocols, one may allocate to an entity as many ports as it needs for setting up parallel associations with another entity. The reader may wonder at this point why this subsection is titled *association names*, while only context identifiers have been discussed. Indeed a context is the only embodiment of an association, and either entity sees one facet. An association is similar to a coin, only one side is visible. Either entity is only concerned with its own side. Thus context identifiers are sufficient to name associations for the participating entities.

It may be necessary to identify associations for external entities, e.g., a monitoring system. The network-wide C-name of an association is simply the pair of its context identifiers C-names.

B. Error Control

When information is moved over some distance, it must use transmission media, such as wires, optical fibers, microwaves, satellites. Occasionally, noise, improper tuning, physical damage, or human interference, corrupt or shut off signals used for conveying messages. Thus mechanisms are necessary to detect and hopefully, recover from transmission errors. The same concern applies when transmission is effected through more complex systems, e.g., packet networks, which purport to carry out their own internal error control. End-to-end service quality may be considerably improved over that of a nonerror-controlled system, but residual errors are often encountered that can be traced to interference problems, installation changes, operation, management, hardware and software failures, etc.

Information exchanged between two entities appears as bit strings, sent and received in byte size blocks. Bit integrity may be assumed with a probability of one bit in error in 10^{10} to 10^{11} bits, which is customarily taken as satisfactory. This results from the properties of error detecting codes carried along with data [79]. Actually, these figures could be made as low as required by using any of several error detecting and correcting codes. However, there may still occur losses or duplication of blocks, which are not trapped by error detecting codes.

In order to assure block integrity, they are labeled with unique identifiers. As long as a receiver keeps track of correctly received identifiers, it can discard duplicates. Blocks received are acknowledged by returning their identifiers. After an agreed delay, called a *time out*, the sender retransmits unacknowledged blocks. This simple scheme, called Automatic Repeat Request (ARQ) is one of the most commonly used in communication protocols [8].

For simplification and efficiency in handling identifiers, blocks are numbered sequentially by the sender. This gives the receiver the ability to detect missing blocks, and possibly speed up transmission by signalling the sender, rather than waiting for time out. Multiple blocks may be acknowledged with a single acknowledgment, if the assignment of identifiers is sequential.

Cycle: Identifiers cannot be unique indefinitely, however, since they would require an ever increasing number of bits. Practically an upper limit is set after which the old identifier numbers are reused. Two different blocks or acknowledgments carry the same identifier if they are separated by one cycle, or any integral number of cycles. Obviously, an undetected error would arise if a duplicate block created within cycle N-1 were received in lieu of the proper block transmitted within cycle N. The lifetime $T \cdot ID$ of an identifier is the sum of: maximum network transit delay for a block; maximum response time for the receiver to generate an acknowledgment; maximum network transit delay for its acknowledgment.

The cycle period must be longer than $T \cdot ID$ at the maximum block sending rate [64]. As a safeguard the minimum cycle period might be chosen to be $2T \cdot ID$.

Example: Let us assume that the maximum transit delay of a transmission network is 1 s, and that the receiver response time is negligible. This would yield $T \cdot ID = 2$ s, thus the minimum cycle period is 4 s. If blocks may contain as little as 200 bits at a sending rate of 50 kbits/s, the cycle may contain up to 1000 identifiers. Thus the identifier field must be at least 10 bits. Communication paths involving satellites and high bit rates require either large identifier fields or larger block sizes [77].

Recovery: A protocol may rely on an underlying service for automatic error recovery. As will be seen in the following, the virtual terminal protocol assumes that a transport protocol at the lower level provides for virtually error free transmission. However, some errors cannot be recovered in case of major system failure. Therefore, protocols must provide for checkpoints; i.e., the possibility to backtrack up to a point in the past from which activities may be restarted safely. Taking a checkpoint involves some context saving. This must be done at each end of the communication when the two corresponding activities are in synchronized states. This aspect will be covered in a later part of the paper.

C. Flow Control

The amount of information transmitted by a particular source might exceed the capacity of the receiver or the capacity of an intermediate transmission system. A crude solution could be to discard traffic when it cannot be absorbed. This form of flow control may be used as a last resort to prevent deadlocks, but it cannot be considered as an efficient tool.

An objective of flow control is to maintain traffic within limits compatible with the amount of available resources [68] This objective could be achieved simply by putting drastic limits on senders rate, but throughput degradation would be significant. In practice, flow control is a set of policies attempting to optimize the use of the system resources, while keeping data rates close to their nominal values. The whole subject is rather complex and is still an important research topic [15], [24], [68], [70].

This paper focuses on two mechanisms commonly used as flow control building blocks. When two entities exchange information, they must be able to regulate each other's flow. Two basic mechanisms are used: stop and go, and credit.

Stop and Go: The receiver either accepts or rejects all traffic, except some short signalling messages. This technique is common with data link control procedures, such as BSC or HDLC [44]. The receiver has to take into account the transit delay necessary to carry stop and go signals to the sender. When this delay is large compared to the transmission delay of a block, a certain amount of traffic may still be flowing for a while. Also, when the receiver can resume traffic, the transit delay of a go signal translates into idle time. In short, this technique is simple in implementation, and works satisfactorily on terrestrial links. It becomes ineffectual for end-to-end flow control on satellite links, or across store-and-forward systems [77].

Credits: We assume the sender is not allowed to transmit unless he has received from the receiver an indication of the amount of traffic that it can accept. These are known as credits. The use of credits completely protects the receiver to the extent that it can set aside enough resources for the credited traffic. However, the receiver may decide to give more credits than its available resources, in the case where transit delays or sender response times are long enough to schedule additional resources in the meantime.

By mutual agreement, an error control scheme may be used as a simple flow control scheme, if acknowledgments are taken to mean credit for one or several more blocks. This is the case in HDLC [44].

Integrity: A flow control scheme might fail if signalling messages were not protected against loss and duplication: e.g., if a go signal were lost, the sender might be programmed to remain silent forever. If credits were duplicated, excessive traffic might be sent. Several techniques are used to protect the signalling messages.

1) Individual signalling messages may be acknowledged.

2) Signalling information may be embedded (piggybacked) in the control field of blocks sent in the reverse direction, when traffic is bidirectional.

3) Redundant signalling may be used: losses are automatically recovered by a subsequent message and duplicates are harmless.

The credit scheme introduced initially in the CYCLADES [65] transport protocol [82] is of the latter type. It is now commonly used in other protocols.

Interrupts: A flow control snare of concern is that it may work too well. If, on occasion a receiver stops accepting traffic for an exceptional period of time, the sender may want to send a command calling for some reinitialization or cancellation of the receiving activity. This could become impossible due to a blockage in the normal input channel at the receiver. A way around this impediment is a mechanism allowing special "interrupt" messages to bypass the flow control system in order to reach the receiver under such circumstances. This is equivalent to a secondary channel or out-of-band channel being available.

Interrupt messages may in turn generate excessive or unwarranted traffic and must be submitted to some form of flow control. Since the use of these control messages is assumed to contribute a small percentage of the total traffic, the flow control schemes adopted for interrupt messages may be rather simplistic, as throughput efficiency is not at a premium (e.g., the protocol may allow only one interrupt in transit between any two entities).

D. Synchronization

As is apparent from the previous sections, entities involved in communications must remember a certain number of parameters, or state variables, e.g., associations, message numbers, credits, delays, etc. As a whole, this information is termed protocol context. Context information is created at system generation, or as a result of explicit context setting commands, or on an incremental basis during traffic exchange. Due to transmission delays and response time of the machinery at each end of a communication, contexts evolve asynchronously. However, it is occasionally necessary to make sure that both sides of a protocol context are simultaneously in a well-defined state, e.g., at initialization, resetting checkpointing, closing, etc. This is termed synchronization. Indeed, one should keep in mind that contexts maintained at each end of an association are actually local views of the same machinery. Discrepancies between both views are due to message transit delays, and are normally transient. Putting a protocol into a quiescent state should automatically leave consistent contexts after a certain delay. Any persisting discrepancy between contexts is an error condition. Synchronization is intended to guarantee context consistency, or else to report an error.

The challenge in distributed systems is that there is no ideal observer capable of freezing each entity in a well-defined state. Only received messages can give clues about the correspondent state as it were at some point in the past. When error control context is not yet set up or has been destroyed, messages exchanged are not protected against loss and duplication. Therefore, context building must work even in adverse environment.

Some Paradoxes:

1) No state can be held totally certain. Indeed, the knowledge of the states of a distributed system is aquired via messages, which require some transit delay. There is no guarantee that states are reported correctly or remain stable. One can only assume that components keep working properly with a certain probability.

2) Certainty is not a Boolean variable. What is certainty when it is distributed? E.g., two entities A and B must start working together. Each of them steps through preliminary tests, such as:

Am I ready?
Is the other ready?
Does the other know that I am ready?

If no other entity is monitoring the situation, the decision to get started relies entirely on both A and B, but they may reach different conclusions (yes, no, perhaps), depending on their preliminary tests. Thus there may not be clear cut definition of readiness. However, it may be statistically satisfactory that either entity start working as soon as it finds both of them ready.

3) The last message cannot be critical. If it were, it had better be acknowledged, and it would no longer be the last one. But, why should one transmit a nonessential message? Some possible reasons are to improve efficiency, to introduce symmetry, to decrease uncertainty. Indeed, uncertainty may be decreased in the sense that certainty is only reached through incremental steps. Thus one more message may bring the entity into a desired state, instead of going to a recovery procedure.

This intriguing problem of uncertainty is ubiquitous in distributed systems. Not everything can ever be ascertained, but general solutions have been proposed in the literature [63] for securing some critical states.

Example: Let us take a seemingly simple example. How can two corresponding activities determine that they both have terminated? A practical situation could be mutual data base updating. Let us call the two activities A and B. One might think of the state diagram in Fig. 7.

In state 1, A is active. When finished, it sends OVER to B and goes to state 2, where it waits for OVER from B. At this point A is certain that they both are finished, and goes to state 3; i.e., vanishes.

Unfortunately, this does not always work because B may be finished before A and send OVER first. The diagram in Fig. 8 appears to be better, as it covers the two cases. It also covers a third case, when both A and B are finished first! Or so they think. Then both A and B send OVER and go to state 3. Both receive OVER from the other, and they both are finished. As long as every message sent is correctly delivered, the diagram in Fig. 8 is foolproof; but this assumption can only be made if we can rely on an underlying transport service that is 100 percent reliable (see Section V-B).

On the other hand, it is instructive to assume the underlying transport service is not 100 percent reliable and that, on occasion, messages may be lost, duplicated, or delivered out of sequence. In that case, the diagram in Fig. 8 is not

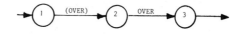

(OVER) : message sent   OVER : message received

Fig. 7. Partial synchronization.

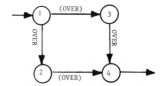

Fig. 8. Simple synchronization.

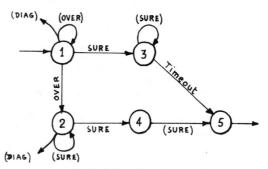

Fig. 9. Safe synchronization.

guaranteed to work either because A may get stalled in state 3, if the OVER message from B has been lost. What could A do? Going to state 4 and vanish is unsafe, since B might have not yet reached state 1 after all and it might need A's cooperation to complete its task. On the other hand, going back to state 1 and repeating OVER would not help, if B had already gone to state 4 and had vanished.

A possible solution is in Fig. 9. No activity can vanish until the other has reached state 2. In case one activity gets hung up in state 2, it knows that the other has at least reached state 1, and possibly state 5. If any activity is hung up in state 1 or 2, it initializes a diagnostic procedure, which is specific to the environment. If there is no remedy, they may be uncertain about the other's state, but both of them have achieved their task completely.

This example should not be taken as a paragon for all synchronization schemes. It only intends to illustrate the nature of the difficulties encountered and of possible solutions.

Guidelines: The complexity of synchronization introduces added logic and delays (message transit delays and timeouts). One should not introduce more complexity into a system than is strictly necessary. Before designing a scheme, one should state explicitly:

What is to be achieved?
What is to be prevented?
What interferences are expected or allowed from
  the environment?

When the same set of operations is repeated, e.g., context initialization, it may occur that messages of several successive instances get intermixed due to duplicates or variable transit delays. Additional protection is required for discriminating messages pertaining to each instance. Practical schemes introduce quiescent states or numbering cycles long enough for all stranded messages to be discarded naturally [32]. Nontrivial solutions have been proposed in the literature [20], [76]. They in turn are sufficiently complex and constraining to create reliability problems. Simple and provable solutions are yet to be discovered.

The design of safe synchronization schemes is far from trivial. A typical approach consists of representing the logic at each end with state diagrams, and then analyzing all possible conditions leading to stable states. Usual pitfalls are loops, deadlocks, or inconsistent stable states. This is illustrated in the literature [4]. It turns out that even simple diagrams are difficult to analyze exhaustively by hand due to the multiplicity of abnormal conditions. Automated validation techniques are now becoming available [25].

IV. DISTRIBUTED SYSTEM ARCHITECTURE

A. Motivation

Within human organizations as well as industrial processes, information is naturally distributed. To date, cost effectiveness considerations have generally led to centralized automatic data processing in order to take advantage of economies of scale and to minimize communications costs. The picture is drastically changing with the advent of microelectronics and packet switching. Technology now makes it possible to fragment data storage and processing capability according to the natural structure of human or industrial organizations and for information to flow along its natural paths within the structure (Fig. 10).

But technology alone is not enough, and a distributed system must be organized and structured in such a way that its different parts can cooperate efficiently and the whole system can evolve according to the evolution of the organization. This is the purpose of the system *architecture*, the organizational chart of a distributed system, which often reflects the organization of the parent enterprise itself (Fig. 11).

Up to this point, we have described a distributed system in terms of the relations (protocols and interfaces) between its various parts. There are various ways to structure a distributed system to achieve these relations. In this section, we emphasize the relational aspects of a distributed system and review the basic structuring techniques. By the term *reference architecture* we refer to the set of relations (and their definitions) to which the various parts of a distributed system must conform. Each element of the system can only see an external manifestation of these relations. It cannot see their internal implementation or structure within the other elements. We refer to this filtered view of the system as its *visibility* relative to that part, or simply as its visibility. In this section, we introduce the concept of visibility in a reference architecture and its application.

A central aspect of distributed systems is the structure of interactions between logical entities, just as the structure of relations between divisions, departments, and individuals is central to an organization.

Interactions between logical entities make sense only if they contribute to a common objective. A protocol defines for each entity how it must interact with the other entities in order to achieve its objective. Commonly, the name of a

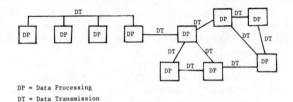

DP = Data Processing
DT = Data Transmission

Fig. 10. Physical structure of distributed systems.

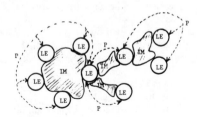

LE = logical Entity
IM = Interaction Means
P = Protocol

Fig. 11. Logical structure of distributed systems.

protocol is matched to the function carried out by interacting entities (e.g., transport protocol, file transfer protocol).

B. Structuring Techniques

Almost any distributed system architecture is based on a small set of simple structuring techniques, viz. multiplexing, switching, wrapping, cascading, and assembling, which are described in this portion of the paper.

Multiplexing: In most cases, resources in distributed systems must be shared between different activities. For instance, in a packet switching network, lines, node processors, and memory are shared between all users. Rather than requesting all activities to coordinate themselves directly when sharing resources, the network can provide statistical multiplexing mechanisms which take care of allocating resources to activities dynamically when they need them. (The allocation itself can be organized as a distributed activity.) This technique, already used in any multiprogrammed system permits activities to proceed as if they were independent from each other (see Fig. 12). Each activity interacts only with the network and its multiplexing mechanisms.

Switching: Interaction is usually restricted to pairs of entities. More generally, however, the action of one entity may be directed to a subset of the entities with which it may interact. Therefore, each action must be associated with a switching parameter which can be interpreted by the underlying mediating system (interaction means), in order to route the action to the proper destination(s) (see Fig. 13). In most cases, the action is directed to one entity only, i.e., each entity must be associated with one address. Other types of switching may also be used such as "all entities" (broadcast) or "any entities with such and such characteristics" (associative switching). Switching permits us to define independently interactions within subsets of entities [54].

Wrapping: A pair of entities interacting according to a specific protocol may be used to provide other entities with a more powerful interaction means (see Fig. 14). When applied recurrently, a hierarchical structure is obtained and we use the term wrapping to indicate how it was created

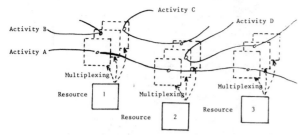

Fig. 12. Multiplexing resources between activities.

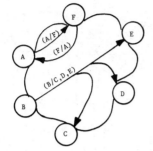

Fig. 13. Addressing actions.

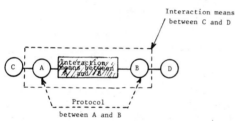

Fig. 14. Wrapping assembly.

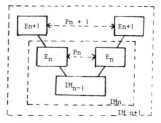

Fig. 15. Hierarchical assembly.

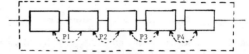

Fig. 16. Cascading assembly.

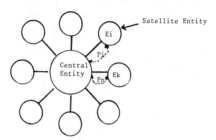

Fig. 17. Star assembly.

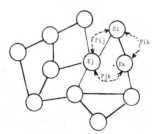

Fig. 18. Meshed assembly.

(see Fig. 15). The essential feature of wrapping is that each protocol layer communicates with its counterpart at the other end of the interaction means. Of course switching and multiplexing techniques can be used in conjunction with wrapping in order to provide interaction means between any pair of higher level entities, as well as alternate paths between them.

Cascading: Cascading consists of forming a linear string of entities, each one interacting only with its neighbors. The resulting cascade can in turn be used as an interaction means (see Fig. 16). A common form of cascade is found between successive nodes in a packet switching network where all protocols between neighbors are identical in the recurring cascade. Again switching and multiplexing techniques can be used in conjunction with cascading in order to form multi-path networks.

Assembling:

Star assembly: Entities can be assembled into a star configuration where a central entity interacts individually with satellite entities (see Fig. 17). There is no direct interaction between satellites entities. The protocol of the star is decomposed into protocols (usually identical) between each satellite and the central entity, which also performs the coupling of all these protocols.

Meshed assembly: Entities can be assembled into a meshed network where each entity interacts with its neighbors (see Fig. 18). The protocol of the meshed network is defined by protocols (usually identical) between neighbors and by their coupling inside each entity. This assembling technique is frequently used to organize route adaptation as a distributed activity in packet-switching networks.

C. Reference Architecture and Visibility

As explained in Section IV-A, defining the architecture of a distributed system consists essentially of specifying interactions between logical entities which compose the system, i.e., their protocols. The design and implementation of any specific entity is only constrained to behave externally according to the system protocols. Any other functional or technical aspects of its design or implementation are independent choices which will not be explicitly visible to the other entities, except perhaps as performance variations. This essential characteristic of a reference architecture is its visibility to the other entities. The architecture serves as a reference for relations between entities.

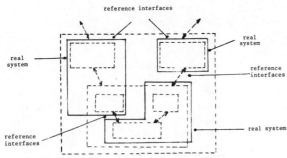

Fig. 19. Reference architecture and reference interfaces.

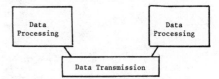

Fig. 20. Data processing wraps data transmission.

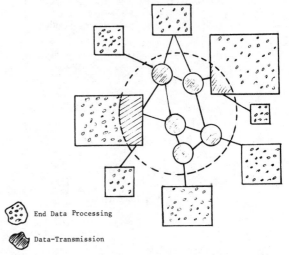

End Data Processing

Data-Transmission

Fig. 21. Border between data transmission and end data processing.

Visibility is essential to the architecture of heterogeneous networks [84], [85]. It allows systems from different manufacturers to interwork according to common rules, even though they may have different internal structures.

Protocols between systems are exercised only at physical or logical interfaces between them. The only points at which the physical embodiment of the protocols are visible are at these interfaces. The architecture must include the definition of relations through such interfaces.

These interfaces (or visibility points) must be properly located within each equipment for them to serve their original purpose. For example, if one manufacturer's equipment is to be replaced by another's, the physical interface to that equipment should be identical with a visibility point. The interfaces at which a system conforms with the architecture are known as *reference interfaces* (see Fig. 19). The reference interfaces occur only at the boundaries of protocol layers and, when used in conjunction with the distributed system protocols, form the visibility points of the reference architecture.

D. Layered Architecture

The concept of visibility can in turn be applied recursively to the reference architecture itself in order to provide maximum flexibility in its own evolution; when a set of entities are assembled using wrapping, cascading, or another technique. The resulting entity can then be considered as a whole; i.e., without any assumption on its internal structure. This entity will in turn be assembled with other entities into a higher level entity, etc.

The resulting architecture is often referred to as "layered architecture;" at each step of its assembly process a new higher layer is added which views the former (lower) layers as a single entity without any assumption about its internal structure. Under this assumption, successive layers are said to be independent. This technique is very similar to the one used in structured programming where each module is defined functionally, as seen from outside, without regard to its internal structure.

E. Computer Network Architecture

The concepts and techniques developed in the previous sections are applicable to any distributed system (e.g., small network of identical microprocessors, an organization, as well as a large heterogeneous resource sharing computer network). In the specific case of computer networks, the constraint of preexisting computers and transmission systems must be taken into account. Most computer networks make use of the same basic reference architecture with three major

protocol layers: data-transmission, end-to-end transport control, and application control [2], [7], [16], [19], [26], [53], [55], [58], [65], [71].

Data Transmission: Data transmission facilities are often provided by common carriers or PTT's. These are physical systems which will interact with data processing systems in order to let them communicate. The reference architecture for computer networks must therefore identify a data-transmission entity used as an interaction means between data-processing entities (Fig. 20).

The data-transmission part of the network may also use computers internally, as in packet switching, but these computers act as intermediate nodes (i.e., they store and forward packets to the next node).

In the reference architecture, the data-transmission entity ends where the end data-processing systems begin. In case a computer is used both as an end data-processing system and as an intermediate node for other data-processing systems, the border between data-transmission and data-processing is located inside the computer itself (see Fig. 21).

End-to-End Transport Control: In most computer systems, several independent processes run simultaneously. In that environment, they happen to share local resources. Cooperation within a distributed network takes place between individual processes rather than between systems as a whole. The network conventions must cater to interprocess communication (i.e., define how processes access each other and how information exchanged between processes is passed from one computer to another through the data-transmission system, how errors are detected and corrected, etc.).

The reference architecture must therefore include a transport layer, on top of the data-transmission entity. Thus each

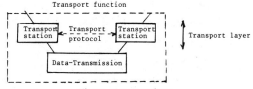

Fig. 22. Transport layer.

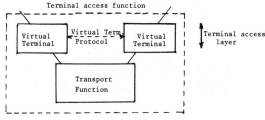

Fig. 23. Terminal access layer.

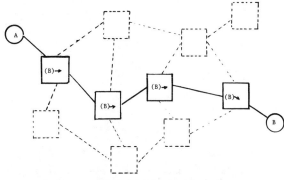

Fig. 24. Cascade of nodes built by routing tables.

end data-processing system will include a transport control entity, often referred to as a transport station. Transport stations interact through the data-transmission entity, according to an end-to-end transport protocol, (see Fig. 22). This assembling is of the "wrapping" type.

The rationale for identifying this transport component is that its function is not specific to one application but can be used by any of them. In other words, all kinds of applications can share the transport components. They add their own sets of conventions (or protocols) to define the meaning as well as the format of data they exchange through the transport function.

As long as the transport layer has not been defined and built, a network cannot exist, since processes cannot communicate, but the transport layer in itself is not sufficient. It just allows resources to get connected and must be augmented with conventions on the use of resources. This type of convention might be referred to as application control.

Application Control: It is clear that conventions for the use of all resources within a network need not, and cannot, be defined and implemented at one time. It is also obvious that some resources are not widely shared, at least at the beginning of the network life. It can be observed that the pressure from users for an application control protocol to be defined is strictly related to the actual amount of sharing of the corresponding resource.

It seems to be a general rule today that the initial requirement on any large network is for a set of conventions allowing people to use the same terminal to access a variety of services. Thus the next layer in the basic reference architecture (above the transport layer) is a terminal layer, which provides higher layers (application programs and human users) with a terminal oriented communication facility (Fig. 23). Terminal-like entities interacting through the transport entity are often referred to as virtual terminals while the protocol between them is termed a virtual terminal protocol.

Later on, when the users start using the network, they are gradually led to compare equivalent services provided by different computer systems and tempted to use the best of each. The pressure then builds up for further compatibility:
1) Users want to move their data from one computer system to another (e.g., they may want to use a text editor on one sys-

tem to prepare a program and ship it to another system for execution). This leads to define extensions of the basic architecture to include file transfer or more generally file management.

2) Users are frustrated with meaningless differences between control languages of each system (e.g., different log-in procedures) and this leads to standardizing the most widely shared procedures between human users and data processing systems.

As sharing develops, extensions to the reference architecture and corresponding high-level protocols will be defined (e.g., for distributed data-base management, etc.). High-level protocols are addressed in the paper by Sproul and Cohen in this issue [72].

V. The Design of Protocols

This chapter attempts to characterize protocols in the various layers of the basic reference architecture. We develop one particular model of network operation for ease in exposition.

A. Data Transmission

Many kinds of data-transmission facilities are used for computer networks (telephone lines, satellite links, etc.) and the most widely used computer networks nowadays are based on packet switching. A packet switching network is basically a meshed assembly of minicomputers, called nodes, interacting through transmission lines. The major activity of a packet switching network is to transmit packets from a source to a destination. Routing tables in each node indicate which line is the next step to each destination. In other words, routing tables determine for each source-destination pair which cascade of nodes will be involved (see Fig. 24). Each node is shared (multiprogrammed) between a number of entities (processes); therefore, the various cascades may be considered to be independent.

Along a cascade, packets are transmitted from node to node (store and forward), using some line control procedure. Route adaptation is usually organized as a meshed-assembly with one entity in each node (see Fig. 25). Neighboring entities exchange information they have on the state of the network components (lines and nodes). Each entity integrates the information from its neighbors and passes the result to all neighbors.

This meshed distributed activity is responsible for determining best routes and updating routing tables accordingly; thus, reducing the problem of reliable packet switching to "packet forwarding" along dynamically predetermined cascades. A similar propagation mechanism may also be used for conges-

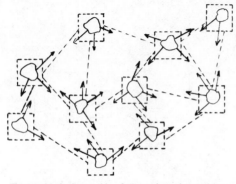

Fig. 25. Meshed assembly of route adaptation entities.

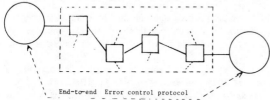

Fig. 26. End-to-end error control performed on top of a cascade.

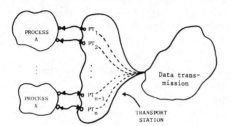

Fig. 27. Port names and process names.

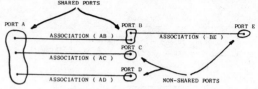

Fig. 28. Ports and association.

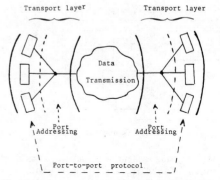

Fig. 29. Port-to-port protocol on top of addressing.

tion control as in CIGALE [66], [68]. If a node or a line fails, route adaptation will immediately determine a new route, but some packets may have been lost. Error detection and recovery is usually performed end to end on top of the cascade (see Fig. 26), either within the data-transmission layer [10] or in the transport layer [2], [26], [28], [39], [58], [65], [75], [82].

Broadcast packet networks exhibit another kind of distributed organization [1], [29], [58]. See the papers by Kahn *et al.* and Jacobs *et al.* in this issue for a further discussion of packet radio and packet satellite systems [45], [46].

B. Transport Protocol

The transport layer must provide the higher level protocols with a reliable interprocess communication facility (see Section IV-C).

The transport protocol comprises a set of tools, along with operating procedures, which allow processes to set up associations, and transfer blocks of information from one process domain to another's.

Naming—Ports and Liaisons: In order to communicate processes must be given *C*-names by the transport layer. A most common scheme is to use hierarchical port names composed of a transport station number and a local port number.

⟨Port Name⟩ :: = ⟨Transport Station #⟩⟨Local Port #⟩.

In each system, processes/resources are mapped locally into port names (see Fig. 27).

When two processes engage in a conversation, an association between ports called liaison must be identified. The simplest way to identify a liaison is to use a pair of port names [13], [26], [28], [82]. This allows a process to participate simultaneously in several associations (see Fig. 28). Another method [9] consists of restricting ports to a single association and thus uses a single-port name to identify the association. A contention problem arises when several processes want to access the same (sharable) resource. Additional protocol machinery [4] must then be provided [31], [36], [61].

It is generally agreed that switching should form the basic sublayer in the transport layer, with error control, flow control, etc., on top (see Fig. 29). This permits the transport protocol to be decomposed into protocols between pairs of ports (see Section III-A) and reduces its complexity. In this case, each packet is related to an association and handled at each end in the context of this association. For transaction-oriented applications, the association between ports may be restricted to the exchange of one message only. It is then simply identified with the pair of port names. This type of association has been included in [13], [28], [42], [82]. It is often referred to as Lettergram or Datagram facility.

Messages—Letters and Telegrams: The transport layer usually provides processes with transportation of two kinds of messages [13], [14], [28], [42], [82]. 1) variable length blocks, often called letters, intended for regular exchange of data [82]. 2) Small pieces of data (e.g., 8 bits), often called telegrams intended to transfer interrupt-like signals, and thus not submitted to flow control on letters.

Letters may be longer than packets accepted by the data-transmission facility, and thus, transport protocols can include fragmentation of letters into packets and their subsequent reassembly upon arrival in the transport station of the destination data-processing system. Fragments are usually sequentially numbered within the letter [82] (i.e., fragments refer

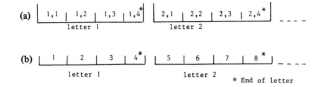

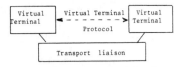

215

Fig. 33. Protocol between a pair of virtual terminals.

Pouzin and Zimmerman: A Tutorial on Protocols

Fig. 30. Fragments numbering. (a) Fragments numbering within letters.
(b) Fragments numbering within a liaison.

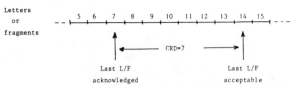

Fig. 31. Flow control with credit.

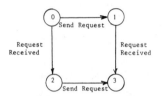

Fig. 32. Symmetric rendez-vouz synchronization.

to letters), or within the liaison (i.e., letters refer to fragments) as in [14] (see Fig. 30).

Error Control: End-to-end error control on lettergrams and telegrams is usually based on individual acknowledgments, while liaisons make use of more refined schemes. In any case, pieces of information must be uniquely identified with cyclically reused names. It is often considered that the (0, 1) numbering is sufficient for Telegrams, while letters or fragments require a longer cycle, e.g., 16 bits [13], [14].

Flow Control: Most transport protocols control flow with a credit scheme. Credits indicate the number of letters [42], [82], or fragments [14] which the receiver is prepared to accept beyond the last acknowledged letter or fragment (see Fig. 31). This scheme is preferred to a stop and go scheme which is ineffectual across packet switching nets because of the round-trip delay. Another advantage of the credit scheme is that the corresponding control information (ACK #, Credit) is self-contained; i.e., it does not refer to previous control information. The information in each new (ACK #, Credit) message replaces any preceding information, thus providing automatic error control in case of loss or duplication [68].

Another scheme used in [9], where each flow control message refers to the preceding ones (incremental credits) has proved to cause deadlocks in case of loss.

Synchronization: Liaisons must be initialized, possibly reinitialized, and terminated. To this effect, symmetric protocols are often used (see Fig. 32) in order to avoid contention when both ends happen to initiate synchronization at the same time.

C. Virtual Terminal Control

The virtual terminal layer provides higher layers in the basic reference architecture, with a terminal oriented communication facility [23], [27], [41], [59], [81]. The virtual termi-

nal protocol defines a set of procedures and messages formats which may be interpreted for controlling a large variety of real terminals.

Naming: Terminal oriented communications take place between pairs of correspondents (terminal-program, terminal-terminal, or program-program). It can thus be nested in the protocol of an association, e.g., a liaison. Hence, the virtual terminal protocol can use ports and liaisons names provided by the transport layer. This permits the virtual terminal protocol to be defined as a protocol between a pair of virtual terminals (see Fig. 33).

Error Control: Virtual terminal belongs to end data-processing systems, where transport stations are also located and the virtual terminal layer relies upon end-to-end transport control performed by the transport layer.

Data Representation: Terminal-oriented communications are based on text representation, i.e., a data structure made of pages, lines, and character positions which may be filled with characters. The data structure may be more complex and be associated with access control functions (e.g., protected zones on data-entry terminals). The virtual terminal protocol must incorporate conventions for data representation and access to the data structure, such as character code, addressing, etc.

Dialogue Control: The data structure, e.g. for display on the terminal, can be accessed by both ends. The virtual terminal protocol must therefore comprise mechanisms to coordinate accesses from both correspondents. A widely used scenario consists of using some sort of semaphore representing the right to write in the data-structure, usually called the "turn," which is passed to the other end when the owner of the turn has finished writing. This type of dialogue is often referred to as "alternating dialogue."

Access to the data-structure is normally done on the basis of messages [27], [41], [59], [81], rather than characters [23]. This reflects the evolution of terminals from a character at a time to message mode, the former mode being totally inadequate on store and forward networks.

Synchronization: Most virtual terminal protocols include some kind of purge mechanism which permits the data-structure to be reinitialized without reinitializing the whole protocol (equivalent to a clear display operation) [41], [59]. This type of mechanism uses telegrams to initiate the operation even if letters are blocked due to flow control in the transport layer.

Negotiation of Options: A characteristic of terminals is the variety of functions they may provide. For this reason, any virtual terminal protocol includes a negotiation mechanism which permits both correspondents to agree on the virtual terminal characteristics which they are going to use [41], [59]. In most virtual terminal protocols conventions on default characteristics permit negotiation to be avoided in simple cases.

D. File Transfer/Management Protocol

The file transfer protocol should provide a set of tools intended for handling complete files across computer systems.

Example of functions to be performed are: finding, copying, deleting, renaming, or creating.

The potential complexity of file transfer/management is due to heterogeneity of file names, structure, access procedures, and data representation. Examples of such protocols can be found in [34], [60].

Another characteristic of file transfer protocols is that they must often include their own error recovery procedures on top of the transport layer. The reason for this is that file transfer is an automatic process which cannot rely on the intervention of a human operator. In addition, the duration of a file transfer may be such that the probability of failure of one computer system is not negligible. The recovery procedure is based on periodical check-pointing involving end-to-end synchronization (see Section III-D). In case of failure, the transfer is restarted from the last check-point in the transfer.

VI. PROTOCOL AND PERFORMANCE INTERPLAY

Good systems, like good meals, do not come out simply by putting together good ingredients. Choosing a proper set of tools and parameters depends on a number of factors.

1) Efficiency—What is to be optimized? Bandwidth, throughput, transit delay?

2) Complexity—Do we want to reduce implementation costs, core size, execution time?

3) Service quality—What error rates, delay variations, manual interventions, are tolerable?

4) Service parameters—What are the requirements for message length, traffic density, response time, number of terminals?

5) Transmission parameters—What are line speeds, error rates, propagation delays?

6) Overall architecture—What is the proper distribution of functions within system layers?

These are just a sampling of considerations which must be taken into account in the design of a network architecture. In practice, a number of parameters or system components are given as constraints. In a sense, this is usually useful, as it narrows the set of options, and often simplifies design decisions.

The reader should not expect to find in this paper off-the-shelf recipes for building protocols. Rather, we shall attempt to point out how typical protocol ingredients interplay with one another, and how they relate to some of the factors mentioned earlier.

A. Acknowledgments

Simple tranmission protocols use an alternation scheme:

Sender	Receiver
Send block $\longrightarrow$	Receive block
Receive ACK $\longleftarrow$	Send ACK
etc.	etc.

During the acknowledgment round-trip time, the sender is blocked. If this time is relatively very short, throughput degradation may be negligible. This is the case with unidirectional data flows on terrestrial circuits. However, data flows are also bidirectional. ACK's compete with data blocks in the reverse direction. Even if ACK's are given priority, they must wait for the end of transmission of the current block, Hence, delayed ACK's translate into lost bandwidth. How much is lost depends on the characteristics of the reverse traffic, but reverse traffic also needs ACK's which are again delayed by competing data blocks. Finally, simple analysis

Fig. 34. Bidirectional traffic.

shows that both traffics become paced with one another [30], as shown in Fig. 34. Block B_i is acknowledged by A_i, while block B'_i, is acknowledged by A'_i.

A remedy could be separate acknowledgment channels. This would decouple the two opposite data flows, at the cost of added hardware, software, and bandwidth. Some communication systems use this scheme in the form of bidirectional modems, which provide a low bandwidth return channel.

A more frequently used technique for improving throughput is to use a protocol allowing transmission of a number of blocks before waiting for ACK. This is termed anticipation, and is used in the more refined error control schemes.

Sender	Receiver
Send block B1 $\longrightarrow$	Receive block B1
Send block B2 $\longrightarrow$	Receive block B2
Send block B3 $\longrightarrow$	Receive block B3
Send block B4 $\longrightarrow$	Receive block B4
Receive ACK A4 $\longleftarrow$	Send ACK A4.

Blocks are numbered, and acknowledging block B4 implies acknowledgments of all previously sent blocks. There is a significant gain in bandwidth at the cost of larger buffering on the sender side. Indeed the sender must keep copies of all unacknowledged blocks in order to be able to send them again if they have not been received correctly.

There is a limit to the number of blocks that can be queued, waiting for ACK's. Sender and receiver may agree on a maximum, but in any case it must be lower than the numbering cycle to avoid ambiguities.

Long blocks transmitted in one direction may still delay ACK's to the point of exhausting the maximum amount of block numbers in the other direction when the blocks are short [22]. A countermeasure could be to increase the numbering cycle, but more buffers would be required for unacknowledged blocks.

Piggybacking ACK's (i.e., embedding ACK's in the control field of data blocks in the reverse direction) yields additional throughput, as it saves the overhead of sending distinct ACK messages. However, the disadvantage is increased ACK delay, since the ACK field cannot be processed until a complete block has been received and validated. This effect is more pronounced when block sizes on each direction are heavily unbalanced [33], [37]. It has also been proposed to place ACK's in a trailer rather than in the header of a block [37].

B. Retransmission

Transmission errors are usually corrected by retransmission, Strategies may be selective or sequential [8]. Selective retransmission is executed on receiver's requests for individual blocks. The advantage is to keep retransmission to a strict minimum. On the other hand, the receiver must manage blocks pending predecessors and deliver them to processing in a correct sequence. This strategy is suitable with random errors when sequential retransmission is too costly. Satellite transmission paths fall into this category [77].

Sequential retransmission may be executed on either side with requests starting back from the first erroneous block and proceeding sequentially. Due to the delay introduced in triggering retransmission, a number of blocks following an erroneous block may have been correctly received, but will be retransmitted. An advantage of sequential retransmission is that it is straightforward to implement. It is suited to bursty errors, which may effect several blocks in a row, when ACK round-trip delay is small. If round-trip delay and numbering cycles are large, a substantial amount of correctly received traffic may have to be retransmitted. This would be acceptable with low block error rates say less than 10^{-5}. Otherwise, lost bandwidth is important. This case applies to terrestrial lines (very short propagation delay) and also to letters exchanged over a liaison across a packet network (very low error rate). Broadcast channel (radio or satellite) utilization depends critically on retransmission strategies. The simplest one is the ALOHA [1], [3] scheme, in which colliding blocks are retransmitted after a random delay. The maximum utilization of the channel is 18 percent. A number of more sophisticated schemes have been analyzed, and new ones are generated every year [7]. Under a number of assumptions, and at the cost of increasing protocol complexity, maximum channel utilization may reach or exceed 80 percent [48], [49].

C. Flow Control

Traffic will be lost if no buffer is available to accept incoming blocks. When the data transmission system includes intermediate storage, a receiver may defer accepting traffic for a certain period of time. However, it is vulnerable to network countermeasure, such as the discard of traffic. Thus the question is: How many buffers must be set aside for the expected traffic? This is a classical problem in real-time systems design. A number of books cover this subject [57]. The system designer has access to analytic models based on queuing theory, simulation, and measurement of live traffic. Assuming that hypotheses about real traffic are valid, results are customarily histograms relating load factors and probability of traffic overflow. Unless buffers, table entries, CPU cycles, are overallocated, chances are not totally negligible that occasionally traffic will have to be delayed or discarded. This is acceptable if retransmission is possible, as long as the percentage of bandwidth used up by retransmission remains low enough. Here the tradeoff is between buffer and bandwidth.

A second question is: how to control the sender's flow? As seen earlier, the use of stop and go mechanism is simple, because it requires little or no strategy. However, stop-and-go messages have to compete with data blocks on two-way channels, hence they may be delayed. Store and forward transmission systems add even more delay between sender and receiver. An evaluation should be made of the probability of excess arrival, or idle time. This is related to block rate, block size, and transit delay characteristics. Occasional loss of bandwidth may be acceptable, up to a point.

Better control is offered with credit schemes in the sense that it is possible to restrict senders to the exact amount of buffer space available at the receivers end. However, in networks comprising a very large number of low activity terminals, the probability of a terminal transmitting a block is small, hence buffers would be mostly squandered. In such conditions it is customary to overbook buffers, as seats on airplanes. As a result, overflow may occur with certain probability. Flow control is still useful, but only as a second line of defense.

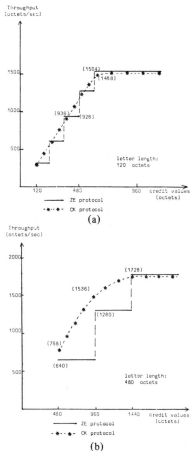

Fig. 35. Flow control and throughput.

Another situation is high volume traffic between computers, such as in the case of file transfer for batch processing. Traffic loss or idle time would degrade service performance. There the question is: What credit policy would assure a certain throughput rate? Experience is still scarce in this area. Modeling and simulation provide interesting insights, and may help tuning flow control parameters in a specific network. It is shown that throughput increases almost linearly with credit values, up to a maximum determined by the block size (see Fig. 35, excerpted from [50]). In this figure, the curve ZE refers to the protocol by Zimmerman and Elie in [82] and CK refers to the protocol by Cerf and Kahn described in [14].

From these results it might be tempting to increase the block size by multiplexing blocks of several parallel data streams into larger blocks. Results from simulation show that the effect is a decrease of throughput of each data stream up to a cross point corresponding to large credit values (see Fig. 36 excerpted from [50]).

D. Protocol Overhead

The influence of the major protocol functions has been covered in the previous sections. There remains an odd lot of considerations that do not come easily under meaningful headings. We lump them together under the term overhead.

When defining message formats one is faced with the choice of fixed versus variable fields. Fixed formats mean economic

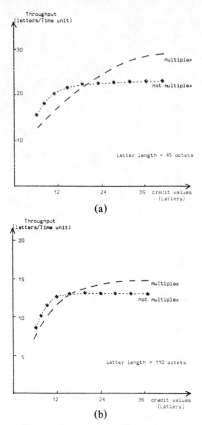

(a)

(b)

Fig. 36. Multiplexing and throughput.

Error Rate: Performance may change only slightly over large variations in error rate. However, there are thresholds past which degradation is drastic. Quite often the threshold may be adjusted in varying some protocol parameters: block length, anticipation window, retransmission policy, acknowledgment policy. If this is not sufficient, it may be necessary to use a different protocol, e.g., forward error correction instead of or in addition to retransmission if the ratio of retransmitted packets exceeds say 10 percent of all transmitted packets.

Transit Delay: Delay is a major factor in tuning properly error and flow control schemes for getting maximum throughput. In particular, large transient variations may be more detrimental than a long but stable delay.

Protocols require that some assumptions be made about maximum transit delay. If these assumptions are not valid, performance may be degraded, or worse, undetected errors may occur if the transit delay exceeds the cycle over which the error control identifiers apply.

Blocks Out of Sequence: A block getting out of sequence by one or a few positions is rarely a burden, when this is anticipated in the protocol. It would be a different matter if some blocks could be out of sequence by, say, 100 or more positions, as this would require considerable buffering to be available at the final destination.

Duplicates: Duplicates arise due to retransmission and create no hardship when they are anticipated and the number of them remains below a few percent. However, the coupling of duplicates with excessive transit delays may be disastrous, since duplicates could appear to be valid messages.

Traffic Patterns: Although one of the least known aspects, it has been clearly recognized that performance is strongly dependent on traffic characteristics: message length, distribution, message rates, competing traffic.

Resource Management: This is the dual of traffic patterns, since any message in transit ties up some resources. It is certainly the most critical area in protocol performance [68].

VII. STANDARDIZATION

A. The Rationale for a Standard

Standardization means agreement on reducing the number of ways of doing the same thing. Diversity may be justified when it stems from substantially different constraints or objectives. Nevertheless, it is fair to recognize that in the computing field diversity is often the result of adherence to local or individual standards. A considerable proportion of human skill and financial resources are devoted to reinventing the wheel and converting from one standard to another. Indeed, not everybody is in favor of standardization, unless the standards are theirs. Hence, we observe a standards war; PL/1 against Cobol, EBCDIC against ASCII, X.25 against HDLC, etc. Thus the term standard should not be taken at face value. In practical terms, standardization does not eliminate diversity and never will. It is only a patient trimming process weeding out excess diversity.

With the policy of limited diversity comes a number of advantages.

1) Techniques may be better understood and improvements can be concentrated on smaller set.

2) Engineering and maintenance costs may be reduced.

3) Individual suppliers and products are likely to support a wide range of these techniques and thus the choice of available systems may be greatly increased for many applications.

parsing, but also unused bits. There is also the tradeoff between the number of different message headers, and the number of control fields within each header. Choices should be based on assumptions about the percentage and average size of each type of message in the total traffic. Measurements of ARPANET pointed out that the user bits going end to end above the transport protocol (NCP) [9] represented 24 percent of all the bits carried through the packet net [47] not counting internal service traffic. The remaining 76 percent were headers, trailers, ACK's, etc. Indeed, a large proportion of real traffic is made of one-character packets, which results from a characteristic of some of the hosts on the network rather than the network itself.

The latter example illustrates perhaps undesirable aspect of layered systems. When layers are carefully designed as independent system components, each one introduces its own layer of overhead without regard to what happens elsewhere. The overall combined overhead can be quite large.

E. Protocol Sensitivity

In various places we have mentioned interplays between protocols and environmental factors. A summary of the most common interplays is introduced here for easier reference.

Error Type: A protocol can only cope with errors for which it has been designed. For example, X.25, a standard interface for public packet networks [11], [12], contains no mechanism for the detection of errors generated by the packet network. It relies entirely on error signalling from the network.

There are also counterarguments.

1) Individual techniques are often tailored to an application and too much generality may be more expensive.

2) Suppliers may not favor increased compatibility when they control a certain market.

3) Past investment may rule out major changes.

4) Innovation could be stifled.

Thus standardization is a subtle game; what techniques should be standardized and when?

C. Major Existing Standards

A complete review of existing standards may be found in the literature [18]. Here we shall only focus on a few major standards that are most relevant from a data processing and data communication standpoint.

CCITT-V24 (alias RS 232) Modem Interface for telephone circuits up to 20 kbits/s. This is the most widely used interface for data transmission. It is data and procedure transparent, but without capability for switched circuit signalling.

CCITT-X.21 Interface for switched digital circuits. It covers the whole range of speeds, from 600 bits/s up to megabits per second range, although there are few switched digital networks which operate at higher speeds. It is completely transparent to data and procedures. The dialing phase (connection setup) is based on electrical signalling, rather than control messages. This is a definite shortcoming due to an early design, at a time when telephone administrations had not yet anticipated packet switching.

CCITT-X.25 Interface for public packet networks [11], [12]. It works on the principle of virtual circuits carrying packets in sequence, and provides for cascading flow control between source and destination. It will presumably be gradually adopted by equipment manufacturers as an alternative to switched circuit or multipoint line interfaces.

There is, however, some dissatisfaction about the complexity of this interface, which was drafted up in a remarkable hurry. Various design flaws and ways to interpret the standard are still matters to be resolved [4]. A major deficiency, especially for the level of complexity of X.25, is the lack of mechanism assuring end-to-end integrity of data flows. This will likely prevent X.25 from becoming the workhorse of data transmission, except where an end-to-end protocol is also provided.

CCITT-X.3, X.28, X.29 (alias PAD) [12]. These standards are a package intended for the remote handling of asynchronous ASCII terminals across public packet networks. X.28 is the interface with the terminal and its human user, X.29 is the interface between the remote handler and an application program (or an access method) located in a computer, X.29 requires the use of X.25.

The Packet Assembler Disassembler (PAD) does not fit within the reference architecture introduced earlier. Rather, it appears to be derived from a centralized network structure (Fig. 37), in which a physical circuit has been replaced with a virtual one. In this context, the PAD is equivalent to a remote terminal handler in a star network (Fig. 38).

ISO-HDLC: High-level Data Link Control (HDLC) [43], [44], is a data link control procedure worked out within the International Standard Organization (ISO) [43], [44]. It was designed primarily for multipoint circuits, at a time when packet switching was not anticipated. Since then it was revamped for handling quasi-symmetric point-to-point circuits.

Although HDLC contained most of the capabilities desirable for virtual circuit interface, it was not chosen by CCITT.

Fig. 37. Central handler.

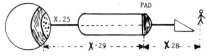

Fig. 38. Remote handler.

It only appears at the link level (physical line) of X.25, while the packet level (virtual circuit) is controlled by a similar but different procedure.

Presently, studies are underway for the definition of a public data network interface (Frame Mode DTE) using HDLC.

C. Future Standard Development

On the computer network scene, early initiatives have been taken by groups of experimenters for reaching some consensus on internetwork protocols. Gradually, these pre-standards seep into the industrial world, and may become a basis for future standards. Most of this groundwork is carried out within research communities developing around experimental computer networks (ARPANET [71], CYCLADES [65], EIN [2]). Their international forum is the Working Group WG 6.1 of the International Federation of Information Processing (IFIP). This group has produced proposals for several basic layers of the reference architecture: datagram format [40], transport protocol [42], virtual terminal protocol [41]. These protocols (or equivalent variants) have been validated through experiments, simulation, and analysis. Higher level protocols, such as file transfer, remote job entry, are still being investigated, and not enough experience is available for reaching a large consensus.

A new subcommittee entitled "Open Systems Interconnection" has been set up by ISO in 1977 (ISO/TC97/SC16) for the purpose of standardizing architecture and protocols required by heterogeneous networks. Hopefully, this may give a serious boost to computer network standards.

The following is a brief summary of recent trends in protocol evolution.

Protocols of the transmission layer are dependent on the technology; circuits, radio, satellite. Circuit characteristics do not change very much, thus line procedures tend to stabilize, however, optical fibers will bring a quantum jump in transmission bandwidth. The utilization of radio and satellite transmission is undergoing evolution [6], [45], [46], [48], [49].

So far, packet-switching standards are limited to virtual circuits (X.25), which may be appropriate for terminal-to-host communication, but datagrams are more effective for broadcast type applications. The controversy [67] between datagrams and virtual circuits should settle down if public networks offer both services. A datagram standard is under study within CCITT and could possibly be agreed upon by 1980.

Transport protocols are now reasonably well understood functionally, but performance tradeoffs still require further investigation [73], [74]. New transmission media using broadcast techniques [29], [45], [46], [58], will likely foster more sophisticated types of associations 1-to-N or M-

to-N, which present protocols do not handle. This may bring elegant solutions to some problems of distributed data base, for which point-to-point protocols are rather awkward.

Virtual terminal protocols of the first generation handled Teletype-like terminals. Some early commercial developments are now under way with a second generation designed for handling programmable displays. A period of maturation will be necessary for better understanding the concepts and practical requirements brought about by the rapid expansion of programmable terminals.

All other protocols are still in the domain of research and experiment.

The convergence of personal computing and CB ratio should not take much time to spawn a new sociological phenomenon. The evolution of protocols in this mass communication context is a fascinating and open question.

It certainly would appear rational to define standards for network architecture, protocols, and interfaces. Existing networks will remain alive much longer than their individual components. What makes networks obsolete are neither their components, which can be replaced, nor their protocols, which can be improved, but the services they offer. It is only when pressure builds for new services which cannot be accommodated on an existing network, that a decision is made to build a new network offering both existing and new services. The life cycle of a network is about 10 years.

The advent of cheap microsystems is pushing industry into a reassessment of its computing resources. A new life cycle is starting with distributed architectures. This opportunity may be auspicious for new standards to succeed. The market scene is a cornucopia of new products from a host of suppliers. Common protocols are the only viable solution towards heterogeneous systems [83], [85]. There is presently a definite concern among users and suppliers for a basic set of standard protocols within data processing systems, not just for transmission interfaces. If this trend is not thwarted by powerful and conflicting interests, standardization should become active in those areas identified previously as the most vital layers of a distributed resource sharing system, viz. transport protocol, virtual terminal, resource management (tasks, files, peripherals), and data base access.

VIII. Conclusions

There is some similarity between the evolution of programming languages and the evolution of protocols, except that they are separated by about twenty years. Initially, programming languages were invented to meet practical needs. Compilation was a matter of brute force. Thereafter, came models, grammars, theories, and theorems. Then structured programming and lastly proofs of correctness.

Like programming languages, protocols realize abstract machines. Messages and their headers are operands and instruction sets of low level languages designed to be used by machines rather than people. The first generation of protocols, basically those which were developed for the ARPANET, was rather difficult. The second generation triggered by CYCLADES attempted a more precise definition in ALGOL like procedures which were simulated before implementation [84].

Thereafter came models and formal representations [7], [21], [36], [38], [51], [62], [75]. Recently, some verification techniques have been proposed [5], [25]. Unfortu-nately, the body of knowledge accumulated with programming languages does not transfer very well to protocols. It seems that the reason is mainly the introduction of asynchrony and noise in the network environment. These protocols must perform in an environment of uncontrolled parallelism and mutual suspicion, which is a far cry from the relative neatness of modern programming languages.

Occasionally, one attempts to name a few great inventions in computer history, the transistor, the core memory, the microprocessors. Objects have always been better candidates than concepts. However, protocols will rank among the major innovations of the 1970's. Not because they introduce new techniques or new applications, but for the sake of their conceptual potential.

For the first time the design of computer systems puts the emphasis not on the internal management of resources, but on communication between resources of different systems. The very idea of a network architecture defined by an international standard is a novel concept in itself.

What we are witnessing is the emergence of an abstract construction as a standard, which in time will become the model from which commercial systems are derived. Again, the similarity with programming languages is striking. But the implications of a common network architecture are far significant. Within such a framework, any set of processes around the world could start exchanging information. The final stage is a distributed world machine. Naturally, there are other than technical implications worth investigating [52].

We may think this goal is not achievable, because technology is changing so fast that concepts will be obsolete almost as fast as they are formulated. It is precisely this problem that a layered architecture can solve. Layers and protocols that are dependent on specific technologies may have to change, but the architecture should remain and evolve.

Acknowledgment

As a finale to this technical coverage of the protocol field, we would like to mention a fundamental nontechnical aspect intimately related to computer protocols. Reaching international consensus among multiple research communities is a remarkable achievement in human protocols. Building a worldwide environment of mutual respect and friendship turns out to be a prerequisite to a common vision of a system architecture. We are intensely grateful to countless friends for the enrichment gained throughout years of protocol saga.

A special mention should be made of the help provided by the referees, particularly R. Kahn and V. Cerf, who contributed a major effort towards improving the structure and the readability of this paper.

References

[1] N. Abramson, "The ALOHA system—Another alternative for computer communications," *AFIPS-FJCC*, vol. 37, pp. 281-285, Nov. 1970.

[2] D. L. A. Barber, "A European informatics network: Achievement and prospects," in *Proc. ICCC* (Toronto, Ont., Canada), pp. 44-50, Aug. 1976.

[3] R. Binder, N. Abramson, F. Kuo, A. Okinaka, and D. Wax, ALOHA packet broadcasting: A retrospect," *AFIPS-NCC*, vol. 44, pp. 203-215, 1975.

[4] G. V. Bochmann, "Notes on the X.25 procedures for virtual call establishment and clearing," *ACM Computer Commun. Rev.*, vol. 7, no.4, pp. 53-59, Oct. 1977.

[5] G. V. Bochmann and J. Geosei, "A unified method for the specification and the verification of protocols," in *Proc.* IFIP Congress, pp. 229-234, Aug. 1977.

[6] F. Borgonovo and L. Fratta, "A new technique for satellite broad

cast channel communications," presented at the 5th ACM-*IEEE Data Commun. Symp.*, pp. 21–24, Sept. 1977.

[7] J. Bremer and A. Danthine, "Communication protocols in a network context," *ACM Sigcom*, in *Proc. Sigops Interface Workshop on Interprocess Communication* (Santa Monica, CA), p.6, Mar. 1975.

[8] H. Q. Burton and D. D. Sullivan, "Errors and error control," *Proc. IEEE*, vol. 60, no. 11, pp. 1293–1301, Nov. 1972.

[9] S. Carr, V. Cerf, and S. Crocker, "Host-to-host protocol in the ARPA computer network," *AFIPS SJCC*, pp. 589–597, May 1970.

[10] P. M. Cashin, "DATAPAC network protocols," in *Proc. ICCC* (Toronto, Ont., Canada), pp. 150–155, Aug. 1976.

[11] CCITT-X.25-*Orange Book*, vol. VIII-2, pp. 70–108, 1977.

[12] CCITT-Provisional recommendations X.3, X.25, X.28, and X.29 on packet switched data transmission services, p. 100, Oct. 1977.

[13] V. Cerf, A. McKenzie, R. Scantlebury, and H. Zimmermann, "Proposal for an international end-to-end protocol," *ACM Sigcom Comp. Commun. Rev.*, vol. 6, no. 1, pp. 68–89, Jan. 1974.

[14] V. Cerf and R. Kahn, "A protocol for packet network intercommunication," *IEEE Trans. on Commun.*, vol. 22, no. 5, pp. 637–648, May 1974.

[15] A. Chatterjee, N. G. Georganas, and P. K. Verma, "Analysis of a packet switching network with end-to-end congestion control and random routing," in *IEEE Proc* (Toronto, Ont., Canada), pp. 488–494, Aug. 1976.

[16] G. E. Conant and S. Wecker, "DNA: An Architecture for heterogeneous computer networks," in *Proc. ICCC* (Toronto, Ont., Canada), pp. 618–685, Aug. 1976.

[17] I. W. Cotton, "Computer network interconnection: Problems and prospects," Nat. Bur. Stand., Compu. Sci. Special Publication, 500–6 Apr. 1977.

[18] I. W. Cotton and H. C. Folts, "International standards for data communications: A Status report," in *Proc. 5th ACM-IEEE Data Commun. Symp.*, pp. 4.26–4.36, Sept. 1977.

[19] S. D. Crocker, J. F. Heafner, R. M. Metcale, and J. B. Postel, "Function oriented protocols for the ARPA computer network," *AFIPS-SJCC*, vol. 40, pp. 271–279, May 1972.

[20] Y. Dalal, "More on selecting sequence numbers, "*IFIP-WG 6.1*, protocol note no. 4, Oct. 1974.

[21] A. Danthine and J. Bremer, "An axiomatic description of the transport protocol of CYCLADES," presented at the Conf. Computer Networks and Teleprocessing, Aachen, Germany, 1976.

[22] A. Danthine and E. Eschenauer, "Influence on packet node behavior of the internode protocol," *IEEE Trans. Commun.*, vol. COM-24, pp. 606–614, June 1976.

[23] J. Davidson, N. Hathaway, J. Postel, N. Mimno, R. Thomas, and D. Walden, "The ARPANET TELNET protocol: Its purpose, principles, implementation, and impact on host operating system design," in *5th ACM-IEEE Data Commun. Symp.*, pp. 4.10–4.18, Sept. 1977.

[24] D. W. Davies, "Flow control and congestion control," presented at COMNET '77, John Von Neumann Soc., Budapest, Hungary, pp. 17–36, Oct. 1977.

[25] J. D. Day, "A bibliography on the formal specification and verification of computer network protocols," in *Proc. Computer, Network Protocols Symp.* (Liege, Belgium), p. 3, Feb. 1978.

[26] Digital Equipment Corp., NSP specifications, version 2, p. 87, Nov. 1977.

[27] A. Duenki and P. Schicker, "Virtual terminal definition and protocol," *ACM Sigcom Comput. Commun. Rev.*, vol. 6, no. 4, pp. 1–18, Oct. 1976.

[28] EIN-An end-to-end protocol for EIN, EIN Rep. 76 002, p. 22, Jan. 1976.

[29] D. J. Farber and K. C. Larson, "The system architecture of the distributed computer system," in *Proc. Communications Syst. Symp. Computer-Communications Networks and Teletraffic*, NY, pp. 21–27, Apr. 1972.

[30] G. Fayolle, E. Gelenbe, and G. Pujolle, "An analytic evaluation of the performance of the "Send and Wait" protocol," *IEEE Trans. Commun.*, vol. COM-26, 3, pp. 313–319, Mar. 1978.

[31] E. Feinler and J. Postel, eds., *ARPANET Protocol Handbook*, Network Info. Center, SRI International, for the U.S. Defense Communication Agency, Jan. 1978.

[32] J. G. Fletcher and R. W. Watson, "Mechanisms for a reliable timer-based protocol," in *Proc. Symp. Computer Network Protocol* (Liege, Belgium), p. 17, Feb. 1978.

[33] E. Gelenbe, J. Labetoulle and G. Pujolle, "Performance evaluation of the HDLC," in *Proc. Computer Network Protocols Symp.* (Univ. Liese, Liese, France), p. 7, Feb. 1978.

[34] M. Gien, "A file transfer protocol (FTP)" in *Proc. Computer Network Protocols Symp.* (Univ. Liege, Liege, Belgium), p. 7, Feb. 1978.

[35] J. Glories, The SITA story, SITA Monogr. 4, p. 26, May 1973.

[36] M. G. Gouda and E. G. Manning, "Protocol Machines: A Concise formal model and its automatic implementation," presented at ICCC, Toronto, Ont., Canada, pp. 346–350, Aug. 1976.

[37] J. L. Grange and P. Mussard, "Performance measurement of line control protocols in the CIGALE network," in *Proc. Computer Network Protocols Symp.* (Univ. Liege, Liege, Belgium), p. 13, Feb. 1978.

[38] J. Harangozo, "An approach to describing a data link level protocol with a formal language," in *Proc. 5th ACM-IEEE Data Communication Symp.*, pp. 4.37–4.49, Sept. 1977.

[39] W. S. Hobgood, "The role of the network control program in systems network architecture," *IBM Syst. J.*, vol. 15, no. 1, pp. 39–52, 1976.

[40] IFIP-Basic Message Format for Inter-Network Communication, ISO/TC97/SC6 N1281, p. 7, Apr. 1976.

[41] IFIP-WG 6.1 Proposal for Standard Virtual Terminal Protocol, Doc. INWG Protocol # gi or ISO/TC97/SC16 N23, p. 56, Feb. 1978.

[42] IFIP-WG 6.1-Proposal for an Internetwork End-to-End Transport Protocol, ISO/TC 97/SC 16 N24, p. 46, Mar. 1978.

[43] *ISO-High Level Data Link Control* –Frame Structure, IS 3309, 1976.

[44] *ISO-High Level Data Link Control*–Elements of Procedure, IS 4335, 1977.

[45] I. M. Jacobs, R. Binder, and E. V. Hoversten, "General purpose packet satellite networks," this issue, pp. 1448–1467.

[46] R. E. Kahn, S. A. Gronemeyer, J. Burchfiel, R. C. Kunzelman, "Advances in packet radio technology," this issue, pp. 1468–1496.

[47] L. Kleinrock, W. E. Naylor, and H. Opderbeck, "A study of line overhead in the ARPANET," *Commun. Ass. Comput Mach.*, vol. 19, no. 1, pp. 3–13, Jan. 1976.

[48] L. Kleinrock and F. Tobagi, "Random access techniques for data transmission over packet switched radio channels," *AFIPS-NCC*, vol. 44, pp. 187–201, May 1975.

[49] S. Lam and L. Kleinrock, "Packet switching in a multi-access broadcast channel: Dynamic control procedures," *IEEE Trans. Commun.*, vol. COM-23, no. 9, Sept. 1975.

[50] G. LeLann and H. LeGoff, "Advances in performance evaluation of communication protocols," in *Proc. ICCC* (Toronto, Ont., Canada), pp. 361–366, Aug. 1976.

[51] G. LeLann, "Distributed system—Toward a formal approach," in *Proc. IFIP Congr.*, pp. 155–160, Aug. 1977.

[52] J. C. R. Licklider and A. Vezza, "Applications of information networks," this issue, pp. 1330–1346.

[53] J. H. McFadyen, "Systems network architecture: An overview," *IBM Syst. J.*, vol. 15, no. 1, pp. 4–23, 1976.

[54] J. M. McQuillan, "Enhanced message addressing capabilities for computer networks," this issue, pp. 1517–1527.

[55] J. M. McQuillan and D. C. Walden, "The ARPA network design decisions," *Comput. Networks*, vol. 1, no. 5, pp. 243–291, Aug. 1977.

[56] E. G. Manning, E. Gelenbe, and R. Mahl, "A Homogeneous Network for Data Sharing—Modeling and Evaluation. New York: Amer. Elsevier, Aug. 1974, pp. 345–353.

[57] J. A. Martin, *Systems Analysis for Data Transmission*. Englewood Cliffs, NJ: Prentice-Hall, 1970.

[58] R. Metcalfe and D. Boggs, "ETHERNET: Distributed packet switching for local computer networks," *Commun. Ass. Comput. Mach.*, vol. 19, no. 7, pp. 395–404, July 1976.

[59] N. Naffah, "High level protocol for alphanumeric data-entry terminals," to be published in *Comput. Networks*.

[60] N. Neigus, "File transfer protocol," in *Proc. ARPA-NIC*-17759, p. 50, July 1973.

[61] J. Postel, "Official initial connection protocol," in *Proc. ARPA-NIC*-7101, p. 5, June 1971.

[62] —, "A graph model analysis of communication protocols," UCLA-ENG-7410, 1974.

[63] L. Pouzin, "Network architectures and components," in *Proc. 1st European Workshop on Comput. Networks* (Arles, France), IRIA ed., pp. 227–265, Apr. 1973; also, IFIP-WG 6.1, INWG no. 49.

[64] —, "Network protocols," NATO International Advanced Study Institute on Computer Communication Networks, University of Sussex, Brighton, Nordhoff ed., pp. 231–255, Sept. 1973.

[65] —, "Presentation and major design aspects of the CYCLADES computer network," in *Proc. 3rd ACM-IEEE Commun. Symp.*, Tampa, FL., pp. 80–87, Nov. 1973.

[66] —, "CIGALE, the packet switching machine of the CYCLADES computer network," in *Proc. IFIP Congress* (Stockholm, Sweden), pp. 155–159, Aug. 1974.

[67] —, "Virtual circuits vs. Datagrams-technical and political issues," Presented at the *AFIPS National Comput. Conf.*, NY, pp. 483–494, June 1976.

[68] —, "Flow control in data networks: Methods and tools," in *Proc. ICCC* (Toronto, Ont., Canada), pp. 467–474, Aug. 1976.

[69] —, "Names and objects in heterogeneous computer networks,"

in *Proc. 1st Conf. European Cooperation in Informatics* (Amsterdam, The Netherlands), pp. 1–11, Aug. 1976.

[70] E. Raubold and J. Haenle, "A method of deadlock-free resource allocation and flow control in packet networks," in *Proc ICCC* (Toronto, Ont., Canada), pp. 483–487, Aug. 1976.

[71] L. G. Roberts and B. D. Wessler, "Computer network development to achieve resource sharing," *AFIPS-SJCC*, vol. 36, pp. 543–549, May 1970.

[72] R. F. Sproull and D. Cohen, "High level protocols," this issue, pp. 1371–1386.

[73] C. Sunshine, "Interprocess communication protocols for computer networks," Stanford Electron. Lab., TR 105, 258 pp., Dec. 1975.

[74] ——, "Factors in interprocess communications protocol efficiency for computer networks," *AFIPS-NCC*, pp. 571–576, June 1976.

[75] *Proc. Symposium on Network Protocols* (Liege, Belgium), Feb. 1978; This is a collection of 36 papers containing a considerable amount of significant information on recent work.

[76] R. Tomlinson, "Selecting sequence numbers," IFIP WG 6.1, protocol note, no. 2, Sept. 1974.

[77] K. C. Traynham and R. F. Steen, "SDLC and BSC on satellite links: A performance comparison," *ACM Sigcom Comput. Commun. Rev.*, vol. 7, no. 4, pp. 3–9, Oct. 1977.

[78] L. Tymes, "TYMNET, a terminal oriented communications network," *AFIPS-SJCC*, pp. 211–216, June 1971.

[79] A. J. Viterbi, "Error control for data communication," *ACM Sigcom Comput. Commun. Rev.*, vol. 6, no. 1, pp. 27–37, Jan. 1976.

[80] D. Walden, "Host-to-host protocols," INFOTECH State of the Art Rep., Network Systems and Software, Maidenhead, England, pp. 287–316, 1975.

[81] H. Zimmermann, "Terminal and access in the CYCLADES computer network," in *Proc. International Computing Symp.* (Juan-les-Pins), North-Holland/American, Elsevier edit., pp. 97–99, June 1975.

[82] ——, "The CYCLADES end-to-end protocol," in *Proc 4th ACM-IEEE Data Commun. Symp.* (Quebec, Canada), pp. 7.21–7.26, Oct. 1975.

[83] ——, "High level protocols standardization: Technical and political issues," in *Proc. ICCC* (Toronto, Ont., Canada), pp. 373–376, Aug. 1976.

[84] ——, "The CYCLADES experience: Results and impact," in *Proc. IFIP Congr.* (Toronto, Ont., Canada), pp. 465–469, Aug. 1977.

[85] H. Zimmermann and N. Naffah, "On open systems architecture," presented at the ICCC, Kyoto, Japan, pp. 669–674, Sept. 1978.

Throughput in the ARPANET—Protocols and Measurement

LEONARD KLEINROCK, FELLOW, IEEE, AND HOLGER OPDERBECK, MEMBER, IEEE

Abstract—The speed at which large files can travel across a computer network is an important performance measure of that network. In this paper we examine the achievable sustained throughput in the ARPANET. Our point of departure is to describe the procedures used for controlling the flow of long messages (multipacket messages) and to identify the limitations that these procedures place on the throughput. We then present the quantitative results of experiments which measured the maximum throughput as a function of topological distance in the ARPANET. We observed a throughput of approximately 38 kbit/s at short distances. This throughput falls off at longer distances in a fashion which depends upon which particular version of the flow control procedure is in use; for example, at a distance of 9 hops, an October 1974 measurement gave 30 kbit/s, whereas a May 1975 experiment gave 27 kbit/s. The two different flow control procedures for these experiments are described, and the sources of throughput degradation at longer distances are identified, a major cause being due to a poor movement of critical limiting resources around in the network (this we call "phasing"). We conclude that flow control is a tricky business, but in spite of this, the ARPANET throughput is respectably high.

I. INTRODUCTION

THE ARPANET, which was the world's first large-scale experimental packet-switching network, needs little introduction; it has been amply documented (see, for example, [5] and the extensive references therein). Our interest in this paper is to describe the message-handling protocols and some

experimental results for the achievable throughput across the ARPANET. These experiments were conducted at the UCLA Network Measurement Center (NMC) and show that the network can support roughly 38 kbit/sec between HOST computers which are a few hops apart; for more distant HOST pairs, the throughput falls off to a level dependent upon the particular version of message processing used, as discussed in detail below.

An earlier NMC experiment reported upon the behavior of actual user traffic in the ARPANET (and also described the NMC itself) [4]. More recent NMC experiments identified, explained, and solved some deadlock and throughput-degradation phenomena in the ARPANET [11] and also measured the effect of network protocols and control messages on line overhead [4]. The experiments reported upon herein consisted of throughput measurements of UCLA-generated traffic (using our PDP 11/45 HOST in a dedicated mode) which was sent through the ARPANET to "fake" HOST's at various topological distances (hops) from UCLA. Each experiment ran for 10 min during which time full (8-packet) multipacket traffic was pumped into the ARPANET as fast as the network would permit. Both throughput (from the UCLA HOST to the destination HOST) and delay (as seen by the UCLA HOST) were measured, along with some other statistics described below.

This paper is organized as follows. We describe the message-handling procedure for multipacket messages in Section II, identify the limitations this procedure imposes on the throughput in Section III, and then quantitatively report upon the October 1974 throughput experiments in Section IV. The issue of looping in the adaptive routing procedure and its erratic effect on throughput is discussed in Section V. Some recent changes to the message-processing procedure are described in Section VI, and in Section VII we describe some of its faults, their correction, and the experimentally achieved throughput as of May 1975, using this new procedure.

II. HANDLING OF MULTIPACKET MESSAGES

In this section, we describe the details for handling multipacket messages in the ARPANET as of October 1974[1]; it was at this time that the initial set of throughput experiments reported here was conducted. This discussion will permit us to identify throughput limitations and to discuss system bottlenecks.

We are interested in the transmission of a long data stream which the ARPANET accepts as a sequence of messages (each with a maximum length of 8063 data bits). Each such message in this sequence will be a "multipacket" message (a multipacket message is one consisting of more than one 1008-bit packet). To describe the sequence of events in handling each multipacket message we refer to Fig. 1 (which gives the details for a data stream requiring only *one* full multipacket message for simplicity). A message is treated as a multipacket message if the HOST–IMP interface has not received an end-of-message indication after the input of the first packet is completed (shown as point *a* in Fig. 1). At this time, transmission of the remaining packets of this message from the HOST to the IMP is temporarily halted until the message acquires some network resources as we now describe. First, the multipacket message must acquire a message number (from the IMP) which is used for message sequencing (point *b*); all messages originating at this IMP and heading to the same destination IMP share a common number space. Next, an entry in the *pending leader table* (PLT) must be obtained as shown at point *c*. The PLT contains a copy of the leader of all multipacket messages that are currently being handled by the source IMP. Among other things, the function of the PLT is to construct the packet headers for the successive packets of the multipacket message. Such an entry is deleted and released when the RFNM (the end-to-end acknowledgment whose acronym comes from "ready-for-next-message") is received from the destination IMP. The PLT is shared by messages from all HOST's attached to the same IMP and used for all possible destinations.

After the PLT entry has been obtained by the multipacket message, a table is interrogated to find out whether there are eight reassembly buffers reserved for this source IMP at the desired destination IMP. If this is not the case, a control message REQALL (request for allocation) is generated and sent from the source IMP (also shown at point *c*) to the destination IMP which requests an allocation of these buffers. The protocol is such that this REQALL steals the acquired message number and the PLT entry for its own use at this time. This request is honored by the destination IMP as soon as it has eight buffers available (point *d*). To report this fact a subnet control message ALL (allocate) is returned to the source IMP, thus delivering the 8-buffer allocation. Since the previously acquired

[1] This is the message-handling procedure referred to as "version 2" in [5].

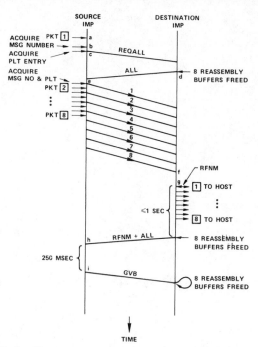

Fig. 1. The sequence of events for one multipacket transmission.

message number and PLT entry have been used, a new message number and a new PLT entry will have to be obtained for the multipacket message itself. (Had 8 reassembly buffers been reserved in the first place, this would have shown at the source IMP by the presence of an unassigned ALL and the steps from *c* to *e* would not have occurred). Only when all these events have taken place can the first packet begin its journey to the destination IMP and can the input of the remaining packets be initiated, as shown at point *e*.

When all packets of the multipacket message have been received by the destination IMP (point *f*), the message is put on the IMP-to-HOST output queue. After the transmission of the first packet to the HOST (point *g*), the RFNM for this message is generated at the destination IMP (also point *g*) to be returned to the source IMP. This RFNM prefers to carry a "piggy-backed" ALL (an implicit reservation of 8 buffers for the next multipacket message) if the necessary buffer space is available. If not, the RFNM will wait for at most 1 s for this buffer space. In case the necessary 8 reassembly buffers do not become available within this second, the RFNM is then sent without a piggy-backed ALL. (We show the case where the buffers do become available in time and so the ALL returns piggy-backed on the RFNM).

After the reception of the RFNM at the source IMP (point *h*), the message number and the PLT entry for this message are freed and the source HOST is informed of the correct message delivery. In case the RFNM carries a piggy-backed ALL, the allocate counter for the proper destination IMP is incremented. This implicit reservation of buffer space is returned to the destination IMP if some HOST attached to the source IMP does not make use of it within the next 250 ms (shown at point *i*); the cancellation is implemented as a control message

GVB (giveback) which is generated at the source IMP. If, however, the next multipacket message to the same destination IMP is received from any source HOST within 250 ms, this message need only acquire a message number and a PLT entry before it can be sent to the destination IMP, and need not await an ALL.

Thus we see that three separate resources must be obtained by each multipacket message prior to its transmission through the net: a message number, a PLT entry, and an ALL.[2]

III. THROUGHPUT LIMITATIONS

Let us now identify the limitations to the throughput that can be achieved between a pair of HOST's in the ARPANET. First we consider the limitations that are imposed by the hardware. The line capacity represents the most obvious and important throughput limitation. Since a HOST is connected to an IMP via a single 100-kbit/s transmission line, the throughput can never exceed 100-kbit/s. If there is no alternate routing in the subnet, the throughput is further limited by the 50-kbit/s line capacity of the subnet communication channels. (The issue of alternate routing is discussed later.)

The processing bandwidth of the IMP allows for a throughput of about 700 and 850 kbit/s for the 316 and 516 IMP's, respectively [6]. Therefore the IMP's can easily handle several 50-kbit/s lines simultaneously. The processing bandwidth of the HOST computers represents a more serious problem. Severe throughput degradations due to a lack of CPU time have been reported in the past [1], [2], [12]. However, these reports also indicate that the degradations are in many cases caused by inefficient implementations of higher level protocols [4]. Therefore, changes in these implementations have frequently resulted in enormous performance improvements [13]. To avoid throughput degradations due to a CPU-limited HOST computer for our throughput experiments, we used a PDP 11/45 minicomputer at UCLA whose only task was to generate 8-packet messages as fast as the network would accept them.

Let us now discuss what throughput limitations are imposed on the system by the subnet *flow control procedure*. As discussed above, there are two kinds of resources a message must acquire for transmission: buffers and control blocks (specifically message numbers and table entries). Naturally, there is only a finite number of each of these resources available. Moreover, most of the buffers and control blocks must be shared with messages from other HOST's. The lack of any one of the resources can create a bottleneck which limits the throughput for a single HOST. Let us now discuss how many units of each resource are available and comment on the likelihood that it becomes a bottleneck. This discussion refers to the ARPANET as of October 1974.

In October 1974, a packet was allowed to enter the source IMP only if that IMP had at least four free buffers available. At that time, the total number of packet buffers in an IMP with and without the very distant HOST (VDH) software was, respectively, 30 and 51. This meant that an interruption of message input due to buffer shortage could occur only in the unlikely event that the source IMP was heavily engaged in handling store-and-forward as well as reassembly traffic.

The next resource the message had to obtain was the message number. There was a limitation of only four message sequence numbers allocated per source IMP-destination IMP pair. This meant that all source HOST's at some source IMP *A* which communicated with any of the destination HOST's at some destination IMP *B* shared the same stream of message numbers from IMP *A* to IMP *B*. This possible interference between HOST's and the fact that there were only four message numbers which could be used in parallel meant that the message number allocation could become a serious bottleneck in cases where the source and destination IMP were several hops apart. (This was the major reason for the recent change to the message processing procedure which has recently been implemented; see Section VI).

After a message number was obtained, the multipacket message had to acquire one of the PLT entries of which there was a shared pool of six. Since the PLT is shared by all HOST's which are attached to the source IMP and used for all possible destinations, it also represents a potential bottleneck. This bottleneck can easily be removed by increasing the number of entries permitted in the PLT. However, the PLT also serves as a flow control device which limits the total number of multipacket messages that can be handled by the subnet simultaneously. Therefore, removal of the throttling effect due to the small size PLT may introduce other congestion or stability problems. A corresponding consideration applies to the message number allocation.

The number of simultaneously unacknowledged 8-packet messages is further limited by the finite reassembly space in the destination IMP. In October 1974, a maximum of 34 buffers was available for reassembly (for IMP's without the VDH software). This meant that at most four 8-packet messages could be reassembled at the same time (leaving space for at least two single-packet messages). (The reassembly space must of course be shared with all other HOST's that are sending messages to the same destination IMP.) It may therefore become another serious throughput bottleneck.

From the above discussion, we know that even if there is no interference from other HOST's there cannot be more than four messages in transmission between any pair of HOST's due to the message number limitation. This restriction decreases the achievable throughput in the event that the line bandwidth times the round trip time is larger than four times the maximum message length. Fig. 2 depicts this situation. The input of the first packet of message *i* is initiated at time *a* after the last packet of message *i* − 1 has been processed in the source IMP. After the input of this first packet is complete, the source IMP waits until time *b* when the RFNM for message *i* − 4 arrives. Shortly after this RFNM has been processed (at time *c*) the transmission of the first packet over the first hop and

---

[2] The procedure just described extracts a price for the implementation of its control functions. This price is paid for in the form of overhead in the packets as they are transmitted over the communication channels, in the packets as they are stored in IMP buffers, in control messages (IMP-IMP, IMP-HOST, HOST-HOST), in measurement and monitoring, etc. We refer the reader to [4] for the effect of this overhead on the line efficiency.

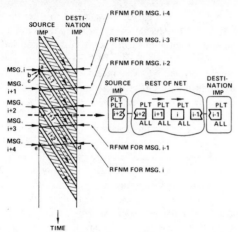

Fig. 2.   The normal sequence of multipacket messages.

the input of the remaining packets from the HOST is initiated. At time *d*, all packets have been reassembled in the destination IMP, the first packet has been transmitted to the destination HOST and 8 reassembly buffers have been acquired by the RFNM which is then sent (with a piggy-backed ALL) to the source IMP. The RFNM reaches the source IMP at time *e* and thereby allows the transmission of message $i + 4$ to proceed. In this figure we also show a snapshot of the net at the time slice indicated by the dashed arrow. We show four messages (each with their own ALL and PLT): $i + 2$ is leaving the source IMP, both $i + 1$ and i are in flight, and $i - 1$ is entering the destination IMP. We also see the two unused PLT entries in the source IMP. The possible gaps in successive message transmissions represent a loss in throughput and can be caused by the limitation of four messages outstanding per IMP pair; this manifests itself in the next (fifth) message awaiting the return of a RFNM which releases one of the message numbers.

We have not yet mentioned the interference due to other store-and-forward packets which can significantly decrease the HOST-to-HOST throughput. This interference causes larger queueing times and possibly rejection by a neighbor IMP. Such a rejection occurs if either there are 20 store-and-forward packets in the neighbor IMP or if the output queue for the next hop is full. (There is an allowed maximum of 20 store-and-forward packets per IMP and of 8 packets for each output queue.) A rejected packet is retransmitted if no IMP-to-IMP acknowledgment has been received after a 125 ms timeout.

We now turn to a brief discussion of alternate routing and its impact on our throughput experiments. By alternate routing we refer to the possibility of sending data over two (or more) completely independent paths from source to destination. The shorter (shortest) path (in terms of number of hops) is usually called the primary path, and the longer path(s) are called secondary (tertiary, etc.) or alternate paths. For reliability reasons there should always be at least one alternate path available in a properly operating network. It turns out in the ARPANET that alternate paths are rarely used if they are longer than the primary path by more than two hops. The reason for this comes from the way the delay estimate is calculated, updated, and used by the routing procedure and from

the way the output queues are managed. Each hop on the path from source to destination contributes four (arbitrary) units to the delay estimate. Each packet in an output queue between source and destination contributes one additional delay unit to the delay estimate. Since the length of the output queues is limited to 8 packets, one hop can therefore increase the delay estimate by at least 4, and at most, 12 units. Thus the minimum and maximum delay estimates over a path of *n* hops are, respectively, $4n$ and $12n$ delay units. Packets are always sent over the path with the smallest current delay estimate. From this, it follows that an alternate path is never used if it is more than three times longer (in terms of hops) than the primary path. Thus, for a primary path of length *n*, alternate routing is possible only over paths of length less than $3n$ hops. Let us assume that all the channels along the primary and alternate secondary path have the same capacity and that there is no interfering traffic. If we send as many packets as possible over the primary path, these packets usually will not encounter large queueing delays because this stream is fairly deterministic as it proceeds down the chain. This means that the delay estimate increases only slightly, although all of the bandwidth is used up. Therefore a switch to an alternate path occurs only if that path is slightly longer than the primary path. In the case of interfering traffic, the output queues will grow in size, and therefore a switch is more likely to occur. Such a switch to an alternate path may therefore help to regain some of the bandwidth that is lost to the interfering traffic. It has already been pointed out in [7] that, even if primary and secondary paths are equally long, at most a 30 percent increase in throughput can be achieved. This is due to the restriction of a maximum of 8 packets on an output queue and the fact that the frequency of switching between lines is limited to once every 640 ms (for heavily loaded 50-kbit/s lines). Thus the back-logged queue of 8 packets on the old path will provide overlapped transmission for only $8 \times 23.5 = 188$ ms of the total of 640 ms between updates (the only times when alternate paths may be selected). The relatively slow propagation of routing information further reduces the frequency of switching between the primary and secondary path. This discussion shows that alternate paths have only a small effect on the maximum throughput that can be achieved. However, the alternate paths are of great importance for the reliability of the network.

IV. THROUGHPUT EXPERIMENTS

The October 1974 throughput experiments produced the results shown in Fig. 3. Here we show the throughput (in kilobits/second) as a function of the number of hops between source and destination. Curve *A* is for the throughput averaged over the entire 10-min experiment; curve *B* is the throughput for the best block of 150 successive messages. Note that we are able to pump an average of roughly 37–38.5 kbit/s out to 5 hops[3]; it drops beyond that, falling to 30 kbit/s at 9 hops due largely to transmission gaps caused by the 4-message limitation. Also, the best 150-message throughput is not much

---

[3]This indicates an approximate efficiency of 75 percent on the 50-kbit/s lines. See [4] for a detailed description of line efficiency.

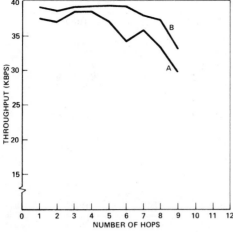

Fig. 3. ARPANET throughput (October 1974).

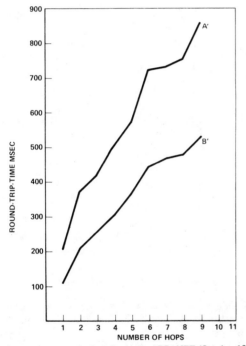

Fig. 4. Average round-trip delay in the ARPANET (October 1974).

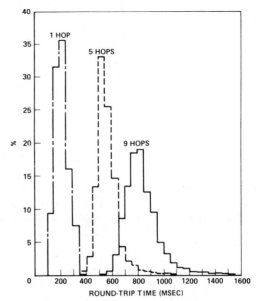

Fig. 5. Histogram of round-trip delay.

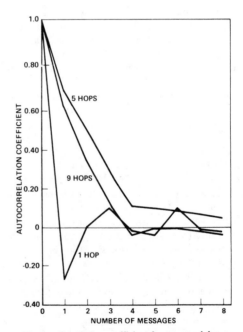

Fig. 6. Correlation coefficient for message delay.

better than the overall average, indicating that we are almost achieving the maximal performance most of the time. In Fig. 4, we show the corresponding curves for the average round-trip delays (as seen by the UCLA PDP 11/45 HOST) as a function of source–destination hop distance; that is, curve A' is for the average and curve B' is for the best 150 successive messages. Note that the average delay for n hops may be approximated by $200 + 90(n-1)$ ms. The measured histogram for delay is given in Fig. 5 for hop distances of 1, 5, and 9. Some of the large delays shown in this figure are caused by looping as explained in the next section. Of further interest is the autocorrelation coefficient of round-trip delay for successive messages in the network; this is shown in Fig. 6. Note

that message delay is correlated out to about 3 or 4 successive messages.

V. LOOPING

The observation of occasional very long network delays recently led to an investigation of this phenomenon. The results showed that at times there was extensive looping in the subnet, i.e., packets were tossed back and forth between neighboring nodes many times and thus did not reach their destination until the loop was removed through the adaptive routing procedure. In what follows we describe how the ARPANET tries to avoid loops and why this procedure may fail in certain cases.

Let us consider a net with the following linear topology.

The exact topology between nodes D and A is immaterial for our discussion; node X is that node in the "rest of the network" which sends routing updates to node A. In this kind of configuration, nodes B and C should always send packets for node D to their left-hand neighbor (nodes A and B, respectively). We adopt the following notation to be used in the three following examples.

$B \to C$ means that node B sends a routing message to node C. $d/1/A$ means that the overall delay estimate to node D is d units, the local delay over the best delay line to a neighbor is 1 unit, and A is the name of the best delay neighbor.

Table I describes an example of how loops can occur if no loop prevention procedure is used. The reader should review the earlier discussion which describes how delay estimates are formed.

Initially, the local delays in IMP's A, B, and C are zero, and the delay estimates to IMP D are, respectively, $d - 4$, d, and $d + 4$ delay units (row 0). Assume now that a sudden increase in traffic between node A and node D causes the delay estimate in A to be increased by 9 units (row 1). This fact is reported to B (row 2). Since C has not yet been informed of the sudden increase in traffic, it sends the old delay estimate to B. This causes B to consider C its best delay neighbor for IMP D (row 3); a loop between IMP's B and C has now been created! This loop remains effective until B tells C about the new situation (row 4), C reports back to B (row 7), and finally A's routing message causes B to switch back to A as its best delay neighbor (row 10). Since routing messages are sent every 640 ms, the loop persists for 640 to 1280 ms in this example.

To prevent the occurrence of this kind of loop in the ARPANET, a line hold-down mechanism was implemented [8]. The function of this mechanism was to continue using the best delay path for up to 2 s (ignoring the estimated delay from nonbest delay neighbors) whenever the delay estimate on this path increased by more than 8 delay units. The argument put forth in favor of this hold-down strategy was that, at times of sudden change, a node cannot be sure that its neighbors have already been informed of this change. Therefore, it should ignore the delay estimates from all but the best delay neighbor for some time until the information on the sudden change has propagated through the net.

Table II shows how this hold-down mechanism prevents the loop in the previous example. The hold-down of a line is indicated by an exclamation mark (!). Note that IMP's A and B start holding down their line to the best delay neighbor since their delay estimate gets worse by more than 8

TABLE I

	Routing	IMP A	IMP B	IMP C
0	initially	$d - 4/0/X$	$d/0/A$	$d + 4/0/B$
1	$X \to A$	$d + 5/4/X$		
2	$A \to B$		$d + 9/0/A$	
3	$C \to B$		$d + 8/0/C$	
4	$B \to C$			$d + 13/1/B$
5	$X \to A$	$d + 5/4/X$		
6	$A \to B$		$d + 8/0/C$	
7	$C \to B$		$d + 17/0/C$	
8	$B \to C$			$d + 21/0/C$
9	$X \to A$	$d + 5/4/X$		
10	$A \to B$		$d + 9/0/A$	

TABLE II

	Routing	IMP A	IMP B	IMP C
0	initially	$d - 4/0/X$	$d/0/A$	$d + 4/0/B$
1	$X \to A$	$d + 5/4/X!$		
2	$A \to B$		$d + 9/0/A!$	
3	$C \to B$		$d + 9/0/A!$	
4	$B \to C$			$d + 14/1/B!$
5	$X \to A$	$d + 5/4/X!$		
6	$A \to B$		$d + 9/0/A!$	
7	$C \to B$		$d + 9/0/A!$	
8	$B \to C$			$d + 13/0/B!$
9	$X \to A$	$d + 5/4/X!$		
10	$A \to B$		$d + 9/0/A!$	

TABLE III

	Routing	IMP A	IMP B	IMP C
0	initially	$d - 4/0/X$	$d/0/A$	$d + 4/0/B$
1	$X \to A$	$d + 1/2/X$		
2	$A \to B$		$d + 5/0/A$	
3	$X \to A$	$d + 5/4/X$		
4	$A \to B$		$d + 9/0/A$	
5	$C \to B$		$d + 8/0/C$	
6	$B \to C$			$d + 13/1/B!$
7	$X \to A$	$d + 5/4/X$		
8	$A \to B$		$d + 8/0/C$	
9	$X \to A$	$d + 5/4/X$		
10	$A \to B$		$d + 8/0/C$	
11	$C \to B$		$d + 17/0/C!$	

delay units (rows 1 and 2). This causes IMP B to ignore the lower delay estimate received from IMP C and the loop is thereby prevented.

Since the decision of whether or not to initiate a line hold-down depends solely on the delay difference between consecutive routing messages, the hold-down mechanism is sensitive to the frequency at which routing messages are sent. The routing message frequency, however, is a function of line speed and line utilization. Therefore it is quite possible, for example, that A sends two routing messages to B before B sends one routing message to C. Table III gives an example of the occurrence of a loop which is due to the fact that routing messages on different lines are sent at different frequencies. In this case B does not initiate a line hold-down when it receives the routing information from A since the delay difference is always smaller than 8 (rows 2 and 4). Therefore, B switches the best delay path from A to C when it receives C's routing message (row 5), i.e., a loop has again been created! In row 6 we see a hold-down at C. The situation becomes even worse when B receives the next routing message from C (row 11). Now B initiates a line hold-down in the wrong direction! This means that the B-C loop cannot be removed for almost 2 s because B ignores further delay estimates if received from A.

We call the occurrence of a loop whose existence is extended because of line hold-down a "*loop trap*." These loop traps have been observed repeatedly by the UCLA Network Measurement Center [9]. When such a loop trap occurs, packets are exchanged between neighbors up to 50 times before they can continue their travel to the destination IMP. We believe that these loop traps represent a major reason for the observation of occasional very long network delays during our throughput experiments.

Recently, the criterion for initiation of a hold-down was changed in such a way that it is now independent of the frequency at which routing messages are sent. As a result, we have not been able to detect loop traps in this modified system. Naylor has studied the problem of eliminating loops completely, and he presents a loop-free routing algorithm in [10].

VI. RECENT CHANGES TO MESSAGE PROCESSING

Some of the problems with the subnet control procedures described in Section III have recently led to a revision of message processing in the subnet.[4] In particular, message sequencing is now done on the basis of HOST-to-HOST pairs and the maximum number of messages that can be transmitted simultaneously in parallel between a pair of HOST's was increased from 4 to 8. Let us now describe the details of this new scheme.

Before a source HOST A at source IMP S can send a message to some destination HOST B at destination IMP D, a message control block must now be obtained in IMP S and IMP D. This message control block is used to control the transfer of messages. It is called a transmit block in IMP S and a receive block in IMP D. The creation of a transmit block–receive block pair is similar to establishing a (simplex) connection in the HOST-to-HOST protocol. It requires an exchange of subnet control messages that is always initiated by the source IMP. The message control blocks contain, among other things, the set of message numbers in use and the set of available message numbers.

After the first packet has been received from a HOST, the source IMP checks whether or not a transmit block-receive block pair exists for the transfer of messages from HOST A to HOST B. If HOST A has not sent any messages to HOST B for quite some time, it is likely that no such message control block pair exists. Therefore, source IMP S creates a transmit block and sends a subnet control message to destination IMP D to request the creation of a receive block. When IMP D receives this control message, it creates the matching receive block and returns a subnet control message to IMP S to report this fact. When IMP S receives this control message, the message control block pair is established.

A shortage of transmit and/or receive blocks will normally cause only an initial setup delay. Currently, there are 64 transmit and 64 receive blocks available in each IMP. This means, for example, that a HOST can transmit data to 64 different HOST's simultaneously, or that two HOST's, attached to the same destination IMP, can each receive messages from 32 different HOST's simultaneously, etc. Since 64 message blocks is a rather large number, it is unlikely that this is a limiting resource.

The remaining resources are acquired in the following sequence: message number, reassembly buffers, and PLT entry. Since there are now 8 message sequence numbers which are allocated on a sending HOST–receiving HOST pair basis, a HOST is allowed to send up to 8 messages to some receiving HOST without having received an acknowledgment for the first message. For multipacket traffic this is more than enough because there are still only 6 entries in the PLT. Suddenly, therefore, the PLT has become a more prominent bottleneck than it used to be in the old message processing procedure when only four messages per IMP pair could exist.

Note that a multipacket message tries to obtain the reassembly buffers before it asks for the PLT entry. (This sequence for resource allocation can lead to difficulties as is described in Section VII.) In case there is no reassembly buffer allocation waiting at the source IMP, then as before, the message number and the PLT entry are used to send a REQALL to the destination IMP.

VII. THROUGHPUT FOR THE NEW MESSAGE PROCESSING

In February 1975, we repeated the throughput measurements of October 1974 to determine what effect the new message processing procedure had on the maximum throughput. Since the subnet had grown in size, we were able to measure the throughput as a function of hop distance up to 12 hops. The measured throughput in February 1975 with the new message processing procedure was significantly less than the throughput that was achieved in October 1974. For paths with many hops, the decrease in throughput was almost 50 percent. The observed throughput degradation was not due to a sudden surge of interfering traffic. Investigation of this performance degradation revealed the following two causes which explain in part the observed decrease in throughput: processing delays in 316 IMP's and an effect which we refer to as "phasing." Recent measurements show that the 316 IMP's in the ARPANET are becoming a major bottleneck. For the 316 IMP's, the queueing delays in the input or processing queue are, on the average, larger than the queueing delays in the output queue. The average queueing delay in the processing queue is about 10 ms. This is more than 5 times as much as the corresponding queueing delays in the 516 IMP. The cause for this increase in processing delay can be found in the more extensive processing which is done at a higher priority level.

A processing delay of 10 ms appears to be within acceptable limits. However, this is only an average number. In particular cases, we observed queueing delays in the input queue of several hundred milliseconds. In addition, it is not clear what

---

[4] This is the "version 3" procedure in [5].

second-order effects these long processing delays have on a system that was originally designed to be limited by line band-width.

Phasing, the second (and more subtle) cause for the through-put degradation, is due to the sending of superfluous REQALL's! A REQALL is called superfluous if it is sent while a previous REQALL is still outstanding. This situation can arise if message i sends a REQALL but does not use the ALL returned by this REQALL because it obtained its reassembly buffer allocation piggy-backed on a RFNM for an earlier mes-sage (which reached the source IMP before its requested ALL). The sending of superfluous REQALL's is undesirable because it unnecessarily uses up resources. In particular, each REQALL claims one PLT entry. Intuitively, it appears to be impossible that more than four 8-packet messages could be outstanding at any time since there is reassembly buffer space for only four such messages (34 reassembly buffers). If, however, the buffer space that is freed when message i is reassembled causes an ALL to piggy-back on an RFNM of message $i - j$ $(j \geqslant 1)$, then the RFNM for message i may queue up in the destination IMP behind $j - 1$ other RFNM's! Thus only four messages really have buffer space allocated. In addition to these four, there are other outstanding messages which have already reached the destination IMP and which have RFNM's waiting for buffer space (i.e., waiting for piggy-backed ALL's).

The sending of more than four 8-packet messages is initially caused by the sending of superfluous REQALL's. The PLT entries which were obtained by these REQALL's are later used by regular messages. When the PLT is full, further input from the source HOST is stopped until a PLT entry becomes avail-able (this results in the inefficient use of transmission facil-ities). Thus we have a situation where our HOST uses all six entries in the PLT for the transmission to a destination HOST.

Fig. 7 graphically depicts the kind of phasing we observed for almost all transmissions over more than 4 hops. Let us briefly explain the transmission of message i. At time a the last packet of message $i - 1$ has been accepted and input of the first packet of message i is initiated. This first packet is re-ceived by the source IMP at time b. Since there is a buffer allocation available (which came in piggy-backed on the RFNM for message $i - 7$) no REQALL is sent. However, the PLT is full at time b. Therefore, message i must wait until time c when the RFNM for message $i - 6$ frees a PLT entry and message i may then proceed. At time d all 8 packets have been accepted by the source IMP. The first and eighth packet are received by the destination IMP at times e and f, respectively. The sending of the RFNM for message i is delayed until the RFNM's for messages $i - 3$, $i - 2$, and $i - 1$ are sent. The buffer space that is freed when message $i + 3$ reaches the destination at time g is piggy-backed on the PFNM for message i which reaches the source IMP at time h. This effect may be seen in Fig. 7 by observing the time slice picture while message i is in flight. Here we show messages as rectangles and RFNM's as ovals. Attached to RFNM's and messages are the ALL and PLT resources they own. We see the four ALL's owned by messages $i - 1, i, i + 1$ and by the RFNM for message $i - 5$; we see the six PLT's owned by messages $i - 1, i$ and by the RFNM's for

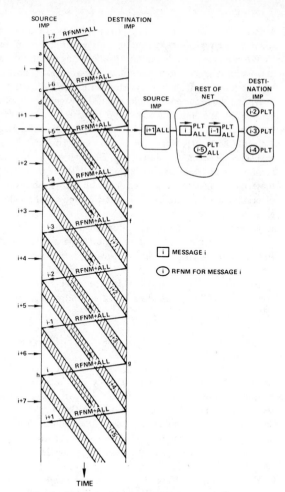

Fig. 7. Phasing and its degradation to throughput.

messages $i - 5, i - 4, i - 3, i - 2$. Message $i + 1$ cannot leave the source IMP since it is missing a PLT; most of the PLT's are owned by RFNM's who are foolishly waiting for piggy-backed ALL's which are not critical resources at the source IMP (message $i + 1$ has its ALL!). The trouble is clearly due to a poor phasing between PLT's and ALL's.

The phasing described above was observed for destination IMP's without the VDH software. For VDH IMP's, which can only reassemble one message at a time (10 reassembly buffers), a different kind of phasing was observed which resulted in even more serious throughput degradations! In this case, a situation is created in which a REQALL control message is sent for every data message. The 6 PLT's are assigned to 3 REQALL's and 3 data messages. Fig. 8 depicts this situation. The first packet of message i is transmitted from the source HOST to the source IMP between times a and b. Since there is no buffer allocation available, the source IMP decides to send a REQALL. However, all the PLT's are assigned and therefore the sending of the REQALL message is delayed until time c when the reply to an old REQALL (for message $i - 3$) delivers an ALL and a PLT. At this time, the PLT entry is immediately stolen by the

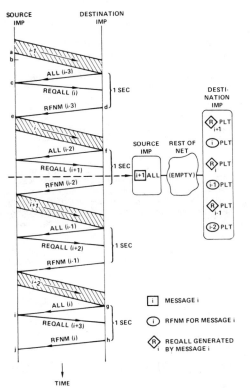

Fig. 8. Phasing when only one multipacket message can be assembled.

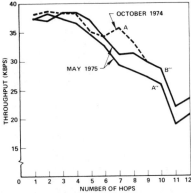

Fig. 9. ARPANET throughput (May 1975).

delayed REQALL (generated for message i). Note that at this point, message i gets the necessary buffer allocation but it cannot be sent to the destination because the PLT is once again full! Only when the RFNM for message $i - 3$ times-out after 1 s (time d) and is received by the source IMP (time e) without a piggy-backed ALL does a PLT entry become free for use by message i. At time f, all 8 packets have been received by the destination IMP. The sending of the RFNM for message i is now delayed by several seconds because the replies for messages $i - 2$, $i - 1$ and for two previous REQALL's must be sent first (see the time-slice given by the dashed line in Fig. 8 which shows REQALL's as diamonds and is taken during the 1-s time-out when nothing is moving in the net). At time g, the ALL control message responding to REQALL (i) is sent to the source IMP, and 1 s later at time h the RFNM for message i times-out. The RFNM is finally received for message i by the source IMP at time j.

The phasing in the case of destination IMP's with VDH software results in throughput degradations by a factor of 3. This large decrease is due to the fact that the system is stalled for almost 1 s while the source IMP has the buffer allocation but no PLT entry; during this delay, the destination IMP, which can free a PLT entry by sending an RFNM, is waiting for the buffer allocation to use as a piggy-back. There are two obvious ways to avoid this undesirable phasing of messages. First, one can avoid sending superfluous REQALL's which are the underlying causes of the phasing. Secondly, one can avoid the piggy-backing of allocates on RFNM's as long as there are other

replies to be sent. This second method was suggested and implemented by BBN.

In Fig. 9 we show some more recent throughput measurements made in May 1975 (after the phasing fix). As in Fig. 3, we show the throughput as a function of hop distance, with curve A'' displaying the throughput averaged over the 10-min experiment and curve B'' displaying the throughput for the best (maximum throughput) 150 consecutive messages. Curve A from Fig. 3 (October 1974) is included for a comparison of the two throughput experiments. We note that the throughput with the new message processing procedure is inferior to that in October 1974, although it is far better than that which we observed in February 1975 prior to the phasing fix.

VIII. CONCLUSIONS

In this paper we have described procedures for, limitations to, and measurement of throughput in the ARPANET. We identified some sources of throughput degradation due to the latest message processing procedure and displayed performance measurements after some of these problems were corrected. Here, as with many other deadlocks and degradations, it is rather easy to find solutions once the fault has been uncovered; the challenge is to identify and remove these problems at the design stage.

The ARPANET experience has shown that the building of a modern data communications network is an evolving process which requires careful observation and evaluation at each step along the way. Although the ARPANET was the first large-scale experimental packet-switched net and therefore underwent regular changes (as one would expect in any pioneering experiment) we foresee a continuing need for system evaluation.

The function of network measurements should not only be to test the initial configuration and make sure that it behaves according to specification. Indeed the rapid growth of these networks and the necessary changes in hardware and software make it extremely important to constantly reevaluate the total system design by means of analysis and measurements. This is the only guarantee for detecting performance problems as they arise and for acting accordingly before users experience degraded service.

REFERENCES

[1] G. Hicks and B. Wessler, "Measurement of HOST costs for transmitting network data," ARPA Network Information Center, Stanford Research Institute, Menlo Park, CA, Request for Comments 392, Sept. 1972.

[2] R. Kanodia, "Performance improvements in ARPANET teletransfer from multics," ARPA Network Information Center, Stanford Research Institute, Menlo Park, CA, Request for Comments 662, Nov. 1974.

[3] L. Kleinrock and W. E. Naylor, "On measured behavior of the ARPA network," in *AFIPS Conf. Proc.*, vol. 43, pp. 767–780, 1974.

[4] L. Kleinrock, W. E. Naylor, and H. Opderbeck, "A study of line overhead in the ARPANET," *Commun. Ass. Computing Machinery*, vol. 19, pp. 3–13, Jan. 1976.

[5] L. Kleinrock, *Queueing Systems, Vol. II: Computer Applications.* New York: Wiley, 1976.

[6] J. M. McQuillan, W. R. Crowther, B. P. Cosell, D. C. Walden, and R. E. Heart, "Improvements in the design and performance of the ARPA Network," in *AFIPS Conf. Proc.*, vol. 41, pp. 741–754, 1972.

[7] J. M. McQuillan, "Throughput in the ARPA network—Analysis and measurements," Bolt, Beranek and Newman, Inc., Cambridge, MA, Rep. 2491.

[8] —, "Adaptive routing algorithms for distributed computer networks," Rep. 2831, Bolt, Beranek and Newman, Inc., Cambridge, MA, 1974.

[9] W. E. Naylor, "A status report on the real-time speech transmission work at UCLA," Network Speech Compression Note 52, Dec. 1974.

[10] —, "A loop-free adaptive routing algorithm for packet switched networks," in *Proc. 4th Data Communications Symp.*, Quebec City, P.Q., Canada, Oct. 1975, pp. 6-1-6-11.

[11] H. Opderbeck and L. Kleinrock, "The influence of control procedures on the performance of packet-switched networks," *Nat. Telecommunications Conf.*, San Diego, CA, Dec. 1974.

[12] B. Wessler, "Revelations in network HOST measurements," ARPA Network Information Center, Stanford Research Institute, Menlo Park, CA, Request for Comments 557, Aug. 1973.

[13] D. C. Wood, "Measurement of the user traffic characteristics," ARPA Network Information Center, Stanford Research Institute, Menlo Park, CA, Network Measurement Note 28, May 1975.

A High Throughput Packet-Switched Network Technique Without Message Reassembly

ROY DANIEL ROSNER, MEMBER, IEEE,
RAYMOND H. BITTEL, MEMBER, IEEE, AND
DONALD E. BROWN

Abstract—A new packet-switching network technique is described which, while utilizing certain aspects of the ARPANET technology, introduces a substantially different technique for handling traffic which is longer than a single packet in length. The technique is keyed to a common-user network environment, where a wide variety of subscriber types, ranging from computers to simple terminals, are to be serviced. Subscribers in most cases would be remotely located from the network switching nodes. By splitting the buffering between the originating and destination nodes and by essentially eliminating the segment reassembly process, substantial reductions in on-line buffering can be achieved, while still maintaining short response times for interactive messages and large bandwidths for long data exchanges.

In this paper we describe the network operational concepts and traffic flow for various subscriber types, show specific examples and timing diagrams for message flows, and present a comparative analysis of the buffer sizing, throughput, and delay for this new technique compared to the well-known ARPANET technique of packet switching.

I. INTRODUCTION AND BACKGROUND

Despite the strong technical similarities between the store-and-forward switching of message traffic and the packet switching of computer-derived data packets, the development of packet switching through the efforts of the Defense Advanced Research Projects Agency (ARPA) has heralded a vast new interest in data communications network development. The development and implementation of the ARPA Intercomputer Network (ARPANET), together with the prolific documentation provided by its researchers, have provided a technical basis for what is actually a new subspecialty of communications technology.

The farsighted leadership provided by the ARPANET researchers in the field of computer communications has resulted in a tendency for many specialists to equate packet switching with the actual ARPANET implementation. The point has been made previously [1] that packet switching as a concept refers to the mode of handling data-oriented communications, but it is not limited to the particular distributed network concept embodied in the experimental (and now operational) ARPANET. This fact has now become more widely recognized as evidenced by numerous new projects [2]–[4] which, while packet-switch based, depart substantially from the ARPANET technology.

This concise paper describes a packet-switching technique which minimizes the effects of certain problems encountered during the early operation of ARPANET, but which were later corrected as described by McQuil-

len *et al.* [5]. More importantly, the described technique is inherently designed to provide service to a wide variety of subscriber types, ranging from large computers to simple terminals, all of which may be remote from the switching nodes. It provides a consistently short response time for interactive (I/A) traffic, while maintaining a moderately large bandwidth to handle high-volume (bulk data) users. By using buffering very efficiently, it requires significantly less buffering at the switch nodes than do present packet-switching techniques, and as far as has been determined, appears to be free of potential lockups and critical race situations, especially the problems associated with message reassembly.

II. THE DATA NETWORK TOPOLOGY

In order to describe the new network technique and traffic flow concept, a few words about the user environment and topology are needed. The concept is based on a large-scale nationwide network, serving a large number of diverse users. It is configured in the context of a nationwide network utility where, in general, the switching nodes are a public resource (though they may be privately owned) and the user sources and sinks (computers, terminals, and other digital input devices such as facsimile) are remote from the switching nodes. To be feasible, the network has to be economical both as a switched network facility and as a direct replacement for private-line, point-to-point services. There should also be minimal user impact in terms of protocols, procedures, and communications overhead in order to make the service attractive.

Key in the description of the user environment are the operating speeds and required throughputs. While the capability of large instantaneous bandwidths (such as the nominal 50 kbit/s capability provided by the ARPANET) is very attractive, both for experimentation and for certain direct computer-to-computer applications, the majority of present data communications takes place at speeds below 9600 bits/s. In fact, current commercial practices classify speeds below 300 bits/s as low speed, 300 to 2000 bits as medium speed, and anything above 2400 bits/s as high speed. Furthermore, the environment of a data communications utility requires in most cases (since the users are remote from the switching nodes) conventional modem access line connections for network access. This leads to a continuation of the practice of achieving a relatively good match between the user's true throughput requirement and his line speed, despite the fact that the common-user network may have a much higher node-to-node throughput capability.

The full protection of error control, message acknowledgment (ACK), and flow control has to be applied to the user-to-node access line as well as between the nodes of the network. Finally, the very bimodal distribution of user traffic is recognized. As pointed out in various studies, most recently by Kleinrock and Naylor [6], typical user data messages are either very short (20 to 600 bits) or very long (bulk messages ranging from 10^4 to 10^7 bits). No single message length in the midrange seems to offer a suitable compromise for the definition of a single typical or average data message length.

The network being considered [1] has an environment of 27 500 subscribers and 3×10^{11} bits (300 000 Mbits) of daily traffic. However, the technique described is applicable in a network of much smaller proportions.

III. THE BASIC NETWORK MODE OF OPERATION

The technique described in this concise paper is designed to provide communications support for the data transactions associated with both terminal-to-computer and computer-to-computer interactive activity, and long message transactions associated with remote job entry (RJE), data base update or exchange (DBU), file transfer (FT), and point-to-point narrative message exchanges. The general traffic flow through the network is to a large extent similar to the current implementation of the ARPANET, with the fundamental difference that all user messages are limited to a single 2000-bit (nominal) packet. Network resources, processing time, and switch node buffering are not expended in the reassembly of user messages. At the same time, however, protocol features are included which to a high degree of confidence, insure the properly sequenced arrival of successive packets (subunits of long messages) and which enable relatively high throughputs to be attained. Finally, as will be analytically shown, these fundamental changes in the operation of the packet-switched network lead to substantial reductions of required buffering for similar congestion probabilities compared to present techniques, or conversely, for similar buffering capabilities will result in sharply reduced buffer overflow (congestin) probabilities.

In the following discussion a segment is the unit of interface between the user and network node, and consists of a leader followed by the text bits. Long data messages and transactions may have to, in general, be broken into more than one segment. A packet is the mode of interface between the network nodes, and consists of a header followed by the text bits. The text length of the segment and packet are the same for this network concept (as opposed to the normal ARPANET mode of operation where a segment may be up to eight times as long as a packet). However, due to differences between the header and leader structures, the total sizes of the segment and packet are not identical.

A. Computer-to-Computer Traffic

In the specific case of computer-to-computer traffic, the source computer blocks transactions into a series of text units of as many as 1952 bits. Each of these blocks

contains control and cyclic redundancy code (CRC) bits. A leader is placed at the beginning of each block, forming a segment which is sent to the originating switch where the CRC is validated. If the CRC indicates that the segment was received without error, the originating switch removes the leader, constructs a header and places it as the beginning of the text block, and transmits header and block as the first packet to the next (or terminating) switch, retaining the packet in storage until the succeeding node has acknowledged its receipt. As soon as the current packet begins output from the originating switch, the source computer is permitted to input another segment on the same logical channel. However, the subsequent segment is held at the originating switch, pending receipt of the "in delivery ACK" (IDA) from the destination switch. It is assumed that there will be buffer space at the destination regional switch for a very high percentage of single-packet transactions, since the probability of a full buffer at the destination switch is very small.

When the terminating switch receives the packet, it assigns buffer space for the transaction in the termination buffer (access line output buffer), forms the segment from the packet by replacing the header with the leader, and links the segment to the termination buffer. When the segment actually begins output from the terminating switch, the IDA is sent back to the originating node.

Upon receipt of the IDA, the source switch releases all accountability for that segment, and immediately transmits the next segment which the source computer may have given to the source switch on that logical channel. When a segment is received and acknowledged by the destination computer, a delivery ACK in the form of a "request for next segment" (RFNS) is transmitted from the destination node to the originating computer. By using a two-level ACK exchange (the IDA between switching nodes and the destination to originator RFNS), the delay time of the originating access line can be avoided, and, in effect, the "pipeline" from source to destination kept full. If, for example, the round-trip network delay from source switch to destination switch is 150 ms, then this technique would keep the access pipeline full for up to a 15 kbit/s access line, which is expected to serve most of the users.

To a large extent network congestion is reduced through the combination of several features. The reduced holding time of buffers at the destination switch due to the fact that message reassembly is not performed leads to a higher probability that buffer space will be available for arriving packets for a given quantity of buffer space. If a queue of traffic exists for a particular destination, no additional traffic will be permitted into the net bound for that destination since the IDA is not returned to the originating nodes until the currently queued packets actually begin output to the destination. This effect also helps prevent the congestion which could result from a high-speed source (such as a computer) transmitting into the net at a much higher rate than the destination (such as a teleprinter terminal) can receive, thus holding a full buffer for an inordinately long time.

B. Low-Speed Terminal-to-Computer Traffic

In the case of low-speed terminal-to-computer traffic, the low-speed source terminal constructs a character-oriented leader and transmits it to the originating switch as a service request, with control characters that earmark it as either a single-packet message or an I/A session. When the originating switch receives and verifies the leader, it activates a pseudohost program used to process character-oriented transactions and acknowledges receipt of the leader to the source terminal. The terminal then transmits the block of text to the switch in character format. The pseudohost program routine of the originating switch converts the characters into a bit stream, generates a header, and transmits them as a packet to the next or terminating switch. At the terminating switch the received packet is assigned to the destination buffer. The terminating switch constructs a destination leader and transmits the leader and character text to the destination computer. When this computer has received the packet and validated the CRC, it transmits an ACK to the terminating switch. When the ACK is received, the terminating switch generates and transmits an RFNS to the originating switch. When the originating switch receives the RFNS, it transmits an ACK to the source terminal, and for a single-segment transaction deactivates the pseudohost program (or for an I/A session waits for the next transaction or the end of session control signal).

For multisegment transactions, the source terminal again constructs a character-oriented leader and transmits it as a service request to the originating switch. When the originating switch receives and verifies the leader, it activates the pseudohost program and generates a service request to the terminating switch, to establish the availability of the destination subscriber. If the destination subscriber is available, the terminating switch sends a ''go ahead'' (GA) to the originating switch. When the originating switch receives the GA, it instructs the source terminal to transmit its text to the originating switch in character format. The originating switch receives the text and builds packets by converting them to bit format and by generating a packet header. The originating switch then sends the packets, one at a time, using the procedure described previously for computer-to-computer exchanges, receiving an ACK for each before sending the next. At the terminating switch, packets are assigned to the termination buffer and then delivered to the destination computer. When the entire transaction has been received and acknowledged by the destination, the terminating switch generates and transmits a RFNS to the originating

switch which sends an ACK to the source terminal. The source terminal is then able to send additional text, or an indication that no further traffic follows which causes the originating switch to deactivate the pseudohost program. The terminating switch will retain destination buffer space for a predetermined timeout period for additional text of the transaction. Buffer space will be released before timeout occurs if the source terminal indicates to the originating switch that no further traffic follows.

C. Computer-to-Low-Speed Terminal Traffic

In the specific ease of computer-to-low speed terminal traffic, for single segment transactions, the source computer constructs a leader and transmits the leader and text to the originating switch. The originating switch validates the CRC, constructs a header, and sends the header and text as a packet to the terminating switch. When the header and text are received, the terminating switch assigns them to the destination buffer, activates the pseudohost program routine, and determines the status of the destination terminal. If the terminal is ready to receive, the pseudohost program performs bit-to-character translation, generates the character leader, and transmits the leader, text bits, and, after the last segment of the transaction, an end-of-transaction sequence to the destination terminal. The terminating switch then generates and transmits an RFNS to the originating switch. The RFNS is relayed to the source computer and is interpreted as an ACK of delivery and completion of transaction. After a predetermined time interval following the transmission of the RFNS, the terminating switch releases the destination buffer allocation for the message.

For multisegment transactions, the source computer constructs a leader and transmits the leader and first 1952 text bits to the originating switch. There the CRC is validated, a header is constructed, and the header and text are transmitted as a packet. The terminating switch receives the packet, assigns it to the destination buffer, activates the pseudohost program, and determines the status of the destination terminal. If the terminal is ready to receive, transmission is started and an IDA is transmitted to the originating switch. The remaining segments of the transaction can now be transmitted, one at a time, to the originating switch, and then on to the terminating switch, according to the procedure discussed for computer-to-computer transactions. When the entire transaction has been received at the terminating switch, and all required ACK's have been received, the pseudohost program, which performs bit-to-character translation and generates a character-oriented leader, will hold termination buffer available for subsequent transactions. This space will be released if the next segment is not received within a predetermined period of time.

In general, the input line rate for a computer is very high compared to the output line rate for the typical I/A or Q/R terminal. In this case, the procedure of always keeping a segment ready for transmission in the originating switch does not reduce significantly the overall segment delivery time (e.g., a 300 bit/s terminal will require over 6 seconds to receive a single packet, compared to a transnetwork time of less than $\frac{1}{2}$ second). Thus, except in the case of high-speed terminals, the procedure should require receipt of the RFNS at the source computer before additional segments are transmitted to the originating switch.

Bit-oriented intelligent terminals are presumed to have buffer capability and limited "functional intelligence." As such they should follow the same procedures as for computer-to-computer exchanges, except that in general they will be capable of supporting only a single logical channel to any other destination. Thus, the input and output streams to such terminals will represent sequences of transaction segments, all bound in order for a particular destination. However, it is possible for intelligent terminals to interleave transaction segments bound for different destinations, but each terminal-destination pair represents a unique message flow, each one over a single different logical channel.

In summary, the logical data network flows for various combinations of data subscribers follows in many aspects the procedures and protocols of the ARPANET. The key distinctions which lead to substantial buffering advantages include using single packet per segment sizing, delivering packets as received, and using in-delivery ACK's in order to pipeline packets to maintain throughput.

IV. SEGMENT AND PACKET DESIGN

The basic elements of communications exchange for the data communications network concepts are the data segment and the data packet. The segment, composed of a leader followed by text, is the element entered by the subscribers into the network for subsequent delivery to some other network subscriber. The packet, composed of a header followed by text, is the fundamental element handled and processed by the switch nodes of the network. The packet contains all of the data fields contained in the segment, and in addition makes provision for carrying various elements needed by the switches to maintain flow and control in the network.

Choosing the proper or optimum size for the packets involves a complex optimization of many factors, including the traffic message length statistics, the line error rate of the trunks (block retransmission rate), time delay objectives, and line throughput requirements. The choice is further complicated by the fact that variable length packets are used; that is, only the maximum packet length P_M is specified. A packet may actually contain from H up to P_M bits, where H is the number of header bits in the packet. Thus, a packet may contain between zero and P_M-H text bits, since certain control or ACK packets may actually contain the header only. Thus, any mathematical optimization of P_M is clearly dependent upon H, as well as the distribution of full and partially full packets.

If it is presumed that the number of bits in the header is fixed by the minimum header information needed by the network at a value H, then the following facts are relevant to the choice of P_M, or the maximum packet length:

1) The longer P_M becomes, the smaller the average functional overhead ratio (H/P_M) for the packet-switching overhead.

2) The longer P_M becomes, the more likely an error will occur during a single-packet transmission time, and the more likely a packet retransmission will be needed. This can reduce channel throughput if P_M is too large for the channel error rate.

3) Increasing P_M increases the average network delay. The delay for full packets is proportional to their length, and the delay for less than full packets is increased by having to wait longer in queue if they happen to be in queue behind full packets.

4) It is desirable to make P_M sufficiently large so that a very high percentage of all data transactions can be accommodated in a single packet. This leads to minimum overhead and reduced complexity in user-to-user protocol.

Using these considerations, a maximum packet length of about 2000 bits was found to be optimum for line efficiency of typical circuits when the existence of rather poor access circuits is presumed. It has been shown [7] that for the ARPANET the optimum packet length to maximize throughput is about 4000 bits, but the network efficiency is not compromised significantly for lengths between 1000 and 8000 bits. (The analysis of [7] did not account for the fact that many of the user messages are much shorter than 4000 bits.) Typical modem access circuits generally operate in a more error-prone environment, compared to the higher quality 50 kbit/s ARPANET trunk circuits, thus leading to a smaller optimum (2000 bits). Further, it is estimated that a packet length of 2000 bits would accommodate approximately 60 percent of the data transactions based on currently measured transaction statistics with only bulk data, long query-response transactions, and some narrative/record messages not accommodated in a single packet. However, it may not always be possible to meet a 1-second delay constraint for a full 2000-bit packet of data, but the delay constraint of less than 1-second delay for 600-bit transactions can be achieved, despite the presence of full 2000-bit packets in the network.

The maximum-size user message segment, the segment being the element of interchange between the subscribers and the network, could theoretically be either the same size as the packet or longer. If longer, it is logical that it would be a length equal to an integer multiple of the maximum packet length. In the ARPANET, for example, the segment at 8000 text bits is eight times as long as the 1000-bit packets. The ARPANET experience to date shows that even with a 1000-bit maximum packet size, a very small fraction (less than 1 percent) of all transactions use multiple packets, and of those that do use multiple packets (long file transfers) most are much longer than 8000 bits (single segment). The conclusion is simply that the ARPANET parameters do not provide a universal solution for meeting all user needs, and that other designs may be better for specific applications.

In the common-user network concept, all users utilize standard communications access lines, in a *non*-error-free environment. Error control, using CRC and limited-length transmission blocks, is required on access lines as well as on the node-to-node trunks. As previously discussed, a maximum block size of about 2000 bits on access lines is attractive. The transmission block on access lines is thus logically the segment, and thus the basic text (and therefore information) content of the segment is made identical to that of the packet. They differ in the overhead structure, however, with the leader and header being significantly different.

The major concern with use of segments and packets of the same size is the sequencing of packets at the destination node, and assurance that segments are delivered in sequence to the destination user. The flow procedure which permits only three segments outstanding in the network at a time (one in delivery to the destination user, one in flow to the destination switch, and one in access to the originating switch) assures that segments can arrive only in the same sequence in which they were originated. To provide a positive check on sequencing and to protect against lost or duplicated packets, the packets should, as part of the host-to-host leader information, include a host-generated sequence number. Throughput is not materially affected by the host having to originate four times as many segments as if the segment length were 8000 bits, because the host would have to block and error control every 2000 bits anyway to utilize the communications access line.

V. TRAFFIC FLOW EXAMPLES

The timing sequences in Figs. 1, 2, and 3 illustrate how the protocol concept described in the preceding sections would work for three typical cases. The figures are drawn under the assumption that on the average the typical source-to-destination path employs one intermediate, or tandem, node. While this is probably a poor assumption under the ARPANET topology, studies [8] have shown that large networks can economically employ tandem nodes in a hierarchical configuration in order to assure that the average number of hops per source-destination node pair is on the order of two. In such a topology, using 50 kbit/s trunks, the average time for a full (2000-bit) packet to transit from the originating

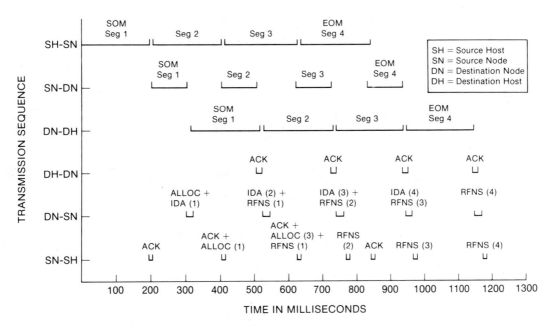

ACK = Error Check Acknowledgement Only
ALLOC = Source Node is ready for next segment on this logical channel
IDA = Segment #n is now is process of output to destination host
RFNS = Destination Host has acknowledged receipt of segment #n

Fig. 1. Approximate traffic flow timing sequence—two 9600 bit/s users.

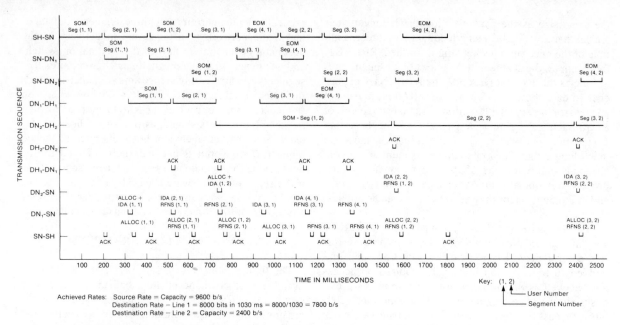

Fig. 2. Approximate user traffic flow timing sequence—one 9600 bit/s user sending to a 9600 bit/s destination and a 2400 bit/s destination.

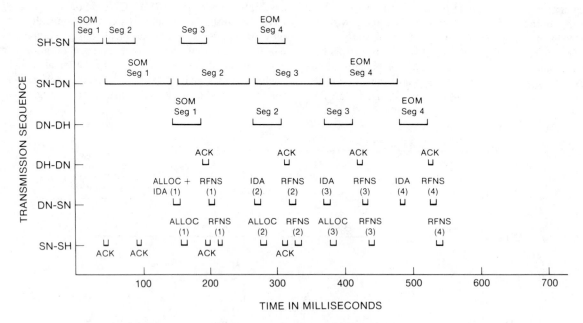

Fig. 3. Approximate traffic flow timing sequence—two 50 kbit/s users.

node to the destination node would be on the order of 100 ms (40 ms on each of two links plus 20 ms of processing and queuing time).

In Fig. 1, the case of two 9.6 kbit/s subscribers is shown, which would be typical of a small computer- (or intelligent terminal) to-computer environment. Fig. 2 shows the case of a single 9.6 kbit/s subscriber (such as a computer) generating traffic for two different destinations, one served at 9.6 kbits/s and the other at 2.4 kbits/s. Fig. 3 shows the high-speed computer-to-computer case with two 50 kbit/s subscribers. In each example, the transmission sequences are shown for the source host to source node (SH-SN), source node to source host (SN-SH), source node to destination node (SN-DN), destination node to source node (DN-SN), destination node to destination host (DN-DH), and destination host to destination node (DH-DN). For simplicity, both terminal and computer subscribers are termed "hosts." In each case, the example is drawn for a message or transaction consisting of four packets. Clearly, the flow control of this technique for single-packet messages is sufficiently similar to the ARPANET as to need no further detailed description. The major advantage of this technique, as elaborated on in the next section, is that, because of the very efficient use of buffer space in the switches, new transactions will be able to enter the network at any time with a high probability of acceptance and delivery.

The figures show the general flow of information through the network, together with the various sequences of ACK's and control messages. The specific implementation of the protocols would have to be developed in significantly more detail in order to actually implement an operating system, but the basic operating principles have been examined by many people who are very familiar with various packet-switching techniques, and the advantages as described appear to be well founded. As seen from both Figs. 1 and 2, 9600 bit/s access lines are readily filled to capacity, for long transfers, both on input and output, while 50 kbit/s access lines are filled to about 50 percent capacity (on a single logical computer-to-computer channel), even for the case of a tandem connection. More sophisticated host-to-host protocols could be implemented which would use multiple logical channels between two hosts having large quanities of traffic to exchange, such that throughputs nearly equal to the access line rates could be achieved.

Various proposals have been advanced to decrease the network overhead [9] by grouping control messages between nodes, and further increasing throughput [10] by positively controlling more than one packet "in flight" between the source and destination nodes by using network-supplied packet sequence numbers. However, since the purpose of this paper is to present the basic network concept, together with its major advantages, further description of possible operational sophistication is reserved for future work.

VI. ANALYSIS OF DELAY AND BUFFER SIZING

A major area of concern in the design of a packet network protocol is the determination of the buffering needed to support the packet "reassembly" function at a destination switch. In general, switch congestion, leading to traffic restriction from the network, is due primarily to the lack of reassembly or termination buffering.

A. The Network Model

The reassembly buffer can be sized from a knowledge of the network and access area delays that a packet incurs, and from knowledge of the traffic and network statistics. The single most important factor that impacts network delays is the network protocol. The protocol influences how the packets are to be sent through the network and what management strategy is to be used to control the packet flows. This directly impacts the delay in getting information through the network. In turn, this delay inversely affects the network throughput which affects the required buffering. Hence, these factors, once determined, contain all the information required to size the reassembly buffer.

The network model used in this analysis consists of an origination switch, a tandem switch, and a destination switch connected by transmission lines of known data rate. The switch processors are assumed to be identical. The path length for this model is 2 links. It is assumed that the access area transmission line speed is a known constant. The traffic is assumed to be statistically characterized by an independent Poisson process with parameter λ, the arrival rate. The switch processors are assumed to serve the arriving packets on a first-come-first-served basis, and the service time is assumed to be exponentially distributed with parameter μ, the mean service rate. It is assumed that only a finite amount of reassembly buffer is available and that packets, if blocked at the switch will be delayed and not cleared.

The traffic in the network is based on a particular requirements model as shown in Fig. 4, but the compar-

COMPUTERS AND TERMINALS

Average Computers/Switch	36
Average Terminals/Switch	324

AVERAGE COMPUTER TRAFFIC

TYPE	RATE (NO./HR)	AVERAGE LENGTH (BITS)	BITS/HR.
Queries	241	600	1.446×10^5
Responses	257	6000	15.42×10^5
Bulk	31	500000	155.0×10^5
Message	42	20000	8.4×10^5

AVERAGE TERMINAL TRAFFIC

TYPE	RATE (NO./HR)	AVERAGE LENGTH (BITS)	BITS/HR.
Queries	144	600	0.864×10^5
Responses	—	—	—
Bulk	14	500000	70.0×10^5
Message	12	20000	2.4×10^5

Fig. 4. Busy hour traffic model.

ative results of the analysis can be generalized over a broad range of user requirements. The network model shows each switching node serving 36 computers and a total of 324 terminals connected to the switching node via access lines ranging from 150 bits/s to 9600 bits/s. Fig. 4 shows the assumed traffic characteristics for the computer and terminal subscribers.

Because only the reassembly buffer being sized, only messages of length greater than some specified minimum length (one full packet) are considered as contributing to the traffic statistics. Messages of shorter lengths do not generate a need for reassembly space. Other traffic to be considered is that generated as a function of the assigned protocol with respect to ACK's and message requirements for buffer allocation.

Two different switch-to-switch protocols are compared as to their effect on buffering. Protocol I assumes that each message, as distinguished from a packet or segment, initiates one (and only one) request for buffer allocation (RFA) at the destination switch. The buffer allocation is held for the entire message, whether single or multiple segment. For Protocol I, it is assumed that the segments are multipacket. The allocation is released to the buffer pool only after the complete message has been received by the destination host and a positive host ACK for the last segment of the message has been received by the destination switch. In the network, each packet is acknowledged on a switch-to-switch basis. An RFNS is generated by the destination switch, after it has correctly received the host ACK for the previous segment. This RFNS is transmitted back to the originating switch indicating that the destination switch is ready for the next segment of data. This multipacket-per-segment technique is analyzed to provide a baseline comparison to which the single-packet-per-segment technique is to be compared.

Protocol II is a radical departure from the multipacket-segment concept. Here, one packet equals one segment. There are no multipacket segments. In the network, each packet is acknowledged on a switch-to-switch basis. An RFNS per packet is returned to the originating switch. It is transmitted only after the packet has been correctly received by the destination switch. In Protocol I, a buffer reassembly allocation was required, but Protocol II does not require a reassembly buffer, only an output queue buffer to the destination subscribers.

A number of cases as shown in Table I are now examined. The traffic is assumed to be statistically characterized by an independent Poisson process with density function

$$f_n(t) = \begin{cases} (\lambda t)^n (e^{-\lambda t})/n!, & n = 0,1,2,\cdots \\ 0, & \text{otherwise} \end{cases} \quad (1)$$

where λ is the arrival rate. The switch service times are assumed to be exponentially distributed with density function

$$f(t) = \begin{cases} \mu e^{-\mu t}, & t > 0 \\ 0, & \text{otherwise} \end{cases} \quad (2)$$

where μ is the mean service rate. It is assumed as a characteristic of the network that any information units which enter the system will not be discarded if they find the switch busy. That is, blocked calls are delayed, not cleared. However, the buffers at the destination switch may be full, implying that no RFNS requesting the next message will be sent. This is a form of flow control. The network must be designed to allow information units which enter the system to reach their destination. The parameters λ and μ are determined from the traffic statistics and processor characteristics.

TABLE I
PACKET AND SEGMENT SIZES

Protocol	Segment Size	Packet Size	Packets/ Segment
IA	8K	1K	8
IB	8K	2K	4
IC	16K	2K	8
IIA	1K	1K	1
IIB	2K	2K	1

B. Delay

To exercise the protocols, the network delay and the access delay must be calculated. The network delay consists of the processor queuing delay, the processing delay, and the transmission delay for the message text, ACK, and control packets. The access delay includes the access-transmission delay, the destination-host processing delay, and the destination-switch ACK-processing delay.

The switch processor appears as a single server to the input queue and the delay experienced by the arriving packets, excluding the time being served, is computed [11], [12] as

$$E(t_w) = \frac{E(n)E(t_s^2)}{2(1 - p)} \quad (3)$$

where E denotes expectation, t_s is the service time for an information unit, n is the number of arrivals per second, and p is the traffic intensity where $p = E(n)E(t_s)$. With the assumed statistics, $E(n) = \lambda$ and $E(t_s^2) = 2E^2(t_s) = 2/\mu$. Equation (3) then becomes

$$E(t_w) = \frac{\lambda}{\mu(1 - p)} = \frac{p^2}{\lambda(1 - p)}. \quad (4)$$

The total expected waiting time in the queue is the sum of the expected queuing time plus one service time. This expected waiting time at the input buffer is

$$E_1(t_q) = E(t_s) + E(t_w) = \frac{1}{\mu(1 - p)}. \quad (5)$$

The expected switch processor delay can be computed from a knowledge of the computational load placed on the switch by the processing of one information packet or segment [5]. If D is the packet length in bits, S is the software overhead in bits, P is the number of packets in a segment of B bits, R is the I/O transfer rate, V is the overhead for various periodic functions (primarily routing computation), and C is the program processing time per packet, then the expected processor delay is

$$E(t_{sw}) = \left[PC + \frac{2(B + PS)}{R} + C + \frac{2S}{R} \right] (1 + V). \quad (6)$$

It is assumed that a piggyback ACK scheme has been employed for packet ACK's [5].

The output delay is a function not only of the traffic statistics and service time, but also of the number of output transmission lines. From [12], assuming an exponential service time, the delay incurred at the output queue is given by

$$E_2(t_q) = \frac{K}{M\mu(1 - p)} + \frac{1}{\mu} \quad (7)$$

where M is the number of parallel transmission lines, μ is the information unit transmission rate, p is the traffic intensity defined as $p = \lambda/\mu M$, and K is a constant defined by

$$K = \frac{\dfrac{(Mp)^M}{M!}}{\dfrac{(Mp)^M}{M!} + (1 - p) \displaystyle\sum_{i=0}^{M-1} \dfrac{(Mp)^i}{i!}}. \quad (8)$$

The transmission line delay for an information unit is given by

$$E(T_D) = \frac{1}{\mu R_T} \quad (9)$$

where $1/\mu$ is the expected number of bits/information unit, and R_T is the transmission rate of the lines.

In the access area, the processing delay is defined by (3) and the transmission delay is given by (9), with suitable changes in parameter values.

The network delay D_N can be found by summing the individual delays. Given the average path length of 2, then

$$D_N = 2E(T_D) + 3[E_1(t_q) + E(t_{sw}) + E_2(t_q)] \quad (10)$$

where $E_1(.)$ and $E_2(.)$ are obtained from (5) and (7), respectively. This formulation corresponds somewhat to Kleinrock's work [13]–[15] where he considered the processor as a single server queuing model. In the formulation given here, the structure of the processor has been used to break up the switch into an input queue, a processor as a single server queue, and an output queue. With this arrangement, the influence of piggy-back versus no piggyback ACK's can be studied in greater detail.

To incorporate the effect of an error environment into the delay calculation, a binary symmetric channel (BSC) has been used [7]. The only effect on the delay calculation, as given in (10), is a multiplicative factor $1/1 - P_r$ on $E(T_D)$, since only the transmission efficiency is directly affected. P_r is the probability of retransmission.

In order to compute the amount of reassembly buffering required, the delays affecting the buffer size must be calculated. It is important to note that these delays are a function of the protocol used. These delays appear to the destination switch as the time interval for which the buffer must be held in order to service the arriving packets. This time interval is the buffer holding time t_h. The buffer holding time is determined from the summation of all the delays in the network from the time the first packet of a segment effectively makes the buffer reservations, until the last segment of the message is acknowledged by the destination host. Of course, for Protocol II, only a single packet per segment is used so the buffer holding time is just the delivery delay to the destination plus the network delay to hold the buffer awaiting the arrival of the next packet.

The buffer pool appears to its destination switch as a multiserver of unknown number. Under the stated assumptions, the formulation is Erlang C. The only unknown parameter is the blocking probability of the buffer, or equivalently, the probability of buffer overflow P_{of}. Specifying a value of P_{of} allows calculation of the number of buffers given the arrival rate and holding time. To evaluate the number of buffers N, (11) is to be solved for N.

$$P_{of} = \frac{\dfrac{E^N}{N!} \left(\dfrac{N}{N - E} \right)}{\dfrac{E^N}{N!} \left(\dfrac{N}{N - E} \right) + \displaystyle\sum_{i=0}^{N-1} \dfrac{E^i}{i!}} \quad (11)$$

where E is the number of Erlangs of traffic, $E = \lambda t_h$. Computer programs have been written to solve (11) for N, given P_{of} and E. Note that the protocol enters the calculation only through t_h and therefore E. Thus the protocol influence on buffering is established.

TABLE II
PROTOCOL I DELAYS

Protocol	Average Network Delay (per message, in seconds)	Average Access Delay (seconds)	Average Buffer Holding Time (seconds)
IA	2.81	2.08	4.89
IB	2.73	2.07	4.80
IC	2.71	4.11	6.82

VII. ANALYTICAL RESULTS

Evaluation of (10) leads to an estimate of the network delays. Under Protocol I, employing multiple packet segments, the delays were estimated for cases IA, IB, and IC (Table I) as shown in Table II. The network backbone rate is assumed at 56 kbits/s and the average access rate is 4.8 kbits/s.

Protocol II, the new protocol concept described in this paper, was analyzed for 1000- and 2000-bit packets, one packet equaling one segment. The delays for a packet are shown in Table III.

Note that the network delay in Protocol I is less than eight times that of Protocol II. This is due to the fact that additional RFNS's are required to control the packet flows; that is, more control information is required for Protocol II.

The reassembly buffer size is a function of the holding time t_h and the arrival rate λ, the holding time being based on the sum of the network and access delays. For Protocol I, Fig. 5 gives the probability of buffer overflow as a function of the reassembly buffering for a backbone transmission rate of 56 kbits/s. Of importance here is the fact that for 16 000-bit segments, the buffers are multiples of 16 000 bits in length, not 8000 bits. Buffer size is given in terms of the total number of bits of buffering.

For Protocol II, Fig. 6 gives the buffer overflow probability. For the case where 2000 bits per packet has been assumed, the required buffering is given in 2000-bit buffers. This last case is particularly important because in doubling the length of a packet, an arrival has associated with it 2000 bits. The average holding time nearly doubles. This says that the traffic intensities for both packet sizes are nearly equal yet the total buffering for the 2000-bit packet case is nearly double that of the 1000-bit packet case. This is not true in the analysis for Protocol I, with respect to case IC, where the segment length was doubled to 16 000 bits. In this case the arrival-holding time product was greater than in the other two cases.

Table IV summarizes the required buffering results for a $P_{of} = 10^{-4}$.

Protocol II requires the least amount of buffering, showing a reduction of nearly 60 percent in the required online reassembly buffering (Case IB relative to IIA). This is expected since the buffer holding times are small in comparison with the other protocols, and also, there is no delay due to reassembly. Note that the buffer required for Protocol II does not depend on the delay that a packet incurs in the network. Rather, it depends only on the switch-processing time and access delay. Hence, the traffic intensity is reduced and the buffer size is reduced correspondingly.

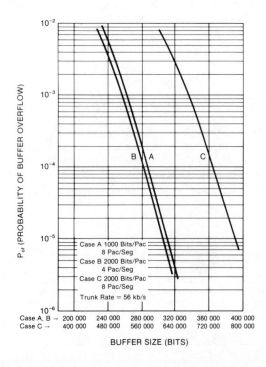

Fig. 5. Probability of buffer overflow versus required buffering—Protocol I.

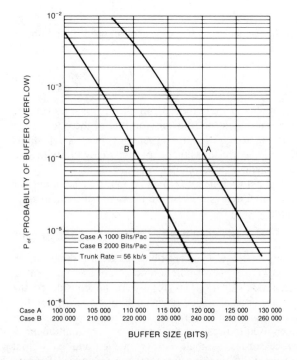

Fig. 6. Probability of buffer overflow versus required buffering—Protocol II.

From the analysis performed, Protocol I, cases A and B, and Protocol II appear to be the most useful candidates for a digital data network. The major advantages for Protocol II, in comparison to the others, are the required amount of buffering, and buffer utilization. However, the design goal of a digital communications system must be kept in mind. If the goal is to optimize the backbone network without consideration for the access area (i.e., as in the ARPANET where the hosts and nodes are collocated), then Protocol I is preferred since it maximizes throughput and network efficiency at the expense of buffering. If the goal is to efficiently utilize access area transmission as well as the backbone network, then Protocol II is preferable because delivery in the access area is simplified (no reassembly is required). Also, smaller packets can more easily compete for delivery with multipacket messages than they can using single segment/multipacket messages. If the delay constraints are flexible then Protocol IIB is the desired one. However, this decision is based on the amount and cost of online buffering.

TABLE III
PROTOCOL II DELAYS

Protocol	Average Network Delay (per message)	Average Access Delay	Average Buffer Holding Time (seconds)
IIA (1000 bits)	0.047	0.287	0.334
IIB (2000 bits)	0.084	0.542	0.626

TABLE IV
BUFFER REQUIREMENTS

Protocol	Buffer (bits) $T_R = 56$ kbits/s
IA	288×10^3
IB	280×10^3
IC	728×10^3
IIA	121×10^3
IIB	222×10^3

As a final note, it may be argued that the statistical assumptions about the processor are invalid. Rather than being exponentially distributed, the processor service time may be constant. From queuing theory, it is known that in most circumstances, an exponential assumption on the service time gives a worst-case analysis. In the analysis presented here, this is true. Thus, the results presented give conservative estimates of the delays, throughput, efficiencies, and buffering. However simple the topological structure of the network model, the basic results should not change in a more complicated topology.

VIII. CONCLUSIONS

A detailed description has been presented of a concept for a digital data network which can operate in the realistic environment of a backbone network serving diverse users who access the network by means of communication lines of limited bandwidth. The technique described significantly reduces the required buffering (by as much as 60 percent, compared to current ARPANET technology) at the network nodes, thus keeping the network nodes small in size, while causing only minimal reduction in network throughput. As this technique is further developed, some rather simple modifications will increase throughput for long messages by allowing for some additional complexity in the protocol procedures.

ACKNOWLEDGMENT

The authors would like to thank the many people who have contributed ideas that have been incorporated in the techniques described in this concise paper. Dr. H. A. Helm, D. L. Kadunce, and W. P. Hawrylko of DCA; E. P. Brown of Data System Analysts, Inc.; Dr. D. Walden and Dr. J. McQuillan of Bolt Beranek and Newman, Inc.; and Dr. L. Kleinrock of UCLA have all been extremely helpful in developing these concepts.

REFERENCES

[1] R. D. Rosner, "A digital data network concept of the defense communications system," in *Proc. Nat. Telecommun. Conf. (NTC'73)*, Atlanta, Ga., 1973, p. 22-C1.

[2] R. Despris, "RCP, the experimental packet switched data service of the French PTT," in *Proc. Int. Conf. Comput. Commun. (ICCC'74)*, Stockholm, Sweden, 1974, p. 171.

[3] K. Hirota, M. Kato, and Y. Yoshida, "A design of a packet switching system," in *Proc. Int. Conf. Comput. Commun. (ICCC'74)*, Stockholm, Sweden, 1974, p. 151.

[4] D. J. Pearson and D. Wilken, "Some design aspects of a public packet switched network," in *Proc. Int. Conf. Comput. Commun. (ICCC'74)*, Stockholm, Sweden, 1974, p. 199.

[5] J. M. McQuillan, W. R. Crowther, B. P. Cosell, D. C. Walden, and F. E. Heart, "Improvements in the design and performance of the ARPA networks," in *1972 Fall Joint Comput. Conf., AFIPS Conf. Proc.*, vol. 41. Montvale, N. J.: AFIPS Press, 1972, p. 741.

[6] L. Kleinrock and W. E. Naylor, "On measured behavior of the ARPA network," in *Proc. Nat. Comput. Conf.*, vol. 43. Montvale, N. J.: AFIPS Press, 1974, p. 767.

[7] R. M. Metcalfe, "Packet communications," rep. MAC TR-114, Dec. 1973.

[8] R. D. Rosner, "Large scale network design considerations," in *Proc. Int. Conf. Comput. Commun. (ICCC'74)*, Stockholm, Sweden, 1974.

[9] E. P. Brown, private communication.

[10] J. M. McQuillan and D. Walden, private communication.

[11] F. S. Hillier and G. J. Lieberman, *Introduction to Operation Research*. San Francisco, Calif.: Holden Day, 1972.

[12] J. Martin, *Systems Analysis for Data Transmission*. Englewood Cliffs, N. J.: Prentice-Hall, 1972.

[13] L. Kleinrock, *Communication Nets*. New York: Dover, 1964.

[14] ——, "Analytic and simulation methods in computer network design," in *1970 Spring Joint Comput. Conf., AFIPS Conf. Proc.*, vol. 36. Montvale, N. J.: AFIPS Press, 1970, pp. 569–579.

[15] H. Frank *et al.*, "Computer communication network design—Experience in theory and practice," in *Spring Joint Comput. Conf., AFIPS Conf. Proc.*, vol. 40. Montvale, N. J.: AFIPS Press, 1972, pp. 255–270.

[16] G. D. Cole, "Computer network measurements—Techniques and experiments," Univ. of California, Los Angeles, rep. UCLA-ENG 7165, Oct. 1971.

THE NETWORK CONTROL CENTER FOR THE ARPA NETWORK

A. A. McKenzie, B. P. Cosell, J. M. McQuillan, M. J. Thrope
Bolt Beranek and Newman, Inc.
Cambridge, Massachusetts, U.S.A.

Abstract

The ARPA Network allows dissimilar, geographically separated computers (Hosts) to communicate with each other by connecting each Host into the network through an Interface Message Processor (IMP); the IMPs themselves form a subnetwork that can be thought of as a distributed computation system. To detect failures in this system each IMP automatically and periodically examines itself and its environment and reports the results to the Network Control Center (NCC), at Bolt Beranek and Newman Inc., for action. The NCC computer, like any other Host, can itself fail without affecting network integrity; further, the NCC central processor can easily be replaced, in case of failure, by any standard IMP.

The present paper briefly describes the NCC hardware; discusses such software issues as NCC-related routines in the IMPs, data-collection and interpretation mechanisms, line status determination, IMP status and program reloading, and Host and line throughput; details NCC operations (manning, problem-handling procedures, track record); and summarizes overall NCC experiences and future plans.

I. Introduction

Almost four years ago the Advanced Research Projects Agency of the Department of Defense (ARPA) began the implementation of a new type of computer network. The ARPA Network provides a capability for geographically separated computers, called *Hosts*, to communicate with each other via common-carrier circuits. The Host computers typically differ from one another in type, speed, word length, operating system, etc. Each Host is connected into the network through a small local computer called an *Interface Message Processor (IMP)*; each IMP is connected to several other IMPs via wideband communication lines. The IMPs, all of which are virtually identical, are programmed to store and forward messages to their neighbor IMPs based on address information contained in each message.

In a typical network operation a Host passes a message, including a destination address, to its local IMP. The message is passed from IMP to IMP through the network until it finally arrives at the destination Host. An important aspect of this operation is that the path the message will traverse is not determined in advance; rather, an IMP forwards each message on the path it determines to be best, based on its current estimate of local network delay. Since the path choices are determined dynamically, IMPs can take account of circuit or computer loading (or failures) in an attempt to insure prompt delivery of each message.

In three years the network has expanded from 4 to over 25 IMPs and is still growing. Early work on the ARPA Network is described in some detail in a set of papers presented at the 1970 Spring Joint Computer Conference[1-5]. Additional work is described in a paper presented at the 1972 SJCC[6].

An interesting aspect of the IMP subnetwork (i.e., the set of IMPs and communication lines) is that it can be considered a distributed computation system. Each IMP performs its own tasks relatively independently of its neighbor IMPs; nevertheless all IMPs are cooperating to achieve a single goal — reliable Host-Host communication — and in some cases, for example, the dynamic path selection mentioned above, each IMP coopperates with its neighbors in making reliable delay estimates for various path choices.

In any distributed computation system it is likely to be difficult to detect component failures quickly; the difficulty is increased in the IMP subnetwork by the wide geographic separation of components. For this reason we chose at the outset to incorporate automatic reporting functions in the IMPs as an aid to failure diagnosis. Each IMP is programmed to examine itself and its environment periodically and to report the results of these examinations to a central mediating agent. This agent has the function of collecting the (possibly conflicting) IMP reports, determining the most likely actual state of the network and, in the case of failures, initiating repair activity. The mediation function is performed by the Network Control Center (NCC) located at Bolt Beranek and Newman Inc. (BBN) in Cambridge, Mass. The mediating agent is the NCC computer, which is attached, as a normal Host, to the BBN IMP. It should be noted that although the NCC computer is an *important* component of the network it is not an *essential* component; as with any other Host it can fail without disturbing overall network integrity.

The NCC computer is concerned primarily with the detection of line failures and IMP failures. In addition, the NCC computer monitors the volumes of Host traffic and line traffic; these are parameters which can give advance warning of network elements whose capacity may need to be increased and which can be used for site usage accounting. Finally, the NCC computer keeps track of other data, such as switch settings and buffer usage, for each IMP; these data are frequently helpful in diagnosing IMP failures.

The remainder of this paper describes the operation of the Network Control Center. Section II describes the NCC hardware located at BBN and Section III provides details of the overall software operation. Section IV discusses the manual procedures followed by NCC operators and technical staff in diagnosing and correcting network malfunctions. In Section V we have provided typical summaries of the types of information collected at the NCC in recent months, and mention some anticipated changes in NCC operation.

II. NCC Hardware

The central site NCC hardware consists of two packages, a central processor with 12K of 16-bit memory, a real-time clock and a "special Host interface", and a special set of hardware designed specifically for NCC functions. The current CPU is a Honeywell 316 computer; this choice provides two important advantages. First, the "special Host interface" required for connection to the network is exactly identical to the "standard Host interface" already designed as part of the IMP, thus reducing the implementation cost. Second, because all special hardware has been kept modular and external to the CPU package, if the NCC computer goes down for an extended period, it can be replaced by any standard IMP (Model 316, Model 516, or Terminal IMP). This is significant because we frequently have several IMPs on site in preparation for field delivery; thus the potential for substitution of the NCC machine is of practical value.

The special NCC hardware consists of two dial-up line controllers, a half-duplex Teletype I/O interface, and hardware associated with a panel of 32 display lights, a programmable audible alarm, and 16 control switches. All of this equipment is housed in a separate cabinet along with the required power supplies. When necessary, it can be simply connected to the I/O bus of an alternate CPU.

Two Model 35 ASR Teletypes handle most of the input and output functions. One, attached through the Teletype interface in the special hardware package, is dedicated to a print-only logging function while the other, the NCC computer's standard console Teletype, serves both as a report printer and as the primary source of operator input. Input can also be provided through the 16 control switches, and other output is given via the 32 display lights and the alarm. The dial-up line controllers are reserved for possible future use. The external I/O equipment is duplicated at nearby locations for the convenience of NCC personnel.

III. Software Operation

The IMP subnetwork consists of three principal classes of components: 1) a collection of wide-band common carrier data lines, 2) a set of IMP processors, 3) IMP system software. Network performance can be affected by failures in any of the components in each of these classes. Therefore, in conjunction with our construction of the network, we had to develop procedures for quickly detecting and repairing component failures within any of these classes. In this section we will describe the software used to assist in detecting such failures.

NCC-Related Software in the IMPs

A basic assumption, which underlies the NCC effort, is that the most effective way of detecting failures is to have each IMP periodically compile a report on the status of its local environment and forward this report through the network to a mediating agent, the NCC. This agent has the task of collecting and integrating the reports from all of the IMPs to build up a global picture of the current state of the network. The data generation within each IMP is performed by two routines: a timing routine which controls the periodic execution of the report routine, and the report routine itself.

The timing routine used is the IMP's statistics mechanism. This mechanism establishes a network-wide synchronized clock which it uses to coordinate the execution of a set of self-measurement (statistics) routines which have been incorporated into the IMP. The bulk of the statistics routines are concerned with factors such as measuring IMP bandwidth capacity and storage utilization, etc. One of the statistics routines, however, is the "Trouble Reports" routine, which provides data to the Network Control Center.

The Trouble Reports routine, when initiated by the timing routine, interrogates various parts of the IMP system to determine which lines are alive, which Hosts are up, etc. It formats that information into a message which is forwarded to the NCC's data collection mechanism. Since space is at a premium in the IMP system, the routine does no pre-processing of the information; it is merely collected and forwarded.

In addition to the statistics and reporting packages, each IMP contains a small debugging package, DDT. DDT is a simple command interpreter capable of such functions as examining and modifying a memory word, clearing a block of memory, searching memory for a particular stored value, etc. DDT is structured so that it can be driven remotely through the network, returning any responses back through the network. The remote use of DDT is important to many NCC operations.

Each IMP contains several routines which perform such NCC-related actions as "looping" data sets and line interfaces, testing Host interfaces, etc. ("Looping" is the interconnection of circuit elements such that all transmissions from an IMP are returned to that IMP rather than being sent to the IMP at the other end of the line.) DDT is used to initiate and terminate these routines by modifying words in IMP memory which contain their parameters, including an enable/disable bit. For example, one routine monitors a word which, when changed to a line interface number, loops the appropriate interface. This particular ability is vital to isolating the malfunction when a line goes down, so that we know whether to notify telephone company personnel to fix the line, or to notify Honeywell field engineering to repair the interface.

NCC Development

While the data generation scheme and (in large part) the data actually collected have remained invariant during the development of the NCC, the data collection/interpretation mechanisms at BBN have undergone steady evolution. In the first versions, while the network was small, the data were sent as ASCII text which was typed out on the BBN IMP console Teletype; personnel at BBN periodically scanned the typescript to determine if anything noteworthy was happening in the network. Since the collection was being done on a Teletype, a low bandwidth device, space within the message was at a premium; however, since a person was required to read it and make sense out of it, the format had to be intelligible. The only way to balance these factors was to omit the collection of much interesting data.

As the network became larger and more reliable, the proportion of status messages which said anything other than "everything's still OK here" decreased, thus making the location of the messages which required action on our part more difficult. The scheme we developed to make the location of critical messages somewhat easier consisted of having each IMP: 1) send us a status message every 15 minutes and 2) examine its status every minute and send an additional message at that time if it detected a *change* in status. Since these routines were being driven by the synchronized clock of the statistics package, the effect of this scheme was that every 15 minutes we would receive a block of "checkin" reports, one from each IMP; interspersed between these "checkin" blocks on the typescript there would be an occasional "change" report.

This setup functioned tolerably for some time, but eventually several factors combined to make it unwieldy. First, the number of IMPs in the network was constantly increasing, so that the amount of typescript which had to be scanned in order to determine what was happening in the network became overwhelming. Second, outside organizations became increasingly interested in receiving monthly reports on IMP and line performance; the prompt and accurate compilation of these reports by hand became more and more difficult. Third, there was pressure to take statistics on line usage and Host traffic in order to obtain advance warning of network elements whose usage was approaching saturation and to investigate accounting algorithms for network usage. All of these factors led us to install a Host on the BBN IMP which is dedicated to monitoring network performance and doing much of the bookkeeping required for our reports.

With a separate Host dedicated to monitoring the network, we were able to abandon ASCII text format in favor of binary format, and to expand the reports to include more internal status information as well as statistics on Host traffic and

line usage. We were also able to increase the frequency of reporting to once a minute for the "checkin", and to send "change" reports as soon as changes are detected. We also worked, and are still working, on the knotty problem of formalizing the heuristics which are used to integrate the (often conflicting) reports from the individual IMPs. The following paragraphs discuss several of the problem areas of greatest interest.

Line Status

For its own routing purposes, each IMP is continually measuring the quality of each of its data lines. Every half-second it sends a thousand-bit status message on each line and expects to receive a similar message from its neighbors. Each status message includes the number of the IMP which originated it. When an IMP receives a status message from one of its neighbors correctly, it marks its next status message to that neighbor with an acknowledge bit. Thus an IMP's receipt of a status message with the acknowledge bit set indicates that the line is in good condition. Conversely, whenever a half-second interval elapses and the IMP does not receive a status message with an acknowledgment of its own previous message, it counts an error on that line.

In conjunction with this acknowledgment scheme, an important system debugging feature is the ability to "loop" lines for test purposes. Each line is nominally a pair of independent one-way circuits, one in each direction. "Looping" is the interconnection of these circuits such that one end is disconnected and the other end receives its own transmissions. A line can be looped in one of three places: either inside the IMP's line interface or at the local data set (under program control), or at the remote data set (manually). The IMP system, by checking the origin of status messages, can detect looped lines.

Using its line error count and detection of looped lines, an IMP can make a simple usable/unusable decision, for its own purposes, for each of its lines. A line can, however, be "network unusable" for a variety of reasons (the IMP at the other end is down, the interface on the local IMP is broken, the line itself is broken, etc.) and at the NCC we must be able to distinguish amongst them in order to initiate the appropriate repair procedure. Therefore we supplement the IMP's report of whether it thinks the line is up, down, or looped with the IMP number of the IMP on the other end of the line, the total number of status messages sent on the line, and the total number of status messages received on the line (whether their acknowledge bit was set or not). The NCC takes the 3-way division from the IMP (up, down, or looped) and incorporates into it a 2-way division (status messages coming in or not) to form a 5-way breakdown of line status as seen by the IMP at one end of a line: up, down without errors (unusable but with status messages without errors being received), down with errors, looped, and "no information" (the IMP has not reported to the NCC recently). Every minute, for each line in the network, the NCC takes the latest status for

McKenzie et al: Network Control Center for the ARPA Network

each end of the line and determines the state of the line according to the decision rules shown in Table 1. Whenever any line's state changes, a message is printed in the log.

The IMPs are essentially synchronized with regard to the generation of status messages; furthermore, status messages constitute a known constant traffic load on each line. Therefore, for lines whose state is declared up, a measure of line quality is given by the fraction formed by dividing the number of status messages correctly received by the number of status messages sent, since only line errors (detected by checksum hardware) will cause status messages to be incorrectly received. This fraction is printed in the log whenever the numerator differs from the denominator by more than one and the fraction is neither zero nor one. Thus we are alerted to line failures before the lines become completely unusable.

Since the IMPs have been designed to infer the network's topology dynamically, they are not directly concerned with the common carrier data lines; rather, they are interested only in which portions of the network they can access through a particular line interface. NCC personnel, however, must deal with the actual lines. A report from an IMP that the line connected to interface

2 has become unusable is not useful unless we can determine which line is actually connected to that interface. Toward this end, the NCC maintains a connectivity table which contains, for each line in the network, the IMP numbers for the IMPs at each end and the interface numbers that that line should be connected to. The NCC types a message in the log whenever it determines that a line has been moved from its nominal interface or when a report for a line not contained in the connectivity table is received.

IMP Status and Program Reloading

The NCC is faced with a difficult problem in attempting to determine that an IMP is no longer functioning. Since a broken IMP can't send us a message indicating that it's broken, we must infer this condition from the absence of its "check-in" messages. In the past, this decision was made after a scan of the typescript and the observation that the IMP had not checked in "for a while". The current NCC system declares an IMP dead when it has not reported for three minutes. Because of the effects of problems like network partitioning, this is an inadequate test for actually determining whether the IMP is up or down, but it does alert our personnel to the need for further diagnosis. For example, all lines to the IMP may

STATUS FROM HIGH NUMBER IMP

		UP	DOWN NO ERRORS	DOWN WITH ERRORS	LOOPED	NO INFORMATION
STATUS FROM LOW NUMBER IMP	UP	UP	IN LIMBO	IN LIMBO	IN LIMBO	UP
	DOWN NO ERRORS	IN LIMBO	DOWN	DOWN ON HIGH END	LOOPED ON HIGH END	UNKNOWN
	DOWN WITH ERRORS	IN LIMBO	DOWN ON LOW END	DOWN	LOOPED ON HIGH END	UNKNOWN
	LOOPED	IN LIMBO	LOOPED ON LOW END	LOOPED ON LOW END	LOOPED ON BOTH ENDS	LOOPED ON LOW END
	NO INFORMATION	UP	UNKNOWN	UNKNOWN	LOOPED ON HIGH END	UNKNOWN

The terms are defined as follows:

HIGH NUMBER IMP   - IMP with higher network address

LOW NUMBER IMP   - IMP with lower network address

UP   - The line is usable for both IMPs.

DOWN   - The line is unusable for both IMPs.

DOWN ON ONE END   - One IMP can transmit to the other, but not vice versa.

LOOPED ON ONE END - The line is looped as seen by one IMP, but not as seen by the other.

LOOPED ON BOTH ENDS - The line is looped as seen by each IMP.

IN LIMBO   - Conflicting reports from the two IMPs.

UNKNOWN   - Insufficient data to make a decision on the line's state.

TABLE 1: DECISION RULES FOR LINE STATE

be down, rather than the IMP being down.

In the rare case of network-wide failures it is often difficult to determine which IMP triggered the network failure, much less what caused that IMP to fail. Nevertheless, personnel at the NCC must attempt to make these determinations. To assist them, each status report that an IMP sends to the NCC contains a snapshot of the IMP's environment. The snapshot information is used to determine if the IMP is experiencing a transient or getting into some kind of trouble. This information includes the version number of the program running in the IMP, the storage utilization and the amount of free storage left in the IMP, the state of the sense switches and the memory protect switch (to detect unauthorized tampering), a list of the statistics programs which are enabled, a list of which Hosts are up, and an indication of whether tracing is enabled. The NCC logs any change in reported status and, in the event of a network failure, we attempt to correlate environmental data for individual IMPs with the network failure as a whole.

Since all IMPs run the same program, we built a small "bootstrap" routine into the IMP which, when initiated, sends out a request for a core image on a line selected either by parameter or at random. When any IMP receives such a request it returns a copy of its entire (running) program as a single message. The bootstrap routine then checks incoming messages on the selected line for correct length and checksum; when the core image is successfully received it is initialized and started. If an incorrect core image is received, the bootstrap routine sends another request. This facility provides a quick and easy way to obtain a fresh copy of the IMP system. Since the bootstrap resides in the protected memory sector, and thus is nearly always intact, site personnel are almost never required to handle IMP system paper tapes when an IMP requires reloading.

Program reloading can be initiated remotely by commanding DDT to execute a transfer to the bootstrap. Without the remote reloading ability, the only way to distribute a revision of the system would be to mail out paper tapes of the new program to each site, and then schedule a time, with personnel available at each site, to load and start the new version. With the remote reloading ability, however, we merely load the new version into the BBN IMP, direct BBN's neighbors to reload from us, then direct their neighbors to reload from them, and so on until the new version is propagated through the entire network. In fact, propagation of a new program release can be accomplished by one person in a few minutes, rather than requiring a month of planning and several hours of work by a nationwide "team". Also, since the procedure doesn't require assistance from site personnel, it can be scheduled to occur at a time when network usage is extremely low (typically very early morning, a time when site assistance would be most difficult to arrange), thus minimizing the loss of network availability.

Host and Line Throughput

With the change from ASCII to binary reporting and the consequent easing of bandwidth limitations on the NCC we have been able to take initial steps toward building an accounting facility for network usage. Toward this end, the IMP measures the amount of use each Host has made of the network in eight categories. The eight categories are the combinations of the following parameters: transmissions from and to the Host, inter-site and intra-site transmissions, and packet and message traffic. Thus, the eight categories are:

1. inter-site messages sent
2. inter-site messages received
3. inter-site packets sent
4. inter-site packets received
5. intra-site messages sent
6. intra-site messages received
7. intra-site packets sent
8. intra-site packets received

The IMP counts data transmissions only; control messages and RFNMs (destination-to-source message acknowledgments) are not included. The NCC tabulates Host traffic data from all the IMPs in the network. At the end of each hour it copies this table into a second table and then clears the first to obtain a clean "snapshot" for the hour, which is then printed on the report Teletype. This table is also added into a daily table which is printed at midnight every day.

In order to be able to better predict when lines may become overloaded, we also keep track of the line utilization in the network. The IMP measures the line throughput by counting the number of successfully acknowledged packets. The NCC accumulates these line throughput data for each line and types them out with the Host throughput at the end of each hour, and at the end of the day.

Visual and Audible Alarms

Although the computerization of the NCC virtually eliminated extraneous typescript, it was still desirable to free the NCC personnel from having to regularly check the typescript to determine whether action was required. We therefore attached a set of lights and an audible alarm to the NCC. The NCC maintains two sets of "virtual light" display information: which IMPs are alive and which lines are functioning. The NCC staff can select either of these sets for output in the physical display lights. This provides for a quick visual survey of the state of the network.

Whenever a line breaks, or an IMP stops working, the alarm is sounded and the virtual light for that IMP or line is flashed. This minimizes the time for the NCC personnel to notice and take some corrective action, while at the same time freeing them from having to watch the lights or log to achieve this rapid response to network failures.

IV. NCC Operation

The Network Control Center staff consists of five computer operators and several regular BBN technical staff members. The operators are familiar with the operation of the NCC machine and, to a certain extent, with the diagnosis and resolution of network problems. The technical staff members are both hardware and software specialists, most of whom have participated in the design and implementation of the network from the outset.

As the network developed and the role of the NCC increased and became more clearly defined, a fairly comprehensive scheme for manning evolved. It became clear very early that 9 to 5 coverage with informal arrangements to contact staff members at home was insufficient. A dedicated Network Control Center telephone line was installed, with computer operators acting in a monitoring capacity to direct inquiries and problems to available staff members. This has become the single contact telephone for Host site personnel, the telephone company and Honeywell field engineering.

The present NCC program, with its detailed log, allows the operators to assume first order responsibility for network operation. Operators now man the NCC 24 hours a day, 168 hours a week; technical staff coverage is normally close to 50 hours a week, with additional "at-home" availability of key personnel. Routine chores are handled by the operators and only more complicated situations are referred to staff members. After hours, the operators attempt to contact specific staff members at home in the event that a problem arises or a phone call is received which the operators cannot properly field. In specific rare cases, such as attempting to pin down an obscure hardware malfunction, a problem may be preserved (i.e., not fixed) until a staff member can investigate. Even outside regular working hours, most problems are resolved, or at least under control, within a few hours.

There are several different means of handling problems, depending on severity and type. For routine controlled situations (such as IMP preventative maintenance and scheduled repair, scheduled Host testing, and scheduled line test and repair) the NCC operators coordinate the activities of the Honeywell field engineering, Host site, and telephone company personnel involved. We have established the policy that the state of an IMP or a line is not to be intentionally modified without first seeking the permission of the NCC. We insist upon strict adherence to this policy in order to prevent a deferrable outage from occurring during an unscheduled failure and thereby jeopardizing network integrity.

The alarm calls attention to IMP or line outages. The display lights, in conjunction with the log and the ability to obtain a quick printout of network status, usually make it fairly easy for the operators to determine what has failed. In the case of an indicated IMP failure, the operator on duty calls the IMP site, verifies that the IMP has failed, gathers some rudimentary information as to the type of failure, and enlists the aid of site personnel to bring the machine back on the network. If this is not possible, technical staff members are called in to investigate further. If a hardware malfunction is indicated, Honeywell field engineering is alerted to repair the problem.

At present, IMP maintenance and repair are carrier out under contract by Honeywell field engineering. Coverage is prime shift with guaranteed 2-hour response time. When circumstances warrant, however, the NCC will request extended coverage for repair or for standby backup. Most repairs are completed by the end of the day they are reported.

In the case of an indicated line failure, the operator performs a series of checks to confirm that the line has actually failed. This is necessary since some IMP failures appear in the log as line failures (the converse is also true). Diagnosis is performed from the NCC by using IMP DDT to test the terminal equipment. If a line failure has isolated a site from the NCC, the operator will contact site personnel and direct them in performing the tests for him. When a line problem has been confirmed, the operator notifies the appropriate telephone company office, frequently supplying considerable detail.

Each line is maintained and tested from a private line office at one end. Manned around the clock, these offices are equipped with test facilities for finding and repairing line problems. Unless there is a manpower or access problem related to local facilities, line failures usually are corrected within a few hours of the initial report. Maximum repair time is normally about a day.

For many NCC activities the cooperation of the sites is essential. Site personnel aid the NCC in the diagnosis of a variety of problems, help in recovering from IMP failures, and take local responsibility for the IMP. Their assistance is particularly useful when investigating obscure hardware and software malfunctions.

Our relationship with the organizations involved in network maintenance has been good. Honeywell field engineering, telephone company, and site personnel have a high regard for the conclusions reached in our problem analysis. This believability has been fostered by a good track record; in at least 75% of all failure reports to Honeywell or the telephone company, an actual problem has been detected. Line problems and many IMP problems are usually fully diagnosed and dispatched to the appropriate maintenance group within half an hour. Some more subtle IMP problems, however, occasionally require gathering data over a number of failures before a conclusion can be drawn.

V. Experience and Future Plans

A great deal of additional work is done with the NCC's log and summary reports in order to produce monthly reports on network status and usage for ARPA and other interested parties. Since the NCC machine has no secondary storage capability we are unable to accumulate monthly summary information on that machine; instead the daily summaries and log information must be used as input to manual preparation of Host traffic reports and IMP Down

Month	Average Line Outage	Average IMP Down	# of Nodes	Average Host Inter-site Output (packets/day)
September 1971	.59%	3.27%	18	51,386
October	1.66%	1.77%	18	95,930
November	1.65%	5.50%	18	116,515
December	3.21%	3.95%	19	107,896
January 1972	1.02%	1.92%	19	172,037
February	1.23%	2.73%	19	224,668
March	1.36%	4.00%	23	240.144
April	.88%	2.86%	25	362,064
May	1.11%	2.57%	25	505,639
June	.41%	.97%	29	807,164

TABLE 2: SUMMARY OF NETWORK OPERATION

and Line Outage summaries. A certain amount of judgment is used in the latter two summaries; several outages which the technical staff feels are due to a single cause are normally combined into a single (longer) reported outage. Table 2 provides summary information for an actual ten-month period of network operation.

Until early this year the Host traffic summaries were produced manually from the NCC's hard-copy summary reports. The NCC now punches a paper tape (on the report Teletype) of all daily summary information and this is used as input to computer programs which produce the reports more rapidly and accurately. Eventually, when experience indicates that several of the network's service Hosts are reliably up around the clock, we expect to have the NCC transmit all of the summary information through the network for storage and later manipulation. This will enable us to more easily provide answers to interesting questions such as:

- What are the peak hours of network use and what is the peak-to-average traffic ratio?

- What percentage of network traffic do single-packet messages constitute, and how does this percentage vary from Host to Host?

- What is the ratio of weekday use to weekend use?

- What percentage of line capacity is used during peak hours, on the average, and during weekends?

Although the data needed to answer these questions are available now, the data manipulation required constitutes a prohibitive manual burden. Thus, the installation of an NCC computer lifted one bandwidth limitation only to reveal another. In an attempt to deal with this new problem, we are planning to experiment with an additional Host which was recently added at BBN. This is the machine which is currently being used "off line" to process the paper tapes mentioned above.

Another change which is under consideration is automated single-point reporting of line problems. The NCC program, after appropriate automated line testing, could report confirmed line failures directly to a Teletype at some telephone company central location via one of the dial-up line controllers. Telephone company personnel would then direct this report to the appropriate office for test and repair.

Finally, certain of the NCC command options will be made available to other organizations (such as the ARPA office) via one of the dial-up line controllers. This will be primarily to allow access to information on the overall state of the network, particularly the up/down status of the IMPs and lines.

Acknowledgment

This work was sponsored by the Advanced Research Projects Agency under Contract No. DAHC15-69-C-0179.

References

1. "Computer Network Development to Achieve Resource Sharing," L.G. Roberts and B.D. Wessler, Proc. 1970 SJCC.

2. "The Interface Message Processor for the ARPA Computer Network," F.E. Heart, R.E. Kahn, S.M. Ornstein, W.R. Crowther, and D.C. Walden, Proc. 1970 SJCC.

3. "Analytic and Simulation Methods in Computer Network Design," L. Kleinrock, Proc. 1970 SJCC.

4. "Topological Considerations in the Design of the ARPA Computer Network," H. Frank, I.T. Frisch and W. Chou, Proc. 1970 SJCC.

5. "Host-Host Communication Protocol in the ARPA Network," C.S. Carr, S.D. Crocker, and V.G. Cerf, Proc. 1970 SJCC.

6. "The Terminal IMP for the ARPA Computer Network," S.M. Ornstein, F.E. Heart, W.R. Crowther, H.K. Rising, S.B. Russell, and A. Michel, Proc. 1972 SJCC.

Introduction to Satellite Communications

This section:

- introduces the technology of satellite communications, emphasizing their unique broadcast properties

- provides a broad overview of the trends and prospects for further improvement of satellite communications

- traces the evolutionary history of satellite communications and discusses the major technical challenges involved in their future development

- shows how the desirable characteristics of a satellite communications system can be employed to provide a general-purpose data communications network serving a wide variety of users.

The words LIVE VIA SATELLITE flashed across millions of television screens nearly every day are probably the most vivid reminder of the success the space program has had since the early 1960s. Satellite-based communication has added many dimensions to the media coverage of worldwide events, allowing them to offer "live", real-time transmission of sports events and instantaneous news coverage. Not quite as visible to much of the world's population is the huge growth in high-quality international telephone service. It is particularly significant that the growth in the availability of this service has not been limited to highly developed, highly industrialized nations; equally high-quality service is available to all developing nations as well. In fact, high-quality telecommunications in the less-developed nations, generally provided by satellite-based systems, is considered to be fundamental to the growth of such nations.

With the exception of satellite distribution of television programming, most applications of satellite technology since its inception have viewed the satellites, in functional terms as a "giant cable in the sky." By and large, satellite-based communications have been carried out on a point-to-point basis, even though many

INTELSAT Generation	Time Frame	Circuits	Design Lifetime	On-Orbit Cost	Annual Cost per Circuit
I	1965–1967	240	1.5 years	$8.2 million	$22,800
II	1967–1968	240	3 years	$8.1 million	$11,300
III	1968–1971	1200	5 years	$10.5 million	$1,800
IV	1972–1975	4000	7 years	$28 million	$1,000
IVA	1975–1980	6000	7 years	$26 million	$600
V	1981–1984	15,000	8 years	$34 million	$280
VA and beyond	Late 1980s	50,000+	10 years	$40 million or less	$80 or less

Figure V-1   Satellite Cost Estimates

pairs of points can be simultaneously served by a single satellite. It was only in the very late 1970s when such applications as demand-assignment satellite-based networks increased that the capability of satellites was recognized as going far beyond the "giant cable in the sky."

COST AND CAPACITY TRENDS

Figure V-1 shows the trends in the capacity and cost of communications satellites over the years since commercial services were first initiated. These figures reflect the costs of the spacecraft itself, including the launch costs, but does not include the cost of the terrestrial portions of a satellite network. With the greater use of the space shuttle as part of the launching process, cost projections into the late 1980s will probably turn out to be quite conservative. Organizations involved in designing and developing satellites have not yet begun to capitalize on the capabilities of the space shuttle, primarily because most of the shuttle launch capacity during its first five to seven years has been reserved for "conventional" satellites or evolutionary improvements in their design. By *conventional,* we mean a satellite with dimensions and shape that could be launched by expendable rocket-launch vehicles rather than by the shuttle. It will probably be 1987 or later before satellites are launched that fully utilize the large weight and physical dimensions the space shuttle makes possible.

SATELLITE SYSTEM CONSIDERATIONS

Figure V-2 shows past and projected costs for satellite communications on an overall system basis including, in addition to the cost of the space segment, the cost of the earth stations, and local user-to-earth station interconnect costs. This figure includes the cost of the earth station, or a portion of an earth station if it is shared among may users. For a static user located close to the earth station, the cost of interfacing the user devices with the station may be inconsequential, but for a mobile user, or a user fairly far from his serving earth station, the cost can be substantial. Finally, the users' end-devices—the voice terminals, video displays, keyboards, facsimile terminals, or any other user devices with specialized technical characteristics—add to the cost of a communications system.

Figure V-2   Past and projected costs: satellite circuit component.

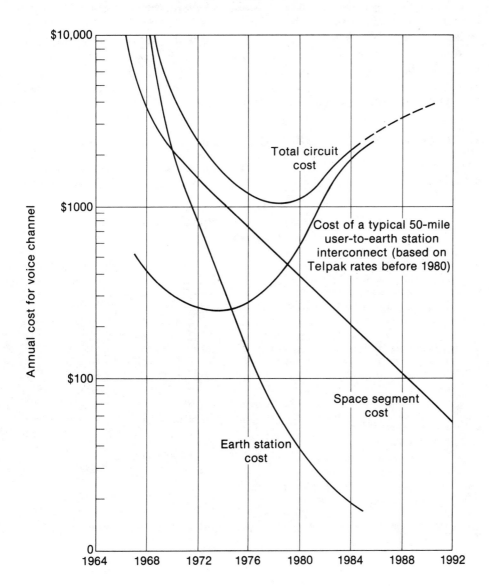

The cost approximations illustrated in Figures V-1 and V-2 were based upon costs referenced to equivalent voice channels of the space-borne portion of the system. The derivation of individual voice channels over the satellite can be achieved in a number of ways, including time-division multiplexing (TDM) and frequency-division multiplexing (FDM), as well as more complex techniques. TDM provides a fundamentally digital mode of operation of the satellite link whereas FDM divides the satellite link into a large number of individual analog channels. Using an analog voice channel to carry digital information, such as computer data, requires modems (modulators-demodulators). Such a channel can then accommodate a data rate of approximately 9600 bits per second. However, as digital technology becomes more cost-effective with the operation of the satellite links based on time-division multiplexing, each voice channel equates to a digital bit rate of 64,000 bits per second. Some systems use lower rate techniques or voice-compression techniques, resulting in equivalent rates of 32,000 bits per second or, in some instances, as low as 16,000 bits per second. In any case, a system based on TDM is highly capable of efficiently integrating high-speed digital data with voice and other communication services.

THE ULTIMATE SATELLITE-BASED SYSTEM

Even if used with only modest overall efficiency, modern satellite-based systems provide enormous total capacity. Figure V-3, assuming a system based on the

Figure V-3 Estimate of total capacity of a satellite system

INTELSAT V—equivalent capacity
17,000 voice circuits @ 64,000 bps/channel

$$\text{total aggregate capacity} = 17 \times 64 \times 10^6$$

$$= 1.088 \times 10^9 \text{ bps}$$
$$\times\ 3600 \text{ secs/hour}$$
$$\times\ 24 \text{ hours/day}$$

$$\overline{94 \times 10^{12} \text{ bits/day}}$$

$$\frac{94 \times 10^{12} \text{ bits/day}}{220 \times 10^6 \text{ persons}} = 427,000 \text{ bits per day per person}$$

$$\frac{427,000 \text{ bits per day per person}}{8 \text{ bits/letter} \times 5 \text{ letters/word} \times 275 \text{ words/page}} = 38.8$$

or approximately 39 pages per day for every person in the United States!

parameters of the INTELSAT V satellite, shows how the information capacity available to each user can be derived. (This derivation assumes that a single satellite devotes all of its capacity to a single distributed network, and that the capacity of the satellite is shared equally by 220,000,000 people over a broad geographic area.) What this derivation shows is that a single hypothetical satellite system provides enough capacity for each of the 220 million men, women, and children in an area the size of the United States to transmit nearly 40 pages of information over the system every day. In addition, by virtue of the satellite mode of operation, all of that information can be received by every other user within the network.

EVOLUTION OF SATELLITE TECHNOLOGY

The four papers included in this section, taken as a group, trace the evolution of satellite technology, further elaborating on the ideas thus far introduced. These papers primarily deal with the overall technology; packet operation plays only a small part in the discussion. (The papers that are more specifically devoted to packet-satellite operation are contained in Part VI.)

The paper by N. Abramson and E.R. Cacciamani, Jr. provides an excellent overview of the satellite technology as viewed from the vantage point of the mid-1970s. As pioneers in this technology, the authors provide insight into the future directions in satellite-based networks, and many of their ideas are already in everyday practice. The principal emphasis of this paper is on the rapidly declining cost of the earth stations and the broadcast property of satellites that permits information transmitted from a single location to be simultaneously received at an unlimited number of different destinations. These ideas foreshadowed the use of satellites for (1) direct-to-home television broadcasting and (2) support of major computer systems using packet broadcast communications systems to provide the bulk of terminal and computer-based interconnections. The first idea is under rapid development by a number of new companies, whereas the second was at the core of the system developed by IBM, Comsat, and Aetna, which became Satellite Business Systems (SBS).

The October 1979 issue of the *IEEE Transactions on Communications* was a special issue devoted to satellite communications. The paper by F.E. Bond and P. Rosen introduced that issue, and provides, within a few short pages, an outstanding synopsis of the complex issues—technical, socio-economic, and political—that influence the evolution of the state of the art. This paper also introduces the relationship between the choice of frequency bands, the international assignment of those bands, and the major services assigned to each band. The authors discuss the role of specialized systems (devoted to a narrow set of customer applications), military systems, and systems employing entirely new techniques (such as on-board switching, high-power spot beams, multibeam antennas) and provide references to other articles/reports that further develop these ideas.

The paper by W.L. Pritchard provides additional coverage of historical trends in satellite communications, and specifically deals with domestic programs in many different countries around the world. Pritchard describes plans for various systems, many of which have come into service since this paper was written. A compact synopsis of the major technical characteristics of more than 20 different systems are given in the form of tables, along with a complete discussion of the various access and sharing techniques that are in use or planned. The paper concludes with a discussion of launch, power, orbital, and spacecraft design considerations.

The last article in this section, by I.M. Jacobs, R. Binder, and E.V. Hoversten describes the application of satellite technology to a highly efficient integrated network supporting data and speech. The article describes the user characteristics and networking techniques that influence the design of a general-purpose satellite network, and the role that packet operation plays in such a network. The authors introduce a class of resource assignment protocol known as Priority Oriented Demand Assignment (PODA) which responds to the recognition that a general-purpose network has to exhibit different performance characteristics for the wide variety of user services being supported. The major advantages of non-channelized resource-sharing techniques that generally use some form of packetized operation are also outlined.

Detailed descriptions of the internal controls and protocols within the General Purpose Satellite Network are also provided, specifically with relationship to system transparency, message sequencing and integrity, addressing, flow control, conferencing, internetworking, and all the other considerations introduced by the protocol articles in Section IV. Of paramount interest and value in this article is the authors' description of the actual implementation of many of these ideas in the Atlantic Packet Satellite Experiment, carried on in cooperation with the Advanced Research Projects Agency of the U.S. Department of Defense and the ARPANET. Actual network performance is measured and contrasted with theoretical performance derived for PODA and similar assignment structures.

The papers in this section comprise a broad overview of networking via satellite. While the power of achieving that networking using packetized operation is alluded to, it is not explicitly derived or addressed. The relationship between distributed satellite networking and packetized operation will be fully explained and derived in the papers in Section VI, all of which draw on the basics established by the papers in this section.

Satellites: not just a big cable in the sky

High bit rates, high reliability, low costs, and versatility are features of satellite communications systems

NORMAN ABRAMSON, FELLOW, IEEE, AND

EUGENE CACCIAMANI, PhD

In 1975, the pace of development of communication satellite systems has accelerated markedly. NASA successfully launched three commercial communication satellites from Cape Canaveral during 1974 but, during 1975, from eight to ten such launches are anticipated. Fueling this development are the lower costs and special capabilities that satellite links offer communication users; particularly data communication users. For example, long-haul service provided by satellite is usually more cost-effective than comparable service provided by purely terrestrial links. This is especially true of specialized higher data rate services. Today there is little, if any, terrestrial capacity available to provide long-haul transmission of data at rates of one million bits per second and higher. To provide new services at these rates in most situations will require both large inflexible investments and new construction of terrestrial facilities, which will require long implementation times. Satellite systems can be quickly and inexpensively installed to establish circuit connections wherever the user desires and at the data rates that he needs.

Generally speaking, highly reliable data channels can be achieved using satellite systems. This is particularly true if the service can be provided directly to the user without intervening terrestrial carrier links. Since the system consists of a single transponder, the channel characteristics are well known and can be controlled within certain limits. Unlike terrestrial circuits, which can experience fading due to microwave propagation path anomalies as well as "bursty errors" due to telephone exchange equipment switching, the satellite circuit is affected primarily by additive, white, Gaussian noise. Hence, performance can be controlled by adjusting earth station transmit power, and further performance enhancement can be achieved using forward error-correction techniques.

Some important communications satellite capabilities have yet to be exploited. For example, commercial satellite channels are usually employed as point-to-point links between two earth stations, just as if these stations were linked by a direct cable connection. But a satellite is not just a big cable in the sky; a single satellite channel has a broadcast character that allows it to transmit and receive signals from any earth station in its antenna pattern. This flexibility provides the satellite communication channel with particularly powerful capabilities for use in a digital network composed of a large number of small earth stations. If a destination address is attached to each message transmitted by an earth station in such a system, then it is no longer necessary to devote a separate channel to each pair of earth stations that wish to communicate. Once this flexibility is appreciated, it becomes possible to employ either novel forms of communication systems architecture to allocate voice circuits dynamically (in seconds), or packet broadcasting to allocate data packets dynamically in microseconds. The extent to which these satellite communications capabilities actually come into use depends, in the first place, on their availability.

Standard and special satellite services

Both the U.S. domestic carriers and INTELSAT now offer a variety of "standardized" tariffed services that the user can order from a "shopping list." COMSAT, as the U.S. representative to, and member of, INTELSAT, acts as a carrier's carrier, providing satellite circuits to the U.S. international carriers, interfacing with them at the COMSAT-operated earth sta-

---

More and more satellites aloft

The commercial use of communication satellites began with the successful launch of INTELSAT I in April 1965. INTELSAT consists of a consortium of 90 member countries that operates a network of geostationary communication satellites positioned over the Atlantic, Pacific, and Indian Oceans. Since each satellite covers approximately one third of the surface of the earth, such a network is adequate for almost complete earth coverage. Since 1965, INTELSAT has designed three new generations of communication satellites; the latest, INTELSAT IV, was first put into operation in March 1971.

In addition to the INTELSAT system, several national or regional communication satellite systems have been or soon will be launched. By early 1975, five other systems were operational—Molniya (U.S.S.R.), Anik (Canada), and Westar (U.S.A.)—with their own satellites, and RCA and American Satellite Corporation employed the Anik and Westar satellites for their space segment. Other communication satellites expected to be launched during this decade are two additional U.S. systems (in competition with Westar), a European system involving nine stations, an Iranian system, a Japanese system, an Arabian system, an Indonesian system (perhaps involving Malaysia and the Philippines), a Brazilian system, and an Indian system. An international maritime communications satellite (MARISAT) is being launched in 1975, and an international aeronautical satellite (AEROSAT) is scheduled for the late 1970s.

---

tions. Services in operation today include: video transmission on demand; 4-kHz voice circuits using both preassigned FM/FDMA transmission and the SPADE demand-assigned digital transmission system; teletype; telex; voice band data up to 9600 bits per second; and digital data circuits operating at 50 000 and 56 000 bits per second.

The U.S. domestic carriers, American Satellite Corporation, Western Union, and RCA, can interface at the user premises using terrestrial interconnections from their respective earth stations to the user's facility. In some specialized situations, these carriers may locate their earth station at the user's premises.

Representative tariff rates for domestic services are shown in Table I and the trend INTELSAT satellite tariffs have been taking over the last few years is shown in Table II.

In addition to these more or less standardized services, satellite carriers develop and provide *new* services to meet evolving user requirements previously not easily met using terrestrial lines. The carriers' ability and willingness to do this is of considerable importance to the consumer of data communication services. Specialized tariff services can be developed dynamically as requirements arise, and tariffs can be filed with the U.S. Federal Communications Commission to cover these specific applications on a case-by-case basis.

One example of such new services is the Government network installed in 1974 by the American Satellite Corporation. It consists of five dedicated-user earth stations on U.S. Government facilities. The system is used to transmit and receive five 1.344-Mb data streams over long- and medium-range distances (from 3000 km down to 450 km) to connect remote data collection terminals to a primary computer facility.

This year, American Satellite will install a dedicated broadcast network for Dow Jones. This service will enable digital broadcast of the *Wall Street Journal* to remote printing plants for regional distribution. The initial service will be at 150 000 bits per second and, combined with facsimile data compression equipment provided by Dow Jones, a complete "master" newspaper page can be transmitted in about three minutes. Eventually, when the system data rate is increased to 1.344 Mb/s, the transmission time will be about one minute per page.

Another proposed service includes off-shore oil rig earth terminals such as the Cities Service terminal. Exxon also has a proposed requirement for a gyrostabilized tracking station on board its oil exploration ships. This proposed voice and data service will link the ships to Exxon's central computer located in Houston, Tex.

In the near future, the costs for satellite communications can be expected to decrease. The reasons for this probable trend lie in changes now underway in satellite system design.

Satellite system costs

The cost of the satellite itself is often thought to be the dominant cost in the overall satellite communication system. Actually, in the present generation of systems, the cost of the satellite is rapidly becoming less significant, taking third place behind earth sta-

tion cost (including multiplexing equipment), and the cost of distribution from the earth station to the user.

Cost estimates for the space segment of a single, full-duplex voice channel on the INTELSAT IV satellite have been estimated at about $600 per year. Comparable cost for the Anik or the Westar satellite is about $300 per voice circuit per year.

From the point of view of satellite communication systems, digital transmission systems are generally more efficient in transmitting both voice and data than are traditional analog FM/frequency-division multiplex systems. In transmitting data, the satellite digital voice circuit is the equivalent of at least six voice circuits in a conventional analog-transmission land line system. A single conventional circuit may achieve a data rate of 9600 bits per second under carefully controlled conditions, but up to 64 000 bits per second of digital data can be transmitted over the equivalent satellite digital voice channel. For example, a 50 000-b/s data circuit, with a tariff comparable to that of a single voice channel and operating over the Pacific Ocean INTELSAT IV satellite, has linked the ALOHA system in Honolulu with the ARPA Computer Network in North America and Europe since December 1972.

The cost of an earth station for a satellite communication system is more difficult to determine than

I. Representative U.S. domestic tariffs in February 1975 (satellite, ground links, and local loops)

Between	And	Satellite Single-Channel Rates	AT&T Single-Channel Rates
Chicago	New York	$ 620	$ 760
Chicago	Los Angeles	820	1674
Dallas	Los Angeles	820	1231
New York	Los Angeles	1120	2300
Washington	San Francisco	1120	2292

II. Satellite rates—U.S. to Europe

Effective Dates	Monthly Half-Circuit Rates
INTELSAT charges to COMSAT	
6/27/65	$ 2 667
1/ 1/66	1 667
1/ 1/71	1 250
1/ 1/72	1 080
1/ 1/73	930
1/ 1/74	750
1/ 1/75	705
COMSAT charges to carriers	
6/27/65	$ 4 200
4/ 4/67	3 800
7/ 1/71	2 850
Carriers' charges to customers	
6/27/65	$10 000
10/ 1/66	8 000
10/ 1/67	6 500
8/ 1/68	6 000
4/ 1/70	4 750
8/15/71	4 625

Note: Carrier, COMSAT, and INTELSAT rates are for circuits to the mid-point (i.e., half-circuit rates). Carrier rates include terrestrial haul from COMSAT earth station to New York.

the cost of the satellite. In analyzing the cost of an earth station, it is necessary to specify the communication capabilities desired of the station. For example, almost all the stations first built in the INTELSAT system were capable of transmitting and receiving color television signals while simultaneously handling voice and data traffic. Furthermore, they were designed to have this capability with satellites transmitting considerably less power than the current generation of satellites. These earth stations typically consist of a 30-meter antenna, cryogenically cooled receivers, and tracking capabilities of questionable utility. The cost of such an installation has been from about $3 000 000 to $5 000 000. The cost estimates for the smaller and simpler earth stations used in the Canadian and U.S. domestic systems have been from $1 000 000 to $2 000 000 for full television and voice-data capabilities. If we consider earth stations with only data communication capabilities, even simpler installations are possible. Earth stations have been installed by U.S. domestic satellite carriers for specialized data applications ranging from $100 000 to $400 000 per station, and even lower costs have been projected for future stations.

Some of the factors that contribute to the lower cost of these earth station installations include: smaller diameter antennas, uncooled low-noise receivers, low-power transmitters, and simplified communication hardware requirements. In many instances, forward-acting error-correction techniques can be effectively added to the system at little additional cost providing significant improvement in performance. Earth station parameters can be adjusted to provide a cost-effective earth station based on capital investment in the earth station and recurring space segment charges.

Increasing equipment redundancy requirements can significantly increase earth station costs. Redundancy implies that additional components are implemented to provide backup in the event of failure of on-line components. User reliability requirements will determine the need for redundancy. For example, assuming unmanned earth station operation, the expected reliability of a nonredundant earth station is 99.4 percent versus 99.9 percent for a redundant system. However, the cost of adding redundant components can amount to a significant portion of the total earth station cost—as much as 40 percent of the total cost of a simple single-data-link earth station. Thus, the user must be careful not to overspecify his reliability requirements.

Other user requirements will affect costs by defining earth stations to be duplex or simplex operating. A simplex operating earth station (receive-only or transmit-only) requires correspondingly less equipment than does the duplex station (receive and transmit) and, hence, is less costly. A typical example of a simplex operating system could be a point-to-multipoint newspaper publishing system whereby the paper is distributed from a central point to a number of printing plants equipped with receive-only terminals. Television and radio distribution networks would be another example of a simplex earth station network.

Reduced costs in earth station design for data communications can be achieved by reduced multiplexor requirements. In some cases, this function (when required) can be replaced by a program in the computer connected to the earth station. This latter possibility is particularly attractive because of the flexibility and added capability of a computer network employing packet-broadcasting architecture. Further decreases in costs, beyond those discussed in this article, can be projected for newer-generation satellites as well as for present-generation satellites, with advanced data transmission-multiplexing techniques.

General distribution earth stations can require extensive ground interconnection circuits to reach the user. This is particularly undesirable to the data user since it can add significantly to the overall service cost. However, many small data-terminal earth stations can be located at the user's premises, thus reducing the ground interconnect costs to zero and insuring high end-to-end performance. The "dedicated user" Government data network mentioned previously, where all stations are located at the user's premises, operates with an overall bit error probability of less than 10^{-8} for 1.344-mb/s data service. These ongoing developments in equipment and systems promise cost reductions for satellite services in the near future, and we can expect them to be followed by more dramatic changes.

As for the future . . .

With the hindsight provided by the ten years since the launch of INTELSAT I, it is easy to forget the revolutionary nature of the satellite telecommunications medium. And it is easy to forget that there was considerable uncertainty before 1965 about incorporating geostationary communication satellites into the existing telecommunications plant. With these uncertainties in mind, satellite communication advocates have tended to minimize the differences between satellite channels and earthbound channels. Actually, until the present time, satellite channels have been used almost exclusively as replacements for cable or microwave channels. This situation now appears to be changing as the special capabilities of satellite channels are becoming more widely recognized.

Because of the greater flexibility of digital data, compared to video or voice signals, the impact of the unique capabilities of satellites is likely to be most visible in satellite data communication systems providing unconventional services. In the U.S., approval for such unconventional data communication services will be necessary from the Federal Communications Commission. But this agency has indicated that it is receptive to proposals that take fuller advantage of the capabilities of satellite channels.

Dynamic channel-sharing

One of these unconventional techniques—the ability to allocate dynamically satellite channel capacity among a number of mobile stations—will be exploited in the planned MARISAT and AEROSAT systems. The first MARISAT satellite is scheduled for launch in 1975 and will be employed for voice and data channels to ships in the Atlantic using L-band frequencies (1.6 GHz). MARISAT will be operated by COMSAT General Corp. and will initially provide one voice channel and 44 teletype channels.

An interesting feature of the MARISAT program is

Communication satellite basics

At an altitude of 36 000 km, a satellite in a circular orbit will have a period of 24 hours. If such a satellite is placed in orbit over the equator, it will appear stationary from any point on the earth. Such a satellite is called geostationary. Three geostationary satellites positioned at roughly equidistant points around the equatorial orbit are sufficient for essentially full communications coverage of the entire earth.

Geostationary satellites are employed as radio repeaters for communication purposes. A signal from one earth station is transmitted up to the satellite (the uplink) and the signal is retransmitted down from the satellite to another earth station (the downlink). In order that the signal transmitted by the satellite not interfere with the signal received by the satellite, the uplink and the downlink signals use different frequency bands. Most of the present generation of geostationary satellites use the 4-GHz band for the downlink and the 6-GHz band for the uplink.

Since the characteristics of most commercial satellite channels match quite well the classical Shannon model of a channel with additive white Gaussian noise, we may calculate the channel capacity by means of the Shannon formula

$$C = W \log (1 + S/N)$$

where C is the channel capacity in bits per second, W is the channel bandwidth in Hz, and S/N is the signal-to-noise power ratio at the receiver. From this equation, we see that the capacity of a satellite to transmit messages depends upon both the bandwidth available and the total effective radiated power of the satellite. But the capacity increases rapidly with the bandwidth while increasing much more slowly with the radiated power.

The upper limits of bandwidth available in the 4-GHz and 6-GHz bands assigned to communication satellites have been reached in present systems, and additional bandwidth will only be obtained by shifting to higher frequencies—frequencies that may require more expensive equipment. The effective radiated power, however, can be expected to increase in the next generation of communication satellites. While this will not lead to large increases in satellite capacity, it will lead to a dramatic decrease in the size and complexity of ground systems necessary to receive satellite signals and it will lead to greater flexibility in satellite communication systems architecture.

Since INTELSAT I was launched in April 1965, almost all of the more than 100 earth stations used in the INTELSAT system have included a 30-meter-diameter antenna, sophisticated tracking and control equipment to point this large antenna, and extremely sensitive, cooled receivers to detect the weak signal from the satellite. Such ambitious earth stations may have been desirable for use with INTELSATs I, II, and III, but they seem overdesigned for use with INTELSAT IV and its successors.

Earth stations designed for use in national and regional satellite systems put into operation in the early 1970s generally employ simpler uncooled amplifiers and smaller diameter antennas—typically 10 to 15 meters. The European satellite system plans to employ antennas with diameters no greater than 15 to 18 meters, and it seems likely that most earth stations in that system will have even smaller diameter antennas. As the diameter of an earth station antenna is decreased, the beam width of its radiation pattern increases, thus leading to the possibility that the signal transmitted to one satellite may interfere with the signal to an adjacent satellite. At the presently planned spacing of geostationary satellites around the equator: it appears that this may limit the minimum antenna diameter to from 3 to 5 meters for the 6-GHz band now used.

---

that during the first few years of operation much of the satellite channel capacity will be employed in a separate U.S. Navy communications network in the UHF band (240–400 MHz). As the civilian maritime needs increase, the satellite power will be shifted from the UHF transmitter to the L-band transmitter in the satellite, and when the Navy has completely phased out its use of MARISAT, the capacity available to the civilian maritime system will be nine times the initial capacity.

Scheduled channel sharing

The MARISAT and AEROSAT services will use unscheduled sharing of the satellite space-segment portion of a communication system. It is also possible to design special data communication systems that share the space segment in a scheduled manner. For example, a banking or credit card system might be designed with small earth stations located at each of ten or twenty regional centers and those centers might transmit financial data to a national center on a regularly scheduled basis over the same channel.

Land lines or microwave systems could also be used for such systems, of course, but a satellite data service has certain properties that can make it more attractive. A land-based system will usually have to locate the central processing facility near the geographical center of the system in order to minimize leased or dial-up line charges. In a satellite system, the central facility can lie anywhere within the antenna pattern of the satellite. Data transmitted from outlying units may be rerouted to a backup central facility with a minimum of effort, even if the backup facility is located at some considerable distance from the primary center.

The tradeoff of earth station costs versus space segment costs for a scheduled channel-sharing satellite data service is different from the tradeoff in the case of conventional fixed point to fixed point satellite channels. Since with channel-sharing there are many earth stations making use of the same satellite channel resources, the earth stations may be small and inexpensive while satellite channel capacity can be quite freely used in order to minimize the total system cost. In fact, as the number of possible nodes in such a satellite data communication network increases, and as the fraction of the satellite capacity used by each node decreases, the overriding consideration in the system design becomes the design of the many inexpensive earth stations of the network.

Digital broadcasting

The future systems discussed so far employ the traditional point-to-point architecture of ground-based data communications, although the receive or transmit points may be reconfigured in order to utilize the

Multibeam antenna developments

The waveguide-lens spacecraft antenna pictured on *Spectrum*'s cover this month is capable of generating multiple beams that collectively cover the whole earth disk or, instead, generate several isolated beams as small as 2 degrees in diameter.

The antenna was built by Lockheed Missiles and Space Co. for proposed use on the phase III spacecraft of the Defense Satellite Communications System, DSCS-III, the next generation of U.S. military communications satellites.—*Howard Falk*

space segment fully. But, as was pointed out, a satellite is basically a broadcast communications medium, connecting any point in its electromagnetic view to all other points. This property may most easily be exploited in the distribution of identical data from one point to several other points. Examples of such systems are news or financial wire services or the distribution of national newspapers and magazines to regional printing plants—or even to plants located in individual cities. A fully redundant earth station employing a 5-meter antenna dish, with receive-only capabilities at a data rate of 1.344 b/s, could be built at a cost of about $150 000. At this data rate, it takes only one minute to transmit a newspaper page. The advent of such a nationwide system would make it economically feasible to have only a few nationwide newspapers with separate sections produced in each city for local news.

Packet broadcasting

It is a relatively small step from the digital broadcasting concept to packet broadcasting. This is a method of operation that allows efficient transmission of data from many earth stations to one earth station, or even from many to many. Certain kinds of data communication systems operate under extreme conditions of traffic variability. For example, terminal time-sharing networks, data base inquiry systems such as those for airline reservations or stock market quotations, and computer networks with file-transfer capabilities, all tend to operate with high peak-to-average data rates.

In these situations, it is possible to allow each earth station in the data network to share a single high-capacity satellite channel. Data are buffered in each earth station to form packets—typically these are 1000 bits in length. Each packet, together with a header containing address and control information, is transmitted over the common shared satellite channel at the maximum rate of the channel. In such a packet broadcasting data network, it is not necessary to con-

trol or synchronize the burst transmissions from the separate earth stations. The header for each packet is received by all the earth stations in the system and the packet is accepted by the station or stations with the proper address. If the average data-rate of each station is low enough, and if there are not too many earth stations, the probability that two packets will be transmitted at the same time—and thus interfere with each other—will be low. If such interference does occur, the two earth stations that transmitted the packets can detect the interference. Fortunately, in a satellite channel, it is possible to monitor your own transmissions. Each station can then repeat the lost packets until a successful transmission occurs.

Packet broadcasting by satellite was first demonstrated in 1973 in a NASA experiment involving the University of Alaska, NASA Ames Research Center, and the ALOHA system at the University of Hawaii. Satellite packet broadcasting experiments using large INTELSAT earth stations (in Etam, W.Va., and Goonhilly, England) and the Atlantic Ocean INTELSAT IV are planned to begin in September 1975.

Packet broadcasting systems are an attractive possibility from the point of view of efficient use of space segment satellite resources, since the satellite transponder need only provide power during the short burst when it is transmitting a packet. Thus, a satellite transponder operating in a packet broadcasting mode at a duty cycle of 10 percent would only transmit 10 percent of its rated power. Alternatively, such a transponder could be adjusted to provide 10 dB more power during its transmission of packets, while keeping its average power output fixed.

An earth station employed for packet broadcasting needs no multiplexor since this function can be carried out by a rather simple software module in the computer system receiving the data. A packet broadcasting earth station using a single transponder at a peak data rate of 1 Mb on a Westar class satellite could be built for about $30 000–$50 000. Such a system could easily handle the data traffic generated by 10 000 computer terminals in a time-sharing or data base inquiry system.

These cost figures for packet broadcasting earth stations are in the same range as, or lower than, the cost of many present-day peripheral devices for medium- and large-scale computer systems. This fact suggests the possibility that such an earth station might very well be marketed as just another computer peripheral device by a large computer manufacturer. In the 1980s, when ordering a large-scale computer, one may well be asked to consider system configurations that include a packet broadcasting earth station with a 5-meter antenna on the roof of the computer center.

Introduction and Overview

FREDERICK E. BOND, FELLOW, IEEE, AND PAUL ROSEN, FELLOW, IEEE

INTRODUCTION

THE explosive growth of communications satellite use and the perceived potential of the medium for novel applications has generated intense interest in both the government and private sectors. The future use of this remarkable communications medium will affect national policy, international geopolitical developments, industrial growth, public service functions, and quality of life. The directions of future developments in the field are manifold (if uncertain), depending on complex interrelationships among new technologies, socioeconomic factors and domestic and international institutional issues.

CURRENT SYSTEMS

Satellite communications systems currently in use or due to be implemented shortly comprise about 50 operating satellites and a population of over 2,000 terminals. Virtually all the spacecraft used for worldwide connectivity (other than some Soviet systems) are in the geostationary orbit. The frequency bands in use range from a few hundred MHz to 30 GHz. Table 1 lists the bands currently in use or allocated and the "popular" band designations generally applied in the literature. Some feeling for the ubiquity of communications satellites is given by Long (pp. 1538–1544), showing both the present occupancy of satellites in the geostationary orbit alone as well as requested reservations for future positions in longitude for each of the major frequency bands. Many varieties and sizes of terminals are in use, ranging from fixed stations with antennas as large as 28 m in diameter to relatively simple mobile equipments on aircraft, ships, and military vehicles. The systems are used to communicate an immense variety of information including teletype, voice, facsimile, computer-to-computer data and video.

Current systems can be characterized by the nature of the relevant user communities and the services provided. INTELSAT, the oldest system, provides multichannel trunking service on a wholesale basis to international carriers as well as leased transponders to individual nations for regional or domestic systems. The Western hemisphere domestic systems (TELESAT, WESTAR, RCA SATCOM, and COMSTAR) offer a variety of private-line, leased services directly to industry, government, or other carriers. Domestic systems have also been established in Europe, Indonesia, and Japan. The

TABLE 1
FREQUENCY BANDS GENERALLY EMPLOYED FOR SATELLITE COMMUNICATION

Popular Designation	Range in MHz		Type Service (Satellites)
Military UHF (in US & Possessions)	240.0 – 328.6 and (Up & Down)		Mobile – Satellite [1] (LES –5,6,8 and 9; MARISAT, FLEETSAT)
L	1535.0 – 1543.5	Down	Maritime Mobile Satellite (MARISAT)
	1636.5 – 1645.0	Up	
	1542.5 – 1558.5	Down	Aeronautical Mobile – Satellite (AEROSAT)
	1644.0 – 1660.0	Up	
C	3700 – 4200	Down	Fixed – Satellite (INTELSAT series, DOMSATs)
	5925 – 6425	Up	
X or Military SHF (in US & Possessions)	7250 – 7750	Down	Fixed – Satellite [2] (DSCS, NATO, GALS)
	7900 – 8400	Up	
Ku	10,950 – 11,200 & 11,450 – 11,700	Down	Fixed – Satellite (INTELSAT V, planned DOMSATs)
	14,000 – 14,500	Up	
Ka	17,700 – 21,200	Down	Fixed – Satellite [3] (Japanese CS)
	27,500 – 31,000	Up	

(1) Shared with terrestrial communications

(2) 450 MHz shared with terrestrial communications – for example, radio relay systems.

(3) Upper 1 GHz for Military use in US & Possessions.

MARISAT system, which is to be replaced by INMARSAT, provides ship-shore communication service in the Atlantic, Pacific and Indian Oceans.

The U.S. Department of Defense operates military systems which serve both strategic and tactical needs. Great Britain operates a military system (SKYNET), and NATO employs a system serving critical command and control needs of the Organization and its member nations.

Other systems of relevance include the Soviet counterparts of the international, domestic and military systems (INTERSPUTNIK, MOLNIYA/STATIONAR). Finally, there are the domestic systems of the lesser developed or emerging nations such as Algeria, which utilize transponder assets of international systems to provide previously non-existent internal communications.

EVOLUTION

In the early stages of development, satellite communications systems focused on high volume, multi-channel voice transmission, employing terminals with very large antennas and extremely sensitive receivers to compensate for satellite power limitations. The satellites, designed to serve a relatively homogeneous user population, were analogously uniform in technical characteristics, growing in capacity as the market demanded. Mobile services materialized slowly, using modest terminals for both commercial and military maritime applications. "Thin route" applications also develped slowly, appearing first as single-channel-per-carrier for sparsely populated INTELSAT members, and then with increasing momentum in a variety of domestic satellite (DOMSAT) applications. Domestic distribution of video channels for cable TV and the Public Broadcasting Service is a recent but very rapidly growing DOMSAT application.

The momentum of expanded satellite applications for commercial long-haul service has been carried recently into more sophisticated concepts formed around digital communications and direct user terminals at the customer's location. One example of such service, exploiting information processing technology as an alternative to conventional common carrier switched systems, is offered by Satellite Business Systems, Inc. Another is the announced intention of the Xerox Corporation to offer a graphic communications service with on-site terminals using a combination of microwave terrestrial and satellite links. Holiday Inn, Inc. has disclosed its desire to tie the hotel chain affiliates together with roof-top satellite terminals for entertainment and business communications. Bell Telephone Labs has discussed concepts involving high-speed switching of narrow satellite antenna beams for improved satellite service for long-haul, high-volume trunk systems, providing efficient service to localities with a wide range of traffic requirements (Acampora, pp. 1406-1415).

Early military systems also started with large terminals and simple satellites, but military systems are developing with a different technical emphasis from commercial systems. The differences have been driven by communications system requirements exemplified by service to a larger number of mobile terminals, an inhomogeneous user mix (terminal sizes, data rates, etc.), and the need for rapid reconfiguration. Military requirements for which there are no civilian counterparts are physical and electronic survivability. The effect of these uniquely military needs is seen in different steps taken in accommodating new technology (Cummings, pp. 1423-1435) and in more complex system designs.

The application of satellite communications in the lesser developed nations has been viewed as a unique opportunity, with a minimum of technical transition problems or institutional issues to settle. The new medium has proven useful at three levels: to link the nations with the outside world with high-grade trunk service (via INTELSAT, for example); to connect major internal population centers (by leasing an INTELSAT transponder or employing its own regional or domestic system), and to provide distributed communications to rural areas or islands.

PROSPECTS FOR THE FUTURE

In addition to the applications described above, satellite communications is also viewed as having great potential for systems of health care delivery, law and order preservation, and for providing a multitude of other services furnished by local, state, and federal governments, including lower cost education. Other possibilities include: electronic mail delivery between major postal centers and eventually to individual locations, television conferencing as an alternative to business travel, and direct (home) broadcast for education or entertainment. These concepts represent major technical, socio-economic, and institutional prospects for change and

challenge, with the knottiest technical problems anticipated in the peripheral terrestrial processing equipment and customer distribution links rather than the satellite communications links.

When we restrict our focus to the satellite communications problems per se, the largest impediments to achieving these somewhat lofty objectives in providing new services appear to be crowding of frequency spectrum and positions in orbit, high costs of satellites and terminals, and contention among the national and international carriers over regulatory issues.

TECHNICAL OPPORTUNITIES

Of all the constraints inhibiting future growth in both commercial and military systems, two highly leveraged technical problems are those associated with frequency/orbit congestion and system costs stemming from sophisticated spacecraft and proliferated terminals (Jeruchim, pp. 1544-1550; Ito, pp. 1551-1558).

The seriousness of the congestion problem can be sensed by noting that the useful parts (i.e. coverage) of the geostationary circle are essentially full at C band and are expected to be saturated in a few years at K_u band. Also, the military UHF and SHF bands are essentially saturated because large portions of the bands are shared with terrestrial links. While the regulatory aspects of frequency management (including orbit locations and power density limits) will be addressed at the 1979 World Administrative Radio Conference (WARC) in Geneva, there are several technical approaches to the relief of spectrum and orbit congestion, with varying degrees of benefit and risk (Long, pp. 1538-1544).

One of the obvious solutions to frequency crowding is to develop the higher frequencies, e.g., the 20-30 GHz band which has an allocated width about seven times greater than the lower bands, and the 43/45 GHz band which has even greater usable width. Although a substantial amount of propagation data is now on hand, residual uncertainties about rain attenuation and depolarization, with the need for site diversity for very high-circuit availability, are still a matter of concern (Fugono, pp. 1381-1391; Tsukamoto, pp. 1392-1405; Cummings, pp. 1423-1435).

Another solution, not mutually exclusive, is the achievement of large-scale frequency reuse at the lower frequencies by means of large, multi-beam antennas in space together with suitable on-board switching matrices. System realization of these techniques will require significant technology advances.

There are short-term solutions with less risk and payoff. These approaches, which work to maximize the utility of available spectrum and orbit positions, are exemplified by more efficient modulation signaling or polarization diversity techniques to communicate more bits per sec per Hertz (Lin-shan Lee, pp. 1504-1512) and by improved multiple access or allocation techniques to ensure more efficient channel occupancy and source compression for voice and imagery (Horstein, pp. 1441-1448; Aein, pp. 1436-1441; Inukai, pp. 1449-1455; Taylor, pp. 1484-1496; Balagangadhar, pp. 1467-1475). New types of data services, using data packet techniques for efficiency, create new considerations in satellite design (DeRosa, pp. 1416-1422; Lam, pp. 1456-1466). These kinds of improvements imply changes in satellite or ground signal processing (or both); hence implementation is often encumbered by system transition, compatibility, and interoperability problems.

There are several approaches for countering the escalating costs of spacecraft and launches. For spacecraft, general-purpose, modular "buses" to carry payloads have been proposed to reduce R & D costs; and the reusable Space Shuttle is expected to reduce launch costs significantly in the long term. Although it seems logical to develop concepts for standardized Shuttle loads, progress in this direction appears slow. Other approaches for reduced satellite cost are aimed at more efficient primary and secondary power sources and more efficient power amplifying devices. Modulation and coding/multiple access techniques which produce more bits per watt of transmitter power (for a fixed error rate) or per Hz of bandwidth can also reduce spacecraft cost per unit of communications capacity (Morihiro, pp. 1512-1518).

A futuristic approach to significant cost reduction in the space segment is the use of large (e.g., 10,000 kg and 100 m), multi-purpose platforms which claim to exploit economy of scale. They are also proposed as a means of offering relief from orbit congestion by concentrating many satellite functions at one location. The concept involves several diverse communication (or other) payloads with multibeam antennas, on-board switching, and processing—all served by a common structure, power subsystem, attitude control, telemetry and command, etc. The use of large aperture multibeam antennas permits frequency reuse and hence relief from frequency congestion. The large platform may require the assembly of several Shuttle loads in low earth orbit followed by low-power, long-duration propulsion to synchronous altitude. Sophisticated fabrication/design techniques for zero-gravity antenna structures may lessen the size impact on shuttle capacity. Cost and reliability considerations dictate developing a capability for on-orbit replenishment, repair and upgrading.

For satellite communications systems serving a large number of terminals, terminal cost is a driving factor. For example, despite spectrum congestion problems, the military UHF band was selected for initial DoD mobile communications satellite programs, primarily because simple aircraft blade antennas as well as relatively simple radio equipment could be used for airborne terminals. In general, terminal costs rise dramatically if a continuously tracking antenna must be used, especially for mobile services. This conflict between uncongested spectrum and related equipment cost/simplicity has slowed the rate of development of both aeronautical and land mobile services and will be an important factor in shaping future development of components and systems for direct user-to-user service.

While considerable money has been spent on terminal development and construction, there are no clear indices pointing the way to design and construction processes for cheap terminals. Some argue that the best approach is "complexity inversion," where the satellite EIRP is large enough and satellite signal processing complex enough to permit adequate

performance with small, relatively simple terminals operating at higher frequencies.

Still another approach for minimizing system costs and (indirectly) relieving spectrum congestion is one which is more administrative than scientific. This involves the aggregation of user community requirements with the goal of reducing system complexity and increasing efficiency by standardizing services, channels, data rates, equipments, etc. The INTELSAT organization and the domestic carriers have followed this process in their system planning and realization. The subordination of individual national systems (DOMSATs) into aggregated regional systems has not yet occurred.

As the range of possibilities for satellite communications continue to grow and the resources (spectral bandwidth and number of orbit locations) continue to shrink, there is an obvious need for generalizing the aggregation and standardization process to cover a wider range of user communities. For example, in the Department of Defense, the Military Satellite Communications Systems Office of the Defense Communications Agency attempts to maximize efficiency of satellite use by aggregating requirements reflecting similar DoD user needs. The necessity for this kind of activity is also recognized by the public service community which advocates the acquisition of a satellite communications system for use by state and local governments, educational institutions, medical groups, etc. The National Telecommunications and Information Agency has been designated as the focal point for both federal non-military government requirements and the public service requirements.

Enforced requirements, aggregation and standardization are exemplary of a class of issues inherent in telecommunications regulation. In this case there are varying opinions among communities. There is a general feeling in the private sector that the proper arena for settling such issues is in the marketplace; and there is agreement in some parts of the U.S. government with this viewpoint. But a completely free telecommunications systems market can be self-defeating when competition forces specialization, restricting the customer to a narrower range of services than might have been potentially available. On the other hand, regulation can discourage the influx of new private capital for investment in high risk technology with long range payoffs because of regulated, fixed rates of return and established operating turf.

ADDITIONAL READING AND ACKNOWLEDGMENTS

We have attempted to keep the introduction to the Special Issue short, hoping to provide a perspective not only of the Special Issue content but of the more comprehensive satellite communications world. Because of the introductory brevity and because the Issue is necessarily finite, we append a reading list segmented into major interest categories, which we hope will be helpful to the interested reader.

I would not have missed the opportunity to function as Editor of the Special Issue. Having been educated, however, I hasten to add that I would not seek to repeat the experience soon. While my education has been improved specifically, it has also been broadened immensely by the generosity of the people who helped form the issue and administer the paper review process. For the latter, I thank Al Weinrich and Bart Batson, two of the three Associate Editors. I want to express my special gratitude to the third Associate Editor, Ron Sherwin, whose work not only contributed to the quality and texture of the issue, but helped me immeasurably in surviving the process.

REFERENCES

Current and Near-Term Systems

Lipke, D. W. et al "MARISAT–A MARITIME SATELLITE COMMU–NICATIONS SYSTEM," *COMSAT Tech. Review* Vol 7, #2 Fall 1977, pp. 351-391.

Briskman, R. D., "THE COMSTAR PROGRAM," *COMSAT Tech. Review* Vol 7, #1 Spring 1977, pp. 1-34.

Abutaleb, G. E. A. et al, "THE COMSTAR SATELLITE SYSTEM," *COMSAT Tech. Review* Vol 7, #1 Spring 1977, pp. 35-83.

Edelson, B. I., "GLOBAL SATELLITE COMMUNICATION," *Scientific American*, Feb 1977, Vol 236 #2, pp. 58-73.

Brook, A. W., "RCA SATCOM SYSTEM," *RCA Engineer* Vol 22 #1, June-July 1976, pp. 42-49.

Keigler, J. E., "RCA SATCOM-MAXIMUM COMMUNICATION CAPACITY PER UNIT COST," *RCA Engineer*, Vol 22 #1, June-July 1976, pp. 50-57.

Keigler, J. E., "RCA SATCOM; AN EXAMPLE OF WEIGHT-OPTIMIZED SATELLITE DESIGN FOR MAXIMUM COMMUNICATION CAPACITY," *IAF ACTA Astronautica*, Vol 5, March-April 1978.

Lee, D. J., "SYSTEM PERFORMANCE OF AMERICA'S FIRST DOMESTIC COMMUNICATION SATELLITE-WESTAR," *EASCON 74, Conference Proceedings.*

Van Cleve, John W., "OPERATION AND CONTROL OF AN INTEGRATED SATELLITE-TERRESTRIAL NETWORK," ICC 77.

Barnal, J. D., and Zitzmann, F. R., "THE S.B.S DIGITAL COMMUNICATIONS SATELLITE SYSTEM," EASCON 77, Washington, D.C., Sept. 26-28, 1977.

Frequency and Orbit Management

MASTER INTERNATIONAL FREQUENCY REGISTER OF THE IFRB, Nov 8, 1978.

NTIA MANUAL FOR FREQUENCY ALLOCATION FOR SATELLITE COMMUNICATION.

Withers, D. J., "EFFECTIVE UTILIZATION OF THE GEOSTATIONARY ORBIT FOR SATELLITE COMMUNICATION," *Proc. IEEE*, Vol 65 #3, March 1977, pp 308-317.

Evolution

Pritchard, W. L., "SATELLITE COMMUNICATION–AN OVERVIEW OF THE PROBLEMS AND PROGRAMS," *Proc IEEE,* Vol 65 #3, March 1977, pp. 294-307.

Bond, F. E. and Curry, W. H., Jr., "THE EVOLUTION OF MILITARY SATELLITE COMMUNICATIONS SYSTEMS," *Signal Magazine,* March 1976, pp. 39-44.

Bargellini, P. and Edelson, B., "PROGRESS AND TRENDS IN COMMERCIAL SATELLITE COMMUNICATIONS," IAF XXVI Congress, Lisbon, Portugal, Sep 21-27, 1975.

Technology and Trends

Van Trees, H. L., Hoversten, E. V., and McGarty, T. P., "COMMUNICATION SATELLITES LOOKING TO THE 1980'S, " *IEEE Spectrum,* Dec. 1977, pp 43-51.

Bargellini, P. and Edelson, B., "PROGRESS AND TRENDS IN COMMERCIAL SATELLITE COMMUNICATIONS," presented at IAF XXVI Congress, Lisbon, Portugal, Sept 21-27, 1975.

Van Trees, H. L., "FUTURE INTELSAT SYSTEM (1986-1993) PLANNING," *Progress in Astronautics and Aeronautics,* AIAA Vol 54, pp 95-141.

Anderson, Howard and Kutnick, Dale, "WHAT IS ELECTRONIC MAIL?", *Telecommunications,* Vol 12 #11, Nov 1978, pp 31-54.

Propagation at Higher Frequencies

Hogg, D. C. and Chu, Ta-Shing, "THE ROLE OF RAIN IN SATELLITE COMMUNICATIONS," *Proc. IEEE,* Vol 63 #9, Sept 1977, pp 1308-1331.

Crane, R. K., "PREDICTION OF THE EFFECTS OF RAIN ON SATELLITE COMMUNICATION SYSTEMS," *Proc IEEE* Vol 65 #3, March 1977, pp 456-474.

Large Platforms

Edelson, B. I., and Morgan, W. L., "ORBITAL ANTENNA FARMS," *Progress in Astronautics and Aeronautics,* AIAA, Sept 1977.

Fordyce, S. and Jaffee, L., "FUTURE COMMUNICATION CONCEPTS; THE SWITCHBOARD IN THE SKY," *Satellite Communication,* February 1978, pp 22-26 and March 1978, pp 22-27.

Regulation and Technology

Jones, Erin Bain, *EARTH SATELLITE TELECOMMUNICATIONS SYSTEMS AND INTERNATIONAL LAW,* Encino Press, 1970.

Galloway, Jonathan F., *THE POLITICS AND TECHNOLOGY OF SATELLITE COMMUNICATIONS,* D. C. Heath, 1972.

Magnant, Robert S., *DOMESTIC SATELLITE: AN FCC GIANT STEP,* Westview, 1977.

Satellite Communication—An Overview of the Problems and Programs

WILBUR L. PRITCHARD, FELLOW, IEEE

Abstract–This paper introduces the subject of satellite communication in its broadest aspects, recounts its history and discusses the principal technical problems. It is primarily communications-oriented but relevant spacecraft and launch considerations are summarized. Tabular summaries of the world's satellite communication programs are given.

I. INTRODUCTION

OMMUNICATION satellites have several important characteristics. One is certainly the availability of bandwidths exceeding anything previously available for intercontinental communications. Although overland transmission of high-quality TV pictures by microwave radio relays or cable has been possible for some years, trans-Atlantic TV transmission took place for the first time only after the first active communication satellite had been put in orbit. Intercontinental relaying of TV programs, now commonplace, is done exclusively by satellites.

Another, perhaps the most important of all, is the unique ability to cover the globe. In the future, it is likely that cables will use much higher carrier frequencies, probably as high as the optical region of the spectrum. If so, a multitude of TV channels or their equivalent could be transmitted from one continent to another without satellites. However, a cable still has two fixed ends and there must be a connection between every pair of points to be in communication. Satellite systems offer, in this respect, a flexibility that cannot now be duplicated. Furthermore, this flexibility applies not only to fixed points on earth, but also to moving terminals, such as ships at sea, airplanes, and space vehicles.

With communication satellites, then, instant and reliable contact can be established rapidly between any points on earth, in addition to, and well beyond, the capabilities of available land lines, microwave line-of-sight relay systems, and other techniques. Satellites are the elements of a communications revolution analogous to that in transportation resulting from the airplane.

II. HISTORY

Although the origins of the whole idea of satellite communication are obscure, there is no question that the synchronous, or more accurately, geostationary, satellite was first proposed by Arthur C. Clarke in an article in *Wireless World* entitled, "Extraterrestrial Relays." He recognized the potential for rocket launches based on the German V2 work during the war and also the conspicuous advantage of the geostationary orbit. Prophetically, his proposal was for the use of these satellites for FM voice broadcast rather than for telephone service. Interestingly enough, Clarke also foresaw the use in space of electric power generated by panels of solar cells. Implementation of his idea still had to wait for the Space Age (Sputnik 1957) and solid-state technology.

Thirty-one years have passed since his prophecy, and there are now 22 satellite communication programs with satellites in orbit or under active construction. There are another score of programs with earth stations only using the satellites of others. A word of tribute to his exceptional vision is certainly in order.

A. The Early Years

Moon reflections for radar and communication purposes were repeatedly demonstrated in the late forties and early fifties. In July 1954, the first voice messages were transmitted by the US Navy over the earth-moon path. In 1956, a US Navy moon relay service was established between Washington DC, and Hawaii. This circuit operated until 1962, offering reliable long-distance communication limited only by the "availability" of the moon at the transmitting and receiving sites. Power used was 100 kW, with 26-m diameter antenna at 430 MHz.

A metallized balloon of the correct size, launched by a rocket and placed in orbit can be used as a scatterer of electromagnetic waves generated by an earth transmitter. Part of the energy can be picked up by receiving stations at any point on earth from which the balloon is visible, thus obtaining a passive communications satellite system.

Through the joint action of Bell Telephone Labs, NASA, and JPL, the ECHO experiment was performed. Successful communications across the US were first established in early August 1960, between Goldstone, CA, and Holmdel, NJ, at frequencies of 960 MHz and 2290 MHz. The ECHO "balloon," in an inclined orbit at 1500-km altitude, was visible to the unaided human eye.

Later in the same month, the first trans-Atlantic transmission occurred between Holmdel, NJ, and a French receiving station [1]. This project alerted the entire world to the prospect of the new medium of communications although the specific method was never exploited commercially.

Although passive satellites have infinite capability for multiple-access communications, they are gravely handicapped by the inefficient use of transmitter power. In the ECHO experiment, for instance, only one part in 10^{18} of the transmitted power (10 kW) is returned to the receiving antenna. Since the signal has to compete with the noise coming from various sources, special low-noise receivers must be used. Luckily, the invention of the maser in 1954 and its successive development, permitted the construction of very low-noise receivers (with temperatures in the neighborhood of 10 which, used with horn-reflector receiving antennas having an aperature of about 43 m^2, made possible the transmission of teletype, voice, and pictures).

The advantage of passive satellites is that they do not require sophisticated electronic equipment on board. A radio beacon transmitter might be required for tracking, but in general, neither elaborate electronics, nor, with spherical satellites, attitude stabilization is needed. Such simplicity, plus the lack of space-flyable electronics in the late fifties, made the passive system attractive in the early years of satellite communications.

As soon as space-flyable electronics became available, it was obvious that passive systems would be replaced by active satellites. The mathematics of the inverse square law for active satellites (versus the radar-like inverse fourth-power law applicable to passive satellites) are overwhelmingly in favor of the former.

The relative disadvantages of a passive system increase with orbital altitude and the on-board power availability of the active satellite. After the early experimental trials, all subsequent satellite communication experimental and operational systems have been of the active type, and there is nothing to indicate that the situation is likely to change.

It is interesting to note that the first active US communication satellite was a broadcast satellite. SCORE, launched on December 18, 1958, transmitted President Eisenhower's Christmas message to the world with a power of 8 W at a frequency of 122 MHz. SCORE was a delayed-repeater satellite receiving signals from earth stations at 150 MHz; the message was stored on tape and later retransmitted. The 68 kg payload was placed in rather low orbit (perigee 182 km, apogee 1048 km).

The communications equipment was battery powered and not intended to operate for a long time. After 12 days of operation, the batteries had fully discharged and transmission stopped.

B. The Experimental Years

Aside from early space probes like Sputnik, Explorer, and Vanguard, as well as the SCORE and Courier projects, which were early communication satellites of the record and retransmit type, the major experimental steps in active communication satellite technology were the Telstar, Relay, and Syncom projects.

Project Telstar is the best known of these probably because it was the first one capable of relaying TV programs across the Atlantic. This project was begun by AT&T and developed by the Bell Laboratories, which had acquired considerable knowledge from the early work of John R. Pierce and his associates, and from the work with the ECHO passive satellite. The first Telstar was launched from Cape Canaveral on July 10, 1962. It was a sphere of approximately 87 cm diameter, weighing 80 kg. The launch vehicle was a Thor-Delta rocket which placed the satellite into an elliptical orbit with an apogee of 5600 km, giving it a period of 2-½ h.

Telstar II was made more radiation resistant because of experience with Telstar I, but otherwise, it was identical to its predecessor. It was successfully launched on May 7, 1963.

The power of Telstars I and II was 2.25 W provided by a TWT, with an RF bandwidth of 50 MHz at 6 and 4 GHz. Both satellites were spin-stabilized. The overall communication capability was 600 voice telephone channels, or one TV channel. To overcome the low carrier-to-noise ratio available in the down-link, receivers at the earth stations used FM feedback in order to obtain an extended threshold. Even though the Telstar system was superbly engineered, it was designed as an experiment and was not intended for commercial operation. Among

other things, the orbit used made it only visible for brief periods. A project with similar objectives, Project Relay, was developed by the Radio Corporation of America under contract to NASA. It was similarly successful.

In early 1962, the President sent proposed legislation to Congress to start the commercial exploitation of these successes. After extensive hearings on the Bill, the US Congress passed the Communications Satellite Act of 1962, which led to the establishment of the Communications Satellite Corporation in 1963.

On August 20, 1964, a significant event occurred when agreements were signed by 11 sovereign nations which resulted in the establishment of a unique organization—the International Telecommunications Satellite Consortium, known as INTELSAT. This new organization was formed for the purpose of designing, developing, constructing, establishing, and maintaining the operation of the space segment of a global commercial communications satellite system.

C. The Commercial Era

Commercial communications by satellite began officially in April 1965, when the world's first commercial communication satellite, INTELSAT I (known as "Early Bird"), was launched from Cape Kennedy. It was decommissioned in January of 1969 when coverage of both the Atlantic and Pacific was accomplished by two series of satellites, INTELSAT's II and III. Interestingly enough, Early Bird was planned to operate for only 18 months. Instead, it lasted four years with 100 percent reliability.

The fully mature phase of satellite communications probably is best considered as having begun with the installation of the INTELSAT IV into the global system starting in 1971. These spacecraft weigh approximately 730 kg in orbit and provide not only earth coverage but also two "pencil" beams about 4° in diameter which can be used selectively to give spot coverage to Europe and North and South America. INTELSAT IV is a spinning satellite, as were its predecessors, but the entire antenna assembly, consisting of 13 different antennas, is stabilized to point continually toward the earth. Two large parabolic dishes form the two spot beams. Each satellite provides about 6000 voice circuits, or more, depending upon how the power in the satellite is split between the spot beams and the earth coverage beam. INTELSAT IV can carry 12 color TV channels at one time.

D. Military Satellites

The first military satellites, the DSCS-I, were launched by the US Air Force in June of 1966. These launches were interesting because 8 satellites were launched simultaneously. Finally, about 30 satellites of a very simple spinning type and without station-keeping were placed in near synchronous orbits. Some are still in operation today. The DSCS-II system was initiated several years ago and constitutes the present US military system although it has had both spacecraft and launch vehicle failures. DSCS-III is being planned.

III. CATEGORIES OF SYSTEMS

There are some 42 satellite communication systems in the world today, 22 of which include both satellite and terrestrial equipment (See Table I). By satellite system, we mean one which is in active operation or one for which the equipment is

TABLE 1
LIST OF PROGRAMS

Program	Class						Coverage		Status				First Operational Date — Actual or Planning	Frequency Bands (MHz)
	Fixed	Mobile	Broadcast	Experimental	Military	Proposed	Regional	Domestic	Operational	Under Construction	In planning	Under Construction		
1 INTELSAT	X						X		X				1965	5925 – 6425 up 3700 – 4200 down (C Band)
2 USSR	X						X	X	X				1965	5725 – 6225 up 3400 – 3900 800 – 1000 down
3 TELESAT – Canada	X							X	X				1973	5927 – 6403 3702 – 4178
4 Western Union	X							X	X				1974	5925 – 6425 3702 – 4178
5 RCA SATCOM	X							X	X	X			1974	Same as WESTAR
6 AMERICAN SATELLITE CORP *	X							X	X				1974	WESTAR Transponder
7 COMSAT/ATT (COMSTAR)	X							X	X				1976	5925 – 6425 3700 – 4200
8 ESA (OTS/ECS)	X						X				X		1977	14152.5 – 14192.5 11490 – 11530 14242.5 – 14362.5 11580 – 11700 14455.0 – 14460 11792.5 – 11797.5
9 ALGERIA*	X							X	X				1975	INTELSAT Transponder
10 INDONESIA	X							X	X				1976	5925 – 6425 3700 – 4200
11 NORWAY/NORTH SEA	X						X			X			1975	INTELSAT Transponder
12 SBS	X							X			X		1978	14000 – 14500 11700 – 12200 Ku Band
13 PHILIPPINES*	X							X				X	1976	5925 – 6425 3700 – 4200
14 INDIA	X	X						X				X	?	C or Ku 2500 B'cast
15 ARAB SYSTEM	X						X					X	1980-90	C Band 2500 B'cast
16 COMSAT GENERAL (MARISAT)		X					X	X					1975	Satellite – Ship Satellite – Shore 300 – 312 1638.5 – 1642.5 up 6174.5 – 6425 up 248 – 260 1537 – 1541 down 3945.4 – 4199.0 down
17 ESA/COMSAT GENERAL (AEROSAT)		X					X				X		1979	Satellite – Plane Satellite – Ground 1645 – 1660 131.425 – 131.975 up 5000 – 5125 up 1543.5 – 1558.5 125.425 – 125.975 down 5125 – 5250 down
18 ESA (MAROTS)		X					X				X		1977	Satellite – Ship Satellite – Shore 1641.5 – 1644.5 up 14490 – 14500 up 1540 – 1542.5 down 11690 – 11700 down
19 DMCO (INMARSAT)		X					X					X	1980+	Satellite – Ship Satellite Shore L Band C or Ku Band
20 EUROPEAN BROADCASTING UNION			X				X					X	1980+	Ku Band
21 FEDERAL REPUBLIC OF GERMANY			X					X				X	~1985	Ku Band
22 SYMPHONIE				X			X	X	X				1975	5926 – 6424 up 3700 – 4200 down
23 ATS-6				X				X	X				1974	Wide variety of experiments in K, C, S, UHF, and VHF Bands
24 JAPAN – COMMUNICATION	X			X				X			X		1977	5925 – 6425 3700 – 4200 30 000 20 000
25 JAPAN – BROADCAST			X	X				X					1978	14 000 12 000
26 CTS				X				X		X			1975	14 000 12 000
27 SIRIO				X				X		X			1977	17009 – 17782 11331 – 11862
28 DSCS	X	X			X		X		X	X			1966	7900 – 8400 7250 – 7750
29 SKYNET	X	X			X			X	X				1969	7976 – 8005 7257 – 7286
30 NATO	X	X			X			X	X				1970	7975 – 8162 7250 – 7437
31 PLTSATCOM	X	X			X			X	X				1977	Satellite – Ship Satellite – Shore 225 – 400 7900 – 8400
32 LES SERIES	X	X	X		X		X		X				1965	Wide variety of experiments at UHF and K Bands
33 ANDEAN NATIONS*	X					X	?	?				X	?	INTELSAT Transponder
34 ARGENTINA*	X					X		X	X			X	?	INTELSAT Transponder
35 AUSTRALIA	X					X		X	X			X	1980's	C Band, Ku, S Band
36 BRAZIL	X	?		?		X		X			X		1978	5925 – 6425 3700 – 4200
37 DENMARK*	X					X		X	X			X	?	INTELSAT Transponder
38 IRAN	X	?		?		X		X	X			X	?	--
39 MALAYSIA*	X					X		X	X			X	1975	INTELSAT Transponder
40 UNITED KINGDOM								–					–	--
41 SYSTEMS USING INTELSAT*	X	X						X	X	X		X	Various	INTELSAT Transponder
42 ASEAN NATIONS	X					X	?	X				X		INTELSAT Transponder

* Earth Stations Only

being built under funded contract. There are literally dozens of other systems under study in various phases. They range, in likelihood of implementation, from remote possibilities to being on the verge of realization.

We can characterize and categorize satellite systems in a variety of manners, either by their technical characteristics or by their operational use. The latter seems more natural and, indeed, is instructive in the sense that after having categorized the satellites operationally, we can examine the diversity of technical methods used to achieve similar operational results.

The first category of system is certainly international civil telecommunications, of which we have only 2 examples today: the highly successful INTELSAT and the inchoate STATSIONAR of the Soviet Union.

The second category is that of regional and domestic satellites, principally for civil telecommunications but occasionally for the distribution of television programs. In this class are: Anik, Westar, Satcom, Comstar, Palapa (Indonesia), and Molniya. For military communications (either to fixed or mobile terminals), we have 4 systems: NATO, Skynet, DSCS, and FLTSATCOM.

There are as yet no operational direct broadcast systems although there are several experimental ones in that category. It is likely that even operational ones will be joined with telecommunications services.

And, finally, still in the operational rather than the experimental category, we have those systems planned specifically for communication with mobile terminals: MAROTS, MARISAT, and AEROSAT.

The experimental programs cover a wide range of purposes, from experimenting with operational problems and proving spacecraft technology to acquiring propagation data in frequency bands of potential interest, especially above 12 GHz. In this category, we have: ATS-6, Canadian Technology Satellite, the Japanese Communication Satellite, Japanese Broadcast Satellite, OTS, Symphonie, SIRIO, and the Lincoln-Laboratory Series (LES). The gross technical features of these systems are summarized in Table II.

IV. MAIN TECHNICAL CONSIDERATIONS

In order to discuss the various ways in which the different systems have chosen to cope with the particular technical problems, we must examine, at least briefly, the technical and operational problems of satellite communications. These problems cover virtually the entire spectrum of physics and engineering. It is our intent here to look particularly at those that are special and even unique to satellite communication.

Clearly, a whole range of problems results from the necessity to insert the satellite into the desired orbit, normally geostationary, to control it remotely, and to maintain it in good operating condition. Even if satellite communications were to consist of passive reflectors, the later problems alone would be obviously not negligible. The wide bandwidths now easily available in the use of satellites in order to be exploited properly lead to another group of problems; many of which, of course, are familiar in connection with terrestrial microwave relay.

Most important of all, we have that category of problem unique to satellite communication that arises from the necessity and desirability of exploiting the geometric availability of a geostationary satellite to any point over almost a third of the earth's surface. Before this convenience can be realized, it is necessary to choose a system of multiple access. In a very real sense, we can call this *the* problem of satellite communications.

The problems in the three categories are not independent of each other. We will discuss, at least briefly, the most important interrelations among them.

A. The Communication Link

The exploitation of the wide bandwidth and the multiple access problem, and the related question of choosing appropriate modulation systems, are all best examined by starting with the elementary equation for the performance of a communication link.

The carrier power received at the earth station is given by

$$C = \frac{P_T G_T A_R}{4\pi R^2 L_i}, \quad A_R = \frac{G_R \lambda^2}{4\pi}, \quad \text{space loss} = \frac{4\pi R}{\lambda^2} \quad (1)$$

where

P_T satellite transmitter power
G_T transmitter antenna gain (over isotropic)
A_R earth station antenna effective area $= G_R \lambda^2 / 4\pi$
λ wavelength
R distance from satellite to earth station
L_i incidental loss

The noise density N_0 is equal to kT_s where k is Boltzmann's constant and T_s is the equivalent system temperatures defined so as to include antenna noise and thermal noise generated at the receiver. Let us further define effective isotropic radiated power (EIRP) as equal to $P_T G_T$ and a dimensionless spaceloss equal to $(4\pi R/\lambda)^2$. It is routine to show

$$\frac{C}{N_0} = \frac{EIRP}{L_s L_i} \frac{G_R}{T_s} \frac{1}{k} \quad \text{dBHz.} \quad (2)$$

Communications engineers usually make these calculations in dB by taking 10 times the $\log_{10}$ of both sides of (2). Care should be taken since some terms are dimensionless and some others are not. (e.g., (C/N_0) is in dBHz, Li will be in dB, and EIRP in dBw.)

Even more interestingly, the transmitter antenna gain G_T is inversely proportional to the solid coverage angles which in turn depends on the terrestrial area to be covered A_{cov}

$$A_{cov} \left(\frac{C}{N_0}\right) \sim P_T \frac{A_R}{T_s} \frac{1}{L_i}. \quad (3)$$

Equation (3) has been written so as to highlight the most basic aspects of space communication. On the left side of the equation are the desired parameters C/N_0 and the area to be covered on the earth. On the right side, by using the reference simple relations, we have succeeded in eliminating all the terms other than the transmitted power in the satellite, the effective physical size of the receiving antenna, the dissipative losses, and the system temperature.

The carrier-to-noise density (C/N_0) can be written as $(C/N)B$ where B is the noise bandwidth and (C/N) is the desired carrier-to-noise ratio. (C/N) typically has a threshold value, the achievement of which will permit substantial improvements in demodulated signal-to-noise ratio. It will also be a function of bit-error rate in digital systems and in general, depends on the modulation used. It has values between 8 and 13 dB for most space communication systems. We now can write

$$A_{cov}B \sim P_T \frac{A_R}{T_s} \frac{1}{L_i} \left(\frac{C}{N}\right)^{-1}. \quad (4)$$

Equations (3) and (4) are both fundamental to appreciating the essential problems of space communications. Although they have been written for a down-link, the same equations apply to the up-link from earth station to satellite if appropriate parameters are substituted.

Either one or both links may determine the overall performance. It is usually assumed that the noise from all sources (including intermodulation to be discussed later) is additive and it is then easy to show that

$$\left(\frac{C}{N_0}\right)_T^{-1} = \left(\frac{C}{N_0}\right)_U^{-1} + \left(\frac{C}{N_0}\right)_D^{-1} + \left(\frac{C}{N_0}\right)_I^{-1} \quad (5)$$

where the subscripts U, D, I, and T apply to the carrier-to-noise density ratios as calculated for the up-link, down-link, equivalent intermodulation noise, and total, respectively.

Note that in both (3) and (4) certain terms do *not* appear, e.g., satellite altitude, carrier frequency, and earth station (G/T).

The carrier frequency has an important second-order effect since clearly the available transmitter power, receiver temperatures, etc., does depend on it's state of the art at this frequency, but the dependence is not a function of the communication performance equation. The commonly used figures of merit EIRP and G/T are not basic and should be used with caution. EIRP has an implied coverage and G/T an implied carrier frequency and when used as figures of merit, this must be remembered.

The implications of the term C/N_0 are fundamental and account for its frequent use in communications systems engineering. If one examines Shannon's expression for the information capacity of a channel and allows the bandwidth to approach infinity as a limit, it turns out that the information that can be transmitted in a channel is proportional to C/N_0.

TABLE II
CHARACTERISTICS

SYSTEM	SATELLITE MASS – STABILIZATION	LAUNCH VEHICLE	MODULATION AND MULTPLE ACCESS	EARTH STATION ANTENNA DIAMETER	CAPACITY PER SATELLITE
1. INTELSAT IV	700 KG. after apogee motor firing	ATLAS–CENTAUR	FM/Video FDM/FM/FDMA	29.5 M. for Std. A 13 M. for Std. B	7500 channels + SPADE + TV 12,000 channels + SPADE + TV
INTELSAT IV–A	790 KG. after apogee motor firing S	"	"	29.5 M.	
2. U.S.S.R. – MOLNIA	~ 1,000 KG. B (elliptical orbit)	A–Z–ε and SL–12	FM	12.M. 25 M.	1 TV channel + unspecified telephones
3. TELESAT – Canada	272.2 KG. S	DELTA 1914	Single carrier FM Multi carrier FM Single channel/ carrier, Delta modu- lation PSK TDMA	Heavy Route – 30M. Network TV – 10.1 M. Northern 10.1 M. Tele- communications Remote TV 8.1M./ 4.7M. Thin Route 8.1M./ 4.7M.	12 transponders of 36 MHz bandwidth
4. WESTERN UNION – (WESTAR)	297 KG. S	DELTA 2914	SSB/FM/FDM Single & Multiple Access Video, SCPC: PSK/TDM/TDMA	15.5M.	12 Video channels (one way) or 14,400 FDM Voice channels (one way)
5. RCA SATCOM	461 KG. B	DELTA 3914	FDMA: FDM/FM & PCM/PSK for voice data, 4 PSK for digital data, FM for monochrome or color TV TDMA: PCM/PSK for voice/ data	10M.	24 Video channels w/34 MHz bandwidth 9000 channels/transponder
7. COMSAT/ATT– (COMSTAR)	750 KG. S	ATLAS–CENTAUR	FDM/FM Digital Transmission 4PSK	30M.	28,800 one-way tele- phony channels or >1,000 mb/s data
8. ESA (OTS/ECS)	324 KG. B	DELTA 3914	4-Phase PSK FM Video TDMA	Eurobeam A 13M. + Spot Beam Eurobeam B 3M.	1–120 MHz Transponder 1–40 MHz Transponder 1–5 MHz Transponder
10. INDONESIA	300 KG. S	DELTA 2914	FDM/FM Multiple carriers per transponder SCPC/FM demand-assigned	9.8M. 7.3M. 4.0M.	
16. COMSAT GENERAL (MARISAT)	326.6 KG. on station S	DELTA 2914	Voice: SCPC–FM Data: 2-Phase coherent PSK	1.22M., mobile term- inals 12.8M., shore term- inals	9 voice channels (both ways) 110 teleprinter channels (both ways)
17. ESA/COMSAT GENERAL (AEROSAT)	470 KG. B	DELTA 3914	Voice: NBFM, PDM and/or VSDM Data: PSK/FSK	Low gain airborne antennas	5–80 kHz comm. ch. for ground-to-air and surveillance 15–40 kHz comm. ch. for air-to-ground 2–80 kHz comm. ch. for ground-to-ground 1–400 kHz or 10 MHz experimental channels
18. ESA (MAROTS)	466 KG. B	DELTA 3914	FDM TDM FDMA TDMA	About 1M.	Shore-to-ship: Up to 50 voice/high-speed data channels Ship-to-shore: Up to 60 voice/high-speed data channels Shore-to-shore: Up to 3 voice/high-speed data channels

If the RF power budget, as determined from the above questions, yields a particular C/N_0 and if a particular bandwidth B is available from a frequency allocation point of view, then C/N_0 is quite simply divided by the bandwidth to determine the carrier-to-noise ratio (C/N) available for detectability with the particular modulation system. For a fixed (C/N) and modulation system, B is proportional to the number of channels. Thus the number of channels in a given coverage can depend only on the transmitter power, receiver antenna size, and system temperature as before. Multiple-beam antennas reduce the coverage and hence increase the number of channels for a constant total power. If the power is divided among the beams proportionally, then B remains constant for each beam and the total bandwidth available per satellite increases by the number of beams.

Increasing capacity by increasing B through frequency reuse so as to avoid the limitations of (4) costs considerably in spacecraft antenna size and complexity. A channel may be available only in a particular beam and must be switched if it is desired in another. The switching can be done slowly, mechanically, as is now done in INTELSAT IV and will be done in INTELSAT V. Increased capacity is obtained at the expense of increased transponder complexity—there will be hundreds of switches in INTELSAT V—and a loss of flexibility that is acceptable because of the traffic patterns for a worldwide fixed system. The

22.	SYMPHONIE	230 ± 5 KG.	B	DELTA 2914	Analog and Digital	16M. 12M.	1200 one-way telephony or 2 color TV channels
23.	ATS-6	1356 KG.	B	TITAN III-C	FM, Video	25.9M. 3M. (Various)	2 Video channels at 2.6 GHz or 1 Video channel at UHF (860 MHz) C-band transponder has 40 MHz bandwidth 1.5 GHz transponder has 12 MHz bandwidth
26.	CTS	350 KG.	B	DELTA 2914	FM Video FM Sound broadcast 10 ch. (FDM) of FM duplex voice	10 - 0.91M. 8 - 2.43M. 2 - 3.05M. 2 - 9.14M.	1 TV Channel 1 sound broadcast channel 10 duplex voice channels
27.	SIRIO	188 KG.	S	THOR-DELTA	PCM-PSK, 2-phase for voice (narrow band communications) FM or Digital for TV (wideband communi- cations)	14.5M. for stations in Italy 14M. for stations in Finland 12M. for stations in USA Various smaller sizes down to 1.2M. for shipboard terminal	12 - 100 kHz telephone channels, total 1.5 MHz band- width or 1 - 4 MHz baseband for TV
28.	DSCS II	500 KG.	S	TITAN III-C	Stage 1a of program: FDMA & CDMA Stage 1b of program: FDMA & CDMA Stage 1c of program: FDMA & CDMA, phasing into TDM/PCM Stage 2 of program: TDMA	18.2M. for largest terminals (Fixed) 0.8M. for smallest transportable terminals (airborne)	1300 duplex voice channels or 100 Mb/s data Total of 410 MHz of transponder bandwidth
29.	SKYNET	232 KG.	S	DELTA 2313 (SKYNET 2A & 2B)	CDMA in 20 MHz channel FDMA in 2 MHz channel	I II 12.8M. 12.2M. III IV 6.4M. 6.4M. V "SCOT" 1.8M. 1.1M.	1 - 20 MHz channel 1 - 2 MHz channel 24 (2400 bps) data channels or 280 voice channels
30.	NATO	340 KG.	S	DELTA 2914	FDMA/FDM (clear mode) CDMA (jamming mode)	12.8M.	
31.	FLTSATCOM	862 KG.	B	ATLAS-CENTAUR			9 UHF and 1 SHF uplink 10 UHF downlink Each UHF has 25 kHz bandwidth
32.	LES SERIES	450 KG.	B	TITAN III-C	DPSK downlink QPSK conferencing link 8-ary FSK forward uplink 8-ary MFSK, hopped at 200/sec.	1.2M. for ABNCP terminal (Lincoln Labs) 92 CM. for airborne terminal, AN/ASC-22 46 dM. for Navy term- inal	36-38 GHz: 10 kb/s, DPSK, to other LES satellites 20 kb/s, DPSK, to ABNCP terminal 8 ary FSK forward uplink: QPSK conferencing uplink 50 K-ary symbols/sec from ABNCP terminal 75 b/s to Navy (shipboard) terminal UHF: 50 8-ary symbol/sec to aircraft 100 8-ary symbol/sec from aircraft

switching can also be electronic and rapid, such as would be the case in a time-division switched satellite in which multiple access was achieved by burst transmissions from each earth station that would be switched in the satellite to the appropriate beam.

The multiple-beam configuration is suitable to high-traffic fixed systems, such as INTELSAT and large-area regional systems. It is less appropriate in domestic systems, such as Canada, Indonesia, and the United States.

When very high carrier frequencies are used because of increased crowding in the more desirable bands, the dissipative and scattering losses and lower available powers will probably force multiple-beam operation even where it might otherwise be undesirable.

Broadcast satellites, especially those for areas covering several time zones, will use them since the same programs do not necessarily have to go simultaneously to different areas. The European Broadcasting Union plans, when brought to fruition, envision multiple beams for Europe.

Satellite connections require up-links and down-links. Normally, since it is relatively easy to supply high transmitter powers and antenna gains at earth stations, the performance is determined by the down-link. The limitations of satellite power and the necessity for covering the appropriate terrestrial area limit the overall performance. The problem is complicated by the necessity for multiple access and the resulting possibilities for intermodulation noise in nonlinear transponders. In the case of small terminals, such as in mobile and data gathering systems, the limitation is often in the up-link.

As we see from (3), the ultimate limit in the down-link performance is the transmitted power in the satellite, and this limitation cannot be avoided for an assigned bandwidth and the requirement for terrestrial coverage.

Satellite power is directly translatable into weight in the spacecraft and, even more pointedly, into cost. Although the carrier frequency is not a first-order problem in satellite system planning, it has many very important second-order effects. External sources of noise, such as the galaxy and propagation through the ionosphere and atmosphere, are generally frequency dependent.

B. Multiple Access

To exploit the unique geometric properties of wide-area visibility and multiple connectivity that go with satellites, the various communications links using it must be separated from each other. This can be accomplished in several ways.

1) Space-Division Multiple Access (SDMA): One is to use different antenna beams and separate amplifiers within the satellite. This is SDMA. Flexibility is only possible at the expense of complications within the satellite, increased weight, and occasional operational difficulties.

2) Frequency-Division Multiple Access (FDMA): A second basic way, and the one in most common use, is that of using different carrier frequencies for each transmitting station. This is FDMA and permits many stations to use the same transponder amplifier until finally the overall noise level limits the capacity of that amplifier. Multiple carriers in any nonlinear amplifier produce intermodulation products which raises the apparent noise level. This requires a "back-off" of drive on the amplifier in order to reduce this intermodulation noise. The carrier level received is less and thus the effect of thermal noise generated in the earth station receiver is increased. This reduction in drive must thus be optimized. Even optimized, the effect is not trivial and the reduction in capability of a transponder over that it would have if all the available information was multiplexed on a single carrier frequency can be as much as 6 dB. Nevertheless, FDMA is the most popular technique for commercial communication satellites. It is efficient if one is not power limited, and it is the natural expansion of terrestrial communication methods.

FDMA can be implemented in two ways. One is to multiplex, in the conventional terrestrial manner, many channels on each carrier that is transmitted through the satellite. Another is to use a separate carrier frequency for each telephone or baseband channel within the satellite. If many carriers are used, the intermodulation problem is still more serious. On the other hand, it does approach, asymptotically, a limiting level that is usually acceptable. This single-channel-per-carrier approach has particular advantages in systems where there are many links to be made, each one having only a few circuits to be handled at any one time. Normal multiplexing is very convenient terrestrially but may be economical only if each carrier has traffic, for example, in a group of 12 channels or more.

Both systems are in extensive use today. INTELSAT uses both systems, the SPADE system being a single-channel-per-carrier multiple-access system. Canada, Indonesia, Algeria, to mention a few, use single-channel-per-carrier systems. The modulation for single-channel-per-carrier systems is a separate decision and in use today, we have PCM, Delta modulation, and narrow-band FM. The arguments as to which is best are rather complicated and are discussed elsewhere in this issue.

3) Time-Division Multiple Access (TDMA): The next basic method is TDMA, in which each earth station is assigned a time slot for its transmission, and all the earth stations use the same carrier frequency within a particular transponder. In terms of total satellite performance, this is the superior method because the intermodulation noise is eliminated and there is an increase in capacity. The required back-off is much less, just that required to achieve acceptable spectrum spreading. The price paid is a considerable increase in complexity of the ground equipment. It does seem as if the long-term trend will be toward more and more TDMA since it fits naturally with the digital communications systems that are so rapidly proliferating terrestrially, not only for data transmission but more and more for digitized voice.

Various experimental TDMA systems in the 6 Mbit/s to 60 Mbit/s range have been built and tested by INTELSAT and others. Their efficiency advantage over FDMA can be illustrated by comparing the approximate channel capacities of an INTELSAT IV global beam transponder operating with standard INTELSAT 30 m earth stations, using TDMA and FDMA, respectively. Assuming 10 accesses, the typical capacity using FM/FDMA is about 450 one-way voice channels [2]. With TDMA, using standard 64 kb/s voice frequency PCM encoding, the capacity of the same transponder is approximately 900 channels. If Digital Speech Interpolation (DSI) is used to process the PCM bit streams, the capacity is further increased to about 1800 channels.

A TDMA system went into commercial operation on Telesat, Canada's system, starting in May 1976. Numerous other TDMA systems are planned for regional and domestic satellite systems throughout the world.

This trend to digital systems both terrestrially and via satellite is reinforced by the ease with which the TDMA methods can be combined with SDMA by switching transmission bursts from one antenna beam to another depending on their ultimate destination. This notion of time-division switching, although not yet exploited in any satellite, seems inevitable for the reasons stated in connection with the discussion on the link equation. It is efficient in its exploitation of both the satellite power and the frequency spectrum, and both these resources are in short supply. The price paid is increasing complexity. That seems less and less of a price considering the awesome technology of large-scale integration and microcomputers.

Time-division switching will be a major factor in communication satellite technology. A satellite-switched TDMA system (SS-TDMA) using a microwave switch matrix of redundant design shows an increase of over 30 percent in available capacity over FDMA/TDMA [3] (separate frequency bands, each carrying TDMA). The satellite-switched TDMA concept uses a single 400 MHz channel, as distinguished from the FDMA/TDMA system, which uses 5 channels of 80 MHz. Its keying rate is 300

MBd/s, rather than 60 MBd/s. Four-phase PSK is used, as with FDMA/TDMA. A total channel capacity of 39 700 is achieved by SS-TDMA, compared with 29 870 for TDMA/TDMA. Note that this time-division switching must be done in nanoseconds so as to connect successive bursts to different spot beams. Diodes of the p-i-n type and similar solid-state switches will be necessary and are under development along with extensive ancillary logic circuitry.

4) Code-Division Multiple Access (CDMA): The final basic method of multiple access is that of CDMA, called occasionally "spread-spectrum multiple access." In either case, the idea is the same. The transmission from each earth station is combined with a pseudo-random code so as to cause the transmission to occupy the entire bandwidth of the transponder. The station for whom the transmission is intended has a duplicate of this pseudo-random code and by cross-correlating techniques can extract it from the "noise level" created by the simultaneous use of many other stations.

It has considerable advantage in military systems because the spread-spectrum technique must be used anyway to harden the satellite receiver against possible jamming and the pseudo-random sequences are necessary to provide cryptographic security.

The use of such crypto and anti-jam systems provides automatic multiple access. In a sense, it is free. The difficulty is that it is not nearly so efficient an exploitation of the resources of power and frequency spectrum as is even the FDMA system, not to mention TDMA. In addition, it requires extra equipment at both ends of the link.

Nevertheless, it is used and will continue to be used for military systems. The possibility of its limited use in commercial systems may appear as satellite users become increasingly concerned with the possibilities of both malicious interference and unauthorized listening. Users of satellite systems for commercial data transmission of the kind envisioned in domestic US systems may well be the first to consider at least the crypto secure aspects of these methods.

C. Multiplexing

Multiplexing is the process of combining a number of information-bearing signals into a single transmission band. This is either a terrestrial or satellite problem and is not to be confused with the related multiple-access question. Theoretically almost any sequence of terrestrial modulation—terrestrial multiplexing, carrier modulation to the satellite, multiple-access system—can be used. For instance, the standard INTELSAT, Telesat, DSCS-1, and Molniya systems use single-sideband AM and frequency-division multiplexing on the ground, FM to the satellite, and separate carrier frequencies for each earth station. In abbreviation, this system is SSB/FDM/FM/FDMA. The proposed TDMA referred to earlier would be described as PCM/TDM/QPSK/TDMA. The SPADE single-channel-per-carrier system is written as PCM/QPSK/FDMA. The most common terrestrial multiplex method in use is frequency-division multiple (FDM), which is used throughout the world. Frequency-division systems include:

a) single-sideband suppressed carrier (SSC or SSB);
b) single-sideband transmitted carrier (SSTC);

c) double-sideband suppressed carrier (DSSC);
d) double-sideband transmitted carrier (DSTC).

Most terrestrial and space systems use SSB, although some short-to-medium-haul systems use other techniques.

Time-division multiplexing (TDM) is becoming of increasing interest in satellite communications. Time-division systems can use many modulation systems, such as pulse-amplitude modulation (PAM) and pulse-duration modulation (PDM). By far the most important for satellite communication are pulse-code modulation (PCM) and delta modulation (DM). Within these headings there are variations, such as differential PCM and variable-slope delta modulation. The tradeoffs are complex.

Although FDM goes naturally with FDMA, and TDM with TDMA, nevertheless hybrid systems are entirely conceivable and will be used; e.g., a FDM-Master Group Codec (coder-decoder) has recently been designed for use in the Telesat TDMA system [4].

A low-loss multiplexer for satellite earth terminals has been developed to eliminate the broadband high-power transmitter and thereby improve satellite earth station reliability and efficiency [5]. The 5925- to 6425-MHz frequency band is divided into 12 contiguous channels, each 36 MHz wide. Each channel is amplified with a separate air-cooled TWTA. Channels can be added by using modular units consisting of two 3-dB quadrature filters. Time delay and amplitude responses are connected with waveguide equalizers placed before the TWT, thereby avoiding the equalizer loss in the high-power TWT output.

These units are expected to find wide application in small, unmanned earth terminals. Successful implementation of multiplexer and equalization circuits has demonstrated the practicality of the modular transmitter as an alternative to single, large, high-power transmitters currently used in satellite earth stations.

D. Demand Assignment (DA)

Earth stations having continuous traffic over a given number of channels use preassigned channels. However, many channel requirements, as in any communications plant, are of a short-term nature, so a channel and terminal equipment economy technique known as demand assignment is used.

Increased space segment efficiency in a fully variable DA network arises from the fact that all channels are pooled and may be used by any station, according to its instantaneous traffic load. This may be contrasted with a system using preassignment in which all channels are dedicated, i.e., both ends of the channel are fixed. With this system, when traffic to a particular destination is light, the utilization is poor. Also, for a given system traffic load, the blocking probability for a system employing preassignment is higher than for a system employing DA. This occurs because some number of channels are "locked in" to a particular link. In a system employing DA, unused channels may be made available to other users. Conversely, for a given blocking probability, the number of channels required to pass a given amount of traffic in a preassigned system is greater than in a DA system. The lighter the traffic per destination, the greater the advantage of the DA system.

DA offers two main advantages when compared to preassigned systems: 1) more efficient utilization of the space segment; 2) more efficient utilization of terrestrial interconnect facilities. Corollary advantages are more direct service (the need for "via" or transit routing is eliminated) and a consequent possible slight increase in communications quality on such links because of the elimination of tandem connections.

There are many possible forms which a DA communications system may take, and there are various ways of categorizing their makeup. If, for example, both ends of all channels in the system are undedicated so that any station may use any channel, then the system is termed "fully variable." On the other hand, if blocks of channels are reserved to an originating station or a destination station, but are still used only on demand, then it is a "semi variable" system.

Another way of categorizing DA systems is as follows: when carriers (in FDMA systems) or bursts (in TDMA systems) are assigned on demand, the approach may be termed DA multiple access (DAMA); when channels on existing carriers (FDMA) or time slots in existing bursts (TDMA) are assigned on demand, then the approach may be referred to as baseband DA (BDA).

Various combinations of these approaches may be created, the choice depending on traffic characteristics and on user requirements. For example, if a network has a multiplicity of users but only a few large earth stations, then a BDA approach may be most suitable. If there are a great many earth stations in a network, each with low traffic requirements, a fully variable DAMA approach would seem best. In an application where priority control of access is vital, the greater restriction of a semi-variable system may be advantageous, since it would enable a certain number of channels incoming to each station or each of several stations to be reserved for priority traffic.

It is also quite possible and perhaps desirable to mix approaches within the same system. The choice depends solely on the user's requirements.

DAMA systems, of which the INTELSAT SPADE system is an example, are characterized by a per-user access technique. In the SPADE system, for example, each user accesses the satellite with a single-channel-per-carrier (SCPC). Other possible approaches to DAMA are single-channel-per-burst (SCPB) TDMA and CDMA. With these approaches, the carrier (FDMA or CDMA) or burst (TDMA) is not transmitted until it is required. The arrival at the ground station via a terrestrial line of a call request results in the establishment of the carrier or the burst for the duration of the call. As stated earlier, these techniques are attractive in systems in which there are a large number of users whose individual traffic requirements are small.

The INTELSAT network uses a common TDM signalling channel on which all stations call each other and keep track of the available channels which are seized in turn on a first-come-first-served basis. This avoids the politically awkward problem of central controlled systems and the choice of country in which to locate such control—Canada and Algeria, for domestic service only, use central common control and avoid the expense of increased equipment complexity necessary to avoid it.

BDA may be implemented using digital speech interpolation (DSI) or speech-predictive encoding (SPEC). In the SPEC type system, each voice channel is monitored, and the present sample value for each channel is compared with the previous sample value for the same channel. If the samples differ by an amount which exceeds some threshold, (i.e., the present sample is not predictable from the previous sample) the present sample is transmitted. If the difference is less than or equal to the prediction threshold, the sample is not transmitted. By making the prediction threshold adaptive, the system can be made able to respond to rapid changes in traffic loading.

In application, both SPEC and digital TASI (the digital counterpart of the well-known Bell system for intersyllabic channel sharing developed for trans-Atlantic cable use) exhibit approximately the same advantage but SPEC is superior in its freedom from speech clipping, simple algorithms, lower complexity, and lower cost.

Using this approach, a 2:1 reduction may be achieved in the bit rate required to transmit a group of voice channels. It is important to note that such economies may be achieved only when at least 30 voice channels are processed as a group. One of the advantages of SPEC is its graceful and slight degradation in the face of an overload condition.

E. On-Board Processing

Increases in circuit reliability that have accompanied advances in solid-state technology allow significant increases in the complexity of satellite on-board circuitry; consequently, designers now can give serious consideration to advantages of on-board processing previously considered too unreliable.

Although on-board processing is being done effectively in the case of earth-resources and weather satellites involved in large-scale data gathering missions, it has not been attempted with commercial communications satellites since their chief purpose has been the provision of a link between points on the ground for the unaltered transmission of voice, data, and television.

In military satellites, up-link signals accompanied by unwanted interference can be converted to baseband, processed for interference removal, and then remodulated on a down-link carrier. This at least prevents the repeating of the interference and the nonlinear "capture" effect of a strong signal in the transponder. The increased noise level must be dealt with by other methods.

Experiments are now being designed in which packets of information sent using TDMA will be sorted on-board the satellite and transmitted via one of the several spot beams. This is the previously discussed time-division switching. On-board DA, the "switchboard in the sky," also is expected to become important toward the end of the coming decade.

Since satellites historically have been power limited, while earth stations tend to be bandwidth limited, it is conceivable that a future system may be designed in which the up-link channels are transmitted using single sideband (amplitude modulation) to minimize bandwidth, while the down-link channels are transmitted using PCM/FM to minimize power required from the satellite. Other combinations of up-link and down-link modulation and multiple-access systems can also be conceived for various optimization plans. Such arrangements would also be accomplished with demodulation and remodulation on board the satellite.

F. Higher Frequencies

With the launch of CTS, Sirio, and other satellites in 1976, the 14/12 and 18/12 GHz bands will enter active use for satellite communication. The Japanese also plan use of these frequencies in their BS and CS satellites. Still higher frequencies are under active consideration, not only because of spectrum crowding in the lower frequency bands, but because of the desire for broader bandwidths to accommodate higher data rates than are now being sent commercially.

COMSAT Laboratories has used IMPATT amplifiers providing about one watt output at 19 GHz (16-dB gain, 700-MHz bandwidth) and at 28.5 GHz (21-dB gain, 1000-MHz bandwidth) for the AT&T Domestic Satellite Propagation Experiment [6].

Efforts to obtain a 1 to 2 Gbit/s data-transmission capability have lead to interest in the 60-GHz band for privacy and interference protection from the high oxygen absorption in the earth's atmospheric blanket; and in the 94-GHz band, which is the shortest wavelength atmospheric window beyond the infrared. Millimeter-wave travelling wave tubes can deliver kilowatts of power in the 50- to 100-GHz range, but they use very large solenoids or permanent magnets. For the space segment, tubes of lower power (e.g., 60 W) have been developed using periodic magnet focusing systems based on samarium cobalt magnet material.

Work has been done at Hughes [7, pp. 4-1–4-6] and the Air Force [7, pp. 4-7–4-12] at 10.6 μm, at which wavelength N_2HeCO_2 lasers can be built with good efficiency. The 10.6 μm band is also being explored by AIL [8].

Work on coherent optical links is being done by TRW [9], by the Air Force [10], and by NASA. Common-carrier relay represents one possible use for optical links, but circuit reliabilities are marginal because of weather conditions. Another severe problem with optical links is the extremely narrow beamwidth, which would require mutual autotracking from the satellite and earth stations to keep a beam pointed properly. Such links may eventually be useful as supplements to saturated long-haul facilities.

Inter-Satellite Relays: Communications between earth stations that are not both visible to the same satellite require either the complexities and time delays of double-hop transmission or a link directly from one satellite to another. Because most of the paths between geostationary satellites would not involve transmission through the atmosphere, which would attentuate them, work on intersatellite relays has concentrated on the use of millimeter waves and optics, because of the small aperture requirements when using such wavelengths.

Wavelengths under consideration for such relays are 5 mm (60 GHz), which is highly attenuated by the oxygen absorption of the earth's atmosphere blanket but otherwise unaffected, and the optical wavelengths of 10.6 and 0.53 μm. At 10.6 μm, highly efficient N_2HeCO_2 laser sources are available, while the 0.53-μm wavelength takes advantage of the simple detection properties of photomultipliers and the availability of energy from doubled Nd: YAG lasers.

A major difficulty for intersatellite laser links is the acquisition and tracking of the two widely separated space packages. Laboratory tests [7] by the Air Force have achieved pointing errors less than 1.2 μm rad peak-to-peak.

Apertures in the 1- to 2-m range and beamwidths of tenths of degrees are achieved in the millimeter (e.g., 60-GHz) systems, while apertures on the order of 25 cm are used for the optical systems. This 10:1 difference, despite a 10^4:1 wavelength difference, results from the facts that: 1) the noise levels at millimeter wavelengths are lower by more than two orders of magnitude: and 2) higher efficiency power generation can be used for millimeter waves at a level at least an order-of-magnitude higher than for lasers.

The principal issue with respect to millimeter-wave systems concerns their relative weight. Systems weighing on the order of 100 kg, drawing 300 W of prime power and having 2-m apertures, appear to be feasible.

Weights of laser transceivers are projected at less than 90 kg as a result of the relatively small apertures and higher laser efficiency which can be used effectively at this wavelength (10.6 μm).

The chief areas for research and development for intersatellite links, in addition to beam stabilization and system weight, are:

a) at 0.53 μm, electrically powered transmitter efficiency and reliability;

b) at 10.6 μm, the internal laser modulator and its driver electronics;

c) at 60 GHz, the reduction of receiver noise through passive cooling techniques.

The first test of an intersatellite relay will take place using LES-8 and LES-9 in the 36- to 38-GHz band.

G. Antennas

At geostationary altitude, the earth subtends an angle of approximately 18°. This, plus the limited power available on board satellites, makes the concentration of RF output into narrow beams (e.g., ≤ 18°) important. However, beamwidth is inversely proportional to antenna diameter, which is constrained by the space available within the fairing of the launch vehicle. Furthermore, attempts to obtain very small beamwidths (e.g., ≤ 1°) may be thwarted by spacecraft attitude-control precision limitations (it is difficult and costly to point antennas to a high degree of accuracy) or by antenna reflector imperfections. One way of alleviating the problem of fairing size is by the use of an antenna that can be deployed in space, as was done on ATS-6, where a 9.1-m diameter antenna was contained in a torus of 2.0-m outside diameter prior to deployment.

Multiple antenna beams are increasing in importance because of the need to concentrate energy toward different parts of the world simultaneously. They are also attractive from the viewpoint of frequency reuse, i.e., transmitting different message groups on the same frequencies, but beaming the groups simultaneously in different directions toward different parts of the earth. A single antenna reflector can provide multiple beams by the use of feeds offset from the focal point. Separate reflectors, however, provide better efficiency and less crosstalk.

Omnidirectional antennas serve a useful purpose for telemetry and command during the launch and orbital injection phases of a spacecraft's life, but once the spacecraft's attitude becomes stabilized correctly, omni-antennas generally serve only for back-up purposes.

The polarization of an antenna's beam is governed by the polarization of its feeds. (Polarization refers to the orientation of the electric vector of the radiated field.) Polarization may be linear or circular. Two linear polarizations (vertical and horizontal) or two circular polarizations (left-hand and right-hand) can be used to achieve isolation of transmitted and received beams from one another, or for the transmission of two separate message groups in a given frequency band.

1) Polarization: Tests on frequency reuse via orthogonal polarization have been sufficiently successful that the COMSTAR satellites launched starting in 1976 have dual linear polarizations with a polarization isolation of 33 dB. The frequency plan calls for the transponder frequencies on orthogonal polarizations to be interleaved. The RCA Satcom is the first satellite to use dual polarization.

Following successful commercial operation of Comstar, as well as similar operation planned for INTELSAT IV-A, F-2, and F-3, it has been predicted [12] that the widespread use of dual polarization as a means of obtaining added channel capacity in the already crowded 6 and 4 GHz bands. For example, INTELSAT V will use both the present INTELSAT polarization and polarization orthogonal to it.

H. Orbits

To appreciate the various tradeoffs made in the satellite communications systems, it is necessary to look briefly at the various orbits in which communication satellites can be placed, how they get there, what the ensuing spacecraft problems are, and how they affect the possibilities for transponder design.

The period of an orbiting satellite is given by

$$T = 2\pi \sqrt{\frac{A^3}{\mu}} \qquad (6)$$

where A is the semi-major axis of the eclipse and μ is the gravitational constant 3.99×10^5 km^3/s^2. For a circular orbit to have a period equal to that of the earth's rotation—a sidereal day 23 h, 56 min, 4.09 s—an altitude of 35803 km is required. In the equatorial plane, this satellite will remain fixed relative to any point on earth to be "geostationary." In other planes at this altitude, it will describe figure eights daily relative to the earth. The geostationary orbit is indeed delightful from many points of view. An earth station can work with a single satellite, or several with multiple antenna beams, without the need for frequent hand-over characteristics of nonstationary satellites.

Three satellite locations can be configured to permit covering almost the entire earth. Nevertheless, it does have some disadvantages. It is a difficult orbit to get into and it does not provide coverage of the polar regions. The civilized parts of the globe are overwhelmingly within the coverage area of geostationary satellites and the latter limitation has not been serious to date. Nevertheless, future marine and aeronautical systems may want to communicate to the far northern and southern latitudes. Certain other application such as data gathering and military communication may also have the same need. When one considers that orbits meeting this requirement, such as the medium-altitude polar, also permit the injection of much greater payloads into orbit, it may be that the future will see such orbits used for satellite communication.

Ten years ago there was concern that the combination of time delay and echo inevitably present on a synchronous altitude link with hybrid two-wire to four-wire transformers would impair intelligibility noticeably. This has simply not been a serious problem except when the required echo suppressors are defective. It is no longer a consideration by system planners if voice only is used. Data transmission with long time delay places special requirements on error-correction protocols. The automatic repeat-request (ARQ) error-detecting system that requires retransmission must have a block length chosen to optimize the throughput. This block length is sensitive to both the round-trip delay and the noise-bit-error rate, normally very low in a satellite link compared to terrestrial links. In tandem connections involving the bit-error rates of a mediocre terrestrial link with the time delay of a satellite connection, the overall throughput can be poor. We may expect that satellite links more and more will use forward error-correction codes that

require no retransmission and thus the time delay again will be of little significance [11]–[15].

Another orbit of interest is that of the Soviet Union's Molniya used for their domestic communication satellites. It is uniquely tailored to the coverage requirements of the far northern latitudes while avoiding the payload handicaps of a launch site at these latitudes. A highly elliptical 12-h orbit with apogee over the northern hemisphere is used for far northern coverage.

Normally the major axis of any elliptical orbit, called the line of apsides, rotates slowly because of the nonspherical or "oblate" earth. There is one angle of orbit inclination in which the effects cancel and this angle is about 62°. A 12-h period orbit at this angle and with apogee of the ellipse over the northern hemisphere is reasonably convenient for northern coverage. It is also an easy orbit in which to launch payloads from sites at northern latitudes. The geometry of launches states that any orbit inclination less than the latitude of the launch site (for instance equatorial) requires a turn or "dog leg." The loss in useful payload can be quite noticeable for far northern sites. This undoubtedly contributed to the Soviet decision to use an inclined orbit from launch sites above 45°N and the French decision to locate its launch facilities at Kourou, French Guiana—almost on the equator. This inclined orbit system gets northern coverage and high payloads in orbit at the expense of multiple satellites and stringent backing and "hand-over" problems. It is not as convenient as a synchronous system for most applications.

I. Spacecraft

Several aspects of spacecraft design deserve discussion since they affect the communication performance in varying ways. They are attitude control, primary power sources, and propulsion.

Once a communications satellite is on station, its attitude must be held fixed so that its antenna beams are always directed as desired. Effects such as gravity gradient (the difference in gravitational attraction caused by the difference in distance to the earth's center of mass of different parts of the spacecraft), the earth's magnetic field, solar radiation pressure and uncompensated motion of internal motors, gear trains, and lever arms all constitute disturbing forces acting on the spacecraft. All but the internal torques are quite small but continuous, whereas the internal torques, although large, are of short duration.

The simplest form of stabilization is that of spinning the satellite in orbit at a rate of 30 to 100 rpm. This makes the satellite act as a gyro wheel with a high angular momentum. The satellite's angular-momentum vector provides attitude "stiffness." However, it requires that the antennas be "despun", i.e., located on a relatively low-inertia platform spinning in the opposite direction so that the net effect is a stationary antenna beam relative to the earth. A bearing and power transfer assembly then couples the spinning and despun portions of the spacecraft. Spin stabilization also means that a given solar cell is effectively illuminated by the sun only $1/\pi$ of the time, thus causing the primary power to be only $1/\pi$ of the value it would have been if the cells were not spinning.

Rather than spinning a substantial fraction of the satellite, angular momentum can be provided by using a fly wheel spinning about the pitch axis and mounted inside. In this case, the entire satellite is the "despun portion."

This trend in dual-spin designs is toward despinning a larger percentage of the satellite's mass. This trend will continue as multiple beam antennas become more common. Systems such as INTELSAT V and DSCS-III will have severe requirements of this kind.

The question arises of when the stabilization system is no longer to be classified "dual-spin" but rather "three-axis with spinning drum providing angular momentum." One possible definition of dual-spin stabilization is that the spinning portion of the satellite performs functions other than providing angular momentum.

As solar arrays and antennas become very large (10 m on a side, or in diameter), the problem of adequately balancing solar disturbing torques becomes difficult, and full three-axis stabilization becomes necessary. More and more satellite designs are of this type.

1) Attitude Control: A comparison of dual-spin versus three-axis stabilization is instructive. The following three points explore dual-spin advantages relative to three-axis stabilization.

a) Simpler attitude-sensing system: Scanning is provided by the spinner, and the spin momentum eliminates the need for direct measurements of yaw angle.

b) Minimum number of jet thrusters: The propulsion system obtains ullage control (i.e., the feeding of propellents to the nozzles) from the centrifugal force of the spinner; a minimum number of jet thrusters are required and the same relatively high thrust level can be used for station keeping as well as attitude control.

c) Attitude "Stiffness:" The spinning momentum creates attitude stiffness that reduces the effects of torques which are created within the spacecraft and also prevents a rapid accumulation of attitude error as a result of environmental torques. Ground command thus has enough time to provide compensation. This attitude "stiffness" also can be used for attitude control during an apogee motor burn (this also applies to a three-axis system, but to a lesser degree).

The following four points explore disadvantages of dual-spin relative to three-axis stabilization.

d) Vulnerability: A single catastrophic bearing-failure mode can cause a total telecommunications outage with dual-spin stabilization. Vulnerable slip-rings and brushes, and binding of the despin bearings can cut off communications, thus rendering the satellite useless. Furthermore, power losses associated with transferring RF signals increase with frequency, and redundant encoders/decoders may have to be used on both sides of the mechanical despin mechanism.

e) Spacecraft diameter limitations: A spinning body, to be stabilized about a desired axis, should have a higher stable shape than a pencil, for example. If the spacecraft diameter is limited by the launch vehicle fairing, then this constraint is very serious.

f) Nutational Instability: Mechanical damping is needed on the despun platform to compensate for nutational instability (i.e., "coning") that results from an unfavorable ratio of spin-to-lateral moments of inertia and by energy dissipation from fuel sloshing in the tanks in the spinning portion of the spacecraft.

g) Power: More solar cells are needed for a given power when mounted on a rotating drum, resulting in a weight and cost penalty. This factor is increasing in importance because of the need for more RF power output from any single antenna,

the need for more channels, the use of higher frequencies with their lower efficiency transmitters, and more onboard data processing and automation.

Some general considerations are: reliability of the three-axis design is decreased by the more complex attitude-sensing system which it requires, but the sensing system can be made redundant; and dual-spin reliability is degraded by the platform despin system, which cannot easily be made redundant.

Spacecraft costs for the two design approaches appear to be comparable.

2) Primary Power

a) Solar Cells: Primary power for communication satellites mostly is obtained by the use of silicon solar cells. They may be fixed to the spacecraft body, or mounted so that they can be oriented continuously for maximum solar energy.

During the equinox seasons, a geostationary satellite will be eclipsed by the earth. This means that the satellite will be in the dark for up to 70 min per day, depending on the inclination of the orbit and the number of days before or after equinox. To maintain operation during such periods, communication satellites depend upon internal batteries, usually consisting of nickel-cadmium cells, although nickel hydrogen and other technologies are improving swiftly. The batteries represent a major tradeoff among weight, power, and performance.

To avoid the solar-cell battery limitations, consideration has been given to the use of nuclear cells for powering satellites. Either radio isotope thermoselectric generations (RTG) or nuclear reactor powered turbines can be used. A kg of U^{235} could supply 2.5 MWh of energy even at a 10 percent conversion efficiency. With a half-life of 10^8 years, it would outlast most spacecraft.

The advantage of the nuclear supply over solar power systems is that no solar orientation is required nor is any battery needed. However, heavy shielding is required to protect the payload from radiation. This disadvantage has caused solar cells to continue to be the preferred primary power source for communications satellites. Nuclear fuel handling continues to present safety problems both during manufacture and in the event of launch malfunctions. The safer fuels, such as Plutonium, Curium (Cm^{244}), etc., are very expensive. Strontium (Sr^{90}), although much cheaper and with a convenient half-life of 25 years, is very dangerous to handle.

b) Propulsion: After launch, one or two types of propulsion are required. Satellites launched by Thor-Delta or Atlas-Centaur launch vehicles inject into transfer orbit only and require the use of an apogee kick motor for injection of the satellite into geostationary orbit. The weight of this apogee motor and its propellant is typically equal to that of the weight of the rest of the spacecraft.

Spacecraft launched by Tital III-C "direct injection" launch vehicles do not require a separate apogee kick motor, the functions of orbit circulation and inclination removal being performed by the launch vehicle itself.

Because of anomalies in the earth's gravitational field and the perturbing effects of the sun and moon, all spacecraft require a small propulsion system for station keeping. Changes in longitudinal position may be desired from time to time and also require propulsion.

Hydrazine is very popular as a monopropellant because it has high density for storage, low molecular weight and high specific impulse. It is dense, storable, and catalytic; i.e., it needs no oxidizer but dissociates on its own.

The change in velocity of a spacecraft Δr that can be achieved (e.g., for station keeping or apogee-kick purposes) is

$$\Delta v = v_e \ln M_0/M_b \tag{7}$$

where v_e = exhaust velocity, M_0 = mass of spacecraft plus hydrazine, and M_b = mass of spacecraft (all hydrazine burned).

The exhaust velocity is related to I, the specific impulse, by the expression

$$v_e = gI \tag{8}$$

where g, the acceleration due to gravity (at the earth's surface), is 9.8 m/s^2. I, specific impulse, is measured in seconds and is a property of the propellant.

By equating molecular kinetic energy to 1/2 kT per degree of freedom à la Boltzman, it is seen that velocity is proportional to the square root of the absolute temperature and inversely proportional to the square root of the molecular weight of the propellants; thus, the importance of high temperature and low molecular mass is readily seen. Equation (7) can be used for sizing apogee-kick engines and hydrazine station-keeping systems. Its important attribute is the "logarithm." This makes the increase in velocity changing ability of any propulsion system insensitive to increases in propellant weight carried. The efficient way to improve the capability is through high specific impulses, that is, higher escape velocities for the propellant molecules. It explains the great attractiveness for future work of ion engines where the particles are accelerated to high velocities electronically. Specific impulses of several thousand seconds are easily achieved. They are still experimental, but one can expect their use during the next decade. The correction of latitude in synchronous orbit, because of the perturbing effects of the gravity of the sun and moon, require a Δv capability of perhaps 100 m/s/yr over a long period—a high value and a natural for ion engines. Longitudinal corrections because of a noncircular earth are very much smaller—perhaps 5 m/s/yr—and would probably continue to be made by hydrazine engines. Even they will probably be improved by various techniques, the most promising of which seems to be heating the hydrazine thermally.

c) Engine Type and Propellant: The choice of engine type and propellant is another major tradeoff area. Accuracy in station keeping simplifies the earth station tracking problems, but at the expense of hundreds of kilograms of propellant in spacecraft of the INTELSAT class.

Ion engines [16], because of the high exhaust velocity provided by electronic acceleration, offer hope of large reductions in propellant requirements but their technology is still not mature enough to be acceptable to operational spacecraft designers.

J. Launch

The delivery of a communication satellite to its geostationary position takes place in four steps:

    a) ascent;
    b) parking orbit;
    c) transfer orbit;
    d) insertion into final orbit.

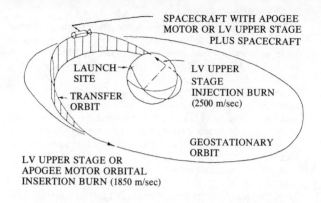

Fig. 1. Typical profile of a geostationary equatorial mission.

The spacecraft mass that can be placed into geostationary orbit is maximized by injecting the spacecraft into a transfer orbit at an equatorial crossing. This means that the spacecraft with its second and third states must coast in the parking orbit until the right time for the injection burn, which uses both the second and third stages and accelerates the spacecraft to 36 700 km/h.

Fig. 1 shows the geometry and events for the transfer orbit and orbital insertion phases of a geostationary mission. It shows the transfer-injection burn occurring at the second equatorial crossing, where the launch vehicle places the spacecraft into a transfer orbit with its apogee equal to geosynchronous altitude.

Actually, to achieve geostationary orbit, another velocity impulse is required at the apogee of the transfer orbit to remove the orbital inclination caused by the launch site latitude and to make the final orbit circular. This last velocity increment can be obtained from the launch vehicle upper stage or from the spacecraft [17]. Current communication satellites launched from Cape Canaveral insert themselves into final orbit by use of a solid propellant moor. The Titan III-C, however, has an upper stage called the Transtage which performs both the transfer orbit and the final orbit injection.

Cape Canaveral in the US is used for launches in which use of the rotation of the earth is desired in order to increase the velocity of the vehicle, i.e., for eastward launches. Most communications satellite launches take place here. The Western Test Range (WTR) is used mainly for southernly launches into near-polar orbits.

The latitude of Cape Canaveral (nearly 29°N) places it at a disadvantage for launches into geostationary equatorial orbit compared with sites closer to the equator. Accordingly, the European Space Agency (ESA) is building an Ariane launch facility at its Kourou, French Guiana, launch site, which is at approximately 5°N latitude. Other near-equatorial sites are at Sriharikota and Thumba (Trivandrum) India and San Marco, an Italian mobile platform base off the coast of Kenya.

Launch vehicles available for satellite communication, especially to geostationary orbit, fall in several groups. The most important to date is that group putting spacecraft into synchronous transfer or low orbit only, e.g., the Thor-Delta in its many versions, Atlas-Centaur, and the Titan-Agena. The Titan III-C brings the spacecraft directly to synchronous orbit, without requiring the use of an apogee-kick motor. This is a very convenient method for the spacecraft designer since he does not have to design for the apogee kick and transfer orbit, but it is expensive.

On the horizon is a new vehicle being developed in France, the Ariane, which will be in the first class but with payload capability almost equal to Atlas Centaur. It will go from Kourou with all the advantages of an equatorial launch site.

Even more interesting will be the NASA shuttle. It will permit very large and complicated satellites to be placed in 200 km parking orbits, but it will be necessary to transfer them to the ultimate operational orbit, normally geostationary. Ultimately an auxiliary vehicle called the ''tug'' will be developed to do this transfer in a recoverable fashion.

Without the tug vehicle to do this, it will be necessary to provide both perigee and apogee stages on the satellite itself, and this will permit launching about one-quarter of the parking-orbit weight into the geostationary orbit.

The economic and operational tradeoffs are extraordinarily complicated. At this moment, it does, indeed, seem as if this may be an efficient and economic way of launching geostationary payloads although the final decision will depend on the total number of shuttle launches. It seems quite possible to design restartable liquid engines, or a combination of liquids and solids, that will transfer the satellites from parking to geostationary orbit efficiently.

Besides the ability to check a spacecraft before putting it into synchronous transfer and after it has experienced the worst in launch environment, the shuttle will have another feature of particular interest to communication satellite designers. It will permit the use not only of much heavier spacecraft but also a bigger spacecraft physically. Notably the diameter of spacecraft can go up to about 5 m. Current spinning satellite designs are seriously hindered by the limit of about 3 m on the diameter which forces large, high-capacity spacecraft to be long and slender. As mentioned previously, this makes them inherently unstable dynamically, and requires sophisticated damping in order to prevent catastrophic nutation.

An increase in diameter from 3 m to 5 m will increase the desired moment of inertia by almost 3 times and make the spacecraft a good deal more stable. In addition, recent developments of solar-cell technology also favor the continued use of spinning satellites because it again permits more primary power for a particular diameter.

V. CONCLUSIONS

In a sense this paper, as a survey of the satellite communication field, is its own conclusion. One need only glance at Table I, a list of the world's programs in all categories, to realize that as an industry, satellite communication has arrived. Table II lists more detail on those programs that include satellites. A complete listing of the characteristics of all those systems would occupy hundreds of pages [12] but we have tried to excerpt those characteristics that epitomize each system. The aggregate serves to make one appreciate the variety of programs already in existence and to make predictions of the future safe, in the sense that the magnitudes will clearly increase and risky in the sense that there are so many diverse possibilities.

With 94 nations participating and some 80 percent of the world's overseas traffic going by satellite, INTELSAT's role is clear. Yet this is only a small part—domestic traffic and services to mobile platforms will probably represent the greatest part of satellite traffic ten years from now. The satellites will continue to exploit the solid-state revolution so as to permit increasingly complicated spacecraft and communications services. One can expect on board message switching and processing in great quantities. The military organizations of the US and other countries and groups will expand their satellite activities so as to exploit the spectacular tactical possibilities. Digital technology will predominate in future development, but FM and FDMA will be around for a long time. Broadcast satellites, long possible technically but involved institutionally and sensitive politically, will slowly come into their own. Antenna techniques to restrict useful signal levels to within a national boundary will be developed. The ability of NASA's shuttle to launch large payloads—and of continuous rather than quantized sizes—will make the exploitation of all the techniques easier.

R&D programs of all kinds will continue—in orbit and on the ground so as to foster the continued development of a mature technologically oriented industry.

Any kind of extrapolation of the past ten years leads to a predicted activity for the next ten that is staggering.

ACKNOWLEDGMENT

The author is grateful to Horizon House-Microwave, Inc., of Dedham, MA, for permission to quote freely from their study entitled ''Communications Satellite Systems Worldwide, 1975–1985.'' Much of the tabular material and some text has been extracted from this study.

REFERENCES

[1] *Bell Syst. Tech. J.*, July 1961.
[2] D. G. Gabbard and P. Kaul, ''Time-division multiple access,'' in *Eascon Rec.*, pp. 179–188, 1974.
[3] T. Muratami, ''Satellite-switched time-domain multiple access,'' in *Eascon Rec.*, pp. 189–196, 1974.
[4] H. Kaneko, Y. Katagiri, and T. Okada, ''The design of a PCM Master-Group Codec for the Telesat TDMA system,'' in *Conf. Rec., Int. Conf. Communications*, vol. 3, pp. 44-6 and 44-10, June 1975.
[5] R. W. Gruner and E. A. Williams, ''A low-loss multiplexer for satellite earth terminals,'' *COMSAT Tech. Rev.*, vol. 5, no. 1, pp. 157–177, Spring 1975.
[6] R. Sicotte and M. Barett, ''Centimeter-wave IMPATT amplifiers for space communications,'' in *Eascon Record*, p. 412, 1974.
[7] J. D. Barry, *et al.*, ''1000 megabits per second intersatellite laser communications system technology,'' in *Conf. Rec., Inter. Conf. Communications*, vol. 1, pp. 4-1–4-6, 4-7–4-12, June 1975.
[8] B. J. Peyton and A. J. D. Nardo, ''Systems measurements of a CO_2 laser communications link,'' in *Conf. Rec., Int. Conf. Communications*, vol. 3, pp. 4-13–4-17, June 1975.
[9] N. Killen, ''Digital communications performance using both temporal and spatial modulation,'' in *Conf. Rec., Int. Conf. Communications*, vol. 3, pp. 4-18–4-22, June 1975.
[10] W. L. O'Hern, Jr., ''Applications of optical communications in space,'' in *Conf. Rec. Int. Conf. Communications*, vol. 3, pp. 4-28–4-32, June 1975.
[11] Heller and Jacobs, ''Viterbi Decoding for satellite and space communications,'' *IEEE Trans. Commun. Technol.*, vol. COM-19, pp. 835–848, Oct. 1971.
[12] W. L. Pritchard *et al.*, *Communications Satellite Systems Worldwide, 1975–1985*. Dedham, MA: Horizon-House Microwave, 1975.
[13] Cacciamani and Kim, ''Circumventing the problem of propagation delay on satellite data channels,'' in *Data Communications*
[14] Cohen and Germano, ''Gauging the effect of propagation delay and error rate on data transmission systems,'' *Telecommun. J.*, vol. 37, pp. 569–574, 1970.
[15] A. R. K. Sastry, ''Performance of hybrid error control schemes on satellite channels,'' *IEEE Trans. Commun.*, vol. COM-23, pp. 689–694, July 1975.
[16] B. A. Free, ''Chemical and electric propulsion tradeoffs for communication satellites,'' *COMSAT Tech. Rev.*, vol. 2, no. 1, p. 123, 1973.
[17] J. R. Mahon, ''Launch vehicles for communication satellites,'' presented at the Nat. Electronics Conf., Chicago, IL, Oct. 1974.

General Purpose Packet Satellite Networks

IRWIN MARK JACOBS, FELLOW, IEEE, RICHARD BINDER, MEMBER, IEEE, AND
ESTIL V. HOVERSTEN, MEMBER, IEEE

Invited Paper

Abstract—The use of satellite communication techniques to provide integrated data network and point-to-point and conference speech services is discussed. The concept of a General Purpose Packet Satellite Network (GPSN) is introduced in terms of its requirements, and consideration is given to techniques that satisfy these requirements. The class of Priority Oriented Demand Assignment (PODA) algorithms is defined and compared with other packet-oriented demand assignment algorithms. PODA is shown to be well suited to the GPSN application. Networking and access protocol issues are considered in the context of a GPSN. The Atlantic Packet Satellite Experiment, an ongoing experimental program which is developing packet satellite technology, is described in some detail.

I. Introduction

BROADCAST communication satellites have a unique potential to support general-purpose packet communications efficiently, including both data and voice, among a large and diverse network of users. Here we consider the techniques of realizing this potential despite certain limitations, including long transmission delay (quarter-second round trip to a geostationary satellite) and the possible constraints on full connectivity imposed by differing size earth stations, local weather conditions, and interference.

Packet communications is a natural choice for the transmission of digital data, including messages and voice, among the many users of a broadcast satellite. Time-division multiple-access (TDMA) techniques permit a flexible and economic sharing of the satellite resource. To use TDMA, the data from each station must be segmented into bursts for time-shared transmission through the satellite. The addition of packet header information to each burst imposes a generally small additional overhead to that already required for phase, time, and frame acquisition but permits switching functions to be readily accomplished. Each station need only monitor the downlink traffic for headers to determine reliably which traffic to accept. Multidestination traffic and infrequent message traffic are both readily accommodated.

In this paper, we specifically address certain aspects of the design of a general-purpose packet satellite network (GPSN). We assume that a GPSN may be required to function simultaneously both as a transit network with attached internetwork gateways and as a local network supporting directly attached hosts and terminals of differing capabilities and requirements. Due to the cost of even a small satellite earth

station, several hosts, terminals, and/or gateways may be interfaced to a single earth station, and hosts as well as gateways may perform a concentrating function. The GPSN is designed to support such activity efficiently, characterized by significant traffic volume generated by one source station and addressed to possibly many destination stations, without degrading the quality of service to a station generating only occasional packets. While doing so, a GPSN should make effective use of the global information available via the broadcast medium to control congestion and to allocate capacity during periods of overload.

It follows that a GPSN must accommodate packets with a wide range of lengths, allowing a short packet containing a message with one or a small number of characters to coexist efficiently with a packet containing a long message (segment) or perhaps several host-multiplexed messages.

It is desirable that a GPSN be able to handle messages with differing priorities in an integrated fashion to accommodate users willing to pay more to assure desired service, military users, and users with flexible requirements, and to support special system and host-to-host control packets, providing in an in-band fashion what appears to be an out-of-band communication path.

A GPSN should be able to handle messages with differing desired delay specifications in order to accommodate short response time uses economically, such as interactive terminal communications and voice, medium response time uses such as mail, transaction oriented devices, and facsimile, and long response time uses such as file transfers. Other desirable properties include a maximum holding time specification, allowing the maximum network lifetime of a message to be controlled by the user, and a choice of message delivery reliability.

One implication of the long propagation time to a geostationary satellite is that special handling is required for a class of volatile periodic traffic referred to throughout this paper as stream traffic and typified by voice, for which the maximum acceptable delay is only slightly larger than the minimum propagation time, or for which the allowable variance in packet interarrival time is small. Such traffic is assumed to be characterized by a message length and interpacket interval which is provided to the GPSN as part of a stream setup.

Finally, since satellite communications is essentially a broadcast medium, the GPSN can and should provide a multidestination addressing capability to permit hosts to provide message broadcasts and voice conferencing service conveniently.

We do not address the problem of whether or not a GSPN should support datagram, virtual circuit, or various host protocols at its external interfaces but instead consider the GPSN only out to a protocol level that provides, in effect, a datagram service with holding time, reliability, priority, and stream options.

Higher level protocol modules that support end-to-end error control, segmentation and reassembly, ordering, duplicate detection and elimination and call set up and teardowns could be located all or in part within the GPSN or external to it without significantly affecting the design choices. Passage of control information across the datagram interface is considered.

This paper is organized into seven sections. Section II considers the nature of the satellite medium in some detail, with emphasis on the techniques of providing multiple channels and the impact on earth station costs and flexibility. Section III examines the central issue of allocating the GPSN satellite channel capacity on demand in an efficient manner. This discussion leads in Section IV to the specification of a demand assignment algorithm denoted priority oriented demand assignment (PODA). In Section V some network issues of a GPSN using PODA are addressed.

The development of the GPSN techniques has been carried out as part of the Atlantic Packet Satellite Program sponsored by the Defense Advanced Research Projects Agency (DARPA). Program objectives include 1) the development and demonstration of packet satellite technology, 2) the determination of the potential role of packet satellite technology in the provision of communication services by establishing basic service capabilities, interface and internetting requirements, and the interaction between the service capabilities and the satellite system, and 3) the investigation of the utility of packet satellite service from the user viewpoint. As part of the program, an experimental network, the Atlantic Packet Satellite Network (SATNET), has been created as a testbed for the concepts. SATNET is discussed in Section VI along with some initial measurement results. Conclusions are presented in Section VII.

II. BROADCAST SATELLITES

The basic characteristics and constraints associated with the use of satellites to provide digital communication networks are considered in this section. The objective is to provide a framework and some background for the discussion of the GPSN which follows. The discussion here is brief; the references contain more detail for the interested reader [1], [2].

A. The Medium

Fig. 1 is a block diagram model of a satellite communication system. The satellite communication link has several important characteristics from the user point of view: data rate, error rate, and propagation delay. Link data rates can range from 75 bits/s to tens or hundreds of megabits per second depending on the application and choice of system parameters. For fixed-power and bandwidth parameters on the RF link through the satellite, a tradeoff is possible between the link data rate and its error performance; decreases in the data rate typically permit better error performance. Forward error correction techniques provide an efficient way to improve error performance with fixed power by reducing data rate or increasing bandwidth [3]. Nominal bit error probabilities on satellite communication links are in the 10^{-4}–10^{-8} range. The round-trip propagation delay (to and from a geosynchronous satellite) is approximately 0.25 s and can vary on the order of milliseconds.

The satellite can be viewed as a centralized power and bandwidth resource which must be shared among a number of communication links. Techniques for sharing, i.e., for allowing multiple access of the satellite by a number of earth stations (and carriers from the same earth station), are considered in the next subsection.

An important characteristic of a satellite communication system is its coverage. This refers to the portion of the earth's surface from which an earth station can access the satellite and from which earth stations can receive the satellite's transmissions. In the case of a geostationary satellite, the former is determined by the satellite's receiving antenna(s) and the latter by the satellite's transmitting antenna(s). In the case of multiple satellite antennas, the processing within the satellite is also important in determining the connectivity. A satellite can provide broadcast capability at any given time to all earth stations within its transmission coverage area. The combination of the multiple access and broadcast capabilities has an important networking implication, i.e., it is possible to form the earth stations into a fully connected one-hop network. These capabilities are also important in facilitating conferencing applications.

B. Channelization

Division of the overall satellite resources (power and bandwidth) among the various earth station accesses (earth station uplink transmissions) occurs at three levels: built-in satellite channelization due to use of multiple transponders operating in different frequency bands, application of basic multiple-access techniques, and dynamic sharing of a channel or group of channels among subgroups of the accesses via the use of demand-assignment techniques. The first two levels are emphasized here with the third, a major topic of this paper, considered in detail in Sections III and IV.

Most present day satellites contain a set of independent transponders. Each transponder contains an input filter to limit reception to a selected region of the uplink frequency band (e.g., 40 or 80 MHz) followed by a frequency translator, power amplifier, and output filter to limit transmission to a specified region of the downlink frequency band. Thus the overall satellite power and bandwidth is divided up on a frequency basis. Typical transponders can support maximum overall data rates of 60–120 Mbits/s for communications to large earth terminals (e.g., 30-m antennas) and perhaps 10–20 Mbits/s with small earth terminals (e.g., 5-m antennas) via the use of coding techniques. Transponder channelization, which sometimes can be modified via switches controlled by ground command, is used to reduce interference problems, remove some access coordination requirements, and provide a means of keeping many satellite failures from being catastrophic to the total communication capability. This channelization, unless coupled with appropriate satellite processing, does result in some loss of flexibility and efficiency relative to handling changing traffic demands, primarily because it constrains connectivity.

Within a given transponder, independent channels (resource allocations) may be provided by the use of orthogonal (or almost orthogonal) signal structures [4]. The basic channelization (multiple-access) techniques are frequency division multiple access (FDMA), time division multiple access (TDMA), and code division multiple access (CDMA). The discussion here concentrates on FDMA and TDMA and their application to satellite networking.

For FDMA [5], channelization is achieved by dividing the total transponder bandwidth among the various access subgroups and confining the transmission of each access to an allocated subband. The transponder is operated in a near linear mode so that the power obtained by each access is approximately proportional to its uplink power. This implies power control on each access and operation of the satellite

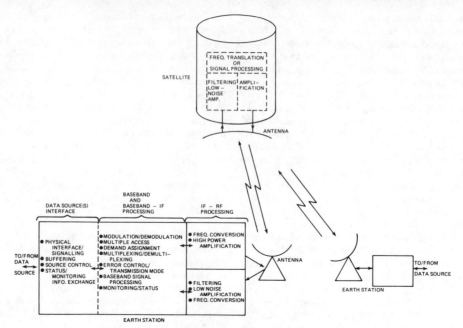

Fig. 1. Block diagram model of satellite communication system.

power amplifier in the linear mode, with some resulting loss in efficiency. Nonlinearities in the power amplifier lead to inter-modulation noise which reduces the capacity (number of accesses) that can be supported by the transponder. FDMA requires no real time coordination among the accesses and can be used to transmit either analog or digital signals. It is economically attractive for point-to-point trunking applications supporting high-duty-cycle high-data-rate accesses such as might occur if many users with traffic for the same destination are multiplexed, and for some point-to-point applications where the total traffic from an earth station is small and/or low-duty cycle.

In the case of TDMA [6], [7], channelization is achieved by temporal sharing of the entire allocated bandwidth and power among the various accesses. For successful message transmission, only one earth station transmits through the satellite at any given time, and the need for power control is minimal. This temporal sharing usually involves the use of a frame structure and implies a requirement for global (network) timing among the earth stations and the use of burst, as con-trasted to continuous, communication techniques. Because of the burst nature, TDMA techniques are associated with the transmission of digital information. Transmit timing at an earth station depends on a global time reference, established either explicitly by a reference station or leader or implicitly, and the propagation delay from the earth station to the satel-lite, typically measured by the station listening to its own transmissions (ranging). The accuracy achieved for the timing will determine the guard times required.

TDMA modems must acquire frequency, phase, bit timing, and bit framing synchronization for each receive burst. This acquisition is usually aided by appending an appropriate bit pattern, a preamble, at the beginning of each transmitted data burst. The number of bits that could be transmitted in the time required for the preamble is an important factor in de-termining efficiency. Typical TDMA systems expend 100–200 bit times in the preamble (including the start of message

sequence used for phase ambiguity resolution and bit framing synchronization), but techniques exist that can be used to re-duce this number. Among these are parallel processing to ac-quire frequency and bit timing simultaneously, exploitation of partial burst-to-burst coherence from a given transmitting station by the use of stored acquisition initial conditions de-rived from the previous burst(s) from that station, use of sampling and post-processing techniques so that some or all of the acquisition operation can be performed on the data bits themselves, and, in some system situations, the use of frequency and/or time precorrection at the transmitter to minimize or eliminate the need for additional correction and, hence, acquisition at the receiver.

Both FDMA and TDMA systems may accommodate varying propagation conditions as well as earth stations with different receiving capabilities (on the order of 10–15 dB) via the use of variable redundancy coding techniques and/or simple modulation changes, e.g., using binary-phase shift keying (BPSK) rather than quaternary-phase shift keying (QPSK). In the case of TDMA, a single burst may contain traffic destined for stations with different receive rates.

C. Networking

The multiple-access and broadcast capabilities of a satellite can be used to form a network from a set of cooperating earth stations in a number of ways. Some realizations involve fixed capacity allocations on links connecting the earth stations; other realizations involve dynamic sharing to provide the ca-pacity and connectivity required by fluctuating traffic de-mands. In systems which utilize dynamically changing con-nectivity or capacity, the response time required to change the connectivity or capacity allocations strongly influences the properties of the network realization. The economics and properties of various network realizations are also affected by whether or not the network occupies an entire transponder or must share the transponder with other users.

Three basic types of satellite network realizations illustrate some of the options and tradeoffs. These realizations are dedicated-channel, dedicated-uplink/broadcast-downlink, and shared-uplink/shared-downlink. In a dedicated-channel network realization, fixed-data-rate point-to-point links connect each pair of stations in the network (assuming a fully connected network). For an N-node network, this realization requires $N(N-1)$ full-duplex links, with each link sized to handle the average point-to-point traffic between the two connected stations while satisfying a suitable utilization or traffic delay constraint.

In a dedicated-uplink/broadcast-downlink realization, each station in the network has a dedicated fixed-data-rate uplink which can be received by all the other stations in the network. This realization requires N-broadcast simplex links for an N-station network, with each link sized (subject to a suitable utilization or traffic delay constraint) to handle the average traffic originating at the transmitting station. Each station multiplexes its own originating traffic on its simplex link to transmit this traffic to the other stations in the network.

In the shared-uplink/shared-downlink realization, a single broadcast communication channel is dynamically shared among all the stations using demand assignment techniques to provide the required connectivity and link capacity.

The choice between the use of TDMA and FDMA depends strongly on the network traffic matrix and the method of realization. A shared-uplink/shared-downlink network, for example, is highly constrained by the use of FDMA. The choice also depends on the other traffic, if any, carried by these stations and whether or not the network must share a transponder with other traffic. As network size increases, TDMA becomes more cost effective than FDMA in providing one-hop connectivity because less equipment (e.g., modems) is required at the ground stations, although each piece of equipment is more expensive. TDMA allows more flexibility than FDMA in establishing and adjusting individual link data rates to the average traffic demands.

When TDMA techniques are used, there is a direct tie between the flexibility available in the TDMA frame and the possible network realizations. If the bursts within the frame are fixed or slowly varying in both length and assignment to specific earth stations and the bits within each burst are also essentially permanently assigned to specific destination earth stations (corresponding to fixed-capacity simplex links from the station transmitting the burst to the various destination stations), only dedicated-channel networks can be formed. If burst lengths within the frame are fixed and allocated to specific stations but the bit assignments to various destinations within the individual bursts can be dynamically varied to match the real-time traffic demands of the transmitting station, then dedicated-uplink/broadcast-downlink types of network realizations can be supported. If the entire frame can be flexibly demand assigned, i.e., there are not fixed burst allocations to specific stations, then the shared-uplink/shared-downlink realization is possible. This paper concentrates on techniques to realize fully connected networks via shared-uplink/shared-downlink TDMA techniques.

FDMA network realization techniques readily accommodate the cases where the network must share the transponder with other users. For example, in SATNET the shared-uplink/shared-downlink network realization is a demand-assigned TDMA system nested in one FDMA-derived channel. (The PODA algorithm discussed in Section IV could, with minor modification, simultaneously demand assign several FDMA-derived channels.) Networks can also be realized in sub-channel(s) of a higher data-rate TDMA system, e.g., a portion of each frame could be allocated to the network. If the transponder is shared via FDMA techniques, there is a potential interference penalty and a loss of efficiency because of the need for linear transponder operation. If the transponder is shared via TDMA techniques there is some loss of system flexibility and a potential cost impact because of the requirement to operate at the higher burst rate.

D. Earth Stations

Fig. 1 shows the basic subsystems of a satellite earth station, namely, antenna, IF–RF processing, baseband and baseband–IF processing, and interface. The subsystem requirements, which depend on both the application and the satellite characteristics, play a major role in determining earth station costs and, hence, the economic feasibility of system concepts which involve large numbers of earth stations.

The receiving capability of an earth station is typically expressed in terms of the parameter G/T, the gain of the receiving antenna divided by the receiving system noise temperature. The receiving antenna gain is determined primarily by its diameter, measured in units of the downlink wavelength, while the noise temperature is normally dominated by the quality of the low-noise amplifier. Larger values of G/T require larger antennas and/or higher quality low-noise amplifiers with a subsequent increase in cost. The earth station's transmitting capability is primarily determined by the transmit gain of the antenna, which is proportional to the antenna area in units of the uplink wavelength squared, and the size of the high-power amplifier, which is a major cost component.

Several other considerations are important in the selection of antenna size [8]. The requirements to avoid interfering with and interference from other satellites which are located close in the orbital arc (e.g., $3-5°$) to the desired satellite place a lower limit on antenna size. On the other hand, the system concept, e.g., user premises or rooftop antennas, may require that the antenna size not exceed some maximum value. In the commercial satellite communications bands at 6/4 GHz and 14/12 GHz, antenna sizes from 1 to 30 m have been successfully employed, with the small antennas used in very special low-capacity applications. Many applications in these frequency bands will probably use 5-m-type antennas with costs in the $10–15 000 range during the next few years.

Fig. 1 lists major functions that are performed in the baseband processing and data-source(s) interface subsystems. In a particular system (application), one or more of these functions might be absent or present in a very simplified form, and some additional functions might be required [9]. It is particularly important to note that the earth station may be serving as a concentrator for a number of data sources/sinks. If the station is to support multiple links, portions of the equipment associated with the functions shown in Fig. 1 may have to be duplicated. The amount and type of equipment duplication required will be strongly influenced by the multiple-access technique employed, e.g., the number of modems required.

With the present generation of satellites, 5-m antennas and burst rates in the low megabit range, earth station costs from IF through RF, including the antenna, are in the $100 000 range [10], [11]. This provides a target figure for the cost of the interface and baseband processing equipment in this data-rate range, assuming as a goal that these costs should at most equal the IF–RF costs. New satellite technology can be expected to reduce the IF–RF costs significantly, which further motivates trying to hold the baseband costs in the $50 000 range.

III. Demand Assignment Approaches

A number of different approaches have evolved in recent years for performing satellite demand assignment. These approaches can be grouped into those designed to handle circuit-switched voice traffic and those designed primarily for packet-switched data traffic. The second group can be segregated further into three categories: random access, implicit reservation, and explicit reservation.

A. Circuit-Oriented Systems

Two early systems designed for circuit-switched voice traffic are SPADE [12] and MAT-1 [13]. In the SPADE system, transponder capacity is divided into subchannels using FDMA, with all except one of these subchannels dynamically assigned to requesting stations. The remaining subchannel is used as a TDMA order wire, with each station permanently assigned one time slot per frame. When a new call request arrives at a station's terrestrial interface, the station sends an allocation request for a pair of subchannels from the FDMA pool, which, if available, establishes a full duplex circuit between the requesting station and the call destination station. When a call is terminated, one of the stations releases the circuit using its TDMA time slot.

In the MAT-1 approach all circuits are derived from the transponder using TDMA channelization. Each subchannel consists of a time slot defined to contain 8 bits of PCM digital voice data, with a frame size of 125 μs (64 kbits/s per subchannel). The subchannels within the frame are partitioned into groups, with each station assigned a group. The number of subchannels in each group is reallocated periodically so that stations with heavy demand can have a larger share of the total.

Both the SPADE and MAT-1 systems are efficient for the voice calls they were designed to service because these calls are typically long compared to the time required to make new circuit allocations. If each circuit were used for bursty data traffic, however, then these systems would not appreciably improve channel efficiency over the case of no demand assignment. Thus new techniques were derived to provide satellite capacity demand assignment for packets.

B. Packet Systems

1) Random Access: An early packet approach makes use of the simple random-access principle developed for ground-based radio [14]. One form is called unslotted ALOHA in which no global timing synchronization is used. In this approach each station which has a packet to send simply transmits it, subject to certain constraints discussed below. If it overlaps in time at the transponder with one or more transmissions from other stations, a conflict occurs, and the packets are not received correctly in the downlink. The broadcast property of the channel allows each sending station to monitor the success or failure of its transmissions and to queue unsuccessful packets for retransmission after the maximum station-to-station propagation-plus-transmission time has elapsed.

The need for constraints on sending a packet arises from two sources. First, a stability problem in general exists for this type of system, since the feedback loop formed by

successive conflicts-retransmissions can cause a runaway condition in the absence of a control technique. Secondly, when station traffic is sufficient to form a queue of new arrivals, transmissions from this queue must also be regulated so as not to increase the probability of a conflict with other stations. For optimum performance, these constraints need to be varied dynamically according to the state of the channel and station queue lengths [15].

The maximum channel efficiency achievable using unslotted ALOHA for a population of about twenty or more stations, each with the same traffic rate, is $\frac{1}{2e}$ or about one sixth of the available capacity [16]. The introduction of slotting, in which each station begins its transmissions at the beginning of a fixed-size slot time, can increase the efficiency bound to $1/e$ for constant-length packets [17]. Either of these bounds can be exceeded if some stations have a significantly higher traffic rate than others—this effect, known as "excess capacity," can result in an efficiency approaching one for certain limiting cases at the expense of many retransmissions per packet (and thus large delays) for stations with small rates [14], [18].

2) Implicit Reservations: A second approach to packet demand assignment is the use of a frame concept to permit implicit reservations with the slotted ALOHA approach. The frame provides a basis for allowing stations with high traffic rates to have one or more slots in each frame for their exclusive use, removing them from the random-access contention occurring in the remaining slots. Reservation-ALOHA [19] utilizes this principle with implicit reservation-by-use allocation. Control is distributed in that each station executes an identical assignment algorithm based on global information available from the channel. Stations maintain a history of the usage of each channel slot for one frame duration; when a station uses a contention slot successfully, the slot is assigned to that station in each successive frame until it stops using it. The frame duration must be at least one round-trip propagation time or "hop" in this scheme. Some other approaches to this mixing of assignment and contention within a frame are given in [20].

3) Explicit Reservations: A third category of packet demand assignment consists of explicit reservation systems. A portion of the channel time is used (as an order wire) by stations to make an explicit reservation for one or more messages. The reservations may be sent in separate subframe(s) distinct from the frame time in which messages are transmitted, or they may be combined with message transmissions ("piggybacked") or both. In addition, the assignments of message transmission times resulting from the reservations may be made centrally by one station (central control), distributedly by all stations (distributed control), or by a mixture of both techniques.

In an early scheme proposed in [21], slotted ALOHA is used to make reservations in a separate subframe with the remaining frame time divided into slots equal to the maximum data packet size. A simple distributed control technique honors each reservation, which could be for up to eight packets, on a first-come-first-served (FIFO) basis. A station must remember only the total number of outstanding reservations and the slots at which its own reservations begin. To allow a station to detect out-of-sync conditions and to acquire the correct reservation count, each station sends in its data packet transmissions what it believes to be the total reservation count. For optimum performance the ratio of data to reservation subframe sizes can be slowly varied according to average traffic loads, with the entire frame used for making reservations whenever the total reservation count is zero.

A second approach to explicit reservations uses a combination of contention and fixed assignment to make the reservations [22]. A fixed assignment frame structure is used in which each station is assigned a slot in the frame permanently. In this slot, a station can both send a data packet and make a (piggyback) reservation using bits in the data packet header. Each station with packets to send always uses its permanently assigned slot but, in addition, can share with other busy stations the slots not in use by idle stations.

Distributed control is again used to perform the dynamic allocation of idle station slots; each station maintains a table of total unserviced reservations heard from each station, with each idle station slot allocated on a round-robin basis to the next station having a nonzero table entry. When an idle station becomes busy, it always first sends in its permanently assigned slot, deliberately creating a conflict if the slot was dynamically assigned to another station. The conflict is assumed to be detected by all stations, causing them to stop using the slot for dynamic allocations. The station can then begin using its slot to send data and also to make reservations for other packets in its queue, allowing it to participate in the dynamic allocations. To maintain synchronization and to allow initial acquisition, one of the stations broadcasts a copy of its table at infrequent intervals.

A third reservation approach, a modified version of the second approach above, uses only fixed assignments to make reservations [30]. In this system, implemented in an early phase of the Atlantic Packet Satellite Experiment and called Reservation-TDMA (R-TDMA), a separate reservation subframe is used with each station permanently assigned a slot for sending reservations. The remaining frame time is divided into data packet slots in the same manner as above; each station is permanently assigned one data slot per data frame, with the idle station slots in each data frame dynamically allocated according to the table of outstanding reservations. However, synchronization is acquired and maintained by having each station send its own reservation table entry in its reservation slot rather than by one station sending all of its table entries. The number of data frames within a reservation frame (where the latter is the time between successive groups of reservation slots) is chosen to optimize performance based on considerations similar to those in the first system above.

C. Shortcomings for GPSN

Although each of the approaches described in Sections III-A and III-B exhibits some desirable performance characteristics for certain traffic contexts and user requirements, no one system satisfies the requirements of a GPSN. These requirements include:

1) efficient use of satellite bandwidth,
2) satisfaction of multiple delay constraints,
3) multiple priority levels,
4) variable message lengths,
5) stream traffic,
6) different station receiving rates,
7) fairness,
8) efficient message acknowledgments,
9) robustness.

A requirement which is absent from this list is a small processing bandwidth at each station. While this was in fact assumed as a design constraint for most or all of the systems described above, present technology trends make it clear that this is no longer a significant cost factor, at least for a general-purpose network with a channel rate of a few megabits per second or less. Higher channel rates can also be accommodated with reasonable processor cost by the use of appropriate techniques to be discussed below.

We now discuss some of the shortcomings of the above systems relative to the GPSN requirements.

1) Efficiency Versus Delay: As is well-known for queuing systems with random arrivals, shorter delays must be achieved at the expense of lower server efficiency and conversely. This tradeoff also applies to the set of geographically dispersed queues (i.e., stations) being serviced by the single satellite transponder; the tradeoff is complicated in this case by the use of the transponder bandwidth to coordinate the dispersed queues and by the quarter-second propagation delay inherent in this coordination.

The simple ALOHA approaches were designed in particular to achieve the shortest possible satellite channel delay for interactive data users, while achieving efficiencies which were high for the assumed bursty traffic relative to the use of dedicated channels. The quarter-second or more delay required to detect a conflict and retransmit, however, means that an ALOHA satellite channel must be operated at only a fraction of its $\frac{1}{2e}$ or $1/e$ maximum possible efficiency if short delays are desired.

The framing approach represented by the Reservation-ALOHA system does provide one-hop delays to stations which have acquired one or more slots, while allowing potentially high efficiency. However, this is accomplished by sacrificing the demand assignment property of the system; that is, the more frequently slots are switched among different stations the more the system reverts to simple ALOHA channel behavior.

The explicit reservation approaches in general involve at least two hops of delay if distributed assignments are used and three hops if the assignments are centralized. (Exceptions can occur for particular systems at certain traffic loads; for example, the last two reservation systems described in Section III-B provide a one-hop minimum delay when no other stations are using the channel at the time access is desired.) If the total reservation overhead per message is sufficiently small, however, efficiency can approach one.

As has been shown previously in [21], this general efficiency-delay behavior makes it clear that, to the extent that satellite capacity cost dominates total system cost, minimum delays of less than about one second (for short messages) are expensive to achieve. This expense is in fact significantly greater if the other requirements of a GPSN are to be satisfied.

2) Delay Classes, Delay Variance, and Priorities: The random-access systems do not easily satisfy arbitrary sets of delay and priority requirements; in fact, the excess capacity effect referred to earlier tends to produce results opposite from those typically desired. A technique for using multiple bursts for more urgent messages is described in [25]. Another approach is to use longer waiting times before transmitting less urgent messages; for certain satellite systems which allow capture, the use of different uplink power levels has also been suggested. In all of these approaches, however, a relatively large efficiency penalty is incurred. A second problem inherent in random-access schemes is the high variance resulting from the probabilistic transmission conflicts and randomized waiting times.

The particular reservation schemes described above also do not provide much flexibility for different delay classes. The FIFO strategy of [21] does not allow assignments of outstanding reservations according to their urgency, while the fixed-assignment data slot structure and round-robin dynamic assignments of [22] introduce similar restrictions.

The delay variance of the explicit reservation schemes depends strongly on whether random access is used to make reservations. If it is, the larger variance associated with the ALOHA technique applies to the reservation-making component of total delay, whereas reservations made using exclusively assigned channel time will have a significantly smaller variance. For a reservation scheme in which assignments are made according to delay class and priority, these statements are of course strictly true only for the highest priority traffic; lower priority messages may have a very-large variance at times of heavy loading by higher priority traffic.

3) Variable Message Lengths: When a high variance in message lengths exists, all of the systems described except unslotted ALOHA become inefficient due to their use of fixed-length data slots. If the slots are chosen to match the largest lengths, slot times are underutilized by shorter messages; if shorter slot sizes are used, more overhead per message results, and some slot time is still wasted. While unslotted ALOHA does not have these particular problems, its overall efficiency is still considerably lower than that achievable by the reservation systems (and, in fact, its maximum capacity is somewhat degraded by the variable lengths [37]).

4) Stream Traffic: None of the above systems provide a mechanism for allocating channel time for streams. While Reservation-ALOHA and the explicit reservation systems which use fixed assignments do provide recurring uncontended time slots to some or all stations, the recurrence interval is constrained by the system frame size rather than being chosen to match the desired stream repetition rate. Further, stations in general cannot obtain the number of slots needed for their differing distinct streams.

5) Different Receiving Rates: As discussed in Section II, different stations may require different receive data rates at a given time. Since the differing rates are achieved by varying the amount of coding and/or changing from QPSK to BPSK, the impact of this requirement on the data portion of the channel is essentially the same as variable message lengths. In addition, the ability to obtain control information directly by monitoring the channel is adversely affected, since the channel is no longer operated in a 'pure' broadcast mode. The explicit reservation schemes and Reservation-ALOHA, which all use a distributed assignment technique, would thus have to be significantly modified. This is also true for the unslotted and slotted ALOHA schemes, if transmission control depends on each station explicitly monitoring the total channel loading.

6) Fairness: Fairness refers to the ability of the system to provide the same quality of service to competing stations when desired. The simple ALOHA approaches can potentially provide this by applying the same transmission control criteria in each station [23]. The Reservation-ALOHA approach is most vulnerable to operation in an unfair state, since (without additional controls which drive it towards the simple ALOHA mode) a single station can capture all available slots for an indefinite time.

The slotted ALOHA reservation system in Section III-B is subject to the same fairness considerations as the simple ALOHA systems, since reservations are serviced in the order made. On the other hand, the round-robin assignment strategy of the second two systems, coupled with the exclusively assigned time for making reservations, does guarantee fairness. (Although contention signaling is used at times in the first of these schemes for initial access, the contention is always limited to one retransmission for all stations.)

7) Efficient Message Acknowledgments: The need to send acknowledgments and other control information from stations which otherwise have no traffic to send introduces certain problems for some of the systems described. In particular, the absence of fixed time assignments in the simple ALOHA, Reservation-ALOHA, and slotted ALOHA reservation systems means that acknowledgment packets must contend with data or reservation packets. This contention further reduces channel efficiency and delay performance; it also has a significant impact on the source buffering required in each station, since the waiting time for an acknowledgment is subject to contention-induced retransmissions. This is also true to some extent for the explicit reservation system using contention signaling for idle stations in [22]. Of the systems described in Section III-B, only the system using fixed assignments for reservations avoids these problems, since acknowledgments (and other control information) can be sent along with the desired reservation information in the assigned slot occurring in each reservation frame.

8) Robustness: The systems depending on distributed control, i.e., with each station independently executing an identical assignment algorithm based on global information, are especially sensitive to local receive errors at each station, the degree of sensitivity depending on the amount of information required and the particular techniques used. The simple ALOHA schemes are probably the least sensitive to noise when reasonably loaded, since contentions normally dominate local errors. The exclusive use of a slot in the Reservation-ALOHA scheme is subject to disruption due to local errors if these are not distinguished from conflicts, as is the dynamic allocation of slots in the first round-robin reservation scheme.

More generally, the explicit reservation schemes using distributed assignment are subject to varying degrees of disruption whenever a reservation is received with errors by any station. If *a priori* knowledge of the time at which new reservation information should be received is available, a station can know when it has not received new information correctly, allowing it to suspend its transmissions before possibly disrupting use of the channel by others. This is done, for example, in the system of Section III-B which uses fixed assignment reservation slots exclusively.

If contention reservations are used and conflicts cannot be distinguished from local errors, then detection of locally missed reservations is in general not possible until subsequent information is sent in the channel; the latter cannot of course take place until at least one hop after the missed information was received at the other stations. In the slotted ALOHA reservation system of [21], for example, a station cannot discover it has missed a reservation until it hears another station's total count received one hop or more later.

In either explicit reservation approach, if the channel data rate and/or local noise levels are sufficiently high, system operation may be badly degraded. Systems which utilize centralized, as opposed to distributed, control are less susceptible to this degradation.

D. Synthesis of Demand Assignment Techniques

In view of the diverse traffic requirements of a GPSN, provision of low-delay (less than 1 s) datagram service is in conflict with good system efficiency. A reasonable decision is to deemphasize low-delay datagram service. With this decision the selection of an explicit reservation approach to demand assignment is natural, with short delays (on the order of one hop) provided for stream traffic by the use of repetitively allocated channel time. Short delays for certain datagram traffic can still be provided, although perhaps at more cost, by use of the stream capabilities (i.e., by setting up circuits). Given this context, the following observations are relevant.

1) Assuming adequate station processor bandwidth, the desired properties of a GPSN such as multiple-delay classes, priorities, stream traffic, and fairness can be accommodated by including appropriate parameters within an explicit reservation and employing an appropriate scheduling algorithm.

2) The advantages of ALOHA access for bursty traffic from large-station populations can be combined with the desirable properties of a preassigned channel allocation for making reservations by using a slotted-ALOHA subframe for access when idle and the headers of scheduled message transmissions for subsequent piggybacked reservations.

3) The use of an integrated assignment scheme that combines distributed and centralized techniques can provide the high-system availability and lower delays of the distributed approach, while also resolving the control problems introduced by different station receiving rates and the need for robustness.

4) The acknowledgment transmission problem can be reasonably resolved by scheduling channel time when necessary as part of the message-scheduling function, without introducing additional channel overhead.

In the next section, we describe a system specifically designed for a GPSN based on the above principles.

IV. PRIORITY-ORIENTED DEMAND ASSIGNMENT (PODA)

A. Basic Structure

PODA [24]–[26] is a demand assignment protocol designed to efficiently satisfy the overall requirements of a GPSN, including support of both datagram and stream traffic, multiple-delay classes and priorities, and variable message lengths. Its design represents an integration of both circuit and packet-switched demand assignment and control techniques.

1) Explicit reservations are used for datagram messages, generally resulting in delays of at least two satellite hops.

2) A single explicit reservation is used to set up each stream, with subsequent packets automatically scheduled at predetermined intervals thus experiencing substantially one-hop delays.

3) Very-high reliability, when desired, is provided by the use of scheduled acknowledgments and packet retransmissions.

4) High availability and improved performance are achieved by the use of distributed control techniques.

5) Robustness and mixed receiving rate operation are achieved by the integrated use of both distributed and centralized control techniques.

The design described here assumes TDMA channelization, although the PODA approach could be used to allocate channel capacity derived by FDMA or CDMA channelization techniques.

The particular techniques used in PODA are based on the synthesis of the demand assignment techniques discussed in Section III. Channel time is divided into two basic subframes, an information subframe and a control subframe. The information subframe is used for scheduled datagram and stream transmissions, which also contain control information such as reservations and acknowledgments in their headers. In normal operation, the control subframe is used exclusively to send

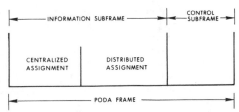

Fig. 2. PODA channel time.

reservations which cannot be sent in the information subframe in a timely manner.

The subframes are shown in Fig. 2, in which the information subframe is seen to be further divided into a section for centralized assignments and a section for distributed assignments. Details on the use of these sections and the other aspects of PODA are described in the following subsections.

B. Control Subframe

The control subframe must be used for initial reservation access by a station which has no impending scheduled transmissions, and may also need to be used at times to make reservations for traffic whose urgency does not allow waiting for the next opportunity to use the header of a scheduled message.

The access method used for this subframe depends on the particular characteristics of the system in which PODA is being used. If the total number of stations is small, then fixed assignments of one slot per station are typically used within the control subframe, and the system is referred to as FPODA. For large populations of low duty-cycle stations, mixed station populations, or situations in which traffic requirements are not well known, random access (slotted ALOHA) is used and the system is referred as CPODA. Combinations of fixed assignment and random access in the control subframe are also possible. In either case subframe time is divided into fixed-sized slots equal to the (constant) length of the control packets.

The slotted ALOHA transmissions in CPODA create potential stability and fairness problems in the control subframe. Since the desirable operating point for contention subframe loading is considerably less than $1/e$ for best overall system performance, however, instability problems are reduced relative to a typical ALOHA channel. In addition to the usual slotted ALOHA control techniques, the following mechanisms enhance stability and fairness.

1) The size of the control subframe is varied according to current scheduling requirements. If no reservations are outstanding, the control subframe occupies almost the entire CPODA frame; as loading increases, the control subframe is reduced to a value determined by the urgency of the messages being scheduled until a minimum value is reached.

2) New reservation transmissions are constrained at all stations allowing only higher priority traffic to compete in the contention subframe when the channel is heavily scheduled.

CPODA does have a potential fairness problem relative to FPODA, due to the ability of heavy-traffic stations to obtain continued use of the channel via reservations made in their scheduled transmissions. This is reasonably resolved, however, by enlarging the control subframe when excessive conflicts are occurring (temporarily increasing the delay of messages waiting to be scheduled).

C. Reservations and Scheduling

All datagram reservations received successfully in the satellite downlink are entered into a scheduling queue, maintained by all stations performing distributed scheduling. These entries are ordered according to reservation urgency, which is a function of potential lateness (relative to the user-specified delay class) and priority. Reservations with the same urgency are further ordered to provide fairness to the stations involved. Channel time in the information subframe is then assigned to messages according to the latest ordering of reservations in the scheduling queue.

A reservation for a stream is made only once, at the beginning of the stream use, and is retained at each station in a separate stream queue when received in the downlink. Each stream reservation contains information defining the stream repetition interval, desired maximum delay relative to this interval, and priority. Whenever the interval starting time is near, a reservation is created for that stream's next message and entered into the scheduling queue; the reservation urgency is calculated according to the stream queue information, and is treated the same as datagram reservation urgency. Thus all channel time requests, whether for datagrams or streams, are ordered according to a single set of rules. In particular, high-priority datagram traffic can preempt lower priority stream allocations for an indefinite period, possibly resulting in termination of the streams.

1) Reservation Efficiency: To maximize channel efficiency while satisfying urgency constraints, each reservation in the scheduling queue may actually be for a group of several distinct datagrams. The maximum burst-length request allowed per group reservation represents a tradeoff between satisfying system urgency constraints for competing stations, which argues for a smaller maximum, and reducing relative burst overhead by using a larger maximum. This burst overhead, discussed in Section II, can be quite significant when the total burst size is small.

Of course, a station must have several messages queued in order to make a single group reservation for them. Furthermore, at certain times a station may have more than one message group being scheduled within the same frame. Overhead can be reduced further by scheduling these message groups contiguously to allow a single burst transmission. Thus reservation efficiency and burst efficiency both cause small messages to be packed into a single burst when possible, as shown in Fig. 3. This packing slightly defeats the scheduling rules based on urgency and fairness ordering within the frame, but the resulting increase in delay is not very significant for channel data rates above 50 kbits/s. (A PODA frame time is about 300 ms at this channel rate.)

A potential burst fragmentation problem can result from the use of longer group reservations, since their use increases the probability that bursts will not fit neatly into the information subframe. This problem is not significant, since if the unused time is small it can be used as part of the control subframe, while if large a burst can be partially sent in the time available, with its remainder sent at the beginning of the next information subframe. If this process results in fragmentation of a message within the burst, reassembly is performed prior to error detection. The need for fragmentation decreases with increased channel rates, since the burst times become smaller while the PODA frame size need not be reduced proportionally to maintain good delay performance.

2) Centralized Assignments: To achieve the reduced delay of distributed control (at least one hop less than centralized control) while also making centralized assignments for stations unable to participate in the scheduling, the information subframe is partitioned as shown in Fig. 2, and an integrated con-

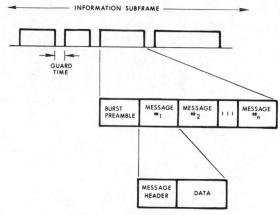

Fig. 3. Information subframe bursts.

trol technique used. Since the centralized assignments cannot be acted on until the assigned stations are informed, they are scheduled for a frame at least one hop in the future. Thus the centralized assignment section of the information subframe in Fig. 2 represents assignments made at least one hop in the past, while the distributed assignment section represents assignments being made from the scheduling queue in the current frame.

Note that all stations which perform distributed scheduling can also make the centralized assignments, so that the high system availability resulting from distributed control is maintained. To avoid introducing unnecessary burst overhead in sending the assignments, they are sent by a station transmitting a scheduled burst in the distributed assignment section. A separate control burst is thus necessary for this function only when no messages are otherwise scheduled for the distributed assignment section.

D. Scheduling Synchronization

As noted in Section III, the use of distributed scheduling introduces potential synchronization problems due to local receiving errors at each station performing scheduling. To prevent increased reservation errors in PODA due to the higher error rates of long data bursts, all global control information is sent in a separately check-summed header in each burst.

The synchronization acquisition and maintenance process consists of three states: 1) the initial acquisition state, 2) the out-of-sync state, and 3) the in-sync state, with the transition into or out of a particular state based upon consistency checks. Each station compares its own scheduling decisions against the transmissions actually taking place in the channel; if a discrepancy is noted, the station detecting it does some further checks and, if necessary, adjusts its scheduling queue.

When a station is first turned on, it enters the initial acquisition state in which it has no information about the state of the scheduling queue in the other stations. In this state it listens for new reservations and observes the packets being transmitted in the channel but does not itself send any data messages or reservations. Since each reservation has a maximum lifetime in the scheduling queue, a station can build up a reservation list compatible with other stations within a reasonable period. When a certain number of correct consistency checks have been made, the station enters the out-of-sync state.

In the out-of-sync state, the station schedules the received reservations and monitors for consistent schedulings, but it does not send any data packets or make reservations. If several consecutive schedulings are made within a certain period of time, the station enters the in-sync state.

While in the in-sync state, the station can send reservations and scheduled messages. Whenever it detects a transmission in the channel which it has not scheduled for the same time, it readjusts its scheduling queue using the header information contained in the transmission. If a certain number of consecutive inconsistent checks are detected within a given time period, the station enters the out-of-sync state, preventing possible subsequent disturbance to transmissions made by other stations due to its own incorrectly sent data packets.

A more detailed description and quantitative evaluation of this approach is given in [26].

E. Acknowledgments

The global scheduling mechanism can be used to achieve efficient message acknowledgment transmissions for high reliability traffic. All scheduling stations note the destination address in each message header sent in the information subframe. If the destination station has a pending reservation which will allow transmission of the acknowledgment within the latter's time constraint, the reservation is expanded to allow time for the acknowledgment. If no reservations are pending, a reservation is created and entered into the scheduling queue to allow the acknowledgment to be sent in a separate burst. Each message header is check-summed separately from the rest of the message, maintaining the low reservation error rate required for global scheduling synchronization.

This method minimizes both acknowledgment channel overhead and source message buffering required in stations, since the acknowledgments are sent at the earliest opportunity. Stations not participating in scheduling are informed of their scheduled acknowledgment times the same as for their scheduled message transmissions. These stations thus have a minimum message buffering time of five hops for reliable datagram traffic sent to another station also not doing scheduling: three-hops minimum for transmission of the message, one-hop minimum for the destination to be informed of its acknowledgment scheduling time, and one-hop minimum to send the acknowledgment.

The other extreme is represented by a scheduling station sending to another scheduling station, which is a minimum of three hops. Note that the minimum buffering time possible in any system using positive acknowledgments is at least two hops, so that the completely distributed scheduling case does not increase buffering requirements significantly.

F. Channel Efficiency and Processor Cost

The satellite channel efficiency achievable with PODA is clearly limited by the average control subframe size and overhead used for reservations in the information subframe. The latter is of course strongly dependent on the average size of bursts being reserved, becoming small as the burst size increases. The relative overhead is not large even for short bursts, however, since the number of bits needed to specify a reservation is small or at most comparable to other burst overhead components such as the modem acquisition preamble.

The average control subframe size represents a tradeoff between channel efficiency and improved delay and fairness performance, as discussed above. This choice can be easily effected by "tuning" certain parameters appropriately, according to the particular system constraints being satisfied.

An important aspect of PODA, relative to other demand assignment approaches, is its station processing requirements. This becomes a significant consideration for high-data-rate channels, in which the processor bandwidth required to schedule channel transmissions may be quite high. The use of centralized control provides a reasonable solution to total system cost, however, since stations receiving assignments need only enough processor bandwidth to handle their local traffic rates. The scheduling can be performed by a small number of stations with large processor power, the actual number dependent on system availability criteria.

A second method which can maintain reasonable processor cost while providing distributed scheduling performance at high channel rates is partitioning. In this approach channel time is partitioned into slowly changing regions, with each station assigned to one or possibly a few of these regions for transmission purposes. Thus each station performs distributed scheduling only for its region, although it can still receive messages sent in any region. This approach trades channel efficiency for the reduced delay obtainable by distributed scheduling, since the slowly changeable partitions cannot allocate total channel time as flexibly as the centralized approach.

V. Network Issues

A number of issues arise in the design of a GPSN in addition to the demand assignment choices discussed in the preceding sections. These issues include the variety of message services provided and the ways in which they are quantified: addressing, flow and congestion control, and support of special network applications such as conferencing and internetwork traffic. Of particular interest are the ways in which one-hop broadcast connectivity and PODA algorithms impact these areas.

A. Message Services

A number of message properties are of concern for both datagrams and streams: priority, delay class, holding time, length, error reliability, duplicates, and sequencing.

1) Priority and Delay Class: While the meaning of maximum acceptable delay (delay class) is clear, the appropriate interpretation of message priority in conjunction with delay class is implementation dependent. A reasonable interpretation of priority is the following: messages of a given priority level should see a class of service independent of the traffic loads imposed by messages of lower priority. In practice, efficiency considerations may dictate that this rule be violated at times. For example, if a high-priority message with a desired delay of 5 s enters the network while a lower priority message with a desired delay of 1 s is queued already, it is more desirable to deliver the lower priority message first if the network is uncongested at the time. However, traffic conditions could change while delivering the lower priority message, causing the higher priority message to be delayed beyond its specified delay class. Thus unless priority receives absolute consideration, it is possible that the above rule will be violated.

In general, one would like to use priority to arbitrate the allocation of resources when the latter are insufficient to meet total momentary demand but otherwise satisfy the delay constraints of all messages in the most efficient manner. A way of accomplishing this is to order messages in system queues according to a function of each reservation's priority and delay class. By using the global time synchronization property of the satellite network, a timestamp can be created for each message as it enters the network and used to track the message age relative to its delay class. By choosing the unit of time appropriately for the age measure, messages can be ordered according to their age in this time unit until they are within a single unit of exceeding their delay, after which they are ordered by priority. Thus when messages are in danger of being delivered late, priority takes over to guarantee service to those with the greatest overall urgency.

2 Holding Time: The network global time property can also be used to allow users to specify a maximum holding time for each message. This holding time specifies the maximum lifetime of the message within the network before discarding and serves several purposes. It provides a convenient mechanism for preventing undeliverable messages from congesting system buffers; it makes acquisition of PODA scheduling synchronization easier, since reservations in the scheduling queue have a finite lifetime; and it allows user-level protocols to operate more efficiently.

An example of this last factor occurs when end-to-end retransmissions are being used, since message holding time can be selected for compatibility with the user-level retransmission timeouts to minimize duplicates arising from retransmissions within the satellite network. A second example is the use of the network for volatile traffic such as speech, in which the loss of a speech packet is more desirable than its late delivery.

3) Message Lengths: As pointed out in preceding sections, satellite channel burst length is an important parameter for efficient channel allocation. The maximum allowed datagram message sizes entering the network can be longer than the maximum burst size, however, since the message can be sent in several bursts.

Allowable datagram length thus depends primarily on the acceptable internal network buffering penalty. Longer lengths reduce the need for hosts or gateways to fragment messages into smaller segments, which reduces both host access processing overhead and the relative overhead in each segment due to repetition of host-to-host or fragmentation-reassembly headers. If the allowed network length is such that the message must be sent in several separate bursts over the satellite channel, however, then the network must fragment and reassemble the message before delivery to the destination host. Because of retransmissions due to channel errors and the long retransmission times involved, buffers could be tied up for significantly longer times than if reassembly is not required.

Thus optimum network efficiency is achieved if satellite channel considerations allow burst lengths comparable to those required for efficient host access. For a channel bit error rate of 10^{-6}, a reasonable upper bound on message length is on the order of 8000–16000 bits for error control, since the packet retransmission rate due to channel errors is less than 1 percent at these lengths. For channel data rates greater than about 50 kbits/s, the scheduling delays introduced by these lengths are also acceptable relative to the total delays incurred and are insignificant at rates above 500 kbits/s.

Finally, since technology trends indicate a continued decrease in memory costs, internal fragmentation-reassembly may still be a reasonable choice, where necessary, relative to the less efficient use of satellite capacity.

4) Reliability: The acknowledgment-retransmission strategy used in the PODA protocol provides a probability of message delivery with undetected errors of typically 10^{-13} or less. However, this is achieved at considerable buffering expense due to the need for the source station to hold the message until the acknowledgment is received. This strategy also introduces the possibility of duplicate message delivery due to missed acknowledgments and of out-of-sequence delivery when retransmissions are needed.

If traffic consists of volatile speech, or if a user-level protocol performs end-to-end error control, then it may be desirable to suppress retransmissions originating within the satellite network. This is easily accomplished for datagrams by use of a bit in each datagram header or for streams by specifying a bit in the stream setup message. When this choice is made, the resulting reliability is determined by the satellite channel performance. If the burst is not acquired or the message header (which is separately check-summed) contains errors, the message is not delivered. If only data errors have occurred, the message can be marked and delivered if requested.

A more difficult problem involves the question of whether an acknowledgment should be returned to the source station when the message is actually delivered to the destination, either in place of the channel acknowledgment or in addition to it. If destination buffers cannot be preempted, then the probability of message errors or loss in the destination station is that due to processor and memory failures and is normally very small. If buffer preemption is allowed, the loss probability could be quite high depending on message priority and system loading. If the channel acknowledgments were delayed until message delivery, considerably longer retransmission timeouts and message numbering overhead would result. Since the delivery time is not known globally, the return of a separate delivery acknowledgment would require significant additional channel overhead. The choice thus depends on whether the cost of system buffering warrants use of preemption and on the degree of reliability required by network users.

5) Duplicates and Sequencing: The use of retransmissions for high-reliability traffic raises two associated issues: 1) should the network detect and discard duplicates, and 2) should the network deliver all messages of a given priority-delay class to the destination in the order received from the source?

Duplicate detection requires that each station retain information in its memory which allows unique identification of each message received from the satellite channel in order to decide whether a message currently being received has been received in the past. This information must span the total possible time in which a particular message may be retransmitted, which in the worst case is given by the maximum allowed holding time. Note that the numbering modulus required for this function is considerably greater than for channel acknowledgment identification, since the latter need only span the maximum time required to send a single-message transmission and its acknowledgment. (For either function, however, the uniqueness of satellite transmission time and global time synchronization among all stations provides an efficient numbering basis.)

For a channel data rate of 1 Mbits/s, 1000-bit average message lengths, and a 2 min maximum holding time, over 100 000 message numbers are involved in the duplicate time span. A reasonable memory requirement can be achieved, however, by remembering only the numbers of messages not received correctly and checking correctly received retransmissions against this list. While this scheme involves typically less than 1 percent of all messages in the checking and in the total list length, it is subject to a probability that the memory space allocated for this task might be exceeded during periods of increased error rates and has complications for stations constrained to low receiving rates.

Sequencing can impose significant increases in the amount of destination buffering required, since a large number of messages may be received between retransmissions over the satellite channel. The sequencing function also implies message numbering which spans the maximum message lifetime, as in duplicate detection. Since the resources required for sequencing and duplicate detection become more reasonable at levels of decreased traffic concentration, these functions appear best left to higher level protocols. Note that a virtual circuit service could be provided by the network by use of, say, an X.25 interface for network access, with sequencing and duplicate detection for each virtual circuit performed between the X.25 interface and the GPSN's internal datagram interface.

B. Addressing

The choice of a logical addressing scheme is particularly well-suited to a satellite network. Each host is assigned a unique address to be used for host-to-host communication, with tables maintained in each station which define an internal mapping between these logical addresses and their associated station and host physical port [32]. Stations must check the table whenever they have a message to send or when a message is received from the satellite channel.

Although processing overhead can be kept small, a more significant factor is the amount of storage required for the logical address table, since the need to send each message at the destination station's (dynamically changeable) receiving rate requires that each station have a table of all host addresses. For a network containing a few thousand hosts, this still results in a reasonable storage requirement while also keeping processing small for address checking. For example, in a 16-bit word processor, 4 K words of core memory allows a simple indexed reference for up to 4000 hosts, with a 12-bit address field used in each message header.

Logical addressing has a further advantage especially well-suited to a broadcast network—a single address can also represent any arbitrary subset of hosts. All that is required is that each station with one or more hosts belonging to the subset maintain a table entry for the group address indicating which of its host ports are members. Recognition is then done on group addresses in each received message in the same way as for individual host addresses.

C. Flow and Congestion Control

1) Host Access Control: The global information inherent in PODA channel scheduling provides an effective means for also performing network congestion control. As the number of reservations in the scheduling queue at each priority level reaches an amount sufficient to keep the channel continuously utilized, acceptance of lower priority messages into the network is reduced or cut off. Stations performing scheduling have immediate access to this information and can apply it directly to their host input traffic; stations receiving centralized assignments can be informed of the congestion parameters after a one-hop delay. The available information on the local buffer conditions at the source and destination stations is also used to regulate host input.

The existence of multiple priorities and delay classes dictates the use of a selective flow control mechanism for host input,

allowing particular messages to be refused while those following can be accepted. A number of strategies are available to accomplish this, ranging from sending explicit reservation requests prior to sending the message itself to simply sending each message which is queued and retrying those which are refused. An efficient approach which does not increase delays in the host-station interface is to send information to hosts on a regular basis which defines the current acceptance state of each message priority and delay class, allowing hosts to withhold messages not being accepted by the network. This information is prefixed to messages being sent to the host to keep overhead low, with explicit status messages sent only in the absence of other traffic.

Messages which have acceptable priority and delay class may still need to be refused for other reasons, such as local buffer congestion or destination blocking. In these cases a refusal can be returned to the host, and the host subsequently informed when the message is acceptable. A method which maintains interface efficiency and minimizes internal book-keeping is to define a small number of categories and have the station assign each refused message to a particular category. A bit is used to indicate the current acceptance–refusal status of each category, and these bits are sent regularly with the priority-delay status information.

2) Internal Flow Control: To optimize use of buffers and channel capacity, it is desirable to suppress acceptance and scheduling of messages destined for stations which do not have buffers available. This situation can arise because of a steady-state flow from one or more sources which exceeds the destination host's average acceptance rate, or because of short-term variations introduced by temporary destination blocking. The central problem here is how to pass the necessary information across the satellite channel to the source stations in a timely manner without using significant channel capacity.

The applicable strategies range from simply discarding the message at the destination station upon arrival and not acknowledging it to maintaining information about the flow to each destination and the destination's buffer status as part of the PODA global scheduling function. In between these extremes are techniques such as broadcasting varying amounts of information about current buffer status and/or sending predictive information in place of an acknowledgment when arriving messages are discarded. The choices are complicated by the quarter-second minimum feedback time which, if comparable to the time constant of flow variances introduced by destination blocking, could cause feedback to degrade rather than improve performance. While certain conditions such as long-term blocking are readily handled by simple status information transmission, the more general problem is not yet well understood.

D. Special Network Applications

1) Dynamic Addressing: The broadcast connectivity of the satellite network makes it especially attractive for applications requiring multidestination messages such as in speech conferencing. The group address capability discussed in Section V-B provides a particularly efficient mechanism for this. To allow it to be used for large numbers of conferences involving arbitrary sets of hosts, however, a dynamic addressing protocol in required.

This addressing protocol requires participation by both the hosts and the network, since stations must have the address in their tables for proper delivery of messages received over the satellite channel, and the hosts which are to be members of the address must be made known to the network. To allow efficient use of the address space, which is constrained by available table space in stations and by the address field size in messages, the network should perform the address allocation. While some addresses might be allocated permanently to certain sets of hosts for on-going functions such as mail, in temporary applications such as conferencing the address would be released when the conference or other use is terminated.

Two distinct ways in which host membership can be established for a group address are 1) having one host supply a list of all members, or 2) allowing each host to join or leave the group through messages exchanged with its local station. In either method information needs to be broadcast by stations to the other stations to effect the allocation or release of an address and appropriate station table entries. Note that the existence of different station receiving rates requires that all participating stations establish the lowest receiving rate required for the address in their tables.

2) Stream Conferencing: The above group addresses can be used for either datagrams or streams in a GPSN. For a PODA system, there is an important difference between these two uses. In the datagram case, messages which might be entering the network at the same time from different hosts will be scheduled at different times and sent to their destinations independently of each other. In the stream case, messages entering at the same time will be sent by their respective stations at the same burst time being globally scheduled for the stream, resulting in destructive interference at the satellite.

Thus the sharing of a stream requires a higher level coordination protocol to be provided either by the network or by the hosts. Note that datagram messages can be sent using the group address to support this coordination simultaneously with the stream messages, providing an 'out-of-band' signaling mechanism for the stream.

3) Internetwork Traffic: The use of a GPSN in conjunction with other networks raises a number of unresolved issues. Among these are the ways in which the diversified message services of a GPSN and those provided by other networks can be indicated uniformly without unreasonable message overhead, whether capabilities such as dynamic addressing and stream setups should be manipulated directly from other networks or by programs residing locally to support users on other networks, and the type and frequency of information which should be exchanged with gateways connected to the GPSN. This last topic is of particular importance to gateways performing broadcast addressing or dynamic routing in situations where alternate routes exist. The potentially large and rapid variation in delay and bandwidth characteristics possible in a PODA system could cause problems for dynamic routing algorithms, making it important for timely status information to be provided by the network.

VI. THE ATLANTIC PACKET SATELLITE EXPERIMENT

The Atlantic Packet Satellite Experiment serves as the development vehicle and experimental test-bed for the overall Packet Satellite Program. The experiment includes both the experimental satellite network, called SATNET, and the supporting development and measurement activities and facilities. A schematic representation of the facilities involved in the Atlantic Packet Satellite Experiment is shown in Fig. 4 with the elements of SATNET enclosed by a solid line.

The experiment has a number of participants and sponsors. The participating organizations are Bolt Beranek and Newman, Inc. (BBN), the Communications Satellite Corporation (COMSAT), the LINKABIT Corporation, M.I.T. Lincoln Laboratory, and the University of California at Los Angeles (UCLA) in the U.S., the University College London in England, and the Norwegian Defense Research Establishment in Norway. Technical coordination is the responsibility of LINKABIT. The project is jointly sponsored by the DARPA, the British Post Office (BPO), and the Norwegian Telecommunications Authority (NTA), with the participation of the Defense Communications Agency (DCA) and the U.S. Air Force Space and Missile Systems Organization (SAMSO).

A basic philosophy of the experimental program and the SATNET development efforts has been to achieve a flexible experimental environment which could serve as a measurement vehicle for a spectrum of specific system realizations and applications. Further, development and measurement emphasis has been placed on those system aspects which are least understood and most in need of feasibility demonstration, e.g., distributed rather than central control with PODA algorithms, with the goal of obtaining a good understanding of the benefits and limitations of these features.

A. Experimental Facilities

As shown in Fig. 4, SATNET is composed of four earth stations which communicate with each other over a shared channel derived from the Atlantic INTELSAT IV-A satellite. The earth stations located at Etam, WV, U.S., Goonhilly Downs, England, and Tanum, Sweden, are INTELSAT Standard A earth stations with approximately 30-m antennas and G/T values of 40.7 dB. The earth station located at COMSAT Laboratories in Clarksburg, MD, is physically smaller, with a G/T value of approximately 29.7 dB (i.e., its receiving capability is inherently 11 dB less than the INTELSAT stations) [27]. It is included in SATNET solely for experimental purposes and is not involved with the passage of operational traffic. While the hardware facilities of SATNET are limited to four earth stations, the design activities assume that potential future networks might include hundreds of earth stations.

1) Communication Channel: A 38-kHz channel is shared among the earth stations via the demand-access algorithms which are under test. This channel is one of the 800 possible frequency division multiplexed channels in the global SPADE transponder of the Atlantic INTELSAT IV-A satellite [28]. This full-period assigned channel is operated at nominal power levels, supporting 64-kbits/s data transmission with a bit error probability on the order of 10^{-6}–10^{-7} to an INTELSAT Standard A earth station.

2) Earth Station Equipment: Fig. 5 shows, in block diagram form, the major subsystems located at each of the earth stations. These subsystems implement the various functions shown in Fig. 1. The earth station RF-transmission, RF-reception, and RF–IF-conversion equipment is conventional and, at the large stations, is shared with other channels which access the SPADE transponder.

Initially, simple interface equipment was used to adapt modems designed for demand-assigned voice-activated applications to this packet communication application. After some experience, it became clear that this initial system did not adequately support the experimental and operational requirements. The original modem and interface equipment are being replaced with equipment specifically designed for the

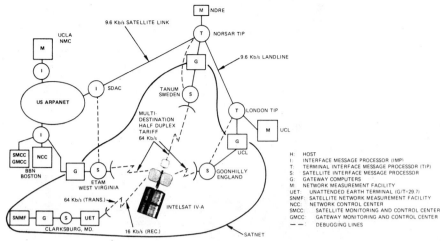

Fig. 4. Schematic of Atlantic Packet Satellite Experiment.

Fig. 5. Earth station equipment associated with the Atlantic Packet Satellite Experiment.

packet satellite application, i.e., the PSP Terminal. (This configuration is scheduled to become operational early in 1979.) This self-contained terminal consists of redundant burst modems and associated frequency selection equipment interface equipment which will support a variety of transmission modes, and appropriate equipment to support both local and remote test and monitoring functions.

The digital 32-kilosymbol per second microprocessor-based modems used in the PSP Terminal are designed to provide very rapid burst acquisition, process both BPSK and QPSK packets with near optimum error rate performance, and automatically provide certain performance and monitoring information, e.g., the signal-to-noise ratio and the frequency offset with each packet reception. The original development effort for these modems was motivated by the stringent signal-to-noise ratio requirements (design input E_b/N_o = 3.5 dB) at the Clarksburg earth station [29].

The interface equipment allows control over the bit patterns (preambles) used for acquisition and bit framing. It supports a number of packet transmission modes. In particular, the data rate for a packet can be 16, 32, or 64 kbits/s or a combination of these rates, e.g., a variable portion of the front end of a packet can be transmitted at 16 kbits/s with the remainder at 64 kbits/s. These rates are achieved with the 32-kilosymbol per second modem by using appropriate modulation (BPSK or QPSK) and coding (uncoded or rate one-half convolutional encoded) combinations. The interface includes a Viterbi decoder.

The 16-kbits/s transmission modes are fundamental to the operation of the Clarksburg earth station as part of the experimental network. This station can easily transmit at rates up to 64 kbits/s, but its nominal receiving rate is limited to 16 kbits/s. Through the use of coding, its error-rate performance at 16 kbits/s can be made as good as the large station performance at 64 kbits/s; the reduced data rate increases the available

energy per bit by 6 dB and the coding gain reduces the energy per bit required to achieve the same bit error performance by more than 5 dB. (The sum of these two factors more than compensates for the inherent 11-dB deficiency of the Clarksburg station.)

The interface equipment also provides both a local and remote capability to control certain system parameters, e.g., preamble lengths, and to support certain monitoring (both on- and off-line) and testing functions. Local control and certain built-in packet generation and reception equipment are available for use by earth station personnel for troubleshooting.

The network protocols are implemented in the Satellite Interface Message Processor (Satellite IMP), a Honeywell 316 minicomputer with 32 K words (16-bit) of memory. This hardware is essentially the same as the ARPANET IMP hardware except for an additional 16 K words of memory and interface changes to support satellite burst mode operation and ranging. This equipment is interfaced to the PSP Terminal by a full-duplex high-speed data path, and two asynchronous character-oriented paths for interfacing certain test, monitoring, and interface control functions.

In the original experimental configuration, the Satellite IMP's interfaced directly with ARPANET IMP's and the software was essentially an augmented version of the IMP software, including the additions required to implement the desired satellite channel protocols [30]. During the course of the experiment, SATNET has been physically (and logically) separated from the ARPANET, while retaining the ability to transmit ARPANET traffic, by the addition of Gateway computers (see Fig. 4), and the software has evolved to the form shown in Fig. 6. Note that the Satellite IMP implements global timing control, the demand-access protocols, input and output to the satellite channel PSP Terminal, internal network protocols (e.g., control of information flow), the SATNET side of the host access protocol, certain measurement capabilities, software to control and receive data from the PSP Terminal test and monitoring functions, and software to allow interaction with the SATNET monitoring and control center. The internal measurement capabilities consist of fake hosts within each Satellite IMP's which can be activated to serve as artificial traffic sources and

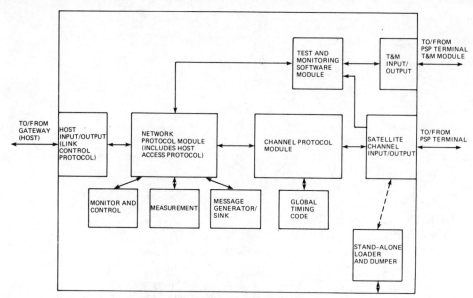

Fig. 6. Schematic of Satellite IMP software.

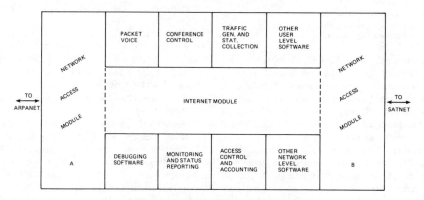

NOTE: IN THE CASE OF THE SMALL STATION, TRANSLATION GATEWAY A
CONNECTS DIRECTLY TO A MEASUREMENT COMPUTER. (IBM 360/65)

Fig. 7. Elements of the gateway.

sinks and other fake hosts which can be activated to collect and transmit cummulative statistics on specified performance parameters (e.g., delay).

3) Network Interface and Data Source: The Gateways which interface SATNET to ARPANET act as host computers on each of the connected networks. Thus the line designating SATNET in Fig. 4 is drawn through the middle of these Gateways. For experimental convenience, a Gateway is also attached to the Clarksburg Satellite IMP. As shown in Fig. 4, this Gateway is attached to another collocated computer (an IBM 360/65) which is used for data collection at the Clarksburg site and certain SATNET application experiments.

The Clarksburg site is isolated (i.e., it has no ARPANET connection). This site provides a realistic experimental model of isolated sites which must be monitored and maintained solely via their satellite connectivity. Except at Clarksburg, the Gateways are physically remote (tens to hundreds of miles) from the Satellite IMP's, and the connection is via 50 kbits/s lines. These Gateways are located at BBN in Cam-

bridge, MA, UCL in London, England, and NDRE in Kjeller, Norway.

The Gateways in the Atlantic Packet Satellite Experiment have several functions, including 1) the separation of the ARPANET and SATNET so that SATNET performance can be separately optimized and measured while still using certain ARPANET facilities for traffic generation and data collection, 2) providing a realistic environment for internetwork experiments involving traffic flow between ARPANET and SATNET, 3) supporting certain experimental applications, e.g., packet voice conferencing, and 4) providing the facilities to emulate various traffic sources and sinks and to measure source-to-sink performance. Note that the Gateways are the only external data sources to SATNET. These gateway functions are realized in a PDP 11 minicomputer with 128 K words (16 bit) of memory. (The large memory size is due to the need to support user programs in the gateway computer for experimental purposes.)

Fig. 7 shows a representation of the various elements of gateway software. The basic gateway function, the interconnection of SATNET and ARPANET, is implemented by two network access modules which interface and terminate the respective network protocols and an internet module which performs certain required network-independent functions, e.g., routing. The other modules indicate various applications programs which are or could be supported by the Gateway. In particular, an extensive traffic generation and statistics collection capability has been implemented in the Gateways. For the purposes of this experiment, the software required to format digital speech into packets and to control conference applications of packet speech also resides in the Gateways, as does appropriate debugging and monitoring software.

4) System Control and Monitoring: The initial SATNET configuration relied exclusively on ARPANET monitoring and control facilities made possible by the presence of the ARPANET IMP program in the initial Satellite IMP software releases. An independent capability has since been developed to provide remote monitoring and control of SATNET through its present gateway connectivity to ARPANET. This is accomplished by the SATNET Monitoring and Control Center (SMCC) and Gateway Monitoring and Control Center (GMCC) programs which run on an ARPANET Tenex host at BBN. These programs provide an interactive user interface which allows network control center personnel to exercise appropriate monitoring and control, while also allowing experimenters to have remote access to the same monitoring information.

The existing SMCC capabilities are undergoing two significant developments. The first involves extending the SMCC software to remotely utilize, via interaction with appropriate software in the Satellite IMP, the test and monitoring capabilities built into the PSP Terminal, and the second involves the evolution of monitoring and control techniques which exploit the one-hop broadcast properties of the network. A goal is to utilize the satellite connectivity for transmission of the monitoring and control information with minimal reliance on terrestrial connectivity between the various SATNET nodes and the SMCC.

5) Measurement Control and Data Collection: Facilities also have been developed to facilitate measurement control and data collection. Interactive control programs have been developed for both Satellite IMP's and Gateways to perform tasks such as turning traffic generators on and specifying desired statistics collection parameters. The programs allow the experimenter to remotely specify, initiate, and terminate an experiment and to direct the data to an appropriate collection site(s).

B. Measurement Activities/Objectives

The experimental activities which use the facilities discussed above can be grouped into four broad categories: channel-oriented, network-oriented, user-oriented, and applications experiments/demonstrations.

1) Channel-Oriented: The channel-oriented measurement activities are concerned with measurements of channel and equipment performance in the broadcast packet satellite environment, with the objectives of developing an understanding of system requirements and achieving and demonstrating reliable system performance. These activities include the efforts required to develop appropriate system monitoring capabilities (SMCC, GMCC, and hardware test and monitoring capabilities). An appreciable effort has been devoted continuously to this area because reliable operation and a knowledge of the actually achieved channel performance are required to support the other measurement activities. The channel performance is monitored on an essentially real-time basis, and this information is available to both the SMCC and the experimenters.

2) Network-Oriented: These experimental activities are concerned with measuring the performance of the various demand assignment algorithms and the related portions of the SATNET protocol, including the various control strategies. The network-oriented measurements primarily involve the use of the artificial message generators and statistics collection programs located in the Satellite IMP's. In most cases, simulation programs have been written to gain insight into the expected behavior, establish the parameter ranges which seem to be of most interest, and provide some results for comparison with the measurement results. Comparisons between the measurement results and analytical and simulation predictions have been used to verify the Satellite IMP software implementations. The goal of these measurements is to characterize the performance of the various channel protocols as a function of traffic and other system parameters and to provide a mechanism for optimizing certain system parameters.

The initial Satellite IMP software implemented the fixed-TDMA (F-TDMA), S-ALOHA, and R-TDMA protocols. An extensive measurement program has been carried out with these protocols [31]. In the case of the S-ALOHA protocol, software was developed which allowed each physical Satellite IMP to emulate up to 10 additional Satellite IMP's, greatly increasing the realism of the measurement. Stability controls for the S-ALOHA protocol were also implemented and the resulting performance measured [23].

Later versions of the Satellite IMP software include the FPODA and CPODA protocols and F-TDMA. This Satellite IMP software supports channel protocol and network protocol experiments involving point-to-point and broadcast datagram and stream traffic. The mixed packet transmission modes required to incorporate the Clarksburg earth station into the network are included, along with the various forms of integrated control discussed in Section IV.

Experiments to exploit all of these capabilities are underway. In addition to extensive PODA measurements with stations of uniform size (receiving capability) and various mixes of datagram and stream traffic, measurements of the performance of a network with mixed stations (different receiving capabilities) operating under the various control strategies are being performed.

3) User-Oriented: The user-oriented measurement activities are concerned with obtaining quantitative information about SATNET's capability to support certain types of applications, such as file transfers. These measurements are performed using the traffic generation and statistics collection programs located in the Gateways, and permit evaluation of end-to-end SATNET performance.

The Gateway traffic generators are quite sophisticated and can reasonably model interactive, file transfer, and stream applications. The majority of these measurement activities commenced during the Spring and Summer of 1978.

4) Applications Experiments/Demonstrations: A set of applications experiments and demonstrations is currently ongoing. The primary goal is to demonstrate the use of SATNET for providing service to certain applications and, in some

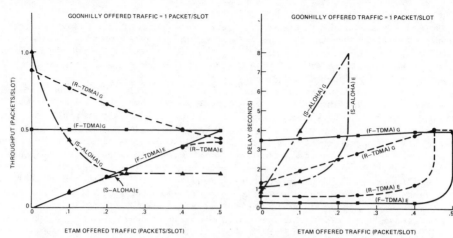

Fig. 8. Throughput and delay performance of different demand-assignment protocols under unbalanced load conditions. The subscripts indicate the station where the traffic originated (after Fig. 7 and 8 of [31]).

cases, to assess the viability of the application or technique. In most cases, the desired results will be primarily qualitative rather than quantitative information, e.g., the effects of the transmission delays on the application. These experiments will be useful in demonstrating and assessing the applicability of the SATNET type of technology to a GPSN.

One of the most important applications is packet speech. The digitized speech produced by a voice compression device (linear predictive coder) operating at bit rates of 2.4, 3.6, or 4.8 kbits/s is packetized and transmitted over SATNET using the stream traffic handling capability of PODA. Both point-to-point and conference applications are supported with the conferencing control performed by distributed conference controllers located in the gateways.

Other applications experiments/demonstrations concern the use of the X.25 access protocol as a SATNET access protocol, the use of SATNET to provide communication between elements of an extended local network or a set of identical remotely located local networks, and the use of SATNET to support interactive traffic.

C. Preliminary Results

SATNET has evolved significantly since its beginning in the fall of 1975 as problems have been discovered and corrected, and capabilities have been added. The development process has been somewhat longer and required more effort than was originally anticipated, in part due to the problems inherent in maintaining an operating network while simultaneously performing development. On the other hand, the development of capabilities, equipment, and measurement and monitoring tools has been influenced strongly by the experience and understanding of requirements gained as the experiment has progressed. In this sense, SATNET has been an effective tool in the development of packet satellite technology.

A significant, although implicit, accomplishment of the experiment has been to demonstrate the effective utilization of a packet-switched network, i.e., the ARPANET, as a central vehicle for effectively coordinating a large project among a geographically dispersed group of people and for remotely controlling experiments and collecting data, maintaining and developing software, and monitoring system (including channel) performance.

The experience to date has provided certain valuable insights into the design of a future GPSN. It has helped to refine and emphasize the requirements that must be met by various subsystems, e.g., modems, Satellite IMP's, and Gateways, in this type of system. The early experience before the introduction of Gateways established the desirability of logically isolating a satellite-based packet network from terrestrially based networks. This is required to permit independent optimization of the satellite network protocols and to decouple the software evolution of the networks. The early SATNET measurements showed that a requirement to be consistent with the terrestrially based ARPANET store-and-forward protocols could significantly degrade SATNET performance. The early experience also indicated the need for comprehensive system and subsystem performance monitoring in this type of distributed system where certain failure modes involve either coupled interactions among various subsystems (e.g., packet errors caused by the receive hardware can result in the loss of transmission synchronization for the station) or among the various earth stations (e.g., with some modem designs, out-of-specification transmission parameters at one station can impair the reception of packets transmitted by the other stations). The monitoring requirement is, of course, particularly acute during the experimental and development stages, but it is believed that it will be equally important to the reliable operation of a fully developed system. This belief has resulted in appreciable SATNET development work in this area, and the fully implemented system has significant remote monitoring and control capability.

The early measurements have concentrated on channel protocol as opposed to network experiments. Some of these measurements have been reported in [31]. Fig. 8 is illustrative of the type of results obtained. This figure shows the throughput and delay for the various protocols in a two station measurement in which the offered traffic at Goonhilly is held constant at one packet per slot, and the offered traffic at Etam is varied. These curves illustrate the general conclusion that a reservation approach (R-TDMA) exhibits better performance than F-TDMA and S-ALOHA over a significant range of traffic values.

Fig. 9 is indicative of the CPODA measurements that have been performed and addresses one of the most important

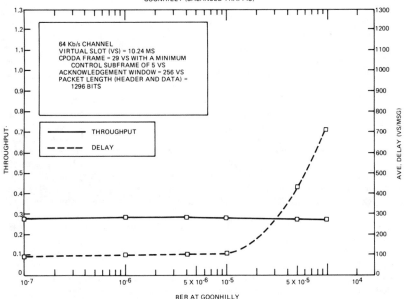

Fig. 9. CPODA performance for Goonhilly traffic versus bit error rate (BER) at Goonhilly. (The measurements, which were performed by N. Hsu at BBN, involved a balanced traffic flow between the Etam and Goonhilly earth stations with artificial noise introduced only at Goonhilly. The delay and throughput results for the Etam traffic versus the Goonhilly BER are very similar to those shown above. Measurements for an unbalanced traffic flow have shown similar delay results, i.e., a break point at a BER of approximately 10^{-5}.)

issues about the CPODA algorithm with distributed control, namely, its performance in the presence of local (single-station) channel errors [26]. These channel errors, to the extent that they affect the control information in a packet, tend to cause a loss of scheduling synchronization among the earth stations. The results of Fig. 9, obtained using reasonable but not necessarily optimum parameters in the scheduling synchronization algorithm, show that the performance of the 64-kbits/s system is not adversely affected until the bit error rate is on the order of 10^{-5}, which corresponds to a probability on the order of 10^{-3} that the control information in a packet is not valid. The bit error requirements on higher data-rate channels will be more stringent, but on the basis of these results distributed control should be feasible for channel rates into the megabit range.

It is believed that the measurement results obtained so far and the additional results which will be obtained from the measurements currently in progress will provide an important source of information for the design of future packet satellite networks.

VII. SUMMARY AND CONCLUSIONS

Basic packet satellite technology will receive continued attention in the future, with particular emphasis on judging its suitability for various applications and on assessing a number of technical, economical, and operational issues, e.g., the impact of new satellite communication technologies and detailed studies of cost and charging policies. For example, in the U.S., DARPA and DCA plan to utilize a high-data-rate domestic packet satellite channel, in part to support multiple-user packet voice experimentation. In addition to the technology and study activities, it is likely that some operational utilization of the technology will occur in the next few years.

A. Potential Applications

The type of packet satellite technology discussed in this paper has the potential to be utilized in a number of communication systems. Because of the general-purpose features, which can be tailored to special applications or provide service to a mixed group of users, the potential applications include military command and control systems, military logistics and administrative communication systems, integrated domestic systems operated by commercial carriers, specialized private networks operated by or for large business organizations, and transit networks which interconnect other national or domestic networks. It is likely that in some applications packet satellite techniques will be combined with terrestrial packet technology, either packet radio [33] or now conventional terrestrial packet networks [34], to form mixed-media networks which exploit the best properties of each.

The time scale of the demand for certain types of services is expected to play an important role in establishing the development pace of nonmilitary applications of packet satellite technology. The growth rate of the demand for transaction-based and distributed data-base services, in general, and the growth of certain specific applications, such as electronic mail, teleconferencing, certain forms of data collection and dissemination, and interconnection of local networks serving large companies, can be expected to have a significant impact. The economics of voice digitization and compression techniques and the results from extensive and detailed satellite packet voice experimentation will also be significant factors in the pace and extent of packet satellite technology application.

B. Future Satellite Communications Technology

Satellite communication technologies which are currently under development and expected to be implemented in the

1980's have the potential to influence the economic viability of various packet satellite applications and system concepts [2]. The developments with the largest potential for significant impact include: 1) the use of higher transmission frequencies (e.g., 14/11 and 30/20 GHz), 2) multibeam satellites, 3) satellite on-board processing, and 4) intersatellite links.

These developments are expected to permit a more flexible tradeoff between transmission and earth station costs. Higher transmission frequencies will permit some reductions in antenna size, as well as significantly reduce the restrictions on earth station siting due to frequency clearance problems. Multibeam satellites can be used to obtain frequency reuse, provide higher energy densities at the earth with a resulting reduction in the G/T required by the receiving earth station, reduce the transmitting antenna gain (size) and/or power amplifier size required at the earth station, and provide interference rejection [35]. These multibeam satellite capabilities come at a cost of some loss in flexibility and connectivity unless some form of on-board processing is employed. On-board processing replaces the simple frequency translation operation used in all of today's commercial satellites with varying levels of signal processing to allow the satellite to play a more active role in the allocation of its resources. The level of on-board processing provided can vary from the use of a microwave switch matrix to provide dynamic connectivity between the beams of a multibeam satellite to a full processing capability which allows signal dependent processing (e.g., switching) and includes storage capability for buffering. Intersatellite links are two-way communication paths between satellites. They have the potential, while maintaining or reducing the cost of the system's earth segment, to provide flexibility in the space segment implementation (e.g., to use a cluster of small low-cost satellites or to use low-altitude satellites) without causing a connectivity penalty.

These technological developments can be expected to have some impact on the system organizations and operational procedures considered in this paper. While most of the basic ideas of the packet satellite technology discussed above will still be applicable, there are likely to be some changes in detail and emphasis. Because the high-frequency bands are subject to local propagation effects, rain losses in particular, the capability to use sufficiently dynamic integrated control techniques to maintain continuous communication capability (perhaps at reduced rates) to all stations, without requiring redundant antennas, is expected to be an important feature [36]. The availability of an adequate level of on-board processing would permit channel scheduling to be implemented in the satellite with the potential for some cost reduction at the earth stations; demand assignment strategies which involve reservation techniques (such as the PODA techniques) could use central scheduling without incurring an additional one-hop delay. The use of intersatellite links in a packet satellite network may require that certain gateway functions (e.g., routing) be implemented in the satellites.

C. Implementation and System Issues

There are certain implementation and system issues, in addition to future satellite communication technology trends, that require further study and experimental verification. Some of these issues will be important in specific system designs, while others may have more general applicability and economic impact. These include:

1) the evolution of packet satellite technology to high-data-rate systems and the feasibility and desirability of combining the hardware which realizes its functions. Special processor architectures to support PODA-like algorithms, and other protocol functions on high-data-rate channels are likely to be desirable. A potential realization is special-purpose high-speed hardware elements acting under microprocessor control to obtain the speed while retaining appreciable flexibility. There is also likely to be significant economic and operational advantage to combining a number of the earth-station functions into an integrated hardware realization, as opposed to the largely functional packaging that is currently employed. For example, it may make sense to realize the modem, interface, Satellite IMP, and portions of the Gateway functions in a single integrated hardware realization.

2) the degree to which general purpose packet satellite networks versus networks specialized to specific applications are desirable. There normally are operational economic advantages to networks providing integrated services, but the initial implementation cost is usually larger. Certain applications may exist which are best satisfied by tailored systems which emphasize the provision of the required characteristics at minimal implementation cost. Some small networks with relatively concentrated traffic may require fewer flexibility features, e.g., FPODA with central control rather than CPODA with distributed control may be appropriate.

3) the assessment of future traffic types and message sizes. The relative importance of various traffic sources, e.g., digital voice, facsimile, word-processing systems, transaction-based terminals, etc., and the corresponding traffic characteristics will be important in detailed system design and sizing. For example, it is expected that future packet systems will have longer message lengths than current systems because local rather than remote echoing will be used on interactive applications, with a significant reduction in the number of single-character messages.

4) the verification that providing the processing required to implement complex demand assignment and network protocols is cost effective. A basic assumption in the work reported here is that the cost of processing power (at the earth stations or in the satellite) will decrease relative to the cost of satellite bandwidth (communication capacity). The relatively complex algorithms considered in this paper might have been considered too complicated and costly to implement a few years ago. However, with modern computer technology a different view of complexity is required. While the complexity of an algorithm, including built-in test features, does influence the initial design costs, software complexity need not degrade maintainability or greatly increase the cost of producing additional systems. SATNET has demonstrated the feasibility of reliably implementing the packet satellite technology in a low-data-rate system, but certain economic and high-data-rate questions remain to be answered and verified.

5) the assessment and further study of certain network access and flow control issues (e.g., the use of preemption as part of buffer-allocation strategies and the amount of explicit station-to-station flow control required), certain aspects of the internetwork use of the capabilities (e.g., broadcast) that packet satellite networks can provide, and design procedures for integrating packet satellite and terrestrial packet technology to form mixed-media networks.

D. Conclusion

While certain technical and system questions require further study, and detailed application experiments and market and cost studies are not yet complete (underway), it seems clear that packet satellite technology has significant promise for providing a variety of data communication services, and that it will be used for at least some applications in the 1980's.

ACKNOWLEDGMENT

Many persons involved in the Atlantic Packet Satellite Experiment contributed significantly to the ideas presented in this paper. We particularly wish to cite major contributions to PODA made by Lin-nan Lee and Nai-Ting Hsu.

REFERENCES

[1] W. L. Pritchard, "Satellite communications—An overview of the problems and programs," *Proc. IEEE*, vol. 65, pp. 294-307, Mar. 1977.

[2] H. L. VanTrees, E. V. Hoversten, and T. P. McGarty, "Communications satellite technology in the 1980's," *IEEE Spectrum*, pp. 42-51, Dec. 1977.

[3] J. Heller and I. M. Jacobs, "Viterbi decoding for satellite and space communications," *IEEE Trans. Commun. Technol.*, vol. COM-19, pp. 835-848, Oct. 1971.

[4] J. G. Puente, W. G. Schmidt, and A. M. Werth, "Multiple access techniques for commercial satellites," *Proc. IEEE*, vol. 59, pp. 218-229, Dec. 1971.

[5] J. L. Dicks and M. P. Brown, Jr., "Frequency division multiple access (TDMA) for satellite communication systems," in *EASCON 74 Con. Rec.*, pp. 167-178, 1974.

[6] G. D. Dill, "TDMA, the state of the art," in *EASCON 77 Con. Rec.*, pp. 31-5A-31-5I, 1977.

[7] P. P. Nuspl, K. E. Brown, W. Steenaart, and B. Ghicopoulos, "Synchronization methods for TDMA," *Proc. IEEE*, vol. 65, pp. 434-444, Mar. 1977.

[8] R. C. Davis, F. H. Esch, L. Palmer, and L. Pollack, "Future trends in communications satellite systems," *Acta Astron.*, Mar./Apr. 1978.

[9] I. M. Jacobs, "Integration of demodulation, decoding, buffering, and control for TDMA demand assignment systems," in *Proc. 2nd Digital Satellite Conf.*, Paris, France, pp. 287-296, Nov. 1972.

[10] L. Palmer, "Economic trends in digital transmission by satellite," in *Proc. 5th Data Communications Symp.* (Snowbird, UT), pp. 2-30-2-36, Sept. 1977.

[11] L. Cuccia and C. Hellman, "Status report: The low-cost low-capacity earth terminal," *Microwave Syst. News*, June/July 1975.

[12] B. Edelson and A. Werth, "SPADE system progress and application," *COMSAT Tech. Rev.*, vol. 2, no. 1, pp. 221-242, Spring 1972.

[13] W. Schmidt *et al.*, "MAT-1: INTELSAT's experimental 700-channel TDMA/DA system," in *Proc. INTELSAT/IEEE Int. Conf. Digital Satellite Communications*, Nov. 1969.

[14] N. Abramson, "Packet Switching with satellites," in *Proc. AFIPS Conf.*, vol. 42, June 1973.

[15] S. Lam and L. Kleinrock, "Packet switching in a multi-access broadcast channel: Dynamic control procedures," *IEEE Trans. Commun. Technol.*, vol. COM-23, pp. 891-905, Sept. 1975.

[16] N. Abramson, "The ALOHA system—another alternative for computer communications," in *AFIPS Proc. FJCC Conf.*, vol. 37, Fall 1970.

[17] L. Roberts, "ALOHA packet system with and without slots and capture," *ACM SIGCOM Computer Commun. Rev.*, vol. 5, no. 2, Apr. 1975.

[18] L. Kleinrock and S. Lam, "Packet-switching in a slotted satellite channel," in *Proc. AFIPS Conf.*, vol. 42, June 1973.

[19] W. R. Crowther, R. Rettberg, D. Walden, S. Ornstein, and F. Heart, "A system for broadcast communication: Reservation-ALOHA," in *Proc. 6th Hawaii Int. Syst. Sci. Conf.*, Jan. 1973.

[20] F. C. Schoute, "Decentralized control in computer communication," Ph.D. dissertation, Dep. Eng., Harvard Univ., May 1977.

[21] L. Roberts, "Dynamic allocation of satellite capacity through packet reservation," *Proc. AFIPS Conf.*, vol. 42, June 1973.

[22] R. Binder, "A dynamic packet switching system for satellite broadcast channels," in *Proc. ICC 75*, San Francisco, CA, June 1975.

[23] M. Gerla, and L. Kleinrock, "Closed loop stability controls for S-ALOHA satellite communication," in *5th Data Commun. Symp.* (Snowbird, UT), pp. 2-10-2-19, Sept. 1977.

[24] I. Jacobs *et al.*, "CPODA—A demand assignment protocol for SATNET," in *5th Data Commun. Symp.* (Snowbird, UT), pp. 2-5-2-9, Sept. 1977.

[25] I. Jacobs and L. Lee, "A priority-oriented demand assignment (PODA) protocol and an error recovery algorithm for distributively controlled packet satellite communication network," in *EASCON 77 Con. Rec.*, pp. 14-1A-14-1F, 1977.

[26] N. Hsu and L. Lee, "Channel scheduling synchronization for the PODA protocol," in *Proc. ICC 78* (Toronto, Canada), June 1978.

[27] L. Pollack and W. Sones, "An unattended earth terminal for satellite communications," *COMSAT Tech. Rev.*, vol. 4, no. 2, pp. 205-230, Fall 1974.

[28] E. R. Cacciamani, "The SPADE system as applied to data communications and small earth station operation," *COMSAT Tech. Rev.*, vol. 1, no. 1, pp. 171-182, Fall 1971.

[29] C. Heegard, J. A. Heller, and A. J. Viterbi, "A microprocessor-based PSK modem for packet transmission over satellite channels," *IEEE Trans. Commun. Technol.*, vol. COM-26, pp. 552-564, May 1978.

[30] R. Weissler, R. Binder, R. Bressler, R. Rettberg and D. Walden, "Synchronization and multiple access protocols in the initial Satellite IMP," in *Proc. Fall COMPCON 78*, (Washington, DC), Sept. 5-8, 1978.

[31] M. Gerla, L. Nelson, and L. Kleinrock, "Packet satellite multiple access: Models and measurements," National Telecommun. Conf. Rec., pp. 12:2-1-12:2-8, 1977.

[32] J. McQuillan, "Enhanced message addressing capabilities for computer networks," this issue, pp. 000-000.

[33] R. E. Kahn, J. Burchfiel, S. Gronemeyer, and R. Kunzelman, "Advances in packet radio technology," this issue, pp. 000-000.

[34] R. E. Kahn, "Resouce-sharing computer communications networks," *Proc. IEEE*, vol. 60, pp. 1397-1407, Nov. 1972.

[35] E. W. Matthews, W. G. Scott, and C. C. Han, "Multibeam antennas for data communication satellites," in *Proc. 5th Data Commun. Symp.* (Snowbird, UT), pp. 2-20-2-29, Sept. 1977.

[36] D. C. Hogg and T. S. Chu, "The role of rain in satellite communications," *Proc. IEEE*, vol. 63, pp. 1300-1331, Sept. 1975.

[37] M. Ferguson, "A study of unslotted ALOHA with arbitrary message lengths," presented at *4th Data Commun. Symp.*, Quebec City, Canada, Oct. 1975.

SECTION SIX

Packet Satellite Networks

This section:

- provides comprehensive treatment of the theory of broadcast channel operation where user data is packetized by "intelligent" network interface devices

- shows how adding a structure such as time synchronization to a packet broadcast channel materially improves channel efficiency

- demonstrates how capacity reservation techniques that use the satellite's own transmission channel permits broadcast satellite operation to reach nearly 100% efficiency.

The articles in the previous section provided a general introduction to the technology of satellite communications, showing the evolution of techniques, facilities, and capacity. In recent years, as a result of the technological improvements and competitive activity in satellite communications, the availability of capacity has rapidly increased. With new systems being planned by at least six major primary suppliers, further rapid increases are essentially guaranteed. Early success of the Shuttle Transportation System will further enhance the opportunity to develop and launch bigger, higher-capacity satellites, at reduced unit costs. In addition, several organizations, including the U.S. National Aeronautics and Space Administration (NASA) are aggressively developing still newer technology—the use of frequency bands at 20 and 30 GHZ; adaptive power control to compensate for attenuation effects at these frequencies; satellite on-board switching and processing; high-intensity, localized, downlink spot beams; and low-cost, small-size user earth stations. The success of these ideas, as indicated by the development of a prototype system to be launched in 1988, will provide the pattern for satellites in the 1990s and, compared to today's most advanced systems, will increase the cost-effectiveness of satellites by at least another order of magnitude. Furthermore, these ideas, combined with several "direct-to-home" satellite television broadcast systems, will make direct satellite access particularly attractive to small-user concentrations.

Yet, regardless of its relative cost and availability, the use of satellite facilities for distributed networking will require careful engineering to ensure cost-effectiveness. A fairly large amount of total information will have to be aggregated at concentration points so that the satellite facilities can be heavily loaded at or near capacity during the busiest parts of the day. At the same time, the ability to apply the satellite resources based on specific user demand allows the flexibility to accommodate demands in cases where geographic and occupancy patterns change rapidly.

Fully distributed satellite-based networks will result from combining large space-borne capacity with local distribution using packet broadcasting and local area networks. The articles in this section are devoted to the general concept of packet broadcasting, with principal emphasis on satellite channel operation. The reprints in the next section will show how local area networking—that is, the clustering of many users within a few miles of each other—benefits from the use of packet broadcasting.

Figure VI-1 shows the overall concept of the generalized distributed network. Highly efficient, dynamic techniques will be used to tie many communications users together in a single local area. In some cases local areas will be tied together with conventional communications facilities and, in other cases, with some of the newer techniques such as cellular radio, microwave distribution systems, or fiber optical systems. Over longer distances, satellite communications will be used to tie local distribution systems to each other. By proper design of the interface protocols and operational procedures, end user-to-user communications can take place almost as if the users had direct, individual access to the broadcast channel. Thus, as satellite capacity increases and satellite power requirements decrease, users will be able to change to more direct satellite access. In addition, users in low-concentration areas—that is, areas remote from high-capacity user concentrations, will gain direct and more economical access to all other users in the system.

BROADCASTING FUNDAMENTALS

The term "broadcasting" has traditionally meant the transmission of information from a single transmitter to a multitude of receivers. Radio and television broadcasting, of course, are the most common examples. In broadcast-based communications networks, whether for voice or data communications, a single centralized facility can converse with any of many possible destinations by means of self-identified information "bursts" or messages. Such messages are broadcast over a common transmission medium. Since each message is uniquely identified, each message is received by the proper addressee. Because the signals radiated from a communications satellite can be received equally over a very broad geographic area, such satellites make an ideal medium for highly distributed communications networks via broadcasting.

The inverse process—sending the data from any one of the dispersed users to the central location—can also be achieved by transmission of self-identified messages. Because of the convergence of many locations' traffic upon a single desti-

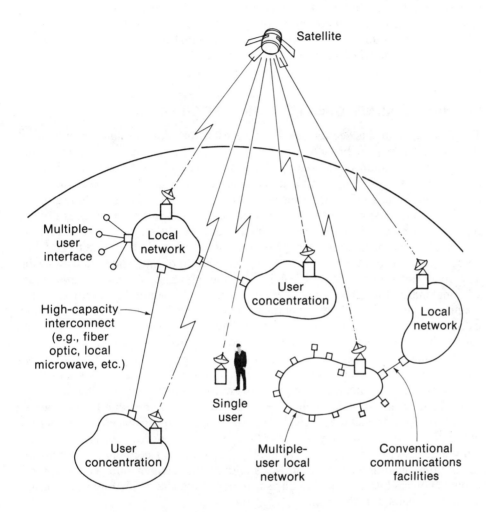

**Figure VI-1   Generic Distributed
Communications Network**

nation, we might refer to this form of transmission as *narrowcasting*. In a more general sense, narrowcasting occurs when any one of many system users transmits information destined to one, or a small subset, of the total user community. When many dispersed users are narrowcasting messages over a shared channel system, the probability of messages being transmitted simultaneously and causing mutual interference is significantly high.

The resolution of the problems of simultaneous transmissions (often referred to as *collisions*) depends upon the fact that users can individually assess the success or failure of their last transmission attempt. The one-quarter-second delay caused by the transmission propagation delay to the synchronous satellites 22,300 miles above the equator makes it easy for individual users to make this assessment, particularly when each packet sent has a time duration of less than one-quarter of a second.

DEFINING CHANNEL OPERATION AND EFFICIENCY

The first paper in this section, by N. Abramson, presents a unified treatment of these types of packet broadcasting (or narrowcasting) channels. The paper describes the operation of the basic ALOHA channel—that is, one in which any user, at any time, can transmit one packet's worth of data into the channel. By listening to the channel, each user is able to assess the success or failure of the previous transmission attempt. Transmission attempts will be successful if users transmit alone during a time interval equal to one packet's length, whereas failure will result if two or more users happen to transmit at nearly the same time. Upon detecting a collision, the user waits a random amount of time and then tries again. The uncoordinated operation of such a channel gives apparently low efficiency, with maximum throughput of only 18.4% of the basic channel rate. However, the channel assigns none of its capacity to inactive users and there is no channel overhead or multiplex equipment—which renders such a channel highly desirable for serving very large numbers of dispersed, low duty-cycle users.

Starting from the concept of the basic ALOHA channel, Abramson's paper shows how such a channel is implemented and describes the refinements to channel operation that materially improve overall channel efficiency. These refinements include time-synchronizing the users into fixed slots within the channel so that users either transmit at precisely the same time and cause a collision over the entire packet length, or they do not collide at all. Although this procedure essentially doubles channel capacity (to about 37% of the basic channel rate), it does require timing equipment at each user station. Other refinements generally applicable to nonsatellite channels, such as listening to the channel before transmitting, are also explained in this paper.

Other aspects of packet broadcast operation discussed in this paper are the effects on network operation that users have when they command a larger than equal share of the total capacity. Another area of interest concerns the spatial properties associated with terrestrial (non-satellite) packet networks. In such networks, users closer to the central stations have a distinct advantage in signal strength compared to users who are farther away.

The paper by L. Kleinrock and S.S. Lam presents the specific results for the slotted, or time-synchronized, channel in detail. This paper is not only the first to present these results, but it provides considerable insight into the mathematical techniques used in analyzing broadcast channels of this general type.

Finally, the paper by L. Roberts introduces and analyzes the ultimate concept in packet broadcast operation—a technique using the channel itself to broadcast very short reservation messages from each user to all other users. Each user maintains a table of all channel reservations and schedules requests for present or future idle time on the channel. Because only a small percentage of overall channel capacity is used by the reservation mechanism, very high channel efficiencies result. In addition, the operation of the reservation mechanism makes it particularly easy to mix the traffic demands of large and small users sharing the capacity of the same broadcast channel.

The three papers in this section provide the theoretical basis of broadcast operation using packetized information. Many of these concepts have direct applicability in non-satellite-based (terrestrial) networks. The major difference between these two operational environments is in the ratio of their propagation delay to packet length, often referred to as "a". For a satellite-based system, "a" is large, typically an integer value in excess of 10. For terrestrial networks, on the other hand, "a" is generally small, 0.1 or less. In functional terms, a small "a" value means that the information heard in the channel at any given time is essentially accurate; i.e., it should be considered in making the decision of whether or not to transmit. If the channel appears to be empty and "a" for all users is small, then it is probably safe to transmit. When "a" is large, however, as it is on a satellite channel, hearing the channel empty at one moment simply means that it was empty one-quarter second earlier; i.e., it contributes no useful information to the decision to transmit.

In the next section are a number of papers that better define the operation of local-area networks and include a number of practical applications finding widespread application in such systems as the Ethernet local-area network. The combination of the techniques of satellite-based broadcast networks and local-area networks that lead to highly integrated, highly distributed telecommunications architectures are discussed in subsequent sections.

The Throughput of Packet Broadcasting Channels

NORMAN ABRAMSON, FELLOW, IEEE

Abstract—Packet broadcasting is a form of data communications architecture which can combine the features of packet switching with those of broadcast channels for data communication networks. Much of the basic theory of packet broadcasting has been presented as a byproduct in a sequence of papers with a distinctly practical emphasis. In this paper we provide a unified presentation of packet broadcasting theory.

In Section II we introduce the theory of packet broadcasting data networks. In Section III we provide some theoretical results dealing with the performance of a packet broadcasting network when the users of the network have a variety of data rates. In Section IV we deal with packet broadcasting networks distributed in space, and in Section V we derive some properties of power-limited packet broadcasting channels, showing that the throughput of such channels can approach that of equivalent point-to-point channels.

I. INTRODUCTION

A. Packet Switching and Packet Broadcasting

THE transition of packet-switched computer networks from experimental [1] to operational [2] status during 1975 provides convincing evidence of the value of this form of communications architecture. Packet switching, or statistical multiplexing [3], can provide a powerful means of sharing communications resources among large number of data communications users when those users can be characterized by a high ratio of peak to average data rates. Under such circumstances, data from each user are buffered, address and control information is added in a "header," and the resulting bit sequence, or "packet," is routed through a shared communications resource by a sequence of node switches [4], [5].

Packet-switched networks, however, still employ point-to-point communication channels and large multiplexing switches for routing and flow control in a fashion similar to conventional circuit switched networks. In some situations [6]–[10] it is desirable to combine the efficiencies achievable by a packet communications architecture with other advantages obtained by use of broadcast communication channels. Among these advantages are elimination of routing and network switches, system modularity, and overall system simplicity. In addition, certain kinds of channels available to the communications systems designer, notably satellite channels, are basically broadcast in their structure. In such cases use of these channels in their natural broadcast mode can lead to significant system performance advantages [11], [12].

B. Outline of Results

Packet broadcasting is a form of data communications architecture which can combine the features of packet switching with those of broadcast channels for data communication networks. Much of the basic theory of packet broadcasting has been presented as a byproduct in a sequence of papers with a distinctly practical emphasis. In this paper we provide a unified presentation of packet broadcasting theory.

In Section II we introduce the theory of packet broadcasting as implemented in the ALOHA System at the University of Hawaii; also in Section II we explain a modification of the basic ALOHA method, called slotting. In Section III we provide some theoretical results dealing with the performance of a packet broadcasting channel when the users of the channel have a variety of data rates. In Section IV we deal with packet broadcasting networks distributed in space, and present some incomplete results on the theoretical properties of such networks. Finally, in Section V we derive some properties of power limited packet broadcasting channels showing that the throughput of such channels can approach that of equivalent point-to-point channels. This result is of importance in satellite systems using small earth stations since it implies that the multiple access capability and the complete connectivity (in the topological sense) of packet broadcasting channels can be obtained at no price in average throughput.

II. PACKET BROADCASTING CHANNELS

A. Operation of a Packet Broadcasting Channel

Consider a number of widely separated users, each wanting to transmit short packets over a common high-speed channel. Assume that the rate at which users generate packets is such that the average time between packets from a single user is much greater than the time needed to transmit a single packet. In Fig. 1 we indicate a sequence of packets transmitted by a typical user.

Conventional time or frequency multiplexing methods (TDMA or FDMA) or some kind of polling scheme could be employed to share the channel among the users. Some of the disadvantages of these methods for users with high peak-to-average data rates are discussed by Carleial and Hellman [13]. In addition, under certain conditions polling may require unacceptable system complexity and extra delay.

In a packet broadcasting system the simplest possible solution to this multiplexing problem is employed. Each user transmits its packets over the common broadcast channel in a completely unsynchronized (from one user to another) manner. If each individual user of a packet broadcasting chan-

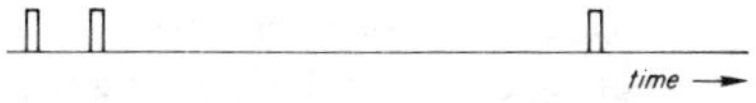

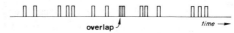

Fig. 1.   Packets from a typical user.

Fig. 2.   Packets from several users on an ALOHA channel.

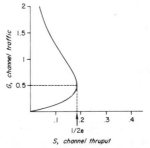

Fig. 3.   Channel throughput versus channel traffic for an ALOHA channel.

nel is required to have a low duty cycle, the probability of a packet from one user interfering with a packet from another user is small as long as the total number of users on the common channel is not too large. As the number of users increases, however, the number of packet overlaps increases and the probability that a packet will be lost due to an overlap also increases. The question of how many users can share such a channel and the analysis of various methods of dealing with packets lost due to overlap are the primary concerns of this paper. In Fig. 2 we show a packet broadcasting channel with two overlapping packets. Since the first packet broadcasting channel was put into operation in the ALOHA System radio-linked computer network at the University of Hawaii [6], they have been referred to as ALOHA channels.

B. ALOHA Capacity

A transmitted packet can be received incorrectly or lost completely because of two different types of errors: 1) random noise errors and 2) errors caused by packet overlap. In this paper we assume that the first type of error can be ignored, and we shall be concerned only with errors caused by packet overlap. In Section II-D we describe several methods of dealing with the problem of packets lost due to overlap, but first we derive the basic results which tell us how many packets can be transmitted with no overlap.

Assume that the start times of packets in the channel comprise a Poisson point process with parameter λ packets/second. If each packet lasts τ seconds, we can define the normalized channel traffic G where

$$G = \lambda\tau. \tag{1}$$

If we assume that only those packets which do not overlap with any other packet are received correctly, we may define $\lambda' < \lambda$ as the rate of occurrence of those packets which are received correctly. Then we define the normalized channel thruput S by

$$S = \lambda'\tau. \tag{2}$$

The probability that a packet will not overlap a given packet is just the probability that no packet starts τ seconds before or τ seconds after the start time of the given packet. Then, since the point process formed from the start times of all packets in the channel was assumed Poisson, the probability that a packet will not overlap any other packet is $e^{-2\lambda\tau}$, or e^{-2G}. Therefore

$$S = Ge^{-2G} \tag{3}$$

and we may plot the channel throughput versus channel traffic for an ALOHA channel (Fig. 3).

From Fig. 3 we see that as the channel traffic increases, the throughput also increases until it reaches its maximum at $S = 1/2e = 0.184$. This value of throughput is known as the capac-

ity of an ALOHA channel, and it occurs for a value of channel traffic equal to 0.5. If we increase the channel traffic above 0.5, the throughput of the channel will decrease.

C. Application of an ALOHA Channel

In order to indicate the capabilities of such a channel for use in an interactive network of alphanumeric computer terminals, consider the 9600 bits/s packet broadcasting channel used in the ALOHA System. From the results of Section II-B we see that the maximum average throughput of this channel is 9600 bits/s times $1/2e$, or about 1600 bits/s. If we assume the conservative [14] figure of 5 bits/s as the average data rate (including overhead) from each active[1] terminal in the network, this channel can handle the traffic of over 300 active terminals and each terminal will operate at a peak data rate of 9600 bits/s. Of course, the total number of terminals in such a network can be much larger than 300 since only a fraction of all terminals will be active and a terminal consumes no channel resources when it is not active.

D. Recovery of Lost Packets

Since the packet broadcasting technique we have described will result in some packets being lost due to packet overlaps, it is necessary to introduce some technique to compensate for this loss. We may list four different packet recovery techniques for dealing with the problem of lost packets. The first three make use of a feedback channel to the packet transmitter and the repetition of lost packets, while the fourth is based on coding.

1) Positive Acknowledgments (POSACKS): Perhaps the most direct way to handle lost packets is to require the receiver of the packet to acknowledge correct receipt of the packet. Each packet is transmitted and then stored in the transmitter's buffer until a POSACK is received from the receiver. If a POSACK is not received in a given amount of time, the transmitter can repeat the transmission and continue to repeat until a POSACK is received or until some other criterion is met. The POSACK can be transmitted on a sepa-

---

[1] A terminal is defined as active from the time a user transmits an attempt to log on until he transmits a log off message.

rate channel (as in the ALOHANET [6]) or transmitted on the same channel as the original packets (as in the ARPA packet radio system [15]). An error detection code and a packet numbering system can be used to increase the reliability of this technique.

2) Transponder Packet Broadcasting: Certain communication channels—notably communication satellite channels—transmit packets on one frequency to a transponder which retransmits the packets on a second frequency. In such cases all units in a packet broadcasting network can receive their own packet retransmissions, determine whether a packet overlap has occurred, and repeat the packet if necessary. This technique has been employed in ATS-1 satellite experiments in the Pacific Educational Computer Network (PACNET) [16] and in the ARPA Atlantic INTELSAT IV packet broadcasting experiments [17].

3) Carrier Sense Packet Broadcasting: For ground-based packet broadcasting networks where the signal propagation time over the furthest transmission path is much less than the packet duration, it is feasible to provide each transmission unit with a device to inhibit packet transmission while another unit is detected transmitting. A carrier sense capability can increase the channel throughput, even if these conditions are not met, when used in conjunction with other packet recovery methods. Carrier sense systems have been analyzed by Tobagi [18] and by Kleinrock and Tobagi [19]. A comprehensive yet compact analysis of such systems is provided in [42].

4) Packet Recovery Codes: When a user employs a packet broadcasting channel to transmit long files by breaking them into large numbers of packets, it is possible to encode the files so that packets lost due to broadcasting overlap can be recovered. It is clear that some of the existing classes of multiple burst error-correcting codes [20] and cyclic product codes [21] can be used for packet recovery in transmissions of long files. It is also clear that these codes are not as efficient as possible for packet recovery and that considerable work remains to be done in this area.

E. Slotted Channels

It is possible to modify the completely unsynchronized use of the ALOHA channel described above in order to increase the maximum throughput of the channel. In the pure ALOHA channel each user simply transmits a packet when ready without any attempt to coordinate his transmission with those of other users. While this strategy has a certain elegance, it does lead to somewhat inefficient channel utilization. If we establish a time base and require each user to start his packet only at certain fixed instants, it is possible to increase the maximum value of the channel thruput. In this kind of channel, called a slotted ALOHA channel, a central clock establishes a time base for a sequence of "slots" of the same duration as a packet transmission [41]. Then when a user has a packet to transmit, he synchronizes the start of his transmission to the start of a slot. In this fashion, if two messages conflict they will overlap completely, rather than partially.

To analyze the slotted ALOHA channel, define G_i as the probability that the ith user will transmit a packet in some slot. Assume that each user operates independently of all other users, and that whether or not a user transmits a packet in a given slot does not depend upon the state of any previous slot. If we have n users, we can define the normalized channel traffic for the slotted channel G where

$$G = \sum_{i=1}^{n} G_i. \tag{4}$$

Note that G may be greater than 1.

As before, we can also consider the rate at which a user sends packets which do not experience an overlap with other user packets. Define $S_i \leq G_i$ as the probability that a user sends a packet and that this packet is the only packet in its slot. If we have n users, then we define the normalized channel throughput for the slotted channel S where

$$S = \sum_{i=1}^{n} S_i. \tag{5}$$

Note that S is less than or equal to 1 and $S \leq G$.

For the slotted ALOHA channel with n independent users, the probability that a packet from the ith user will not experience an interference from one of the other users is

$$\prod_{\substack{j=1 \\ j \neq i}}^{n} (1 - G_j).$$

Therefore we may write the following relationship between the message rate and the traffic rate of the ith user:

$$S_i = G_i \prod_{\substack{j=1 \\ j \neq i}}^{n} (1 - G_j). \tag{6}$$

If all users are identical, we have

$$S_i = \frac{S}{n} \tag{7}$$

and

$$G_i = \frac{G}{n} \tag{8}$$

so that (6) can be written

$$S = G \left(1 - \frac{G}{n}\right)^{n-1} \tag{9}$$

and in the limit as $n \to \infty$, we have

$$S = Ge^{-G}. \tag{10}$$

Equation (10) is plotted in Fig. 4 (curve labeled "slotted ALOHA"). Note that the message rate of the slotted ALOHA channel reaches a maximum value of $1/e = 0.368$, twice the capacity of the pure ALOHA channel.

This result for slotted ALOHA channels was first derived by Roberts [41] using a different method.

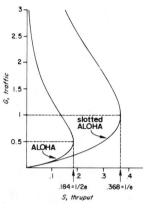

Fig. 4.   Traffic versus throughput for an ALOHA channel and a slotted ALOHA channel.

III. PACKET BROADCASTING WITH MIXED DATA RATES

A. Unslotted Case: Variable Packet Lengths

In Section II we were concerned with the analysis of ALOHA channels carrying a homogeneous mix of packets. If some channel users have a higher average data rate than others, however, the high rate users must either transmit packets more frequently or transmit longer packets. In this section we shall analyze the unslotted ALOHA channel when carrying packets of different lengths, and we shall analyze the slotted ALOHA channel when the probability of transmitting in a given slot varies from user to user.

Let us assume an unslotted ALOHA channel with two different possible packet durations, τ_2 and τ_1. Assume $\tau_2 \geqslant \tau_1$, and therefore we refer to the two different length packets as long packets and short packets, respectively. Assume also the start times of the long packets and short packets form two Poisson point processes with parameters λ_2 and λ_1 packets/second, and that the two Poisson point processes are mutually independent. Then we can define the normalized channel traffic for those packets of duration τ_i:

$$G_i = \lambda_i \tau_i, \qquad i = 1, 2. \tag{11}$$

Again assume that only those packets which do not overlap with any other packet are received correctly and define $\lambda_i' < \lambda_i$ as the rate of occurrence of those packets of duration τ_i which are received correctly. Define the normalized throughput of packets of duration τ_i as

$$S_i = \lambda_i' \tau_i, \qquad i = 1, 2. \tag{12}$$

Since we assumed two independent Poisson point processes, the probability that a short packet will be received correctly is

$$\exp\left[-\lambda_1(2\tau_1) - \lambda_2(\tau_1 + \tau_2)\right] \tag{13}$$

and if we define

$$G_{12} \triangleq \lambda_1 \tau_2 \tag{14a}$$

$$G_{21} \triangleq \lambda_2 \tau_1 \tag{14b}$$

(13) becomes

$$\exp\left[-2G_1 - G_{21} - G_2\right]. \tag{15}$$

Therefore

$$S_1 = G_1 \exp\left[-2G_1 - G_{21} - G_2\right] \tag{16a}$$

and, by a similar argument, the throughput of long packets is

$$S_2 = G_2 \exp\left[-G_{12} - G_1 - 2G_2\right]. \tag{16b}$$

For any given values of λ_1 and λ_2 we may calculate G_1, G_2, G_{12}, and G_{21}; substitution of these values into (16a) and (16b) will allow calculation of the throughputs S_1 and S_2. Therefore (16a) and (16b) may be used to define an allowable set of throughput pairs (S_1, S_2) in the (S_1, S_2) plane.

To determine the boundary of this region we define

$$\alpha \triangleq \frac{\tau_2}{\tau_1}. \tag{17}$$

Note that $\alpha \geqslant 1$. We may rewrite (16a) and (16b) in terms of α, the ratio of long packet duration to short packet duration:

$$S_1 = G_1 \exp\left[-2G_1 - \left(1 + \frac{1}{\alpha}\right)G_2\right] \tag{18a}$$

$$S_2 = G_2 \exp\left[-(1 + \alpha)G_1 - 2G_2\right]. \tag{18b}$$

The boundary of the set of allowable (S_1, S_2) pairs in the (S_1, S_2) plane is defined by setting the Jacobian

$$J = \left| \frac{\partial S_i}{\partial G_j} \right|, \qquad i, j = 1, 2 \tag{19}$$

equal to zero. A simple calculation shows that the Jacobian is zero when

$$G_2 = \frac{1 - 2G_1}{\left(\dfrac{(\alpha - 1)^2}{\alpha} G_1 + 2\right)}. \tag{20}$$

Note that this checks for $G_1 = 0$ and for $\alpha = 1$.

We need only substitute this expression for G_2 into (18a) and (18b) to obtain two equations for S_1, the short packet throughput, and S_2, the long packet throughput, in terms of the single parameter G_1; and as G_1 varies from 0 (all long packets) to 1/2 (all short packets), we will trace out the boundary of the achievable values of throughput in the (S_1, S_2) plane. These achievable throughput regions are indicated for several values of α in Fig. 5.

The basic conclusion of this analysis is that the total channel throughput can undergo a significant decrease if all packets are not of the same length. Thus if the two different

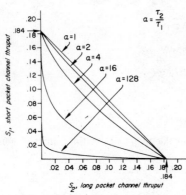

Fig. 5.   Achievable throughput regions in an unslotted ALOHA channel.

packet lengths differ by a large factor, it is often preferable to break up long packets into many shorter packets as long as the overhead necessary to transmit the text in each packet is small. Ferguson [23] has generalized these results to show that channel throughput is maximized over all possible packet length distributions with fixed length packets.

In view of this discouraging result, we might conclude that an inhomogeneous mix of users inevitably leads to a decrease in the maximum value of channel throughput. Surprisingly, this conclusion is not warranted, and we shall show in Section III-B that a mix of users of varied data rates can lead to an increase in the maximum values of channel throughput.

B. Slotted Case: Variable Packet Rates

In the section we shall consider a slotted ALOHA channel used by n users, possibly with different values of channel traffic G_i. From (6) we have a set of n nonlinear equations relating the channel traffics and the channel throughputs for these n users:

$$S_i = G_i \prod_{\substack{j=1 \\ j \neq i}}^{n} (1 - G_j), \qquad i = 1, 2, \cdots, n. \tag{21}$$

Define

$$\alpha = \prod_{j=1}^{n} (1 - G_j); \tag{22}$$

then (21) can be written

$$S_i = \frac{G_i}{1 - G_i} \alpha, \qquad i = 1, 2, \cdots, n. \tag{23}$$

For any set of n acceptable traffic rates $G_1, G_2, \cdots, G_n$, these n equations define a set of channel throughputs $S_1, S_2, \cdots, S_n$ or a region in an n-dimensional space whose coordinates are the S_i. In order to find the boundary of this region, we calculate the Jacobian:

$$J = \left| \frac{\partial S_j}{\partial G_k} \right|, \qquad j, k = 1, 2, \cdots, n. \tag{24}$$

Since

$$\frac{\partial S_j}{\partial G_k} = \begin{cases} \prod_{\substack{i=1 \\ i \neq j}}^{n} (1 - G_i), & j = k \\ -G_j \prod_{\substack{i=1 \\ i \neq j,k}}^{n} (1 - G_i), & j \neq k \end{cases} \tag{25}$$

after some algebra we may write the Jacobian as

$$J = \alpha^{n-2} \begin{vmatrix} (1 - G_1) & -G_1 & -G_1 \\ -G_2 & (1 - G_2) & -G_2 & \cdots \\ -G_3 & -G_3 & (1 - G_3) \\ & \vdots & \end{vmatrix}$$

$$= \alpha^{n-2} [1 - G_1 - G_2 - \cdots - G_n]. \tag{26}$$

Thus the condition for maximum channel throughputs is

$$\sum_i G_i = 1. \tag{27}$$

This condition can then be used to define a boundary to the n-dimensional region of allowable throughputs $S_1, S_2, \cdots, S_n$.

Consider the special case of two classes of users with n_1 users in class 1 and n_2 users in class 2:

$$n_1 + n_2 = n. \tag{28}$$

Let S_1 and G_1 be the throughputs and traffic rates for users in class 1, and let S_2 and G_2 be the throughputs and traffic rates for users in class 2. Then the n equations (21) can be written as the two equations

$$S_1 = G_1 (1 - G_1)^{n_1 - 1} (1 - G_2)^{n_2} \tag{29a}$$

$$S_2 = G_2 (1 - G_2)^{n_2 - 1} (1 - G_1)^{n_1}. \tag{29b}$$

For any pair of acceptable traffic rates G_1 and G_2, these two equations define a pair of channel throughputs S_1 and S_2 or a region in the (S_1, S_2) plane.

From (27) we know that the boundary of this region is defined by the condition

$$n_1 G_1 + n_2 G_2 = 1. \tag{30}$$

We can use (30) to substitute for G_1 in (29a) and (29b) and obtain two equations for S_1 and S_2 in terms of a single parameter G_2. Then as G_2 varies from 0 to 1, the resulting (S_1, S_2) pairs define the boundary of the region we seek. These achievable regions are indicated for various values of n_1 and n_2 in Figs. 6 and 7.

The important point to notice from Figs. 6 and 7 is that in a lightly loaded slotted ALOHA channel, a single large user can transmit data at a significant percentage of the total channel data rate, thus allowing use of the channel at rates well above the limit of $1/e$ or 37 percent obtained when all users have the same message rate. A throughput data rate above the $1/e$ limit

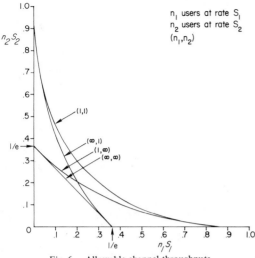

Fig. 6. Allowable channel throughputs.

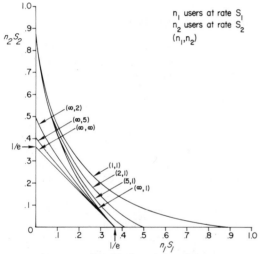

Fig. 7. Allowable channel throughputs.

$$S_1 = G_1{}^2 \tag{31a}$$

$$S_2 = (1 - G_1)^2. \tag{31b}$$

2) For $n_2 \to \infty$:

$$S_1 = G_1(1 - G_1)^{n_1 - 1} \cdot \exp\left[-(1 - n_1 G_1)\right] \tag{32a}$$

$$S_2 = (1 - n_1 G_1)(1 - G_1)^{n_1 - 1} \cdot \exp\left[-(1 - n_1 G_1)\right]. \tag{32b}$$

3) For $n_1 = n_2 \to \infty$:

$$S_1 = \frac{G_1}{e} \tag{33a}$$

$$S_2 = \frac{1 - G_1}{e}. \tag{33b}$$

Additional details dealing with excess capacity and the delay experienced with this kind of use of a slotted ALOHA channel may be found in [11] and [25]. A different view of the use of a slotted packet broadcasting for different sources may be found in [43].

IV. SPATIAL PROPERTIES OF PACKET BROADCASTING NETWORKS

A. Packet Repeaters

In this section we deal with certain spatial properties of packet broadcasting networks. Not long after the initial units of the ALOHA System went into operation, it was realized that the range of the network could be extended beyond the range of a single radio link in the network (about 200 km) by the use of packet repeaters. A packet repeater operates in much the same manner as a conventional radio repeater with one major exception. Since radio transmission in a packet broadcasting network is intermittent, a packet repeater can receive a packet and retransmit that packet in the same frequency band by turning off its receiver during a retransmission burst. Thus a packet repeater can sidestep many of the frequency allocation and spatial cell problems [26] of conventional land-based repeater networks.

The use of packet repeaters leads to the consideration of packet broadcasting networks with more than one central station distributed over very large areas. Users transmit a packet, and if the packet cannot be received directly by its destination, it is forwarded to its destination by one or more packet repeaters according to some routing algorithm [27]. The study of such networks has led to the analysis of two communication theory issues related to the performance of the networks: 1) capture effect and 2) the distribution of packet traffic and packet throughput in space.

B. Capture Effect

Up to this point we have analyzed packet broadcasting channels under the pessimistic assumption that if two packets overlap at the receiver, both packets are lost. In fact, this assumption provides a lower bound to the performance of

has been referred to as "excess capacity" [24]. Excess capacity is important for a lightly loaded packet broadcasting network consisting of many interactive terminal users and a small number of users who send large but infrequent files over the channel. Operation of the channel in a lightly loaded condition, of course, may not be desirable in a bandwidth-limited channel. For a communications satellite where the average power in the satellite transponder limits the channel, however, operation in a lightly loaded packet-switched mode is an attractive alternative. Since the satellite will transmit power only when it is relaying a packet, the duty cycle in the transponder will be small and the average power used will be low (see Section V).

Finally, we note that it is possible to deal with certain limiting cases in more detail, to obtain equations for the boundary of the allowable (S_1, S_2) region.

1) For $n_1 = n_2 = 1$:

Upon using (30) in (29), we obtain

real packet broadcasting channels, since in many receivers the stronger of two overlapping packets may capture the receiver and may be received without error. Metzner [40] has used this fact to derive an interesting result, showing that by dividing users into two groups—one transmitting at high power and the other at low power—the maximum throughput can be increased by about 50 percent. This result is of importance for packet broadcasting networks with a mixture of data and packetized speech traffic.

In order to include the effect of capture in a packet broadcasting network, we consider a distribution of packet generators over a two-dimensional plane and a single packet broadcasting receiver which receives packets from these generators [41]. The receiver then may be viewed as a "packet sink" and the packet generators as a distribution of "packet sources" in the plane. We assume that the rate of generation of packets in a given area depends only on r, the distance from the packet sink, and is independent of direction θ.

Then we may define a traffic density and a throughput density analogous to the normalized traffic G and normalized throughput S defined in Section II-B.

$G(r)$ = normalized packet traffic per unit area at a distance r.

$S(r)$ = normalized packet thruput per unit area at a distance r.

The traffic due to all packet generators in a differential ring of width dr at a radius r is

$$G(r)\, 2\pi r\, dr. \tag{34}$$

We assume that packets from different users are generated so that the packet starting times of all packets generated in the differential ring constitute a Poisson point process. Then since the sum of two independent Poisson processes is a Poisson point process, if users in different rings are independent, the start times of all packets generated in a circle of radius r also constitute a Poisson point process, and the total traffic generated by all users within a distance r of the center is

$$\int_0^r G(x)\, 2\pi x\, dx. \tag{35}$$

If we assume that a packet from a user at a distance r from the center will be received correctly unless it is overlapped by a packet sent from a user at a distance ar or less ($a \geqslant 1$), then using the results of Section II-B the probability that such a packet will be received correctly is

$$\exp\left[-4\pi \int_0^{ar} G(x)x\, dx\right] \tag{36}$$

Any packet generated from a packet source in the circle of radius ar shown in Fig. 8 will interfere with packets generated from a source in the circle of radius r. A packet generated outside the circle of radius ar will not interfere with packets generated from a source in the circle of radius r.

We can relate the normalized packet throughput to the normalized packet traffic in the usual way:

Fig. 8.   Regions of interfering packets.

$$2\pi r S(r)\, dr = 2\pi r G(r) \exp\left[-4\pi \int_0^{ar} G(x)x\, dx\right] dr$$

or

$$S(r) = G(r) \exp\left[-4\pi \int_0^{ar} G(x)x\, dx\right]. \tag{37}$$

If we take a derivative of (37) with respect to r and use (37) to substitute for the exponential, we get

$$S'(r)G(r) = G'(r)S(r) - 4\pi r a^2 S(r)G(r)G(ar). \tag{38}$$

We have not found a general solution of (38) for relating $S(r)$ to $G(r)$ in the presence of capture. We have been able to analyze two special cases, however.

C. Two Solutions

In the first of these special cases we assume a constant traffic density $G(r)$. We can then show that the throughput density $S(r)$ has a Gaussian form, due to the fact that those packets generated further from the receiver will be received correctly less frequently than those packets generated close to the receiver.

In the second special case analyzed we assume a constant packet throughput density $S(r)$ and perfect capture ($a = 1$). Under these assumptions, the packet traffic density will increase as the distance from the receiver increases. We show that there exists a radius r_0 such that the packet traffic density is finite within a circle of radius r_0 around the receiver, while the packet traffic density becomes unbounded on the circle of radius r_0.

For the important case of a packet broadcasting channel distributed over some geographical area and using a packet retransmission policy (Section II-D), this result has an interesting interpretation. In such a situation any packet transmitted from a terminal located within the circle of radius r_0 will be received correctly with probability one (after a finite number of retransmissions), while the expected number of retransmissions required for a packet transmitted from a terminal further from the center than r_0 will be unbounded. Thus there exists a circle of radius r_0 such that terminals transmitting from within this circle can get their packets into the central receiver, while terminals transmitting from outside this circle spend all their time retransmitting their packets in vain. We call r_0 the Sisyphus distance of the ALOHA channel.

1) Constant Packet Traffic Density: Assume the density of normalized packet traffic is constant over the plane

$$G(r) = G_0 \tag{39}$$

and define the distance r_1 as the radius of a circle within which the total packet traffic is unity:

$$\pi r_1{}^2 G_0 \triangleq 1. \tag{40}$$

Then (38) reduces to

$$S'(r) = -\frac{4ra^2}{r_1{}^2} S(r) \tag{41a}$$

with the boundary condition

$$S_0 = G_0 \tag{41b}$$

so that the packet throughput density is

$$S(r) = G_0 \exp\left[-2a^2\left(\frac{r}{r_1}\right)^2\right] \tag{42}$$

and the total normalized packet thruput from a circle of radius r is

$$S = \int_0^r S(r')2\pi r' \, dr'$$

$$= \frac{1}{2a^2}\left\{1 - \exp\left[-2\left(\frac{ar}{r_1}\right)^2\right]\right\} \tag{43}$$

and

$$\lim_{r \to \infty} S = \frac{1}{2a^2}. \tag{44}$$

Note that a total throughput which can be supported by a single packet sink with "perfect capture" ($a = 1$) is equal to one half.

2) Constant Packet Throughput Density: Another case of interest where we have found a solution for (38) is that of constant packet throughput density in the plane. Assume

$$S(r) = S_0 \tag{45}$$

over the region in the plane where $S(r)$ and $G(r)$ are bounded.
Then (38) becomes

$$G'(r) = 4\pi r a^2 G(r)G(ar). \tag{46}$$

For the case of $a = 1$ (perfect capture), (46) becomes

$$G'(r) = 4\pi r G^2(r). \tag{47}$$

with the boundary condition

$$G(0) = S_0 \tag{48}$$

so that

$$G(r) = \frac{S_0}{1 - 2\pi r^2 S_0}. \tag{49}$$

Note that the normalized packet traffic per unit area is finite

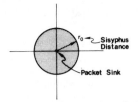

Fig. 9. Region of constant packet throughput S_0 for a single packet sink.

for

$$0 \leqslant r < r_0 \tag{50}$$

where

$$r_0 \triangleq [2\pi S_0]^{-1/2} \tag{51}$$

and r_0 is the Sisyphus distance mentioned in Section IV-C. Note that the Sisyphus distance also has the property that

$$\pi r_0{}^2 S_0 = \frac{1}{2}. \tag{52}$$

As in the previous case, the total packet throughput which can be supported by a single packet sink operating with perfect capture is one half.

V. PACKET BROADCASTING WITH AVERAGE POWER LIMITATIONS

A. Satellite Packet Broadcasting

In previous sections we have analyzed the performance of packet broadcasting channels and compared the performance of these channels to that of conventional point-to-point channels operating at the same peak data rate. Such a comparison is of interest in the case of channels limited by multiple access interference rather than noise, since an increase in the transmitted power of such channels will not lead to improved performance. But just as the average data rate of a packet broadcasting channel can be well below its peak data rate when it is operated at a low duty cycle, the average transmitted power of a packet broadcasting channel can be well below its peak transmitted power.

In this section we analyze the throughput of a packet broadcasting channel when compared to that of a conventional point-to-point channel of the same average power. This analysis is of interest in the case of satellite information systems employing thousands of small earth stations. For a satellite system the fundamental limitation in the downlink is the average power available in the satellite transponder rather than the peak power. Our results show that in the limit of large numbers of small earth stations, the packet throughput approaches 100 percent of the point-to-point capacity. Thus the multiple access capability and the complete connectivity (in the topological sense) of an ALOHA channel can be obtained at no price in average throughput. Furthermore, since our results suggest the use of higher peak power in the satellite transponder (while the average power is kept constant), the small earth stations may use smaller antennas and simpler receivers and modems than would be necessary in a conventional system.

In existing satellite systems the TWT output power in each transponder cannot be varied dynamically. In such systems the advantages implied by our analysis may be realized by frequency-division sharing a single transponder among several voice users and a single channel, operating in an ALOHA mode or some other burst mode, and occupying a frequency band equivalent to one or more voice users. The type of operation implied by our analysis also suggests investigation of high peak power satellite burst transponders (perhaps employing power devices similar to those used in radar systems) for use in information systems composed of large numbers of ultra-small earth stations.

B. Burst Power and Average Power

The capacity of a satellite channel can be calculated by the classical Shannon equation

$$C = W \log \left(1 + \frac{P}{N} \right) \tag{53}$$

where C is the capacity in bits (if the log is a base two logarithm), W is the channel bandwidth, P is the average received signal power at the earth station, and N is the average noise power at the earth station. Equation (53) expresses the capacity of the satellite channel under the assumption that the transponder transmits continuously.

If the channel is used in burst mode the transponder will emit power only when a data burst occurs, and the average power out of the transponder will be less than the burst power. Let D be the ratio of the average power transmitted to the power transmitted during a data burst. For a linear transponder D will equal the channel traffic G, and for a hard-limiting transponder D will equal the duty cycle of the channel. For both the unslotted and slotted ALOHA channel the duty cycle is $1 - e^{-G}$. Thus for a linear transponder[2]

$$D = G, \tag{54a}$$

while for a hard-limiting transponder

$$D = 1 - e^{-G}. \tag{54b}$$

Note that in the case of a hard-limiting transponder with small values of channel traffic, the duty cycle approaches that of a linear transponder.

If we retain P as the notation for the average signal power received at the earth station, the power received during a data burst will be P/D. Thus (53) should be modified in two ways.

[2] Our analysis is of significance only for $G < 1$. The analysis is formally correct, however, for all G, even though the designation of the power transmitted during bursts as "peak power" becomes inappropriate for the linear transponder case when $G > 1$. (In such a situation the "peak power" is less than the average power.)

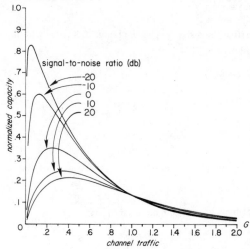

Fig. 10.   Linear transponder; unslotted channel.

1) We replace W by SW to account for the fact that the channel is only used intermittently.

2) We replace P in (53) by P/D to keep the average power of the channel fixed at P.

We should note that when we make these changes, we are assuming that the packet length of the system is long enough so that the asymptotic assumptions which are used to derive (53) still apply. In practice, this is not a problem.

With these two changes then, we have four different cases.

1) Unslotted channel, linear transponder:

$$C_1 = Ge^{-2G} W \log \left(1 + \frac{P}{GN} \right). \tag{55a}$$

2) Unslotted channel, limiting transponder:

$$C_2 = Ge^{-2G} W \log \left(1 + \frac{P}{(1 - e^{-g})N} \right). \tag{55b}$$

3) Slotted channel, linear transponder:

$$C_3 = Ge^{-G} W \log \left(1 + \frac{P}{GN} \right). \tag{55c}$$

4) Slotted channel, limiting transponder:

$$C_4 = Ge^{-G} W \log \left(1 + \frac{P}{(1 - e^{-G})N} \right). \tag{55d}$$

We have calculated the normalized capacities C_i/C for $i = 1, 2, 3, 4$ for different values of P/N, the signal-to-noise ratio of the earth station when the transponder operates continuously. The normalized capacities are plotted in Figs. 10, 11, 12, and 13 for P/N equal to -20, -10, 0, 10, and 20 dB. Of particular interest in these curves is the fact that the highest values of C_i/C occur just where we would want them to occur—for small values of channel traffic (G) and for small earth stations (low P/N). In the limit we have (for a fixed value of G)

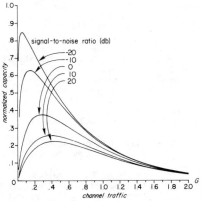

Fig. 11.   Limiting transponder; unslotted channel.

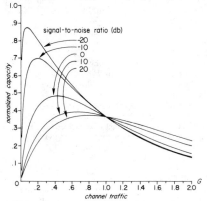

Fig. 12.   Linear transponder; slotted channel.

$$\lim_{\frac{P}{N}\to 0} \frac{C_i}{C} = \frac{S}{D}, \qquad i = 1, 2, 3, 4 \tag{56}$$

so that

1) unslotted channels, linear transponder

$$\lim_{\frac{P}{N}\to 0} \frac{C_1}{C} = e^{-2G} \tag{57a}$$

2) unslotted channels, limiting transponder

$$\lim_{\frac{P}{N}\to 0} \frac{C_2}{C} = \frac{Ge^{-2G}}{(1 - e^{-G})} \tag{57b}$$

3) slotted channel, linear transponder

$$\lim_{\frac{P}{N}\to 0} \frac{C_3}{C} = e^{-G}$$

4) slotted channel, limiting transponder

$$\lim_{\frac{P}{N}\to 0} \frac{C_4}{C} = \frac{Ge^{-G}}{(1 - e^{-G})} \tag{57d}$$

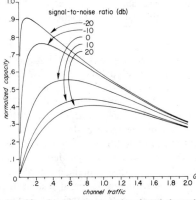

Fig. 13.   Limiting transponder; slotted channel.

and in all cases

$$\lim_{G\to 0} \lim_{\frac{P}{N}\to 0} \frac{C_i}{C} = 1. \tag{58}$$

Thus this multiplexing technique allows a network of small inexpensive earth stations to achieve the maximum value of channel capacity, at the same time providing complete connectivity and multiple access capability.

VI. BACKGROUND AND ACKNOWLEDGMENT

The term "packet broadcasting" was first coined by Robert Metcalfe in his Ph.D. dissertation [28]. As is often the case with simple ideas, the concept of combining burst transmission and Poisson user statistics to provide random access to a channel has occurred independently to a number of investigators. The first attempt at an analysis of such a system of which I am aware is contained in an internal Bell Laboratories memorandum by Schroeder [29], suggested by an earlier paper by Pierce and Hopper [30]. Two other early related papers were written by Costas [31] and Fulton [32]. Of course, a theoretical analysis is not necessary in order to build such a system, and anyone who has sat in a taxi listening to the staccato voice bursts of a radio dispatcher and a set of taxi drivers sharing a single voice channel will recognize the operation of a voice packet broadcasting channel using a carrier sense protocol. And even after an analysis is available, the concept of packet broadcasting may be suggested without reference to the theory [33].

The first papers analyzing packet broadcasting in the form implemented in the ALOHA System [6] assumed fixed packet throughput and a retransmission protocol as described in Section II-D-1). This approach leads to a number of questions involving optimum retransmission policy [28], the behavior of the channel with a finite number of users [39], stability of the channel [13], and transmission of long files by means of various reservation schemes [34], [44]. A comprehensive treatment of these as well as other interesting packet broadcasting questions may be found in Kleinrock [42]. In this paper we have taken a different approach by assuming a given packet traffic rather than throughput. With such a starting point, the questions mentioned above do not assume key importance in the theory, although their practical importance is not diminished.

Much of the theory of packet broadcasting was developed in two working groups sponsored by the Advanced Research Projects Agency of the Department of Defense. These groups circulated a private series of working papers—the ARPANET Satellite System notes (ASS notes) and the Packet Radio Temporary notes (PRT notes)—where many of the theoretical results described or referenced in this paper appeared for the first time. Unfortunately, the several references to ASS notes in papers subsequently published in the open literature may have produced some confusion in the minds of those trying to trace the references. Among the most significant of the ASS note and PRT note results was the first derivation of the capacity of a slotted ALOHA channel and the first analysis of the use of the capture effect in packet broadcasting, both by Larry Roberts. That note has since been republished in the open literature [41].

The results of Section III-A dealing with two different packet lengths were suggested by an ASS note written by Tom Gaarder, and the results of Section III-B dealing with the excess capacity of a slotted channel were suggested by an ASS note written by Randy Rettberg. Other problems which were first analyzed in ASS notes or PRT notes but not emphasized in this paper include various packet broadcasting reservation systems [22], [35], [36], carrier sense packet broadcasting [18], [19], and questions dealing with packet routing and protocol issues in a network of repeaters [37]. The reader interested in theoretical network protocol questions should also see Gallagher [38], although this work did not originate in an ASS note or PRT note.

The first system to employ packet broadcasting techniques was the ALOHA System computer network at the University of Hawaii in 1970. Subsequently, packet repeaters were added to the network and packet broadcasting by satellite was demonstrated in the system. Some of the people involved in the implementation and development of the system were Richard Binder, Chris Harrison, Alan Okinaka, and David Wax.

The historical relevance of [29] and [32] was pointed out to me by Joe Aein, to whom I am indebted, in spite of my embarassment at having forgotten I was thesis supervisor on the second of these papers.

REFERENCES

[1] L. G. Roberts and B. D. Wessler, "Computer network development to achieve resource sharing," in *1970 Spring Joint Comput. Conf., AFIPS Conf. Proc.*, vol. 36. Montvale, NJ: AFIPS Press, 1970, pp. 543–549.

[2] L. G. Roberts, "Data by the packet," *IEEE Spectrum*, vol. 11, pp. 46–51, Feb. 1974.

[3] W. W. Chu, "A study of asynchronous time division multiplexing for time-sharing computer systems," in *1969 Spring Joint Comput. Conf., AFIPS Conf. Proc.*, vol. 35. Montvale, NJ: AFIPS Press, 1969, pp. 669–678.

[4] F. Heart, R. Kahn, S. Ornstein, W. Crowther, and D. Walden, "The interface message processor for the ARPA computer network," in *1970 Spring Joint Comput. Conf., AFIPS Conf. Proc.*, vol. 36. Montvale, NJ: AFIPS Press, 1970, pp. 551–567.

[5] H. Frank, I. Frisch, and W. Chou, "Topological considerations in the design of the ARPA computer network," in *1970 Spring Joint Comput. Conf., AFIPS Conf. Proc.*, vol. 36. Montvale, NJ: AFIPS Press, 1970, pp. 581–587.

[6] N. Abramson, "The ALOHA system—Another alternative for computer communications," in *1970 Fall Joint Comput. Conf., AFIPS Conf. Proc.*, vol. 37. Montvale, NJ: AFIPS Press, 1970, pp. 281–285.

[7] R. E. Kahn, "The organization of computer resources into a packet radio network," in *Nat. Comput. Conf., AFIPS Conf. Proc.*, vol. 44, May 1975, pp. 177–186.

[8] N. Abramson and E. R. Cacciamani, Jr., "Satellites: Not just a big cable in the sky," *IEEE Spectrum*, vol. 12, pp. 36–40, Sept. 1975.

[9] I. T. Frisch, "Technical problems in nationwide networking and interconnection," *IEEE Trans. Commun.*, vol. COM-23, pp. 78–88, Jan. 1975.

[10] R. M. Metcalfe and D. R. Boggs, "ETHERNET: Distributed packet switching for local computer networks," *Commun. Ass. Comput. Mach.*, to be published.

[11] N. Abramson, "Packet switching with satellites," in *Nat. Comput. Conf., AFIPS Conf. Proc.*, vol. 42, 1973, pp. 695–702.

[12] R. D. Rosner, "Optimization of the number of ground stations in a domestic satellite system," in *EASCON'75 Rec.*, Sept. 29–Oct. 1, 1975, pp. 64A–64J.

[13] A. B. Carleial and M. E. Hellman, "Bistable behavior of ALOHA-type systems," *IEEE Trans. Commun.*, vol. COM-23, pp. 401–410, Apr. 1975.

[14] P. E. Jackson and C. D. Stubbs, "A study of multiaccess computer communications," in *1969 Spring Joint Comput. Conf., AFIPS Conf. Proc.*, vol. 34. Montvale, NJ: AFIPS Press, 1969, pp. 491–504.

[15] T. J. Klein, "A tactical packet radio system," in *Proc. Nat. Telecommun. Conf.*, New Orleans, LA, Dec. 1975.

[16] K. Ah Mai, "Organizational alternatives for a Pacific educational computer-communication network," ALOHA Syst. Tech. Rep. CN74-27, Univ. Hawaii, Honolulu, May 1974.

[17] R. Binder, R. Rettberg, and D. Walden, *The Atlantic Satellite Packet Broadcast and Gateway Experiments*. Cambridge, MA: Bolt Beranek and Newman, 1975.

[18] L. Kleinrock and F. A. Tobagi, "Packet switching in radio channels: Part I—Carrier sense multiple-access modes and their throughput-delay characteristics," *IEEE Trans. Commun.*, vol. COM-23, pp. 1400–1416, Dec. 1975.

[19] F. A. Tobagi and L. Kleinrock, "Packet switching in radio channels: Part II—The hidden terminal problem in carrier sense multiple-access and the busy-tone solution," *IEEE Trans. Commun.*, vol. COM-23, pp. 1417–1433, Dec. 1975.

[20] R. T. Chien, L. R. Bahl, and D. T. Tang, "Correction of two erasure bursts," *IEEE Trans. Inform. Theory*, vol. IT-15, pp. 186–187, Jan. 1969.

[21] N. Abramson, "Cyclic code groups," *Problems of Inform. Transmission*, Acad. Sci. USSR, Moscow, vol. 6, no. 2, 1970.

[22] L. G. Roberts, "Dynamic allocation of satellite capacity through packet reservation," in *Nat. Comput. Conf. AFIPS Conf. Proc.*, vol. 42, June 1973, pp. 711–716.

[23] M. J. Ferguson, "A study of unslotted ALOHA with arbitrary message lengths," in *Proc. 4th Data Commun. Symp.*, Quebec, Canada, Oct. 7–9, 1975, pp. 5-20–5-25.

[24] R. Rettberg, "Random ALOHA with slots-excess capacity," ARPANET Satellite Syst. Note 18, NIC Document 11865, Stanford Res. Inst., Menlo Park, CA, Oct. 11, 1972.

[25] N. Abramson, "Excess capacity of a slotted ALOHA channel (continued)," ARPANET Satellite Syst. Note 30, NIC Document 13044, Stanford Res. Inst., Menlo Park, CA, Dec. 6, 1972.

[26] L. Schiff, "Random-access digital communication for mobile radio in a cellular environment," *IEEE Trans. Commun.*, vol. COM-22, pp. 688–692, May 1974.

[27] H. Frank, R. M. Van Slyke, and I. Gitman, "Packet radio network design—System considerations," in *Nat. Comput. Conf., AFIPS Conf. Proc.*, vol. 44, May 1975, pp. 217–232.

[28] R. M. Metcalfe, "Packet communication," Rep. MAC TR-114, Project MAC, Massachussetts Inst. Technol., Cambridge, July 1973.

[29] M. R. Schroeder, "Nonsynchronous time multiplex system for speech transmission," Bell Lab. Memo., Jan. 19, 1959.

[30] J. R. Pierce and A. L. Hopper, "Nonsynchronous time division with holding and random sampling," *Proc. IRE*, vol. 40, pp. 1079–1088, Sept. 1952.

[31] J. P. Costas, "Poisson, Shannon, and the radio amateur," *Proc. IRE*, vol. 47, p. 2058, Dec. 1959.

[32] F. F. Fulton, Jr., "Channel utilization by intermittent transmitters," Tech. Rep. 2004-2, Stanford Electron. Lab., Stanford Univ., Stanford, CA, May 12, 1961.

[33] K. D. Levin, "The overlapping problem and performance degradation of mobile digital communication systems," *IEEE Trans. Commun.*, vol. COM-23, pp. 1342–1347, Nov. 1975.

[34] R. Binder, "A dynamic packet-switching system for satellite broadcast channels," in *Conf. Rec., Int. Conf. Commun.*, vol. III, June 1975, pp. 41-1-41-5.

[35] S. S. Lam and L. Kleinrock, "Packet switching in a multiaccess broadcast channel: Dynamic control procedures," *IEEE Trans. Commun.*, vol. COM-23, pp. 891–904, Sept. 1975.

[36] L. Kleinrock and S. S. Lam, "Packet switching in a multiaccess broadcast channel: Performance evaluation," *IEEE Trans. Commun.*, vol. COM-23, pp. 410–423, Apr. 1975.

[37] I. Gitman, "On the capacity of slotted ALOHA networks and some design problems," *IEEE Trans. Commun.*, vol. COM-23, pp. 305–317, Mar. 1975.

[38] R. G. Gallager, "Basic limits on protocol information in data communication networks," *IEEE Trans. Inform. Theory*, vol. IT-22, pp. 385–398, July 1976.

[44] W. Crowther, R. Rettberg, D. Walden, S. Ornstein, and F. Heart, "A system for broadcast communication: Reservation ALOHA," in *Proc. 6th Hawaii Int. Conf. Syst. Sci.*, Western Periodicals, Jan. 1973, pp. 371–374.

[39] M. J. Ferguson, "On the control, stability, and waiting time in a slotted ALOHA random-access system," *IEEE Trans. Commun.*, vol. COM-23, pp. 1306–1311, Nov. 1975.

[40] J. J. Metzner, "On improving utilization in ALOHA networks," *IEEE Trans. Commun.*, vol. COM-24, pp. 447–448, Apr. 1976.

[41] L. G. Roberts, "ALOHA packet system with and without slots and capture," *Comput. Commun. Rev.*, vol. 5, pp. 28–42, Apr. 1975.

[42] L. Kleinrock, *Queueing Systems, Volume 2: Computer Applications.* New York: Wiley, 1976, pp. 360–407.

[43] I. Gitman, R. M. Van Slyke, and H. Frank, "On splitting random accessed broadcast communication channels," in *Proc. 7th Hawaii Int. Conf. Syst. Sci.–Suppl. on Comput. Nets*, Western Periodicals, Jan. 1974, pp. 81–85.

Packet-switching in a slotted satellite channel*

by LEONARD KLEINROCK and SIMON S. LAM

University of California
Los Angeles, California

INTRODUCTION

Imagine that two users require the use of a communication channel. The classical approach to satisfying this requirement is to provide a channel for their use so long as that need continues (and to charge them for the full cost of this channel). It has long been recognized that such allocation of scarce communication resources is extremely wasteful as witnessed by their low utilization (see for example the measurements of Jackson & Stubbs).[1] Rather than provide channels on a user-pair basis, we much prefer to provide a single high-speed channel to a large number of users which can be shared in some fashion; this then allows us to take advantage of the powerful "large number laws" which state that with very high probability, the demand at any instant will be approximately equal to the sum of the average demands of that population. In this way the required channel capacity to support the user traffic may be considerably less than in the unshared case of dedicated channels. This approach has been used to great effect for many years now in a number of different contexts: for example, the use of graded channels in the telephone industry,[2] the introduction of asynchronous time division multiplexing,[3] and the packet-switching concepts introduced by Baran et al.,[4] Davies,[5] and finally implemented in the ARPA network.[6] The essential observation is that the full-time allocation of a fraction of the channel to each user is highly inefficient compared to the part-time use of the full capacity of the channel (this is precisely the notion of time-sharing). We gain this efficient sharing when the traffic consists of rapid, but short bursts of data. The classical schemes of synchronous time division multiplexing and frequency division multiplexing are examples of the inefficient partitioning of channels.

As soon as we introduce the notion of a shared channel in a packet-switching mode then we must be prepared to resolve conflicts which arise when more than one demand is simultaneously placed upon the channel. There are two obvious solutions to this problem: the first is to "throw out" or "lose" any demands which are made while the channel is in use; and the second is to form a queue of conflicting demands and serve them in some order as the channel becomes free. The latter approach is that taken in the ARPA network since storage may be provided economically at the point of conflict. The former approach is taken in the ALOHA system[7] which uses packet-switching with radio channels; in this system, in fact, *all* simultaneous demands made on the channel are lost.

Of interest to this paper is the consideration of satellite channels for packet-switching. The definition of a packet is merely a package of data which has been prepared by a user for transmission to some other user in the system. The satellite is characterized as a high capacity channel with a fixed propagation delay which is large compared to the packet transmission time (see the next section). The (stationary) satellite acts as a pure transponder repeating whatever it receives and beaming this transmission back down to earth; this broadcasted transmission can be heard by every user of the system and in particular a user can listen to his own transmission on its way back down. Since the satellite is merely transponding, then whenever a portion of one user's transmission reaches the satellite while another user's transmission is being transponded, the two collide and "destroy" each other. The problem we are then faced with is how to control the allocation of time at the satellite in a fashion which produces an acceptable level of performance.

The ideal situation would be for the users to agree collectively when each could transmit. The difficulty is that the means for communication available to these geographically distributed users is the satellite channel itself and we are faced with attempting to control a channel which must carry its own control information. There are essentially three approaches to the solution of this problem. The first has come to be known as a pure "ALOHA" system[7] in which users transmit any time they desire. If, after one propagation delay, they hear their successful transmission then they assume that no conflict occurred at the satellite; otherwise they know a collision occurred and they must retransmit. If users retransmit immediately upon hearing a conflict, then they are likely to conflict again, and so some scheme must be devised for introducing a random retransmission delay to spread these conflicting packets over time.

The second method for using the satellite channel is to "slot" time into segments whose duration is exactly equal to the transmission time of a single packet (we assume constant length packets). If we now require all packets to begin their transmission only at the beginning of a slot, then we enjoy a

* This research was supported by the Advanced Research Projects Agency of the Department of Defense under Contract No. DAHC-15-69-C-0285.

gain in efficiency since collisions are now restricted to a single slot duration; such a scheme is referred to as a "slotted ALOHA" system and is the principal subject of this paper. We consider two models: the first is that of a large population of users, each of which makes a small demand on the channel; the second model consists of this background of users with the addition of one larger user acting in a special way to provide an increased utilization of the channel. We concern ourselves with retransmission strategies, delays, and throughput. Abramson[8] also considers slotted systems and is concerned mainly with the ultimate capacity of these channels with various user mixes. Our results and his have a common meeting point at some limits which will be described below.

The third method for using these channels is to attempt to schedule their use in some direct fashion; this introduces the notion of a *reservation system* in which time slots are reserved for specific users' transmissions and the manner in which these reservations are made is discussed in the paper by Roberts.[9] He gives an analysis for the delay and throughput, comparing the performance of slotted and reservation systems.

Thus we are faced with a finite-capacity communication channel subject to unpredictable and conflicting demands. When these demands collide, we "lose" some of the effective capacity of the channel and in this paper we characterize the effect of that conflict. Note that it is possible to use the channel up to its full rated capacity when only a single user is demanding service; this is true since a user will never conflict with himself (he has the capability to schedule his own use). This effect is important in studying the non-uniform traffic case as we show below.

SLOTTED ALOHA CHANNEL MODELS

Model I. Traffic from many small users

In this model we assume:

(A1) an infinite number of users* who collectively form an independent source

This source generates M packets per slot from the distribution $v_i = \text{Prob}[M = i]$ with a mean of S_0 packets/slot.

We assume that each packet is of constant length requiring T seconds for transmission; in the numerical studies presented below we assume that the capacity of the channel is 50 kilobits per second and that the packets are each 1125 bits in length yielding $T = 22.5$ msec. Note that $S'_0 = S_0/T$ is the average number of packets arriving per second from the source. Let d be the maximum roundtrip propagation delay which we assume each user experiences and let $R = d/T$ be the number of slots which can fit into one roundtrip propagation time; for our numerical results we assume $d = 270$ msec. and so $R = 12$ slots. R slots after a transmission, a user will either hear that it was successful or know that it was de-

stroyed. In the latter case if he now retransmits during the next slot interval and if all other users behave likewise, then for sure they will collide again; consequently we shall assume that each user transmits a previously collided packet at random during one of the next K slots, (each such slot being chosen with probability $1/K$). Thus, retransmission will take place either $R + 1, R + 2, \ldots$ or $R + K$ slots after the initial transmission. As a result traffic introduced to the channel from our collection of users will now consist of new packets and previously blocked packets, the total number adding up to N packets transmitted per slot where $P_i = \text{Prob}[N = i]$ with a mean traffic of G packets perr slot. We assume that each user in the infinite population will have at most one packet requiring transmission at any time (including any previously blocked packets). Of interest to us is a description of the maximum throughput* rate S as a function of the channel traffic G. It is clear that S/G is merely the probability of a successful transmission and G/S is the average number of times a packet must be transmitted until success; assuming

(A2) the traffic entering the channel is an independent process

We then have,

$$S = Gp_0 \tag{1}$$

If in addition we assume,

(A3) the channel traffic is Poisson

then $p_0 = e^{-G}$, and so,

$$S = Ge^{-G} \tag{2}$$

Eq. (2) was first obtained by Roberts[11] who extended a similar result due to Abramson[7] in studying the radio ALOHA system. It represents the ultimate throughput in a Model I slotted ALOHA channel without regard to the delay packets experience; we deal extensively with the delay in the next section.

For Model I we adopt assumption A1. We shall also accept a less restrictive form of assumption A2 (namely assumption A4 below) which, as we show, lends validity to assumption A3 which we also require in this model. Assume,

(A4) the channel traffic is independent over any K consecutive slots

We have conducted simulation experiments which show that this is an excellent assumption so long as $K < R$.
Let,

$$P(z) = \sum_{i=0}^{\infty} p_i z^i \tag{3}$$

$$V(z) = \sum_{i=0}^{\infty} v_i z^i \tag{4}$$

---

* These will be referred to as the "small" users.

* Note that $S = S_0$ under stable system operation which we assume unless stated otherwise (see below).

Using only assumption A4 and the assumption that M is independent of $N - M$, we find [10] that $P(z)$ may be expressed as

$$\left[\frac{p_1}{K}(1 - z) + P\left(1 - \frac{1 - z}{K}\right)\right]^K V(z)$$

If, further, the source is an independent process (i.e., assumption A1) and is Poisson distributed then $V(z) = e^{-S(1-z)}$, and then we see immediately that,

$$\lim_{K \uparrow \infty} P(z) = e^{-G(1-z)}.$$

This shows that assumption A3 follows from assumptions A1 and A4 in the limit of large K, under the reasonable condition that the source is Poisson distributed.

We have so far defined the following critical system parameters: S_0, S, G, K and R. In the ensuing analysis we shall distinguish packets transmitting in a given slot as being either newly generated or ones which have in the past collided with other packets. This leads to an approximation since we do not distinguish how many times a packet has met with a collision. We have examined the validity of this approximation by simulation, and have found that the correlation of traffic in different slots is negligible, except at shifts of $R + 1$, $R + 2, \ldots, R + K$; this exactly supports our approximation since we concern ourselves with the most recent collision. We require the following two additional definitions:

q = Prob[newly generated packet is successfully transmitted]

q_t = Prob[previously blocked packet is successfully transmitted]

We also introduce the expected packet delay D:

D = average time (in slots) until a packet is successfully received

Our principal concern in this paper is to investigate the trade-off between the average delay D and the throughput S.

Model II. Background traffic with one large user

In this second model, we refer to the source described above as the "background" source but we also assume that there is an additional single user who constitutes a second independent source and we refer to this source as the "large" user. The background source is the same as that in Model I and for the second source, we assume that the packet arrivals to the large user transmitter are Poisson and independent of other packets over $R + K$ consecutive slots. In order to distinguish variables for these two sources, we let S_1 and G_1 refer to the S and G parameters for the background source and let S_2 and G_2 refer to the S and G parameters for the single large user. We point out that the identity of this large user may change as time progresses but insist that there be only one such at any given time. We introduce the new variables

$$S = S_1 + S_2 \qquad (5)$$

$$G = G_1 + G_2 \qquad (6)$$

S represents the total throughput of the system and G represents the traffic which the channel must support (including retransmissions). We have assumed that the small users may have at most one packet outstanding for transmission in the channel; however the single large user may have many packets awaiting transmission. We assume that this large user has storage for queueing his requests and of course it is his responsibility to see that he does not attempt the simultaneous transmission of two packets. We may interpret G_2 as the probability that the single large user is transmitting a packet in a channel slot and so we require $G_2 \leq 1$; no such restriction is placed on G_1 (or on G in Model I).

We now introduce a means by which the large user can control his channel usage enabling him to absorb some of the slack channel capacity; this permits an increase in the total throughput S. The set of packets awaiting transmission by the large user compete among each other for the attention of his local transmitter as follows. Each waiting packet will be scheduled for transmission in some future slot. When a newly generated packet arrives, it immediately attempts transmission in the current slot and will succeed in capturing the transmitter unless some other packet has also been scheduled for this slot; in the case of such a scheduling conflict, the new packet is randomly rescheduled in one of the next L slots, each such slot being chosen equally likely with probability $1/L$. Due to the background traffic, a large user packet may meet with a transmission conflict at the satellite (which is discovered R slots after transmission) in which case, as in Model I, it incurs a random delay (uniformly distributed over K slots) plus the fixed delay of R slots. More than one packet may be scheduled for a future slot and we assume that these scheduling conflicts are resolved by admitting that packet with the longest delay since its previous blocking (due to conflict in transmission or conflict in scheduling) and uniformly rescheduling the others over the next L slots; ties are broken by random selection. We see, therefore, that new packets have the lowest priority in case of a scheduling conflict; however, they seize the channel if it is free upon their arrival. The variable L permits us a certain control of channel use by the large user but does not limit his throughput. We also assume, K, $L < R$. Corresponding to q and q_t in Model I, we introduce the success probabilities q_i and q_{it} ($i = 1, 2$) for new and previously blocked packets respectively and where $i = 1$ denotes the background source and $i = 2$ denotes the single large source. Finally, we choose to distinguish between D_1 and D_2 which are the average number of slots until a packet is successfully transmitted from the background and large user sources respectively.

RESULTS OF ANALYSIS

In this section we present the results of our analysis without proof. The details of proof may be found in Reference 10.

Model I. Traffic from many small users

We wish to refine Eq. (2) by accounting for the effect of the random retransmission delay parameter K. Our principal result in this case is

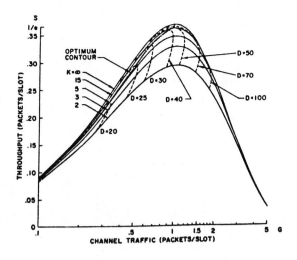

Figure 1—Throughput as a function of channel traffic

traffic G for various values of K. We note that the maximum throughput at a given K occurs when $G = 1$. The throughput improves as K increases, finally yielding a maximum value of $S = 1/e = .368$ for $G = 1$, $K = $ infinity. Thus we have the unfortunate situation that the ultimate capacity of this channel supporting a large number of small users is less than 37 percent of its theoretical maximum (of 1). We note that the efficiency rapidly approaches this limiting value (of $1/e$) as K increases and that for $K = 15$ we are almost there. The figure also shows some delay contours which we discuss below. In Figure 2, we show the variation of q and q_t with K for various values of G. We note how rapidly these functions approach their limiting values as given in Eq. (11). Also on this curve, we have shown Roberts' approximation in Eq. (10) which converges to the exact value very rapidly as K increases and also as G decreases.

Our next significant result is for packet delay as given by

$$D = R + 1 + \frac{1 - q}{q_t}\left[R + 1 + \frac{K - 1}{2}\right] \quad (12)$$

We note from this equation that for large K, the average delay grows linearly with K at a slope

$$\lim_{K \uparrow \infty} \frac{\partial D}{\partial K} = \frac{1 - e^{-G}}{2e^{-G}}$$

Using Eq. (11), we see that this slope may be expressed as $G - S/2S$ which is merely the ratio of that portion of transmitted traffic which meets with a conflict to twice the throughput of the channel; since $G - S/2S = \frac{1}{2}(G/S - 1)$, we see that the limiting slope is equal to $\frac{1}{2}$ times the average number of times a packet is retransmitted. Little's well-known result[12] expresses the average number ($\bar{n}$) of units (packets in our case) in queueing system as the product of the average arrival rate ($S_0 = S$ in our case) and the average time in system (D). If we use this along with Eqs. (7) and (12), we get

$$\bar{n} = SD = G\left[R + 1 + \frac{K - 1}{2}\right] - S\left[\frac{K - 1}{2}\right] \quad (13)$$

In Figure 1 we plot the loci of constant delay in the S, G plane. Note the way these loci bend over sharply as K in-

$$S = G\frac{q_t}{q_t + 1 - q} \quad (7)$$

where

$$q = \left[e^{-G/K}\frac{G}{K}e^{-G}\right]^K e^{-S} \quad (8)$$

and

$$q_t = \left[\frac{1}{1 - e^{-G}}\right][e^{-G/K} - e^{-G}]\left[e^{-G/K} + \frac{G}{K}e^{-G}\right]^{K-1}e^{-S} \quad (9)$$

The considerations which led to Eq. (7) were inspired by Roberts[11] in which he developed an approximation for Eq. (9) of the form

$$q_t \cong \frac{K - 1}{K}e^{-G} \quad (10)$$

We shall see below that this is a reasonably good approximation. Equations (7–9) form a set of non-linear simultaneous equations for S, q and q_t which must be solved to obtain an explicit expression for S in terms of the system parameters G and K. In general, this cannot be accomplished. However, we note that as K approaches infinity these three equations reduce simply to

$$\lim_{K \uparrow \infty} \frac{S}{G} = \lim_{K \uparrow \infty} q = \lim_{K \uparrow \infty} q_t = e^{-G} \quad (11)$$

Thus, we see that Eq. (2) is the correct expression for the throughput S only when K approaches infinity which corresponds to the case of infinite average delay; Abramson[8] gives this result and numerous others all of which correspond to this limiting case. Note that the large K case avoids the large delay problem if T is small (very high speed channels).

The numerical solution to Eqs. (7–9) is given in Figure 1 where we plot the throughput S as a function of the channel

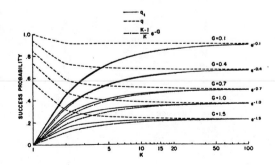

Figure 2—Success probabilities as a function of retransmission delay

creases defining a maximum throughput $S_{\max}(D)$ for any given value of D; we note the cost in throughput if we wish to limit the average delay. This effect is clearly seen in Figure 3 which is the fundamental display of the tradeoff between delay and throughput for Model I; this figure shows the delay-throughput contours for constant values of K. We also give the minimum envelope of these contours which defines the optimum performance curve for this system (a similar optimum curve is also shown in Figure 1). Note how sharply the delay increases near the maximum throughput $S = 0.368$; it is clear that an extreme price in delay must be paid if one wishes to push the channel throughput much above 0.360 and the incremental gain in throughput here is infinitesimal. On the other hand, as S approaches zero, D approaches $R + 1$. Also shown here are the constant G contours. Thus this figure and Figure 1 are two alternate ways of displaying the relationship among the four critical system quantities S, G, K, and D.

From Figure 3 we observe the following effect. Consider any given value of S (say at $S = 0.20$), and some given value of K (say $K = 2$). We note that there are two possible values of D which satisfy these conditions ($D = 21.8$, $D = 161$). How do we explain this?* It is clear that the lower value is a stable operating point since the system has sufficient capacity to absorb any fluctuation in the rate S_0. Suppose that we now

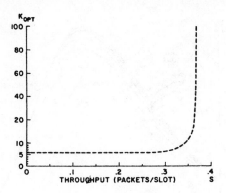

Figure 4—Optimum K

slowly increase S_0 (the source rate); so long as we do not exceed the maximum value of the system throughput rate for this K (say, $S_{\max}(K)$), then we see that $S = S_0$ and the system will follow the input. Note that $S_{\max}(K)$ always occurs at the intersection of the $G = 1$ curve as noted earlier. However, if we attempt to set $S_0 > S_{\max}(K)$, then the system will go unstable! In fact, the throughput S will drop from $S_{\max}(K)$ toward zero as the system accelerates up the constant K contour toward infinite delay! The system will remain in that unfortunate circumstance so long as $S_0 > S$ (where now S is approaching zero). All during its demise, the rate at which new packets are being trapped by the system is $S_0 - S$. To recover from this situation, one can set $S_0 = 0$; then the delay will proceed down the K contour, round the bend at $S_{\max}(K)$ and race down to $S = 0$. All this while, the backlogged packets are being flushed out of the system. The warning is clear: one must avoid the knee of the K contour. Fortunately, the optimum performance curve does avoid the knee everywhere except when one attempts to squeeze out the last few percent of throughput. In Figure 4, we show the optimum values of K as a function of S. Thus, we have characterized the tradeoff between throughput and delay for Model I.

Model II. Background traffic with one large user

In this model the throughput equation is similar to that given in Eq. (7), namely,

$$S_i = G_i \frac{q_{it}}{q_{it} + 1 - q_i} \quad i = 1, 2 \tag{14}$$

the quantities q_{it} and q_i are given in the appendix. Similarly the average delays for the two classes of user are given by

$$D_1 = R + 1 + \frac{1 - q_1}{q_{1t}} \left[R + 1 + \frac{K - 1}{2} \right] \tag{15}$$

$$D_2 = R + 1 + \frac{1 - q_2}{q_{2t}} \left[R + 1 + \frac{K - 1}{2} \right]$$

$$+ \frac{L + 1}{2} \left[E_n + \frac{1 - q_2}{q_{2t}} E_t \right] \tag{16}$$

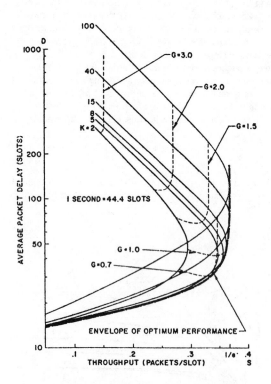

Figure 3—Delay-throughput tradeoff

---

* This question was raised in a private conversation with Martin Graham (University of California, Berkeley). A simulation of this situation is reported upon in Reference 13.

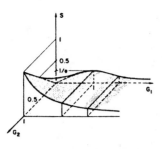

Figure 5—Throughput surface

where E_n and E_t are given in the appendix. It is easy to show that as K, L approach infinity,

$$q_1 = q_{1t} = e^{-G_1}(1 - G_2) \qquad (17)$$

$$S_1 = G_1 e^{-G_1}(1 - G_2) \qquad (18)$$

$$q_2 = q_{2t} = e^{-G_1} \qquad (19)$$

$$S_2 = G_2 e^{-G_1} \qquad (20)$$

$$S = (G - G_1 G_2)e^{-G_1} \qquad (21)$$

where we recall $G = G_1 + G_2$ and $S = S_1 + S_2$. From these last equations or as given by direct arguments in an unpublished note by Roberts, one may easily show that at a constant background user throughout S_1, the large user throughput S_2 will be maximized when

$$G = G_1 + G_2 = 1 \qquad (22)$$

This last is a special case of results obtained by Abramson in Reference 8 and he discusses these limiting cases at length for various mixes of users. We note that,

$$\frac{\partial S}{\partial G_2} = e^{-G_1}(1 - G_1) \qquad (23)$$

$$\frac{\partial S}{\partial G_1} = -e^{-G_1}(G - G_1 G_2 - 1 + G_2) \qquad (24)$$

In Figure 5 we give a qualitative diagram of the 3-dimensional contour for S as a function of G_1 and G_2. We remind the reader that this function is shown for the limiting case K, L approaching infinity only. From our results we see that for constant $G_1 < 1$, S increases linearly with G_2 ($G_2 < 1$). For constant $G_1 > 1$, S decreases linearly as G_2 increases. In addition, for constant $G_2 < \frac{1}{2}$, S has a maximum value at $G_1 = 1 - 2G_2/1 - G_2$. Furthermore, for constant $G_2 > \frac{1}{2}$, S decreases as G_1 increases and therefore the maximum throughput S must occur at $S = G_2$ in the $G_1 = 0$ plane.

The optimum curve given in Eq. (22) is shown in the S_1, S_2 plane in Figure 6 along with the performance loci at constant G_1. We note in these last two figures that a channel throughput equal to 1 is achievable whenever the background traffic drops to zero thereby enabling $S = S_2 = G_2 = 1$; this corresponds to the case of a single user utilizing the satellite channel at its maximum throughput of 1. Abramson [8] discusses a variety of curves such as those in Figure 6; he considers the generalization where there may be an arbitrary number of background and large users.

In the next three figures, we give numerical results for the finite K case; in all of these computations, we consider only the simplified situation in which $K = L$ thereby eliminating one parameter. In Figure 7 we show the tradeoff between delay and throughput similar to Figure 3. (Note that Figure 5 is similar to Figure 1.) Here we show the optimum performance of the average delay $D = S_1 D_1 + S_2 D_2/S$ along with the behavior of D at constant values of K and $S_1 = 0.1$ (note the instability once again for overloaded conditions). Also shown are minimum curves for D_1 and D_2, which are obtained by using the optimum K as a function of S. If we are willing to reduce the background throughput from its maximum at $S_1 = 0.368$, then we can drive the total throughput up to approximately $S = 0.52$ by introducing additional traffic from the large user. Note that the minimum D_1 curve is much higher than the minimum D_2 curve. Thus our net gain in channel throughput is also at the expense of longer packet delays for the small users. Once again, we see the sharp rise near saturation.

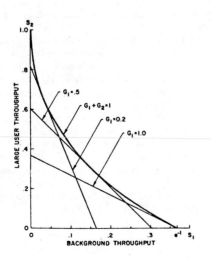

Figure 6—Throughput tradeoff

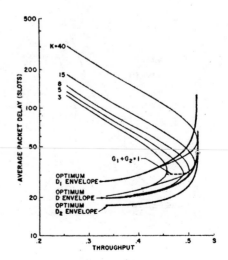

Figure 7—Delay-throughput tradeoff at $S_1 = 0.1$

In Figure 8, we display a family of optimum D curves for various choices of S_1 as a function of the total throughput S. We also show the behavior of Model I as given in Figure 3. Note as we reduce the background traffic, the system capacity increases slowly; however, when S_1 falls below 0.1, we begin to pick up significant gains for S_2. Also observe that each of the constant curves "peels off" from the Model I curve at a value of $S = S_1$. At $S_1 = 0$, we have only the large user operating with no collisions and at this point, the optimal value of L is 1. This reduces to the classical queueing system with Poisson input and constant service time (denoted $M/D/1$) and represents the *absolute optimum performance* contour for any method of using the satellite channel when the input is Poisson; for other input distributions we may use the $G/D/1$ queueing results to calculate this absolute optimum performance contour.

In Figure 9, we finally show the throughput tradeoffs between the background and large users. The upper curve shows the absolute maximum S at each value of S_1; this is a clear display of the significant gain in S_2 which we can achieve if we are willing to reduce the background throughput. The middle curve (also shown in Figure 6 and in Reference 8) shows the absolute maximum value for S_2 at each value of S_1. The lowest curve shows the net gain in system capacity as S_1 is reduced from its maximum possible value of $1/e$.

CONCLUSIONS

In this paper we have analyzed the performance of a slotted satellite system for packet-switching. In our first model, we have displayed the trade-off between average delay and average throughput and have shown that in the case of traffic consisting of a large number of small users, the limiting throughput of the channel $(1/e)$ can be approached fairly closely without an excessive delay. This performance can be

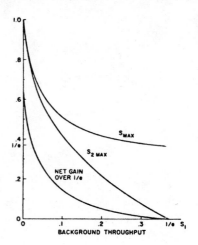

Figure 9—Throughput countours

achieved at relatively small values of K which is the random retransmission delay parameter. However, if one attempts to approach this limiting capacity, not only does one encounter large delays, but one also flirts with the hazards of unstable behavior.

In the case of a single large user mixed with the background traffic, we have shown that it is possible to increase the throughput rather significantly. The qualitative behavior for this multidimensional trade-off was shown and the numerical calculations for a given set of parameters were also displayed. The optimum mix of channel traffic was given in Eq. (22) and is commented on at length in Abramson's paper.[8] We have been able to show in this paper the relationship between delay and throughput which is an essential trade-off in these slotted packet-switching systems.

In Roberts' paper[9] he discusses an effective way to reserve slots in a satellite system so as to predict and prevent conflicts. It is worthwhile noting that another scheme is currently being investigated for packet-switching systems in which the propagation delay is small compared to the slot time, that is, $R = d/T \ll 1$. In such systems it may be advantageous for a user to "listen before transmitting" in order to determine if the channel is in use by some other user; such systems are referred to as "carrier sense" systems and seem to offer some interesting possibilities regarding their control. For satellite communications this case may be found when the capacity of the channel is rather small (for example, with a stationary satellite, the capacity should be in the range of 1200 bps for the packet sizes we have discussed in this paper). On the other hand, a 50 kilobit channel operating in a ground radio environment with packets on the order of 100 or 1000 bits lend themselves nicely to carrier sense techniques.

In all of these schemes one must trade off complexity of implementation with suitable performance. This performance must be effective at all ranges of traffic intensity in that no unnecessary delays or loss of throughput should occur due to complicated operational procedures. We feel that the slotted satellite packet-switching methods described in

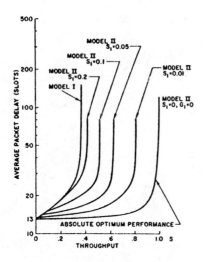

Figure 8—Optimum delay-throughput tradeoffs

this paper and the reservation systems for these channels described in the paper by Roberts do in fact meet these criteria.

REFERENCES

1. Jackson, P. E., Stubbs, C. D., "A Study of Multi-access Computer Communications," Spring Joint Computer Conf., *AFIPS Conf. Proc.*, Vol. 34, 1969, pp. 491–504.
2. Syski, R., *Introduction to Congestion in Telephone Systems*, Oliver & Boyd, Edinburgh and London, 1960.
3. Chu, W. W., "A Study of Asynchronous Time Division Multiplexing for Time-Sharing Computer Systems," Spring Joint Computer Conf., *AFIPS Conf. Proc.*, Vol. 35, 1969, pp. 669–678.
4. Baran, P., Boehm, S., and Smith, P., "On Distributed Communications," series of 11 reports by Rand Corp., Santa Monica, Calif., 1964.
5. Davies, D. W., "The Principles of a Data Communication Network for Computers and Remote Peripherals," *Proc. IFIP Hardware*, Edinburgh, 1968, D11.
6. Roberts, L. G., "Multiple Computer Networks and Inter-Computer Communications," *ACM Symposium on Operating Systems*, Gatlinburg, Tenn., 1967.
7. Abramson, N., "The ALOHA System—Another Alternative for Computer Communications," Fall Joint Computer Conf., *AFIPS Conf. Proc.*, Vol. 37, 1970, pp. 281–285.
8. Abramson, N., "Packet Switching with Satellites," these proceedings.
9. Roberts, L. G., "Dynamic Allocation of Satellite Capacity through Packet Reservation," these proceedings.
10. Kleinrock, L., Lam, S. S., Arpanet Satellite System Notes 12 (NIC Document #11294); 17 (NIC Document #11862); 25 (NIC Document #12734); and 27 (NIC Document #12756), available from the ARPA Network Information Center, Stanford Research Institute, Menlo Park, California.
11. Roberts, L., Arpanet Satellite System Notes 8 (NIC Document #11290) and 9 (NIC Document #11291), available from the ARPA Network Information Center, Stanford Research Institute, Menlo Park, California.
12. Kleinrock, L., *Queueing Systems: Theory and Applications*, to be published by Wiley Interscience, New York, 1973.
13. Rettberg, R., Arpanet Satellite System Note 11 (NIC Document #11293), available from the ARPA Network Information Center, Stanford Research Institute, Menlo Park, California.

APPENDIX

Define $G_s \triangleq$ Poisson arrival rate of packets to the transmitter of the large user

$$= S_2[1 + E_n + E_2(1 + E_t)] \quad (A.1)$$

The variables q_i, q_{it} ($i = 1, 2$) in Eqs. (14–16) are then given as follows (see Reference 10 for details of the derivations):

$$q_1 = (q_0)^K (q_h)^L e^{-s} \quad (A.2)$$

$$q_{1t} = (q_0)^{K-1} q_{1c} (q_h)^L e^{-s} \quad (A.3)$$

where

$$q_0 = e^{-G_1/K} + \frac{1}{K}[(1 - e^{-G_s})(e^{-G_1} - e^{-G_1/K})$$

$$+ G_1 e^{-(G_1 + G_s)}] \quad (A.4)$$

$$q_h = \begin{cases} (G_s + 1)e^{-G_s} & L = 1 \\ \dfrac{1}{L - 1}(Le^{-G_s/L} - e^{-G_s}) & L \geq 2 \end{cases} \quad (A.5)$$

$$q_{1c} = \frac{1}{1 - e^{-(G_1 + G_s)}}\left[e^{-G_1/K}\left(1 - \frac{1 - e^{-G_s}}{K}\right) - e^{-(G_1 + G_s)}\right] \quad (A.6)$$

Let us introduce the following notation for events at the large user:

SS = scheduling success (capture of the transmitter)
SC = scheduling conflict (failure to capture transmitter)
TS = transmission success (capture of a satellite slot)
TC = transmission conflict (conflict at the satellite)
NP = newly generated packet

Then,

$$q_2 = \frac{r_n + r_s E_n}{1 + E_n} \quad (A.7)$$

$$q_{2t} = \frac{r_t + r_s E_t}{1 + E_t} \quad (A.8)$$

where

$E_n \triangleq$ average number of SC events before an SS event conditioning on NP $= \dfrac{1 - a_n}{a_s}$ (A.9)

$E_t \triangleq$ average number of SC events before an SS event conditioning on TC $= \dfrac{1 - a_t}{a_s}$ (A.10)

The variables a_i, r_i ($i = n, t, s$) are defined and given below:

$$a_n \triangleq \text{Prob}[SS/NP] = \left(\frac{q_0}{q}\right)^K (q_h)^L \left(\frac{1 - e^{-S_2}}{S_2}\right) \quad (A.11)$$

$$r_n \triangleq \text{Prob}[TS/SS, NP] = q^K e^{-S_1} \quad (A.12)$$

$$a_t \triangleq \text{Prob}[SS/TC] = \frac{1}{K}\frac{1 - (q_0/q)^K}{1 - q_0/q} \quad (A.13)$$

$$r_t \triangleq \text{Prob}[TS/SS, TC] = q^{K-1} q_{2c} e^{-S_1} \quad (A.14)$$

$$a_s \triangleq \text{Prob}[SS/SC] = \left(\frac{q_0}{q}\right)^K \frac{q_{sc}}{L}\frac{1 - (q_h)^L}{1 - q_h} \quad (A.15)$$

$$r_s \triangleq \text{Prob}[TS/SS, SC] = q^K e^{-S_1} \quad (A.16)$$

where

$$q = e^{-G_1/K} + \frac{G_1}{K} e^{-(G_1 + G_s)} \quad (A.17)$$

$$q_{2c} = \frac{e^{-G_1/K} - e^{-G_1}}{1 - e^{-G_1}} \quad (A.18)$$

$$q_{sc} = \frac{1}{G_s - 1 + e^{-G_s}}\left(\frac{L}{L - 1}\right)^2 \left[G_s\left(1 - \frac{1}{L}\right)e^{-G_s/L}\right.$$

$$\left. - e^{-G_s/L} + e^{-G_s}\right] \quad (A.19)$$

Dynamic allocation of satellite capacity through packet reservation

by LAWRENCE G. ROBERTS

Department of Defense
Arlington, Virginia

INTRODUCTION

If one projects the growth of computer communication networks like the ARPANET[1,2,3,4] to a worldwide situation, satellite communication is attractive for intercommunicating between the widespread geographic areas. For this variable demand, multi-station, data traffic situation, satellites are uniquely qualified in that they are theoretically capable of statistically averaging the load in total at the satellite rather than requiring each station or station-pair to average the traffic independently. However, very little research has been done on techniques which permit direct multi-station demand access to a satellite for data traffic. For voice traffic statistics, COMSAT Laboratories has developed highly efficient techniques; the SPADE[5] system currently installed in the Atlantic permitting the pooled use of 64KB PCM voice channels on a demand basis, and the MAT-1[6] TDMA (Time Division Multiple-Access) experimental system. Both systems permit flexible demand assignment of the satellite capacity, but on a circuit-switched basis designed to interconnect a full duplex 64KB channel between two stations for minutes rather than deliver small blocks of data here and there. This work forms the technical base for advanced digital satellite communication, and provides a very effective means for moving large quantities of data between two points. However, for short interactive data traffic between many stations, new allocation techniques are desirable.

TRAFFIC MODEL

In order to evaluate the performance of any new technique for dynamic assignment of satellite capacity and compare it with other techniques, a complete model of the data traffic must be postulated. Given the model, each technique can be analyzed and its performance computed for any traffic load or distribution. Although it is difficult to fully represent the complete variation in traffic rates normal in data traffic, the following model describes the basic nature of data traffic which might arrive at each satellite station from a local packet network.

There are Poisson arrivals of both single packets (1270 bits including the header) and multi-packet blocks (8 packets) at

each station. The overall Poisson arrival rate for both is L with a fraction F of single packets and the remainder multi-packets. For simplicity, the arrival rates at all stations are stationary and equal. This is not completely representative of normal data traffic but for the assignment techniques of interest, non-stationary and unequal arrival rates will produce nearly identical performance to the stationary case. Techniques which subdivide the satellite capacity in a preassigned manner would be seriously hurt by non-stationary traffic rates but the poor performance of these systems will be demonstrated, at least in part, by their inability to handle Poisson packet arrivals effectively. The average station traffic in packets per second is:

$$T = L(F + 8(1 - F)) \qquad (1)$$

The destination of this traffic is equally divided between all of the other stations.

For a truly reliable data communications network, each packet or block should be acknowledged as having been correctly received. Positive error control using acknowledgments and retransmissions is very important for data traffic. Thus, acknowledgment traffic must be added to the station traffic. To achieve rapid recovery from errors there must be one small packet (144 bits) sent for each packet or block sent. This traffic is administrative overhead and will not be counted when computing the channel utilization.

The analytic results presented later in the paper are all for equal arrival rates for single packets and multi-packets (F = .5). Other values of F have been examined as well as cases where the input traffic contains small (144 bit) data packets as well. The detailed effect of these variations is not sufficiently pronounced to consider here, however. For comparing techniques the equal arrival distribution is quite representative.

ARPANET experience indicates that the data traffic one can expect is proportional to the total dollar value of computer services being bought or sold through the network. The total traffic generated by one dollar of computer activity is about 315 packets, half going each way.[3] Thus, $200K/year of computer activity within a region produces 2KB of traffic, of which IKB is leaving the region. Within the next few years it is probable that the computer services exchanged internationally will be between $50K/country and $2M/country which suggests that the traffic levels, T, to consider are from .25KB to 10KB. For domestic satellite usage the dollar flow would be far greater than this if the regions are ones like the east and west coast. However, if small stations become economically attractive, the individual user complexes or computer sites will have traffic levels within this range. Therefore, several of the analytic results presented are for a station traffic of T = 1KB. This corresponds to one packet or multipacket arriving every 4.5 seconds, on the average. It is extremely important to note the infrequency of this, considering that the block must be delivered within less than a second. Even at 10KB, with arrivals every .45 seconds, each arrival must be treated independently, not waiting for a queue to build up if rapid response is to be maintained. Only after the individual traffic exceeds 50KB is there significant smoothing and uniformity to the station's traffic flow. Thus, it is quite important to devise techniques which do not depend on this smoothing at each station if stations with under $10M of remote computer activity are to be served economically.

CHANNELIZED SATELLITE TRANSMISSION TECHNIQUES

FDM—*FULL interconnection*

The most common technique in use today is for each pair of stations which have traffic to lease a small full duplex data channel directly. If this technique were used for a large net of N stations, it would require N (N − 1) half duplex channels, each large enough to provide the desired delay response. The total satellite bandwidth required is the sum of the N(N − 1) individual requirements plus 2KHz* per channel (minimum) for guardbands. However, since the channels are dedicated, variable packet sizes can be handled and the small acknowledgments fit in efficiently.

FDM—*Store and forward star*

Since it is clearly very costly for full interconnection, store and forward is an obvious alternative. With short, leased ground lines, the ARPANET very effectively uses this technique, but since each hop adds at least .27 sec due to the propagation delay, it is important to minimize the number of hops. Thus a star design is probably as good an example of this technique as any. The total number of channels for a star is N − 1. The delay is the two hop total plus any switch delay (herein presumed zero and of infinite capacity).

TDMA

Since all stations could theoretically hear all the transmissions, a store and forward process is really unnecessary if each packet has an address and its destination can receive it. Further, the guardbands required for FDM can be eliminated if Time Division Multiple Access techniques are used. Instead, an 80 bit start up synchronization leader is required. This increases the small acknowledgment packets to 225 bits and the normal packets to 1350 bits, a 7.6 percent overhead. For this type of data traffic a strict alternation of time slot ownership between the stations was evaluated. All slots are the same size, 1350 bits, except for small acknowledgment packets which are packed in at the necessary intervals. Thus, each station has one Nth of the channel capacity and can use it freely to send to any station. Each station must examine all packets for those addressed to itself. To adapt to unequal or non-stationary traffic levels, there are many techniques[6] for slowly varying the channel split.

ALOHA

Instead of preassigning time slots to stations and often having them be unused, in the ALOHA system they are all freely utilized by any station with traffic. When there are many stations this reduces the delay caused in waiting for your own slot, but introduces a channel utilization limit of 36 percent to insure that conflicts are not too frequent. When conflicts do occur the sum check clearly indicates it and both stations retransmit. A very complete treatment of this technique is presented in the papers by Abramson[7] and Kleinrock and Lam.[8] For the comparison curves presented here, an approximation to the precise delay calculation was used and the possibilities of improved performance due to excess capacity were ignored. Thus, the ALOHA results are slightly conservative.

RESERVATION SYSTEM

In order to further improve the efficiency of data traffic distribution via satellite, the following reservation system is proposed. As with TDMA and ALOHA the satellite channel is divided into time slots of 1350 bits each. However, after every M slots one slot is subdivided into V small slots. The small slots are for reservations and acknowledgments, to be used on a contention basis with the ALOHA technique. The remaining M large slots are for RESERVED data packets. When a data packet or multi-packet block arrives at a station it transmits a reservation in a randomly selected one of the V small slots in the next ALOHA group. The reservation is a request for from one to eight RESERVED slots. Upon seeing such a reservation each station adds the number of slots requested to a count, J, the number of slots currently reserved. The originating station has now blocked out a sequence of RESERVED slots to transmit his packets in. Thus, there is one common queue for all stations and by broadcasting reservations they can claim space on the queue. It is not necessary for any station but the originating station to remember which space belongs to whom, since the only requirement is that no one else uses the slots.

Referring to Figure 1, a reservation for three slots is transmitted at t = 0 so as to fall in an ALOHA slot at t = 5. If a

---

* Two KHz is the minimal possible channel separation determined by oscillator stability for current INTELSAT IV equipment based on a private communication with *E.* Cacciamani, COMSAT Laboratories. Actual guardbands in use are wider.

conflict occurs, the originating station will determine the sum check is bad at t = 10 and retransmit the reservation. However, if it is received correctly at t = 10 and assuming the current queue length is thirteen, the station computes that it can use the slots at t = 21, 22 and 24. It does this by transmitting at t = 16, 17 and 19. By t = 30 the entire block of three packets has been delivered to their destination. If no other reservations have been received by t = 19 the queue goes to zero at this point and the channel reverts to a pure ALOHA state until the next valid reservation is received.

Reservations

To maintain coordination between all the stations, it is necessary and sufficient that each reservation which is received correctly by any station is received correctly by all the stations. This can be assured even if the channel error rate is high by properly encoding the reservation. The simplest strategy is to use the standard packet sum check hardware, and send three independently sumchecked copies of the reservation data. A reservation requires 24 bits of information and with the sum check is 48 bits. Three of these together with the 80 bit sync sequence made a 224 bit packet. Given this size for the small slot and 1350 bits for the large slot, we can pack six reservations in the large slot space; therefore, V = 6. If the channel error rate is 10^{-5} and there are 1000 stations, the probability that one or more of the stations will have errors in all three sections is approximately 1000 (48 × $10^{-5})^3$ or 10^{-7}. With a 1.5MB channel this is one error every three days, a very tolerable rate considering the only impact is to delay some data momentarily. If the reservation were not triplicated, however, the probability of an error is .48, sufficient to totally confuse all the stations.

Channel states

There are two states, ALOHA and RESERVED. On start up and every time thereafter when the reservation queue goes to zero, the channel is in the ALOHA state. In this state, all slots small and the ALOHA mode of transmission is used. Reservations, acknowledgments and even small data packets can be sent using the 224 bit slots. However, the first successful reservation causes the RESERVED state to begin. Let us define Z to be the channel rate in large slots per second and R to be the number of large slots per round trip (R = .27Z). Then, considering time as viewed from the satellite, the data packets associated with the first reservation should be transmitted so as to start R + 1 large slots after the reservation. To avoid confusion, M is kept constant for the entirety of each RESERVED state but it is allowed to change each time the state is entered. The initial reservation which starts the state contains a suggested new value for M. This value is used until the state terminates. The determination of M will be considered later.

Channel utilization

The traffic of small packets (reservations, acknowledgments) is twice the overall arrival rate (NL) since every data block requires a reservation and an acknowledgment. If we assume that the arrival rate of these small packets is independent of the state (a good approximation since they are fully independent at both low and high traffic levels where the average duration of one of the states is short compared to R), then:

Small Slot Channel Utilization in ALOHA State:
$$S_1 = 2NL/ZV \quad (2)$$
Small Slot Channel Utilization in RESERVED State:
$$S_2 = 2NL(M + 1)/ZV \quad (3)$$

The channel utilization for large slots must be computed as if the channel were always in the RESERVED state since the ALOHA state is a *result* of the non-utilization of the reserved slots, not the cause. Thus:

Large Slot Channel Utilization: $S_3 = BNL(M + 1)/MZ$

Where, average block size:
$$B = F + 8(1 - F)$$
$$B = 4.5 \quad (4)$$

For the ALOHA transmissions, the channel utilization is related to the actual transmission rate (G) by the relation (see references 7 and 8):

ALOHA State: $\quad S_1 = AG_1 e^{-G1}$
RESERVED State: $\quad S_2 = AG_2 e^{-G2}$

These relations must be solved for G by iteration since S is the known quantity. The correction constant, A, depends on the retransmission randomization technique and R, but is always between .8 and 1.0. As a result of these relations the maximum useful ALOHA throughput is S = A/e. An empirically derived approximation* to A used for this analysis was (K = retransmission randomization period in slots):

$$A = \frac{K - 1}{K} \text{ where } K = 2.3 \sqrt{R}$$

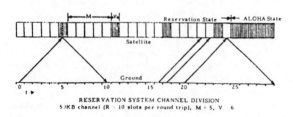

RESERVATION SYSTEM CHANNEL DIVISION
50KB channel (R - 10 slots per round trip), M - 5, V 6

Figure 1

---

* For an accurate and more detailed solution to the effect of a fixed retransmission delay, refer to Reference 8.

The average fraction of time the system is in the RE-SERVED state is equal to the large slot channel utilization, S_3, since that is the fraction of time the reserved packets are being sent. Thus, if we compute the delay for the small packets in both states a weighted average can be taken, using S_3, to obtain the average delay.

ALOHA State:

Initial queueing delay: $\quad W_1 = \dfrac{G_1/N}{2V(1 - G_1/N)}$

Retransmissions: $\quad H_1 = \dfrac{1 - Ae^{-G_1}}{Ae^{-G_1}}$

Small Packet Delay: $\quad D_1 = \dfrac{R + 1.5/V + W_1 + H_1}{Z}$

$$\times \; \frac{(R + W_1 + 1/V + K/2V)}{Z}$$

RESERVED State:

Initial queueing delay: $\quad W_2 = \dfrac{(M + 1)G_2/N}{2V(1 - G_2/N)}$

Retransmissions: $\quad H_2 = \dfrac{1 - Ae^{-G_2}}{Ae^{-G_2}}$

Small Packet Delay: $\quad D_2 = \dfrac{R + 1.5V + M/2 + W_2}{Z}$

$$\frac{+ \; H_2(R + W_2 + 1/V + \dfrac{K(M + 1)}{2V})}{Z}$$

Now, the overall average small packet delay can be determined:

Overall Small
Packet Delay: $\quad D^2 = D_1(1 - S_3) + D_2 S_3 \qquad (5)$

Large packet and block delay

For the reserved packets, the delay has three components; the reservation delay (D_s), the central queueing delay and the transmission-propagation delay of the packet or block. For a block of B packets where the general load is the defined traffic distribution the delay is:

Average Delay
for reserved: $\quad D_r = \dfrac{ZD_s + R + B\dfrac{(M + 1)}{M} + \dfrac{YS_3(M + 1)}{2M(1 - S_3)}}{Z}$

$$(6)$$

Where: $\quad Y = 7.2$ packets (second moment of block size/ avg. block size)

and: $\quad B = 4.5$ packets (average block size)

An optimal value for M can now be determined numerically for any given channel and traffic load. However, this value is not very critical at low channel loading factors. It is only when the channel is operating near peak capacity that M affects the delay more than a few percent. Since M cannot be changed rapidly it is desirable to set M to the value which optimizes the channel capacity and thereby minimizes the delay at peak load. For peak capacity, both the small and large slot portions of the channel in the RESERVED state should be fully loaded. This occurs when $S_2 = A/e$ and $S_3 = 1$. Doing this and solving equations (3) and (4) for the arrival rate, L, gives us:

$$L = \frac{ZVA}{2eN(M + 1)} = \frac{ZM}{BN(M + 1)}$$

Solving for M:

$$M = \frac{AV}{2e}B \; \text{rounded up to nearest integer}$$

for $B = 4.5$, $V = 6$: $\quad M = 5$

If this peak capacity value for M is always used the delay is within 10 percent of optimal and the system is quite stable. As can be seen, the only traffic parameter M depends on is B, the average block size. M can be adjusted by the stations if the channel is monitored and the fractions of each type of packet sent are measured. From these fractions it is easy to determine M.

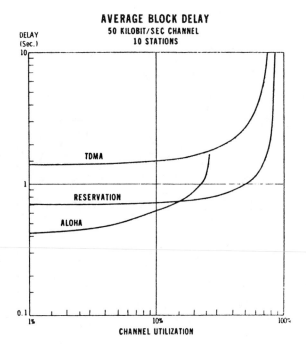

AVERAGE BLOCK DELAY
50 KILOBIT/SEC CHANNEL
10 STATIONS

Figure 2

Now it is possible to determine the delay given the traffic distribution (F,B), number of stations (N), and input arrival rate (L). One common way to examine performance is by plotting delay versus the channel utilization for a fixed channel. The channel utilization, C, is the ratio of the good data delivered to the new channel speed:

Channel Utilization: $C = NLB/Z$

Figure 2 shows the delay vs. C for the TDMA, ALOHA, and Reservation techniques. The traffic distribution is as previously defined; half single packets and half blocks of eight.

This type of presentation is not the best for deciding what technique to use for a specific job, but it does show the general behavior of the systems for a fixed channel size, as the traffic load varies.

In order to really compare the cost of the various techniques to do a certain job, it is necessary to set the traffic level, number of stations, and the delay permissible. Then, for each technique, the channel size required to achieve the delay constraint can be searched for. To make the presentation more meaningful the cost of this channel per megabit of traffic can then be determined using as a price basis the current tariffed price of the 50KB INTELSAT IV channel (45 KHz) used in the ARPANET between California and Hawaii.

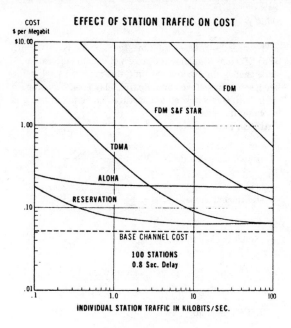

Figure 4

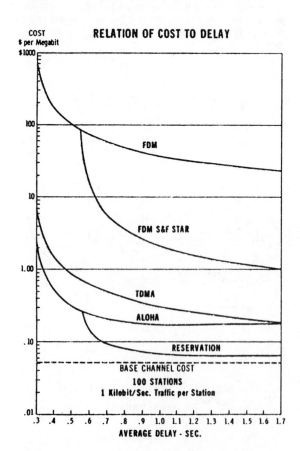

Figure 3

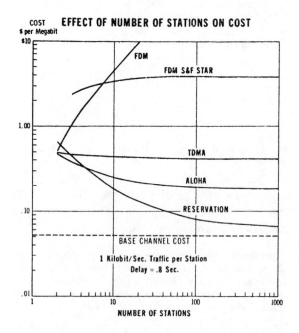

Figure 5

It is presumed that any bandwidth could be purchased for the same price per KHz. Converting the cost to dollars per megabit permits easy comparison with the cost in the current ARPANET where distributed leased line capacity can be achieved for $.10 per megabit.

Figures 3, 4 and 5 show communications cost as a function of the three variables; delay, traffic and number of stations. Examining Figure 3 it is clear that if a delay of less than two round trips (.54 sec) is required, the ALOHA system is superior. However, the cost for .4 sec service is over 6 times that of .8 sec service (using the reservation technique). It is also clear that delays of more than .8 sec are not necessary and save very little money. Figure 4 shows that as the individual station traffic is increased to 50KB or higher, TDMA becomes almost as good as the reservation system since there is sufficient local averaging of traffic. Similarly, at this same traffic level FDM—Store and Forward achieves its maximum efficiency but due to sending each packet twice its asymptotic cost is twice that of TDMA or Reservation. These traffic levels for each station are unrealistically high, however, and the flat performance of ALOHA and the Reservation System is vastly preferable since the cost of data communications to small stations is the same as for large stations. Finally, Figure 5 shows the effect of adding stations to the net. With FDM the cost grows out of bonds quickly whereas the reservation technique improves its efficiency until the total traffic from all stations exceeds 100KB. Below 5KB total traffic ALOHA is superior, but this is not a very important case. For large numbers of stations at 1KB traffic per station and .8 seconds delay the reservation system is 3 times cheaper than ALOHA, 6 times cheaper than TDMA, and 56 times cheaper than FDM Store and Forward.

CONCLUSIONS

The reservation technique presented here is one of several techniques which have been developed recently to take full advantage of the multi-access capabilities of satellites for data traffic.[9,10] Both the ALOHA technique and the reservation system depend for their efficiency on the total multi-station traffic rather than the individual station traffic as does TDMA and FDM Store and Forward. The performance improvement reflects this with the reservation system being up to 10 times as efficient as TDMA for small station traffic levels. The worst possible technique for data traffic is pure FDM links between each station pair since this is only efficient if all *pairs* of stations have 50KB of traffic, driving the cost out of bounds for normal usage. The reservation system is also a factor of 3 more efficient than the ALOHA system and for large (100KB) traffic levels achieves almost perfect utilization of the channels.

REFERENCE

1. Roberts, L. G., Wessler, B., "Computer Network Development to Achieve Resource Sharing," *AFIPS Spring Joint Computer Conference Proceedings*, pp. 543–549, 1970.
2. Heart, F., Kahn, R. E., Ornstein, S., Crowther, W., Walden, D., "The Interface Message Processor for the ARPA Computer Network," *AFIPS Spring Joint Computer Conference Proceedings*, pp. 551–567, 1970.
3. Roberts, L. G., "Network Rationale: A 5-Year Reevaluation," *Proceedings of COMPCON 73*, February 1973.
4. Kahn, R. E., "Resource-Sharing Computer Communications Networks," *Proceedings of IEEE*, pp. 1397–1407, November 1972.
5. Cacciamani, E. R., "The Spade System as Applied to Data Communications and Small Earth Station Operation," *COMSAT Technical Review*, Vol. 1, No. 1, Fall 1971.
6. Schmidt, W. G., Gabbard, O. G., Cacciamani, E. R., Maillet, W. G., Wu, W. W., "MAT-1: INTELSAT's Experimental 700-Channel TDMA/DA System," *INTELSAT/IEE International Conference on Digital Satellite Communications Proceedings*, London, November 25–27, 1969.
7. Abramson, Norman, "Packet Switching With Satellites," *These proceedings*.
8. Kleinrock, Leonard, Lam, Simon S., "Packet-Switching in a Slotted Satellite Channel," *These proceedings*.
9. Crowther, W., Rettberg, R., Walden, B., Ornstein, S., Heart, F., "A System for Broadcast Communication: Reservation—ALOHA," *Proceedings of the Sixth Hawaii International Conference on System Sciences*, 1973.
10. Binder, Richard, *Another ALOHA Satellite Protocol*, ARPA Satellite System Note 32, NIC 13147.

SECTION SEVEN

Local-Area Networks

This section:

- provides a comprehensive, generalized introduction to local-area networks, the outgrowth of the bus architectures of distributed computer systems

- introduces packet radio technology where packet broadcasting is limited to a relatively small geographic area

- gives a complete mathematical and operational analysis of a packet broadcast channel in a localized area where it is possible to sense the status of the channel before transmitting

- shows how various techniques can be applied to actually lay out the configuration of the network components in a local geographic area

- presents operational data on one of the most widely implemented local-area networks, the Ethernet, showing that the theoretically high efficiencies are actually achieved.

A basic premise in developing and implementing telecommunications networks is that the communications resources are relatively expensive and, therefore, must be deployed cautiously and utilized efficiently. In order to use communications-transmission facilities efficiently, processing power, in the form of switches, protocol-driven interfaces, and "intelligent" terminals, are generally employed. Applied to the basic transmission facility, the processing capability allocates and shares the total resources among the large community of potential users.

Figure VII-1 shows the cost trends for computing and communications using packet switching technology. The curves in this figure, based on ARPANET experience, represent the rough costs for processing one million bits of data and for transmitting one million bits of data across the United States. In 1971 the cost of processing a million bits of data and the cost of transmitting a million bits were

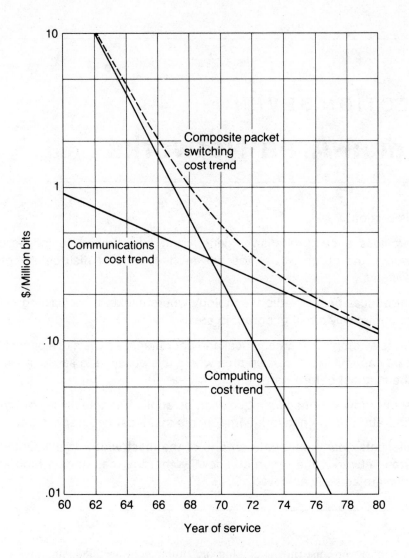

**Figure VII-1   Packet-switching cost-performance
trends. Cost for processing and transmitting
1 million bits of data across the United States.
© 1974, IEEE.**

about equal, at approximately 35¢. By the late 1970s, however, computers had become less expensive and the cost of transmission, on an incremental cost basis, was at least ten times that of computation. Without some major breakthrough in transmission costs, this differential will be about 100 times by the mid-1980s.

If, on the other hand, we look at communications costs on a local rather than nationwide basis, the situation changes dramatically. Extrapolating from Figure VII-1, we arrive at a 1985 processing cost per megabit of roughly $0.001, or a tenth of a cent. If, for example, we can connect all of our users in a local area together with about $25,000 worth of cable and connectors, use the system about eight hours a day for a ten-year lifetime, and put about three million bits per second through the local cable, the allocated, incremental communication cost would also be about $0.001 per megabit. In this case, the cost of transmission and the cost of processing are approximately equal. Thus, in many cases, the advantages of processing over communications are less evident in local networks used over a limited geographical area. When this occurs, using transmission media efficiently without overburdening the processors with a great deal of overhead requires a balanced approach. Many of the techniques discussed in earlier sections, including carrier sense multiple-access and capacity reservation, meet this requirement well. These approaches permit efficient utilization of channels, a large degree of network structure and interface independence, and a relatively low demand on the complexity of communications and interface processors.

TYING LOCAL NETWORKS TO NATIONAL NETWORKS

Local networks can play a very important role in allowing processor-based devices to efficiently utilize nationwide, long-distance networks. At the boundary between local and long-distance facilities, a variety of techniques not only achieve efficient transmission of user data, they also permit continued operation of nationwide facilities under anomalous conditions or system failures, fairly allocate the network resources, and accommodate the interface needs of various user devices.

National and international standards provide the features and protocols necessary for utilizing such networks with similar or dissimilar equipment over broad geographic distances. As shown in Figure VII-2, user end-devices are tied together by relatively simple, low-cost local networks that use interface protocols to minimize the complexity of the local-network interface. One device on the local network, the *internetwork gateway,* concentrates the interfacing of the long-distance networks in a single device. The internetwork gateway can accept data from any of the locally connected devices, add the long-distance network protocols, overhead, and controls, and transmit efficiently through the public switched networks to similar but distant internetwork gateways. After receiving the information at the distant end, the internetwork gateway converts it back to the simple local-network protocol (which may not be the same as that in the local network at the originating end) for delivery to the destination device.

USING LOCAL NETWORKS TO REDUCE COSTS

To explain the economic motivation for using local networks, let us return to Figure VII-1. The economic tradeoff situation illustrated in the figure was limited because the extrapolated trends were based on the technology of the early 1970s. The possibility of direct interfacing of user concentrations, however, particularly via satellite technology, alters the cost trends for basic transmission services. This change would modify Figure VII-1 as shown in Figure VII-3 where total network cost trends, with and without local network influence, are depicted. It is assumed

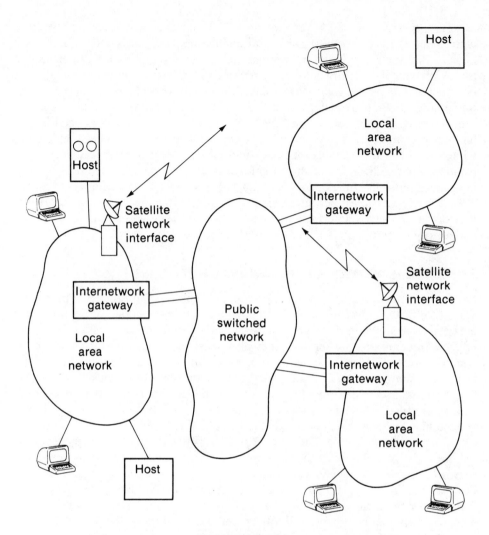

Figure VII-2   Local networks in a long-distance environment.

that, in the absence of a local network, individual user devices would have to directly interface with satellite facilities through shared earth stations. For most users this would necessitate relatively expensive, short-distance, carrier-provided terrestrial interconnections. On a per-voice channel, or, more significantly, on a per-megabit-of-capacity basis, these facilities are much more expensive than the satellite capacity itself. Tying many users together through local networks permits the use of an earth station dedicated to the total requirements of the local network, thereby eliminating the need for costly carrier-provided short-distance communications to shared earth stations. Use of the local network, where unit costs approximate that of the processing elements themselves, permits any limitations on the system's cost to be driven by the cost of the long-haul satellite facilities rather than by short-distance, carrier-provided interconnections.

In addition to their effect on the raw transmission costs, local networks significantly impact the number of individual interface protocols that need to be employed, simplifying those interfaces and allowing the local interconnection of a

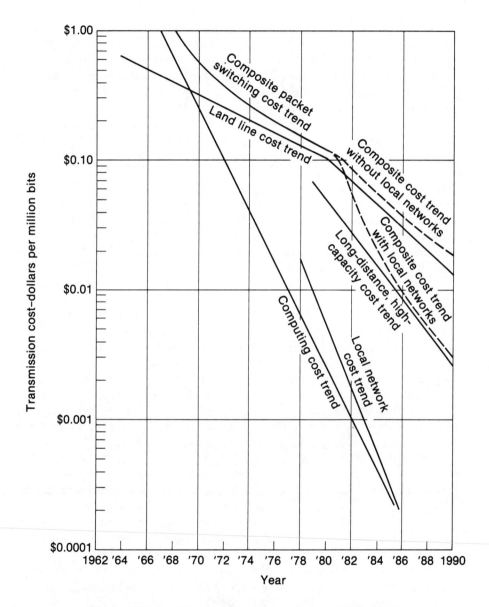

Figure VII-3   Total network cost trend (projected to 1990).

wide variety of devices. They also simplify the design and analysis of the overall network structure by allowing the design of the network to be partitioned into separate pieces and attacked with different, simpler design techniques than those needed to solve the long-haul network design problem.

CONSIDERATIONS IN DESIGNING LOCAL NETWORKS

The papers in this sections treat all of the major issues associated with local-area network design. The introductory paper by D.D. Clark, K.T. Pogran, and D.P. Reed separates the design question into two major issues. The first deals with the organization of the hardware elements—the low cost of raw-transmission capacity places further emphasis on hardware complexity, and care is required to keep the associated hardware and processing costs comparably low. The other issue deals with the design and structure of protocols. Protocol is an important factor because protocols that are suitable for expensive, long-delay, long-distance facilities are probably not directly applicable to the very high-speed, low-delay, local connections.

Because much of the technology of local networks originates with the bus architectures used in large computers, the authors quickly draw the distinction between the computer bus and the local network. For the most part, this distinction lies in the autonomy of the various users in a local network in contrast to the much tighter functional relationship among the components of a computer tied through the bus. The attributes of the star, ring, and bus structure are then discussed, and the authors relate cost, complexity, reliability, and performance.

This paper also discusses the approaches to the hardware interconnection and interface between user devices, such as terminals and computers, and the various local network structures. The relationships between the interface hardware, software, and protocols are developed. A discussion of network interconnection, gateways, and management conclude this paper.

While the Clark, Pogran, and Reed paper presumes a physical/electrical interconnection among the local area network components, principally using a cable, bus, or direct wiring, R.E. Kahn, S.A. Gronemeyer, J. Burchfiel, and R.C. Kunzelman address the range of issues dealing with implementation of local networks by means of radio-based interconnections. The primary differences between radio-based networks and cable-based local networks are that radio systems generally cover a considerably larger geographic area than cable-based systems and the transmission over this larger area is prone to errors arising from propagation effects, noise, low signal level, interference, and other factors, in addition to collision-induced errors, which are of primary concern on satellite or cable-based systems. The authors of the second paper in this section introduce these issues, indicate the directions taken in order to accommodate these potential problems, and provide considerable detail on the actual implementation of an experimental packet radio local network operating in the San Francisco Bay Area.

The next paper in this section, by L. Kleinrock and F.A. Tobagi, presents, in two parts, the technical and mathematical basis on which most of the present random-access local networks have been built. Part I introduces a number of different user models as well as a variety of different protocols and decision processes that the users can implement to determine the conditions under which they transmit, wait, and retransmit their packets. The concept of *persistent* and *nonpersistent* users is introduced; persistent users constantly monitor a busy channel, waiting for it to become vacant before making a decision to transmit. Because two persistent users would collide with probability of unity as soon as a busy channel became vacant, a transmit probability "p" is introduced to reduce the probability of a collision. The authors derive the complex relationships between traffic characteristics, propagation delay, packet length, and transmission probability. The basic assumption in Part I of this paper is that all users can hear each other equally well. Part II of the paper extends the analysis to the case where a number of terminals are hidden from each other because of propagation effects or other location-dependent effects. The authors propose solving this condition by using part of the basic channel capacity to designate the "busy" or "idle" condition of the input channel.

This paper by Kleinrock and Tobagi is the basis of techniques classed as CSMA—Carrier Sense Multiple Access. Researchers at XEROX Corporation expanded this technique by adding the ability to detect collisions while they are occurring. When possible, particularly on cable-based systems, each user is directed to stop transmitting as soon as a collision is detected. In this way the system capacity heretofore wasted to collisions is reduced and overall performance enhanced. The expanded concept, CSMA/CD—Carrier Sense Multiple Access with Collision Detection—is the basis for Ethernet and IEEE standards for local networks.

The final paper of this section, by J.F. Shoch and J.A. Hupp, details the experimental measured performance of the Ethernet, and shows that actual performance between 83% and 96% can be achieved.

Also included in this section is a paper by P.V. McGregor and D. Shen that deals with the problem of locating the various facilities of a local network within the intended service area. The intent of the local network concept is to capitalize on both low-cost transmission facilities as well as low-cost processor-driven interface devices. However, the fact that there are frequently a large number of relatively low-cost elements means that careful attention must be given to their number, cost, and location. By properly choosing the location of the processing nodes of a network, overall costs can be minimized. Handling all nodes at the same time is often impossible, even with the assistance of a large computer, and the authors of this paper propose and test a heuristic algorithm that groups and partitions the total nodes into a number of solvable subsets.

The first paper in Section VIII provides a more general discussion of the problem of network topological design, especially in the context of nationwide systems. The algorithmic approach described in McGregor and Shen's paper however, provides initial insight into the facility location problem.

An Introduction to Local Area Networks

DAVID D. CLARK, MEMBER, IEEE, KENNETH T. POGRAN, MEMBER, IEEE, AND DAVID P. REED

Invited Paper

Abstract—Within a restricted area such as a single building, or a small cluster of buildings, high-speed (greater than 1 Mbit/s) data transmission is available at a small fraction of the cost of obtaining comparable long-haul service from a tariffed common carrier. Local area networks use this low-cost, high-speed transmission capability as the basis for a general-purpose data transfer network. There are two basic issues in local area network design. First, how should the hardware realizing the network be organized to provide reliable high-speed communication at minimum cost? With the low cost of the raw transmission capability, care is required to keep the associated hardware costs correspondingly low. Second, what protocols should be used for the operation of the network? While many protocol problems are common to local area networks and long-haul networks such as the ARPANET, new protocols are required to exploit the extended capabilities of local area networks. This paper addresses these two basic issues. It also considers the interconnection of local area networks and long-haul networks and presents a case study which describes in detail the host computer interface hardware required for a typical local area network.

I. INTRODUCTION

AS ITS NAME IMPLIES, a local area network is a data communication network, typically a packet communication network, limited in geographic scope.[1] A local area network generally provides high-bandwidth communication over inexpensive transmission media. This paper discusses what local area networks are, their structures, the sorts of protocols that are used with them, and their applications. It also discusses the relationship of local area networks to long-haul networks and computer system I/O buses, as well as the impact of these networks on the field of computer communications today.

A. Components of a Local Area Network

Like any other data communication network, a local area network is composed of three basic hardware elements: a *transmission medium*, often twisted pair, coaxial cable, or fiber optics; a *mechanism for control* of transmission over the medium; and an *interface* to the network for the host computers or other devices—the *nodes* of the network—that are connected to the network. In addition, local area networks share with long-haul packet communication networks a fourth basic element: a set of software *protocols*, implemented in the host computers or other devices connected to the networks, which control the transmission of information from one host or device to another via the hardware elements of the network. These software protocols function at various levels, from low-level *packet transport* protocols to high-level application protocols, and are an integral part of both local area networks and their close relatives, long-haul packet communication networks. This combined hardware–software approach to com-

munication serves to distinguish networks, as discussed in this paper, from other arrangements of data communication hardware.

B. Relationship of Local Area Networks to Long-Haul Networks

1) The Evolution of Networking: Local area networks share a kinship with both long-haul packet communication networks and with I/O bus structures of digital computer systems; their structure and protocols are rooted in packet communication, while their hardware technology derives from both networks and computer busses. Local area networks arose out of the continuing evolution of packet communication networks and computer hardware technology. Packet communication techniques have become well known and widely understood in the nine years since development of the ARPANET was begun. Meanwhile, computer hardware has come down in price dramatically, giving rise to environments where, within a single building or a small cluster of buildings, there may be one or more large mainframe computers along with a number of mini-computers, microprocessor systems, and other intelligent devices containing microprocessors. Local area networks evolved to meet the growing demand for high data rate, low-cost communication among these machines.

2) Geographic Scope; Economic and Technical Considerations: Fig. 1 illustrates the geographic scope spanned by long-haul packet networks, local area networks, and computer system busses. Long-haul packet networks typically span distances ranging from meters[2] to tens of thousands of kilometers (for intercontinental packet networks); bus structures used in computer systems range from those of microprocessor systems, which can be as short as 1–10 cm, to those used in large-scale multiprocessor systems, which can be as much as 100 m in length. As Fig. 1 indicates, local area networks span distances from several meters through several kilometers in length.

The first local area networks evolved in environments in which the distances to be spanned by the network were within the range of inexpensive high-speed digital communication technologies. Today, the relationship has been turned around, so that the distance range of local area networks is governed by the distance over which these inexpensive techniques can be used. The result is high-data-rate networks in which the cost of transmission and the cost of control of transmission is very low compared to the costs associated with traditional

---

[1] Conventional Packet communications networks, not limited in geographic scope, are referred to in this paper as "long-haul" networks.

[2] Long-haul packet communication network technology has been used in environments that could be served more effectively, and at lower cost, by local area network technology. This local area use of long-haul network technology, indicated by the shaded area of Fig. 1, is due to the fact that long-haul packet communication technology has been available commercially for several years, while local area network technology has not.

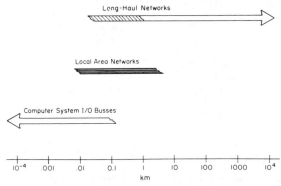

Fig. 1. Geographic range of computer communication networks and I/O buses. The shaded area of the long-haul network bar indicates the distance range for which that technology has been used in the past, but which could be better served, in both cost and performance, by emerging local area network technology.

data communication networks, providing some unique opportunities conventional long-haul networks do not afford.

For long-haul networks, the cost of communication is high. Wide-band common carrier circuits, satellite circuits, and private microwave links are expensive. Long-haul packet communication networks commonly employ moderately expensive (i.e., 50 000-dollar) minicomputers as packet switches to manage and route traffic flow to make the most effective use of the network communication links, delivery of packets to their proper destinations. The geographic characteristics of local area networks yield economic and technological considerations that are quite different. Inexpensive, privately owned transmission media can be used. For example, simple twisted pair can support point-to-point communication in the 1–10-Mbit/s range over distances on the order of a kilometer between repeaters. Coaxial cable, such as low-loss CATV cable, can support either point-to-point or broadcast communication at similar data rates over comparable distances. Typically, base-band signaling is used to place digital signals directly on the medium, rather than by modulation of a carrier. Because the hardware needed to drive and control these transmission media is inexpensive, there is little motivation or need to employ computing power to make the most effective use of the available bandwidth. On the contrary, it is quite reasonable to provide additional bandwidth, or to use a greater fraction of the existing bandwidth, if by doing so some other network cost—either hardware or software—can be reduced.

3) New Opportunities: The economic and technical characteristics of local area networks engender new applications of networking techniques and provide some unique opportunities to simplify traditional networking problems. Many of the constraints that long-haul networks impose on models of communication over a computer network are not present in local area networks. One example is broadcast communication, such as that used in the Distributed Computer System developed at the University of California at Irvine [1].

The high-bandwidth and low-delay attributes of local area networks make possible distributed multiprocessor systems utilizing the sort of information sharing between processors commonly associated with multiprocessor systems sharing primary memory. Local area networks can also be used to provide a central file system for a group of small computers which do not have their own secondary storage; it is even

possible to use such a central file system, accessed over the local area network, for swapping or paging—an application made especially attractive by the fact that the cost of local area network interface hardware for a typical minicomputer can be less than the cost of a "floppy disk" drive and its associated controller.

The high bandwidth of local area networks can be exploited to simplify the control structure of communication protocols by removing any motivation to minimize the length of control or overhead information in a packet. Fields of packet headers in local area networks can be arranged to simplify the processing involved in creating or interpreting the packet header, using as many bits as are necessary. There is little need to use "shorthand" techniques often found in the protocols of long-haul networks which necessitate additional table lookups by the receiver of a message. Simplicity also extends to other aspects of local area network protocols, such as schemes for allocation of network bandwidth, flow control, and error detection and correction.

It should be emphasized that local area networks are *not* an off-the-shelf, plug-in panacea for all local area computer communication needs. For the distance range over which they operate, the technology of local area networks holds the promise of doing for computer communications what the hardware innovations of the last five years have done for computing power: they can bring down the cost of high-bandwidth communication and make possible new applications. But they cannot by themselves solve the "software problem," for with low-cost hardware, the costs of software development will dominate the cost of any system development using local area network technology.

C. Interconnection with Other Networks

While some local area networks now in use or under construction are "stand-alone" networks, not connected to other networks, the trend is toward interconnection of local area networks with long-haul networks.

Interconnection can be motivated either by economics or simply by the needs of users of the hosts of a local area network. For example, a local area network can provide an economical means of connecting a number of hosts within a small area to one or more long-haul packet networks. The savings thus obtained is most obvious in the situation where a number of local host computers are to be connected to more than one long-haul network; each computer, rather than being directly connected to every network (an "*M*-by-*N* problem"; see Fig. 2(a)), can be connected only to the local area network, and one host (called the *gateway*) can be connected between the local area network and each of the long-haul networks (Fig. 2(b)). This cost savings can be worthwhile even in a situation in which local hosts are to be connected only to a single long-haul network, for two reasons; first, host interface hardware for local area networks can be less expensive than that for long-haul networks; and, second, only a single port to the long-haul network, rather than one port for each local host, is required.

Examples of motivation for interconnection based on users' needs are as follows.

a) A computer-based mail system, in which messages to and from users of the several hosts of a local area network can be exchanged via a long-haul network.

b) Access to specialized computing resources, occasionally required by the hosts of a local area network but too expensive

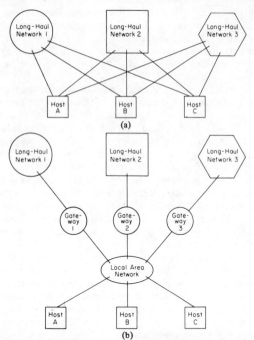

Fig. 2. The "$M \times N$ Problem" and a local area network as one solution to it. In (a), each of three hosts at a particular site is to be connected to three long-haul networks; each host must implement the communication protocols for, and be equipped with a hardware interface to, all three networks; there may be nine different interfaces and nine protocol implementations in all. In (b) each host needs only one hardware interface and one protocol implementation; the gateway machines each handle communication between one long-haul network and the local area network.

to maintain locally; they can be accessed by users of local area network hosts on a fee-for-service basis via a long-haul network.

c) Communication between local area networks maintained by a company at each of its major locations.

The interconnection of a local area network to a long-haul network presents problems, as well as benefits. At some point, the protocols used within the local area network must be made compatible with those of the long-haul network(s). Compatibility can be achieved either by adopting the protocols of a long-haul network for the local area network, or by performing appropriate protocol transformation on messages as they pass through the gateway. Care must be taken with the former approach to ensure that needed capabilities of the local area network are not sacrificed for the sake of ease of the network interconnection. These issues are discussed in greater detail in Section VI of this paper.

II. WHY ISN'T A LOCAL AREA NETWORK MERELY A "BIG BUS"?

A. Distinctions Between Local Area Networks and Computer Busses

One distinguishing feature of local area networks is the geographic restrictions that permit them to utilize low-cost but very high-bandwidth transmission media. That characterization can also apply to the bus structure of a computer. How, then, is a local area network different from a "big bus"? The significant differences are not, as one might initially expect, topological, technological, or geographic. Rather, the distinction is a philosophical one. A computer bus is usually conceived as

connecting together the components of a single computer system; it is difficult to imagine a computer continuing to perform any sort of useful action in the absence of its bus. In contrast, a network is understood to connect together a number of autonomous nodes, each capable of operating by itself in the absence of the network.

1) Defensiveness: This philosophical distinction manifests itself in the management and control strategies of a network, which are far more defensive than the equivalent strategies of a bus are required to be. For example, the reliability issues surrounding a bus are somewhat different from those surrounding a local area network. While both should be reliable, it is usually not critical that a bus continue to work if one if one of the devices attached to it has failed; it is usually acceptable to have the system halt momentarily until the failing device can be manually disconnected from the bus. In contrast, it is presumed that a network will continue to operate despite arbitrary failures of one or more nodes.

The handling of traffic overloads is another example of the defensive nature of a network. One must anticipate, when designing a network, that independently initiated transfers will occasionally demand more bandwidth than the network has available, at which time the network itself must mediate gracefully between these conflicting demands. There is usually no such concern for a computer bus. The problem of insufficient capacity on a computer bus to transfer all the information required is generally solved by reconfiguration of the hardware of the system, or through use of a different programming strategy.

2) Generality: The intended generality of a computer bus and a network is a second philosophical distinction between the two. For example, the protocols that control communication on a local area network are often designed with the explicit intention that messages can be exchanged between a local network and a long-haul network, an idea that is usually missing from the addressing and control structure of a computer bus. For another example, networks usually transmit variable size messages, while buses often transfer single, fixed-size words.

Another sort of generality that serves to distinguish a bus from a network is the nature of the interface that each provides to the nodes attached to it. A computer bus often has a specialized interface, oriented towards the addressing and control architecture of a particular computer. A network, on the other hand, usually attempts to provide an interface equally suitable for a wide variety of computers and other devices. In this respect, the interface is often less efficient than a bus interface, but is easier to implement for arbitrary devices. Fraser describes one plausible specification for a general network interface [2].

3) Minor Distinctions: Current realizations of networks and busses suggest other differences which are much less relevant. Busses are often even more "local" than our definition of a local area network. A computer bus is often highly parallel, with separate control, data, and address lines; networks tend to carry this information serially over a single set of lines. On the other hand, the idea of a computer bus which is completely serial is very attractive in the design of microprocessor systems, both to reduce component pinout and to eliminate problems of skew on parallel lines.

B. The IEEE Instrumentation Bus as a Border-Line Case

The IEEE Instrumentation Bus [3] is a good example of a communication medium that lies on the boundary between a network and a bus. In certain respects, this bus resembles a local area network, since it has a general interface capable of interconnecting a variety of instruments and computers, each of which operates with a certain degree of autonomy. Philosophically, however, the Instrumentation Bus is indeed very much a bus, since its specification is clearly based on the assumption that all of the devices connected to a particular bus are intended to operate harmoniously as one system to perform a single experiment under the control of one particular experimenter. The experimenter, not the bus itself, is expected to ensure that the capacity of the bus is not exceeded, and to detect and remove failing nodes which disable the bus. Also, the addressing structure of the Instrumentation Bus, as defined, is not extendable, so that the idea of connecting several of these busses together to make a larger network is difficult to realize. This limitation on addressing may prove a hindrance to experimenters who wish to use the Instrumentation Bus as a component of a larger interconnected array of computers and experimental equipment.

III. TOPOLOGIES AND CONTROL STRUCTURES FOR LOCAL AREA NETWORKS

The introduction of this paper identified three hardware components of a local area network: the transmission medium, a mechanism for control, and an interface to the network. This section will discuss the first two of these, which together provide the lowest-level functionality of the network, the ability to move messages from place to place in a regulated manner.

A. Network Topology

Network topology is the pattern of interconnection used among the various nodes of the network. The most general topology is an unconstrained graph structure, with nodes connected together in an arbitrary pattern, as illustrated in Fig. 3. This general structure is the one normally associated with a packet-switched network; its advantage is that the arrangement of the communication links can be based on the network traffic. This generality is a tool for optimizing the use of costly transmission media, an idea which is not germane to local area networks. Further, this generality introduces the unavoidable cost of making a routing decision at each node a message traverses. A message arriving at a node cannot be blindly transmitted out of all the other links connected to that node, for that would result in a message that multiplied at every node and propagated forever in the network. Thus each node must decide, as it receives a message, on which link it is to be forwarded, which implies a substantial computation at every node. Since this general topology is of no significant advantage in a local area network, and does imply a degree of complexity at every node, local area network designers have identified a variety of constrained topologies with attributes particularly suited to local area networks. We shall consider three such topologies: the *star*, the *ring*, and the *bus*.

1) The Star Network: A star network, illustrated in Fig. 4(a), eliminates the need for each network node to make routing decisions by localizing all message routing in one central node. This leads to a particularly simple structure for each of the other network nodes. This topology is an obvious choice if the normal pattern of communication in the network conforms to its physical topology, with a number of secondary nodes communicating with one primary node. For example, the star is an obvious topology to support a number of terminals communicating with a time-sharing sys-

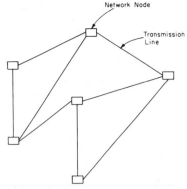

Fig. 3. Unconstrained topology. Each node receiving a message must make a routing decision to forward the packet to its final destination.

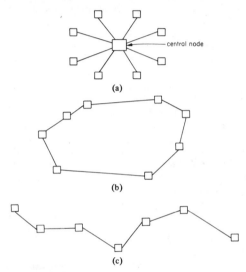

Fig. 4. Examples of constrained topologies. (a) The star. (b) The ring. (c) The bus.

tem, in which case the central node might be the time-sharing machine itself.

If, however, the normal pattern of communication is not between one primary node and several secondary nodes, but is instead more general communication among all of the nodes, then reliability appears as a possible disadvantage of the star net. Clearly, the operation of the network depends on the correct operation of the central node, which performs all of the routing functions, and must have capacity sufficient to cope with all simultaneous conversations. For these reasons, the central node may be a fairly large computer. The cost and difficulty of making the central node sufficiently reliable may more than offset any benefit derived from the simplicity of the other nodes.

2) Ring and Bus Networks: The ring and bus topologies attempt to eliminate the central node on the network, without sacrificing the simplicity of the other nodes. While the elimination of the central node does imply a certain complexity at the other nodes of the net, a decentralized network can be constructed with a surprisingly simple structure of the nodes. In the ring topology, illustrated in Fig. 4(b), a message is passed from node to node along unidirectional links. There are no routing decisions to be made in this topology; the sending node simply transmits its message to the next node in the ring, and the message passes around the ring, one node at a time, until it reaches the node for which it is intended. The only routing requirement placed on each node is that it be able to recognize, from the address in the message, those messages intended for it. Similarly, in the bus structure pictured in Fig. 4(c), there are no routing decisions required by any of the nodes. A message flows away from the originating node in both directions to the ends of the bus. The destination node reads the message as it passes by. Again, a node must be able to recognize messages intended for it.

B. Network Control Structures

Both the ring network and the bus network introduce a problem, not immediately apparent in the star net, of determining which node may transmit at any given time. The mechanism for control, the second component of the network as listed in the introduction, performs this determination. This task is not difficult in the star network; either the central node has sufficient capacity to handle a message for every node simultaneously, or it may poll each of the other nodes in turn to determine if that node wishes to transmit. Both the ring and the bus topology, lacking any central node, must use some distributed mechanism to determine which node may use the transmission medium at any given moment.

1) Daisy Chain, Control Token, and Message Slots: There are a variety of control strategies suitable for the ring topology, based on the general idea that permission to use the net is passed sequentially around the ring from node to node. In what is often called a *daisy chain network*, dedicated wires are used to pass the control information from one node to the next. Alternatively, the control information may be a special bit pattern transmitted over the regular data channel of the ring. For example, the network for the Distributed Computing System uses an 8-bit *control token* that is passed sequentially around the ring [4]. Any node, upon receiving the control token, may remove the token from the ring, send a message, and then pass on the control token. A third strategy for ring control is to continually transmit around the network a series of *message slots*, sequences of bits sufficient to hold a full message. A slot may be empty or full, and any node, on noticing an empty slot passing by, may mark the slot as full and place a message in it. This strategy was described by Pierce [5], and has been used in the Cambridge Net [6] and in the network described by Zafiropulo and Rothauser [7]. This technique is not completely decentralized, since one node must initially generate the slot pattern.

2) Register Insertion: Another control strategy particularly suited for the ring topology is called *register insertion*. In this technique, a message to be transmitted is first loaded into a shift register. The network loop is then broken and this shift register inserted in the net, either when the net is idle or at the point between two adjacent messages. The message to be sent is then shifted out onto the net while any message arriving during this period is shifted into the register behind the message being sent. Since the shift register has then become an active component of the network, no further messages can be sent by this node until the register is switched back out of the ring. This can only be done at a moment when there is no useful message in the register. One obvious way to remove the register from the network is to allow the message transmitted by the node to pass all the way around the network

and back into its shift register. At the moment when the message is again contained in the register, both message and register can be simultaneously removed from the network. If this technique is not used, or if the message is damaged and does not return intact back to its original sender, it is necessary to wait for the network to become idle before removing the buffer from the network.

The performance characteristics of the register insertion technique are rather different from those of the previously discussed techniques, since the total delay encountered in the network is variable, and depends on the number of messages currently being sent around the net. Further, a message to be sent is inserted in front of, rather than behind, a message arriving at a node. One analysis [8] indicates that the register insertion network may, under certain circumstances, have better delay and channel utilization characteristics than either the control token or message slot strategy. The register insertion strategy is complex, however, especially in the technique for removing the register from the network when the transmitted message does not return.

The register insertion technique was initially described by Hanfer *et al.* [9], in a paper that discusses a variety of operating modes that can be achieved using this technique. In the initial description of the Cambridge Network [10], register insertion was proposed as a control strategy, because it ensured a fair share allocation of network bandwidth. Since a node that has transmitted one message cannot transmit a second until the first message has passed completely around the ring, and the node has removed its register from the network, every other node has a chance to send one message before a given node may transmit a second. However, circulating message slots were finally chosen as the control mechanism of the Cambridge Network. Slots, if suitably employed, ensure fair share allocation, and the slot mechanism reduces the number of components whose failure can disrupt network operation, since there is no buffer in series with the net. Thus the slot scheme seemed more reliable. The register insertion technique is currently being used as the control strategy in the Distributed Loop Computing Network (DLCN) described by Liu and Reames [11].

3) Contention Control: A bus topology also requires a decentralized control strategy. One very simple control strategy that has been used for bus networks is *contention*. In a contention net, any node wishing to transmit simply does so. Since there is no control or priority, nothing prevents two nodes from attempting to transmit simultaneously, in which case a *collision* occurs, and both messages are garbled and presumably lost. The contention control strategy depends on the ability of a node to detect a collision, at which point it waits a random amount of time (so that the same collision will not recur), and then retransmits it message. Assuming that network traffic on the average consumes only a small percentage of the available bandwidth, the number of collisions and retransmissions will be reasonably small. The essential local area network characteristic of inexpensive bandwidth makes this strategy well suited to a local network. The bandwidth wasted in order to keep the channel utilization low is a small price to pay in return for the very simple mechanisms that must be implemented at each node: a timer capable of generating a random distribution, and some means of detecting collisions.

A variety of strategies have been used to detect collisions. The first use of a contention packet network was the ALOHA

Net [12], not a local area network, but a network using radio transceivers to connect together computer terminals to a computer center on the island of Oahu. The techniques for detecting a collision in this network was very simple: the transmitting node started a timer when it transmitted the message, and if an acknowledgment for the message had not been received when the timer expired, the message was retransmitted. The disadvantage of this collision detection scheme is that it leads to a very low theoretical upper limit on the percentage of channel capacity which can be utilized without causing the network to overload with retransmission traffic [13].

A strategy which greatly increases the maximum effective transmission capacity of the network is to listen before transmitting, which changes the whole pattern of network operation. A collision will now occur only if two nodes attempt to transmit at nearly the same instant, because if one node has started sufficiently in advance of the other so that its signal has propagated over the transmission medium to that other node, the other node will hear that signal and will refrain from transmitting.

A further embellishment of this idea is to listen, *while* transmitting as well. This permits colliding nodes to detect the collision much more promptly than if they detected the collision only by noticing the absence of an acknowledgment. This strategy not only reduces the delay caused by a collision, it makes the transmission medium available sooner, as well, since colliding nodes can cease transmitting as soon as they detect a collision. The strategy of listening while transmitting is not suitable for terrestrial radio, because the transmitter overloads the receiver, but is quite reasonable when transmitting over wire or cable. This strategy is used in the ETHERNET, developed at the Xerox Palo Alto Research Center [14].

C. Combinations of Topology and Control Structure

We have identified three network topologies: the star, the ring, and the bus topology, and three control strategies: ring control, contention control, and centralized control. It is important to note that any control strategy can be used with any topology. Several interesting combinations are described in the following paragraphs.

A variation on the use of a control token, suitable for a bus topology, is described by Jensen [15]. Every node is provided with a list of the order in which each may send, since the bus itself imposes no natural order. A special signal on the bus causes every node to move to the next entry on the list. The node named by that entry may send a message if it has one, after which it must in turn send the special signal. A mechanism is required to recover synchronism should one or more nodes miss the special signal, so that the lists get out of step. This scheme has the interesting advantage that some interfaces can be entered in the list more often than others, so that they receive a larger proportion of the bandwidth.

A bus topology using a daisy chain ring control strategy is a common means of implementing a computer bus. An example is the UNIBUS architecture of the Digital Equipment Corporation PDP-11 [16].

A ring topology controlled by a contention strategy produces a network with some very promising attributes, being explored in an experimental contention ring currently under development at the Laboratory for Computer Science of the Massachusetts Institute of Technology. In a bus topology contention network, collisions most commonly occur immediately following the end of a message, for at that moment all of the nodes that have refrained from transmitting during the previous message simultaneously attempt to seize the bus. In a ring topology contention net, the unidirectional flow of messages from node to node provides a natural ordering of all nodes wishing to transmit at the end of a previous message. Thus the contention ring will experience a much lower collision rate for a given degree of channel utilization. Further, in a contention ring it is very easy to implement the concept of listening while transmitting in order to detect collisions promptly. One way of operating a ring topology is for the transmitting node to place the message on the ring and also to remove the message from the ring. The message flows all the way around the ring; it is not removed by the recipient. A collision is detected when a transmitting node discovers that the message it is removing from the ring is different from the one it is sending.

It is also possible to devise a ring network with centralized control. Such a network was proposed by Farmer and Newhall [17]; the SPIDER network, described by Fraser [18] also has this structure. Line control protocols such as SDLC [19] use centralized control to regulate both a ring topology and multidrop line, a topology that somewhat resembles a bus.

D. Reliability Characteristics of Ring and Bus

The chief motivation for the ring and the bus topology was to avoid a potential reliability problem with the central node of the star. Reliability considerations arise both from the topology and the control strategy of the network. The contention control strategy has an inherent reliability advantage over the ring control strategies described above, for in any ring control strategy there is some entity, be it a control token or an explicit signal on a wire, which is passed from node to node to indicate which node currently has the right to transmit. The control strategy must always take into account the possibility that a transient error will destroy this entity. For example, a control token may be destroyed by a noise burst on the transmission medium. Therefore, any ring control strategy must be prepared to restart itself after a transient error by regenerating the permission to send and bestowing it uniquely upon one of the nodes. Unfortunately, with a completely decentralized control strategy, it is very difficult to determine with certainty that the control entity has been lost, and it is even more difficult to decide which node should take it upon itself to recreate the control entity. Thus one must either use some sort of contention scheme to deal with error recovery, as is done in the Distributed Computing System network, or have a single node provide a centralized mechanism for restart, as is done in the Cambridge Network.

In contrast, almost any transient failure in a contention control network has exactly the same effect as a collision, and is thus dealt with automatically. If a message is garbled it must be retransmitted, but no long-term failure of the network results. This is one of the very appealing attributes of the contention control strategy. On the other hand, contention control does require that the recipient detect a garbled message, and be able to request a retransmission if the original transmitter has for some reason failed to discover that the message was garbled. Thus higher level protocols must provide mechanisms which ensure reliable recovery. In fact, attention to reliability at higher levels is required regardless of the control strategy of the network; this is not a particular disadvantage of the contention strategy.

The topology of a network, as well as its control strategy, influences its reliability. The ring topology requires that each node be able to selectively remove a message from the ring or pass it on to the next node. This requires an active repeater at each node, and the network can be no more reliable than these active repeaters. The active repeater must also play a role in implementing the control strategies used on a ring, for almost without exception the control strategies depend on the ability of a node to modify a message is it passes by. For example, in the control token strategy, the token is removed from the network by modifying its last bit as it passes by, so that it is no longer recognized as a control token. Thus there can be a significant amount of logic at each node whose failure disables the network. The repeater portion of a ring network interface should, therefore, be made very reliable, with a reliable power supply, to reduce its probability of failure. In the Cambridge Network, repeaters are powered from the ring, so that the network is immune to loss of local power at any node. Another technique that enhances the reliability of the network is to provide a relay at each repeater that can mechanically remove it from the network in the event of a failure, including a local power failure.

The bus topology, on the other hand, does not require the message to be regenerated at each node. The bus is a passive medium, with each node listening. A node can fail without disrupting the bus so long as it fails in a manner that presents a high impedance to the bus. A bus network node that is designed to operate in this way is described later in this section.

This analysis suggests that a network constructed with a bus topology could be more reliable than one constructed with a ring topology. While both have active elements whose failure can disrupt the network, the bus components must fail in a particular way, so as to drive and disrupt the bus, whereas almost any failure of the ring repeater will disable the ring. However, not all the reliability issues favor the bus topology. If, for example, the transmission medium is subjected to a catastrophic disruption, such as a lightning strike or an errant cross connection to the power lines, one can expect all electronic components connected to the medium to be destroyed. In the case of a ring, this will be one set of line drivers and one set of receivers. In the case of the bus, every node may be damaged. While in both cases the net will be disabled, the bus may require much longer to repair. Practical experience, although somewhat limited and not well reported in the literature, suggests that with care *both* topologies can be made sufficiently reliable that the possibility of failure can essentially be ignored in a practical system. The various factors previously discussed may tend to favor a passive bus or an active repeater design in some particular case. However, we believe that the most significant factor in hardware reliability is the quality and care in the engineering design. In this context it is worth noting that certain design problems present in a bus do not arise in a ring (Several such problems are discussed in the case study of a bus interface in the next section.). Thus it may be somewhat easier to design a ring than a bus.

E. Patterns of Data Flow; Addressing

So far we have identified three criteria for choosing one of the three constrained network topologies: simplicity, reliability, and the need for a network control strategy. There is one additional criterion of importance—the patterns of data flow that each topology will support. Message flow in the arbitrary graph structure discussed at the beginning of this section is inherently point-to-point. That is, a message inserted into the network flows over some subset of the available links and is routed to some eventual destination. There is an alternative pattern, in which any message placed in the network is automatically broadcast to all nodes, each node examining the address on the message and copying it as appropriate. As we will discuss later, many applications of a local area network can benefit greatly from the ability to send messages in a broadcast mode. Thus there is an advantage to a topology that naturally supports broadcast at the low level.

The bus topology is inherently a broadcast medium, as there is no way to selectively route a message along the bus. The ring topology can either be used in a point-to-point mode or a broadcast mode. In the point-to-point mode, the message is transmitted around the ring until it reaches the recipient, who then removes it. In broadcast mode, the message passes completely around the ring and is removed by the original sender. The star configuration, also, can operate as either a point-to-point or a broadcast network.

The possibility of operating a network in the broadcast mode raises the question of what addressing mechanism is used on the network to identify the recipient of a message. The simplest addressing strategy, a pre-assigned, wired-in, fixed size address for each node, is easily implemented, but precludes the use of multidestination addressing. Instead, it forces all entities communicating over the network to be aware of the low-level routing of messages. An alternative to fixed addressing is associative addressing, so called because the address recognition mechanism in a node's network interface is implemented using an associative memory. In the simplest form of associative addressing, each interface contains a set of addresses for which the interface is a destination. Attached to a broadcast network, the interface listens to each packet as it is transmitted, and picks up those packets that are addressed to one of the addresses contained in the interface. A multidestination packet can be sent by using an address recognized by several interfaces. It is also possible to move a destination from one node to another transparently with this scheme. These features may be used to great advantage in the design of distributed system software, as discussed by Mockapetris and Farber [20]. Two disadvantages of this scheme are 1) its implementation requires more complex address recognition hardware containing host-loadable memory for the addresses, and 2) it is more difficult to determine the cause of failure when messages are not delivered.

Mockapetris has designed a more general associative addressing scheme [21] that allows host interfaces to match subfields of the message destination address against subfields of addresses contained in the host interface name table. For example, see Fig. 5. Host A will receive all messages addressed to destinations with the first four bits zero. Host B will receive all messages with the first two bits zero, and the fifth bit one. The additional power gained by this scheme is the ability to specify large classes of destinations with a small amount of memory in the interface.

The comparison of point-to-point and broadcast transmission in a ring topology raises the interesting question of how a message is removed from the ring. In a point-to-point mode, each node must examine the message before deciding whether to remove it. Thus each node must buffer enough of the message to see the destination address before passing the message

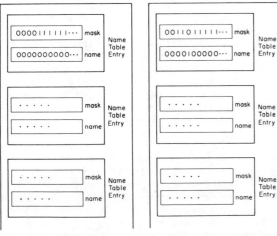

Host A Name Table   Host B Name Table

Fig. 5. Name Tables for Host Associative Addressing. Zeros in host A's first name table entry's mask select the first four bits of destination addresses for comparison. Destination addresses whose first four bits match the first entry's name are thus accepted by host A. Zeros in host B's first name table entry's mask select the first, second, and fifth bit for comparison. Destination addresses of the form 00XX1XXX ··· are thus accepted by host B.

on. Considerable delay will be introduced by that buffer unless a special addressing technique is used which allows this decision to be made in 1-bit time. In contrast, it is possible to build a broadcast net in which the message can be removed without examining it first. In the network for the Distributed Computing System, there is one control token, so there is one message on the network at any time. Thus a transmitting node knows that the next message it sees will be its own, and can remove it without examination. In the Cambridge Network, several message slots circulate on the net, so the transmitting node must do something slightly more complicated to determine which slot contains the returning message to be removed. In fact, the technique used is to count the number of slots going by. When the correct number has passed, the next slot is marked as empty.[3] Again, no buffering is required. In fact, if techniques such as this are used for message removal, it is possible to build a network in which the only delay at each node is due to gate propagation delays.

F. Transmission Media

Most local area networks now in use or being designed use bit-serial transmission over either coaxial cable or twisted pair. The geographic limitations of local area networks are precisely those encountered when attempting to send high-speed digital information over such wire or cable. The goal of simplicity has led to the use of baseband signaling, the simplest class of modulation schemes, as the means of encoding data. Thus if one is interested in achieving the highest bit rates or the longest distances, one will choose coaxial cable because of its more uniform impedance characteristics. On the other hand, it is much easier to splice new nodes in between two existing nodes if a twisted pair is used.

[3] To determine the total number of slots in the ring, each node makes use of a unique pattern occuring only once on the sequence of slots circulating on the ring, counting the number of slots between arrivals of this pattern.

One promising candidate for local area data transmission is cable television (CATV) technology. CATV makes tremendous bandwidth available; its wide-spread utilization tends to make CATV system components low in cost. In many cases, CATV may already be installed, and a network can be produced using some of the channels of the existing equipment. The MITRE Corporation has developed MITRIX, a CATV network with a bus topology and centralized control [22], and an alternative with decentralized contention control [23].

Fiber optics is another promising candidate for local area data transmission. Transmission of signals for a small number of kilometers (but significantly longer than possible using wire or cable) using bit rates between 1 and 20 Mbit/s appears to be a fairly simple task for fiber-optic technology; other characteristics of optical fibers such as their high noise immunity and inherent ground isolation make the technology even more tempting. However, transmission using fiber-optic technology is inherently unidirectional, which seems to eliminate the bus topology at the present time. An interesting and challenging problem is the design of a high-speed bus topology contention network using fiber optics as the transmission medium.

Radio broadcast has been demonstrated for a local network using packet switching [24]. Other ideas have even been suggested such as communication between computers using light signals reflected off a mirrored ceiling or a blimp. In general, the physical transmission medium for a local area network should be reliable, simple, inexpensive, high-speed, noise-free, and physically robust. It should also be easy to install, maintain, and reconfigure. There is room for further creativity in this area.

IV. The Nature of Hosts and their Interfaces

The hardware of a local area network is keyed to high performance at low cost. In a decentralized local area network, the interface hardware associated with a host generally provides all the transmission control and address recognition circuitry that is required. Because of the desire to connect low-cost minicomputer and microprocessor systems to local area networks, there is a good deal of motivation to make the host interface hardware as inexpensive as possible. Ultimately, a good portion of a host interface for a local area network could be implemented as a single large-scale integrated circuit (LSI) chip.

A. Hardware Structure of a Host Interface

Generally, the host interface hardware for a network may be viewed as having two parts: a *network-oriented* part that performs whatever transmission control functions are required for the network, and a *host-specific* part that fits into the I/O structure of a particular type of host computer and controls the exchange of data between the host and the network-oriented portion of the interface.

The simple architecture and control structures of local area networks aid in reducing the complexity and cost of the network-oriented portion of the interface. However, the situation can be quite different for the host-specific portion of the interface, as microprocessor systems, minicomputers, and large-scale systems present a wide range of I/O interface complexity. Interfaces for microprocessor systems tend to be the least complex, because of the simple bus structures of microprocessor systems, and the availability of LSI peripheral interface circuits. Large-scale systems tend to require the most complex interfaces, as one might expect. Interfaces for minicomputer systems tend to be more complex than one might expect at first glance, largely due to the need for a "direct-memory access" type of interface required by a high-bandwidth peripheral device, which the host interface for a local area network is.

B. Approaches to Attachment of a Host to a Network

Over the entire range of computer systems, from microprocessors to large-scale systems, there is little difference between the complexity of the host-specific portion of an interface for a local area network and for a long-haul network. With the reduced complexity of the network-oriented portion of a local area network interface, and the low cost of transmission medium of a local area network, domination of the cost of attachment of a host computer to a network shifts from network-related costs, in the case of long-haul network, to host-related costs, for a local area network. This shift has two major effects: first, it becomes practical—that is, economically justifiable—to connect microprocessor systems to a network, and, second, it causes those responsible for the attachment of large-scale hosts to a local area network to examine their approach very carefully.

One approach to the interfacing of large-scale systems to long-haul networks that has become popular among those who do not wish to develop specialized hardware interfaces (until packet network interfaces become standard offerings of large-scale system vendors) or modify vendor-supplied operating system software is to interconnect a packet network to the system via a *front-end processor*, usually a minicomputer, which has an appropriate packet network interface and which connects to the large-scale host system in a way that mimics a standard method of attachment to the system, such as a group of remote interactive terminal lines, or a remote job entry (RJE) port. With such an attachment, the host is usually limited to utilizing only that portion of the network's functional capabilities which correspond to those of the standard attachment being mimicked. This front-end approach is less satisfactory for attachment of a local area network to a large-scale host, for, although a large-scale host system may be even better able to utilize the high data rate offered by the local area network than a minicomputer or microprocessor, actual data rates available through standard interfaces mimicking RJE or interactive terminal ports are meager by comparison. In addition, the protocols and applications used with and envisioned for local area networks are less well matched to standard interactive and RJE ports than are those of long-haul packet networks. In short, more of the potential of the local area network is lost through front-ending.

Although development of specialized hardware and software to interface a large-scale host system to a local area network may initially be the more expensive path to follow, it is likely to be the most fruitful in the long run; for, with properly designed interface, the high-speed local area network and the large-scale host system are particularly well matched. Such an interface is virtually a necessity in applications in which the large-scale host serves as a central data repository, as a specialized or centralized information processing resource, or in tightly coupled distributed processing systems.

C. Case Study of a Local Network Interface

We have discussed a number of alternatives for the topology, control strategy, and interface hardware of a local area net-

work. We shall now examine a particular local network interface, both to see how the various design issues were resolved in the particular unit, and to gain an overall perception of its complexity. Our example is the Local Network Interface, or LNI, originally developed at the University of California at Irvine and now being used as part of the local area network under development at the Laboratory for Computer Science of the Massachusetts Institute of Technology [24]. This interface can be made to operate a control token ring network, a contention ring network, or a contention bus network with only small modifications. These varied capabilities will enable us to perform experiments comparing these different network control strategies in the same operational environment. By examining the modifications needed to achieve each of these operational modes it is possible to perceive the similarities and differences required to support each of them.

1) Host-Specific Part: The structure of the local network interface (LNI) follows the general plan outlined in Section IV-B above, comprising a host-specific part and a network-oriented part. The first implementation of the LNI is for a Digital Equipment Corporation PDP-11 minicomputer host; therefore, the host-specific portion of the initial LNI is a full-duplex direct memory access (DMA) interface connected to the PDP-11 UNIBUS [16]. With this interface, the LNI can transfer data to or from the memory of the PDP-11 without the intervention of the processor. Although there are simpler forms of I/O interfaces for the PDP-11; namely, programmed I/O and interrupt-driven I/O, the data rate of the LCS Network (initially, 1 Mbit/s, with an eventual goal of 4–8 Mbit/s) requires a DMA interface to ensure that the PDP-11 processor will be available for tasks other than servicing data transfers to and from the LNI.

a) Registers: A PDP-11 full-duplex DMA interface is a surprisingly complex device. As implemented for the LNI, it contains ten registers directly addressable by the PDP-11 processor, including two 16-bit and two 18-bit counter registers:

Command
Status
Transmitted Data
Received Data
Transmit Address (lower 16 bits)
Transmit Address (upper 2 bits)
Receive Address (lower 16 bits)
Receive Address (upper 2 bits)
Transmit Byte Count
Receive Byte Count.

The Transmitted and Received Data Registers allow the LNI to be used in an interrupt-driven or programmed I/O mode for testing purposes. The Command Register is a path to flip-flops in the network-oriented portion of the LNI which controls its operation; setting bits in the Command Register initializes transmission of messages over the network, enables receipt of messages, etc. Similarly, the Status Register is a path to flip-flops in the LNI which indicate the success, failure, or other status of the LNI.

b) Input/output transactions: In a typical DMA transaction, the PDP-11 processor loads the memory address of the first byte to be transmitted into the Transmit Address register, and loads the Transmit Byte Count Register with the number of bytes to be transmitted. The processor then sets a bit in the

Command Register to initiate a transmit operation. The DMA interface requests memory cycles of the PDP-11, transferring data bytes from PDP-11 memory into a first-in-first-out (FIFO) buffer in the LNI. Once the buffer is full, further transfers from the PDP-11 take place only when actual transmission over the network has begun, as data are shifted out of the FIFO buffer. The setup of the DMA for receipt of data from the network is similar: the Receive Address and Receive Byte Count Registers are set, and the LNI enabled for input. Data are transferred into the PDP-11 only when a message addressed to this network host begins to arrive. Good programming practice suggests that an input operation should *always* be pending, and, unless complex I/O buffer strategies are used, the input byte count register should be set to permit receipt of a message of the maximum expected length (a "full packet"). A DMA operation terminates either when the byte count goes to zero, or upon a signal from the network-oriented portion of the LNI (e.g., a complete received message contained fewer bytes than the maximum initially set in the Byte Count Register). Terminations are generally signalled to the PDP-11 via an interrupt, although interrupts may be inhibited, under the control of a bit in the command register.

c) Advantages of full-duplex operation: The full-duplex nature of the interface makes it possible to initiate a transmit operation while a receive operation is pending. This is important for a network interface, especially a local network interface, as the receiver of a message has no control over when that message will arrive. A host may initiate a transmit DMA operation, with the network-oriented portion of the interface awaiting an opportunity to begin actual transmission of the message over the network, when a message may arrive on the network addressed to that host. If the host cannot perform a receive DMA transaction while the transmit operation is in progress, the message addressed to the host will be lost (unless the network-oriented portion of the interface can buffer entire messages).

d) Interfaces for other host computers: The PDP-11 DMA interface described here is typical of the host-specific part of a local area network interface for minicomputers, and for some microprocessor systems as well. DMA interfaces for microprocessors are available as LSI chips, although they are generally half-duplex interfaces, and two would be required for the full-duplex operations described here. Later versions of the MIT-UCI LNI will provide host interfaces for the PDP-10 and LSI-11; the PDP-10 interface will be a program–interrupt, I/O bus interface, while the LSI-11 bus interface will be similar to the PDP-11 UNIBUS interface described here.

2) Network-Oriented Part: Like any interface for a local area network, the network-oriented portion of the LNI (shown in Fig. 6) performs four basic functions:

1) Control of transmission: It observes transmissions on the network, determining when it may send a message of its own.

2) Control of reception: It observes transmissions on the network, looking for incoming messages. Incoming messages addressed to this host are transferred to the host-specific portion of the interface.

3) Address recognition: It determines whether or not a message detected by the interface is addressed to this host.

4) Signal conditioning: It provides appropriate transformations between the logic-level signals of the interface and signals appropriate to the network transmission medium. Examples are: differential voltage signals on twisted pair; bipolar pulses on coaxial cable; light pulses on optical fibers.

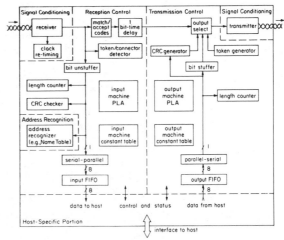

Fig. 6. Block diagram of the Local Network Interface (LNI). The five major components described in the text are separated by dashed lines; the Signal Conditioning Section appears in two parts, at the upper left and upper right. The arrows depict data flow; control, vested in the Programmed Logic Arrays (PLA's) of the input and output machines, is not indicated.

In the LNI control of transmission and control of reception are each achieved with a sequential state machine composed of a state counter, field-programmable logic array (FPLA), header field length and data length counters, and programmable read-only memories (PROM's) which serve as constant tables providing the lengths of header fields. The actions taken by these machines, and their sequence, can be changed by reprogramming the FPLA. The lengths of the various fields of a packet header can be changed by reprogramming the constant table PROM's. The initial version of the LNI implements a ring network similar to the original UC-Irvine DCS ring network. When an LNI is not transmitting a message, it serves as a repeater, retransmitting, one bit-time later, each bit it receives. It recovers clock from the incoming signal and synchronizes its transmit clock to the recovered clock.

When the transmission control section of the LNI has a message to send, it waits for the passage of the *control token*, a particular bit pattern, on the ring. Since the LNI introduces only one bit-time of delay into the ring, the token detector circuit must itself be a small sequential machine. The transmission control section inverts the last bit of the token to transform it into a *connector* which indicates to each LNI on the ring that a message follows. The transmission control mechanism then ceases repeating bits it has received, and instead transmits its own message. The transmit clock is decoupled from the recovered clock, and the transmitting LNI sets the timing of the ring.

The *output machine* of the transmission control section of the LNI follows each field of the packet header as it is transmitted, noting in particular the *length* field of the packet. After all the data of the packet have been transmitted, the output machine outputs a 16-bit cyclic redundancy checksum (CRC) followed by a match/accept field and a new control token. While this is taking place, the *input machine* of the reception control section is following the message, which has traveled around the ring and returned to the transmitting LNI. It verifies the received checksum, as it would for any message it would receive for this host, and passes the received checksum to the output control machine, to verify that it is identical to the one transmitted. This, together with the fact that the input machine detects extraneous tokens or connectors appearing in the middle of a message, provides a means of detecting ring errors due to faulty LNI's which may have begun to transmit a message without waiting for a token.

When the LNI is *not* transmitting a message of its own, the input machine monitors data arriving on the ring and passes them to the output machine to be repeated. When the token detector detects a *connector*, the input machine begins to follow the fields of the message. The destination address fields are passed, a bit at a time, to the address recognition section of the LNI. In addition, the reception control section of the LNI assembles the bits into 8-bit bytes and passes them to a FIFO buffer. When the entire destination address field has been received, the address recognition section indicates whether or not the message is for the host. If it is, the input machine signals the host specified part of the LNI that the data in the FIFO may be passed to the host; if not, it clears the FIFO and does not load subsequent data bits into the FIFO.

The input machine compares the checksum which follows the data of the received message with the checksum it has computed; checksum errors are reported to the host, if the message was addressed to the host. Following the checksum, the machine modifies the match/accept bits to indicate that the message was received. Neither the checksum nor the match/accept bits are placed into the FIFO to be passed to the host.

The address recognition section of the LNI is a sophisticated associative memory *name table* [21] which can be loaded by the host. It is an essential element of the UC-Irvine Distributed Computing System concept, in which the name table is loaded with the names of processes currently located in the host. The LNI name table, and the destination address fields of packets on the ring, contain masks as well as names, to facilitate the addressing of classes or groups of processes. From the point of view of the LNI as an example of a local network interface, the name table contains more sophistication than is necessary: it could be replaced by a simple mechanism which recognizes a single address, either wired into the unit or encoded in a PROM.

With either the name table or the single address recognizer, address recognition is done on a bit-by-bit basis, since that is how the data are presented to the address recognition circuit by the reception control section. The address recognition circuit indicates to the reception control section, after all the address bits have been processed, whether the address fields match—whether the message is addressed to this host.

The signal conditioning section of the LNI is quite straightforward; a simplified version of it is shown in Fig. 7. The transmission medium of the ring network is a shielded twisted pair; differential signals are placed on the pair by the transmitter of one LNI, and current flow on the pair is detected by high-speed optocouplers placed across the pair at the next LNI, which also serve to terminate the pair. Two optocouplers are used, connected across the pair in opposite directions, one to detect current flow in each direction. During the first half of each bit-time, the transmitter output driving each side of the pair is low, so that no current flows in the pair and neither optocoupler turns on; during the second half of the bit-time one transmitter output goes high if the value of the data bit is 1, and the other goes high

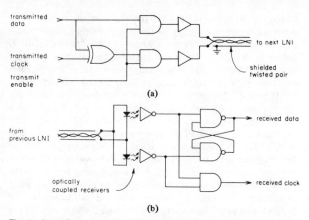

Fig. 7. Simplified schematic diagram of the Signal Conditioning Section of the LNI. (a) The transmitter circuit. (b) The receiver circuit.

if the bit to be transmitted is a 0. Thus the direction of current flow in the pair during the second half of a bit-time indicates the value of the data bit; the current flow is detected by one or the other of the optocouplers, one coupler turning on to indicate a 1, and the other turning on to indicate a 0. The couplers drive a simple latch, the output of which presents recovered data to the reception control section of the LNI, with a half-bit-time of delay. The logical OR of the outputs of the two couplers yields the recovered clock: 0 for the half-bit-time the pair is quiesient, and 1 for the half-bit-time during which there is current flow on the pair.

3) Modifying the LNI for Other Types of Networks: The claim was made earlier in this section that the structure of the LNI was representative of the structure of interfaces for various types of local area networks, and that the LNI could readily be modified to operate either a contention bus network or a contention ring network. We now investigate how this can be done, to further illuminate the nature of local area network interfaces.

Little change needs to be made in the host-specific part of the LNI, as the nature of data interchange with the host remains the same. Different types of networks may require somewhat different command bits, or report different status bits, but, since the most of the bits of command and status registers of the host-specific part of the LNI are but paths to and from flip-flops in the LNI's network-oriented portion, these changes have little impact on the host-specific part of the LNI.

The several sections of the network-oriented portion of the LNI are affected to varying degrees. Least affected is the address recognition section; its function remains the same: to examine the bits of the address field of an arriving message to see if it addressed to this host. Although the DCS ring network and the LNI have been described in the literature as containing a name table associative memory, and contention bus networks such as the Xerox PARC ETHERNET have been described as having a fixed address recognition mechanism, these are design decisions based on the intended initial applications of the network technology, rather than choices dictated by the nature of the technology itself.

The changes made to the transmission control section of the LNI are the most illustrative. In the ring network previously described, the LNI output machine section must wait for a token to pass before initiating the transmission of a message. In a contention bus or contention ring network, the output machine may transmit only when the network is quiet. The "token present" signal is replaced by a "network quiet" signal. In the ring network, the reception control section signals the transmission control section if it detects another token in the midst of its receipt of the message the transmission control section sent; this has its analogue in the collision detection capability of the contention network. In both cases, the LNI must abort transmission of its message and take corrective action. In the ring network this is an error condition, an exception; more than one control token is present in the ring. In the contention network, a collision is an expected event. Both situations can be handled by the LNI reporting the event to host software, which can attempt to restart a token on the ring, in the ring network case, or apply a retransmission backoff algorithm in the contention network case.

A better solution for the contention network is to modify the transmission control section to execute a simple retransmission backoff algorithm in hardware. This requires that the entire message remain accessible to the transmission control section without host intervention. The FIFO buffer cannot be used in this situation; a complete packet buffer which is not erased until the message has been successfully transmitted is an appropriate alternative.

Two features of the ring network LNI's transmission control section are not needed in the contention bus network version: the data repeater which passes bits from the receive side of the LNI to its transmit side when the LNI is not transmitting a message, and the token generator which places a new token or connector onto a quiescent ring. Of course, the connector is a brief sequence of bits, and there is no good motivation to delete it from the beginning of messages transmitted by the contention bus version of the LNI. In fact, retention of the connector at the head of a message results in fewer changes to the input machine of the LNI. It can use its token/connector detector to signal the beginning of an incoming message. Its function remains the same, for the most part; extra connectors detected in the middle of a message indicate a collision, just as they do for the ring network version. However, in the contention bus network, because bits are not repeated from one LNI to another, there is no way to set the match/accept bits for the benefit of the transmitting LNI, and the match/accept field of the message cannot be used.

The signal conditioning section of the LNI undergoes an interesting transformation. For a contention ring network, of course, the signal conditioning section remains the same. However, for a contention bus network, the logic levels of the LNI must be converted to appropriate signal levels and waveforms for the coaxial cable of the bus. This is done in a two-step process. First, a cable transceiver is added to the configuration. To minimize impedance mismatches, reflections, etc., the transceiver is located immediately adjacent to the network cable, and is often packaged separately from the LNI.[4] It is connected to the cable either directly, or via

<hr>

[4] This has become common practice in local area networking; the networking transmission medium is generally *not* brought into the racks, equipment bays, etc., of a host computer where it would be subject to accidental disconnection and other physical abuse that could disrupt the entire network. Instead, the connection point for a host is designed to be physically stable: a box on the wall, above a false ceiling, etc.

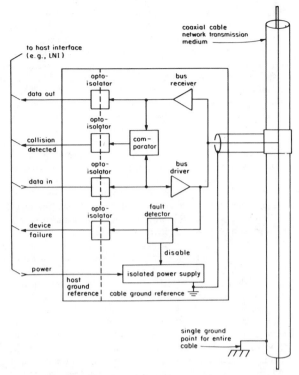

Fig. 8. A typical bus transceiver. The opto-isolators and isolated power supply permit the drivers and receivers to be referenced to cable ground; the cable, in turn, is grounded at only one point along its length, eliminating problems that would result if each transceiver tied the cable to local host ground.

a short stub cable attached to the main cable via a tap. Second, since the transceiver is located adjacent to the network bus cable, and the LNI is located next to its host, an appropriate transmission scheme must be selected to span the intervening distance. For distances up to 30 ft or so, "single-ended" drivers and receivers will suffice. For better reliability, greater distances, or both, differential signals over a shielded twisted pair can be used—just as in the transmission medium of the ring network itself. So, the signal conditioning section of the original LNI can be modified to interconnect the LNI and the cable transceiver.

4) The Cable Transceiver: The care taken in the design of a cable transceiver for a contention bus network will strongly influence the overall reliability and performance of the network. Therefore, we conclude our case study by examining a hypothetical contention bus cable transceiver, shown in Fig. 8, that is similar to one designed and built for the CHAOS Network at the MIT Artificial Intelligence Laboratory; it is typical of transceivers built for various contention bus networks.

The cable transceiver performs the following functions:

1) transmission (cable driving);
2) reception;
3) power and ground isolation;
4) collision detection;
5) transceiver fault detection ("watchdog").

The first three of these constitute part of the signal conditioning function described previously.

The basic design principle of the transceiver is that it must present a high impedence to the bus except when it is transmitting and actually driving the bus. This is essential to the operation of the contention bus network; a large number of receivers on the bus must not present impedence lumps or in any way interfere with a transceiver which is actively transmitting.

The receiver must be able to detect and properly receive signals from the most distant point on the bus; in addition, it must be able to detect a colliding signal while its companion transmitter is itself driving the bus. This requirement impacts the choice of an encoding scheme for data transmitted on the bus. A number of data encoding schemes can be used, all of which require that the transmitter be able to place the transmission medium in two distinct states. At first glance, it might seem that *three* states could be used: the quiescent, high-impedance state, to indicate that no transmission is in progress, and two active driver states, for example $+V$ and $-V$. However, with two active driver states, when two or more network nodes attempt to transmit simultaneously, the cable will be driven to different voltage levels at different points. This has two effects. First, it places a severe load on drivers. Second, it makes the detection of a colliding signal more difficult than it needs to be. On the other hand, if the transceiver drives the cable to some voltage to represent one signaling state, and represents the other signaling state by *not* driving the cable, the problem of overloaded drivers is eliminated, and the task of collision detection is greatly simplified. Collision detection is accomplished looking at the bus during the transmitter's quiescent state. Any signal present during that time must come from another transceiver, and constitutes a collision. The transceiver can detect an incoming signal with 20-dB attenuation, which corresponds to about 1 km of the particular cable used.

The transceiver must be able to cope with ground potential differences at the various network hosts. Isolation is accomplished by high-speed optocouplers and an isolated power supply which enables the major circuit elements of the transceiver to be referenced to cable ground, rather than local host ground. Finally, the fault detection, or watchdog circuit examines the output of the driver to guard against transceiver failures which drive the bus and disrupt the network. The signaling states used by the transceiver result in the driver being quiescent approximately 50 percent of the time; if the driver remains on steadily for several bit-times, it is deemed to be faulty, and the fault detector disconnects its power, which, of course, returns the driver to its high-impedance state.

5) Complexity of the Local Network Interface: In its present form, the LNI comprises about 350 TTL SSI and MSI integrated circuits, apportioned as follows:

PDP-11 full-duplex DMA	100
Name table controller	25
Name table cells (8 provided)	90
Network-oriented portion	120
Test and diagnostic	15
Total	350

The count of 120 chips for the network-oriented portion of the LNI, excluding the associative name table, is well within the capabilities of current large-scale integration. As the field of local area networking matures, and standards are arrived at, it is likely that integrated circuit manufacturers will add local area network controllers to their product lines, to take their place alongside other LSI data communication chips which are already available, making high-performance local area network technology available at a very reasonable cost.

V. PROTOCOLS FOR LOCAL AREA NETWORKS

As in long-haul networks, local area network protocols can be divided into two basic levels—low-level protocols and high-level protocols. At each level, the characteristics of local networks impact effects on protocol design and functionality.

A. Low-Level Protocols

The term *low-level* protocol identifies the basic protocols used to transport groups of bits through the network with appropriate timeliness and reliability. The low-level protocols are not aware of the meaning of the bits being transported, as distinct from higher level application protocols that use the bits to communicate about remote actions. Two aspects of local area networks have a very strong impact upon the design of low-level protocols. First, the high performance achievable purely through hardware technology enables the simplification of protocols. Second, low-level protocols must be designed to take advantage of and preserve the special capabilities of local networks, so that these capabilities can be utilized, in turn, by higher level application protocols. We will explore these two issues in this section.

1) Simplicity: Local area networks must support a wide variety of hosts, from dedicated microprocessors to large time-sharing systems. The existence of extremely simple hosts (such as microprocessor-based intelligent terminals, or even microprocessor printer controllers) leads to a desire for simple, flexible, low-level protocols that can be economically implemented on small hosts, while not compromising the performance of large hosts. Supporting a variety of hosts also leads to a difficult software production and maintenance problem that can be ameliorated somewhat by having a protocol that is simple to implement for each new kind of host. Although quite a variety of hosts has been attached to long-haul networks such as the ARPANET, the problem of software development has not been too severe, since each individual host in such environments usually has a software maintenance and development staff. In the local area network context where a variety of computers are all maintained by a small programming staff, the arguments for simplicity in protocol design are far stronger in our view.

In a long-haul network, complexity results from strategies that attempt to make as much of the costly network bandwidth as possible available for transport of high-level data. The costs of a local area network are concentrated instead in the host interfaces, the hosts themselves, and their software. Two factors lead to the simplicity of low-level local area network protocols.

a) Unrestricted use of overhead bits: Bandwidth is inexpensive in a local area network; there is little motivation to be concerned with protocol features designed to reduce the size of the header or overhead bits sent with each message. This is in contrast to protocols developed for networks making the more conventional assumption that bandwidth is expensive. For example, the ARPANET NCP host-to-host protocol [26] initiates a connection using a 56-bit (net, host, socket) identifier for the destination, but then goes through a negotiation so that instead of sending this 56-bit value on subsequent messages, a 32-bit (net, host, link) value can be sent instead. It is not clear whether this conservation of bits is appropriate even in a long-haul network; in a local area network, where bandwidth is inexpensive, it is clearly irrelevant. Other examples of ways in which extra header space can be used to simplify processing include:

1) having a single standard header format with fields in fixed locations, rather than having optional fields or multiple packet types; field extraction at the host can be optimized, reducing processing time;

2) using addresses that directly translate into addresses of queues, buffers, ports, or processes at the receiver without table lookup.

b) Simplified flow control, etc.: The low transmission delay inherent in local area networks, as well as their high data rate, can eliminate the need for complex buffer management, flow control, and network congestion control mechanisms. Consider, for example, flow control: the problem of assuring that messages arrive at the recipient at the rate it can handle, neither too fast, so that its buffers overflow, nor too slow, so that it must wait for the next message after processing the previous one. In a long-haul network, a receiver typically allocates to the transmitter enough buffer space for several messages following the one currently processed by the receiver, so that messages can be placed in transit well before the receiver is ready to process them. Considerable mechanism is required to keep the sender and the receiver properly synchronized under these circumstances. In a local area network, the delay will typically be low enough for a much simpler flow control mechanism to be employed. For example, one can use the very simple strategy of not sending a message until the recipient has explicitly indicated, by a message in the other direction, that it is ready for it. In contrast, a network using communication satellites has such a high transmission delay that very complex predictive flow control algorithms must be used to obtain reasonable data throughput.

It is crucial to understand that other factors may obviate these simplifications. While the data rate and delay characteristics of a local area network can render it essentially instantaneous, its speed cannot eliminate the intrinsic disparity that may exist between the capabilities of two hosts that wish to communicate with each other. These disparities may not show up when the two hosts are communicating through a long-haul network whose characteristics are so constraining that the principal problem is dealing with the restrictions of the network. While protocols for local area networks need not include mechanisms designed to cope with the limitations of the network itself, it is still necessary to design protocols with sufficient generality to cope with disparities between the capabilities of machines wishing to communicate through the network. Such disparities include:

1) mismatch between the rate at which hosts can generate and absorb data;

2) host delay between the time a packet is received and the time it is successfully processed and acknowledged;

3) amount of buffer space available at the sender and the receiver.

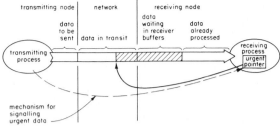

Fig. 9. The urgent pointer mechanism. By transmitting a new, larger value of the urgent pointer, a pointer into the data stream, a sender can indicate the data buffered in the sender, network; and receiver are holding up data that must be processed quickly. The receiver can then adjust his use of the data stream flow control to process the buffered data until the urgent data is processed. The shaded area indicates the location of potentially urgent data specified by a particular urgent pointer value.

Further, considerable effort may be required to modify host software to provide a suitable interface to the network. If one were to consider the simple flow control mechanism mentioned earlier, where a message is sent in the reverse direction requesting transmission of each message as it is needed, one would discover that in many cases the scheme was unworkable, not because the network introduced intolerable delays, but because the hosts communicating with each other themselves introduced excessive delay. In a large host with a time-shared operating system, for example, the real time that elapses from the time a message is received, one or more processes are scheduled in response to this message, and that process runs, to the time a message is sent in response, could well run into a large number of milliseconds, milliseconds during which the other host is forced to wait.

c) Example of protocol simplification: The low-level protocol initially proposed for the Laboratory for Computer Science Network at MIT is an example of the sort of protocol that results when simplicity of mechanism is a primary design goal. The Data Stream Protocol (DSP) was based on the Transmission Control Protocol (TCP) used in internetworking experiments sponsored by the Defense Advanced Research Projects Agency [27], but evolved from original TCP due to the continuing desire to simplify the protocol features, packet formats, and implementation strategies. Most of these simplifications have subsequently been incorporated into the TCP.

One specific example is the mechanism used to signal *interrupts* and other urgent messages that are logically part of the sequence of data in a virtual circuit. The basic model is that the sender occasionally wants to signal the receiver that all data in the stream preceding the signal (buffered somewhere in the network) must be scanned immediately in order to respond promptly to some other important signal. A mechanism is provided whereby a pointer into the data stream is maintained at the receiver, which can be moved, when the sender chooses, to point to a more recently transmitted piece of data. This pointer, called the *urgent pointer*, can be used to indicate the point in the data stream beyond which there is no more urgent data. (See Fig. 9.) The urgent pointer can be implemented in two ways, depending upon the nature of the host receiving the message. In the case of a simple (e.g., microprocessor) host dedicated to a task that processes the incoming stream as it arrives, the host need not process the urgent pointer, since by design, all data, urgent or not, are processed as quickly as possible. In contrast, on a large time-shared host, data need not be processed until either

a) the process to receive the data is scheduled and requests input, or b) the urgent pointer points to data not already received by the process. In case b) an interrupt is sent to the receiving process, indicating that data should be read and processed until the urgent pointer is past. The corresponding mechanism in TCP required that a host be capable of understanding and responding to a special interrupt signal in the data stream, even if the signal had no meaning to the host in its particular application of TCP. The urgent pointer, then, is a simple mechanism that meets the needs of sophisticated host implementations without placing an excessive burden on unsophisticated hosts.

2) Special Capabilities: The other aspect of low-level protocols for local area networks to be discussed is the manner in which protocols must be structured to take advantage of, and provide to higher levels, the unique capabilities of local networks. Conventional low-level protocols have provided a function best characterized as a bidirectional stream of bits between two communicating entities—a *virtual circuit*. The virtual circuit is implemented by a process that provides sequenced delivery of packets at the destination. While a virtual circuit is one important form of communication, two others easily provided by a local network are very useful in a variety of contexts. These are *message exchange* communication, where the packets exchanged are not viewed as being members of a sequence of packets but are rather isolated exchanges, and *broadcast* communication in which messages are sent not to one particular recipient but to a selected subset of the potential recipients on the network.

a) Message exchange: A typical example of a message exchange is the situation in which one message asks a question and another provides the answer. For example, if there are a large number of services provided by nodes connected to a local net, it is disadvantageous to maintain, on every node, a table giving all of the addresses of these, for whenever a change is made in the network address of any service, every node's table will need to be revised. Rather, it may be advantageous to maintain, as a network service, a facility which will take the name of a desired entity and give back its network address. Clearly, the pattern of communication with this service is not one of opening a connection and exchanging a large number of messages, but instead is a simple two-message exchange, with a query of the form "What is the address of such and such a service?" and a reply of similarly simple form. While a virtual circuit *could* be used for this exchange, it is unneeded and uses excessive resources.

b) Broadcast: The example given above demonstrates the need for a broadcast mechanism. If the service described above is intended to provide the address of network services, how can we find the address of this service itself? An obvious solution is to broadcast the request for information. The query then takes the form "Would anyone who knows the address of such and such a network service please send it to me?" There are many other examples, some apparently trivial but nonetheless very useful, for support of broadcast queries in a local network. A microprocessor with no calendar clock may broadcast a request for the time of day. A new host attached to the network for the first time may broadcast a message announcing its presence, so that those who maintain tables may discover its existence and record the fact. Broadcast mechanisms in the low-level protocols can also be quite useful in implementing higher level protocols for such applications as document distribution to multiple host nodes, and for speech and video conference calls.

Why are these alternative models of communication not commonly found in traditional networks? The first, and perhaps most important reason is that long-haul networks have not been extensively exploited for applications in which computers directly query other computers with individual, self-contained queries. Instead, the major use of long-haul networks has been for long-term, human-initiated interactions with computers, such as direct terminal use of a remote computer, or long-term attachments of remote job entry stations. Such human interactions usually involve many message exchanges between sender and receiver, so that the extra delay and cost of initial setup of a virtual circuit is insignificant—perhaps even recovered by reducing redundant information in each message. As new applications such as distributed data base systems become more important, these alternative models will become important in long-haul networks, but long-lived connections between terminals and host computers continue to dominate the usage.

The second reason is precisely that discussed in the previous section concerning the relative simplicity of protocols for local area networks—a variety of functions performed in conventional networks are very difficult to understand except in the context of a sequence of ordered messages (a virtual circuit) exchanged between two nodes. For example, flow control is normally handled in network protocols by placing an upper bound on the number of messages which may be flowing at any one time between the sender and the receiver. This concept has meaning only in the restricted case where the sender and the receiver are a well-identified pair exchanging a sequence of messages. There is no obvious equivalent of flow control that can be applied to situations where sender and receiver communicate by sending arbitrary unsequenced messages, or where a sender broadcasts to several receivers. Similarly, if efficiency requires use of the shorthand version of an address for communication between the sender and the receiver, this clearly implies that the sender and the receiver have negotiated this address, and agree to use it over some sequence of messages. Again, this idea makes no sense if communication is isolated in unsequenced messages.

Another problem that is traditionally handled in the context of a sequence of messages is the acknowledgment to the sender that the receiver has correctly received a message. If messages are sequenced, acknowledgment can be very easily done by acknowledging the highest member of the sequence that has been successfully received. If messages bear no relationship to each other, then each must be identified uniquely by the sender, and acknowledged uniquely by the receiver. This increases the complexity and overhead of acknowledgment. However, in most cases where message exchange communication is the appropriate underlying communication model, no acknowledgment mechanism is required of the low-level protocol at all. For example, if a microprocessor system asks the time of day, it is not at all necessary to acknowledge that the query has been successfully received; the receipt of the correct time is sufficient acknowledgment. Similarly, a request for a network address is acknowledged by a return message that contains the desired address. Depending on a low-level acknowledgment message to handle all failures can be dangerous, for it may lead to the practice of assuming that acknowledgment of receipt of a message implies that the message was processed at a high level.

In the broadcast context, it is difficult to formulate a useful definition of acknowledgment that can be supported by a low-level protocol. What does it mean to say that a broadcast message has been successfully received? By one of the possible recipients? By all of the possible recipients? One appropriate strategy is to rely on the high-level application to deal with these problems as a part of its normal operation, rather than have the low-level protocol concern itself with issues of flow control or acknowledgment at all.

3) Protocol Structure: Based on the previous observations, a two-layer structure is a very natural one for low-level protocols in a local area network. The bottom layer should provide the basic function of delivering an addressed message to its (one or many) destinations. This level corresponds to the concept of a *datagram* network [28]. It should also take on the responsibility of detecting that a message has been damaged in transit. To this end it may append a checksum to a message and verify the checksum on receipt. However, this layer probably should not take on the responsibility of ensuring that messages are delivered, and delivered in the order sent, since different applications have different needs and requirements for these functions. The first layer might be implemented entirely in hardware; however, if the packet size, addressing structure, or routing topology of the hardware is not sufficient to provide adequate message size, process addressing, or broadcast selectivity, some software help will be needed to make up the difference.

Above this first layer should be made available a variety of protocols. One protocol should support a virtual circuit mechanism, since a virtual circuit is definitely the appropriate model for a great deal of the communication that will go on in any network, local or otherwise. As alternatives to the virtual circuit protocol, there should be mechanisms for sending isolated messages, for message exchange communication, and additional alternatives to provide support for message models other than the ones we have discussed here. For example, transmission of digitized speech requires a communication model with some but not all of the attributes of the virtual circuit; in particular, reliability is of less concern than timeliness of arrival.

B. Applications of Local Area Networks; Higher Level Protocols

In the previous section we considered low-level protocols for a local area network. These protocols exist, of course, to support higher level protocols, which, in turn, support user applications. In this section we will consider a number of applications for which local area networks are suited.

1) Access to Common Resources: The model of computing most common over the last few years is that of a large centralized computer, with the only remote components being terminals and, perhaps, a few other I/O devices. Line control protocols such as SDLC [19] were created to serve this sort of arrangement. A simple but very important application of a local area network is to generalize this picture very slightly to include more than one central computer. As the total workload grows to exceed the capacity of a single machine, a common solution is to procure a second machine, and to divide the applications and workload between the two. The communication problem to be solved in this arrangement is simple but critical—to allow an individual terminal to have access to both of the central machines. A local area network can solve this problem, and provide some additional capabilities as well. For example, if the central facility has specialized I/O devices such as plotters or microfilm writers, they

can be placed on the local area network and made accessible to both central machines—an advantage if a device is expensive and is not heavily enough loaded to justify having one for each computer. Further, I/O devices can be placed remote from the central site but convenient to users; for example, a line printer can be placed near a cluster of users.

This pattern of sharing among several computers can be expanded to include more than just I/O devices. In fact, the network can be used to move computations from one machine to another in order to spread the computing load equally. The high speeds available in the local area networks make this sort of load leveling much more practical than do the bandwidths traditionally available on long-haul networks.

2) Decentralized Computing: A wide variety of new uses for a local area network arises if the computing power available is not strongly centralized. Let us consider the alternative of a computing environment consisting of a large number of relatively small machines, each dedicated to a small number of users or a small number of tasks. In the extreme, we can look to the future and imagine the day when each user has a computer on his desk instead of a terminal. Such a completely distributed computing environment by no means eliminates the need for an interconnecting network, for users will still need to exchange information. Data files containing the results of one person's computation will need to be shipped through the local area network to be used as input to other tasks. Users will wish to communicate with each other by exchanging computer mail, as is now done over the ARPANET [29]. Users will still want access to specialized resources which cannot be provided to each user, resources such as large archival storage systems, specialized output devices such as photo typesetters, or connection points to long-haul networks. All of these features can be made available through the local area network.

3) Protocol and Operating System Support: The applications outlined in the previous paragraph can be supported by high-level protocols very similar to the ones already in existence in the ARPANET: TELNET for logging into a remote system through the network, and File Transfer Protocol for exchanging data between machines [26]. When one examines how these protocols might be modified to take advantage of the special attributes of a local area network, for example, its higher speed, one discovers that the problem is not one of modifying the protocols, but of modifying the operating system of the hosts connected to the network so that the services available through the network appear to be a natural part of the programming environment of the operating system. The File Transfer Protocol in the ARPANET, for example, is usually made available to the user as an explicit command which he may invoke to move a file from one machine to another. As part of this invocation he may be required to identify himself at the other machine, and give explicit file names in the syntax of the local and the foreign machine, describing exactly what action he wishes to perform.

This particular view of file transfer has two disadvantages. First, there is a lot of overhead associated with moving a file. Much of the delay in moving the file seen by the user has nothing to do with the time required to send the data itself through the network, but is rather the time spent establishing the connection, identifying the user at the other site, etc. Second, the file system on the local computer understands nothing about the existence of files accessible through the network. No matter how sophisticated the local file system is, in terms of keeping track of the various files that the user cares about, it requires explicit user intervention in order to reach through the network and retrieve a file from another machine. The use of a high-speed local area network will not eliminate any of these problems, but will instead make even more obvious to the user the overhead that the protocol imposes on the transfer of data. Clearly, what is needed is a further integration of the local area network into the file system and user authentication mechanism of the individual operating systems, so that interchange of information between the various machines can be done with less direct user intervention. Some attempts have been made to do this within the context of the ARPANET. RSEXEC is an example of a protocol which makes files on various TENEX operating systems in the ARPANET appear to the user to exist on a single machine [30].

The design of operating system structures to take full advantage of the capabilities of local area networks represents the current edge of research in this area. Examples of operating systems that incorporate a high-speed local area network into their architecture are the Distributed Computing System [31], the Distributed Loop Operating System [11], and MININET [32].

VI. Interconnection of Local Area Networks with other Networks

A. Motivation for Interconnection

As was mentioned earlier, a local area network will be only a part of the overall communication system used by the hosts attached to it. A very important use of the local area network can be to provide an interconnection between hosts attached to a local area network and other networks such as long-haul packet-switched networks and point-to-point transmission links. The advantage of this method of interconnection is reduced cost, by taking advantage of the fact that connection of a host to a local area network is relatively inexpensive. Instead of connecting all machines directly to the long-haul network, one can connect all the host computers to the local area network, with one machine, the *gateway*, connected to both the local area network and the long-haul network.

B. Protocol Compatibility

There are two pitfalls that should be avoided when planning for the interconnection of a local area network with a long-haul network. On the one hand, long-haul networks currently cannot provide all of the functions that local area networks can. If a local area network is initially designed to serve only the function of connecting hosts to a long-haul network, the protocols of the local network may be designed to serve only the needs of communicating with the long-haul network, and may not support the other functions that make a local area network especially attractive. On the other hand, if a local network is initially designed with no thought given to the possibility that it may be interconnected with another network, the protocols designed for it may lack the necessary generality. For example, the addressing structure used on the local area network may not be able to express destinations outside the local network. In either case, the only after-the-fact solution is to implement a second set of protocols for the local area network, so that different protocols are used for intercommunication with long-haul networks and for local services. This proliferation of protocols is undesirable,

as it adds to the cost of software development associated with each new host added to the local area network. To avoid these pitfalls, it is important that all the functions a local area network is to provide must be considered from the very inception of the design of the network, and the protocols for the network must be designed to support that entire range of functionality.

Fortunately, initial experiments with protocols for local area networks suggest that a uniform approach to protocol design can support both specialized local network functions and interconnection with other networks, provided that both functions are envisioned from the start. Although the protocols used in the local area network must be made slightly more general to handle the internetworking situation, there is no interference with the realization of the purely local network functions. For example, a more general address field must be used to specify the destination of a message, but the only overhead implied if this same addressing structure is used for purely local messages is additional bits in the message to hold a presumably larger address. Since bandwidth is inexpensive, the bits "wasted" on this larger address are presumably irrelevant.

A slightly more difficult problem, one that is still being studied, is the problem of speed matching between the local area network and the long-haul network. As this paper has characterized the difference between local nets and long-haul nets, it is reasonable to presume that the local network will have a much higher data rate. If a host sends a large number of packets into the local area network with an ultimate destination to be reached through the long-haul network, the packets may arrive at the gateway much faster then the gateway can pass them to the long-haul network. Some mechanism will be required to prevent the gateway from exhausting its buffer space. The speed matching problem is not unique to the gateway between the local area network and the long-haul network; it occurs any time two networks of differing speed are connected together. (The problem may be more extreme here, though, due to the greater speed difference that can be encountered between local area and some long-haul networks. Satellite networks with speeds comparible to local networks are quite conceivable, yet are a long-haul technology.) A general discussion of the problems of internetworking, and some proposed solutions can be found in a companion paper by Cerf and Kirstein in this issue [33].

At the next higher level of protocol, one finds facilities that support various communications models, such as virtual circuits, broadcast, and message exchange. In interconnecting to a long-haul network we are chiefly forced to deal with a virtual circuit model, since that is the only pattern of communication usually supported by commercial long-haul networks. Here, it is appropriate to use a virtual circuit protocol in the local area network as similar as possible to that used in the long-haul network, so that translation between the two is easy. Although there is not as much practical experience available in the area of network interconnection as could be desired, it appears that one can develop a virtual circuit protocol for a local area network that is a compatible subset (in the sense of using compatible packet formats and control algorithms) of a suitable long-haul virtual circuit protocol. This means that it is not necessary to implement two complete virtual circuit protocols, one for internal local network use and the other for communication out through the local net. It leaves unanswered the question of how the additional features, such as complex flow control, buffering, and speed matching required for the long-haul protocol should be implemented. One approach would be to implement them in every host that desires to communicate over the long-haul network; this implies a programming burden for every machine. An alternative would be to implement the additional functions in the gateway machine that interconnects the local area network to the long-haul network. This would add considerable complexity to the gateway, for it will have to cope with such problems as the speed differential between the two networks without having the benefit of the flow control mechanisms normally used for this purpose in the long-haul network. At this time, it is not clear whether the gateway can assume the entire responsibility for augmenting a local network virtual circuit protocol with the functions required for communication through a long-haul network.

It would be advantageous to make sure the local area network protocols are also compatible with other communication models, such as single message exchange or selective broadcast, that may become available on commercial long-haul networks in the future. However, this presupposes that the long-haul networks attached to the local area network use a two-layer low-level protocol implementation such as that described for the local area network, and if the long-haul networks do use such an implementation, that they provide an interface that allows direct use of the datagram layer. Many current long-haul networks do not provide that interface.

VII. THE SUBNETWORK CONCEPT

Resting midway between the monolithic, single-technology, local area network and the internetworking environment is an approach to local area networking that we term the *subnetwork concept*, which provides for a mix of network technologies within a uniform addressing and administrative structure.

A. General Approach

A local area network can be composed of a collection of subnetworks, possibly implemented with various network technologies and perhaps with various transmission rates, but using identical software protocols, compatible packet sizes, and a single overall homogeneous address space.[5] These subnetworks are interconnected by *bridges*, which are midway in complexity between the repeaters used in a multisegment contention bus network (ETHERNET) and the gateway processor used between networks in an internetworking environment. This general structure is indicated in Fig. 10. A bridge links two subnetworks, generally at a location at which they are physically adjacent, and selectively repeats packets from each of them to the other, according to a "filter function."[6] In addition, since they buffer the packets they repeat, they can perform a speed-matching function as well.

B. Advantages of Subnetworking

The subnetworking concept enables a variety of technologies and data rates to be utilized in a single local area network, each to its best advantage. For example, a network could

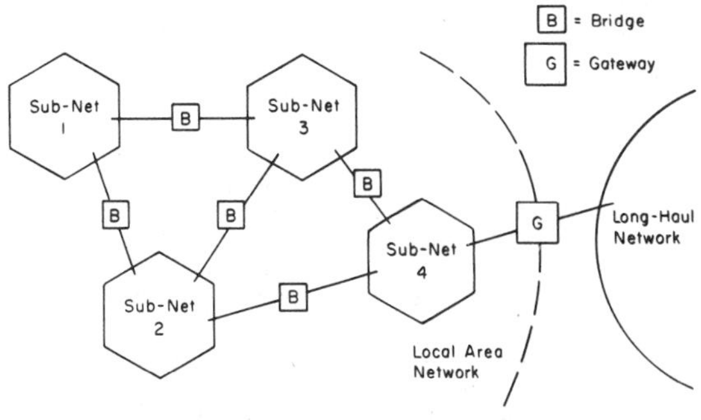

---

[5] The subnetwork concept, as we describe it, is a generalization of an approach suggested by Pierce [5] for use with multiple loops or rings.
[6] The concept of the filter function is introduced in the "filtering repeaters" described by Boggs and Metcalfe [14].

Fig. 10. The subnetwork concept. Here, a local area network is composed of a number of subnetworks, linked in some fashion by bridges. The subnetworks, though of differing technologies, share one address space, and the same protocols are used over the entire network. Thus, the bridges can be simpler than the gateway which connects the local area network to the long-haul network. Viewed externally, from outside the dashed line in the figure, the local area network appears to be monolithic.

be constructed with a contention bus subnetwork, perhaps using coaxial cable originally installed for CATV, and with a ring subnetwork, using twisted pair which can be easily installed in a crowded laboratory environment. These two subnetworks could be of different data rates; the bridge between the two will handle the speed difference between them.

Subnetworking also provides an orderly means for handling growth in traffic. Local area networks perform best, providing high throughput with low delay, when they are not heavily loaded. As traffic on a local area network grows with time, if a higher speed technology is not available, it may be desirable to split the network into two or more interconnected subnetworks. Since the bridges which interconnect the subnetworks are selective in their repeating of packets "across the bridge," not all packets from a subnetwork will flow to all other subnetworks, and the traffic density on each subnetwork will be less than that of the original monolithic network. If the partitioning of the hosts into subnetworks can be done along the lines of "communities of interest," such that a group of hosts with high traffic rates among themselves but with substantially lower traffic rates to other hosts are placed in the same subnetwork, traffic across the bridges will be minimized, and a greater fraction of all packets will stay within their subnetwork of origin.

C. Bridges

A bridge, depicted in Fig. 11, contains:

two network interfaces, one appropriate to each of the two subnetworks it interconnects,
a limited amount of packet buffer memory, and
a control element, which implements an appropriate filter function to decide which messages to "pull off" one subnetwork and buffer until it has an opportunity to retransmit it to the other subnetwork.

The topology of the subnetworks interconnected by a bridge determines the complexity of its filter function. A bridge with a simple filter function can be implemented using a finite state machine as its control element; a complex filter function, which may involve a periodic exchange of information among bridges on the network to determine correct routing, may require the capabilities of a microprocessor [34].

A bridge *must* buffer packets since, upon receiving a message from one subnetwork which it decides to repeat to the other

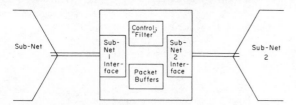

Fig. 11. The structure of a bridge. A bridge would most naturally be located at a point where the two subnetworks it interconnects have been made physically adjacent.

subnetwork, it must wait for an opportunity to transmit on that subnetwork, according to the control structure of that subnetwork. Packet buffers also aid a bridge in handling instantaneous cross-bridge traffic peaks during which the traffic offered by one subnetwork exceeds the available capacity of the other. This situation can arise if the bridge interconnects subnetworks of dissimilar data transmission rates, or subnetworks of drastically different traffic densities. However, if the sustained cross-bridge traffic offered is greater than the target subnetwork can handle, the bridge must discard packets. This is an acceptable course of action, as local area network protocols are generally prepared to handle lost packets.

D. Transparency

The subnetwork structure of a local area network should be transparent, both to the hosts on the local area network and to the "outside world"—other networks to which the local area network may be connected via gateways. A host on the local area network wishing to transmit a packet to another need have no knowledge of whether that host is on the same subnetwork, in which case the packet will be received by the destination host directly, or whether the destination host is on another subnetwork, in which case the packet is retransmitted by one or more bridges. In particular, no ordinary data packets *are ever addressed* to a bridge; rather, packets are simply addressed to their destination hosts, and may be picked up by a bridge and passed along through other subnetworks, finally reaching their destinations. This is a key distinction between subnetworking, with bridges, and internetworking, with gateways: in internetworking, a host about to transmit a packet must realize that the host to which it is addressed is on a different network. The sending host must transmit the message in a local network "wrapper" to an appropriate gateway, which "unwraps" it, performs protocol conversions, if any, packet fragmentation, etc., as necessary, and then transmits the message into the other network. In subnetworking, protocols are identical over all subnetworks, and packet sizes are compatible, so that neither protocol conversion nor fragmentation takes place in the bridges. Finally, as was mentioned above, a packet is directly addressed to its destination host, not to a bridge, for hosts do not know that the local area network is composed of subnetworks.

E. Impact on Network Characteristics

Splitting a local area network into subnetworks has little impact on the key characteristics of the network. From the point of view of the users and hosts of the network, addressing is affected only slightly, if at all. Bridges must determine whether or not a packet should be picked up for retransmission; one way to aid bridges in this determination is to include

a subnetwork field in the address of each host. Other routing techniques which have no impact at all on addressing (such as complete table look-up of host addresses by the bridges) can be implemented, although usually at the expense of greater complexity within the bridges.

Splitting a local area network into subnetworks should have no effect on the protocols of the network. One exception is if a particular subnetwork technology provides a hardware acknowledgment of delivery of a packet (as in the DCS Ring Network) [2]; this acknowledgment may only indicate successful receipt by a bridge. However, not all network technologies provide hardware acknowledgments, and, in a network of mixed technologies, host-to-host acknowledgments will generally be provided by software protocols. Traffic is, of course, affected by subnetworking in a positive way. Splitting a local area network into subnetworks in a judicious way can minimize the overall traffic of the network; bottlenecks can be eliminated by using higher bandwidth technologies for affected subnetworks.

F. The Long-Distance Bridge

There are situations in which it is necessary to interconnect two subnetworks of a local area network which cannot be brought physically adjacent to one another so that an ordinary bridge may be connected between them. An example of this would be a local area network on a university campus, with a major research laboratory across town. The laboratory may be beyond the range of a twisted-pair ring network or a coaxial cable contention bus network; or it may be within range, but it may be impossible for the university to install its own cables between them.[7] The off campus research laboratory can be given its own subnetwork, connected to the main campus subnetwork via a specialized *long-distance bridge*.

A long-distance bridge is made up of two *half-bridges* at either end of a suitable full-duplex point-to-point communication link, such as a high-bandwidth common carrier circuit, an optical link, or a private microwave link (Fig. 12). Some other network technology such as packet radio could be used to derive this point-to-point link as desired.[8] Each half-bridge contains an appropriate interface to its subnetwork, packet buffers, and a controller. In addition to its filtering function, the controller of a half-bridge regulates the flow of data over the communication link between the two halves of the bridge. Of course, it is possible that the bridge communication link may be of lower bandwidth than the two subnetworks it interconnects. Additional packet buffers at each half-bridge can help to smoothe out traffic peaks, but if the communication link is a bottleneck, the long-distance bridge must discard packets just as an ordinary bridge does when it is overloaded.[9]

---

[7] Although common carriers such as the Bell System operating companies are moving in the direction of leasing wire pairs for transmission of digital signals with customer-provided equipment, these circuits are not intended for use at the high bandwidth of local area networks, and are generally routed through central offices rather than point-to-point.

[8] Although we do not discuss it further in this paper, there is an interesting philosophical issue whether the intervening network should be viewed in the internetworking context using gateways or as a point-to-point link within a single bridge.

[9] If the bottleneck created by the communication link of a long bridge is severe, the local area network advantages of high-bandwidth communication with low delay will be forfeited.

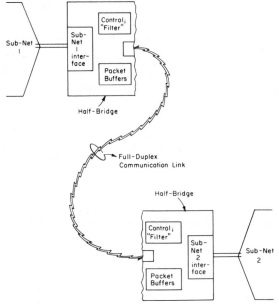

Fig. 12. The "long bridge." In this case, the two subnetworks cannot be made physically adjacent, so a half-bridge is attached to each, and a full-duplex communication link is employed to interconnect the two half-bridges. The control and filter functions, and the packet buffers, are replicated in each half-bridge.

VIII. CONCLUSION

The utilization of a technological innovation often occurs in two stages. In the first stage, the innovation is exploited to perform better the same tasks that were already being performed. In the second stage, new applications are discovered, which could not be reasonably performed or even forseen prior to the innovation. Local area networks are now on the threshold of this second stage. While there is still much room for creativity in improving the innovation itself—reducing the cost of the network interface and increasing its speed and convenience—the real challenge lies in identifying new sorts of applications that a local area network can make possible.

Current trends in hardware costs encourage abandonment of a single large computer in favor of a number of smaller machines. This decentralization of computing power is, for many applications, a natural and obvious pattern. In many information processing applications, for example, the information itself is distributed in nature, and can most appropriately be managed by distributed machines. Distributed applications can only be constructed, however, if it is possible to link their machines together in an effective manner. Subject to their geographical limitations, local area networks offer a very effective and inexpensive way to provide this interconnection. The greatest impact of local area networks will come with the development of operating systems that integrate the idea of distribution and communication at a fundamental level.

The impact of local area networks on the decentralization of computing is sociological as well as technological. Operational control of centralized computers has traditionally been vested in the staff of a computer center. The trend toward decentralized computing greatly increases the autonomy of individual managers in the operation of their machines, and appears to reduce the need for a centralized staff of computer managers. The communication capability made available by local area networks will serve to bind these decentralized machines together into a unified information processing resource. The effectiveness of this resource can be measured by the degree of coherence it achieves, which, in turn, depends upon the care and foresight put into the design of the local area network and the development of standards for communication at all levels. It is in the identification of areas in which standards are needed, and in their development, that the staff of the "computer center" of the future will find its work.

REFERENCES

[1] D. J. Farber and K. C. Larson, "The system architecture of the distributed computer system—The communications system," presented at the *Symposium on Computer Networks* (Polytechnic Institute of Brooklyn, Brooklyn, NY, Apr. 1972).
[2] A. G. Fraser, "On the interface between computers and data communications systems," *Commun. Ass. Comput. Mach.*, pp. 31–34, July 15, 1969.
[3] IEEE Instrumentation and Measurements Group, *IEEE Standard Digital Interface for Programmable Instrumentation*, IEEE Standard 488, 1975.
[4] D. C. Loomis, "Ring communication protocols," University of California, Department of Information and Computer Science, Irvine, CA, Tech. Rep. 26, Jan. 1973.
[5] J. R. Pierce, "Network for block switching of data," *Bell Syst. Tech. J.*, vol. 51, pp. 1133–1143, July/Aug. 1972.
[6] A. Hopper, "Data ring at computer laboratory, University of Cambridge," in *Computer Science and Technology: Local Area Networking*. Washington, DC, Nat. Bur. Stand., NBS Special Publ. 500-31, Aug. 22–23, 1977, pp. 11–16.
[7] P. Zafiropulo and E. H. Rothauser, "Signalling and frame structures in highly decentralized loop systems," in *Proc. Int. Conf. on Computer Communication* (Washington, DC), IBM Res. Lab., Zurich, Switzerland, pp. 309–315.
[8] G. Babic and T. L. Ming, "A performance study of the distributed loop computer network (DCLN)," in *Proc. Computer Networking Symp.*, (National Bureau of Standards, Gaithersburg, MD, December 15, 1977), pp. 66–76.
[9] E. R. Hafner *et al.*, "A digital loop communication system," *IEEE Trans. Commun.*, p. 877, June 1974.
[10] M. V. Wilkes, "Communication using a digital ring," in *Proc. PACNET Conf.* (Sendai, Japan, August 1975), pp. 47–55.
[11] M. T. Liu and C. C. Reames, "Message communication protocol and operating system design for the distributed loop computer network (DLCN)," in *Proc. 4th Annu. Symp. Computer Architecture*, pp. 193–200, Mar. 1977.
[12] N. Abramson, "The ALOHA system," University of Hawaii Tech. Rep. No. B72-1, Jan. 1972; also *Computer Communication Networks*. Englewood Cliffs, NJ: Prentice-Hall, 1972.
[13] R. M. Metcalfe, "Packet communication," M.I.T., Project MAC, Tech. Rep. 114, Cambridge, MA, Dec. 1973.
[14] D. R. Boggs and R. M. Metcalfe, "Ethernet: Distributed packet switching for local computer networks," *Comm. Ass. Comput. Mach.*, vol. 19, no. 7, pp. 395–404, July 1976.
[15] E. D. Jensen, "The Honeywell experimental distributed processor—An overview," *Computer*, Jan. 1978.
[16] Digital Equipment Corporation, *PDP-11 Processor Handbook*. Maynard, MA: Digital Equipment Corporation, 1973.
[17] W. D. Farmer and E. E. Newhall, "An experimental distributed switching system to handle bursty computer traffic," in *Proc. ACM Symp. Problems in the Optimization of Data Communication Systems* (Pine Mountain, GA, Oct. 1969), pp. 31–34.
[18] A. G. Fraser, "A virtual channel network," *Datamation*, pp. 51–56, Feb. 1975.
[19] *IBM Synchronous Data Link Control General Information*, GA27-3093-0, File GENL-09, IBM Systems Development Division, Publications Center, North Carolina, 1974.
[20] P. Mockapetris and D. J. Farber, "Experiences with the distributed computer system," submitted to the *J. Distributed Processing*, 1978.
[21] P. Mockapetris, "Design considerations and implementation of the ARPA LNI name table," Univ. California, Dep. Information and Computer Sci., Tech. Rep. 92, Irvine, CA, Apr. 1978.
[22] D. G. Willard, "A time division multiple access system for digital communication," *Comput. Des.*, vol. 13, no. 6, pp. 79–83, June 1974.

[23] N. B. Meisner *et al.*, "Time division digital bus techniques implemented on coaxial cable," in *Proc. Computer Networking Symp.* (National Bureau of Standards, Gaithersburg, MD, Dec. 15, 1977).

[24] R. E. Kahn, S. A. Gronemeyer, J. Burchfiel, and R. C. Kurnzelman, "Advances in packet radio technology," this issue, pp. 1468–1496.

[25] P. Mockapetris *et al.*, "On the design of local network interfaces," *Informat. Process.*, vol. 77, pp. 427–430, Aug. 1977.

[26] *ARPANET Protocol Handbook*, Network Information Center, SRI International, Menlo Park, CA, NIC 7014, revised Jan. 1978.

[27] V. Cerf and R. Kalin, "A protocol for packet network interconnector," *IEEE Trans. Commun.*, vol. COM-25, No. 1, pp. 169–178, May 1974.

[28] L. Pouzin, "Virtual circuits vs. datagrams—Technical and political problems," in *AFIPS Conf. Proc.* (National Computer Conf., June 1976), p. 483.

[29] D. H. Crocker *et al.*, "Standard for the format of ARPA network text messages," ARPA Network RFC 733, NIC 41952, Nov. 21, 1977.

[30] R. H. Thomas, "A resource sharing executive for the ARPANET," *AFIPS Conf. Proc.*, vol. 42 (Nat. Computer Conf. and Exposition, 1973), pp. 155–163.

[31] D. J. Farber and F. H. Heinrich, "The structure of a distributed computer system—The distributed file system," in *Proc. Int. Conf. on Computer Communication* (Washington, DC, 1972), pp. 364–370.

[32] E. G. Manning and R. W. Peebles, "A homogeneous network for data sharing communications," Computer Communications Network Group, University of Waterloo, Waterloo, ON, Tech. Rep. CCNG-E-12, Mar. 1974.

[33] V. G. Cerf and P. T. Kirstein, "Issues in packet network interconnection," this issue, pp. 1386–1408.

[34] S. L. Ratliff, "A dynamic routing algorithm for a local packet network," S.B. thesis, M.I.T., Department of Electrical Engineering and Computer Science, Cambridge, MA, Feb. 1978.

Advances in Packet Radio Technology

ROBERT E. KAHN, SENIOR MEMBER, IEEE, STEVEN A. GRONEMEYER, MEMBER, IEEE, JERRY BURCHFIEL, MEMBER, IEEE, AND RONALD C. KUNZELMAN, MEMBER, IEEE

Invited Paper

Abstract—Packet radio (PR) is a technology that extends the application of packet switching which evolved for networks of point-to-point communication lines to the domain of broadcast radio. It offers a highly efficient way of using a multiple access radio channel with a potentially large number of mobile subscribers to support computer communication and to provide local distribution of information over a wide geographic area. We discuss the basic concepts of packet radio in this paper and present the recent technology and system advances in this field. Various aspects of spread spectrum transmission in the network environment are identified and our experience with a testbed network in the San Francisco Bay area is discussed.

I. INTRODUCTION

AN EXCITING set of developments has taken place during the last few years in the field of digital radio networks. The advantages of multiple access and broadcast radio channels for information distribution and computer communications have been established and several experimental digital radio networks are now in operation. Packet radio is a perfect example of the rapid technological progress which has been achieved. It utilizes packet-switched communications and is particularly important for computer communications in the ground mobile network environment. In this paper, we provide an overview of the basic concepts of packet radio and discuss the recent technology and system advances in this field. We also address the closely related subjects of signaling in the ground mobile radio environment and the advantages of spread spectrum technology in the multiple access network environment. This paper is intended to be tutorial with emphasis on the current state-of-the-art in packet radio.

The rapid growth in packet communications which has taken place during the last decade following the successful development of the ARPANET [1], [2] is directly related to the increasing demand for effective telecommunication service to handle computer communications [3], [4]. Only in this time frame did it become cost effective to utilize minicomputers, and later microprocessors as packet switches in a large scale network [5]. In a packet-switched network, the unit of transmission is called a packet. It contains a number of data bits, and is usually of variable length up to a maximum of a few thousand bits. A packet includes all the addressing and control information necessary to correctly route it to its destination.

Packet switching was originally designed to provide efficient network communications for "bursty" traffic and to facilitate

computer network resource sharing. It is well known that the computer traffic generated by a given user is characterized by a very low duty cycle in which a short burst of data is sent or received followed by a longer quiescent interval after which additional traffic will again be present. The use of dedicated circuits for this traffic would normally result in very inefficient usage of the communication channel. A packet of some appropriate size is also a natural unit of communication for computers. Processors store, manipulate, and transfer data in finite length segments, as opposed to indefinite length streams. It is therefore natural that these internal segments correspond to the computer generated packets, although a segment could be sent as a sequence of one or more packets. Computer resource sharing techniques which exploit the capabilities inherent in packet communications are still primarily in the research stage, but significant progress has already been achieved in this area [6].

The packet switching concept incorporates individual per-packet processing by each switch or node in the network such that incremental network capacity can be dynamically allocated by a node immediately after it receives a packet. Each packet wends its way from node to node through the network until eventually it arrives at the final destination and is delivered. The transit time through the network is typically a fraction of a second. Due to the low duty cycle of individual user traffic, allocating a portion of the network capacity for that user in advance (e.g., to provide a dedicated circuit) could also lead to very inefficient use of the network resources even if the allocation is valid over a very short time.

An essential attribute of any network is its ability to provide full connectivity among all network participants. Full connectivity implies that any set of computers can communicate subject only to the overall network performance and administrative limitations. Specific performance parameters or constraints such as throughput, delay, cost and reliability are usually quite important to the critical mass of interconnected subscribers. Within a given network environment, delay, throughput, and cost are intricately related because lower delay usually means higher data rates which, in turn, implies higher throughput and greater cost. Of course, the delay can never be reduced below the speed of light propagation delay.

Packet radio [7] is a technology that extends the original packet switching concepts which evolved for networks of point-to-point communication lines to the domain of broadcast radio networks. The rapid development in this area has been greatly stimulated by the need to provide computer network access to mobile terminals and computer communications in the mobile environment. Packet radio offers a highly efficient way of using a multiple access channel, particularly with mobile subscribers and large numbers of users with bursty traffic. During the early 1970's, the ALOHA project at the University of Hawaii demonstrated the feasibility of using packet broadcasting in a single-hop system (see reference [8], [9]). The Hawaii work led to the development of a multihop multiple access packet radio network (PRNET) under the sponsorship of the Advanced Research Projects Agency (ARPA). The PRNET is a fundamental network extension of the basic ALOHA system and broadens the realm of packet communications to permit mobile applications over a wide geographic area. The use of broadcast radio technology for local distribution of information can also provide a degree of flexibility in rapid deployment and reconfiguration not currently possible with most fixed plant installations. Although the original impetus for packet radio development was and still is largely based on tactical military computer communication requirements [10], the basic concept is applicable to an extremely wide range of new and innovative computer communication applications never before possible in any practical way.

In addition to the strong ARPANET and ALOHA system influences, three technical developments in the early 1970's were largely responsible for the evolution of packet switching to the radio environment. The first was the microprocessor and associated memory technology which made it possible to incorporate computer processing at each packet radio network node in a form that was compatible with mobile usage and portable operation. The second was the reduction to practice of surface acoustic wave (SAW) technology which can perform matched filtering (to receive wide-band radio signals) on a very small substrate of quartz or similar piezo-electric material. The third development was purely conceptual and involved an awareness within the computer and communications communities of the importance of "protocols" in the development of network management strategies. It was upon those three pillars that the technical approach to the PRNET was founded.

Packet radio network technology will be essential for military and other governmental needs as terminals and computer systems become pervasive throughout essentially all aspects of their operations. Initially, the needs for radio-based computer communications are expected to be prevalent in training, on or near the battlefield, and for crisis situations. The first operational systems, even if of limited availability, are most likely to be deployed for use in one of these areas where a higher relative cost of providing the advanced capability can be tolerated. Within the civilian sector, there is also a strong need for terminal access to information in the mobile environment, but the cost of service to the user (e.g., personal terminal, tariffs) will dictate when such capabilities should be publicly provided. We expect to see a considerable increase in the usage of civilian terminals and microcomputers "on the move" during the early 1980's but, in contrast to the military environment, these applications are expected to involve relatively simple equipment, reduced capabilities and lower costs.

All users in a packet radio network are assumed to share a common radio channel, access to which is controlled by microprocessors in the packet radios. In contrast to a CB radio channel in which contention for the channel is directly controlled by the users (who at best can do a poor job of scheduling the channel), the packet radio system decouples direct access to the channel from user requests for channel access. Within a fraction of a second, the microprocessors can dynamically schedule and control the channel to minimize or avoid conflicts (overlapping transmissions) particularly when the transmissions are very short. The use of computer control for channel access can lead to very efficient system operation relative to other more conventional manual methods of access control [11].

In recent years, the subject of efficient spectrum utilization has received increasing attention. A special issue of *IEEE Transactions on Electromagnetic Compatibility* on spectrum management [12] addresses this topic in considerable detail. This subject is not addressed further in this paper, other than to note that because of its capability for dynamic allocation of the spectrum, packet radio is a particularly good choice to obtain efficient utilization for bursty traffic. The ability to achieve effective usage of the spectrum will be a central factor along with cost in determining the ultimate viability of radio based networks for local distribution of information.

In this paper, we discuss the basic concepts of packet radio and present the recent technology and system advances. In Section II, we indicate various capabilities and services which a packet radio network might provide. In Section III, we consider the problem of signaling over a ground radio channel with all its attendant environmental factors. In Section IV, we discuss the basic operation of a packet radio network with underlying emphasis on the network elements and system protocols. In Section V, we discuss several advanced system capabilities for operation and control of the packet radio system. In Section VI, we separate out a subject of particular interest, namely spread spectrum transmission in the network environment. Although a packet radio system need not employ spread spectrum, there are several noteworthy attributes arising from its use. The experimental packet radio network (PRNET) under development by the ARPA is discussed in Section VII. The final section contains conclusions.

II. CAPABILITIES AND SERVICES OF A PR NETWORK

A primary objective of a packet radio network is to support real-time interactive communications between computer resources (hosts) connected to the network and user terminals (e.g., terminal–host, host–host, and terminal–terminal interactions). In order to satisfy this objective, the network should provide certain basic capabilities and services which can be grouped roughly into two categories: those which are always or automatically provided by the network and those which a user may select based on his application. The former category includes such capabilities as network transparency, area coverage/connectivity, mobile operation, internetting, coexistence, throughput with low delay, and rapid deployment. The last category includes error control options, routing options, addressing options, and services for various tactical applications.

We identify here a few of these basic packet radio network services and capabilities. While the list is not intended to be exhaustive, those items on it are all major factors of interest.

We assume that computer resources (hosts) need to be connected with each other and with individual users who might access data bases, manipulate files, run programs or write and execute programs to run on remote hosts. The packet radio network merely provides a high throughput, low delay means of interconnection for the (potentially mobile) community of users. Many of these operations will be interactive, with a computer response to a remote user entry being desired in real-time. Although the primary objective of the net is to provide service to computer communication traffic, other types of service, such as might be required for real-time speech, can be accommodated along with the capability for end-to-end security based on packet encryption techniques.

A. Transparency

The basic internal operation of the network should be transparent to the user. We use this term to mean that all user data presented to the net should be delivered to the destination without modification of the information content in any way. Only the data to be delivered, and the necessary control and addressing information should be required of the user as input. All other aspects of routing, reliable delivery, protocols and network operation should be handled by the network itself. Only in the case of communication difficulties should the user or user process be advised of internal network status. Transparency is desirable in order to allow the network to regulate and optimize its internal flow of traffic in a global manner without unnecessary constraints applied by users. The users, in turn, need not be concerned with the activities taking place in the net or their effect on network operations, but need only specify the services desired.

B. Area Coverage and Connectivity

Area coverage with full connectivity should be provided. For ground mobile radio, network diameters on the order of 100 miles are appropriate, but the system architecture should allow the geographic area of coverage to be expanded at the expense of increased end-to-end delay across the network. All valid traffic originators within the net must be provided connectivity with all other valid receivers subject only to the overall reliability and performance of the system. The network need have no prior knowledge of which users may wish to connect to which other users or resources in the net. This is particularly important (and necessary) for the mobile subscribers.

C. Mobility

The system should support mobile terminals and computers at normal vehicular ground speeds within the area of coverage. Packet radios in mobile applications must satisfy reasonable size, weight, and power consumption constraints.

D. Internetting

The packet radio network structure should be capable of internetting in such a way that a user providing a packet address in another net can expect his network to route the associated packet to a point of connection with the other net or to an intermediate (transit) net for forwarding. Similarly, arriving internet packets should also be routed to the local user.

E. Coexistence

Radio frequency characteristics of the packet radio system should allow coexistence with existing users of a chosen frequency band. This could provide a greater degree of spectrum sharing, particularly among similar systems, and may facilitate the introduction of the technology in new geographic locations.

F. Throughput and Low Delay

The capacity of the packet radio system should allow for variable length packet sizes up to a few thousand information bits, and provide delivery of packets with delays on the order of 0.1 s in nets of 100 mi area coverage size. Parameters on this order are required to provide the real-time interactive services, and to accommodate efficient data transfers. With 100 kbits/s signaling rates, a maximum packet size might be a few thousand bits.

G. Rapid and Convenient Deployment

Deployment of the packet radio net should be rapid and convenient, requiring little more than mounting the equipment at the desired location. No alignment procedure should be required, and in most applications omni-directional antennas would be used, thus eliminating the need for antenna alignment. Once installed, the system should be self-initializing and self-organizing. That is, the network should discover the radio connectivity between nodes and organize routing strategies on the basis of this connectivity and on the source/destination data of traffic presented to the net. Packet radios should be capable of unattended operation.

H. Error Control

Data integrity is crucial for most computer applications. Error control should be provided by the network, so that packets delivered to a user with undetected errors occur less frequently than about one in 10^{10} packets. This is a critical requirement for computer communications, since even one undetected error in a large file may render it useless or cause troublesome and unpredictable problems during subsequent use of such a file. While detection of errors is essential, choices exist in dealing with the detected errors. In some cases, error detection and retransmission may be used, while in other environments, more sophisticated forward error correction technology must be used in order to maintain satisfactory throughput and delay when communicating through land mobile radio channels.

I. Routing Options

The network should support efficient communication between any pair of users and the capability for users to broadcast a packet to a subset or to all users on the net. Land mobile radio traditionally has been used for point-to-point voice communications. Dispatching services, walkie talkies, and, recently, CB radio all have supported a broadcast mode of operation as well. These capabilities could be requested by a "type of service" field provided by the user in the packet header.

J. Addressing Options

The network should provide a capability for addressing a subset of the network participants and for efficiently establishing communication among them. This might be used for real-time conferencing or to support message delivery to a list of addressees with the minimum number of network transmissions. Logical network connectivity is a necessary foundation upon which protocols for these services can be built.

K. Tactical Applications

In tactical military applications, the RF waveform used by packet radios should provide resistance to jamming, spoofing, detection, and direction finding. In many cases, waveforms with these capabilities lead naturally to the capability for position location and relative navigation. With the addition of a communication security function, the packet radio net would then provide an integrated communication, navigation, and identification system for secure tactical use.

III. Signaling in the Ground Radio Environment

Packet radio technology is applicable to ground-based, airborne, seaborne, and space environments. In this paper, we focus on ground-based networks which encounter perhaps the most difficult environment in terms of propagation and RF connectivity. Ground radio links, particularly when mobile terminals are involved, are subject to severe variations in received signal strength due to local variations in terrain, man-made structures and foliage. In addition, reflections give rise to multiple signal paths leading to distortion and fading as the differently delayed signals interfere at a receiver [13]. As a result of these phenomena, RF connectivity is difficult to predict and may abruptly change in unexpected ways as mobile terminals move about. An important attribute of a packet radio system is its self-organizing, automated network management capability which dynamically discovers RF connectivity as a function of time for use in packet routing. The multipath phenomena also provide a strong motivating factor for the use of spread spectrum waveforms in packet radio systems. In the paragraphs which follow, we first bound (roughly) the radio frequency choices which are most appropriate for ground-based radio networks. We then discuss the characteristics of propagation path loss, multipath effects, and man-made noise at these frequencies, and conclude with a discussion of spread spectrum signaling [14] and its applicability to ground-based packet radio systems.

A. Frequency Band

The operational characteristics of the radio frequency band have a major impact on the packet radio design. The lowest and highest frequencies which can be used for a packet radio system are determined primarily by considerations of bandwidth and propagation path loss (and the associated RF power generation requirement) respectively.

The required systems bandwidth effectively determines the lowest desirable radio frequency in two ways. Practical, cost-effective radio equipment is difficult to achieve if the ratio of RF bandwidth to RF center frequency is much larger than about 0.3. This lower bounds the range of acceptable RF center frequencies. In practice, a center frequency well in excess of this lower bound is also desirable if the received signals would otherwise have too wide a multipath spread (e.g., due to sky wave phenomena at HF). For a packet radio system to deliver 2000 bit packets through a network with delays on the order of a tenth of a second, the data rate of the system must be in the range of a few hundred kilobits per second, which implies RF bandwidths of a few hundred kilohertz. From an implementation point of view, then, the RF center frequency should be at least a few megahertz, or in the lower high-frequency (HF) band extending from 3 MHz to 30 MHz. Propagation in the HF band can provide long distance communication due to sky wave reflections from the earth's ionosphere, but the propagation suffers from noticeable multipath spreading of the signal which, as will be described later in the section, limits the data-rate of signals which can be used. Multipath spreading in the very-high-frequency (VHF) band from 30 MHz to 300 MHz, where line-of-sight propagation

dominates, is typically reduced to a few microseconds as compared to the millisecond spreads encountered at HF, and data rates on the order of a hundred kilobits can be supported. Multipath fading and distortion are still a problem at VHF, particularly for terminals which are mobile or do not operate with radio line-of-sight. However, diversity techniques or the spread spectrum signaling techniques discussed later can overcome these difficulties.

The upper limits on usable radio frequencies for packet radio are primarily established by propagation path loss. As the operating frequency rises to about 10 GHz, absorptive losses due to the atmosphere and rain rapidly increase, and the resulting radio range is reduced accordingly. In general, packet radio systems must use closely spaced relays in order to provide adequate area coverage at these frequencies. The cost of providing a dense relay population may be acceptable if the distribution of users is also dense and if packet radios co-located with the users can provide the relay function. For most applications, however, 10 GHz is a practical upper limit for a useful radio frequency in a ground-based packet radio system. We conclude, then, that practical packet radio systems should use radio frequencies in the upper VHF band, in the ultra-high-frequency (UHF) band from 300 MHz to 3 GHz, and in the lower portion of the super high frequency (SHF) band from 3 GHz to 30 GHz.

An additional factor which must be considered for operational systems is the authorization to radiate packet radio transmissions. The VHF and UHF bands are already heavily allocated. The use of spread spectrum signals potentially could allow coexistence of a packet radio system with existing users of some frequency band. However, this is a relatively new concept from the regulatory point of view, and significant technical issues would have to be resolved to establish the feasibility of coexistence.

The discussion of propagation, multipath, and background noise which follows focuses on the UHF band, although, qualitatively, the phenomena discussed apply to the VHF and SHF bands as well. Later sections of the paper describe an experimental packet radio which operates at 1710–1850 MHz in the upper UHF band.

B. Propagation Characteristics

Packet radio network operations would be greatly simplified if all radios were sited such that a radio line-of-sight path existed to nearby radios. Link design procedures for such paths are well understood, and RF connectivity within the network would then be fixed and reliable. Such stringent restrictions on siting are not reasonable from the user point of view, however. Many users of a packet radio network will have to operate from facilities previously established without consideration of radio propagation. Use of packet radio in a mobile environment would be almost useless if siting were required for reliable operation.

The minimum theoretical path loss is achieved on a radio link in free space (i.e., a vacuum), where received signal strength decreases as the inverse square of link range. For a ground radio link, the path loss of free space may be approached on a link having a radio line-of-sight path, although even under this desirable condition diffraction and multipath phenomena can greatly reduce received signal strength.

When a radio line-of-sight path does not exist on a given link, one can still speak of sited or non-sited terminals, due to

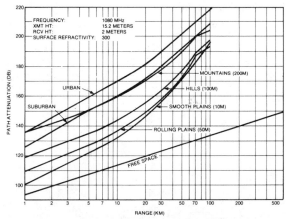

Fig. 1. Path loss versus range. 15-m transmitter height.

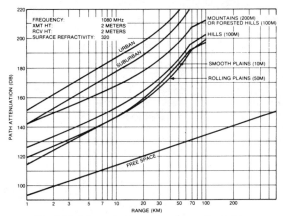

Fig. 2. Path loss versus range. 2-m transmitter height.

radios will encounter larger path losses [35], [36]. These factors lead to large variations in achievable radio range among users and make RF connectivity difficult to predict in a large, mobile user community. The objective of packet radio net design is to overcome this difficulty without placing undue restrictions on allowable user locations. In general, this requires automated network management procedures capable of sensing the existing RF connectivity in real-time and instantly exploiting this connectivity for network control and packet routing.

In addition to the wide variations in path loss, the ground–mobile, nonsited radio channel is subject to the effects of multipath propagation. When several differently delayed versions of the radio signal arrive at a receiver, constructive and destructive interference results. For stationary users, the effect of this phenomenon is that additional attenuation of the signal may be observed when the receiver is located at a point on the ground where the signal interference is destructive. Nulls on the order of tens of decibels may be observed. When communicating users are in motion, or when a multipath component arises from a moving reflector, received signal strength fading is observed as a function of time. The rate of fading is proportional to the velocity of user motion. Movement by a mobile user of only a few meters can cause received signal strength to drop below the threshold of the receiver, thus effectively disabling the link. Radio connectivity to a mobile terminal may change dramatically even for small displacements, and for continuous motion, signal strength may fluctuate above and below the receiver threshold several times during the reception of a packet, causing several short bursts of errors in the data or even loss of synchronization altogether.

We assume the modulation technique used by the packet radio results in a transmitted signal which is structured as a sequence of identifiable segments called symbols, where each symbol is one of a finite set of waveforms. In a simple binary modulation technique, one of two symbols is selected for each transmitted bit, depending on the value of the bit. In a spread spectrum system, the set of symbols may change with time, but a binary system would still choose each transmitted symbol from the set of two, which is in use at that particular time. Typically, a receiver processes the arriving waveform symbols one at a time, making a decision on each one as to which of the finite set has been received. The existence of multipath signal components affects the reliability with which symbol decisions can be made by causing symbol distortion and intersymbol interference. Intersymbol interference occurs when a symbol is overlapped by the delayed components of adjacent symbols. Such interference can lead to lower limits on symbol error probability which cannot be improved by increasing the signal to additive noise ratio on the radio link. When simple modulation techniques, such as phase-shift keying (PSK), are used, the symbol rate must be low enough that only a small portion of the symbol is overlapped by multipath signals of adjacent components. More sophisticated receivers (e.g., using adaptive equalization) can improve performance by suppressing the multipath components, but in order to do so they must rapidly obtain good estimates of the channel impulse response, which can be very difficult if not impossible to achieve. Spread spectrum signals are capable of reducing intersymbol interference effects, as described in Section III-D.

One way to break through the intersymbol interference barrier and achieve megabit rates with binary signaling is to use an

the strong influence of shadowing by local terrain and objects and of the elevation of the antenna above the ground. A sited terminal is one which has been located to avoid surrounding obstacles and whose antenna has been elevated to the maximum extent possible, while a terminal operating from a moving vehicle would generally be nonsited.

Average path attenuation exceeds that of a free space radio link by a significant amount in the ground radio environment, depending on the type of terrain and the elevation of the radio antenna. The curves in Fig. 1 and 2 show path loss as a function of link range for a frequency near 1 GHz, and illustrate these dependencies for two different transmitter heights. These curves are typical of propagation of UHF, and the variation of mean path loss as a function of frequency is typically much less than the variations due to terrain at a particular frequency. For example, the mean path loss from 700 MHz to 2000 MHz varies about 8 dB, while path loss at a 20-km range is seen from the figure to exceed that of free space by 25 to 80 dB depending on terrain and antenna heights. Furthermore, the path loss in urban and suburban areas, where many area coverage packet radio net applications might occur, is more severe than that of most natural terrain. The curves shown reflect average values of path loss which apply to a link of a given length which is randomly selected without regard to user siting. Well sited radios will typically encounter less path loss than shown in the curves, while poorly sited

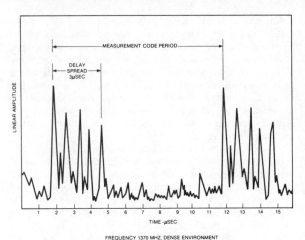

Fig. 3. Received multipath energy.

estimator-correlator type of receiver, such as typified by the RAKE structure [15] and its successors [16], [17]. This type of receiver, which is relatively complex, first estimates the channel impulse response and then adjusts its matched filters accordingly to match the channel characteristics. An alternative approach is to use M-ary, rather than binary, signaling to increase the bit rate to the desired value while still keeping the symbol rate sufficiently low to avoid the intersymbol interference problem. For example, a maximum rate of 250K symbols per second is reasonable to expect with a 4-μs multipath spread; thus four bits per symbol will be needed to achieve a 1-Mbit transmission rate. In this example, the symbol alphabet must contain at least 16 signals, each of 4-μs duration. The M-ary receiver may be relatively complex compared with more conventional schemes such as binary modulation.

In addition to causing intersymbol interference, multipath distorts individual symbols by smearing the received signal components over time. Another view of this effect can be seen in the frequency domain. If signal components arrive with differential propagation delay spread dt seconds, nulling of RF signal components occurs at intervals on the order of $B_c = 1/dt$. If the spectral width of the signal exceeds the value B_c, called the channel coherence bandwidth, notches occur in the received signal spectrum spaced B_c Hz apart. A matched filter receiver designed to match the undistorted (or unnotched) symbol spectrum will thus suffer a loss in detection performance since it is no longer matched to the actual received signal. Either the symbol rate must be restricted so that the signal spectrum falls mainly within the channel coherence bandwidth or other techniques such as spread spectrum must be used to suppress the distortion.

Fig. 3 illustrates the received multipath energy as a function of delay for an urban propagation path at 1370 HMz as measured with a spread spectrum test waveform. Note that significant delay components span a 3-μs interval. The figure illustrates the presence of several major multipath components for each symbol. Ideally, without multipath, the received energy would be concentrated at the beginning of each symbol in a single pulse of short duration. Thus a simple binary modulation system such as phase shift keying would begin to suffer significant intersymbol interference and symbol distortion at data rates exceeding a few hundred kilobits per second.

C. Man-Made Interference

Man-made interference in the RF frequency band includes both intentional and unintentional interference. Resistance to intentional interference (jamming) is of utmost importance in tactical military applications and strongly affects the details of waveform design and system complexity. Unintentional interference results from sources such as automobile ignition spark discharges, arc-welders, electric trains, radars, and ac power distribution systems. These sources of interference can be characterized as impulsive in nature. Such interference is often generated at relatively low signal strength levels, but in some environments, such as urban areas, the density of interference sources is high enough that packet radios will be near enough to some of them to encounter impulsive interference levels which are 60 to 80 dB above the thermal noise level of the receiver. Measurements near major roads and freeways [13] indicates that packets of 10-ms duration would typically suffer interference from several impulses of this magnitude during reception in such areas. Packet radios in such an environment might therefore experience a few bit errors in almost every packet received, and would require forward error correction in order to maintain system throughput.

D. Spread Spectrum Waveform Design Considerations

The packet radio signaling waveform must be designed to perform well with respect to both the natural environment and the induced environment arising from both intentional and unintentional interference including system self-interference arising from the multiple access/random access nature of the packet radio system. We have already noted that the limitations on signaling rate due to multipath can be reduced by using spread spectrum techniques. In addition, spread spectrum provides rejection of interference and the ability to coexist with other signals in the RF band. For these reasons, we consider spread spectrum signaling and identify the performance attributes of two major types of spread spectrum signals which make them well suited for packet radio applications. We do not consider non-spread signaling waveforms in this paper, nor other bandwidth expansion techniques such as FM. It should be noted that spread spectrum is, in reality, nothing more than a particular form of low rate coded transmission and that higher data rates are possible to achieve in principle through the use of more sophisticated systems which couple coding techniques directly with bandwidth expansion. However, engineering and implementation of these coded systems can be quite difficult, particularly at bandwidths in excess of a hundred megahertz. The following discussion illustrates that a variety of spread spectrum systems are well suited to deal with the RF environment, both natural and man-made.

The most commonly used forms of spread spectrum waveforms are direct sequence pseudo-noise (PN) modulation, frequency hopped (FH) modulation, and hybrid combinations of the two. A typical form of PN modulation is illustrated in Fig. 4(a). A source produces binary data at a rate of R bits per second. A pseudo-random generator produces a stream of binary "chips" at nR chips per second where n is an integer which is one or more orders of magnitude greater than unity. The term "chip" is used to distinguish the PN stream, or chip pattern, from the data stream. Each data bit is modulo two added to a sequence of n chips to form a "PN modulated" data stream which is then input to a modulator in order to

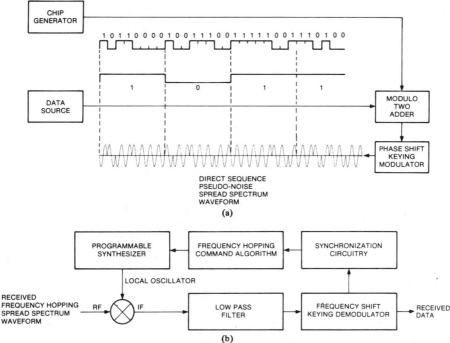

Fig. 4. (a) Pseudonoise spread spectrum modulation. (b) Psuedonoise spread spectrum demodulation.

convert it to an analog form suitable for transmission over the radio channel. The phase shift keying modulator illustrated in the figure is one way of accomplishing this digital-to-analog conversion. A typical form of demodulation is shown in Fig. 4(b). The effect of PN modulating the data stream is to increase the digital rate going into the PSK modulator from R bits per second to nR chips per second. Consequently, the occupied RF bandwidth of the resulting waveform is increased by a factor of n.

Several benefits are received in return for this increased bandwidth.

1) By using matched filter or correlation techniques, the signal-to-noise ratio is improved by a factor called the processing gain which is equal to $10 \log(n)$ dB in the above system. This improvement is realized with respect to both interference and receiver thermal noise as long as the interference in the RF channel is not highly correlated with the chip sequence used to encode a particular bit. In tactical systems, it is necessary to continuously change the chip pattern with time, so that a jammer cannot mimic or replay the waveform to appear as a highly correlated signal at the receiver.

2) The wide-band PN waveform can provide the ability to separate the various multipath signal components using correlation or matched filter techniques. Once separated, these signal components can be combined to reduce signal fading over time and improve signal-to-noise ratio. In addition, a suitably wideband signal will experience frequency selective fading only over small portions of the band. The received energy is therefore relatively constant over time and the overall communication reliability is quite similar to that of a frequency diversity system.

3) Because the signal is spread over a wider bandwidth than would be required for transmission of the signal without PN modulation, the spectral density of the signal is reduced for a constant transmitted power level. This factor coupled with the pseudo-random nature of the waveform gives the system a lower electromagnetic profile.

4) If the PN chip sequence used changes with each bit, the waveform will have a strong capture property, as explained in Section VI. Capture is the ability of a receiver to correctly receive one packet in the presence of other interfering packets. The capture property greatly enhances multiple access efficiency.

5) By using different PN chip patterns, various groups of users can coexist in the same area with greatly reduced interference. This "code division multiple access" (CDMA) capability again depends on the ability to receive one chip pattern while rejecting others as noise by using matched filter or correlation techniques.

6) Intersymbol interference may be suppressed even if the same chip pattern is reused for each symbol provided that the multipath spread is only a fraction of a symbol duration. If the multipath spread exceeds a symbol duration, intersymbol interference can still be suppressed if chip patterns are changed from symbol to symbol.

Fig. 5 illustrates one of many variations of a frequency hopped system. Data from the source is input directly to a modulator, in this example a frequency-shift keyed (FSK) modulator is assumed, to produce a signal which might ordinarily be converted to a fixed RF frequency for transmission. In this case, however, a programmable synthesizer is used as a local oscillator (LO) to dynamically select the RF frequency for transmission. The LO pseudorandomly hops among n frequencies over time as determined by a suitable algorithm. As with PN modulation, the bandwidth used by the system is n times that used without frequency hopping and the wider bandwidth of a FH system yields certain advantages in return for expanded bandwidth.

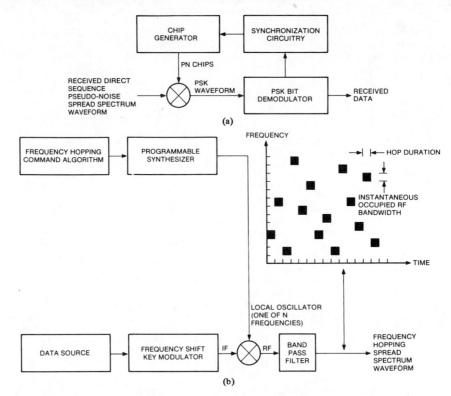

Fig. 5. (a) Frequency hopped spread spectrum waveform generation. (b) Frequency hopped spread spectrum reception.

1) The degree of interference rejection afforded by a FH system depends on the rate with which the RF carrier frequency changes and on auxiliary techniques such as forward error correction. The basic concept is that data may be received in error when transmission occurs on frequencies subject to strong interference, but that data correctly received on other frequencies subject to weaker interference will allow error correction to restore the lost data. In tactical systems, it is often required that the hopping rate of the system be sufficiently fast that a jammer cannot detect the frequency being used and jam this frequency while it is still being used because of the time delay accumulated by the signals during propagation over normal operational distances. If the interference is distributed uniformly over all n hopping frequencies, then, as in the PN system, the system may realize an average signal-to-noise ratio improvement of $10 \log(n)$ dB by filtering the instantaneously correct RF band from the total RF bandwidth.

2) FH systems do not inherently suppress multipath fading effects. However, if the average frequency spacing between successive hops exceeds the channel coherence bandwidth, diversity reception may be implemented by sending data once on each of two or more successive hopping frequencies.

3) The coexistence properties of FH are different from those of a PN system. A PN system has reduced spectral density. A FH system also has reduced spectral density, when the spectrum is averaged over many hops. However, the shorter term spectral density in any portion of the band will be relatively high during the $1/n$ fraction of the time that this portion of the band is selected for transmission. As a result,

FH systems may interfere with other systems which are sensitive to low duty cycle interference due, for example, to loss of synchronization when the FH signal falls in-band.

4) The capture properties of an FH system are very effective, since the filtering process in an FH receiver may be able to handle interference over a wider dynamic range than is possible in a PN system.

5) If they use different hopping patterns, FH techniques can support overlayed CDMA networks in a common area thus yielding increased throughput in the FH band.

6) If each RF hop is used to transmit exactly one data symbol, FH systems can suppress intersymbol interference. Symbol distortion, however, may still occur if the signaling rate is high enough that the short-term signal spectrum significantly exceeds the channel coherence bandwidth. It is usually necessary to use a noncoherent signaling technique in FH ground radio systems, since the medium may destroy signal coherence from symbol to symbol due to frequency hops that exceed the channel coherence bandwidth.

Combinations of FH and PN techniques can result in a waveform with desirable attributes of both techniques. For example, on each RF hop, one could transmit one of M PN sequences in order to send $\log_2 M$ bits of information. Such a technique could achieve diversity reception, based on suppression of multipath components and reduced symbol distortion, but a receiver for this type of system can be complex.

IV. BASIC SYSTEM CONSIDERATIONS

In this section we introduce the basic packet radio system concepts along with a description of the key elements and how they work. We describe two key routing options (point-to-point and broadcast) and the protocols which support them. Packet radio combines the use of time division multiple access with radio broadcasting to form a powerful, self-organizing store-and-forward network. We begin with a discussion of the multiple-access channel and its control.

A. The Multiple Access Channel

A multiple access (MA) channel is one which two or more users may nominally share at the same time. A simple form of MA channel is obtained by partitioning the channel into separate nonoverlapping frequency subbands with each user assigned to a separate subband. Another form of MA channel is obtained by scheduling each user's transmission to occur in short nonoverlapping intervals in time. In the first case, known as Frequency Division Multiple Access (FDMA), each user has access to a dedicated portion of the channel at all times. In the second case, known as Time Division Multiple Access (TDMA), each user has access to the whole channel for only a portion of the time. In these two cases, the signaling waveforms are orthogonal in frequency and time, respectively. With a properly designed receiver, orthogonal signals should not interfere with each other in reception. Other desirable forms of multiple access are possible, such as the use of spread spectrum waveforms. Spread spectrum signals may overlap in both time and frequency but use signaling structures that allow a receiver to separate one signal from the others using correlation or matched filtering techniques.

Although relatively simple to implement, a disadvantage of FDMA is that the use of separate dedicated frequency bands between each pair of users does not generally result in efficient use of the frequency spectrum or provide cost effective interconnection strategies for interconnecting multiple users. Only in the event that each pair of users can make almost full use of their frequency subband will this strategy be at all efficient.

We assume each user has a radio receiver that is capable of receiving only one frequency subband at a time and refer to him as a single channel user. The simultaneous use of many radios or receivers per user for increased connectivity is simply not cost effective and the protocols required to support their use would be inelegant and unworkably difficult in practice. If single channel user A is communicating with single channel user B at a given time, then user C cannot talk to user A or B unless they all share the same channel. Inaccessability of this form is clearly undesirable in the network environment.

Due to the burst nature of its transmissions, a TDMA system requires time synchronization of its transmissions to achieve nonoverlapping bursts. For this reason, a TDMA system is more complex to implement than FDMA, and some form of local or global control is typically needed to schedule the transmissions so they are nonoverlapping in time. However, an important advantage is the connectivity which results from the fact that all receivers listen on the same channel while all senders in a TDMA system transmit on the same common channel at different times.

A common radio channel provides a cost-effective way to achieve complete network connectivity between subscribers and efficient spectrum utilization can be achieved through statistical multiplexing of many traffic sources onto the channel using TDMA. Each user is guaranteed to be able to communicate with every other user when he transmits at his scheduled transmission times. In principal, a very large number of users each with very low duty cycle can simultaneously and efficiently share the channel. In order to achieve efficient performance in practice though, a control mechanism is required for access to the channel.

B. Controlling the MA Channel

A variety of theoretical and experimental studies have been carried out to determine the most effective techniques for sharing a MA channel. One of the simplest techniques, known as pure ALOHA was designed for very low duty cycle applications and involves no control other than the ability to detect overlapping packet transmissions (conflicts) when they occur and to randomly reschedule these unsuccessful transmissions at a later time. This scheme is normally implemented using a positive acknowledgment and timeout procedure based on packet checksums. Packets are simply transmitted randomly in time according to the underlying arrival process [18]. A time slotted version of random access is also possible. In this case, accurate timing must be maintained on the channel such that the radios can synchronize their transmissions to begin only at the time slot boundaries. A packet transmission occurs within a time slot with some underlying probability derived from the arrival process. This access method is often referred to as slotted ALOHA [19].

A pure ALOHA system is more efficient than a system consisting of many subchannels dedicated to a collection of very low duty cycle users, but the expected number of conflicts increases as the pure ALOHA channel loading increases until the channel reaches its maximum throughput at $1/2e$ (about 18 percent) of peak channel capacity (peak instantaneous transmission rate) for fixed length packets and Poisson arrivals. For this same assumed type of traffic, slotted ALOHA achieves twice the maximum throughput of random access (or pure ALOHA). A slotted system may appear advantageous for reasons of efficiency, but the system is also more vulnerable in this case, since packet synchronization preambles which are critical for packet acquisition begin in predictable locations in time, and this information could be exploited by an adversary in tactical military applications. Both pure ALOHA and slotted ALOHA can become unstable and some form of stability control is desirable in operation.

One of the more efficient control techniques for the ground radio MA channel is Carrier Sense Multiple Access (CSMA). In CSMA, each sender first senses the channel, and then transmits a packer only if the channel is idle. If the channel is determined to be in use, the transmission is rescheduled at a later time when the same procedure will be invoked. Various elaborations on the CSMA scheme offer the possibility of achieving 80–90 percent utilization of the channel with low end-to-end transmission delay per packet [20]–[23].

In a slotted carrier sense system, each transmitter senses the channel during the beginning of each slot. This elementary observation has further implications when we address the network aspects of spread spectrum in Section VI.

C. System Structure

When critical parts of the system are geographically distributed and often unattended (or unprotected), special provisions are required to ensure the overall network integrity. One way to protect against compromise due to tampering with radios in the field is to control all functions which can have system-wide implications from one or more protected locations. For example, a fully distributed routing algorithm such as that used in the ARPANET would not be suitable for use with unprotected radios, since a small modification at one radio could totally affect the network routing and performance (e.g., imagine that a radio declared itself to be the shortest path to everywhere). Aspects of the network protocols (such as the radio acknowledgment procedures) which must be performed by each radio would be distributed among all radio elements. However, all network control protocols which can have global effect are specifically initiated by one or more entities in the network called *stations*. The resulting network control thus takes the form of a two level hierarchical system. The normal mode of operation utilizes a single station or multiple station. However, a stationless mode is also possible. The implications of stationless operation are discussed later.

The functions of a station are associated with global management of the radio net [24]. Generally speaking, each station is aware of all operational radios in the network. The stations discover the existence of new radios waiting to enter the net and determine when other radios have departed. The station determines the route to each of these radios and plays an active role in initializing, organizing, and maintaining the operational network. In particular, all routes are assigned by the station to minimize PR cost and complexity. PR's are not required to store information about every other PR and terminal device in the network.

One of the requirements for controlling the PRNET is assessing the reliability of radio links between PR's and using the information to assign good routes. A primary source of link information is the PR neighbor table whose entries are collected by each radio, summarized, and regularly sent to the station along with other status information. For example, each radio reports which other radios it can hear along with raw or processed information for the station to determine the quality of the transmission path between these radios. The station then deduces the overall connectivity of the network (we assume topologically rather than topographically) and determines good routes to itself from each of the radios in its subset. The station then distributes to each radio in its subset the route from that radio to the station. This process is known as *labeling*. The neighbor table is maintained by each PR whether or not an operational station is present and can be used in a stationless mode if necessary.

We assume that a set of radios distributed throughout a geographic area, which we call the backbone, provides a carrier-like packet communication network service to the users. These backbone radios, known as repeaters, receive packets from nearby users and relay them. The repeaters also accept packets from other nearby repeaters for relaying. This extends the range of the system beyond line of sight. Users communicate with each other via a common frequency band [25].

For military operation, where a separate backbone network might be infeasible to deploy, each user's radio might be equipped to support not only his own traffic but that of other designated users. That is, the user's radio may also have to "double up" as a repeater, to support network traffic. In this case, we do not identify a separate backbone repeater network *per se*, since it would be indistinguishable from the network of user packet radios.

D. Store-and-Forward Operation

An individual packet radio unit is a small piece of electronic equipment which consists of a radio section and a digital section which controls the radio [26]. The radio section contains the antenna, RF transmitter/receiver, and all signal processing and data detection logic associated with modulation and demodulation. The digital section contains a microprocessor controller plus semiconductor memory for packet buffering and software. The radio and digital sections are connected by a high speed interface (see Fig. 6). For each transmitted packet, the digital unit selects the transmit frequency (normally fixed), data rate, power, and time of transmission. In addition, it performs the packet processing to route the packet through the network. In a half duplex mode of operation, a radio may be transmitting or receiving, but not both simultaneously. In the remainder of this paper we assume that each radio operates as a half duplex transceiver in the common frequency band.

Normal store-and-forward operation within the network takes place as follows. A user generated packet with associated addressing and control information in the packet header is input to the digital section of his packet radio, which adds some network routing and control information and passes the packet to the radio section for transmission to a nearby repeater which is identified within the packet. Upon correct receipt of the packet, the nearby repeater processes the header to determine if it should relay the packet, deliver it to an attached device, or discard it. Several nearby repeaters may actually hear the packet, but only one repeater (which we call the next downstream repeater) will typically be identified to relay it. The other repeaters, will discard the packet. The packet will then be relayed from repeater to repeater through the backbone (in a store-and-forward fashion using the procedure described above) until it arrives at the final repeater which broadcasts it directly to the user's packet radio. At each repeater, the packet is stored in memory until a positive acknowledgment is received from the next downstream repeater or a time-out occurs. In the latter case, the packet will be retransmitted. Each packet is uniquely identified by a set of bits in its header called the Unique Packet Identification (UPI). The relaying of a packet by omnidirectional broadcast can also serve as a reverse acknowledgment (an echo acknowledgment) to the previous upstream repeater as long as the system is nonspread or a common spread spectrum code is used by the radios.

The propagation delay between radios in the ground radio environment is typically on the order of a small fraction of a millisecond (it could be as low as a few microseconds). As a result, the minimum time a packet must be stored in each digital unit awaiting a positive acknowledgment when it was correctly received downstream is little more than the time to transmit the packet and for the next downstream repeater to quickly return the acknowledgment. Each digital section therefore needs to contain enough memory to store at most a few packets (e.g., 4–6). Due to the half-duplex operation, a repeater which has just transmitted a packet should, under normal conditions, immediately enter receive mode to listen for the acknowledgment.

If, instead, another packet should arrive during the time-out period at the packet radio waiting for the acknowledgment (due to the half duplex operation, the radio must be in receive mode to receive the acknowledgment), the radio will not accept

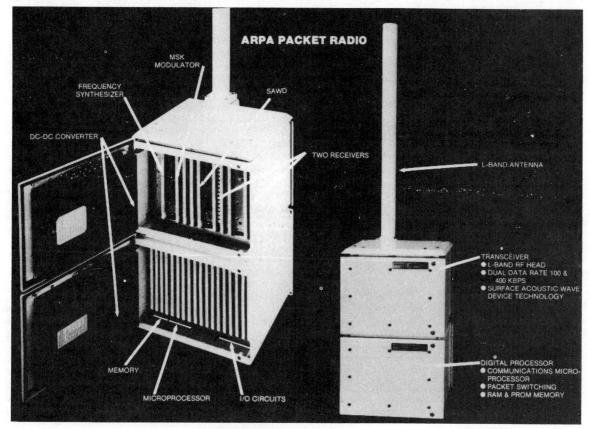

Fig. 6. An experimental packet radio.

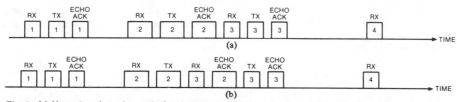

Fig. 7. (a) Normal packet relay cycle (carrier sense mechanism may have delayed Packet 3 until completion of acknowledgment for Packet 2). (b) Alternative packet relay cycle (carrier sense mechanism delays echo acknowledgment for Packet 2 until packet 3 is received.

it if there are no unoccupied buffers in its memory. The half-duplex operation is keyed to a nominal cycle of packet transmission, acknowledgment receipt, new packet receipt, followed then by its transmission and acknowledgment (see Fig. 7). By providing only a few buffers in each packet radio, efficient system performance can be achieved with half-duplex transceivers[27]. A large backlog of packets never accummulates within the individual radios.

Along with mobile operation, rapid deployment and portability are two of the essential attributes of the packet radio technology. Reliable system operation is achieved under these constraints in a number of ways. Each packet radio unit shares a single wideband channel with omnidirectional antennas and extensive siting or alignment procedures are therefore unnecessary in practice. Good connectivity can be maintained with mobile terminals as long as a line of sight path exists. We assume that with omnidirectional antennas (in the azimuthal plane) communication is possible with a multiplicity of radios within approximate line of sight propagation range. It is further assumed that the network traffic matrix cannot be estimated accurately or is expected to change very rapidly. In this case, dynamic allocation of a common frequency band is a very good strategy.

For the following discussion, we refer to the operation of an experimental packet radio, in which a transmitted packet has the structure shown in Fig. 8. It consists of a 48 bit preamble followed by a variable length header (typically 96–144 bits), followed by the text and a 32 bit checksum.

The packet preamble is used by the radio section of the receiver for several purposes. The first few bits are used to detect the carrier energy and to set the automatic gain control (AGC) to compensate for differing signal strengths of the arriving packets. Correct reception of the packet is totally dependent upon acquisition of the preamble. The next few

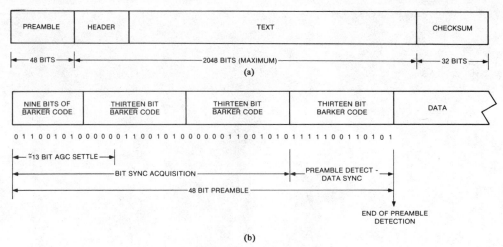

Fig. 8. Structure of a transmitted packet. (a) EPR packet format. (b) EPR packet preamble detail.

bits are used to acquire bit timing. Following these, the next set of bits is used to acquire packet timing (identify the end of the preamble and the start of the header). Both the header and text are delivered from the radio section to the digital section which knows the header format and can therefore determine the exact start of the text. The error control bits consist of a checksum appended by the transmitter and checked by each receiver. After checking, the error control bits are stripped off by the radio section as was the preamble before it. The digital section checks a status register in the interface to determine if the packet is correct.

The per packet error rate before error control is usually much higher in the ground radio environment than over a leased line. A packet of duration 5–10 ms will typically experience one or more bit errors due to automobile ignition noise and other sources of interference. In addition, mobile terminals experience signal strength fading while in motion due to shadowing and destructive multipath interference which were described in Section III. As a result, and depending on the environmental conditions and type of error control in existence at the time, it is possible that a packet will not be successfully relayed even after several attempts. Two actions are appropriate for the system to take in this case. The packet may be dynamically rerouted via an alternate repeater, or an end-to-end retransmission strategy may be invoked external to the network. In the latter case, the packet may either arrive back at the same repeater at a later time when the original problem is no longer present, or a different route may be provided from source to destination. In the meantime, critical network resources (e.g., buffer, channel) are available for more productive use.

Each radio has an identifier which we shall call its selector.[1] The selectors play a central role in the network routing and control procedures. We discuss the point-to-point and broadcast routing options below. The more advanced network control functions are discussed in the next section.

[1] For the purposes of this paper, we assume the selectors are unique and preassigned, but this is a simplifying restriction which is adopted for clarity in exposition. Nonunique selectors can, in fact, be used as effectively when the probability of nonuniqueness in any given area is sufficiently low.

E. Point-to-Point Routing

In the point-to-point routing procedure, a packet originating at one part of the network proceeds directly through a series of one or more repeaters until it reaches its final destination. The point-to-point route (which consists of an ordered set of selectors) is first determined by a station which is the only element in the net that knows the current overall system connectivity. Having determined a good point-to-point route, where should the station send the point-to-point routing information? One possibility is for it to distribute the information to the individual repeaters along the point-to-point route. In this case, each succeeding packet would only require some form of source and/or destination identifier but would not have to carry the entire route in its header. Alternatively, the station can send it directly to the digital section of the sender's (or receiver's) packet radio. In this case, each packet originating at that radio could then contain the entire set of selectors in its header. However, this choice may have a significant impact on the network efficiency and ultimately its extendability since the selectors would contribute overhead to the packet and, at most, only a small finite set of them could be carried along.

A far more attractive choice is for the sender's or receiver's digital unit to take on the responsibility for setting up the route that was specified by the station. In particular, it transmits a route setup packet (along the designated route) which contains the entire ordered set of selectors and which is used to create appropriate routing entries in the specified repeaters. Regardless of how the routing entries are finally created within the repeaters along the point-to-point route, it may still be desirable to carry along within each data packet the selector for the next downstream repeater, or even the next few repeaters. The latter strategy may have significant operational as well as performance advantages as is discussed further in Section V.

In the event that one or more stations are available in the net and a point-to-point route fails (e.g., an intermediate repeater fails) the existing traffic on that route can be diverted to the station for forwarding to the final destination. The station must recompute a good route to the destination before it can forward the packet. Alternate routing is per-

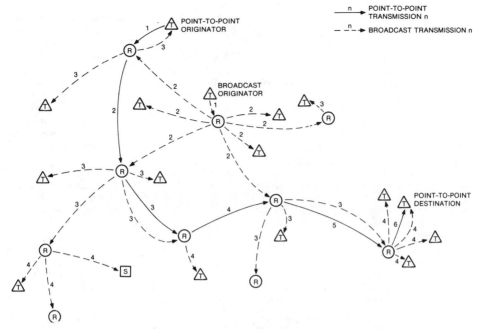

Fig. 9. Point-to-point and broadcast routing.

formed by the PR, when necessary, using parameters assigned by the station. Multiple receivers are allowed to forward a packet around a failed temporarily inaccessable repeater.

As network conditions change (terminal movement, repeater failure or recovery, changes in hop reliability, and changes in network congestion) routes will be dynamically reassigned by the station to satisfy the minimum-delay criteria. In general, only major status changes will result in dynamic routing changes. Hosts and terminals attached to the PRNET remain unaware of (do not participate in) the station's assignment and reassignment of routes; these routes are stored in their attached PR. Establishment of a point-to-point route is an automatic process, transparent to the host or terminal initiating the communication. If the source device's PR does not already have a good point-to-point route to the desired destination, it forwards the incoming packets to some station. The station reroutes the packets to their destination, and in parallel, computes a good point-to-point route from the indicated source PR to the destination. It then deposits this route in the source PR, and later packets are routed directly to their destination instead of being routed through a station.

F. Broadcast Routing

Broadcasting to all radios in the network is another important mode of system operation. In this mode, every repeater in the network keeps a short list of unique packet identifiers (UPI's) for previously broadcast packets that it recently received and retransmitted. A broadcast packet is identified by a bit in its header. If a repeater receives a broadcast packet whose UPI is already on its list, it will discard it. Otherwise it will accept the packet if correct and retransmit it. In the broadcast mode, each packet radiates away from the source radio as in a wavefront type of propagation. The use of a list of UPI's in each repeater prevents any packet from returning to portions of the

net through which it (or any of its copies) has already passed.

Broadcasting is not a particularly efficient mode of operation for two party communications, but it is a very robust way to distribute packets to all parts of the net. This type of service may also be unreliable in the sense that not all radios may actually receive every broadcast packet and end-to-end procedures must be invoked if reliable delivery is essential.

A broadcast routed packet will generally be intended for actual delivery to all users or a subset of users (e.g., conferencing). However, a user could also use this routing option if, for example, there was no working station on the net to supply a point-to-point route to a single intended recipient. In this case, the destination address is used to identify the intended recipient so that it will accept the packet and the others will discard it. Point-to-point routing and broadcasting to all users are illustrated in Fig. 9. The circles indicate repeaters, the triangles represent user terminals and the square boxes denote stations.

G. Mobile Operation

All elements of the packet radio network can be mobile, as within a fleet of ships, or certain elements can be fixed while others are in motion. In general, we assume that stations and repeaters move relatively slowly, if at all, so that topological changes in the backbone are not frequent events. Normally, user terminals will also move slowly enough that assigned point-to-point routes will remain effective for at least a few minutes, if not longer. However, certain user terminals may move sufficiently rapidly relative to the other elements in the system that the use of point-to-point routes would not be a practical choice. A particularly significant aspect of broadcasting is that only the identity of the intended recipients need be known to the sender and not their actual location or the routing. Thus broadcasting is a useful technique to bypass the need for control of rapidly changing routes. A station is assumed to use broadcast routing to communicate with the few "high speed" users.

Fundamental performance limits are imposed on the system by the presence of mobile terminals. The need to update routes in real-time as users move, and the time in which the system can adapt to environmental changes along the mobile users path, will determine a maximum allowable speed for mobile terminals which use point-to-point routes.

A significant problem associated with the use of point-to-point routes in mobile operation arises from the repeater currently servicing a mobile vehicle's traffic. The radio link can typically disappear and reappear on a time scale of a few seconds as the vehicle moves past various obstructions. The reliance upon summary reports from each radio (e.g., twice a minute) provides a simple way for station to track the mobile units. However, loss of communication between the mobile unit and the repeater handling its traffic would only be detected by the station after a lapse on the order of a minute, after which the station could assign a different point-to-point route for that mobile unit. The following procedure will help to speed up the process. Whenever a mobile unit loses contact with its next downstream repeater, it broadcasts a distress message, and all repeaters within range will forward it to the station. The station can then assign a new point-to-point route and quickly restore communication with the mobile unit.

H. Reliable Delivery Mechanisms

The inherent undependability of a mobile radio channel requires an end-to-end protocol to provide reliable operation. For internal network control traffic (nonuser traffic), where perhaps one outstanding unacknowledged packet is sufficient, a highly efficient but specially designed protocol suffices. For intranet user traffic, or internet operation, a more flexible and higher performance protocol is desirable.

We assume an end-to-end error detection and retransmission technique to be used in the network for reliable delivery of individual packets. Each source/destination pair on the network could utilize an end-to-end protocol such as described in [38], which also supports internetworking. In this case the user's terminal could be equipped with a microprocessor-based device known as a Terminal Interface Unit (TIU) which performs the end-to-end protocol, and any local support for the terminal (e.g., local echoing, formatting). The TIU interfaces directly to the digital unit of a packet radio.

Within the PRNET, stations and radios need to communicate control packets reliably. For example, the regular reports from each radio to the station are used to validate the radio's continued availability. Without an effective recovery procedure the station could declare a perfectly good radio to be out of order and remove it from service if several of its reports were lost consecutively. Similarly, parameter change packets from the station to the radio should be delivered reliably since these are used to set dynamically important radio parameters, such as the retransmission interval or the maximum number of allowed retransmissions. The Station-PR Protocol (SPP) provides the reliable delivery mechanism.

V. ADVANCED SYSTEM FUNCTIONS

The basic elements and operation of a packet radio network were described in the previous section. The more advanced aspects of system operation and control are briefly highlighted in this portion of the paper. We begin by discussing the way in which the network is initialized and the techniques used to introduce new radios and stations into the net. We then discuss the need for multiple stations in a single radio net and describe the basic multistation operation.

The use of stations has clear efficiency advantages when elements of the network are mobile or otherwise prone to rapidly changing fluctuations in communications reliability or connectivity. However the network should support a stationless mode of operation in which no operational stations are available. A method is discussed for implementing good routes under the stationless mode. Finally, we discuss several alternatives for achieving multi-destination routing.

A. Network Initialization and Control.

Consider a collection of geographically distributed packet radios, each of which is powered on and capable of communicating packets to some subset of radios within line of sight propagation range. We assume for the moment that no operational stations are present. Under these circumstances, each radio periodically (e.g., every 30 s for fixed radios and every 5 s for mobile radios) broadcasts a "radio-on packet" (ROP) announcing its existence and containing selected status and identification information from its digital unit. This information is also sent out over its host interface. Some set of radios within range will hear this ROP, will note the event in its tables along with the measured strength of the received signal, and will discard the ROP. There are several ways a PR can determine the quality of a radio link using ROP's. For the purposes of this paper, we simply assume that each ROP contains a cumulative count of the number of packets the PR has received from every other PR it can hear. Upon receipt of an ROP, a radio can determine its connectivity and the percentage of packets successfully communicated on each link. In this fashion, each radio is able to determine the set of local radios with which it can communicate reliably.

If there was an operational station on the net, the PR's would send summary ROP's directly to the station at appropriate times (on a point-to-point route) to convey the current labeling information and also the neighbor table information. This is done both periodically and upon detection by the PR of a possibly significant change in some link. To coordinate transmit counts with receive counts (which is essential for knowing what percentage of traffic on the link is heard), and to get the information to the station, PR's include their transmit counts in each summary ROP, and each receiving PR includes the matching receive count in the summary ROP before forwarding it to the station. All traffic counts are cumulative modulo a very large number.

If the radio is not aware of the existence of a station, it does not transmit any summary ROP's. In this case, it will support the stationless mode of operation described later in this section.

When a new radio is powered on, it will begin broadcasting ROP's, be recognized by neighboring PR's and quickly enter the system. If a radio moves, its new connectivity will be determined locally by the ROP mechanism. If a radio is powered down, fails, or can no longer communicate with any radios, its new connectivity state can also be determined locally (the lack of connectivity can be inferred by the other radios if they keep a record of the previous state or set of states). The ROP mechanism is designed to insure that each radio directly reports its own status and identification information to other radios within range and to the station.

B. Station Entry

When a station is first connected to an operational packet radio, it will soon hear an ROP from that radio over the host interface after which the station will promptly "label" the radio. The labeling process consists of first determining and then supplying the radio with a route (i.e., a set of selectors) to the station. In this first step, the task is trivial since the route to the station is via the host interface. In principle, once labeled, the packet radio periodically will continue to send its ROP's over the host interface to the station along with the less frequent summary ROP's. From the status information in these ROP's, the station will learn which radios are in direct communication range of its radio.

We adopt the following terminology for ease in explaining the labeling process. The radio directly connected to the station is said to be at level 0 with respect to its station. All radios in direct communication range of the level 0 radio and not directly attached to the station are said to be at level 1 with respect to the station. Similarly, a radio in communication range of a level n-1 radio which is not itself already at level n-1 or less is said to be at level n. A radio may be at different levels with respect to different stations, and its levels may change during mobile operation. A level simply indicates the minimum number of radio hops to the station and does not otherwise affect packet routing.

Having labeled its level 0 repeater, and learned of the level 1 repeaters, the station then reliably labels the level 1 repeaters one at a time using the SPP protocol over the radio channel. The level 1 repeaters then begin sending status reports to the station which prompt the station to label the level 2 repeaters and so forth. Various implementation steps can be taken to insure that labeling occurs quickly and that the station is not inundated with status reports during the labeling process or afterwards. If the order of events is reversed and a packet radio is attached to an already operational station, the sequence of actions taken is identical.

In the case where a station is connected to a network which has already been labeled by one or more other stations, the sequence of actions taken by the station is also identical. Each radio has table space (labeling slots) for storing routes to several stations and by design the number of slots in each radio limits the number of stations that are allowed to label it. The route from a radio to a station is supplied by that station via the labeling process. Two stations which have labeled a common repeater are known as *neighbors*. Summary ROP's, which are sent by each radio to the stations that have labeled it, include sufficient information about the current labeling slot entries for each station to learn which other stations are its neighbors. No direct station-to-station coordination is required to acquire this status information.

One having labeled a radio, the station must relabel the radio within a given time or the labeling slot entry will expire. These entries are timed out (relatively slowly) by the radio, and the age of each entry is reported in its status reports. A station will always fail to successfully label a repeater which has no available labeling slots, but it can refresh an existing entry of its own at any time. A slot whose entry has expired is not erased by the radio (the route is not normally used either) but it may be overwritten by another station if no other labeling slots in the radio are free. In principle, a radio could be provided with diverse routes to a given station for applications such as mobile hand-off.

This portion of the multi-station design achieves the following goals.

1) It provides complete redundancy among multiple, functionally identical stations in such a way that "hot switchover" is provided, i.e., other stations will automatically and immediately assume the responsibilities of any station which fails.

2) It shields all end devices from any knowledge about stations. In particular, no terminal or host device needs to know which stations are currently operational, where they are, or even how many of them are in operation. This comment also applies to the single station design.

3) It keeps PR's simple for reasons of reliability and economy.

As we shall see below, stations may collaborate in establishing routes between users which are within the jurisdiction of different stations, for graceful handover of responsibility for mobile terminals, and for supporting expansion of the PRNET to several hundred stations and several thousand PR's. After a brief discussion of renewal points and routing, we discuss the multiple station operation in greater detail.

C. Renewal Points and Routing

A renewal point is a PR along the route of a packet where the route (as specified in its header) may be altered. In point-to-point routing, the header contains fields which identify the next few designated repeaters along the path to a specified destination. Every repeater on a point-to-point route can act as a renewal point where these fields are rewritten. Two reasons for identifying more than just the next repeater in the header are to allow a detour via an alternate repeater in the event of failures along the designated path and to allow some repeaters not to serve as renewal points. The detour must be such that it eventually rejoins the designated path.

To function as a renewal point for a point-to-point route, a repeater must have a renewal table containing the next few designated repeaters for that route. When a packet arrives at the next downstream repeater for relaying, its routing fields are rewritten in the header according to the current renewal table entries. To conserve table space, each repeater maintains, at most, one table entry for each source/destination PR pair. In addition, the table must identify the last few upstream repeaters on the path so that the source can be notified in the event of communication failures at any point of the path.

A packet may also contain an entire path of route selectors in the text of the packet. This case may be distinguished by a special bit in the packet header which indicates that the text contains all the route selectors. Such a packet is known as a route setup packet and is used to initialize or refresh the renewal table entries in each repeater. Upon receipt of a route setup packet, a repeater extracts the renewal table information (normally a few entries) from the entire list of selectors in the text and writes it into the renewal table. The contents of a route setup packet are normally inserted by the digital unit of the destination PR. Any packet may be a route setup packet, subject only to the maximum packet length constraints of the network. A route setup packet may also contain data.

D. Multiple Station Operation

A single station in a network with a relatively small number of radios (tens to hundreds) is sufficient to organize it efficiently and to control its operation effectively. If the station fails, one of two cases will arise. The backup case is a station-less mode in which communication is still possible through the net but with a lowered overall utilization. This case is discussed in Section V-E. below. In the normal case of multiple station operation, the system control responsibilities are jointly assumed and if one station fails the others will temporarily take over its functions with little or no degradation in system performance. Thus, a major incentive behind multi-station operation is efficient and reliable system performance in the event of station failure.

However, a single operational station is no longer sufficient if the number of radios becomes too large (thousands or more) or if its capacity is otherwise exceeded. A set of distributed multiple stations are required in that case to maintain the system performance without degradation. Multiple station operation is necessary to support the setup and maintenance of routes through many geographically distributed repeaters under the jurisdiction of different stations and, in particular, to handle mobile terminals at the two ends of these long routes. Various alternatives for multiple station operation are currently under investigation. One of these alternatives is presented below.

Each station individually competes for control of radio resources by attempting to label radios which report free labeling slots or expired entries. If there are more labeling slots per radio than stations which are attempting to label them, a maximum number of radios will be labeled by each station. In general, a station has a limit on the maximum number of radios it is allowed to label, as well as on the maximum traffic loading it can operationally support. Usually there will be many fewer stations than radios and the station locations will be effectively distributed throughout the net (rather than concentrated) to distribute the station related traffic and to allow greater station coverage of radios in the event that each station cannot label all radios in the net.

We do not treat the case where the density of stations is sufficiently high or the number of labeling slots sufficiently small that there are more stations than slots per radio. In this situation, the stations would greedily compete for the labeling slots and some stations would be prevented from acquiring a full set of radios in the labeling process. We also assume that stations do not coordinate with each other to adjust the apportionment of radios to stations. By the nature of the labeling process, stations can only label contiguous sets of radios. If a particular entry in a repeater at level $n-1$ should somehow be allowed to expire by its station, and that labeling slot is overwritten by another station, then other radios at level n and higher may also be cut off completely from the station's labeling process. In this case, the labeling slots in these other radios will shortly expire as well.

In the remainder of this paper, we shall assume each radio has a sufficiently large number of labeling slots to accommodate any number of stations which attempt to label it. We also assume that the net if sufficiently large so that each station cannot label every radio. From the summary ROP's, each station knows the identity of its neighboring stations and a point-to-point route to each radio labeled by itself and a neighbor. If a source and destination are both within the control of a single station, that station can handle the route assignment. Otherwise, each station must also collaborate with its neighbors to find good routes.

It is an open question as to whether stations should communicate directly with other than their neighbors, except via a broadcast technique which requires no state information to be kept about the existence of all other stations or how to get to them. One can visualize a logical store-and-forward network of stations interconnected by point-to-point radio routes (the station-to-station routes can also be provided by other media) using distributed adaptive routing. Participation in such a logical network may complicate the station's taks and will increase the steady state traffic load somewhat on both the station and the radio channel. We only consider the case where neighboring stations communicate directly with each other. The use of a logical network of stations may be a viable alternative if the overall system performance without it is deficient in some significant way.

We net describe routing through the multistation environment. Each station is assumed to know which radios it has labeled, but must inquire of other stations to learn the whereabouts of other radios (and their users). This inquiry is assumed to take place upon request by a user via a broadcast to all stations via neighboring stations. The inquiry process also provides the destination station with a set of selectors for a point-to-point route to the destination station from the originating user (if fixed) or station (for mobile users). If both users are highly mobile, a point-to-point route will be established between the two end stations which will individually handle the final distribution. If both users are fixed, the destination station will choose the last few selectors from among the radios it has labeled to obtain a point-to-point route directly between the two end users (with no intervening stations) and will supply it to the destination user. The actual point-to-point route setup is initiated by the destination user in this case, rather than by the destination station as was the case for mobile users. If one user is fixed and one user is highly mobile, the resulting point-to-point route will be between the fixed user and the remote station which will handle local distribution for the mobile user.

The route selection process takes place as follows. A user's packet radio generates a packet for a destination outside the control of his station and his radio routes it to an appropriate local station (e.g., the closest one according to some metric). The station converts this packet into several distinct route finding packets which it sends to each neighboring station via some repeater jointly labeled by both stations. The conversion involves adding to the packet the station ID and a list of selectors from the user to the jointly labeled repeater. When the packet is received by the neighboring station, it checks to see if the specified destination user is under its control. If not, it again converts the packet into several distinct packets by adding its own ID and a list of selectors from the original jointly labeled repeater to another repeater jointly labeled by a station not previously visited by the packet. If the packet arrives at a station which has just previously handled the same request via another route, it will be discarded by that station. In this way, one or more route finding packets will eventually arrive at a station which has labeled the destination user and will contain a composite list of selectors. The destination station then passes a complete list of selectors to the destination user's packet radio which initiates the route setup procedure described earlier in this section. This process is illustrated in Fig. 10. The route finding path is shown as a series of dotted lines and the point-to-point route is shown as a series of solid lines.

Once having provided a point-to-point route between two

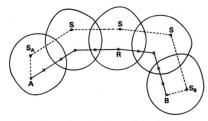

Fig. 10. Route finding and route setup with multiple station operation.

radios under its control, a station does not need to retain detailed state information about that route, regardless of whether the route is part of a larger multistation route or used by two radios under its control. It may be appropriate for the station to retain certain summary information, however, such as the cumulative number of point-to-point routes it has provided, or the number of renewal table entries which were made in each repeater over the last minute, etc.

If a radio fails while a packet is enroute (e.g., repeater R in Fig. 10), and this disrupts a point-to-point route, local alternate routing around the failed radio will take place if possible. The particular station in control of the failed radio will attempt to take a locally corrective measure using only the routing selectors contained in the en route packet (for transit traffic) and the destination ID for nontransit traffic. In this context, transit traffic consists of packets which are passing through the domain of that particular station en route to their final destination. Nontransit traffic consists of packets destined for a radio that has been labeled by that station.

If the locally corrective measure fails (remember there is no state information about specific point-to-point routes in the station), an error message will be returned to the source and the original route setup process will be reinitiated. In this process, certain packets may be discarded, so the end-to-end host protocol must be prepared to recover from this situation.

E. Stationless Operation

The main difference in operation when no stations are present is that each radio initially relies on broadcast routing to all radios in the net in order to communicate with any destination. Broadcasting is a stable and effective way to communicate, particularly with mobile terminals, but it is inherently very inefficient (We assume that distributed adaptive routing cannot be used). For stationless operation, it is highly desirable for the radios to determine acceptable point-to-point routes, even if they only remain acceptable for short periods of time. In a technique that is very similar to multistation route finding, broadcasting can be used to find point-to-point routes which may be usable if the radio links are relatively stable. The repeater and terminals in the network cooperate with each other to discover and set up routes in three phases.

1) Route Discovery: A route finding packet is broadcast from the source PR when its attached device attempts to communicate with a destination for which a route is not yet known. This packet contains the source selector, the desired destination selector, and a sequence number which insures uniqueness. Any PR which hears the route finding packet appends its own selector to the data field, stores the information which uniquely identifies the packet, increments a "hop" count in the packet and rebroadcasts it. This PR will then ignore any subsequent copies it hears of this packet, so the broadcast proceeds outward eventually reaching all PR's in a "wavefront" type of propagation. If the hop count exceeds a maximum value, that packet will be discarded.

If the route finding packet reaches a PR connected to the desired destination, a successful route to the destination is contained in it. Of course, the destination may receive a number of such packets over different routes. With some probability, the route finding packet may also never reach the destination. Radio networks are inherently lossy and no guarantee of delivery can be given. The user can try again or give up. The forwarding of the packet also involves an incremental route delay computation. A delay field is initialized to zero by the source PR; each PR forwarding the packet adds in an estimate of the delay the packet incurred in traversing the previous hop. This estimated delay is derived from the PR's neighbor table, which estimates reliability (and hence expected delay, through the hop retransmission algorithm) for the hop from each neighbor.

Once the route finding packet arrives at the destination PR, it contains the estimated round-trip delay for a route which works in both directions. The destination PR will wait long enough to receive most of these packets (which may have travelled over different routes), will select the route with minimum delay, and will store this as the best route.

2) Route Setup Phase: For efficient channel utilization, a short routing field in the header is desirable for normal data packets. Since the path discovered above may involve a very large number of selectors, it is clear that not all routing information can be stored in the data packets themselves. Some of the routing information must be stored in the intermediate PR's along the route. This is accomplished by a route setup procedure which is nominally identical to that which is used in the operational station case. A route setup packet is sent from the digital unit of the destination. PR which traverses the route specified in the selected route. This packet causes renewal table entries to be written in the intermediate PR's (in a table indexed by source and destination). Once this route setup packet arrives back at the source PR, the entire route is set up in both directions and we proceed to the final phase.

3) Normal Data Traffic: Normal data traffic is sent using the short routing field in the header to specify the next few hops. The source PR and each intermediate PR along the route overwrite this route information with their stored routing information for the destination specified in the packet, giving it a fresh route for the next few hops each step of the way. As before, the choice of a few next hops rather than a single next hop provides the flexibility of permitting alternate routing if the next PR in the route is failing.

There are four major differences in the functioning of a stationless network and a network with operational stations: First, the station collects connectivity information so that it can supply a route without requiring a broadcast search procedure involving all radios; this economizes on channel use and contention. Secondly, the station can compare all possible routes when making its choice, not just the routes which happened to be traversed successfully on a single attempt; this results in a better route selection. Third, the station can detect changes in connectivity quality and issue better routes as appropriate without interrupting on-going communication. Finally, the station can perform global congestion control, changing PR parameters in a coordinated fashion to relieve congestion in different areas of the network. The stationless

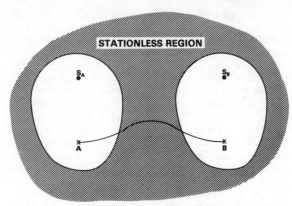

Fig. 11. Communication across a stationless region.

mode of PRNET operation is primarily envisioned as a fail-soft (degraded) mode of operation rather than the normal mode.

There may be instances in which an initial deployment strategy calls for radios to be made operational before stations and where a stationless mode of operation would be very useful. It is even possible for a set of radios under station control to be within range of another set of radios operating in stationless mode and to communicate with each other. Normally, the station would attempt to label the other set of radios, if it could accommodate more radios. If, however, it cannot label them for whatever reason, communication can still take place between the two sets by a concatenation of the station routing with the stationless routing. We do not address this area further in this paper, but if two or more station controlled sets of radio are connected via a stationless region, they can communicate by this same procedure (see Fig. 11).

F. Multidestination Routing

Many applications require the ability to communicate identical information to more than one destination (e.g., conferencing, data base updating, etc.). This capability could be achieved by sending multiple copies of the original information, one copy to each destination. The alternative, a multidestination routing capability could be provided for efficiency and low delay. We view broadcasting as a special case of multidestination routing where all radios (rather than a subset) actively participate.

The inherent broadcast mode of the radio channel (the "ether") provides an opportunity for efficient multidestination routing which is not present in dedicated-channel networks. In particular, the "fan-out" required to transmit a message from its source to multiple destinations can be provided by the inherent fan-out of the radio channel. Each transmission may be simultaneously received by many network nodes at no additional cost in radio channel utilization. Alternative strategies for multipoint routing include: Minimal Spanning Tree Routing, Source-Based Routing, and Multi-Address Routing. In the first case, one minimal spanning tree is defined for the network, and all traffic flows on that tree to the selected destinations. In the second case, each radio derives its own minimal spanning tree with itself at the root. In the third case, a list of addresses is used to selectively route packets through the network [28].

A source-based routing scheme is a good choice for broadcast applications and is generally preferable to the use of minimal spanning tree routing. The multiaddress routing scheme is preferred for conference applications involving only a subset of the radios.

VI. Network Aspects of Spread Spectrum

The use of spread spectrum waveforms in a packet radio system is motivated largely by the desire to achieve good performance in the fading multipath channels resulting from nonsited mobile system users, by the need for coexistence with other systems and for antijamming capabilities in tactical applications. Once a spread spectrum waveform is adopted, however, a number of other benefits, which in themselves suggest the use of spread spectrum, are also available, such as a strong capture capability to enhance access efficiency, the potential for an integrated position location feature, and the ability to operate links, nets, and subnets on pseudo-orthogonal waveforms using CDMA.

These capabilities are not received free of cost, however. Use of a fixed spread spectrum waveform adds a modest amount of system complexity, while a time varying spread spectrum waveform requires some degree of network time synchronization to be established and may require added system protocols to effect the distribution of variables to control waveform generating algorithms.

The use of spread spectrum, although desirable for many applications, is not an *a priori* requirement for a packet radio system. It is worthwhile to distinguish those aspects of network operation which are primarily due to the presence of spread spectrum from those which are not. Consequently, we have identified six spread spectrum network concepts for treatment in this section to separate them from the other more generic packet radio concepts.

A. Synchronization

Because packet radio networks operate with random access transmissions on a common radio channel, a receiver cannot anticipate the sender of any particular packet or its exact time of arrival. The receiver must use the initial portion of the packet (preamble) to detect the arrival of the packet, adjust any automatic gain control loops, and acquire synchronization with the packet symbol and waveform structure to the degree needed to receive the remainder of the packet successfully. In systems which use a spread spectrum waveform which varies with time, packet radios must acquire and maintain a common network time of day, so that the proper reference waveforms can be provided to the receiver for reception of a packet at any given time.

The design of a packet preamble and the corresponding receiver system suitable for achieving rapid synchronization to a received packet is a function of the particular spread spectrum waveform structure being used. Significant design differences exist depending upon whether PN, FH, or other waveforms are used, upon whether or not the waveform used is a function of time and, in the case of a fixed waveform, on its repetition period. A typical design is discussed in Section VII which has been implemented in an experimental packet radio network.

1) Unsynchronized Network: In designs where a fixed PN or FH spread spectrum pattern is reused for each packet, packet radio networks require no common time of day synchronization. If the repetition period of the fixed pattern is chosen correctly, all of the benefits of a spread spectrum waveform may be realized, with the exception of the antijamming properties needed for military applications. The pattern period should exceed the maximum multipath spread expected on the channel in order to suppress intersymbol interference. For example, the PN waveform used to estimate the multipath channel depicted in Fig. 3 had a period three times the multipath spread, and allowed the channel multipath profile to be estimated without overlap. It is desirable that the pattern not repeat during the duration of a maximum length packet, so that overlapping packets interfere with each other to the minimum extent. In addition, a strong capture property is thereby realized. For ease of packet acquisition and synchronization, the fixed pattern should always be restarted from the same point at the beginning of each packet.

2) Code Slotted Network: In tactical applications, the waveform pattern must not be reused if the system antijamming properties are to be retained. To coordinate the waveform changing process, network time is defined in terms of discrete intervals called slots. Each slot is uniquely identified by a number. Slots are numbered sequentially, beginning at some specified time. Slot number may be viewed as the most significant digits of the system time of day. During each slot a spread spectrum pattern is defined which is unique to that slot and is used both to generate the waveform for any packet transmission beginning in that slot and as a reference input to the programmable matched filter receiver for incoming packets.

Since packet radios may operate in a random access mode, several radios could transmit overlapping packets in a given slot with the same pattern. If the slot duration is short enough, the probability of overlapping packets with identical patterns is small. This advantage from the self interference and antijamming points of view is obtained at the expense of increased difficulty of maintaining system synchronization with very small slots. Typically, one might choose a slot duration which is larger than the average direct path propagation delay by one or two orders of magnitude to facilitate synchronization and to minimize packet loss due to disagreement between transmitter and receiver as to the current slot number and corresponding valid waveform. For a typical ground radio application, this implies a slot length of a few milliseconds.

Having made a correspondence between slot number and spread spectrum pattern, a choice still remains as to the points in time at which packet transmissions may begin. Two modes, termed slotted and nonslotted, are of interest. In a slotted system, transmission is allowed to begin only at a specified point in each slot, and a guard time must be provided in the slot to allow the packet to propagate beyond normal radio range prior to the end of the slot. In this mode, the packet duration is somewhat less than a slot. Individual users make the decision independently to transmit in a given slot in random access fashion. This independence of the access contention process from slot to slot has advantageous consequences in the carrier sense mechanism described later in this section.

In contrast to a slotted system, a nonslotted system allows users to transmit at any point within a slot, and the duration of a packet transmission may exceed the duration of a single slot. The nonslotted mode may reduce system delay since the radio does not have to wait for a given time within the slot to transmit. We conjecture that this may also provide increased efficiency compared to slotted operation due to the presence

of capture. We note that this conjecture is the opposite of what is known for nonspread systems without capture. In particular, the ALOHA and slotted ALOHA [19] techniques have channel efficiencies of $1/2e$ and $1/e$, respectively. These efficiencies, however, are based on the assumption, called the zero capture model, that no packets are received correctly if they are overlapped in time at a receiver, which is not the case for spread spectrum waveform reception.

B. Capture

By the term capture, we mean the ability of a receiver to successfully receive a packet (with nonzero probability) even though part or all of the packet arrives at the receiver overlapped in time by other packets. The basic mechanism for capture is the ability of the receiver to synchronize with and lock on to one packet and subsequently reject other overlapping packets as noise. A system with perfect capture is one in which the first arriving packet at an idle receiver is captured with probability one, even if another packet arrives after a vanishingly small delay.

A system cannot achieve perfect capture for a number of reasons. If the preamble portions of packets overlap, the AGC and packet synchronization process is more likely to fail, causing loss of one or both packets. If the relative time of arrival of two packets using the same pattern is less than one chip in a PN system or less than one frequency hopping interval in an FH system, successful reception of any one of the packets is unlikely. In general, there is a vulnerable period at the beginning of a packet, denoted Tc and called the capture interval, during which collision with the same portion of another packet results in the loss of both. The magnitude of Tc varies both with the waveform structure and the complexity of the receiver synchronization circuitry which can be tolerated in a given implementation. Minimizing Tc increases the capture probability.

A second major reason for imperfect capture is a result of relative received signal strength. If a packet is captured, but a later arriving packet having sufficiently greater signal strength overlaps this packet, the captured packet may be destroyed. The stronger packet will in effect jam the earlier packet being received by overwhelming the interference rejection capability of the system. The likelihood that such unintentional self-jamming occurs depends upon the distribution of users in the network area and the distribution of signal strength ratios observed at a receiver. With a typical radio implementation and assuming a uniform distribution of users in an area the probability of such unintentional jamming can be quite low. One approach to further reducing this self-interference effect is to use transmit power control. By using cooperative network protocols, each radio can measure the signal strengths of received packets and send this information back to the transmitting radio. Using this feedback, received signal strengths may be roughly normalized by adjusting transmit power and system self-interference may be reduced. The control problems associated with the conflicting requirements of this procedure appear to be complex.

The capture phenomenon exhibits different characteristics in slotted and nonslotted systems. In slotted systems, the capture interval of the packet T_c arrives in some period of duration Tu near the beginning of the slot. If all users transmit at the same point in the slot, packets arrive in order of increasing range from the receivers. This leads to discrimination against the more distant users. Packet radios may implement a randomizing function which tends to distribute arrivals uniformly over the uncertainty interval T_u. If this is done, range discrimination is decreased, and capture performance improves in proportion to Tu/Tc. Unfortunately, a portion of the slot, Tu, has been lost to the communication function as a result. This loss, however, may be more than compensated for by the increase in capture performance.

In a nonslotted system, Tu is equal to the slot duration. Thus, Tc/Tu is minimum and the capture probability would be expected to increase. Offsetting this factor is the potential for a packet in a nonslotted system to be interfered with by packets from adjoining slots as well as from the slot in which the transmisssion began. It is not clear at this point whether capture performs best in a slotted or a nonslotted system, or under what conditions, and this is expected to be a subject for both theoretical and experimental study.

C. Receiver Addressed Waveforms

In the code slotted system described earlier, there is a requirement for a waveform generator which produces a different spread spectrum pattern for each slot. If one views the mechanization of this requirement as a device which has the slot number as an input and the pattern as an output, it is easy to generalize the concept to include several input parameters and an output which is unique for each possible set of inputs. An important potential use of such a mechanism is to use the unique identifier of the next intended receiver as one input, thus generating a waveform which is associated with a particular slot and a particular receiver. While in the receive mode, each radio would use its own unique identifier to generate the reference pattern for its receiver. One benefit of this procedure is that contention at a given receiver is reduced to only those radios trying to send a packet to that particular receiver in the same slot. Another benefit is that the processing overhead (and even the incremental routing overhead) can be reduced since only those packets intended for a given receiver are actually received by it. In networks using point-to-point routing, increased network throughput may result.

A number of difficulties arise, however, if receiver-addressed waveforms are used in place of a common code by all radios. Echo acknowledgment techniques must be replaced by separate acknowledgments if a hop-by-hop acknowledgment protocol is used. Network synchronization may require more overhead transmissions, since a simple broadcast on a common code pattern no longer is received by all PR's in radio range. The discovery of new RF connectivity also becomes more difficult without a common code pattern. Potential solutions to these problems include the use of dual receive channels to operate simultaneously with both a receiver-addressed pattern and a common code pattern or a time shared partition of slots into a receiver addressed portion and a common code portion. The first solution requires additional hardware, while the second solution may reduce the efficiency of the shared radio channel.

D. Carrier Sense

The carrier sense multiple access (CSMA) technique provides one way to significantly reduce the number of conflicts due to self-interference on the channel. It achieves this capability by inhibiting transmissions which would otherwise cause con-

tention to occur. The capture effect described earlier provides an alternative method to deal with contention. In a way, CSMA captures the channel, while spread spectrum allows the waveform to be captured at a receiver. Carrier sense is, in reality, a misnomer in the spread spectrum context. A better term might be "spread spectrum packet acquisition," or SSPA. However, we will use the term carrier sense to refer to SSPA for compatibility with terminology commonly used in the literature.

In carrier sense operation, a packet radio operates in the receive mode until the last possible instant prior to the onset of a packet transmission. If channel activity is present, the transmission is inhibited and rescheduled for later transmission. Perfect channel sensing is impossible in actual application due to finite propagation delay. Nevertheless, carrier sense can provide a very efficient access mechanism [11, 29].

Early models of carrier sense assumed that packet radios would implement the sensing function by measuring in-band RF energy or by observing an in-lock status indicator associated with packet synchronization circuitry. In high interference environments, however, sensing RF energy may result in false activity sensing and unnecessary transmission delays. In tactical environments, the observation of bit synchronization status may also result in false sensing if the same waveform is continually reused in the system, since this would allow a jamming transmitter to hold off the other packet radios in an area. In code slotted systems, these false sensing problems can be avoided, since new waveforms are used in each slot. A number of remaining issues related to CSMA in a spread spectrum system are addressed below.

1) Carrier Sense Versus Capture: Even though a packet radio system operates with a strong capture mechanism, it may still be desirable to use carrier sense to limit contention for a number of reasons. In a slotted system, sensing a packet on the channel often indicates that the potential recipients of a packet to be transmitted are committed to receiving the packet which is already on the channel. This is particularly true if receiver addressed waveforms are used, and sensing is performed using the waveform of the intended receiver as a reference.

While the capture mechanism provided by a spread spectrum waveform reduces the adverse effects of contending packets, contention may still have undesirable effects. Battery operated packet radios may waste energy unnecessarily if they transmit when it would have been possible to sense that the intended receiver was probably busy and delay the transmission. Transmitting densely overlapped packets may not result in zero capture, but the code division multiple access and anti-jamming performance of the system will be reduced. Thus, even if a strong capture mechanism exists, the use of carrier sensing procedures may allow the original margin against interference to be retained. In a slotted system the presence of carrier sense will normally increase system efficiency as well.

2) Sensing Strategy: Since the spread spectrum patterns in a code slotted system are constantly changing, carrier sensing can only take place practically during the preamble portion of the packet. Two alternatives are apparent. A packet radio may detect preambles and continue to receive a packet once detected. Thus, collisions with that particular packet are avoided, but any packet which arrives at the packet radio subsequent to capture of the first packet cannot be detected since the preamble will be missed. After the end of the first packet, the channel will probably still be occupied with the end of the second packet, but, unfortunately, this will be invisible to the radio. Alternatively, if the packet radio is in a sensing mode with all local packet buffers full with packets waiting to be transmitted but with no outstanding acknowledgments, the packet radio may choose simply to sense preambles only. If a preamble is detected, transmission is prohibited for some period of time and the remainder of the detected packet is intentionally dropped. The receiver then returns to the preamble sensing mode once again in search of an idle channel interval. If during this time another preamble is sensed, the transmission is further prohibited. The difficulty with this procedure is that the location of the end of a sensed packet is unknown, due to a variable packet length. Thus, the minimum delay prior to re-attempting transmission may have to be set for a maximum length packet, adding unnecessarily to packet delivery delay.

The sensing strategies discussed above can be applied directly to the cases of slotted and nonslotted operation. However, the results of the sensing strategy are more precise in slotted operation, since packet transmission and reception must end before a slot boundary. The slot size would generally be selected to accommodate the maximum packet size plus an allowance for propagation delay and the uncertainty interval T_u if used. Thus, the slot is normally capable of accommodating only one packet per slot for a given receiver, and the sensing process may be restarted at the beginning of each slot without danger of missing a preamble due to overlapping prior transmissions.

E. Position Location

The wide bandwidth of a PN spread spectrum waveform or a hybrid PN/FH waveform allows good time resolution for measurement of packet time of arrival (TOA). In a code slotted system, a certain degree of network synchronization must be maintained to support the communication function. This degree of synchronization can be further refined to allow multiple TOA or pseudoranging measurements to be made very accurately at each packet radio with respect to other radios, and these measurements can be used in a multilateration position location procedure. Periodically, each packet radio could broadcast its current estimate of its location in order to provide adjacent units with additional data to improve the pseudoranging procedure. For typical ground-based mobile terminals, these position messages would only need to be broadcast a few times per minute, so that the load on the channel due to these messages would be a small fraction of total capacity.

In addition to the obvious applications to mobile user navigation or position reporting, knowledge of geographic relative positions and velocity vectors of packet radios could aid significantly the process of point-to-point route establishment, mobile terminal route evaluation, and other network procedures. However, the value of this tool has yet to be established.

F. ECCM Performance

A code slotted packet radio network can be designed to provide an effective electronic counter countermeasure (ECCM) capability. If the valid waveform for any given slot cannot be predicted by a network nonparticipant, jammers are constrained to operate as noncoherent rather than coherent noise generators. Jammers are also unable to "spoof" the network

with false traffic if previously used waveforms are rapidly discarded in favor of new waveforms valid in the very next slot. The decreased spectral density of spread sprectrum signals, lack of waveform predictability, use of multiple access and random access, and co-located CDMA networks all contribute to a decreased susceptibility to timely detection, direction finding, and intercept of signals by nonnetwork participants. All of these properties could be important in tactical applications of packet radio.

VII. EXPERIMENTAL PACKET RADIO DEVELOPMENTS

Although significant analytical progress has been achieved, a complete mathematical analysis of a multiple access, multihop radio network is not yet possible. Computer simulations having reasonable detail are being developed, and show great promise, but their practicality and degree of realism have yet to be validated. Thus, experimentation with a real network is still a primary tool of the network architect.

Toward this end, the Advanced Research Projects Agency (ARPA) initiated in 1973 a theoretical and experimental packet radio program. The initial ARPA program objective was to develop a geographically distributed network consisting of an array of packet radios managed by one or more mini-computer based "stations," and to experimentally evaluate the performance of the system. The first packet radios were delivered to the San Francisco Bay area in mid-1975 for initial testing and a quasi-operational network capability was established for the first time in September 1976, shortly after the prototype station software was developed. Approximately 25 radios are currently available for use and about 50 radios are expected to be available by the end of 1979. As of this writing (June 1978), the packet radio network has been in daily operation for experimental purposes for almost two years. During that period, nearly three dozen major demonstrations of the network were scheduled and successfully carried out. The location of the major elements of the packet radio tested during 1977 is shown in Fig. 12.

The packet radio equipment currently in use in the Bay Area testbed is designed to support the development and evaluation of fundamental network concepts and techniques. This initial radio equipment was designated the Experimental Packet Radio (EPR), and is briefly described below. A major new development in 1978 has been the completion of an Upgraded Packet Radio (UPR) which is similar in architecture to the EPR but which has, in addition, the ECCM features necessary to verify the viability of packet radio concepts in tactical military applications. Selected features of the EPR and UPR designs are described in this section.

Functions provided within the station software installed in 1977 included: network routing control; a gateway to other networks; a network measurement facility which collects, stores, and delivers experimental statistics from any network components; a debugging facility which supports examining and depositing the contents of memory in the PR units; an information service which assists in locating and connecting to people currently using the PRNET; and an experiment configuration control module. Of these functions only the network routing control must reside within the station. All the other functions can reside in separate hosts attached to the PRNET. This permits convenient expansion of the quantity and quality of services provided to users of the PRNET. At the same time the station software itself may be kept small and simple permitting economical replication for high redundancy and

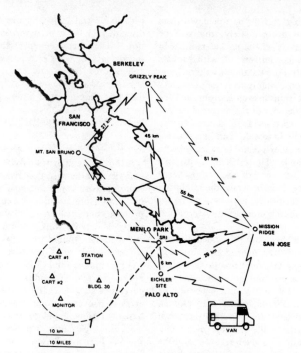

Fig. 12. Location of major elements of the packet radio testbed during 1977.

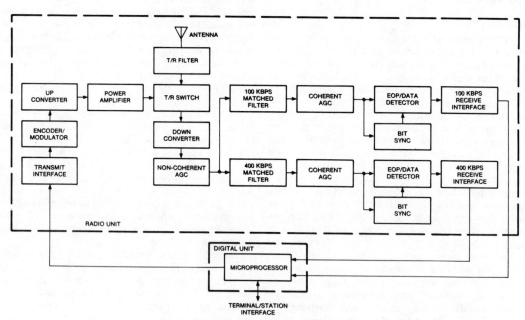

Fig. 13. Experimental packet radio configuration.

reliability of network operation.

The major objectives of the experimental testbed are the development and evaluation of efficient network protocols and operational techniques and the test use of these capabilities by selected communities of real users. Toward this end, techniques have been developed to allow a variety of debugging and performance measurement operations to be carried out

in the network, and these capabilities are also described in this section.

A. An Experimental Packet Radio Unit

The EPR design which has been implemented for use in the testbed is functionally configured as in Fig. 13 (also see Fig. 6). An in-depth discussion of the implementation issues and applied technology relevant to the EPR is available [26]. The description provided here focuses on the operation and evolution of the EPR in the packet radio test bed.

Each EPR consists of a radio unit, which transmits and receives packets, and a microprocessor-based digital unit, which controls the radio and provides packet header processing (e.g., for routing of packets between nodes). An EPR may operate as a repeater, or may be connected to a user's host computer or terminal, or to a station. The interface between the user equipment and the EPR digital unit is the portal through which packets enter and leave the network.

The EPR radio unit operates with a fixed PN spread spectrum pattern which, for simplicity in implementation, is identical for each transmitted bit. Two transmission data rates are available, 100 and 400 kbits/s, with corresponding spread spectrum patterns of 128 and 32 chips per bit, respectively. The 100-kbits/s rate is used for links with potentially large multipath spreads because the fixed bit length PN chip pattern does not provide the ability to discriminate against intersymbol interference. The radio unit operates in a half duplex mode. When a packet is transmitted, the preamble, header and text are read from microprocessor memory under direct memory access (DMA) control. The radio unit completes the packet format previously illustrated in Fig. 8 by adding a 32 bit cyclic redundancy checksum (CRC), then differentially encodes the data, and adds (modulo two) the appropriate PN chip pattern for the selected data rate. The resulting PN modulated stream is then applied to a minimum shift keying (MSK) modulator, and the signal is up-converted to a selected 20 MHz portion of the 1710–1850 MHz band, power amplified, and transmitted through an azimuthally omnidirectional antenna.

When not transmitting, the EPR remains in the receive mode. An arriving packet proceeds through RF amplification, downconversion, IF amplifier and wide-band (noncoherent) automatic gain control (AGC) functions. Because the PN chip patterns used for the 100 and 400-kbits/s data rates are chosen for low cross correlation performance, two parallel receive chains following the IF amplifier/AGC can be simultaneously active. A fixed surface acoustic wave (SAW) device is used to match filter the PN modulated waveform two bits at a time in a differential detector. The outputs of the differential detector (i.e., the correlation pulses) are used to set a narrowband (coherent) AGC, and after a phase locked loop bit sync circuit has settled and the end of preamble (EOP) bit pattern has been detected, the differentially detected data in the active channel is passed to microprocessor memory under DMA control. The packet preamble design illustrated in Fig. 7 allows AGC settling, bit sync acquisition, and EOP detection (data sync) to occur prior to arrival of the first data bit. The microprocessor executes the appropriate protocol software to determine whether the received packet should be relayed, delivered to an attached user or station, or discarded.

The EPR digital unit presently uses a National IMP-16 microprocessor with 4096 16-bit words of RAM and 1024 words of PROM. The PROM contains the processor operating system and all DMA I/O routines. Network protocols and packet buffers are stored in RAM. Eventually, the protocol routines will also be stored in PROM, but the evolving nature of the software in an experimental system requires the use of RAM at present. Protocols currently implemented are the channel access protocol (CAP), the reliable station to PR protocol (SPP), a statistics gathering feature called CUMSTATS, and a debugging package called X-RAY. CAP is responsible for the primary EPR function of transferring packets to or from the adjacent EPR on a route through the network. CAP is responsible for monitoring the hop-by-hop echo acknowledgment process, retransmission of nonacknowledged packets, invoking alternate routing procedures, and determining packet disposition. CAP currently implements pure ALOHA, CSMA, and a variant of pure ALOHA in which random transmission is deferred until the end of an on-going reception process so as not to needlessly discard an arriving packet.

SPP is an end-to-end protocol which is used for reliable delivery of network monitor and control packets, such as "labeling packets" sent to the EPR. SPP uses CAP as its interface to the network. Similarly, an external user device which submits a packet to the EPR uses CAP to launch the packet into the network, but is responsible for any user end-to-end protocols which might be required to be layered on top of CAP. Currently, approximately 1000 words of memory are devoted to the EPR operating system, 600 words to the DMA I/O routines, 2500 words to CAP, SPP, and CUMSTATS, and 900 words to packet buffers.

B. An Upgraded Packet Radio Unit

1) Motivation: The EPR was designed to allow experimentation and verification of the majority of basic packet radio protocols and concepts. The need exists, however, for a limited number of packet radios having enhanced capabilities in order to demonstrate the feasibility and advantages of packet switching radio networks in ground-mobile tactical environments. The upgraded packet radio (UPR) design which has been developed in response to this need differs from the EPR design primarily in its enhanced electronic counter measures (ECCM), low probability of intercept and position location capabilities.

In the tactical environment, reuse of the same PN sequence would result in vulnerability to jamming and waveform spoofing. In the UPR, a PN pattern, which varies on a bit-by-bit basis, is used to spread spectrum modulate each bit. As a result, the UPR must have a programmable matched filter to receive the PN modulated waveform, and the UPR network must maintain a degree of synchronization among network elements to enable the receiver to generate the same PN sequence as the sender. Timing is provided by means of an accurate time of day clock maintained by all the UPR's. In addition to varying the PN pattern used, a higher spread factor (or number of chips per bit) is used in the UPR than in the EPR. The data rate in the UPR is approximately the same as in the EPR, but the corresponding bandwidth of the UPR is larger—approximately 140 MHz (i.e., 1710–1850 MHz). An electroacoustic convolver is used in the UPR to provide the wideband, programmable matched filtering element. This convolver is a single passive device which currently occupies a few cubic inches and is equivalent to many conventional matched filters. Further discussion of the operation of the convolver is contained in reference [30], [31].

Reliable packet transport in the EPR system relies on error detection and retransmission strategies. These techniques are inadequate in a tactical environment because they allow very simple jamming strategies to force large number of packet retransmissions with a low attendant throughput. The UPR is provided with a forward error correction (FEC) mechanism based on convolutional encoding and sequential decoding which operates in combination with error detection and packet retransmission techniques.

2) Organization and Basic Operation of the UPR: The func-

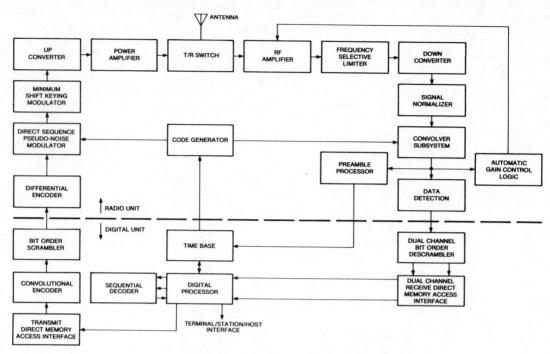

Fig. 14. Upgraded packet radio configuration.

tional organization of the UPR is illustrated in Fig. 14. As in the EPR, the UPR digital unit interfaces with user terminal equipment by means of a direct memory access (DMA) channel and with the radio unit by means of DMA channels under processor control. In addition, the UPR processor controls the local time of day clock through addressable registers and supervises the activity of the sequential decoder which operates into and out of the processor memory by means of three DMA channels.

Network synchronization is achieved through the use of time stamped packets. Transmitted packets contain their time of transmission as determined from the local time of day clock. When packets are received, the time of arrival is stored in memory. The processor uses this data to develop timing corrections for the local time base in accordance with network timing protocols.

System time is divided into basic units called slots which have a 5-ms duration. For experimental purposes, however, the slot duration is adjustable. Slots are uniquely identified by sequentially numbering them. In order for two network UPR elements to communicate, their time of day clocks must agree on the current slot number. There is a one-to-one correspondence between a slot number and a PN code sequence used to transmit a packet such that when a UPR initiates a packet transmission in a particular slot, the PN code sequence used is the one corresponding to that slot. The transmission is allowed to begin at any point in the slot and the duration of the transmission many be many slots. The same strategy is used at a UPR which is waiting to receive a packet. At the beginning of each time slot, the time base delivers a new slot number to the PN code generator. This slot number completely determines the PN pattern used to receive any packet whose reception begins during that slot. Regardless of the time of arrival of the

packet within the slot, or its length, the PN code sequence used is completely defined by the slot in which reception begins. Thus, the UPR implements a code slotted system with nonslotted transmission. The hardware is also capable of slotted transmission for experimental purposes.

When a packet is to be transmitted, the processor activates a DMA channel to control and monitor the transmission. Under DMA control, the packet is read from the processor memory, convolutionally encoded with a constraint length 24 code, and loaded into a buffer prior to scrambling (bit order permutation). The packet data is read from the buffer bit by bit in pseudo-random order, differentially encoded, and passed to the spread spectrum modulator where each data bit is modulo two added to each chip of the PN chip sequence used to encode that bit. The PN modulated chip sequence is then passed to an MSK modulator, implemented with a SAW device, and having an IF output at 300 MHz. This signal is up-converted to 1780 MHz, amplified to 10 W, and fed to the azimuthally omnidirectional antenna. Fig. 15 shows the basic UPR packet and preamble format. In the discussion above, only the header and text bits of the packet are read from the processor memory. The preamble and postamble bits are supplied by the code generator circuitry, and are used in combination by the receiver to determine the receive data rate and coding format of the packet.

The UPR operates in a half duplex mode. When the UPR is not transmitting, received packets pass from the antenna through a number of RF amplifier/automatic gain control (AGC) stages. The signal is then processed by two frequency selective limiter (FSL) stages which provide an adaptive notching mechanism for narrow-band interference. After down conversion to 300 MHz, the signal is processed by a signal normalizer which tries to normalize wideband interference to the same level as the desired signal prior to spread spectrum processing. The normalizer is a signal processing device which operates at IF. When used in conjunction with the spread spectrum processing gain, it can nullify the effects of a single wideband source of interference. The UPR can also be operated with an adaptive antenna array to null wideband jammers.

The convolver subsystem provides the spread spectrum matched filtering function. During the preamble of the packet, the convolver outputs are further processed by a preamble processor where packet detection and synchronization functions are performed. During the remainder of the packet, the convolver outputs are processed by the data detection circuitry which makes hard binary bit decisions and provides a matched filter function for the postamble sequence. Received packets are buffered in one of the receiver's descramblers prior to bit reordering and storage of the packet in processor memory under control of a DMA channel. Two receive descrambler/DMA channels are provided to allow reception of two successive packets with minimum interpacket arrival time. When a packet has been received and stored in memory, the processor initializes DMA transactions with the sequential decoder to decode the packet. The decoding process thus takes place "off-line" in that additional packets may be received in memory while previous packets are still being decoded. As soon as the packet header is decoded, it may be submitted to network protocol processing to determine disposition of the packet.

3) Spread Spectrum Code Selection and Generation: One of the fundamental differences between the EPR and UPR equipment is the use in the UPR of a direct sequence spread spectrum waveform which changes from bit to bit. The selection of the code and an efficient means for its generation was an important consideration in the UPR development. The attributes of the code which influenced the choice were as follows. In order to provide good antijamming and antispoofing properties, it should be impossible to predict the PN sequence which will be used to encode some future data bit based on an observation of the PN code used for encoding past data. The PN code sequences of bit duration should also have good autocorrelation properties to facilitate detection and synchronization of the packet. Code autocorrelation becomes even more important in view of the position location subsystem being developed for the UPR which relies on accurate time of arrival measurement for its operation. The PN code sequences used to encode bits should have low cross-correlation. This attribute is important since it provides a capture mechanism for the channel access technique and will support multiple simultaneous packet transmissions using code division multiple access (CDMA). The equipment required to generate the code should require low power and occupy low volume. The PN chip stream rate is quite high, and low power is therefore of particular concern. In the current application, it is desirable to have a code generation algorithm which is easily time reversible, since the convolver reference chip stream is the reverse of the modulator chip stream over a bit. Any code generator can be made reversible by running it in advance, storing the chips in a memory, and reading them in reverse order. However, a reversible code algorithm may result in savings of power and hardware.

A code generation technique having the desired attributes was selected from several candidates for implementation in the UPR. For brevity, a code pattern will refer to the length of PN chip sequence used to spread spectrum encode one data bit. In the initial versions of the UPR, a finite set of approxi-

mately 2000 code patterns are being used. For any pattern in this set, its chip-by-chip binary complement is also in the set. This feature makes it more difficult for an unauthorized user to extract the data from a packet. For any given transmitted packet, a sequence of codes is constructed from this finite code set according to a nonlinear secure algorithm. The sequence constructed depends on the slot number during which transmission begins and certain other parameters used by the algorithm. Use of this technique allows a good algorithm operating at low power and low speed to specify a high speed chip stream having the desired unpredictability to an observer outside of the system. In addition, the autocorrelation and cross-correlation properties desired are a function only of the finite set of codes selected. For the UPR equipment, the code set was constructed from the Gold codes [32], which provide the desirable property of reversibility and which can be implemented in a memory structure using parallel operations at low speed and low power consumption prior to a final high speed parallel to serial operation.

4) Error Control: The UPR employs convolutional encoding and sequential decoding for forward error correction in conjunction with the error detection and retransmission techniques used in the EPR [33]. As indicated in Fig. 15(a), only the header and text of the packet are encoded. The convolutional encoder is capable of encoding at three rates: 1/4, 1/2, and 3/4. Two code rates applied to each coded packet. A lower rate code is used for the first portion of the packet and a higher rate is used for the larger second portion. The length of the first portion is under software control and need not be identically the header. The code rate combinations allowed in the header and text, respectively, are (1/4, 1/2), (1/2, 3/4), and 1/2, 1), where code rate 1 indicates that the encoder is bypassed. The UPR processor submits the two distinctly coded portions of a received packet to the sequential decoder as separate tasks. This allows the processor to decode one or more packet headers quickly so that header protocol processing may take place while the longer portions of the packet are being decoded.

If the decoding time of the second portion becomes excessive, the decoding can be aborted by the processor and the original packet can be saved in the encoded state pending the arrival of a retransmission. If the first portions, which are more heavily encoded, of both the original and the duplicate can both be decoded, the two packets can be identified as duplicates. Since the encoded second portions of the two packets are identical, except for their error sequences, they can be combined to form a virtual binary erasure channel. The sequential decoder is designed to operate on either binary symmetric or binary erasure channel data. The result of this procedure is that if the first portions of the packets were decodable at, say, rate 1/4 as separate binary symmetric channel inputs, then the combined second portions will, with high probability, be successfully decoded as a rate 1/2 binary erasure channel input.

It should be pointed out that sequential decoders operate best with random channel errors. For this reason, a bit order randomization across the entire packet is performed after encoding, and the inverse operation is performed prior to decoding. The randomization algorithm is a function of both slot number and packet length. As a result, a jammer cannot impose the same error pattern in two copies of a packet, and burst errors in the channel appear as random errors to the decoder. This may also significantly increase the power of the binary erasure combining technique.

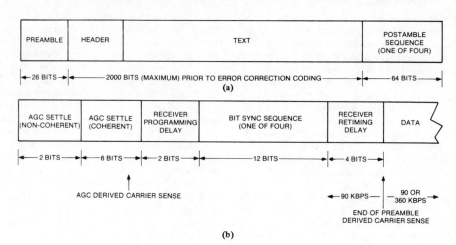

Fig. 15. (a) UPR packet format. (b) UPR packet preamble detail.

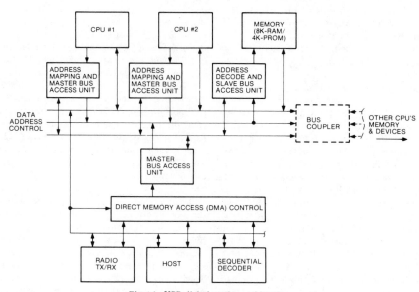

Fig. 16. UPR digital section architecture.

As indicated in Fig. 15, there are four distinct preamble bit sync patterns and four distinct postamble patterns. Part of the purpose of these patterns is to inform the receiving UPR as to which forward error correction encoding format applies to the packet. The other information imparted is the bit rate subsequent to the preamble (90 kbits/s or 360 kbits/s) and whether or not the bit order descrambling function should be bypassed. The latter feature is used in conjunction with an uncoded packet format which uses only a cyclic redundancy checksum for error detection, as was the case in the EPR.

5) Processor Architecture: The processor being used in the UPR design is considerably more powerful than that used in the original EPR. The added capability is driven both by the added functions which the UPR processor must perform, and the results of network tests in the EPR test bed which indicate that system throughput can be processor-limited rather than channel limited. Since efficient multiple access use of the limited RF spectrum resouce is a key program objective, the

UPR processor should provide sufficient processing power to allow maximum use of the channel.

Fig. 16 shows the general architecture of the UPR digital section. The central processing units (CPU's) are Texas Instruments 9900 microprocessors. These microprocessors have the capability for implementing a multiprocessor architecture by providing the appropriate monitor and control signals for bus sharing. Analysis showed that for the UPR application two CPU's could most efficiently share a bus directly, but that for larger numbers of CPU's higher net efficiency was obtained using bus couplers similar to Texas Instrument's TI-Line architecture. The UPR will initially be configured with the dual CPU processor, and will have the capability to add additional processing power using bus couplers and additional CPU and memory hardware. The dual CPU processor increases the processing power available to the UPR to support additional functions such as position location, network synchronization, and distribution of spread spectrum keys. A picture of the UPR is shown in Fig. 17.

Fig. 17. An upgraded packet radio.

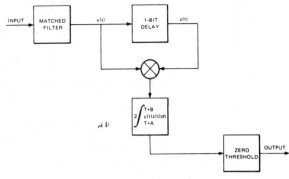

Fig. 18. DPSK receiver using post-detection integration.

The basic architecture of the multiprocessor digital unit is compatible with the EPR radio section and the latest EPR units are being constructed using the dual processor design. These units will provide the added processing power desired at high traffic levels in the EPR test bed.

6) Mobile Operation: An *L*-band radio operating from a mobile platform in a nonsited, ground environment encounters severe fading due to multipath signal components. The EPR equipment was originally designed to lock onto a single multipath component during the packet preamble and to track this component for the remainder of the packet. This was a simple and reliable approach for initial experiments which did not involve mobile platforms. When operating from a platform moving at sufficient velocity, however, fading of the acquired multipath component could occur rapidly enough that portions of the packet would be lost. In order to combat this fading phenomenon, a diversity mechanism based on post-detection integration of multipath components is provided in the UPR. The simplicity of the mechanism has allowed the EPR equipment to be retrofitted with similar circuitry, so that reliable mobile operation is currently available in the EPR test bed. A block diagram of the receiver is shown in Fig. 18.

C. Network Management and Operation

At the present time two packet radio experimental networks are operating: an experimental testbed network covering much of the San Francisco Bay Area; and a local distribution network in the Boston Area, which is used for station software development.

1) Network Monitoring and Control: A centralized network management facility (NMF) has been developed for managing and operating the Bay Area experimental testbed. It is somewhat similar to the ARPANET network control center (NCC) [39] in that it collects and displays relevant PRNET status information on a continuous basis. The facilities available at the NMF include two stations and a network monitoring system employing an Interdata 70 and an EPR to monitor radio traffic (during test and debugging only). From the NMF, the network can be debugged, the status of the network can be monitored, tests and measurement experiments can be run, and faults can be detected, diagnosed, and isolated.

2) Debugging the Network: All elements of the packet radio network have been designed to be debugged remotely under test as well as operational conditions. The memory of any packet radio's microprocessor can be remotely examined or altered through the use of the X-RAY debugger by a person at the station. The X-RAY process is routinely used to alter operating parameters in the packet radios (such as power output, frequency, timing, and protocol values) and to examine or alter program code. Its operation can also be automated so that other station processes can utilize it directly.

The PRNET is normally connected to the ARPANET. This connection is accomplished using a gateway [34] process, co-located with the network station processor, to communicate with an ARPANET IMP [2]. The station can then be remotely debugged from an authorized ARPANET host using a cross-internetwork debugger known as X-NET. By using internet protocols to access the station's X-RAY process, even the radios can be remotely debugged from the ARPANET. The user's terminal interface unit (TIU) also supports both forms of remote debugging. This feature has proved essential to the Bay Area PRNET development in that station software developers located in Boston and packet radio software developers in Texas can remotely participate in network debugging as the need arises, and new software versions can be conveniently installed from remote development sites as frequently as required.

3) System Monitoring: Once initialized, each packet radio in the network periodically announces its existence by transmitting to the station summary ROP's which contain neighbor tables and other status information. Similarly, terminal devices periodically send summary TOP's (terminal-on packets), which serve much the same function as their counterpart summary ROP's.

Both the station and the network monitor make extensive use of summary ROP's and TOP's. The station maintains a connectivity matrix based on the information contained in the ROP's for assigning routes. Current network connectivity may be displayed at the station upon request, and all state changes for nodes and links may be time stamped and logged. When active, the independent network monitoring system also listens to ROP's, and maintains a table of the last time that ROP's and TOP's were heard, for each packet radio or terminal interface unit ID. Thus, the exact time of failure of any network element can be obtained—even if a component of the station fails.

D. Test and Experimentation

This section describes the measurement capabilities that have been designed-in as an integral part of the experimental network. Measurement facilities have been built into the PR and station software. They provide for the collection and delivery of measurement data over the radio channel in real-time while an experiment is being run; after the experiment has been completed, the data are reduced and analyzed at a remote site. Operating software in the PRU's, TIU's, and station performs the collection of measurement data and uses the system protocols for delivery of this data to a measurement file located at the station. No additional hardware is needed to make measurements on throughput, delay and several other parameters, and the design goal of the facilities is that the taking of measurements should have a minimal impact on PRNET performance. A discussion of PR measurement techniques is given in Tobagi *et al.* in this issue [29].

From the station, parameters in each PR and terminal device in the network can be set remotely, selected elements can be halted, if appropriate, the collection of statistical data from selected devices may be enabled, traffic sources may be turned on or off, and data collection may be initiated. At the conclusion of a measurement run, the data can be automatically spooled over the ARPANET to a remote site (e.g., UCLA) for analysis.

The four primary measurement tools that have been developed are: cumulative statistics (CUMSTATS), snapshots, pickup packets, and neighbor tables, CUMSTATS consist of a variety of activity counters in each node. Snapshots periodically record the disposition of packet buffers and other node resources. Pickup packets are "crates" that start out empty at a traffic source, and pick up information at each node they traverse enroute to their destination, thus providing a trace of their history. Neighbor tables are a table of counts of packets received from each "neighbor" PR in range. The location of these built-in measurement tools are as follows.

1) TIU measurement software—provides sources and sinks of controlled traffic streams; generates and collects pickup packets; collects end-device CUMSTATS; and periodically sends collected data to station measurement process. End-device cumstats collected consist of packet activity counters, retransmission histograms, and end-to-end acknowledgment time delay spectra.

2) PRU measurement software—collects subnet CUMSTATS and snapshots; enters local data into pickup packets; and periodically sends collected data to station measurement process. PRU CUMSTATS include counters for packets transmitted, packets received, packets in error, and retransmission histograms.

3) Station measurement software—controls experiments and collects the resulting measurement data. Internal packet-handling processes (such as the forwarder and gateway) also collect cumstats which are written on the measurement file. Network connectivity, labeling, and route updates are all written on the measurement file as they occur.

4) Off-line measurement software—the final destination of the PRNET measurement data is the UCLA 360/91 computer. The data are sent from the station over the ARPANET and are stored at UCLA, for use by several analysis programs.

These facilities as described provide an advanced set of tools for network experimentation. Controlled experiments can be run, data collected and transmitted to a general-purpose machine for reduction and analysis, all using elements that are a built-in, permanent, part of the network under test. Today these capabilities are an accepted part of experimental networks; in the future, it is likely that many operational networks will have similar capabilities as well.

VIII. CONCLUSIONS

Packet radio has emerged as a viable technology for both fixed and mobile computer communications. The elements of a PRNET are already small enough (the repeater is about a cubic foot) to envision the day when a complete packet radio terminal will be about the size of a pocket calculator. A packet radio is still too expensive for widespread usage outside the military. However, the application of recent advances in integrated circuit technology and sophisticated signal processing techniques is expected to lead to low cost radios within the next five to ten years.

The cost of digital technology and advanced signal processing is steadily decreasing. It is very reasonable to expect that digital radio technology will be married with the personal computer technology and that a very powerful personal network technology will emerge which will be capable of revolutionary impact on both the communications and computing fields.

In the space environment, the packet radio technology is a potential candidate for organizing multiple satellites into a store-and-forward network. This architecture would entail the provision of significantly more processing within each satellite than is typical of existing satellites.

Packet radio will most likely play a major role in achieving local distribution of information, particularly when the source or destination can be mobile. Numerous cost, fabrication, and operational issues are being pursued and various regulatory bodies are evaluating the possible role of packet radio in achieving efficient spectral utilization for computer communications. However, the basic technology has now been demonstrated and it seems likely that packet radio will play a significant future role in computer communications and the local distribution of information.

ACKNOWLEDGMENT

Many of our colleagues working on the packet radio program have contributed to the ideas presented in this paper. In particular, we would like to acknowledge contributions from ARPA, Rockwell International/Collins, Bolt, Beranek and Newman, Inc., SRI International, UCLA, MIT Lincoln Laboratory, Texas Instruments, and Linkabit Corporation.

REFERENCES

[1] L. G. Roberts and B. D. Wessler, "Computer network development to achieve resource sharing," *AFIPS Proc. SJCC*, pp. 543–549, 1970.

[2] F. E. Heart, R. E. Kahn, S. M. Ornstein, W. R. Crowther, and D. C. Walden, "The interface message processor for the ARPA computer network," *AFIPS Proc. SJCC*, pp. 551–567, 1970.

[3] R. E. Kahn, "Resource-sharing computer communication network," *IEEE Proc.*, vol. 60, pp. 1397–1407, Nov. 1972.

[4] L. G. Roberts, "The evolution of packet switching," this issue, pp. 1307–1313.

[5] L. G. Roberts, "Data by the packet," *IEEE Spectrum*, Feb. 1974.

[6] R. E. Millstein, "The national software works: A distributed processing system," *Proc. ACM Nat. Conf.*, Seattle, WA. 1977.

[7] R. E. Kahn, "The organization of computer resource into a packet radio network," *IEEE Trans. Commun.*, vol. COM-25, no. 1, pp. 169–178, Jan. 1977.

[8] N. Abramson, "The Aloha system—another alternative for computer communications," in *AFIPS Conf. Proc.*, vol. 37, FJCC, pp. 695–702, 1970.

[9] R. Binder *et al.*, "ALOHA packet broadcasting—A retrospect," in *AFIPS Conf. Proc.*, vol. 44, pp. 203–215, 1975.

[10] T. Klein, "A tactical packet radio system," in *NTC Proc.*, New Orleans, LA, Dec. 1975.

[11] L. Kleinrock and F. Tobagi, "Packet switching in radio channels: Part I—Carrier sense multiple access modes and their throughput-delay characteristics," *IEEE Trans. Commun.*, pp. 1400–1416, Dec. 1975.

[12] Special Issue on Spectrum Management, *IEEE Trans. Electromagn. Compat.*, vol. EMC-19, no. 3, May 1978.

[13] D. L. Nielson, "Microwave propagation measurements for mobile digital radio applications," EASCON-77 Proc., Sept. 1977, pp. 14-2A to 14-2L.

[14] R. C. Dixon, "Spread spectrum systems," New York: John Wiley and Sons, 1976.

[15] G. L. Turin, "Communication through noisy, random multipath channels," IRE Convention Record, vol. 4, part IV, 1956, pp. 154-166.

[16] R. Price and P. G. Green, Jr., "A communication technique for multipath channels," IRE Proc., vol. 46, Mar. 1958, pp. 555-570.

[17] V. Charash, "A study of multipath reception with unknown delays," Ph.D. dissertation, Dep. Electrical Engineering and Computer Science, University of California, Berkeley, Dec. 1973.

[18] N. Abramson, "Packet switching via satellite," in AFIPS Conf. Proc., 1973, pp. 703-710.

[19] L. Kleinrock and S. Lam, "Packet switching in a slotted satellite channel," in AFIPS Conf. Proc., 1973, pp. 703-710.

[20] L. Kleinrock and M. Scholl, "Packet switching in radio channels: New conflict free multiple access schemes for a small number of data users," ICC Proc., 1977, pp. 22.1-105 to 22.1-111.

[21] L. Kleinrock and F. Tobagi, "Carrier sense multiple access for packet switched radio channels," ICC Proc., June 1974, pp. 21B-1 to 21B-7.

[22] L. Kleinrock and M. Scholl, "Packet switching in radio channels: New conflict free multiple access schemes," to appear in IEEE Trans. Communications.

[23] J. Capetanakis, "The multiple access broadcast channel: Protocol and capacity considerations," Ph.D. dissertation, M.I.T., Aug. 1977.

[24] J. Burchfiel, R. Tomlinson, and M. Beeler, "Functions and structure of a packet radio station," AFIPS Conf. Proc., vol. 44, 1975, pp. 245-251.

[25] H. Frank, I. Gitman and R. VanSlyke, "Packet radio system—Network considerations," AFIPS Conf. Proc., May 1975, pp. 217-231.

[26] S. Fralick and J. C. Garrett, "Technological considerations for packet radio networks," in AFIPS Conf. Proc., pp. 233-243, 1975.

[27] F. Tobagi and L. Kleinrock, "On the analysis of buffered packet radio systems," in Proc. 8th HICCS, Honolulu, HI, Jan. 1976, pp. 42-45.

[28] Y. Dalal, "Broadcast protocols on packet-switched computer networks," Ph.D. dissertation, Stanford University, Digital Systems Laboratories, technical report no. 128, Apr. 1977.

[29] F. A. Tobagi, M. Gerla, R. W. Peebles, and E. Manning, "Modeling and measurement techniques in packet communication networks," this issue pp. 1423-1447.

[30] J. Cafarella, J. Allison, W. Brown, and E. Stern, "Programmable matched filtering with acoustic–electric convolvers in spread spectrum systems," in 1975 Ultrasonics Symp. Proc., IEEE CAT. #75, CHO 994-4su, pp. 205-208.

[31] —, "Acousoelectrical convolvers for programmable matched filtering in spread spectrum systems," Proc. IEEE, vol. 64, May 1976.

[32] R. Gold, "Optimal binary sequences for spread spectrum multiplexing," IEEE Trans. Inform. Theory, pp. 617-621, Oct. 1967.

[33] J. Wozencraft and L. Jacobs, Principles of Communication Engineering. New York: John Wiley and Sons, 1966.

[34] V. G. Cerf and P. T. Kirstein, "Issues in packet network interconnection," this issue, pp. 1386-1408.

[35] Y. Okamura et al., "Field strength measurements and its variability in VHF and UHF land mobile radio services," in Communication Channels: Characterization and Behavior, B. Goldberg, Ed. New York: IEEE Press, 1976.

[36] A. Longley and P. Rice, "Prediction of tropospheric radio transmission over irregular terrain: A computer method," ESSA Tech. Report ERL 79-ITS 67, 1968.

[37] I. M. Jacobs, R. Binder, and S. V. Hoversten, "General purpose packet satellite networks," this issue, pp. 1448-1467.

[38] V. G. Cerf and R. E. Kahn, "A protocol for packet network intercommunications," IEEE Trans. Commun., vol. COM-22, pp. 637-648, May 1974.

[39] A. McKenzie, "The ARPA Network control center," in Proc. Fourth Data Commun. Symp., Quebec City, Canada, pp. 5-1-5-6, Oct. 1975.

Packet Switching in Radio Channels: Part I—Carrier Sense Multiple-Access Modes and Their Throughput-Delay Characteristics

LEONARD KLEINROCK, FELLOW, IEEE, AND FOUAD A. TOBAGI

Abstract—Radio communication is considered as a method for providing remote terminal access to computers. Digital byte streams from each terminal are partitioned into packets (blocks) and transmitted in a burst mode over a shared radio channel. When many terminals operate in this fashion, transmissions may conflict with and destroy each other. A means for controlling this is for the terminal to sense the presence of other transmissions; this leads to a new method for multiplexing in a packet radio environment: carrier sense multiple access (CSMA). Two protocols are described for CSMA and their throughput-delay characteristics are given. These results show the large advantage CSMA provides as compared to the random ALOHA access modes.

I. INTRODUCTION

LARGE COMPUTER installations, enormous data banks, and extensive national computer networks are now becoming available. They constitute large expensive resources which must be utilized in a cost/effective fashion. The constantly growing number of computer applications and their diversity render the problem of *accessing* these large resources a rather fundamental one. Prior to 1970, wire connections were the principal means for communication among computers and between users and computers. The reasons were simple: dial-up and leased telephone lines were available and could provide inexpensive and reasonably reliable communications for short distances, using a readily available and widespread technology. It was long recognized that this technology was inadequate for the needs of a computer-communication system which is required to handle bursty traffic (i.e., large peak to average data rates). For example, the inadequacies included the long dial-up and connect time, the minimum three-minute tariff structure, the fixed and limited data rates, etc. However, it was not until 1969 that the cost to switch communication bandwidth dropped below the cost of the bandwidth being switched [1]. At that time, the new technology of packet-switched computer networks emerged and developed a cost/effective means for connecting computers together over long-distance high-speed

lines. However, these networks did not solve the *local* interconnection problem, namely, how can one efficiently provide access from the user to the network itself? Certainly, one solution is to use wire connections here also. An alternate solution is the subject of this paper, namely, ground radio packet switching.

We wish to consider broadcast radio communications as an alternative for computer and user communications. The ALOHA System [2] appears to have been the first such system to employ wireless communications. The advantages in using broadcast radio communications are many: easy access to central computer installations and computer networks; collection and dissemination of data over large distributed geographical areas independent of the availability of preexisting (telephone) wire networks; the suitability of wireless connections for communications with and among mobile users (a constantly growing area of interest and applications); easily bypassed hostile terrain; etc. Perhaps, this broadcast property is the key feature in radio communication.

The Advanced Research Projects Agency (ARPA) of the Department of Defense recently undertook a new effort whose goal is to develop new techniques for packet radio communication among geographically distributed, fixed or mobile, user terminals and to provide improved frequency management strategies to meet the critical shortage of RF spectrum. The research presented in this paper is an integral part of the total design effort of this system which encompasses many other research topics [3]–[9].

Consider an environment consisting of a number of (possibly mobile) users in line-of-sight and within range of each other, all communicating over a (broadcast) radio channel in a common frequency band. The classical approach for satisfying the requirement of two users who need to communicate is to provide a communication channel for their use so long as their need continues (line-switching). However, the measurements of Jackson and Stubbs [10] show that such allocation of scarce communication resources is extremely wasteful. Rather than providing channels on a user-pair basis, we much prefer to provide a single high-speed channel to a large number of users which can be shared in some fashion. This, then, allows us to take advantage of the powerful "large number laws" which state that with very high probability, the demand at any instant will be approximately equal to the sum of the average demands of that population. We wish to take advantage of these gains due to resource sharing.

Of interest to this paper is the consideration of radio channels for packet switching (also called packet radio channels). A packet is merely a package of data prepared by one user for transmission to some other user in the system. As soon as we deal with shared channels in a packet-switching mode, then we must be prepared to resolve conflicts which arise when more than one demand is simultaneously placed upon the channel. In packet radio channels, whenever a portion of one user's transmission overlaps with another user's transmission, the two collide and "destroy" each other. The existence of some acknowledgment scheme permits the transmitter to determine if his transmission was successful or not. The problem we are faced with is how to control the access to the channel in a fashion which produces, under the physical constraints of simplicity and hardware implementation, an acceptable level of performance. The difficulty in controlling a channel which must carry its own control information gives rise to the so-called random-access modes. A simple scheme, known as "pure ALOHA," permits users to transmit any time they desire. If, within some appropriate time-out period, they receive an acknowledgment from the destination, then they know that no conflicts occurred. Otherwise, they assume a collision occurred and they must retransmit. To avoid continuously repeated conflicts, some scheme must be devised for introducing a *random* retransmission delay, spreading the conflicting packets over time. A second method for using the radio channel is to modify the completely unsynchronized use of the ALOHA channel by "slotting" time into segments whose duration is exactly equal to the transmission time of a single packet (assuming constant-length packets). If we require each user to start his packets only at the beginning of a slot, then when two packets conflict, they will overlap completely rather than partially, providing an increase in channel efficiency. This method is referred to as "slotted ALOHA" [11]–[13].

The radio channel as considered in this paper is characterized as a wide-band channel with a propagation delay between any source-destination pair which is very small compared to the packet transmission time.[1] This suggests a third approach for using the channel; namely, the carrier sense multiple-access (CSMA) mode. In this scheme one attempts to avoid collisions by listening to (i.e., "sensing") the carrier due to another user's transmission.[2] Based on this information about the state of the channel, one may

think of various actions to be taken by the terminal. Two protocols will be described and analyzed which we call "persistent" CSMA protocols: the nonpersistent and the *p*-persistent CSMA. Below, we present the protocols, discuss the assumptions, and finally establish and display the throughput-delay performance for each.

II. CSMA TRANSMISSION PROTOCOLS AND SYSTEM ASSUMPTIONS

The various protocols considered below differ by the action (pertaining to packet transmission) that a terminal takes after sensing[3] the channel. However, in all cases, when a terminal learns that its transmission was unsuccessful, it reschedules the transmission of the packet according to a randomly distributed retransmission delay. At this new point in time, the transmitter senses the channel and repeats the algorithm dictated by the protocol. At any instant a terminal is called a *ready terminal* if it has a packet ready for transmission at this instant (either a new packet just generated or a previously conflicted packet rescheduled for transmission at this instant).

A terminal may, at any one time, either be transmitting or receiving (but not both simultaneously). However, the delay incurred to switch from one mode to the other is negligible. Furthermore, the time required to detect the carrier due to packet transmissions is negligible (that is a zero detection time is assumed).[4] All packets are of constant length and are transmitted over an assumed noiseless channel (i.e., the errors in packet reception caused by random noise are not considered to be a serious problem and are neglected in comparison with errors caused by overlap interference). The system assumes noncapture (i.e., the overlap of any fraction of two packets results in destructive interference and both packets must be retransmitted). We further simplify the problem by assuming the propagation delay (small compared to the packet transmission time) to be identical[5] for all source–destination pairs.

We first consider the *nonpersistent CSMA*. The idea here is to limit the interference among packets by always rescheduling a packet which finds the channel busy upon arrival. More precisely, a ready terminal senses the channel and operates as follows.

1) If the channel is sensed idle, it transmits the packet.

2) If the channel is sensed busy, then the terminal schedules the retransmission of the packet to some later time according to the retransmission delay distribution. At this new point in time, it senses the channel and repeats the algorithm described.

A slotted version of the nonpersistent CSMA can be

---

[1] Consider, for example, 1000-bit packets transmitted over a channel operating at a speed of 100 kbits/s. The transmission time of a packet is then 10 ms. If the maximum distance between the source and the destination is 10 mi, then the (speed of light) packet propagation delay is of the order of 54 μs. Thus the propagation delay constitutes only a very small fraction ($a = 0.005$) of the transmission time of a packet. On the contrary, when one considers satellite channels [13] the propagation delay is a relatively large multiple of the packet transmission time ($a \gg 1$).

[2] Sensing carrier prior to transmission is a well-known concept in use for (voice) aircraft communication. In the context of packet radio channels, it was originally suggested by D. Wax of the University of Hawaii in an internal memorandum dated Mar. 4, 1971.

[3] Each terminal has the capability of sensing carrier on the channel. The practical problems of feasibility and implementation of sensing, however, are not addressed here.

[4] The detection time is considered negligible for relatively wide-band channels (100 kHz). In Part II [19] the detection time on the "busy-tone" narrow-band channels (on the order of 2 kHz) will be accounted for in the analysis.

[5] By considering this constant propagation delay equal to the largest possible, one gets lower (i.e., pessimistic) bounds on performance.

considered in which the time axis is slotted and the slot size is τ seconds (the propagation delay). All terminals are synchronized[6] and are forced to start transmission only at the beginning of a slot. When a packet's arrival occurs during a slot, the terminal senses the channel at the beginning of the next slot and operates according to the protocol described above.

We next consider the *p-persistent CSMA* protocol. However, before treating the general case (arbitrary p), we introduce the special case of $p = 1$.

The *1-persistent CSMA* protocol is devised in order to (presumably) achieve acceptable throughput by never letting the channel go idle if some ready terminal is available. More precisely, a ready terminal senses the channel and operates as follows.

1) If the channel is sensed idle, it transmits the packet with probability one.

2) If the channel is sensed busy, it waits until the channel goes idle (i.e., persisting on transmitting) and only then transmits the packet (with probability one—hence, the name of 1-persistent).

A slotted version of this 1-persistent CSMA can also be considered by slotting the time axis and synchronizing the transmission of packets in much the same way as for the previous protocol.

The above 1-persistent and nonpersistent protocols differ by the probability (one or zero) of not rescheduling a packet which upon arrival finds the channel busy. In the case of a 1-persistent CSMA, we note that whenever two or more terminals become ready during a transmission period (TP), they wait for the channel to become idle (at the end of that transmission) and then they all transmit with probability one. A conflict will also occur with probability one! The idea of randomizing the starting time of transmission of packets accumulating at the end of a TP suggests itself for interference reduction and throughput improvement. The scheme consists of including an additional parameter p, the probability that a ready packet persists ($1 - p$ being the probability of delaying transmission by τ seconds). The parameter p will be chosen so as to reduce the level of interference while keeping the idle periods between any two consecutive nonoverlapped transmissions as small as possible. This gives rise to the *p-persistent CSMA*, which is a generalization of the 1-persistent CSMA.

More precisely, the protocol consists of the following: the time axis is finely slotted where the (mini) slot size is τ seconds. For simplicity of analysis, we consider the system to be synchronized such that all packets begin their transmission at the beginning of a (mini) slot.

Consider a ready terminal. If the channel is sensed idle, then: with probability p, the terminal transmits the packet; or with probability $1 - p$, the terminal delays the transmission of the packet by τ seconds (i.e., one slot). If at this new point in time, the channel is still detected

idle, the same process is repeated. Otherwise, some packet must have started transmission, and our terminal schedules the retransmission of the packet according to the retransmission delay distribution (i.e., acts as if it had conflicted and learned about the conflict).

If the ready terminal senses the channel busy, it waits until it becomes idle (at the end of the current transmission) and then operates as above.

III. TRAFFIC MODEL: ASSUMPTIONS AND NOTATION

In the previous section, we identified the system protocols, operating procedures, and assumptions. Here we characterize the traffic source and its underlying assumptions.

We assume that our traffic source consists of an infinite number of users who collectively form an independent Poisson source with an aggregate mean packet generation rate of λ packets/s. This is an approximation to a large but finite population in which each user generates packets infrequently and each packet can be successfully transmitted in a time interval much less than the average time between successive packets generated by a given user. Each user in the infinite population is assumed to have at most one packet requiring transmission at any time (including any previously blocked packet).

In addition, we characterize the traffic as follows. We have assumed that each packet is of constant length requiring T seconds for transmission. Let $S = \lambda T$. S is the average number of packets generated per transmission time, i.e., it is the input rate normalized with respect to T. Under steady-state conditions, S can also be referred to as the channel throughput rate. Now, if we were able to perfectly schedule the packets into the available channel space with absolutely no overlap or gaps between the packets, we could achieve a maximum throughput equal to 1; therefore we also refer to S as the *channel utilization*. Because of the interference problem inherent in the random nature of the access modes, the achievable throughput will always be less than 1. The maximum achievable throughput for an access mode is called the *capacity* of the channel under that mode.

Since conflicts can occur, some acknowledgment scheme is necessary to inform the transmitter of its success or failure. We assume a positive acknowledgment scheme[7]: if within some specified delay (an appropriate time-out period) after the transmission of a packet, a user does not receive an acknowledgment, he knows he has conflicted. If he now retransmits immediately, and if all users behave likewise, then he will definitely be interfered with again (and forever!). Consequently, as mentioned above, each user delays the transmission of a previously collided packet by some random time whose mean is $\bar{X}$ (chosen, for example, uniformly between 0 and $X_{\max} = 2\bar{X}$). The traffic

---

[6] In this paper, the practical problems involved in synchronizing terminals are not addressed.

[7] The channel for acknowledgment is assumed to be separate from the channel we are studying (i.e., acknowledgments arrive reliably and at no cost).

offered to the channel from our collection of users consists not only of new packets but also of previously collided packets: this increases the mean *offered* traffic rate which we denote by G (packets per transmission time T) where $G \geq S$.

Our two further assumptions are the following.

Assumption 1: The average retransmission delay $\bar{X}$ is large compared to T.

Assumption 2: The interarrival times of the point process defined by the start times of all the packets plus retransmissions are independent and exponentially distributed.

It is clear that Assumption 2 is violated in the protocols we consider. (We have introduced it for analytic simplicity.) However, in Section V, some simulation results are discussed which show that performance results based on this assumption are excellent approximations, particularly when the average retransmission delay $\bar{X}$ is large compared to T. Moreover, in the context of slotted ALOHA it was analytically shown [14] in the limit as $\bar{X} \to \infty$, that Assumption 2 is satisfied; furthermore, simulation results showed that only the first moment of the retransmission delay distribution had a noticeable effect on the average throughput-delay performance.

So far, we have defined the following important system variables: S (throughput), G (offered channel traffic rate), T (packet transmission time), $\bar{X}$ (average retransmission delay), τ (propagation delay), and p (p-persistent parameter). Without loss of generality, we choose $T = 1$. This is equivalent to expressing time in units of T. We express $\bar{X}$ and τ in these normalized time units as $\delta = \bar{X}/T$ and $a = \tau/T$.

IV. THROUGHPUT ANALYSIS

We wish to solve for the channel capacity of the system for all of the access protocols described above. This we do by solving for S in terms of G (as well as the other system parameters). The channel capacity is then found by maximizing S with respect to G. S/G is merely the probability of a successful transmission and G/S is the average number of times a packet must be transmitted (or scheduled) until success. In Section V, we discuss delay and give the throughput-delay tradeoff for these protocols.

This analysis is based on renewal theory and probabilistic arguments requiring independence of random variables provided by Assumption 2. Moreover steady-state conditions are assumed to exist. However from the (S,G) relationships found below one can see that steady state may not exist because of inherent instability of these random-access techniques. This instability is simply explained by the fact that when statistical fluctuations in G increase the level of mutual interference among transmissions, then the positive feedback causes the throughput to decrease to 0. Nevertheless, the results are useful for the following reasons.

1) They are meaningful for a finite (and possibly long) period of time. (Simulations supporting these analytic results showed no saturation over the simulated period of time when $\bar{X}$ was large enough; see Section V.)

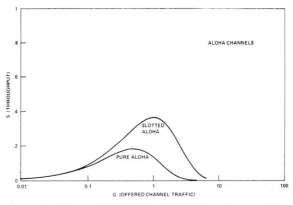

Fig. 1. Throughput in ALOHA channels.

2) In finite population cases, stable situations are possible for which steady-state results prevail over an infinite time horizon. (See [14] and [16].)

3) Control procedures have been prescribed for the slotted ALOHA random access [14] which stabilize unstable channels, achieving performance very close to the equilibrium results.

A. ALOHA Channels

In the *pure ALOHA* access mode, each terminal transmits its packet over the data channel in a completely unsynchronized manner. Under the system and model assumptions (mainly Assumption 2), we have

$$S = GP_s$$

where P_s is the probability that an arbitrary offered packet is successful. A given packet will overlap with another packet if there exists at least one start of transmission within T seconds before or after the start time of the given packet (i.e., over a "vulnerable" interval of length $2T$). Using the Poisson traffic assumption, Abramson [2] first showed that

$$S = Ge^{-2G}. \tag{1}$$

Thus, we see that pure ALOHA achieves a maximum throughput of $1/(2e) = 0.184$ (at $G = 1/2$).

In the *slotted ALOHA*, if two packets conflict, they will overlap completely rather than partially (i.e., a vulnerable interval only of length T). The throughput equation then becomes

$$S = Ge^{-G} \tag{2}$$

and was first obtained by Roberts [12] who extended Abramson's result in (1). With this simple change, the maximum throughput is increased by a factor of two to $1/e = 0.368$ (at $G = 1$). In Fig. 1, we plot the throughput S versus the offered traffic G for these two systems. From these results, it is all too evident that a significant fraction of the channel's ultimate capacity ($C = 1$) is not utilized with the ALOHA access modes; we recover a major portion of this loss with the CSMA protocols, as we now show.

B. Nonpersistent CSMA

The basic equation for the throughput S is expressed in terms of a (the ratio of propagation delay to packet transmission time) and G (the offered traffic rate) as follows:

$$S = \frac{Ge^{-aG}}{G(1 + 2a) + e^{-aG}}. \quad (3)$$

Proof: G denotes the arrival rate of new and rescheduled packets. All arrivals, in this case, do not necessarily result in actual transmissions (a packet which finds the channel in a busy state is rescheduled without being transmitted). Thus, G constitutes the "offered" channel traffic and only a fraction of it constitutes the channel traffic itself. Consider the time axis[8] (See Fig. 2)[9] and let t be the time of arrival of a packet which senses the channel idle and such that no other packet is in the process of transmission. Any other packet arriving between t and $t + a$ will find (sense) the channel as unused, will transmit, and hence will cause a conflict. If no other terminal transmits a packet during these a seconds (the "vulnerable" period), then the first packet will be successful.

Let $t + Y$ be the time of occurrence of the last packet arriving between t and $t + a$. The transmission of all packets arriving in $(t, t + Y)$ will be completed at $t + Y + 1$. Only a seconds later will the channel be sensed unused. Now, any terminal becoming ready between $t + a$ and $t + Y + 1 + a$ will sense the channel busy and hence will reschedule its packet. The interval between t and $t + Y + 1 + a$ is called a *transmission period* (TP). Note that there can be at most one successul transmission during a TP. Define an *idle period* to be the period of time between two consecutive TP's (also called busy periods in this simple case). A busy period plus the following idle period constitute a cycle. Let $\bar{B}$ be the expected duration of the busy period, $\bar{I}$ the expected duration of the idle period, and $\bar{B} + \bar{I}$ the expected length of a cycle. Let U denote the time during a cycle that the channel is used without conflicts. Using renewal theory arguments, the average channel utilization is simply given by

$$S = \frac{\bar{U}}{\bar{B} + \bar{I}}. \quad (4)$$

The probability that a TP is successful is simply the probability that no terminal transmits during the first a seconds of the period and is equal to e^{-aG}. Therefore

$$\bar{U} = e^{-aG}. \quad (5)$$

The average duration of an idle period is simply $1/G$. The average duration of a busy interval is $1 + \bar{Y} + a$, where $\bar{Y}$ is the expected value of Y.

[8] The reference time axis considered in this and subsequent proofs is the transmitter's time. Shifting all transmissions by τ seconds will give a description of events on the station's time axis. Any time overlap in transmission on the station's time axis results in packet interference.

[9] In this and other figures, a vertical arrow represents a terminal becoming ready.

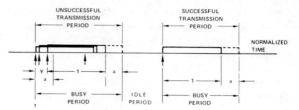

Fig. 2.   Nonpersistent CSMA: Busy and idle periods.

The distribution function for Y is

$$F_Y(y) \triangleq \mathrm{Pr}\{Y \le y\} = \mathrm{Pr}\{\text{no arrival occurs in an interval of length } a - y\}$$
$$= \exp\{-G(a - y)\}, \quad (y \le a). \quad (6)$$

The average of Y is therefore given by

$$\bar{Y} = a - \frac{1}{G}(1 - e^{-aG}). \quad (7)$$

Applying (4) and using the expressions found for $\bar{U}$, $\bar{B}$, and $\bar{I}$, we get (3).   Q.E.D.

It is easy to prove that the throughput equation for the *slotted* nonpersistent CSMA is given by

$$S = \frac{aGe^{-aG}}{(1 - e^{-aG}) + a}. \quad (8)$$

Note that for both cases we have

$$\lim_{a \to 0} S = G/(1 + G). \quad (9)$$

This shows that when $a = 0$, a throughput of 1 can theoretically be attained for an offered channel traffic equal to infinity. S versus G for various values of a is plotted in Fig. 3.

C. 1-Persistent CSMA

The throughput equation for this protocol is given by

$$S = \frac{G[1 + G + aG(1 + G + aG/2)]e^{-G(1+2a)}}{G(1 + 2a) - (1 - e^{-aG}) + (1 + aG)e^{-G(1+a)}}. \quad (10)$$

Proof: Consider Fig. 4 and again let t be the time of arrival of a packet which senses the channel to be idle with no other packet in the process of transmission. In this protocol, any packet arriving in the interval $[t + a, t + Y + 1 + a]$ will sense the channel busy and hence must wait until the channel is sensed idle (at time $t + 1 + Y + a$) at which time they will *all* transmit! The number of packets accumulated at the end of TP is the number of arrivals in $1 + Y$ seconds. If this total is equal to or greater than two, then a conflict occurs in the next TP with probability 1.

Define a *busy period* to be the time between t and the end of that TP during which no packets accumulate. De-

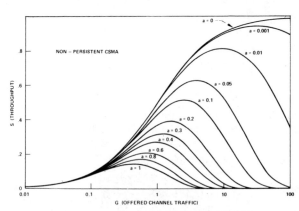

Fig. 3. Throughput in nonpersistent CSMA.

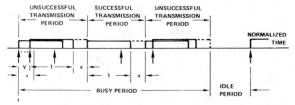

Fig. 4. 1-persistent CSMA: TP's, busy periods, and idle periods.

fine an *idle period* to be the period of time in which the channel is idle and no packets are present awaiting transmission. A busy period plus the following idle period constitute a *cycle*.

Let $\bar{B}$ be the expected duration of the busy period, $\bar{I}$ the expected duration of the idle period, and $\bar{B} + \bar{I}$ the expected length of a cycle.

Let us now consider the transmission of an arbitrary packet. Three situations must be considered.

1) If the packet arrives to an idle system, then its transmission is successful if and only if no packets arrive during its first a seconds; its probability of success is therefore e^{-aG}.

2) If the packet arrives during the first a seconds of a TP, then its probability of success is 0.

3) If the packet arrives during the channel busy period (excluding the first a seconds of the TP), then it is successful (in the next TP) if and only if it is the only packet to arrive during this TP and no packets arrive during *its* first a seconds. To calculate its probability of success, we observe that a TP is of random length equal to $1 + a + Y$ where Y is a random variable. Let B' denote the time during a cycle that the channel is in its busy period excluding the first a seconds of each TP. B' is a sequence of segments of random length $1 + Y \triangleq Z$ separated by periods of a seconds. Knowing that a packet arrives in B', this packet is more likely to arrive in a longer segment Z than in a shorter one (due to the "paradox of residual life" [17]). Let $\hat{Z}$ denote the segment in which the arrival occurred, and $\hat{q}_0$ (derived below) be the probability that no arrival occurs in $\hat{Z}$; the probability of success of the packet is therefore $\hat{q}_0 e^{-aG}$.

Only cases 1) and 3) contribute to a successful transmission. Let $\bar{B}'$ be the expected value of B'. From renewal theory arguments, the probability that an arrival finds the channel idle [case (1)] is given by $\bar{I}/(\bar{B} + \bar{I})$, and the probability that an arrival finds the channel in situation 3) is $\bar{B}'/(\bar{B} + \bar{I})$; then the probability of success of the packet is given by

$$P_s \triangleq \Pr \{\text{success}\} = \frac{\bar{I}}{\bar{B} + \bar{I}} e^{-aG} + \frac{\bar{B}'}{\bar{B} + \bar{I}} \hat{q}_0 e^{-aG}. \quad (11)$$

The determination of $\bar{I}, \bar{B}, \bar{B}'$, and $\hat{q}_0$ follows.

Since the traffic is Poisson, it is clear that the average idle period is given by

$$\bar{I} = 1/G. \quad (12)$$

For $\bar{B}, \bar{B}'$, and $\hat{q}_0$ we must first obtain some intermediate results as follows. The distribution function for Y and its average are given in (6) and (7), respectively. The Laplace transform of the probability density function of Y, defined as

$$F_Y^*(s) \triangleq \int_0^\infty e^{-sy} \, dF_Y(y),$$

is given by

$$F_Y^*(s) = e^{-aG} + \frac{G(e^{-as} - e^{-aG})}{G - s}. \quad (13)$$

Let us now find the distribution of the number of packets accumulated at the end of a TP.

Let

$$q_m(y) \triangleq \Pr \{m \text{ packets accumulated at end of TP} \mid Y = y\}$$

and

$$q_m = \int_0^a q_m(y) \, dF_Y(y).$$

Let $Q(z)$ denote the generating function of q_m defined by

$$Q(z) \triangleq \sum_{m=0}^\infty q_m z^m.$$

The number of packets accumulated at the end of a TP is equal to the number of packets arriving during a period of time equal to $1 + Y$. Let m_1 denote the number of packets arriving in $T = 1$, and m_2 the number of packets arriving in Y. Let $Q_1(z)$ and $Q_2(z)$ denote the generating functions of the probability distributions for m_1 and m_2, respectively. Since the arrival process is Poisson, the random variables m_1 and m_2 are independent and the generating function $Q(z)$ of q_m, where $m = m_1 + m_2$, is given by

$$Q(z) = Q_1(z)Q_2(z).$$

We have [17]

$$Q_1(z) = \exp \{G(z - 1)\}$$

and

$$Q_2(z) = F_Y^*(G(1 - z)).$$

From (13) we get

$$Q(z) = \exp\{G(z-1)\} \exp\{-aG\}\left[1 + \frac{\exp\{aGz\}-1}{z}\right]$$

(14)

We may invert this explicit expression for $Q(z)$; in particular we find that the probability of zero packets accumulated at the end of a TP is

$$q_0 = Q(z)\mid_{z=0} = \exp\{-G(1+a)\}[1+aG]. \quad (15)$$

To find the average busy period, we let Y_i denote the random variable Y defined above corresponding to the ith TP in a busy period. All Y_i, $i = 1,2,\cdots$, are independent and identically distributed. It is easy to see that the number of TP's in a busy period is geometrically distributed with mean $1/q_0$. Conditioned on the fact that we have exactly k TP's in the busy period and that $Y_i = y_i$ for $i = 1,2,\cdots,k$, the average busy period is

$$\bar{B}(y_1,y_2,\cdots,y_k) = k(1+a) + y_1 + y_2 + \cdots + y_k.$$

Therefore, by removing the conditions on k and Y_i, we get $\bar{B}$ as

$$\bar{B} = \cdots\int_{y_i=0}^{a}\cdots\int_{y_1=0}^{a}\sum_{k=1}^{\infty}[k(1+a) + y_1 + \cdots + y_k]$$

$$\cdot q_0(y_k)\prod_{i=1}^{k-1}(1-q_0(y_i))\,dF_{Y_1}(y_1)\cdots dF_{Y_i}(y_i)\cdots.$$

It is easy to see that by inverting the order of summation and integration, the contribution of the term $k(1+a)$ reduces to $(1+a)/q_0$ and the contribution of the generic term y_j simply reduces to $\bar{Y}(1-q_0)^{j-1}$. Finally, we have

$$\bar{B} = \frac{1+a}{q_0} + \sum_{j=1}^{\infty}\bar{Y}(1-q_0)^{j-1} = \frac{1+a+\bar{Y}}{q_0}. \quad (16)$$

Since the average number of TP's is $1/q_0$, from the distribution of B' we have

$$\bar{B}' = \frac{1+\bar{Y}}{q_0}. \quad (17)$$

In (11) for P_s, it remains only to compute $\hat{q}_0$. The probability density function of $Z = 1 + Y$ is easily obtained from the distribution of Y. From (6), the probability density function of Y can be expressed as

$$f_Y(y) = \exp\{-aG\}u_0(y) + G\exp\{-aG\}\exp\{Gy\},$$

$$0 \le y < a$$

where $u_0(y)$ is the unit impulse at $y = 0$. Thus we have

$$f_Z(x) = \exp\{-aG\}u_0(x-1) + G\exp\{-aG\}$$

$$\cdot \exp\{G(x-1)\}, \quad 1 \le x \le 1+a.$$

The probability density function of $\hat{Z}$ is given by [17]

$$f_{\hat{Z}}(x) = \frac{xf_Z(x)}{\bar{Z}}$$

$$= \frac{e^{aG}}{1+\bar{Y}}u_0(x-1) + \frac{Gxe^{-aG}e^{G(x-1)}}{1+\bar{Y}},$$

$$1 \le x \le 1+a.$$

Finally, the probability that no arrival occurs (from our Poisson source) in the interval $\hat{Z}$ is simply

$$\hat{q}_0 = \int_{x=1}^{1+a}\exp\{-Gx\}f_{\hat{Z}}(x)\,dx$$

$$= \frac{\exp\{-G(1+a)\}}{1+\bar{Y}}[1+aG(1+a/2)]. \quad (18)$$

Using our expressions for $\bar{I}$, $\bar{B}$, $\bar{B}'$, and $\hat{q}_0$ in (12), (16), (17), and (18), respectively, we immediately obtain from (11)

$$P_s = \frac{\dfrac{1+\bar{Y}}{q_0}e^{-aG}\hat{q}_0 + \dfrac{1}{G}e^{-aG}}{\dfrac{1+a+\bar{Y}}{q_0} + \dfrac{1}{G}}.$$

Substituting the expressions obtained for q_0, $\hat{q}_0$, and $\bar{Y}$, and recalling that $S = GP_s$, we have finally established (10).

Q.E.D.

Slotted 1-persistent CSMA: Let us now consider the *slotted* version of 1-persistent CSMA. The throughput equation for this case is given by

$$S = \frac{G\exp\{-G(1+a)\}[1+a-\exp\{-aG\}]}{(1+a)(1-\exp\{-aG\}) + a\exp\{-G(1+a)\}}$$

(19)

Proof: In this slotted version, as in slotted ALOHA, if two packets conflict, they will overlap completely. The length of a TP is always equal to $1 + a$. (We have assumed that the packet transmission time is an integer multiple of the propagation delay.)

Since the traffic process is an independent one (Assumption 2), the number of slots in an idle period is geometrically distributed with a mean equal to $1/(1-e^{-aG})$. Thus the average idle period is given by

$$\bar{I} = \frac{a}{1-e^{-aG}}. \quad (20)$$

Using a similar argument, we find that the average busy period is given by

$$\bar{B} = \frac{1+a}{\exp\{-G(1+a)\}}. \quad (21)$$

Let $\bar{U}$ again denote the expected time during a cycle that the channel is used without conflicts. In order to find $\bar{U}$ we need to determine the probability of success over each

TP in the busy period. The probability of success over the first TP is given by

$$\text{Pr} \{\text{success over first TP}\} = \text{Pr} \{\text{only one packet arrives during the last slot of the preceding idle period/some arrival occurred}\}$$

$$= \frac{aGe^{-aG}}{1 - e^{-aG}}.$$

Similarly we have:

Pr {success over any other TP}

$$= \frac{G(1+a) \exp \{-G(1+a)\}}{1 - \exp \{-G(1+a)\}}.$$

The number of TP's in a busy period is geometrically distributed with a mean equal to

$$\exp \{G(1+a)\} \triangleq 1/q_0,$$

thus

$$\bar{U} = \frac{aG \exp \{-aG\}}{1 - \exp \{-aG\}} + \left(\frac{1}{q_0} - 1\right)$$

$$\cdot \frac{G(1+a) \exp \{-G(1+a)\}}{1 - \exp \{-G(1+a)\}}. \quad (22)$$

Applying (4) and using the expressions found for $\bar{U}$, $\bar{I}$, and $\bar{B}$, we get (19). Q.E.D.

The ultimate performance in the ideal case ($a = 0$), for both slotted and unslotted versions, is

$$S = \frac{Ge^{-G}(1+G)}{G + e^{-G}}. \quad (23)$$

For any value of a, the maximum throughput S will occur at an optimum value of G. In Fig. 5 we show S versus G for the nonslotted version of 1-persistent CSMA for various values of a.

D. p-Persistent CSMA

For a given offered traffic G and a given value of the parameter p, we can determine the throughput S as

$$S(G,p,a) = \frac{(1 - e^{-aG})[P_s'\pi_0 + P_s(1 - \pi_0)]}{(1 - e^{-aG})[a\bar{t'}\pi_0 + a\bar{t}(1 - \pi_0) + 1 + a] + a\pi_0} \quad (24)$$

where P_s', P_s, $\bar{t'}$, $\bar{t}$, and π_0 are defined in the following proof in (37), (34), (36), (30), and (25), respectively.

Proof: Consider a TP and assume that some packets arrive during the period as shown in Fig. 6. These packets sense the channel busy and accumulate at the end of the TP, at which point they randomize the starting times of their transmission according to the randomizing process described in Section II. This randomization creates a random delay before a TP starts, called the initial random

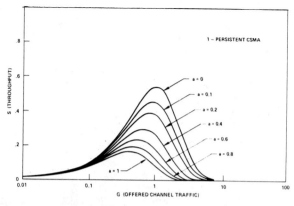

Fig. 5.   Throughput in 1-persistent CSMA.

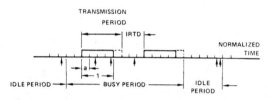

Fig. 6.   *p*-persistent CSMA: TP's, busy periods, and idle periods.

transmission delay (IRTD), during which time the channel is "wasted." If, at the start of a new TP, two or more terminals decide to transmit, then a conflict will certainly occur. All other packets which have delayed their transmission by τ seconds will then sense the channel busy and will have to be rescheduled for transmission by incurring a retransmission delay δ. Thus, at the expense of creating this IRTD, we greatly improve the probability of success over a TP.

Consider Fig. 6 in which we observe two TP's separated by an IRTD. One can also define busy periods and idle periods in much the same way as before. An idle period is that period of time during which the channel is idle and no packets are ready for transmission. A busy period consists of a sequence of transmission periods such that some packets arrive during each transmission period *except* the last one. Let $\mathfrak{I}_i$ denote the ith TP of a busy period. In order to find the channel utilization, we once again apply (4), which requires identifying and determining the average busy and idle periods, the gaps between TP's, as well as the condition for success over each TP. This we do as follows.

Recall that we require the system to be (mini-) slotted (the slot size equal to a, the normalized propagation delay) and all transmissions to start at the beginning of a slot. Here again we consider the transmission time of a packet to be an integer number $1/a$ slots (recall $T = 1$). Let $g = aG$; g is the average arrival rate of new and rescheduled packets during a (mini) slot.

We first determine the distribution of the number of packets accumulated at the end of a TP. Let N denote this number and let $\pi_n \triangleq \text{Pr} \{N = n\}$. According to the protocol described in Section II, only those packets arriving during a TP will accumulate at the end of that TP. Therefore, by Assumption 1, we have

$$\pi_n = \frac{[(1+a)G]^n}{n!} \exp\{-(1+a)G\}, \qquad n \geq 0. \quad (25)$$

To find the distribution of the IRTD between two successive TP's in the same busy period, we condition $N = n$ and we let t_n be the number of slots elapsed until some packet is transmitted. Let $q = 1 - p$. It is easy to see that

$$\Pr\{t_n > k\} = q^{(k+1)n} \prod_{j=1}^{k}\left[\sum_{m=0}^{\infty} \exp\{-g\}\frac{g^m}{m!}q^{m(k-j+1)}\right]$$

$$= q^{(k+1)n} \prod_{j=1}^{k} \exp\{g(q^{k-j+1}-1)\}$$

$$= q^{(k+1)n} \exp\left\{g\left(\frac{q(1-q^k)}{p}-k\right)\right\} \quad (26)$$

and, therefore, for $k > 0$ we have

$$\Pr\{t_n = k\} = \Pr\{t_n > k-1\} - \Pr\{t_n > k\}$$

$$= q^{kn}[1 - q^n \exp\{-g(1-q^k)\}]$$

$$\cdot \exp\left\{g\left(\frac{q(1-q^{k-1})}{p}-(k-1)\right)\right\} \quad (27)$$

and for $k = 0$,

$$\Pr\{t_n = 0\} = 1 - q^n. \quad (28)$$

The average IRTD is given by

$$\bar{t}_n = \sum_{k=0}^{\infty} \Pr\{t_n > k\}. \quad (29)$$

Removing the condition on N, we get

$$\bar{t} = \sum_{n=1}^{\infty} \bar{t}_n \cdot \frac{\pi_n}{1 - \pi_0}. \quad (30)$$

$\bar{t}$ is the average gap between two consecutive TP's in a busy period.

In order to find the probability of success over a TP $\mathfrak{I}_i$ one has to distinguish two cases: $i = 1$ and $i \neq 1$. We first treat the second case, $i \neq 1$. Given $N = n$, define[10]:

$P_s(n)$ — probability of success over $\mathfrak{I}_i$

L_n — the number of packets present at the starting time of $\mathfrak{I}_i$

$L_n - n$ — merely the number of packets arriving during the gap t_n.

By the Poisson assumption we have

$$\Pr\{L_n = l/t_n = k\} = \frac{(kg)^{l-n}}{(l-n)!}e^{-kg}, \qquad l \geq n. \quad (31)$$

[10] The quantities $P_s(n)$, P_s, and L_n need no index i since they are identical for all $\mathfrak{I}_i$, $i \neq 1$.

Removing the condition on t_n,

$$\Pr\{L_n = l\} = \sum_{k=1}^{\infty} \frac{(kg)^{l-n}}{(l-n)!}e^{-kg}\Pr\{t_n = k\}$$

$$+ (1-q^n)\delta_{l,n}, \qquad l \geq n \quad (32)$$

where $\delta_{i,j}$ is the Kronecker delta. The probability of success over $\mathfrak{I}_i$ is equal to the probability that none of the L_n transmit over $\mathfrak{I}_i$:

$$P_s(n) = \sum_{l=n}^{\infty} \frac{lpq^{l-1}}{1-q^l}\Pr\{L_n = l\}. \quad (33)$$

Removing the condition on N, we get

$$P_s = \sum_{n=1}^{\infty} P_s(n)\frac{\pi_n}{1-\pi_0}. \quad (34)$$

For the probability of success over $\mathfrak{I}_1$ we note that the number of packets present at the beginning of a busy period, denoted by N', is the number of packets arriving in the last slot of the previous idle period. We then have

$$\pi_n \triangleq \Pr\{N' = n\}$$

$$= \frac{g^n}{n!}\frac{e^{-g}}{1-e^{-g}}, \qquad n \geq 1. \quad (35)$$

Given $N' = n$, let t_n' denote the first initial random transmission delay of the busy period, and $P_s'(n)$ denote the probability of success over $\mathfrak{I}_1$. The distribution of t_n' and its average $\bar{t}_n'$ are the same as for t_n [(27) and (29)]. $P_s'(n)$ is the same as $P_s(n)$ [see (33)]. Removing the condition on N', we get

$$\bar{t}' = \sum_{n=1}^{\infty} t_n'\pi_n'. \quad (36)$$

$$P_s' = \sum_{n=1}^{\infty} P_s'(n)\pi_n'. \quad (37)$$

It remains to compute $\bar{B}$, $\bar{U}$, and $\bar{I}$. It is clear that the number of TP's in a busy period is equal to m with probability $\pi_0(1-\pi_0)^{m-1}$.

Consider a busy period with m TP's. Let N_i denote the number of packets accumulated at the end of the ith TP. We know that $N_m = 0$, and that all other $N_i \geq 1$ are independent and identically distributed random variables. Conditioned on the fact that $N_i = n_i, i = 1, \cdots, m-1$, the average busy period is given by

$$\bar{B}_m(n_1,\cdots,n_{m-1}) = a\bar{t}' + \sum_{i=1}^{m-1} a\bar{t}_{n_i} + m(1+a). \quad (38)$$

The expected time, during the busy period, that the channel is used without conflicts is given by

$$\bar{U}_m(n_1,\cdots,n_{m-1}) = P_s' + \sum_{i=1}^{m-1} P_s(n_i). \quad (39)$$

On the other hand, we know that

$$\Pr\{N_i = n_i\} = \frac{\pi_{n_i}}{1 - \pi_0}, \qquad n_i \geq 1, \; i = 1, 2, \cdots, m - 1. \tag{40}$$

Therefore, removing the conditions $N_i = n_i$ in (38) and (39), we get

$$\bar{B}_m = a\bar{l}' + (m - 1)a\bar{l} + m(1 + a) \tag{41}$$

$$\bar{U}_m = P_s' + (m - 1)P_s \tag{42}$$

and removing the condition on m we get

$$\bar{B} = \sum_{m=1}^{\infty} \bar{B}_m \pi_0 (1 - \pi_0)^{m-1} = a\bar{l}' + \frac{a\bar{l}(1 - \pi_0) + 1 + a}{\pi_0} \tag{43}$$

$$\bar{U} = P_s' + \frac{1 - \pi_0}{\pi_0} P_s. \tag{44}$$

The idle period is geometrically distributed with mean $1/(1 - e^{-g})$; its average is:

$$\bar{I} = \frac{a}{1 - e^{-g}}. \tag{45}$$

Finally, using (4) and substituting for $\bar{B}$, $\bar{U}$, and $\bar{I}$ the expressions found in (43), (44), and (45), respectively, we get the throughput S; it is a function of G, p, and $a = 1/T$ and is expressed as

$$S(G,p,a) = \frac{P_s' + \frac{1 - \pi_0}{\pi_0} P_s}{a\bar{l}' + a\bar{l}\frac{1 - \pi_0}{\pi_0} + \frac{1 + a}{\pi_0} + \frac{a}{1 - e^{-g}}} \tag{46}$$

which reduces to (24). Q.E.D.

In order to evaluate $S(G,p,a)$, a PL/1 program was written and run on the IBM 360/91 of the Campus Computing Network at UCLA. For small values of p ($0.01 \leq p \leq 0.1$), the numerical computation as suggested by (24) becomes time consuming and requires an extremely large amount of storage. Fortunately some approximations have been found useful which lead to a closed-form solution for the throughput (see the derivations of $S'(G,p,a)$ in Appendix A).

Special case $a = 0$: Let us now consider the special case $a = 0$. For finite G, $g = aG = 0$. Equation (26) becomes

$$\Pr\{t_n > k\} = q^{(k+1)n}.$$

The average IRTD is then given by (29), and is expressed as

$$\bar{l}_n = \sum_{k=0}^{\infty} \Pr\{t_n > k\} = \frac{q^n}{1 - q^n}.$$

It is important to note that $\bar{l}_n$ is finite, so is $\bar{l}$. On the other hand the idle period given in (45) becomes

$$\bar{I} = \frac{1}{G}.$$

Since $\bar{l}$ and $\bar{l}'$ are finite, by letting $a \to 0$ in (46) we get

$$S(G,p,a = 0) = \frac{P_s' + \frac{1 - \pi_0}{\pi_0} P_s}{\frac{1}{\pi_0} + \frac{1}{G}}. \tag{47}$$

To compute P_s we have to get back to (31) through (34). With $a = 0$ we have

$$\Pr\{L_n = l/t_n = k\} = 1, \qquad l = n$$

and

$$\Pr\{L_n = n\} = 1.$$

Therefore

$$P_s(n) = \frac{npq^{n-1}}{1 - q^n} \tag{48}$$

and

$$P_s = \sum_{n=1}^{\infty} \frac{npq^{n-1}}{1 - q^n} \frac{\pi_n}{1 - \pi_0} \tag{49}$$

where

$$\pi_n = \frac{G^n}{n!} e^{-G}. \tag{50}$$

By the same token, we see from (35) that

$$\pi_1' = \frac{ge^{-g}}{1 - e^{-g}} \xrightarrow[g \to 0]{} 1$$

and that

$$P_s' = P_s'(1) = 1.$$

With these considerations, the throughput is given by

$$S(G,p,a = 0) = \frac{G[\pi_0 + (1 - \pi_0)P_s]}{G + \pi_0} \tag{51}$$

where P_s and π_n are given in (49) and (50), respectively. When $p = 1$, we have, from (48),

$$P_s(1) = 1$$
$$P_s(n) = 0, \qquad n > 1$$

and therefore

$$P_s = \frac{Ge^{-G}}{1 - e^{-G}}.$$

Equation (51) then becomes

$$S(G, p = 1, a = 0) = \frac{G(1 + G)e^{-G}}{G + e^{-G}}$$

which is (and should be) identical to the 1-persistent CSMA when $a = 0$ [see (23)]. Let us now consider $p \to 0$. Since $1 - q_n \approx np$, (48) then becomes

$$P_s(n) = q^{n-1}$$

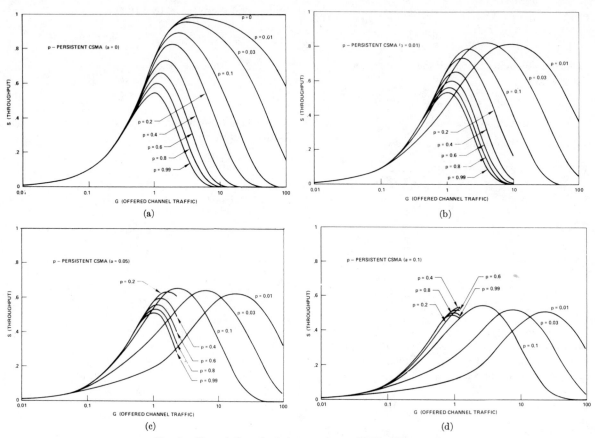

Fig. 7.   Channel throughput in p-persistent CSMA. (a) $a = 0$.
(b) $a = 0.01$. (c) $a = 0.05$. (d) $a = 0.1$.

and

$$P_s = \sum_{n=1}^{\infty} \frac{q^{n-1}G^n e^{-G}}{n!(1 - e^{-G})}$$

$$= \frac{qe^{-G}(e^{qG} - 1)}{1 - e^{-G}}.$$

In particular, $p \to 0$ gives $P_s(n) \to 1$, for all n, and $P_s \to 1$. In this limit the throughput is then given by

$$S(G, p \to 0, a = 0) \to \frac{G}{G + e^{-G}} \qquad (52)$$

which shows that a channel capacity of 1 can be achieved when $G \to \infty$.

For each value of a, one can plot a family of curves S versus G with parameter p [as shown in Fig. 7 (a)–(d)]. The channel capacity for each value of p can be numerically determined at an optimum value of G. In Fig. 8 we show the channel capacity as a function of p, for $a = 0$, 0.01, 0.05, and 0.1. We note that the capacity is not very sensitive to small variations of p; for $a = 0.01$, it reaches its highest value (i.e., the channel capacity for this protocol) at a value $p = 0.03$. When $p = 1$, the (slotted)

p-persistent CSMA reduces to the slotted 1-persistent CSMA. Indeed we can check that, when $p = 1$, (24) reduces to (19), since P_s, $\bar{l}$, and $\bar{l}'$ then become

$$P_s = \frac{aGe^{-G}}{1 - e^{-aG}}$$

$$\bar{l}' = \bar{l} = 0.$$

E. Performance Comparison and Sensitivity of Capacity to the Parameter a

To summarize, we plot in Fig. 9 for $a = 0.01$, S versus G for the various access modes introduced so far and thus show the relative performance of each, as also indicated in Table I.

While the capacity of ALOHA channels does not depend on the propagation delay, the capacity of a CSMA channel does. An increase in a increases the vulnerable period of a packet. This also results in "older" channel state information from sensing. In Fig. 10 we plot, versus a, the channel capacity for all of the above random-access modes. We note that the capacities for nonpersistent and p-persistent CSMA are more sensitive to increases in a, as compared to the 1-persistent scheme. Nonpersistent CSMA drops below 1-persistent for larger a. Also, for large a,

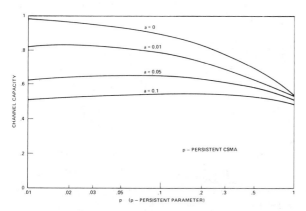

Fig. 8. *p*-persistent CSMA: effect of *p* on channel capacity.

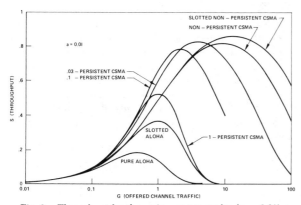

Fig. 9. Throughput for the various access modes ($a = 0.01$).

TABLE I
CAPACITY C FOR THE VARIOUS PROTOCOLS CONSIDERED ($a = 0.01$)

Protocol	Capacity C
Pure ALOHA	0.184
Slotted ALOHA	0.368
1-Persistent CSMA	0.529
Slotted 1-Persistent CSMA	0.531
0.1-Persistent CSMA	0.791
Nonpersistent CSMA	0.815
0.03-Persistent CSMA	0.827
Slotted Nonpersistent CSMA	0.857
Perfect Scheduling	1.000

slotted ALOHA (and even "pure" ALOHA) is superior to any CSMA mode since decisions based on partially obsolete data are deleterious; this effect is due in part to our assumption about the constant propagation delay. (For *p*-persistent, numerical results are shown only for $a \leq 0.1$. Clearly, for larger a, optimum *p*-persistent is lower-bounded by 1-persistent.)

V. DELAY CONSIDERATIONS

A. Delay Model

In the previous section, we analyzed the performance of CSMA modes in terms of maximum achievable throughput. We now introduce the expected packet delay D de-

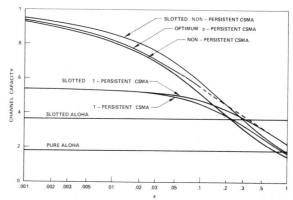

Fig. 10. Effect of propagation delay on channel capacity.

fined as the average time from when a packet is generated until it is successfully received.

Our principal concern in this section is to investigate the tradeoff between the average delay D and the throughput S.

As we have already stated, for the correct operation of the system, a positive acknowledgment scheme is needed. If an acknowledgment is not received by the sender of a packet within a specified time-out period, then the packet is retransmitted (incurring the random retransmission delay X, introduced to avoid repeated conflicts). For the present study, we assume the following.

Assumption 3: The acknowledgment packets are always correctly received with probability one.

The simplest way to accomplish this is to create a separate channel[11] (assumed to be available) to handle acknowledgment traffic. If sufficient bandwidth is provided to this channel overlaps between acknowledgment packets are avoided, since a positive acknowledgment packet is created only when a packet is correctly received, and there will be at most one such packet at any given time. Thus, if T_a denotes the transmission time of the acknowledgment packet on the separate channel, then the time-out for receiving a positive acknowledgment is $T + \tau + T_a + \tau$, provided that we make the following assumption.

Assumption 4: The processing time needed to perform the sumcheck and to generate the acknowledgment packet is negligible.

Assumption 2 further simplifies our delay model by implicitly assuming that the probability of a packet's success is the same whether the packet is new or has been blocked, or interfered with any number of times before; this probability is simply given by the throughput equation, i.e.,

$$P_s = \frac{S}{G} = \frac{\text{throughput}}{\text{offered traffic}}.$$

Bearing these assumptions in mind, we can write the delay equations for each of the previous access modes.

[11] The reader is referred to [16] for a study of the effect of acknowledgment traffic on channel throughput when acknowledgment packets are carried by the same channel.

As an example let us consider the ALOHA mode. Let R be the average delay between two consecutive transmissions (i.e., a retransmission) of a given packet. R consists of the transmission time of the packet, the transmission time of the acknowledgment packet, the round-trip propagation delay, and the average retransmission delay, that is

$$R = T + \tau + T_a + \tau + \ddot{X}.$$

Using our normalized time units, we have

$$R = 1 + 2a + \alpha + \delta \qquad (53)$$

where $\alpha = T_a/T$. Since $(G/S - 1)$ is the average number of retransmissions required, the average delay is given by

$$D = \left(\frac{G}{S} - 1\right)R + 1 + a. \qquad (54)$$

(Special attention must be devoted to the CSMA modes in which packets may incur pretransmission delays, and in which all arrivals do not necessarily correspond to actual transmissions. The delay equations and their derivations are given in Appendix B.)

Let us begin with some comments concerning the above delay equations. First, G/S as obtained from the throughput equations rests on two important and strong Assumptions 1 and 2; namely, that we have an independent Poisson point process and that δ is infinite, or large compared to the transmission time (in which case delays are also large and unacceptable). On the other hand, δ cannot be arbitrarily small. It is intuitively clear that when a certain backlog of packets is present, the smaller δ is, the higher is the level of interference and hence the larger is the offered channel traffic G. Thus, $G = G(S,\delta)$ is a decreasing function of δ such that the average number of transmissions per packet, $[G(S,\delta)]/S$, decreases with increasing values of δ, and reaches the asymptotic value predicted by the throughput equation. Thus, for each S, a minimum delay can be achieved by choosing an optimal δ. Such an optimization problem is difficult to solve analytically, and simulation techniques have been employed in our evaluations below.[12]

Before we proceed with the discussion of simulation results, we compare the various access modes in terms simply of the average number of transmissions (or average number of schedulings[13]) G/S. For this purpose, we plot G/S versus S in Fig. 11 for the ALOHA and CSMA modes, when $a = 0.01$. Note that CSMA modes are superior in that they provide lower values for G/S than the ALOHA modes. Furthermore, for each value of the throughput, there exists a value of p such that p-persistent is optimal. For small values of S, $p = 1$ (i.e., 1-persistent) is optimal. As S increases, the optimum p decreases.

---

[12] We have been able to solve the problem analytically in the case of the nonpersistent CSMA when we are in presence of a large population but with a finite number of users; all conclusions obtained from simulation in Section V-B have been verified by the analysis. For this the reader is referred to reference [16].

[13] For the nonpersistent and p-persistent CSMA, G measures the offered channel traffic and not the actual channel traffic. G/S represents, then, the average number of times a packet was scheduled for transmission before success.

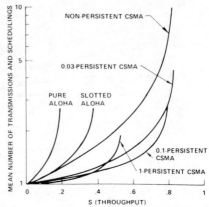

Fig. 11. G/S versus throughput ($a = 0.01$).

B. Simulation Results

The simulation model is based on all system assumptions presented in Section II. However, we relax Assumptions 1 and 2 concerning the retransmission delay and the independence of arrivals for the offered channel traffic. That is, in the simulation model, only the newly generated packets are derived independently from a Poisson distribution; collisions and uniformly distributed random retransmissions are accounted for without further assumptions.

In general, our simulation results indicate the following.

1) For each value of the input rate S, there is a minimum value δ for the average retransmission delay variable, such that below that value it is impossible to achieve a throughput equal to the input rate.[14] The higher S is, the larger δ must be to prevent a constantly increasing backlog, i.e., to prevent the channel from saturating. In other words, the maximum achievable throughput (under assumed stable conditions) is a function of δ, and the larger δ is, the higher is the maximum throughput.

2) Recall that the throughput equations were based on the assumption that $\ddot{X}/T = \delta \gg 1$. Simulation shows that for finite values of δ, $\delta > \delta_0$, but not too large compared to 1, the system already "reaches" the asymptotic results ($\delta \to \infty$). That is, for some finite values of δ, Assumption 2 is excellent and delays are acceptable. Moreover, the comparison of the (S,G) relationship as obtained from simulation and the results obtained from the analytic model exhibits an excellent match. Simulation experiments were also conducted to find the optimal delay; that is, the value of $\delta(S)$ which allows one to achieve the indicated throughput with the minimum delay.

Finally, in Fig. 12[15] we give the throughput-minimum delay tradeoff for the ALOHA and CSMA modes (when $a = 0.01$). *This is the basic performance curve.* We conclude

---

[14] Such behavior is characteristic of random multiple-access modes. Similar results were already encountered by Kleinrock and Lam [13] when studying slotted ALOHA in the context of a satellite channel.

[15] In Fig. 12, the curve corresponding to slotted ALOHA is obtained from the analytical model developed in [13] successfully verified by simulation.

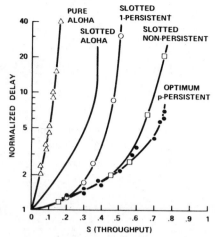

Fig. 12. Throughput-delay tradeoffs from simulation ($a = 0.01$).

that the optimum p-persistent CSMA provides us with the best performance; on the other hand the performance of the (simple) nonpersistent CSMA is quite comparable.

VI. SUMMARY AND DISCUSSION

We have introduced and evaluated the new CSMA mode and have shown it to be an efficient means for randomly accessing packet switched radio channels which have a small ratio of propagation delay to packet transmission time. Just as with most "contention" systems, these random multi-access broadcast channels (ALOHA, CSMA) are characterized by the fact that the throughput goes to zero for large values of channel traffic. At an optimum traffic level, we achieve a maximum throughput which we define to be the system capacity. This and the throughput-delay performance were obtained by a steady-state analysis under the assumption of equilibrium conditions.

However, these channels exhibit unstable behavior at most input loads as shown by Kleinrock and Lam [18]. In this last reference, the dynamic behavior and stability of an ALOHA channel are considered; quantitative estimates for the relative stability of the channel are given, which indicate the need for special control procedures to avoid a collapse. Optimal control procedures have been found [14], [15] and similar procedures are necessary for CSMA as well, since it can be shown [16] that CSMA exhibits similar unstable behavior.

Throughout the paper, it was assumed that all terminals are within range and in line-of-sight of each other. A common situation consists of a population of terminals, all within range and communicating with a single "station" (computer center, gate to a network, etc.) in line-of-sight of all terminals. Each terminal, however, may *not* be able to hear all the other terminals' traffic. This gives rise to what is called the *"hidden-terminals"* problem. The latter badly degrades the performance of CSMA as shown in Part II of this paper [19]. Fortunately, in a single-station environment, the hidden-terminal problem can be elim-

inated by dividing the available bandwidth into two separate channels: a busy tone channel and a message channel. As long as the station is receiving a signal on the message channel, it transmits a busy tone signal on the busy tone channel (which terminals sense for channel state information). The CSMA with a busy tone under a nonpersistent protocol has been analyzed. It is shown to provide a maximum channel capacity of approximately 0.65 when $a = 0.01$ for a channel bandwidth W of 100 kHz (modulated at 1 bit/Hz); when $W = 1$ MHz and $a = 0.01$, the channel capacity is 0.71 [19]. These values compare favorably with the capacity of 0.815 for nonpersistent CSMA with no hidden terminals.

APPENDIX A

SMALL p APPROXIMATIONS IN p-PERSISTENT CSMA

We claim, for small p, that $S(G,p,a)$ may be approximated by

$$S'(G,p,a)$$

$$= \frac{(1 - e^{-aG})[\hat{P}_s{}'\pi_0 + \hat{P}_s(1 - \pi_0)]}{(1 - e^{-aG})[a\hat{\bar{t}}'\pi_0 + a\hat{\bar{t}}(1 - \pi_0) + 1 + a] + a\pi_0} \tag{A1}$$

where $\hat{P}_s{}'$, $\hat{P}_s$, $\hat{\bar{t}}$, and $\hat{\bar{t}}'$ are defined hereafter in the proof.

Proof: We show here that, with some approximations, we can get a closed-form solution for the throughput when p has *small values* ($p < 0.1$). These approximations are validated by comparing the results obtained in this section with those obtained from Section IV-D for $p = 0.1$.

For the distribution of idle time between two TP's, we have from (26)

$$\Pr\{t_n > k\} = q^{(k+1)n} \exp\left\{g\left(\frac{q(1 - q^k)}{p} - k\right)\right\}. \tag{A2}$$

When p is small, we may make the following approximation (actually a lower bound):

$$q^k = (1 - p)^k \simeq 1 - kp \tag{A3}$$

and therefore we may rewrite (A2) as

$$\Pr\{t_n > k\} \simeq q^{(k+1)n}e^{-kpg} = q^n[q^n e^{-pg}]^k. \tag{A4}$$

Let $t_{n>}{}^*(z)$ and $t_n{}^*(z)$ be the generating functions defined by

$$t_{n>}{}^*(z) \triangleq \sum_{k=0}^{\infty} \Pr\{t_n > k\}z^k \tag{A5}$$

$$t_n{}^*(z) \triangleq \sum_{k=0}^{\infty} \Pr\{t_n = k\}z^k. \tag{A6}$$

We have

$$t_{n>}{}^*(z) = q^n \sum_{k=0}^{\infty} (q^n e^{-pg}z)^k = \frac{q^n}{1 - q^n e^{-pg}z}. \tag{A7}$$

Since

$$\Pr\{t_n = k\} = \Pr\{t_n > k - 1\} - \Pr\{t_n > k\}, \quad k > 0$$

and

$$\Pr\{t_n = 0\} = 1 - \Pr\{t_n > 0\},$$

we have

$$t_n^*(z) = 1 + (z - 1)t_{n>}^*(z) = 1 + \frac{q^n(z - 1)}{1 - q^n e^{-pg}z}. \quad (A8)$$

The averages defined in (29) can now be written as

$$\bar{t}_n = \left.\frac{\partial t_n^*(z)}{\partial z}\right|_{z=1} = \frac{q^n}{1 - q^n e^{-pg}}. \quad (A9)$$

Equation (30), which defines $\bar{t}$ as $\sum_{n=1}^{\infty} \bar{t}_n \pi_n/(1 - \pi_0)$, does not lead to a closed-form expression. Instead, we replace $\bar{t}$ by $\hat{\bar{t}}$, which is defined as

$$\hat{\bar{t}} = \frac{C}{1 - Ce^{-pg}} \quad (A10)$$

where $C = \sum_{n=1}^{\infty} q^n \pi_n/(1 - \pi_0)$. ($\hat{\bar{t}}$ is smaller than $\bar{t}$ since $\bar{t}_n = q^n/(1 - q^n e^{-pg})$ is a convex function of q^n.)

C can be expressed as

$$C = \frac{\exp\{-(1 + a)pG\} - \pi_0}{1 - \pi_0} = \frac{\pi_0^p - \pi_0}{1 - \pi_0} \quad (A11)$$

and therefore,

$$\hat{\bar{t}} = \frac{\pi_0^p - \pi_0}{1 - \pi_0 - (\pi_0^p - \pi_0)e^{-pg}}. \quad (A12)$$

To find the probability of success over TP 3_i, $i \neq 1$, we first define the following generating functions:

$$L_n^*(z) \triangleq \sum_{l=n-1}^{\infty} \Pr\{L_n = l\}z^l \quad (A13)$$

$$L_n^*(z/k) \triangleq \sum_{l=n-1}^{\infty} \Pr\{L_n = l/t_n = k\}z^l. \quad (A14)$$

It is clear that

$$L_n^*(z/k) = \exp\{kg(z - 1)\}z^{n-1}. \quad (A15)$$

Removing the condition on k, we get

$$L_n^*(z) = \sum_{k=0}^{\infty} L_n^*(z/k) \cdot \Pr\{t_n = k\}$$

$$= z^{n-1} \sum_{k=0}^{\infty} \exp\{kg(z - 1)\} \cdot \Pr\{t_n = k\}$$

$$= z^{n-1} t_n^*(\exp\{g(z - 1)\})$$

$$= \frac{q^n(\exp\{g(z - 1)\} - 1)z^{n-1}}{1 - q^n \exp\{-pg\}\exp\{g(z - 1)\}} + z^{n-1}. \quad (A16)$$

The probability of success $P_s(n)$, defined in (33), is now simply expressed (since $1 - q^l \approx lp$) as

$$P_s(n) = L_n^*(q)$$

$$= q^{n-1} - \frac{(1 - e^{-gp})q^{2n-1}}{1 - q^n e^{-2gp}}. \quad (A17)$$

Here again, (34) defines

$$P_s = \sum_{n=1}^{\infty} P_s(n) \cdot \frac{\pi_n}{1 - \pi_0}$$

which does not lead to a closed-form expression. Instead, we replace P_s by $\hat{P}_s$, which is defined as

$$\hat{P}_s = \frac{C}{q} - \frac{(1 - e^{-gp})C'}{q(1 - Ce^{-gp})} \quad (A18)$$

where C is as expressed in (A11), and

$$C' = \sum_{n=1}^{\infty} q^{2n} \frac{\pi_n}{1 - \pi_0} = \frac{\pi_0^{1-q^2} - \pi_0}{1 - \pi_0}. \quad (A19)$$

Finally, $\hat{P}_s$ can be expressed as shown in the following equation:

$$\hat{P}_s = \frac{\pi_0^p - \pi_0}{q(1 - \pi_0)} - \frac{(1 - \exp\{-gp\})(\pi_0^{1-q^2} - \pi_0)}{q(1 - \pi_0) - q\exp\{-2gp\}(\pi_0^p - \pi_0)}. \quad (A20)$$

The quantities $\hat{\bar{t}}'$ and $\hat{P}_s'$ are readily obtained from (A12), and (A20), respectively, by replacing

$$\pi_0 \triangleq \exp\{-G(1 + a)\}$$

by the quantity e^{-g}. The substitution of $\hat{P}_s$, $\hat{\bar{t}}$, $\hat{P}_s'$, and $\hat{\bar{t}}'$ for P_s, $\bar{t}$, P_s', and $\bar{t}'$, respectively, in (46) provides us with a closed-form solution for $S(G,p,a)$ when p is small.

In Table II, we compare for $p = 0.1$ the "exact" results obtained from Section IV-D to those obtained by the approximation; note that the closed-form solution is quite satisfactory for $p < 0.1$.

APPENDIX B

DELAY EQUATIONS

A. Nonpersistent CSMA

In this case, the average delay R between two successive sense points of the same packet is

$$R = \begin{cases} 1 + \alpha + 2a + \delta, & \text{if the packet is transmitted} \\ \delta, & \text{if the packet is blocked.} \end{cases} \quad (B1)$$

Let P_b be the probability that an arrival gets blocked (i.e., senses the channel busy). We have

$$1 - P_b = \frac{a + 1/G}{\bar{C}}$$

$$= \frac{1 + aG}{1 + G(1 + a + \bar{Y})}. \quad (B2)$$

Under the traffic independence assumption, the rate of

TABLE II
COMPARISON OF RESULTS FOR THROUGHPUT S OBTAINED FROM THE
EXACT ANALYSIS (24) AND RESULTS OBTAINED FROM THE
APPROXIMATION (APPENDIX A) WHEN $p = 0.1$

	$a = 0.01$		$a = 0.05$	
G	Exact	Approximate	Exact	Approximate
0.1	0.098	0.098	0.095	0.094
0.2	0.192	0.192	0.179	0.178
0.3	0.279	0.279	0.252	0.251
0.4	0.358	0.358	0.316	0.314
0.5	0.428	0.428	0.370	0.367
0.6	0.490	0.490	0.417	0.413
0.7	0.544	0.544	0.457	0.453
0.8	0.589	0.590	0.490	0.486
0.9	0.628	0.630	0.519	0.515
1.0	0.661	0.663	0.543	0.539
1.1	0.689	0.691	0.563	0.560
1.2	0.711	0.714	0.580	0.578
1.3	0.730	0.733	0.594	0.593
1.4	0.745	0.749	0.606	0.605
1.5	0.757	0.761	0.616	0.616
1.6	0.766	0.771	0.624	0.625
1.7	0.773	0.778	0.630	0.632
1.8	0.778	0.784	0.635	0.638
1.9	0.781	0.787	0.639	0.643
2.0	0.783	0.790	0.642	0.647
2.1	0.784	0.791	0.644	0.649
2.2	0.784	0.791	0.645	0.651
2.3	0.783	0.790	0.646	0.653

actual transmissions is given by

$$H = G(1 - P_b).$$

Since $(H/S) - 1$ represents the average number of actual retransmissions per packet, the average delay D is therefore

$$D = (H/S - 1)[1 + \alpha + 2a + \delta][(G - H)/S]\delta + 1 + a \quad (B3)$$

where G/S is given by the nonpersistent CSMA throughput equation (3).

If we choose to treat all packet arrivals in a uniform manner, we may assume that when a packet is blocked, it behaves as if it could transmit, and learned about its blocking only T_a seconds after the end of its "virtual" transmission. With this simplification, the delay equation is

$$D = (G/S - 1)(1 + 2a + \alpha + \delta) + 1 + a \quad (B4)$$

thus introducing an additional delay equal to (GP_b/S) $[1 + \alpha + 2a]$.

B. 1-Persistent CSMA

Unlike the ALOHA channel, a packet on a CSMA channel incurs an additional pretransmission delay r, if upon its arrival, that packet detects the channel busy. Recall that the probability of finding the channel busy is given by (see Section IV-C)

Pr {a packet finds the channel busy}

$$= \frac{\bar{B} - a/q_0}{\bar{B} + \bar{I}} = \frac{1 + \bar{Y}}{q_0(\bar{B} + \bar{I})} \quad (B5)$$

where $\bar{B}$, $\bar{I}$, $\bar{Y}$, and q_0 are given in (16), (12), (7), and (15), respectively.

Under the condition that the packet found the channel busy, the average waiting time until the channel is detected idle (i.e., until the end of the TP) is simply equal [17] to $\overline{Z^2}/2\bar{Z}$ by the Poisson assumption. The second moment of Z is simply given by

$$\overline{Z^2} = \overline{(1 + Y)^2} = 1 + 2\bar{Y} + \overline{Y^2}.$$

From the distribution of Y given in (6) we then have

$$\overline{Z^2} = 1 + a^2 + 2(1 - 1/G)\bar{Y}. \quad (B6)$$

Therefore the average pretransmission delay $\bar{r}_1$ can be easily expressed as

$$\bar{r}_1 = \frac{1 + a^2 + 2(1 - 1/G)\bar{Y}}{2(1 + \bar{Y})}$$

$$\cdot \text{Pr \{the packet finds the channel busy\}}$$

$$= \frac{1 + a^2 + 2(1 - 1/G)\bar{Y}}{2q_0(\bar{B} + \bar{I})}. \quad (B7)$$

Finally, the expected packet delay is

$$D = (G/S - 1)(1 + 2a + \alpha + \delta + \bar{r}_1) + \bar{r}_1 + 1 + a \quad (B8)$$

where G/S is given by the 1-persistent CSMA throughput equation (10).

C. p-Persistent CSMA

Similar to the special case of 1-persistent CSMA, a packet in this general scheme incurs an initial delay which we denote by r_p. In order to compute its expected value $\bar{r}_p$, one must consider the following situations.

1) An arbitrary packet, upon arrival, will find the channel idle with probability $\bar{I}/(\bar{B} + \bar{I})$, in which case its average initial wait is $a\bar{l}'$.

2) An arbitrary packet, upon arrival, will find the channel in the first IRTD (first t' seconds) of a busy period with probability $a\bar{l}'/(\bar{B} + \bar{I})$. In this case, its average initial delay is $a\overline{l'^2}/2\bar{l}'$.

3) An arbitrary packet, upon arrival, will find the channel in the remaining part of a busy period with probability $(B - a\bar{l}')/(\bar{B} + \bar{I})$, in which case the average initial wait is $\overline{(1 + a + at)^2}/2(1 + a + a\bar{l})$.

Therefore

$$\bar{r}_p = \frac{\bar{I}}{\bar{B} + \bar{I}} a\bar{l}' + \frac{a\bar{l}'}{\bar{B} + \bar{I}} \cdot \frac{a\overline{l'^2}}{2\bar{l}'} + \frac{B - a\bar{l}'}{\bar{B} + \bar{I}} \cdot \frac{\overline{(1 + a + at)^2}}{2(1 + a + a\bar{l})}. \quad (B9)$$

Treating all transmissions and schedulings uniformly (by introducing artificial delays due to "virtual" transmissions and acknowledgment), the expected delay can simply be expressed as

$$D = (G/S - 1)[1 + 2a + \delta + \bar{r}_p] + 1 + a + \bar{r}_p \quad (B10)$$

where G/S is given by the p-persistent CSMA throughput equation (24).

REFERENCES

[1] L. G. Roberts, "Data by the packet," *IEEE Spectrum*, vol. 11, pp. 46–51, Feb. 1974.

[2] N. Abramson, "The ALOHA System—Another alternative for computer communications," in *1970 Fall Joint Comput. Conf., AFIPS Conf. Proc.*, vol. 37. Montvale, N. J.: AFIPS Press, 1970, pp. 281–285.

[3] R. E. Kahn, "The organization of computer resources into a packet radio network," in *Nat. Comput. Conf., AFIPS Conf. Proc.*, vol. 44. Montvale, N. J.: AFIPS Press, 1975, pp. 177–186.

[4] L. Kleinrock and F. Tobagi, "Random access techniques for data transmission over packet-switched radio channels," in *Nat. Comput. Conf., AFIPS Conf. Proc.*, vol. 44. Montvale, N. J.: AFIPS Press, 1975, pp. 187–201.

[5] R. Binder, N. Abramson, F. Kuo, A. Okinaka, and D. Wax, "ALOHA packet broadcasting—A retrospect," in *Nat. Comput. Conf., AFIPS Conf. Proc.*, vol. 44. Montvale, N. J.: AFIPS Press, 1975, pp. 203–215.

[6] H. Frank, I. Gitman, and R. Van Slyke, "Packet radio system—Network considerations," in *Nat. Comput. Conf., AFIPS Conf. Proc.*, vol. 44. Montvale, N. J.: AFIPS Press, 1975, pp. 217–231.

[7] S. Fralick and J. Garrett, "Technological considerations for packet radio networks," in *Nat. Comput. Conf., AFIPS Conf. Proc.*, vol. 44. Montvale, N. J.: AFIPS Press, 1975, pp. 233–243.

[8] J. Burchfiel, R. Tomlinson, and M. Beeler, "Functions and structure of a packet radio station," in *Nat. Comput. Conf., AFIPS Conf. Proc.*, vol. 44. Montvale, N. J.: AFIPS Press, 1975, pp. 245–251.

[9] S. Fralick, D. Brandin, F. Kuo, and C. Harrison, "Digital terminals for packet broadcasting," in *Nat. Comput. Conf., AFIPS Conf. Proc.*, vol. 44. Montvale, N. J.: AFIPS Press, 1975, pp. 253–261.

[10] P. E. Jackson and C. D. Stubbs, "A study of multi-access computer communications," in *1969 Spring Joint Comput. Conf., AFIPS Conf. Proc.*, vol. 34. Montvale, N. J.: AFIPS Press, 1969, pp. 491–504.

[11] N. Abramson, "Packet switching with satellites," in *Nat. Comput. Conf., AFIPS Conf. Proc.*, vol. 42. Montvale, N. J.: AFIPS Press, 1973, pp. 695–702.

[12] L. Roberts, "ARPANET Satellite System," Notes 8 (NIC Document 11290) and 9 (NIC Document 11291), available from the ARPA Network Information Center, Stanford Research Institute, Menlo Park, Calif.

[13] L. Kleinrock and S. Lam, "Packet-switching in a slotted satellite channel," in *Nat. Comput. Conf., AFIPS Conf. Proc.*, vol. 42. Montvale, N. J.: AFIPS Press, 1973, pp. 703–710.

[14] S. Lam, "Packet switching in a multi-access broadcast channel with application to satellite communications in a computer network," School of Eng. and Appl. Sci., Univ. of California, Los Angeles, rep. UCLA-ENG 7429, Apr. 1974.

[15] S. Lam and L. Kleinrock, "Dynamic control schemes for a packet switched multi-access broadcast channel," in *Nat. Comput. Conf., AFIPS Conf. Proc.*, vol. 44. Montvale, N. J.: AFIPS Press, 1975, pp. 143–153.

[16] F. Tobagi, "Random access techniques for data transmission over packet switched radio networks," Ph.D. dissertation, Comput. Sci. Dep., School of Eng. and Appl. Sci., Univ. of California, Los Angeles, rep. UCLA-ENG 7499, Dec. 1974.

[17] L. Kleinrock, *Queueing Systems, Vol. I, Theory; Vol. II, Computer Applications.* New York: Wiley Interscience, 1975.

[18] L. Kleinrock and S. S. Lam, "Packet switching in a multi-access broadcast channel: Performance evaluation," *IEEE Trans. Commun.*, vol. COM-23, pp. 410–423, Apr. 1975.

[19] F. A. Tobagi and L. Kleinrock, "Packet switching in radio channels: Part II—The hidden terminal problem in carrier sense multiple access and the busy tone solution," *IEEE Trans. Commun.*, this issue, pp. 1417–1433.

Packet Switching in Radio Channels: Part II—The Hidden Terminal Problem in Carrier Sense Multiple-Access and the Busy-Tone Solution

FOUAD A. TOBAGI AND LEONARD KLEINROCK, FELLOW, IEEE

Abstract—We consider a population of terminals communicating with a central station over a packet-switched multiple-access radio channel. The performance of carrier sense multiple access (CSMA) [1] used as a method for multiplexing these terminals is highly dependent on the ability of each terminal to sense the carrier of any other transmission on the channel. Many situations exist in which some terminals are "hidden" from each other (either because they are out-of-sight or out-of-range). In this paper we show that the existence of hidden terminals significantly degrades the performance of CSMA. Furthermore, we introduce and analyze the busy-tone multiple-access (BTMA) mode as a natural extension of CSMA to eliminate the hidden-terminal problem. Numerical results giving the bandwidth utilization and packet delays are shown, illustrating that BTMA with hidden terminals performs almost as well as CSMA without hidden terminals.

I. INTRODUCTION

THE USE of packet switching in a multiple-access broadcast radio channel for communication between terminals and a central station was presented in Part I [1]. We also introduced and analyzed a new random-access mode, the carrier sense multiple-access mode (CSMA), as a means of multiplexing a large number of terminals communicating with the station over the shared radio channel.[1] Briefly, CSMA consists of reducing the level of interference (caused by overlapping packets) in the random multiaccess environment by allowing terminals to sense the carrier due to other users' transmissions; based on the information gained in this way about the state of the channel (busy or idle), the terminal takes an action prescribed by the particular CSMA protocol being used (in particular, a terminal never transmits when it senses the channel busy). In Part I we described and analyzed three protocols referred to as: 1-persistent; nonpersistent; and p-persistent CSMA. The evaluation of performance of the various protocols obtained there (Part I) was based on the assumption that all terminals are in line-of-sight (LOS) and within range of each other. However there are many situations in which this is not true, forcing us to

relax the above assumption here. Two terminals can be within range of the station but out-of-range of each other; or they can be separated by some physical obstacle opaque to UHF signals. Two such terminals are said to be "hidden" from each other. It is evident that the existence of hidden elements in an environment affects (degrades) the performance of CSMA. In this paper we first attempt to gain some insight about this effect. This is the subject of Section II. (For simplicity, we restrict our study to 1-persistent and nonpersistent CSMA protocols only.)

Second, in this paper, we consider a solution to the hidden-terminal problem which we call the busy-tone multiple-access (BTMA) mode. This is the subject of Section III in which we give 1) a description of the operation of BTMA under a nonpersistent protocol, 2) an analysis to determine the throughput-delay characteristics along with the effect of various system parameters, and 3) a discussion of some numerical results.

II. THE HIDDEN-TERMINAL PROBLEM

Below, we shall describe the model and define an adequate representation for hidden-terminal configurations. Then we proceed with the analysis for throughput and delay. Finally we shall consider some examples to which we apply analytical and simulation techniques.

A. The Model

We assume an environment consisting of a large number of terminals communicating with a single station over a shared radio channel. All terminals are in line-of-sight and within range of the station but not necessarily with respect to each other. All other system assumptions introduced in Part I hold true. The total traffic source will be approximated by an independent Poisson source with an aggregate mean packet generation rate of λ packets/s.

We characterize a terminal configuration with hidden elements as follows. Let $i = 1,2,\cdots,M$ index the M terminals in the population. By definition, terminal i "hears" (is connected to) terminal j if i and j are within range and in line-of-sight of each other. To represent the connections among terminals, we use an $M \times M$ square matrix $\boldsymbol{M}$ such that the element m_{ij} is

$$m_{ij} = \begin{cases} 1, & \text{if } i \text{ hears } j \\ 0, & \text{otherwise.} \end{cases}$$

---

[1] Throughout the paper, the reader is assumed to be familiar with the results and terminology introduced in [1].

Since two terminals that hear the same subset of the population behave similarly, it is advantageous to partition the population into several groups (say $N, N \leq M$) such that all terminals within a group hear exactly the same subset of terminals in the population. This partitioning is easily formed by collecting all terminals with identical rows or columns in $\boldsymbol{M}$ into one group. We now define a "hearing" graph with N nodes and make a one-to-one correspondence between the nodes of the graph and the N groups just obtained. A link between two nodes k and l represents the fact that group k and group l hear each other, and this is easily determined by the fact that there exists a terminal i in group k that hears a terminal j in group l. This procedure provides us with the minimum number of groups describing the configuration. Let $h(i)$ be the set of groups that group i can hear. In the sequel, we shall isolate the case of independent groups from the general case of dependent groups. The former is characterized by the absence of links in the hearing graph.

We shall further assume that each group i consists of a large number of users who collectively form an independent Poisson source with an aggregate mean packet generation rate λ_i packets/s such that $\sum_{i=1}^{N} \lambda_i = \lambda$.

As in Part I, we characterize the traffic as follows. Let

B. Analysis

We wish to answer the following basic questions of interest.

Question 1: Given an input pattern $\mathfrak{U}$, what is the channel capacity $C(\mathfrak{U})$? An equivalent question is: Is a given set of input rates $\mathfrak{S}(\mathfrak{U})$ achievable or does it saturate the channel?

Question 2: For a given achievable set of input rates $\mathfrak{S}(\mathfrak{U})$, what is the relative performance of the various groups?

We shall first treat the simple case of independent groups for both the 1-persistent and nonpersistent CSMA, then we shall proceed with an approximate analysis for the dependent groups case under a nonpersistent protocol.

Independent Groups Case: We first recognize that S_i/G_i is merely the probability of success of an arbitrary packet from group i. This quantity is a function of the traffic vector $\mathcal{G}$. By expressing S_i/G_i for each i in terms of $\mathcal{G}$, we obtain a set of equations relating the components of $\mathfrak{S}$ to the components of $\mathcal{G}$. For a given $\mathcal{G}$ and under the system and model assumptions stated above, the probability of success of an arbitrary packet from group i is given as follows.

1-Persistent CSMA:

$$P_{s_i} = \frac{S_i}{G_i} = \frac{[1 + G_i + aG_i(1 + G_i + aG_i/2)]}{(1 + aG_i) \exp\{-G_i(1 - 2a)\}} \prod_{j=1}^{N} \frac{(1 + aG_j) \exp\{-2G_j\}}{G_j(1 + 2a) - (1 - \exp\{-aG_j\}) + (1 + aG_j) \exp\{-G_j(1 + a)\}} \tag{1}$$

$S_i = \lambda_i T$. Under steady-state conditions, S_i is the throughput of group i. Let $S = \lambda T = \sum_{i=1}^{N} S_i$; S is the total throughput and utilization of the channel. Let $\mathfrak{S} = (S_1, S_2, \cdots, S_N)$. Let $\mathfrak{U} = (u_1, u_2, \cdots, u_N)$ where $u_i = (S_i/S)$; $\mathfrak{U}$ describes a direction in N-dimensional space. The *capacity* of the channel along the direction $\mathfrak{U}$ is defined as

$$C(\mathfrak{U}) = \max_{0 \leq S \leq 1} S$$

such that the set of inputs determined by the vector $S\mathfrak{U}$ is achievable. In other words, $C(\mathfrak{U})$ is the maximum achievable throughput or maximum attainable channel utilization, when for all i, the input source of group i constitutes a fraction u_i of the total input source. In addition, let G_i denote the mean offered traffic rate (per T seconds) of group i. Let $\mathcal{G} = (G_1, G_2, \cdots, G_N)$ and $G = \sum_{i=1}^{N} G_i$.

We determined in Part I [1] the necessity of introducing a random retransmission delay X with mean $\bar{X}$ to avoid repeated conflicts. We shall further assume here that $\bar{X}$ is the same for all groups and that Assumptions 1 and 2 of Part I still hold true, as follows.

Assumption 1': $\bar{X}$ is large compared to the transmission time T, so that the interarrival times of the point processes defined by the start times of all the (new) packets plus retransmissions from group i are of independent increments and exponentially distributed, with mean interarrival time $1/G_i$.

Nonpersistent CSMA:

$$P_{s_i} = \frac{S_i}{G_i} = \exp\{G_i(1 - 2a)\}$$

$$\cdot \prod_{j=1}^{N} \frac{\exp\{-G_j(1 - a)\}}{G_j(1 + 2a) + \exp\{-aG_j\}}. \tag{2}$$

Proof: By definition, a packet transmission is said to be *i-successful* if the packet is free from interference caused by packets from group i. A packet transmission is said to be *totally successful* if and only if it is i-successful for all i; $i = 1, 2, \cdots, N$.

Consider first the 1-persistent CSMA case. An arbitrary packet from group i is successful if the following two mutually independent conditions are satisfied.

$\mathcal{C}_1$: The packet transmission is i-successful.

$\hat{\mathcal{C}}_2$: The packet is j-successful, for all $j \neq i$.

($\mathcal{C}_1$ and $\mathcal{C}_2$ are independent since we are dealing with independent groups.)

Consider for each group i a time line which exhibits packet transmissions from group i only (see Fig. 1). We observe on time line i an alternate sequence of busy and idle periods as defined in [1]. Moreover, because of the independence among groups (completely disconnected graph), this sequence is completely determined by the traffic rate G_i. Condition $\mathcal{C}_1$ is satisfied with a probability equal to the probability of success of a packet in 1-persistent CSMA without hidden terminals when the traffic rate is G_i. It is given by [1]

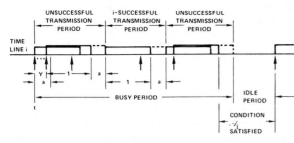

Fig. 1. 1-persistent CSMA: time line i.

same property,[2] we have:

Pr {the tagged packet starts transmission during the last

a seconds of a busy period on time line j} = $a/(\bar{I}_j + \bar{B}_j)$.

In this event, the probability that no packets from group j start transmission during the transmission time of the tagged packet is the probability of no transmission from group j in an interval $(1 + x - a)$ with x uniformly distributed over the last a seconds of the busy period and is given by

$$\text{Pr}\,\{\mathcal{C}_1\} = \frac{[1 + G_i + aG_i(1 + G_i + aG_i/2)]\exp\{-G_i(1 + 2a)\}}{G_i(1 + 2a) - (1 - \exp\{-aG_i\}) + (1 + aG_i)\exp\{-G_i(1 + a)\}}. \tag{3}$$

Consider now the time line j corresponding to group j, $j \neq i$. Here again, we observe an alternate sequence of busy and idle periods denoted by B_j and I_j, respectively, completely determined by the rate G_j. The average busy and idle periods are expressed as [1]

$$\bar{B}_j = \frac{1 + a + \bar{Y}_j}{q_{0,j}} \tag{4}$$

$$I_j = \frac{1}{G_j} \tag{5}$$

$$\int_0^a \exp\{-G_j(1 + x - a)\}\,\frac{dx}{a}$$

$$= \frac{1}{aG_j}\left[\exp\{-G_j(1 - a)\} - \exp\{-G_j\}\right].$$

On the other hand, the probability that the tagged packet starts transmission during an idle period is $\bar{I}_j/(\bar{I}_j + B_j)$ and the probability that no packets from group j start transmission during its transmission time is $\exp\{-G_j\}$. Since the groups are independent, we then have

$$\text{Pr}\,\{\mathcal{C}_2\} = \prod_{j \neq i} \frac{(1/G_j)[\exp\{-G_j(1 - a)\} - \exp\{-G_j\}] + \bar{I}_j\exp\{-G_j\}}{\bar{I}_j + \bar{B}_j}$$

$$= \prod_{j \neq i} \frac{(1 + aG_j)\exp\{-2G_j\}}{G_j(1 + 2a) - (1 - \exp\{-aG_j\}) + (1 + aG_j)\exp\{-G_j(1 + a)\}}. \tag{8}$$

where

$$\bar{Y}_j = a - \frac{1}{G_j}(1 - \exp\{-aG_j\}) \tag{6}$$

and

$$q_{0,j} = (1 + aG_j)\exp\{-G_j(1 + a)\}. \tag{7}$$

Our tagged packet is j-successful if and only if the following two conditions are satisfied.

$\mathcal{A}_j$: The start of transmission of the tagged packet does not occur during any transmission period (with the exception of the last a seconds of the last transmission period of the busy period [see Fig. 1]).

$\mathcal{B}_j$: No packet from group j starts transmission during the transmission time of the tagged packet.

We know that, by assumption 1′, the arrival of an arbitrary packet represents a random look in time. In 1-persistent CSMA, the start of transmission of the packet may not correspond to the arrival of the packet since the packet may incur a pretransmission delay in case the channel is sensed busy at its arrival time. However, by assuming that the start of transmission possesses the

Conditions $\mathcal{C}_1$ and $\mathcal{C}_2$ being mutually independent, we have

$$P_{s_i} = \frac{S_i}{G_i} = \text{Pr}\,\{\mathcal{C}_1\}\,\text{Pr}\,\{\mathcal{C}_2\}$$

in which we substitute the expressions found above to get (1).

The proof for (2) is exactly identical to the one above, in which we have the following expressions for the various quantities [1]:

$$\bar{B}_j = 1 + a + \bar{Y}_j$$

$$\bar{I}_j = \frac{1}{G_j}$$

$$\bar{Y}_j = a - \frac{1}{G_j}(1 - \exp\{-aG_j\})$$

---

[2] Such an assumption is not needed in the nonpersistent CSMA case since packets will not incur pretransmission delays. In that case the analysis will be exact. The comparison between results obtained from simulation of the "Poisson" model, to be discussed in the following section, and those obtained by this analysis (with $a = 0.01$) shows that the effect of this assumption is not noticeable.

$$\Pr\{\mathcal{C}_1\} = \frac{\exp\{-aG_j\}}{G_i(1 + 2a) + \exp\{-aG_i\}} \qquad (9)$$

and

$$\Pr\{\mathcal{C}_2\} = \prod_{j \neq i} \frac{\exp\{-G_j(1 - a)\}}{G_j(1 + 2a) + \exp\{-aG_j\}} . \qquad (10)$$

Therefore

$$P_{si} = \frac{S_i}{G_i} = \exp\{-G_i(1 - 2a)\}$$

$$\cdot \prod_{j=1}^{N} \frac{\exp\{-G_j(1 - a)\}}{G_j(1 + 2a) + \exp\{-aG_j\}} . \qquad \text{Q.E.D.}$$

Thus we obtain a set of equations relating the components of the input vector $\mathcal{S}$ to the components of the traffic vector $\mathcal{G}$ of the form

$$\frac{S_j}{G_i} = f_i(G_1, G_2, \cdots, G_N). \qquad (11)$$

For a given input vector $\mathcal{S}$, we can numerically solve for G_i, $i = 1, \cdots, N$. This we do by writing (11) in the form

$$G_i = S_i / f_i(G_1, \cdots, G_N) \qquad (12)$$

and by solving the set of equations iteratively, starting with the initial values $\mathcal{G} = \mathcal{S}$. If the iterative procedure results in a (finite) traffic vector $\mathcal{G}$ then the input vector is feasible. (We do not claim we can prove existence and unicity of the solution but it has been our experience that $\mathcal{G} = \mathcal{S}$ is a good starting solution and simulation results always agreed with the results obtained by the iterative procedure whenever a finite solution could be reached. Lack of convergence is assumed whenever a certain preset maximum number of iterations is exceeded.) Thus, the convergence of the iterative procedure determines the feasibility of the input vector $\mathcal{S}$ and the final values G_i/S_i; $i = 1, 2, \cdots, N$ give the average number of transmissions and schedulings a packet from group i undertakes before success. This will be our measure of relative performance of the various groups. Some simple examples are treated in the following section.

Before we proceed with the case of dependent groups, we consider here the particular case in which the N independent groups are identical, i.e.,

$$S_i = s, \qquad \forall i$$

$$G_i = g, \qquad \forall i.$$

Equations (1) and (2) reduce to, respectively,

$$\frac{S}{G} = \frac{s}{g} = \frac{1 + g + ag(1 + g + ag/2)}{(1 + ag)\exp\{-g(1 - 2a)\}} \left[\frac{(1 + ag)\exp\{-g\}}{g(1 + 2a) - (1 - \exp\{-ag\}) + (1 + ag)\exp\{-g(1 + a)\}} \right]^N \qquad (13)$$

$$\frac{S}{G} = \frac{s}{g} = \exp\{g(1 - 2a)\} \left[\frac{\exp\{-g(1 - a)\}}{g(1 + 2a) + \exp\{-ag\}} \right]^N . \qquad (14)$$

Naturally, if we let $N \to \infty$, $s \to 0$, and $g \to 0$ such that $Ns = S$ and $Ng = G$, we expect the CSMA mode to reduce

to the ALOHA access mode. Indeed it is easy to see from (13) and (14) that for both protocols

$$\frac{S}{G} \underset{N \to \infty}{\longrightarrow} e^{-2G}$$

the probability of success of a packet in ALOHA mode!

The Case of Dependent Groups: The dependence among the groups renders the determination of the sequence of idle and busy periods relative to a group, say i, rather difficult. Moreover this sequence is a function of the entire set $\{G_j, j \in h(i)\}$. Thus, no tractable analysis is yet available for the 1-persistent CSMA mode. However, an *approximate* model is presented here for the nonpersistent CSMA protocol. For this we make some fairly strong assumptions of statistical independence among the groups as well as exponential distributions for the interpoint times of various processes. (We claim that such assumptions are particularly valid when the load on the channel is low). Simulation techniques are considered in the next section allowing us to verify the validity of the approximate models.

Consider again the time axis on which we represent packet arrivals and packet transmissions. Time line i, relative to group i, is obtained by deleting from the time axis all packet transmissions belonging to groups other than i; that is, time line i exhibits packets from group i only. As discussed above, we can observe on time line i an alternate sequence of idle and busy periods (see Fig. 2). The simplicity in studying this protocol is mainly due to the fact that any busy period consists of a single transmission period [1]. In nonpersistent CSMA, when a terminal becomes ready it senses the channel. An arrival corresponds then to a *sense point*. A sense point will result in an actual transmission if the channel is sensed idle, otherwise the sense point (or arrival) is said to be *blocked*. Furthermore, by definition a sense point is said to be *j-unblocked* if the sense point is not blocked by packet transmissions from group j.

Consider time line i for example and let G_i, as before, denote the *total* rate of sense points generated by group i. The point process defined by these sense points is assumed to be of independent increments and Poisson (see Section II-A). Let G_i' be the rate of sense points which are j-unblocked for all j; $j \neq i$; i.e., G_i' is the rate of sense points which did not find the channel busy because of the transmission of a packet from group j; $j \neq i$. Obviously, j must then belong to the subset $h(i)$. The independence assumption that we make at this point can be stated as follows.

Assumption 2': The point process defined by the unblocked sense points relative to group i is independent of

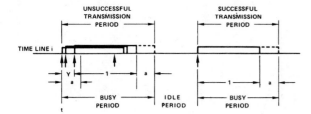

Fig. 2. Nonpersistent CSMA: busy and idle periods on time line i.

the state (busy or idle) of any time line j; $j \neq i$, is of independent increments, and is Poisson. That is, the point process is completely determined by the rate G_i'.

We recognize that this statement would be true only if all groups were independent. Nevertheless it is valid when the system is lightly loaded (at any rate, simulation is used to check the validity of results obtained under these assumptions); if $P_b{}^i$ is the probability that a sense point on time line i is blocked by packet transmissions from and group j; $j \neq i$, then we can write

$$G_i' = G_i(1 - P_b{}^i).$$

The introduction of Assumption 2' simplifies the problem yielding *approximate* relationships between the various quantities defined so far.

Under the model assumptions and the additional Assumption 1', the relationship between the components of S and the components of G is given by the following system of equations:

$$S_i = G_i \frac{\prod_{j \notin h(i)} \exp\{-aG_j'\} \prod_{k \notin h(i)} \exp\{-G_k'(1 - a)\}}{\prod_{l=1}^{N} [G_l'(1 + 2a) + \exp\{-aG_l'\}]} \quad (15)$$

where

$$G_i' = G_i \prod_{j \in h(i); j \neq i} \frac{1 + aG_j'}{G_j'(1 + 2a) + \exp\{-aG_j'\}}. \quad (16)$$

Proof: Consider time line i on which we observe an alternate sequence of busy and idle periods (Fig. 2). By Assumption 2', this sequence is completely determined by the rate G_i'. The average busy and idle periods can then be expressed as [1]

$$\bar{B}_i = 1 + \bar{Y}_i' + a$$

$$\bar{I}_i = \frac{1}{G_i'}$$

where

$$\bar{Y}_i' = a - \frac{1}{G_i'}(1 - \exp\{-aG_i'\}).$$

By the Poisson assumption, an arbitrary sense point represents a random look in time. The probability that an arbitrary sense point from group i is j-unblocked, $j \neq i$, is the probability that a random look at time line j falls either in an idle period or during the first a seconds of

a busy period of time line j and is expressed as

$$\frac{\bar{I}_j + a}{\bar{B}_j + \bar{I}_j} = \frac{1 + aG_j'}{G_j'(1 + 2a) + \exp\{-aG_j'\}}.$$

By the independence assumption we then establish (16).

$$G_i' = G_i \prod_{j \in h(i); j \neq i} \frac{1 + aG_j'}{G_j'(1 + 2a) + \exp\{-aG_j'\}}.$$

Consider now an arbitrary (unblocked) sense point from group i. For this to result in a totally successful transmission, the following conditions must be simultaneously satisfied:

α: The sense-point corresponds to the start of an i-successful transmission.

$\mathcal{B}$: The tagged sense point (which is j-unblocked, $\forall j \in h(i)$; $j \neq i$) occurs during an idle period of time line j, $\forall j \in h(i)$; $j \neq i$.

$\mathcal{C}$: There are no arrivals from any group $j \in h(i)$, $j \neq i$ during the first a seconds of the tagged transmission.

$\mathcal{D}$: The tagged packet is k-successful, for all $k \notin h(i)$.

It is easy to see that, as in (9), we have:

$$\Pr\{\alpha\} = \frac{\exp\{-aG_i'\}}{G_i'(1 + 2a) + \exp\{-aG_i'\}}. \quad (17)$$

On the other hand, knowing that the tagged packet is j-unblocked, $j \in h(i)$; $j \neq i$, we have

$$\Pr\{\mathcal{B}/\text{packet is } j\text{-unblocked}, j \in h(i), j \neq i\}$$

$$= \prod_{j \in h(i); j \neq i} \frac{1/G_j'}{1/G_j' + a}$$

$$= \prod_{j \in h(i); j \neq i} \frac{1}{1 + aG_j'} \quad (18)$$

and

$$\Pr\{\mathcal{C}\} = \prod_{j \in h(i); j \neq i} \exp\{-aG_j'\}. \quad (19)$$

Moreover, similarly to (10), we have

$$\Pr\{\mathcal{D}\} = \prod_{k \notin h(i)} \frac{\exp\{-G_k'(1 - a)\}}{G_k'(1 + 2a) + \exp\{-G_k'\}}. \quad (20)$$

By the independence assumption, we have

$$S_i = G_i' \Pr\{\alpha\} \cdot \Pr\{\mathcal{B}\} \cdot \Pr\{\mathcal{C}\} \cdot \Pr\{\mathcal{D}\}.$$

Using the expressions found in (17), (18), and (20) and the expression for G_i' given by (16), we get (15). Q.E.D.

Before we proceed with the examples, let us consider again the case of N independent and identical groups to which we apply (15) and (16). In this case, $h(i) = \{i\}$; $\forall i = 1, 2, \cdots, N$. Denoting as before $G_i = g$, $S_i = s$; $\forall i$, and $Ns = S$, $Ng = G$, (16) reduces to $G_i' = G_i = g$; $\forall i$, and (15) reduces to

$$\frac{s}{g} = \frac{S}{G} = \exp\{g(1 - 2a)\} \left[\frac{\exp\{-g(1 - a)\}}{g(1 + 2a) + \exp\{-ag\}}\right]^N$$

which is identical to (14).

C. Examples

In the present section, we consider some examples to which we apply the analytical results found in Section II-B. Simulation techniques are also used 1) whenever the analysis is intractable, and 2) to check the validity of the assumptions on which the *analysis* was based. The simulation model is based on the same system assumptions as in Section II-A. Among these, in particular, we assume that the input processes for the various groups are Poisson. However the assumptions pertaining to the characterization of the offered traffic (see Assumption 1′) and the independence assumptions introduced for analytic tractability are all relaxed. For the various examples simulated, the comparison of results obtained from simulation and the results obtained from the analytic model match very well. In the present section, we also draw various conclusions about the effect hidden terminals have on the performance of CSMA. For the following examples and numerical results, we restrict ourselves to $a = 0.01$.

Independent Groups Case: A Symmetric Configuration: We have already considered the example in which the population is partitioned into N groups of equal size. The (S,G) relationship is given by (13) for 1-persistent CSMA and by (14) for the nonpersistent protocol.

In this example, for each terminal there exists a fraction β of the population which is hidden, namely $\beta = [(N-1)/N]$ (>0.5). The channel capacity for various values of N is plotted in Fig. 3. Note that the channel capacity experiences a drastic decrease between the two cases: $N = 1$ (no hidden terminals, $\beta = 0$) and $N = 2$ ($\beta = 0.5$). For $N \geq 2$, slotted ALOHA performs better than CSMA.[3] This decrease is more critical for the nonpersistent CSMA than for the 1-persistent CSMA as shown in the figure. For $N > 2$, the channel capacity is rather insensitive to N and approaches pure ALOHA for large N, as was shown in Section II-B.

Independent Groups Case: Complementary Couple Configuration: The previous example did not show the effect of a small fraction of the population being hidden from the rest. In this example the population consists of two independent groups ($N = 2$) of unequal sizes such that $\mathfrak{u} = (\alpha, 1 - \alpha)$; that is,

$$S_1 = \alpha S$$

$$S_2 = (1 - \alpha)S.$$

Equations (1) and (2) are readily applicable. The channel capacity is plotted versus α for both CSMA protocols in Fig. 4. Here again we note that the capacity decreases *rapidly* as α increases from 0. This decrease is much more critical for the nonpersistent than for the 1-persistent.

In answering Question 2 we note that a good measure of delay is given by G_i/S_i, the average number of transmissions and scheduling of a packet until success. We note that the larger group (i.e., the group with the higher aggregate input rate) performs much better than the smaller one. (See [2] for further details.)

[3] Recall that a is assumed constant and equal to 0.01.

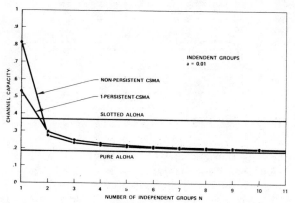

Fig. 3. Independent groups case—channel capacity versus the number of groups.

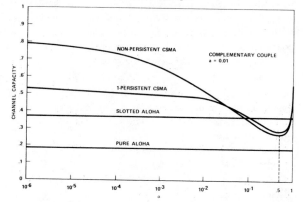

Fig. 4. Complementary couple configuration—channel capacity versus α.

Dependent Groups Case: A Symmetric Configuration: Let us now consider the situation in which the population is partitioned into N groups of equal size such that for each group all but one group are within sight. The graph representation of such a configuration is shown in Fig. 5. Obviously, this situation falls within the case of dependent groups, and corresponds to the instance in which, for each terminal, there is a fraction β of hidden terminals such that $\beta < 0.5$, namely $\beta = (1/N)$. Simulation techniques have been employed to study this instance under a 1-persistent CSMA protocol. In Fig. 6, we show the relationship between the total throughput S and the total traffic G for various values of β. We note again that the higher the value of β is, the smaller is the channel capacity. We further note that for a given achievable throughput S, the larger β is, the larger the traffic rate G is and hence the smaller is the probability of success of a packet, and the larger its average number of transmissions is.

Dependent Groups Case: The "Wall" Configuration: Consider a uniform distribution of terminals over a circular area, the station being located at the center. All terminals are within range of each other, but the presence of a "wall" (hill) as displayed in Fig. 7(a) causes some terminals to be hidden from others. A terminal T_0 at an angle α_0 ($\alpha_0 < 180°$) from the wall can only hear terminals in the region

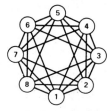

Fig. 5.   The hearing graph of a symmetric dependent groups configuration.

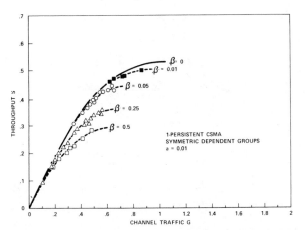

Fig. 6.   Symmetric dependent groups configuration—throughput versus channel traffic.

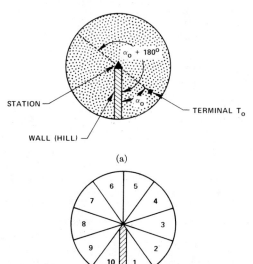

Fig. 7.   The wall configuration.

$(0, \alpha_0 + 180°)$. To study a continuous problem such as this, we examine the discrete approximation obtained by breaking up the uniform population into small sectors, each sector considered then as a group. The smaller the sectors are, the more precise the approximation becomes. The problem is solved numerically below for $N = 10$; that is, we partition the population into 10 equal sectors as shown in Fig. 7(b). In the latter, we note that, for each

group i, there exists a group $\bar{i}$ diametrically opposite such that a terminal in group i can only hear a fraction of the terminals in group $\bar{i}$. Therefore, the solution to the continuous problem can be bounded by the following two cases.

Case 1—$\bar{i} \notin h(i)$: This gives a lower bound on CSMA performance; that is a lower bound on channel capacity or upper bound on the channel traffic and on the average number of transmissions and schedulings.

Case 2—$\bar{i} \in h(i)$: This gives an upper (i.e., optimistic) bound on the performance of CSMA.

The hearing graph representations corresponding to the two cases are shown in Fig. 8. The analytic results and simulation results were compared for the case $\bar{i} \notin h(i)$ for values of $a = 0.01$ and 0.1 (see [2]). The close match again validates the analytical model. The results obtained by the latter are shown in Figs. 9 and 10. In Fig. 9 we plot the total throughput S versus the aggregate rate G. The channel capacity is shown to be bounded by $0.37 \leq C \leq 0.44$ ($a = 0.01$). The existence of the wall decreased the channel capacity by a factor approximately equal to 2. Similarly, in Fig. 10 we plot the upper and lower bounds on the overall average number of transmissions and schedulings per packet.

Remark: The set of equations relative to the dependent groups case given in Section II-2 provides us with an approximate solution based on the independence Assumption $2'$. Simulation results agree with those obtained by the model for the examples considered so far. However, there are cases where the independence assumption is not satisfied and the model inapplicable. Consider, for example, the symmetric configuration depicted in Fig. 5. Assume $a = 0$ for simplicity and let $g' = G_i'$, $g = G_i$, and $s = S_i$; $\forall i$. From (15) and (16) we have

$$g' = g \left(\frac{1}{1 + g'} \right)^{N-2}$$

$$s = g \frac{e^{-g'}}{(1 + g')^N}.$$

If we now let $N \to \infty$ $s \to 0$, $g \to 0$, $g' \to 0$ such that $Ns = S$, $Ng = G$, and $Ng' = G'$, then we get

$$S = G' = Ge^{-S}. \tag{21}$$

Such a result is certainly wrong in the limit since, for this particular case, we expect to reach the nonpersistent CSMA result with no hidden terminals, namely [1]

$$S = \frac{G}{1 + G}. \tag{22}$$

This is basically due to the independence assumption introduced above. However, (21) and (22) are equivalent in the case of low channel utilization, since then approximating e^{-S} by $1 - S$, (21) reduces to (22).

In summary, for the various particular configurations that were considered, it is to be noted that the performance of carrier sense is badly degraded by the existence

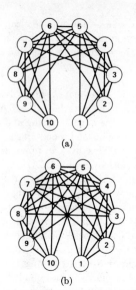

Fig. 8. The wall configuration—the hearing graphs for the ten sectors discrete approximation. (a) $\bar{\imath} \notin h(i)$. (b) $\bar{\imath} \in h(i)$.

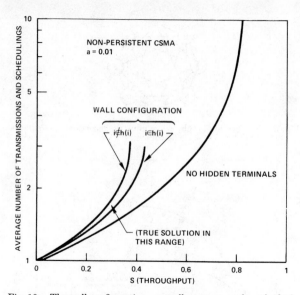

Fig. 10. The wall configuration—overall average number of schedulings and transmissions.

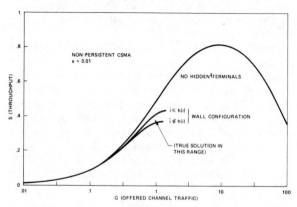

Fig. 9. The wall configuration—throughput versus offered channel traffic.

of hidden elements. Fortunately, in a single-station environment, the hidden-terminal problem can be eliminated by dividing the available bandwidth into two separate channels: a busy-tone channel and a message channel. This solution is the subject of the following section.

III. BUSY-TONE MULTIPLE-ACCESS (BTMA)

A. System Operation and Protocol.

The operation of BTMA rests on the assumption that the station is, by definition, within range and in line-of-of-sight of all terminals. The total available bandwidth is to be divided into two channels: a message channel and a busy-tone (BT) channel.[4] As long as the station senses (terminal) carrier on the incoming message channel it transmits a (sine wave) busy-tone signal on the busy-tone channel. It is by sensing carrier on the busy-tone channel

---

[4] The busy-tone concept in the context of packet radio was first suggested by Fralick [3].

that terminals determine the state of the message channel. The action pertaining to the transmission of the packet that a terminal takes (again) is prescribed by the particular protocol being used. In this section, we shall restrict ourselves to the nonpersistent protocol because of its simplicity in analysis and in implementation, as well as its relatively high efficiency as shown in Part I. In CSMA [1], the difficulty of detecting the presence of a signal on the message channel when this message used the entire bandwidth was minor and therefore was neglected. It is not so when we are concerned with the (statistical) detection of the (sine wave) busy-tone signal on a narrow-band channel. The detection time, denoted by t_d, is no longer negligible and must be accounted for. The nonpersistent BTMA protocol is similar to the nonpersistent CSMA protocol [1] and corresponds to the following. Whenever a terminal has a packet ready for transmission, it senses the busy-tone channel for t_d seconds (the detection time) at the end of which it decides whether the busy-tone signal is absent (in which case it transmits the packet); otherwise it reschedules the packet for transmission at some later time incurring a random rescheduling delay. At this new point in time, it senses the busy-tone channel and repeats the algorithm. In the event of a conflict, which the terminal learns about by failing to receive an acknowledgment from the station, the terminal again reschedules the transmission of the packet for some later time, and repeats the above process.

Of interest to this paper is first, the determination of the channel capacity under a nonpersistent BTMA protocol and second, the throughput delay characteristics of the latter. The total available bandwidth being the limiting resource, the problem then reduces to selecting the system parameters in order to achieve the best system performance. For this analysis we make the same assumptions as in Part I. While the effect of noise is assumed to be negligible on the message channel, we do account for it in the (narrow-band) busy-tone channel (See Section III-B). τ is the one-way propagation delay to (and from) the station. Each packet is of constant length requiring T_m seconds for transmission on the message channel. We let $S_m = \lambda T_m$. Let ψ be the fraction of the bandwidth assigned to the busy-tone channel. The overall channel utilization S is

$$S = (1 - \psi) S_m.$$

Let γ denote the mean offered traffic rate. (This is the rate of sense points since each arrival corresponds in this protocol to sensing the busy-tone channel before taking an action.) In Section III-B the problem of detecting a sine wave signal on a narrow-band channel is examined and the effect of various system parameters is characterized.

B. Signal Detection

The detection of the busy-tone signal is the problem of detecting a signal of known form in the presence of noise. The useful signal is a given function with some unknown parameters, namely, phase and amplitude.[5] However the observation (detection) time is usually small compared to the "fluctuation time" of these parameters, and the unknown phase and amplitude can be regarded as constant.

The problem of detecting a signal in a background of white Gaussian noise is a classical statistical problem involving the choice of one hypothesis from two mutually exclusive hypotheses. This has been extensively studied in the literature [4]. The quality of the decision can be characterized by the following two probabilities:

D Probability of correct detection (in presence of the signal)

F Probability of incorrect detection or false alarm.

Again let t_d be the observation time, i.e., the width of the window over which the channel is observed in making the decision. If μ_{t_d} is the signal-to-noise ratio (SNR)[6] when the signal is present over the entire window t_d, and r_* is the optimum threshold of the statistical detector, then for the usually assumed Rayleigh channel, it can be shown that [4]

$$F = \exp\{-r_*^2 / 2\mu_{t_d}\}$$
$$r_* = (-2\mu_{t_d} \log F)^{1/2}. \tag{23}$$

That is, for a given observation time t_d and a desired false alarm probability F, the optimum threshold r_* can be determined from (23).

In this case (i.e., when the *useful* signal is present over the entire window), the probability of correct detection is given by [4]

---

[5] Because of the mobility of terminals, the signal fluctuates. Thus we assume it to be of unknown amplitude. In the case of fixed terminals, we may idealize the problem to be that of detecting a signal with known amplitude but unknown phase.
[6] Ratio of the signal energy to the noise energy in the time-frequency window.

Fig. 11. Block diagram of the busy tone signal detector.

$$D_{t_d} = \exp\left(-\frac{r_*^2}{2\mu_{t_d}(1 + \mu_{t_d})}\right)$$
$$= F^{1/(1+\mu_{t_d})}. \tag{24}$$

Equation (24) rests on the fact that the SNR of the received useful signal during the observation time is actually μ_{t_d}. In the following paragraph we investigate the (transient) situations where the signal may be present during only a fraction of the observation window.

Transient Behavior: The detector at the receiver consists mainly of a filter, an integrator, and a threshold decision box (see Fig. 11). Assume the step response of the busy-tone detect filter is exponential [3]; the amplitude at time t for the output of the busy-tone filter is then[7]

$$A_{BT}(t) = A_{\max}(1 - e^{t/k}) \tag{25}$$

where $A_{\max}$ is the maximum amplitude and k is the filter time constant. If we assume [3] that the same peak power is used for the busy tone as for the message on the message channel, then

$$A_{\max} = A_m$$

where A_m is the amplitude of the message signal on the message channel. Since the energy of a signal is the integral of its squared amplitude (which equals $(A^2 t/2)$ for a sinusoid of amplitude A and duration t), we define the SNR as the ratio of the signal energy to the noise-energy and express it as

$$\mu = \frac{A^2}{2N_0 W}$$

where W is the bandwidth of the channel under consideration and N_0 is the (assumed white) noise power density.

Let μ_m be the SNR of the message on the message channel required for suitable operation (typically $\mu_m = 10$). Then from the last equation, it is clear that

$$A_{\max} = A_m = (2N_0 W_m \mu_m)^{1/2}$$

and

$$A_{BT}(t) = [2N_0 W_{BT}\mu(t)]^{1/2} = (2N_0 W_m \mu_m)^{1/2}(1 - e^{-t/k}).$$

Taking the time constant k to be one half of the inverse of the busy-tone channel bandwidth and recalling that ψ is the fraction of the total bandwidth W assigned to the busy-tone channel, we have the following function defining the SNR $\mu(t)$ on the busy-tone channel:

$$\mu(t) = \mu_m \frac{1 - \psi}{\psi}(1 - \exp\{-2\psi Wt\})^2. \tag{26}$$

---

[7] At time $t = 0$, the filter is just turned on, and the signal is assumed to be present.

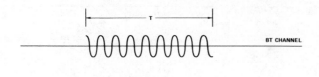

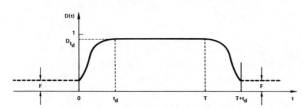

Fig. 12. $D(t)$ for an isolated busy tone signal of length T.

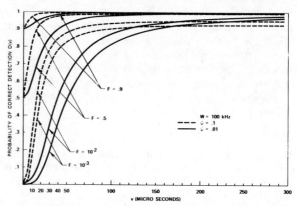

Fig. 13. Probability of correct detection $D(v)$.

Particularly

$$\mu_{t_d} = \mu(t_d) = \mu_m \frac{1-\psi}{\psi}(1 - \exp\{-2\psi W t_d\})^2. \quad (27)$$

As the starting time of a busy tone is unpredictable, the busy tone may not be present during the entire window of t_d seconds. The resulting SNR is a function of the time u over which the busy-tone signal is present [see (26)]. By the same token, the probability of proper detection is also a function of the time over which the busy-tone signal is present.

Consider now a signal starting at $t = 0$ and terminating at $t = T$. Let $D(t)$ be the probability of correct detection at time t after having observed the channel over t_d seconds (t is the time at which the decision is made). $D(t)$ is determined by [4]

$$D(t) = F^{\{1/[1+\mu(u)]\}} \quad (28)$$

where

$$u = \begin{cases} t, & \text{if } 0 \le t \le t_d \\ t_d, & \text{if } t_d \le t \le T \\ T - t + t_d, & \text{if } T < t \le T + t_d. \end{cases}$$

For $t > T + t_d$, the probability of false alarm is F. $D(t)$ is sketched in Fig. 12. A detailed graph of $D(t)$ for $0 < t < t_d$ can be seen in Fig. 13 in which we plot, for various values of F and ψ, the function

$$D(v) = F^{\{1/[1+\mu(v)]\}}, \quad v \in [0, \infty]$$

where $\mu(v)$ is given in (26). ($D(v)$ here is the probability of correct detection if the useful signal is present over v seconds.) For very large v, namely when $v \to \infty$, the probability of correct detection reaches an asymptotic value equal to

$$D(v) \xrightarrow[v \to \infty]{} F^{(1+[(1-\psi)/\psi]\mu_m)^{-1}}.$$

As usual, the larger F is, the better is the probability of correct detection. It is interesting to note that when v in-

creases from 0, $D(v)$ increases quite rapidly reaching its asymptotic value for relatively small v. Moreover, this increase is faster for larger busy-tone bandwidth (larger ψ). A more complex situation occurs when two busy-tone signals are separated by a gap shorter than t_d. The window now can overlap over two busy-tone signals. Let $t = 0$ be the time at which the first signal terminates and t_g be the gap between the two signals. $D(t)$ for various values of t_g is sketched in Fig. 14. In the following section further approximations are introduced to avoid dealing with these complex situations. The approximations are checked to have a negligible effect on the evaluation of the system performance.

C. Throughput Analysis

Let us first summarize the notation in use:

τ One-way propagation delay.

b_m Number of bits per packet.

W Total bandwidth available.

ψ Fraction of W assigned to the busy tone channel.

T_m Transmission time of a packet on the message channel;

$$T_m = \frac{b_m}{(1-\psi)W}.$$

γ Rate of the offered channel traffic (in packets/s).

G Normalized rate of offered channel traffic (i.e., $G = \gamma T_m$).

t_d Detection time.

F The false alarm probability.

$D(t)$ Probability of correct detection given the signal is present for t seconds.

We wish to solve for the channel capacity, given the system parameters $F, \psi, W, b_m, \tau, t_d$. This we do by solving for S in terms of γ and other system parameters. The channel capacity is then found by maximizing S in respect to γ.

Contrary to the CSMA modes the fraction of the population which decides to transmit is a function of time. The analytical approach consists (as in Part I [1]) of identifying the busy and idle periods and of determining the

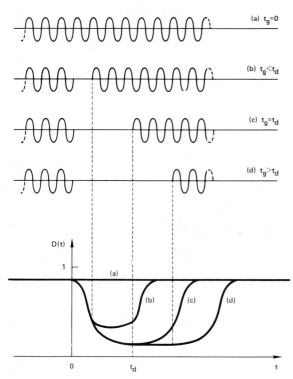

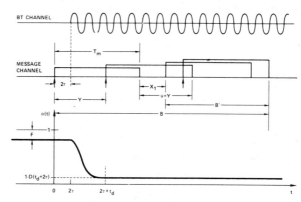

Fig. 15. BTMA—$\alpha(t)$ at the start of a busy period.

Fig. 14. $D(t)$ for various values of the gap t_g between two consecutive busy tone signals.

condition for a successful transmission over the busy period.

As stated above, to keep the analysis simple, some approximations will be made yielding a lower bound on throughput. Corresponding numerical results are presented and discussed in the following section.

Under stationary conditions and the model assumptions, a lower bound on the channel utilization S_l is given by

$$S \geq S_l = \frac{b_m}{W} \frac{\exp\left(-\gamma m(0, T_m)\right)}{\bar{B} + \bar{I}} \tag{29}$$

where the quantities $\bar{B}, \bar{I}, m(0, T_m)$ are defined in the following proof in (42), (46), and (31), respectively.

Proof: Since a terminal senses the busy tone channel for t_d seconds before deciding to transmit, an arrival is not effective, as far as the channel operation is concerned, until t_d seconds following its arrival time. To view the occurrence of events in this access mode more easily, we consider the Poisson arrival process to be shifted t_d seconds in time, bearing in mind that the terminal has already observed the channel for t_d seconds. The busy-tone signal emitted by the station is, as seen by the terminals, shifted in time by 2τ, a round-trip delay to the station (time until the station hears the transmission) and back (time until the busy tone from the station is heard by the other terminals).

If we let γ (packets/s) denote the offered traffic, the rate of transmission of packets at any time t is $\alpha(t)\gamma$ such that

$$\alpha(t) = \begin{cases} 1 - D_{t_d}, & \text{if the busy-tone signal is present during the entire observation window} \\ 1 - F, & \text{if the busy-tone signal is absent during the entire window} \\ 1 - D(v), & \text{for the case when the busy-tone signal is only present over } v \text{ s,} \\ & v \leq t_d \end{cases}$$

where $D(t)$ is the probability of correct detection at time t and F is the false alarm probability. That is, at any time t there is a fraction $\alpha(t)$ of the population which decides to transmit.

Contrary to CSMA, an arrival in the middle of an ongoing transmission has a nonzero probability $(1 - D(t))$ of transmitting; thus, a busy period (period of time over which the channel is continuously used) can now exceed $T_m + 2\tau$.

Let t_{P1} be the time at which the transmission of the first packet of a busy period starts. The busy-tone signal of such a busy period starts 2τ seconds later, at time $t_{P1} + 2\tau$. The time dependent *packet transmission rate* $\alpha(t)\gamma$ depends on the length of the gap between the start of the busy-tone signal corresponding to the current busy period and the end of the busy-tone signal corresponding to the previous busy period. Let us assume $t_g > t_d + 2\tau$ for the present time. (This corresponds to the worst yet "cleanest" case.) Without loss of generality, let $t_{P1} = 0$. From the previous section, we can easily derive $\alpha(t)$ as follows (see Fig. 15):

$$\alpha(t) = \begin{cases} 1 - F & 0 < t \leq 2\tau \\ 1 - D(t), & 2\tau < t \leq 2\tau + t_d \\ 1 - D(t_d + 2\tau), & 2\tau + t_d < t \leq T_m \end{cases} \tag{30}$$

where

$$D(t) = F^{\{1/[1+\mu(t)]\}}$$

$$\mu(t) = \mu_m \frac{1-\psi}{\psi} \left[1 - \exp\left\{-2\psi W(t-2\tau)\right\}\right]^2.$$

Let

$$m(x_1, x_2) = \int_{x_1}^{x_2} \alpha(t) \, dt. \tag{31}$$

The first packet in the busy period is successful if there is no arrival from the nonhomogeneous Poisson arrival process with rate $\alpha(t)\gamma$ during the entire transmission time of the packet. The probability of success is given by

$$P_s = \exp\left(-\gamma m(0, T_m)\right). \tag{32}$$

To find the channel throughput, we use a "cycle" analysis [5]. For this, we must calculate the average busy and idle periods.

Let us first find the average busy period. Let Y be the random variable defining the time until the last arrival during the packet transmission time T_m (see Fig. 15). It is easy to see that:

$$\Pr\{Y \le y\} = \Pr\{\text{no arrival in } (y, T_m)\}$$

$$= \exp\left[-\gamma m(y, T_m)\right]$$

$$\Pr\{Y = 0\} = \exp\left[-\gamma m(0, T_m)\right]. \tag{33}$$

The Laplace transform $Y^*(s)$ is defined by

$$Y^*(s) \triangleq \int_0^{T_m} \exp\{-sy\} \, d \Pr\{Y \le y\}.$$

From this we obtain

$$Y^*(s) = \Pr\{Y = 0\}$$

$$+ \int_0^{T_m} \gamma \exp\{-sy\} \exp\{-\gamma m(y, T_m)\} \, dm(y, T_m)$$

$$= \exp\{-\gamma m(0, T_m)\}$$

$$+ \int_0^{T_m} \gamma \alpha(y) \exp\{-sy\} \exp\{-\gamma m(y, T_m)\} \, dy. \tag{34}$$

Let

$$D \equiv D(t_d + 2\tau) = D_{t_d} \tag{35}$$

$$\Delta = 1 - D \tag{36}$$

$$\Phi = 1 - F. \tag{37}$$

The busy period is equal to:

$$\begin{cases} T_m + Y, & \text{if no arrival occurs in the period} \\ & \text{of time } (T_m, T_m + Y) \\ T_m + X_1 + B', & \text{otherwise} \end{cases}$$

where (see Fig. 15) 1) X_1 is the random variable defining the time elapsed from $t = T_m$ up to the first arrival in

$(T_m, T_m + Y)$, and 2) B' is the length of a busy period when the Poisson arrival process is homogeneous with rate $\Delta\gamma$. (B' is the "sub-busy" period (see [5]) created by the first arrival in $(T_m, T_m + Y)$. Beyond $t = T_m$, $\alpha(t) = \Delta$.)

Conditioning on $Y = y$, the average busy period is given by

$$\bar{B}_y = (T_m + y)e^{-\Delta\gamma y} + (T_m + \bar{X}_{1y} + \bar{B}')(1 - e^{-\Delta\gamma y})$$

$$= T_m + ye^{-\Delta\gamma y} + (\bar{X}_{1y} + \bar{B}')(1 - e^{-\Delta\gamma y}).$$

The expected value $\bar{X}_{1y}$ (defined as the expected value of X_1 conditioned on $Y = y$) is derived as follows. First, we have

$$\Pr\{X_1 \le x / \text{some arrival occurred}, Y = y\} = \frac{1 - e^{-\Delta\gamma x}}{1 - e^{-\Delta\gamma y}}.$$

Therefore,

$$\bar{X}_{1y} = \frac{1}{1 - e^{-\Delta\gamma y}} \int_0^y (e^{-\Delta\gamma x} - e^{-\Delta\gamma y}) \, dx$$

$$= \frac{1}{1 - e^{-\Delta\gamma y}} \left[\frac{1}{\gamma\Delta}(1 - e^{-\Delta\gamma y}) - ye^{-\Delta\gamma y}\right]$$

$$= \frac{1}{\Delta\gamma} - \frac{ye^{-\Delta\gamma y}}{1 - e^{-\Delta\gamma y}}. \tag{38}$$

The expected busy period $\bar{B}'$ is given by

$$\bar{B}' = \frac{1}{\Delta\gamma}\left(e^{\Delta\gamma T_m} - 1\right). \tag{39}$$

Indeed, we have

$$B' = \begin{cases} T_m, & \text{with probability } e^{-\Delta\gamma T_m} \\ X_2 + B', & \text{with probability } 1 - e^{-\Delta\gamma T_m} \end{cases}$$

where X_2 is the time elapsed from the start of transmission of a packet until the start of transmission of the first overlapping packet. $\bar{X}_2$ is obtained from $\bar{X}_{1y}$ (37) by simply setting $y = T_m$.

$$\bar{X}_2 = \frac{1}{\Delta\gamma} - \frac{T_m e^{-\Delta\gamma T_m}}{1 - e^{-\Delta\gamma T_m}}. \tag{40}$$

Therefore,

$$\bar{B}' = T_m e^{-\Delta\gamma T_m} + (\bar{X}_2 + \bar{B}')(1 - e^{-\Delta\gamma T_m})$$

$$= T_m e^{-\Delta\gamma T_m} + \frac{1 - e^{-\Delta\gamma T_m}}{\Delta\gamma}$$

$$- T_m e^{-\Delta\gamma T_m} + \bar{B}'(1 - e^{-\Delta\gamma T_m})$$

$$= \frac{1}{\Delta\gamma}(e^{\Delta\gamma T_m} - 1).$$

The average busy period $\bar{B}_y$, conditioned on $Y = y$, is given by

$$\bar{B}_y = T_m + ye^{-\Delta\gamma\nu} + \frac{1 - e^{-\Delta\gamma\nu}}{\Delta\gamma} - ye^{-\Delta\gamma\nu}$$

$$+ \frac{1}{\Delta\gamma}(e^{\Delta\gamma T_m} - 1)(1 - e^{-\Delta\gamma T_m})$$

$$= T_m + \frac{1}{\Delta\gamma}e^{\Delta\gamma T_m}(1 - e^{-\Delta\gamma\nu}). \tag{41}$$

Removing the condition on Y, we get

$$\bar{B} = \int_0^{T_m} \bar{B}_y \, d\Pr\{Y \leq y\}$$

$$= T_m + \frac{1}{\Delta\gamma}e^{\Delta\gamma T_m}(1 - Y^*(\Delta\gamma)). \tag{42}$$

Let us now calculate the average idle period. Let $t = 0$ now denote the *end* of the busy period. It is easy to see that the transmission rate, which we denote by $\alpha'(t)\gamma$, is defined by (see Fig. 16)

$$\alpha'(t) = \begin{cases} 1 - D(t_d + 2\tau) = \Delta, & 0 \leq t \leq 2\tau \\ 1 - D'(t), & 2\tau \leq t \leq 2\tau + t_d \\ 1 - F = \Phi, & t \geq 2\tau + t_d \end{cases} \tag{43}$$

where

$$D'(t) = F^{\{1/[1+\mu'(t)]\}}.$$

Denoting by I the idle period, we have

$$\Pr\{I > z\} = \Pr\{\text{no arrival occurs in }(0,z)\}$$

$$= \exp\left[-\gamma m'(0,z)\right] \tag{44}$$

where

$$m'(z_1,z_2) = \int_{z_1}^{z_2} \alpha'(t) \, dt. \tag{45}$$

The average idle period is given by

$$\bar{I} = \int_0^\infty \exp\{-\gamma m'(0,z)\} \, dz. \tag{46}$$

The exact expression[8] for the channel throughput depends on the distribution of the gap t_g. Indeed, $D(t)$ as well as $\alpha(t)\gamma = [1 - D(t)]\gamma$ depends on t_g as shown in Fig. 14. However, to keep the analysis fairly tractable, we choose to give a lower bound on channel throughput by considering $\alpha(t)\gamma$ as determined in the worst case $(t_g \gg t_d)$, that is as given in (30). Under this condition, the first transmission in the busy period is successful with probability

$$P_s = \exp\left(-\gamma m(0,T_m)\right)$$

and the expected period of time $\bar{U}$ that the channel is

---

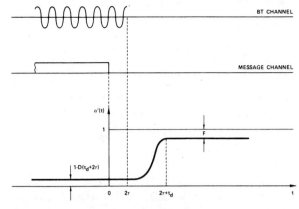

Fig. 16. BTMA—$\alpha'(t)$ at the start of an idle period.

transmitting packets without interference during a cycle (defined as a busy period plus an idle period) is

$$\bar{U} = T_m \exp\left[-\gamma m(0,T_m)\right].$$

Therefore, the lower bound on throughput is given by

$$S_l = (1 - \psi)\frac{T_m \exp\left[-\gamma m(0,T_m)\right]}{\bar{B} + \bar{I}}$$

$$= \frac{b_m}{W}\frac{\exp\left[-\gamma m(0,T_m)\right]}{\bar{B} + \bar{I}}. \qquad \text{Q.E.D.}$$

"Upper bound" on S: To check the validity of the lower bound provided by (29), we consider the fraction f of busy periods for which the gap t_g is less than $t_d + 2\tau$. Indeed, it is for these cases that $\alpha(t)$ is overestimated by the expression given in (30), and could be underestimated by $1 - D_{t_d} = \Delta$.

The probability that the gap t_g is less than $t_d + 2\tau$ is readily given by

$$f = \Pr\{t_g \leq t_d + 2\tau\} = 1 - \exp\{-\gamma m'(0,t_d + 2\tau)\}. \tag{47}$$

The smaller f is, the closer is the lower bound on S to the exact expression. On the other hand, an "upper bound"[9] on S is obtained by underestimating $\alpha(t)$ by Δ for a fraction f of the busy periods. That is, a lower bound on the expected busy period is obtained by

$$\bar{B}_l = T_m + \frac{1}{\Delta\gamma}e^{\Delta\gamma T_m}[1 - fY_1^*(\Delta\gamma) - (1 - f)Y^*(\Delta\gamma)] \tag{48}$$

where $Y_1^*(s)$ is the Laplace transform of Y when $\alpha(t) = \Delta\gamma$, and is given by

$$Y_1^*(s) = \gamma\Delta\frac{e^{-sT_m} - e^{-\gamma\Delta T_m}}{\gamma\Delta - s} + e^{-\gamma\Delta T_m} \tag{49}$$

---

[8] Exact under the provision that a is constant.

[9] With respect to the approximation concerning the gap t_g.

so that

$$Y_1^*(\Delta\gamma) = e^{-\gamma\Delta T_m}[1 + \gamma\Delta].$$

An upper bound on the probability of success of the first transmission in a busy period is obtained by

$$P_{s_u} = f \exp\{-\gamma\Delta T_m\} + (1-f)\exp\{-\gamma m(0, T_m)\}.$$

The upper bound on throughput is then given by

$$S_u = \frac{B_m}{W} \frac{f\exp\{-\gamma\Delta T_m\} + (1-f)\exp\{-\gamma m(0, T_m)\}}{\bar{B}_l + \bar{I}}. \tag{50}$$

Limit when $t_d \to 0$: When $t_d \to 0$, the channel capacity reduces to

$$S = (1-\psi)\frac{1}{2e}. \tag{51}$$

Proof: We realize that in the limit $(t_d \to 0)$, the problems caused by the transient behavior are insignificant (nonexistent when $t_d = 0$). We see then from (24) that

$$D_{t_d} = F^{\{1/[1+\mu_t d]\}} \to F$$

since $\mu_{t_d} \to 0$. Therefore, $\alpha(t)$ is constant and equal to $\Phi = \Delta = 1 - F$; (34) becomes:

$$Y^*(s) = \exp\{-\Phi\gamma T_m\}$$
$$+ \Phi\gamma \exp\{-\Phi\gamma T_m\}\left(\frac{\exp\{(-s+\Phi\gamma)T_m\} - 1}{\Phi\gamma - s}\right)$$

yielding

$$Y^*(\Delta\gamma) = \exp\{-\Phi\gamma T_m\}(1 + \Phi\gamma T_m).$$

Therefore, $\bar{B}$ as given by (42) becomes

$$\bar{B} = \frac{1}{\Phi\gamma}\exp\{\Phi\gamma T_m\} - \frac{1}{\Phi\gamma}.$$

Similarly, we can see that

$$\bar{I} = \frac{1}{\Phi\gamma}.$$

Substituting in (29) we get

$$S = (1-\psi)\Phi\gamma T_m \exp\{-2\Phi\gamma T_m\}$$

yielding a channel capacity equal to $(1-\psi)(1/2e)$ under the optimum condition $\Phi\gamma T_m = 1$.　Q.E.D.

D. Numerical Results and Discussion

The expressions given in (29) and (50) relate the throughput S to the offered channel traffic γ. When all the system parameters have fixed values, the *information capacity* of the channel is defined as the maximum achievable throughput. This throughput is obtained at an optimum value of the traffic γ and results in infinite packet delays. To obtain finite delays, we must reduce the throughput below the capacity.

The design problem in BTMA consists of maximizing the channel capacity (under the nonpersistent protocol) by properly selecting the design variables ψ, F, and t_d when the number of bits per packet, b_m, and the total available bandwidth W are assumed to be given. Because of the complicated form of the expressions for S, numerical optimization techniques are used. Below, we first discuss the restrictions on the input data and the accuracy of the approximations. Then we give the numerical results in the form of curves. The subsequent curves depict the changes in system performance due to variations in the design parameters. The various tradeoffs which influence the performance of BTMA are also discussed.

Let us first discuss some restrictions and approximations. To reduce the dimensionality of the problem, and to provide an easy comparison with the previously analyzed CSMA protocols we restrict ourselves to the following:

τ (maximum propagation delay) $= 100 \ \mu s$[10]

$\mu_m = 10$

$\dfrac{b_m}{W} = 10^{-2}$ seconds.[11]

We consider two cases for b_m and W.

Case I: $b_m = 1000$ bits; $W = 10^5$ Hz.
Case II: $b_m = 10\,000$ bits; $W = 10^6$ Hz.

It is important to note that, for the same μ_m, Case II requires higher transmitting power than does Case I; the following curves also show that Case II offers a channel capacity higher than that offered by Case I.

Along with the numerical computation of S_l and S_u we computed the probability f that the idle period is smaller than $t_d + 2\tau$. The probability f never exceeded a few percent (less than 0.04) and the two estimates on throughput are *very close* to each other. (As an example, in Table I we give the values of f encountered for various values of the system parameters.) Therefore all numerical results given in the sequel will only correspond to the lower bound S_l.

For $F = 10^{-3}$ and various values of ψ we plot in Fig. 17 the channel capacity versus the observation window t_d. Similar curves can be plotted for other values of F. For each couple (F, ψ), the channel capacity reaches its maximum at some optimum value of t_d. This optimum is explained by the fact that the larger t_d is, the better is the probability of correct detection D_{t_d} when the signal is present during the entire window. However, the larger t_d is, the longer the idle period will be, as it can be seen from Fig. 16. The effect is reversed as t_d gets smaller.

Note that when the observation window shrinks to 0, the capacity of the channel decreases to $(1-\psi)(1/2e)$

---

[10] This corresponds to a maximum distance of about 20 miles. The ratio of propagation delay to transmission time of a packet, denoted by a, is in all cases less than (but very close to) or equal to 0.01.
[11] The bandwidth is assumed to be modulated at 1 bit/Hz·s.

TABLE I
ACCURACY OF THE APPROXIMATIONS

	γ	S_l	S_u	f
$F = 10^{-3}$ $\psi = 10^{-2}$ $t_d = 7 \times 10^{-4}$	10	0.0897	0.0897	0.0008
	100	0.4580	0.4581	0.0084
	400	0.6245	0.6269	0.0333
	500	0.6201	0.6238	0.0414
	600	0.6084	0.6137	0.0495
$F = 10^{-2}$ $\psi = 10^{-2}$ $t_d = 7 \times 10^{-4}$	10	0.0890	0.0890	0.0006
	100	0.4586	0.4587	0.0067
	500	0.6411	0.6440	0.0334
	600	0.6338	0.6380	0.0400
	700	0.6222	0.6279	0.0465
$F = 10^{-1}$ $\psi = 10^{-2}$ $t_d = 5 \times 10^{-4}$	10	0.0818	0.0818	0.0004
	100	0.4443	0.0444	0.0045
	500	0.6635	0.6653	0.0226
	600	0.6625	0.6651	0.0271
	700	0.6565	0.6599	0.0315
$F = 0.5$ $\psi = 10^{-2}$ $t_d = 5 \times 10^{-4}$	10	0.0473	0.0473	0.0002
	100	0.3206	0.3206	0.0021
	500	0.6318	0.6322	0.0106
	1000	0.6807	0.6825	0.0210
	2000	0.6406	0.6476	0.0417
$F = 0.7$ $\psi = 10^{-2}$ $t_d = 5 \times 10^{-4}$	100	0.2249	0.2249	0.0012
	500	0.5511	0.5512	0.0062
	1000	0.6540	0.6546	0.0125
	2000	0.6830	0.6855	0.0248
	3000	0.6583	0.6639	0.0370
$F = 0.9$ $\psi = 10^{-2}$ $t_d = 5 \times 10^{-4}$	1000	0.4692	0.4692	0.0041
	2000	0.6018	0.6020	0.0083
	3000	0.6552	0.6558	0.0124
	4000	0.6779	0.6790	0.0165
	5000	0.6858	0.6876	0.0206

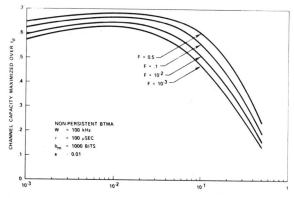

Fig. 18. BTMA—channel capacity (maximized over t_d) versus ψ.

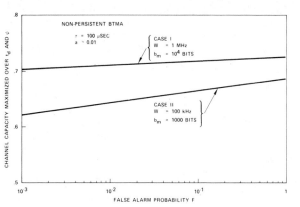

Fig. 19. BTMA—channel capacity (maximized over t_d and ψ) versus F.

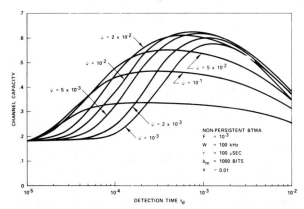

Fig. 17. BTMA—channel capacity versus observation window t_d.

(capacity provided by the pure ALOHA access mode [1], [6]) as shown in the previous section. Qualitatively speaking, $t_d \to 0$ reduces to very bad detection (i.e., $D \to 0$), and terminals behave in a pure ALOHA mode.

In Fig. 18, we plot for various F, the maximum capacity of the channel (maximized over t_d) versus ψ. We note here that the maximum capacity is not very sensitive to *small* variations of ψ. However there is a certain range of ψ which yields the best performance. For those values of F considered in the graph ($F = 10^{-3}, 10^{-2}, 10^{-1}, 0.5$), the optimum ψ is in the range $[10^{-2}, 2 \times 10^{-2}]$.

In Fig. 19, we plot the capacity (maximized over ψ and t_d) versus F. Note that for both Case I and Case II,

the capacity of the channel is a logarithmic function of F. The ultimate performance ($\simeq 0.68$ for Case I and $\simeq 0.72$ for Case II) is obtained for $F \to 1$. However, the channel capacity is not very sensitive to variations of F. The effect of F can be explained as follows: For fixed values of ψ and t_d the larger F is, the better is the quality of correct detection $D(t)$ (see, for example, Fig. 13). Thus, for a particular value of γ, the channel time wasted by interference is smaller; the channel idle time, however, which is a function of $(1 - F)\gamma$ is larger; thus a tradeoff exists between idle time and time wasted by interference. The overall performance is not easily expressed. The plot corresponding to Case II exhibits the same linear behavior on the semilogarithmic graph, but acheives a larger capacity.[12]

To compare the delay performance of BTMA for various values of the system parameters, we first consider the quantity G/S, the average number of transmissions and schedulings that a packet incurs before successful transmission.

---

[12] The larger the bandwidth is, the better is the correct detection $D(v)$ [see (26)]. Thus larger W provides larger channel capacity. However, we note from Figs. 15 and 16 that the channel capacity is always bounded from above by the capacity of CSMA with propagation delay equal to 2τ.

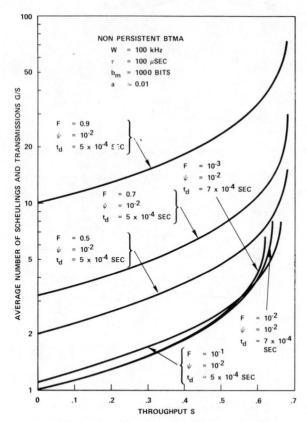

Fig. 20. BTMA—average number of schedulings and transmissions.

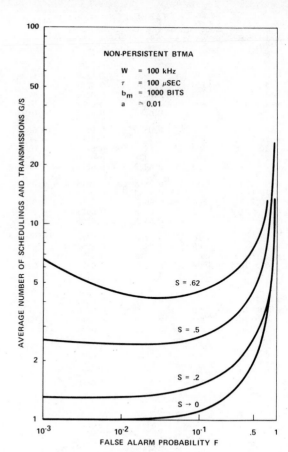

Fig. 21. BTMA—G/S versus F for constant S.

In Fig. 20 we plot, for each value of F, G/S versus S for those values of ψ and t_d yielding the maximum channel capacity. (Strictly speaking, we should plot G/S versus S for all pairs (ψ, t_d) and then draw their lower envelope. However, the difference between this lower envelope and the plotted curve corresponding to the optimum ψ and t_d for maximum capacity is so minor that we restrict ourselves to the latter.)

Note that for each value of S there exists a value of F minimizing G/S. However for relatively small values of S (not too close to the saturation point of the channel) we note that the higher the probability of false alarm F is, the larger is G/S. An explanation can be given by the following fact: when $G \rightarrow 0$ and $S \rightarrow 0$, the terminal incurs an average number of schedulings and transmissions equal to $(1/1 - F)$. This is shown on Fig. 20 at $S = 0$. (In some cases such as in the curve corresponding to $F = 0.9$, we have $\lim_{S \rightarrow 0} (G/S)$ slightly larger than $(1/1 - F)$; this is due to the fact that the curve does not correspond to the lower envelope.) To best compare the effect F has on G/S, we plot in Fig. 21 G/S versus F for constant S. Thus, we show that for $S < 0.5$ and $10^{-3} < F < 10^{-1}$, G/S is small and fairly insensitive to F, and that for $F > 10^{-1}$ it increases rapidly with increasing values of F. For larger values of throughput, the choice of F is more

critical. A good operating point should then be in the flat part of the curves.

G/S, as a measure of delay, can be of importance since the complexity of the equipment and the implementation of the protocol can be directly related to the number of schedulings and transmissions that a packet incurs. For example, at each scheduling the terminal has to generate a random number determining the scheduling delay. Of even more importance in evaluating the performance of such a system is the determination of the actual packet delay, defined as the time lapse since the packet is first generated, until the time it is successful. As discussed in Part I [1], the mathematical determination of packet delays is fairly complex, and simulation techniques are employed. For various values of F ($F = 10^{-3}$ and $F = 0.5$), by selecting the optimum system parameters (ψ, t_d) with respect to channel capacity, we simulated the BTMA mode. In Fig. 22, we plot the throughput-minimum-delay[13] curve for these values of F. It is to be noted that, even though G/S can be significantly affected by F, the minimum delay is insensitive to F. However, for each value of S there exists a value of F which provides the lowest

[13] Delay is minimized with respect to $\bar{X}$. (See [1].) In BTMA, the larger F is, the larger is G/S. The minimum delay is obtained for very small values of $\bar{X}$.

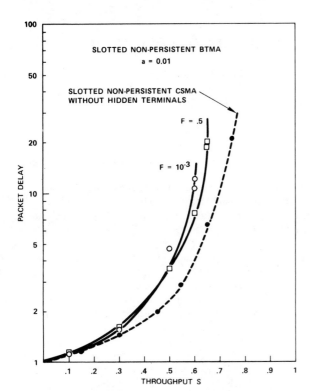

Fig. 22. BTMA—throughput-delay tradeoffs ($a = 0.01$).

0.66]. These capacities are obtained for the optimum values of ψ lying in the range $[10^{-2}, 2 \times 10^{-2}]$ and the corresponding optimum values of t_d. Similar results are readily obtainable for other values of W.

TABLE II
Capacity C for the Various Protocols Considered ($a = 0.01$)

Protocol	Capacity C
Pure ALOHA	0.184
Slotted ALOHA	0.368
1-persistent CSMA	0.529
Slotted 1-persistent CSMA	0.531
Nonpersistent BTMA	
$\quad W = 100$ kHz	0.680
$\quad W = 1000$ kHz	0.720
0.1-persistent CSMA	0.791
Nonpersistent CSMA	0.815
0.03-persistent CSMA	0.827
Slotted nonpersistent CSMA	0.857
Perfect scheduling	1.000

Thus the nonpersistent BTMA constitutes a fairly good solution to the "hidden terminal" problem, providing a maximum channel capacity of approximately 0.68 when $a = 0.01$ and $W = 100$ kHz, and 0.72 when $a = 0.01$ and $W = 1$ MHz, as compared to 0.82 for nonpersistent CSMA with no hidden terminals (0.85 for slotted nonpersistent CSMA with no hidden terminals). It should be noted that while in CSMA sensing the presence of a transmission involves a one-way propagation delay $a = 0.01$, BTMA requires a round-trip delay $2a = 0.02$ to perform the same operation. Furthermore, the performance of the nonpersistent BTMA is insensitive to the precise setting of the various system parameters ψ, t_d, and F.

We summarize the above results in Table II (CSMA capacities assume no hidden terminals) for $a = 0.01$.

delay. By comparing the lower envelope of these throughput-delay curves to the curve corresponding to the nonpersistent CSMA without hidden terminals, we note the relatively good performance of BTMA.

IV. CONCLUSION

The hidden-terminal problem seriously degrades the performance of CSMA. To eliminate this problem in single-station environments, the use of a busy-tone channel has been considered. In this paper, the nonpersistent BTMA mode has been analyzed and the approximations made in the analysis were shown to have a very minor effect in evaluating the channel performance. The channel capacity (optimized over ψ and t_d) and the packet delays are not very sensitive to F; but when G/S is considered, we see that (unreasonably) large values of F may have a significant effect even when $S < 0.5$ and that the choice of F is more critical for $S > 0.5$. For $W = 100$ kHz and $b_m = 10^3$ bits (Case 1), and $a \lesssim 0.01$, in order to keep G/S low, F should lie in the range $[10^{-3}, 10^{-1}]$. For this range of F the channel capacity lies in the range [0.62,

REFERENCES

[1] L. Kleinrock and F. Tobagi, "Packet switching in radio channels: Part I—Carrier sense multiple access modes and their throughput delay characteristics," *IEEE Trans. Commun.*, this issue, pp. 1400–1416.
[2] F. Tobagi, "Random access techniques for data transmission over packet switched radio networks," Comput. Sci. Dep., School Eng. Appl. Sci., Univ. Calif., Los Angeles, UCLA-ENG 7499, Dec. 1974.
[3] S. Fralick, Stanford Res. Inst., Menlo Park, Calif., private communication.
[4] L. A. Wainstein and V. D. Zubakov, *Extraction of Signals from Noise.* Englewood Cliffs, N. J.: Prentice-Hall, 1962.
[5] L. Kleinrock, *Queueing Systems, Vol. I, Theory, Vol. II, Computer Applications.* New York: Wiley Interscience, 1975.
[6] N. Abramson, "The ALOHA system—Another alternative for computer communications," in *1970 Fall Joint Computer Conf., AFIPS Conf. Proc.*, vol. 37. Montvale, N. J.: AFIPS Press, 1970, pp. 281–285.

Network Design: An Algorithm for the Access Facility Location Problem

PATRICK V. Mc GREGOR, MEMBER, IEEE, AND DIANA SHEN

Abstract—In any network where a large number of widely dispersed "users" share a limited number of "resources," the strategy for access will play a large part in determining the cost and performance of the network. In this paper we consider a topological design aspect of the access problem. In particular, we consider the problem of locating "access facilities," or concentration points, to obtain an economic connection of users to resources.

The problem is formulated as the locating of generic access facilities (GAF's) to obtain an economic connection of nodes (users) to a resource connection point (RESCOP). The nodes may be connected through multipoint lines, but with a constraint on the number of nodes which may share a single line. The GAF's are constrained in capacity, expressed as the number of nodes they can support, and have a cost associated with them.

The basic solution technique presented is a heuristic algorithm characterized by the following four steps.

1) Simplify the problem to a point-to-point problem by replacing clusters of nodes by single "center-of-mass" (COM) nodes.

2) Partition the reduced set of COM nodes by applying an Add algorithm, resulting in one of the COM nodes selected as a GAF site.

3) Select one of the original nodes as a real GAF site in each partition by examining the original nodes closest to the COM node selected in the Add algorithm, and selecting the best.

4) Apply a line-layout algorithm to each partition, with its selected GAF site serving as the central node.

I. INTRODUCTION

IN any network where a large number of widely dispersed "users" share a limited number of "resources," the strategy for access will play a large part in determining the cost and performance of the network. The users may include time-sharing terminals, terminals used for message transfers, remote automatic sensing devices (such as might be found in an environment-monitoring situation), manned sensing stations, and several others. The resources may be as sophisticated as many heterogeneous computers tied together in a packet-switching high-level subnet, such as in the ARPANET, or as simple as a single computer processing data received from automatic remote sensing devices. An almost endless number of user and resource combinations, both covering and extending the range described above, appear possible. Effective, economical user access will depend on the development of hardware to facilitate access, protocols to ensure satisfactory operation, and topological design techniques to efficiently utilize the hardware. In this paper we consider a topological

design aspect of the access problem. In particular, we consider the problem of locating "access facilities," or concentration points, to permit economical connection of users to resources.

The access-facility location problem can be formulated in many ways, and can appear at several different levels within the same network design problem. This is typified by a network such as the ARPANET where users access the resources through a packet communications subnet. If the resource locations and user locations are fixed, the problem becomes one of locating the packet switches for user access to the subnet to minimize the cost, subject to capacity, performance, and reliability constraints. When the locations of these facilities are fixed, the problem may appear at the level of using concentrators or multiplexers as access facilities for connecting the users to these new resources (i.e., the user-access packet switches). Because the facility-location problem can be posed for various levels of the same design problem, with devices performing different "functions" depending on context, we will pose the problem in terms of a generic access facility (GAF) and a generic resource to which users and/or GAF's are to be connected, called a resource connection point (RESCOP). We will also use the term "node" as a generic replacement for "user." Thus, the general problem is one of locating GAF's to most economically allow connections of a given set of nodes to a given set of RESCOP's. In this paper we present a heuristic design algorithm to solve this problem.

For simplicity, the algorithm is presented for the problem with only one RESCOP. Following presentation of the algorithm, its extension to the multiple RESCOP's problem is considered. The network design constraints allow multipoint line connection of nodes to GAF's or the RESCOP. This permits economy in line cost, but results in a far more complex design problem than that obtained when only point-to-point connections are allowed. The problem is formally stated below.

Given:

A	set of node sites $i = 2, \cdots, N$
Site 1	RESCOP
H	set of possible GAF sites
w_{max}	line capacity (nodes/line)
c_{max}	GAF capacity (nodes/GAF)
$d_i(j)$	cost of connecting site j to site i
charge	cost of GAF,

find the low-cost network design subject to the following constraints: 1) no line may be connected to more than w_{max} nodes; and 2) no GAF may serve more than c_{max} nodes.

In this formulation the set H of possible GAF sites is defined separately from the set A of nodes. In most practical problems the possible GAF sites are limited to a subset of the nodes selected on considerations of maintenance, rental space, access by trained company personnel, security, etc. However, it is quite feasible for situations to occur where the possible GAF sites are in fact disjoint from the nodes (such as in a commercial time-sharing operation where GAF's must be at company-provided locations but all terminals are at customer locations), partially overlap with the set of nodes, are a proper subset of the nodes, are the same as the nodes, or have the nodes as a proper subset. For simplicity, we have chosen to present the algorithm for the case where the nodes are the possible GAF sites, thus dealing with only one set, A. In a later section the algorithm will be also shown to handle easily the other cases noted above. Without loss of generality, we have chosen to assume the cost of connecting a GAF at node i to the RESCOP to be the same as the cost of connecting node i directly to the RESCOP.

In Section II, related problems and solution techniques are described, and in Section III previous approaches to this problem are considered. In Section IV, a simplified description of our approach is given. Results of experiments to determine the performance of the algorithm are given in Section V, and extensions to more general problems are given in Section VI. A formal description of the algorithm is presented in the Appendix.

II. RELATED PROBLEMS AND SOLUTION TECHNIQUES

There are many problems that are related to the GAF-location problem, including warehouse-location problems, clustering problems, and partitioning problems. In this section, we discuss these problems, their solution techniques, and the applicability of these techniques to the GAF-location problem.

A. Warehouse-Location Problems

The warehouse-location problem may be briefly defined as the determination of the number, location, and capacity of source sites in order to minimize the cost of satisfying a set of shipping requirements under a given cost matrix [7]. Various formulations of this problem have been treated with numerous solution techniques [8], [10], [12], [16], [17], [27], [29], [32], [36], [38], [39]. The approaches described below are particularly reflective of the GAF-location problem.

The basic approach to obtaining algorithms for the exact solution of the warehouse-location problem is to formulate them as mixed integer programs and to solve these programs by decomposing them into a master problem (which is an integer program) and a series of subproblems (which are linear programs). Efroymson and Ray present a branch-and-bound algorithm solution technqiue for the problem when it is formulated in a manner analogous to the GAF-location problem, but with a point-to-point connection constraint and no GAF capacity constraint [11]. Gray presents exact solution techniques based on mixed integer programming for several problem variations, including that of Efroymson and Ray but with a GAF capacity constraint [19]. Relative to the general GAF-location problem, these approaches have two drawbacks: they do not easily handle the multidrop line case,

and for large problems, the computational requirements are prohibitive.

The computational requirements for obtaining exact solutions to these problems have given rise to many heuristic approaches. Among the more successful is the "Add'" algorithm introduced by Kuehn and Hamburger [26]. The approach consists of two basic parts, a main procedure which locates warehouses one at a time until no additional warehouses can be added without increasing cost, and a bump-and-shift routine which tries to improve the initial solution by removing and by interchanging warehouses. The main procedure begins with all "customers" directly connected to a central warehouse. Each possible warehouse location is then evaluated by determining the cost-reduction which would be achieved by placing a warehouse at the location. The location giving the greatest reduction is then selected for the first placement of a warehouse. With the new warehouse in place, the processs is repeated for next location. When no further cost-reduction is achieved, the main procedure yields to the bump-and-shift routine. This routine is designed to check on the solution in two ways: 1) any warehouse which is no longer economical because some of its customers have been shifted to newly located warehouses is eliminated (bumped) and its customers reassigned, and 2) attempts are made to interchange each remaining site with others in the territory it serves.

Several variations of the Add algorithm are possible, including examination of only the largest demand points as potential warehouse locations [26] and problem reduction by permanent assignment of nodes to each new site as selected. Such variations are generally trades between design improvement and program complexity and execution time. Relative to the GAF-location problem, an Add approach has the drawback of not handling the multidrop line case efficiently; evaluating each site with a complete line layout would be too computationally costly. In addition, the main procedure is based on addition of one new GAF at a time, but there is little reason to think that the location of the best single GAF is also one of the best locations for the two GAF's, etc.

The "Drop" algorithm [13] is basically the reverse of the Add algorithm. Warehouses are initially assumed to be at all possible locations, and customers are assigned to multiple warehouses. The procedure is to then eliminate the warehouse whose elimination most reduces the cost. The process is repeated until no further reduction is possible. The difficulties here are directly analogous to those of the Add algorithm.

B. Clustering Problem

Basic aspects of the GAF-location problem can be posed in a clustering context. The clustering problem may be thought of as detecting inherent separations between subsets of a given point set where the separations may be in terms of several different measures. In the GAF-location problem, one might expect some measure to be available such that nodes clustered under this measure would most appropriately share a common line or be connected to the same GAF. Many clustering techniques exist [1], [3], [18], [28], [31], [34], [35] with numerous measures that may be applicable to the GAF-location problem. Among the more promising are those which attempt to cluster points on a plane [21], [41]. Zahn [41] attempts to identify gestalt clusters (two-dimensional point sets naturally perceived as separate groupings) by connecting the points with a minimum spanning tree and deleting relatively long branches to form components or clusters. Jarvis and Patnik [21] offer a similarity measure based on shared, near neighbors to generate "nonglobular" clusters. This technique and that of Zahn have much in common. However, neither of the approaches appears to offer more than insight to the complexity of the GAF-location problem.

C. Partitioning Problems

The GAF-location problem may be viewed as that of partitioning a set of elements in a way that optimizes some objective function defined on the set of all partitions. The partitioning may be in terms of nodes sharing a common line or in terms of nodes connected to the same GAF, and the objective function is simply network cost. There are several formulations to the partitioning problem in terms of set theory [5], [15] and graph theory [22], [25], [24]. Perhaps the most interesting of these formulations, for the GAF-location perspective, is the formulation by Roach [33] in which the objective function is to minimize the maximum within-cluster distance. Although the solution technique offered is one of forming reduced subproblems and is somewhat improved over integer programming, it is still far too complex computationally to be of interest in the GAF-location problem.

III. PREVIOUS APPROACHES

The GAF-location problem has received considerable attention in the data communication network literature [2], [4], [6], [9], [14], [20], [37], [40]. The approaches have almost all been heuristic in nature, and most have dealt only with the less general version of the problem where only point-to-point connections are permitted. For this simpler problem, two attractive approaches seem to be a direct adaptation of the Add algorithm of the warehouse-location problem, and a "graceful" drop algorithm based on link removal rather than GAF removal [2]. The former seems to be considerably more efficient computationally, but the latter seems to yield, in general, slightly better results. An approach based on combinatorial optimization over certain selected subsets of the network has been offered by Greenburg [20] and has the attraction of partitioning the main problem into subproblems in an iterative manner, and reoptimizing the partitions based on global considerations. However, the basic heuristic used in the partition process does not restrict the size of partition elements, the procedure for optimization is based on enumeration, and the approach is again for point-to-point problems.

The general GAF-location problem has been attacked by Woo and Tang [40] with an algorithm that approximates the problem with a simpler point-to-point problem, in which the point-to-point cost of connecting two nodes is a weighted average of the direct connection cost and the shared cost of connecting through a minimum spanning tree. We call this the average tree-direct (ATD) algorithm. After the problem simplification, an Add algorithm is applied to make site selections. Recall that in the Add algorithm, all candidate sites are

evaluated for savings with the best selected for insertion as a given site in the next iteration. The procedure is iterated until no new site is found to give positive savings. A rigorous proof of the following theorem, which justifies the termination condition for the Add algorithm, has been reported [40].

Theorem: Let L, L_1, and L_{12} be optimal network assignments corresponding, respectively. to GAF sets $(j_0, j_1, \cdots, j_i)$, $(j_0, \cdots, j_i, j_{i+1})$, and $(j_0, \cdots, j_i, j_{i+1}, j_{i+2})$, and let $C(1)$, $C(L_1)$, and $C(L_{12})$ be their respective costs. Then $C(L) - C(L_1) \geqslant C(L_1) - C(L_{12})$.

Note that this theorem does not say that the network cost is a convex function of the number of GAF's, but rather that the cost is convex over the iterations of adding a new GAF to an existing set of GAF's without reoptimization of the existing GAF locations. In fact, counterexamples are easily produced in which the cost is not convex over the number of GAF's when the given number of GAF's are optimally located in each case.

The ATD algorithm described above appears to be the best algorithm that is currently in the literature for the general GAF-location problem. We will use it as the basis for comparative evaluation of the algorithm we present next.

IV. DESCRIPTION OF GENERAL APPROACH

There are many possible formulations for the GAF-location problem, and many possible approaches to finding acceptable solutions. We present here a heuristic approach to the problem, as formulated in Section I.

The object is a total network design. Because the approach is heuristic, no promise is made of optimality, only feasibility in terms of the given constraints. The constraints used in this formulation were chosen because they are often appropriate in reality, and are particularly simple. The approach is also reasonable for various other constraints. These will be discussed later. GAF's are used only where they appear to be beneficial, and consequently for some node arrangements, designs will result which use no GAF's.

The general approach is characterized by the four steps discussed below with the numbers in parentheses used in the text referring to the corresponding blocks in the flow chart shown in Fig. 1. A formal description of the algorithm is given in the Appendix.

A. Simplification by Clustering

Simplification is achieved when the problem is reduced in size and converted to a point-to-point formulation. To accomplish this, clusters of nodes are replaced by single nodes. The clusters are intended to reflect natural groupings of nodes that can be most appropriately approximated by single nodes at their center of mass. The clusters are limited in size by the line constraint, and thus also reflect possible groupings of nodes to share a line.

The clusters are formed by "rolling snowballs" in a rather "balanced" fashion. First, the two nodes closest together are selected (2). If these two nodes can be put in the same cluster (i.e., their being joined does not violate the line constraint) (3), they are replaced by a single node at their center-of-mass

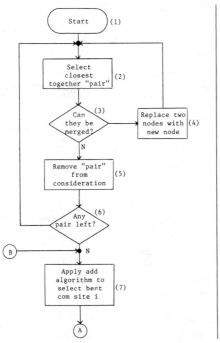
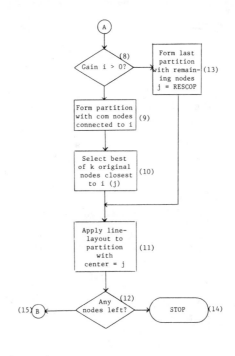

Fig. 1. Flow chart of approach.

(4), called a COM node. The "weight" of the nodes is simply the number of nodes contained in the cluster they represent, with all weights initially equal to one.

If the two nodes cannot be merged, the next closest together "pair" will be considered. As pairs of nodes are identified which cannot be merged, they are removed from further consideration in the clustering process (5), although the individual nodes may reappear as members of other pairs. The clustering process continues until no pair of nodes can effectively be merged (6). An example of the clustering process is shown in Fig. 2. At this point, the original set of nodes has been replaced by the set of nodes representing the clusters. This set is smaller in number by a factor slightly less than the line constraints, and the relative costs of connecting the nodes in a cluster to different sites can be approximated by the point-to-point form. An example of the reduced set, with associated weights for the nodes, is shown in Fig. 2.

B. Partitioning

The Add algorithm examines the benefit of placing a GAF at each COM node. The benefit is determined by iteratively associating with each COM node the other COM nodes which give the greatest cost benefit by being connected to the GAF instead of the RESCOP, subject to the GAF capacity constraint. After the assignment of COM nodes to the GAF, a heuristic estimate is made of the cost benefit. The estimate can be expressed as

$$r_i = \sum_{j \in B_i} s_{ij} * w_j - d_1(i) * \gamma$$

where

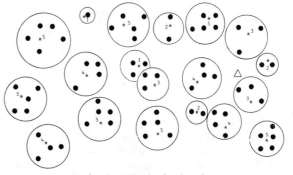

Fig. 2. Simplification by clustering.

r_i	estimate of relative benefit of GAF at node i,
B_i	set of nodes associated with GAF at node i,
s_{ij}	savings of connecting node j to GAF at i instead of to RESCOP,
w_j	weight of node j
$d_1(i)$	cost of connecting GAF at i to RESCOP,
γ	parameter used to emphasize proximity of GAF to RESCOP.

The COM which, as a GAF site, has the greatest heuristic benefit estimate is selected as the best (7). The simple point-to-point cost gain of this COM node is then checked (8). If the cost gain is positive, the selected COM node and all those COM nodes assigned to the selected location in the Add algorithm are partitioned from the remaining nodes to form a separate subproblem (9), as shown in Fig. 3. If the cost gain is not positive, it is concluded that the best GAF site is not cost-effective, and all the remaining COM nodes are then assigned

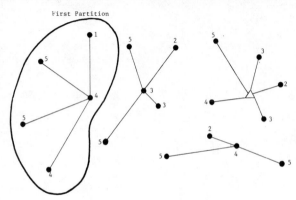

Fig. 3. Partition by add algorithm.

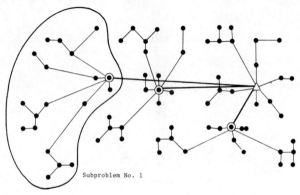

Fig. 5. Line layout.

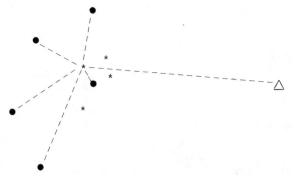

Fig. 4. Local optimization.

to the RESCOP (13). This forms a last partition element to be treated as a subproblem.

C. Local Optimization

The output of the partition process is a particular subset of COM nodes identified as deserving a GAF, and a particular member of the subset identified as the appropriate site for the GAF. However, the COM nodes are representatives of clusters, and in reality some particular original node site must be selected for the GAF. To select this site, each of the k nodes closest to the selected COM node are evaluated, as shown in Fig. 4, with $k = 3$.

The same measure is used in the evaluation as is used in the COM node selection. The node with the greatest heuristic estimate of benefit is then selected as the actual GAF site (10). This gives a partition of the nodes into a subproblem complete with an actual node site chosen for the GAF.

D. Line Layout

In order to apply a line-layout algorithm to the subproblem formed above, it is first necessary to replace all the COM nodes by the actual nodes they represent. This gives a partition of actual nodes with an actual node selected as the GAF site. Thus, the subproblem in the partition is now a classic case of a multidrop line layout in a centralized network. A line-layout algorithm is then applied to the partition (11), giving a result as shown in Fig. 5. The iteration is completed by removing the nodes in the partition from further consideration in the overall

problem. This removal is an elementary variation of the Add algorithm in which design effectiveness is traded for execution time, with the removal reducing the number of nodes to be considered in the Add algorithm at each iteration, and simplifying the line-layout problem by reducing it to subproblems. Such a trade has been made to allow cost-effective experimentation with the algorithm. Other variations of the Add algorithm oriented towards design effectiveness (such as by incorporating postprocessing) may be appropriate for actual design cases.

The removal of the partitioned nodes from further consideration marks the end of an iteration of the Add algorithm phase. If there are no nodes remaining to be considered (12), the design is complete (14) with each node connected (via a multidrop line) to a GAF. If there are nodes remaining to be considered (12), the next iteration of the Add algorithm phase is commenced (15). The iterations continue until all nodes are partitioned and connected to GAF's or until no additional GAF is estimated to beneficial (8). In this later case, the remaining nodes are assigned to the RESCOP (13), and the line-layout algorithm is applied (11) to complete the design as shown in Fig. 5.

V. PERFORMANCE RESULTS

The COM algorithm is a heuristic approach to a rather complex problem. In order to evaluate its performance, we have implemented the algorithm and applied it to a number of problems with randomly positioned nodes. The results of these experiments are reported below. For all experiments, the implementation was in Fortran, and the experiments conducted on a CDC 6600 computer. Each of the problems were of the simple type described earlier, with constraints on the number of nodes per line and nodes per GAF, and on preset RESCOP site to which all nodes must be connected, either through GAF's or directly through their multipoint lines. The cost function was a rate of $0.50/mile and $40/drop as a monthly charge. This corresponds to a prorated Telpak rate. In each experiment, the values chosen for the program parameters were held fixed for all cases, no attempt was made at "fine tuning."

In order to have a comparison basis for the evaluation, we

TABLE I
PERFORMANCE OF COM ALGORITHM

EXPERIMENT	RESCOP	NUMBER OF NODES - COST OF DESIGNS (COM/ATD)							
		50		100		200		400	
1. 4 nodes/line 20 nodes/GAF $200/GAF	1	12754	13409	20348	21332	34353	34443		
	2	13373	13215	20809	22745	37411	38011		
	3	12885	13503	21371	22000	36845	37556		
2. 4 nodes/line 20 nodes/GAF $1500/GAF	1	14005	15385	25846	25955	44275	45794		
	2	16136	15757	25888	27366	48005	48160		
	3	15485	17403	26601	27205	47337	47821		
3. 10 nodes/line 50 nodes/GAF $15000/GAF	1			18393	20881	32023	33258	55406	57438
	2			18765	21419	32107	33816	58520	
	3			18706	22148	32184	34046	56966	
Avg. Exec. Time (Sec.)				2.1	9.4	8.2	67.7	36	275

implemented another heuristic algorithm for this problem that was the best we could find in the literature. This was the approach of converting the problem to a point-to-point problem by forming a cost matrix based on the average of the cost of connecting two nodes through the minimum spanning tree and the cost of connecting them directly, as reported by Woo and Tang [40]. The implementation of this algorithm [which we call the average tree-direct (ATD) algorithm], was also in Fortran on a CDC 6600. The same elementary Add algorithm as used in the COM implementation was used in the ATD implementation (this version is different from that used in the implementation reported by Woo and Tang). The COM and ATD implementations used the same line-layout procedure. The ATD algorithm also allows parametric fine tuning. However, again the values chosen for the parameters were based on "optimizing" the design of a pilot problem. Our experience with the pilot problem supported the observation by Woo and Tang that the design results are rather insensitive to variations in the parameters around the optimum values. In the implementation of each algorithm, the same programming techniques were used for efficiency (and inefficiency). More recent implementations have used additional insight to the COM procedure to substantially improve its execution efficiency.

A. Uniformly Randomly Distributed Nodes

The first series of experiments were performed on problems where the nodes were uniformly randomly distributed over a 2000 × 3000 mi rectangle. The algorithms were applied to problems of 50, 100, 200, and 400 nodes. Three RESCOP sites were considered, one in the center (1), one in the corner (2), and one at the midpoint of a side (3). The results of the experiments are shown in Table I. The COM algorithm shows an average cost advantage of approximately five percent in comparison to the ATD algorithm ([ATD − COM]/ATD), and produced the lower cost design in 23 out of the 25 experiments. A typical COM design is shown in Fig. 6. The average

costs of the designs produced by the COM algorithm as a function of the number of nodes is shown in Fig. 7. Note that, as would be expected, the larger capacity constraints not only give lower cost designs, but also a less rapidly growing cost curve.

B. Randomly Distributed Based on Population

In most real problems, the nodes will not be uniformly distributed over a nice rectangular region. In order to pose a more realistic problem, nodes were located throughout the United States in a random manner based on population density. The model used in distributing the nodes is described in detail in a report [30].

In each experiment, the RESCOP site was Atlanta, the GAF cost was $2500/month, and the line constraint was 10 nodes/line. With a constraint of 100 nodes/GAF, the ATD algorithm produced a design with a cost of $41 508.50/month, and the COM algorithm produced the design shown in Fig. 8, with a cost of $39 418.50/month. In this case the COM algorithm produced a design lower in cost by five percent.

As a final experiment, we applied the algorithms to the same problems as described above, but with no constraint on the GAF capacities. The designs produced by the two algorithms were very similar. The COM algorithm produced a lower cost design by only 0.50 percent. This result, coupled with the others, suggests that the COM algorithm is perhaps more sensitive to GAF capacity. In fact, in the 27 comparison cases examined, the COM algorithm had fewer GAF's in 18 cases, and the same number in the remaining 9 cases. On the average, it used 21 percent fewer GAF's ($[N_{ATD} - N_{COM}]/N_{ATD}$).

C. Computation Time

A significant factor for a design algorithm is its efficiency, i.e., the computer time it requires. In order to appraise this attribute of the COM algorithm, the execution time for the algorithm was measured on several problems. The times are

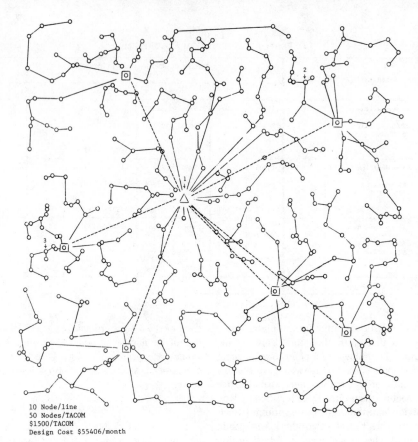

10 Node/line
50 Nodes/TACOM
$1500/TACOM
Design Cost $55406/month

Fig. 6.   Typical COM design experiment 3; 400 nodes.

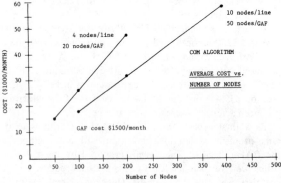

Fig. 7.   COM algorithm cost performance.

this comparison, based on log-log scale plot of the two curves, the execution time of the COM algorithm for these problems may be approximated by the function $t = (2 \times 10^{-4})N^2$, and the ATD algorithm by the function $t = (10^{-4})N^{2.5}$.

D. Performance Interpretation

The basic execution time advantage for the COM algorithm results from its problem-simplification strategy. The merging process can be implemented with only slightly more complexity than a basic Kruskal minimum spanning tree algorithm [25]. However, the ATD algorithm involves considerable computation to determine the equivalent tree cost of connecting each pair of nodes in addition to generation of a minimum spanning time. Furthermore, as noted above, the results of the simplification is not only conversion to a point-to-point problem, as is true for both algorithms, but also reduction of the number of nodes to be used in the Add phase for the COM algorithm, whereas the ATD algorithm has no such reduction.

The Add algorithm variation used in the implementations compared above is oriented towards execution efficiency with some sacrifice of design effectiveness. However, the basic strategy of the algorithms is, first, problem simplification by conversion to a point-to-point problem (also with a reduced number of nodes in the COM case) and, then, application of an

only for the GAF-selection portion of the problem, and not the line-layout portion. For comparison purposes, the execution times for the ATD algorithm were also measured in the same way. The same basic strategies for efficiency were used in each implementation. For the problems with constraints of 10 nodes/line, 50 nodes/GAF, and GAF cost of $1500/month, the average results are shown in Table I. Curves portraying these results are shown in Fig. 9. It would appear that the COM algorithm is substantially more efficient. To quantify

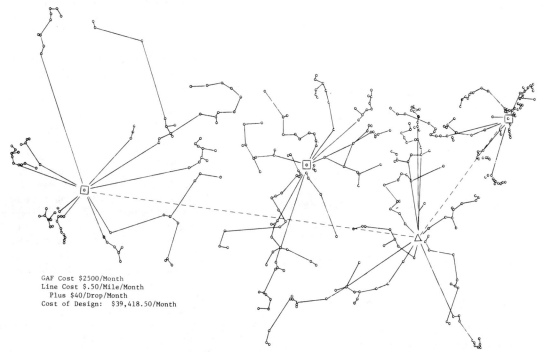

GAF Cost $2500/Month
Line Cost $.50/Mile/Month
   Plus $40/Drop/Month
Cost of Design: $39,418.50/Month

Fig. 8.   COM algorithm applied to 400 nodes distributed in a random manner based on population.

Add algorithm for site selection. Thus, the second step is independent of the first, and effectiveness-oriented versions of the Add algorithm could easily be substituted for the elementary version described above. In fact, this has been done in application of the COM algorithm to design of the National Airspace Data Interchange Network (NADIN) for the Federal Aviation Administration. This network has approximately 350 terminal sites evenly distributed over the coterminous United States. Application of the ATD and COM algorithms with the elementary Add approach gave typical comparison characteristics. Based on these results, a more sophisticated Add approach was used with the COM algorithm to generate a final design.

VI. GENERALIZATIONS AND EXTENSIONS

The GAF-location problem has been posed in its most basic form, and an algorithm, with associated performance results, has been presented which appears to be an effective approach to the problem. Several generalizations and extensions of the problem and algorithm are now outlined.

A. Line Constraints

Variation: Constraints on multiple node attributes.

Approach: Attributed vector V_l and constraint function for merging,

$$C(V_i, V_j) = \begin{cases} 1, & i \text{ can be merged with } j \\ 0, & i \text{ cannot be merged with } j. \end{cases}$$

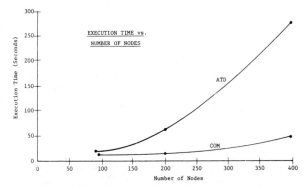

Fig. 9.   Execution time of COM and ATD algorithms.

B. GAF Constraints

Variation: Constraints on multiple node attributes.
   Approach: As in Section **VI-A**.

C. Possible GAF Sites

1) Variation: Small number of possible GAF sites.
   Approach: During Add phase, evaluate actual possible GAF sites (i.e., members of H) rather than COM nodes.
2) Variation: Large number of possible GAF sites.
   Approach: During Add phase, evaluate k nodes in H rather than in A.
3) Variation: Possible GAF sites subset of nodes.
   Approach: Flag COM nodes containing possible GAF sites and evaluate only these during Add phase.

D. Multiple Capacity GAFs

Variation: Different models of GAF with different costs.

Approach: Evaluate each model at each site in order of increasing capacity.

E. Staging

Variation: Smaller GAF's can be connected to larger GAF's.

Approach: Locate larger GAF's first and then recast problem for smaller GAF's.

F. RESCOP Variations

1) Variation: No nodes can be connected to RESCOP.

Approach: Cost of GAF added to cost of connecting a node to the RESCOP and terminate condition changed to "no nodes left."

2) Variation: Limited number of nodes can be connected to RESCOP.

Approach: First GAF preselected at RESCOP and problem continues as in Section VI-F1) above.

3) Variation: Multiple RESCOP's.

Approach: Evaluate on basis of connection to closest RESCOP.

4) Variation: Determine best location of RESCOP's.

Approach: Assign each COM node a RESCOP including cost which can be eliminated by connection to another node.

5) Variation: Require greater connectivity.

Approach: In each iteration, select k most economical nodes as usual and then evaluate in detail.

G. Very Large Networks and Time Reductions

Variation: 5 000–50 000 nodes.

Approach: Introduce sparse graph by superimposing a grid over the network and considering only neighbors in adjacent boxes for interconnection—merge on a box-by-box basis in the clustering phase.

VII. CONCLUSION

A new algorithm has been presented for the design of multidrop networks that may incorporate GAF's to economically connect nodes (users) to RESCOP's. Experiments with the algorithm indicate that it is both effective and efficient. Extention of the basic algorithm to handle more general problems was shown to be easily accomplished.

Research is continuing on extending the concepts reported here to the integrated design of large (5 000–50 000) hierarchial networks with various levels of access facilities.

APPENDIX

THE CENTER-OF-MASS (COM) ALGORITHM

In this Appendix, we give a precise description of the COM algorithm. The algorithm is formally stated at the end of the section. The first part of the section is devoted to interpreting the formal statement.

The formal statement begins with a list of the relevent definitions, parameters, constants, constraints, and cost function. Note that the constraints are of the simple capacity variety, and the cost function for connecting two nodes is simply the Euclidean distance between the nodes. It will be clear after examination of the algorithm that other constraints and cost functions are equally usable.

In the initialization phase (Step 0), A is formed as the set of $N - 1$ nodes of interest, where the RESCOP (node 1) is not included. The RESCOP is not considered eligible for a GAF as nodes may connect to it directly. The set DO is the set of all nearest neighbor distances. For each node i a list L_i is kept of the real nodes represented by the node. Initially, the list for each node contains only the node itself. After two nodes merge, the list for the new node will contain the members of the lists of the two nodes that merged, thereby keeping track of the represented nodes. The weight assigned to each node is initially set equal to one. Thus, the constraints are simply on the number of nodes/line and the number of nodes/GAF. For each node i, its nearest neighbor j is represented as n_i. The average "nearest neighbor distance" d_{avg} is used to set a maximum distance d_{max} over which nodes may be merged. The parameter α determines this maximum distance. The reason for establishing such a maximum distance is based on the clustering objective and procedure, as discussed below.

The clustering objective is to replace natural groupings of nodes by a single node. The procedure is to iteratively merge the closest feasible pair of nodes. As this procedure nears completion, most nodes are representative of near capacity clusters, and consequently, cannot be merged. Thus, the shortest distance between feasible nodes may become quite large. Merging of these nodes would not reflect the objective of natural groupings. The maximum distance is designed to prevent such mergings.

After determining a maximum allowable distance, the set D is formed as the set of allowable nearest neighbor distances.

The merging of two nodes is accomplished in Step 1 of algorithm. Prior to entering this step, it is known that all nodes with their nearest neighbor distance contained in D can be feasibly merged with their nearest neighbor. When Step 1 is entered after the initilization process, this is certainly true, or else the problem is point-to-point in form, and not multipoint. Following Step 1 is an update procedure designed to ensure that this is true when the step is reentered.

The first task in the merge process is selection of the minimum nearest neighbor distance. Then a new node is formed with its location at the center-of-mass (COM) of the two nodes being merged, where the weights used in the COM calculation are simply the weights assigned to each node. The list of real nodes represented by the new node is formed by merging the lists of the two old nodes. The old nodes are then removed from further consideration by deleting them from the set A. The new node will be added to the set A later. The weight of the new node is simply the sum of the weights of the old nodes. Note that with the initial weights all set equal to one, the weight is simply the number of real nodes represented by the new node.

After a new node is formed, its nearest neighbor must be found, and all nodes which had one of the old nodes as a nearest neighbor must also have new nearest neighbors found. This is accomplished in the update procedure (Step 2). $D1$ is the subset of D containing all the distances for which the nearest neighbors are not one of the two nodes just merged in the previous step. $A2$ is the set of all nodes whose nearest neighbors were one of the two nodes just merged, plus the new node; i.e., the set of all nodes for which new nearest neighbors are to be found. The new node is added to the set A, making it again the set of all nodes which are candidates for merging. The nearest feasible neighbor distances are then found for the nodes which need new distances, i.e., the members of $A2$. This is the set $D2$. $D3$ is the subset of $D2$ containing distances less than the maximum allowed distance. For each node i, which has a new feasible, allowable distance neighbor j, n_i is defined as j. The set D of all allowable nearest neighbor distances is then formed by combining the unchanged distances $D1$ with the new distances $D3$. Note that all members of D are defined for pairs of nodes which are feasible to merge. If D is not empty, no further merging is possible, and the Add process is entered (Step 3.)

The first task of the Add algorithm is the iterative examination of each node for the possible benefit of locating a GAF at its site. This task is accomplished in Step 3. The savings achieved by connecting a node j to a GAF at i versus the RESCOP (node 1) is defined as s_{ij}. For each node i, nodes are iteratively associated with the node in order of decreasing savings, with the maximum savings node associated first. The iterations continue until no further savings are possible, or until the capacity constraint prevents any further associations. When the iterative process is terminated, the benefit of placing a GAF at the node site is evaluated on the basis of the associated nodes. Thus, for each node i the iterative process is initiated by forming Λ as the set of candidate nodes for connection to i, and B_i as the list of nodes actually to be associated (initially empty). T is the running sum of weights of the associated nodes, used to check the capacity constraint.

Part A) is the iterative process of association. If all candidates have been examined, then the evaluation, Part B), is commenced. If there are still unchecked candidates, select the one with the greatest savings. If the savings are not positive, no further savings are possible, and commence the evaluation. Otherwise, remove the selected node from the candidate set to ensure that it is not selected a second time, and then determine if the association is consistent with the constraint. If not go on to the next candidate. If the association is feasible, add the node to the association list, and add its weight to the running sum. If the sum is equal to the capacity, no further associations are possible, and the evaluation process is commenced. If the sum is less than the capacity, additional associations may be made, and thus, return to consider the next candidate.

Part B) is the evaluation process. First, the point-to-point savings obtained by placing a GAF at the node are determined. This is simply the sum of all the individual savings found for the associated nodes, minus the cost of connecting the GAF to the RESCOP and the cost of the GAF. Then a relative multipoint benefit is calculated. This is a heuristic measure of the node as a site for a GAF considering the multipoint nature of the problem. The measure is calculated as the sum of the weighted savings minus an emphasized cost of connecting the GAF to the RESCOP. The weights serve to move GAF's towards the bigger clusters, and the emphasis parameter γ serves to move GAF's towards the RESCOP.

After evaluating each node as a possible GAF site, the one with the greatest heuristic measure of benefit is selected (Step 4). If the point-to-point savings for this site are not positive, then it is predicted that an additional GAF will not be cost-effective, and the final stage, Step 5, is then entered. This stopping condition has its foundation in the theorem stated earlier which characterizes the costs resulting from the intermediate steps of the Add algorithm as a convex function of the addition of new GAF's. If the selected site is found to be cost-effective on a point-to-point basis, then the process of forming a subproblem is commenced.

First, the set of COM nodes associated with the selected site is removed from further consideration in the main problem. Then a partition element P is formed as the set of real nodes represented by all the COM nodes associated with the selected GAF site. Since a real node must be selected for the actual GAF site, a local optimization procedure is used to evaluate possible real node sites. The set K is formed as the k real nodes closest to the COM node selected as the GAF site. A heuristic measure z_l of relative cost for each of these nodes is then evaluated, and the node with the minimum cost measure is selected. Thus, a partition element of real nodes and selected GAF site has been formed as a subproblem for the line-layout algorithm. If additional COM nodes remain, the next iteration of the Add algorithm is commenced. Otherwise, we are done.

When COM nodes still exist in the consideration set, A, but GAF's are predicted to not be cost-effective, then the remaining nodes are associated with the RESCOP site. The line layout for the real nodes represented by these COM nodes is the only remaining problem, and is handled in Step 5.

COM ALGORITHM

Definitions

A	Set of nodes.
L_i	Set of nodes associated with $i \in A$.
x_i, y_i	Coordinates of node i.
$d_i(j)$	Cost of connecting j to i.
w_i	Weight assigned to node i.

Parameters

γ	Used to emphasize proximity of possible GAF sites to RESCOP.
α	Used to limit distance over which nodes may be merged.
k	Number of real nodes nearest to the merged node which are to be examined as possible GAF sites.

Constants

charge Cost of GAF.
w_{max} Line capacity.
c_{max} GAF capacity.
node 1 RESCOP.

Constraints

1) For a line shared by the nodes $i = 1, 2, \cdots, M$ to be feasible

$$\sum_{i=1}^{M} w_i \leqslant w_{max}.$$

2) For a GAF serving the nodes $i = 1, 2, \cdots, M$ to be feasible

$$\sum_{i=1}^{M} w_i \leqslant c_{max}.$$

Cost Function

$$d_i(j) = \| i, j \| = \sqrt{(x_i - x_j)^2 + (y_i - y_j)^2}.$$

Step 0–Initialization

$$A = \{i \mid i = 2, 3, \cdots, N\}$$

$$D0 = \{d_i(j) \mid k_i(j) = \min_{\substack{l \in A \\ l \neq i}} d_i(l), i = 2, 3, \cdots, N\}.$$

For $i = 2, 3, \cdots, N$,

$$L_i = \{i\}$$

$$w_i = 1.$$

For each $d_i(j) \in D0$, $n_i = j$,

$$d_{avg} = \frac{1}{N-1} \sum_{D0} d_i(j)$$

$$d_{max} = \alpha * d_{avg}$$

$$D = \{d_i(j) \mid d_i(j) \in D0, d_i(j) \leqslant d_{max}\}.$$

Step 1–Merge

$$d_l(n_l) = \min_D d_i(n_i).$$

Form a new node k with

$$x_k = \frac{(w_l * x_l + w_{n_l} * x_{n_l})}{w_l + w_{n_l}}$$

$$y_k = \frac{(w_l * y_l + w_{n_l} * y_{n_l})}{w_l + w_{n_l}}$$

$$L_k = L_l \cup L_{n_l}$$

$$A = A - \{l, n\}$$

$$w_k = w_l + w_{n_l}.$$

Step 2–Update

$$D1 = D - \{d_i(n_i) \mid n_i \in \{l, n_l\}\}$$

$$A2 = \{i \mid i \in A, n_i \in \{l, n_l\}\} \cup \{k\}$$

$$A = A \cup \{k\}$$

$$D2 = \left\{ d_i(j) \mid i \in A2, d_i(j) = \max_{\substack{l \in A \\ l \neq i \\ w_i + w_l \leqslant w_{max}}} d_i(l) \right\}$$

$$D3 = \{d_i(j) \mid d_i(j) \in D2, d_i(j) \leqslant d_{max}\}.$$

For all i, j such that $d_i(j) \in D3$, $n_i = j$,

$$D = D1 \cup D3.$$

If $D \neq \emptyset$, then go to Step 3.

Step 3–Evlauate Each Site

Let $s_{ij} = d_j(1) - d_j(i)$. For each i,

$$\Lambda = A$$

$$B_i = \emptyset$$

$$T = 0.$$

1) If $\Lambda = \emptyset$, then go to 2). Select l such that $s_{il} = \max_{j \in \Lambda} s_{ij}$. If $s_{il} \leqslant 0$, then go to 2).

$$\Lambda = \Lambda - \{l\}.$$

If $w_l + T > C_{max}$, then go to 1).

$$B_i = B_i \cup \{l\}$$

$$T = T + w_l.$$

If $T = C_{max}$, then go to 2). Go to 1).

2)

$$s_i = \sum_{J \in B_i} s_{ij} - d_1(i) - \text{charge}$$

$$r_i = \sum_{j \in B_i} s_{ij} * w_j - d_1(i) * \gamma.$$

Step 4—Select Best

$$r_l = \max_{i \in A} r_i.$$

If $S_l \leqslant 0$, then go to Step 5.

$$A = A - B_l$$

$$P = \bigcup_{j \in B_l} L_j$$

$$K = \{i_n \mid i_n \in P, n = 1, \cdots, k, d_l(i_n) \leqslant d_l(i_m), i_m \notin K\}$$

$$Z_l = \min_{i \in k} \left[\sum_{j \in B_l} d_i(j) * w_j + d_1(i) * \gamma \right]$$

$$i_c = l.$$

Do line layout on P with $i_c = $ GAF site. If $A \neq \emptyset$, then go to Step 3. Else stop.

Step 5—Finish

$$P = \bigcup_{j \in A} L_j$$

$$i_c = 1.$$

Do line layout on P with $i_c = $ RESCOP site. Stop.

ACKNOWLEDGMENT

The authors wish to thank their colleagues at Network Analysis Corporation for many useful comments, and in particular I. Gitman for suggesting many improvements in the manuscript.

REFERENCES

[1] J. G. Auguston and J. Minker, "An analysis of some graph theoretic cluster techniques," *J. Ass. Comput. Mach.*, vol. 17, pp. 571–578, 1970.

[2] L. R. Bahl and D. T. Tang, "Optimization of concentrator locations in teleprocessing networks," in *Proc. Symp. Comput. Commun. Networks and Tele-Traffic*, Polytechnic Inst. of Brooklyn, Brooklyn, NY, Apr. 4–6, 1972.

[3] R. E. Booner, "On some clustering techniques," *IBM J.* pp. 22–32, Jan. 1964.

[4] J. Cahit and R. Cahit, "Topological considerations in the design of optimum teleprocessing tree networks," in *Conf. Rec., Nat. Telecommun. Conf.*, Atlanta, GA, Nov. 26–28, 1973, pp. 37F1–37F7.

[5] N. Christofides and S. Korman, "A computational survey of methods for the set covering problem," *Management Sci.*, vol. 21, Jan. 1975.

[6] W. Chou and A. Kershenbaum, "A unified algorithm for design multidrop teleprocessing networks," in *Proc. Data Networks: Analysis and Design, 3rd Data Commun. Symp.*, St. Petersburg, FL, Nov. 13–15, 1973, pp. 148–156.

[7] L. Cooper, "Location-allocation problems," *Oper. Res.*, pp. 331–343, 1962.

[8] L. Cooper and C. Drebes, "An approximate solution method for the fixed charge problem," *Naval Res. Logist. Quart.*, vol. 14, pp. 101–114, Mar. 1967.

[9] D. R. Doll, "Topology and transmission rate considerations in the design of centralized computer-communication networks," *IEEE Trans. Commun. Technol.*, pp. 339–344, June 1971.

[10] P. S. Dwyer, "Use of completely reduced matrices in solving transportation problems with fixed charges," *Navy Res. Logist. Quart.*, vol. 13, pp. 289–315, Sept. 1966.

[11] M. A. Efroymson and T. L. Ray, "A branch-bound algorithm for plant location," *Oper. Res.*, May–June 1966.

[12] J. Elizinga and D. W. Hearn, "Geometrical solutions for some minimax location problems," *Oper. Res.*, May–June 1966.

[13] E. Feldman, F. A. Lehner, and T. L. Ray, "Warehouse location under continuous economies of scale," *Management Sci.*, vol. 12, pp. 670–684, May 1966.

[14] H. Frank and W. Chou, "Topological optimization of computer networks," *Proc. IEEE*, vol. 60, pp. 1385–1397, Nov. 1972.

[15] R. S. Garfinkel and G. L. Nemhauser, "The set-partitioning problem: Set covering with equality constraints," *Oper. Res.* pp. 848–856, Oct. 1968.

[16] A. J. Goldman, "Optimal center location in simple networks," *Oper. Res.*, pp. 212–221, Aug. 1970.

[17] —, "Minimax location of a facility in a network," *Oper. Res.*, pp. 407–418, Nov. 1971.

[18] J. C. Gower and G. J. S. Ross, "Minimum spanning trees and single link cluster analysis," *Appl. Statist.*, vol. 18, pp. 54–64, 1969.

[19] P. Gray, "Exact solution of the site selection problem by mixed integer programming," Stanford Res. Inst., Menlo Park, CA, tech. rep. 5205-28, 1967.

[20] D. A. Greenberg, "A new approach for the optimal placement of concentrators in a remote terminal communications network," in *Conf. Rec., Nat. Telecommun. Conf.*, Atlanta, GA, Nov. 26–28, 1973, pp. 37D-1–37D-7.

[21] R. A. Jarvis and E. A. Patrick, "Clustering using a similarity measure based on shared near neighbors," *IEEE Trans. Comput.*, vol. C-22, pp. 1025–1034, Nov. 1973.

[22] P. A. Jensen, "Optimum network partitioning," *Oper. Res.*, pp. 916–931, July–Aug. 1971.

[23] B. W. Kerninghan and S. Lin, "An efficient heuristic procedure for partitioning graphs," *Bell Syst. Tech. J.*, pp. 291–307, Feb. 1970.

[24] B. W. Kernighan, "Optimal sequential partitions of graphs," *J. Ass. Comput. Mach.*, vol. 18, pp. 34–40, Jan. 1971.

[25] A. Kershenbaum and R. Van Slyke, "Computing minimum spanning trees efficiently," in *Proc. ACM Annual Conf.*, Aug. 1972, pp. 518–527.

[26] A. A. Kuehn and M. J. Hamburger, "A heuristic program for locating warehouses," *Management Sci.*, vol. 9, pp. 643–666, July 1963.

[27] F. Levy, "An application of heuristic problem solving to accounts receivable management," *Management Sci.*, vol. 12, pp. 245–254, Feb. 1966.

[28] R. F. Ling, "A computer generated aid for cluster analysis," *Commun. Ass. Comput. Mach.*, vol. 16, pp. 355–365, June 1973.

[29] A. S. Manne, "Plant location under economies of scale-decentralization and computation," *Management Sci.*, vol. 11, pp. 213–235, Nov. 1964.

[30] Network Analysis Corporation, *The Practical Impact of Recent Computer Advances on the Analysis and Design of Large Scale Networks, Third Semiannu. Tech. Rep.*, June 1974.

[31] E. A. Patrick and F. P. Fisher, II, "Cluster mapping with experimental computer graphics," *IEEE Trans. Comput.*, vol. C-18, pp. 987–991, Nov. 1969.

[32] C. Revelle, D. Marks, and J. C. Liebman, "An analysis of private and public sector location models," *Management Sci.*, vol. 16, July 1970.

[33] C. Roach, "An optimization algorithm for cluster analysis," *Rand Corp.*, pp. 4878, Aug. 1972.

[34] G. Salton, A. Wong, and C. S. Yang, "A vector space model for automatic indexing," *Commun. Ass. Comput. Mach.*, vol. 18, pp. 613–620, Nov. 1975.

[35] D. H. Schwartzmann and J. J. Vidal, "An algorithm for determining the topological dimensionality of point clusters," *IEEE Trans. Comput.*, vol. C-24, pp. 1175–1182, Dec. 1975.

[36] R. W. Swain, "A parametric decomposition approach for the soultion of uncapacitated location problems," *Management Sci.*, vol. 21, Oct. 1974.

[37] D. T. Tang, "Network optimization for teleprocessing systems," in *Proc. 5th Annu. Southeast. Symp. Syst. Theory*, Duke University, Durham, NC, Mar. 22–23, 1973.

[38] M. B. Teitz and P. Bart, "Heuristic methods for estimating the

generalized vertex median of a weighted graph," *Operations Res.*, pp. 955–961, July 1967.

[39] R. R. Trippi, "The warehouse location formulation as a special type of inspection problem," *Management Sci.*, vol. 21, pp. 986–988, May 1975.

[40] L. S. Woo and D. T. Tang, "Optimization of teleprocessing networks with concentrators," in *Conf. Rec., Nat. Telecommun. Conf.*, Atlanta, GA, Nov. 26–28, 1973, pp. 37C1–37C5.

[41] C. T. Zahn, "Graph theoretical methods for detecting and describing gestalt clusters," *IEEE Trans. Comput.*, vol. C-20, pp. 68–86, Jan. 1971.

Measured Performance of an Ethernet Local Network

by John F. Shoch and Jon A. Hupp

February 1980

© Xerox Corporation 1980

Abstract: The Ethernet communications network is a broadcast, multi-access system for local computer networking, using the techniques of *carrier sense* and *collision detection*. Recently we have measured the actual performance and error characteristics of an existing Ethernet installation which provides communications services to over 120 directly connected hosts.

This paper is a report on some of those measurements -- characterizing "typical" traffic characteristics in this environment, and demonstrating that the system works very well. About 300 million bytes traverse the network daily; under normal load, latency and error rates are extremely low and there are very few collisions. Under extremely heavy load -- artificially generated -- the system shows stable behavior, and channel utilization remains above 97%, as predicted.

CR Categories: 3.81.

Key words and phrases: Local computer networks, Ethernet system, carrier sense multiple access with collision detection (CSMA/CD), packet switching, measurements.

1. Introduction

One of the most attractive architectures for a local computer network is a shared, multi-access bus with distributed control. This approach was originally articulated in the design of the Ethernet communications network, in which all of the host machines share access to a single, passive coaxial cable [Metcalfe & Boggs, 1976].

A number of interconnected Ethernet system installations have been in operation for several years, providing communications service for many hosts and supporting a wide variety of applications. The networks have always functioned very well, and probably represent the largest collection of local networks in use today. The intent of the present study was to further our understanding of this system by characterizing the actual behavior of one of these installations in normal use: traffic volume, packet sizes, error rates, efficiency, and more. In addition, we have constructed a test environment that allows us to artificially generate extreme amounts of traffic -- well beyond the expected operating conditions -- in order to help assess performance under heavily loaded conditions.

Successful use of a local computer network depends on much more than the specific communication medium: applications are using higher level protocols, including a complete architecture for internetwork communication among many different systems [Boggs, *et al.*, 1980]. Those topics, however, are beyond the range of the current study, which is directed primarily at the behavior of the Ethernet system itself. A preliminary version of these results was reported in [Shoch & Hupp, 1979]; more detailed discussion will be found in [Shoch, 1980].

In the sections which follow, we review the Ethernet local network, describe the experimental environment and test methodology, measure the system under normal operating conditions, and then measure the behavior with artificially high loads.

2. Review of the Ethernet principles

The approach used in the Ethernet system traces its roots back to the radio-based Aloha packet switching network, developed at the University of Hawaii [Abramson, 1970]. In the Aloha network, terminals equipped with packet radios shared a common multi-access radio channel. There was, however, no central controller allocating access to the channel; instead, a *random access* procedure was used, in which each terminal independently decided when to transmit. With this simplest *pure Aloha* scheme, two terminals may transmit at the same time producing a collision; after a collision, each terminal will have to wait for a random interval and then retransmit. The attraction of this approach is the elimination of any central control; as the load increases, however, the maximum possible utilization of a pure Aloha channel is only 18% [Abramson, 1977].

Unlike the pure Aloha scheme, when an Ethernet station wishes to send a packet a *carrier sense* mechanism is first used, forcing the station to *defer* if any transmission is in progress. If no other station is transmitting, the sender can begin immediately, with zero latency; otherwise, the sender waits until the packet has passed before transmitting.

It is possible that two or more stations will sense that the channel is idle and begin transmissions simultaneously, producing a *collision*. Each sender, however, continues to monitor the cable during transmission, and can provide *collision detection* when the signal on the cable does not match its own output. In that case, each station stops transmitting, uses a *collision consensus enforcement* procedure to ensure that all other colliding stations have seen the collision, and then stops. A retransmission is then scheduled for some later time.

To avoid repeated collisions, each station waits for a random period of time before retransmitting. To avoid overloading the channel, thus making it unstable, the range of this retransmission interval is increased under times of heavy load, using a *binary exponential backoff* algorithm.

Taken together, these mechanisms represent the Ethernet random access procedure, also referred to as *carrier sense multiple access with collision detection*, or CSMA/CD. This access procedure can be applied to any suitable broadcast channel, such as radio, twisted pair, coaxial cable, fiber optics, diffuse infrared, or others. Coaxial cable is well suited for use in constructing a local computer network -- a system designed to provide high-bandwidth communications for machines within a distance of approximately 1 kilometer.

The shared component of an Ethernet local network consists of a single coaxial cable, typically strung in a meandering fashion through a building, perhaps in the ceiling or under a raised floor. As shown in Figure 1, individual computers, or *stations*, connect to the cable with the use of a CATV-style *tap*; a small *transceiver* is connected at the tap, with a cable running down to the *interface* which might be located in the station. The use of a passive medium, and the lack of any active elements in the shared portion, combine to help provide a very reliable and flexible system. This approach also provides for easy reconfiguration of stations: machines can be moved -- disconnected from one point and reattached at another -- without any need to take down the network. (For further details on the design of the Ethernet system, see [Metcalfe & Boggs, 1976; Metcalfe, *et al.*, 1977; Shoch, 1979, 1980; and Crane & Taft, 1980].)

All of these mechanisms are combined to increase the probability that an Ethernet system will correctly deliver a client's packet. The network gives its best effort, but -- like any other low-level communications subsystem -- the Ethernet communications network cannot provide a 100% guarantee of delivery. In a complete network architecture, suitable *packet protocols* will be layered on top of this local network, to provide further reliability.

3. Experimental environment

The principles of the Ethernet system were first realized in the *experimental* or *prototype* system developed at the Xerox Palo Alto Research Center: using regular coaxial cable, running at 2.94 Mbps. There are now many of these networks in use in locations throughout the United States, tied together with the Pup internetwork architecture [Boggs, *et al.*, 1980].

The particular local Ethernet network installation chosen for these measurements is one of the oldest in existence: it spans about 1800 feet (550 m.) and connects over 120 machines. These machines include large numbers of single-user stand-alone computers (mainly the Alto computer [Thacker, *et al.*, 1980]), two time-sharing servers running the Tenex operating system, numerous shared printers and file servers, as well as several internetwork gateways.

Applications which use the network include file transmission to the printers, file transfer to one of the storage systems, specialized multi-machine programs, access to shared data bases, remote diagnostics, down-line loading of programs, terminal access to the time-sharing machines, and others.

To conduct the measurements, a series of specialized test and monitoring programs has been built to assess the behavior of the network. The broadcast nature of an Ethernet system makes it particularly well suited for the passive collection of measurements: an individual station can sense the state of the cable and -- by operating its interface in a *promiscuous* mode -- can receive all of the packets passing by. (We thus avoid many of the Heisenberg problems: no processing is done at the source or destination to collect statistics, and no communications bandwidth is used to report them.)

Most of the measurements in the next two sections were collected using this passive technique, watching packets that go by on the cable.

4. Reliability and packet error rates

One of the major objectives of any local network is to provide reliable communications facilities, reflected both in the continued availability of the network itself, and in the lowest possible error rate as seen by individual hosts. Overall reliability was one of the most important considerations in the development of the Ethernet network. Thus, the only shared component in the communications subsystem is the passive coaxial cable; the shared portion has no electronics, no active components, no switches, no power being supplied through the line, and no power supplies.

In practical operation, there have been very few system-wide failures; almost all of these are caused by human error of some sort -- for example, removing the terminator at the end of the cable. On one occasion, a lightning strike generated a large surge in the building ground, disabling a number of improperly installed transceivers. Modified installation procedures and a minor circuit change have now significantly improved lightning resistance.

These system-wide failures are easy to control, and overall reliability is very high. Individual packets, however, are subject to damage when being transmitted, and may arrive in error. The design of the Ethernet system does not attempt to guarantee 100% error-free operation, and applications may use suitable error control procedures, such as packet acknowledgments and retransmissions. But, for sustained high performance, one would certainly want to minimize the occurrence of packet errors.

There are several mechanisms available to detect and discard a damaged packet. When a station receives a packet, the interface checks that an integral number of 16-bit words has been received and that the 16-bit cyclic redundancy checksum (CRC) is correct. If either of these conditions is not met, the packet is considered damaged and is normally discarded.

With a passive listener set to receive every packet on the net, our current machines initially reported one damaged packet in about 6000, but there was wide variance among machines. Further examination has indicated that most of these errors reflect problems in the receiver section of the interface itself, and the packets were actually well-formed on the cable. Experimentation with a revised interface has produced normalized packet error rates of about 1 in 2,000,000 packets. Since these error rates are so low, packets received in error have been excluded from all of the results reported below.

We should note that there is a much more significant cause of lost packets which has nothing to do with network performance: failure to keep a receiving interface running properly. Some networking systems have proposed use of an interface which will receive a packet and then have a lengthy *refractory period* or *blind spot* before the receiver is restarted; these devices have a long *receiver-to-receiver turnaround time*. During these intervals, any packets sent to this station may be correct on the cable, but entirely missed by the station. Interface turnaround times require careful attention in the design of network equipment (see [Shoch, 1980]); but this is a problem beyond the scope of the local network itself.

5. Performance characteristics under normal load conditions

5.1. Overall traffic characteristics

In normal use, this single Ethernet system carries about 2.2 million packets in a 24-hour period, totalling almost 300 million bytes. (For comparison, this traffic roughly corresponds to about half of the volume carried through the Arpanet on an average day.) Not surprisingly, the load is very light at night and is heaviest during the regular daytime hours, with a slight dip around lunchtime; see Figure 2, showing network load over a 24-hour period, sampled over fairly long intervals (six minutes each).

5.2. Utilization

Again measured over a full 24-hour day, this traffic represents an extremely modest usage of the net: ranging from about 0.60% to as much as 0.84%. During shorter periods, however, maximum utilization in the busiest interval is much higher: about 3.6% over the busiest hour, 17%

over the busiest minute, and 37% in the busiest second. Figure 3 represents the load observed over four minutes on the Ethernet network, with individual samples summed over one second intervals; note that the full range of this figure would be contained within just one sample in the previous figure.

These results help to verify our design assumption that computer communications applications tend to produce a "bursty" pattern of requests; what we observe on the shared Ethernet channel is the aggregation of some number of independent sources of "bursty" traffic.

5.3. Packet length

Packets sent through the system exhibit a bimodal distribution of packet length: most of the packets are short ones (containing terminal traffic, acknowledgments, etc.), but most of the total volume is carried in the large packets (often containing file transfer traffic) -- see Figure 4. This is similar to some of the measurements reported for the Arpanet [Kleinrock & Naylor, 1974], but very different from the distributions frequently used in analytical models of networks.

To some extent, the upper and lower limits on the length of packets represent artifacts of the various implementations. Almost all of the traffic consists of Ethernet packets carrying encapsulated internetwork packets, or Pups [Boggs, et al., 1980]. Thus, the minimum sized packet with no data would usually include the Ethernet and Pup headers, while acknowledgments and packets with only 1 or 2 bytes of data would be just a bit larger. At the other extreme, software considerations usually impose an effective upper limit on the size of a packet; depending on the particular system, this ranges from about 200 to 540 bytes.

The mean packet length is about 122 bytes, and the median is about 32 bytes.

5.4. Source-destination traffic patterns

On a given day, over 120 hosts use this Ethernet installation -- nearly every machine known to be connected. The extent of this utilization ranges from a fraction of a percent to over 25% of the observed network traffic. Nearly 1300 different source-destination pairs communicate during this period; on the average, each host sends packets to more than 10 other hosts, but some of this traffic is concentrated to and from specialized servers (see the following section).

Figure 5 shows the source-destination traffic matrix for a fairly typical day. A heavily used server both sends and receives packets from many other hosts; each server appears in the figure as a pair of broken lines, intersecting on the diagonal and roughly symmetric about that point.

This matrix only indicates sources and destinations on the local Ethernet system, but some of these servers are gateways to other networks. Thus, they represent the path to some larger number of internetwork addresses, and there are many more internetwork source-destination pairs.

5.5. Server traffic patterns

As suggested from the source-destination traffic matrix, many of the packets are going to or from specialized servers: time-sharing systems, gateways, information servers, file systems, and printers.

Over one typical day, these identifiable servers sent about 69% of the packets, and received about 73%.

5.6. Inter-packet arrival times

The distribution of inter-packet arrival times over a 24-hour period is shown in Figure 6. The mean inter-packet arrival time is 39.5 ms., with a standard deviation of 55.0 ms. and a median of 8.5 ms.

A fair portion of the current traffic consists of request/response transactions with a server. With low utilization, it is not uncommon to have this exchange take place with no intervening packets being transmitted; some of the spikes in the inter-packet arrival time distribution represent the turnaround times of these servers.

Figure 7 indicates the cumulative distribution of inter-packet arrival times: it shows that 50% of the packets are followed by the next packet within 10 ms., 90% of the packets are followed by the next packet within 64 ms., and 99% within 183 ms.

5.7. Latency and collisions detected by a sender

With the current level of traffic, it is not surprising that most attempts to send a packet succeed on the first transmission -- there are no collisions and no need to backoff and retransmit.

To measure this result, a test program has been run which periodically wakes up and tries to transmit a packet; under normal load, it indicates that 99.18% of the packets make it out with zero latency, 0.79% of the packets are delayed due to deference, and less than 0.03% of the packets are involved in collisions.

5.8. Overhead

In addition to useful data, there are several forms of overhead which impact Ethernet communications: headers and error checking on every packet, as well as entire packets carrying acknowledgments, routing tables, or other ancillary information.

The overhead due specifically to Ethernet encapsulation is only 4 bytes; this represents about 3% of an average length packet. The standard internetwork protocol header consumes another 22 bytes, or about 18% of an average packet. Thus, data fields of all packets represent about 79% of the bits carried. It is worth noting that overhead accounts for less than 5% of the maximum length packet used for high-volume data transfers.

To get a precise measure of the actual data bits sent by user protocols one would need to exclude acknowledgment packets, and perhaps model both ends of the end-to-end process. As an approximation, however, we can obtain a reasonable estimate by measuring the amount of data carried inside data packets of the most common protocols. This will generally provide a conservative estimate of user traffic, counting as overhead all of the packets used to set up a connection, error packets, boot loader packets, and all of the specialized packet types exchanged by other user protocols. (This procedure may also double-count the data in any packets which are actually retransmitted by the source; in general, that number is very small.)

Using this alternative approach, about 69% of all the bits carried through the Ethernet system have been classified as user data, leaving a total of about 31% encompassing all forms of overhead.

5.9. Intranet vs. internet traffic

About 72% of all of the packets seen have both their source and destination on this network -- they are intranet packets. About 28% of the packets are internet traffic, coming from or going to another network via an internetwork gateway. The presence of the two large time-sharing systems on this particular network accounts for much of this traffic, as users access these hosts from distant locations. In addition, it is possible for a local Ethernet system to serve as a transit network for traffic originating on one network and destined for another; with the current topology, however, we see few such "transit packets" on this particular network.

6. Performance characteristics under high load conditions

The foregoing discussion has considered network performance under normal operating conditions, reflecting the current demands placed upon the network by the existing set of applications. Further growth of new systems will increase the load on the net, and the system should certainly be able to accommodate short term bursts of very high load. One of the considerations in designing an application would be the performance of an Ethernet system as the load increases dramatically.

Initial modelling of the Ethernet system approach [Metcalfe & Boggs, 1976] has indicated that the system should continue to function very well as the load goes up. The following measurements serve to verify that prediction.

In a high load situation, each individual host can be set to generate packets at a specified rate, generally less than the channel capacity, and independent of the rate at which these packets are actually sent through the channel. The total offered load then represents the sum of these individual offered loads, and may be less than 100% of the channel capacity, or more than 100% (in which case some of the transmit queues overflow, and packets are lost). This load can be varied by changing the number of hosts, or the load generated per host. (In section 7 below, we explore a bit further the special case in which each host can be viewed as a continuously queued source, always trying to transmit a packet, and generating a new one as soon as this one is sent. Thus, each host can be viewed as attempting to present a load of 100%. In this case, it is only the number of hosts attempting to transmit which is of interest; it is not meaningful to talk about increasing the load generated by each host.)

Measurements of interest include the actual level of channel utilization, the stability of the system as the load increases (does the total network utilization dramatically decrease as the load increases?), and the fairness with which the channel is shared.

To help gauge this behavior under high load, and to help stress the capabilities of the system, we have constructed test programs which generate artificially high levels of traffic. Using a special control program, we use the network to find idle machines and then load them with a test program which can be set to produce a specified offered load to the network. Statistics are accumulated as to the number of packets successfully sent and at the end of a test period, after all the test machines are idle, the controlling machine can reach out and collect these statistics.

When a test program generates a packet for transmission, it is placed upon an output queue; actual transmission is performed by an output process, which also handles retransmissions due to collisions. The presence of the queue allows the sender to tolerate short term variations in the availability of the channel. If the total offered load exceeds 100%, however, it is clear that some of the traffic being generated will never be transmitted successfully, and it merely gets dumped off of the output queue.

One can start with a modest offered load, and then increase that level by either adding more machines or increasing the load being offered by each host. With ideal scheduling, one would like to see the total channel utilization increasing with the total offered load, up to 100% (see Figure 8). Beyond 100% load -- under *very* heavy load -- we would like to see the channel utilization remain at 100%, representing full use of the available capacity.

Few real system can perform this well. A pure Aloha channel gets 18% utilization at best and a slotted Aloha channel can realize 37% utilization. Both Aloha mechanisms can become unstable: utilization declines significantly as offered load increases [Abramson, 1977]. In general, one tries to use a control procedure that will maximize the utilization, while remaining stable as the load increases.

Actual test runs were made on the regular Ethernet system described above, usually at night, when there was very little other traffic. The high load results reported below were for full-size data packets (512 bytes), typical of most of the volume moved through the net.

6.1. Maximum utilization

As the total offered load increases from 0% to 90%, channel utilization matches it perfectly: all of the traffic gets out correctly. As the offered load moves above 90%, the channel utilization flattens out at a level above 96% (see Figure 9). This experiment has been run with even greater loads, as high as 9000% (well beyond any reasonable operating region!); utilization continued to increase slightly, approaching 98% -- just 2% short of the ideal case.

6.2. Stability

As Figure 9 indicates, an Ethernet system under high load shows no instability: the throughput curve does not decline as the total offered load increases.

6.3. Fairness

The Ethernet control discipline is also very fair in its allocation of channel capacity: at 100% offered load (10 hosts at 10%/host) we observe a total utilization of 94% with individual throughputs ranging from 9.3 to 9.6%.

7. Performance characteristics with continuously queued sources

The preceding section described a general model in which both the load per host and the number of hosts could be varied, producing total offered load in the range of 0% to 150%. In this section we examine a bit more closely the special case in which every host is continually queued, trying to generate a load of 100% of the channel rate.

This corresponds to the analytical model developed in the original paper on the Ethernet communications network (see [Metcalfe & Boggs, 1976]); the actual measurement results are of use in trying to verify this model. In the ideal case, a single machine continuously transmitting would produce a utilization of 100%; with perfect scheduling, the addition of other machines would not cause the total utilization to decrease. But we do not have perfect knowledge nor scheduling, and as more users queue up to transmit, collisions take place, and the utilization decreases. The analytical results have predicted, however, that for reasonably large packet sizes the distributed control algorithm of the Ethernet would only cause the total utilization to decrease by a small percentage (see Figure 10a).

In practice, we can only approximate these test conditions, since the channel cannot be driven continuously by one host (due to non-zero transmitter-to-transmitter turnaround times). A single machine transmitting large 512 byte packets, for example, can offer a 95% load, but with smaller packet sizes the effective load becomes even lower. With the current population of machines, the most interesting (and feasible) measurements use from 5 to 64 hosts.

As predicted, the total utilization starts out very high, and decreases somewhat as additional hosts are added. The test results for various combinations of packet length and number of hosts are summarized in the table below (also see Figures 10a and 10b):

Ethernet utilization with continuously queued sources

Q = number of hosts, each generating 100% load

Packet length P (in bytes)

Q	512	128	64	6	4
5	97%	95%	94%	72%	----
10	97%	91%	89%	68%	58%
32	97%	90%	83%	64%	56%
64	97%	92%	85%	61%	54%

These measured results are very similar to the results predicted in the model used in the original paper on the Ethernet system. As expected, Ethernet utilization increases with packet size. For full-size data packets -- 80% of the typical traffic -- total utilization remains above 97%, even with 64 continuously queued hosts.

With small packets, any time lost to collisions and collision resolution becomes larger compared to the packet size, and total utilization decreases. Four bytes represents the smallest packet we can send through the Ethernet network, and utilization would decrease further if the packets were even smaller. Indeed, with even smaller packets the packet transmission time should approach the *collision interval* and network utilization approaches $1/e$, the maximum efficiency of a slotted Aloha network [Abramson, 1977].

7.2. Stability

As the table above shows, the Ethernet system remains stable even under extreme overload conditions. In experiments with as many as 90 hosts sending medium- to large-size packets (each offering up to a 95% network load) Ethernet utilization remains high, and shows no signs of suddenly decreasing or becoming unstable.

7.3. Fairness

With more than one continuously queued source, some of the traffic cannot be accommodated, and each station can get only some fraction of its nominally desired bandwidth (100% per host). In the full range of tests the system exhibits very good fairness to all of the machines. With 90 hosts sending, the average utilization per host is 1.1%, ranging from about .9% to 1.3%.

8. Concluding remarks

From these results, we can now begin to verify some important hypotheses about the performance of an Ethernet local network:

1. The error rates are very low, and very few packets are lost.

2. Under normal load, transmitting stations rarely have to defer and there are very few collisions. Those few collisions are resolved very quickly. Thus, the access time for any station attempting to transmit is virtually zero.

3. Under heavy load there are more collisions, but the collision detection and resolution mechanisms work well, and channel utilization remains very high -- almost 100%. In addition, channel sharing is quite fair.

4. Even under extreme overload, the Ethernet channel does not become unstable. (There have been several proposals for complex control schemes which have claimed higher utilization than the Ethernet approach, fairer allocation of the channel, or better performance [Shoch, 1979]. In light of the above results, these alternatives offer increased complexity with little potential benefit.)

With all of these characteristics, an Ethernet system remains a particularly attractive choice as an architecture for local communication.

9. Acknowledgments

The original design of the Ethernet system is due primarily to Robert M. Metcalfe and David R. Boggs; they have both been of great help in undertaking this project. We are also indebted to Edward A. Taft for his help in conducting this work.

10. References

[Abramson, 1970]
N. Abramson, "The Aloha system -- another alternative for computer communications," *AFIPS Conference Proceedings (FJCC 1977)*, 37, 1970, pp. 281-285.

[Abramson, 1977]
N. Abramson, "The throughput of packet broadcasting channels," *IEEE Transactions on Communications*, com-25:1, January 1977, pp. 117-128.

[Boggs, *et al.*, 1980]
D. R. Boggs, J. F. Shoch, E. A. Taft, and R. M. Metcalfe, "PUP: An internetwork architecture," *IEEE Transactions on Communications*, com-28:4, April 1980.

[Crane & Taft, 1980]
R. C. Crane and E. A. Taft, "Practical considerations in Ethernet local network design," *Proc. of the 13th Hawaii International Conference on Systems Sciences*, January 1980, pp. 166-174.

[Kleinrock & Naylor, 1974]
L. Kleinrock and W. E. Naylor, "On measured behavior of the ARPA Network," *AFIPS Conference Proceedings (NCC 1974)*, 43, June 1974, pp. 767-780.

[Metcalfe & Boggs, 1976]
R. M. Metcalfe and D. R. Boggs, "Ethernet: Distributed packet switching for local computer networks," *CACM*, 19:7, July 1976, pp. 395-404.

[Metcalfe, *et al.*, 1977]
R. M. Metcalfe, D. R. Boggs, C. P. Thacker, and B. W. Lampson, *Multipoint data communication system with collision detection*, United States Patent No. 4,063,220, December 13, 1977.

[Shoch, 1979]
J. F. Shoch, *An annotated bibliography on local computer networks (1st edition)*, Xerox Parc Technical Report SSL-79-5, and IFIP Working Group 6.4 Working Paper 79-1, October 1979.

[Shoch, 1980]
J. F. Shoch, *Design and performance of local computer networks*, in press, to appear 1980.

[Shoch & Hupp, 1979]
J. F. Shoch and J. A. Hupp, "Performance of an Ethernet local network -- a preliminary report," *Local Area Communications Network Symposium*, Boston, May 1979, pp. 113-125. Revised version presented at the *20th IEEE Computer Society International Conference (Compcon '80 Spring)*, San Francisco, February 1980.

[Thacker, *et al.*, 1979]
C. P. Thacker, E. M. McCreight, B. W. Lampson, R. F. Sproull, and D. R. Boggs, *Alto: A personal computer*, Xerox Parc Technical Report CSL-79-11, August 1979. To appear in Siewiorek, Bell, and Newell (Eds.), *Computer Structures: Readings and Examples, 2nd edition*.

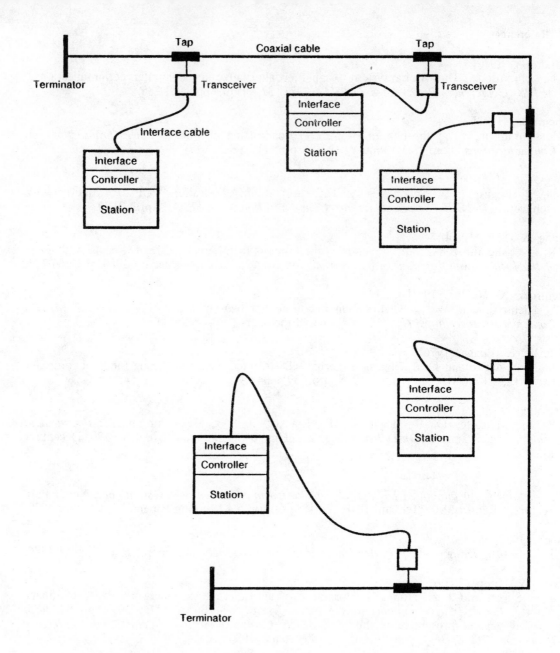

Figure 1: An overview of an Ethernet communications network (after [Metcalfe & Boggs, 1976]).

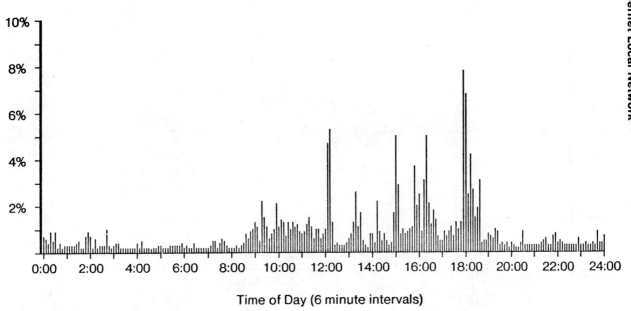

Max Load This Period = 7.9%
Min Load This Period = 0.2%
Average Load This Period = 0.8%

Figure 2: Ethernet load on a typical day, sampled over 6 minute intervals.

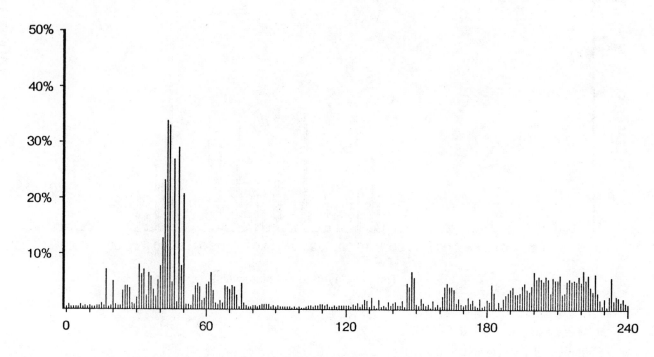

Sampled during 1 second intervals

Max Load This Period = 32.4%
Min Load This Period = 0.2%
Average Load This Period = 2.7%

Figure 3: Ethernet load over a 4 minute period, sampled over 1 second intervals.

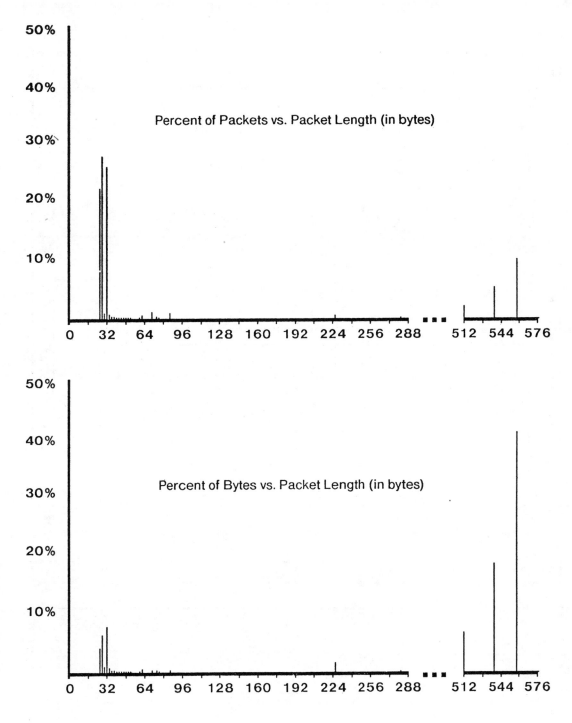

Figure 4: Distribution of packet lengths.

Source host number (octal)

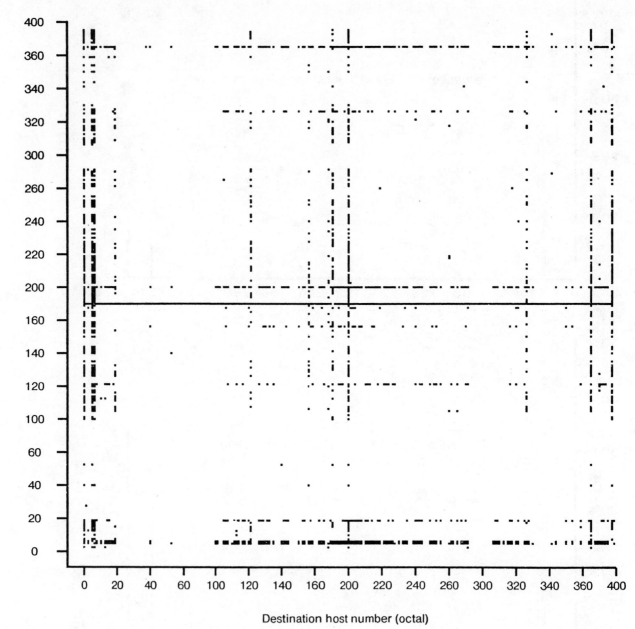

Destination host number (octal)

Figure 5: Source-destination traffic matrix.

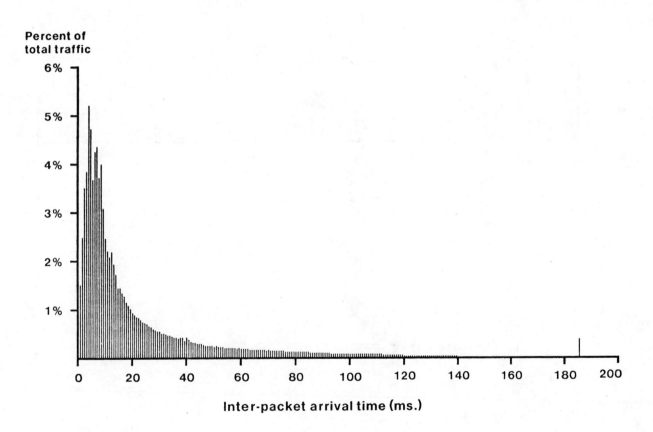

Figure 6: Distribution of Inter-Packet arrival time.

**Percent of
total traffic**

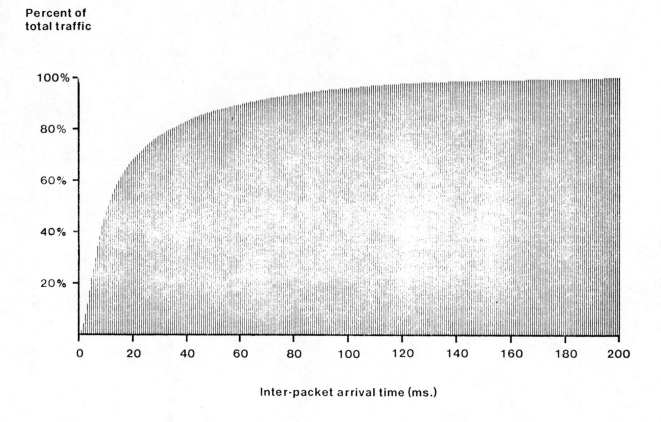

Inter-packet arrival time (ms.)

Figure 7: Cumulative inter-packet arrival times.

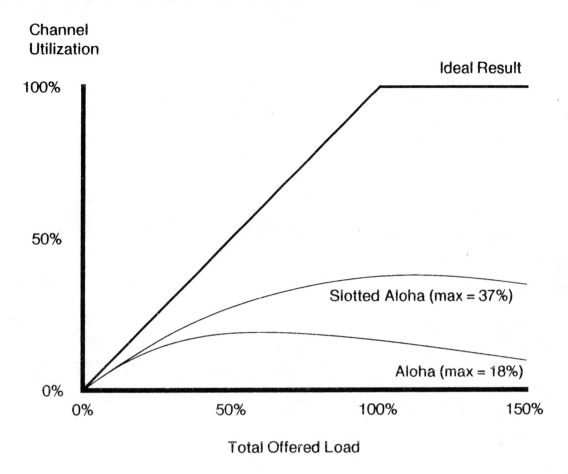

Figure 8: Utilization for several Aloha schemes.

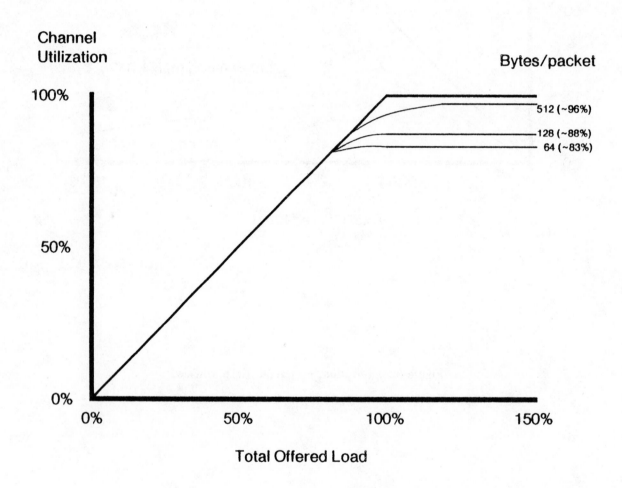

Figure 9: Measured utilization of the Ethernet network under high load.

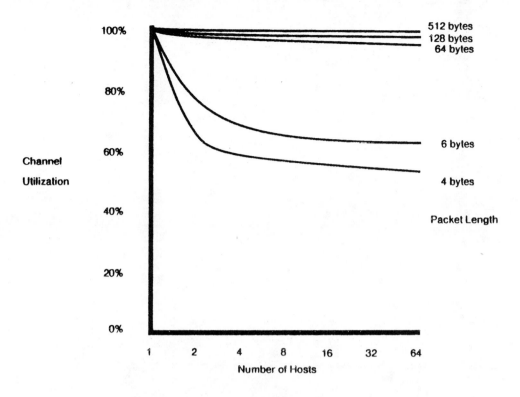

Figure 10a: Predicted utilization with continuously queued sources
(recomputed from [Metcalfe & Boggs, 1976]).

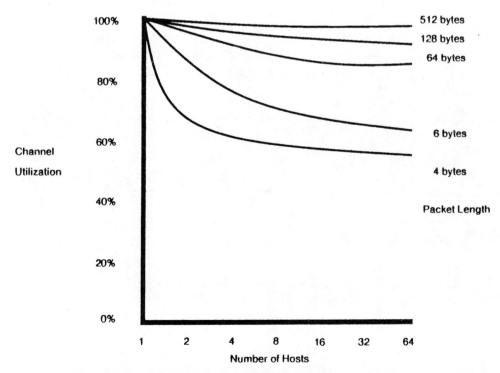

Figure 10b: Measured utilization with continuously queued sources.

Combining Voice and Data into Large-Scale Satellite/Terrestrial Networks

This section:

- integrates many of the separate ideas and technologies of the previous sections into fully integrated nationwide and worldwide telecommunications systems

- shows various considerations and approaches to designing large-scale networks

- introduces various techniques for combining satellite facilities with carefully selected terrestrial facilities both to improve the network delay performance and to optimize cost

- discusses a number of different ways to achieve the best features of circuit switching and packet switching in common facilities and networks

- presents the results of a comprehensive tradeoff study that shows the economic impact of various design choices as they affect a nationwide network

- summarizes some of the approaches being taken by the major common carriers to evolve to integrated voice/data networks.

Thus far, each section of the book has dealt with one major aspect of networks where each portion of the technology could be used, almost by itself, to achieve highly efficient, worldwide telecommunications. Yet, it is clear, that no one approach, be it satellite or terrestrial, packet-switched or circuit-switched, bus-connected or loop-connected, could be best for all situations. The papers presented in this section illustrate a number of approaches to combining and integrating the various techniques to achieve better efficiency, lower cost, and greater service capability.

Up to this point, much of the discussion has centered on the implementation of distributed data communications networks. Though voice communications were not specifically excluded from the discussions, the data rates of the applications used in the discussions and in the examples were clearly directed to the distributed data communications user. A number of the papers in this section deal more explicitly with the integration of voice and data services within a common network. This integration is evolving in many ways within the world's common carrier networks. Though there are many different technical implementations, the general class of such networks has come to be known as *integrated services digital network,* or ISDN.

A fundamental precursor to ISDN is the conversion of the basic analog voice signals into a digitized bit stream. A wide range of techniques are employed to achieve this voice digitization process, techniques that result in effective voice data rates from as low as 2400 bits per second up to 120,000 bits per second, or more. While low rates are most desirable from a transmission resources point of view, the cost of the devices that implement the analog-to-digital conversion tend to rise rapidly in price as the output data rate is decreased. In addition, the widespread use of Pulse Code Modulation, PCM, operating at 64,000 bits per second throughout most of the world's common carrier networks has created this technique essentially as a de facto standard. PCM, using 8000 samples per second, converts each sample into an 8-bit codeword, is robust, is only minimally affected by bit errors on the link, and provides such an accurate representation of the analog voice channel that any signal format capable of passing through the voice channel will accurately pass through the PCM channel.

Recent advances in coding techniques and activity detection reduce the effective channel bit rate for voice users. "Activity detection" is simply the recognition that there are frequent, natural pauses in the flow of human speech. Long pauses occur while one person listens to another. Shorter pauses occur between thoughts, between sentences, between words, and even between syllables of words. On a microscopic timescale of short-data packets, speech pauses are frequent and long enough in duration to accommodate a number of data packets without, apparently, interrupting normal speech. Advanced processing techniques are beginning to capitalize on this phenomenon in the configuration and integration of systems using activity detection. Using this approach can reduce the effective bit rate of a voice user by a factor of three or four and significantly reduce the total bit transfer rate required in an integrated system.

HANDLING VOICE AS IF IT WERE DATA

The ultimate goal of digitizing voice, and processing the digitized voice bit stream by using activity detection, is to achieve a "bursty" data stream that can be handled in a generalized data network. The network need not then be concerned with recognizing any distinction between voice and data services. More logically, certain users of the network, voice users as well as interactive data users, would likely demand a highly reliable, low-delay service. At the same time large-volume data transmissions, facsimile transmission, document distribution systems and the like, would utilize a high-bandwidth, lower-cost service, although such high-capacity service may occasionally suffer some measurable network delays.

The ability to fully integrate voice with data is of interest not just from the viewpoint of economy and efficiency, but also because many new services are being developed where the combination of the two formats is an inherent feature. Voice mailbox systems, where messages can be left in the voice of the caller yet stored in digital computers, are already in widespread use. Voice answerback systems, where callers can make simple data inquiries into a system by using the telephone keypad and getting their response from a computer-synthesized voice, are currently employed in many financial and business applications. In other words even those papers in this section that seem to relate specifically to computer networks, cover the techniques that are applicable to voice- and data-integrated networks.

TOPOLOGICAL ASPECTS OF LARGE-SCALE NETWORKS

In designing a large communications network, selecting the best mix of lines and switches is, by itself, a major challenge. The communications circuits available, from a very large number of suppliers, differ widely in capacities, data rates, tariffs, bulk discounts, service periods, and the like. The paper by M. Gerla and L. Kleinrock attempts to cut through the maze by developing computer-based algorithms to attack the fundamental problem of minimizing resources. Given the node locations and traffic estimates of busy hours, the models attempt to minimize the total line cost by optimizing the network topology, channel capacities, and routing techniques. The designs also have to take into account constraints on overall capacity, user delay, and reliability. Even stated as simply as this, achieving a mathematical solution is considerably difficult, primarily because of the highly non-linear cost/capacity relationships for transmission facilities. The authors introduce various approaches to dealing with these nonlinearities, and provide very useful models for large-scale networks that can be used nationwide.

The paper by R.D. Rosner deals with a somewhat different set of topological issues. Earlier examination of the operation of alternative protocols within a packet-based network revealed a tight interrelationship between the operation of the protocols and the overall packet transit time through the network. When transit times become excessive, severe limitations are imposed on network performance, particularly with respect to user throughput and message delay. One of the best ways to control this degradation of protocol performance is to limit the number of network nodes or switching centers along the user-to-user path. Hierarchical networks, which employ a high degree of interconnectivity among the higher level (tandem) nodes, can achieve these effects, although they generally incur a penalty in terms of the total network resources (lines and circuit miles) required. In this paper, this effect is examined over a wide range of parameters and network structures. The results provide an optimistic picture: the resource penalty appears to be generally quite small, even in large networks. Thus, the use of hierarchical network designs leads to efficient systems that exhibit very desirable operational characteristics.

COMBINATIONS OF SATELLITE AND TERRESTRIAL CONNECTIVITY

There are many ways in which to combine the capabilities of satellite and terrestrial packet networks to exploit the advantages of both. In packet network use, satellites have the very desirable property of relatively high capacity and low incremental cost, whereas terrestrial packet networks have substantially less inherent user-to-user delay. The paper by D. Huynh, H. Kobayashi, and F.F. Kuo analyzes a unique approach to combining these capabilities. Their concept involves combining a high-capacity satellite network, operated in the ALOHA or Slotted ALOHA mode, with a relatively thin-line, low-capacity terrestrial network. Operation would normally proceed over the satellite channel; however, when a packet collision is detected by two users, the retransmission attempt would automatically be made over the terrestrial network. This idea has two desirable features: First, retransmission packets would not clutter the ALOHA satellite channel, thus reducing the possibility of unstable behavior under heavy traffic loading; second, delays in the system would not accumulate as rapidly because the retries of collided packets would not suffer the one-quarter-second satellite delay. In other words, retried packets would be only slightly delayed compared to packets that might have gotten through successfully on their first attempt.

Interestingly, in terms of improving cost-effectiveness, the benefits of using this technique are only marginal—largely because a thin-line terrestrial packet network is significantly more expensive than the equivalent additional capacity contained within the satellite system. However, under a number of other conditions it appears that very favorable cost-performance will be obtained by using this system and others that combine satellite and terrestrial packet operation. satellite and terrestrial packet operation.

The fourth paper in this section, by R.D. Rosner, takes a look at the conditions needed to optimize the balance of satellite and terrestrial connectivity within a combined voice and data network. Unlike the situation discussed in the previous paper, the premise here is that the individual user has no alternative in its basic route choice through the network. The attempt in this analysis is to route as much traffic as possible through the satellite system. However, the optimization of resources involves selecting the proper number of earth stations used in the network and the amount of residual connectivity needed to connect each user to the nearest satellite earth station. Because user-to-earth station connectivity requires terrestrial facilities, some user-to-user traffic can be diverted to the terrestrial facilities as well. This paper derives "families" of system-cost curves based on earth station and terrestrial circuit costs. The results show the potential for great cost savings in a large nationwide network and they foreshadow the applications currently in use by a number of commercial networks.

HYBRID SWITCHING TECHNIQUES

Although circuit-switching techniques have been traditionally applied to voice networks, and packet switching has been principally applied to modern data networks, the technology of each has progressed sufficiently that each approach can be adapted to either service. In fact, without predefining voice or data as the principal user, it is generally agreed that circuit-switching techniques are best used for information sources that transmit a continuous stream of data. These include large computer file transfers, full-motion video signals, coded and compressed facsimile transmission, and some graphics applications. Packet switching, on the other hand, is best for users who send their data on a bursty or interruptible basis, such as is done for interactive data, short data files, and even conversational speech. Since both types of user demand are likely to coexist in common-user networks, various techniques have evolved, or been suggested, that combine circuit and packet switching in a common switching structure.

In their paper, N. Keyes and M. Gerla discuss commercial implementation of their product bearing the registered trademark, PACUIT SWITCH, which combines packet and circuit switching into a single system. Their approach uses a master frame, or master super packet, which carries the traffic of continuous, noninterruptible, and bursty users. A preassigned portion of each successive interswitch packet would be assigned to the continuous users, thereby creating a low-delay, high-capacity path. The remaining space in each packet would accommodate the information of the other, lower-data-rate, interruptible users.

The paper by C.J. Weinstein, M.L. Malpass, and M.J. Fischer presents a mathematical analysis of an adaptive frame structure that is quite similar to the interswitch data blocks used in the PACUIT switch. The authors present both analytical and simulation results to show the performance improvement achievable with a frame structure that can be adapted to the changing demands of the users. Included in their analysis are the specific effects of such a structure on voice and data packets mixed over a single system.

INTEGRATED SERVICES DIGITAL NETWORKS—ISDN

Given the variety of methods by which voice and data traffic can be combined into common networks by using different forms of switching technology, the analysis and tradeoff study described by I. Gitman and H. Frank provides useful insight into the economic issues. This study, funded by the U.S. Department of Defense posits a number of different switching technologies for implementing a moderately large, nationwide backbone network. Even when subjected to a wide range of assumptions, packet switching emerges as the most cost-effective approach for a new network to be implemented from the ground up.

On the other hand, I. Dorros, in the paper following, shows how the Bell System (American Telephone and Telegraph Company) plans to integrate data and other digital services into its largely voice-dominated network. By combining the very high capacity, interswitch digital trunks with intelligent nodal controllers, a variety of services can be integrated. Much of the ability to achieve this level of integration stems from AT&T deployment of a specialized packet network used to carry the high-speed signalling information among the network's backbone circuit switches.

Finally, in the last paper in this section, S.W. Johnston and B. Litofsky provide more detail on how the end-to-end implementation of 56 KB/S services will be achieved in the common carrier voice networks. In addition to the features described in the previous paper, a key feature of this service offering will be the ability to pass the 56 KB/S data directly over most subscriber loops by using a time-compression multiplexing technique, which allows the simple two-wire loop to carry independent 56 KB/S data streams in both directions at the same time.

As you study the papers in this section, remember that they are but a sampling of thousands of ideas that eventually will lead to the integration of many services into combined satellite/terrestrial networks. Ultimately, networks structured according to that shown in Figure VIII-1 will evolve. As competition in the industry increases and the time of capital recovery decreases, such innovative ideas will flourish.

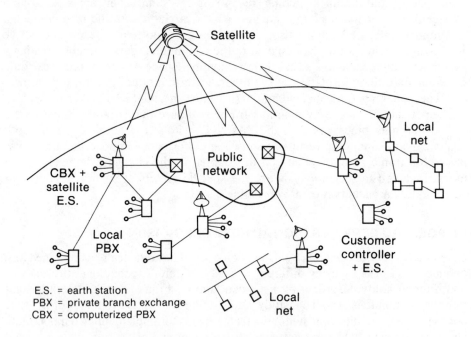

E.S. = earth station
PBX = private branch exchange
CBX = computerized PBX

Figure VIII-1   An integrated distributed communications architecture.

On the Topological Design of Distributed Computer Networks

MARIO GERLA, MEMBER, IEEE, AND LEONARD KLEINROCK, FELLOW, IEEE

Abstract—The problem of data transmission in a network environment involves the design of a communication subnetwork. Recently, significant progress has been made in this technology, and in this article we survey the modeling, analysis, and design of such computer-communication networks. Most of the design methodology presented has been developed with the packet-switched Advanced Research Projects Agency Network (ARPANET) in mind, although the principles extend to more general networks.

We state the general design problem, decompose it into simpler subproblems, discuss the solutions to these subproblems, and then suggest a heuristic topological design procedure as a solution to the original problem.

I. INTRODUCTION

MANY stand-alone computer systems were configured and put into operation long before anyone seriously analyzed their performance (a procedure which sometimes led to embarrassing failures). In contrast, the field of computer-communication networks is at once both unusual and fortunate in that great care has gone into analysis and design techniques prior to system implementation. In this paper we wish to survey some of the recent mathematical techniques which have been found useful in the topological design and performance evaluation of computer-communications networks. Most of the procedures we describe below were first developed in the process of designing the Advanced Research Projects Agency Network (ARPANET) [3], [6], [7], [13], [15], [21], [24], [25], [31], [32], [43], [44], [46], [47], but were later applied to the design of a large variety of Government and commercial distributed data networks.

Many of the early computer networks were constructed mainly to provide access to a *centralized* computer service from a large number of remote users. Such centralized networks have a tree structure, with the computer located at the root of the tree and the terminals located at the nodes. The communication lines are shared among several terminal users by means of multidrop, multiplexing, and concentration techniques. Considerable research effort has been spent on the minimum cost design of these centralized computer networks, and a vast literature is now available [2], [4], [12].

In the pursuit of more efficient computer configurations, it was recognized in the late 1960's that the utilization of existing computer systems could be improved by connecting them

together as a resource-sharing network [32]. Among the shared resources we include: computer power (for load sharing); specialized hardware; specialized software; and data files. This type of network differs from the former in that the computer resources are distributed among the nodes, rather than accumulated in a central node; this configuration is here referred to as a *distributed* computer network. One of the first examples of a distributed network is represented by the ARPANET, a recent configuration of which is shown in Fig. 1.

In distributed networks traffic demands can arise between any two nodes of the network, and not only between terminals and the "central" node. Consequently, better cost-effectiveness and performance are achieved with topological configurations which present a higher degree of connectivity than the centralized tree structures [20].

Computer network users share not only processing facilities, but also communication facilities [30]. The cost-effective configuration and use of communication channels is the main concern of this paper. Conventional line-switching techniques, as used by the common carrier switching network, in which a dedicated path is established for each conversation, are inefficient for computer communications in a bursty mode (e.g., terminal-to-computer conversations). In fact, with the present technology, the time required for establishing and clearing (disconnecting) the path is much longer (on the order of seconds) than the average intercomputer conversation. A more efficient solution for line sharing and speed conversion in a bursty data communication environment is provided by the packet-switching (P/S) technique [27]. Each message is segmented into packets at the source; these packets, instead of traveling along paths reserved in advance, adaptively find their way through the network independently, in a store-and-forward fashion. More than one route between source and destination may be used; such routes can be thought of as pipelines, along which several packets may travel simultaneously, interleaved with other packets corresponding to different source-destination pairs. It is through this pipelining and interleaving that large improvements in network thoughput and message delay are achieved.

The design of distributed P/S networks is substantially different and more complex than the design of centralized networks. The presence of more than one route between each origin and destination in the distributed case requires the solution of complex routing and capacity allocation problems; furthermore, the use of P/S techniques requires the analysis of the relationship between packet delay and line and buffer utilization.

Several algorithms have been proposed for the design of distributed networks. Some of the algorithms are of a heuristic nature; some others are based on more rigorous mathematical

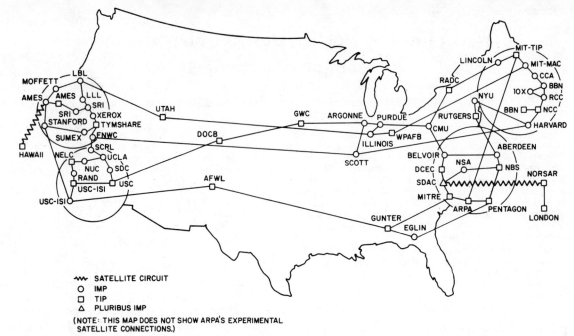

Fig. 1.   ARPA geographical map, June 1976.

programming, queueing, and network flow concepts. In this paper we present a survey of the mathematical programming and network flow approaches that are available for the design of packet networks. We also describe the most common heuristics and compare them to the mathematical programming approach based on the criteria of computation time and solution accuracy.

II. THE DESIGN PROBLEM

Several different formulations of the design problem can be found in the literature [27]; generally, they correspond to different choices of performance measures, of design variable, and of constraints [20], [29]. Here, we select the following very general formulation:

Given	Node locations
	Peak-hour traffic requirements between node pairs
Minimize	Total line cost
Over the design variables	Topology
	Channel capacities
	Routing policy
Subject to	Link capacity constraints
	Average packet delay constraints
	Reliability constraint

Other common formulations of the design problem are the following: 1) minimize average packet delay given the network cost; 2) maximize network throughput given cost and admis-

sible average delay. It is shown in Gerla [20] that all these formulations are closely related and that the solution techniques that apply to our general formulation also apply to the other problems.

In the following sections, we first introduce a network model and discuss the relations between performance measures, input parameters, design variables, and constraints that appear in the general design problem. Then we define and solve (in various degrees of completeness) three design subproblems (capacity assignment, flow assignment, flow and capacity assignment) which are derived from the general problem formulation by fixing some of the design variables. Finally, we study the solution of the general problem (i.e., including the topological design) in which the three above-mentioned problems appear as essential subproblems.

III. MODELING AND ANALYSIS

A. The Model

In a P/S network, packets are transmitted through the network using a store-and-forward technique [26], [27]. That is, a packet traveling from source node s to destination node d is received and "stored" in queue at any intermediate node k, while awaiting transmission, and is sent "forward" to node p, the next node on the route from s to d, when channel (k,p) permits. Even when this channel is free, the packet must first be received fully in node k before transmission to node p may be started. Given the destination d and the present node k, the selection of the next node p is made by a well-defined decision rule referred to as the routing policy. A routing policy is said to be fixed (or static) if a predetermined fraction of the packets arriving at k and directed to d is sent to each output queue; it is said to be adaptive if the selection of the output channel at each node depends on some estimate of current network traffic [27].

Traffic requirements between nodes arise at random times and the size of the requirement is also a random variable. Consequently, queues of packets build up at the channels and the system behaves as a stochastic network of queues. For routing purposes, packets are distinguished only on the basis of their destination [17], [20], [27]; thus, messages having a common destination can be considered as forming a "class of customers." The P/S network, therefore, can be modeled as a network of queues with n classes of customers where n is the number of different destinations [29].

B. Delay Analysis

A vital performance measure for a computer-communication network is the average source-to-destination packet delay T, defined as follows [27]:

$$T = \sum_{\substack{j,k \\ j \neq k}} \frac{\gamma_{jk}}{\gamma} Z_{jk} \tag{1}$$

where

γ_{jk} average packet rate flowing from *source j* to *destination k*

Z_{jk} average packet delay (queue and transmission) from j to k

$$\gamma = \sum_{j,k} \gamma_{jk}. \tag{2}$$

A straightforward application of Little's result [28], [29] to the network of queues model leads to the following very useful expression for T:

$$T = \sum_{i=1}^{b} \frac{\lambda_i}{\gamma} T_i \tag{3}$$

where b is the number of links (arcs), λ_i is the average traffic rate, and T_i is the average queueing plus transmission delay on link i. This expression, established by Kleinrock in 1964 [27], is very general (as general as Little's result!), and extremely simple. Unfortunately, we are not able in general to evaluate λ_i and T_i. However, if we make the following assumptions: 1) external Poisson arrivals; 2) exponential packet length distribution; 3) infinite nodal storage; 4) fixed routing; 5) error-free channels; 6) no nodal delay; and 7) independence between interarrival times and transmission times on each channel, then the evaluation of (3) can be carried out analytically [26], [27]. In fact, the network of queues reduces to the model first studied by Jackson [23], in which each queue behaves as an independent $M/M/1$ queue [28]. Thus, the average delay T_i on channel i is given by

$$T_i = \frac{1}{\mu C_i - \lambda_i} \tag{4}$$

where

$1/\mu$ average packet length (bits/packet)

C_i capacity of channel i (bits/s)

λ_i average packet rate on channel i (packets/s).

The average rates λ_i are easily computed from the routing tables and the traffic requirement matrix [20], [27].

By substituting (4) into (3) and letting f_i be the average bit rate on channel i (bits/s), we obtain the following expression for T:

$$T = \frac{1}{\gamma} \sum_{i=1}^{b} \frac{f_i}{C_i - f_i}. \tag{5}$$

Although the delay expression (5) is sufficiently accurate for most design purposes, it is possible to obtain expressions which correspond more accurately to measured results. Kleinrock proposed in [25] the following very general formula, which includes propagation and nodal processing delay:

$$T = \frac{1}{\gamma} \sum_{i=1}^{b} \lambda_i [T_i + P_i + K_i] \tag{6}$$

where

P_i propagation delay in channel i (seconds)

K_i nodal processing time in the node at which channel i terminates (s/packet).

The term T_i depends on the nature of the traffic and on the packet length distribution.

Finally, assuming a general packet length distribution, with mean $1/\mu$ and variance σ^2, we obtain [26]:

$$T_i = \frac{1}{\mu C_i}(1 - \beta) + \frac{\beta}{\mu C_i - \lambda_i} \tag{7}$$

where $\beta = (1 + \mu^2 \sigma^2)/2$. An even more detailed model has been recently proposed and is discussed in [31].

The validity of the above assumptions and approximations has been tested and verified through simulation and measurements by many authors in a variety of applications [17], [25], [27], [31]. The results indicate that the model is robust.

In this paper, without loss of generality, we limit our considerations to the delay expression in (5). Most of the techniques described in the sequel can be easily extended to more elaborate delay expressions. Furthermore, the solution of some of the design problems seems to be rather insensitive to the introduction of additional details in the delay formula [20].

C. The Communications Cost

With C_i the capacity of link i, we let $d_i(C_i)$ be the cost of leasing capacity value C_i for link i. The communication cost D is defined as

$$D = \sum_{i=1}^{b} d_i(C_i). \tag{8}$$

The availability and effectiveness of the design algorithms depends rather critically upon the form for $d_i(C_i)$. In most applications $d_i(C_i)$ is a discrete variable; however, for computational efficiency it is often convenient to approximate the discrete costs with continuous costs during the initial optimization phase, and to discretize the continuous values during a refinement phase [25].

D. Traffic Requirements

Average (busy-hour) traffic requirements between nodes can be represented by a requirement matrix $R = \{r_{jk}\}$, where r_{jk} is the average transmission rate from source j to destination k. In some cases, we define the requirement matrix as $R = \rho\overline{R}$, where $\overline{R}$ is a known basic traffic pattern and ρ is a variable scaling factor usually referred to as the traffic level.

In general, R (or $\overline{R}$) cannot be estimated accurately *a priori*, because of its dependence upon network parameters (e.g., allocation of resources to computers, demand for resources, etc.) which are difficult to forecast and are subject to changes with time and with network growth. Fortunately, the analysis of several different traffic situations has shown that the optimal design is rather insensitive to traffic pattern variations [1], [20]. This insensitivity property, which seems to be typical of distributed networks, justifies the use of traffic averages for network design.

E. Routing Policy

In designing network topologies, one generally assumes fixed routing, since fixed routing is easy to describe (by means of routing tables, for example) and allows the direct evaluation of channel flows and average delay as a function of routing tables and traffic requirements. Adaptive routing, on the other hand, is complex to describe, and requires simulation to evaluate channel flows and delay. Furthermore, it was shown that *at steady state,* flow patterns and delays induced by good adaptive routing policies are very close to those obtained with optimal fixed policies [18]. This fact suggests that network configurations optimized with fixed routing, are also (near) optimal for adpative routing operations [27].

F. Link Flows

The routing policy and the traffic requirements uniquely determine the vector $f \triangleq (f_1, f_2, \cdots, f_b)$ where f_i is the average data flow on link i. The evaluation of f is straightforward in the case of fixed routing; it can, at this point, only be obtained by simulation if the routing is adaptive.

Conversely, not any generic vector f corresponds to a realizable routing policy and requirement matrix. If it does, then f is a multicommodity (MC) flow for that particular requirement matrix. An MC flow results from the sum of single commodity flows f^{jk} ($j,k = 1,2, \cdots, n$) where f^{jk} is the average flow vector generated by packets with source node j and destination node k, and n is the number of nodes. Clearly, each single commodity flow of the MC flow must separately satisfy nonnegativity and flow conservation constraints.

G. The Capacity Constraint

The presence of capacity constraints $f \leqslant C$ (where $C = (C_1, C_2, \cdots, C_b)$) makes the design problem in Section II a constrained MC flow problem. From the delay expressions (4)

and (7), we notice that if the link flow approaches the link capacity, then the delay approaches infinity, thus violating the delay constraint [16]. Therefore, if both capacity and delay constraint must be satisfied the *capacity constraint is implied by the delay constraint and can be disregarded.*

H. Reliability

Links and nodes in a real network can fail with nonzero probability, thus interrupting some communication paths. It is important to evaluate the overall network reliability in the presence of such failure probabilities.

Several reliability measures have been proposed for computer-communication networks. Among them we mention: the probability of the network being connected (PC); and the fraction of communicating node pairs (FR). Such measures must be verified during the design phase.

Van Slyke and Frank developed very efficient techniques for the evaluation of PC and FR [36]. The techniques, however, are based on simulation and are too time-consuming to be included in a global design algorithm.

Roberts and Wessler [32] proposed as a reliability measure, the two-connectivity of the network (i.e., two node-disjoint paths available between each node pair). This measure is easy to include as a constraint in the topological design. Furthermore, it is adequate for networks with a relatively small number of nodes (on the order of 20–40) and relatively small component failure probability (on the order of 0.01).

For larger networks (or higher failure rates), stronger constraints must be applied to the network topology (e.g., three-connectivity, no long chains, etc.) in order to obtain adequate reliability.

This concludes our model description. In the next four sections, we discuss some of the important design problems and their solutions.

IV. THE LINK CAPACITY ASSIGNMENT (CA) PROBLEM

A. Problem Formulation

The CA problem can be formulated as follows:

Given	Topology Requirement matrix R Routing policy (and therefore link flow vector $f = (f_1, f_2, \cdots, f_b)$)
Minimize	$D = \sum_{i=1}^{b} d_i(C_i) \qquad (9)$
Over the design variables	$C = (C_1, C_2, \cdots, C_b)$
Subject to	$f \leqslant C$
	$T = \dfrac{1}{\gamma} \sum_{i=1}^{b} \dfrac{f_i}{C_i - \lambda_i} \leqslant T_{\max}$

The optimal assignment of capacities to a distributed network with arbitrarily fixed routes is not very interesting as a stand-alone problem, since routing plays a determinant role in the optimization of network performance. Rather, the CA is of practical interest as subproblem of more general optimization problems.

The technique used for the solution of the CA problem depends on the nature of the cost-capacity functions $d_i(C_i)$. In the following, we present optimal and suboptimal algorithms for linear, concave, and discrete costs.

B. Linear Costs

Assuming that $d_i(C_i) = d_i C_i + d_{i0}$ where d_{i0} is a positive start-up cost, the optimal solution is obtained using the method of Lagrange multipliers [27]. In particular, the optimal capacity for channel i is given by

$$C_i = f_i + \frac{\sum_{j=1}^{b} \sqrt{d_j f_j}}{\gamma T_{\max}} \sqrt{\frac{f_i}{d_i}} \qquad (10)$$

and the minimum cost D is given by

$$D = \sum_{i=1}^{b} \left[d_i f_i + d_{i0} + \frac{\left(\sum_{j=1}^{b} \sqrt{d_j f_j}\right)^2}{\gamma T_{\max}} \right]. \qquad (11)$$

C. Concave Costs

The concave case can be solved iteratively by linearizing the costs and solving a linearized problem at each iteration [20]. The method leads, in general, to local minima. However, for the very important case of a power law cost function (i.e., $d_i(C_i) = d_i C_i^{\alpha} + d_{i0}$ where $0 \leq \alpha \leq 1$), Kleinrock showed that there exists a unique local minimum [25]. The iterative procedure therefore yields the optimal solution in the power law case.

D. Discrete Costs

The optimal solution to the discrete CA problem can be obtained with a dynamic programming (DP) technique. The DP algorithm, described by Gerla in [20], requires an amount of computation which, in most applications, is close to $(b)^2$. Another suboptimal technique for the solution of the discrete CA problem is the Lagrangian decomposition (LD). The LD method, first developed by Everett [9], and subsequently improved by Fox [10] and Whitney [37], is suboptimal in the sense that it determines only a subset of the set of optimal solutions corresponding to various values of the parameter $T_{\max}$. The amount of computation required by LD is slightly more than linear with respect to the number of arcs b.

The delay versus cost plot in Fig. 2 shows DP and LD solutions for a large range of values of $T_{\max}$ for a discrete CA application relative to a 26-node ARPANET topology [20]. The circles correspond to LD solutions, whereas the union of circles and dots corresponds to DP solutions. As a property of the LD method, the LD solutions belong to the convex envelope of the global set of optimal DP solutions.

V. THE ROUTING PROBLEM

A. Problem Definition

The routing problem is here defined as the problem of finding the fixed routing policy which minimizes the average delay T. A possible formulation of the problem is the following:

Given	Topology
	Channel capacities $\{C_i\}$
	Requirement matrix R

Minimize

$$T = \frac{1}{\gamma} \sum_{i=1}^{b} f_i \left[\frac{1}{C_i - f_i} + \mu(P_i + K_i) \right] \qquad (12)$$

Over the design variable $\qquad f = (f_1, f_2, \cdots, f_b)$

Subject to
a) f is a multicommodity flow satisfying the requirement matrix R
b) $f \leq C$

From formulation (12), we notice that the routing problem is a convex MC flow problem on a convex constraint set; therefore, there is a unique local minimum, which is also the global minimum and can be found using any downhill search technique [5].

Several optimal techniques for the solution of MC flow problems are found in the literature [8], [35]; however, their direct application to the routing problem proves, in general, to be cumbersome and computationally not efficient. Consequently, considerable effort was spent in developing heuristic suboptimal routing techniques [5], [17], [33]. Satisfactory results were obtained and computational efficiency was greatly improved. However, all of these techniques are affected by various limitations and may fail in some pathological situations.

A new downhill search algorithm, called flow deviation (FD), was recently developed by Fratta, Gerla, and Kleinrock [16]. The FD algorithm finds the optimal solution and is computationally as efficient as the heuristics. To place the FD algorithms in the proper perspective we first introduce the most popular among the heuristic algorithms—the "minimum link" algorithm—and compare it to the FD algorithm.

B. The Minimum Link Algorithm

We begin by giving an outline of the heuristic algorithm reported in [5].

:Algorithm:

Step 1: For a given source j and destination k, determine all paths Π_{jk} with the minimum number of intermediate nodes. Such paths are called "feasible" paths.

Step 2: Choose, among the feasible paths, the least utilized path (or the path with maximum residual capacity).

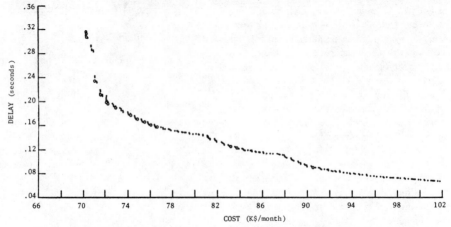

Fig. 2. Delay versus cost for discrete capacity assignments on 26-node 30-link topology.

Step 3: Route the requirement γ_{jk} along such a path.

Step 4: If all source destination pairs have been processed, stop; otherwise select a new pair and go to 1.

Step 1, repeated for all node pairs, corresponds to the evaluation of all shortest paths between all pairs of nodes, assuming unitary link length. Such a computation requires from $(n)^2$ to $(n)^2 \log (n)$ operations, depending on network connectivity. It can be shown that the total amount of computation required by the algorithm has proportion between $(n)^2$ and $(n)^2 \log (n)$.

The minimum link algorithm is conceptually simple and computationally very efficient. Its major drawback is that of being rather insensitive to queueing delays and therefore possibly far from optimum in heavy traffic situations.

C. The Flow Deviation Algorithm

Before introducing the FD algorithm, we mention the following properties of the optimal routing solution.

Property 1: The set of MC flows f satisfying the requirement matrix R is a convex polyhedron. The extreme points of such a polyhedron are called "extremal flows" and correspond to shortest route policies. A shortest route policy is a policy that routes each (j,k) commodity along the shortest (j,k) path, evaluated under an arbitrary assignment of lengths to the links. To each such assignment there corresponds an extremal flow and conversely. Any MC flow can be expressed as a convex combination of extremal flows [16].

Property 2: For a given MC flow f, let us define link length as a function of link flow of the form $l_i \triangleq \partial T/\partial f_i$. Let ϕ be the shortest route flow associated with such link lengths and let $f' = (1 - \lambda)\phi + \lambda f$ be the convex combinations of ϕ and f minimizing the delay T. If $T(f') = T(f)$, then f is optimal.

Property 2 provides a way of finding a downhill direction, if it exists. Based on such property, we may now state the FD algorithm, as follows.

Algorithm:

Step 0: Let $f^{(0)}$ be a starting feasible flow. (A starting feasible flow can be obtained using a modified version of the FD algorithm [16].) Let $p = 0$.

Step 1: Compute $\phi^{(p)}$, the shortest route flow corresponding to $l_i^{(p)} = [\partial T/\partial f_i]_{f=f^{(p)}}, \forall i = 1, \cdots, b$.

Step 2: Let $\bar{\lambda}_p$ be the minimizer of $T[(1 - \lambda)\phi^{(p)} + \lambda f^{(p)}]$, $0 \leqslant \lambda \leqslant 1$. Let $f^{(p+1)} = (1 - \bar{\lambda}_p)\phi^{(p)} + \bar{\lambda}_p f^{(p)}$.

Step 3: If $|T(f^{(p+1)}) - T(f^{(p)})| < \epsilon$, stop: $f^{(p)}$ is optimized to within the given tolerance. Otherwise let $p = p + 1$ and go to Step 1. A geometric representation of the FD algorithm is given in Fig. 3.

Step 1 is the most time-consuming operation of the algorithm, and requires an amount of computation between $(n)^2$ and $(n)^2 \log (n)$. Therefore, the amount of computation required by the minimum link algorithm and the FD algorithm are comparable. A typical central processing unit (CPU) requirement is from 2 to 4 seconds for a 30-node application on a large computer.

VI. THE CAPACITY AND FLOW ASSIGNMENT (CFA)

A. Problem Formulation

The CFA problem can be formulated as follows:

Given	Topology
	Requirement matrix R
	Cost-capacity functions $d_i(C_i)$

Minimize
$$D(C) = \sum_{i=1}^{b} d_i(C_i) \qquad (13)$$

Over the design variables $\quad f, C$

Such that
a) f is an MC flow satisfying the requirement matrix R
b) $f \leqslant C$
c) $T(f, C) = \dfrac{1}{\gamma} \displaystyle\sum_{i=1}^{b} \dfrac{f_i}{C_i - f_i}$
$\leqslant T_{\max}$

The CFA problem requires the simultaneous optimization of routes and line capacities. The existence of a huge number

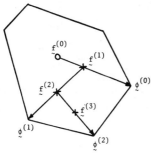

Fig. 3.   Geometrical representation of the FD algorithm. Characters with tildes beneath are boldface in text.

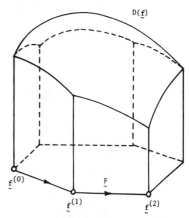

Fig. 4.   Concave objective function $D(f)$. Characters with tildes beneath are boldface in text.

of local minima makes the exact solution difficult to obtain. Therefore, we discuss here only suboptimal techniques.

B. Linear Costs.

In the linear cost case we can obtain, for a given f, a closed form expression of the optimal C in terms of f [see (9)]. In particular, the total cost D can be expressed in terms of f only, and problem (13) can be reformulated as follows:

Given	Topology Requirement matrix R
Minimize	$D(f) = \displaystyle\sum_{i=1}^{b} \left[d_i f_i + d_{i0} \right.$ $\left. + \dfrac{\left(\displaystyle\sum_{j=1}^{b} \sqrt{d_j f_j} \right)^2}{\gamma T_{\max}} \right]$ (14)
Over the design variable	f
Such that	f is an MC flow satisfying R

It can be shown that $D(f)$ is concave over the convex polyhedron of feasible multicommodity flows [20]. This implies that there are in general several (in fact, enormous numbers of) local minima corresponding to some corners of the polyhedron, i.e., corresponding to some extremal flows (see Fig. 4). The FD method described in the previous section can still be applied, in a properly modified form; however, it leads to local minima. More precisely, the FD algorithm performs a local search on extremal flows, until it finds a local minimum. The modified FD algorithm is next introduced.

FD algorithm (for concave objective function):

Step 0: Let $f^{(0)}$ be a feasible starting flow. Let $p = 0$.

Step 1: Let $f^{(p+1)}$ be the extremal flow corresponding to the following definition of equivalent lengths:

$$l_i^{(p)} = [\partial D/\partial f_i]_{f_i=f_i^{(p)}}, \qquad \forall i = 1, 2, \cdots, b.$$

Step 2: If $D(f^{(p+1)}) \geqslant D(f^{(p)})$, stop: $f^{(p)}$ is a local minimum. Otherwise, let $p = p + 1$ and go to 1.

The convergence follows from the fact that there are only a finite (albeit large) number of extremal flows and repetitions are not allowed because of the stopping rule 2.

From (14) the equivalent length l_i defined in Step 1 has the following expression:

$$l_i = d_i \left[1 + \frac{\displaystyle\sum_{j=1}^{b} \sqrt{d_j f_j}}{\gamma T_{\max}} \cdot \frac{1}{\sqrt{d_i f_i}} \right]. \qquad (15)$$

Notice that $\lim_{f_i \to 0} l_i = \infty$. This implies that whenever the flow (and therefore the capacity) of link i is reduced to zero at the end of an iteration, flow and capacity will remain zero for all subsequent iterations, since the incremental cost of restoring the flow on a link is proportional to l_i which in this case is infinity. In other words, uneconomical links tend to be eliminated by the algorithm. This link elimination property, first observed by Yaged [38], can be utilized in the topological design, as shown in the following section.

The solution obtained with the FD algorithm is a local minimum which depends on the selection of the starting flow $f^{(0)}$. In order to obtain a more accurate estimate of the global minimum, several locals are usually explored, starting from randomly chosen flows.

C. Concave Costs

For concave channel costs there is no closed-form expression of D in terms of f. However, it has been shown by Gerla that $D(f)$ is concave over f, and that the FD algorithm can be still applied to obtain local minima [20].

An application of the FD method to the topology shown in Fig. 5 is next introduced. Channel capacities are available only in discrete sizes (see Table I); therefore, discrete channel costs are approximated with continuous power law costs, to apply the concave cost version of the FD algorithm.

A uniform traffic requirement of 1 kbit/s was assumed between all node pairs. Fifty different local minima were obtained with the FD method using randomly generated

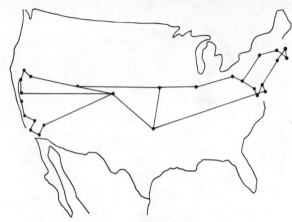

Fig. 5. ARPA-like topology with 26-nodes and 32 links.

TABLE I
LINE COST (TELEPAK RATES)

CAPACITY [K BITS/S]	TERMINATION COST [$/MONTH]	MILEAGE COST [$/MONTH/MILE]
9.6	650	.40
19.2 (2 x 9.6)*	1,300	.80
19.2	850	2.50
50.0	850	5.00
100.0 (2 x 50)*	1,700	10.00
230.4	1,350	30.00

TABLE II
DISTRIBUTION OF LOCAL SOLUTIONS

D[$/MONTH]	NO. OF SOLUTIONS
88,400 - 88,500	8
88,500 - 88,600	11
88,600 - 88,700	11
88,700 - 88,800	19
88,800 - 90,000	1

starting flows. The distribution of the local solutions in Table II shows that all the solutions fall in a very narrow range. Due to the random procedure used to select starting flows, we may conjecture that the optimal solution is also close to this cost range [20].

The execution time required to generate the 50 local minima was about 50 seconds on an IBM 360/91 (about 1 second per local solution).

D. Discrete Costs

For the discrete cost problem one of the following heuristic approaches may be used.

Approach 1: Solve, iteratively, a routing problem with fixed capacities, followed by a discrete CA problem with fixed flows, until a local minimum is obtained.

Approach 2: Interpolate discrete costs with continuous, concave costs. Solve the corresponding concave CFA problem.

Adjust the continuous capacities to the smallest feasible discrete values. Reoptimize the flow assignments by solving a routing problem.

Four algorithms are next introduced, the first three following Approach 1 and the last following Approach 2.

1) Minimum Link Assignment [13]:

Step 1: Apply the minimum link routing algorithm, as described in Section V-B, to assign link flows.

Step 2: Allocate to link i the smallest feasible capacity C_i, such that

$$C_i > f_i, \qquad \forall i = 1, \cdots, b.$$

Step 3: Reoptimize the routing, so as to maximize throughput.

2) Bottom Up Algorithm [20]:

Step 1: Assign minimum available capacities to the links.

Step 2: Maximize the throughput ρ at $T \leqslant T_{max}$, using the FD algorithm.

Step 3: If $\rho \geqslant \bar{\rho}$ (where $\bar{\rho}$ is the required traffic level) stop; we have a feasible suboptimal solution. Otherwise, increase the capacity on the "most utilized" link and go to Step 2.

3) Top Down Algorithm [20]:

Step 1: Assign maximum available capacities.

Step 2: Maximize ρ at $T \leqslant T_{max}$, using the FD algorithm.

Step 3: If $\rho < \bar{\rho}$ (where $\bar{\rho}$ is the required traffic level) stop; routing and capacities of the previous iteration represent the feasible suboptimal solution. Otherwise, decrease the capacity on the "least utilized" channel and go to Step 2.

4) Discrete Capacity (Dis Cap) Algorithm [20]:

Step 1: Interpolate discrete costs with continuous, power law costs.

Step 2: Solve the continuous CFA problem and find a local minimum (f, C).

Step 3: Keeping f fixed, solve a discrete CA problem.

Step 4: Keeping C fixed, solve a routing problem.

Step 5: If cost D did not decrease between two successive iterations, stop. Otherwise, go to Step 3.

In order to evaluate the effectiveness of the above heuristics, the three latter algorithms were applied to the network shown in Fig. 5, and the results compared. The difference in cost between the best and the worst solution was less than 3 percent; furthermore, two of the three solutions (Top Down and Dis Cap) were identical! Considering the fact that the methods are conceptually very different, the narrow cost range of the solutions implies that they are close to optimum.

Execution time for each of the three algorithms (Top Down, Bottom Up, and Dis Cap) was about 90 seconds on an IBM 360/91.

VII. TOPOLOGICAL DESIGN

A. Problem Formulation

The topological problem can be generally designed as follows:

Given	Requirement matrix R
Minimize	$$D(A, C) = \sum_{i \in A} d_i(C_i)$$ where the set of arcs A specifies the topology
Such that	a) f is an MC flow satisfying the requirement matrix R b) $f \leqslant C$ c) $T = \dfrac{1}{\gamma} \sum_{i \in A} \dfrac{f_i}{C_i - f_i} \leqslant T_{\max}$ d) The set A must correspond to a 2-connected topology

There exists no efficient technique for the exact solution of this topological problem. Several heuristics, however, have been proposed and implemented and are discussed below.

B. The Branch X-Change (BXC) Method [13], [34]

This method starts from an arbitrary topological configuration and reaches local minima by means of local transformations (a local transformation, often called branch X-change, consists of the elimination of one or more old links and the insertion of one or more new links).

The BXC method has found applications in a variety of topological problems (natural gas pipelines [11], minimum cost survivable networks [34], centralized computer networks [12], etc.). In particular, BXC has been applied to the topological design of distributed computer networks [13]. The algorithm described in [13] is iterative and each iteration consists of three main steps, as follows.

Step 1: Local transformation. A new link is added and an old link is deleted in such a way that two-connectivity is maintained.

Step 2: Capacities and flows are assigned to the new topological configuration using the minimum link assignment described in Section VI, and cost and throughput are evaluated. If there is a cost-throughput improvement, then the topological transformation from Step 1 is accepted. Otherwise, it is rejected.

Step 3: If all local transformations have been explored, stop. Otherwise, go to Step 1.

C. Concave Branch Elimination (CBE) Method [20], [38]

The CBE method can be applied whenever the discrete costs can be reasonably approximated by concave curves [20]. The method consists of starting from a fully connected topology, using concave costs and applying the FD algorithm described in Section VI until a local minimum is reached. Typically, the FD algorithm eliminates uneconomical links, and strongly reduces the topology. Once a locally minimal topology is reached, the discrete capacity solution can be obtained from the continuous solution with the techniques discussed in Section VI. Since two-connectivity is required, the FD algorithm is terminated whenever the next link removal violates this constraint; the last two-connected solution is then assumed to be the local minimum. In order to obtain several local minima, and therefore several different topological solutions, the FD algorithm is applied to several randomly chosen initial flows.

D. Other Methods

Both the BXC and CBE methods have some shortcomings. For example, the BXC method requires an exhaustive exploration of all local topological exchanges and tends to be very time consuming when applied to networks with more than 20 or 30 nodes. The CBE method, on the other hand, can very efficiently eliminate uneconomical links, but does not provide for insertion of new links. In order to overcome such limitations, new methods derived from BXC or CBE have been recently proposed and are now being investigated.

The cut-saturation method, discussed in [22], can be considered as an extension of the BXC method, in the sense that, rather than exhaustively performing all possible branch exchanges, it selects only those exchanges that are likely to improve throughput and cost. In particular, at each iteration: a routing problem is solved; the saturated cut (i.e., the minimal set of most utilized links that, if removed, leaves the network disconnected) is found and a new link is added across the cut; then the least utilized link is removed. The selection of the links to be inserted or removed depends also on link cost.

The concave branch insertion method, discussed in [20], identifies and introduces links which provide cost savings under a concave cost structure. The method can be efficiently combined with the CBE method, to compensate for the inability of the latter to introduce new links.

In some applications with very irregular distributions of node locations, or with constraints which are difficult to formulate analytically (e.g., no chains longer than m hops; connectivity higher than 2, etc.), network design can be greatly enhanced using man–computer interaction. To this end, interactive design programs have been developed in which the network designer can observe (and eventually correct) the topological transformations performed by the computer and displayed iteration after iteration on a graphic terminal.

In general, the selection of the appropriate algorithm will depend on the cost-capacity structure, on the presence of additional topological constraints, on the degree of human interaction allowed and, finally, on the tradeoff between cost and precision required by the particular application.

E. Bounds

In the development and evaluation of topological design algorithms, lower bounds are investigated for the following purposes: 1) appraisal of the accuracy of heuristic algorithms; and 2) development of optimal algorithms based on branch-and-bound type approaches. Lower bounds are obtained by relaxing the topological connectivity constraint and by approximating the discrete cost-capacity curves with lower envelopes (see Fig. 6). Linear lower envelopes lead to *linearized bounds,* while concave envelopes lead to *concave bounds.*

1) Linearized bounds: The linearized bounding problem is generally formulated as a CFA problem (see Section VI) with

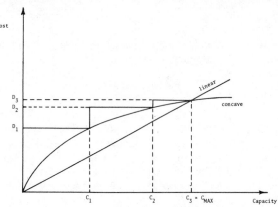

Fig. 6.   Linear and concave lower envelope.

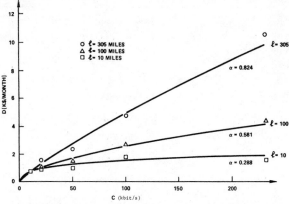

Fig. 7.   Concave approximations to link costs, for various link lengths (Telpak tariff assumed).

linear line costs and fully connected topology. The direct solution of the bounding problem is difficult because of the concavity of the objective function $D(f)$ [see (14)]. Rather, the objective $D(f)$ is further bounded as follows:

$$D(f) = \sum d_i f_i + \left(\sum \sqrt{d_i} f_i\right)^2 \Big/ \gamma T_{amx}$$

$$\geqslant \sum d_i f_i + \left(\sum \sqrt{d_i} f_i\right)^2 \Big/ \gamma T_{max} C_{max} = D_{LB}(f)$$

(16)

where C_{max} = max admissible link capacity option.

The lower bound $D_{LB}(f)$ in (16) is *convex*. Thus, a lower bound to the topological problem is obtained by minimizing the convex objective $D_{LB}(f)$ using the FD method.

The procedure as defined above applies to the case in which no link (and link capacity) is preassigned; but can be extended to applications in which a set of links is assigned *a priori*, and new links must be added in order to meet the requirements (e.g., network expansion problem) [39].

The linearized bound can also be applied in branch-and-bound (B-B) algorithms [39]. To this end, recall that at each step of a B-B algorithm a bound is required on the cost of a partially specified topology with a set n_a of assigned links, a set n_p of potential links, and a set n_e of excluded links. This bounding problem is similar to the topological design with some preassigned links, and can be approached with the linearized bounding procedure previously mentioned.

2) Concave bounds: Linearized bounds are simple and exact. However, they are often too loose, especially if line cost versus capacity shows strong economy of scale, or more generally, the cost-capacity structure cannot be accurately bounded with a linear envelope. In such cases, concave bounds lead to better results.

Unfortunately, the presence of concave link costs makes the solution of the bounding problem difficult, since the objective $D(f)$ cannot be expressed in closed form (see Section VI-C). One possible (but complex) approach consists of formulating linear or convex bounds for $D(f)$, and then solving the problem exactly with the FD method. A simpler approach,

which we follow, consists of finding an *approximate* solution to the bounding problem with the technique indicated in Section VI-C. The lower bound is then derived from the approximate solution taking into account the accuracy of the solution method. For example, if D is the cost of the approximate solution and ϵ is the relative accuracy, the lower bound is $D_{LB} = D(1 - \epsilon)$.

VIII. APPLICATIONS

We now evaluate the efficiency of some of the heuristic techniques as applied to the topological design of a proposed 26-node ARPANET configuration (see Fig. 5). Capacity options vary from 9.6 to 230.4 kbit/s; discrete cost-capacity functions as well as concave approximations are shown in Fig. 7. Delay requirement is $T_{max} = 200$ ms. Traffic demands are uniformly distributed between node pairs. Several levels of throughput requirement in the range from 400 to 700 kbit/s are considered.

The suboptimal solutions obtained with different techniques are displayed in a throughput versus cost diagram in Fig. 8. For each technique several solutions were generated at different throughput levels. Since each technique typically generates several locally optimal solutions, only the non-dominated solutions were shown in Fig. 8. (Note: Solution A is defined to be dominated by Solution B if B has lower cost and better performance than A.) Figs. 9, 10, and 11 display some typical topologies obtained with BXC, cut-saturation, and CBE, respectively.

From Fig. 8, it is noticed that different techniques lead to solutions which fall in a narrow cost-throughput range. The resulting topological structures, on the other hand, may vary considerably from technique to technique, as can be seen by comparing cut-saturation and CBE solutions in Figs. 10 and 11, respectively. Cost and throughput of the two solutions are approximately the same, but cut-saturation yields about 30 links while CBE yields about 60 links. We note that the marginal cost [dollars/(bit/s)/month] varies over a moderate range for these three procedures.

These facts lead us to conjecture that there are a large number of low-cost solutions which may correspond to very

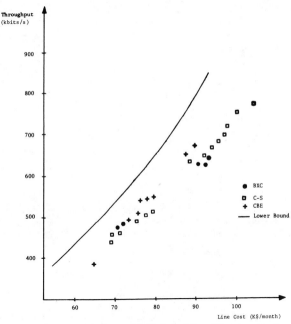

Fig. 8. Heuristic solutions and lower bound for a 26-node network design.

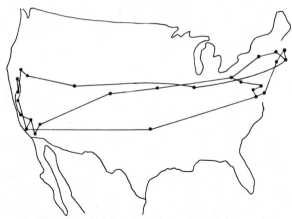

Fig. 9. Example of BXC solution. All links—50 kbits/s; throughput—480 kbits/s; cost—$72 K/month; cost/bit—$0.15/(bit/s)/month.

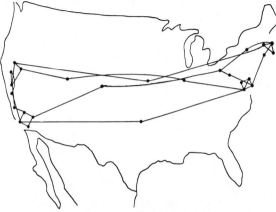

Fig. 10. Example of C-S solution. All links—50 kbits/s; throughput—650 kbits/s; cost—$83 K/month; cost/bit—$0.135/(bit/s)/month.

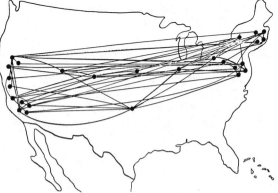

Fig. 11. Example of CBE solution. Line speeds—9.6 and 19.2 kbits/s; throughput—700 kbits/s; cost—$89 K/month; cost/bit—$0.127/(bit/s)/month.

TABLE III
CPU TIME PER NONDOMINATED SOLUTION FOR A 26-NODE
NETWORK APPLICATION

BXC	25 seconds
Cut-Saturation	3 seconds
CBE	4 seconds

different topological configurations and capacity assignments. Thus well-constructed heuristics can easily identify good solutions, although the optimal solution may be elusive and extremely difficult to obtain.

In order to evaluate the computational efficiency of the various techniques, we compare in Table III, the CPU times required in the 26-node application. The runs were made on a CDC 6600 computer.

Cut-saturation and CBE methods require comparable amounts of CPU time to generate a nondominated solution while the BXC method is much more time consuming. The experience, based on a variety of topological design problems, has demonstrated that cut-saturation is consistently more efficient than BXC. This result was expected since cut-saturation can be viewed as a refinement of BXC.

As for the near optimality of the heuristics, we conjecture that since suboptimal solutions obtained with different starting topologies and different techniques all fall within a narrow cost-performance range, they are within the same range from the optimal solution as well. From the results in Fig. 8, we may, therefore, conjecture that the accuracy of the heuristics is within 5 percent.

A more rigorous appraisal of near optimality is obtained by comparing suboptimal solutions with lower bounds. For this purpose, we calculated both linearized and concave bounds for the 26-node application.

The linearized bound was used to evaluate the near optimality of the cut-saturation solution shown in Fig. 10 which is constructed exclusively with 50 kbit/s capacity options. To obtain a reasonably tight linearized bound, we generated a large number of low-cost trees (the optimal topology must contain at least a tree) and evaluated the bound on the optimal completion of each tree so as to satisfy the design requirements, following the procedure indicated in Section VII-E1. The lowest cost over all completed trees is $72K/month, 20 percent below the cut-saturation solution.

To obtain tighter bounds we used the concave bounding approach and approximated the discrete costs with concave curves as shown in Fig. 7. We then solved the associated concave problem for various throughput levels. The resulting bound is represented by the solid curve in Fig. 8. The bound is 15 percent below the cut-saturation solution.

IX. CONCLUSION

In this paper we have surveyed some of the mathematical programming and network flow techniques that have been found useful for the design of distributed computer-communication networks. We reported some exact algorithms for the capacity assignment (CA) and routing problems, and discussed efficient suboptimal techniques that are available for the capacity and flow assignment (CFA) problem. As for the topology design, we showed that good suboptimal solutions can be obtained with the concave branch elimination (CBE) method in the case of smooth (i.e., small economy of scale) cost-capacity characteristics and relatively simple topological constraints; or with the cut-saturation method in those applications that have only a few link capacity options and require some man-machine interaction to verify and satisfy nontrivial topological constraints.

Some bounding techniques were presented, and were used to evaluate the near optimality of the topological design algorithms. The difference between the cost of the suboptimal solutions and the lower bound is typically less than 15 percent. This degree of accuracy is deemed adequate for most topological designs, especially considering that user requirements cannot be predicted with much accuracy before network implementation, or tend to change during the life of the network.

Although the present algorithms seem adequate for most applications, there is a need for better bounds and (possibly) exact topological design solutions to measure and compare the near optimality of the various techniques. More work should be devoted also to the analysis of algorithm complexity (expected and worst-case time and storage requirements) to have a better concept of the applicability and limits of the various techniques, and to identify the areas where research effort should be expended [48].

So far, we have discussed and solved a rather simple network design problem, with fixed switch locations, pure terrestrial topology, average delay constraints, etc. More general design problems can be formulated, requiring for example, the selection of backbone switch number and location; the use of both terrestrial and satellite facilities; the consideration of different traffic applications (e.g., bulk transfers, interactive traffic, digitized voice) with different priorities and performance requirements; the use of more general performance constraints (e.g., maximum end-to-end delay instead of average delay); the integration of circuit and packet-switching

techniques, etc. Further work is required to extend the existing algorithms to handle the more general cases. Some preliminary results in this direction are reported in [41] for the mixed terrestrial and satellite design and in [42] for the selection of backbone nodes.

REFERENCES

[1] "Analysis and optimization of store-and-forward computer networks," NAC 4th Semiannu. Tech. Rep. for the ARPA Project, Defense Documentation Center, Dec. 1971.

[2] L. R. Bahl and D. T. Tang, "Optimization of concentrator locations in teleprocessing networks," presented at Symp. Computer-Communications Networks and Teletraffic, Polytechnic Inst. Brooklyn, Brooklyn, NY, Apr. 4-6, 1972.

[3] S. Carr, S. Crocker, and V. Cerf, "HOST-HOST communication protocol in the ARPA network," in *Conf. Rec., 1970 Spring Joint Computer Conf., AFIPS Conf. Proc.*, vol. 36. Montvale, NJ: AFIPS Press, 1970, pp. 589-597.

[4] W. Chou and A. Kershenbaum, "A unified algorithm for designing multidrop teleprocessing networks," in *Data Networks Analysis and Design, 3rd Data Communication Symp.*, Nov. 1973, pp. 148-156.

[5] W. Chou and H. Frank, "Routing strategies for computer network design," presented at Symp. Computer-Communications Networks and Teletraffic, Polytechnic Inst. of Brooklyn, Brooklyn, NY, Apr. 4-5, 1972.

[6] G. C. Cole, "Computer network measurements: Techniques and experiments," School of Eng. and Appl. Sci., Univ. of California, Los Angeles, rep. UCLA-ENG-7165, 1971.

[7] S. Crocker, J. Haefner, R. Metcalfe, and J. Postel, "Function-oriented protocols for the ARPA computer network," in *Conf. Rec., 1972 Spring Joint Comput. Conf., AFIPS Conf. Proc.*, vol. 40. Montvale, NJ: AFIPS Press, 1972, pp. 271-279.

[8] G. B. Danzig, *Linear Programming and Extensions.* Princeton, NJ: Princeton Univ. Press, 1963.

[9] H. Everett, III "Generalized Lagrange multipliers method for solving problems of optimal allocation of resources," *Operations Res.*, vol. II, pp. 399-418, 1963.

[10] B. Fox, "Discrete optimization via marginal analysis," *Management Sci.*, vol. 13, pp. 210-216, Nov. 1966.

[11] H. Frank, B. Rothfarb, D. Kleitman, and K. Steiglitz, "Design of economical offshore natural gas pipeline networks," Office of Emergency Preparedness, Washington, DC, rep. R-1, Jan. 1971.

[12] H. Frank, I. T. Frisch, W. Chou, and R. Van Slyke, "Optimal design of centralized computer networks," *Networks*, vol. 1, pp. 43-57, 1971.

[13] H. Frank, I. T. Frisch, and W. Chou, "Topological considerations in the design of the ARPA computer network," in *Conf. Rec., 1970 Spring Joint Comput. Conf., AFIPS Conf. Proc.*, vol. 36. Montvale, NJ: AFIPS Press, 1970.

[14] H. Frank, M. Gerla, and W. Chou, "Issues in the design of large distributed networks," presented at the IEEE Nat. Telecommun. Conf., Atlanta, GA, Nov. 1973.

[15] H. Frank, R. E. Kahn, and L. Kleinrock, "Computer communication network design—Experience with theory and practice," in *Conf. Rec., 1972 Spring Joint Comput. Conf., AFIPS Conf. Proc.*, vol. 40. Montvale, NJ: AFIPS Press, 1972, pp. 255-270.

[16] L. Fratta, M. Gerla, and L. Kleinrock, "The flow deviation method: An approach to store-and-forward computer-communication network design," *Networks*, vol. 3, pp. 97-133.

[17] G. L. Fultz and L. Kleinrock, "Adaptive routing techniques for store-and-forward computer communication networks," presented at IEEE Int. Conf. Commun., Montreal, Canada, June 14-16, 1971.

[18] M. Gerla, "Deterministic and adaptive routing policies in packet-switched computer networks," presented at the ACM-IEEE 3rd Data Commun. Symp., Tampa, FL, Nov. 13-15, 1973.

[19] M. Gerla, H. Frank, and W. Chou, "Computational considerations and routing problems for large computer networks," presented at IEEE Nat. Telecommun. Conf., Atlanta, GA, Nov. 1973.

[20] M. Gerla, "The design of store-and-forward networks for computer communications," Ph.D. dissertation, School of Eng. and Appl Sci., Univ. of California, Los Angeles, Jan. 1973.

[21] F. E. Heart, R. E. Kahn, S. M. Ornstein, W. R. Crowther, and D. C. Walden, "The interface message process for the ARPA computer network," in *Conf. Rec., 1970 Spring Joint Comput. Conf., AFIPS Conf. Proc.*, vol. 36. Montvale, NJ: AFIPS Press, 1970, pp. 551–567.

[22] "Issues on large network design," Network Analysis Corp., Glen Cove, NY, ARPA rep., Jan. 1974.

[23] J. R. Jackson, "Networks of waiting lines," *Operations Res.*, vol. 5, pp. 518–521, 1957.

[24] R. E. Kahn and W. R. Crowther, "A study of the ARPA computer network design and performance," Bolt, Beranek, and Newman, Inc., rep. 2161, Aug. 1971.

[25] L. Kleinrock, "Analytic and simulation methods in computer network design," in *Conf. Rec., Spring Joint Comput. Conf., AFIPS Conf. Proc.*, vol. 36. Montvale, NJ: AFIPS Press, 1970, pp. 568–579.

[26] —, "Models for computer networks," in *Proc., Int. Conf. Commun.*, Boulder, CO, June 1969, pp. 21-9-21-16.

[27] —, *Communication Nets: Stochastic Message Flow and Delay.* New York: McGraw-Hill, 1964.

[28] —, *Queueing Systems: Volume I, Theory.* New York, Wiley-Interscience, 1975.

[29] —, *Queueing Systems: Volume II, Computer Applications.* New York: Wiley-Interscience, 1976.

[30] —, "Resource allocation in computer systems and computer-communication networks," presented at IFIP Cong., Aug. 1974.

[31] L. Kleinrock and W. Naylor, "On measured behavior of the ARPA network," in *Conf. Rec., Nat. Comput. Conf., AFIPS Conf. Proc.*, vol. 43. Montvale, NJ: AFIPS Press, 1974, pp. 767–780.

[32] L. G. Roberts and B. D. Wessler, "Computer network development to acheive resource sharing," in *Conf. Rec., 1970 Spring Joint Comput. Conf., AFIPS Conf. Proc.*, vol. 36. Montvale, NJ: AFIPS Press, pp. 543–599.

[33] D. J. Silk, "Routing doctrines and their implementation in message switching networks," *Proc. Inst. Elec. Eng.*, vol. 116, pp. 1631–1638, Oct. 1969.

[34] K. Steiglitz, P. Weiner, and D. J. Kleitman, "The design of minimum cost survivable networks," *IEEE Trans. Circuit Theory*, pp. 455–460, Nov. 1969.

[35] J. A. Tomlin, "Minimum cost multi-commodity network flows," *Operations Res.*, vol. 14, pp. 45–47, Jan. 1966.

[36] R. Van Slyke and H. Frank, "Network reliability analysis: Part I," *Networks*, vol. 1, pp. 279–290, 1971.

[37] V. K. M. Whitney, "Lagrangian optimization of stochastic communication systems models," presented at MRI Symp. Comput. Commun. Networks, Brooklyn, NY, Apr. 1972.

[38] B. Yaged, Jr., "Minimum cost routing for static network models," *Networks*, vol. 1, pp. 139–172, 1971.

[39] "The practical impact of recent computer advances on the analysis and design of large scale networks," Network Analysis Corp., Glen Cove, NY, 3rd Semiannu. Tech. Rep., Defense Documentation Center, June 1974.

[40] M. Gerla, H. Frank, W. Chou, and J. Eckl, "A cut-saturation algorithm for topological design of packet-switched communications networks," in *Proc. Nat. Telecommun. Conf.*, San Diego, CA, Dec. 2–4, 1974.

[41] M. Gerla, W. Chou, and H. Frank, "Cost-throughput trends in computer networks using satellite communications," in *Proc. Int. Conf. Commun.*, Minneapolis, MN, 1974.

[42] W. Hsieh, M. Gerla, P. McGregor, and J. Eckl, "Locating backbone switches in a large packet network," presented at Int. Conf. Comput. Commun., Montreal, Canada, 1976.

[43] L. Kleinrock and H. Opderbeck, "Throughput in the ARPANET—Protocols and measurement," in *Conf. Rec., Network Structures in an Evolving Operational Environment, 4th Data Commun. Symp.*, Quebec City, Canada, Oct. 1975, pp. 6-1-6-11.

[44] —, "The influence of control procedures on the performance of packet-switched networks," in *Proc. Nat. Telemetering Conf.*, Dec. 1974.

[45] L. Kleinrock and F. Kamoun, "Data communications through large packet-switching networks," presented at 8th Int. Teletraffic Cong., Australia, Nov. 1976.

[46] L. Kleinrock, W. E. Naylor, and H. Opderbeck, "A study of line overhead in the ARPANET," *Commun. Ass. Comput. Mach.*, vol. 19, pp. 3–12, Jan. 1976; also in *Proc. IIASA Conf. Comput. Commun. Networks*, Oct. 1974, pp. 87–109.

[47] L. Kleinrock, "ARPANET lessons," in *Conf. Rec. IEEE Int. Conf. Commun.*, Philadelphia, PA, June 14–16, 1976, pp. 20-1-20-6.

[48] A. Aho, J. Hopcraft, and J. Ullman, *Design and Analysis of Computer Algorithms.* Reading, MA: Addison-Wesley, 1974.

LARGE SCALE NETWORK DESIGN CONSIDERATIONS

©Rosner, R D, Defence Communications Agency, Reston, Virginia, USA

ABSTRACT

Various network structures have been studied to meet the
potential data network requirements of the Defense
Communications System. Having selected a packet switched
approach for further investigation both distributed and
hierarchical network structures were examined. The impact
of different network structures, connectivities, link
counts and traffic distribution were studied to arrive at
fairly general results useful in the design of large data
networks. These results are described together with the
application of a combination analytical-heuristic approach
to the design of large data networks.

1. INTRODUCTION

Significant research has been undertaken over the past
few years concerned with the design of communication networks oriented toward computer and data communications
(1-4). Networks which were studied fall into two broad
categories. One category is the design of centralized
networks servicing a large number of relatively low
speed remote terminals from one (or two) large central
(time shared) computer sites. The other being the
design of distributed networks servicing a number of
geographically dispersed highly concentrated traffic
sources in a computer resource sharing environment.
Techniques have been developed to generate optimium
network structures for a fixed traffic flow matrix, given
the processor node locations and convex cost functions.
The optimal solution of network structure and capacity
for non-continuous cost functions (those that appear in
present transmission tariff structures) or for networks
with unspecified concentrator or nodal locations has
been shown to be mathematically unsolvable, but various
heuristic techniques have been developed for finding
local optimium solutions.

In this paper a different approach is taken toward network
design, oriented toward satisfying a widely distributed
mix of high and low density users. Network optimization
is made a secondary consideration, subject to the design
of a basic network topology and connectivity which will
meet the nominal data flow requirements, will meet other
network performance objectives, and will be relatively
insensitive to changes in, or inaccurate forecasts of, user
traffic flow requirements. The attributes of both distributed and hierarchical network structures are
discussed. The impact of different network connectivities,
link counts, and traffic distributions are examined to
arrive at fairly general results useful in the design of
large data networks.

2. BACKGROUND - THE DATA NETWORK DESIGN PROBLEM FOR THE FUTURE DEFENSE COMMUNICATIONS SYSTEM (DCS)

In an earlier paper (5) a digital data network concept
for the future DCS was described. This concept involves
the integration of a packet switched data communications
subnetwork of the future integrated digital DCS. In
this paper the problem of designing the data subnetwork
is attacked as if it were a stand alone data network,
but subject to the following design constraints:

    A. Network switching node locations are restricted
to the 70 CONUS AUTOVON (Automatic Voice Network) switch
locations.

    B. Network traffic eminates from approximately 2400
geographically dispersed user locations with sites varying
from single low speed terminals to those encompassing 10
to 20 computers plus 40 or more high and low speed remote
terminals.

    C. A nominal design load of 4.4×10^{10} bits in the busy
hour (44,000 Megabits of data) in a total of 1,600,000
data transactions.

D. A geographic distribution of the data traffic flow is estimated by extrapolation of currently measured traffic flow and by relating flow to computer CPU core size.

E. Concentration is performed in local areas such that 16, 32, 64, or 96 kb/sec of transmission capacity is used depending on the data traffic load presented by that local area. In general a local area represents a military installation, government office, or industrial site of a major contractor.

F. In order to simplify the design model, no attempt to use multi-drop lines is made. All lines originate and terminate in two distinct locations only.

Subject to these constraints, it was desired to design an efficient, survivable, reliable data communications network. Because only a very approximate estimate of the traffic magnitude and flow distribution was known (which is the general case for all network designs) the design procedure was not formally oriented to determining an optimium network design. Optimium networks are fictions of mathematics, and cannot occur in practice. The actual network traffic flow, in quantity and distribution, is a dynamic parameter, which changes from second to second, and at most matches the design optimium for an infinitesimal time. What is demonstrated here is the application of analytical and heuristic techniques to develop a data network concept which is efficient (though not necessarily optimium) flexible, and relatively insensitive to the dynamics of a changing user community.

3. THE STRUCTURE OF THE DATA NETWORK

Switched Communications networks have classically been implemented using two basic network structures. AUTOVON, the Automatic Voice Network of the U.S. Department of Defense, and the ARPA Computer network are examples of non-hierarchical or distributed switched communications networks, where each switch of the network is functionally identical and has similar structure and connectivity. The public telephone network is the prime example of a hierarchical switched network where switches at various levels or stages of the hierarchy are given greater power, authority, connectivity and concentration.

In developing the future Defense Communications System (DCS) data network alternatives were identified which encompassed both distributed and hierarchical techniques. Various tradeoffs were examined to help identify the relative advantages and disadvantages of each, and the final recommendations indicated strong preference for a modified hierarchical structure which was somewhat of a compromise between the two extremes, to achieve most of the advantages of each.

One of the most pressing concerns of the future DCS data subnetwork is the achievement of the short end-to-end delay times essential to computer related traffic. In addition, the future DCS data network should utilize simple routing and processing functions at the switches, provide concentration points for store and foreword message (S/F) functions, and should provide reliable, easily implemented network control and monitoring functions. A hierarchical configuration can be readily structured to achieve consistently short end-to-end delays, achieve efficient channel utilization, and provide the other features required by the future data network. However, compared to distributed networks concepts, the hierarchical network is potentially more vulnerable, and may require more links and channel miles to achieve the required connectivity and capacity.

The two basic structures are illustrated in Figure 1 and Figure 2. In Figure 1 the general distributed (non-hierarchical) network is shown, with a representative connectivity among the functionally identical switching nodes. The route between any pair of switches may be direct or may have to employ one (or more) intermediate tandem switches. The routing could be predetermined as in AUTOVON, or could be adaptive and determined in real time as in the ARPA network.

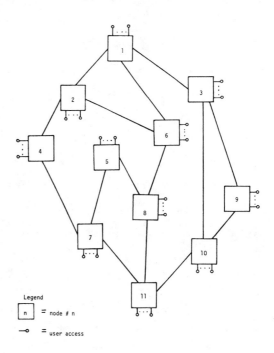

Legend

| n | = node # n
—o = user access

FIGURE 1
EXAMPLE OF BASIC DISTRIBUTED NETWORK

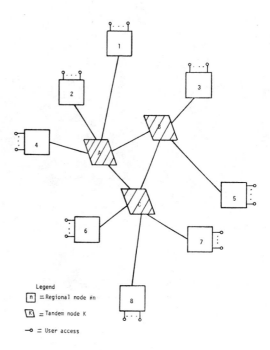

Legend

| n | = Regional node #n
| K | = Tandem node K
—o = User access

FIGURE 2
EXAMPLE OF BASIC HIERARCHICAL NETWORK

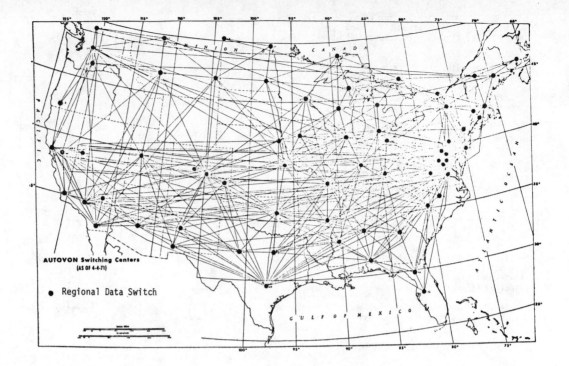

FIGURE 5
DISTRIBUTED CONNECTIVITY, 362 LINKS

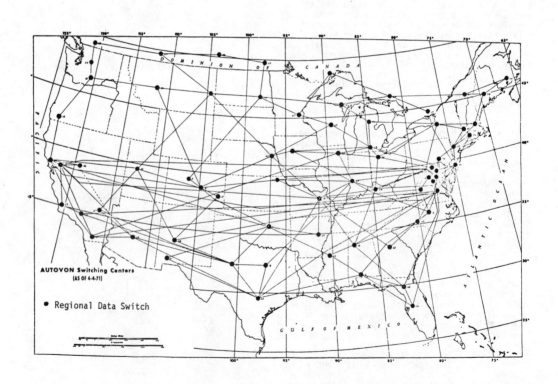

FIGURE 6
DISTRIBUTED CONNECTIVITY, 149 LINKS

In Figure 2 the purely hierarchical network is shown, where different levels of switching exist. The lower level (regional) switches serve the subscribers and are connected to the higher level (tandem) switches. A regional switch may be connected to more than one tandem (dual homed) switch for redundancy or survivability. The high level of tandem switches are highly connected to each other, but since n(n-1)/2 links are required to fully connect tandem switches they are not generally fully connected in large networks in practice. The key is that the principal routes for network traffic always utilize the tandem switches, except for that traffic turned around locally at the regional switches. The complexities of routing are then confined primarily to the tandem switches, while the overhead and processing of the user interface is confined to the regional switches. The result is that the network functions are partitioned and individual switches are simpler in design and operation.

The concentration, and thus the vulnerability of the hierarchical configuration of Figure 2 is considered unacceptable for a future DCS data subnetwork, but the use of a modified hierarchy, shown in Figure 3 can overcome the deficiencies of the true hierarchical structure.

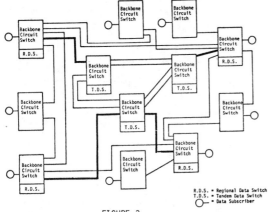

FIGURE 3
MODIFIED HIERARCHICAL NETWORK CONCEPT

In figure 3 the basic data subnetwork for the future DCS is shown in its context as an integrated voice and data communications system. This concept introduces within the non-hierarchical network of DCS circuit switches an overlay of a modified hierarchical data network. All major nodes shown in the sample network are functionally identical with respect to the part of each box labeled "Backbone Circuit Switch." A subset of the switches has appended to them a functional addition called a regional data switch. Ideally this regional data switch would be an additional software function within the same processor as the circuit switch, but could in fact be a parallel processor or added small computer. Each regional data switch would be connected, via the circuit switch matrix to at least two other regional data switches. The tandem data switches would be a relatively small number of switches arranged in a fully connected mesh, so that each one has a direct channel to every other one. Each regional data switch would have a line directly to at least one tandem data switch. The tandem data switch would perform tandem switching as well as provide a concentration point, where, for example the store and forward functions for message traffic provided by today's AUTODIN can be located. The mode of operation of the regional and tandem data switches would be a packet switched mode, with an adaptive connectivity which could be readily modified via the circuit switched connections. The principal route through the network for a data message will be from data subscriber to the nearest regional data switch. If the first regional switch can reach the destination regional switch directly, it is so routed. If this is not possible, then the message, in packet form, is passed up to the tandem data switch. This insures that no more than four data switches will ever be needed in the end-to-end chain between two subscribers, and thus, keeps

the problem of minimizing network delay a tractable one which is one of the essential concerns of the data subnetwork design. In case of loss or failure of one or more tandem data switches, the more circuitous routes employing only regional data switches would provide backup at longer than average delays. By routing the traffic via the distributed arrangement of regional nodes in case of failures, the redundancy and survivability disadvantages of the hierarchical network are alleviated. For convenience, the connectivity arrangement illustrated by figure 3 will henceforth be referred to as a modified hierarchical configuration.

4. COMPARISON METHODOLOGY OF NETWORK STRUCTURES

In order to quantify the question of network structure a comparison on the basis of channels, channel miles and network delay was performed on representative data networks. An analytic store and forward network performance model was used together with the traffic model developed for future data flows to determine network connectivity and sizing parameters for the modified hierarchical and distributed network configurations.

Quantitative comparisons of different network topologies and structures were made using a very fast analytic network synthesis program. The program does not optimize the topology, but given the basic network connectivity matrix and traffic flow matrix, sizes the links and flows to the minimum requirements needed to meet the traffic flow and delay requirements. Because the program operates so fast (analyzes and sizes a 70 node network in 5 to 10 seconds) many runs can be made using different structures, topologies, and traffic flow requirements to determine the most desirable network configuration.

The model assumes that the basic network comprises a specified number of nodes. These nodes are connected by a set of links, each of which comprises a number of channels. Figure 4 shows how traffic is processed through a typical node.

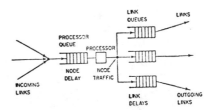

FIGURE 4
FUNCTIONAL FLOW DIAGRAM OF A NODE

Traffic from the incoming links is first delayed at the node processor. It is assumed that the processor performs a variety of functions, such as determining the priority class of the message, placing the message on the correct outgoing link, etc. If the processor is busy when a message arrives, it is stored and processed on a first-come, first-served basis. Once on the links, messages are again stored when blocked and then transmitted on a first-come, first-served basis.

When analyzing a network with priority classes, the following queue discipline is assumed. The processor accepts customers on a first-come, first-served basis, regardless of priority class. On the links, service is given to the highest priority class first. That is, when a channel becomes free, waiting messages with the highest priority are transmitted on a first-come, first-served basis. Once on a channel, a message may not be preempted by messages with a higher priority level.

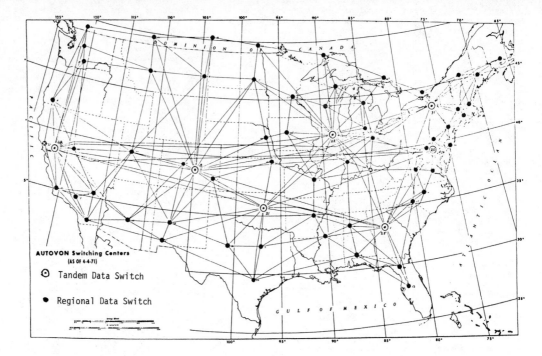

FIGURE 7
HIERARCHY I, 224 LINKS

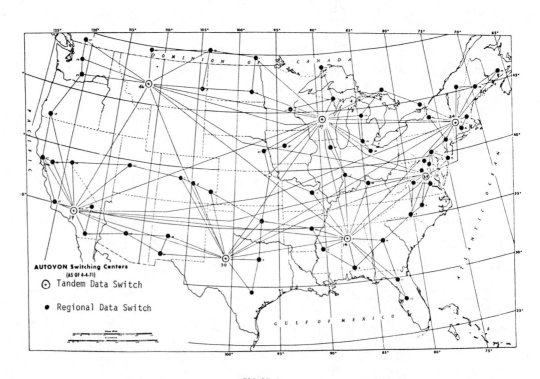

FIGURE 8
HIERARCHY II, 149 LINKS

The model assumes that the number of nodes and the connectivity are fixed; that is, traffic from node A to node B has one and only one path in the network. For convenience, the program includes an algorithm which, given the connectivity and the distance between each of the nodes, determines a shortest path routing table for the network.

The general philosophy of the model and its mathematical justification stem from a theorem in queueing theory by P. J. Burke (6). This theorem states that in a store-and-forward Markovian queue with N parallel channels (Poisson arrivals and exponential service times), the output process has the same probability distribution as the input process. Further, the behavior of the traffic at a given node in a tandem array is independent of the behavior at any other node in the array. Since the routing is fixed, and because of the well-known result that the sum of independent Poisson processes is a Poisson process, the input to each node is a Poisson process. If it is assumed that the interarrival times of messages to the network can be described by a negative exponential distribution, and that the message length is also distributed exponentially, then Burke's Theorem allows analysis of the network on a nodal basis. At each node, delays may be determined from known results in queueing theory.

The shortest path is determined for each node pair, based on the connectivity of the network and the internodal distance matrix. The loading at each switch is easily found by adding the traffic loads of all node pairs whose routes pass through the switch. Once the traffic loading at each node is known, the model calculates the delay for all messages passing through the node. Alternatively, the links leading out of the node may be sized to ensure a given average delay at the node.

Aside from the obvious uses in the behavioral and synthesis mode, the model may be used in sensitivity analyses of various system parameters. Further, it may evaluate the survivability and availability of a given network under a particular scenario. The feasibility of various routing doctrines can also be determined by the model. An advantage of the model is that it can rapidly analyze a large network under variations in configurations and parameters.

The projected future data and narrative traffic of the DCS was used to drive the model, and both actual projected as well as uniform distributions of the data traffic were assumed. The current AUTODIN and AUTOVON statistics were used in the formation of a 70x70 to/from traffic matrix representing the data traffic flow in the future DCS from regional switch node to regional switch node. In the case of the uniform traffic distribution, it was assumed that 1/70th of the total data traffic originated at each switch, and that the traffic was equally destined for every other switch (i.e., each entry of the 70x70 traffic matrix is equal to 1/4900th of the total traffic). This assumption implies no a priori knowledge of the network traffic flow or user distribution, and can be shown to represent an upper bound on the line capacities required for a given total traffic flow. This provides a useful measure of how important (or unimportant) an accurate geographical traffic projection has to be in order to be able to perform meaningful system studies and tradeoffs.

The operation of the synthesis procedure of the analytical queuing model used in the studies required the preliminary establishment of the node-to-node network connectivity, and the model then sized the links on the network and computed point to point and average delays and throughputs. The starting point for the model was established using the final results of voice network optimization runs, which used a total of 362 links to connect the 70 CONUS switching nodes. A computer run was made with that connectivity at a line rate of 4 kb/sec to get an initial network configuration employing 362 links. Because of the large number of links relative to the amount of data traffic, the results of the model run yielded a large number of links with only a few data channels on them. The number of links used was then reduced by eliminating

all links with three or fewer channels yielding a connectivity of 224 links, and then further reduced by eliminating all links with five or fewer channels yielding a 149 link connectivity. Finally an 84 link connectivity was used which represented the (nearly) minimal spanning tree for the network to establish a lower bound on the connectivity. In the 149 and 224 link cases each regional node was connected to at least two other regional nodes, and in the modified hierarchical case at least one tandem node.

For design study purposes, computer runs were then made with the various network connectivities, for various bit rates, with and without the hierarchical tandem switch nodes, and for uniform and actual projected traffic distributions. The number of tandem nodes in the hierarchical cases was fixed at seven, because that required only a modest number (21) links to fully interconnect them. However, this number should be the subject of future studies. Four of the many different network structures studied are shown in Figures 5 through 8. Figures 5 and 6 illustrate two 70 node distributed networks, while Figures 7 and 8 illustrate two different modified hierarchical structures.

5. RESULTS OF STRUCTURE COMPARISONS

The results of comparing many different network structures are shown in a group of figures which follow. They graphically portray some of the results of the computer runs in summary curves.

In Figures 9 and 10 the number of channels and channel miles respectively are plotted versus channel bit rate for both distributed and modified hierarchical network configurations composed of 149 and 224 links. It can be seen that at any bit rate there is a fairly large spread in the number of channels between the alternatives, but the number of channel miles for the 149 and 224 link cases was tightly grouped around a single value. As the connectivity gets thinner, the number of channels for the distributed and modified hierarchical networks tend to converge. This is further illustrated by Figure 11 and Figure 12 in which the number of channels and channel miles are plotted against the number of links used in the 70 node network for both the modified hierarchical and distributed network cases at a channel bit rate of 16 kb/sec.

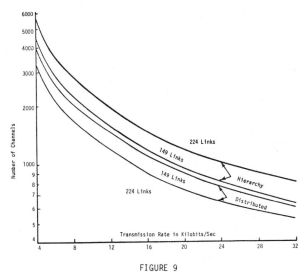

FIGURE 9
NUMBER OF CHANNELS VS BIT RATE DISTRIBUTED
AND 7 NODE HIERARCHY NETWORKS

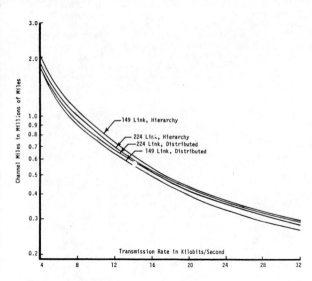

FIGURE 10
CHANNEL MILES VS BIT RATE FOR DISTRIBUTED
AND 7 NODE HIERARCHY NETWORKS

In Figure 11 the number of channels is shown to rapidly diverge as the number of links increases from the 84 link case, with the homogeneous network showing a maximum in the number of channels in the area of 150 links. The number of channels required by the modified hierarchical network increases rather significantly as the number of links increases, and there was not enough data developed to indicate the presence of a maximum point for the number of channels in the modified hierarchical network.

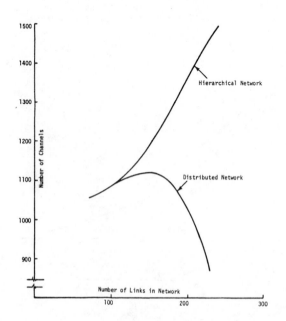

FIGURE 11
NUMBER OF CHANNELS VS NUMBER OF LINKS

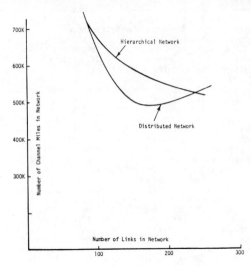

FIGURE 12
NUMBER OF CHANNEL MILES VS NUMBER OF LINKS

The plots of channel miles for each network configuration (Figure 12) show a much closer and regular comparison of the modified hierarchical and distributed network configurations. The relative difference between the number of channel miles is generally within 10 percent, with the distributed network showing a minimum of channel miles in the region of 175 links, while the modified hierarchical network appears to be reaching a minimum outside the range of the configurations studied (beyond 300 links).

It is difficult to try to make firm conclusions, or to extrapolate very much more significance from the information of Figures 9, 10, 11 and 12 because of the relatively limited number of cases studied as well as the limitations of the queuing model used in the analysis. However, the implications are strong that the advantages of the modified hierarchical network structure in terms of network routing, delay, delay variance, and system management, control, and implementation phasing, has to be achieved with an increase in both the number of channels, and number of channel miles used in the network. In the region of interest in terms of the number of links employed (e.g. 150 to 200 links) the modified hierarchical network appears to require about 15% more channels and 10% more channel miles of backbone connectivity than the distributed network case. As shown in another paper (7) this could reflect a cost differential of approximately $2 million annually, the cost effectiveness of which is yet to be proven. However, as noted earlier, the performance advantages of the modified hierarchical network appear to be of such significant value that pending further investigation, tentative judgment is made that the relatively small increase (about 5% on a total system basis) in system cost is warranted and the modified hierarchical network described is being pursued.

The delay aspects of the network were investigated using the same model as was used for the channel characteristics. In Figure 13, the average delay is plotted as a function of bit rate for two classes of traffic for the 149 node modified hierarchical network configurations. Class 1 traffic is considered the high priority traffic, and represents the interactive and computer-to-computer data traffic. Class 1 traffic amounts to 1/3 of the total data traffic flow. Class 2 traffic representing the bulk and narrative traffic and some longer duration query/response traffic accounts for 2/3 of the total traffic and is given second priority in handling via the network. Delay times of one second or better were required for the Class 1 traffic and 30 seconds or better for the Class 2 traffic. In Figure 13 it is seen that in all cases the required times were exceeded, and that at the higher bit rates delays of one second or better were achieved for both classes of traffic.

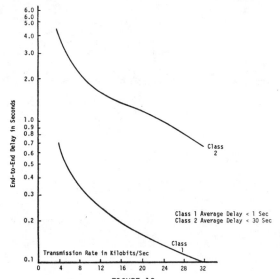

FIGURE 13
END-TO-END DELAY VS BIT RATE OF CLASS 1 AND 2
TRAFFIC IN A 149 LINK, 7 NODE MODIFIED
HIERARCHICAL NETWORK

The implication of Figure 13 is clear. Once the packet switched network has sufficient capacity to handle the busy hour traffic without the creation of infinite queues and can provide a significant portion of the traffic rapid response, then all traffic for which it was sized achieves an excellent grade of service as measured by network delay. This is a direct consequence of the packet size, where the largest component of the end-to-end delay is the actual physical transmission time of the packet. All work done to date has held the packet size fixed at 1000 bits including overhead, but studies related to optimization of the packet size based on actual message lengths and line error statistics are required. As the packet size is decreased it is expected that the end-to-end delay would similarly decrease but at the same time overhead as a percentage of total useful transmitted data would increase. Therein lies the potential for an optimization study based on packet length. The principal reason for improvement in network delays for class 2 traffic, as the bit rate is increased is related to the operation of the network queuing model. The model computes the integer number of channels that any given link needs in order to carry the required throughput. Any "fractional" channels are always rounded upward since "fractional" channels cannot physically exist. As the bit rate increases, the throughput capability represented in the rounding up of the fractional channels increases, so that network then performs even better than required, since network performance is computed based on the actual network configuration composed of discrete channels. The excess capacity, achieved by the channelization process is what permits the excellent network performance for both class 1 and class 2 traffic. The higher the bit rate per channel, the more excess capacity is available above the minimum throughput required to meet class 1 delay. The implication is that if one full channel were removed from each link, either the class 1 or class 2 delay requirement (or both) would not be met.

In Figure 14, the achieved network delay for class 1 and class 2 traffic for the 16 kb/sec case is plotted for various network connectivities. As can be seen in each case the achieved average delay initially increases as the number of links increase, reaches a maximum, and then decreases. This effect is due to the fact that the total end-to-end delay is composed of essentially three parts, the transmission delay (e.g., the time to send the 1000 bits of the packet down the 16 kb/sec line) on each link of the end-to-end path, the processing time at each switch in the path, and the queuing delay at each switch. (Propagation delay is ignored as negligible, which is not a valid assumption if satellite links are used.) Since

the delay elements are associated with the number of links that are required between the end points of a message, the fewer average number of links the shorter the average network delay. As the number of available links in the network increases, the likelihood of reaching the destination node with no intervening nodes increases for any node pair thus reducing the average delay. However, as the number of links get very small (approaching a minimal tree as in the 84 link case) the capacity of network is concentrated in a relatively small number of large cross sections, which leads to very small (virtually nonexistent) queuing delays, which more than offset the increase in the average number of nodes and links traversed. Thus the achieved delays show somewhat of a reduction as the network gets very thin.

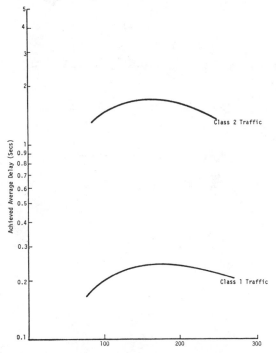

FIGURE 14
NUMBER LINKS IN MODIFIED HIERARCHICAL NETWORK

The last aspect of the network structure analysis is shown in Figures 15 and 16. Here, some of the previous plots for channels and channel miles are repeated, but in these plots curves are shown for both uniform as well as non-uniform traffic flow. As indicated previously, the uniform traffic case handles the same total volume of traffic but assumes no knowledge of the distribution of the traffic among the various nodes. In most cases, agreement between two different non-uniform traffic cases, taken at random, would likely be even closer. It is thus implies that a very precise forecast of actual distribution of network traffic is not nearly as important as the total traffic carried in performing design and tradeoff studies dealing with the large DCS data network. While it is valuable to establish a nominal distribution of traffic in order to simulate varying traffic flows, the fact that changing geo-political factors are likely to alter the node-to-node traffic needlines is shown to be a second order consideration in performing sizing and tradeoff studies. The dynamic nature of the user community is recognized however, and the actual configuration of the network does depend on the eventual user needlines at the time of implementation. The use of the circuit switched network to provide for the dynamic alteration of the network configuration in (nearly) real time further minimizes the impact of loading variations on design studies.

In a companion paper (7) various aspects of the cost

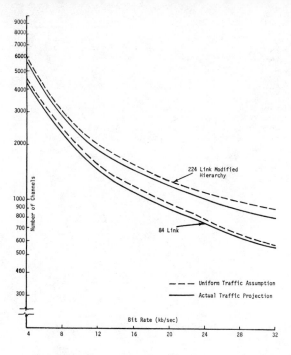

FIGURE 15
NUMBER OF CHANNELS VS BIT RATE

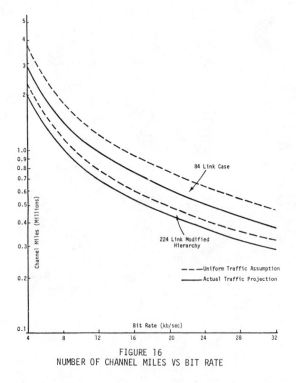

FIGURE 16
NUMBER OF CHANNEL MILES VS BIT RATE

This report has covered a variety of aspects of the development of a large data network suitable for the data requirements of future DCS. The following conclusions are warranted at this time:

1. A modified hierarchical configuration of tandem and regional data nodes compared to a distributed network topology can achieve the desired network objective at minimal increase in channel miles and complexity. This is in general agreement with other results((8, 9) which show that for reliability reasons, large data networks should be implemented using hierarchical network structures.

2. A packet switched, modified hierarchical network, sized to meet a very rapid speed of service for part of the design load, provides a surprisingly good speed of service for the total traffic load.

3. Good networks can be designed using simple design tools when special constraints have to be met.

4. Network design and tradeoff studies can be made in confidence even though exact traffic statistics, flows, and node to node distributions are not known. The most important parameter, total design data flow can be made a simple parameter of the analysis when rapid design programs are used.

7. ACKNOWLEDGEMENT

The author gratefully acknowledges the excellent work of Dr. Martin J. Fischer who developed and programmed the network analysis program, and William Cohen and Michael J. DeFrancesco who assisted in much of the data reduction. All are employees of the Defense Communications Agency.

8. REFERENCES

1. M. Gerla, "The Design of Store-and-Forward (S/F) Networks for Computer Communications", UCLA Engineering Report Number UCLA-ENG-7319 (ARPA) January 1973, University of California, Los Angeles.

2. R. M. Metcalfe, "Packet Communication", MIT Engineering Report Number MAC-TR-114 December 1973, Massachusetts Institute of Technology, Cambridge, Mass. 02139.

3. G. L. Fultz, "Adaptive Routing Techniques for Message Switching Computer Communication Networks", UCLA Engineering Report Number UCLA-ENG-7252 (ARPA), July 1972, University of California, Los Angeles.

4. H. Frank and W. Chou, "Topological Optimization of Computer Networks", Proceedings of the IEEE, Vol 60, No. 11 November 1972, Pp. 1385-1396.

5. R. D. Rosner, "A Digital Data Network Concept for the Defense Communications System", Conference Record of the National Telecommunications Conference, NTC '73, Cat No 73 CHO 805-2NTC November 1973, Pp 22C1-22C6.

6. P. J. Burke, "The Output of a Queueing System", Operations Research, Vol 4 (1956) Pp. 699-704.

7. G. J. Coviello and R. D. Rosner, "Cost Considerations for a Large Data Network", '74 ICCC, These proceedings.

8. H. Frank and R. M. VanSlyke, Reliability Considerations in the Growth of Computer Communications Networks", Conference Record of the National Telecommunications Conference, NTC '73, Cat No 73 CHO805-2NTC November 1973, Pp. 2201-2205.

9. H. Frank, M. Gerla, and W. Chou, "Issues in the Design of Large Distributed Computer Communication Networks", Conference Record of tee National Telecommunications Conference, NTC '73, Cat No 73 CHO805-2NTC November 1973, Pp 37A1-37A9.

Optimal Design of Mixed-Media Packet-Switching Networks: Routing and Capacity Assignment

DIEU HUYNH, MEMBER, IEEE, HISASHI KOBAYASHI, SENIOR MEMBER, IEEE, AND FRANKLIN F. KUO, FELLOW, IEEE

Abstract—This paper considers a mixed-media packet-switched computer communication network which consists of a low-delay terrestrial store-and-forward subnet combined with a low-cost high-bandwidth satellite subnet. We show how to route traffic via ground and/or satellite links by means of static, deterministic procedures and assign capacities to channels subject to a given linear cost such that the network average delay is minimized. Two operational schemes for this network model are investigated: one is a scheme in which the satellite channel is used as a slotted ALOHA channel; the other is a new multiaccess scheme we propose in which whenever a channel collision occurs, retransmission of the involved packets will route through ground links to their destinations. The performance of both schemes is evaluated and compared in terms of cost and average packet delay tradeoffs for some examples. The results offer guidelines for the design and optimal utilization of mixed-media networks.

I. INTRODUCTION

IN recent years two major packet-switched communication techniques are becoming increasingly important in the design of large computer communication networks: one technique is to store-and-forward packets over terrestrial communication links; the other technique is to transmit packets over a random multiaccess radio or satellite channel.

Up to the present, the studies and implementations have been centered on networks utilizing solely either terrestrial [e.g., Advanced Research Projects Agency Network (ARPANET), National Physical Laboratory Data Network [1], etc.] or satellite (e.g., ALOHANET) links. Recently ARPA has augmented its terrestrial network with packet satellite communication between the U.S. and the United Kingdom via INTELSAT IV using satellite interface message processors (SIMP's) built by Bolt Beranek and Newman Inc. (BBN) [2]. This multiple-access broadcast system was initiated in September 1975 and is expected to include four ground stations shortly. TELENET Communication Corporation, one of the new value-added carriers [3], announced a plan to offer public

packet-switched data service in which initially terrestrial and eventually satellite links will be available. Therefore, it is of great interest and importance to investigate, or at least to extend the current knowledge to cover, such a mixed-media packet-switched computer communication network.

Our mixed-media network model consists of a terrestrial store-and-forward packet-switching network, referred to here as the ground subnet, and a multiaccess/broadcast satellite which, together with the associated SIMP's, forms the satellite subnet. The store-and-forward ground subnet can be implemented to provide a low delay by using, for example, leased lines of low error rate. This, however, makes the network necessarily expensive. The satellite subnet, on the other hand, is subject to the intrinsic propagation delay of about 0.26 seconds, but its cost per channel bandwidth is substantially less than that of the ground links. Therefore, the combination of high-delay low-cost satellite subnet that operates in contention mode and a low-delay high-cost ground subnet that operates in queueing mode into an overall system model presents many interesting problems.

It is our goal in this paper to examine a number of key issues in the design of the proposed mixed-media packet-switching network. Assuming that network topology and traffic characteristics are given, we concentrate on the following problems in the present paper: 1) routing of packets via ground or satellite links; 2) capacity assignments for ground and satellite channels; and 3) retransmission strategies.

Routing procedures have been investigated using various approaches [4]-[7]. The routing we will consider is a deterministic procedure, which optimizes the overall average packet delay given a set of link capacities and message traffic characteristics. This is the approach taken by Kleinrock [4], Felperin [5], Fultz [6], and Cantor and Gerla [7] in their studies on optimal deterministic routing for a terrestrial store-and-forward network; their results will be used here to obtain the optimal routing for our mixed-media network model. A capacity assignment problem in communication networks was first formulated by Kleinrock [4], who assumed the linear cost model and a continuum of channel capacities. We solve our capacity assignment problem using the same approach and find tradeoffs between cost and overall average packet delay.

We also investigate two operational schemes for satellite channels: one is a scheme in which the satellite channel is used as a slotted ALOHA channel as discussed by Kleinrock and Lam [8], [9]; the other is a new multiaccess scheme we propose, in which no retransmissions are attempted via satellite channel; whenever a channel collision occurs, retransmission of the involved packets will route through ground links to their destinations. These two different schemes will be compared in terms of such performance measures as overall average delay of a packet, maximum allowable traffic loads, capacity requirements, etc.

The proposed network model consists of the following.

1) A set of store-and-forward IMP-like devices interconnected by capacity-limited ground channels (a connected, distributed subnet). For the sake of reliability this subnet is at least two-connected.

2) A set of SIMP-like devices directly connected to satellite ground stations. These SIMP's are usually geographically scattered and relatively far apart from each other. IMP's and SIMP's all have buffering and scheduling capabilities.

3) A multiaccess/broadcast satellite transponder linking all SIMP's in a star configuration. The SIMP's together with the broadcast satellite channel will be referred to as a satellite subnet.

In this network model we assume that the network is regionalized. That is, the network is partitioned into regions. Each region contains a SIMP and a number of IMP's. An IMP can only access its regional SIMP. In this study we assume that the number of SIMP's and their locations are given *a priori*. A SIMP is usually colocated with an IMP at some node. The regionalization of IMP's is determined by the closeness of an IMP to a SIMP in terms of the number of hops and the distances between them. Such a structure is shown in Fig. 1.

In the following, we first review previously known results which are important to our studies, and at the same time we develop those analyses which are pertinent to our network model but have not been considered before. We then proceed on to formulate and solve the related design problems.

II. SATELLITE SUBNET MODEL

The satellite subnet consists of a set of SIMP's which are linked together via a satellite multiaccess/broadcast channel. Each SIMP is equipped with buffering and scheduling capabilities. The satellite channel model we will use is based upon the ALOHA technique of random-access synchronous time-division multiplexing [10], [11]. We assume throughout this study that the satellite channel is time-slotted [8], but as to multiaccessing methods, we consider two different schemes: one is the slotted ALOHA, and the other is a new scheme which we name multiaccess satellite with terrestrial retransmission (MASTER).

Scheme I: Slotted ALOHA

In our satellite subnet the user population consists of a set of SIMP's, and we assume that each SIMP is provided with sufficiently large buffer capacity that new arrivals will never be blocked. Similarly a SIMP is capable of transmitting a new packet, even when previously sent packets are outstanding due to collisions.

Lam [9] has made an extensive study of slotted ALOHA proposed for a satellite communication system. His result on a finite population model without blocking is given in terms of

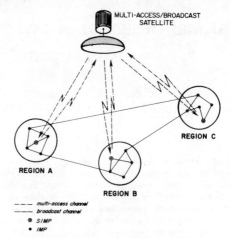

Fig. 1.   Proposed network model.

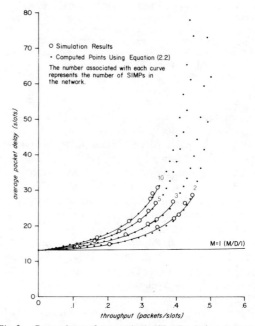

Fig. 2.   Comparisons of curves obtained by simulations in Lam [9] and by curve-fitting (2).

throughput–delay curves as shown in Fig. 2. These curves were obtained using simulation rather than by analysis and assumed equal input rates from all users and uniform retransmission and rescheduling delays. A rescheduling delay results from a scheduling conflict in which one or more packets are scheduled to transmit in a slot. The scheduling conflict is resolved by first sending the high-priority packet and rescheduling the other packets for later slots. An important result from his simulation study is that when the satellite system has $M = 10$ users, its throughput–delay tradeoffs approximate those of an infinitely many user population.

If there were only one SIMP transmitting, i.e., $M = 1$, then there should be no contention in the channel. Thus the delay over the satellite subnet consists of the propagation delay and the average queueing delay; the latter component can be derived from the result of an $M/D/1$ queueing system:

$$\tau_s = \tau_{min} + \frac{1}{\mu_s C_s} + \frac{\lambda_s}{2(\mu_s C_s - \lambda_s)} \cdot \frac{1}{\mu_s C_s} \qquad (1)$$

where

τ_s	average packet delay (seconds)
τ_{min}	0.26 second: delay time due to propagation delay
C_s	satellite channel capacity (bits/s)
λ_s	total input rate to (hence throughput rate from) the satellite channel (packets/s)
$1/\mu_s$	packet length (bits/packet).

For $M > 1$ we found that a simple modification of (1) is very satisfactory to those curves obtained by simulation in a manner as indicated by Fig. 2.

$$\tau_s = \tau_{min} + \frac{1}{\mu_s C_s} + \frac{b(M)\lambda_s}{2(\mu_s C_s a(M) - \lambda_s)} \cdot \frac{1}{\mu_s C_s} \qquad (2)$$

where $a(M)$ represents the degradation factor of channel capacity due to collision. From (1) and the known result on an infinite population model we have $a(1) = 1$ and $\lim_{M \to \infty} a(M) = 1/e$. The other parameter $b(M)$ gives another degree of freedom for us to fit the curves in the vertical direction. The set of parameters found are tabulated in Table I.

Scheme II: MASTER

A possible drawback of the ALOHA scheme as applied to the satellite subnet is that each time a packet needs a retransmission, its delay time must be increased at least by the 0.26 seconds due to the inherent propagation delay of the satellite channel. Furthermore if the message traffic rate exceeds some critical level, the number of backlog packets will rise sharply, and will further increase the occurrence of collisions. This positive feedback effect will possibly result in an avalanche of collisions: the channel utilization will rise up to almost 100 percent, yet the throughput will suddenly drop to virtually zero. This "instability" problem of the slotted ALOHA is discussed by Kleinrock and Lam [9].

This observation had led us to propose a new scheme, in which all retransmissions by SIMP's are carried out through the ground subnet, and retransmitted packets are rerouted from the originating SIMP's to their destination IMP's directly via ground without going through the destination SIMP's. We will refer to this technique as MASTER. In this scheme, there will be no outstanding packets in the satellite subnet, and no packet will experience more than one roundtrip propagation delay over the satellite channel. If the ground subnet is not as congested as the satellite subnet, then we not only improve the overall delay response, but also ensure the stability of the system.

Since MASTER has never been analyzed previously, let us consider the throughput-delay tradeoffs here. Suppose that the satellite subnet has M SIMP's. Let λ_σ [packets/s] be the average message rate from SIMP σ to the satellite channel,

TABLE I
PARAMETERS OF (2)

# of SIMPs M	a(M)	b(M)
1	1	1
2	.531	3.059
3	.528	4.674
5	.494	5.871
10	.489	7.219

$\sigma = 1, 2, \cdots, M$. Then the probability q_σ, that SIMP σ attempts a packet transmission in a given time slot, is

$$q_\sigma = \frac{\lambda_\sigma}{\mu_s C_s}, \qquad \sigma = 1, 2, \cdots, M \qquad (3)$$

where C_s [bits/s] is the capacity of multiaccess satellite channel and $1/\mu_s$ [bits/packet] is the packet size as defined earlier. We define random sequences $X_\sigma(h)$ and $Y_\sigma(h)$ by

$$X_\sigma(h) = \begin{cases} 1, & \text{if SIMP } \sigma \text{ transmits in slot } h \\ 0, & \text{otherwise} \end{cases} \qquad (4)$$

and

$$Y_\sigma(h) = \begin{cases} 1, & \text{if SIMP } \sigma \text{ successfully transmits in slot } h \\ 0, & \text{otherwise} \end{cases} \qquad (5)$$

for $\sigma = 1, 2, \cdots, M$, and $h = 1, 2, 3, \cdots$. Certainly, $X_\sigma(h) = 1$ if $Y_\sigma(h) = 1$, but not vice versa. If $X_\sigma(h) = 1$ and $Y_\sigma(h) = 0$, then the attempted transmission is a failure due to channel collision.

From the definitions of (3) and (4) it follows that

$$P[X_\sigma(h) = 1] = q_\sigma \qquad (6)$$

and

$$P[X_\sigma(h) = 0] = 1 - q_\sigma \triangleq \bar{q}_\sigma \qquad (7)$$

for $\sigma = 1, 2, \cdots, M$, and all h.

Furthermore, by assuming that the input message sequences from different SIMP's are statistically independent, we readily obtain the following properties for the sequences $Y_\sigma(h)$.

$$P[Y_\sigma(h) = 1] = q_\sigma \prod_{\tau \neq \sigma} \bar{q}_\tau \triangleq q_\sigma P_\sigma \qquad (8)$$

where P_σ represents the probability of success of any attempt transmission from SIMP σ.

The throughput S_0 [packets/slot time] of the satellite channel is therefore given by

$$S_0 = \sum_{\sigma=1}^{M} P[Y_\sigma(h) = 1] = \sum_{\sigma=1}^{M} q_\sigma P_\sigma. \qquad (9)$$

Since the input (or offered) traffic is

$$S_I = \sum_{i=1}^{M} q_\sigma \qquad (10)$$

the difference between S_I and S_o,

$$S_I - S_0 = \sum_{\sigma=1}^{M} q_\sigma (1 - P_\sigma), \tag{11}$$

is the portion of the packet flows that are rerouted via the ground subnet.

If we assume that message flows from regional IMP's to a SIMP can be characterized by a Poisson process, and that nodal processing delays can be very small compared with channel queueing delays, then the average delay T_σ [s/packet] experienced by a packet successfully traveling from SIMP σ through the satellite channel is given by

$$T_\sigma = \tau_{\min} + \frac{1}{\mu_s C_s} + \frac{1}{2\mu_s C_s} \cdot \frac{\lambda_\sigma}{\mu_s C_s - \lambda_\sigma} \tag{12}$$

where the last term is the average waiting time obtained from an $M/D/1$ queueing model.

The overall average packet delay over the satellite channel τ_s [s/packet] can then be obtained by averaging T_σ's:

$$\tau_s = \frac{1}{\lambda_s} \sum_{\sigma=1}^{M} \lambda_\sigma T_\sigma$$

$$= \tau_{\min} + \frac{1}{\mu_s C_s} + \frac{1}{2\lambda_s} \sum_{\sigma=1}^{M} \frac{q_\sigma}{1 - q_\sigma} \tag{13}$$

where

$$\lambda_s = \sum_{\sigma=1}^{M} \lambda_\sigma \text{ [packets/s]} \tag{14}$$

is the total input rate to the satellite channel. Note that the condition for the individual queue to be stable is $\lambda_\sigma < \mu_s C_s$, $\sigma = 1, 2, \cdots, M$, rather than $\lambda_s < \mu_s C_s$. This is because those packets which require retransmissions are not placed on the satellite channel and rerouted to the ground.

For the purpose of this paper (13) is all we need. For further information concerning MASTER, the reader is referred to a companion paper by the authors [13].

III. GROUND SUBNET MODEL

The ground subnet is a store-and-forward distributed network. In accordance with the two different operational schemes of the satellite subnet, we discuss the following two cases for the ground subnet model. In Scheme I (i.e., slotted ALOHA), under which the satellite subnet uses slotted ALOHA without blocking, the ground subnet, has only originally assigned ground traffic (i.e., only one type of traffic). In Scheme II (i.e., MASTER), under which SIMP's divert all retransmissions to the ground net, IMP's can either add these retransmission to their originally assigned packet flows on a first-come, first-served (FCFS) basis, or give these retransmission packets a higher priority, by holding off, but without preempting, regular packets flow. Under the FCFS case, the operations of the ground subnet associated with

Scheme I and Scheme II are the same. Under the priority scheduling scheme, the IMP's should be equipped with a facility to manipulate traffic with two priority classes.

For a queueing network with general structure, a closed-form analytic solution has been obtained only under the Markovian assumption: that is, the origination of messages from sources (IMP nodes in our model) should be characterized by Poisson processes, and the service times (message lengths) are independent and identically distributed (iid) with exponential distributions. The Poisson assumption has been statistically validated in several empirical studies of data traffic (see [14], [15]). The length of packets in our ground subnet is assumed to be variable with a possible constraint on the maximum allowable packet size. As long as the coefficient of variation (i.e., the ratio of standard deviation to mean) of the packet size distribution is not far from unity, the exponential assumption should be quite reasonable. However, the length of a packet, once generated at the originating IMP, should remain unchanged throughout the entire transmission over different links within the network; the independence assumption stated above certainly is not realistic in this regard, but it seems that this assumption is not so critical for the analysis of such performance measures as throughput and the average delay. It may well be a poor assumption if one is interested in estimating the delay of a particular packet or in the delay distribution rather than just the average value. As for further discussions on these assumptions the reader is referred to Kleinrock [4], [16] who reports the validation of such a model based on simulation studies.

Under the Markovian assumptions made above, we can apply the well-known decomposition theorem due to Jackson [17] and its recent extension by Baskett *et al.* [18] and Kobayashi and Reiser [19]. Hence the average delay T_l [s/packet], that a packet experiences due to queueing and transmission over the link l, is

$$T_l = \frac{1}{\mu C_l - \lambda_l} \tag{15}$$

in which λ_l [packets/s] is the total message flow over link l, C_l [bits/s] is the capacity of that link, and $1/\mu$ [bits/packet] is the average length of a packet.

A remark is in order concerning (15). This formula holds not only under the FCFS scheduling rule but under a large class of disciplines, which are often referred to by the name of "work-conserving" queueing disciplines [20]. This class includes the case with two class priority scheduling described earlier. Furthermore, it has been shown recently [19] that in order for (15) to apply, the message-routing behavior can be any stochastic or deterministic one, as long as it can be characterized by a Markov chain of some order. Furthermore, we can assume arbitrary number of different classes, as long as their routing behavior is concerned (we still have to assume, however, that all messages are coming from the common distribution, which is an exponential distribution of mean $1/\mu$. This last property allows us to assume different routing patterns depending on the origin of messages. A deterministic split traffic (bifurcated) routing strategy to be discussed later is certainly a particular case of general routing behavior to which the simple formula (15) applies. Note that the traffic rates $\{\lambda_l\}$ over the individual links will differ depending on the routing pattern or algorithm to be chosen.

IV. ROUTING

With a mixed-media network the issue of routing is a major concern. With two possible courses to choose from—one via satellite and one via ground—the issue is to choose the set of routes so as to minimize the overall average network delay. The tradeoffs to consider are these: the satellite channel has an inherent minimum delay of 0.26 seconds. However, satellite capacity is less costly than ground channel capacity for medium to long distances, and therefore more satellite capacity is available at less cost than comparable facilities on the ground. The ground channels are inherently faster than the satellite channel, but because of capacity limitations, are subject to queueing delays, which combined with the store-and-forward nodal processing delays, may, in heavy traffic situations, result in larger overall delays than the satellite delays. To summarize, satellite channels have greater delays but also more cost effective channel bandwidth than ground channels.

The routing procedure used in the ARPANET is a distributed adaptive algorithm in which each node has a routing table which is periodically updated with minimum distance estimates from its immediate neighbor [6]. In our case study to follow, we have chosen a deterministic split traffic routing strategy [6] which because it allows traffic to flow on more than one path between a given source-destination node pair, gives a better balance than a fixed routing procedure. We wish to emphasize, however, that our analytical model does not require any specific routing algorithm, but can accommodate any that are static and can be modeled mathematically.

Suppose we are given a network of N nodes with a specified topology which includes a satellite channel of capacity C_s [bits/s], a set of ground links capacities C_l [bits/s] ($l = 1,2, \cdots, L$), and a demand matrix $[\gamma_{ij}]$, where γ_{ij} [packets/s] is the average rate of messages originated at node (IMP) i and destined for node (IMP) j, $i,j = 1, 2, \cdots, N$. The routing problem is to optimally assign the traffic demand γ_{ij}'s along different paths of the network so that the resulting overall average packet delay is minimized. Note that the link traffic rate λ_l ($l = 1,2, \cdots, L$) defined in the preceding section is uniquely determined by the demand matrix $[\gamma_{ij}]$, the routing rule, and retransmission strategy.

Let us define the *traffic splitting factor* g_{ij} as the fraction of the traffic, originated at node i and destined for node j, which goes through the ground subnetwork. It is clear that $g_{ij} = 1$ if IMP's i and j are in the same region. We define $\bar{g}_{ij}$ by

$$\bar{g}_{ij} = 1 - g_{ij}.$$

This fraction ($\bar{g}_{ij}$) of the traffic is first routed to the regional SIMP of IMP i, sent through the satellite channel to the regional SIMP of IMP j, and finally directed to the destination IMP j. Of courses this sequence of steps takes place only when

the transmission over the satellite channel is successful. If a collision takes place, two different courses of action follow, depending on the scheme assumed; in case of the ALOHA channel (i.e., Scheme I), the collided packets will attempt retransmission through the satellite channel, whereas in the MASTER channel (i.e., Scheme II), these packets will be rerouted to the ground subnet.

In both the slotted ALOHA and MASTER schemes, the overall average delay T for a packet traveling from its origin to destination is given by

$$T = \frac{1}{\gamma} \left[\sum_{l=1}^{L} \lambda_l T_l + \sum_{\sigma=1}^{M} \lambda_\sigma T_\sigma \right] \tag{16}$$

where

$$\gamma = \sum_{ij} \gamma_{ij} = \text{total traffic rate in the network.} \tag{17}$$

The definition of λ_l and λ_σ, and derivations of the average delay T_l and T_σ were already discussed in Sections II and III. In [21, appendix II] we derive the expressions for λ_l and λ_σ in terms of $\{g_{ij}\}$ and $\{\gamma_{ij}\}$.

Having derived the expression for the overall average packet delay, we can now state formally our routing problems as follows: given network configuration, traffic demand matrix $[\gamma_{ij}]$, and link capacities C_l's and C_s,

$$\min_{\{g_{ij}\}} T \qquad \text{subject to } \{0 \leqslant g_{ij} \leqslant 1, \text{ for all } i,j\}. \tag{18}$$

In the system with slotted ALOHA channel (i.e., Scheme I), the average delay T_σ of those packets sent from SIMP σ ($\sigma = 1,2, \cdots, M$) over the satellite subnet is equal to τ_s of (2), if the traffic demand and routing are such that the input rates from all the SIMP's into the satellite channel are well balanced, i.e.,

$$\lambda_\sigma \cong \frac{\lambda_s}{M}, \qquad \sigma = 1, 2, \cdots, M. \tag{19}$$

Those packets which go through the ground link l will experience the average delay of T_l given by (15). Thus the overall average delay of a packet during its entire travel in the net is, from (2), (15), and (16),

$$T = \frac{1}{\gamma} \sum_{l=1}^{L} \frac{\lambda_l}{\mu C_l - \lambda_l} + \frac{\lambda_s \tau_{\min}}{\gamma}$$

$$+ \frac{\lambda_s}{\gamma \mu_s C_s} \left[1 + \frac{b(M)\lambda_s}{2(\mu_s C_s a(M) - \lambda_s)} \right]. \tag{20}$$

For the scheme with MASTER channel (Scheme II) the average delay T_σ for packets from SIMP σ is given by (12). The average delay that a packet receives in its transmission over the ground link l is given by T_l of (15), irrespective of the scheduling rule chosen. Therefore, by substituting these equations into (16) we obtain the following expression for the overall delay which a packet receives in the net under Scheme II:

$$T = \frac{1}{\gamma} \sum_{l=1}^{L} \frac{\lambda_l}{\mu C_l - \lambda_l} + \frac{\lambda_s \tau_{\min}}{\gamma}$$

$$+ \frac{1}{\gamma \mu_s C_s} \sum_{\sigma=1}^{M} \lambda_\sigma \left\{ 1 + \frac{\lambda_\sigma}{2(\mu_s C_s - \lambda_\sigma)} \right\}. \qquad (21)$$

In (21), the traffic rate λ_l over the ground link l includes not only the originally assigned ground traffic, but also the traffic due to those packets which are rerouted from SIMP's after unsuccessful transmissions over the satellite channel.

For the class of deterministic and probabilistic routing rules we discussed earlier, we can show that both T of (20) and that of (21) are *convex* functions of g_{ij}, $i, j = 1, 2, \cdots, N$. The reason is as follows: as discussed earlier the traffic rates λ_l ($l = 1, 2, \cdots, L$), λ_σ ($\sigma = 1, 2, \cdots, M$), and λ_s are all linear with respect to g_{ij}, and T's of (20) and (21) are both convex functions with respect to λ_l, λ_σ, and λ_s. Because of the convexity, the set of $\{g_{ij}\}$ that minimize T can be found by any optimization procedure. In numerical computations of the case study problems we used Box's COMPLEX optimization algorithm [22]. This method is a sequential search technique which has proven effective in solving problems with nonlinear objective functions subject to nonlinear inequality constraints. It has an advantage over gradient methods in that no derivatives are required. It also tends to find the global optimum due to the fact that the initial set of starting points are randomly scattered throughout the feasible region. Moreover, its rate of convergence has been shown to be better than Rosenbrock's algorithm [22]. In [21, appendix III] we show the flow chart and give a brief description of this optimization procedure.

Example 1: Consider the network with eight nodes (IMP's) and twenty links shown in Fig. 3. In this net there are two regions consisting of nodes {1,2,3,4} and nodes {5,6,7,8}, respectively, and the regional SIMP's are located at nodes 1 and 7. The traffic demand matrix is assumed to be uniform with $\gamma_{ij} = 20$ [packets/s] for all $i \neq j$ and $\gamma_{ii} = 0$ for all $i = 1, 2, \cdots, 8$, and the average packet length is assumed to be 512 bits on all ground channels. The packet length on the satellite channel is fixed and equals 1 kbit. The ground link capacities are all assumed to be 50 kbits/s, i.e., $C_l = 5 \times 10^4$ [bits/s] for all $l = 1, 2, \cdots, 20$, and the satellite capacity to be $C_s = 1.5 \times 10^6$ [bits/s].

The ground subnet routing we used in this example is the split traffic routing (or alternate routing) which is based on the minimum number of hops required to transmit packets from a given source node to a destination node. For example, if we want to send packets from IMP 1 to IMP 8 via the ground net, the minimum number of hops between IMP's 1 and 8 is four, and there are four alternate paths of four hops: they are path a) $1 \rightarrow 3 \rightarrow 5 \rightarrow 7 \rightarrow 8$; path b) $1 \rightarrow 3 \rightarrow 5 \rightarrow 6 \rightarrow 8$; path c) $1 \rightarrow 3 \rightarrow 4 \rightarrow 6 \rightarrow 8$; and path d) $1 \rightarrow 2 \rightarrow 4 \rightarrow 6 \rightarrow 8$. At any node along the paths selected above, if there are two links of the selected paths emanating from the node, then the traffic rate is bifurcated equally on each of these two links. For instance the traffic coming into IMP 3 will be split into links ④ and ⑤ equally. Using this ground subnet routing algorithm

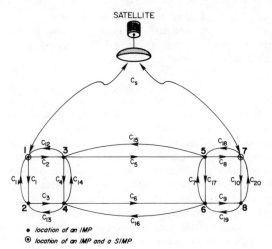

Fig. 3. Network model with two SIMP's and eight IMP's.

- location of an IMP
- ⊙ location of an IMP and a SIMP

TABLE II
ROUTING INDEXES FOR TRAFFIC FROM REGION 1
TO REGION 2

for ALOHA

		destination		
	5	6	7	8
1	1	0	0	0
2	.855	.681	0	.706
3	1	1	0	.315
4	1	1	0	1

origin

T = .229 seconds

for MASTER

		destination		
	5	6	7	8
1	.889	0	0	0
2	1	.995	0	.394
3	1	1	.025	.692
4	1	.817	0	.805

origin

T = .326 seconds

we find that the traffic assignments between IMP's 1 and 8 are 1/8 of the total traffic $\gamma_{1,8}$ over path a), 1/8 over path b), 1/4 over path c), and 1/2 over path d).

After obtaining these split factors for each node pair and adding them up appropriately, we obtain the link traffic λ_l for each $l = 1, 2, \cdots, 20$. Because of the symmetry of both the network topology and traffic demand matrix, balanced traffic is maintained throughout the net by use of the ground subnet routing algorithm discussed above.

With the above data as input to the routing optimization program, the routing indexes g_{ij} for both ALOHA and MASTER are computed and are given in Table II. Also shown in Table II are the minimum overall average delays of 0.229 seconds for the network using ALOHA and 0.326 seconds for the network using MASTER. ALOHA outperforms MASTER. This is not too surprising; here satellite capacity is much larger than the ground capacities ($C_s = 1.5 \times 10^6$ bits/s $> C_l = 5 \times 10^4$ bits/s). In addition, the ground capacities are shared by intraregional traffic. (Recall that $g_{ij} = 1$ if IMP's i and j reside in the same region.) Thus, the optimum average delay points fall inside the region favoring ALOHA. At the end of this paper, we show how this difference can be narrowed by a reassignment of the capacities subject to a fixed linear cost.

In Table II, only routing indices for traffic from Region 1 to Region 2 are shown. Because of the symmetry of the network and demand matrix, the routing indices for traffic from Region 2 to Region 1 are symmetrical to those from Region 1 to Region 2. For instance, for network using MASTER scheme (lower table), $g_{82} = g_{28} = 0.394$; $g_{72} = g_{18} = 0$; etc.

V. CAPACITY ASSIGNMENTS

In the packet-routing studies of the previous section we assumed that the ground link capacities and the satellite channel capacity are given. Now we proceed to the capacity assignment problem; that is, we wish to minimize the total average message delay T under the total budget constraint. We assume as given the topology of the network (including SIMP locations), and the demand matrix $[\gamma_{ij}]$. Furthermore we assume, in the present section, that the routing indexes $[g_{ij}]$ are given; hence the link traffic λ_l, $l = 1, 2, \cdots, L$ and the satellite channel traffic λ_s are also known. (The last assumption will be removed in Section VI which will discuss the joint optimization problem.)

We can formulate the capacity assignment problem as

$$\begin{array}{c} \text{minimize } T \\ \{c_l\}, c_s \end{array} \quad \text{subject to } \sum_{l=1}^{L} b_l C_l + b_s C_s \leqslant B \quad (22)$$

where T is given by (20), i.e.,

$$T = \frac{1}{\gamma}\left[\sum_{l=1}^{L} \lambda_l T_l + \sum_{\sigma=1}^{M} \lambda_\sigma T_\sigma\right]. \quad (23)$$

The parameter b_s [cost/bit·s^{-1}] of (22) is the cost of the satellite channel per unit traffic rate [bit/s]; similarly b_l is the cost factor of link l, $l = 1, 2, \cdots, L$.

The optimization problem defined by (22) and (23) is quite difficult if the capacities must be chosen from a discrete set of options. Following Kleinrock [4] we assume that the capacities of ground links and satellite channel are continuous variables and use analytic procedures involving Lagrange multipliers similar to that used by Kleinrock in his capacity assignment problem [4]. We now outline our solutions for the two different schemes.

Scheme I: Slotted ALOHA

The (continuous) capacity assignment problem is to minimize T of (20) subject to the following equality constraint:

$$B = \sum_{l=1}^{L} b_l C_l + b_s C_s. \quad (24)$$

By using the well-known Lagrange multiplier method, we could obtain closed form expressions for the optimal values of C_l and C_s. This would involve, however, solving fourth-order algebraic equations, which would result in rather complicated solutions.

The solution can be greatly simplified, however, if we slightly modify the result of (20) which we write $\tilde{T}$

$$\tilde{T} = \frac{1}{\gamma}\sum_{l=1}^{L} \frac{\lambda_l}{\mu C_l - \lambda_l} + \frac{\lambda_s \tau_{\min}}{\gamma}$$

$$+ \frac{\lambda_s}{\gamma\mu_s C_s}\left[\frac{b(M)}{2} + \frac{b(M)\lambda_s}{2(\mu_s C_s a(M) - \lambda_s)}\right]'$$

$$= \frac{1}{\gamma}\sum_{l=1}^{L} \frac{\lambda_l}{\mu C_l - \lambda_l} + \frac{\lambda_s \tau_{\min}}{\gamma} + \frac{a(M)\lambda_s}{2\gamma(\mu_s C_s a(M) - \lambda_s)}. \quad (25)$$

The difference between T of (20) and its approximation is

$$T - \tilde{T} = \left(1 - \frac{b(M)}{2}\right)\frac{\lambda_s}{\gamma\mu_s C_s}, \quad (26)$$

which is negligibly small compared with the second term of (25) since

$$\mu_s C_s \tau_{\min} \gg 1 \quad (27)$$

which holds practically for any value of C_s in the range of our interest. Minimization of $\tilde{T}$ involves only second-order algebraic equations, and the optimal set of capacities are given by

$$C_l = \frac{\lambda_l}{\mu} + \frac{B - B_c}{b_l} \cdot \frac{\sqrt{f_l}}{\sum_{l=1}^{L}\sqrt{f_l} + \sqrt{f_s/2}}, \quad l = 1, 2, \cdots, L \quad (28)$$

and

$$C_s = \frac{\lambda_s}{\mu_s a(M)} + \frac{B - B_c}{b_s} \cdot \frac{\sqrt{f_s/2}}{\sum_{l=1}^{L}\sqrt{f_l} + \sqrt{f_s/2}} \quad (29)$$

where

$$B_c = \sum_{l=1}^{L} \frac{b_l \lambda_l}{\mu} + \frac{b_s \lambda_s}{\mu_s a(M)} \quad (30)$$

$$f_l = \frac{b_l \lambda_l}{\gamma\mu} \quad (31)$$

and

$$f_s = \frac{b_s \lambda_s}{\gamma\mu_s}. \quad (32)$$

We can interpret the above result as follows. The first term λ_l/μ of (28) is the minimum required capacity of link l. If C_l were less than this critical value the link l would become unstable. Similarly, the first term of (29) is the lower bound of the satellite capacity, below which the ALOHA channel will completely collapse. The constant f_l defined by (31) represents the cost of the link l per unit level of the total traffic γ. Similarly f_s is that of the satellite channel. The value B_c defined by (30) represents the critical budget below which the network cannot accommodate the given traffic level γ. According to (28) and (29), the remaining budget $B - B_c$ is distributed among all the links and satellite in proportion to $\sqrt{f_l}$, $l = 1, 2, \cdots, L$, and $\sqrt{f_s/2}$, respectively.

The substitution of (28) and (29) into (25) leads to

$$\min \tilde{T} = \frac{\lambda_s \tau_{\min}}{\gamma} + \frac{\left(\sum_{l=1}^{L} \sqrt{f_l} + \sqrt{f_s/2}\right)^2}{B - B_c} \qquad (33)$$

which is a hyperbola function of B.

Scheme II: MASTER

We now minimize T of (21) with the same budgetary constraint, i.e., (24). Because of the reason similar to that discussed in Scheme I, we attempt to minimize, instead, the following objective function $\tilde{T}$:

$$\tilde{T} = \frac{1}{\gamma} \sum_{l=1}^{L} \frac{\lambda_l}{\mu C_l - \lambda_l} + \frac{\lambda_s \tau_{\min}}{\gamma}$$

$$+ \frac{1}{\gamma \mu_s C_s} \sum_{\sigma=1}^{M} \lambda_\sigma \left[\frac{1}{2} + \frac{\lambda_\sigma}{2(\mu_s C_s - \lambda_\sigma)} \right]$$

$$= \frac{1}{\gamma} \sum_{l=1}^{L} \frac{\lambda_l}{\mu C_l - \lambda_l} + \frac{\lambda_s \tau_{\min}}{\gamma} + \frac{1}{2\gamma} \sum_{\sigma=1}^{M} \frac{\lambda_\sigma}{\mu_s C_s - \lambda_\sigma}. \qquad (34)$$

The difference

$$T - \tilde{T} = \frac{1}{2\gamma \mu_s C_s} \sum_{\sigma=1}^{M} \lambda_\sigma = \frac{\lambda_s}{2\gamma \mu_s C_s} \qquad (35)$$

is again negligibly small compared with the second term of (34); their ratio is equal to the ratio of one half of packet time $1/2\mu_s C_s$ to the propagation time $\tau_{\min} = 0.26$ seconds. The modified function (34) is still not suitable to optimization; the last term consists of M components ($\sigma = 1, 2, \cdots, M$) each of which has the common value of C_s in the denominator. Thus, in order to obtain a closed form solution for optimal solution, one has to deal with an algebraic equation of degree $2M$. By further assuming that the traffic rates from SIMP's are well balanced, we use the approximation

$$\lambda_\sigma = \frac{\lambda_s}{M}, \qquad \sigma = 1, 2, \cdots, M. \qquad (36)$$

Then the objective function $\tilde{T}$ (34) reduces to that of (25) except that $a(M)$ is to be replaced by M. Hence the optimal sets of C_l and C_s are

$$C_l = \frac{\lambda_l}{\mu} + \frac{B - B_c{}^*}{b_l} \cdot \frac{\sqrt{f_l}}{\sum_{l=1}^{L} \sqrt{f_l} + \sqrt{f_s/2}} \qquad (37)$$

and

$$C_s = \frac{\lambda_s}{\mu_s M} + \frac{B - B_c{}^*}{b_s} \cdot \frac{\sqrt{f_s/2}}{\sum_{l=1}^{L} \sqrt{f_l} + \sqrt{f_s/2}} \qquad (38)$$

where the critical budget $B_c{}^*$ is now

$$B_c{}^* = \sum_{l=1}^{L} \frac{b_l \lambda_l}{\mu} + \frac{b_s \lambda_s}{\mu_s M}. \qquad (39)$$

The minimized $\tilde{T}$ possesses the same form as that of (33):

$$\min \tilde{T} = \frac{\lambda_s \tau_{\min}}{\gamma} + \frac{\left(\sum_{l=1}^{L} \sqrt{f_l} + \sqrt{f_s/2}\right)^2}{B - B_c{}^*}. \qquad (40)$$

Although the two schemes discussed above give similar results [(33) versus (40)], their performance comparison is not so simple as it may appear. If all the traffic rates λ_l, $l = 1, 2, \cdots, L$ and λ_s were common for both cases, then clearly Scheme II should achieve a better performance than Scheme I, since $B_c{}^* < B_c$ (recall that the values $a(M)$ obtained in Section II are all smaller than unity, hence smaller than M). The assumption made above, however, never holds. Even when the same set of routing indexes $\{g_{ij}\}$ and the ground subnet routing rule are applied to both of the schemes, the ground traffic λ_l, $l = 1, 2, \cdots, L$ are different, though the values of λ_s will be the same. In Scheme II the link traffic λ_l includes not only the originally assigned portion but the traffic rerouted from SIMP's, which will increase $B_c{}^*$. Note also the values f_l are different under the two different schemes, because of the difference in λ_l.

Since the link traffic is dependent on the routing indices g_{ij}, we cannot separate the capacity assignment problem from the routing problem.

VI. JOINT OPTIMIZATION OF ROUTING AND CAPACITY ASSIGNMENTS

In the previous two sections we treated the problems of routing and capacity assignments separately. The optimal solution of one problem, however, depends on the solution of the other; thus we need to seek joint optimization of the problems. Let us denote by g the set of routing indexes $\{g_{ij}; 1 \leqslant i, j \leqslant N\}$, and by C the set of link capacities $\{C_l, 1 \leqslant l \leqslant N, C_s\}$. Similarly let λ be the set of link and channel traffics $\{\lambda_l, 1 \leqslant l \leqslant L, \lambda_s\}$ which, in turn, is a function of g. We can write our objective function T as

$$T = T(\lambda(g), C) \qquad (41)$$

and our joint optimization problem as

$$\minimize_{g, C} T \qquad \text{subject to } 0 \leqslant g \leqslant 1$$

$$\text{and } \sum_{l=1}^{L} b_l C_l + b_s C_s \leqslant B.$$

It seems quite unfeasible to attempt to minimize the above function directly with respect to g and C. We instead seek the optimum solution g^* and C^* according to the iterative numerical procedure schematically shown in Fig. 4. Brief discussions of each block follow.

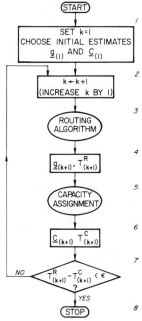

Fig. 4. Flow chart for joint optimization of routing and capacity assignment.

Blocks 1 and 2: Set the iteration step index k equal to one, and choose the initial estimates $g_{(1)}$ and $C_{(1)}$. For the specified ground subnet routing algorithm and the operational scheme of satellite channel (i.e., ALOHA channel or MASTER channel), $\lambda_{(1)}$ is computed from the given demand matrix $[\gamma_{ij}]$ for the chosen $g_{(1)}$. With this initialization, enter the iterative cycle by increasing the index k to $k + 1$.

Blocks 3 and 4: For the parameters $C_{(k)}$ and $\lambda_{(k)}$, the routing algorithm of Section IV yields the optimum routing indices $g_{(k+1)}$ and the corresponding performance index $T^R_{(k+1)}$.

Blocks 5 and 6: With the updated link traffic $\lambda_{(k+1)} = \lambda(g_{(k+1)})$, we then use the analytic solution method of Section V to find the optimum capacity assignment $C_{(k+1)}$, and the corresponding delay $T^C_{(k+1)}$.

Blocks 7 and 8: The average packet delay $T^C_{(k+1)}$ is an improvement over $T^R_{(k+1)}$ within the same cycle step $(k + 1)$. Hence it follows always that $T^C_{(k+1)} \leq T^R_{(k+1)}$. If the improvement is less than a sufficiently small parameter $\epsilon > 0$ (*a priori* set value), then we judge the minimum point has been achieved, and stop the whole procedure. Otherwise go back to Block 2 and iterate the same cycle.

From the above description, we see that each iteration is a two-stage optimization; in the first stage the set of link capacities is fixed, and T is minimized over the set of traffic splitting factors, whereas in the second stage, the set of traffic splitting factors is fixed and T is minimized by the choice of link capacities. Since T is a convex function with respect to both C and g, the optimum solution can be achieved within the error determined by ϵ. In fact, for the numerical examples to follow, with $\epsilon = 10^{-5}$ seconds, each optimum point is obtained after 10–20 iterations depending on how close the

starting point is to the optimum solution; the speed of convergence is quite reasonable.

Example 2: As an example of this joint optimization, let us again consider the network model of the routing example (Example 1), depicted in Fig. 3.

In solving the capacity assignment problem we need to know the total allowable budget and unit cost of ground and satellite channel capacities. We assume the unit cost of the ground channel capacities are identical, so we can normalize all costs by the unit cost of ground channel capacity. In obtaining the following results, we further assume a ratio of 1 to 10 for satellite and ground channel capacity costs, i.e., 1 unit cost of ground channel capacity equals 10 unit cost of satellite channel capacity. Our program is general enough so that we can assume any ratio between satellite capacity and ground capacity costs. Perhaps a more realistic ratio to use today is a 1 to 3 proportion. The input data consist of the original demand matrix, initial values of $C_l = 50$ kbits/s and $C_S = 1500$ kbits/s, and an assumed total budget B in ground channel cost units and $b_l = 1$ and $b_s = 0.1$.

We have performed a number of runs with different values of budget B and the results are given in Fig. 5. We see that the throughput–delay tradeoff curves, after the removal of the assumption made in Section V that the link traffic rates are fixed, are again hyperbolic. As shown in Fig. 5, network under ALOHA tends to perform better than network under MASTER. Recall that we have observed this fact in the routing example. As seen in Fig. 5 for the 2-SIMP case, the difference in the delays is reduced as compared with the results in the previous routing example (27 percent relative to MASTER's delay value there in comparison with 21 percent here). Nevertheless, the results are still in favor of ALOHA. Notice that the delay differences between ALOHA and MASTER are negligible for budget ranging from $B = 1\,300\,000$ cost units to $B = 1\,700\,000$ cost units. As B drops below $1\,300\,000$ cost units, the delay curve of MASTER departs from that of ALOHA. A reasonable explanation for these outcomes is as follows. When the budget B is sufficiently large, one has money to spend on the more expensive ground channel capacities; therefore, most traffic go through the lower delay ground subnet. (Recall that the satellite subnet-delay is at least 0.26 seconds.) Since satellite channel traffic is small, the difference between ALOHA and MASTER disappears; their performances are thus similar. When budget becomes small, the ground subnet alone cannot support the entire traffic load, more and more traffic is diverted to the satellite channel, and the distinction between MASTER and ALOHA becomes apparent. Since we have assumed a 10 to 1 cost ratio in favor of satellite channel, more capacity is assigned to the satellite subnet and less to the ground subnet. As we know, the larger the satellite channel capacity, the better ALOHA performs, whereas the smaller the ground channel capacities, the poorer MASTER performs. Altogether these make ALOHA outperform MASTER.

Example 3: To further clarify the problem let us consider an 8-region network shown in Fig. 6 with same configuration as the previous example, but each node now has an IMP and a SIMP. It is a network with 8 IMP's and 8 SIMP's. We assume

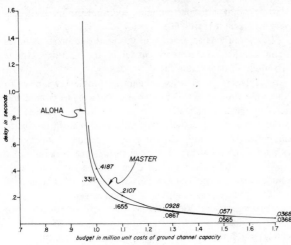

Fig. 5. Delay–cost tradeoffs for network with eight IMP's and two SIMP's.

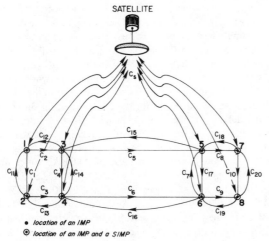

Fig. 6. Network model with eight IMP's and eight SIMP's.

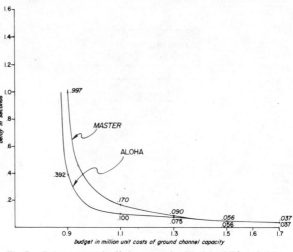

Fig. 7. Delay–cost tradeoffs for network with eight IMP's and eight SIMP's and ten to one cost ratio.

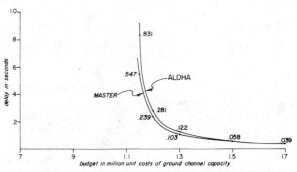

Fig. 8. Delay–cost tradeoffs for network with eight IMP's and eight SIMP's and three to one cost ratio.

that the demand matrix, mean packet lengths, and unit-cost ratio remain the same as before. For ALOHA, however, $a(M)$ and $b(M)$ are now estimated to be 0.492 and 6.724, respectively. Results for this network model are depicted in Fig. 7. It is seen that this network model has generally smaller delay than the previous one. This is because of the addition of 6 more SIMP's.[1] The network under ALOHA is again superior to that under MASTER, and the difference is more substantial. This is a general trend: as the number of SIMP's is increased, more traffic goes through the satellite channel, which draws more capacity from the total, and ground subnet gets less of its share. This puts MASTER in a disadvantageous position relative to ALOHA. For the same network (shown in Fig. 6)

[1] Unfortunately, our mathematical model does not include the cost of IMP's and SIMP's to reflect this addition. Another way of looking at this problem is: the cost of the SIMP's has been averaged out and included in the cost unit b_s; so by retaining the same cost ratio, we essentially assume that the effective cost of the satellite channel in this example is even lower than that in the previous case.

and parameters, if we now reduce the cost ratio to 3 : 1, we obtain different results which are depicted in Fig. 8. As seen in Fig. 8, again, when the budget $B > 1\,500\,000$ cost units, no distinction exists between ALOHA and MASTER, but, when $B < 1\,500\,000$ cost units, MASTER is slightly better than ALOHA, since now the cost ratio is advantageous for ground subnet, ground channels get more of their share of the total capacity, and excessive capacity in the ground subnet favors MASTER.

CONCLUSIONS

In this paper we have presented some of the important design issues for mixed-media packet-switching networks. Satellite packet switching has considerable promise for low-cost high-bandwidth data communications. However, there is inherent high delay in satellite links which does not appear in ground links. Therefore, a mix of the two communication media seems to offer the best of both worlds. In this paper we have examined a number of tradeoffs which offer guidelines for the design and optimum utilization of mixed-media networks. We have introduced a new communication scheme called multiaccess satellite with terrestrial retransmission

(MASTER) with the hope that it would offer significant advantages over slotted ALOHA. Results of this paper show that MASTER does work better than ALOHA under certain circumstances, but not always so. The capacity assignment, determined by the cost ratio of ground and satellite channels, determines which system is better. Also, the greater sensitivity to small changes in traffic of ALOHA may make MASTER the better system for many applications.

The fact that neither of the two schemes is clearly dominant suggests that a mixture of both might be the best system. When a channel collision occurs in such a system, retransmissions can go either again through the satellite subnet, as in ALOHA, with some probability δ, or through the ground subnet, as in MASTER, with probability $1 - \delta$. The probability δ would be chosen to minimize the average packet delay. This idea will be explored further.

In this paper we have not included the cost of IMP's and SIMP's in our model. We have also not explored the possibility of sending network control information along the ground and using the satellite for bulk data transmission. Another logical extension of the MASTER scheme will be to use the satellite channel on a reservation basis, such as suggested by Crowther *et al.* [23], Roberts [24], and Binder [25], but use the ground channel to set up reservations. We plan to explore these ideas in a subsequent paper.

REFERENCES

[1] D. L. A. Barber and D. W. Davies, "The NPL data network," in *Proc. Conf. Laboratory Automation,* Novosibirsk, U.S.S.R., Oct. 1970; also *NPL Com. Sci. T.M. 47 (NIC 14671),* Oct. 1970.

[2] S. Butterfield, R. Rettberg, and D. Walden, "The satellite IMP for the ARPA network," in *Proc. 7th Hawaii Int. Conf. System Sciences–Subcon. Comput. Nets,* Jan. 1974, pp. 70-73.

[3] B. D. Wessler and R. B. Hovey, "Public packet-switched networks," *Datamation,* pp. 85-87, July 1974.

[4] L. Kleinrock, *Communication Nets: Stochastic Message Flow and Delay.* New York: Dover, 1964.

[5] K. D. Felperin, "Interactive techniques for evaluation of command-control store-and-forward net performance," Stanford Res. Inst., Stanford, CA, Tech. Note TN-CDS-1, 1969.

[6] G. L. Fultz, "Adaptive routing techniques for message switching computer-communications networks," Ph.D. dissertation, Dep. Comput. Sci., Univ. of California, Los Angeles, available as rep. UCLA-ENG-7252, July 1972.

[7] D. G. Cantor and M. Gerla, "Optimal routing in a packet-switched computer network," *IEEE Trans. Comput.,* vol. C-23, pp. 1062-1069, Oct. 1974.

[8] L. Kleinrock and S. S. Lam, "Packet-switching in a slotted satellite channel," in *AFIPS Nat. Comput. Conf. Proc.,* vol. 42, June 1973, p. 703.

[9] S. S. Lam, "Packet-switching in a multi-access broadcast channel with application to satellite communication in a computer network," Ph.D. dissertation, Dep. Comput. Sci., Univ. of California, Los Angeles, available as rep. UCLA-ENG-7429, Apr. 1974.

[10] N. Abramson, "The ALOHA System," in *Computer-Communication Networks,* N. Abramson and F. F. Kuo, Eds. Englewood Cliffs, NJ: Prentice-Hall, 1973, pp. 501-518.

[11] F. F. Kuo and N. Abramson, "Some advances in radio communications for computers," in *Dig. Papers–COMPCON '73,* San Francisco, CA, Feb. 1973, pp. 57-60.

[12] L. G. Roberts, "ALOHA packet system with and without slots and capture," Stanford Res. Inst., Stanford, CA, ARPANET Satellite Syst. Note 8 (NIC 11290), June 1972.

[13] D. Huynh, H. Kobayashi, and F. F. Kuo, "Design issues for mixed media packet switching networks," in *Proc. Nat. Comput. Conf., AFIPS Conf. Proc.,* vol. 45. Montvale, NJ: AFIPS Press, 1976, pp. 541-549.

[14] E. Fuchs and P. E. Jackson, "Estimates of distribution of random variables for certain computer communications traffic models," in *Proc. ACM Symp. Optimization of Data Communications Systems,* Oct. 1969, pp. 202-225.

[15] P. A. W. Lewis and P. C. Yue, "Statistical analysis of series of events in computer systems," in *Statistical Computer Performance evaluation,* W. Freiberger, Ed. New York: Academic, 1972, pp. 265-280.

[16] L. Kleinrock, "Performance models and measurements of the ARPA computer network," in *Proc. NATO Advanced Study Inst. Computer Communication Networks,* R. L. Grimsdale and F. F. Kuo, Eds. Noordhoff International, 1975, pp. 63-88.

[17] J. R. Jackson, "Job shop-like queueing systems," *Management Science,* vol. 10, p. 131, Oct. 1963.

[18] F. Baskett, K. M. Chandy, R. R. Muntz, and F. G. Palacios, "Open, closed, and mixed networks of queues with different classes of customers," *J. Ass. Comput. Mach.,* vol. 22, pp. 248-260, Apr. 1975.

[19] H. Kobayashi and M. Reiser, "On generalization of job routing behavior in a queueing network model," IBM Thomas J. Watson Res. Center, Yorktown Heights, NY, res. rep. RC 5679, Oct. 1975.

[20] L. Kleinrock, "A conservation law for a wide class of queueing disciplines," *Naval Res. Log. Quart.,* vol. 12, pp. 181-192, 1965.

[21] D. Huynh, H. Kobayashi, and F. F. Kuo: "Optimal design of mixed-media packet-switching networks: Routing and capacity assignment," Univ. of Hawaii, Honolulu, ALOHA Syst. Tech. Rep. B76-3, Mar. 1976.

[22] M. J. Box, "A new method of constrained optimization and a comparison with other methods," *Comput. J.,* p. 42, Aug. 1965.

[23] W. Crowther *et al.,* "A system for broadcast communications: Reservation ALOHA," in *Proc. 6th Hawaii Int. Conf. System Sciences,* Jan. 1973, pp. 371-374.

[24] L. G. Roberts, "Dynamic allocation of satellite capacity through packet reservation," in *AFIPS Nat. Comput. Conf. Proc.,* vol. 42, p. 711, June 1973.

[25] R. Binder, "A dynamic packet switching system for satellite broadcast channels," Univ. of Hawaii, Honolulu, ALOHA Syst. Tech. Rep. B74-5, Aug. 1974.

OPTIMIZATION OF THE NUMBER OF GROUND STATIONS IN A DOMESTIC SATELLITE SYSTEM

ROY DANIEL ROSNER

Defense Communications Engineering Center, Reston, Virginia

Abstract

In developing system concepts for large communications systems, a traditional viewpoint of a
backbone network with a user defined access line
connectivity to the backbone has been consistently
retained. Satellites have been considered as a
high quality, highly reliable source of very long
haul circuits, and viewed as if they were simply
"cables in the sky." A different perspective is
taken in this study in the interest of developing
an advanced system concept which can effectively
utilize demand assignment satellites. In this concept, satellites can take the principal role in all
connectivity, including most of the access network
in addition to the main backbone. In fact, carrying
the concept to its natural conclusion leads to the
elimination of a network backbone as the term is
used today.

This study shows that the potential impact of satellites on developing new communication concepts
can be substantial since it appears that cost savings
of between 200 and 800 million dollars over a ten
year period appear possible with systems employing
hundreds of earth stations as compared to totally
terrestrial systems, or systems employing just a
few satellite earth stations. The exact values, of
course, are highly dependent upon the cost trends
for future satellite earth stations. It is concluded that advanced concepts for the future should
make more widespread use of satellite facilities
than has been recommended to date.

Introduction

In developing system concepts for large communication systems, a traditional viewpoint of a backbone network with a user defined access line connectivity to the backbone has been consistently retained.
Satellites have been considered as a high quality,
highly reliable source of very long haul circuits,
and viewed as if they were simply "cables in the
sky." A different perspective is taken in this
study in the interest of developing an advanced
system concept, which can effectively utilize demand
assignment satellites. In this concept, satellites
can take the principal role in all network connectivity, including most of the access network in
addition to the main backbone. In fact, carrying
the concept to its natural conclusion leads to the
elimination of a network backbone as the term is
used today.

The study is presented in terms of a demand
assignment domestic satellite system serving a
diverse set (over 2500) of geographically dispersed
user concentrations. In the absence of a satellite
system the users would be connected using leased
terrestrial access lines between themselves and
large backbone switches, and leased terrestrial
trunks between the switches. Using the satellite
system, the costs associated with the leased
terrestrial facilities can be avoided, thus offsetting the cost of the satellite spacecraft and
earth stations, and leading to a cost optimization
based on the number of earth stations.

It should be pointed out that this analysis does
not have   validity in the context of an established common carrier who is providing incremental
increase in network capacity through a domestic
satellite system. It is aimed at the situation,
where, for example, a large private user communications network could potentially fulfill all of
its capacity requirements with satellites. Another
example would be the case of new domestic satellite
carrier, with a charter to serve users anywhere in
the U.S. (not just in the large metropolitan centers)
and who is faced with the decision of leasing his
user to ground station lines from the established
carriers, or using a large number of earth stations,
placed, in many cases, at the users' locations.

The Satellite Concept

The concept investigated in this study is based on
the potential availability of a new class of high
capability communications satellites with sufficient
effective radiated power (ERP) margins to support a
large number of relatively small, simple, and
inexpensive ground terminals. The satellites and
earth stations would constitute the substance of
the communication network rather than provide supplemental capacity to meet special or long haul
requirements. The costs of the satellite facilities
would then be offset by large reductions in conventional (terrestrial) communication facilities;
judicious replacement of terrestrial facilities by
satellite facilities would achieve an overall cost
minimization.

The degree of replacement (satellite for terrestrial)
would naturally be governed by the technical capabilities and economics of the satellite network
components compared with the equivalent terrestrial
components. In an ideal case, if the number of
satellite ground stations employed was the same as
the number of backbone switches, then the entire

terrestrial backbone could be eliminated. If, then, the number of earth stations could be economically increased by the number of access areas, the entire terrestrial network, both backbone and access, could be eliminated. Individual satellite ground stations would then perform the concentration, switching, and transmission functions and the need for backbone switches would disappear. The results of this study shows that the optimum level occurs between these two extremes and it would be cost effective to put satellite earth stations at hundreds of user locations.

The Analytical Model

A simplified topological model of a hypothetical communication system was used for this study. The model approximates the total requirements forecast for the Defense Communications System in the 1985-1990 era, but also employs some gross averaging to create a model which can be readily analyzed.

The model utilizes a communication user requirement profile which was developed previously[1] for data networking studies, and adds to it the required aspects of the voice network user model. All user locations in the model are confined to a rectangular area, approximating the size of the U.S. with dimensions of 2700 by 1500 miles. The area is divided into 70 cells, each cell being 270 miles long and 215 miles wide with a backbone switch in the center. User access areas or user clusters are distributed approximately uniformly throughout the cells. Traffic flow is distributed uniformly among the 70 cells and within each cell, traffic to each user location is proportioned on the basis of its size (in terms of personnel). Large scale system design studies[2] have previously been shown to be sufficiently accurate for comparative system studies, using uniform traffic distributions, to permit the use of this highly simplified modeling technique.

The salient features of the user requirement profile used in this study are shown in Table 1. The backbone model is configured with approximately 7,800 backbone trunks and 5,000,000 backbone channel miles which is representative of a large backbone transmission network, such as that in the CONUS Defense Communication System.

The geographical distribution of the user access areas within a cell is shown in Figure 1. The cell is considered to be nominally centered on a backbone switch location. The large user locations (2 per cell) are presumed to be located an average of 135 miles from the switch. Medium users are typically 70 miles from the switch, and small locations 90 miles. The large number (11 each per cell) of very small and individual locations are assumed distributed relatively uniformly throughout the cell, with average distances to the switch of 80 and 88 miles respectively. The location of the access areas and calculations of access line lengths is shown in Appendix A. The model thus described forms the basis for this satellite transmission tradeoff study.

TABLE 1. SATELLITE STUDY ACCESS AREA MODEL

Access Area Size Category	Large	Medium	Small	Very Small	Individual	Total
No. of locations per cell	2	5	7	11	11	36
Total locations in U.S.	140	350	490	770	770	2,520
Voice Terminal per location	1,835	346	34	6	1	
Access Line Traffic (Erlangs)	74	13.84	1.36	0.24	0.20	
Access Lines Required per location GOS = P.05	80	19	4	2	1	
Average Distance to switch (miles)	135	70	90	90	88	105
Total Lines in U.S.	11,200	6,650	1,960	1,540	770	22,120
Total Channel Miles in U.S.	1,500,000	465,000	176,000	138,500	67,600	2,347,100
Data Traffic in Busy Hour (per location kilobits)	103,530	33,670	8,833	1,516	418	
Equivalent Load in Erlangs @ 16 kb/s lines	3.6	1.17	0.31	0.05	0.01	
Additional Lines per location GOS=P.05	7	4	2	1	1	
Total Additional Lines in U.S.	980	1,400	980	770	770	4,900
Total Additional Channel Miles in U.S.	132,000	98,000	88,000	68,000	67,600	453,600
Total Lines in U.S.	12,180	8,050	2,940	2,310	1,540	27,020
Total Channel Miles in U.S.	1,632,000	563,000	264,000	206,500	135,200	2,800,700

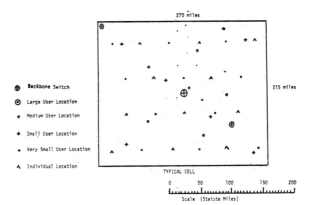

⊕ Backbone Switch
◉ Large User Location
• Medium User Location
+ Small User Location
• Very Small User Location
∧ Individual Location

Figure 1. User Access Areas Geographical Distribution

Study Methodology

A cost optimization analysis was performed on the model of the user environment described in the previous section to determine if there was a concave cost relationship for the total transmission system cost as a function of the number of satellite earth stations. Such a relationship would insure that a system cost minimization could be achieved by the proper selection of the number of satellite earth stations to be utilized in the overall network design.

As the number of earth stations in the network increases, the dominant portion of the total system cost becomes that associated with the ground portion of the satellite system. Recognizing this, and in order to insure feasibility of the concept, very conservative estimates of the cost, performance, and capacity of the space segment of the network were made. Costs were deliberately overestimated and capacities underestimated, resulting in a satellite space segment which is more costly and has more capacity than is really needed. (More accurate estimates of the space segment parameters are included in Appendix B.)

The space segment of the system was increased in size and complexity as the number of ground stations served was increased. Table II indicates the nature of the space segment for different ranges of ground stations supported.

TABLE II. SATELLITE SPACE SEGMENT FOR VARIOUS
GROUND STATION CONFIGURATIONS

Ground Stations	No. of Satellites	Est Sat Weight (#)	Est Cost ($ Mil) Per Sat	No. of Launches	Cost for ($ Mil) Launch
Up to 50	2	1500	15	1	22
50 to 100	4	1500	15	2	22
101 to 350	4	2200	20	2	26
351 to 1050	6	2200	20	3	26

As the number of ground stations increases, the satellites are presumed to require additional weight to permit more sophisticated beam steering and switching of narrow beam antennas, and additional power to service smaller ground stations.

Several quantities of ground stations, over the range of interest of the analysis, are of particular significance. Referring to Table 1 for the user model, values of 70, 210, 560, and 1050 for the number of ground stations, represent, respectively, ground stations at all backbone switches, at all backbone switches and all large locations, at all backbone switches and all large and medium locations, and at all backbone switches and all large, medium and small locations. The study was limited to ground terminals with G/T values no smaller than 18 dB, and thus serving "very small" and "individual" areas directly by satellite was found to be not cost effective. Development of very low cost earth terminals with the capability of one to ten equivalent voice channels could extend the range of application of these concepts to smaller locations such as the "very small" and "individual" locations of this access model used. Moreover, the calculations did not account for the possible rehoming of lines from these locations to satellite ground stations which might be closer than the switch location. Further cost savings could result from this type of rehoming, lowering the total system costs for those cases with a large number of ground stations. The ultimate case, eliminating the backbone switches entirely by the use of direct access-area to access-area communications (the satellite equivalent of a fully connected network) was not considered (though a significant reduction in backbone switching capability could be easily attained with a large number of earth stations).

A uniform channel rate of 16 kb/s was utilized in order to simplify the computations. The 16 kb/s rate provides acceptable voice A/D quality and sufficient data transmission bandwidths for most ADP applications. Cost factors for the terrestrial portions of the network were taken from earlier estimates[3]. While the trend to lower voice A/D

rates (8000 b/s or lower) would significantly reduce the cost of the leased terrestrial facilities, operating at such bandwidths would also significantly reduce the power and data rate requirements (and hence costs) for the replacing satellite system. While it is uncertain at this time which effect would dominate, it appears that the terrestrial costs would reduce more rapidly with decreasing bit rate than would the satellite costs, thus narrowing the advantage of the satellite system employing many ground stations over the terrestrial system. This effect would at most slightly decrease the optimum number of earth stations but is unlikely to change the shapes of the curves shown below.

All cost estimating was done over a ten-year period, nominally 1980 to 1990. Previous projections of digital terrestrial line costs for 16 kb/s service were made as shown in Table III[3,4]. Averaging over the ten-year period results in an average access line cost of $1.40 per mile per month and a trunk cost of $0.70 per mile per month. A present worth factor of 6.54 was used to convert the ten-year lease costs to an equivalent acquistion cost occurring at the beginning of the ten-year period, assuming a 10% annual rate of interest.

TABLE III. ESTIMATED FUTURE COSTS PER CHANNEL
MILE PER MONTH

	1980	1985	1990
Access Lines	$2.00	$1.30	$1.10
Trunks	.90	.70	.50

All costs are subject to a fixed termination charge of $175/month per-channel-end.

The assumed satellite system components consist of a new class of technologically advanced satellites (but not significantly beyond the present state of the art) and basically three different size categories of earth stations. The satellite would operate in a time-division-multiple-access (TDMA) demand-assignment mode and, particularly in the case where a large number of earth stations are employed (50 or more), would use a number of steerable, switchable, narrow-beam antennas. While sufficient capacity appears to be available in the lower SHF band to handle the traffic for all cases, it would probably be desirable to utilize the 20-30 GHz region for handling much of the peak load, especially for some of the larger ground stations where antenna gain and available power could maintain communications in all but the worst of rainstorms.

The ground stations to be employed can be categorized as large, medium, and small, with G/T figures of 40, 30, and 18 dB respectively. This corresponds approximately to 60-, 18-, and 8-foot diameter antennas in the SHF region. Figures 2 and 3, developed from COMSAT data [5], summarize some of the critical considerations in choosing earth station operating parameters. Figure 2 shows that when using TDMA, the relative space segment cost using a 30 dB G/T ground station is about 20% higher than the cost of using a 40 dB G/T station. Conversely, Figure 3 shows the cost of a 40 dB G/T station is about 200% higher than that of a 30 dB G/T station. In satellite systems which employ a relatively small number of earth stations, the costs are heavily space segment dominated, so that saving 20% of the space segment costs outweighs the effects of the much more expensive earth terminals. However, in the cases considered in this study, as the number of earth stations increases the costs become dominated by the earth segment costs, thus making it desirable to use smaller earth stations at the expense of a more sophisticated space segment.

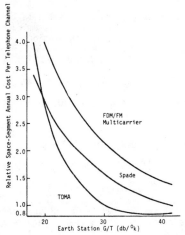

Figure 2. Relative Space-Segment Cost, Spot-Beam Transponder (From [5])

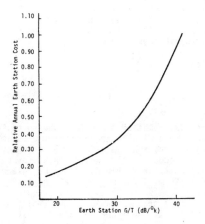

Figure 3. Relative Earth Station Cost (From [5])

This line of reasoning was used in selecting earth station sizes as a function of the number of earth stations employed. For cases where 50 or fewer earth stations were employed, only large (40 dB G/T) stations were used. For cases which utilized between 50 and 560 earth stations (locations up to and including medium subscriber concentrations), only medium (30 dB G/T) size earth stations were employed. For cases where more than 560 earth stations were used, 30 dB G/T stations were used to serve the backbone switches and the large and medium access areas, while 18 dB G/T terminals were used to serve the small access areas. Since the unit costs of earth stations are subject to very wide variations, the study used a large variation in earth terminal costs, and as will be shown later, results are plotted parametrically based on variations of the earth station unit costs.

The last aspect of the study methodology is related to quantity discounts for the earth stations. Satellite earth terminals are generally not bought in significant quantities, but in this study it is postulated that large quantities of earth terminals would be used. The notion of a quantity discount factor is introduced such that d_f is the ratio of the unit cost of an earth terminal bought in large quantity to the cost of a single unit. Thus $0 \leq d_f \leq 1$, and an exponential relationship as a function of the quantity bought (n) is arbitrarily assumed to exist such that:

$$d_n = d_f + (1 - d_f) e^{-n/100} \qquad (1)$$

where n = the number of units purchased. Since even in moderately large quantities (200 or more) earth terminal costs are likely to have a large component of direct labor, it is presumed that the ultimate unit cost eventually levels off. At the same time, a "time constant" of 100 units is used since it would appear to be a sufficient quantity for which an efficient assembly line process could be established, thus realizing a large fraction (about 2/3) of the ultimate cost savings per unit.

Calculations and Results

Using the cost factors and methodology described in the previous sections, cost curves were derived relating the total transmission system cost associated with the user model to the number of satellite earth stations used. The baseline costs for an all-terrestrial implementation were computed from the relationship

$$C_o = C_t + C_a \qquad (2)$$

$$C_o = m_t C_{tm} + 2n_t C_{tt} + m_a C_{am} + n_a C_{at} \qquad (3)$$

where C_o = cost of system with 0 ground stations (i.e., all-terrestrial system)

C_t = trunking costs
C_a = access costs
m_t = trunk channel miles
m_a = access channel miles

C_{tm} = cost per channel mile for trunks
C_{am} = cost per channel mile for access
n_t = number of trunk channels
n_a = number of access channels
C_{tt} = termination cost per trunk end
C_{at} = termination cost per access end.

Ten-year costs were calculated from monthly rates using the factor 78.48 as a multiplier (12 months/ year x 6.54 cost factor relating annual lease costs to ten-year present value costs). For the all-terrestrial system, the parameter values were

m_t = 5,000,000 channel miles
m_a = 2,800,700 channel miles
C_{tm} = \$0.70 x 78.48
C_{am} = \$1.40 x 78.48
n_t = 7800 trunks
n_a = 27,020 access lines
C_{tt} = C_{at} = \$175 x 78.48 per termination.

Substituting these values in equation (3) results in

$$C_o = \$1.168 \times 10^9 \text{ for the ten year period}$$
$$(1980 \text{ to } 1990)$$

For system configurations which employ satellites, the general expression for the system cost is

$$C_N = C_o + C_S - C_A \qquad (4)$$

where N = number of earth stations
C_S = satellite subsystem cost
C_A = terrestrial system cost avoidance as a result of the satellite subsystem.

The cost of the satellite system was derived from the sum of the launch cost, the space segment cost, and the aggregate cost of the ground stations including both acquisition and operation/maintenance. The space segment and launch cost estimates can be obtained directly from Table II, for different values of N. The space segment costs, C_{ss}, are

$2 < N \leq 50$	C_{ss} = \$52 million
$50 < N \leq 100$	C_{ss} = \$104 million
$100 < N \leq 350$	C_{ss} = \$132 million
$350 < N \leq 1050$	C_{ss} = \$198 million.

$$(5)$$

The ground station costs are based on the number of ground stations, size of the ground stations, unit cost, and quantity discount factor. Ground costs, $C_{gs}(N)$, are thus

$$C_{gs}(N) = 1.65NC_g d_n \qquad (6)$$

where C_g is the ground station unit cost, and d_n is given by equation (1). The factor 1.65 accounts for the acquisition cost plus annual costs of 10% of the acquisition cost for O&M, calculated for a ten year present value. Equation (6) has to be used differently over various ranges because of the different size terminals used for different applications. Equation (6) then becomes

$$C_{gs}(N) = 1.65NC_l d_n, \text{ for } 0 \leq N \leq 50, \qquad (6a)$$

$$C_{gs}(N) = 1.65NC_m d_n, \text{ for } 50 < N \leq 560, \qquad (6b)$$

and $C_{gs}(N) = 1.65(560)C_m d_n + (N-560)C_s d_n,$
$$\text{for } 560 < N \leq 1050 \qquad (6c)$$

where C_l is the unit cost of a 40 dB G/T ground station, C_m is the unit cost of a 30 dB G/T ground station, and C_s is the unit cost of an 18 dB G/T ground station.

Combining equations (5) and (6) results in these equations for C_s, the satellite subsystem cost as a function of the number of earth stations

$$C_s = 52 \times 10^6 + 1.65NC_L d_n, \text{ for } 2 \leq N \leq 50, \qquad (7a)$$

$$C_s = 104 \times 10^6 + 1.65NC_m d_n, \text{ for } 50 < N \leq 100, \qquad (7b)$$

$$C_s = 132 \times 10^6 + 1.65NC_m d_n, \text{ for } 100 < N \leq 350 \qquad (7c)$$

$$C_s = 198 \times 10^6 + 1.65NC_m d_n, \text{ for } 350 < N \leq 560 \qquad (7d)$$

and $C_s = 198 \times 10^6 + 1.65(560)C_m d_n + (N-560)C_s d_n,$
$$\text{for } 560 < N \leq 1050, \qquad (7e)$$

where d_n is given by equation (1).

The terrestrial system cost avoidance as a result of using the satellite system (C_A in equation (4)) is derived for the various ranges of interest. Over the range from 2 to 70 ground stations, the satellite system replaces terrestrial trunking facilities only, such that when N = 70 (a satellite earth station at each backbone switch) essentially all terrestrial trunking charges are eliminated. Over this range, the following relationship appears to give a good fit to the estimated trunking cost avoidance:

$$C_A = C_t(1-e^{-N/15}) \qquad (8)$$

where $C_t = m_t C_{tm} + 2n_t C_{tt}$ from equation (3) and $0 \leq N \leq 70$.

This exponential relationship drives the terrestrial trunking cost to nominally zero for N = 70 (99.1% of the trunking costs are avoided), while achieving only a modest reduction in trunking costs for values of N significantly less than fifteen. For example, for N=5, C_A=\$128 million, or about 25% of the total trunking costs. Looking at the actual situation if five ground stations were employed in U.S., about ½ of the trunking mileage costs could be avoided, but no termination costs could be saved since lines would still be required from the switches to the satellite terminals. (Half of the mileage costs is about \$130 million.) It should be emphasized that a large fraction of trunking mileage can be saved using a relatively small number of satellite earth stations because demand assignment satellite systems have the unique property of complete connectivity as new terminals are added. In other words, it is presumed that each satellite earth station can communicate directly with every other earth station, with no tandem switching of traffic to add capacity requirements to the trunking cross sections. Each satellite earth station then provides for a reduction in terrestrial trunking from its vicinity, to the vicinity of all other earth stations simultaneously. As the number of earth stations increases above five, collocation of satellite terminals with switches becomes realistic, leading to savings in both mileage and termination costs. When N=70, all switches are presumed to have collocated satellite terminals leading to a cost avoidance equal to the full trunking cost.

For N > 70, the cost avoidance is equal to C_t (the trunking cost) plus the access line cost to each access area location which has a satellite terminal. The number and average length of the access lines for each size category of access area location can be determined from Table 1.

In the range of $70 < N \leq 210$, access circuits from large user concentrations can be avoided saving 87 access lines of 135 miles average length for each additional satellite terminal. Thus, in the range $70 < N \leq 210$

$$C_A = C_t + 87(78.48)(N-70)\{m_a C_{am} + C_{at}\} \qquad (9)$$

where m_a = 135 miles
C_{am} = \$1.40 per mile per month
C_{at} = \$175 per month

and 78.48 converts monthly costs to ten-year costs. Thus

$$C_A = C_t + 2.476 \times 10^6 (N-70), \text{ for } 70 < N \leq 210. \quad (10)$$

For N=210,

$$C_A = C_t + 2.476 \times 10^6 (140) = 488 \times 10^6 + 346 \times 10^6$$
$$= \$834 \times 10^6$$

$$C_{A_{210}} = \$834 \times 10^6.$$

In the range from $210 < N \leq 560$, the medium size access areas are accommodated, each one avoiding 23 access lines with an average length of 70 miles. Following the same line of reasoning as above leads to

$$C_A = \$834 \times 10^6 + 0.492 \times 10^6 (N-210),$$
$$\text{for } 210 < N \leq 560. \qquad (11)$$

For N=560, C_A = \$1008 \times 10^6.

Similarly, for $560 < N \leq 1050$ accommodating small access areas as well (6 lines at an average length of 90 miles),

$$C_A = 1008 \times 10^6 + 0.1315(N-560), \qquad 560 < N \leq 1050. \quad (12)$$

Equations (8), (10), (11), and (12), together with equations (1) and (7), and the constants

$C_0 = \$1168 \times 10^6$ (all terrestrial costs) and
$C_t = \$489 \times 10^6$ (all trunking costs)

can be substituted in equation (4) to yield the complete set of cost equations over the range of zero to 1050 satellite terminals. These are

$$C_N = 731 \times 10^6 + 1.65 N C_1 [d_f + (1-d_f)e^{-N/100}] + 489 \times 10^6 (e^{-N/15}), \text{ for } 0 \leq N \leq 50 \quad (13a)$$

$$C_N = 783 \times 10^6 + 1.65 N C_m [d_f + (1-d_f)e^{-N/100}] + 489 \times 10^6 (e^{-N/15}), \text{ for } 50 < N \leq 70 \quad (13b)$$

$$C_N = 956 \times 10^6 + N\{1.65 C_m [d_f + (1-d_f)e^{-N/100}] - 2.47 \times 10^6\}. \text{ for } 70 < N \leq 100 \quad (13c)$$

$$C_N = 984 \times 10^6 + N\{1.65 C_m [d_f + (1-d_f)e^{-N/100}] - 2.476 \times 10^6\}, \text{ for } 100 < N \leq 210 \quad (13d)$$

$$C_N = 569 \times 10^6 + N\{1.65 C_m [d_f + (1-d_f)e^{-N/100}] - 0.492 \times 10^6\}, \text{ for } 210 < N \leq 350 \quad (13e)$$

$$C_N = 635 \times 10^6 + N\{1.65 C_m [d_f + (1-d_f)e^{-N/100}] - 0.492 \times 10^6\}, \text{ for } 350 < N \leq 560 \quad (13f)$$

and $$C_N = 432 \times 10^6 + [924 C_m + (N-560)C_s][d_f + (1-d_f)e^{-N/100}] - 0.1315 \times 10^6 (N), \text{ for } 560 < N \leq 1050. \quad (13g)$$

Equations (13a) through (13g) were evaluated for various values of N, C_1, C_m, C_s, and d_f using a straightforward computer program. Some representative results are shown in Figures 4, 5, 6, and 7. In these figures the total transmission system costs, C_N, are plotted as a function of the number of earth stations, N, with the cost of the ground stations (C_1, C_m, C_s) as a parameter. The uppermost curve on each of the figures utilizes ground station cost, which are estimates of the present installed costs of large (40 dB), medium (30 dB), and small (18 dB) ground terminals. Each successively lower curve assumes a less expensive set of costs for the earth stations. It is contended that under the conditions of this study the basic earth station costs should be considerably less than present costs. Some of the reasons for this are that the improved effective isotropic radiated power of the satellites and the improved receiver techniques should eliminate the need for cryogenically cooled receiver front ends, the improved satellite stationkeeping should eliminate (or at least substantially reduce) ground station tracking requirements, and the improved solid state, LSI implementation of digital TDMA techniques in the ground station should be much less expensive than the FDMA channel banks of present ground stations.

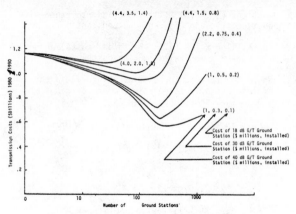

Figure 4. Total Transmission System Costs, $d_f = 1$.

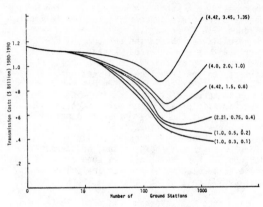

Figure 7. Total Transmission System Costs, $d_f = 0.25$.

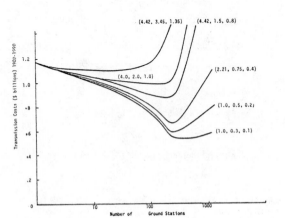

Figure 5. Total Transmission System Costs, $d_f = 0.75$.

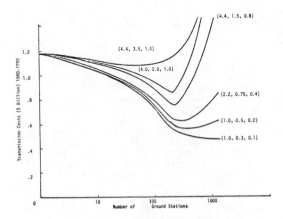

Figure 6. Total Transmission System Costs, $d_f = 0.50$.

Each curve in Figures 4-7 is labeled with a set of three numbers. This triple represents the values of C_l, C_m, and C_s, respectively, used in calculating the curves. The set of curves in Figure 4 is drawn for $d_f=1$, meaning no quantity discount is used, while Figures 5, 6, and 7 are drawn for $d_f=0.75$, 0.5, and 0.25, respectively (discounts of 25%, 50%, and 75% on large quantities of earth stations). It should be pointed out that in the region of the discontinuities of equation (13), particularly where the cost of the space segment changes (N=50, N=100, and N=350), the data on either side of the discontinuity has been smoothed slightly so that the curves would be continuous.

In order to show the relative importance of the various components of the system cost curves, one of the curves of Figure 6 is redrawn as Figure 8. Each of the components of this curve is shown, the cost of the space segment, the cost of the ground segment, and the residual cost of the conventional terrestrial communications facilities. Of interest is the relative size of the various components in the region of the minimum point of the curve (in this case around 200 ground stations). At this point the satellite ground segment costs are more than double the costs of the space segment, which leads to the observation that adding complexity and capability to the satellites would have a leveraged impact on overall system costs if commensurate reductions in ground station costs could be effected. Improvements in satellite EIRP through higher power transmitters or more sophisticated antenna techniques, appears to be the most direct approach to improving the balance between earth and space segment costs, but other technical advances, such as satellite-borne switching and signal processing equipment, adaptive antenna arrays, or use of additional frequency bands and diversity techniques are possibilities. If such a tradeoff is possible, i.e., reducing ground station costs through limited increases in space segment costs, not only would the minimum of the total cost curve be reduced in absolute magnitude, but the location of the minimum in terms of the number of ground stations would move toward a larger number of ground stations.

Figure 9 is drawn using the data from Figures 4 through 7. Here the optimum number of ground stations is plotted as a function of the cost of a medium size (30 dB G/T) ground station, with the discount factor (d_f) as the parameter for each of the different curves. These curves are a smoothed plot of the locus of minimums for each of the curve sets in Figures 4 through 7. The use of the cost of the medium size ground station as the abscissa is based on the cost of the medium station being the dominant cost parameter as seen from equation (13), in the range 70 to 560 ground stations. Since the minimums of the cost curves generally fall in this range, Figure 9 is drawn with the medium stations cost as the independent variable.

From Figure 9 it is observed that if medium size ground stations could be developed to cost in the one to two million dollar range, ground station quantities on the order 100 to 300 would result in optimized system costs. Since the medium size ground terminals would be handling only about 100 16-kb/s voice-equivalent channels, or a nominal 1.5 Mb/s (T-1 Data Rate) bit stream, many experts project potential ground station costs (utilizing TDMA) significantly less than one million dollars. Should this projection be achieved, terminal counts on the order of 300 to 500, or even higher, would be needed to achieve optimization of the system costs.

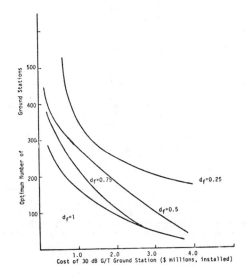

Figure 9. Optimum Ground Station Population

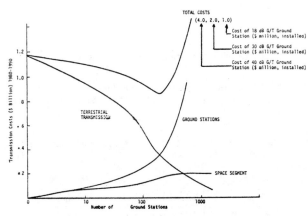

Figure 8. Breakout of Transmission System Costs, $d_f = 0.50$

Summary and Conclusions

This study has shown that the potential impact of satellites on developing new communication networks can be substantial if a fairly large number of ground stations are employed. Cost savings overs a ten year period of between 200 and 800 million dollars (present value adjusted to the beginning of a ten year period) appear possible with systems employing between 100 and 1000 earth stations, the exact values being highly dependent on the cost trends for future satellite earth stations. The optimization of network costs using satellites could evolve over a long period of time, driven by the technology and costs of the earth stations. Using today's costs and no discounts for quantity procurements thirty to fifty earth stations could be employed. As 30 dB G/T earth terminals at the $1 million range evolve (probably 5 to 10 years hence) between 200 and 500 earth stations would be feasible. Ultimately, probably 15 years or more years downstream, 1000 or more earth stations will result in an optimization of total system costs.

It is emphasized that these results are derived using a very simplified model which aggregates the projected traffic into the U.S. modeled as a rectangular area divided into 70 cells. The analysis accounts only for transmission cost savings, even though substantial reduction in switching costs are likely through implementation of this concept. Expansion of the model to include possible switching savings may increase the possible cost savings noted above by an appreciable amount. However, further investigation is warranted in order to quantify these impacts.

Finally, it must be remembered that this entire concept is founded on the premise than an efficient, highly flexible time division multiple access, demand assignment (TDMA/DA) technique will be available to implement a system connectivity and topology based heavily on satellite technology. The "SPADE" techniques developed by COMSAT would provide this capability except that the terminal costs would probably remain at the upper end of the ranges considered because of the complexities of the frequency channelization equipment needed. The "ALOHA" techniques developed by ARPA and University of Hawaii would also meet much of the TDMA/DA requirement, but may suffer from lack of efficiency with respect to achieving high channel loading factors and stable network performance.

The conclusion of this study is that advanced concepts for the future systems should make much more widespread use of satellite facilities than has been recommended to date. To further the definition of this concept, efforts are needed to expand the study results to include leased satellite facilities, particularly leased space segment assets, and to develop multiple access, demand assignment concepts which can achieve the required flexibility of connectivity as well as utilize advancing technology to significantly decrease the ground station costs.

References

[1] R.D. Rosner, "A Digital Data Network Concept for the Defense Communication System," Proceedings of the 1973 National Telecommunications Conference NTC '73, Atlanta, Georgia (November 1973).

[2] R.D. Rosner, "Large Scale Network Design Considerations," Proceedings of ICCC '74, Stockholm, Sweden (August 1974).

[3] G.J. Coviello and R.D. Rosner, "Cost Considerations for a Large Data Network," Proceedings of ICCC '74 Stockholm, Sweden (August 1974).

[4] M. Gerla, "New Line Tariffs and Their Impact on Network Design," Proceedings of the 1974 National Computer Conference, AFIPS Volume 43 (May 1974).

[5] J.L. Dicks, "Domestic and/or Regional Services Through Intelsat IV Satellites," COMSAT Technical Review, Volume 4 No. 1 (Spring 1974).

APPENDIX A

Calculation of Average Distances to Various Area Groups

The distances shown are based on the switch cell model which is shown in Figure 1 of the main text.

	Average Distance (miles)
Large Access Areas - 2 per cell	
1 at 100 miles and 1 at 170 miles	135
Medium Access Areas - 5 per cell	
5 at 70 miles	70
Small Access Areas - 7 per cell	
1 at 30 miles	
2 at 40 miles	
2 at 135 miles	
2 at 125 miles	
$[30 + 2(40) + 2(135) + 2(125)] \times 1/7 = \frac{630}{7} = 90$	90
Very Small Access Areas - 11 per cell	
2 at 140 miles	
2 at 85 miles	
2 at 110 miles	
2 at 100 miles	
1 at 25 miles	
2 at 50 miles	
$[2(140) + 2(85) + 2(110) + 2(100) + 25 + 2(50)] \times 1/11 = \frac{995}{11} = 90$	90
Individual Access Areas - 11 per cell	
2 at 110 miles	
2 at 80 miles	
2 at 140 miles	
2 at 50 miles	
2 at 90 miles	
1 at 30 miles	
$[2(110) + 2(80) + 2(140) + 2(50) + 2(90) + 30] \times 1/11 = \frac{970}{11} = 88$	88

To compute the weighted average distance (D) for all access lines, the number of lines needed to each location size category is used with the 16 kb/s, P.05 sizing values.

$$D = \frac{2 \times 135 \times 87 + 5 \times 70 \times 23 + 7 \times 90 \times 6 + 11 \times 90 \times 3 + 11 \times 88 \times 2}{2 \times 87 + 5 \times 23 + 7 \times 6 + 11 \times 3 + 11 \times 2}$$

$$= \frac{23500 + 8050 + 3780 + 2970 + 1940}{174 + 115 + 42 + 33 + 22}$$

$$D = \frac{40240}{386} = 105 \text{ miles}$$

APPENDIX B

Estimate of Satellite Performance and Capacity Requirements

The following computations are derived from information provided in the DCA Satellite Communication Reference Data Handbook (July 1972), particularly Section 5.6 and Table 5-3.

The required bit rate for each class of user area is computed as follows:

User Area Class		Required Rate
Large	87 channels x 16 kb/s	1.39 Mb/s
Medium	23 channels x 16 kb/s	368 kb/s
Small	6 channels x 16 kb/s	96 kb/s
Switch Node	111 channels x 16 kb/s	1.78 Mb/s

Extrapolating from Table 5-3, the required satellite EIRP can be found for a given data rate and ground station G/T. Since satellite EIRP and ground station G/T can be directly interchanged on a one for one basis in dB, the requirement can be specified for the sum of the EIRP and G/T as a function of the bit rate. The following is derived from factors which assume a value of E_b/N_0 of 10 dB (10 dB normalized signal-to-noise ratio), which corresponds to an error rate of 10^{-5} in an uncoded channel, and a performance margin of 6 dB over all known losses. For a 1 Mb/s channel, a value of G/T and EIRP of 51 dBW is required. The data rates required for each station size are also increased by 50% (nominally) to allow for synchronization losses and time slot guard bands.

Station	Raw Rate	150% of Raw Rate	G/T + EIRP (dBW)
Reference	1 Mb/s	---	51
Large	1.32 Mb/s	2.0 Mb/s	54
Medium	368 Kb/s	0.5 Mb/s	48
Small	96 kb/s	0.15 Mb/s	42.8
Switch Node	1.78 Mb/s	2.7 Mb/s	55.3

In the course of the study, it was assumed that if many earth stations were employed, the small stations would be served by 18 dB G/T ground stations and the others by 30 dB G/T stations.

Thus:

Station	EIRP & G/T	G/T	EIRP	# of Stations	$10 \log_{10}$ of # of Stations	Total EIRP (dBW)
Large	54	30	24	140	21.4	45.4
Medium	48	30	18	350	25.4	43.4
Small	42.8	18	24.8	490	26.9	51.7
Switch Node	55.3	30	25.3	70	18.4	43.7

Remember that dB's are not additive; the EIRP has to be converted back to true power to determine the total requirement:

$$\text{Total EIRP} = \log^{-1}\left(\frac{45.4}{10}\right) + \log^{-1}\left(\frac{43.4}{10}\right) + \log^{-1}\left(\frac{51.7}{10}\right) + \log^{-1}\left(\frac{43.7}{10}\right)$$

$$= 3.46 \times 10^4 + 2.19 \times 10^4 + 1.48 \times 10^5 + 2.34 \times 10^4$$

$$= 22.79 \times 10^4$$

$$\text{EIRP} = 2.279 \times 10^5 \text{ watts}$$

$$\text{EIRP} = 53.6 \text{ dBW}.$$

Assume that the satellites developed to meet this system concept utilize a 20-watt output per transponder, and a narrow beam antenna corresponding to a 3' diameter parabola, at 7 GHz. At 7 GHz, the gain of a 3' dish (55% efficiency) is 34 dB. 20 watts is equal to 13 dBW.

Thus the required EIRP=53.6 dBW = $13+34+10 \log N$ where N is the number of transponders,

$$10 \log N = 6.6 \text{ dB}$$
$$N = \log^{-1}\left(\frac{6.6}{10}\right) = 4.5 \text{ transponders.}$$

Thus fewer than 5 full transponders, at 20 watts output each, operating into 3 foot diameter (equivalent) antennas at 7 GHz, provides sufficient power to meet the requirements of the study, for up to 1050 ground stations (with 6 dB margin on the RF link and 50% rate margin for TDMA overhead). The total bandwidth requirement of this system is (with margin):

$$\text{Rate} = 2 \times 140 + 0.5 \times 350 + 0.15(490) + 2.7 \times 70$$
$$= 280 + 175 + 73.5 + 189 = 717 \text{ Mb/s.}$$

Using a modulation technique with 1.5 bits/Hz achieved capacity, total bandwidth of less than 500 MHz would be required, even if all possible channels were in use simultaneously (which is statistically very unlikely).

Hybrid Packet And Circuit Switching

NEIL KEYES and MARIO GERLA
Computer Transmission Corporation
El Segundo, CA

Many existing data networks were justified after an extensive performance/cost ratio analysis. The justifications were based upon specific user requirements, performance, bandwidth, price and availability, network protocol, and equipment technology and price. The results usually dictated independent networks for each type of traffic (message traffic, inquiry/response traffic, interactive traffic, batch traffic, facsimile traffic, etc.).

During the past two or three years we have experienced drastic changes in the performance/cost ratios of equipment available to mechanize networks, and over the next few years we are certain to experience remarkable decreases in the cost of bulk bandwidth for implementing networks. Because of the decrease in cost of equipment and bandwidth, there has been considerable effort in the past year in analyzing network architectures which can simultaneously support different types of user requirements while maintaining the performance of individual specialty networks.

Protocol is the most salient commitment when choosing the network architecture. Existing trouble-free application programs become important users' assets as compared to developing new programs for network protocol compatibility. With the decreasing cost of hardware and bandwidth, the most important benefit to be derived by having protocol compatibility between the host and the network is the ability to support and address multiple terminals from a single host computer network interface. In order that network users be given the option of an evolutionary rather than a revolutionary change

from existing, trouble-free application programs, to programs providing efficient support of the network protocol, the network must also support transparent operation. For our purposes, transparent support will mean that following the call setup, the users' channel appears as though it were provided by data sets and a private circuit. Thus, the transparent feature allows the initial use of existing programs, without changes.

SYSTEM DESCRIPTION

The M3200^{tm} *PACUIT*® data network provides the ability to simultaneously support circuit switching, *PACUIT* switching, packet switching, error correction, concentration, short delays, etc., because of the interswitch protocol and the hardware architecture. This section contains short descriptions of these designs.

Internode Multiplexing

Time division multiplexing is used to dynamically share trunk bandwidth. Each trunk is framed and divided into nine-bit time slots, and each nine-bit slot can be assigned as required to support the instantaneous bandwidth requirement.[1]

When a circuit-switched channel is provided, a sufficient number of nine-bit time slots are assigned to provide the bandwidth requirements. However, all time slots are assigned to be evenly distributed through the frame so that

the mean delay time in accessing a frame is one-half a user's character time. All bandwidth not assigned for circuit switching is available to support *PACUIT* and packet switching.

The M3200 network is master-clocked, making it possible to drop and insert, patch, and patch-through data on trunks without fear of clock slippage. Master clocking provides the means of circuit switching of transparent synchronous channels and circuit switching of packet data through intermediate nodes. Master framing is then slaved to master clocking.

The ratio of packet/circuit switch bandwidth is dynamically proportioned. *PACUIT*'s are also time division multiplexed. Composite *PACUIT*'s are formed at a constant rate using data from many users. The beginning of a *PACUIT* is identified in the multiplexed trunk data stream by a unique 9-bit frame code.[1] The header following the frame code identifies the number of characters inserted for management and control information before the beginning of the data field. If no data exists from a terminal, only a zero is transmitted for that terminal. If data exists, a byte count is inserted to indicate the number of 8-bit characters which are included for each terminal. The permitted concentration level within the *PACUIT* is established by the network manager. The data field is followed by the CRC, and a full duplex, selective *PACUIT* repeat is used to maintain high performance.

Hardware Architecture

A bus architecture provides the means of transporting data, command, and controls between processing modules within the data switch. Four individual buses are included within the data switch architecture (Figure 1). The shuttle bus is the lowest level and a separate shuttle bus exists for each expansion chassis. All low speed and trunk interface modules are inserted into an expansion chassis serviced by a shuttle bus. Circuit-switched channels can be set up by the minicomputer to exist between trunk interface modules coexisting on the same shuttle bus, thereby having little influence on the throughput of the switch.

All shuttle buses attach to the express bus. Any transfer of data between shuttle buses or between a module on the shuttle bus and a processor must travel over the express bus. The express bus interfaces with a 10-MHz microprocessor called the Switch Processor Unit (SPU). The SPU provides two bus interfaces: the 16-bit express bus and the 18-bit extended maxibus. The extended maxibus interconnects with an 18-bit semiconductor memory, a second 10-MHz microprocessor called the *PACUIT* Processor Unit (PPU), and a unit called the maxibus

extender. The maxibus extender also interfaces with two buses: the 18-bit extender bus and the 16-bit maxibus, where the LSI-2 minicomputer and 16-bit core memory reside.

Confirmed control characters, packet, and *PACUIT* data received from the line interfaces are routed through the express bus to the SPU. Controls are tested to determine if an LSI-2 interrupt is required, and data are transferred to the 18-bit semiconductor memory for future processing by the PPU (incoming *PACUIT*'s are changed into user receive data, and user transmit data are combined into packets and shipped through the SPU to the destination node). *PACUIT* data passing through intermediate nodes are circuit-switched and shuttle/express buses do not affect the processor throughput at these intermediate (transit) nodes.

Additional processor buses can be established by adding additional Switch Processor Units, as shown in Figure 1 for the X.25 processor bus.

Software Architecture

The M3200 Data Switch Operating System (DSOS) was developed to pro-

vide the desired switch functions.[2] The DSOS is a real-time, re-entrant operating system containing eight priority levels. Each level contains one or more tasks under the control of task control blocks. The executive distributes interrupt data to the various levels via queues referred to as level control queues. Each significant interrupt activates this distribution mechanism, as well as causing dispatching to restart at the highest priority level. Thus (within control bounds), the higher the interrupt load, the greater the service given to high priority tasks. Other functions performed by the executive include memory management, clock control, intra-task and inter-task communications, subroutine linkage, error processing, etc.

Identical DSOS software is loaded from a floppy disk into each *PACUIT* data switch, and the configuration information can either be loaded locally or downline loaded from a stand-alone Network Control Center. The Network Control Center has its own Network Management System software. The Network Control Center can monitor and control the entire network. This includes status messages, gathering of statistics, controlling the online configuration, diagnostics, and collecting of billing information.

Addressing

The numbering plan permits closed user groups and a common user's network to coexist. Stations are classified as members of access groups, and inter-access group connecting privileges are under control of the network manager. All fields in the addressing scheme vary in length, providing no significant constraint on the size of the network because of addressing limitations.

Different closed user groups utilize the same physical network resources while each group remains securely separated operationally. Node numbers are assigned for each partitioned network. The actual node/station address is generated algorithmically so that previously assigned addresses do not require changes as the network grows.

Flow Control

Beyond the X.25 flow control, the M3200 network offers a class of service for synchronous subscribers where flow control is invoked for instantaneous bandwidth management. This flow control consists of controlling the clock rate to the synchronous terminal and/or host. Experience has indicated that timeouts are the only restriction in reducing the clock rate to a subscriber.

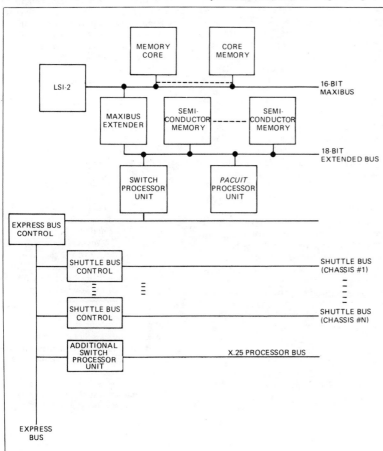

Fig. 1 M3201-2 *PACUIT* Data Switch Hardware Architecture

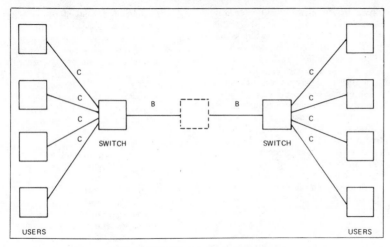

Fig. 2 *PACUIT* **Link Configuration.**

PERFORMANCE

Hybrid Systems

The *TRAN*® hybrid system introduces two novel features with respect to traditional networking approaches: (1) the partitioning of the network into two components, a circuit-switched and a packet-switched component, separated by a dynamically adjustable boundary, and (2) the use of the *PACUIT* protocol for internal packet communications. The two critical issues in evaluating the performance of this system are the effectiveness of the movable boundary between packet and circuit subnetwork, and the advantages of the *PACUIT* protocol with respect to the traditional packet protocol.

We will limit ourselves to the analysis of the *PACUIT* component. First, we evaluate average *PACUIT* delay as a function of message length, terminal speed, link bandwidth, link utilization, and *PACUIT* construction period. We then apply the results of the analysis to a simple network case involving exclusively packet-type traffic (no circuit-switched requirements). For this case, *PACUIT* performance is compared with conventional packet-switched performance, and some general conclusions on the effectiveness of the two schemes are derived.

PACUIT Model

A *PACUIT* link is essentially a full duplex physical circuit maintained between any two nodes in the network, and used to carry data in packet form. The *PACUIT* link is established via time division multiplexing, and may traverse transparently (i.e., without disassembly/reassembly of the packets) one or more intermediate nodes on the

path from source to destination. Error and flow control on the link is provided on a packet basis by an SDLC-like protocol. The bandwidth of a *PACUIT* link is dynamically adjusted, depending on the traffic load on the link.

Traditional packet transport schemes are based on the "private" packet concept (i.e., one packet for each user message), the *PACUIT* scheme is based on the "composite" packet concept (i.e., a packet may multiplex data from several users). Normally, user data that arrived at the entry switch during the interval τ (*PACUIT* construction period) are assembled in a block called a *PACUIT* block. Blocks are transmitted on the *PACUIT* link in a first-come, first-served (FCFS) fashion. User data are packetized "on the fly" without awaiting the complete reception of the message. This implies that the tail of a message may enter the network while the beginning of the message is already traveling toward the destination.

The composite packet scheme proves to be much more economical than the private packet scheme for asynchronous, low speed uses since it permits the transport of small messages or portions of messages, without paying the overhead of private packetization. Composite packet features are now being implemented also by other manufacturers.[4],[5]

The model considered in this section consists of a large population of users statistically multiplexed on a *PACUIT* link of fixed bandwidth B (Figure 2). Users may be connected to the network locally or remotely. User speed is uniformly set to C. Messages are generated at the terminals at random times (Poisson arrival distribution) Message length is exponential. Line errors, propagation delays, and buffer overflows are neglected. *PACUIT* delay

is defined as the time between the arrival of the last character of the message at the entry switch and the arrival of the same character at the exit switch. This definition is consistent with the end-to-end delay definition in conventional packet networks.

PACUIT delay T can be expressed as follows:

$$T = T_L + T_Q + T_T$$

where

T_L = *PACUIT* latency delay, i.e., the time between the arrival of the user data and the next *PACUIT* construction opportunity.

T_Q = *PACUIT* queueing delay, i.e., the delay spent on the *PACUIT* link queue. This delay is incurred when the instantaneous traffic rate exceeds *PACUIT* link bandwidth (the average traffic rate must, of course, be less than the link bandwidth for stability).

T_T = *PACUIT* transport delay, i.e., the delay incurred by transmitting the composite packet over the link.

The evaluation of T_L, T_Q, and T_T may be carried out using queueing theory techniques.[3],[6] Here we limit ourselves to a qualitative comparison of *PACUIT* delay versus traditional packet delay, as a function of network parameters. In the following section we substantiate our findings with a simple network example (Figure 3).

As far as delay performance is concerned, the elements in favor of the *PACUIT* scheme are:

- *No transmission and queueing at intervening nodes on the path from source to destination.* While the traditional packet scheme stores and forwards the packet at each hop, the *PACUIT* scheme transports the packet directly to its destination, without intermediate stops.

- *"On the fly" transmission:* user data are delivered to the destination immediately while they arrive at the source node, without awaiting full reassembly of the user message in the source node.

- *Lower overhead incurred in the transmission of short user packets.* These packets are combined into a composite packet with a common header, instead of being transmitted in private packets.

The elements that tend to favor the traditional packet scheme are:

- *Better trunk bandwidth sharing.* Trunks are shared among all source-destination traffic components, instead of being fragmented into individual source-destination "pipes".

TRAN® is a registered trademark of Computer Transmission Corporation.

- *Lower overhead for transaction-type traffic*. While a connection via *PACUIT* requires a call setup procedure, the delivery of a packet in a traditional packet scheme may be done with no prior setting of resources (datagram mode).

A definitive statement on the superiority of one scheme over the other is not possible. Indeed, the balance of pro's and con's is in favor of packet in some cases and of *PACUIT* in other cases, depending on user requirements and traffic characteristics. Since one of the determining parameters is packet length, in the next section we illustrate the trade-offs between packet and *PACUIT* schemes using a simple network example in which packet length (i.e., the length of the packet submitted by the user to the network) is assumed to be variable.

NETWORK CASE STUDY
PACUIT vs Packet

The network is shown in **Figure 3**. Traffic requirements are uniform between all node pairs. Trunk bandwidth is 9.6 kbps; user speed is 1.2 kbps; and *PACUIT* period is $\tau = 0.1$ s.

Due to the symmetry of topology and traffic requirements, the optimal routing and *PACUIT* bandwidth allocation can be easily determined. One single *PACUIT* link of bandwidth, B = 6.4 kbps, is maintained between each adjacent node pair. For non-adjacent node pairs (namely, (1,4), (2,5), and (3,6), two parallel links of bandwidth 3.2 kbps each are implemented.

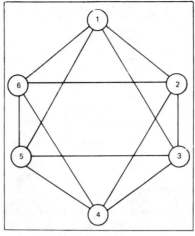

Fig. 3 Network Topology.

For the above implementation we evaluate average network delay (average over all pairs) as a function of message length b and network load ρ (note: ρ is defined as the net trunk utilization, or equivalently, net *PACUIT* link utilization).

We also wish to compare *PACUIT* delay with conventional packet-switching delay. So we assume that user messages are transported through the network in private packets, with 168-bit overhead per packet, a typical value for most experimental and commercial packet networks.

Figure 4 shows delay performance for the two schemes, as a function of message length b, for net trunk utilization = 60%. We note that for b ≤ 1500 bits, *PACUIT* prevails over conventional packet. This is essentially due to the

lower *PACUIT* overhead achieved using composite packets. We also note that b = 300 bits is the threshold below which packet operation cannot yield ≥ 60% trunk efficiency. No practical lower bound exists for *PACUIT*.

For 1500 ≥ b ≥ 2000 bits, *PACUIT* and packet delays grow linearly with b with comparable slopes (note that b = 2000 bit is the typical upper limit on packet size in public and private packet networks; for largest values of b, the message is segmented into smaller blocks at the entry node).

Further analysis of the example network for various network loads shows that *PACUIT* performance is superior to packet performance over the entire range of normal network operations. This result is clearly valid only for the particular topology and traffic requirements considered in this example. For more general configurations, we expect to have ρ and b regions in which *PACUIT* is superior to packet, and vice versa. *PACUIT* will prevail for small message length because of the lower O/H with respect to packet. For large message size, there is a trade-off between "on the fly" benefits and high *PACUIT* overhead. The outcome will depend on the specific network application.

GROWTH POTENTIAL

The means of dynamically allocating time division multiplexing time slots, the modular architecture of both the hardware and software, the addressing options, the call setup procedures, the routing algorithms, and centralized control of the M3200 System provide

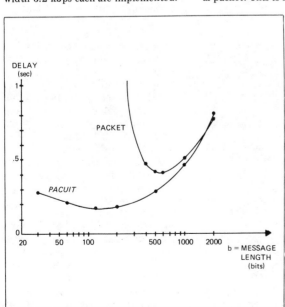

Fig. 4 Delay vs Message Length (load = 60%).

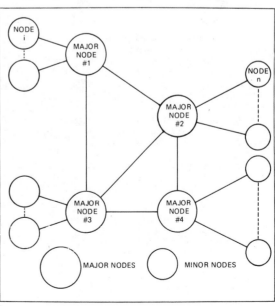

Fig. 5 Example of Two-Level Network.

the basis for considerable growth potential.[2] This section briefly summarizes the ease of growth for an existing integrated network.

Two-Level Networks

To minimize network delays, network processing, and the management problem of instantaneous buffer requirements during fault periods, the M3200 System is designed to form packets and to disassemble packet data at the entrance and the exit node, using circuit switching through all intermediate nodes. If all nodes have terminals talking to terminals in all other nodes, then a very large number of interconnecting paths must exist.

The M3200 System will offer a two-level network architecture consisting of major and minor nodes. A "freeway" will be constructed through the heart of the network consisting of major nodes, and all minor nodes will connect to the major node network. Data from many minor nodes will be collected together at a major node and transmitted over the "freeway" to the major node serving the minor node which serves the addressed subscriber. Each minor node will have a packet link to each major node to which the minor node is directly connected (typically two paths) plus the option of end-to-end packet links wherever a large volume of traffic flows. In Figure 5 subscriber A is served by node i and subscriber B is served by node n. Whenever there exists a large volume of traffic between subscribers served by nodes i and n, a direct packet link can be established between the two nodes using circuit switching through the intermediate major nodes.

Utility Functions

Utility functions can be easily attached to the described network architecture for the purposes of code and/or speed conversion and for message switching. The procedure is to recognize in the address the requirement for utility function support and to route the connection to an available supporting utility function switch. The utility function switch can consist of either a separate switch architecture or an M3200 switch using the basic operating system with many existing software modules replaced with the modules required by the utility functions.

Integrated Voice and FAX

The M3200 hybrid architecture provides eventual support of both voice and FAX transmission. Since composite *PACUIT*'s are formed at a constant rate, a maximum compatibility exists if either voice or FAX is to be supported by packets. The network architecture also provides means of conten-

tion circuit switching through the network where bandwidth is seized only when a signal threshold is exceeded by the voice transmission level or by the initiation of a page transmission.

Network Interconnects

The M3200 System will offer network interconnects through gateways. The addressing remains compatible with the CCITT international numbering plan. All incoming calls from other networks will be prefaced by the Data Network Identification Code (DNIC) to indicate if the call is transit to the M3200 System or terminating within the M3200 System. All calls leaving the M3200 network will do so by placing a "0" in front of the DNIC. The M3200 System will interface at the gateway with the CCITT specification for compatibility (X.7X, X.60, X.70, X.71, etc.).

Net Protocols

Modules under microprocessor control can be inserted into the shuttle bus and programmed for protocol compatibility on a channel basis. For example, the M45402 RS232/V.24 BISYNC/HASP interface module provides a local protocol compatibility for either IBM BISYNC or HASP multileaving at each channel end. The block data are then transmitted via error-corrected *PACUIT*'s between the two data switches serving the channel ends. The result is a BISYNC throughput which is independent of network delays, which is extremely critical if the network trunk is provided via satellite.

REFERENCES

1. Gerla, M. and G. de Stasio, "Integration of Packet and Circuit Transport Protocols in the *TRAN* Data Network," *Computer Network Protocol Symposium*, Liege, Belgium, 13-15 Feb. 1978.
2. Klein, D. M. and S. Frankel, "A Hybrid Circuit- and Packet-Switching System," *Proc. of NTC 76*, Dallas, Texas, Nov. 1976.
3. Kleinrock, L., "Queueing Systems: Theory and Application," Vol. I, 1975, Wiley-Interscience, New York.
4. Vander Mey, J.E., "The Architecnous Time Division Multiplexing for Time-Sharing Computer Systems," *Fall Joint Computer Conference Proceedings, 1969*.
5. Rinde, J., "Tymnet I: An Alternative to Packet Technology," *Proc. of ICC 76*, Toronto, Canada, Aug. 1976.
6. Keyes, N. and M. Gerla, "Report on Experience in Developing a Hybrid Packet and Circuit-Switched Network," *Proc. of ICC 78*, Toronto, Canada, June 1978. ⓉⒼⓈ

competition, whether different regulatory standards and procedures should be established for carriers participating in the handling for domestic PMS and private record services and, if so, the specific rules and regulations which should be adopted. This should include consideration of:

a) whether Western Union should be relieved of any of the regulatory requirements now imposed as a result of its offering of PMS service; and

b) the maximum extent to which the Communications Act of 1934 will permit those markets opened to competition to be deregulated.

The nature and scope of Value Added Network Services, both domestic and international, will be strongly influenced by whether a public service can be offered. The outcome of this inquiry may well affect the viability of VANS and hence the larger issues which have tended to presume that VANS provide needed support, e.g., the automated office.

THE BOTTOM LINE

In summary, for domestic VANS the hardest issue for policy implementation is the determination of the proper standard of review. The *Resale Decision* left this for future consideration after some experience with the actual operation of VANS. Associated with this is the issue of the appropriate methodology by which data — especially cost data — shall be collected. In the international area the basic questions — which technology(s) is the most appropriate, what are the statutory limitations, what is the position of the foreign jurisdictions and, most importantly, where does the public interest lie — have yet to be answered. The FCC is working to obtain answers to these questions but there is no indication as to how long this will take. As is the case with all SCCs, however, the issue most crucial to VANS is the location of the boundary between monopoly and competition. ⓉⒼⓈ

Data Traffic Performance of an Integrated Circuit- and Packet-Switched Multiplex Structure

C. J. WEINSTEIN, M. L. MALPASS, AND M. J. FISHER

Abstract—Results are developed for data traffic performance in an integrated multiplex structure which includes circuit-switching for voice and packet-switching for data. The results are obtained both through simulation and analysis, and show that excessive data queues and delays will build up under heavy loading conditions. These large data delays occur during periods of time when the voice traffic load through the multiplexer exceeds its statistical average. A variety of flow control mechanisms to reduce data packet delays are investigated. These mechanisms include control of voice bit rate, limitation of the data buffer, and combinations of voice rate and data buffer control. Simulations indicate that these flow control mechanisms provide substantial improvements in system performance.

I. INTRODUCTION

An analysis of the performance of an integrated, circuit- and packet-switched multiplexer structure was presented recently by Fischer and Harris [1]. The structure (first advanced by Kummerle [2] and Zafiropulo [3], and refined by Coviello and Vena [4]) is based on a time-slotted frame format where a certain portion of the frame is allocated to circuit-switched calls and the remaining capacity is reserved for data packets. To increase channel utilization, a "movable boundary" feature is included, so that data packets are allowed to use any residual circuit-switched capacity momentarily available due to voice traffic variations. Results indicated significant saving, in terms of channel requirements for the integrated system, as compared to a system with separate circuit- and packet-switched facilities handling the same traffic loads with the same blocking probability and average packet waiting time. However, an error has been discovered in the original derivation of average data packet waiting time for the movable boundary case. New results [5], [6] indicate much larger data packet delays than previously predicted for the case where the fixed capacity reserved for data is smaller than the data traffic load. These results tend to indicate that the advantage of the integrated system in terms of channel utilization is much less than indicated previously. Although it turns out that the basic overload phenomenon discussed here was originally pointed out by Kummerle [2], the importance of presenting these corrected results is underlined by the observation that a number of works subsequent to [1] have presented data delay analysis results which are based directly [7], [8] on [1] or apply different analysis models [9] which yield patterns of packet delay behavior similar to those in [1]. In addition, the results presented in [2] are based on much smaller ratios of call holding time to system frame duration than might be considered typical for voice calls, and therefore the conclusions in [2] are much more favorable than the conclusions here to the advantages of the movable boundary.

However, the results presented here should certainly not be considered in isolation in evaluating integrated system approaches. In particular, the analysis here assumes constant levels of average voice and data traffic load, and shows that a dynamically movable boundary is not very effective in allowing data packets to utilize free capacity due to momentary fluctuations in voice traffic. However, it is also shown here that appropriate flow control mechanisms can appreciably improve data traffic performance with a dynamically movable boundary. In addition, an integrated system will allow effective adaptation to the more gradual variations in voice and data traffic loads that would typically occur over a 24-h period. Finally, other criteria such as flexibility and overall hardware cost must be considered together with efficiency of channel utilization in comparing various system configurations.

II. SIMULATION OF MULTIPLEXER PERFORMANCE

A. Frame Structure

The multiplexing technique under investigation is a time division scheme in which fixed duration frames are partitioned and allocated to the transmission of digitized voice and data packets. The voice component of a frame is further divided into slots allocated to ongoing line switching communications. In general the voice slots can be of varying size depending on the bit rates of different voice users, and the size of data packets is also variable. For simplicity, and in order to allow direct comparison with the earlier analysis [1], the simulation here has assumed a multiplexer frame (taken to be of duration $b = 10$ ms) divided into $S + N$ equal-capacity slots. The nominal bit rate for the voice coders is taken as 8 kbits/s, which is accommodated with an 80 bit slot size and a 10 ms frame period. Voice traffic is allowed to occupy the first S slots, and the remaining N slots are reserved for data packets which are each assumed to occupy one slot. A voice call retains its slot for the duration of the connection; but if all S voice slots are busy when a new call is initiated, that call is blocked. Data traffic is permitted to use any voice slots which may be momentarily free due to statistical fluctuations in the voice traffic.

B. Voice Traffic Simulation Model

Voice traffic is modeled by a Poisson call arrival process of rate λ and exponentially distributed call holding times

Paper approved by the Editor for Computer Communication of the IEEE Communications Society for publication after presentation at the International Conference on Communications, Boston, MA, June 1979. Manuscript received March 12, 1979; revised November 18, 1979. This work was sponsored by the Defense Communications Agency. The views and conclusions contained in this document are those of the contractor and should not be interpreted as necessarily representing the official policies, either expressed or implied, of the United States Government.

C. J. Weinstein and M. L. Malpass are with the Lincoln Laboratory, Massachusetts Institute of Technology, Lexington, MA 02173.

M. J. Fisher is with the Defense Communications Engineering Center, Reston, VA.

with mean μ^{-1} The average number of call arrivals, λb, in a frame interval is generally small enough so that the frame structure can be ignored for voice traffic analysis. Let $n_v(t)$ represent the number of calls in progress at a given time. A simulation of the variation of $n_v(t)$ with time, based on a Markov S-server loss system model has been implemented as follows. Assume that $n_v(t) = k$ at a particular starting time. Then $n_v(t)$ is held at k for a time τ drawn from an exponential distribution with mean holding time τ_k determined as

$$\tau_k = 1/(\lambda + k\mu) \qquad k = 0, 1, \cdots, S - 1 \qquad (1)$$

$$\tau_S = 1/S\mu.$$

After a time τ, $n_v(t)$ is increased to $k + 1$ with probability

$$P_{up}(k) = \lambda/(\lambda + k\mu) \qquad k = 0, 1, \cdots, S - 1 \qquad (2)$$

$$P_{up}(S) = 0$$

or decreased to $k - 1$ with probability $1 - P_{up}(k)$. This process is repeated as often as desired to generate sample functions of $n_v(t)$.

C. Data Traffic Simulation Model

Data packets are assumed to arrive in a Poisson process with rate θ packets/s. Packet size is assumed fixed and equal to the number of bits in one slot. During any frame, the number of slots available for transmission of data packets is $N + S - n_v(t)$. Consider a period of time of duration t_i between the ith and $(i + 1)$st transition in $n_v(t)$, during which $n_v(t) = n_v^{(i)}$; the number, n_+, of data packets arriving during this interval is drawn from a Poisson distribution with mean θt_i. The maximum number of data packets which can be sent out during this time is

$$n_- = (N + S - n_v^{(i)})(t_i/b) \qquad (3)$$

(for simplicity t_i/b is constrained to be an integer). Let the number of data packets waiting for service at the beginning of the ith interval be $n_d^{(i)}$. Then the simulation sets

$$n_d^{(i+1)} = \max(n_d^{(i)} + n_+ - n_-, 0). \qquad (4)$$

If the data queue empties during the interval, then $n_d^{(i+1)}$ may be slightly greater than as given. The slight discrepancy would have negligible effects on the results presented below (particularly for the interesting case of long average data delays), and has been ignored. Note that in the simulation n_d is updated only at times of transition in $n_v(t)$. The average data packet waiting time in the queue is determined as $W_d = \langle n_d \rangle/\theta$, where $\langle n_d \rangle$ is a time average of n_d.

D. Simulation Results

1) Analysis of Sample Run: Sample functions of $n_v(t)$ and $n_d(t)$ are shown in Fig. 1 for $S = 10$, $N = 5$. The average call holding time is taken as $1/\mu = 100$ s, with $\lambda = 0.05$ s^{-1} so that the offered voice traffic is $\lambda/\mu = 5$ Erlangs. The blocking probability is $P_L = 0.018$. The average voice slot occupancy is $\langle n_v \rangle = 4.91$ so that the average number of slots available for data packets is 10.09. The data packet arrival rate is $\theta b = 9.0$ packets/frame, so that, on the average, 89 percent of the data slots are utilized. The plots represent 2500 s of

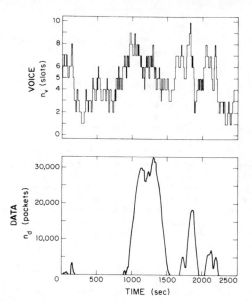

Fig. 1. Sample functions of $n_v(t)$ and $n_d(t)$ for $S = 10$, $N = 5$, $\lambda = 0.05^{-1}$, $\mu = 0.01$ s^{-1} and $\theta = 900$ packets/s ($\theta b = 9$ packets/frame).

real time, during which approximately 230 voice calls either entered or left the system. It is clear that $n_v(t)$ is a highly correlated (in time) random process which exhibits swings of long duration above and below $\langle n_v \rangle$. The error in the previous analysis [1] effectively caused the time correlation of $n_v(t)$ to be neglected, leading to large underestimates of data packet waiting time. During "idle" periods where $n_v(t)$ is low enough so that $\theta b < N + S - n_v(t)$ more than enough capacity is available to handle incoming data traffic and an initially empty data queue $n_d(t)$ will remain essentially empty. But $n_d(t)$ will build up significantly during busy periods where $\theta b > N + S - n_v(t)$. For this example, a long busy period during which $n_v(t) \geqslant 6$ for about 500 s begins at about $t = 900$ s. During this period the data queue builds up to about 30 000 packets. The data queue is eventually emptied during a subsequent idle period of the voice channel, but then builds up again during the next busy period.

2. Average Data Performance Statistics

Average data packet delays as measured by the simulation are depicted in Fig. 2 for $S = 10$, $N = 5$, $\lambda/\mu = 5$, and $\mu = 0.01$. Actually, the average buffer size $\langle n_d \rangle$ was measured, and Little's theorem used to obtain W_d. For example, at $\theta b = 9$, $W_d \approx 10$ s and $\langle n_d \rangle \approx 9000$ packets. In each run, the maximum as well as the average of $n_d(t)$ was measured. Generally (see [6] and Fig. 1) the maximum of n_d, and therefore the required storage allocation, is significantly larger than $\langle n_d \rangle$. The results shown in Fig. 2 (and in Fig. 1) are typical of similar results [6] for a range of the system parameters S, N, λ/μ, and θ.

Results were presented previously [1] for data waiting times for the same values of N, S, and λ/μ as utilized in obtaining Fig. 2. Assuming $b = 10$ ms, the predicted waiting time was $W_d \approx 20$ ms for $\theta b = 9$. This predicted delay is significantly below the simulation results obtained here. The data delay results presented in [2] were based on call

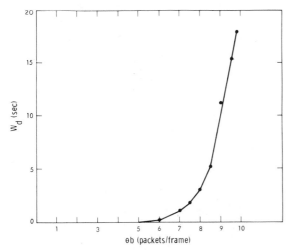

Fig. 2. Average data packet delays as function of data load for $S = 10$, $N = 5$, $\lambda = 0.05$ s^{-1}, $\mu = 0.01$ s^{-1}. Each plotted point represents an average of 4 runs, with each run comprising 5000 transitions in $n_v(t)$ and about 50 000 s of real time.

holding times in the range 5–10 frames, which would correspond to a range of 0.05 to 1.0 s with 0.01 s frame durations. This range of call durations is much shorter than typical voice calls, which accounts to some extent for the much more optimistic conclusions in [2] regarding data packet delays in the movable boundary case.

III. MODIFIED ANALYTICAL MODEL

The previous analysis (see [1, Eqs. (9)–(13)]) effectively treated the number of occupied voice channels and the number of data customers in the system as independent random variables, whereas in fact the two tend to be correlated. This had the effect [5], [6] of treating n_v as a random process which is independent between adjacent 10 ms frames. Actually, as discussed above and illustrated in Fig. 1, n_v changes only with call initiation or termination and is very highly correlated across frame-length time intervals. A different analysis model, based on a single server which switches randomly between voice and data, was applied in [9]. This model, however, assumes the switch position to be a random process which is uncorrelated from frame to frame, and thus also neglects the strong time correlation of the number of voice callers in the system.

If one ignores the effect of the time quantization introduced by the frame structure and assumes exponentially distributed (with mean b) rather than constant data packet lengths, the entire system can be analyzed as a two-dimensional Markov chain, with two classes of customers and priority for voice over data, and no queueing allowed for voice. In [6] a general set of two-dimensional difference equations for the probability of i voice customers and j data packets in the system is written, and a solution procedure is outlined. One can either try to solve the resultant equations directly or use generating functions as was done in [10]. The latter approach was carried out by Chang [11]. For either type of solution, significant computational problems can arise because of the significant differences in the magnitude of the parameters for a normal application. Numerical results have been obtained for the finite-buffer case, and have been shown to match closely

to corresponding finite-buffer simulation results. These results are presented in [6].

For the special case of a single channel ($S = 1$, $N = 0$) and an unlimited buffer, a reasonably simple expression for W_d emerges, in the form

$$W_d = \frac{\rho_2(1 + \rho_1)^2 + \rho_1 \theta/\mu}{\theta_2(1 + \rho_1)(1 - \rho_2 - \rho_1\rho_2)} \tag{5}$$

where $\rho_1 = \lambda/\mu$ and $\rho_2 = \theta b$. The second term of this formula represents the contribution to delay when the channel is not available to the data packets. The factor θ/μ in the numerator of that term represents the expected number of data packets to arrive during an average voice holding time. This factor can be quite large and accounts in large part for the very large data delays noted in the simulation. In fact, the 100 s holding time used in the simulations is somewhat shorter (by about a factor of 2.5) than typically encountered in telephone traffic, so that the results presented above on data delay in Section II are actually optimistic. Applying (5) to the case where $S = 1$, $N = 0$, $\lambda_1 = 0.01$, $\mu_1 = 0.01$, $\lambda_2 = 40$, and $\mu_2 = 100$ results in $W_d \approx 250$ s. This analytic result, based on exponentially distributed packet lengths, is reasonably consistent with the result of a simulation, where for the same set of parameters [6] except for an assumption in the simulation of constant data packet lengths, W_d was estimated at 176 s.

IV. INVESTIGATION OF FLOW CONTROL TECHNIQUES

The conclusion that can be drawn from the simulation and analytic results shown above is that attempts to achieve high channel utilization can lead to data packet queues so large as to overflow any reasonable amount of storage in the multiplexer. Even assuming infinite storage, the mean data delays are large because of the buffer buildup during periods when voice channel occupany is high. The need for a flow-control mechanism is apparent, and two types of flow control have been investigated: 1) voice rate control, where voice coders are assumed to be capable of operating at a variety of rates and a new voice call is assigned (at dial-up) one of several available bit rates, based on the current voice utilization and/or data queue size; and 2) data flow control, where a fixed limit is imposed on the size of the data queue. The simulation parameters for the experiments described below generally correspond to those used in obtaining the results in Fig. 2, except that flow control was introduced. More extensive simulation results are given in [6].

A. Experiments with Voice Rate Control but No Data Flow Control

The idea behind the voice rate control techniques was to cut down the peaks of voice channel utilization by assigning lower bit rates to callers who enter the system when utilization is high. This scheme assumes that each voice user has a flexible vocoder capable of a variety of rates. The following notation is introduced:

R_V = sum of bit rates of active voice users;
C_V = maximum allocated voice capacity;
Q_d = number of packets in the data queue;
R_{new} = bit rate assigned to a new voice caller.
C_V was generally fixed at 80 kbits/s, sufficient to accommodate 10 users at the nominal 8 kbit/s rate. A voice-rate

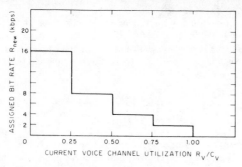

Fig. 3. Example of voice-rate control based on current voice channel utilization. The bit rate assigned to a new user is plotted as a function of the voice channel utilization R_V/C_V.

TABLE I
VOICE/DATA FLOW CONTROL SCHEMES

VOICE/DATA FLOW CONTROL SCHEMES
d1, d2, d3: all impose fixed limit Q_{max} = 150 packets on data buffer size.
d1: R_{new} = constant = 8 kbps
d2: $R_{new} = \begin{cases} 8 \text{ kbps} & R_V/C_V \leq .5 \\ 4 \text{ kbps} & .5 < R_V/C_V \leq .75 \\ 2 \text{ kbps} & .75 < R_V/C_V \leq 1.0 \end{cases}$
d3: $R_{new} = \begin{cases} 8 \text{ kbps} & Q_d = 0 \\ 4 \text{ kbps} & 0 < Q_d \leq Q_{max}/2 \\ 2 \text{ kbps} & Q_{max}/2 < Q_d \leq Q_{max} \end{cases}$

control scheme is defined by a specification of R_{new} as a function of R_v/C_v and/or Q_d. A voice rate control scheme where R_{new} depends only on R_V/C_V is depicted in Fig. 3; other schemes are detailed in [6].

In general, the voice rate control schemes produced some decrease in data delay at a cost in average voice bit rate. However, none of the data performance results [6] was really satisfactory at the high data packet arrival rates. The conclusion was that, although voice rate control alone can enhance data performance, some form of data traffic limitation is necessary to keep delays and queue sizes within reasonable limits. However, the ability of voice users to operate at a variety of rates depending on traffic loads would enhance the voice-traffic handling capability of the system, independent of its effect on data traffic.

B. Experiments with Combined Data Flow Control and Voice Flow Control

The experiments including data flow control involved the simplest form of data traffic limitation, where the data buffer was constrained to a fixed maximum size Q_{max}. Q_{max} was (somewhat arbitrarily) set equal to 150 packets. Data packets were denied entry to the multiplexer when its queue was full. This represents added delay since these packets have to be retransmitted to the multiplexer. If the packets enter the multiplexer directly from a user terminal, then the terminal would retransmit when no acknowledgment was received. Otherwise, a store-and-forward node feeding the multiplexer would handle the retransmission.

Three combinations of data and voice flow control techniques were tested. These will be referred to as $d1$, $d2$, and $d3$, and are summarized in Table I. Thus far, performance of the multiplexer has been displayed as a function of packet arrival rate for convenient comparison with previous results [1]. In order to take into account the effects of packet discard due to data flow control, and to provide a convenient means for comparing systems with different capacity, the results involving data flow control will be presented as a function of utilization (by data packets which are actually transmitted) of the variable capacity available to data due to fluctuations in voice traffic. Variable data capacity is defined as the difference between the total capacity allotted for voice (80 kbits/s) and the average capacity actually utilized by voice. A correspondence between utilization of variable data capacity and packet arrival rate for $d1 - d3$ is shown in Table II. (Note that an arrival rate $\theta b = 9$, as used in obtaining Fig. 1, corresponds to about 70 percent utilization of variable data capacity.) Results for the three schemes are depicted in Figs. 4-6.

TABLE II
PERCENT UTILIZATION OF VARIABLE CAPACITY AVAILABLE
FOR DATA FOR FLOW CONTROL SCHEMES $d1$-$d3$

PERCENT UTILIZATION OF VARIABLE CAPACITY AVAILABLE FOR DATA FOR FLOW CONTROL SCHEMES d1-d3		
Flow Control Scheme	d1	d2	d3
Packet Arrival Rate (⊖b in packets/frame)	Utilization of Variable Data Capacity (percent)		
600	19	18	19
700	39	37	37
800	55	55	55
900	70	71	70
950	78	78	76
1000		84	81
1050		88	87
1100			92

Fig. 4 shows that in all cases, the improvement in data packet delay is dramatic as compared to the situation where no data flow control is imposed. Fig. 5 shows that the percentages of data packets arriving when the queue is full are rather low (below about 70 percent utilization). This implies that the additional delay seen by a user who must retransmit these packets also ought to be modest. Fig. 6 plots average assigned voice bit rate as a function of data capacity utilization. For $d1$ and $d2$, the voice bit rate is independent of data load. For $d3$, the average voice bit rate decreases as a function of data utilization. A comparison among the schemes tested depends on the relative costs assigned to data packet delay versus decreased voice bit rate.

V. CONCLUSIONS

Modified results have been presented for the data packet delay performance of an integrated voice/data multiplexer structure, which showed that excessive data queues and delays

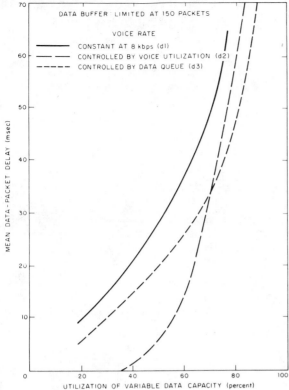

Fig. 4. Mean packet delay as a function of utilization of variable data capacity for schemes $d1$-$d3$.

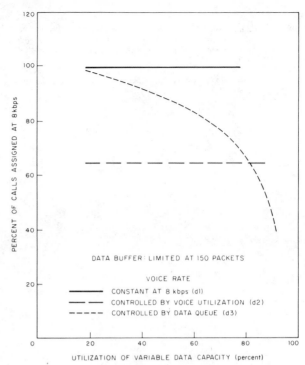

Fig. 6. Percent of calls assigned at 8 kbits/s as a function of utilization of variable data capacity for schemes $d1$-$d3$.

could build up under heavy loading conditions. A variety of voice and data flow control mechanisms to prevent these large delays were investigated. Voice rate control techniques caused modest improvements in data performance, while allowing a potential for handling more voice users. Data buffer limitation permitted substantial decrease in the delay of packets in the multiplexer, at a modest cost in quality of service for voice users and in the percentage of packets held off at the multiplexer input.

ACKNOWLEDGMENT

The authors would like to acknowledge the efforts of W. Hale of the Defense Communications Engineering Center in obtaining numerical results [6] for the finite-buffer case.

REFERENCES

[1] M. J. Fischer and T. C. Harris, "A model for evaluating the performance of an integrated circuit- and packet-switched multiplex structure," *IEEE Trans. Commun.*, vol. COM-24, pp. 198–202, Feb. 1976.

[2] K. Kummerle, "Multiplexer performance for integrated line- and packet-switched traffic," presented at the ICCC'74, Stockholm, Sweden, 1974.

[3] P. Zafiropulo, "Flexible multiplexing for networks supporting line-switched and packet-switched data traffic," presented at the ICCC '74, Stockholm, Sweden, 1974.

[4] G. Coviello and P. A. Vena, "Integration of circuit/packet switching in a SENET (slotted envelope NETwork) concept," in *Conf. Rec. Nat. Telecomunn. Conf.*, New Orleans, LA, Dec. 1975, pp. 42-12–42-17.

[5] "M.I.T. Lincoln Laboratory annual report on network speech processing program," Rep. DDC-AD-A053015, Sept. 30, 1977, pp. 45–54.

[6] C. J. Weinstein, M. L. Malpass, and M. J. Fischer, "Data traffic performance of an integrated, circuit- and packet-switched multiplex structure," M.I.T. Lincoln Lab. Tech. Note 1978-41, Oct. 1978.

[7] M. J. Ross, A. C. Tabbot, and J. A. Waite, "Design approaches and performance criteria for integrated voice/data switching," *Proc. IEEE*, vol. 65, pp. 1283–1295, Sept. 1977.

[8] H. Miyahara and T. Hasegawa, "Integrated switching with variable frame and packet," in *Conf. Rec. Int. Commun. Conf.*, Toronto, Ont., Canada, June 1978, pp. 20.3.1–20.2.5.

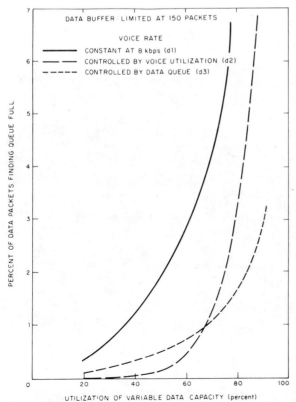

Fig. 5. Percent of packets finding queue full as a function of utilization of variable data capacity for schemes $d1$-$d3$.

520

[9] B. Occhiogrosso, I. Gitman, W. Hsieh, and H. Frank, "Performance analysis of integrated switching communications systems," in *Conf. Rec. Nat. Telecommun. Conf.*, Los Angeles, CA, Dec. 1977, pp. 12.4-1–12.4-13.

[10] U. N. Bhat and M. J. Fischer, "Multichannel queueing systems with heterogeneous classes of arrival," *Naval Res. Logist. Quart.*, June 1976.

[11] L.-H. Chang, "Analysis of integrated voice and data communication network," Ph.D. dissertation, Dep. Elec. Eng., Carnegie-Mellon Univ., Pittsburgh, PA, Nov. 1977.

Economic Analysis of Integrated Voice and Data Networks: A Case Study

ISRAEL GITMAN MEMBER, IEEE, AND HOWARD FRANK, FELLOW, IEEE

Invited Paper

Abstract—This paper describes the results of a study to evaluate alternative switching strategies for future integrated voice and data networks. Three fundamental problems are addressed:

1) the economics of integrating voice and data applications in a common communications system;
2) the comparison of alternative switching technologies for integrated voice and data networks;
3) the cost-effectiveness of alternative voice digitization rates and strategies.

Three broad switching technologies and variations thereof are investigated and compared. The switching technologies are: circuit switching, packet switching, and hybrid (circuit-packet) switching. Each switching technology can accommodate either voice or data applications separately or combined voice and data requirements in an integrated fashion.

Results of studies regarding communication systems options are provided and the sensitivity of the results are tested with respect to traffic variations, cost trends of switching and transmission, and network performance variables. The significant variables which affect the results are identified and quantified.

The intent of this study is to identify and quantify network technologies which demonstrate long-term low operating costs. This is a necessary effort to provide the basis for determining the most cost-effective evolutionary path for future communication systems. It is recognized that transition problems and associated costs may be other important factors determining the ultimate evolutionary path. However, determining these costs was not an objective of this study. Nevertheless, this study provides a framework and a target technology for detailed evolution, planning, and cost analysis.

I. INTRODUCTION

THIS PAPER presents the results of an economic analysis of alternative network strategies based on typical voice and data communications requirements. The study [1] was motivated by the emergence of new communications requirements and by recent advances in computer and communications technologies. These developments created the conditions under which the sophisticated operational techniques considered in the paper have become feasible, quantifiable, and hence worthy of a comprehensive study.

Major Department of Defense networks, for example, are evolving into digital communications systems [2]. Such systems will enable a higher degree of interoperability between the strategic Defense Communication System (DCS) and tactical systems, the provision of digital secure voice for a larger subscriber base, and the opportunity to support secure voice

521

and data applications on a common integrated communications system.

Three fundamental problems are addressed in the paper:

1) the economics of serving voice and data applications on a common integrated communications system;
2) the comparison of alternative switching technologies for integrated voice and data networks;
3) the cost-effectiveness of alternative voice digitization rates and strategies.

Results of studies of communication network options are provided in the paper. The sensitivity of the results are tested with respect to traffic variations, switching and transmission costs and network performance constraints. The significant variables which affect the results are identified and quantified.

Emerging communications requirements and recent advances in communications technology which motivated this study are summarized. An assumption is made that a digital technology will be employed in many future networks, with the analog voice waveform being digitized prior to transmission over the network.

The advantages of digital transmission include: 1) greater immunity to interference, such as noise and cross talk; 2) compatibility and ability to serve voice and data traffic with a common integrated system; 3) capability to apply digital encryption/decryption techniques.

The integration of heterogeneous traffic categories into a common system is desirable because of the economies of scale in switching and transmission available from modern industrial manufacturing processes. Integration also offers the capability to dynamically share transmission and switching facilities and has the capability to accommodate new applications which must access different types of data or voice processes. Most man-machine data communication is extremely bursty in nature, as for example in time-sharing and query-response applications. Indeed, all real-time communications which involve a human in the loop is of necessity bursty in nature. Conversational (real-time) speech is also of a bursty character. Shared transmission and switching facilities allow more efficient utilization of these facilities in bursty environments and consequently provide opportunities for significant cost reductions over the systems of the past.

In the past, most communication systems used dedicated transmission facilities which were not efficiently utilized. The rapid decrease in computer hardware cost over the last twenty years allows modern communication systems to take advantage of the burstiness of the requirements to achieve better utilization of transmission facilities. This can be generally accomplished by dynamically sharing transmission facilities, by using them only when information is being sent, by compression techniques, by error detection and correction techniques, etc.

Three general switching technologies and their variations are evaluated in this paper. The switching technologies considered are: circuit switching; packet switching; and hybrid (circuit-packet) switching. Each switching technology accommodates either independent voice or data applications separately or integrated voice and data requirements. Specific recent developments which motivated this study are as follows.

1) The feasibility of the packet-switching technology for data communications has been successfully demonstrated.
2) Computer and communications cost trends indicate that the price of computing (switching) has been decreasing much faster than the price of communication (transmission). In the recent past, computing cost has decreased by a factor of 10 every five years, while communications cost has decreased at a much slower rate of a factor of 10 over twenty-two years [3]. Furthermore, no spectacular breakthroughs are expected in the cost of communications in the next decade [2].

3) Digital transmission and multiplex equipment for microwave systems is becoming much cheaper than equivalent analog equipment [2]. This supports the assumption of the digital technology used in this paper.

4) Recent advances in low bit rate speech processors have been made. A survey of the state of the art in voice digitization techniques and devices is given in reference [25].

II. System Options Considered

Three broad switching technologies and variations thereof are investigated and compared. The switching alternatives compared are defined below.

A. Circuit Switching Options

1) Traditional Circuit Switching: An end-to-end circuit is established for a pair of voice or data users. The end-to-end transmission facilities are dedicated to users for the duration of use. The circuit is disconnected when either party hangs up.

2) Fast Circuit Switching: Voice and bulk data applications use the traditional circuit-switching concept. For interactive data users, a circuit is established for every message when ready to be sent and then disconnected after transmission. Specifically, the circuit is not dedicated to the user during his idle "think time" period; however the channel capacity not used during circuit setup and disconnection is taken into account. This concept assumes advanced digital switches enabling the set up of a circuit in 140 ms so that delay requirements for interactive data applications can be satisfied. It is likely that future switches will have this capability.

3) Ideal Circuit Switching: This scenario is almost the same as fast circuit switching except that circuit setup and disconnection are assumed to occur in zero time. Hence no channel capacity is wasted during setup and disconnection. While ideal circuit switching is not physically realizable, it is considered in order to obtain a lower bound on transmission cost for the circuit-switching technology.

B. Hybrid Switching Options

Hybrid (circuit-packet) switching has been considered in many previous studies [10]–[12], [14] as a desirable technology for integrating voice and data applications. In such switching schemes, circuit switching is used for voice, and packet switching is used for data applications. This arrangement was partially motivated by studies [23], [24] which demonstrated that circuit switching is cost-effective for traffic characterized by long holding time, while packet switching is cost-effective for bursty traffic, characterized by short messages and long pauses between messages. However, "short" and "long" are relative terms which depend upon the technology realizing the switched system. A traffic source is considered bursty in this study if the time to establish and release an end-to-end circuit in the network plus the message transmission time is shorter than the time interval until the next message generation.

Switching and transmission facilities are dynamically shared between traffic using both circuit-switched and packet-switched

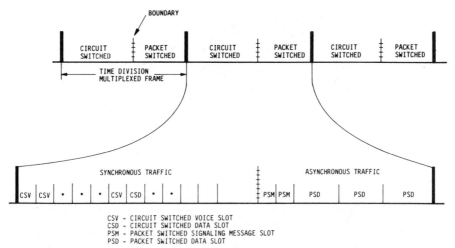

CSV - CIRCUIT SWITCHED VOICE SLOT
CSD - CIRCUIT SWITCHED DATA SLOT
PSM - PACKET SWITCHED SIGNALING MESSAGE SLOT
PSD - PACKET SWITCHED DATA SLOT

Fig. 1. Exemplary hybrid-switching channel structure: Time Division multiplexed frame.

modes. Voice is accommodated by the circuit-switched mode, interactive data applications are accommodated by the packet-switched mode, and bulk data applications may use either the circuit-switched or packet-switched modes depending on the operating discipline selected.

A communications channel linking two nodes in a hybrid-switched system utilizes a time division multiplexed master frame format. The frame is defined as a constant time interval throughout the entire backbone network. Network links may have different channel capacities, resulting in a different number of bits per frame. For example, a frame of 10 ms duration on a $T1$ carrier (1.544 Mbps) will contain 15 440 bits. A voice circuit using a digitization rate of 8 kbps will require an 80-bit slot in each frame. The absolute frame times in the network need not be synchronized. For example, the start of frame time on different links emanating from a single switching node or the start of frame times on the two directions of a full-duplex link need not be the same. The models used to obtain the quantitative results in this paper are presented in [8], [9]. Fig. 1 shows an exemplary link and frame structure. In the structure shown, voice traffic and data traffic (e.g., bulk data transfer) use the circuit switched subchannel and signalling message traffic and other data communications applications use the packet switched subchannel. The slot in the frame could be of different size depending on the bandwidth requirement in the circuit-switched mode and the packet size and signalling message size in the packet-switched mode. Two options for sharing of transmission capacity are examined.

1) Fixed Boundary Frame Management: The partition of link capacity between circuit-switched and packet-switched traffic is fixed.

2) Moveable Boundary Frame Management: While a boundary is assigned between the packet and circuit transmission capacities, packet-switched traffic can dynamically utilize idle channel capacity assigned to the circuit-switched mode.

C. Packet-Switching Options

Packet switching of voice is a natural way to take advantage of the "burstiness" of speech in a network environment and prevent network resources from being dedicated to speakers during silent periods. Packet speech protocols and experimental results are reported in [15], [16], [19].

Under these options, both voice and data are accommodated by the store-and-forward packet-switched concept. However, different packet sizes and different transport protocols are used for data and speech. Packet voice information flow is shown schematically in Fig. 2. The example shown illustrates the case where voice digitization and encryption are performed at the user site. In the paper, however, we also consider the case where voice is being digitized at the origination backbone node.

The packet voice protocol options considered are as follows.

1) Fixed Path Protocol (FPP): When a voice call originates, a signaling message is propagated to the destination to set up a path for the call. The path setup involves setting appropriate pointers at tandem switching nodes which determine the outgoing link for every input voice packet. No channel capacity is reserved or switch capacity dedicated. Voice packets follow the fixed path. When either party hangs up the path is released.

2) Path Independent Protocol (PIP): In this protocol, no path is set up. Each voice packet is transported to the destination independently of other packets of the same conversation. Packets can use alternate routes as appropriate.

It is noted that other techniques which avoid dedication of transmission facilities to speakers during silence periods have been previously considered. During the past two decades, numerous techniques, both analog and digital, for compressing the conversations of a number of speakers N onto a smaller number of channels M, where typically $N \gg M$, have been developed. The earliest strategy was the Bell System Time Assignment Speech Interpolation (TASI) [17] in which channel capacity is allocated only when appropriate hardware detected that a subscriber was actively speaking. Once the channel is seized, the speaker is given uninterrupted access to the channel. During periods of silence, the channel is relinquished and becomes available to other speakers. Digital variations of the original TASI concept, such as Digital Speech Interpolation (DSI) [22] and Speech Predictive Encoding (SPEC) [21] have also been implemented. The above systems "freezeout" speakers when the number of active speakers temporarily exceeds the available channel capacity. This results in

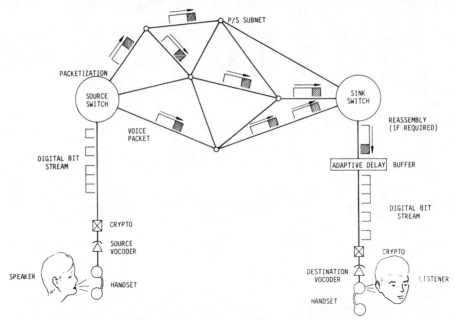

Fig. 2. Exemplary packet voice/data network and information flow.

clipping and segmentation of certain conversations with an associated loss in intelligibility. Refinements of the TASI concept based on digital encoding techniques whereby the bandwidth per active speaker is systematically reduced to accommodate additional speakers have also been implemented. Two such techniques are Adaptive Pulse Code Modulation (APCM) and Variable Rate Adaptive Multiplexing (VRAM), both developed by U.S. Army ECOM. In these systems, when the number of active speakers exceeds the channel capacity during "overflow periods," performance degradation is shared among all speakers by reducing the sampling rate per conversation. Thus, no single speaker suffers excessive degradation. Many variations of speech interpolation with priorities and multiplexing of data in speech idle periods have also been proposed.

III. ASSUMPTIONS AND DISCUSSION

Assumptions include voice and data traffic volumes to be accommodated, the voice digitization rate of active voice sources, switching and transmission cost components, and network performance requirements. The general problem formulation and design methodology are given at the end of this section.

A. The Data Base

The traffic data base used in this study was derived from the present DOD voice traffic on the AUTOVON voice system and a scaled DOD data traffic projected for the AUTODIN II data network. Only the projected traffic volume of AUTODIN II is varied whereas the traffic pattern is assumed unchanged.

The nominal traffic requirements are 2700 erlangs (E), AUTOVON voice traffic, and 36.15 Mbps scaled AUTODIN II data traffic. Data traffic is assumed to be composed of bulk data transfer applications and interactive applications. The nominal volume composition assumes 50 percent bulk and 50 percent interactive. In addition to the nominal data traffic of 36.15 Mbps, systems costs are obtained for data through-

puts of 11.6 Mbps, 86.8 Mbps, and 202.4 Mbps. Moreover, cost and performance for voice load requirements of 675 E and 1350 E (25 percent and 50 percent of the AUTOVON load) are also investigated. The voice digitization rates considered vary from 2.4 kbps to 64 kbps for an active speech source. These combinations of traffic and voice digitization rates result in consideration of digital traffic ranging from 3 percent voice and 97 percent data to 94 percent voice and 6 percent data.

Apart from the user data and voice traffic to be accommodated by the networks studied, the networks must carry a variety of control, protocol and signaling messages. The volumes and pattern of this overhead traffic depend upon the switching technology, communications protocols, and traffic characteristics. For example, the volume and pattern of messages for circuit setup and disconnection depends on the average voice holding time and the routing algorithm for circuit switching. Overhead traffic volumes and patterns are computed by automated network analysis and design tools developed for the study and are based on explicit detailed models of network operation.

B. Cost Models

No attempt is made to predict cost trends of computers or communication facilities. Cost trends depend not only on technology but also on cultural and regulatory factors. In particular, technological developments depend upon demand as much as upon the capability of the technology. For example, the widespread adoption of low voice digitization devices would result in significantly lower per unit costs than those currently predicted. Existing computer and communications costs are used in this paper and the major results are sensitivity tested with respect to large variations in switching or transmission costs.

Cost factors are based on current procurement estimates for tariffed communication lines and hardware. Communication line costs include mileage and termination charges. Hardware cost factors include purchase price, installation, initial support, operations, maintenance, and amortization. Cost factors not considered are network management costs, the security costs of specially cleared switches, operational personnel, and the cost of encryption devices.

1) Transmission Cost Model: The transmission cost model used in the paper is:

$$LC(i) = a_1 [a_2 + D(i)^{\alpha_1}] \left[\frac{C(i)^{\alpha_2}}{a_3} \right] \quad (1)$$

where

$LC(i)$ Total monthly cost for link i [dollars per month];
$D(i)$ Length of link i [miles];
$C(i)$ Channel capacity of link i [kbps].

The parameters a_1, a_2, a_3, α_1, and α_2, are estimated to obtain best fit to specific service offerings. $LC(i)$ takes into account the mileage charge as well as the fixed end charges for a specific capacity offering. The above equation was found to be sufficiently general to represent a large variety of tariff structures. It is assumed that terrestrial lines are used throughout the system.

The Dataphone Digital Service (DDS) offerings are used in this paper; parameter estimation was performed to fit the DDS tariff. The resulting formula is:

$$LC(i) = 0.61 [40.05 + D(i)^{0.873}] \ C(i)^{0.728} \quad (2)$$

2) Switching Cost Model: The switch cost model is a function of processing, memory size, and the cost of channel interfaces. The general formulae used to compute the purchase price of a backbone switching node is:

$$SC(i) = b_1 P(i)^{\beta_1} + b_2 M(i)^{\beta_2} + \sum_j b_{3j} C_i(j) \quad (3)$$

where:

$SC(i)$ Total cost of switch i [dollars];
$P(i)$ Processing capacity of switch i [10^6 instructions/s];
$M(i)$ Memory size [10^6 (32 bit) words];
$C_i(j)$ Channel capacity of outgoing link j.

The parameters b_1, b_2, b_{3j}, β_1, and β_2 are estimated using current switching technology. The parameter b_{3j} represents the cost of the channel interface and is a function of the channel capacity.

The following components are taken into account in computing processing capacity and memory size.

Processing Components
  Operating System Overhead
  Circuit-Switching Rate
  Data Packet-Switching Rate
  Voice Packet-Switching Rate
  Total Character Transfer Rate
  Hybrid Switch Complexity
    Frame Rate
    Circuit and Packet Rates Per Frame
    Moving Boundary Complexity
Memory Components
  Overhead Memory
  Storage for Tables
    Circuit and Packet Routing Tables
    Calls in Progress Tables
  Storage for Store-and-Forward Data

TABLE I
Parameters Used for Switch Processing Capacity

Number of Instructions per Switching Node	Type of Operation
10 000	Circuit switch processing of a signaling message for circuit setup
5000	Message processing for fixed path setup under the fixed path packet voice protocol
600	Data packet processing under packet packet switching and hybrid switching
400	Voice packet processing under the path independent packet voice protocol
100	Packet voice processing for a packet using the fixed path protocol
100	Frame management processing (per frame) in hybrid switching
2	Character transfer processing assuming DMA

Processing capacity is the most significant component of the switch cost. The processing capacity is computed for every switch in the network, taking into account the throughput of various message types and communications protocols. For example, the signaling message rate through a switch depends on the routing algorithm and upon the blocking probability for which the network is engineered. In particular, if a network is engineered for a higher blocking probability, a higher degree of alternate routing results which contributes to a higher signaling throughput requirement and, consequently, to a higher switching cost.

Table I shows key parameters used for computing switch processing capacity. In addition, a 20 percent operating system overhead was assumed, as was a 15 percent higher processing per frame in the hybrid-switching technology for the moving boundary frame management strategy than for the fixed boundary case. Costs of computer systems of major U.S. manufacturers as a function of processing capacity is part of the data used for estimating parameters in the cost formulas. The cost function resulting from the parameter estimation is:

$$SC(i) = 0.35 P(i)^{0.65} + 0.3 M(i)^{0.9} + \sum_j b_{3j} C_i(j). \quad (4)$$

The formuls used for computing switch processing power and memory size are (for convenience, we omit reference to switch i):

$$P = (1 + OVH) \left\{ CPS \cdot NC + PVPS \cdot NPV + PDPS \cdot NPD \right.$$
$$+ CHPS \cdot WPCH \cdot NCH + NF \cdot FPS \cdot \sum_{j=1}^{N \text{ Links}}$$
$$\left. \cdot [1 + \alpha \cdot CPF(j) + \beta \cdot KPF(j)] \right\} \quad (5)$$

where

OVH   operating system overhead factor;
CPS   rate of call set up and disconnection at the switch;
NC   number of instructions for processing a call set up (or disconnection) signalling message, including the identification of the next link (routing);
PVPS rate of voice packet throughput at the switch;
NPV   number of instruction to process a voice packet header, including its routing;
PDPS data packet throughput at the switch;

NPD	number of instructions to process a data packet header;
CHPS	character transfer rate;
WPCH	words per character (in our case $\frac{1}{4}$);
NCH	number of instructions per character transfer;
NF	number of instructions for frame management in hybrid switching;
FPS	rate of frames at the switch (depends on the frame time);
N Links	number of links at the switch;
CPF(j)	average number of circuit-switched slots per frame in the jth outgoing link of the switch;
KPF(j)	average number of packets per frame on the jth outgoing link from the switch;
α	overhead complexity factor in movable boundary hybrid switching per circuit slot;
β	overhead complexity factor in movable boundary hybrid switching per packet

Note that not all terms are nonzero for all switching strategies. For example, the last term, starting with NF, is used only for hybrid switching, and the terms multiplied by α and β are used only for the moveable boundary hybrid-switched strategy. The equation for computing memory size is:

$$M = \text{MOS} + \left\{ \sum_j \sum_k \text{PKPS}(k,j) \cdot \text{PL}(k) \cdot \overline{T}(k,j) \cdot \text{WPB} \right\}$$

$$+ \left\{ \sum_j \sum_k N(k,j)A(k) \right\} + \{\text{CKRT} + \text{PKRT} + \text{CIPS}\} \quad (6)$$

where

MOS	memory requirement for the operating system and communications software;
PKPS(k,j)	kth packet class arrival rate on the jth outgoing link at the particular switch;
PL(k)	average packet size of the kth packet class;
$\overline{T}(k,j)$	the time (in terms of transmission time on link j) that the packet of class k will be stored in the switch;
WPB	words per bit (1/32);
$N(k,j)$	number of slots of VDR_k on link j at the particular switch;
A(k)	storage requirements for a call in progress;
CKRT	circuit routing table size;
PKRT	packet routing table size;
CIPS	table space for calls in progress;

The second term in (6) is the buffer requirement for packets, the third term is for circuit switched calls in progress under the circuit and hybrid switching technologies, and the last term is table space for routing and for recording connections in progress. The packet class refers to a voice packet or one of several data packet classes. The time that a packet of a particular class needs to be stored in the system takes into account the expected time to receive an acknowledgment for classes where an acknowledgment is used; however, no analysis to guarantee an overflow below a given probability was done in determining this value.

As presented above, there are many factors which are taken into account in computing the cost of every switch in the network. Given the flows of various traffic types on network links, (5) and (6) are used to compute the processing power and memory requirement of every switch taking into account the throughput of the various traffic types (signalling messages, voice packets, data packets, calls in progress, etc.) and the particular switching strategy. Finally, (4) is used to compute the switch purchase price, taking into account processing power, memory and channel interfaces. The comparison of switches for different network technologies, in isolation, without considering the network as a whole, is rather difficult and will not provide insight into the comparison of the switching technologies. However, when we compare the cost of the alternative technologies in Section IV, we specifically focus on the cost of switches of the different technologies and explain the reason for the relative costs of such switches.

Apart from current switching and transmission costs, the sensitivity of network cost to component hardware and transmission costs is derived using the following two cost scenarios:

1) switching cost–10 percent of current, transmission cost–current;
2) switching cost–current transmission cost–10 percent of current.

a) Conversion factor: Communications channels are leased, hence their cost is recurring and given in dollars per month. The switch costs are given in purchase price and converted to monthly cost using cost analysis practices of the Defense Communications Agency (DCA). The conversion factor is based on a ten-year amortization plan; it includes installation charge at 20 percent base cost, initial support charge at 67 percent base cost, and a ten-year operation and maintenance cost at 47 percent base cost. It assumes a redundancy factor of 1.5; a capital factor of about 5 percent per year, based on 10 percent annual interest over ten years. Using the above analysis results in a conversion factor 0.0438 from purchase price to monthly cost.

3) Cost of VDR Devices: Low bit rate voice digitizers are of particular interest for future voice systems or for systems under development. Although there exist low VDR devices for commercial use, the purchase prices quoted (e.g., $10 000 for a 2.4 kbps device) are not representative of their potential large volume cost. Furthermore, with recent advances in LSI technology, the cost of such devices is expected to decrease significantly in future years. Since this report addresses systems in the mid 1980's and beyond, the cost of VDR devices is parameterized with analysis oriented towards determining the cost at which VDR devices become economical for various switching strategies and digitization rates. The conversion factor used for switching nodes is used to convert capital cost to recurring monthly cost for voice digitizers.

C. Network Performance

In the circuit and hybrid-switching technologies, voice calls are engineered on a blocking basis and data subscribers on a delay basis. Blocking implies that a certain percentage of voice calls will be rejected by the system via a busy tone because of unavailability of facilities. Data subscribers under the circuit-switching technology are assumed to automatically redial every 10 ms when blocked, until an end-to-end circuit is established. The packet switching network is engineered on a delay basis for voice and data subscribers.

All backbone networks are engineered for nominal:

1 percent	end-to-end blocking for circuit switched voice;
200 ms	end-to-end packet delay for interactive data users and packet voice;
600 ms	end-to-end packet delay for bulk data applications.

The percentage of blocked calls for which the networks are engineered is varied from 1 percent to 10 percent, and the end-to-end packet delay is varied from 200 ms to 1 s; the corresponding network cost is calculated to determine the effect of performance requirement variations. In several cases, packet switching with 50 ms end-to-end delay requirements for voice packets were also examined.

D. Network Structure

The investigation and comparison of switching technologies is developed for backbone networks spanning the U.S. with backbone switching nodes at eight switch locations. The backbone voice traffic corresponding to these locations is determined by assigning the current AUTOVON switch traffic requirements to the eight backbone nodes according to the nearest distance criterion. Network links and their capacities are obtained by automated network design techniques using the minimum cost criterion subject to satisfying network performance requirements. However, a two-connectivity requirement (each node connected to at least two other nodes) is imposed to improve network reliability.

The number of backbone nodes is held constant in the present investigations since previous studies [5] have shown that the optimum number of backbone nodes for comparable throughput levels ranged from five to twelve, and that the cost differences in this range are insignificant. Moreover, well-designed networks with as many as 30 distributed switches have been shown [20] to lead to networks with communication and hardware costs only a few percent above the minimum. Thus, the number of backbone nodes is not a critical issue from a communications efficiency perspective and this number can be determined based on other criteria such as security and survivability.

It should be pointed out that the model used here has been applied only to backbone structure, and the analysis of costs and performance for access lines connecting the backbone to the users was not evaluated. However, see the comments of Section III-E for a discussion of local access issues.

E. Location of Voice Digitization Devices

Investigations are performed under one of the following assumptions.

1) Voice requirements occur in the backbone network in digital form at the bit rate indicated. The location and cost of the digitization process is not considered in this case.

2) Voice is digitized at the origination backbone node.

3) Voice is digitized at subscriber handsets.

The absolute backbone network cost differences between alternative network technologies are independent of the location of the digitization process, and are not affected by the location of the voice digitizers. Under Assumption (2) low bit rate digitizers are provided in the backbone network with the objective of reducing total system cost. This case is applicable for voice and data systems where subscribers do not require end-to-end encryption. An integrated system may consist of some subscribers with digitizers at the handset and other subscribers whose voice signals are digitized in the backbone network. This digitization strategy is also investigated. Assumption (3) relates to the case where all subscriber handsets include voice digitizers. In this case, while total cost of digitizers would increase because of the larger number of units required, savings would be generated by reducing the cost of local access. These savings were not quantified but are

expected to follow similar trends to those observed for the backbone network. This expectation derives from the following argument. For voice digitization at the backbone switch, local access lines would be similar for the circuit, hybrid or packet-switched technologies. Furthermore, if digitization occurs at the telephone (or at the local PBX) at high digitization rates, the local access lines would also be the same. However, as VDR decreases, the natural multiplexing ability of packet switching combined with silence detection at the vocoder would tend to increase the number of conversations achievable via packet switching over a given set of channels and hence decrease the number and cost of the local access lines needed to meet a given voice traffic requirement.

F. General Problem Formulation and Design Methodology

A detailed description of the computational algorithms and computer programs which have been developed during this study [1] and used to obtain the results reported is beyond the scope of this paper. In [26] we present the detailed design methodology for hybrid switched networks. In this section, we briefly state the problem and outline the subproblems which form the design methodology.

Given the following.

1) A set of switching nodes N and their locations.

2) Traffic volume requirements (voice traffic matrix in erlangs and data traffic matrix in kbps).

3) Traffic characteristics, which include
   a) voice digitization rate
   b) average call holding time
   c) average bulk message size
   d) average interactive message size and average idle time between messages.

4) Signalling procedure and signalling message size.

5) Network operation and design parameters, which include:
   a) voice and data switching strategy
   b) communication protocols (e.g., FPP or PIP for packet voice)
   c) routing algorithms
   d) packet sizes for bulk data applications, interactive applications and for voice traffic in the case of packet switching
   e) header size for the various packet sizes
   f) priority structure for the traffic categories.

6) Tolerance values for meeting end-to-end performance requirements.

7) Cost versus capacity and distance for terrestrial communication channels: $LC(i) = f[C(i), D(i)]$, equation (2).

8) Cost versus processing capacity, storage requirements, and channel interfaces for switching nodes: $SC(i) = f[P(i), M(i), C_i(j)]$, equation (4).

9) Cost of voice digitization devices, where appropriate.

Determine the following.

1) The set of links $A = \{1_i\}$.

2) The channel capacity $C(i)$ for each link in A.

3) The processing capacity $P(j)$ and memory requirement $M(j)$ for each switching node.

That *minimize* the total cost

$$\sum_A LC(i) + \sum_N SC(j) + \text{Cost of VDR devices}$$

such that:

1) Traffic volume requirements are accommodated.

2) End-to-end traffic performance requirements are satisfied. (This includes blocking and delay for the various traffic classes as in Section III-C).

3) Average voice call set up and disconnection delays in circuit and hybrid switching strategies are satisfied.

4) 2-connectivity constraint is satisfied.

The design methodology involves iterative application of the solution to the topological problem, link dimensioning and capacity assignment, and the routing problem. The design methods are significantly more complex than those used for either circuit switched network design or packet switched network design. Among the factors contributing to complexity are: existence of several traffic classes with different priorities, several message sizes and routing algorithms, the interaction between the voice and data traffic flows, and the complexity of the switching strategies. This resulted in the necessity to develop and automate numerical optimization techniques because many of the subproblems did not lend themselves to closed form solutions. Some of the optimization techniques developed for the hybrid switching examples are described in [26].

IV. Cost Comparison of Switching Technologies

A. Comparison of Switching Technologies as a Function of Voice Digitization Rate

This section compares the circuit-switching, the packet-switching, and the hybrid (circuit-packet) switching technologies for integrated voice and data applications. The results for each switching technology are first discussed and comparisons are then provided. The comparison in this section assumes that voice requirements are presented to the backbone network in digital form. That is, the cost of the digitization process is not taken into account. Furthermore, the issue of voice quality is not addressed, implying that voice is considered to be at an acceptable quality for each of the digitization rates compared.

Fig. 3 shows the total monthly cost in million dollars for all the switching technologies as a function of the Voice Digitization Rate (VDR). The current prices of switching and transmission are used to obtain the results shown. The VDR value indicates the maximum bit rate of an active voice source and is assumed the same for all sources. Voice is offered in erlangs and converted to bit rate throughput requirements by multiplying the number of erlangs by the voice digitization rate. Hence the VDR impacts both total throughput of the system (in bits per second) and the fraction of digital traffic represented by voice requirements. The total throughput is shown on a second horizontal axis in Fig. 3.

Based on numerous parametric studies it was found that the voice digitization rate is a significant parameter affecting the cost of an integrated voice and data communications system. The cost of digital communications systems carrying voice with VDR as a variable has not previously been exposed. When requirements for secure end-to-end paths are not present, it is cost-effective to convert voice to a low bit rate at the backbone network level. In environments which include some users who require end-to-end security and others who do not, the impact of low VDR is quite significant.

1) Circuit-Switching Technology: Three circuit-switching strategies are compared in Fig. 3: traditional circuit switching, where a circuit at the VDR capacity is dedicated to a pair of users for the duration of the call; fast circuit switching which

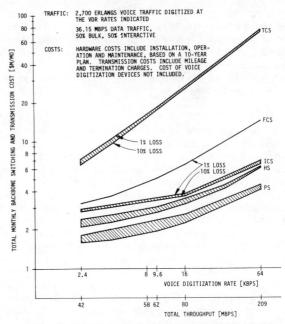

Fig. 3. Monthly backbone switching and transmission cost as a function of voice digitization rate (VDR) and switching technology (TCS-Traditional Circuit Switching; FCS-Fast Circuit Switching; ICS-Ideal Circuit Switching; HS-Hybrid Switching; PS-Packet Switching). Cost range indicate alternative operation scenarios (HS and PS) or blocking probability range (TCS, ICS).

takes advantage of the low duty cycle of interactive data users by establishing a circuit for each message when ready to be sent and disconnecting afterwards (a delay of 140 ms is assumed for circuit setup or disconnection); and ideal circuit switching using the same protocol as fast circuit switching with circuit setup and disconnection requiring zero seconds. While ideal circuit switching is not physically realizable, it is examined to obtain a lower bound on transmission cost for the circuit-switching technology.

The duty cycle of an interactive user is assumed to consist of an average idle (think) time of 10 s followed by an average message size of 1000 bits. This is characteristic of interactive users sending messages "line-at-a-time" rather than "character-at-a-time." The actual duty cycle of interactive users depends upon many factors including user patterns, the computer system and applications environment (e.g., scientific/engineering applications, business applications, etc.), and the I/O and CPU load of the computer system. The results of a study of traffic characteristics for user-computer interactive processes are reported in [18]. Sensitivity to the duty cycle parameters is reported in [1]. Voice calls are engineered on a blocking basis and data calls on a delay basis. An interactive user is assumed to redial automatically at 10 s intervals when blocking occurs.

Among the feasible circuit-switching strategies, fast circuit switching is clearly superior to traditional circuit switching for integrated voice and data networks under current prices of switching and transmission. The cost of traditional circuit-switching increases very quickly with VDR, from $7.28 million per month at 2.4 kbps to $77.18 million per month at 64 kpbs, for 1 percent loss. However, the rate of increase is smaller than that of the VDR value, because voice is only part of the traffic accommodated and because of economy of scale in the purchase of switching and transmission facilities. The absolute cost increase primarily results from an increase in cost of the transmission facilities. Although the circuit setup and disconnection load is independent of VDR, the cost of switching increased from \$0.7 million per month at VDR = 2.4 kbps to \$9.3 million per month at VDR = 64 kbps. This occurs because the expansion of transmission facilities increases two components of the switching cost: the interface cost to communications channels, and the processing resulting from the higher bit transfer rate. Converting these monthly switch costs to purchase prices under the assumption of identical capacity for each of the eight backbone switching nodes yields prices per switch of \$2 million at VDR = 2.4 kbps and \$26.5 million at VDR = 64 kbps.

The switching cost component of fast circuit switches is much higher than for traditional circuit switching. Nevertheless, since fast circuit switching is less sensitive to the VDR parameter than traditional switching, the switching cost increase from VDR at 2.4 kbps to VDR at 64 kbps is from \$2.03 to \$7.7 million per month. These monthly figures correspond to purchase prices per circuit switch of \$5.8 million at VDR = 2.4 kbps and \$22 million at VDR = 64 kbps. The cost of switching nodes is higher in the fast circuit-switching technology than in the traditional circuit-switching technology at low VDR; the reverse is true at high VDR. The transmission cost is the significant cost component contributing to the relatively large difference in the total cost between the traditional and fast circuit-switching technologies.

The end-to-end loss does not have a significant effect on network cost. The cost reduction for traditional circuit switching ranges from 4.1 percent to 9.2 percent when the loss increased by an order of magnitude from 1 percent to 10 percent. There are two main reasons for the relatively small impact of the loss parameter. First, when the network is engineered for a high loss, a higher degree of alternate routing takes place which contributes to an increase in switching cost. Second, for high volume traffic the channel capacity (number of circuits) needed to accommodate the traffic is very high. The efficiency of large trunk groups is such that only a small percentage of additional trunks are needed to reduce the link blocking probability. Coupled with economy of scale in the purchase of transmission facilities, this results in a relatively small loss effect on the cost of transmission facilities.

Two significant observations can be made from the analysis results of the circuit-switching technology.

1) The total cost of circuit-switching network technologies is not sensitive to the loss probability for which the networks are engineered.

2) The cost of the traditional circuit-switching technology is more sensitive to the voice digitization rate than the other network technologies. The increase in the voice digitization rate not only penalizes the voice applications, but also the interactive data applications which under the assumptions made, occupy a voice equivalent channel.

2) Hybrid (Circuit-Packet) Switching Technology: Fig. 3 shows the range of cost as a function of VDR for hybrid-switching strategies. The upper curve corresponds to the case where only voice is circuit switched and data is packet switched. The lower curve corresponds to the case where both voice and bulk data traffic are circuit switched and the interactive data applications are packet switched. The range between the two curves corresponds to varied mixes of bulk and interactive data applications—between 0 percent to 50 percent bulk out of the total 36.15 Mbps data traffic. The strategy with circuit-switching bulk data applications requires further investigation and additional experiments. This would only be of interest in evaluating alternative realizations of hybrid switching; it would not impact the comparison of alternative network technologies. For hybrid switching it is assumed that all data is being packet-switched unless otherwise indicated.

It should be noted that in the hybrid-switching alternative, no attempt was made to utilize speech compression techniques such as those discussed in Section II-C. Further studies to consider such options are currently being considered and could affect the cost comparisons.

The hybrid-switching designs shown were engineered for a voice loss of 1 percent. The packet size for interactive data is 800 bits (including header) and engineered for 200 ms average end-to-end delay; the bulk data packet size is 1200 bits (including header) and engineered for 600 ms average end-to-end delay. Progressive routing (where the entire address information is exchanged between tandem switches and an end-to-end path is established on a link-by-link basis) is assumed for circuit switching and the corresponding signaling messages derived are accommodated as packet-switched data traffic, without priority. For packet switching, adaptive routing is assumed, and appropriate network overheads are taken into account in the designs.

The hybrid-switching technology matches the switching concept to traffic characteristics as traditionally conceived; circuits are dedicated to voice users for the duration of the call. As expected, its backbone network cost is well below that of circuit-switching strategies. The switching cost component is higher than for traditional circuit switching but lower than in fast circuit switching; a special contributing factor to the hybrid switching cost is the switch complexity.

The cost of the hybrid-switching technology increases with VDR, but at a slower rate than in the circuit-switching technologies. At VDR = 2.4 kbps, the total monthly cost is \$2.46 million and at VDR = 64 kbps, the total monthly cost is \$6.35 million; a cost ratio of 2.58. Most of the increase is attributable to the cost of transmission. The rate of increase in total cost is smaller than anticipated because of economy of scale.

The hybrid-switching technology has been examined under variations of data traffic mix (between bulk applications and interactive applications) and under variations in packet size [1]. The analysis of hybrid-switched networks has shown that cost is sensitive to the mix of data traffic applications and leads to the following conclusions.

1) An important aspect of the design and operation of hybrid-switching networks is the partition of traffic between circuit and packet-switching services.

2) The most cost-effective strategy under hybrid switching may result when bulk data are handled in a circuit-switched mode and interactive data applications plus voice are handled in a packet-switched mode.

The last conjecture is motivated by the superior channel utilization obtained by packet-switching voice and interactive

TABLE II
BACKBONE NETWORK COSTS*

VDR	2.4 kbps	8 kbps	16 kbps	64 kbps
Traditional Circuit Switching	$ 7.3M/mo	$ 16.0M/mo	$ 27.0M/mo	$ 77.2M/mo
Fast Circuit Switching	$ 3.3M/mo	$ 4.8M/mo	$ 6.6M/mo	$ 14.7M/mo
Hybrid Switching	$ 2.3M/mo	$ 3.0M/mo	$ 3.5M/mo	$ 6.3M/mo
Packet Switching	$ 1.6M/mo	$ 1.9M/mo	$ 2.3M/mo	$ 4.2M/mo

*Cost of voice digitization devices is not included.

data applications and the reduced switching capacity and channel capacity requirements for packet headers when using circuit switching for bulk data applications.

3) Packet-Switching Technology: Circuit-switching and hybrid-switching technologies require transmission facilities and some switch resources to be dedicated to users; packet switching requires virtually no dedication of resources. Packet switching copes well with traffic burstiness whether data or speech is being communicated. The penalties paid for potentially eliminating all idle capacity are the header overhead appended to each packet and the increased switch processing. Packet speech communication is in an early stage of development; many issues with regard to protocols, error control, routing, etc., are still being studied [15], [16]. Consequently, a variety of protocols and header sizes were examined to obtain a range of cost and performance relationships.

The following strategies were used to obtain the range of costs associated with packet designs shown in Fig. 3:

1) Fixed Path Protocol (FPP) with headers of 48 bits and 96 bits, with and without the compound packet protocol option;

2) Path Independent Protocol (PIP) with headers of 152 bits and 256 bits under the compound packet protocol option.

The protocol type, FPP or PIP, directly impacts switch cost. In FPP, an initial signaling message is propagated to the destination switch to determine a fixed path. It causes the setting of appropriate pointers at tandem (intermediate) switching nodes which determine the outgoing link for every arriving voice packet; however, once the path is established subsequent packets require less processing per switch to use the fixed path. Under PIP there is no path setup; however, the processing per voice packet at each switch is greater than under FPP. The protocol type also impacts the header size—a larger header is needed for PIP—which increases the transmission cost. The compound packet concept provides for the encoding of speech segments from several speakers into the same packet when the speakers have a common destination switch. The technique seeks to reduce packet header overhead while retaining a small speech packetization delay at the origination switch.

The use of compound packets becomes important at low VDR, where a long delay may be required to collect a packet from a single vocoder, and especially significant when using a PIP involving a relatively large header. The use of compound packets affects both transmission and switching cost. The designs developed are based on a model where a speech segment (window) contains 10 ms of speech; a number of segments is associated with the triplet (vocoder VDR, Header Size, Compound Packet Option). The number of speech segments, the vocoder VDR, and the Header Size determine the packet size and overhead. The results reported in Fig. 3 include a range of packet overheads from 3.6 percent (VDR = 64 kbps, 2 segments, $H = 48$) to 44.4 percent (VDR = 2.4

kbps, 5 segments, $H = 96$). Hence, the range covered by the designs anticipates a wide variety of practically engineered packet voice and data networks.

The packet voice and data networks are engineered for an average end-to-end delay of 200 ms for speech and interactive data packets, and 600 ms for bulk data packets (as in the hybrid switching case). It is important to note that no optimization is performed in selecting packet speech size. Consequently the total costs could be somewhat lower than those derived. Also the packet speech size varies with the different VDR values. On the other hand, for some networks involving up to 4 or 5 links in tandem, maintaining an end-to-end delay may require reduction in the per-link delay which could increase cost. In addition to limit loss of packets to say 1 percent, the 99th percentile of delay would be a more appropriate parameter [27]. However, the design for the 99th percentile delay in the presence of other traffic is an intractable network design problem. Hence, the design could be made to satisfy average delay and the results must be verified via simulation and analytical techniques relating the average delay and the delay distribution. Investigations of packet voice delay distributions at destination nodes and processing requirements for smoothing the packet speech stream to the listener are reported in [5]. Furthermore, a preliminary investigation of network cost sensitivity with respect to delay variations (e.g., 50 ms average) led to only small increases in cost (e.g., 1–3 percent).

As a function of VDR, the lower bound of the set of designs shown in Fig. 3 increases from $1.6 million per month at VDR = 2.4 kbps to $4.15 million per month at VDR = 64 kbps; the upper bound increases from $1.86 million per month at VDR = 2.4 kbps to $4.6 million per month at VDR = 64 kbps.

4) Comparison of Switching Technologies: Fig. 3 provides the cost comparison of all switching technologies as a function of the voice digitization rate. The cost comparison leads to the following conclusions.

1) The packet-switching technology is superior to all other switching technologies for the entire range of the voice digitization rate.

2) The traditional circuit-switching technology is inferior to any of the other alternatives.

Table II gives sample costs comparing traditional and fast circuit-switching strategies, the hybrid-switching strategy with circuit-switched voice and packet-switched data, and the packet-switching strategy using the path independent protocol with a 152 bit header. Using the packet-switching cost as the base, the following relationships result:

1) The hybrid-switching technology is more costly than packet switching by 56 percent and 50 percent for VDR = 2.4 kbps and 64 kbps, respectively.

2) The fast circuit-switching technology is more costly than packet switching by 106 percent and 250 percent for VDR = 2.4 kbps and 64 kbps, respectively.

3) The traditional circuit-switching technology is more costly than packet switching by 356 percent and 1740 percent for VDR = 2.4 kbps and 64 kbps, respectively.

5) Switching Cost Components: There exists a tradeoff in telecommunications between communications and computation since more sophisticated processing performed at switching nodes usually results in a reduction of transmission facilities. Different switching concepts demonstrate a different breakdown of switching and transmission cost components at the optimum (minimum cost) design. To reflect this tradeoff, percentage of switch cost is shown in Fig. 4 with voice digitization rate as a parameter. The first observation made is that traditional circuit switching is characterized by the smallest switching cost component and fast circuit switching is characterized by the largest switching cost component. The latter results because of the large amount of processing required to set up and disconnect an end-to-end circuit for every data message of an interactive user. The second observation is that traditional circuit switching is the only technology where the percentage of switching cost increases with VDR. This indicates that the rate of growth of switching cost is higher than that of transmission cost for the range of VDR considered. As indicated before, the switching cost increases with an increase in transmission cost because of the channel interface and character transfer rate components in the switch cost model.

The sensitivity of the switch cost component to variations in VDR is smaller in the circuit-switching strategies than in hybrid switching or packet switching. Specifically, the percentage of switching cost over the VDR range considered varies from:

9–13 percent   for traditional circuit switching;
53–62 percent   for fast circuit switching;
25–55 percent   for hybrid switching;
27–38 percent   for packet switching.

6) Discussion: The results and conclusions of Section IV-A4 are the major results of the paper. The understanding of these results is important in order to enable comprehension of the results of studies that follow. Since the results may initially appear counterintuitive, an explanation follows. The discussion follows questions of reviewers from which the authors greatly benefited.

Two important questions were raised by reviewers of the paper relative to the results shown in Fig. 3. The first question relates to the relatively slow rate of cost increase of the packet and hybrid switching technologies and the relatively high rate of cost increase of the traditional circuit switching technology, when the voice digitization rate increases. The second question relates to the divergence between the cost curves of the traditional circuit switching technology and the fast circuit switching technology when the voice digitization rate increases.

First, it is noted that the rate of increase in all switching technologies is much smaller than the rate of increase in the voice digitization rate. Even for traditional circuit switching, the cost at VDR = 64 kbps is 10.6 times the cost at VDR = 2.4 kbps, whereas the ratio in VDR values is 26.7. There are three reasons for the smaller rate of cost increase than that of the VDR ratio: economies of scale in the purchase of transmission facilities which result from the assumed tariff struc-

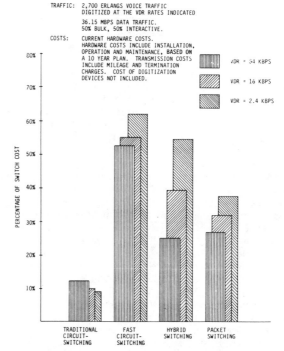

TRAFFIC: 2,700 ERLANGS VOICE TRAFFIC DIGITIZED AT THE VDR RATES INDICATED
36.15 MBPS DATA TRAFFIC. 50% BULK, 50% INTERACTIVE.

COSTS: CURRENT HARDWARE COSTS. HARDWARE COSTS INCLUDE INSTALLATION, OPERATION AND MAINTENANCE, BASED ON A 10 YEAR PLAN. TRANSMISSION COSTS INCLUDE MILEAGE AND TERMINATION CHARGES. COST OF DIGITIZATION DEVICES NOT INCLUDED.

VDR = 64 KBPS
VDR = 16 KBPS
VDR = 2.4 KBPS

Fig. 4. Comparison of switching technologies: switch component cost in backbone network cost.

ture (based on DDS), the fact that voice is only part of the network traffic, and the fact that transmission is only part of the total cost. The reason for the slower cost increase in packet and hybrid switching technologies as compared to traditional circuit switching is because the cost of switches constitutes a higher fraction of the total cost, and because the former technologies also realize a better utilization of transmission facilities as a result of the increase in link channel capacities.

The relatively high rate of cost increase for the traditional circuit switching technology and consequently the divergence observed in Fig. 3 results primarily because of its inefficiency in handling interactive data applications. We recall that an interactive data user is given an end-to-end channel at a bit rate equal to the voice digitization rate, for the duration of the interactive session. Hence, when VDR is increased the waste of channel capacity by interactive data applications increases, because the time for the transmission of an interactive message decreases, whereas the time gap between interactive messages remains constant. That is, the increase in VDR not only penalizes the voice users but primarily penalizes the efficiency for interactive data applications.

The reader may observe that when the voice digitization increases to 64 kbps, the fraction of voice traffic is high (83 percent) leading to a conclusion that traditional circuit switching must be efficient at such conditions. Such a conclusion will be erroneous because although the voice traffic constitutes a high percentage in bits per second it constitutes a small (constant and independent of VDR) percentage of the total traffic in terms of erlangs load offered to the system. The relative efficiency of the alternative network technologies for voice traffic and data traffic, and in particular the inefficiency of traditional circuit switching for interactive data applications, is established in the next section. This will provide additional support and rational to the results shown in Fig. 3.

Finally, we explain the reason for the higher rate of cost increase of the fast circuit switching technology as compared to hybrid switching and packet switching. Recall that in fast circuit switching there is no dedication of channel capacity to interactive users during idle periods and that voice traffic is being served in the same manner as in hybrid switching. The reason for the higher rate of cost increase as a function of VDR is because of the channel capacity wasted during circuit setup and disconnection. The model for circuit setup (and disconnection) used in the paper assumes that an end-to-end circuit is reserved link by link and that a reserved link capacity cannot be used by other traffic; hence such capacity is being idle (thus wasted) during the end-to-end circuit setup (and disconnection). The holding time for the transmission of an interactive message is relatively small and decreasing with an increase in VDR. On the other hand, the channel capacity that is idle (thus wasted) is increasing with VDR because a circuit at the VDR value is being set up. Hence, although the number of circuit setups and disconnections is constant, the wasted capacity is increasing with VDR. It is instructive to indicate that the primary reason for the higher rate of cost increase of fast circuit switching results from serving the interactive data applications. It is also noted that higher link capacities also result in an increase in switch cost, e.g., because of higher channel interface costs.

B. Sensitivity of Comparison of Switching Technologies to Variations in Voice Traffic

The evaluation and comparison of the alternative switching technologies as a function of voice digitization rate has demonstrated a high degree of sensitivity to the voice digitization rate parameter. In particular, the cost of the traditional switching technology increases rapidly with the voice digitization rate while the increase in the cost of the packet switching technology is relatively small. This section compares network costs for different volumes of voice traffic and determines network cost sensitivity to variations in this parameter.

To facilitate the cost comparison, the alternative network technologies were designed for voice and data applications with voice loads of 675 E and 1350 E. These values correspond to 25 percent and 50 percent of the nominal voice traffic volume used for the previous discussed designs. The data traffic is kept constant at 36.15 Mbps with the composition of 50 percent bulk data transfer applications and 50 percent interactive data applications. The voice digitization rate is varied to investigate the compound effect of voice erlang load and VDR variations. The nominal blocking and delay performance requirements and the current price of switching and transmission are used.

Fig. 5 shows backbone network switching and transmission costs for the four network technologies under three voice load scenarios and four values of voice digitization rate. The main conclusion is as follows

The ranking of switching technologies remains the same under variations in voice erlang load, with packet switching the least costly and traditional circuit switching the most costly.

The analysis of cost shown in Fig. 5 provides insight into the efficiency of handling voice and data by the alternative network technologies. When the voice load is reduced from 2700 E to 675 E, the costs of packet switching, hybrid switching, and fast circuit switching decreases as one could have anticipated. On the other hand, the cost reduction of the traditional circuit-switching technology is smaller than what might be expected. A detailed examination of the operation of traditional circuit switching and of the voice-data traffic mix, demonstrates that the small cost reduction results from the inefficiency of traditional circuit switching in handling interactive data applications. It is important to distinguish between the voice-data traffic mix in terms of bits per second on the one hand and in terms of erlangs on the other hand. Voice traffic constitutes a high traffic fraction of total traffic in terms of bits per second, particularly at high voice digitization rates. However, when the data traffic is converted to erlangs for transmission via a circuit-switching network, the data load constitutes the majority of the total traffic volume and thus a relatively large reduction in the voice traffic reduces the erlang load only slightly. Traditional circuit-switching inefficiency results from the dedication of end-to-end circuits to interactive users for the duration of a session, thus inefficiently using transmission capacity during the idle periods. Note from Fig. 5 that the total cost of traditional circuit switching reduced by the small amount of at most 6.2 percent when the voice offered load reduced from 2700 E to 675 E, and the voice digitization rate was varied from 2.4 kbps to 64 kbps. For further corroboration, one can examine the costs of the hybrid switching and fast circuit-switching technologies. These network technologies use a circuit switching concept for voice communications and corresponding subnetworks are designed on the basis of the erlang voice load, similar to the design of a traditional circuit switching network; however, data traffic is handled more efficiently.

Quantitative results for the alternative network technologies over the range of voice and data traffic scenarios and voice digitization rates considered (see Fig. 5) are as follows.

1) Hybrid switching for voice and data are more costly than packet switching by 30 percent–64 percent.

2) Fast circuit switching for voice and data is more costly than packet switching by 94 percent–250 percent.

3) Traditional circuit switching is more costly than packet switching by 356 percent–2916 percent.

C. Unit Cost Comparison of Alternative Network Technologies

This section compares the network technologies in terms of backbone network switching and transmission cost per megabit of traffic. The cost per megabit is equivalent to the cost per kilopacket, assuming 1000 bit packets. This cost is often used to examine economy of scale as a function of traffic load [5] or to examine switching and transmission cost trends. The comparison is provided as a function of voice digitization rate.

The unit costs presented in this section are not directly comparable with previous results (e.g., [3], [5]) because of additional factors taken into account in this paper which include mix of voice and data traffic, voice digitization rate, and the different switching technologies. The reader should proceed with caution in attempting to compare the unit costs presented with prior published results, and pay careful attention to the underlying assumptions given below.

The traffic scenario used for the comparison includes: 2700 E voice traffic and 36.15 Mbps data traffic. The total back-

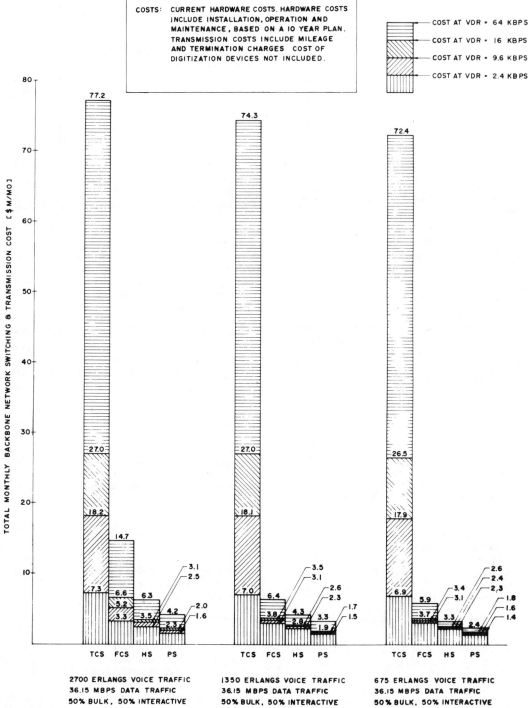

Fig. 5. Monthly backbone switching and transmission cost as a function of voice digitization rate (VDR), traffic scenario and switching technology (TCS—Traditional Circuit Switching; FCS—Fast Circuit Switching; HS—Hybrid Switching; PS—Packet Switching).

bone switching and transmission costs used to derive the units costs are those shown in Fig. 5 as obtained from current prices of switching and transmission. As in [3] it is assumed that the network operates eight hours each working day, 173 h per month. Increases in the operating time decrease proportionally, the unit costs reported below.

Table III gives the unit costs of backbone switching and transmission per megabit of traffic. From the tables over the

TABLE III
UNIT COST OF ALTERNATIVE NETWORK TECHNOLOGIES[a]

Voice Digitization Rate (kbps)	Cost of Backbone Switching and Transmission per Million Bits (in Cents)			
	Network Technology			
	Packet Switching	Hybrid Switching	Fast Circuit Switching	Traditional Circuit Switching
2.4	6.0	9.4	12.4	27.5
9.6	5.2	8.0	13.5	47.1
16	4.7	7.1	13.4	54.6
64	3.2	4.8	11.2	58.8

[a]Traffic scenario: 2700 E voice and 36.15 Mbps data. Current price of switching and transmission.

voice digitization range 2.4 kbps–64 kbps the unit cost ranges per megabit are:

3.2¢– 6.0¢ for packet switching;
4.8¢– 9.4¢ for hybrid switching;
11.2¢–13.5¢ for fast circuit switching;
27.5¢–58.8¢ for traditional circuit switching.

D. Sensitivity of Comparison of Switching Technologies to Switching and Transmission Cost Parameters

The analysis of the switching and transmission cost components, in an optimum design, demonstrates that some switching technologies require substantially more transmission cost than switching cost while others reverse this relationship. Furthermore, the switching or transmission cost components vary as a function of the voice digitization rate. Hence, to obtain confidence in the conclusions resulting from the comparison of switching technologies under nominal (current) switching and transmission cost parameters, it is mandatory to examine whether the conclusions remain valid if the price of switching and transmission varies, since future switching and transmission cost trends will result in cost relationships markedly different from those existing today.

To investigate this problem, integrated voice and data networks are redesigned under all switching technologies assuming different trends in the price of switching and transmission. Apart from current prices, the technologies are investigated under the following two cost scenarios:

1) switching cost—10 percent of current transmission cost—current; and
2) switching cost—current transmission cost—10 percent of current.

Note that these scenarios span two orders of magnitude in the ratio of switching to transmission cost.

The results are shown in Fig. 6 where the total monthly backbone network costs of the four switching technologies are shown under three sets of assumptions (the current costs and the above two scenarios) for VDR values of 2.4 kbps, 16 kbps, and 64 kbps. The costs of backbone networks for a voice digitization rate of 8 kbps are also shown for the current cost scenario. From Fig. 6 one can observe that:

The major conclusion that packet switching for integrated voice and data provides lower cost networks than any of the other technologies remains valid over the entire range of cost scenarios and voice digitization rates considered.

The ranking of the four switching technologies under scenario 1) where the price of switching is 10 percent of the cur-

rent price remains the same as under the current price of switching and transmission. However, the fast circuit-switching technology, which heavily relies on switching, is nearly as cost-effective as the hybrid-switching technology, particularly at low VDR values. This results directly from the reduction in switching cost component and the fact that the complexity of a circuit switch is much lower than that of a hybrid switch.

The ranking of switching technologies only changes under cost scenario 2) where the price of transmission is 10 percent of the current price. Traditional circuit switching costs less than fast circuit switching and hybrid switching for VDR = 2.4 kbps. However the cost difference between traditional circuit switching and hybrid switching is less than 8 percent.

E. Integrated versus Segregated Voice and Data Networks

The objective of this section is to compare integrated versus segregated voice and data communications systems under the four switching technologies as a function of voice digitization rate.

The interest in this comparison is twofold. First, separate voice and data networks using the same switching technology may be developed as part of an evolutionary plan for integrated communication systems. Second, it is of value to examine the cost differences between integrated and segregated systems since the cost differences will indicate the potential savings to be gained by fully integrated systems. Current prices of switching and transmission are used in these investigations. The cost for switching reflects the price of existing computer systems and the cost of transmission is modeled according to current DDS tariff offerings.

Fig. 7 shows the total monthly cost of two separate voice and data networks and an integrated voice and data network for each of the switching technologies as a function of VDR. The cost difference between separate networks and a common network does not demonstrate a consistent trend as a function of VDR. That is, for some switching technologies the cost difference increases as a function of VDR while for others it decreases. The only trend that can be observed from Fig. 7 is that the percentage savings that could be obtained from integrated communications relates to the ranking of switching technologies with the highest percentage savings for packet switching and the lowest for traditional circuit switching.

The major conclusion from these investigations is that:

The backbone network of two separate packet-switching systems (one for voice and one for data) is lower than the cost of an integrated voice and data system under any of the alternative switching technologies, for the entire range of voice digitization rate.

F. Network Economics with Vocoders

The analysis and cost comparison of switching technologies in the preceding sections examined the switching and transmission costs of the backbone network. The location of voice digitization devices, and the cost of the digitization process were not taken into account. This is equivalent to assuming that voice is digitized prior to its processing in the backbone network.

The ranking of network technologies for the backbone is independent of the cost of digitization devices; however a total system cost comparison including the digitization cost will reduce the percentage difference between alternative strategies. The study of alternative locations for voice digitizers is important from both cost and security perspectives. It

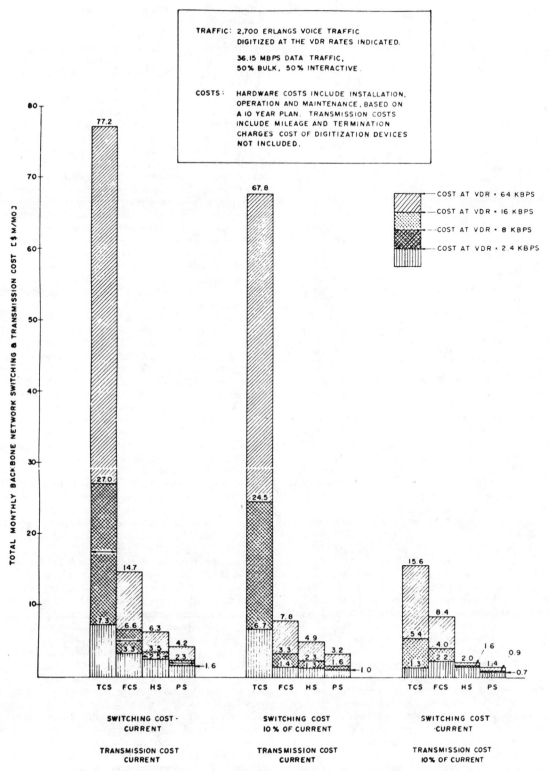

Fig. 6. Monthly backbone switching and transmission cost as a function of voice digitization rate (VDR), component cost, and switching technology (TCS-Traditional Circuit Switching; FCS-Fast Circuit Switching; HS-Hybrid Switching; PS-Packet Switching).

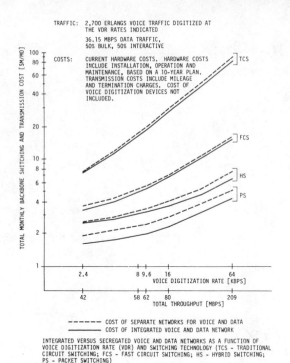

TRAFFIC: 2,700 ERLANGS VOICE TRAFFIC DIGITIZED AT THE VDR RATES INDICATED

36.15 MBPS DATA TRAFFIC,
50% BULK, 50% INTERACTIVE

COSTS: CURRENT HARDWARE COSTS. HARDWARE COSTS INCLUDE INSTALLATION, OPERATION AND MAINTENANCE, BASED ON A 10-YEAR PLAN. TRANSMISSION COSTS INCLUDE MILEAGE AND TERMINATION CHARGES. COST OF VOICE DIGITIZATION DEVICES NOT INCLUDED.

------- COST OF SEPARATE NETWORKS FOR VOICE AND DATA
——— COST OF INTEGRATED VOICE AND DATA NETWORK

INTEGRATED VERSUS SECREGATED VOICE AND DATA NETWORKS AS A FUNCTION OF VOICE DIGITIZATION RATE (VDR) AND SWITCHING TECHNOLOGY (TCS - TRADITIONAL CIRCUIT SWITCHING; FCS - FAST CIRCUIT SWITCHING; HS - HYBRID SWITCHING; PS - PACKET SWITCHING)

Fig. 7. Integrated versus segregated voice and data networks as a function of voice digitization rate (VDR) and switching technology (TCS-Traditional Circuit Switching; FCS-Fast Circuit Switching; HS-Hybrid Switching; PS-Packet Switching).

was demonstrated in previous sections that the switching and transmission cost of the backbone network increases with the voice digitization rate. Hence it is important to investigate the tradeoff between the cost of a low VDR backbone network plus the cost of digitizers to provide the low bit rate, against the cost of a network using high VDR such as PCM (Pulse Code Modulation) at 64 kbps.

This section compares costs of the alternative network technologies including the cost of voice digitizers. The comparison is done under a variety of assumptions of digitization rate, location, and purchase price of voice digitizers. It is noted that differences in costs for terminals where packetization is required, as opposed to those working in a circuit switch environment, while expected to be small, were not evaluated. The following two problems are addressed.

Problem 1: How cost-effective are voice digitization devices (vocoders) in the backbone network? For this problem it is assumed that the voice subscriber operates in an analog environment. The question to be answered is whether it is cost-effective to provide low bit rate voice digitizers in the backbone network to reduce the total system cost (backbone network plus digitizers).

Problem 2: What are the economics of voice digitization devices in handsets? The problem is to determine the total cost, including backbone network cost for a given VDR value and the cost of VDR devices as a function of number of handsets.

Problem 1 is of interest where secure end-to-end digital voice is not a requirement. Hence, the emphasis is strictly on cost-effectiveness. In the DOD environment, for example, a solution whereby the voice digitizers reside only in the backbone network may not be desirable if end-to-end encryption is

sought. In this case a question to be posed is: for a specified total budget, what is the number of handsets which can be equipped with digitizers, of say 2.4 kbps, under the various switching technologies.

To address Problem 2 when voice is digitized at the handset, it is necessary to take into account the savings in the local distribution network which results from the low bit rate voice. Detailed costing of local distribution networks is not examined in this study. Previous studies [5], [6] have shown that the communications cost of local distribution networks (communication channels, multiplexing, concentration) is on the order of 50 percent of the total system cost. Hence the quantitative results under the handset digitization option are derived under the assumption that the communications cost of the local distribution system is equal to the communications cost of the backbone network (switching and transmission) for each of the network technologies compared. This assumption is valid for systems such as AUTOVON which contains a large number of backbone nodes. Projections for the AUTODIN II System which contain eight backbone nodes are also in this range. However, further study should be made of the relationship between local and backbone network costs to fully quantify the tradeoffs.

A digitization strategy whereby some voice requirements are digitized at the handset and others at the backbone network (combination of Problems 1 and 2) is also addressed. This option is attractive when the number of voice subscribers requiring end-to-end encryption is small compared to the total number of subscribers using the network. In such a case, voice digitizers can be provided in the backbone network to reduce total system cost; these digitizers are shared by subscribers with analog local loops. Furthermore, this mixed digitization strategy could be cost-effective as an interim step in the evolution of communications to an all digital system.

1) Voice Digitization in the Backbone Network: The model for providing digitization devices in the backbone network is shown in Fig. 8. A bank of digitization devices is provided at each switching node. As shown, digitization devices at the origination and destination switches are dedicated to the pair engaged in conversation for the duration of the call. No additional digitization devices are used by the pair in tandem (intermediate) switches. The bank of VDR devices is dynamically shared rather than permanently dedicated to subscribers, in order to minimize the number of VDR devices required to achieve the low bit rate backbone network.

The maximum number of digitizers needed in the backbone network for the voice traffic is 15 912. This allows the maximum number of voice connections in the network for the given Erlang traffic. That is, for every point-to-point "circuit" needed for accommodating voice communication, a digitizer is provided at both ends. The above number was assumed for all switching technologies in the results that follow. However, this number can be reduced by providing a smaller number than the maximum (of voice digitizers in backbone nodes), resulting in the possibility of subscriber rejection by the system because of unavailability of digitizers. The rejection event should be made rare, for example, less than 0.2 percent.

Figs. 9 and 10 show the total cost of the backbone network plus the cost of all digitizers needed in the backbone network for all switching technologies, for the 2.4 kbps and 16 kbps cases, respectively, as a function of the purchase price of a digitizer. The cost of the networks with VDR = 64 kbps

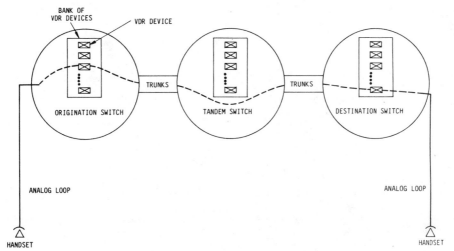

Fig. 8. Model for providing VDR devices in backbone network.

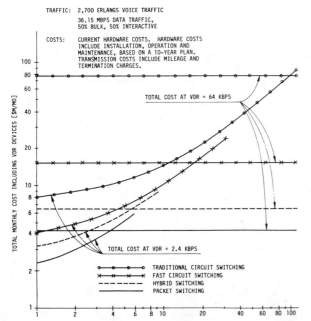

Fig. 9. Cost-effectiveness of 2.4 kbps devices: total costs include cost of switching and transmission of the backbone network and cost of voice digitization devices.

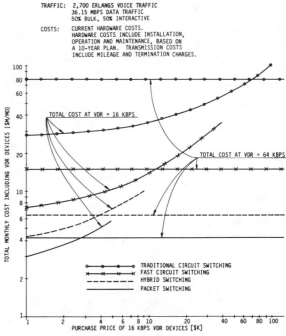

Fig. 10. Cost-effectiveness of 16 kbps VDR devices: total costs include cost of switching and transmission of the backbone network and cost of voice digitization devices.

(PCM) are also shown. The cost of networks with VDR = 64 kbps is nearly constant since these devices are very inexpensive. These so called codecs are available from several manufacturers at a price below $5 per subscriber [7].

It is interesting to observe the price of VDR devices below which the cost of the low bit rate network plus the cost of VDR devices is lower than the cost of the network with VDR = 64 kbps, for the same switching technology. This is the price range within which it is cost-effective to provide low bit rate voice digitizers in the backbone network.

It is apparent that when the switching and transmission costs of a particular technology increase rapidly with VDR, it would

be economical to provide low bit rate voice digitizers in the backbone network at relatively high costs per device. For example, in Fig. 9 one can see that for traditional circuit switching it is cost-effective to provide 2.4 kbps VDR devices in the backbone network (rather than using PCM rate) when the purchase price of such a device is below $98 000. Table IV summarizes the break-even points of the purchase price per voice digitizer below which it is cost-effective to provide these devices in the backbone network.

From Table IV one can conclude that with currently quoted prices for voice digitizers, circuit-switched networks should be

TABLE IV
COSTS BELOW WHICH IT IS ECONOMICAL TO PROVIDE VDR
DEVICES IN THE BACKBONE NETWORK

	2.4 kbps VDR Device	16 kbps VDR Device
Traditional Circuit Switching	$ 98 000	$ 70 000
Fast Circuit Switching	$ 16 500	$ 11 500
Hybrid Switching	$ 5 500	$ 3 900
Packet Switching	$ 3 700	$ 2 800

engineered with low bit rate digitizers in backbone switches. Alternatively, given a digital switching and transmission system for voice at PCM rates, it is cost-effective to incorporate low VDR devices. The break-even point for other switching technologies is also favorable, since the purchase price of low bit rate voice digitizers is expected to be within the indicated price range in the very near future, particularly when purchased in large quantities. Fig. 11 shows the purchase price of a digitization device below which it is cost effective to incorporate such devices in the backbone network for voice digitization rate from 2.4 kbps to 32 kbps.

2) System Cost with Digitizers in Handsets: Fig. 12 shows the total estimated monthly costs for the alternative switching technologies when voice is digitized at 2.4 kbps at subscriber handsets. The total cost shown includes the cost of switching and transmission of the backbone network, the cost of the local distribution system which is assumed to be the same as that of the backbone network, and the total cost of voice digitizers in handsets. The total cost of all network technologies is shown as a function of number of handsets for purchase prices of $2000 and $5000 per 2.4-kbps VDR device.

The total cost differences between the alternative switching technologies decreases when the number of handsets increases. For example, for 50 000 handsets at $2000 per digitizer, the total costs are $7.6M/mo for packet switching, $9.4M/mo for hybrid switching, $11.0M/mo for fast circuit switching, and $19.0M/mo for traditional circuit switching. On the other hand, when the number of handsets is 600 000 or larger, there is a small difference between the alternative switching technologies, because most of the cost would be for purchase of VDR devices in the handsets.

An alternative way to view Fig. 12 is to determine the number of handsets that can be equipped with low VDR devices for a given total cost. For example, for a total cost of $20M/mo and $2000 purchase price per VDR device, the number of handsets ranges from 60 000 for traditional circuit switching to 200 000 for packet switching.

The total system cost for voice digitization varying between 8000 and 16 000 bps is now examined under the option of digitization at the subscriber handsets. Fig. 13 shows the total cost (including the cost of digitization devices at handsets) as a function of number of handsets for the purchase price of $2000 per digitization device. The current price of switching and transmission and the nominal data base were used to obtain these results.

Fig. 13 shows a range of cost for each network technology obtained for a digitization range between 8 kbps (lower bound) to 16 kbps (upper bound). One can observe that for a given number of handsets, the cost differences between

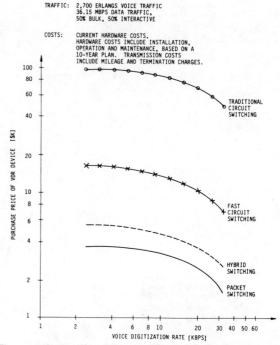

Fig. 11. Cost of a digitization device below which it is cost-effective for incorporation into the backbone networks.

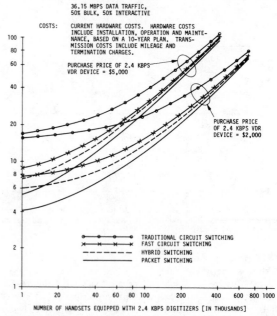

Fig. 12. Comparison of switching technologies with voice digitization at the subscriber handsets: total monthly cost includes backbone network, estimated cost of the local distribution system, and cost of voice digitization devices.

alternative network technologies is larger than in the case of digitization at 2.4 kbps (Fig. 12). Furthermore, the cost differences between digitizing at 8 kbps and digitizing at 16

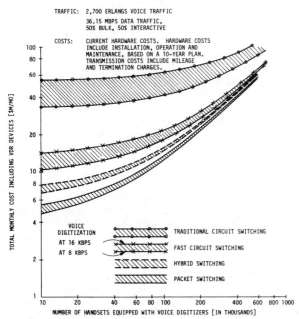

Fig. 13. Comparison of switching technologies with voice digitization between 8 kbps and 16 kbps; total monthly costs include backbone network, estimated cost of voice digitization device in handsets.

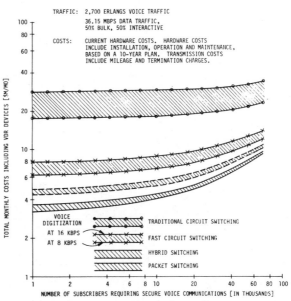

Fig. 14. Comparison of switching technologies under voice digitization at the backbone and in the handsets as a function of number of subscribers requiring secure voice communication. Cost components include: backbone network including cost of digitization devices and cost of digitization devices at the handsets. Purchase price per digitization device assumed $2000.

kbps is smallest under the packet switching technology and largest under the traditional circuit switching technology.

3) Voice Digitization in the Handset and in the Backbone Network: It was demonstrated above that the cost of voice digitization devices becomes a significant component of the total system cost when the number of voice subscribers is large and digitization is done at the handset. For example, at a digitizer price of $2000 and 500 000 subscribers, the cost component for voice digitizers ranges between 74 percent in the traditional circuit-switching technology to 91 percent in the packet-switching technology. End-to-end digital encryption is an important requirement in DOD for subscribers requiring secure voice communications. However, even in such an environment, the fraction of subscribers requiring secure voice communications is expected to be small compared to the total number of voice subscribers.

In previous sections it was demonstrated that it is cost effective to digitize voice at the backbone network at anticipated prices for voice digitization devices. Hence, a digitization strategy whereby voice is digitized at the backbone switches for subscribers not requiring end-to-end digital encryption and digitized at the handset for subscribers requiring secure voice communications appears to be a cost-effective approach for integrating secure and nonsecure voice communications. The results under this strategy are reported below.

Fig. 14 provides the cost comparison of alternative network technologies as a function of number of handsets equipped with voice digitization devices. The values shown in Fig. 14 were obtained using the current price of switching and transmission, the nominal data base of 2700 erlangs and 36.15 Mbps data (50 percent bulk and 50 percent interactive), and a purchase price of $2000 per voice digitization device. The costs are for switching and transmission in the backbone

network, voice digitizers at the backbone switching nodes and voice digitizers at the secire subscriber handsets. The range of handsets with digitizers considered is from 1000 to 70 000. For a given number of handsets, a range of cost is shown for each network technology; this cost range corresponds to the voice digitization bit rate range from 8 kbps (lower value cost) to 16 kbps (higher value cost). The cost of the local distribution system is not taken into account. Under the digitization strategy considered, local distribution can be analog based for all subscribers apart from those requiring end-to-end digital encryption. Hence, the cost of local distribution systems for all network technologies would be equal.

Fig. 14 shows that the cost of the packet switching technology under the mixed digitization strategy is lower than the cost of alternative network technologies. For example, for a voice digitization rate of 8 kbps and 10 000 subscribers requiring secure voice communications, the hybrid switching, fast circuit switching, and traditional circuit switching technologies are more costly than the packet switching technology by 27 percent, 71 percent, and 360 percent, respectively.

A different way of comparing the alternative network technologies is to determine the number of digitization devices that can be supported under a given total budget for the communications system. Fig. 15 shows the number of voice digitization devices that can be supported as a function of the purchase price per VDR device for a voice digitization rate of 8 kbps. Two sets of curves are shown for total monthly budgets of $20 million and $5 million. For example, for a purchase price of $10 000 per VDR device, packet switching can support 41 000 devices and traditional circuit switching can support 9000 devices. When considering the mixed digitization strategy of digitizing in the backbone and at the handsets, it is necessary to provide 15 912 devices in backbone

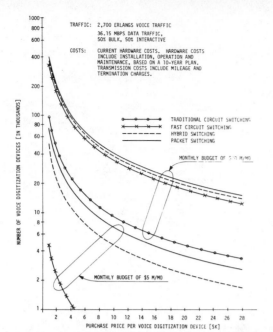

TRAFFIC: 2,700 ERLANGS VOICE TRAFFIC
36.15 MBPS DATA TRAFFIC,
50% BULK, 50% INTERACTIVE

COSTS: CURRENT HARDWARE COSTS. HARDWARE COSTS
INCLUDE INSTALLATION, OPERATION AND
MAINTENANCE, BASED ON A 10-YEAR PLAN.
TRANSMISSION COSTS INCLUDE MILEAGE AND
TERMINATION CHARGES.

TRADITIONAL CIRCUIT SWITCHING
FAST CIRCUIT SWITCHING
HYBRID SWITCHING
PACKET SWITCHING

MONTHLY BUDGET OF $20 M/MO

MONTHLY BUDGET OF $5 M/MO

Fig. 15. Number of voice digitization devices which can be supported by switching technologies as a function of purchase price per digitization device under fixed monthly budget for the backbone network and digitization devices. The voice digitization rate is 8 kbps.

switches. Hence the mixed digitization strategy is not feasible for traditional circuit switching when the purchase price per device is higher than $6000, under a budget of $20 million per month. On the other hand, for other network technologies, the digitization strategy is feasible when the purchase price per device is below $22 000. Assuming a purchase price per digitization device of $4000 and the mixed digitization strategy, the number of secure voice subscribers which can be supported are:

87 000 by packet switching;
81 000 by hybrid switching;
71 000 by fast circuit switching;
 7 000 by traditional circuit switching.

The monthly cost of the traditional circuit-switching backbone network when voice is digitized at 8 kbps is $16 million; hence under a monthly budget of $5 million, no voice digitization device can be supported. The number of devices that can be supported by other network technologies as a function of purchase price per digitization device is shown in Fig. 15. For example, if the purchase price per digitization device is $4000, the number of devices which can be supported is:

 1 100 by fast circuit switching;
11 400 by hybrid switching;
17 700 by packet switching.

Assuming the device cost is invariant with the switching technique used.

Finally it is noted that if voice digitization is performed in backbone switching nodes, the cost of digitization per active voice subscriber is expected to be lower than digitization in the subscriber handset because of the possibility of dynamically sharing processing resources. This possibility was not taken into account in the quantitative results presented.

V. CONCLUSIONS

The major conclusions of this paper are summarized below.

1) On the basis of total backbone network cost (lines and switching) for the specific data bases and requirements that were studied and the cost models assumed, the ranking of switching technologies in increasing cost for integrated voice and data is: packet switching, hybrid (circuit-packet) switching, ideal circuit switching, fast circuit switching, and traditional circuit switching.

2) The ranking of switching technologies remains virtually unchanged under a variety of traffic, cost, and parameter assumptions, with packet switching providing the lowest cost networks for all cases studied. This conclusion is independent of whether voice and data are carried on separate networks or a single integrated network.

3) The backbone network costs of alternatives to the packet-switching technology range from 30 percent to over 1700 percent higher than packet switching. Packet switching remains superior to the other technologies even if switching or transmission costs decrease by a factor of 10.

4) Backbone network cost was found to be insensitive to parameter variations such as: blocking probability (0.01 to 0.1) for which the network is engineered, end-to-end average packet delay (within 200 ms to 600 ms), and priority alternatives. This conclusion holds for each of the alternative network technologies. Additionally, several cases where average packet delay was required to be 50 ms were also examined and costs were found to be 1 percent–3 percent greater than those reported here.

5) Segregated voice and data networks result in only slight cost increases over an integrated voice and data network for all the network technologies considered.

6) Segregated packet systems for voice and data cost less than integrated systems using either the hybrid or circuit-switching technologies.

7) It is cost effective to incorporate low rate digitizers in the backbone network to reduce total cost. The purchase price of a digitizer below which it is cost effective to incorporate such devices in the backbone network differs for the different switching technologies. For traditional and fast circuit switching technologies it is cost effective to incorporate low rate digitizers at presently quoted prices.

VI. DISCUSSION AND FURTHER RESEARCH

The authors have benefited from extensive review and critique of the results presented here in the course of the preparation of this paper and the report [1]. Although none of the results were challenged, the critiques raised important questions relative to the validity of the results under different data bases, model assumptions and design criteria, and in particular, the validity of the results and conclusions when considering extended systems which include local access and satellite channels. For the benefit of the reader, we briefly discuss the major questions that have been raised.

The results of this paper initially appear quite surprising and perhaps nonintuitive. This is the main reason that the study has undergone such detailed scrutiny. Indeed, the results were at first surprising to the authors as well, and an extensive effort of experimentation and sensitivity analyses was necessary to understand and rationalize every single result and conclusion. On the other hand, it is important to point out that, to the authors' knowledge, no study of the magnitude reported has been previously done for a detailed comparison of switching technologies over such a broad range of data bases, cost variations, and network design criteria. Hence, the results do not contradict any previous results. This, however, also presents a difficulty in that it does not enable one to easily verify the results without extensive efforts of modelling, algorithm and program development. The main equations raised by critiques ask about:

1) the impacts on the results and conclusions of variations in switch models and parameters;

2) the impact on the total cost comparison of the switching technologies when the local access cost constitutes a high fraction of the total cost;

3) voice intelligibility when packet switching is extended to local access;

4) the appropriateness of the average delay as a criterion for the design of packet voice networks;

5) the impact on conclusions when silence detection is incorporated into the hybrid switching strategy.

Brief answers follow.

1) The switch models used in the study are sufficiently general and reflect the operation of the switching strategies compared in sufficient detail. Furthermore, the cost of switching nodes, as a function of processing, storage, and channel interfaces reflect current costs of computer systems. Such costs were used in the parameter estimation of the formula. However, since there are no existing switches which realize all the switching strategies compared, it was necessary to assume the number of instructions for processing various functions. These assumptions were based on our knowledge of existing switching nodes and our estimation of the complexity of the particular functions. Nevertheless, one can expect variations in the processing values assumed depending on computer technology, switch architecture, and specific implementation of various functions. We performed some sensitivity studies as a function of several parameters (e.g., assuming 200 instructions per node for a packet speech using the fixed path protocol instead of 100 instructions) with virtually no changes in the results. We believe that changes in these parameters may slightly impact the quantitative comparison of the switching strategies compared but will not change any of the results which rank these technologies or any of the conclusions. The latter is supported by the fact that the major results and conclusions remained unchanged over two orders of magnitude of variations of the ratio of switching to transmission cost. Indeed, one may observe that in many cases, if the cost of switches of the poorer switching alternatives is ignored, the ranking of technologies still remains unchanged.

2) Local access cost can indeed be much higher than that of the backbone network cost in some situations (e.g., small number of backbone nodes). (In fact, if the number of backbone nodes is small, the system may be hierarchical where lower level switching nodes may exist). However, the cost differences observed for the backbone network are valid, independent of the cost of the local access systems. Obviously, full life cycle costs including development should be eventually considered. The ranking of switching technologies in the local access system is expected to be the same as in the backbone system. If the ranking remains the same and the cost of the local access is higher than the cost of the backbone system, the absolute cost differences between the alternative switching technologies in the local access system may be higher than in the backbone system.

3) End-to-end delay for packetized voice, including the local distribution system. This problem needs further investigation. There are several ways to insure intelligible packet voice communication such as:

a) smoothing at the destination for interpacket arrival variances;

b) reducing end-to-end delay constraint in the backbone network to allow some delay in the local distribution system.

An appreciable part of the delay is packetization delay, and this will occur only once, either at the source terminal or nearest concentrator or switch. Of course, it is not mandatory to packetize in the local distribution system. In this case, packetization and possibly digitization starts in the backbone network, and the results should be entirely valid as they stand. Further, a preliminary design with a significantly lower average delay (50 ms) was investigated. This led to an increase in cost for the packet-switching case examined of between 1 percent and 3.2 percent over a range of VDR's, protocols, header sizes, etc. Thus, even with a significantly lower delay constraint, the conclusions (for the cases tested) are still valid.

4) The question relative to the appropriate design criterion for packet speech communication was discussed in Section IV-A3. In general, a 95th or 99th percentile of delay could be a more appropriate criterion than the average delay. However, using such a criterion directly is an intractable network design problem, in particular in the presence of other traffic categories. Hence, the design could be made to satisfy average delay and the results must be verified via simulation and analytical techniques relating the average delay and the delay distribution. Investigations of packet voice delay distributions at destination nodes and processing requirements for smoothing the packet speech stream to the listener are reported in [5].

5) The costs of hybrid switching, while greater than packet switching by 30–64 percent, could be reduced by incorporating speech silence detection methods. Hybrid switch costs, which under our model, are greater than packet switch costs, would then further increase, but line costs should decrease resulting in reduced hybrid switching costs. However, if the appropriate silence detection methods were used, the difference between hybrid and packet switching would probably become a matter of *semantics* rather than technology. That is, the operation of hybrid switching will be quite similar to that of packet switching and the cost differences would depend upon the realization of the two schemes. Additional criteria, apart from cost, will have to be utilized in the comparison, and the modelling of the schemes will have to be more detailed.

The objective of this study was to identify and quantify network technologies demonstrating long-term (1980 and beyond) low operating costs. Since line costs were examined on the basis of tariffs and not costs to a common carrier, it should not be assumed that the conclusions automatically translate to the common carriers environment. Our conclusions relate to the large user who leases tariffed lines and leases or purchases hardware. Furthermore, if the potential cost savings of the packet or hybrid-switching technologies are to be realized, a detailed examination of the transition issues to be encountered in evolving from current circuit-switched voice networks must be performed. The examination of this issue is of importance because the compatibility of existing communications technologies is not a solved problem. To the extent that low rate voice digitization networks are required to interface with higher rate systems (either domestically or abroad) higher near-term costs than the costs projected in the report may be encountered.

While conducting this study, new problem areas were identified. These problems are recommended for further study with the objectives of uncovering the risks in the conclusions and quantitative results, as well as broadening the study into local and regional distribution and more detailed protocol formulations. Among the areas recommended for further investigation are as follows.

1) Further investigations and comparison of hybrid-switching and packet-switching technologies under more detailed protocol scenarios (Although the ranking of these two strategies was consistent throughout the study, the quantitative differences were not extremely large. Furthermore, the hybrid-switching technology may provide a natural evolutionary path for evolving from circuit networks towards a total packet-switching technology.).

2) Investigation and comparison of local distribution strategies for hybrid and packet-switching networks.

3) Investigation of the most appropriate partition between local distribution and backbone networks for hybrid and packet switching.

4) Investigation of postulated evolution strategies from existing systems to integrated voice and data communication systems.

5) Study of alternative concepts for network and message security in an integrated voice and data network.

6) Study of the survivability and reliability of integrated voice and data systems under the packet and hybrid-switching strategies.

It is apparent that the above problem areas are natural and necessary further investigations if additional quantitative information is sought in support of the conclusions of this paper. It is equally apparent that the above problems are extremely difficult.

ACKNOWLEDGMENT

A large number of coworkers at NAC have made contributions to the research summarized in this paper. The authors would particularly like to acknowledge B. Occhiogrosso, W. Hsieh, and K. Schneider who have made substantial contributions.

REFERENCES

[1] Network Analysis Corporation, "Economic analysis of integrated DOD voice and data networks," Final Report, Sept. 1978.
[2] "The evolving defense communications system," Signal, pp. 78-88, Aug. 1977.
[3] L. G. Roberts, "Data by the packet," IEEE Spectrum, pp. 46-51, Feb. 1974.
[4] H. Falk, "Communications," IEEE Spectrum, pp. 42-45, Apr. 1975.
[5] Network Analysis Corporation, "Integrated DOD voice and data networks and ground packet radio technology," Eighth Semiannual Tech. Rep., Feb. 1977.
[6] Network Analysis Corporation, "Local, regional, and large scale integrated networks," Sixth Semiannual Tech. Rep., Feb. 1976.
[7] H. Falk, "Chipping in to digital telephones," IEEE Spectrum, pp. 42-46, Feb. 1977.
[8] I. Gitman, H. Frank, B. Occhiogrosso, and W. Hsieh, "Issues in integrated network design," in Proc. Int. Conf. Communications, pp. 38.1-36-38.1-43, June 1977.
[9] B. Occhiogrosso, I. Gitman, W. Hsieh, and H. Frank, "Performance analysis of integrated switching communications systems," in Proc. Nat. Telecommunications Conf., 1977.
[10] G. Coviello and P. Vena, "Integration of circuit/packet switching by a SENET (Slotted Envelope Network) concept," in Proc. Nat. Telecommunications Conf., New Orleans, LA, pp. 42-12-42-17, Dec. 1975.
[11] M. J. Fischer and T. C. Harris, "A model for evaluating the performance of an integrated circuit and packet switched multiplex structure," IEEE Trans. Commun., pp. 195-202, Feb. 1976.
[12] GTE Sylvania, Inc., "ESD. SENET-DAX interim report," prepared for the Defense Communications Agency, Dec. 10, 1975.
[13] E. Lyghounis et al., "Speech interpolation in digital transmission systems," IEEE Trans. Commun., pp. 1179-1189, Sept. 1974.
[14] J. A. Blackman, "Integrated switching of voice and non-voice traffic," in Proc. ICC, 1976.
[15] A. G. Nemeth, "Behavior of a link in a PVC network," Lincoln Labs, Technical Note 1976-45, Dec. 7, 1976.
[16] "Network speech system inplications of packetized speech," Lincoln Labs, Annual Report, ESD-TR-77-178, Sept. 30, 1976.
[17] K. Bullington and J. M. Frazer, "Engineering aspects of TASI," Bell Syst. Tech. J., Mar. 1959.
[18] E. Fuchs and P. E. Jackson, "Estimates of distributions of random variables for certain computer communications traffic models," Commun ACM, vol. 13, no. 12, pp. 752-757, Dec. 1970.
[19] F. J. McAulay, "A robust silence detector for increasing network channel capacity," in Proc. Int. Conf. Commun., pp. 38.4-54-38.4-56, June 1977.
[20] Network Analysis Corporation, "Alternative network strategies for defense ADP communications," NAC Report No. TR.048.01, July 1975.
[21] J. A. Sciulli and S. J. Campanella, "A speech predictive encoding communication system for multichannel telephony," IEEE Trans. Commun., vol. COM-21, no. 7, pp. 827-835, July 1973.
[22] S. J. Campanella and H. Suyderhoud, "Performance of digital speech interpolation systems for commercial telecommunications," in Proc. NTC '75.
[23] J. Miyahara et al., "A comparative evaluation of switching methods in computer communication networks," in Proc. Int. Conf. Commun., San Francisco, CA, pp. 6-6-6-10, 1975.
[24] R. D. Rosner, "Packet switching and circuit switching: A comparison," in Proc. NTC, 1975.
[25] B. Ochiogrosso, "Digitized voice comes of age: Part I—Tradeoffs, Part II—Techniques," Data Communications, Mar. 1978 (Part I), April 1978 (Part II).
[26] W. Hsieh, I. Gitman, and B. Occhiogrosso, "Design of hybrid switched networks for voice and data," in Proc. Int. Conf. Communications, Toronto, Canada 1978, pp. 20.1.1-20.1.9.
[27] G. J. Coviello, D. L. Lake, and G. R. Redinbo, "System design implications of packetized voice," in Proc. Int. Conf. Communications, Chicago, IL, June 1977.

ISDN

IRWIN DORROS

A challenge and opportunity for the 80's.

WHAT'S AN ISDN?

THE words "Integrated Services Digital Network" or the letters "ISDN" seem to bring on a feeling of well-being among planners in the telecommunications industry. The idea of an ISDN presents an exciting vision.

But what is an ISDN? How do we translate this vision into reality? Its major characteristic is "a public end-to-end digital telecommunications network providing for a wide range of user applications."

We all agree it is evolving; and although some of us would claim that it is already partially here, others believe it will not exist as a ubiquitous network in our lifetime. For most of us, it represents a broad concept that embraces everything good and virtuous about the technology, the architecture, and the service capabilities of the future telecommunications network. Since the ISDN is a vision with a halo, and since it is not yet encumbered by physical existence, it is still a perfect dream with no flaws. In fact, it is still young and amorphous enough to have the potential to correct most of the flaws of today's telecommunication networks of the world.

The challenge of network planners is to fulfill that dream as completely as possible. As usual, when vision and reality meet, there is a crucial period of challenge when basic decisions must be made. Now is such a time: a time when standards are being set, large amounts of capital are being committed, and new industry structures are being mandated not just in Washington, but in other capitals around the world. Let us now examine some ISDN concepts and accomplishments toward meeting the challenges which these concepts present.

The main motivations for the ISDN are the *economies* and *flexibilities* which the integrated nature of the network would foster. The economies come about since many new and emerging services are digital in nature and can be combined with existing services to utilize an integrated transport capability at a significantly lower overall cost than it would take for each service to utilize a separate transport capability. Since we cannot yet envision all the services which the ISDN would carry, we are fortunate that digital transport is, for the most part, insensitive to the services carried, although the traffic characteristics of the services could be quite different.

ISDN ARCHITECTURE

The ISDN concept will enable networks to be planned and constructed before the specific services of the future become defined and shaken out in the marketplace. Using this flexibility, the telecommunications industry can position itself through an appropriate network architecture to handle a wide variety of as yet undefined services. Many of these services will arise from a wide range of information suppliers, other "head end" interactive information points, and numerous special purpose networks which can be accessed through an ISDN. These, and other factors, lead to the general local architecture shown in Fig. 1.

The key ISDN concepts are that the customer will be supplied with a telecommunications transport capacity measured in maximum bit rate at a standardized interface. This bit stream capacity will be provided by the network to a customer's premise in what might be called a "digital pipe." The customer will aggregate the variable bit rate capacity needs of his or her terminals at a control device that interfaces with this network pipe. Various pipe sizes will be available much like service capacity in amperage from power companies today. Packet or circuit switching access will be integrated and be provided on the same pipe. In fact, as technology progresses, the customer will not even need to make a choice in advance between packet and circuit switching; the techniques will merge.

Both the number of multiplexed signals and the bit rates presented by a customer can be variable as long as the total signal rate at the interface does not exceed the maximum bit rate, or size of the digital pipe. There will also be control, or signaling information multiplexed in the digital access pipe to instruct the serving center what to do with portions of the bit stream, that is, how to sort the bit stream out in time and space to reach various services or networks. Within the access network, at either a remote terminal or a serving center, several customers' signals will be multiplexed together. Up to the maximum bit rate available to a customer, no fixed bandwidth is assigned to a customer. A "variable bandwidth assignment" is made each time a customer accesses the network, depending on customer and network needs.

ISDN EVOLUTION

Different perspectives may arise around such issues as: 1) architectural details, 2) interface specifications, 3) how to go about making this happen, 4) what should motivate the

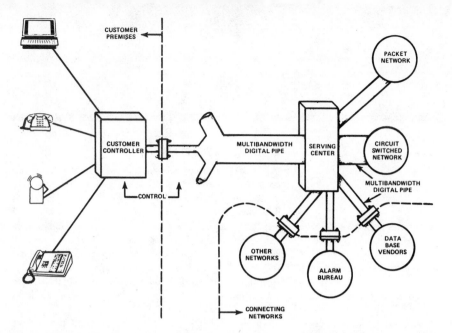

Fig. 1. ISDN Local access.

evolution, 5) how rapidly it will evolve, and 6) what will be the prime services driving the evolution.

One thing is clear. For an integrated services network, voice needs will be dominant for a long time to come and will be the factor which sets the pace of a truly integrated evolution. That is, despite the excitement of new services tending to be driven by nonvoice, voice traffic will be the dominant form of communications for many more years. Thus, the bit stream in a typical digital pipe in the year 2000 will carry a mixture of voice, data, facsimile, and, perhaps, video signals. But, except possibly for video, coded voice is likely to dominate. Packet techniques or low bit rate coding may be used for voice. The architecture must support these possibilities.

Let us now take stock of where we are in achieving the ISDN vision and then look at where we are going and the challenges before us. In the United States, we are approaching the 1000 time division central office switch mark. Most of these are in small local offices, but about 75 are time division toll systems such as No. 4 ESS or No. 3 EAX. In addition, there are thousands of digital PABX's of various sizes which connect into our national communications network. The Bell System will deploy its first regular applications of digital local offices in 1981 using both the DMS-10 and No. 5 ESS.

In transmission, there are nearly 100 million circuit miles of T-carrier and 5 million circuit miles of digital radio in the United States. There are approximately a quarter of a million loops employing digital subscriber carrier. Lightwave is a fast emerging digital transmission technology. There are many customized systems in service throughout the United States. A standard exchange area system entered service this year in Atlanta.

Satellites are handling digits, by both FDMA and TDMA multiplexing of transponders. Some means to carry digital bit streams on FDM systems, such as DUV, are in place and others are being devised to carry digits on radio and on coaxial cable systems.

The Bell System expects to have a 1000 km lightwave system in service in the Boston-Washington corridor by 1984. In addition, plans are being made for a 140 Mbit/s digital transmission system on the L4 coaxial cable route between Oakland and Chicago for service in 1983.

So far, in the United States, the digital data services offered are TWX, Telex, Telenet, Tymnet, Dataphone Digital Service, and various private line satellite-based offerings. SBS is expected to begin this year with a large user private network offering. A Bell System switched digital capability allowing for alternate voice and 56 kbit/s digital data over the public switched network can be available in the early 1980's. This capability takes advantage of the ability of space division ESS's to switch bit streams of 56 kbit/s side-by-side with analog voice signals.

An important capability in an evolving ISDN is common channel signaling. Approximately 25 percent of U.S. intertoll trunks are now utilizing the world's largest packet switched network, the Common Channel Interoffice Signaling (CCIS) network. Extensions to local switches are planned to begin on a limited basis in 1981 and building up thereafter. And a transition to the next vintage, CCITT #7, is being planned and developed to match the expanding signaling needs of an ISDN.

Thus, in the United States, we have been and will continue to introduce the transmission, signaling, and switching piece parts of the ISDN. Although, this introduction must always be tempered by the realities of the economic, operational, and service opportunities involved.

The TransCanada Telephone System is developing a fundamental plan for an ISDN through 1990. A complete family of Canadian switching system designs featuring the DMS switch line supports such a plan. By the mid-80's, about

2 million Canadian lines are expected to be served from digital central offices. Lightwave is already being used on a transmission link into Calgary and by 1984 the Saskatchewan Telecommunications Company plans to install a 3200 km lightwave system carrying both video and voice. In digital services, Canada already has an extensive packet switched network called DATAPAC with interconnections to other countries.

There can be little doubt about the commitment of France to the digitization of its network. The French PT&T was the first to commit its future to digital switching and transmission. They expect to be ordering only digital switches after 1983. France now has E10 digital switches with 1.5 million lines being served. A new generation of switches, the E10B and E12 from CIT-Alcatel and the MT20 and MT25 from Thompson-CSF-Telephones is being introduced. Lightwave transmission is already being used in Paris. France's TRANSPAC data network is in operation and is interconnected into Euronet. Plans are under way for Videotex and Teletex services with a trial of Videotex slated for Velizy. An electronic telephone directory trial is proposed for Ille et Villaine involving approximately 250 000 terminals.

In Great Britain, the Post Office is planning an ISDN, centered around System X and a nationwide lightwave network. West Germany, Japan, Italy, and Sweden are all proceeding towards digital switching and transmission. National data networks, which are interconnected into Euronet, now exist in the major European countries. Videotex and Teletex are also emerging in Europe, Japan, Canada, and the United States.

This quick and incomplete summary of the status of the ISDN evolution around the world gives the flavor of where we are today. The accomplishments which have been mentioned involved the production and deployment of some piece-parts of the ISDN. Each of these has been deployed largely because it was the economic choice for a specific application, such as digital local switch, a digital carrier system, or even a digital service network. The combined capital and operational savings available through digital switching and transmission would continue to expand the deployment of digital piece parts even without an ISDN.

ECONOMIC AND SERVICE SYNERGIES

Utilizing today's base of ISDN piece-parts, we must look for ways to 1) accelerate the prove-in of the lagging piece-parts, namely, long-haul transmission and local loops, and 2) take best advantage of the economic and service synergies of the ISDN.

An illustration of the economic synergy is the prove-in distance for T-carrier over wire pairs. This went from greater than 10 mi when T-carrier was first used in metropolitan areas between analog switches to zero miles when working into a No. 4 ESS because of termination savings. Similar results apply for local digital switches where termination savings encourage the use of digital trunks and digital carrier on subscriber loops. Planners can no longer consider the economic merits of switches or carrier systems alone. Rather, area planning is required to examine trunk, loop, and host and

remote switching systems for an area simultaneously. In addition, since the cost considerations of operating a wire center are comparable to the inital capital costs, life cycle studies of all costs must be the decision vehicle.

Service synergies will become evident as end-to-end digital connectivity grows. Until bits can be transported intact from end-to-end, digital services will not be possible, except by expensive conversions to analog with data sets.

FUNCTIONAL REALIZATION

It is not practical to try to predict all future services. Based on the services contemplated, however, the ISDN functional requirements appear to include qualitatively:

* variety of data speeds or bit rates,
* wide range of holding times,
* wide range of calling rates,
* economic transport of bursty as well as continuous data,
* customer ability to control costs and services,
* sufficiently fast call setup and tear-down time,
* sufficiently low impairment and error rate,
* sufficiently low data transfer delay times,
* various levels of secure transmission, and
* variety of service grades.

Obviously, it is an enormous challenge to plan, design, and evolve to a network meeting such wide-ranging objectives. It is certainly a far cry from the old days of analog voice. Let us test the ISDN vision against this challenge and speculate on what it takes to meet these qualitative functional requirements.

As a basic principle, customer/network interfaces must be simple and few. Once defined, they must remain unchanged for many years. A large variety of interfaces would mean higher costs in manufacturing, provisioning, testing, maintenance, and inventory. It would also result in the loss of flexibility, requiring an interface change when the customer changes from one type of service to another.

In the general ISDN architecture shown in Fig. 1, customer interface equipment (the controller) converts signals into a standard interface protocol. With out-of-band signaling, the network is notified of the service choice and types of the customer signals. At a point internal to the network, the network converts the protocol to achieve transport efficiency and then converts back to the original protocol before reaching the customer/network interface at the other end. For example, given the 64 kbit/s information capacity at the standard interface, a 2.4 kbit/s signal is padded up to the higher bit rate by the customer before entering the network. Redundant bits are stripped off at an appropriate place within the network, such as a loop carrier remote terminal, and the signal is then multiplexed together with other signals to transmit over longer distances over a single channel, such as a 1.5 or a 2 Mbit/s line.

Several variable bit rate customer signals can be packed by the customer into a single bit stream before entering the network. The packing by the customer, while transparent to customer/network interfaces, influences the selection of the interfaces. Within a reasonable cost range, the highest bit rate interface should be selected in order to accommodate a maximum number of services. These are the digital pipe and variable bandwidth concepts discussed earlier.

In the Bell System, a few standard interface alternatives are being studied: a 1.544 Mbit/s interface for large business customers; an undetermined bit rate larger than 1.544 Mbits/s for broadband video; in the neighborhood of 144 kbits/s for small business or residence customers requiring simultaneous voice and high speed data transmission; and 72 or 80 kbits/s for small business and residence customers requiring nonsimultaneous voice and high speed data.

Now let us focus on some of the techniques which may be used in transmission, switching, and signaling to meet the network design challenge. Loop carrier systems, using glass fiber or copper as the transmission medium, may be implemented widely for local distribution. The rapid progress of lightwave technologies may one day make it economical to provide such a medium all the way to a customer's premises. Until then, the remote terminals of these broad-band carrier systems can be located up to 9 kft from a customer's premises and still provide satisfactory voice and data transmission. This distance is selected to allow high-speed digital transmission over the path connecting the remote carrier terminal with the customer premises. At the remote carrier terminal, at the serving center, and at tandem switches, signals from several customers are multiplexed together for more distant transmission. No fixed bandwidth is assigned to customers. Dynamic bandwidth assignment is used to enhance the sharing of loop and trunk facilities.

The integration of a variety of services with divergent traffic characteristics and performance requirements presents a significant challenge to switching. Initially, the network will respond to this challenge with separate packet and circuit switching. Both types of switching are used, since for certain types of traffic, packet switching increases facility fill by concentrating several channels of packetized information into one channel. Eventually, circuit and packet switching may be unified to allow flexible routing and control at each switching node based on the needs of individual calls. At that point, customers need no longer specify the switching technique. The network internally would decide routing at each intermediate node which meets the desired performance requirement and maximizes the utilization of the network facilities.

The concepts of digital pipe and dynamic bandwidth assignment will be implemented in interoffice transmission as well. The concept of a 64 kbit/s trunk will become history. Provisioning will be done on a 1.5 or 2 Mbit/s primary multiplex rate or on an even higher capacity basis. Within the digital pipe, the bandwidth assignment will be done dynamically.

Many future services require sophisticated network signaling control. Some of these require interactions between the network and the terminal, and between network elements and centralized network routing and control databases. New protocols need to be developed to handle the control signal which will grow beyond the simple datagram. The protocols will provide: 1) different control messages to different centralized databases and data terminal equipment, 2) voice announcements and tones for digital telephones, and 3) control information for other terminals such as host computers and CRT terminals. Network control information for call setup, disconnect, and status will be transported over the common channel signaling network.

A major challenge now is to select the appropriate set of protocols and interface standards to support an optimum mix of service needs of the future. This will be of major concern to Study Groups VII, XI, XVI, XVII, and XVIII of the CCITT.

Routing and control databases will be implemented in the network to allow customer control of call establishment such as call forwarding and incoming call screening. A personal number will be available which allows a customer to have calls follow his/her location at a given time, if the customer is faithful in keeping the network informed.

SUMMARY

The Integrated Services Digital Network is truly a vision and a challenge for the 80's and beyond. It will be many years before we get there. Although the ISDN must evolve from existing systems in an orderly manner, in order to achieve full integration, a network architecture fundamentally different from today's is required. This new network architecture is characterized by:

- digital pipe and variable bandwidth assignment in transmission,
- common channel signaling and flexible routing and control at switching nodes,
- centralized intelligence such as routing and control databases, and
- new switch architectures.

The challenge is to plan the details, design the right systems, and continue to prove-in the parts in ongoing deployment decisions. If this is done well, we can reap the benefits of the economic and service synergies on a wide scale by 1990 and everywhere by the year 2000. This is the challenge of ISDN.

Stanley W. Johnston and Barry Litofsky,
Bell Telephone Laboratories

END-TO-END 56 KBPS SWITCHED DIGITAL CONNECTIONS IN THE STORED PROGRAM CONTROLLED NETWORK

The Stored Program Controlled (SPC) telecommunications network is built on the foundation of digital technology for control and signaling. Digital switching and transmission technology is now in widespread use in the SPC network to provide high quality, cost effective telephony. As the SPC network evolves into the future Integrated Services Digital Network (ISDN), this technology will be used to provide a full spectrum of digital telecommunication service capabilities.

A fundamental ingredient of ISDN is the capability for making end-to-end digital connections through the SPC network. In the Bell System, progress in electronic switching and transmission will make possible the first of these all-digital connections in the early 1980s.

Allowing establishment of such connections just like ordinary telephone calls, this Switched Digital Capability (SDC) has the potential to make switched, high-speed digital communications almost as ubiquitous and easy to use as today's voice telephony. SDC resembles, in many respects, DATAPHONE® service on the public telephone network, in that it can provide common user circuit-switched connections which can be used alternately for voice or data. But it also can provide some of the desirable features of DATAPHONE Digital Service (DDS) leased lines, such as a 56 Kbps data rate, digital data transfer, and full-duplex operation.

© 1982 *Business Communications Review*. Reprinted, with permission from Volume 12, No. 1, pp. 17–23. January 1982.

SDC connections could potentially support a wide variety of applications, such as bulk data transfer, high-speed terminal access, computer graphics, high-speed facsimile, teleconferencing, electronic mail, encrypted voice, high fidelity audio, and combined voice/data services. They could also be used as dialed back-up for leased lines, or as supplements to leased lines in peak load periods.

Switched Digital Capability builds on the large base of compatible digital equipment in the SPC network. It takes maximum advantage of existing switching and transmission systems. The required modifications to existing equipment are mostly in software and in plug-in circuits. Only a limited portion of the network need be equipped for digital transmission, since SPC switches can selectively route SDC calls.

Such an approach would minimize the development effort needed, and will permit rapid deployment of the capability. A relatively simple initial capability can be supported, but additional features and subsequent improvements are possibilities.

Network Architecture

A plan for end-to-end digital connectivity must account for all elements of the telecommunications network. Subscriber loops, local switching, exchange transmission, toll switching, and long-haul transmission must all provide digital transparency. Appropriate technology must be implemented in each area in order to insure the integrity of a 56 Kbps bit stream from end-to-end. A plan for initial SDC has been developed which uses No. 1A ESS and No. 4 ESS as switching offices and interconnects them with compatible digital transmission systems (see Figure 1). Experiments at Bell Laboratories have shown that No. 1A ESS has the capability of switching and transmitting high speed digital bit streams, giving it a central role in future digital network capabilities.

Loops

Subscriber access today is almost entirely over metallic loops. Several transmission systems have been reported which could transmit 56 Kbps or higher data rates full-duplex over most unloaded two-wire loops. The most promising digital loop transmission systems are generally of two basic types: "hybrid balance" (including "echo canceler") systems, and "time compression multiplex" (TCM) (or "burst-mode") systems.* For its initial Switched Digital Capability, the Bell System contemplates using a TCM loop transmission system. This system provides relatively good immunity to bridged taps in the loop plant, and is expected to have sufficient range to cover a majority of applications.

For longer loops beyond the range of the TCM system, digital access can be provided through the SLC-96 digital Subscriber Loop Carrier system. SLC-96 is just starting to be used in large quantities in the Bell System, especially in areas with longer than average loops. With new plug-in circuits, it can provide SDC connectivity for selected lines.

Subscriber access for customers remote from an SDC-equipped switching office is discussed below.

Local Switching

Among Bell System local switching offices, the No. 1A ESS will soon serve more lines than any other type of equipment. This very high capacity SPC office is rapidly becoming the dominant line switch in most metropolitan areas, and it is an ideal system for providing initial digital connectivity in local switching.

The No. 1A ESS has a space-division metallic network, which is being used today to switch analog signals. However, its network is capable of carrying 56 Kbps baseband digital signals. When used in this mode, the No. 1A ESS becomes, in effect, a space-division digital switch. This concept is illustrated in Figure 2.

The same technologies available for digital transmission on two-wire metallic loops can be used on the two-wire metallic network of the No. 1A ESS. For an initial SDC a hybrid balance system can be used. It provides relatively good immunity to impulse noise, which can be a problem in the switch environment. Since the transmission method through the switch is different than on the loops, an interface circuit is needed between them. Implementing a single digital transmission method across the switch and on the loops, a future possibility, would eliminate the need for this interface.

---

* B. S. Bosik, "The Case in Favor of Burst-Mode Transmission for Digital Subscriber Loops," Proceedings of the International Symposium on Subscriber Loops and Services, September 15–19, 1980.

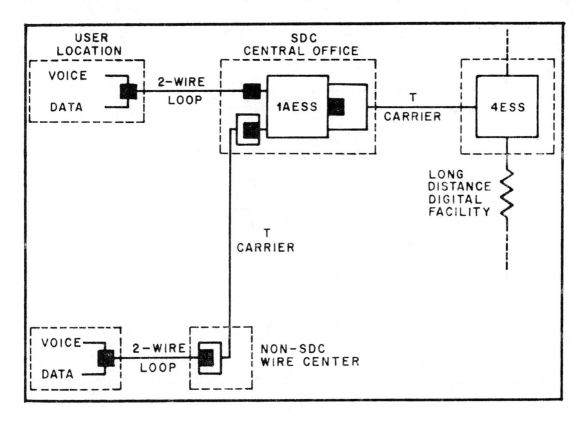

Figure 1. Switched digital capability.

SDC trunks on No. 1A ESS would terminate on the Digital Carrier Trunk (DCT) frame. This currently deployed frame provides integrated channel bank and trunk circuit functions. New plug-in circuits in the DCT can permit selected trunks, used only by SDC, to provide digital connections through the No. 1A ESS.

No. 1A ESSs providing SDC require additional software to handle these special calls, maintenance circuits to test the new capabilities, and a synchronization clock, ultimately tied to the Bell System Reference Frequency, to insure that all equipment used on a digital connection is properly synchronized.

Exchange Transmission

The exchange transmission plant in the Bell System already contains a large portion of digital systems. T-carrier systems are in wide-spread use, especially in metropolitan and suburban areas, and these can be used directly for SDC trunks. Although these trunks would be logically segregated into separate groups, a physical facility (digroup) could contain any combination of SDC trunks, ordinary telephone trunks, and switched special service circuits. Having separate trunk groups for SDC would enable each switch to provide special treatment and routing to incoming calls without the need for traveling class marks.

The same kinds of digital transmission facilities can also be used as remote access lines. A line in one wire center area can thus be given access to SDC through a No. 1A ESS in another wire center. New plug-in circuits for D4 channel banks can provide appropriate interfaces on both ends. In this way, a single local switch equipped for SDC can provide access for selected lines spread over a wide geographical area. This is shown in the lower part of Figure 1.

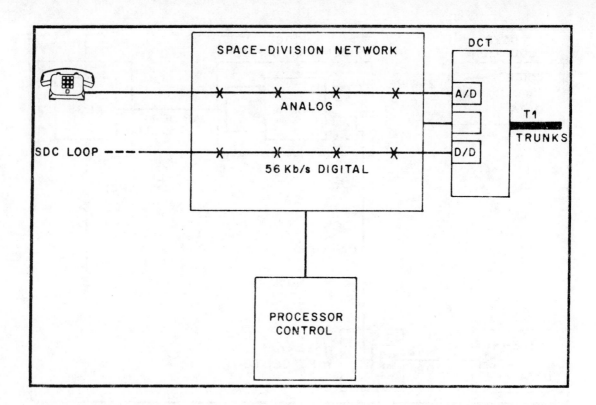

Figure 2. Digital switching in a No. 1A ESS.

Toll Switching

The No. 4 ESS is a very high capacity time-division digital SPC toll office, which is rapidly becoming the backbone of the Bell System long distance network. Since it switches standard 64 Kbps voice channels, it can be readily adapted for end-to-end digital connections. Although No. 4 ESS would be the primary toll switch for an initial SDC, No. 1A ESS can also switch SDC trunk-to-trunk calls using the specially equipped DCT trunks described earlier for its local switching function.

Long-haul Transmission

Although most long-haul voice circuits use analog facilities, a substantial amount of long-distance digital capacity has been built for DDS, among other applications. As demand grows, a wide range of intercity digital systems can be installed as needed. These include Data Under Voice (DUV) and similar systems, digital microwave radio systems, fiber optic systems, digital satellite systems, and digital systems which displace part of an analog carrier system on radio or coaxial cable. By selectivity

disabling or avoiding equipment such as echo cancelers or suppressors which could mutilate the bits, these digital transmission systems will be compatible with a transmission plan for SDC.

Network Functions

SDC network connections can be operated in either a voice or a data mode (see Figure 3). In the voice mode, the connection acts like an ordinary telephone circuit. Voice-band analog transmission would be provided on the metallic loops and across the No. 1A ESS. At that point the voice would be digitized and carried on digital trunks to and through No. 4 ESS. Where appropriate, voice-quality treatment on loops and trunks can be supplied.

In the data mode, on the other hand, an end-to-end 56 Kbps digital channel can be provided. Digital transmission can be provided on the metallic loops and across the No. 1A ESS. Analog-to-digital conversion and voice-quality treatment, if any, would be disabled. The signal format, being digital end-to-end, would eliminate the need for modems.

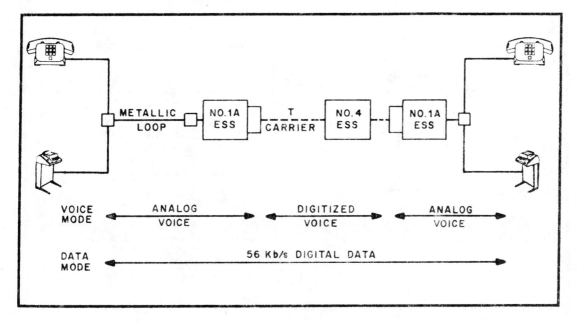

Figure 3. Voice and data modes.

With initial SDC, each call would be set up in the voice mode. This would greatly ease initial implementation, since conventional telephone technology could be used directly. A line would obtain dial-tone in the ordinary manner, and use Touch-Tone dialing. On trunks where CCIS is not yet available, multi-frequency signaling can be used. Existing call progress indications, such as busy-tone, audible ringing or recorded announcements can be used unchanged. Standard ringing would be provided.

Once the called party answers, the connection can be alternated between the voice and the data modes at will. In many applications the users may enter the data mode immediately upon answer, and stay there for the duration of the call, as, for example, in a computer-to-computer file transfer. The call would be originated via an automatic calling unit and automatically answered at the distant end. Upon answer, both ends would be optioned for immediate transition to the data mode. In this way, completely automatic operation will be possible with SDC.

Signaling

The caller would request an SDC connection on a per-call basis by dialing an access code before the standard telephone number. A

line equipped for making digital connections could also be used to make ordinary telephone calls.

Switched digital connections would use standard 64 Kbps channels as the underlying circuits. In the data mode. The end user would have direct access to 56 Kbps, consisting of the first seven bits of each eight-bit byte. The eighth bit would be reserved for signaling and control.

Network signaling to set up SDC connections would use CCIS wherever possible. All No. 4 ESS trunks would use CCIS. Therefore, a No. 1A ESS would require CCIS to directly connect to No. 4 ESS. A No. 1A ESS without CCIS would be required to tandem switch its calls through a No. 1A ESS with CCIS before reaching a No. 4 ESS.

Features

Many switching features already developed in the network could apply directly to SDC. These include Speed Calling, Call Forwarding, INWATS (800 service), direct connect, originate-only, and terminate-only. With Speed Calling, the access code can be included in the stored number if desired. An "SDC calls only" option could also be made available. This would prevent ordinary telephone calls from being originated or terminated on that line.

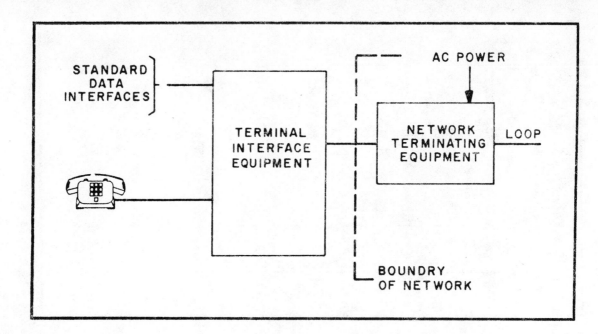

Figure 4. Network interface.

Network Interfaces

A proposed customer-premises equipment arrangement for SDC is illustrated in Figure 4. Network Terminating Equipment (NTE) would terminate the digital transmission system on the loop, and allow testing and measurement of network performance. It would require local AC power to operate in the data mode. The NTE, which represents the network boundary, would provide a simple interface for data, voice and control leads.

Terminal Interface Equipment (TIE) may also be desirable in order to provide standard interfaces for terminal equipment.

Planning for Switched Digital Connections

The Switched Digital Capability can potentially be deployed rapidly in the SPC Network. The key switching systems it relies on — No. 1A ESS and No. 4ESS — are already in nationwide use. The needed transmission systems are either in place or are being rapidly installed. Initially, a few switching systems can serve selected lines across wide areas. Access can be through standard digital transmission systems, such as T-carrier or 56 Kbps DDS based lines where available (DDS presently serves 96 metropolitan areas in the United States.)

To complement the technical design, a comprehensive operations plan for SDC is required which addresses the full range of network operations issues, including installation, testing, maintenance and traffic engineering.

Though an initial SDC can use standard methods to provide simple 56 Kbps circuit switching, many improvements and additions can be envisioned. Among these are faster call set-up using all-digital rather than Touch-Tone® signaling; additional data rates other than 56 Kbps; and advanced calling features.

Switched, end-to-end digital connections is evolving to be a fundamental new capability in the SPC network which can provide the cornerstone for many new digital communications features.

Security and Privacy Protection in Networks

This section:

- introduces both the theory and practical implementation of the *Data Encryption Standard* endorsed by the U.S. National Bureau of Standards

- describes an exciting new concept of public key cryptography wherein each user's encryption key is publicly known—only the decryption key is a secret

- shows a number of different ways to use the Data Encryption Standard in practical communication systems.

Among all of the advantages of distributed, satellite-based communications networks is one large and worrisome problem. If, in fact, information is widely distributed and easily received by many (if not all) users of the network, how can the private information of one user be protected from the unfriendly interception by some other user? Protection techniques have been developed to protect one user's data from reception by some other user. They are particularly easy to use with digital information and provide long-term protection from intentional decoding. Many of the techniques being used for commercial services were developed originally to protect the classified communications of the military services. Recent simplification of many of these techniques, however, together with the cost reductions achievable through mass production of large-scale integrated circuits, makes these systems applicable to essentially all common user systems.

In 1972, the U.S. National Bureau of Standards contracted with IBM to develop a simplified, presumably low-cost, and highly reliable encryption technique that could be applied to nonmilitary systems. After five years of development and testing, the Data Encryption Standard (DES) was adopted by the U.S. National Bureau of Standards. Trying to balance design complexity with reliability and low cost, the DES uses a 64-bit random key (of which 56 bits are actually used in the algorithm). While the relatively short length of the key could leave DES susceptible to a very high-powered, long-term, brute-force decryption attack, the DES does provide a very high level of protection.

One of the other useful features of the DES technique is that it is easily applied to the digital stream associated with a voice-digitization process. For example, numerous satellite-based voice systems are using a single DES box to encode 24 digitized voice channels simultaneously. When applications of this type are employed, the incremental cost of the encryption is very minor indeed, especially when measured in terms of the potential losses that can result from not adequately protecting information.

The paper by H. Bryce describes both the operational characteristics of the DES algorithm and the specific integrated-circuit implementation of the algorithm achieved by Motorola, Inc. Despite the complexity of the overall algorithm, Motorola, as well as other integrated-circuit manufacturers, have been able to confine all of the mathematical encryption/decryption functions to a single microcircuit chip. Combined with other input/output and control circuitry, these devices—using a total equipment volume of only a few cubic inches—can protect digital data streams at rates in excess of 6 million bits per second. The DES device can be incorporated within a data modem, user terminal, or packet-switch interface device.

The paper by M.E. Hellman describes a totally different approach to data protection. The idea behind "public key" cryptography is to try to find encoding algorithms that render it significantly easier to encrypt information than to decrypt it. All users would send information to a specific destination using the same public encryption key, but only the legitimate receiver would have the information needed to easily decrypt the information.

The viability of public-key systems is based upon the theory of trapdoor, or noninvertible, functions. The concept exploits the fact that certain functions are relatively easy to compute in the forward direction but very difficult to compute in the reverse direction. For example, it takes just a few seconds to compute $(123)^3$ and find that it is 1,860,867; it takes considerably longer, however, to determine that the cube root of 94,818,816 is 456. By employing relatively large numbers, such as the product of two 50-digit prime numbers, it is possible to choose an exponent for the encoding process that makes it mathematically not feasible to invert the encoded message unless the receiver also knows the two prime numbers that were used in generating the exponents used in the encoding. And if the number is sufficiently long, it is also not feasible mathematically to factor it back into its two prime components. For example, if a 200-bit-long number is generated by the product of two 100-bit-long prime numbers, it would take a computer with the ability to process 1,000,000 instructions per second more than three days, operating continuously, to factor the number back into its prime components. Without these prime components, it is not feasible to compute the inverse of the encoding exponent. If the numbers were lengthened to 150-bit-long prime numbers, it would take more than three years for the factoring process; and with a 250-bit prime number, more than one million years.

The last paper in this section, by C.M. Campbell, illustrates the various ways that encryption, particularly the DES, can be employed in practical communication systems. He distinguishes between link, node, and end-to-end encryption, and discusses the relative advantages of each.

As the use of distributed data systems becomes more widespread, especially for such sensitive applications as electronic banking, personal data files, medical information, business, financial, and sales information, and certain government information, the use of data encryption and protection will become mandatory. And, as computer theft and fraud perpetrated via networks becomes more frequent, security protection equipment will be increasingly used for digital networks.

The NBS Data Encryption Standard: products and principles

HEATHER BRYCE, Motorola, Inc.

*How the National Bureau of Standards' Data Encryption Standard
does what it does, and what's available to do it*

Valuable information of any kind needs protection from unauthorized access and/or alteration. In the case of computer data, this means more than placing a padlock on a door. Protection requires sophisticated cryptographic techniques, such as those provided by the National Bureau of Standards Data Encryption Algorithm (DES).

Data are particularly vulnerable when being moved from one site to another. Not only are the data subject to unauthorized access, but information may be altered before it reaches its destination. If this happened in a bank, money could be transferred to a thief's account. The theft could go unnoticed for a long time, and when finally discovered, it could be untraceable.

Cryptographic techniques used for this type of data protection scramble, or encrypt, the data before they are sent downline. The data are encrypted by the DES with a variable 56-bit key known only by the sender and intended recipient. Unless the key is known, an intruder cannot unscramble the data (see "The NBS Data Encryption Standard," p.114).

Component, subsystem, system

The DES has been implemented in a single chip, the MC6859 data security device (DSD). The DSD is available alone, incorporated into a board-level product or as the central element in a self-contained, stand-alone module. As the level of integration increases, so does the initial

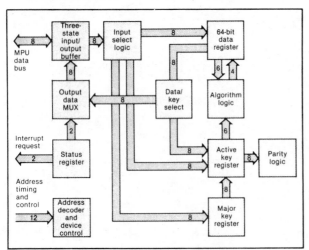

Fig. 1. Data security *device block diagram.*

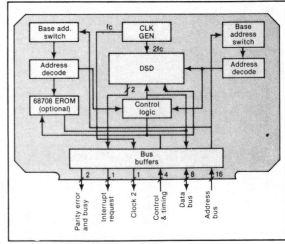

Fig. 2. MGD6800DSM *clock diagram.*

The Info-guard DES4100 series from Codex, for users who do not want to develop their own hardware or software, has the highest level of integration.

price. The price of development time, software and hardware decreases, however, so the trade-offs involved must be weighed.

Purchasing an MC6859 alone initially costs less than buying it as part of a board-level product or stand-alone module. Further, it enables a user to customize hardware and software.

The MC6859 is an NMOS device in a 24-pin package. It requires a single 5V power supply and is directly compatible with the MC6800 μp. Other μps can also control the DSD.

Fig. 1 shows an internal block diagram. Eight bytes are fed in one at a time to fill either the 64-bit data register or the 64-bit active key register. The 64-bit active key register codes the algorithm that scrambles

or unscrambles the 64-bit data block before transmission.

Motorola's Government Electronics Division offers three data security modules (DSMs) for systems integrators who prefer working at the board level. The MGD6800 DSM (Fig. 2) supports the M6800 EXORciser or a Micromodule system. An optional 68708 EROM allows control program storage or encryption key storage. A similar module, the MGD8080DSM, supports an Intel 8080.

A third module, the DES1100DSM, which supports a PDP-11, has a μp on-board and is a self-contained data encryption module.

System-level integration

The Info-guard DES4100 series from Codex, for users who do not want to develop their own hardware or software, has the highest level of integration. The series consists of a stand-alone network security module (NSM) and an optional key management system, consisting of a key generator module and a key loader module. These products (Fig. 3) perform the complete cipher function, require no programming, have minimal effect on system throughput and are transparent to the communication system in which they are used.

The NSM, which performs encryption/decryption, can be installed between a modem and a terminal or a controller in an RS232C network, or between a communications line and a terminal in a current loop (TTY) network. Fig. 4 shows a functional diagram of an NSM-based communication system.

The front panel of the basic NSM desk-top unit measures 7.2 × 6.2 × 15 in. It contains three system alarm LEDs that indicate parity error, cryptographic alarm (indicating the system is in clear text mode and encryption is not being performed) and alarm condition; an illuminated power input switch; and two Underwriters Laboratories-approved security locks. Two dissimilar keys are required for access to a 24-key keypad and a nine-digit hexadecimal display. Partial removal of the front panel activates a tamper switch that zeros the RAM, making key access impossible.

The keypad is used to enter primary and secondary keys and for downline loading of the entered secondary

Fig 3. The Info-Guard *data security system includes this network security module, which performs the NBS Standard. The complete Info-Guard key management system also includes a key generator and a key loader.*

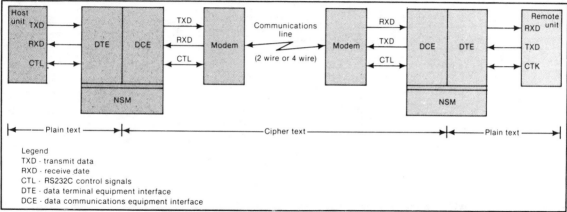

Legend
TXD · transmit data
RXD · receive date
CTL · RS232C control signals
DTE · data terminal equipment interface
DCE · data communications equipment interface

Fig. 4. DES4100 NSM *functional diagram.*

THE NBS DATA ENCRYPTION STANDARD

The National Bureau of Standards' Data Encryption Standard (DES) defines how to encrypt 64 bits of binary data. The result is 64 bits of cipher text. A 56-bit key is used for coding, and because the key is determined by the user, it is at the core of the security. To decipher data, a user would have to know the key used to encipher data. Fig. 5 shows this basic concept.

With 56 bits to define a key, there are more than 70 quadrillion different keys that can be used. This means that there are more than 70 quadrillion variations on the encryption algorithm. However, if a key is obtained by an intruder, the system loses its security.

The DES uses a series of basic cryptographic techniques in the algorithm that make up a lengthy and intricate computation. The flowchart for the enciphering computation is shown in Fig. 6.

First, the 64-bit block of data undergoes a permutation that rearranges the bits according to a specific matrix. After a series of multiple cipher operations, a 64-bit block of data undergoes a final permutation, the "inverse initial permutation," that results in the totally ciphered text of 64 bits. Between the initial permutation and the inverse initial permutation, 16 iterations take place. Each iteration is mechanically the same, but a different key value is used each time.

After the initial permutation, the 64-bit permuted block of data is split into two 32-bit halves. The right half undergoes a permutation that results in 48 bits (Fig. 7). This is because the matrix that specifies how the bits are to be arranged duplicates 16 of the bits.

These 48 bits undergo an exclusive - OR with a 48-bit key value. The key value has been obtained through encipherment on the original 56-bit key. The exclusive - OR operation results in 48 bits that are altered to result in 32 bits. This is accomplished by taking the eight sections of 6 bits each ($8 \times 6 = 48$) and making them eight sections of 4 bits each ($8 \times 4 = 32$), using a 6-bit to 4-bit selection table. The 32 bits are then permuted again to another 32-bit value.

When the right half and the key have been reduced to 32 bits, this value is exclusive - ORed to the until-now-unaltered left half. That completes the first iteration of the computation.

For the second iteration, the result just obtained becomes the right half, and the unaltered right half from the first iteration becomes the left. Then the new left and right halves are ready for the second iteration, a procedure that is repeated 16 times with a different key value used each time. Fig. 8 shows the key schedule calculation giving the 16 key values. Only one key is entered to begin the computation. The selection and protection of this key is of utmost importance in determining the

security of the total system. There are three ways to protect the key. It can be made as random as possible and changed as often as possible, and a key hierarchy of two or more keys can be used. If the key is random, an intruder could not guess its value easily. If the key is changed, an intruder would have only a limited time to use it. A hierarchy of two or more keys is best illustrated by a safe within a safe. The first key is used to obtain the second key. Then the second key allows access to the data.

The Motorola MC6859 supports a two-key hierarchy. A primary key is written to the device and stored. Then a secondary key is written to the device. The secondary key can be encrypted using the DES and the primary key by writing the secondary key to a specific address in the MC6859. When the secondary key has been encrypted, that value will be used as the key for coding the data that follow. Thus, an encrypted secondary key is the key used to code or decode data.

If a key is difficult to obtain, then two keys will be twice as difficult. An intruder must know that a two-key hierarchy is being used, and he must know the values of both keys. A user should store the key values in two different locations so that both cannot be stolen at once.

The two-key hierarchy allows versatility in system design. One guard can generate and control one key, while another guard handles the second key. Systems developed with the MC6859 take advantage of this extra security capability.

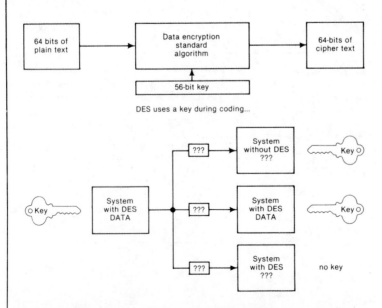

DES uses a key during coding...

Fig. 5. The DES and the key *are required before encrypted data can be deciphered.*

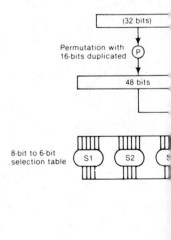

Fig. 7. One *of 16 iterations.*

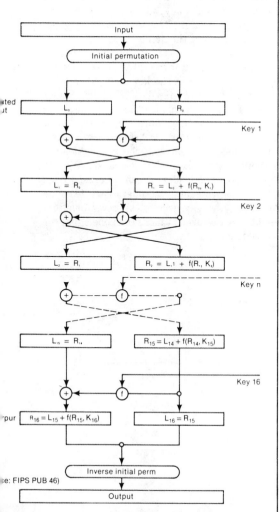

ted
ut

Key 1

$L_1 = R_0$ | $R_1 = L_0 + f(R_0, K_1)$

Key 2

$L_2 = R_1$ | $R_2 = L_1 + f(R_1, K_2)$

Key n

$L_{15} = R_{14}$ | $R_{15} = L_{14} + f(R_{14}, K_{15})$

Key 16

$R_{16} = L_{15} + f(R_{15}, K_{16})$ | $L_{16} = R_{15}$

Inverse initial perm

e: FIPS PUB 46)

Output

Enciphering *computation.(Colored section is detailed in Fig. 7.)*

f of 64-bit permuted block of data

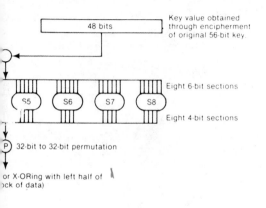

48 bits — Key value obtained through encipherment of original 56-bit key.

Eight 6-bit sections

S5 S6 S7 S8

Eight 4-bit sections

P) 32-bit to 32-bit permutation

or X-ORing with left half of ock of data)

The NSM includes diagnostic routines that are invoked each time the NSM is powered up or when the test switch is pushed.

key. Following installation, a unique set of keys can be installed manually.

The NSM includes diagnostic routines that are invoked each time the NSM is powered up or when the test switch is pushed. Faulty operation is signaled by a test lamp on the back panel. Automatic secondary key generation enables a user to change a secondary key easily and frequently by opening the front panel and entering a few keystrokes on the keypad. This causes an internal random number generator in conjunction with the DES to generate a secondary key. The secondary key can be encrypted and downline-loaded to the receiving station.

Key management

Because the selection, storage and communication of the keys are the most important measures of the security of the system, a key generator and key loader are available to assure the highest level of protection. The key generator randomly generates and stores keys

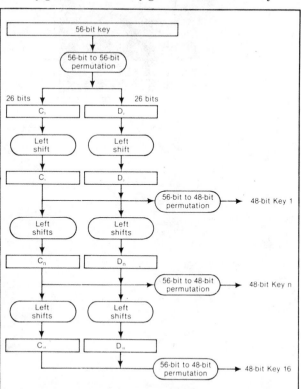

Fig. 8. The key schedule calculation *is less complicated than the enciphering computation. The 56-bit key is permuted and then split into two 26-bit halves. Each half undergoes a 1- or 2-bit left rotation, depending on the iteration. The halves are reconcatenated and undergo another permutation, this time resulting in the 48-bit halve that is used in the enciphering computation. This procedure is repeated 16 times to give a different key value for each of the 16 iterations.*

Automatic secondary key generation enables a user to change a secondary key easily and frequently by opening the front panel and entering a few keystrokes on the keypad.

The key loader *provides an optical link for transferring the generated key to the NSM in the Codex Info-Guard line of NBS-DES data security products.*

and otherwise provides optical control over them. An operator selects keys via a keypad secured by locks on the front of the generator. The operator can select which key to use, but not the actual key itself.

The key loader is connected through an optical interface to the key generator to load the random keys. After the key loader obtains the key, it is taken to an NSM, where it can transmit new keys. The key loader can carry as many as 32 keys at a time. The keys are stored in a nonvolatile semiconductor memory. Power can be maintained as long as the loader case remains intact. If the panels are removed, the key values are lost, thus protecting the key while in transit. When a key is entered from the loader, it can be erased from the loader's memory, keeping intruders from learning or stealing the key. A key usually can be transmitted only once, but it can be used at more than one site if the key generator is used.

An Overview of Public Key Cryptography

Martin E. Hellman

With a public key cryptosystem, the key used to encipher a message can be made public without compromising the secrecy of a different key needed to decipher that message.

I. COMMERCIAL NEED FOR ENCRYPTION

Cryptography has been of great importance to the military and diplomatic communities since antiquity but failed, until recently, to attract much commercial attention. Recent commercial interest, by contrast, has been almost explosive due to the rapid computerization of information storage, transmission, and spying.

Telephone lines are vulnerable to wiretapping, and if carried by microwave radio, this need not entail the physical tapping of any wires. The act becomes passive and almost undetectable. It recently came to light that the Russians were using the antenna farms on the roofs of their embassy and consulates to listen in on domestic telephone conversations, and that they had been successful in sorting out some conversations to Congressmen.

Human sorting could be used, but is too expensive because only a small percentage of the traffic is interesting. Instead, the Russians automatically sorted the traffic on the basis of the dialing tones which precede each conversation and specify the number being called. These tones can be demodulated and a microprocessor used to activate a tape recorder whenever an "interesting" telephone number (one stored in memory) is detected. The low cost of such a device makes it possible to economically sort thousands of conversations for even one interesting one.

This problem is compounded in remote computing because the entire "conversation" is in computer readable form. An eavesdropper can then cheaply sort messages not only on the basis of the called number, but also on the content of the message, and record all messages which contain one or more keywords. By including a name or product on this list, an eavesdropper will obtain all messages from, to, or about the "targeted" person or product. While each fact by itself may not be considered sensitive, the compilation of so many facts will often be considered highly confidential.

It is now seen why electronic mail must be cryptographically protected, even though almost no physical mail is given this protection. Confidential physical messages are not written on postcards and, even if they were, could not be scanned at a cost of only $1 for several million words.

II. THE COST OF ENCRYPTION

Books about World War II intelligence operations make it clear that the allies were routinely reading enciphered German messages. The weakness of the Japanese codes was established by the Congressional hearings into the Pearl Harbor disaster, and while it is less well publicized, the Germans had broken the primary American field cipher.

If the major military powers of World War II could not afford secure cryptographic equipment, how is industry to do so in its much more cost-conscious environment?

Encryption is a special form of computation and, just as it was impossible to build good, inexpensive, reliable, portable computers in the 1940's, it was impossible to build good (secure), inexpensive, reliable, portable encryption units. The scientific calculator which sells for under $100 today would have cost on the order of a million dollars and required an entire room to house it in 1945.

While embryonic computers were developed during the war (often for codebreaking), they were too expensive, unreliable, and bulky for field use. Most computational aids were mechanical in nature and based on gears. Similarly, all of the major field ciphers employed gear-based devices and, just as Babbage's failure indicates the difficulty of building a good computer out of gears, it is also difficult to build a good cryptosystem from gears. The development of general-purpose digital hardware has freed the designers of cryptographic equipment to use the best operations from a cryptographic point of view, without having to worry about extraneous mechanical constraints.

As an illustration of the current low cost of encryption, the recently promulgated national Data Encryption Standard (DES) can be implemented on a single integrated circuit chip, and will sell in the $10 range before long. While some have criticized the standard as not being adequately secure [1], this inadequacy is due to political considerations and is not the fault of insufficient technology.

III. KEY DISTRIBUTION AND PUBLIC KEY SYSTEMS

While digital technology has reduced the cost of encryption to an almost negligible level, there are other major problems involved in securing a communication network. One of the most pressing is key distribution, the problem of securely transmitting keys to the users who need them.

The classical solution to the key distribution problem is indicated in Fig. 1. The key is distributed over a secure channel as indicated by the shielded cable. The secure channel is not used for direct transmission of the plaintext message P because it is too slow or expensive.

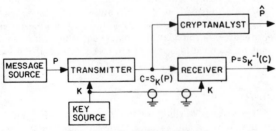

Fig. 1. Conventional Cryptographic System.

The military has traditionally used courier service for distributing keys to the sender and receiver. In commercial systems registered mail might be used. Either way, key distribution is slow, expensive, and a major impediment to secure communication.

Keys could be generated for each possible conversation

and distributed to the appropriate users, but the cost would be prohibitive. A system with even a million subscribers would have almost 500 billion possible keys to distribute. In the military, the chain of command limits the number of connections, but even there, key distribution has been a major problem. It will be even more acute in commercial systems.

It is possible for each user to have only one key which he shares with the network rather than with any other user, and for the network to use this as a master key for distributing conversation specific keys [2], [3]. This method requires that the portion of the network which distributes the keys (known as the key distribution center or node) be trustworthy and secure.

Diffie and Hellman [4] and independently Merkle [5] have proposed a radically different approach to the key distribution problem. As indicated in Fig. 2, secure communication takes place without any prearrangement between the conversants and without access to a secure key distribution channel. As indicated in the figure, two way communication is allowed and there are independent random number generators at both the transmitter and the receiver. Two way communication is essential to distinguish the receiver from the eavesdropper. Having random number generators at both ends is not as basic a requirement, and is only needed in some implementations.

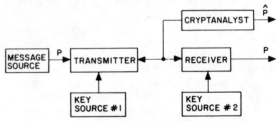

Fig. 2. Public Key Cryptographic System.

The situation is analogous to having a room full of people who have never met before and who are of equal mathematical ability. I choose one other person in the room and, with everyone else listening, give him instructions which allow the two of us to carry on a conversation that no one else can understand. I then choose another person and do the same with him.

This sounds somewhat impossible and, from one point of view, it is. If the cryptanalyst had unlimited computer time he could understand everything we said. But that is also true of most conventional cryptographic systems— the cryptanalyst can try all keys until he finds the one that yields a meaningful decipherment of the intercepted message. The real question is whether we can, with very limited computations, exchange a message which would take the cryptanalyst eons to understand using the most powerful computers envisionable.

A public key cryptosystem [4] has two keys, one for enciphering and one for deciphering. While the two keys effect inverse operations and are therefore related, there must be no easily computed method of deriving the deciphering key from the enciphering key. The enciphering key can then be made public without compromising the deciphering key so that anyone can encipher messages, but only the intended recipient can decipher messages.

The conventional cryptosystem of Fig. 1 can be likened to a mathematical strongbox with a resettable combination lock. The sender and receiver use a secure channel to agree on a combination (key) and can then easily lock and unlock (encipher and decipher) messages, but no one else can.

A public key cryptosystem can be likened to a mathematical strongbox with a new kind of resettable combination lock that has two combinations, one for locking and one for unlocking the lock. (The lock does not lock if merely closed.) By making the locking combination (enciphering key) public anyone can lock up information, but only the intended recipient who knows the unlocking combination (deciphering key) can unlock the box to recover the information.

Public key and related cryptosystems have been proposed by Merkle [5], Diffie and Hellman [4], Rivest *et al.* [6], Merkle and Hellman [7], and McEliece [8]. We will only outline the approaches, and the reader is referred to the original papers for details.

Electronic mail unlike ordinary mail is machine readable and can be automatically scanned for sensitive messages.

The RSA (Rivest *et al.*) scheme [6] is based on the fact that it is easy to generate two large primes and multiply them together, but it is much more difficult to factor the result. (Try factoring 518940557 by hand. Then try multiplying 15107 by 34351.) The product can therefore be made public as part of the enciphering key without compromising the factors which effectively constitute the deciphering key. By making each of the factors 100 digits long, the multiplication can be done in a fraction of a second, but factoring would require billions of years using the best known algorithm.

As with all public key cryptosystems there must be an easily implemented algorithm for choosing an enciphering-deciphering key pair, so that any user can generate a pair, regardless of his mathematical abilities. In the RSA scheme the key generation algorithm first selects two large prime numbers p and q and multiplies them to produce $n = pq$. Then Euler's function is computed as $\phi(n) = (p - 1)(q - 1)$. ($\phi(n)$ is the number of integers between 1 and n which have no common factor with n. Every p^{th} number has p as a common factor with n and every q^{th} number has q as a common factor with n.) Note that it is easy to compute $\phi(n)$ if the factorization of n is known, but computing $\phi(n)$ directly from n is equivalent in difficulty to factoring n [6].

$\phi(n)$ as given above has the interesting property that for any integer a between 0 and $n-1$ (the integers modulo n) and any integer k

$$a^{k\phi(n) + 1} = a \mod n. \qquad (1)$$

Therefore, while all other arithmetic is done modulo n, arithmetic in the exponent is done modulo $\phi(n)$.

A random number E is then chosen between 3 and $\phi(n)$ -1 and which has no common factors with $\phi(n)$. This then allows

$$D = E^{-1} \mod \phi(n) \qquad (2)$$

to be calculated easily using an extended version of Euclid's algorithm for computing the greatest common divisor of two numbers [9, p. 315, problem 15; p. 523, solution to problem 15].

THE RIVEST–SHAMIR–ADLEMAN PUBLIC KEY SCHEME

Design

Find two large prime numbers p and q, each about 100 decimal digits long. Let $n = pq$ and $\psi = (p-1)(q-1)$.

Choose a random integer E between 3 and ψ which has no common factors with ψ. Then it is easy to find an integer D which is the "inverse" of E modulo ψ, that is, $D \cdot E$ differs from 1 by a multiple of ψ.

The public information consists of E and n. All other quantities here are kept secret.

Encryption

Given a plaintext message P which is an integer between 0 and $n-1$ and the public encryption number E, form the ciphertext integer

$$C = P^E \mod n.$$

In other words, raise P to the power E, divide the result by n, and let C be the remainder. (A practical way to do this computation is given in the text of Hellman's paper.)

Decryption

Using the secret decryption number D, find the plaintext P by

$$P = C^D \mod n.$$

Cryptanalysis

In order to determine the secret decryption key D, the cryptanalyst must factor the 200 or so digit number n. This task would take a million years with the best algorithm known today, assuming a 1 μs instruction time.

The information (E,n) is made public as the enciphering key and is used to transform unenciphered, plaintext messages into ciphertext messages as follows: a message is first represented as a sequence of integers each between 0 and $n - 1$. Let P denote such an integer. Then the corresponding ciphertext integer is given by the relation

$$C = P^E \bmod n. \tag{3}$$

The information (D,n) is used as the deciphering key to recover the plaintext from the ciphertext via

$$P = C^D \bmod n. \tag{4}$$

These are inverse transformations because from (3), (2), and (1)

$$C^D = P^{ED} = P^{k\phi(n)+1} = P. \tag{5}$$

As shown by Rivest et al., computing the secret deciphering key from the public enciphering key is equivalent in difficulty to factoring n.

As a small example suppose $p = 5$ and $q = 11$. Then $n = 55$ and $\phi(n) = 40$. If $E = 7$ then $D = 23$ ($7 \times 23 = 161 = 1 \bmod 40$). If $P = 2$, then

$$C = 2^7 \quad \bmod 55 = 18 \tag{6}$$

and

$$C^D = 18^{23} \quad \bmod 55 \tag{7}$$

$$= 18^1 18^2 18^4 18^{16} \tag{8}$$

$$= 18\ 49\ 36\ 26 \quad \bmod 55 \tag{9}$$

$$= 2 \tag{10}$$

which is the original plaintext.

Note that enciphering and deciphering each involve an exponentiation in modular arithmetic and that this can be accomplished with at most $2(\log_2 n)$ multiplications mod n. As indicated in (8), to evaluate $Y = a^X$, the exponent X is represented in binary form, the base a is raised to the 1st, 2nd, 4th, 8th, etc. powers (each step involving only one squaring or multiplication), and the appropriate set of these are multiplied together to form Y.

Merkle and Hellman's method [7] makes use of trap-door knapsack problems. The knapsack problem is a combinatorial problem in which one is given a vector of n integers, a, and an integer S which is a sum of a subset of the $\{a_i\}$. The problem is to solve for the subset, or equivalently, for the binary vector x which is the solution to the equation

$$S = a * x. \tag{11}$$

While the knapsack problem is very difficult to solve in general, there are specific cases which are easy to solve. For example, if the knapsack vector is

$$a' = (171, 197, 459, 1191, 2410) \tag{12}$$

then, given any S', x is easily found because each component of a' is larger than the sum of the preceding components. If $S' = 3798$, then it is seen that x_5 must be 1 because, if it were 0, $a_5' = 2410$ would not be in the sum and the remaining elements sum to less than S'. After subtracting the effect of a_5' from S', the solution continues recursively and establishes that $x_4 = 1$, $x_3 = 0$, $x_2 = 1$, and $x_1 = 0$.

The knapsack vector

$$a = (5457, 4213, 5316, 6013, 7439) \tag{13}$$

does not possess the property that each element is larger than the sum of the preceding components, and the sim-

ple method of solution is not possible. Given $S = 17665$, there is no obvious method for finding that $x = (0,1,0,1,1)$ other than trying almost all 2^5 subsets.

But it "just so happens" that if each component of a is multiplied by 3950 modulo 8443 the vector a' of (12) is obtained. By performing the same transformation on S, the quantity $S' = 3798$ is obtained. It is now seen that there is a simple method for solving for x in the equation

$$S = a * x \tag{14}$$

by transforming to the easily solved knapsack problem

$$S' = a' * x. \tag{15}$$

The two solutions x are the same provided the modulus is greater than the sum of the $\{a_i\}$.

The variables of the transformation (the multiplier 3950 and the modulus 8443) are secret, trap-door information used in the construction of the trap-door knapsack vector a. There is no apparent, easy way to solve knapsack problems involving a unless one knows the trap-door information.

When a is made public anyone can represent a message as a sequence of binary x vectors and transmit the information securely in the corresponding sums, $S = a * x$. The intended recipient uses his trap-door information (secret deciphering key) to easily solve for x, but no one else can do this. Of course the a vector must be significantly longer than that used in this small, illustrative example.

McEliece's public key cryptosystem [8] is based on algebraic coding theory. Goppa codes are highly efficient error correcting codes [10], but their ease of error correction is destroyed if the bits which make up a codeword are scrambled prior to transmission. To generate a public enciphering key, a user first selects a Goppa code chosen at random from a large set of possible codes. He then selects a permutation of the codeword bits, computes the generator matrix associated with the scrambled Goppa code and makes it public as his enciphering key. His secret deciphering key is the permutation and choice of Goppa code.

Key distribution, the secure transmission of keys to the users who need them, is a major problem in securing a communication network.

Anyone can easily encode information (scrambling does not greatly increase the difficulty of encoding since the scrambled code is still linear), add a randomly generated error vector, and transmit this. But only the intended recipient knows the inverse permutation which allows the errors to be corrected easily.

McEliece estimates that a block length of 1000 bits with 500 information bits should foil cryptanalysis using the best currently known attacks.

The other two known methods for communicating securely over an insecure channel without securely transmitting a key are not true public key cryptosystems. Rather, they are public key distribution systems which are used to securely exchange a key over an insecure channel without any prearrangement, and that key is then used in a conventional cryptosystem.

Merkle's technique [5] involves an exchange of "puzzles." The first user generates n potential keys and hides them as the solutions to n different puzzles, each of which costs n units to solve. The second user chooses one of the n puzzles at random, solves it, and sends a test message encrypted in the associated key. The first user determines which key was chosen by trying all n of them on the test message.

The cost to the first user is proportional to n. He must generate and store n keys, generate and transmit n puzzles, and try n keys on the test message. The cost to the second user is also proportional to n because he must solve one puzzle which was designed to have solution cost equal to n.

The cost to an eavesdropper appears to grow as n^2. He can try solving puzzles at random and see if the associated key (solution) agrees with the test message. On the average, he must solve $n/2$ puzzles, each at a cost of n.

Diffie and Hellman [4] describe a public key distribution system based on the discrete exponential and logarithm functions. If q is a prime number and a is a primitive element, then X and Y are in a 1:1 correspondence for $1 \leqslant X, Y \leqslant (q - 1)$ where

$$Y = a^X \mod q \qquad (16)$$

and

$$X = \log_a Y \quad \text{over } GF(q). \qquad (17)$$

While the discrete exponential function (16) is easily evaluated, as in (7) and (8), no general, fast algorithms are known for evaluating the discrete logarithm function (17). Each user chooses a random element X and makes the associated Y public. When users i and j wish to establish a key for communicating privately they use

$$K_{ij} = a^{X_i X_j} \qquad (18)$$

$$= (Y_i)^{X_j} = (Y_j)^{X_i}. \qquad (19)$$

Equation (19) demonstrates how both users i and j use the easily computed discrete exponential function to calculate K_{ij} from their private and the other user's public information. An opponent who knows neither user's secret information can compute K_{ij} if he is willing to compute a discrete logarithm, but that can be made computationally infeasible using the best currently known algorithms [11].

The various public key systems are compared in Section V.

IV. DIGITAL SIGNATURES

Business runs on signatures and, until electronic communications can provide an equivalent of the written signature, it cannot fully replace the physical transportation of documents, letters, contracts, etc.

Current digital authenticators are letter or number sequences which are appended to the end of a message as a crude form of signature. By encrypting the message and authenticator with a conventional cryptographic system, the authenticator can be hidden from prying eyes. It therefore prevents third party forgeries. But, because the authentication information is *shared* by the sender and receiver, it cannot settle disputes as to what message, if any, was sent. The receiver can give the authentication

information to a friend and ask him to send a signed message of the receiver's choosing. The legitimate sender of messages will of course deny having sent this message, but there is no way to tell whether the sender or receiver is lying. The whole concept of a contract is embedded in the possibility of such disputes, so stronger protection is needed.

A true digital signature must be a number (so it can be sent in electronic form) which is easily recognized by the receiver as validating the particular message received, but which could only have been generated by the sender. It may seem impossible for the receiver to be able to recognize a number which he cannot generate, but such is not the case.

While there are other ways to obtain digital signatures, the easiest to understand makes use of the public key cryptosystems discussed in the last section. The ith user has a public key E_i and a secret key D_i. This notation was chosen because E_i was used to encipher and D_i was used to decipher. Suppose, as in the RSA scheme, the enciphering function is **onto**, that is, for every integer C less than n, there exists an integer m for which $E_i(m) = C$. Then, we can interchange the order of operations and use D_i first to sign the message and E_i second to validate the signature. When user i wants to sign and send a message M to user j, he operates on M with his secret key D_i to obtain

$$C = D_i(M) \qquad (20)$$

which he then sends to user j. User j obtains i's public key E_i from a public file and operates with it on C to obtain M

$$E_i(C) = E_i[D_i(M)] = M \qquad (21)$$

User j saves C as proof that message M was sent to him by user i. No one else could have generated C, because only i knows D_i. And if j tries to change even one bit in C, he changes its entire meaning (such error propagation is necessary in a good cryptosystem).

If i later disclaims having sent message M to user j, then j takes C to a "judge" who accesses the public file and checks whether $E_i(C)$ is a meaningful message with the appropriate date, time, address, name, etc. If it is, the judge rules in favor of j. If it is not, the ruling is in favor of i.

Digital signatures have an advantage over written signatures because written signatures look the same, independent of the message. My signature is supposed to look the same on a $100 check as on a $1000 check, so a dishonest recipient can try to alter the check. Similarly, if a photostat of a contract is acceptable as proof, a dishonest person can alter the contract and make a copy which hides the alteration. Such mischief is impossible with digital signatures, provided the signature system is truly secure.

The disadvantage of digital signatures is that the ability to sign is equivalent to possession of a secret key. This key will probably be stored on a magnetic card which, unlike the ability to sign one's name, can be stolen.

V. COMPARISON OF PUBLIC KEY SYSTEMS

This section compares the public key systems which have been proposed. Speed, ease of signature generation, and certain other characteristics can be compared more readily than the all important question of security level. We can compare the security level using the best known methods for breaking each system, but there is the danger that better methods will be found which will change the relative rankings.

If signatures are desired, attention should be directed primarily to the RSA [6] and trap-door knapsack systems [7]. The RSA scheme yields signatures directly. While the trap-door knapsack signature method described in [7] is not direct, Merkle and Reeds have developed a method for generating "high density" trap-door knapsacks which simplify signature generation, and Shamir has recently suggested a direct method for obtaining signatures. Both of these approaches are not yet published.

The $a^{(X_1 X_2)}$ and Goppa code methods do not appear to lend themselves to signatures, but Merkle has developed a puzzle-like technique for generating signatures.

So far, as storage requirements for the public file, the $a^{(X_1 X_2)}$ and RSA schemes are most interesting. Each requires on the order of 500 bits of storage per user. The trap-door knapsack scheme requires on the order of 100 kbits of storage per user, and the Goppa code method requires on the order of a megabit per user. Merkle's puzzle scheme is not really suited to public file storage and rather depends on transmission of public information at the start of each new conversation. The transmitted information must be on the order of a gigabit before significant levels of security are afforded.

Instead of storing each user's public key in a public file (similar to a telephone book), Kohnfelder [12] has suggested having the system give each user a signed message, or certificate, stating that user's public key. The certificate could be stored by the user on a magnetic card, and transmitted at the start of a conversation. This method converts public file storage requirements into transmission requirements. The system's public key would be needed to check the certificate and could be published widely. Protecting the system's secret key might be easy because no one else ever has to use it and it could be destroyed after it was used to certify a group of users.

Computation time on the part of the legitimate users is smallest with the trap-door knapsack method. The $a^{(X_1 X_2)}$ and RSA schemes each require several hundred times as much computation, but are still within reason. Merkle's technique requires even more computation. The Goppa code technique is extremely fast for enciphering, requiring approximately 500 XOR's on 1000 bit vectors, but I have not yet estimated its deciphering requirements.

Turning to security level, Merkle's puzzle method [5] has the advantage of being the most solid method for communicating securely over an insecure channel. That is, it is extremely doubtful that a better method will be found for breaking it. Unfortunately, it is also the least secure using the best known algorithm. Its work factor (ratio of cryptanalytic effort to enciphering and deciphering effort, using the best known algorithms) is only $n^2:n$. Since encryption should cost on the order of $0.01 and cryptanalysis should cost on the order of $10 mil-

lion or more, this ratio needs to be 10^9 or more and corresponds to $n = 10^9$. If all of the enciphering and deciphering effort were in computation, this might be possible in the near future (a $10 microprocessor can execute on the order of 1 million instructions per second), but Merkle's method requires n transmissions as well as n operations on the part of the legitimate users. Current technology therefore limits Merkle's scheme to $n \leq 10\ 000$ which corresponds to approximately 500 kbits of transmission. If fiber optic or other low cost, ultra-high bandwidth communication links become available, Merkle's technique would become of greater practical interest.

Diffie and Hellman's exponentiation method [4] requires the legitimate users to perform an exponentiation in modular arithmetic while the best known cryptanalytic method requires the computation of a logarithm in modular arithmetic. Exponentiation is easily accomplished in at most $2b$ multiplications, much as in (8), where b is the number of bits in the representation of the modulus. Each multiplication can be accomplished with at most $2b$ additions or subtractions, and each of these operations involves at most b gate delays for the propagation of carry signals. Overall, an exponentiation in modular arithmetic can be accomplished in at most $4b^3$ gate delays.

Computation of a logarithm in modular arithmetic is much more complex, and the best currently known algorithm [11] requires $2^{b/2}$ or more operations provided the modulus is properly chosen. Each operation involves a multiplication, or $2b^2$ gate delays. The work factor is therefore exponential in b.

If $b = 500$, then 500 million gate delays are required at the legitimate users' terminals. With current technology this can be accomplished in several seconds, a not unreasonable delay for establishing a key during initial connection. Using $b = 500$ results in the cryptanalyst having to do more than 10^{75} times as much work as the legitimate users, a very safe margin. The real question is whether better methods exist for computing logarithms in modular arithmetic, or if it is even necessary to compute such a logarithm to break this system.

The following table gives the number of operations and time required for cryptanalysis for various values of b assuming a 1 μs instruction time:

b (bits)	100	200	300	500	750	1000
Operations	1.1×10^{15}	1.3×10^{30}	1.4×10^{45}	1.8×10^{75}	7.7×10^{112}	3.3×10^{150}
Time (yrs.)	36	4×10^{16}	5×10^{31}	6×10^{61}	2×10^{99}	1×10^{137}

The storage requirements of this system are small. The public file stores a single b-bit number for each user and only several b-bit words of memory are required at the transmitter and receiver, so that single chip implementation is possible for b on the order of 500.

The RSA system [6] also requires that the legitimate users perform a modular exponentiation, but cryptanalysis is equivalent to factoring a b-bit number. Schroeppel has developed a new, as yet unpublished factoring algorithm which appears to require approximately $\exp\{[\ln(n)\ \ln(\ln n)]^{1/2}\}$ machine cycles where $n = 2^b$ is the number to be factored. The following

THE KNAPSACK PROBLEM

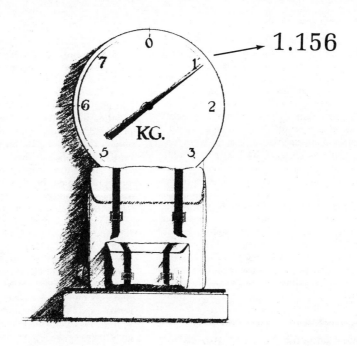

1.156

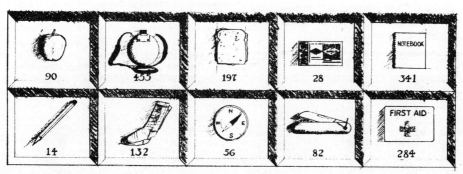

90	455	197	28	341
14	132	56	82	284

The knapsack is filled with a subset of the items shown, with weights indicated in grams. Given the weight of the filled knapsack, 1156 grams, can you determine which of the items are contained in the knapsack? (The scale is calibrated to deduct the weight of the empty knapsack.)

This simple version of the classic knapsack problem generally becomes computationally infeasible when there are 100 items rather than 10 as in this example. However, if the set of weights for the items happens to have some nice properties known only to someone with special "trap-door" information, then that person can quickly decipher the secret information, i.e., a 100 bit binary word that specifies which of the items are in the knapsack.

ART: Jeff Wyszkowski

table gives the number of operations and time to factor a b-bit number again assuming a 1 μs instruction time:

b (bits)	100	200	300	500	750	1000
Operations	$2.8{\times}10^7$	$2.3{\times}10^{11}$	$2.9{\times}10^{14}$	$3.6{\times}10^{19}$	$5.8{\times}10^{24}$	$1.8{\times}10^{29}$
Time	30 s	3 days	9 yr	1 Myr	2 Gyr	$6{\times}10^{15}$ yr

Public file storage for the RSA scheme is reasonable, being several hundred to a thousand bits per user. Memory requirements at the transmitter and receiver are also comparable to the $a^{(X_1 X_2)}$ scheme, so that a single chip device can be built for enciphering and deciphering.

The best known method of cryptanalyzing the trap-door knapsack system requires on the order of $2^{n/2}$ operations where n is the size of the knapsack vector. Enciphering requires at most n additions, so the work factor is exponential. If n is replaced by b, the first table above gives the cryptanalytic effort required for various values of n, so $n \geqslant 200$ provides relatively high security levels. Since each element of the a vector is approximately $2n$ bits long, if $n = 200$, the public storage is approximately 80 kbits/user. Memory requirements at the transmitter and receiver are on the same order.

Both enciphering and deciphering require less computation than either the $a^{(X_1 X_2)}$ or RSA scheme. Enciphering requires at most n additions and deciphering requires one multiplication in modular arithmetic, followed by at most n subtractions.

Until electronic communications can provide an equivalent of the written signature, it cannot fully replace the physical transportation of documents, letters, contracts, etc.

Care must be exercised in interpreting these tables. First, they assume that the cryptanalyst uses the best currently known method, and there may be much faster approaches. For example, prior to the development of Schroeppel's algorithm, the best factoring algorithm appeared to require $\exp\{[2 \ln(n) \ln(\ln n)]^{1/2}\}$ operations. When $b = 200$, that would have predicted that 360 yr, not 3 days, would be required for cryptanalysis. There is the danger that even faster algorithms will be found, necessitating a safety margin in our estimates. As demonstrated by this example, the safety margin is needed in the exponent, not the mantissa.

A similar comment applies to the seemingly higher security level afforded by the $a^{(X_1 X_2)}$ and trap-door knapsack methods when compared to the RSA scheme. For a given value of b the two tables show that the RSA scheme requires much less computation to break, using the best currently known techniques. But it is not clear whether this is because factoring is inherently easier than computing discrete logarithms or solving knapsack problems, or whether it is due to the greater study which has been devoted to factoring.

As computers become faster and more parallel, the time for cryptanalysis also falls. A 1 ns computer with million-fold parallelism might reduce the time estimates given in the tables by a factor of 10^9.

VI. CONCLUSIONS

We are in the midst of a communications revolution which will impact many aspects of people's every day lives. Cryptography is an essential ingredient in this revolution, and is necessary to preserve privacy from computerized censors capable of scanning millions of pages of documents for even one sensitive datum. The public key and digital signature concepts are necessary in commercial systems because of the large number of interconnections which are possible, and because of the need to settle disputes.

A major problem which confronts cryptography is the certification of these systems. How can we decide which proposed systems really are secure, and which only appear to be secure? Proofs are not possible using the currently developed theory of computational complexity and, while such proofs may be possible in the future, something must be done immediately. The currently accepted technique for certifying a cryptographic system as secure is to subject it to a mock attack under circumstances which are extremely favorable to the cryptanalyst and unfavorable to the system. If the system resists such a concerted attack under unfavorable conditions, it is hoped that it will also resist attacks by one's opponents under more realistic conditions.

Governments have built up expertise in the certification area but, due to security constraints, this is not currently available for certification of commercially oriented systems. Rather, this expertise in the hands of a foreign government poses a distinct threat to a nation's businesses. It has even been suggested that poor or nonexistent encryption will lead to international economic warfare, a concern of importance to national security. (There is speculation that this occurred with the large Russian grain purchases of several years ago.)

There is a tradeoff between this and other national security considerations which needs to be resolved, but the handling of the national data encryption standard indicates that public discussion and resolution of the tradeoff is unlikely unless individuals make their concern known at a technical and political level.

REFERENCES

[1] W. Diffie and M. E. Hellman, "Exhaustive cryptanalysis of the NBS data encryption standard," *Computer*, pp. 74–84, June 1977.

[2] D. Branstad, "Encryption protection in computer data communications," presented at the IEEE Fourth Data Communications Symposium, Oct. 7–9, 1975, Quebec, Canada.

[3] *IBM Syst. J., (Special Issue on Cryptography)*, vol. 17, no. 2, 1978.

[4] W. Diffie and M. E. Hellman, "New directions in cryptography," *IEEE Trans. Inform. Theory*, vol. IT-22, pp. 644–654, Nov. 1976.

[5] R. C. Merkle, "Secure communication over an insecure channel," *Common. Ass. Comput. Mach.*, vol. 21, pp. 294–299, Apr. 1978.

[6] R. L. Rivest, A. Shamir, and L. Adleman, "On digital signatures and public key cryptosystems," *Commun. Ass. Comput. Mach.*, vol. 21, pp. 120–126, Feb. 1978.

[7] R. C. Merkle and M. E. Hellman, "Hiding information and signatures in trap-door knapsacks," *IEEE Trans. Inform. Theory*, vol. IT-24, pp. 525–530, Sept. 1978.

[8] R. J. McEliece, "A public key system based on algebraic coding theory," JPL DSN Progress Rep., 1978.

[9] D. E. Knuth, *The Art of Computer Programming, Vol. 2, Seminumerical Algorithms.* Reading, MA: Addison-Wesley, 1969.

[10] E. R. Berlekamp, "Goppa codes," *IEEE Trans. Inform. Theory*, vol. IT-19, pp. 590–592, Sept. 1973.

[11] S. C. Pohlig and M. E. Hellman, "An improved algorithm for computing logarithms over $GF(p)$ and its cryptographic significance, *IEEE Trans. Inform. Theory*, vol. IT-24, pp. 106–110, Jan. 1978.

[12] L. Kohnfelder, *Towards a Practical Public-Key Cryptosystem*, M.I.T. Lab. for Comput. Sci., June 1978.

Design and Specification of Cryptographic Capabilities

Carl M. Campbell

A variety of techniques for communications security are based on the use of a block encryption algorithm such as the DES.

Abstract—Cryptography can be used to provide data secrecy, data authentication, and originator authentication. Non-reversible transformation techniques provide only the last. Cryptographic check digits provide both data and originator authentication, but no secrecy. Data secrecy, with or without data authentication, is provided by block encryption or data stream encryption techniques. Total systems security may be provided on a link-by-link, node-by-node, or end-to-end basis, depending upon the nature of the application.

I. INTRODUCTION

Up to the present, cryptography has been a relatively unknown science, used primarily to secure sensitive governmental communications. However, with the introduction of the Data Encryption Standard (DES) we expect to see cryptography widely applied in data processing systems, especially in digital communications, to provide data security. It is thus essential that the designers of these systems gain an understanding of this new technology.

This paper was originally published in *Computer Science & Technology: Computer Security and the Data Encryption Standard, Proc. Conf. on Computer Security and the Data Encryption Standard*, National Bureau of Standards, Gaithersburg, MD, Feb. 15, 1977.

The author is a Consultant to the Interbank Card Association. He is at 809 Malin Road, Newton Square, PA 19073.

II. USES OF CRYPTOGRAPHY

Cryptography can be used to provide three aspects of data security:

1) Data secrecy.
2) Data authentication.
3) Originator authentication.

The first use of cryptography, data secrecy, is relatively well understood, and will be an important use in an EDP environment.

Data authentication and originator authentication are less understood, but will be very important uses of cryptography in the future. To understand data authentication, assume that A is transmitting data to B. B wants assurance that the data it is receiving is precisely the data which A transmitted. Though conventional error control techniques can protect against communications errors, B is concerned that someone with a sophisticated "active wiretapping" capability may have deliberately modified the data from A, and made the appropriate modifications in any associated error control fields. Cryptographically implemented data authentication provides assurance that the data was received as originated.

Originator authentication is similar to data authentication. This time B requires assurance that it is receiving data from the real A and not from an impostor who may have assumed A's identity. Again, cryptography can provide the solution.

There are an almost unlimited number of ways in which cryptography can be applied. Some applications meet only one or two of the above objectives, and some meet them all.

III. ORIGINATOR AUTHENTICATION

A simple use of cryptography meets only the third objective, originator authentication. In this approach, Fig. 1, each authorized user of a system is given a secret **authorization code**. Each terminal incorporates a cryptographic capability into which he enters this code. The code is **nonreversibly transformed** into another code. This means that, given the transformed code, there is no way to determine the actual code except for an exhaustive trial and error procedure, which is presumed to be nonfeasible if the original code is quite long (approximately 56 bits) and reasonably random. The system's central processor stores, in a manner which may be nonsecure, each user's transformed code. A simple comparison is thus sufficient to authenticate the user.

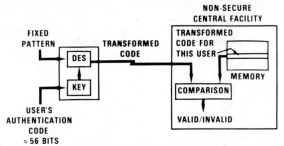

Fig. 1. Non-reversible transformation for user authentication.

Note that this approach does not require a unique terminal key, so imposes no "key management" requirements. Note also that it does not require any on-line cryptographic capability at the central facility.

IV. DATA AUTHENTICATION

A very useful cryptographic technique, **cryptographic check digits**, provides data authentication and can provide originator authentication, but provides no data secrecy. Cryptographic check digits may be likened to parity check digits or to a cyclic redundancy check in that a check field is added to the message by the originator and verified by the recipient. However, unlike a conventional error-control check field, the cryptographic check digit field is generated by a cryptographic algorithm and utilizes a secret key known (desirably) by originator and recipient alone. Thus the field protects not only against accidental garbles, but also against deliberate attempts to modify the transmitted data. Without knowing the secret key, the one attempting such data modification would be unable to make the appropriate changes in the cryptographic check digits field which would be required for his modification to escape detection.

Note that originator authentication is provided if the recipient is certain that only the authorized originator possesses the secret key.

The DES may be used to generate cryptographic check digits, as, for example, is illustrated in Fig. 2. Each group of 64 message bits is passed through the algorithm after being combined with the output of the previous pass. The final DES output is thus a residue which is a cryptographic function of the entire message. All or part of this residue may be used as the cryptographic check digits.

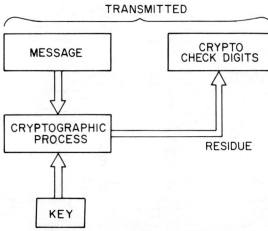

Fig. 2. Generation of cryptographic check digits for authentication of originator data.

Cryptographic check digits alone cannot detect the fraudulent replay of a previously valid message, nor the deletion of a message. To protect against these threats, each transmission of a message must be made unique. One technique is to insert a cryptographically-protected sequence number into the message. Another is to use a different key for each message.

V. DATA SECRECY

Secrecy of transmitted data may be provided by a number of techniques, some providing data authentication and some not. All of the suggested techniques utilize a secret key, and so provide originator authentication if this key is properly controlled.

A. Block Encryption

The Data Encryption Standard is inherently a block encryption algorithm, requiring blocks of precisely 64 bits. Given a plain-text block of 64 bits, a secret key, and the "encrypt" command, the DES algorithm produces 64 cipher bits. Given these 64 cipher bits, the same key, and the "decrypt" command, the algorithm produces the original 64 plain-text bits. Thus, as long as the block size is exactly 64 bits, block encryption with DES is extremely simple.

Short blocks: If the block size is less than 64 bits, these bits must be "padded" (with any fixed or, preferably, variable pattern) to make 64 bits if the algorithm is to be used in its normal block-encryption manner. All 64 of the resulting cipher bits must be transmitted to the recipient even though, for example, only 20 bits of underlying information are present. The recipient block-decrypts these 64 bits, resulting in 64 plain-text bits. All but 20 of these must be discarded, leaving the 20 original information bits.

The use of DES for a block size of less than 64 bits is thus somewhat inefficient, in that the full 64 bits must still be transmitted. Different techniques for using DES are possible, which overcome this disadvantage, but they introduce other disadvantages.

Multi-blocks: Where the block to be encrypted is long, it can be broken up into groups of 64 bit blocks, and each such block encrypted independently. This simple

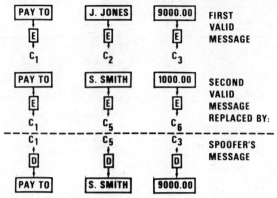

Spoofing. Known pairs of plaintext and ciphertext are used to modify enciphered message.

approach provides secrecy, but it does not provide a high degree of data authentication. For example, assume two block-encrypted messages, one reading: "PAY TO J. JONES $9,000.00" and the second: "PAY TO S. SMITH $1,000.00." If the "$9,000.00" and the "$1,000.00" should each fall precisely within a block, it would be possible to replace the cipher block for "$1,000.00" with that for "$9,000.00" so that when the recipient decrypts the second message it reads: "PAY TO S. SMITH $9,000.00."

This process, by which cipher is manipulated, is called "*spoofing*." Note that the "spoofer" knows corresponding cipher and plain text, but does not know the secret key. His objective is to intercept, modify, and then retransmit the cipher, all in such a manner that his deception is not detected.

Encryption techniques can be devised which prevent "spoofing," but in order to do so it is necessary to introduce something called **garble extension**. This means that if any portion of the cipher becomes garbled (i.e., changed) the decryption by the recipient of a certain amount of subsequent cipher is also garbled.

Fig. 3 illustrates one method by which garble extension, and hence spoofing prevention, can be incorporated into a

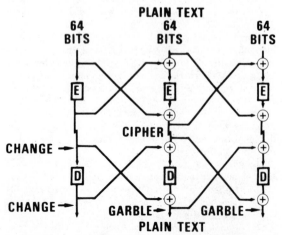

Fig. 3.   Block interconnections to provide "infinite" garble extension. *E* and *D* denote respectively the block encryption and decryption operations.

block encryption system. The "E" boxes perform block encryption, and the "*D*" boxes block decryption. The "+" function indicates exclusive-or. The approach of Fig. 3 provides "infinite" garble extension. That is, any change to the cipher garbles the decryption of all subsequent cipher. Infinite garble extension has the features that the originator can place in the final block a pattern expected by the recipient. If the recipient finds the expected pattern at the end of the message, he is assured that the entire message, regardless of length, was received precisely as originated.

B. Data-Stream Encryption

The term **data-stream** refers to the serial flow (serial by bit, by character, or any other increment) of data, as over a communications line. **Data-stream encryption** refers to the encryption of such data in real-time, for subsequent "data-stream decryption," also in real-time. It is possible to use block encryption for data-stream encryption, but this is not desirable. In DES block encryption, the first bit cannot be encrypted until the 64th bit has been received, so that a block-encryption technique in a data-stream environment inherently imposes a delay of

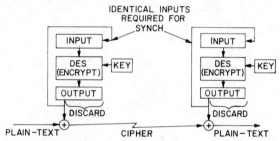

Fig. 4.   Internal feedback configuration—The DES generates a stream of pseudorandom "encrypting" bits.

64 bit times. Block decryption imposes an equal delay. Thus, communications delays would be unacceptably increased were block techniques to be used.

Fortunately, DES can be applied to a data-stream environment so as to minimally impact communications delays. Two such techniques are **internal feedback** and **cipher feedback**.

Internal Feedback: The internal-feedback approach to data-stream encryption uses DES to generate a stream of pseudorandom **encrypting bits**. These bits are exclusive-ored with the plain-text bits to form the cipher bits, as illustrated in Fig. 4. The decryption process operates the same way, with the exact same pseudorandom stream of encrypting bits being generated. Exclusive-oring these bits with the cipher bits then produces the original plain-text bits.

To use DES in this manner, any number of the 64 output (i.e., cipher) bits may be used. For simplicity of explanation, it is assumed that only 1 bit is used, and the other 63 discarded. The selected bit is not only used to encrypt the plain-text data, but is also fed back as the input to DES, and another algorithm cycle initiated. Thus, one algorithm cycle is required per encrypting bit.

To ensure that the decryption process generates the same pseudorandom encrypting bits as does the

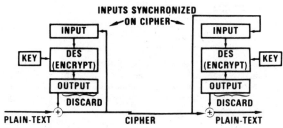

Fig. 5. Cipher feedback configuration. Cipher bits are used to feed the DES input register.

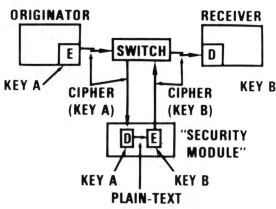

Fig. 7. Node-by-node encryption.

encryption process, the DES input registers of the two devices must commence operation with the same **initial fill**. The process by which this is accomplished is called **crypto synchronization.**

Cipher Feedback: This approach to data-stream encryption is very similar to the internal feedback approach, the difference being that cipher bits, rather than "encrypting bits," are used as the DES input. Note that this approach, Fig. 5, if used in a 1-bit feedback mode, is **self synchronizing** because after 64 bit times, the DES input register of the decryption device will contain the same data as does the input register of the encryption device. Note also that the approach provides garble extension, thus providing anti-spoofing protection.

VI. SYSTEM PHILOSOPHIES

There are three basic approaches to incorporating encryption into a communications system: link-by-link, node-by-node, and end-to-end encryption.

Link-by-link encryption: Fig. 6, is the technique most commonly used today. It may be implemented in a transparent manner with currently available devices, which are placed in series with the circuit between data terminal equipment and data communications equipment. This approach has the disadvantage that it allows all traffic to pass through the CPU of any node in plain-text.

Node-by-node encryption: Fig. 7, is a modified version of link-by-link encryption to overcome this disadvantage. Each link uses a unique key, but the "translation" from one key to the next occurs within a single "security module" which might serve as a peripheral device to the node's CPU. In this way plain-text data does not traverse the node, but exists only within this physically secure module. Note that enough message data must remain encrypted so that the node's CPU can properly route the message.

End-to-end encryption: Fig. 8, requires a **Key Control Center,** located somewhere within the communication system. Each end-point in the system holds a unique **long-term** key, and this center alone holds a copy of each such key. When one end-point wishes to communi-

cate to another, a request to this effect is sent to the Key Control Center. This center then generates a temporary **per conversation** key, encrypts this in the long-term key of the originator, and also in the long-term key of the recipient, and sends the appropriate version to each. The originator decrypts this just-received encrypted temporary key using its long-term key, the recipient does likewise with its long-term key, and the two parties then converse with end-to-end encryption using this temporary key.

VII. PROCUREMENT CONSIDERATIONS

For retrofitting an existing system, link-by-link encryption utilizing transparent link encryption devices is a reasonable approach. DES feedback is a desirable choice for these devices.

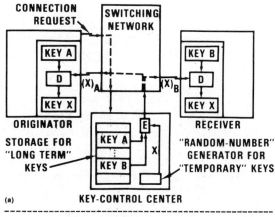

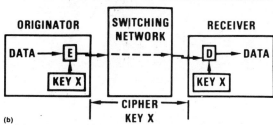

Fig. 8. End-to-end encryption. (a) Connection set-up. (b) Data transfer.

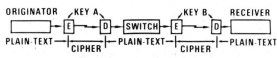

Fig. 6. Link-by-link encryption.

574

For a new system, in which cryptography can be "designed in" rather than "added on," block-encryption techniques should be considered because of their more efficient use of the algorithm, and their absence of initial synchronization requirements. For a transaction-oriented system, in which messages are very short and routed to varying destinations, the node-by-node approach appears preferable because it does not impose any per-conversation overhead for key distribution. However, for a "session"-oriented environment in which conversations may be relatively long, end-to-end encryption appears to be the obvious choice.

REFERENCES

[1] D. K. Branstad, "Encryption protection in computer data communications systems," presented at the 4th *Data Communications Symp.*, Quebec, Canada, Oct. 7–9, 1975.

[2] S. T. Kent, "Encryption-based protection protocols for interactive user-computer communications," Lab. Comput. Sci. Mass. Inst. Technol., Cambridge, May, 1976 Tech. Rep. 162.

[3] D. J. Sykes, "Protecting data by encryption," *Datamation Magazine*, Aug. 1976.

Commercial Services and Systems

This section:

- introduces the major commercial telecommunications carriers operating within the United States

- presents a detailed description of the services of GTE-Telenet, Tymnet, Western Union, and American Satellite, and compares and contrasts satellite and packet-switched services

- shows how packet switching is enhancing international data communications

- describes the operation of Tymnet switching facilities, which provide packet services at the user interface but use a structure quite different from that of most packet networks

- details various ways in which international communications services are being stimulated by the interfacing of many national packet networks.

Throughout the material presented in this book, we have provided a practical view of the concepts of modern, advanced communications technology. We have emphasized those ideas and techniques that, when integrated, will permit the establishment of highly flexible, efficient, and reliable communications. Because the practical applications of these ideas will, for most organizations, entail a combination of user-owned equipment and leased carrier services, it is important to appreciate the range of services available from the world's common carrier companies. Most users will need a mixture of user-owned facilities and services provided by commercial providers and common carriers. The balance between the facilities and services depends upon many factors, including capital investment costs, individual requirements, geographical distribution of users, traffic concentrations, and the availability of commercial services in areas where a particular user may require service. In general, telecommunications users will lease the long-haul transmission media portions of their networks but own and operate some or all of the nodal and processing facilities. If satellite networking techniques are used, the number, deployment, and ownership of satellite earth stations are also considerations.

THE DIVESTITURE OF AT&T

The most dramatic changes in the history of the telecommunications marketplace are in the process of being implemented. As a result of a long-standing antitrust suit between the U.S. Justice Department and the American Telephone and Telegraph Company, AT&T has agreed to divest itself of all of its local operating companies. These 22 wholly owned operating companies, which supply 82% of the total end-user telephones throughout the United States, provide about $\frac{1}{3}$ of AT&T total revenue. The majority of AT&T's revenue derives from its interstate and interexchange facilities that are used by most operating companies, including the 22 AT&T companies, to provide connections for long-distance (toll) traffic. It is only after 1976 that any significant amount of interexchange traffic has flowed over non-AT&T facilities. MCI Corporation initiated the competition, and was followed by others such as Southern Pacific, Satellite Business Systems, International Telephone and Telegraph (ITT) and, more recently, a multitude of service providers and service resellers. Despite this diversification of services, less than 2% of present voice-toll traffic is carried by non-AT&T carriers.

As a result of the AT&T/Justice Department consent decree, as modified following review by the Federal courts, AT&T will operate tariffed interstate facilities and, through an unregulated, arms-length subsidiary, provide competitive, nonregulated equipment and services. Through the arms-length subsidiary, now known as American Bell Incorporated, AT&T will likely be a major force in station equipment, data communications equipment and services, PBX's and local networks, as well as in enhanced value-added and data processing services. Early product offerings of American Bell include ABI Net 1000 service, which provides packet switching and enhanced processor-supported services on a nationwide basis.

Although AT&T will remain an enormous company with huge plant and resources even after divestiture, the consent decree has carefully controlled its ability to dominate the nonregulated markets. In fact, with its equal-access provisions the divested operating companies must give all interexchange carriers equally easy (or in some areas, equally difficult) access to the local subscribers in that area. The net result of new companies and services entering the marketplace will be to enhance competition and stimulate further technological innovation.

DATA COMMUNICATIONS AND SPECIALIZED CARRIERS

Because the earliest competition permitted in the U.S. communications marketplace was for specialized services, particularly data communications services, the data communications carriers have the most mature and reliable services. The following paper by L.C. Shaw introduces 12 companies that provide the overwhelming majority of total data communications services. It is interesting to note that, at the time of this market analysis, AT&T derived only 5% of its revenue from data communications services yet it accounted for about 55% of the total data communications market. With the rapid growth in data communications applications since the publication of this survey, this apportionment of revenue has undoubtedly changed. However, despite the large number of new competitors, AT&T still commands more than 50% of the data communications market. In fact, to a large extent, companies such as MCI, Southern Pacific, and SBS have recently concentrated on developing their voice-services business—in most cases at the neglect of the data communications and specialized services customers.

PACKET AND SATELLITE NETWORKS

Given the importance of both packet switching and satellite facilities in reaching highly distributed communications users, it is useful to examine in somewhat more detail the four largest providers (besides AT&T) of those types of services. The paper by C.A. Taylor and G. Williams examines the packet operations of GTE-Telenet and Tymnet and the satellite offerings of Western Union and American Satellite. This article describes the services and then compares them in terms of availability and cost. Even though the details and numbers associated with all of these services change frequently and dramatically, the overall structure of the service charges remains essentially intact. Thus, the specific information provided on costs and tariffs should be checked against current prices. The aspects of the cost structures and the methods of comparison, however, are still useful.

INTERNATIONAL SERVICES

In contrast to the competitive situation in this country, most countries outside of the United States limit competition, usually by simple virtue of the fact that all telecommunications are provided by a government-owned monopoly. Even in these situations, many countries (Canada, France, the United Kingdom, and Germany) have been very innovative in applying new technology, particularly packet switching, within their national infrastructures. In fact, Canada, France, and the United Kingdom, together with representatives of U.S. carriers, were principally responsible for the rapid development and agreement on the CCITT $\times.25$ packet network interface standard. In the third article in this section, R. Harcharik, of Tymnet Inc., provides some insight into the interoperation of packet networks on a worldwide basis. His discussion specifically relates to the extension of packet services into more than 20 foreign countries. Harcharik emphasizes that the costs for such services are materially influenced by the pricing structure in the distant-end country.

In the following paper, J. Rinde, also of Tymnet, Inc., discusses the unusal way that Tymnet implements the operation of its data communications network. That is, although it adheres to the CCITT X.25 standard for packet-networks, the internal network protocols and operation differ considerably from the protocols employed in the ARPANET and most other commercial packet networks. The entire network is always under the control of a central, supervisory program that establishes the end-to-end path through the network at the beginning of each new logical connection. This path is maintained throughout the data call, although a subsequent call between the same two locations may result in a different path as the routing algorithm optimizes the distribution of traffic. The individual transmission blocks between network switching nodes are, in effect, variable length packets which may consist of multiple logical records from up to 20 different users. Rinde's article explains the operation of this structure, and draws a comparison between this technique and the more conventional packet-switching technique where each packet is exclusively devoted to the data of a single user. Interesting comparisons of network efficiency and overhead result, depending upon user terminal speeds and traffic mix.

While Tymnet, Inc. plays an important role in the international operation of packet networks, many other vendors play their role. We conclude this volume with a paper by A. Trivedi who also discusses the spread, worldwide, of advanced packet services and includes a discussion of services that go beyond basic terminal and computer-oriented services. The potential application of such services to office automation and videotext service represents important growth areas for high-speed, high-accuracy, low-delay digital services.

An interesting sidelight on international data services is the recognition that most countries, particularly in Europe and the Far East, rely more heavily on digital message services than does the United States. While historically operating at very low speeds (50 baud or less), Telex service is the normal mode of international commerce, as opposed to long-distance phone calls so commonly used within the United States. With the increased use of electronic mail systems and the automation of many routine office functions by means of computers and intelligent terminals, the acceptance for digitally based services will quickly evolve. Not only will demand for high-quality packet-switched services grow rapidly but, given the large, intercontinental distances involved, so will the use of distributed satellite-based services. The combination of the two will lead to large, integrated, satellite/terrestrial, voice/data networks—the ultimate mechanism of international commerce and communication.

DATA COMMUNICATION CARRIERS

NAME	DATA COM REVENUE ($K)	% OF TOTAL COMPANY REVENUE	1978 DATA COM REVENUE ($K)	1979 TOTAL REVENUE ($K)	1979 TOTAL NET INCOME (LOSS) ($K)	FISCAL YEAR ENDS
AT&T	2,309,150	5	1,845,000	46,183,000	5,674,000	Dec. 31
GT&E	797,656	8	206,960	9,957,817	612,193	Dec. 31
Western Union Corporation	451,227	62.8	171,900	718,472	4,033	Dec. 31
ITT World Communications Inc.	169,788	100	125,700	169,788	41,707	Dec. 31
United Telecommunications	68,098	3.8	48,800	1,792,078	182,887	Dec. 31
TRT	28,382	100	—	28,382	7,119	June 30
Tymnet	23,900	100	16,500	23,900	2,900	Dec. 31
Continental Telephone	21,370	1.9	14,939	1,347,587	114,296	Dec. 31
Central Telephone	20,000	3.3	16,200	750,500	83,100	Dec. 31
WUI	16,425	15	4,600	110,163	19,656	Dec. 31
COMSAT	16,283	6.2	11,366	262,635	40,185	Dec. 31
RCA	15,819	1	14,680	7,454,600	—	Dec. 31
Graphic Scanning	12,033	43	9,112	27,985	2,201	June 30
SP Communications	9,900	10	4,990	99,000	14,900	Dec. 31
Rochester Telephone	8,064	4.7	7,700	171,574	26,950	Dec. 31
MCI	6,412	5	3,702	128,252	10,925	Mar. 31
American Satellite	5,619	35	3,565	16,057	(4,812)	Dec. 31
FTC Communications	1,100	20	647	5,500	—	Dec. 31

The business of the carriers listed represents essentially all of the data transmission revenue generated in the U.S. or between the U.S. and other nations. Communications is a regulated game, but we know efforts are under way to reduce that regulation which we feel will be reflected in new names appearing on this list in the coming years. For the present, however, it's still a corporate field with the likes of AT&T, GT&E *and* ITT.

AMERICAN TELEPHONE AND TELEGRAPH COMPANY
195 Broadway
New York, NY 10007
(212) 393-9800

The principal business of the Bell System is communications. Most of the work is done through a group of individual operating companies that provide an integrated nationwide network of facilities that can transfer information in virtually any form—voice, data, graphics—between appropriately equipped terminals throughout the country. This business provided AT&T with an annual revenue in excess of $46 billion in 1979. We do not know how much of this is directly attributable to *data* communications. Since Bell refuses to break out even a hint of these figures, we relied on Bell watchers, available data, and gut feelings to reach the estimate that 4% of Bell's revenue comes from data transmission, while another 0.5% comes from tariffed products, including data sets and Teletypes used in data transmission. Our best estimate puts the data com revenue figure at slightly more than $2.3 billion for 1979.

GENERAL TELEPHONE & ELECTRONICS CORPORATION
One Stamford Forum
Stamford, CT 06611
(203) 357-3797

GTE remains number two in the telephone business. GTE Communications Network Systems was established in December 1979 to provide private and public voice, data, and message communications systems tailored to meet the diverse requirements of corporations, government agencies, and other organizations. This group brings together the expertise and resources of three companies to form Communications Network Systems Resources. GTE Telenet, first on the list, is a newly acquired but well-established supplier of public and private data communications network services and systems. GTE Telecommunications Systems is a major supplier of private all-digital switching systems for voice and data. The third company in Communications Network Systems Resources is GTE Information Systems, a supplier of financial data base services for business. GTE's data communications figures for 1979 are impressive. The company claims 8% of the annual revenue is data com related, a figure totaling $797.6 million—up from the $206.9 million data com revenue figure of a year ago.

WESTERN UNION CORPORATION
One Lake Street
Upper Saddle River, NJ 07458
(201) 825-5316

If Western Union Corporation would send itself a telegram it would probably sing "Better luck next year." WUC heads into its 128th year of business following a less-than-successful 127th year of business in 1979. For the corporation as a whole, net income was $4 million, compared with $45.7 million the previous year. The total operating revenue picture looked better with revenues up 11% to $718 million in FY '79. As a holding company, WUC conducts business through subsidiaries with its Western Union Telegraph Company, a communications common carrier and principal holding, engaged primarily in the business of providing telecommunications services to business, government, and the public. In 1979 the corporation's two largest nonregulated subsidiaries—Western Union Data Services and National Sharedata —were unprofitable, and it became clear that substantial additional investment would be required to rehabilitate them. National Sharedata was sold and Data Services reconstructed by reducing the work force by 45% and integrating the functions of sales and maintenance of Data Services' terminals with those of the Telegraph Company. Without these two companies, WUC estimated a 62.8% data com share of the annual revenue, which this year totals $415.2 million.

ITT WORLD COMMUNICATIONS INC.
67 Broad Street
New York, NY 10004
(212) 797-3300

ITT World Communications Inc. is one of 14 groups within the ITT Communications Operations Group (COG). Last year Worldcom extended its record, voice, and data com services into broader markets and inaugurated several services. City-Call service offers savings of up to 40% on long-distance calls between 88 cities in the United States. The Marsplus service links travel agents to reservations computers of participating airlines. Similar services are planned for hotels, car rental agencies, and cruise lines. FaxPak service now provides a nationwide network linking previously incompatible facsimile terminals. Worldcom also offers telex, telegram, leased channel, and other services between the U.S. and other countries. ITT recently announced a new service named Intertext, which it claims is an alternative to telex. Intertext is under the wing of another subsidiary, ITT Domestic Transmission Systems, Inc. Worldcom lists its 1979 annual revenue at $161.8 million, of which 100% is data communications related.

UNITED TELECOMMUNICATIONS, INC.
Box 11315
Kansas City, MO 64112
(913) 676-3232

United Telecommunications offers a fine mix of resources. Its telephone system is the third largest in the country, with 4.5 million telephones installed. The company owns United Computing Systems, a computing service organization, and also has a 24% share in data com manufacturer Rixon. In addition to this, United has a software sales activity, and owns Calma, a manufacturer of interactive graphics equipment. North Supply Company, another affiliated company, distributes telecommunications, electrical, and security and alarm products. This year the data communications line (without Rixon) accounted for 3.8% of the annual revenue, or about $68 million. This is a $20 million increase from last year.

TRT TELECOMMUNICATIONS CORPORATION
1747 Pennsylvania Avenue N.W.
Washington, DC 20006
(202) 862-4556

TRT, a subsidiary of United Brands, Inc., started business in 1903 as Tropical Radio Telegraph Company—the communications link of the United Fruit Company and its fleet of banana ships. In the early '70s TRT was operating one of the primary telex networks, serving the Caribbean and Latin America. In April 1979, TRT acquired Norfield Electronics (now TRT Data Products), which produces corporate message switching systems for data, telex, and TWX. Among the products that TRT recently introduced is the multispeed Stortex, an automated high speed store-and-forward telex system, and the Eltex IIE system, which delivers a 400% capacity increase in switching telex messages. TRT achieved a 24% growth rate over '78, bringing the '79 total revenue figure to $28.4 million.

TYMNET, INC.
20665 Valley Green Drive
Cupertino, CA 95014
(408) 446-6659

The largest U.S. public packet communications network, Tymnet provides local telephone-call access for about 250 U.S. metropolitan areas and, via IRCs, from 28 countries. Approximately 300 host computers are connected to the network. Services include advanced electronic mail system (Ontyme II); a wide variety of financial services to the credit/finance industry; and private network services. Typical devices with which Tymnet interconnects include asynchronous and synchronous terminals, polled terminals, transaction terminals, and synchronous and asynchronous computer interfaces. Network applications include time-sharing, order entry, credit card processing, inventory systems, data base management, and flight planning. Tymnet says that data com is 100% of the annual revenue, which totaled $23.9 million in '79, up from $16.5 million in '78.

CONTINENTAL TELEPHONE CORPORATION
56 Perimeter East
Atlanta, GA 30346
(404) 391-8446

Continental, a small community telephone company that does not handle heavy metropolitan business, is the nation's fourth largest telephone company. In 1979, Continental enhanced its terrestrial communications capabilities by moving into the satellite field. It entered into an agreement to own one-half of American Satellite Corporation's business and signed agreements with two Western Union Corporation subsidiaries to secure the furnishing of available satellite capacity to American Satellite. ConTel Data Services Corporation, the data com arm of the company, provides automatic data processing services in payroll, customer billing, accounting, and processing of long distance revenue for system companies. A major data center is located in each of the company's three regions. It estimates the data com revenue to be 1.9% of the total revenue, or $21.4 million. This is a one-third increase over '78's $14.9 million figure.

CENTRAL TELEPHONE & UTILITIES CORPORATION
O'Hare Plaza
5725 East River Road
Chicago, IL 60631
(312) 399-2749

Central Telephone & Utilities is the fifth largest regulated telephone company in the industry with 1.9 million telephones installed. The biggest portion of its business is in Nevada, followed by Florida and Illinois. Highlighting the company's utility business is the recent acquisition of Digitech Data Industries. CTU estimates its data com revenue at 3.3% of annual revenue. With roughly $4.5 million attributable to Digitech and the manufacturing end of the business, we estimate another $20 million attributable to the carrier portion of the business, up from $16.2 million one year ago.

WESTERN UNION INTERNATIONAL, INC. (WUI)
One WUI Plaza
New York, NY 10004
(212) 363-6400

The family tree for WUI, Inc. now reads: the major subsidiary of Western Union International, which became a wholly owned subsidiary of the Xerox Corporation in 1979. WUI, Inc. is the number three international record carrier behind RCA Globcom and ITT Worldcom. Western Union International, Inc. provides international telecommunications services, including telex, cablegrams, leased channels (both teletype and voice/data), facsimile, datel, data base (DBS), high speed data (50- and 56-kilobit), maritime satellite (Marisat) and satellite television. The company recently began an expansion of data services to additional points. It estimates the data com revenue to be about 15% of annual revenue, or about $16.4 million.

COMSAT
Communications Satellite Corporation
950 L'Enfant Plaza
Washington, DC 20024
(202) 554-6000

Comsat is engaged primarily in providing international, maritime and domestic communications satellite services. It furnishes satellite services to common carrier companies for communications principally between the United States and foreign points. It is the U.S. agent for Intelsat, from which it derives more than half of its revenue; the rest is from Comstar satellites leased to AT&T for U.S. domestic communications. A subsidiary of Comsat General is a partner with subsidiaries of IBM and Aetna Life & Casualty in Satellite Business Systems (SBS). We estimate that the data com revenues are at 6.2%, or $16.3 million, up from '78's mark of $11.4 million.

RCA CORPORATION
30 Rockefeller Plaza
New York, NY 10020
(212) 598-5900
With the sale of RCA Alaska Communications on June 1, 1979, RCA Corporation is left with two communications subsidiaries: RCA Global Communications (Globcom) and RCA American Communications (Americom). Globcom marked its 60th anniversary year hauling in 37% of total international record carrier industry revenues. It offers a broad variety of communications services to more than 200 overseas locations. Americom achieved profitability for the first time in 1979 as demand for satellite services exceeded the capacity available on the company's two orbiting spacecraft. To meet this growing demand RCA launched its third domestic communications satellite on Dec. 6. However, contact with the satellite was lost four days later during a maneuver to place it into a permanent orbit over the equator. RCA plans another launch for June 1981. A new wide-band data service for business and industry—56 Plus—was introduced in 1979. We estimate the total data com revenue to be much less than one percent (.0024%) of the total revenue, or $15.8 million.

GRAPHIC SCANNING CORPORATION
99 West Sheffield Ave.
Englewood, NJ 07631
(201) 569-7711
Deciphering the revenue figures of Graphic Scanning and its communications carrier subsidiary, Graphnet, is not an easy task. The main portion of the corporate business consists of receiving unstructured customer data, compiling and formating it, selecting a route and carrier for its transmission, translating the outgoing message to a format and protocol for whatever terminal or computer is to receive it, and then transmitting it over other carriers' lines. Only a limited amount of the functions performed, 10% Graphics claims, are pure data processing (such as data base update), and no inquiry response is said to be supported. Of the $28 million annual revenue that Graphic Scanning charted last year, we figure slightly more than $12 million is from data com. Other business includes radio paging, using its own subsidiary radio common carrier, and a telecommunications system through another subsidiary, Comnet.

SOUTHERN PACIFIC COMMUNICATIONS COMPANY
One Adrian Court
Burlingame, CA 94010
(415) 692-5600
SP Communications Company (SPC) is a wholly owned subsidiary of the Southern Pacific Company, the train people. The initial SPC network, established in 1970, followed the right-of-way of the existing Southern Pacific railroad system from San Francisco through Dallas. SPC's growth and expansion to a national system required completely new northern routes. In July 1974, SP Communications became the first of the nation's specialized common carriers to offer coast-to-coast private line communications service. It now provides nationwide private line and real service to 80 metropolitan areas, and uses a combination of land-based microwave, broadband cable, and satellite facilities. We estimate SPC's data com revenue to be almost $10 million, or about 10% of annual revenue.

ROCHESTER TELEPHONE CORPORATION
100 Midtown Plaza
Rochester, NY 14646
(716) 325-9871
Rochester is the seventh largest telephone company, with holdings mostly in New York State. The origin of the company was in 1899; in 1921, the company assumed its present name and form. In 1959, Rochester Telephone became the first unaffiliated, independent telephone company to be listed on the New York Stock Exchange. Rochester Telephone Corporation, along with two operating subsidiaries, Highland Telephone and Sylvan Lake Telephone Company, gives regular service to over 326,000 customers. An unregulated subsidiary, Rotelcom Inc., was started in 1978 to provide corporate growth through "new opportunities" in telecommunications; late in 1979 Rotelcom announced establishment of a data communications subsidiary. We estimate this subsidiary to be responsible for about 4.7% of the annual revenue, a figure totaling slightly more than $8 million for 1979.

MCI COMMUNICATIONS CORPORATION
1150 Seventeenth St. N.W.
Washington, DC 20036
(202) 872-1600
MCI, a long distance carrier employing microwave circuits to serve business users, was authorized in 1969 as the first of the specialized common carriers. Its primary business is intercity long distance voice telecommunications. The majority of MCI's data com income is derived from dedicated leased lines arranged exclusively for data transmission. MCI also offers the Execunet service, a one-way, dial-in-dial-out, shared intercity service enabling customers to originate or terminate calls for MCI-served cities via MCI-provided local

business telephones. For calendar year 1979, MCI claims that 5% of its annual revenue was generated in data com—a total of $6.4 million.

AMERICAN SATELLITE CORPORATION
20301 Century Blvd.
Germantown, MD 20767
(301) 428-6000
American Satellite Corporation, a subsidiary of Fairchild Industries, Inc., has been a satellite carrier since 1974. In mid-1979, Fairchild and Continental Telephone Corporation agreed to share jointly in the ownership and operation of the American Satellite Corporation. Toward year-end, Fairchild and Continental entered into agreements with Western Union Corporation to acquire ownership of satellite capacity. ASC concentrates on data transmission as well as commercial broadcasting, fax, teletypewriters, etc. It offers two basic types of service—leased lines through its own satellite, and customer-site dedicated earth stations. We estimate ASC's data com revenue to be 35% of its annual revenue, or about $5.6 million, up $2 million from '78.

FTC COMMUNICATIONS
25 Broad Street
New York, NY 10004
(212) 747-5660
FTCC calls itself the smallest of the five recognized international carriers. Among its offerings are cablegram service to 133 countries, overseas telex message service via its computerized Eltex Switch, and Datel service to Europe. Data com revenues for 1979 totaled $1.1 million.

Considering the alternatives to AT&T

Carol A. Taylor and Gerald Williams,
Auerbach Publishers Inc., Pennsauken, N.J.

Value-added networks and satellite carriers are proliferating. Links using these services may offer advantages in cost, efficiency, and capacity.

The majority of today's data communications networks use AT&T's phone lines to haul their bits. But there are alternatives. Unlike AT&T's facilities, value-added networks typically charge by traffic volume instead of distance traversed. In addition, satellite services can accommodate high-volume data traffic more efficiently. Whether a network has a single link or a thousand, the user should explore these options.

There are two basic approaches to implementing a data communications network. The first is to design a network using leased and/or switched lines offered by common carriers. In so doing, however, the user is responsible for interfacing his equipment to the carrier's lines, checking for transmission errors, and designing software to support the communications. Without extensive knowledge of communications technology, setting up such a network is a formidable task.

The alternative is to select one of the commercial communications networks, which consist of privately owned or leased public facilities that use either terrestrial or satellite transmission technologies. However, choosing the appropriate commercial network for a business is also a complicated undertaking—one that we will examine here.

This article evaluates and compares four of the many commercial network services: two value-added networks (Tymnet and GTE Telenet) and two satellite carriers (American Satellite Corporation and Western Union). The selection of these services is not an en-

dorsement; rather, these networks represent the current technological options. Specifically, GTE Telenet and Tymnet apply different philosophies in their packet-switching networks, and the two satellite services offer varying approaches in establishing satellite communications links.

The relative values of packet-oriented networks and satellite carriers (their interface capabilities, geographic scope, transmission efficiency, integrity of data, and cost) must be considered in light of a company's requirements. The network planner should carefully study the strengths and weaknesses of each commercial network in order to find a cost-effective solution to the organization's communications problems.

As in any decision to procure data processing equipment, the user must examine the practicality of setting up a network. This requires a detailed understanding of both the present and projected information flow within the company. The study should include such statistics as the number of stations in the network, the technology of these stations (interface, programmability, and so on), their physical locations, and the anticipated amount of data traffic.

Armed with this information, the user can choose a network based on three major factors: the technology of the network, its geographic scope, and its cost.

The geographic scope and availability of a network are key considerations in three of these four services. [Only American Satellite Corporation (ASC) with its

1. GTE TELENET TECHNOLOGY: PACKET SWITCHING

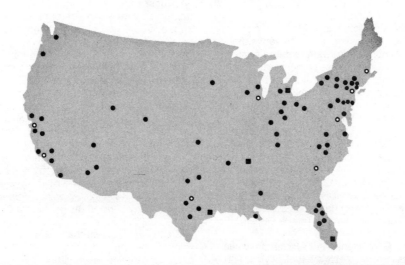

○ CLASS I TCOs

■ CLASS II TCOs

● CLASS III TCOs

Telenet access points

CLASS I TCOs (56-KBIT/S ACCESS)
San Francisco, Boston, Newark, Washington D.C.,
    Atlanta, Chicago, New York, Dallas, Los Angeles

CLASS II TCOs (9.6-KBIT/S ACCESS)
Detroit, Houston, St. Louis, Miami

CLASS III TCOs (1.2-KBIT/S ACCESS)
Ohio: Cincinnati, Akron, Cleveland, Columbus,
    Dayton, Toledo, Youngstown
N.Y.: Albany, White Plains, Buffalo,
    Hempstead, Rochester, Syracuse
Pa.: Philadelphia, Allentown, Pittsburgh
Tex.: Austin, Ft. Worth, San Antonio
Calif.: Colton, Glendale, Sacramento,
    Marina Del Ray, San Carlos, San Diego,
    San Jose, San Pedro, Santa Ana, Ventura
N.C.: Charlotte, Winston-Salem, Raleigh/
    Durham
Conn.: Hartford, New Haven, Stamford
Mass.: Springfield
Fla.: Ft. Lauderdale, Tampa, Orlando,
    St. Petersburg, Jacksonville
Mich.: Ann Arbor
Md.: Baltimore
Ala.: Birmingham
Colo.: Denver
Ind.: Indianapolis
Mo.: Kansas City

CLASS III TCOs (cont.)
Nev.: Las Vegas
Ky.: Louisville
Wis.: Madison, Milwaukee
Tenn.: Memphis, Nashville
Minn.: Minneapolis
La.: New Orleans
Va.: Norfolk, Richmond
Okla.: Oklahoma City, Tulsa
N.J.: Paterson, Trenton
Neb.: Omaha
Ariz.: Phoenix, Tucson
Oreg.: Portland
Utah: Salt Lake City
Wash.: Seattle
Del.: Wilmington

Telenet costs

PUBLIC DIAL-IN PORTS
Standard (to 1.2 kbit/s) $3.25/hr

WATS 15.00/hr

PRIVATE DIAL-IN PORTS
110-4.8K bit/s $160-450/mo

TWX PORTS 210/mo

DEDICATED ACCESS FACILITIES
50-1.8K bit/s $300-380/mo

2.4-9.6 kbit/s 1,100/mo

56 kbit/s 2,100/mo

CHARGE PER THOUSAND PACKETS $0.50

PRIVATE PACKET EXCHANGE
Controller $800/mo

Ports (110-56K bit/s) 60-200/mo

TELENET PROCESSORS
TP 1000 (to 300 bit/s) $240-600/mo

TP 2200 (with no ports) 550/mo

    per asynchronous port 120-200/mo

    per synchronous port 175/mo

TP 4000 (with no ports) 850/mo

    per asynchronous port 120-200/mo

    per synchronous port 175/mo

SDX service and on-site antennae is exempt from this concern. This ASC service is, however, proportionately more expensive than the others.]

If, for example, a customer is a significant distance from a service's access point, it may be very expensive to develop a communications path to enter the network. In addition, some of these carriers offer differing services between particular cities. For example, GTE Telenet supports a maximum transmission speed of 9.6 kbit/s only in certain locations. Let us assume that a company wants to use 9.6-kbit/s transmission from a site in Tampa, Fla. The closest Telenet access point supports only 1.2-kbit/s transmission. Therefore, in order to achieve the desired transmission rate from this site, the company has to access the service at a different point. In this example, the nearest access point which supports 9.6-kbit/s transmission is in Atlanta, and some connection (leased or switched lines) must be established between the Tampa site and the Atlanta access point.

As mentioned, GTE Telenet and Tymnet are value-added networks (VANs). VANs are typically configured around a series of geographically separated nodes, or computers, which control data transmission between a user's terminals and host computer, and often regulate the operation of the network as well. The terminals and host computer are connected to the nodes through switched or leased lines that are usually not supplied by the VANs. As the name implies, VANs provide enhancements to the basic services offered by the common carriers. These can include converting data from one character set to another, adjusting transmission speeds for maximum data flow without compromising the data's integrity, permitting a network device to talk to any other device on that network, routing data along the most efficient path, and reducing transmission costs by multiplexing data from multiple sources onto fewer telephone lines. The VANs generally claim an error rate of approximately one error in 10^8 bits. Services vary, and include packet-switching, message-switching, and electronic mail capabilities. The cost of using VANs is volume-sensitive, whereas common carriers' tariffs are distance-sensitive.

Satellite transmission

Satellite-based transmission services provide high-speed (over 56 kbit/s) communications between sending and receiving stations. Instead of nodes, however, the satellite services use antennae to send and receive signals via geostationary satellites orbiting approximately 22,300 miles above the earth. The user connects his terminals and host computer to the antennae in much the same manner as to a VAN node. Satellite networks are capable of accommodating many types of traffic, including data, voice, alternate data and voice, and video. Satellite facilities provide comparable error rates, and most carriers also claim performance levels of one error in 10^8 bits. Unlike that of VANs, the cost of satellite services is not volume-sensitive, but it includes charges for accessible equipment and channels (bandwidth).

GTE Telenet uses a packet-switching network to provide data communications between customer stations at speeds ranging from 50 to 56K bit/s. It accommodates a wide variety of terminal speeds and protocols, and customer stations communicating over the network do not have to operate at the same speeds or use the same protocols. Within the network, customer data is assembled into packets containing a maximum of 128 characters (1,000 bits). These data packets are then transmitted through the network to the receiving node along the fastest path, transparent to the user. Upon arriving at the destination, the packets are reassembled by the nodes into the data's proper form and sequence.

The actual number of characters in each packet is determined, to a large extent, by three or four operational parameters selected by the customer at the time of network installation. As will be demonstrated later, the efficiency of using the packets is directly related to the cost of this network service.

Customers connect to the GTE Telenet nodes through asynchronous single-port (including private and public dial-in), asynchronous multiple-port, or station-to-station links. Customer stations link with one another, through the network, over virtual connections. Once a virtual connection is established, data is dynamically routed by the network from one node to another along the fastest path. Customer stations attached to an access node through an asynchronous single-port can communicate over a single virtual connection only, so no further addressing is necessary during the session. Stations attached to the network by means of an asynchronous multiple-port can concurrently operate up to 255 independent virtual connections to different customer stations. Station-to-station network links can be set up (through the nodes) between either compatible customer stations operating at the same speed or incompatible stations operating at different speeds. These stations are equipped with either asynchronous or synchronous interfaces and access the network through leased lines.

Over station-to-station links, dissimilar asynchronous stations operate in a network virtual terminal (NVT) mode, with the network handling differences in code, format, and timing. Synchronous stations are connected to the network by station-to-station links that may likewise use single- or multiple-connection ports. When connected to a single-connection port, only synchronous stations with the same protocol can communicate with each other. Although the two stations can function at different speeds, the average data transfer rate from the sending station must be compatible with the receiving station, as the network provides limited buffering. When synchronous stations are attached to the network through multiple-connection ports, the stations can communicate with each other or with asynchronous stations using any standard network speed and standard transmission protocol.

The GTE Telenet service protects data integrity by using acknowledgment packets, which are sent by the receiving node to the transmitting node upon arrival of a packet. If the transmitting node does not receive an acknowledgment or if an error is detected, then the

2. TYMNET

TECHNOLOGY: STORE-AND-FORWARD MESSAGE SWITCHING

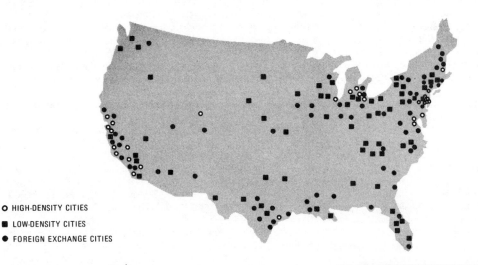

○ HIGH-DENSITY CITIES
■ LOW-DENSITY CITIES
● FOREIGN EXCHANGE CITIES

Tymnet access points

HIGH-DENSITY ACCESS CITIES
Calif.: El Segundo, Los Angeles, Mountain View,
    Newport Beach, Oakland, Palo Alto,
    Riverside/Colton, San Francisco, San Jose/
    Cupertino, Ventura/Oxnard
Colo.: Denver
Ill.: Chicago
Washington D.C.
Mass.: Boston/Cambridge
Md.: Baltimore
Mich.: Ann Arbor, Detroit, Plymouth, Southfield
N.J.: Englewood Cliffs, Lyndhurst, Wayne,
    Newark/Union, Piscataway
N.Y.: New York
Pa.: Philadelphia
Tex.: Houston
Va.: Arlington

LOW-DENSITY ACCESS CITIES
Ala.: Birmingham
Ariz.: Phoenix, Tucson
Calif.: Hayward, Sacramento, San Diego,
    Santa Rosa
Conn.: Darien, Hartford
Ga.: Atlanta
Fla.: Jacksonville, Miami, Orlando, Tampa,
    St. Petersburg
Iowa: Des Moines, Iowa City

LOW-DENSITY ACCESS CITIES (cont.)
Idaho: Boise
Ill.: Freeport, Rockford, Springfield
Mass.: Springfield
Ind.: Indianapolis, South Bend
Minn.: Minneapolis
Kan.: Wichita, Shawnee Mission
Mich.: Jackson, Kalamazoo
Mo.: Kansas City, St. Louis
La.: Baton Rouge, New Orleans
Neb.: Omaha
N.C.: Durham/Raleigh, Winston-Salem
N.J.: Princeton
Nev.: Reno/Carson City
Okla.: Oklahoma City, Tulsa
N.Y.: Buffalo, Corning, Rochester,
    Syracuse, White Plains
Ohio: Akron, Cincinnati, Cleveland,
    Columbus, Dayton
Pa.: Erie, Pittsburgh, Valley Forge
Oreg.: Portland
Tenn.: Chattanooga, Memphis, Nashville
Tex.: Austin, Dallas, El Paso, Midland,
    San Antonio
Utah: Salt Lake City
Wash.: Richland, Seattle
Wis.: Madison, Milwaukee

Tymnet costs

ASYNCHRONOUS INTERFACES (TO 1.2 KBIT/S)
Single user $100/mc
Up to 8 users 1,000–1,250/mo
Up to 16 users 1,500–1,750/mo
Up to 62 users 2,150–2,750/mo

SYNCHRONOUS INTERFACES (TO 4.8 KBIT/S)
Up to 64 users 1,400/mo
Up to 256 users 2,150/mo

CONNECT TIME
High-density (to 4.8 kbit/s) $1–5/hr
Low-density (to 4.8 kbit/s) 4–8/hr
Foreign exchange (to 1.2 kbit/s) 5–6/hr
WATS (to 1.2 kbit/s) 14–15/hr

DEDICATED PORTS (TO 300 BIT/S)
Each, for first 15 ports $475–650/mo
Each, for over 15 . . . . 300–400/mo

USAGE CHARGES (110–300 BIT/S)
First 40 million characters . . . . $.10/thousand
From 40–80 million characters 08/thousand
Over 80 million characters 01/thousand

USAGE CHARGES (1.2–4.8 KBIT/S) . . . $.03/thousand characters

packet is automatically retransmitted.

GTE Telenet services are available throughout the continental U.S. The services are accessible in three classes of transmission speeds, which depend on the location of the Telenet Central Offices (TCO). Class I, Class II, and Class III TCOs have access ports which support maximum speeds of 56, 9.6, and 1.2 kbit/s, respectively. Figure 1 lists the locations of Class I, II, and III TCOs.

The several means of network access include public dial-in ports (which use local telephone exchanges or In-WATS telephone connections), private dial-in ports, leased lines, TWX lines, and foreign exchange channels. There are also dedicated access facilities, which consist of a Telenet processor (TP) and an X.25 software interface. In addition, GTE Telenet recently acquired Cambridge Telecommunications, which had developed many synchronous interfaces (BSC, SDLC, HDLC) for the network. Telenet will most likely tariff these features in the near future.

Interface processors

Three microprocessor-based data communications processors, the TP 1000, TP 2200, and TP 4000, enable communications between the GTE Telenet network and the user's computers and terminals without requiring changes in the installation's hardware or software. Connections between user stations and TP access ports are established by either directly connected cables or leased and dial-up facilities.

The TP 1000 is a host- and terminal-interface concentrator that supports up to 14 asynchronous ports operating at speeds of 75 to 300 bit/s. The TP 2200, a host-interface processor, supports several hundred asynchronous ports at speeds up to 9.6 kbit/s. The TP 4000, which interfaces computers and terminals to Telenet or other X.25-compatible networks, provides protocol-conversion and packet-switching facilities. It accommodates computers and terminal equipment operating at speeds up to 9.6 kbit/s using ASCII, EBCDIC, or correspondence codes, and can be configured with a maximum of 480 ports. As a rule, the TPs are located at the nodes, although they can be placed at the customer's site.

TPs located at a customer's site require a dedicated access facility (DAF) in order to be connected to the network. On-site TPs are attached to DAFs by common carrier lines. With the TP 1000, the DAF operates at 1.2 kbit/s. Two DAFs are needed, though, when the TP 1000 is configured with 14 asynchronous ports. With the TP 2200 or TP 4000, each DAF operates at speeds ranging from 2.4 to 56 kbit/s.

TPs also permit the customer to use GTE Telenet's rotary and private network features. The rotary feature allows customers to designate that certain ports are grouped together, assigned a single network address, and serviced by a single TCO. Thus, a virtual connection can be established between a distant station and the first available port in the group. The private network feature is a control mechanism that lets a customer restrict access to and from his ports. Both features must be specified before start-up of the network.

Additional Telenet offerings are a private packet exchange (PPX) service, Nightline service, and Hot Line Data Service. The PPX service reserves a group of ports in a TCO for a specific customer and performs as the customer's multiplexer, concentrator, or packet switch. It is accessed through dial-in ports, DAFs, or leased access ports. Nightline service offers reduced rates for public dial-in subscribers who use the Telenet network during prescheduled, light-traffic hours. The Hot Line Data Service gives customers the option of a fixed monthly charge in lieu of packet charges for all traffic between a specified pair of leased network ports. Prices and transmission speeds for GTE Telenet's ports and services are listed in Figure 1.

The GTE Telenet service can also be accessed by subscribers outside the continental U.S. Access is presently offered from 30 foreign countries, as well as Puerto Rico and Hawaii. Canadian access is available in 55 Canadian cities though an interconnection between GTE Telenet and Canada's Datapac. Access in Mexico is provided by a connection between GTE Telenet and Teleinformatica in Mexico. Access from Hawaii is through an interconnection between GTE Telenet in the continental U.S. and Hawaii's Dasnet (an X.25 packet-switching network run by the Hawaiian Telephone Company). With the exception of U.S.-to-U.K. service, access to Telenet from abroad is limited to a maximum speed of 300 bit/s. Between the U.S. and the U.K., the full range of services up to a speed of 9.6 kbit/s is available.

GTE Telenet's geographic scope, along with its range of supported transmission speeds and interfaces, offers relatively facile access to most prospective users. Note, however, that the transmission speeds are limited by the class of TCO a customer uses.

A major concern in using GTE Telenet, as will be later discussed, is efficiently using the packets to transmit data. This can be achieved if the user carefully selects the values assigned to several variables. For example, there is a timeout mechanism that determines the maximum amount of time a packet remains at a node while being filled. By optimizing this value to suit his particular needs, the customer can have fairly efficient transmissions.

Tymnet services

Tymnet, another value-added network, uses a store-and-forward message-switching philosophy. Tymnet nodes accept information from several terminals (not necessarily from the same customer), separate the data by destination, and then multiplex the data into full packets. When the packet is full, it is sent to its destination along a preselected path. The receiving node then disassembles the packet and sends the reconstructed information to the user's addressed destination device. A single packet, therefore, can contain characters from several terminals. From the user's viewpoint, messages are sent without regard to packet efficiency. Figure 3 compares the Tymnet and GTE Telenet packet-switching philosophies.

Each packet transmitted over the Tymnet network contains a checksum, a binary sequence used for error

3. Network comparison. In the Telenet network, packetized user data remains intact, although the route taken through the network may vary from packet to packet. In the Tymnet network, data from several users is multiplexed into larger packets, which follow the same route from entry node to destination node.

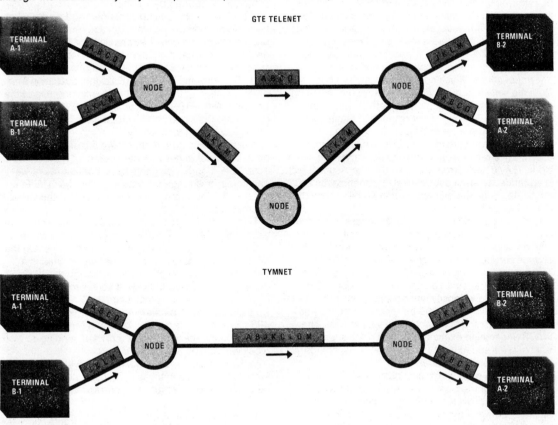

detection. When an error is found, the data packets are retransmitted automatically. If traffic is congested on any preselected path, the Tymnet nodes initiate alternate routing and connection re-establishments.

The Tymnet service is based on what it calls advanced communications technology (ACT), which combines multiprogramming and memory protection. ACT uses two communications tools: the Tymnet Engine, a minicomputer that drives the network nodes, and its software operating system called the Internationally Switched Interface System (ISIS).

The Tymnet Engine, based on a 32-bit microprocessor-complemented minicomputer, accommodates up to 32 high-speed synchronous I/O channels (operating at up to 56 kbit/s) or up to 256 medium-speed asynchronous channels.

The operating system, ISIS, allows the interconnection of dissimilar terminal devices and adjusts to different protocols automatically. Under ISIS, each job slot, or partition, has a specific communications function. These partitions communicate externally through either the network protocol or the protocol of the device it supports. Incoming data is converted to a common internal format, passed to a special software routine

called the dispatcher, transferred to another partition, and then reconverted to the protocol required by the receiving device. ISIS supports asynchronous and synchronous interfaces, as well as the IBM 3270, X.25, X.75, X.27, X.28, HDLC, SDLC, and ADCCP protocols.

Users can access the Tymnet network over dial-up or private lines. Terminals are connected to terminal interface processors, called Tymsats, which match transmission speeds and perform code conversions automatically. A host computer is connected to the network in asynchronous or synchronous mode. The host interface, called Tymcom, is a software module located in the host-end network node. If the connection is synchronous, each Tymcom emulates a bisynchronous cluster controller. Up to four host computers accommodating as many as 256 simultaneous users can be configured with a synchronous Tymcom.

In addition to the network-based services, Tymnet also provides OnTyme, a store-and-forward message-switching service. OnTyme subscribers can create a message, edit it, then send it within a geographically dispersed corporate network or to other Tymnet subscribers. The message, which is held in Tymnet system files, can be either subsequently recalled or stored indefinitely. System files may also hold a user's distribution list, which a user can reference to send a message to multiple accounts. In addition, messages can be retrieved on demand or automatically on a predefined schedule. Customers of OnTyme can check for and retrieve lost messages, and they receive detailed reports of messages sent and received by each user. Every message is assigned a sequence number and stamped with the transmission time and date.

Tymnet nodes, which contain Tymsat, Tymcom, and OnTyme software, are intelligent processors based on general-purpose minicomputers. Both Tymsats and Tymcoms, in addition to their respective functions as terminal and computer interfaces, provide an error detection/retransmission capability which ensures essentially error-free data transmissions over the network. They detect errors by using the checksum, which is appended to each data packet. Tymnet claims an error rate of one undetected error in 40 billion bits of data (4×10^{10}).

Either nationwide or international access to Tymnet is possible. Users may access a high-density access city, a low-density city, a city served by a foreign exchange service, or a city served by WATS facilities. The major difference between the accesses is in the price and the transmission speeds of the ports (Fig. 2). High- and low-density cities support the same transmission speeds (110 to 300 bit/s asynchronous and 2.4 to 4.8 kbit/s synchronous) but vary in cost. The WATS and foreign exchange access ports support transmission speeds of 110 to 300 bit/s asynchronous and 1.2 kbit/s synchronous.

The Tymnet service can also be accessed in over 25 foreign countries, as well as Alaska and Hawaii. Connection is provided under agreements (where applicable) between the U.S. International Record Carriers (IRCs) and the communications administration of each international access country, except Canada. Canadian access to Tymnet is supported through Canada's Datapac network, which offers gateway connections to both Tymnet and Telenet. The Tymnet/Datapac connection enables local access from 57 cities in Canada and 150 cities in the U.S. Access from Italy is achieved through the Direct Access to Remote Databases Overseas (DARDO) provided by Italcable (Italy's IRC) and Tymnet. Access from Hong Kong is provided by an International Database Access Service (IDAS). Both DARDO and IDAS can support simultaneously as many as 64 terminal users to U.S. locations.

VAN comparisons

Tymnet's competitive edge over GTE Telenet results primarily from its geographic scope. Tymnet services many smaller cities that GTE Telenet does not; therefore, it can cover more of a corporation's office locations. When one considers the additional cost of attaching a terminal to a node in a distant city, the proximity of that terminal to an access city (or port) becomes an important factor in selecting a service.

However, it should be pointed out that the two services do not offer the same range of transmission speeds. Whereas GTE Telenet supports transmission speeds of up to 9.6 and 56 kbit/s, the maximum supported by Tymnet is 4.8 kbit/s. This disparity is offset to some degree by the additional compatibilities and protocols supported by Tymnet.

GTE Telenet charges 50 cents per thousand packets transmitted over the network. As Jonathan P. Smith and Pete Moulton demonstrated in their article, "Telenet, Tymnet: What the difference is" (DATA COMMUNICATIONS, October 1978, p. 67), the actual price per character of information in the GTE Telenet system is a direct function of the efficient (or inefficient) use of the packets. With GTE Telenet services, one character might cost the same as 128 characters because of the packet-switching technology.

Assume, for example, that there are eight stations on a GTE Telenet network generating 12 million characters per month. Using this basic configuration, we can compute a range of costs for using the GTE Telenet service, the Tymnet service, and finally, the satellite alternatives.

To provide the cost range for using GTE Telenet, we will assume worst-case and best-case efficiency. In worst-case efficiency, only one character is sent in the original packet, and the return message contains a packet with only one character. Although this inefficient type of traffic is unlikely, it serves to compute a high-end cost figure. To calculate the least-cost figures, we assume that all packets contain the maximum of 128 characters. We further develop a range of prices by computing these costs for a packet sending-to-receiving ratio of 1:1, 1:5, and 1:10.

In their article, Smith and Moulton developed two formulas which aid in these computations. The first determines monthly message, or transaction, volume.

$$\text{Transactions/month} = \frac{\text{Total characters/month}}{\text{Characters/transaction}}$$

A second formula is used to determine the cost of this data traffic.

$$\begin{aligned} \text{Monthly packet cost} = &\ (50\text{ cents}/1{,}000\text{ packets}) \\ &\times (\text{packets/transaction}) \\ &\times (\text{transactions/month}) \end{aligned}$$

Take, for example, the worst case of single-character messages and a 1:1 packet ratio (one packet going out and one packet with one data character coming in). Assuming 12 million characters per month, this equals a total of 6 million transactions per month. Applying the second formula, the cost of using GTE Telenet in this inefficient manner is about $6,000 a month.

With a 1:5 ratio, one character is sent and 513 characters are received (4 full packets $\times$ 128 characters + 1 packet with a single character), yielding about 23,346 transactions per month. The cost of data traffic with this ratio is about $70 per month.

The large disparity between these figures is caused by the fact that the packets received in the second case contain four full packets and, therefore, achieve a significantly better efficiency. A ratio of 1:10 implies that 1,154 characters are used for each transaction (9 full packets $\times$ 128 characters + 2 characters), which results in a total of 10,399 transactions, at a

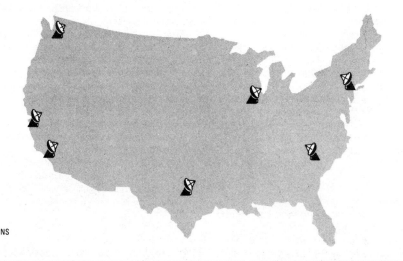

△ EARTH STATIONS

Western Union zones

ZONE 1 CITY PAIRS

Los Angeles or San Francisco to: Atlanta, Boston,
    Buffalo, Baltimore, New York, Cincinnati,
    Cleveland, Columbus, Detroit, Philadelphia,
    Wilmington (Del.), Washington D.C.

Seattle to: Boston, New York, Philadelphia,
    Pittsburgh, Washington D.C.

ZONE 2 CITY PAIRS

Dallas/Ft. Worth to: Baltimore, Boston,
    Buffalo, Los Angeles, Philadelphia,
    Washington D.C., New York, San Francisco

Houston to: Baltimore, Boston, Cleveland,
    Columbus, Dayton, Los Angeles, New York,
    Detroit, Philadelphia, Pittsburgh,
    Washington D.C., Wilmington, San
    Francisco

Kansas City (Mo.) to: Boston, Los Angeles,
    San Francisco, New York

Los Angeles to: Chicago, Indianapolis,
    Milwaukee, St. Louis

San Francisco to: Chicago, Indianapolis,
    Milwaukee, St. Louis

Seattle to: Chicago, Cleveland, Dallas/Ft. Worth,
    Detroit, Kansas City (Mo.), St. Louis,
    Milwaukee

Minneapolis to Boston

ZONE 3 CITY PAIRS

Atlanta to: Baltimore, Boston, Chicago,
    Cleveland, Houston, Dallas/Ft. Worth,
    Detroit, Indianapolis, Kansas City (Mo.),
    New York, Milwaukee, Philadelphia,
    Washington D.C., Wilmington (Del.)

Boston to: Chicago, Cincinnati, Columbus,
    Dayton, Indianapolis, Milwaukee, St. Louis

Chicago to: Baltimore, Dallas/Ft. Worth,
    Houston, New York, Philadelphia,
    Washington D.C., Wilmington (Del.)

Dallas/Ft. Worth to: Cincinnati, Cleveland,
    Columbus, Dayton, Detroit, Indian-
    apolis, Milwaukee, Pittsburgh, St. Louis

Houston to: Cincinnati, Indianapolis,
    Milwaukee, St. Louis

New York to: Columbus, Dayton, Indian-
    apolis, Minneapolis

Philadelphia to: Indianapolis,
    Kansas City (Mo.)

Milwaukee to: Baltimore, New York,
    Philadelphia, Washington D.C.

St. Louis to: Baltimore, New York,
    Washington D.C., Wilmington (Del.)

Seattle to: Los Angeles, San Francisco

Washington D.C. to: Indianapolis, Minneapolis

Western Union costs

BASIC CHANNEL SERVICE (TO 9.6 KBIT/S)
    Zone 1 city pairs $1,100/mo
    Zone 2 city pairs 820/mo
    Zone 3 city pairs 550/mo

TRANSPONDER SERVICE, MONTH-TO-MONTH
    Protected $200,000/mo/transponder
    Unprotected 120,000/mo/transponder
    Unprotected, interruptible 100,000/mo/transponder

TRANSPONDER SERVICE, FIXED-TERM
    Protected $150,000/mo/transponder
    Unprotected 83,500/mo/transponder
    Unprotected, interruptible 66,000/mo/transponder

REMOTE-AREA CHANNEL SERVICE (TO DALLAS/FT. WORTH EARTH STATION)
    Single channel charge $1,900/mo

total monthly cost of about $57.

Best-case situations, because of an implied 100 percent efficiency, are not sensitive to ratios. The price, therefore, is about $47 per month if full packets are always used (also an unlikely occurrence).

Tymnet charges are easier to compute since they are based on character volume. As a result, Tymnet's cost is not sensitive to packet efficiency. The price of Tymnet, when operating at 110 to 300 bit/s, is 10 cents per thousand for the first 40 million characters, 8 cents per thousand for the second 40 million characters, and 5 cents per thousand thereafter. Applying these figures to the previous example, traffic of 12 million characters costs $1,200 per month. Figure 5 provides a comparison of GTE Telenet and Tymnet charges.

Satellite networking

There are, however, alternatives to land-based transmission facilities for communicating voice and data traffic. Two carriers, Western Union and American Satellite Corporation, provide a variety of satellite-based services and charge for the channels used—not for the volume of data sent.

For its satellite transmission, Western Union uses three geostationary satellites, called Westar I, II, and III, and earth stations (which are basically sending and receiving antennae) set up by the company near major population areas in the continental U.S. Customers access the earth stations through extension channels, which could be leased-line, cable, or microwave links. These are provided either by Western Union or by other common carriers.

Every Western Union satellite carries 12 transponders, each capable of receiving and retransmitting data at 50 Mbit/s. A single transponder can support, for example, roughly 600 two-way voice channels or 1,200 one-way voice channels. The satellite transmission service can accommodate alternate data/voice traffic, private-line voice traffic, two-way point-to-point data traffic, wideband data traffic, video transmission, and intercarrier transmission.

Western Union's satellite services were the first of that type to be offered by a common carrier. They can be interconnected with, and accessed by, land-based facilities provided by another common carrier or by one of the international record carriers. The satellites themselves may be accessed by either the customer's or Western Union's earth stations.

Western Union offers satellite services between points in the continental U.S., Puerto Rico, and other U.S. possessions. These locations are served by the seven Western Union earth stations near Atlanta, Chicago, Dallas, Los Angeles, San Francisco, Seattle, and New York. In addition, the company uses more than 8,000 miles of microwave links to connect these earth stations with Western Union communications facilities in other cities. The entire network is controlled by the New York station.

Western Union's Satellite Transmission Service is available between specified pairs of cities. City pairs are combined to form zones, which form the base of Western Union's rate structure for the service (Fig. 4). All cities in the zone pairs are connected to an earth station through Western Union's landline and microwave facilities.

There are two specific types of services: protected and unprotected. Protected service provides redundant transponders that are designated and reserved by Western Union. This permits a customer to access a replacement transponder if the one he is using fails. One replacement transponder is provided for every two transponders the customer is using. Unless otherwise requested by a customer, the reserve transponder is on the same satellite. Unprotected service does not provide backup transponders.

Western Union offers several options to its standard satellite services: Alternate Data/Voice Channel Service, Space Tel Channel Service, Remote Area Channel Service, Synchronous Digital Service, and Wideband Data Channel Service.

The Alternate Data/Voice Channel Service uses one or more full-duplex channels and is offered between Western Union facilities on specific routes and/or between customers located on these routes. The channels are analog and support transmission speeds of up to 9.6 kbit/s on one voice-grade channel, and up to an aggregate speed of 56 kbit/s on 12 voice-grade channels.

The Space Tel Channel Service links a customer and his correspondents through a combination of satellite and terrestrial facilities. The service supports connections between (1) telephone switchboards and telephones, (2) two telephone switchboards, or (3) two telephones. Space Tel channels are zoned (and charged) according to the distance between specified pairs of cities (Fig. 4).

The Remote Area Channel Service is available to customers who maintain their own receive-only earth

5. Costs. *Because the user can regulate the efficiency of his traffic over the Telenet network (through message length), the cost of the service may vary significantly.*

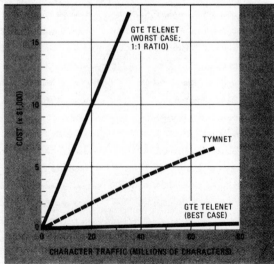

stations located at an installation that connects to Western Union's satellite facilities in Dallas, Tex. This service provides full-duplex, voice-grade channels which accommodate voice, facsimile, or data traffic at 2.4 kbit/s. The customer-provided facilities must both communicate and be compatible with Western Union's transmit/receive equipment in Dallas.

The Synchronous Digital Service supports all-digital data transmissions at speeds of either 9.6 or 56 kbit/s, depending on location. It is available on a limited basis, and users should contact Western Union to determine applicability. The Wideband Channel Service provides leased facilities only to other common carriers in bandwidths of 48 kHz, 240 kHz, and 1.2 MHz. Figure 4 details the pricing for these Western Union services.

On-site satellite service

American Satellite Corporation (ASC) also has satellite data services which combine ASC-owned earth stations and a satellite leased from Western Union. ASC provides commercial services which use either their existing earth stations or those it constructs on or near a user's premises.

Of the various ASC offerings, the most noteworthy is its Satellite Data Exchange (SDX) Service. SDX is a high-resolution, all-digital transmission service that uses private, direct, location-to-location communications links between user sites. With SDX, service is implemented through ASC-designed and installed earth stations complemented by various types of data communications equipment. SDX, through user-based stations, transmits at 1.544 Mbit/s, using a 10-meter antenna earth station, or at midrange speeds of 56 to 115.2 kbit/s, using a 5-meter antenna earth station. There is also a voice-grade data service that employs both SDX earth stations and ASC's commercial network facilities.

ASC offers special equipment to enhance its SDX service. The Satellite Delay Compensation Unit (SDCU) provides conditioned transmission for increased data throughput efficiency and costs $75 per month, in addition to channel costs. The CryptoLine unit guarantees data transmission security and costs an additional $375 per month. The Digital Communications Controller facilitates expansion and lets a customer change the mix of communications services to optimize the available satellite channel bandwidth.

High-speed dedicated service operates at speeds from 56 kbit/s to 1.544 Mbit/s, using the 10-meter antenna, and at speeds from 56 to 115.2 kbit/s with the 5-meter antenna. Both antennae can be placed on a customer's site and provide satellite communications without requiring any additional land-based carrier connection.

The SDCU equipment operates at either 9.6 or 56 kbit/s and virtually eliminates the propagation delay inherent in satellite transmission. The CryptoLine option provides data transmission and traffic flow security to protect against volume and pattern analysis. With CryptoLine, data is scrambled by an electronic variable key located in the unit. Once received, the scrambled data is taken from the receiver and unscrambled. All data scrambling and unscrambling is controlled by the National Bureau of Standards Data Encryption Standard (DES) algorithm. CryptoLine operates at speeds up to 112 kbit/s.

SDX on-site installations include the earth station, a weather-proof electronics shelter, transmit/receive equipment, and one or more communications channels. Full- and half-duplex operation and broadcast mode (one station transmitting to multiple receiving stations) or selective mode (one station transmitting to one receiving station) transmissions are supported.

In addition, ASC provides a controller which interfaces various types of data communications equipment onto a single 56- or 112- kbit/s channel. The unit consists of both SDCU and CryptoLine devices and can handle such varied mixtures as digitized voice, facsimile, digital video, and computer data transmission.

The major advantage of SDX over Western Union's service is that additional earth stations can be easily added to SDX. SDX does not, however, offer the full range of services available through Western Union. The costs for satellite channels are the same as those for Western Union, because ASC uses Western Union satellites for its SDX service. User-site earth stations are generally customized configurations, making detailed pricing very difficult.

VAN versus satellite

Comparing the cost of VANs and satellite services is, by nature, unfair. Because of the wide satellite bandwidth, many firms will choose to use these channels for voice, facsimile, and video, as well as for data. As the volume of traffic and the number of stations increases on the packet-oriented networks, the price of using those networks also rises. If a large number of terminals are located near earth stations, it may be economically preferable to use satellite facilities.

A stigma has been associated with satellites for many years. Often, potential users have feared that they lacked the volume requirements or technical know-how, but carriers like Western Union and ASC are rapidly changing this view. Satellite links are available now to the common user — it is no longer a technology limited to large international corporations.

The capabilities of commercial communications networks cannot be discussed without a look towards the future, because they are rapidly developing. (GTE Telenet recently announced plans to incorporate satellite and microwave links into its network, possibly making dedicated links available to customers.) AT&T's ACS will reportedly provide a universal interface (made possible through intelligent node processors) and will use the existing telephone network. In addition, Xerox's XTEN and Satellite Business Systems' satellite services are both likely to influence existing network offerings because of their potentially favorable prices and geographic coverage. How these services will affect existing commercial networks remains to be seen.

The International Spread of Packet-Switching Networks

BOB HARCHARIK
Tymnet, Inc.
Cupertino, CA

The mid-seventies spawned a new type of data communications service based on technological concepts that are called packet-switching. While there has often been debate over just what does or does not meet the purist's definition of packet switching, the message transmission technologies developed by the Department of Defense's Advanced Research Projects Agency (ARPA) and by Tymshare, Inc., a remote-computing services company, are conceptually similar in design and serve the same basic communications need. They both package data picked up from transmitting terminal devices into synchronous packets which are then moved to their destination through a network of communications circuits and minicomputer nodes. These nodes provide the intelligence to, among other things, insure virtually error-free transmission, alternate routing around facility outages, and transparency between sending and receiving devices. The technical details of how this is accomplished differ between the ARPA and Tymshare sources of the technology and the respective merits of each are often debated.

Two significant public networks have evolved. One is operated by Tymnet Inc., a common-carrier subsidiary of Tymshare created in 1976 after four years of operating the TYMNET network on a joint-use basis. The second network is operated by Telenet, a company founded by the principal researchers of packet switching of ARPA and recently acquired by General Telephone and Electronics. The TELENET network operated nearly four years as an independent company. After seven years of TYMNET and four years of TELENET operations, it is appropriate to review what these two networks have brought to the data communications world.

Tymnet today provides local access to terminals in some 175 US metropolitan areas and Telenet to approximately 100. Together they carry data traffic to and from over four-hundred computers belonging to their subscribers, and tens of thousands of terminals belonging to the users of those computers. Tymnet carries more than a half-billion characters daily through more than 450 network nodes and nearly 125,000 miles of leased communications circuits. At peak loads, Tymnet has approximately 5,000 active circuit terminations.

The success of these networks and their basic technologies has spread around the world. Foreign telephone administrations in several countries have or are constructing their own networks. In addition, the International Record Carriers (IRC's) - ITT Worldcom, Western Union International, RCA Globcom and TRT Communications - have spread the technology to over twenty foreign nations. The IRC's have installed the technology in US gateways and, working together with foreign Postal Telephone and Telegraph (PTT) administrations, have installed packet-switching gateway nodes in the respective foreign countries. The IRC's have accomplished this by purchasing the technology and equipment for the US gateways from Tymnet and Telenet and, under joint arrangements with the PTT's, nodes have been placed in the foreign countries.

The international traffic carried today originates predominantly from terminals, located in foreign countries, and this traffic is destined for US computers connected to Tymnet and Telenet. The foreign terminal user contacts the PTT node which, in turn, establishes contact through the IRC gateways to the Tymnet or Telenet network. These networks complete the access to the US computer as prescribed by the originating terminal user. The communications facilities provide terminal access for users of US data-base and information-services companies and also for the foreign-based operations of companies with central computer operations in the US. Currently, the reverse process is beginning with US terminals contacting foreign computers.

The countries currently operating Tymnet technology nodes and providing these services are as follows: Argentina, Australia, Austria, Belgium, Brazil, Canada, Denmark, France, Germany, Hong Kong, Italy, Japan, Mexico, Netherlands, Philippines, Portugal, Puerto Rico, Singapore, Spain, Sweden, and Switzerland, plus the non-continental US states of Alaska and Hawaii. Access from England similarly originates on Telenet equipment installed by the British Post Office.

Prices are set by the PTT of each respective country. With few exceptions all terminals operate at 300 baud or less. A typical application, using a 300-baud terminal, will result in charges from as low as $25 per terminal hour to as much as $65 or $70 per hour, depending on the country. Arrangements for foreign access are made by the foreign user with the PTT in his country. Of course, arrangements for access to the specific computer must be made independently with the operator of the computer.

The international installations of these gateway services began in early 1977 with the installation of standard Tymnet nodal technology. This is still continuing in some European countries and in South America.

However, many international gateways being ordered and installed today are constructed around Tymnet's new Internally Switched Interface System (ISIS) technology. ISIS is the combination of new hardware and software concepts. The software turns a node into a multipurpose device able to handle local terminals, interface with computers, act as a gateway to other networks public or private, and perform other more specialized functions. The hardware is called the TYMNET Engine.

By installing ISIS nodes, the foreign PTT can not only provide the international gateway services to Tymnet and Telenet but, at the appropriate time, can interconnect to the packet networks evolving in other countries such as England, France, and Japan. Most of these interconnections will be made using various versions of X.75, the CCITT standard for interconnection of such

networks. Although these "standards" vary from one network to another, ISIS was designed to deal with these "varying opinions" on the implementations of network technology.

Through the use of multiprogramming and time-sharing concepts in the microprocessor-based TYMNET Engine, a single node can have one partition handling local traffic, another a gateway to France's Transpac network, another an X.25 connection to the BPO network, and still another operating Tymnet protocol to other similar nodes elsewhere in the country. Also, the node might incorporate a partition containing one or more gateways

INTERNALLY SWITCHED INTERFACE SYSTEM (ISIS)

ISIS Operating System
IRC Gateway
X.75 FRANCE TRANSPAC Gateway
X.75 JAPAN VENUS Gateway
X.25 British Post Office Gateway
Tymnet protocol to other domestic nodes
X.25 Gateway to private network A
X.25 Gateway to private network B
Local terminal handler
Supervisor

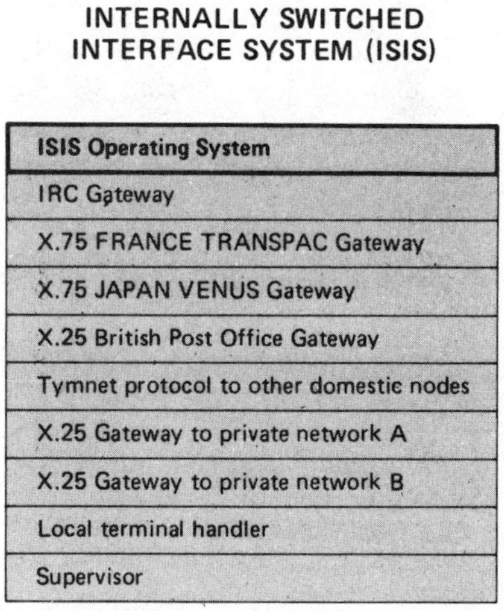

Fig. 1 Internally switched interface system (ISIS).

to private networks belonging to organizations within the country. Figure 1 illustrates the concept. Optionally, the node can obtain central control facilities from Tymnet US or operate its own controlling supervisor.

ISIS switches the traffic between the partitions within the node. Each partitition contains software which translates the external (user) protocol to a common internal (network) protocol. Conversely, when receiving a message, the partition translates the internal protocol to the external protocol it is handling. Adding new protocols for terminals, computers, or other network "standards" is simply a matter of developing the protocol handling program. The time required to do this is typically a few man-months at most. ISIS also includes such tools as a Dynamic Debug package which operates in the node to facilitate the development process and can proceed in a mode that is also handling a production environment.

As late as four years ago these packet switching networks were unknown technological curiosities, even in the US. Today not only are they commonplace in the US data communications environments, but they have also spawned the development of facilities around the world that are rapidly shrinking the time and cost of sharing and disseminating information.

TYMNET I: An Alternative to Packet Switching Technology

by J. Rinde

Tymshare, Inc.
Cupertino, California, U.S.A.

ABSTRACT

TYMNET I is a commercial computer network, designed, implemented and operated by Tymshare, Inc., that has been in operation since 1969. TYMNET I is oriented to terminal users and also permits higher speed host to host communication. A user entering TYMNET I at any point can have a path established to any host connected to the network. An "optimal" path is plotted by a central supervisory program that establishes the path through communication with the nodes along the path.

The physical records transmitted between nodes are of variable length and may consist of multiple logical records whose routing is determined by (Permuter) table entries.

Some comparisons are made between the path technology of TYMNET I and the packet switching technology of ARPANET. The argument is presented that packet switching is a subset of TYMNET technology.

INTRODUCTION

Computer networks are receiving a great deal of attention today as a means of improving human to computer and computer to computer communication. In particular the development by the Advanced Research Projects Agency (ARPA) of a sophisticated, publicly funded computer network, ARPANET, has been widely published in the professional journals. TYMNET I, by contrast, was developed by a small to medium size commercial firm, and only a few articles about its technology have been published. That even these few articles have not been widely read is evidenced by frequent mistaken reference to TYMNET I as a packet switched network. As a result the computing community has come to associate the concept of sophisticated computer networks exclusively with the packet switching technology used by ARPANET.

The next section briefly reviews the concept of packet switching as embodied in the ARPANET implementation. The section following that introduces the TYMNET I path concept. Both these sections employ analogies to help clarify the more detailed explanations. The paper concludes with an examination of the effects of design decisions regarding data grouping (variable in TYMNET I, fixed in ARPANET) and routing (static in TYMNET I, dynamic in ARPANET).

Nomenclature

The computer networks discussed here consist of a collection of computers (nodes; typically minicomputers) interconnected by communication lines (e.g. leased telephone lines), so as to allow alternate paths between nodes in the network. The primary function of a node is to pass information. The nodes do little or no processing of application tasks. Some nodes are connected to host computers (typically a large scale computer that processes application tasks), while others can answer telephones to communicate with a user's terminal. Some nodes may perform both functions. Nodes have buffers that hold data enroute to another node, to an attached host, or to an attached terminal.

A user connects his terminal to a nearby node by a telephone call. The user then arranges to communicate with a host computer connected to some node in the network.

ARPANET was designed to provide ARPA supported computers with a communications link. The primary task of this link was to transmit large quantities of data (e.g. big files) between hosts. To accomplish this task, data traveling through the net is divided into packages, up to one kilobit in size, called "packets". Each such packet contains a destination address. The packet is passed between nodes, each of which decides, based on its global knowledge of the network, to what neighbor node to pass the packet.

ARPANET Analogy

> Though analogy is often misleading,
> it is the least misleading thing we have.
> —Samuel Butler

To see the grouping and routing algorithms more clearly, consider the mythical fly-on-demand airline "ARPA-AIR". ARPA-AIR has a fleet composed of 737's, 747's, and everything in between. ARPA-AIR flies a limited set of routes set by the FAA. When a group of passengers wants to travel from airport X to airport Y, they board the smallest aircraft that will hold them all for immediate departure. The aircraft hops from airport to airport along the assigned FAA route. At each airport the pilot informs the tower of his final destination. The tower (which speaks only to the airports to which it is directly connected by the FAA routes) decides what airport it would be best for the pilot to fly to next. Passengers neither board nor deplane at intermediate stops.

From the analogy we can see that if the group of passengers is large (one or more 747's worth), the larger, more efficient aircraft are used. Note that, at an airport, travelers in a small plane sharing a destination with passengers in another small plane cannot join them to fill a larger plane.

TYMNET I

TYMNET I was developed at the same time as ARPANET, but with different goals in mind. TYMNET I was designed to be a terminal oriented network.

A terminal user typically has a 10 to 30 character per second (cps) terminal, thus limiting the host computer to that data rate. The user to host data rate is likely to average less than 10 or 30 cps and the data flow in either direction is not likely to be continuous throughout the terminal session.

To understand how data is transmitted through TYMNET I, let us start with a user dialing a local number for access. Upon connection he types an identifying character that informs the node what the terminal characteristics are (10 or 30 cps, etc.). The node then sends the message "PLEASE LOG IN:". The user types his account name, optionally the number of the host to which he wishes to connect, and the password for his account name. This information is sent to the TYMNET I network supervisor (a program on an INTERDATA 7/32). The network supervisor verifies, in the Master User Directory (MUD), that the account name is valid and that the password is the correct one for it. If the user has not specified a host number, his standard host number is taken from the MUD. The supervisor then finds the "optimal" path through the network to connect the remote node called by the user to the base node connected to the desired host computer. Optimality is a function of the capacity between nodes, the number of hops (links) to be traveled and the cost of these links. The cost of a path is a measure of the resources used, not a dollar value. Table 1 shows the costs associated with various link speeds plus link overload costs. The algorithms for computing a minimum cost path between two nodes considers all nodes whose cost to the destination node is less than that of the source node and, on the average, half of all nodes with equal cost. On the average, 12ms are used to calculate the optimal path.

TABLE I–Link Costs in TYMNET I

Line speed	cost unoverloaded	cost overloaded one way	cost overloaded both ways
9600 bps	10	26	42
7200 bps	11	27	43
4800 bps	12	28	44
2400 bps	16	32	48

Once an optimal path has been selected, the supervisor program assigns a port in the host computer. It then sends the account name to that port, sends messages to each node along the path to make appropriate entries in the node's permuter tables and then informs the host that the path is complete and that the login may proceed. At this point the supervisor function is complete with regard to this login.

The network supervisor handles all logins (i.e. verifying user names and passwords, and building paths) and maintains network control. When a link goes out or a host or node goes down, the operators at the network control center are informed by the supervisor within seconds. Network operators can instruct the supervisor not to build paths on certain links to allow traffic to fall off before taking a link or node down for maintenance.

One objection to a centrally directed network is the consequences of a supervisor going down. (It should be noted that existing paths are totally unaffected by a

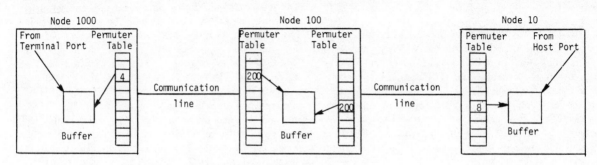

Figure 1—A simplified TYMNET I path

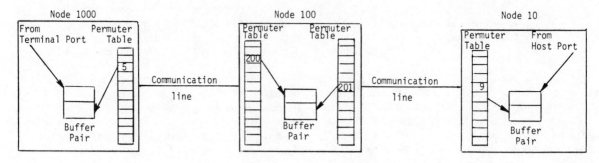

Figure 2—An actual TYMNET I path

supervisor going down.) TYMNET I's solution to the problem is to provide a hierarchy of four supervisors. Only one supervisor is active at a time, and it keeps the others dormant by regularly sending them sleeping pills. If a supervisor goes down, it is likely to be noticed by a network operator who will start another supervisor. However, even without human intervention one of the dormant supervisors will awaken and take control of the network. Although sleep times are staggered, it is possible to have two supervisors trying to take control of the network simultaneously. This situation is handled gracefully when the less dominant supervisor learns of the activities of the more dominant supervisor and goes to sleep. A supervisor can take control of the network in 2 to 2.5 minutes. The TYMNET I supervisor typically runs a week at a time before software or hardware problems force a supervisor move (i.e. a disruption of logins for a few minutes).

To understand what constitutes a path in TYMNET I, we now consider how a node uses its permuter tables. Figure 1 shows a 2 link path. A node has a permuter table for each link (connection to another node). A physical record sent between nodes contains a record header that indicates the record's length, followed by a collection of logical records (see figure 3). As seen in figure 4 each logical record consists of a logical record number, a byte count and that many data bytes.

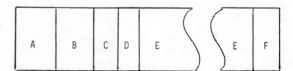

A.   5 bit pattern to identify start of a record
B.   5 bit size (count of 16 bit words, not including checksum)
C.   3 bit record number
D.   3 bit acknowledgement record number
E.   logical records
F.   2 16 bit checksums

Figure 3—TYMNET I internodal physical record

A.   8 bit logical record number
B.   8 bit byte count
C.   As many data bytes as specified by B.

Figure 4—TYMNET I logical record

A TYMNET I path is defined by a set of permuter table entries. Each permuter table entry points to a buffer in the node into which incoming characters are placed or from which characters are removed for shipment. The entry's position in a permuter table corresponds to a logical record number. In figure 1, a record sent from node 1000 to node 100 containing a logical record 2 would cause the data characters from that logical record to be placed in buffer 200. When node 100 builds a record to send to node 10, a logical record 5 would be created containing the data characters in buffer 200. For the path in figure 1 the second permuter table entries for the link between nodes 1000 and 100 must point to buffers in the same path, as must the fifth entries of the permuter tables in nodes 100 and 10. At the ends of the path (node 10 for example) the buffer pointed to by the permuter table entry is a buffer associated with a port (a host port for node 10, a terminal port for node 1000).

To keep the above explanation simple, figure 1 shows only one buffer per node for the path. In actuality each path has a buffer pair in each node, one buffer for each direction. Figure 2 shows this same path as it would actually exist in TYMNET I. The assignment of buffer pairs and permuter table entries in a series of nodes constitutes a TYMNET I path.

TYMNET I Analogy

To simplify the above description, consider the mythical fly-on-demand airline "TYM-AIR". TYM-AIR has a fleet of aircraft including Cessnas, 707's and everything in between. TYM-AIR flies a limited set of routes established by the FAA. To start a series of trips, the tour leader calls a travel agent who arranges the path by calling all the selected airports enroute. When one or more passengers arrive for a given tour, they report to a dispatcher who tells them to what gate to go and gives them a boarding card. A plane is immediately assigned to fly them to the next airport on the itinerary. At the gate, the passengers may meet fellow travelers on different tours who share the next leg of the trip. The aircraft assigned is just large enough to carry all the waiting passengers (up to one 707's worth). Upon arrival at the next airport, the passengers turn in their boarding cards to the dispatcher who gives them new boarding cards and directs them to the proper gate. Those passengers who have reached their destination are so informed by the dispatcher.

From this analogy we see that, at times of heavy traffic, the aircraft used are always full 707's. If small aircraft are being used it means traffic is light. Passengers need not carry the extra "baggage" of knowing what their destination is, but carry a boarding card, which is a much smaller piece of "baggage". Indeed, there need be only one boarding card per tour group on an aircraft. If a large group arrives from the same tour, they will most likely get a 707 to themselves. They may have to share it with a smaller group that arrived sooner. Seating is fair, so that small groups are never shut out by large groups.

COMPARISONS

Data Grouping

Both TYMNET I and ARPANET use a checksum and acknowledgement scheme to insure that records sent between nodes are properly received. These records utilize the most expensive single resource of telecommunications today—line bandwidth.

Bandwidth utilization becomes critical when the quantity of user data to be transmitted approaches the maximum physical bandwidth available. At times of high volume traffic, TYMNET I sends maximum size physical records between nodes. From figures 3 and 4 we can calculate that a 528 bit (maximum size) record with only one logical record has 64 bits of overhead or 12.1%. The worst case is 20 logical records, with one data byte each, giving 368 bits of overhead or 69.7%. An ARPANET packet contains 168 bits of overhead[4] (record header, checksum, destination address, etc.). Thus a full packet (1007 bits) has 16.7% overhead. The worst case is a packet with only one data character (8 bits) having 95.5% overhead. The TYMNET I figures show a lower overhead than ARPANET because TYMNET technology allows grouping of data based on the indivisible unit of the 8 bit byte. This grouping causes maximum size records to be used in times of heavy traffic, allowing the necessary physical record overhead to be distributed over more data bits. In ARPANET the grouping (packet) is arbitrarily set by the sender and remains independent of other traffic flow in the network. Note that full duplex interactive users will tend to generate the worst case in ARPANET, whereas volume senders can help compensate for interactive users in TYMNET I, making the worst case unlikely.

A packet switched network has the property, independent of the routing strategy, that data grouping remains unchanged during travel through the network. Therefore, if TYMNET I were modified to allow only one logical record per physical record, TYMNET I would become a centrally directed, statically routed, packet switched network. Thus it may be said that packet switching, as a data grouping discipline, is a subset of TYMNET technology.

Routing

Paths in TYMNET I are set up upon request and remain static for the duration of the session. The path is optimal based on the condition of the network when the path was requested. Changes of conditions during the session have no effect on the path.

In ARPANET each node maintains a global picture of the network on which to base its adaptive routing decisions. This global picture of the network is maintained by having each node send two lists, every two thirds of a second, to each of its neighbors. First, a list of costs (in arbitrary units) of getting from itself to each destination (i.e. node) in the network; second, a list of the minimum number of links from itself to every destination in the network.[6] Based on these values, each node decides the direction that outbound packets should take.

As traffic increases, the routing information transmitted between nodes competes with user data for the available bandwidth. To see how much bandwidth is used, consider a 160 node network using 8 bits per entry in tables described above. The routing chit chat between nodes would be 3840 bits per second or 160% of the bandwidth of a 2400 bps link! The corresponding worst case overhead for TYMNET I is 5% of a 2400 bps link (see the appendix for calculations). Using 9600 bps links, a 160 node TYMNET I would have a worst case 1.25% bandwidth overhead for routing, compared to 40% for adaptive routing. When physical record overhead is added, the best combined overhead (one logical record per physical record, or full packets) is 13.2% for TYMNET I and 50% for adaptive routing. The worst case (20 logical records per physical record or one character packets) overhead is 70.1% for TYMNET I and 97.3% for adaptive routing. The difference in routing overhead of more than an order of magnitude is not a reflection of different implementations, but shows the inherently high cost of adaptive routing.

Propagation delays in transmitting routing information in ARPANET have caused routing loops (e.g. a packet sent from node A to node B is then sent back to node A). The number of iterations in routing loops has been as high as 50.[4]

Another consequence of ARPANET's routing algorithm is load oscillation. Oscillations arise when a node "A" has two neighbors, "B" and "C", whose cost to some destination "D" is equal. Node "A" will arbitrarily pick one neighbor (say "B") to send all traffic volume to "D". If "A" gets a large volume of traffic for "D" it could cause "B" to become overloaded. "B" would then raise its cost to "D", causing "A" to shift its entire load to "C". Now "B" is no longer overloaded, but "C" becomes overloaded, causing "A" to shift its entire load to "B", etc., etc.

Another problem with adaptive routing is lockup. This condition occurs during heavy load as neighboring nodes accumulate their maximum number of packets and, therefore, cannot accept packets sent by their neighbors.

Adaptive routing also presents a problem for multi-packet messages. ARPANET allows up to 8 packets to be sent at once to permit efficient file transfers. The eight packets may arrive at the destination out of order. The solution to this problem requires reservation of buffers prior to transmission and end to end acknowledgments after any transfer.[4]

The static nature of TYMNET I paths avoids the problems of looping, load oscillation, lockup and the need for end to end acknowledgment, and allows a continuous flow of data. As in ARPANET, there is still a problem of interference by other traffic in the network which may slow the flow in a particular path, or even stop it momentarily. Another possible disturbance is a data rate difference between sender and receiver. A common example is host to terminal communications, in which the host may be sending data at 120 cps to a 10 cps terminal. TYMNET I solves this problem through backpressure. Each node periodically informs its neighbor nodes on which path to permit data to flow. When the size of a buffer in a path gets too large (based on path speed characteristics) the node no longer permits its neighbor to send data for this path. When the buffer size drops the data flow is again permitted. The backpressure information is sent as a bit array, one bit per path, and is passed between nodes every half second. The positive nature of backpressure in TYMNET I prevents a node that is overloaded (and therefore late in sending backpressure information) from being deluged by data from its neighbors. In the host to terminal case above, the backpressure would emanate from the remote and, if need be, propagate back to the base, which will ask the host to stop sending.

The overhead of end to end acknowledgment does provide another service. When a node in ARPANET goes down, the packets within that node are lost. After some time the sender will notice the absence of an acknowledgment and retransmit the packet. Thus a node going down does not necessarily break a connection. This feature is purchased at the cost of an extra trip through the network to acknowledge each message sent, and the cost of keeping a copy of every packet sent until its receipt is acknowledged.

TYMNET I does not provide end to end acknowledgment in the network itself, but does allow the user the option of paying the overhead at the host level if he feels the cost is justified. A TYMNET I node going down loses the information in the node, and all paths terminating or passing through the node are torn down. The network user is now provided with the option of relying on infrequent node outages, or paying the overhead of end to end acknowledgment in the host software. In the case of a terminal user, a node going down in his path will cause PLEASE LOG IN to be typed on his terminal. Upon logging back into a host, such as a PDP10, the user finds his job detached in the state he last saw it. The increasing availability of "smart" terminals opens other possibilities.

For host to host communication, an individual protocol can be implemented at the operating system or user program level to effect end to end acknowledgment. When a program is informed that its path has been torn down, it may request another path to be established and continue its transmission without loss of information.

MORE ABOUT TYMNET I

Security

The TYMNET I supervisor's function of logging in users provides a measure of network security. A host operating system may be assured that only a user with a valid user name and password can establish a path to the host. The host operating system need only verify that the user name is valid to be sure that the user is authorized. The ciphered passwords stored on the dedicated supervisory computers make violation of security difficult.

Orientation

There is an assumption[3] that once a host to host network is provided, network terminal access to hosts is a minor matter of a minicomputer interface. The TYMNET experience indicates that just the opposite is true. Host to host communication in TYMNET I was a natural by-product of a terminal oriented network that required almost no additional code in the nodes or the supervisor.

Whether one starts with a terminal oriented or host to host oriented network, the base to host interface and the internodal communication come first. When interfacing to terminals one is confronted with a vast sea of terminal types. To accommodate many different classes of terminals through special hardware for each class is an expensive and rigid approach. A software solution (used in TYMNET I), though difficult, keeps the capital investment small and allows graceful expansion as new terminal types appear.

The terminal oriented nature of TYMNET I is seen in the mechanisms available to control echo for full duplex terminals. As an example, consider the ability to pass echo control between the remote and the host. To maintain fast echo for the full duplex terminal user, the remote node echoes characters whenever possible. When a special character is typed (e.g. a control character) or if typing occurs during output, the remote cannot be sure what or when to echo. TYMNET I allows a remote, under these circumstances, to pass the responsibility for character echo to the host. A complementary facility exists to return the echo responsibility to the remote without any characters echoing out of order. This is a mechanism that would not naturally develop in a host to host oriented network, and may be difficult to add as an afterthought.

Accessibility

Tymshare accesses its IBM 370's, PDP10's and SDS940's exclusively through TYMNET I. For the latter two, even the operators' consoles are connected through the network.

The efficiency of TYMNET I keeps the cost of access through the network low, while eliminating capital investment in other forms of telecommunication access to Tymshare's hosts.

CONCLUSIONS

TYMNET I has achieved its economic design goals, as evidenced by its rapid expansion even during times of economic recession. TYMNET I now consists of over 160 nodes on two continents. Prime time load is over 1000 simultaneous users. In February 1976 over 3.9 billion data characters were transmitted in 13 million connect minutes.

The use of flexible software multiplexing in the building of physical records permits good utilization of line bandwidth. Inside each node, expanding and contracting buffers for each path provides an efficient utilization of core.

This efficient use of resources has enabled TYMNET I to expand beyond the immediate needs of Tymshare's communication requirements and enabled Tymshare to share network resources at prices far below what a sharer would spend were he to supply his own communication needs. At present TYMNET I connects to 90 host computers in 25 cities.

TYMNET I supports a mixed load of high and low data rates with satisfactory efficiency. As explained in the TYMNET I section, a large data transfer will tend to get its own physical record (just as in ARPANET it gets its own packet). However, like a timesharing system, the network insures that the little user is not permitted to be blocked out in the network. This may cause some delay for the large data transfer but will maintain fast response for the conversational user, while maintaining high throughout in the network. TYMNET I's proportion of high data rate transfers is increasing and the replacement of 2400 bps links with higher speed links is increasing with it.

APPENDIX

TYMNET I's routing overhead comes from two sources. First, nodes report any links that are out by sending a supervisory record every 16 seconds. A supervisory record is 48 bits long and travels as user data. Assuming (worst case) that this is the only supervisory record in the physical record, an additional 16 bits of logical record header make the overhead 64 bits per 16 seconds. This is an overhead of 0.17% of a 2400 bps link. The overhead

from this source on a particular link in the network is
dependent on its position relative to the supervisor and
the out links in the network. Obviously, the links connecting to the supervisory node will have the heaviest
overhead from reports of out links. To have 1% overhead
on a 2400 bps link (it should be noted that links connecting to the supervisory node are typically of higher
speed) requires 6 out links. Each supervisor has at least
8 links. Therefore, 48 links would have to be out throughout the network to cause as high as 1% overhead on the
2400 bps links connecting to the supervisory node. The
overhead on other links in the network would be less.

The second source of routing overhead in TYMNET I
results from supervisory records instructing nodes to
change their permuter tables and the acknowledgment of
those changes. The average path in TYMNET I is just
over three links. For worst case let us assume a four link
path. A total of eight permuter tables must be changed
(two in each intermediate node, one in each end node)
by eight supervisory records. Worst case would require
all these records to leave the supervisor on the same link,
causing 320 bits of overhead (assuming there are no other
supervisory records to share the overhead of the logical
record header). The prime time login rate is one per
second. This gives an overhead on a 2400 bps link of 13%.
Since there are at least 8 links on a supervisory node, the
overhead per link is 1.6%.

After a node changes a permuter table it sends an
acknowledgment to the supervisor. Assuming the acknowledgments returned by each of the 5 nodes arrive in different physical records and that they are the only supervisory records in those physical records (worst case), the
overhead is 19.3%. The average over 8 links is 2.4%. This
brings TYMNET I's worst case average total routing overhead
to 5% of a 2400 bps link with 48 out links in the network.

Acknowledgments

The author owes a great debt to LaRoy Tymes, the
designer of TYMNET, and to Norm Hardy, a constant
source of ideas.

BIBLIOGRAPHY

1. Bell C. G.   More power by networking
                     IEEE spectrum Vol 11 No 2 (Feb '74)
                     pg 40—45

2. Doll D. R.   Telecommunications turbulence and the
                     computer network evolution
                     Computer Vol 7 No 2 (Feb '74) pg 13—22

3. Kimbleton S. R.,   Computer communications net-
   Schneider G. M.   works
                      ACM Computing Surveys Vol 7
                      No 3 (Sept '75) pg 129—173

4. Kleinrock L.,   On measured behavior of the
   Naylor W. E.   ARPA network
                      AFIPS National Computer Con-
                      ference Vol 43 ('74) pg 767—780

5. Kleinrock L.   Throughput in the ARPANET
   Opderbeck H.   Proceedings Fourth data commu-
                      nications symposium (Oct '75)
                      pg 6—1 to 6—11

6. McQuillan J. M.   Adaptive routing algorithms for
                      distributed computers
                      Bolt, Beranek & Newman Inc.
                      report No 2831 May '74 pg
                      157—162

7. Roberts L. G.   Data by the packet
                      IEEE spectrum Vol 11 No 2 (Feb
                      '74) pg 46—51

8. Tymes L.   TYMNET — A terminal oriented
                      communication network
                      AFIPS Vol 38 (Spring '71)
                      pg 211—216

Emerging Services in International Public Packet-Switching Networks

By Ashok Trivedi

Group Leader
New Data Systems Development
ITT World Communications

In the last 15 years, packet switching has evolved from an experimental endeavor upon whose outcome would be based the continuance or cessation of further research and development of the technology itself, to the very basis for the coming information age.

The enlightened application of the packet-switching technology to the emerging requirements of data communications and computer network users is at once both a challenging task as well as an exciting opportunity for the development of new value-added international data services.

It was in August of 1964 that Paul Baran proposed the first fully distributed packet switching system to provide all the communications requirements of the US Air Force. Shortly thereafter, both Donald Davies of the National Physical Laboratory (NPL) in the United Kingdom and Lawrence Roberts of the Advanced Research Projects Agency (ARPA) in the United States began the independent development of store-and-forward packet switching systems. By the early 1970s, both the NPL and ARPA were operating packet switching networks. The ARPANET, in particular, was the focus of a considerable amount of the early research and development of packet networks.

The early 1970s saw the development of the first networks based on the packet technology: SITA, the network for the international air transport carriers; Tymnet, which uses a rather unique flavor of packet transmission; CYCLADES, with its datagram-based communications subnet CIGALE; EIN, the European Informatics Network; and RCP (Reseau a Commutation par Paquets), a virtual circuit based experimental network of the French PTT Administration.

During the early to mid-1970s, public packet switching networks began proliferating in a number of countries of the world.

The development of the ARPANET and the NPL network has given rise directly to two of the public packet switching networks in North America: Telenet, in the US, and Datapac in Canada. The Tymnet network also uses packet transmission principles in the offering of its services. In addition, value-added packet switching networks are also being planned by a number of different organizations, including Satellite Business Systems, Xerox, Graphnet, AT&T, and ITT. In fact, a public packet-switched network for store-and-forward facsimile service between incompatible terminals is now operational in the US. This network, developed by ITT Domestic Transmission Systems, is called Fax-Pak.

Foreign Packet Networks Developing

In France, the experimental RCP network has evolved into the Transpac network, while in the UK, the Experimental Packet Switching System (EPSS) that has been operating for a number of years is now being replaced by a permanent public packet switching network (PSS). Hybrid packet and circuit or message switched networks are operated by the Spanish PTT and by Canadian National-Canadian Pacific Telecommunica-

tions (CNCPT). National data networks have also been developed in the Nordic countries, while Japan's DDX packet-switching network has been developed by Nippon Telephone and Telegraph.

Plans are being made for national public packet networks in many other countries, including West Germany, Australia, The Netherlands, Belgium, Switzerland and Italy.

In all of these countries, there are also related or independent development projects generally under way for providing gateway packet interconnections and services, either with planned nodes or gateway networks. Thus, Kokusai Denshin Denwa (KDD) of Japan is developing their VENUS system, while ITT World Communications of the United States has its Universal Data Transfer Service.

The organization of the communications carriers in the United States has placed the international record carriers (IRCs) in a rather unique position. On the one hand, they are interested in the traditional service offerings which include telex, cable and messages. It is a recognized fact that the international telex network, for example, operates to provide rapid and vital communications among users in over 200 administrations and countries across the world. In the day-to-day interactions and long-term working relationships with various governmental administrations and domestic US organizations, the IRCs have also recognized the need for higher speed communications with high data integrity for an increasing number of applications. Thus, on the other hand, the IRCs have been instrumental in initiating the growth and demand for packet-switching systems and networks around the world. This role of the IRCs, in their competitive yet regulated environment, is a significant factor in the international transfer of modern technology and the evolution of the applications of packet systems and networks around the world.

In 1977, ITT World Communications inaugurated its international gateway packet switching service called the Universal Data Transfer Service (UDTS) to provide communications between dissimilar terminals and computers in the United States and overseas locations. The initial service offerings include access to remote computer data-bases and remote computing facilities. UDTS now provides international packet communications between the United States and over a dozen countries and administrations all over the world.

X.25 Significant Step Forward

With the CCITT ratification of the X.25 protocol for interfacing of packet-mode terminals to public data networks in March of 1976, a very significant step forward was achieved in the standardization of protocols for public data networks. The development subsequently of the CCITT draft recommendation X.75 for international internetworking has added further impetus to the growth of an eventual worldwide network of national public data networks. These standardization achievements of CCITT are viewed by many as of critical importance to the coherent development internationally of public data networks, some writers comparing these achievements as equal in importance to the development of the packet-switching technology itself.

The present and future applications of packet-switched networks are many and varied. For our purposes, we shall categorize the diverse applications of packet switched networks into broad categories.

Electronic message systems provide not only communications (the circuit, or rather, the virtual circuit), but also the intelligence associated with it. Now if A wants to say something to B, he sends an electronic message which will be received by B whenever B is ready to receive his messages.

The reason that packet-switched networks are such a natural choice for this application is simply that by their nature, packet networks already have much of the intelligence necessary to "value add" the message switching application to the basic communications function. And if the communications function happens to make use of virtual circuits, this "value addition" may be even more straightforward.

Electronic mail applications are really a generalization of the "traditional" electronic message systems. In the international arena, these applications are becoming of increasing importance for the rapid and efficient intercommunication among users in different countries. One large set of these is comprised of the intra-company electronic mail applications of multinational company branches located in different countries.

Intra-company and inter-company communications requirements are also spearheading the implementation of systems to provide graphics communications capabilities. In the United States, the first public packet-switched network to offer facsimile transmission between incompatible facsimile terminals has begun operation—this is ITT's Fax-Pak network. This store-and-forward packet network provides distributed intelligence and packetizes and depacketizes transmissions to and from incompatible facsimile terminals in geographically dispersed locations.

Finally, intercommunicating word processing applications are providing the leading edge for the combined text and graphics transmission systems of the future. Once again, the volume-sensitive communications costs afforded by packet-switched networks is providing the motivating force behind the development of this application.

Need for Computer Augmentation

These applications are based on the notion that not only are two-way or n-way communications a user requirement, but of increasing importance is the need for computer-augmentation of such interactions. The telephone conversation of today may serve well for some uses, but speech is not the only commodity that often needs to be exchanged. Thus, the communication of the future must include the means to exchange speech, writing, drawing, graphic logos, photographs, and moving images (television), all of which have been integrated into a set of coherent displays with computer support. The evolution of the plain old telephone set (POTS) conversation towards a system that can support such an enhanced form of communications has already begun, and various intermediate versions of it are already in existence. The use of packet communications is providing the economic cost-effective spur to hasten the process.

Teleconferencing received its first major impetus for development with the decreasing costs of computing and of communications. The second major impetus is clearly the rapidly rising cost of traveling to attend meetings. Not only is the cost of labor relevant here, but so also are the costs of transportation (including fuel) itself, and the cost of the inefficiencies involved in meetings. The inefficiency of a conference or meeting is a function of the following factors,

among others: the number of participants that speak, the length of time of the speeches and, for each of the speakers, the number of participants that are actually interested in what he is saying. There are clear advantages, in these aspects, in teleconferencing where participants "log on" to sessions and make contributions thereto depending upon their own interest and time priorities.

The use of packet-switched networks to provide the communications requirements of the automated offices of the future is extremely likely, and in fact, prototype systems have already begun to appear in a number of intra-company scenarios. Applications within this category have as their driving force the tremendous need for streamlining and modernization of the day-to-day office and inter-office operations of today's business office. The cost of sending a typewritten memorandum from one office to another, even within an organization, has escalated tremendously in recent years. Coupled with the need for frequent corrections and modifications to existing documents, multi-destination transmission, and the inefficiencies of manual filing and retrieval, the office automation development has received an increasing impetus: witness the increasingly common use of word processors in today's offices.

Remote Office Work More Common

As office communications become automated and office work becomes more and more computer-based, it is expected that in the future, remote office work will become increasingly common. The primarily volume-sensitive nature of the cost for packet communications means that the actual physical location of office workers will become increasingly irrelevant. This will of course have its own concomitant benefits which include: the saving of time, energy, and costs in the daily commuting to and from offices; the decreasing number of sick days that should result; and the increasing longevity of the office worker's useful career due to the ability of "working at home."

Other applications in this category include project coordination, augmentation of the managerial roles of problem-solving and decision-making and what has been described as "augmentation of the intellect." In this last application, computers will perform search and compare activities, string and pattern searches, spelling checks, and other informational aids to the human intellect.

These applications are based on the concept of the "electronic market," where buyers and sellers of goods and services can interact. The initial versions of these applications will be based upon the sale of information, while later versions will allow interactive commercial applications.

A number of experiments and preliminary forms of systems which provide some of these capabilities are now in existence. Some of these fall under the category of "teletext" information systems in which the digital information is transmitted in analog form in a television signal, with the television receiver "grabbing" the frame (page) of interest and storing it locally. Others of these initial service applications fall under the category of "viewdata" systems, where the digital code is modulated onto an audiofrequency carrier and transmitted over a telephone channel to the receiving television set. A decoding circuit is attached to the telephone line and the television set. Whereas teletext systems are generally noninteractive, viewdata systems are interactive and thus have a broader scope for use in different applications.

The classification and standardization activities that are being carried out by CCITT for such systems offering new telecommunications services are referred to under the title of "videotex" services. Although we have chosen to broadly categorize videotex under "commercial applications," the nature of these applications varies widely.

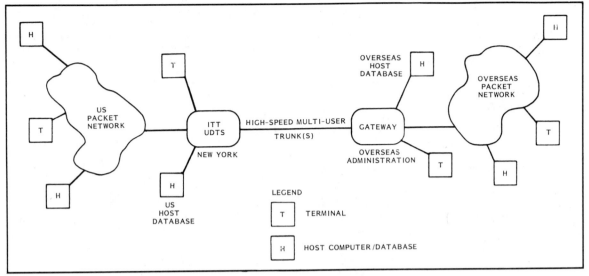

Figure 1. ITT Worldcom's UDTS network provides international packet data services.

Variety of Videotext Applications

Some of the applications of videotext services are: information retrieval for commerce such as price lists, advertising and stock quotations; commercial transactions including hotel, car, airline ticket reservations and purchases, catalog shopping and electronic funds transfer; computer services for education, games, software distribution; and other possible applications.

Examples of videotext systems that are currently being tested, evaluated or offered for limited service include: Prestel in the UK (BPO); Teletel in France; Vista and Telidon in Canada (Bell Canada); Captains in Japan; Bildschirmtext in the Federal Republic of Germany; AT&T and GTE plans or tests in the USA; Green Thumb in the USA (Department of Agriculture); and other systems in Holland, Spain, Hong Kong and the Nordic countries.

These applications of packet-switched networks provide, in many cases, what may be considered to be the basic value-added function of the packet technology. In this sense, we can expect that these applications are the most widespread today, and that is indeed the case. The value-added functions invoked in these applications include: code and speed conversion of unlike devices, terminals and computers; protocol translation; flow control and alternate routing; congestion control; and reliable, virtually error-free data transmission.

As an example of the innovative use of technology, the Universal Data Transfer Service (UDTS) of ITT World Communications provides access to US databases and remote computing facilities on host computers to users in over a dozen countries via data terminals and in over 200 countries via telex terminals. These US host computers may be accessible off the UDTS system itself, or via interconnections to the GTE Telenet and Tymnet packet networks in the US.

There are numerous other applications that have been, are being, or are planned for implementation on packet-switched networks. The reason for grouping them in this major category is a matter of convenience. Moreover, these applications are considered to have a somewhat smaller initial relevance to the international scene than most of the previously discussed applications.

Enhanced Capabilities of IRC Services

As is generally well-known, the traditional services of the international record carriers are based upon telex, channel leases, and international messages (telegrams). ITT World Communications has enhanced the service capabilities of these basic services, and added a number of other services for varying user needs. Store-and-forward telex facilities, for example, are provided by the Timetran service for outbound (out of the United States) traffic, while inbound traffic can utilize the In-Sure service. Various addressing, routing and delivery options are provided. Another service offered is the ARX service that provides automatic, store-and-forward, electronic message switching service for traffic from telex networks, leased channels and Timetran.

The philosophy of ITT Worldcom in providing interservice interconnection or service integration extends most naturally to its Universal Data Transfer Service (UDTS). UDTS is ITT Worldcom's packet-switched international gateway service. One of the primary purposes of UDTS initially is to serve as a network integrator; in this role, it makes compatible the developing new international data networks of the world. In addition to this role, UDTS also provides various packet and non-packet terminal services of its own, and includes the design philosophy to evolve into a network. Interfaces are also planned for the various existing services of ITT Worldcom, thus giving UDTS its interservice interconnection capability (see Figure 1).

The basic subscription traffic classes of UDTS are as follows:

● *Traffic Class R: Real Time Applications*—This traffic class provides real-time communications services between DTEs. The term DTE (Data Terminal Equipment) is used here in the CCITT-defined sense, and represents an intelligent or nonintelligent "terminal" operating in a packet or a non-packet mode.

● *Traffic Class T: Time-Sharing Applications*—This class, which is a subset of Class R traffic, provides communications facilities for database access, remote computing and other computer time-sharing applications.

● *Traffic Class M: Store-and-Forward Applications*—The Class M subscription class provides for the acceptance of an entire message before transmission to its destination address(es). Class M transmissions are basically one-way and non-conversational.

● *Access Facilities*—Asynchronous access to UDTS is designed for line speeds in the range 50 to 9,600 bits per second, while the speeds for synchronous interconnections are in the range of 1,200 to 64,000 bits per second.

● *Code Sets and Protocols*—The code sets that are generally supported on UDTS include: ASCII, EBCDIC, International Alphabet Number 2 (Baudot), International Alphabet Number 5 (upon which ASCII is based), Extended Binary Coded Decimal (EBCD) and Correspondence code (IBM 2741 Selectric). The protocols that are currently supported and planned for future support on UDTS include the following: the CCITT X.25 protocol; the international internetwork protocol X.75; the X.3/X.28/X.29 PAD facility related protocols; the Synchronous Data Link Control (SDLC) protocol; the Binary Synchronous Communications (BSC) protocol; the asynchronous protocols 8A1, 83B3 and 85A1; and the free-wheeling (transmit-at-will) protocol.

● *Interfaces to Other Services*—UDTS is designed to provide direct or indirect interfaces to ITT Worldcom's ARX, Timetran and telex services. As such, the interconnection capability exists to provide a spectrum of different services. In practice, of course, the actual services that are offered are based upon current tariff and regulatory policy provisions and considerations.

Glossary

Access line. The communications circuit between a user device and the network nodes.

ACK. ACKnowledgement. The control message sent between switches or other intelligent network devices to signal the correct receipt of a block of data.

ACS. *A*dvanced *C*ommunications *S*ervice. A new common user data communications service being developed by AT&T.

Active control. Network control technique that has a frequent and real-time impact on the network.

Adaptive directory routing. Routing technique that changes the path between any two points in a network according to current operating conditions.

ADCCP. *A*dvanced *D*ata *C*ommunications *C*ontrol *P*rocedure. The level 2 procedure that assures accurate transmission of data blocks between network nodes or between user devices and the network.

AFIPS. *A*merican *F*ederation of *I*nformation *P*rocessing *S*ocieties. An organization tying together a number of professional societies interested in computers and automated data processing.

ALL or ALLOCATE. A packet switching function that sets aside specific network resources to assure the completion of a particular call or message.

ALOHA. The broadcast multiple access technique that permits users to transmit packets whenever they desire.

Analog channel. A communications channel that responds linearly to changes in the frequency and amplitude of the information transmitted; will accurately represent the input signal over a specified range of parameters.

ANSI. *A*merican *N*ational *S*tandards *I*nstitute. An organization that develops and distributes standards for a wide range of commercial products.

ARPANET. The computer network developed by and for ARPA (the *A*dvanced *R*esearch *P*rojects *A*gency within the U.S. Department of Defense). It has been the basis for much of the technology of packet switching.

ARQ. *A*utomatic *R*epeat re*Q*uest. A mode of operation where a receiving station requests a retransmission of any data block it perceives to be in error.

ASCII.   *A*merican *S*tandard *C*ode for *I*nformation *I*nterchange. The standard code agreed upon by most computer equipment manufacturers for representation of the alphabet, numerals, and a variety of control characters.

Asynchronous.   A form of communications where each transmitted character has self-contained beginning and ending indications, so individual characters can be transmitted at arbitrary times.

AUTODIN (or AUTODIN I).   *AUTO*matic *DI*gital *N*etwork. The store-and-forward message network used worldwide by the U.S. Department of Defense and various other government users.

AUTODIN II.   A complementary system to AUTODIN that uses packet switching to handle computer and ADP information much more efficiently than AUTODIN I.

AUTOVON.   *AUTO*matic *VO*ice *N*etwork. The standard analog private voice communications network operated worldwide for the U.S. Department of Defense.

Bit.   Contraction of *BI*nary Digi*T*. A single symbol, either a one or a zero, which when used in groups represents the numbers, letters, and other symbols of communications. Generally used in groups of 5, 8, or 16.

Blocking.   A phenomenon in a communications network where one user cannot reach another due to any one, or a combination of, network resource limitations.

BPS.   *B*its *P*er *S*econd; sometimes written as B/S or b/s. A measure of the speed with which data communications can move over a line. The prefixes K (for thousand) or M (for million) are often used to represent higher speeds.

Buffer.   Part of a communications processor or switch used to store information temporarily.

Buffer blocking.   Blocking caused by insufficient buffers or by their inefficient use.

Bursty traffic.   Communications traffic characterized by short periods of high intensity separated by fairly long intervals of little or no utilization.

Busy hour.   The single hour in the day of greatest total communications network usage; often used for sizing the resources of the network.

Byte.   A sequence of successive bits, most often a group of eight, handled as a unit in computer manipulation or data transmission.

Call.   A complete, two-way interchange of information between two or more parties in a network, extended over a period of time. It will generally consist of a number of sequential messages or transactions passed over the communications circuits in each direction.

Call arrival rate.   The number of new call originations that occur over a specified interval of time.

Called party.   The destination of a newly originated call in a communications network.

Calling party.   The originator of a new call in a communications network.

Call-second.   The use of a facility during a call over a period of one second.

Capacity.   The ultimate limitation of any resource in a network to hold or move information.

Capture.   A phenomenon of communications whereby the stronger of two signals captures the receiver and remains relatively insensitive to interference effects from the weaker signal.

Carterfone decision.   A landmark decision in federal communications regulation that permitted the attachment of foreign devices—those not supplied by the franchised carrier—to the telephone network. It opened the way for competition in the communications marketplace.

CCITT. *C*onsultive *C*ommittee for *I*nternational *T*elephone and *T*elegraph. An international advisory committee set up under United Nations sponsorship to recommend standards for international communications.

Centralized control. Control of a network from a single centralized point, which receives status information from various locations and makes and disseminates control decisions.

Centralized topology. A form of data communications network where a central computer is accessed by a large number of remote terminals.

Channel. A single physical communications medium capable of moving intelligence from one point to another. Specific physical and electrical parameters generally define its capacity. See **group, link.**

Channel-mile cost. The portion of a communication tariff that refers to the costs associated with the distance between the circuit endpoints. Also called line-haul cost.

Circuit switching. A form of switched network that provides an end-to-end path between user endpoints under the control of the network switches. Often called channel switching.

Collisions. The condition where two packets are transmitted sufficiently close in time that some portion of each intersects and interferes with the other.

Common carrier. An organization, generally franchised by a governmental body, to provide communications services to the general public.

Common user network. A network, generally provided by a common carrier, whereby public users can gain access to one another.

Concentrator. A device that improves the efficiency of a communications circuit by fitting a number of low-speed inputs into a single, higher speed output that has a lower speed than the sum of the individual input speeds.

Congestion. A network condition that causes information to be delayed or interrupted, even though capacity may be available elsewhere in the network.

Contention packet switching. Implementation of packet switching functions by allowing devices to transmit at will into a commonly available channel or medium. A variety of different protocols may be used to resolve interference (contention) resulting from (nearly) simultaneous transmission.

Continuous traffic. Communications traffic that (nearly) completely fills the resource while it is in progress; for instance, the transmission of a television program. Compare **bursty traffic.**

CRC. *C*yclic *R*edundancy *C*heck. A procedure used to insure the correct transmission of a block of data by performing a known mathematical operation on the data at the transmitter and comparing it with the same operation performed at the receiver.

Cross-office time. In reference to a circuit switch, the time from the receipt of customer-dialed digits until the switch establishes the connection to the next switch or destination party.

CRT. Literally, *C*athode *R*ay *T*ube. Used in a generic sense to refer to data terminals that display transmitted and received information on a televisionlike screen.

CSMA. *C*arrier *S*ense *M*ultiple *A*ccess. A method of contention operation whereby the terminals sense the state of the channel before attempting transmission.

CVSD. *C*ontinuously *V*ariable *S*lope *D*elta *M*odulation. A method for converting analog speech into a digital format by transmitting a signal that is proportional only to the difference between two successive samples of the original analog signal.

Datagram. A mode of packet network operation whereby the contents of a single packet are handled as a distinct entity with no functional connection with the preceding or following packets.

DCE. *D*ata *C*ircuit *E*quipment. The device and connections placed at the interface to a network by the network provider, to which the user's equipment **(DTE)** is connected.

DDD. *D*irect *D*istance *D*ial network. The major nationwide long-distance network, provided mainly by the Bell System, that permits users to complete connections to distant users without the assistance of an operator.

DDS. *D*ataphone *D*igital *S*ervice. A Bell System service that provides for direct connection of digital sources to the communications medium.

Decentralized control. Control of a network from multiple points, using locally known information or information provided by distant points via the network itself.

Dedicated line. A communications circuit between two endpoints that is permanently connected and always available.

Delay. As applied to packet switching, the additional time introduced by the network in delivering a packet's worth of data compared to the time the same information would take on a full-period point-to-point circuit.

Directory routing. Technique for routing information through a network based on directories, or instructions, kept in the memory of each switch.

Distributed control.   See **decentralized control.**

Downlink. The transmission path from a satellite to the satellite earth station to which it is transmitting.

DPCM. *D*ifferential *P*ulse *C*ode *M*odulation. A method that permits the conversion of analog information into a digital format, by encoding the difference in amplitude between successive samples.

DSDS. *D*ataphone *S*witched *D*igital *S*ervice. A limited service of the Bell System for high-capacity digital circuits switched among major cities; charged by time and distance of the connections.

DTE. *D*ata *T*erminal *E*quipment. The device, generally belonging to a data communications user, that provides the functional and electrical interface to the communications medium. For instance, a teleprinter, CRT, or computer.

Duplex channel. A communications channel capable of transmitting information in both directions at the same time.

Erlang. A measure of communications traffic intensity representing the full-time use of a communications facility. For example, one erlang represents the traffic that can be carried on a single line used continuously for one hour.

Error detection. The process of using information added to a data transmission to detect the presence of errors in the received information.

ESS. *E*lectronic *S*witching *S*ystem. A generic term for the switching facilities in commercial networks utilizing computerlike processors rather than purely electromechanical switching relays.

Ethernet. A form of contention operation, being commercially deployed by Xerox Corp., used to tie facilities together in local geographic areas.

Fail-soft operation. An operational characteristic whereby failures of individual components only reduce network performance rather than causing loss of service to some users.

Faxpak. A commercial network permitting the interconnection of dissimilar facsimile terminals.

FCC. *F*ederal *C*ommunications *C*ommission. The principal regulatory body in the United States responsible for interstate communications and common carrier services.

FDM. *F*requency *D*ivision *M*ultiplexing. A means whereby a number of separate communications circuits are combined over a common facility.

Flooding. A routing technique whereby copies of the message are transmitted over every possible route to the destination.

Front-end processor. A specialized computer processor, generally used in conjunction with a larger mainframe computer, that interfaces the computer to communications facilities and remote users.

Full-duplex channel. See **duplex channel.**

Full-period. A circuit that is always available to a particular pair of users and is generally paid for on a fixed monthly basis without regard to total usage.

Gateway. A node or switch that permits communication between two dissimilar networks.

Group. A number of communications channels handled as a single entity.

GTE Telenet. The packet switching subsidiary of *G*eneral *T*elephone and *E*lectronics. It provides nationwide common user data communications service via packet switching.

Half-duplex channel. A communications circuit capable of carrying traffic in either direction but only one way at a time.

HDLC. *H*igh-Level *D*ata *L*ink *C*ontrol. A generic name for the digital link control procedure specified by CCITT standards. **ADCCP** is one U.S. implementation.

Header. The initial part of a data block or packet that provides basic information about the handling of the rest of the block.

Hierarchical network. A network composed of various switches with different functions and connectivity operating at various levels of importance in the network.

Hold and forward. Switching technique where each message is held at intermediary switches long enough for them to check the accuracy of the received information before relaying it on to the next switch.

Holding time. The duration of a call in the network, measured from the time connection is established until all requirements for the connection are completed.

Host. An intelligent processor or device, connected to a network, that satisfies the needs of remote users.

IDA. *I*n *D*elivery *A*cknowledgement. A control message used to signal the network that a message is in the process of leaving the network.

IMP. *I*nterface *M*essage *P*rocessor. The name given to the switching nodes in the ARPANET.

Intelligent terminal. A data communications terminal that has sufficient intelligence (processing power) to perform fairly complex interface functions and local formatting and processing.

Interarrival time. A statistical measure of the average time between successive new calls or messages to a network.

Interrupt. A computer operation that temporarily postpones action on a program in progress so that the main processor can service a higher priority function.

Interruptible traffic. Information transfers that, by nature of their content, need not be transmitted in one continuous stream.

ISO. *I*nternational *S*tandards *O*rganization. An international body that standardizes goods and services. ISO works in conjunction with CCITT for standards that impact communications.

Keyboard CRT. A data terminal device that combines a typewriterlike keyboard for data input with a TV-like screen for data output.

Leader. The initial part of a user data block that tells the network the destination and handling of the following data.

Line-haul cost. See **channel-mile cost.**

Link. A physical or electrical connection between two endpoints; for communications purposes may consist of one or more channels.

Logical channel number. A designator of an apparent connection via a packet switched network, by time sharing the channel to the network switch.

Logical multiplexing. The ability of given user to communicate with various network destinations simultaneously, using a single access line, by time sharing the line and using different logical identifiers for each connection in progress.

Loop (local). Telephone terminology which refers to the local connection between a network switch and the subscribers end instrument.

Loop (routing). The undesirable condition in a network where traffic gets routed in a circular path due to an anomaly of the software or address information.

Low duty-cycle. A network user characteristic whereby bursts of transmitted data are separated by relatively long idle periods.

LPC. *L*inear *P*redictive *C*oding. A technique for converting analog speech to digital format that uses information about the vocal tract to synthesize the transmitted speech with a very small number of transmitted bits.

MCI. *M*icrowave *C*ommunications *I*ncorporated. One of the first common carriers licensed to compete with the Bell System for interstate communications services.

M/D/1 queue. Generalized notation of queueing theory, designating a delay process characterized by Poisson (M) arrivals, deterministic (D) message lengths, and a single (1) server.

Message switching. Switching technique where messages are stored in their entirety at each intervening switch.

M/M/1 queue. Generalized notation of queueing theory, designating a delay process characterized by Poisson (M) arrivals, Poisson distributed (M) message lengths, and a single (1) server.

Modem. *MO*dulator-*DEM*odulator. A device that allows digital signals to be transmitted over analog facilities.

MPL. *M*ultischedule *P*rivate *L*ine. A tariff of the Bell System that provides for full-period, point-to-point analog circuits between service locations within the United States.

Multidrop. The data communications analogy of a party line, where various user terminals are connected on a common, shared line.

Multiplexing. See **logical multiplexing.**

Multiplexor. A device that combines a number of low-speed channels into a single higher speed channel.

Multipoint. See **multidrop.**

NCC. *N*etwork *C*ontrol *C*enter. The major control point of a network; collects performance data and issues control commands.

NCP. *N*etwork *C*ontrol *P*rogram. The software that must be executed by a front-end computer in order to interface with a packet switched network in the full packet mode.

Nearest neighbor. Network switches that are directly connected via a single link to a given node.

Node. A point of a network where various links come together; generally containing a switching element used to direct traffic.

Overhead. Information required by a network for its operation, over and above the basic information that is being moved on behalf of the subscribers.

Packet switching. A network technique that divides user messages into relatively short blocks and uses numerous geographically distributed switching nodes, to achieve low end-to-end delay for real-time data traffic.

PACUIT switching. A switching technique, using a transmission frame structure, that combines many of the desirable attributes of packet and circuit switching.

PAD. *P*acket *A*ssembly–*D*isassembly. A packet network–based function that allows terminals with little or no intelligence to interface a packet network.

Passive control. A technique for control over a long-term time frame, based on accumulated statistics on network operation.

PCM. *P*ulse *C*ode *M*odulation. A technique for coding analog signals for transmission on a digital circuit, by sampling the analog signal at regular intervals and converting each sample into a digital codeword.

Persistence. Characteristic of a user who continuously monitors the occupancy of a channel and transmits a packet as soon as he detects that the channel is idle.

Polling. A technique that permits a large number of terminals to share a common channel. A central controller asks each terminal, in turn, to transmit any information it may currently have queued.

POTS. *P*lain *O*ld *T*elephone *S*ervice. The common user voice-based nationwide telephone network.

Protocol. A set of rules and procedures that permit the orderly exchange of information within and across a network.

PVC. *P*ermanent *V*irtual *C*ircuits. A logical connection across a packet switched network that is always in place and available; used to emulate a full-period connection.

Queueing.　Any process that combines elements of storage and delay together with a number of servers. The delay experienced by the users of the process can be estimated on the basis of statistical behavior of the various elements involved.

Random routing.　Routing technique that moves information through the network in a statistically random manner.

REQALL.　*REQ*uest for *ALL*ocation. A control message used in a packet switched network that assigns network resources to the handling of a new call.

Reservation technique.　Any of a number of possible packet broadcast methods requiring that users reserve capacity in advance of transmitting their data.

RFNM.　*R*equest *F*or *N*ext *M*essage. A network control message that confirms reception of a message and indicates that resources are still assigned for the next message that is part of the same call.

RFNS.　*R*equest *F*or *N*ext *S*egment. Similar to **RFNM,** but confirms only a segment of a message, rather than a full message.

RJE.　*R*emote *J*ob *E*ntry. A data terminal used to enter complete jobs for processing at a remote computer location.

Robust.　A network characteristic indicating the ability to operate nearly normally when certain network elements fail.

Routing.　The process of finding a suitable path to move information through the network. See also **adaptive directory routing, directory routing,** and **random routing.**

Routing table.　A set of instructions stored at each switch indicating the path to move a given packet to a given destination.

SBS.　*S*atellite *B*usiness *S*ystems. A partnership of IBM, COMSAT, and Aetna Insurance to provide private network services via satellite communications facilities.

SDLC.　*S*ynchronous *D*ata *L*ink *C*ontrol. IBM's version of the **ADCCP** link control technique.

Segment.　A part of an overall information exchange that is transmitted between the user device and the network. It may be the same length as or longer than a packet, depending on the protocol implementation.

SENET Switching.　*S*lotted *E*nvelope *NET*work Switching. A master frame technique that combines packet and circuit switched functions in a common network.

Slotted channel.　A packet transmission time that has a fixed (rather than random) relationship to allowable transmission times of other packets.

Store and forward.　Switching technique where each message is stored in full at each switch it passes through.

SVC.　*S*witched *V*irtual *C*ircuits. A logical connection across a packet switched network. It is established on an as-needed basis and can provide connection to any other switched user in the network.

Synchronous.　A form of communications where characters or bits are sent in a continuous stream, with the beginning of one contiguous with the end of the preceding one. Separation of one from another requires the receiver to maintain synchronism to a master timing signal.

TADI. *T*ime *A*ssignment *D*igital *I*nterpolation. A digital technique for interleaving data bursts during silent intervals in voice conversations.

Tandem switches. Switches in a network that provide a path between other switching nodes, rather than originating or terminating traffic.

Tariffs. The formalized charges for telecommunications services that are filed and approved by state and federal regulatory organizations.

TASI. *T*ime *A*ssignment *S*peech *I*nterpolation. A technique for carrying a group of voice channels over a physical facility by interleaving conversations in the idle periods of normal voice communications.

TDM. *T*ime *D*ivision *M*ultiplexing. A means whereby a number of separate communications circuits are combined over a common facility by dividing the common facility into discrete time intervals.

Telenet. See **GTE Telenet.**

Telpak. A nationwide AT&T tariff that provided substantial discounts for the lease of 60 or 240 channels in one facility.

Tessellated pattern. A physical arrangement of facilities (such as nodes or earth stations) that follows a regular, repetitive geometric pattern.

Timeout period. The length of time a switch will wait for an expected action (such as an acknowledgement) before it takes unilateral action.

TIP. *T*erminal *I*nterface *P*rocessor. A network switch in the ARPANET that can interface up to 64 user terminals in addition to several computer hosts.

Topology. The physical arrangement of nodes and links to form a network, including the connectivity pattern of the network elements.

Transaction. A computer-based message that represents a complete unidirectional transfer of information between two points on a data network.

Transactional switching. A network switching technique that handles information as discrete entities, rather than providing a fixed end-to-end connection.

Transparent switching. A network switching technique that handles information by providing a logical or physical end-to-end connection.

Trunk. The communications circuit between two network nodes or switches.

Tymnet. A nationwide data service with packet switching-like characteristics, although it is not literally a packet switched network.

Uplink. The transmission path from a satellite earth station to the satellite itself.

VAN. *V*alue-*A*dded *N*etwork. The class of public network that leases facilities in the form of basic transmission from one carrier, adds intelligence, and provides more "valuable" services to end users.

VFCT. *V*oice-*F*requency *C*arrier *T*elegraph. A technique that permits the combination of up to 24 teletype channels over a single voice-frequency channel.

Virtual circuit. A logical connection across a packet switched network that emulates a point-to-point circuit by insuring data integrity, transparency, and data sequence.

WATS. *W*ide *A*rea *T*elephone *S*ervice. A nationwide long distance phone service, where users contract for 10 or 240 hours of use per month, rather than paying for each call individually.

Window.   The major element of the flow control mechanism used to prevent the overload of a packet network. The window size indicates the number of packets a given user can have outstanding (unacknowledged) in the network at any given time.

XTEN.   *X*erox *TE*lecommunications *N*etwork. A proposed nationwide telecommunications offering, originally developed by XEROX Corp., that combines satellite intercity transmission with local microwave distribution.

X.25.   The international standard developed by CCITT that provides the foundation for public packet switching networks.

Index